S0-AKJ-314

studies in physical and theoretical chemistry 25

ADVANCES IN MÖSSBAUER SPECTROSCOPY

studies in physical and theoretical chemistry

Other titles in this series

1 R. Ginell, Association Theory
2 T. Boublik, I. Nezbeda and K. Hlavatý, Statistical Thermodynamics of Simple Liquids and Their Mixtures
3 P. Hobza and R. Zahradník, Weak Intermolecular Interactions in Chemistry and Biology
4 S. Fraga, K.M.S. Saxena and M. Torres, Biomolecular Information Theory
5 A. Vértes, L. Korecz and K. Burger, Mössbauer Spectroscopy
6 H.E. Edwards, S. Navaratnam, B.J. Parsons and G.O. Phillips (Editors), Radiation Biology and Chemistry: Research Developments
7 D.C. Walker (Editor), Origins of Optical Activity in Nature
8 F.J. Comes, A. Müller and W.J. Orville-Thomas (Editors), Spectroscopy in Chemistry and Physics: Modern Trends
9 A. Chełkowski, Dielectric Physics
10 A.G. Walton (Editor), Structure and Properties of Amorphous Polymers
11 S. Trasatti (Editor), Electrodes of Conductive Metallic Oxides
12 B.E. Conway, Ionic Hydration in Chemistry and Biophysics
13 I. Hargittai and W.J. Orville-Thomas (Editors), Diffraction Studies
14 G. Földiák, Radiation Chemistry of Hydrocarbons
15 D.A.J. Rand, G.P. Power and I.M. Ritchie (Editors), Progress in Electrochemistry
16 Z. Hippe (Editor), Data Processing in Chemistry
17 D. Papoušek and M.R. Aliev, Molecular Vibrational-Rotational Spectra
18 G. Náray-Szabó (Editor), Steric Effects in Biomolecules
19 R. Paul, Field Theoretical Methods in Chemical Physics
20 W.B. Person and G. Zerbi (Editors), Vibrational Intensities in Infrared and Raman Spectroscopy
21 R. Carbó (Editor), Current Aspects of Quantum Chemistry 1981
22 Yu. N. Molin (Editor), Spin Polarization and Magnetic Effects in Radical Reactions
23 J. Maruani and J. Serre (Editors), Vibrational Intensities in Infrared and Raman Spectroscopy
24 G. Spach (Editor), Physical Chemistry of Transmembrane Ion Motions
25 B.V. Thosar and P.K. Iyengar (Editors), Mössbauer Spectroscopy: Applications to Physics, Chemistry and Biology
26 E. Wyn-Jones and J. Gormally (Editors), Aggregation Processes in Solution

Studies in Physical and Theoretical Chemistry 25

ADVANCES IN MÖSSBAUER SPECTROSCOPY

Applications to Physics, Chemistry and Biology

Edited by

B.V. Thosar
J.K. Srivastava (Co-Editor)

Tata Institute of Fundamental Research, Bombay, India

P.K. Iyengar
S.C. Bhargava (Co-Editor)

Bhabha Atomic Research Centre, Bombay, India

ELSEVIER SCIENTIFIC PUBLISHING COMPANY
Amsterdam — Oxford — New York 1983

7150-5647

ELSEVIER SCIENTIFIC PUBLISHING COMPANY
Molenwerf 1
P.O. Box 211, 1000 AE Amsterdam, The Netherlands

CHEMISTRY

Distributors for the United States and Canada:

ELSEVIER SCIENCE PUBLISHING COMPANY INC.
52, Vanderbilt Avenue
New York, NY 10017

ISBN 0-444-42186-6 (Vol. 25)
ISBN 0-444-41699-4 (Series)

© Elsevier Science Publishers B.V., 1983
All rights reserved. No part of this publication may be reproduced, stored in a retrieval system or transmitted in any form or by any means, electronic mechanical, photocopying, recording or otherwise, without the prior written permission of the publisher, Elsevier Science Publishers B.V. P.O. Box 330, 1000 AH Amsterdam, The Netherlands

Printed in The Netherlands

QC491
A38
1983
CHEM

FOREWORD

The Resonance Spectroscopy of Gamma-radiation, conventionally known as Mossbauer Spectroscopy, after its discovery a quarter of a century ago has developed into an important and versatile tool for the study of solid state properties.

A wide range of applications became possible, largely due to the instrinsic extreme resolution in energy and also due to the relative simplicity of the necessary instrumentation, which make such measurements available even for small laboratories.

The present volume on "Advances in Mossbauer Spectroscopy: Applications to Physics, Chemistry and Biology" aims at reviewing the experimental method and at describing its applications to problems in the general domain of solid state physics. I do hope that the book can serve as an introductory text to students and to researchers who are new in the field, as well as to provide the specialist with information outside of his immediate field of interest.

It is a characteristic of Mossbauer Spectroscopy that there exists a wide range of possible domains of applications, particularly in the fields of physics, chemistry, biology, medicine and engineering. Even the specialist, therefore, is no longer able to follow all these developments and easily misses interesting and relevant features which may be of pertinence to his own field of interest. The present volume should ease this situation.

Munich, 1982

Rudolf L. Mossbauer

PREFACE

About two and a half decades have passed since the discovery of the recoil-less emission and absorption of low energy gamma rays by nuclei in solids by R.L. Mossbauer; yet the utility of this phenomenon in the investigations of solid state properties continue to increase. By now, Mossbauer spectroscopy has found applications in almost all branches of physical and life sciences. Recent advances show promise in new areas like in the investigations in medical sciences, static and dynamical interactions in important classes of materials (ferroelectric, superconductors, spin glasses, magnetic oxides, etc.) and so on. Whereas excellent reviews and books are available on the basic aspects of the technique of Mossbauer spectroscopy, an effective use of the method requires greater emphasis on the understanding of those quantities (like the parameters of the Spin Hamiltonian of the widely studied S state ion) which have been hitherto treated merely as parameters to be obtained experimentally. Better familarity with the areas in which the method holds great promise is also desirable. This volume has been compiled with the belief that articles with emphasis on such aspects would enable better application of the method. Obviously, a complete coverage cannot be made in a book of this size and hence a selection has been made of certain areas of applications. For example, whereas the interactions of the S-state ions in solids have been covered extensively, certain important areas like Mossbauer study of catalysts, thin films, surfaces, conduction electron relaxation, minerology, archaelogy, environmental samples, etc., could not be included. The choices are restricted due to the size of the volume and of course reflect the bias of the editors.

The first chapter provides an introduction to Mossbauer spectroscopy which is necessary for other chapters as well as discussion of certain aspects not covered elsewhere in the book. It emphasizes the factors which influence the recoilless emission and absorption processes, centre shifts, and give a detailed discussion of the sources of electric field gradient, magnetic hyperfine field, and time dependent hyperfine interactions. This chapter, like other chapters, also contains an exhaustive list of references. The second chapter deals with an essential part of Mossbauer spectroscopy; instrumentation. It deals not only with aspects like drive waveforms, cryogenics, drive systems, etc., but also with recent developments like microprocessor based data acquisition systems and detectors for scattering experiments. The rest of the chapters are of specialised nature but the emphasis on the average is on the theoretical aspects and hence the discussion is directed to research workers in the respective areas. The authors have done extensive research in the areas covered by them. Chapters 4 and 6 cover areas of active research for a long time, but holds good promise for future too. The interpretation of the results on metallic systems has been a difficult one. Reviewing this area is far more challenging in view of the large body of experimental data available on a wide variety of systems, which has been accomplished fruitfully. The physical pictures employed in the discussions in Chapter 6 are helpful in understanding the interactions of the Fe^{2+} ion; static as well as dynamic aspects. In view of the extensive research experience of the authors in this area, their optimism for the great potential of future research in this area should serve as useful advice. The large body of experimental data obtained using rare earth Mossbauer nuclei (^{149}Sm, ^{151}Eu, ^{155}Gd, ^{161}Dy, ^{166}Er, ^{167}Er, ^{169}Tm, ^{170}Yb, etc.) have been discussed and tabulated in Chapter 15. Chapters 7, 8, 9, 12 and 13 can be expected to provide better understanding of the relevance of the parameters normally measured by experimentalists. Chapters 8, 9, 11-13, are strongly interrelated, deal with hyperfine splittings obtainable when relaxation rates become low. Characterisation of the environment of the Mossbauer ion using the spin Hamiltonian makes the method a very valuable tool in areas like biological processes, phase transitions, etc. An understanding of the dependence of the parameters like D,λ, etc., on the interactions of the ion in solids is indeed necessary for fruitful applications. The intensity tensor formulation for dipole transitions relevant to the widely used Mossbauer nuclei, ^{57}Fe, and ^{119}Sn, developed by the author is described in Chapter 5 extensively giving examples for illustration. The use of polarised gamma-rays, thickness corrections and texture effects are also treated. The elaborate process of interpretation of paramagnetic hyperfine spectra obtained when the relaxation rate is low has been covered in Chapter 8. The experience of the author in this area makes the article useful from practical point of view as well. Chapters 3, 9, 10, 14 and 16 cover areas in which distinct possibility exist for future research. The usefulness of Mossbauer experiments on the physics of the co-existence of superconductivity and magnetism is treated in Chapter 10. The application of Mossbauer spectroscopy to the study of spin lattice relaxation has been treated in detail in Chapter 12 and is expected to be of great value in the study of phase transitions, spin glasses and biological systems.

To summarise, this collection of articles would be useful for beginners as well as advanced research workers interested in the areas covered. We have attempted to establish a reasonable coherence among the chapters and to minimise overlaps in discussions in different chapters, though we realise we have not succeeded fully in this effort. To some extent, a little repetition of basic concepts is justified for the sake of completeness. Further, due to practical difficulties, we could not adopt uniformity in notations as well as approach and style used in various chapters. It is hoped that the readers will find the detailed discussion in each of the area covered up to date, and will thus enable them to make further advances in these areas of applications.

We take this opportunity to sincerely thank all the contributors for readily agreeing to join us in this venture in spite of their heavy research commitments. We wish to acknowledge all the authors (and the concerned publishers) who readily permitted the reproduction of their works in this book. We like to make special mention of Mr. A.P. Jadhav who shouldered the responsibility of preparing the manuscripts in the c.r.c. form.

Finally, it is a pleasure for us to bring out this volume at the time of the 25th Year of Mossbauer research.

Bombay, INDIA

P.K. Iyengar
B.V. Thosar
S.C. Bhargava
J.K. Srivastava

CONTENTS

CHAPTER 2

CHAPTER 3

CHAPTER 4

CHAPTER 5

CHAPTER 6

CHAPTER 7

CHAPTER 8

CHAPTER 9

CHAPTER 10

CHAPTER 11

CHAPTER 12

CHAPTER 13

CHAPTER 14

CHAPTER 15

CHAPTER 16

CHAPTER 1

INTRODUCTION TO MOSSBAUER SPECTROSCOPY

J.K. Srivastava*, S.C. Bhargava†,
P.K. Iyengar† and B.V. Thosar*

* Tata Institute of Fundamental Research,
Bombay - 400 005, INDIA

† Bhabha Atomic Research Centre,
Bombay - 400 085, INDIA

1. INTRODUCTION

The discovery of Mossbauer effect was the result of an experiment in which the effect seen was just the opposite of what was expected. Doppler broadening of source and absorber lines due to the thermal motion of atoms is a well-known phenomenon in spectroscopy, in which the widths of the spectral lines increase with the temperature of the source; consequently the cross section for resonance fluorescence increases with temperature. As a part of his graduate work, Mossbauer was trying to decrease nuclear resonance scattering of gamma rays (129 keV) from ^{191}Ir by decreasing the temperature, which contrary to his expectation, resulted in enhanced scattering cross section. This was puzzling at first sight. However, the theoretical work of Lamb (1) dealing with the effect of crystal binding on neutron resonances provided the ready explanation for the effect observed by Mossbauer.

There are indeed many similarities between the discoveries of Raman effect and Mossbauer effect. Both the discoveries involved very simple apparatus and the phenomena involving simple theories should have been looked for earlier. Compton effect had been discovered earlier than the Raman scattering, so the energy change on scattering from a system in which the atoms and molecules are undergoing rapid motion should have been foreseen. Similarly for a long time prior to the discovery of the Mossbauer effect the diffraction of X-rays from crystal lattices and the famous derivation of Debye-Waller factor taking into account the thermal motion of the atoms in the lattice were already known. It is surprising that it never struck anyone earlier that pure elastic scattering could occur when low energy gamma rays are scattered by the bound nuclei. Just as in the case of the Raman effect, the importance of the Mossbauer effect is not just in the discovery of a new phenomenon but due to its numerous applications. It is the universal application of Raman effect to study quantum states of molecules and different aggregates of matter using simple spectroscopic techniques that made the discovery an important milestone in the advancement of physics. Similarly, the Mossbauer effect made it possible to study extremely small energy changes ($\frac{\Delta E}{E} \sim 10^{-13}$) brought about by various types of interactions: gravitational and electromagnetic. It therefore became a technique complementary to other spectroscopies like the nuclear magnetic resonance and electron paramagnetic resonance. The aim of this book is to bring out the various applications of this technique in several areas of condensed matter physics. Though it is impossible to cover every field of application, we have attempted to collect in this book, a good cross section of the various fields of current interest. The articles attempt to give a rather complete coverage of the basic theory, so that fresh applications could be thought of.

This chapter is intended to serve as an introduction necessary for other specialised chapters to follow and covers, though briefly, certain other applications of the method which are important but have not been covered elsewhere in the book.

2. MECHANISM OF MOSSBAUER EFFECT

In a gamma-ray resonance absorption experiment the energy of incident gamma radiation should match exactly with the energy separation between two levels of the absorber nuclei. When the incident gamma ray is emitted from a radioactive source, the resonance condition requires that the energy separation between the two levels in the source nucleus and those in the absorber nucleus should be exactly equal. Therefore both the source and the absorber nuclei must necessarily be identical. The optical resonance fluorescence (2,3) is an example, involving the electronic energy levels of atoms and molecules. Tunable sources of radiation and perturbations to energy levels in the absorber by electric and magnetic fields have been used in a variety of resonance phenomena like the nuclear magnetic resonance (NMR) (4) and electron spin resonance (ESR) (5). In the case of the gamma ray resonance fluorescence, the situation is more similar to that of the optical resonance fluorescence with the difference that nuclear energy states are involved. The intrinsic narrow width of the nuclear energy levels and the ability to obtain the same width of the emitted and absorbed gamma rays even when the nucleus, bound in a solid, is in thermal motion using the Mossbauer effect has provided a unique tool to study hyperfine interactions and similar phenomena involving extremely small energy changes.

Consider the process of emission of a gamma ray from a free nucleus at rest. If E is the energy of the emitted gamma ray, the recoil energy E_R given to the nucleus to conserve momentum is

$$E_R = \frac{P^2}{2M} = \frac{E^2}{2Mc^2} \quad , \quad E = E_o - E_R \, , \qquad (1)$$

P being the momentum given to the nucleus, equivalent to momentum of the gamma photon and M the mass of the emitting nucleus. Similarly, whenever a gamma photon is absorbed, it loses an energy E_R due to the recoil energy imparted to the absorbing nucleus. This is schematically shown in Fig. 1. The line width (Γ) of the absorption (or emission) lines is inversely proportional to the mean life time (τ) of the excited state; $\Gamma\tau = \hbar$, where h is the Planck's constant. For the 14.4 keV transition of ^{57}Fe nucleus, $\Gamma = 4.5 \times 10^{-9}$ eV. For resonant absorption to occur, the emission and absorption lines must overlap appreciably. For nuclear transitions, E_R is in the range of 10^{-2} - 10^{-3} eV for a typical gamma ray energy of 100 keV and for

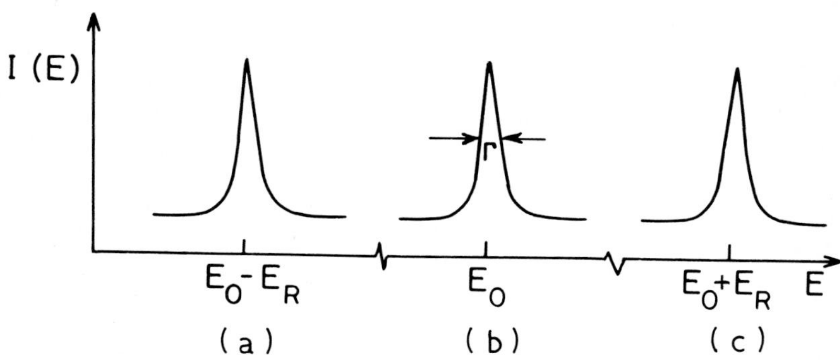

Fig. 1. Schematic representation of gamma ray line shape, (a) when there is recoil energy loss during emission from a free nucleus (E_o - E_R), (b) without energy loss (E_o), (c) energy needed for resonance absorption by a free nucleus (E_o + E_R)

nuclear mass around 100, and Γ is in the range of 10^{-7} - 10^{-8} eV for most of the low lying nuclear levels. Thus resonant absorption of gamma rays is an impossibility when the source and absorber nuclei are free and at rest. For optical transitions, on the other hand, the energy E is in the region of a few eV, $E_R \sim 10^{-11}$ eV for atomic mass around 100, and Γ is $10^{-7} - 10^{-8}$ eV. Thus optical resonance fluorescence is easily possible. The effect of thermal vibrations is to cause a spread in the photon energy of the order of the "Doppler energy" (D) given by (6,7)

$$D \sim 2\sqrt{E_K E_R} \quad , \qquad (2)$$

where E_K is the average kinetic energy of the nucleus due to thermal energy. At normal temperatures for nuclear transitions, D is of the order of 10^{-3} eV which is comparable to or smaller than E_R. Thus, for resonance involving nuclear gamma rays, even the Doppler broadening is not sufficient to make the emission and absorption lines overlap significantly. An order of magnitude increase in temperature could cause an overlap of the emission and absorption lines through Doppler broadening. However, the lines become very broad, the cross section in the region of overlap becomes small and the effect is of no value. It is therefore necessary to compensate for the recoil energy loss by other means. Moon (8) as early as in 1950, attempted to compensate for the recoil energy loss by mounting the radioactive source on the tip of a rapidly spinning rotor to change the energy of the gamma ray by Doppler effect. For ^{198}Au a peripheral linear velocity of the order of 8 x 10^4 cm/sec was required for observing resonance.

In 1958, R.L. Mossbauer made the important discovery (9) that when the emitting (or absorbing) nuclei are bound in a solid, certain fraction of gamma rays are emitted (or absorbed) with negligible energy loss due to recoil. Phenomenologically, this could be understood in the following way:

The chemical binding energy of an atom (nucleus) in a solid is of the order of 10 eV. When the recoil energy E_R is greater than this, the atom (nucleus) gets dislodged from the lattice. The situation is similar to the recoil of a free atom as in a gas. When E_R is smaller than the binding energy but is of the same order of magnitude as the characteristic phonon energies of the solid defined in terms of ω_D, the Debye frequency, or ω_E, the Einstein frequency, the atom (nucleus) will remain in the lattice but dissipate the recoil energy by creation of phonons. Finally when E_R is smaller than the binding energy as well as the characteristic phonon energy, the recoil energy need not be dissipated either in dislodging the nucleus from the lattice site or in heating the lattice. The nucleus can behave as if it is rigidly bound to the solid such that the recoil is taken up by the entire solid. In this case, M in the denominator in Eqn. (1) represents the mass of the entire solid making E_R negligible. When this happens in both the source and absorber, the condition for resonance absorption of gamma rays is satisfied.

3. PROBABILITY FOR OBSERVING MOSSBAUER RESONANCE

The fraction of gamma rays which are emitted (or absorbed) without energy loss due to recoil is related to the lattice dynamical properties of the solid in which the emitter or absorber nucleus is located (10-13). In general, gamma ray emission (or absorption) process involves the nucleus changing its energy state from $|N_i\rangle$ to $|N_f\rangle$ and the crystal lattice simultaneously changing its vibrational state from $|L_i\rangle$ to $|L_f\rangle$. However, there is a finite probability that during the period of the emission or absorption of the gamma ray, the lattice state does not undergo a change. The fraction of such events in which gamma ray emission (or absorption) is recoilless is given by

$$f = \frac{|\langle L_i|e^{i\vec{k}.\vec{X}}|L_i\rangle|^2}{\Sigma_{L_f}|\langle L_f|e^{i\vec{k}.\vec{X}}|L_i\rangle|^2}$$

$$= |\langle L_i|e^{i\vec{k}.\vec{X}}|L_i\rangle|^2, \qquad (3)$$

since the term in the denominator can be shown to be unity. Here $\vec{k}$ is the wave vector of the photon and $\vec{X}$ is the coordinate of the centre of mass of the nucleus. $\vec{X}$ is the sum, $\vec{X}_o + \vec{u}$, $\vec{X}_o$ defining the equilibrium position and $\vec{u}$ representing the displacement from it. Without any loss of generality the origin can be shifted to $\vec{X}_o$. In the harmonic approximation, the above expression can be reduced to (6,7)

$$f = \exp(-k^2 \langle u_k^2 \rangle), \qquad (4)$$

where u_k is the component of the displacement vector $\vec{u}$ of the nucleus in the direction of the gamma ray, and $\langle u_k^2 \rangle = \langle L_i|u_k^2|L_i\rangle$. For a harmonic oscillator corresponding to the i th normal mode of the lattice vibration, the thermal average of u_k^2 is given by

$$\langle u_{k_i}^2 \rangle_{Th} = \frac{\hbar}{M\omega_i}\left[\langle n_i \rangle_{Th} + \frac{1}{2}\right], \qquad (5)$$

where

$$< n_i >_{Th} = \frac{1}{\exp(\frac{\hbar\omega_i}{k_B T}) - 1} .$$

Averaging over all the vibrational modes, the average mean square displacement is given by:

$$< u_k^2 > = \frac{\int < u_{k_i}^2 >_{Th} \, g(\omega_i) d\omega_i}{\int g(\omega_i) d\omega_i} , \quad (6)$$

where $\int g(\omega_i) d\omega_i = 3N$, N being the number of atoms in the lattice. In the usual notation, $g(\omega)d\omega$ denotes the number of lattice modes in the interval ω and $\omega+d\omega$. Thus,

$$f = \exp\{- \frac{k^2\hbar}{3NM} \int (\frac{1}{2} + < n_i >_{Th}) \frac{g(\omega_i)}{\omega_i} d\omega_i\},$$

$$= \exp\{- \frac{2E_R}{\hbar} \frac{1}{3N} \int \left[\frac{1}{2} + \frac{1}{\exp(\frac{\hbar\omega_i}{k_B T}) - 1} \right] \frac{g(\omega_i)}{\omega_i} d\omega_i\}. \quad (7)$$

f is variously referred to as the probability of Mossbauer effect, the recoilless (or recoil-free) fraction, Lamb-Mossbauer factor, f-factor or sometimes loosely as the Debye-Waller factor (13). For calculating f from Eqn. (7), one uses $g(\omega)$ obtained either directly from neutron inelastic scattering experiments (14) or from analyses, based on the various models of lattice vibrations, of the phonon disperson curves obtained using neutron diffraction measurements or using Debye or Einstein models (15,16).

Using Debye model to describe $g(\omega)$, we have (16)

$$f = \exp\left[- \frac{6E_R}{k_B \theta_D} \left\{ \frac{1}{4} + (\frac{T}{\theta_D})^2 \int_0^{\theta_D/T} \frac{x dx}{e^x - 1} \right\} \right], \quad (8)$$

where $x = (\frac{\hbar\omega}{k_B T})$. The integral appearing in Eqn. (8) has been evaluated in many references. The value of f in the limit when the temperature is large or small in comparison to Θ_D is

$$f = \exp \left[- \frac{E_R}{k_B \Theta_D} \left(\frac{3}{2} + \frac{\Pi^2 T^2}{\Theta_D^2}\right) \right] \quad \text{when} \quad T << \Theta_D \quad (9)$$

and $f = \exp(- \frac{6E_R T}{k_B \Theta_D^2})$ when $T \gtrsim \frac{1}{2} \Theta_D$. (10)

It is seen that f is large when $\frac{T}{\Theta_D}$ is small, which means Mossbauer resonance is better observed at lower temperatures. Since E_R is proportional to the square of gamma ray energy it is seen that the Mossbauer fraction is higher for the gamma rays of lower energy. Resonance absorption has been seen up to a gamma ray energy of $\approx$ 190 keV (17,18).

Apart from the above parameters, the line width, related to the lifetime of the excited state of the emitting (or absorbing) nucleus, also plays an important role in determining the feasibility of observing the Mossbauer resonance. If the line width is too small (or too large), the resonance is too sharp (or too broad) to be observed. Mossbauer effect has been observed for excited nuclear state lifetime τ in the range from 10^{-11} to 10^{-5} sec. (17,18). If the lifetime of the level is short, compared to the phonon life time of 10^{-13} sec, the recoil momentum cannot be transmitted to the neighbouring nuclei and hence Mossbauer effect is not observed (19). Very favourable conditions are obtained in the case of 14.4 keV gamma ray transition of ^{57}Fe (Fig. 2(a)) and hence this has been very widely used in experiments.

The Eqn. (8) is valid for a situation where the Mossbauer nucleus is part of a monoatomic lattice (14,16,19). However, when it is introduced as an impurity in a host lattice, the difference in the masses of the impurity and the host atoms and the difference in the force constants between the host-host and host-impurity atoms have to be considered to obtain the mean square displacement of the impurity atom and thus the Mossbauer fraction (f) and its temperature dependence. The function $g(\omega)$ has been theoretically calculated for such a case (20) and the f-factor of impurity atom thus evaluated was compared with the experimental value in many cases (21). The effect of pressure on f-factor has also been studied experimentally. It is found to increase with pressure (14,22).

Fig. 2(b) shows schematically the set-up used in a typical Mossbauer experiment in transmission geometry which is used in most of the Mossbauer studies. The energy of the gamma ray is changed by means of the Doppler effect, using relative motion between the source and absorber, according to the relation $E_v = E_o(1 \pm \frac{v}{c})$, v being the relative velocity. For 14.4 keV gamma radiation from ^{57}Fe, this equation gives an energy shift of 4.8 x 10^{-8} eV for a relative velocity of one mm/sec. Thus the resonance curve is obtained by observing the transmitted counts as a function of relative velocity as shown in

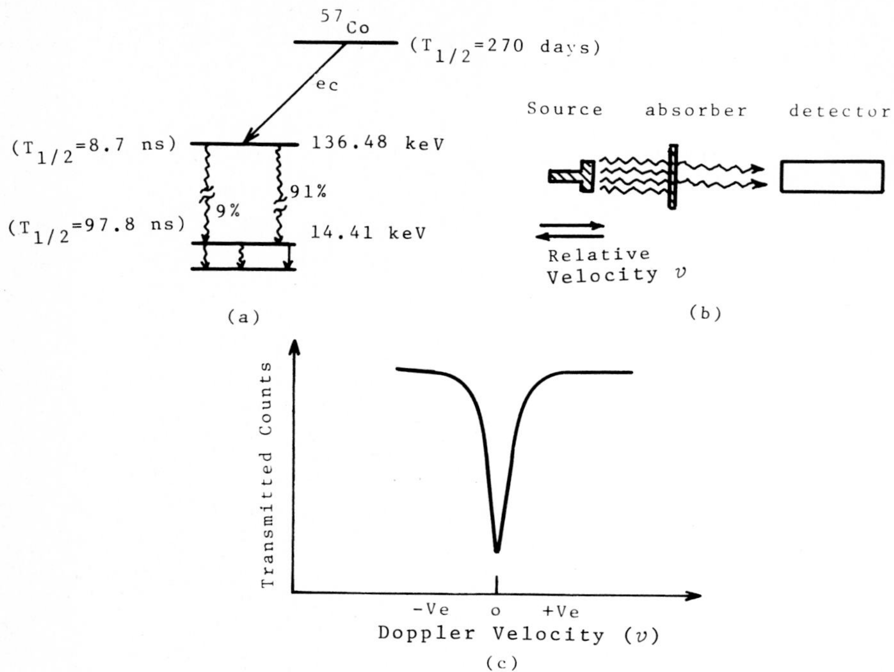

Fig. 2. (a) Energy level scheme of ^{57}Fe nucleus (b) Schematic of experimental arrangement for Mossbauer spectroscopy in transmission mode (c) Mossbauer spectrum in absence of hyperfine interactions.

Fig. 2(c). The resonance can also be observed by counting the scattered gamma rays, or by detecting the internal conversion electrons (13). In general, the Mossbauer spectrum will contain many resonance lines owing to the various hyperfine interactions of the source and absorber nuclei. Details of instrumentation for Mossbauer spectroscopy are described in Chapter 2.

Using certain approximations, it can be shown that the total area under the Mossbauer spectrum, A, is related to the recoilless fractions by (5,23,24),

$$A \propto \frac{\pi}{2} \sigma_0 \Gamma n' f_s f_a \quad , \tag{11}$$

where σ_0 is the maximum cross section, n' the number of the absorber nuclei per unit area, and f_s and f_a are the recoilless fractions for source and absorber respectively. For evaluating f_a from the area of the Mossbauer spectrum, it is necessary to know f_s (and vice versa). In the literature, various methods have been proposed and used for obtaining the absolute value of f_a (23,25-30).

For most of the applications discussed in this book, knowledge of absolute values of f_s and f_a are not essential. However, there have been several attempts to relate experimental values of f_s and f_a to the frequency distribution function $g(\omega)$ obtained using lattice dynamical models. These investigations are not very rewarding in view of more direct methods available for determining $g(\omega)$. Only in cases where impurity modes of Mossbauer nuclei are involved, this technique has some advantages.

4. SHIFTS AND SPLITTINGS OF MOSSBAUER RESONANCE LINES

The single line spectrum shown in Fig. 2(c) is for the case when there is no hyperfine interaction present either in the source or in the absorber. This is, however, a hypothetical situation because the nucleus is always accompanied by the electrons. Due to the electric and magnetic interactions between these electrons (and also other neighbouring charges in the lattice) and the nucleus, known as hyperfine interactions (13), the nuclear energy levels shift and split. Consequently, the observed Mossbauer spectra also show shifts in their position as well as splitting. In order to make the interpretation simpler, Mossbauer measurements are usually done when electric quadrupole and magnetic hyperfine interactions are present either in the source or in the absorber but not in both, though in some cases this may be unavoidable.

4.1 Isomer Shift

The electrostatic interaction between the nuclear charge Ze and the surrounding electrical charges, can be written as (31)

$$H_{el} = \int_{\tau_n} \int_{\tau_e} \frac{\rho_e(\vec{r}_e)\rho_n(\vec{r}_n)d\tau_e d\tau_n}{r} \quad . \qquad (12)$$

$\rho_e(\vec{r}_e)$ is the electronic charge density in the volume element $d\tau_e$ at position $\vec{r}_e$ with the origin at the centre of the nucleus under consideration, $\rho_n(\vec{r}_n)$ is the nuclear charge density in the volume element $d\tau_n$ at $\vec{r}_n$ and $r = |\vec{r}_e - \vec{r}_n|$ is the distance between the volume elements $d\tau_e$ and $d\tau_n$. For the case $r_e > r_n$, $\frac{1}{r}$ may be written as (31,32)

$$\frac{1}{r} = \frac{1}{\sqrt{r_e^2 + r_n^2 - 2r_e r_n \mathrm{Cos}\Theta_{en}}}$$

$$= \frac{1}{r_e} + \frac{r_n}{r_e^2} P_1(\mathrm{Cos}\Theta_{en}) + \frac{r_n^2}{r_e^3} P_2(\mathrm{Cos}\Theta_{en}) + \ldots \qquad (13)$$

where Θ_{en} is the angle between $\vec{r}_e$ and $\vec{r}_n$, and P_ℓ are the Legendre polynomials. When $r_n > r_e$, i.e. for electronic charges within the nuclear volume, $\frac{1}{r}$ has the expansion as given above but with $\vec{r}_e$ and $\vec{r}_n$ interchanged. From Eqns. (12) and (13) we have

$$H_{el} = \int_{\tau_e} \int_{\tau_n} \rho_e(\vec{r}_e)\rho_n(\vec{r}_n)\{\frac{1}{r_e} + \frac{r_n}{r_e^2} \mathrm{Cos}\Theta_{en} + \frac{r_n^2}{r_e^3} \frac{1}{2}(3\mathrm{Cos}^2\Theta_{en} - 1) + \ldots\} d\tau_e d\tau_n \quad . \qquad (14)$$

The first term in Eqn. (14) is the "electric monopole interaction" which arises due to the interaction of the nuclear monopole charge with the electrostic Coulomb potential produced by the surrounding charges. The second and third terms represent the interactions of the nuclear electric dipole and quadrupole moments with the Coulomb field and the field gradient due to surrounding charges, respectively. The second term is zero since the nucleus does not have an electric dipole moment and the terms higher than the third are either zero or negligibly small (31-34).

In obtaining Eqn. (14), we have assumed that the electronic charges are external to the nucleus. However, the electronic charge density is non-zero within the nuclear volume also and contributes to the monopole interaction. Thus, the monopole interaction represented by the first term in Eqn. (14) gets modified to (13):

$$H_{el}^1 = \int_{\tau_n} \{[\int_{\tau_e} \frac{\rho_e(\vec{r}_e)}{r_e} d\tau_e]_{r_n<r_e} + [\int_{\tau_e} \frac{\rho_e(\vec{r}_e)}{r_n} d\tau_e]_{r_e<r_n}\} \rho_n(\vec{r}_n) d\tau_n \quad . \qquad (15)$$

The effect of this interaction is to cause shifts in the energies of the excited and the ground states of the nucleus (Fig. 3a).

If δE is the shift in the energy of the nuclear level, it can be shown (13) from Eqn. (15) that

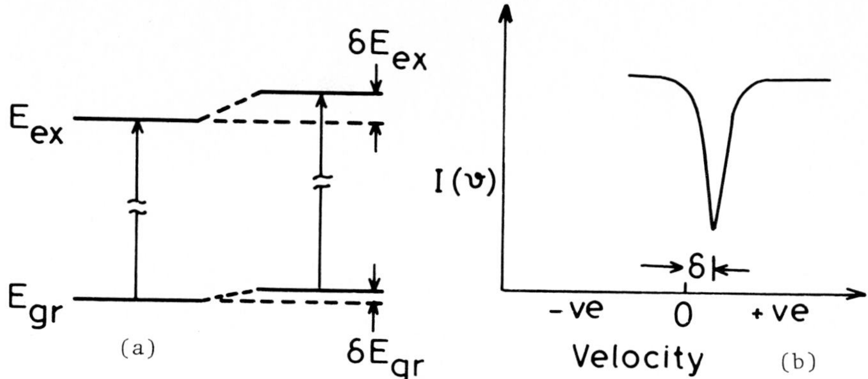

Fig. 3. (a) Shift of nuclear energy levels due to electrostatic monopole interaction. (b) Typical Mossbauer spectrum in presence of isomer shift alone.

$$\delta E = \frac{2\pi}{3} Ze^2 |\Psi(o)|^2 < r_n^2 >, \tag{16}$$

where

$$< r_n^2 > = \frac{\int r_n^2 \rho_n(\vec{r}_n) d\tau_n}{\int \rho_n(\vec{r}_n) d\tau_n} = \frac{1}{Ze} \int r_n^2 \rho_n(\vec{r}_n) d\tau_n \text{ and}$$

$-e|\Psi(o)|^2 = \{\rho_e(r_e)\}_{r_e=0}$ is the charge density at the origin.

The change in the energy of the emitted gamma ray due to this interaction is the net effect of unequal shifts of the ground and excited levels. Similar shift occurs in the absorber as well. Thus, the observed shift is given by

$$\delta = (\delta E_{ex} - \delta E_{gr})_a - (\delta E_{ex} - \delta E_{gr})_s$$

$$= \frac{2\pi}{3} Ze^2 \left[|\Psi(o)|_a^2 - |\Psi(o)|_s^2 \right] \left[< r_n^2 >_{ex} - < r_n^2 >_{gr} \right], \tag{17}$$

where, the suffixes a and s refer to the absorber and source, and suffixes ex and gr refer to excited and ground nuclear states respectively. This is called the "isomer shift" or "chemical shift" and is zero if $|\Psi(o)|_a = |\Psi(o)|_s$ which is the case when the source and absorber nuclei are in chemically identical environment. In a model in which the nucleus is considered as a point charge, the value of $< r_n^2 >$ is zero and hence there is no shift. It can be shown that for a spherical nucleus of finite radius R

$$< r_n^2 > = \frac{3}{5} R^2 . \tag{18}$$

From Eqns. (17) and (18), we get

$$\delta = \frac{2\pi}{5} Ze^2 \left[|\Psi(o)|_a^2 - |\Psi(o)|_s^2 \right] \left[R_{ex}^2 - R_{gr}^2\right] . \tag{19}$$

A typical spectrum in the presence of isomer shift is shown in Fig. 3(b). The Isomer Shift is thus dependent on the detailed nature of the distribution of the electrons in various orbitals in the source and absorber. It therefore gives information about the nature of the chemical bonding.

In general, in any atom the electron wave functions are of the well known form (33-45):

$$\Psi_{n\ell m} = R_{n\ell}(r)\ \Theta_{\ell m}(\theta)\ \Phi_m(\phi) \quad .$$

The s electrons (ℓ=0) are isotropic and the wave function has non-zero value at the origin. On the other hand, non-s electrons have zero density at the origin in the non-relativistic approximation (Fig. 4a). Nevertheless, the non-s electrons indirectly influence the isomer shift through their overlap with s electrons.

Fig. 4(b) shows a plot of radial density distribution for different orbitals. As can be seen, there is considerable overlap of the 3d and 3s charge densities. The chemical shift arises mainly from the change in 3s wave function due to the change in shape and occupancy of valence (3d) orbitals from solid to solid. Similar phenomena occur in other systems also like rare earths, actinides and compounds of 4d, 5d transition elements. A change in the density of valence s electrons or conduction s-electrons influences isomer shift directly. The influence of applied pressure, temperature change, valence fluctuation and phase transitions has been studied using isomer shift measurements (6, 43, 45-47). These measurements have also been used to determine nuclear radii in ground and excited states (6, 43, 48).

The relativistic correction to the wave function at the origin has been tabulated by Shirley (49). Methods like muonic isomer shift have been developed for absolute determination of the electronic charge density at the nucleus (6,50). By correlating the isomer shift and quadrupole splitting in solids (51), presence of charge density waves in some metallic layer compounds has been studied (52).

4.2 Second Order Doppler Shift

This effect, like isomer shift, also results in a shift of the

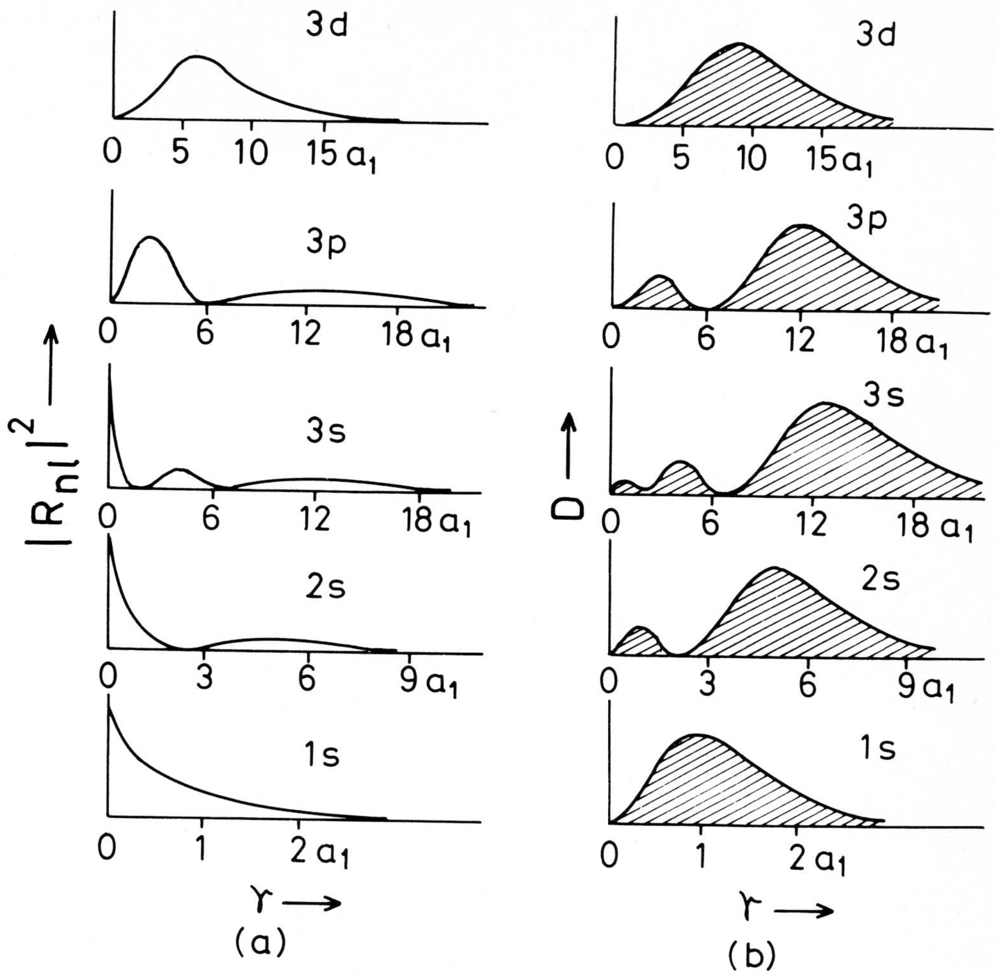

Fig. 4. (a) The radial dependence of $|R_{n,\ell}|^2$*.* a_1 *is the Bohr radius. (b) The radial dependence of* $D = 4\pi r^2|R_{n,\ell}|^2$*. It represents the probability of finding the electron at a distance r from the nucleus.*

centroid of the Mossbauer spectrum. The shift, however, has a strong temperature dependence and this must be taken into consideration while obtaining electronic charge densities from the centre shift. The temperature dependence of this shift is easily obtained from the considerations given below:

Consider two reference frames, one fixed in the laboratory and the other moving with the nucleus. We denote the quantities in the two frames by the subscript lab and 0 respectively. The theory of relativity shows that the frequency of the photon in the two frames of references are related by:

$$\gamma_{lab} = \frac{\gamma_o(1-\frac{v^2}{c^2})^{\frac{1}{2}}}{1-\frac{v}{c}\ \mathrm{Cos}\alpha_{lab}} \ ,$$

where v is the magnitude of the relative velocity of the two reference frames, which is same as the velocity of the nucleus in the frame at rest in the laboratory. α_{lab} is the angle between the direction of the photon emitted and the direction of $\vec{v}$ in the laboratory frame. When $v << c$

$$\gamma_{lab} \approx \gamma_o(1 + \frac{v}{c}\ \mathrm{Cos}\alpha_{lab} - \frac{v^2}{2c^2} + \ldots) \ .$$

During the lifetime of the emitting nucleus, generally in the range from 10^{-7} to 10^{-11} sec, the nucleus performs several oscillations around the equilibrium positions. Consequently $v\ \mathrm{Cos}\alpha_{lab}$ averages out to zero, and v^2 can be replaced by the thermal average $< v^2 >_{Th}$

The contribution of the term $<v^2>_{Th}/2c^2$ is significant and always leads to $\gamma_{lab} < \gamma_o$. (Higher terms are negligible.) The observed shift is resultant of the shifts occurring in emission and absorption processes and thus can be positive or negative. For a harmonic oscillator of mass M

$$M<v^2>_{Th} = 3(<n>_{Th} + \frac{1}{2})h\omega \ , \qquad (20)$$

where $<n>_{Th} = \dfrac{1}{\exp(\frac{\hbar\omega}{k_BT}) - 1}$ is the occupation number of the harmonic oscillator, and $<v_x^2>_{Th} = <v_y^2>_{Th} = <v_z^2>_{Th} = \frac{1}{3} <v^2>_{Th}$ for an isotropic oscillator. Averaging over all the modes of lattice vibrations,

$$<v^2>_{Th} = \frac{\int<v^2>_{Th}\ g(\omega)d\omega}{\int g(\omega)d\omega} \ . \qquad (21)$$

Thus the relative frequency shift is

$$\frac{\Delta\gamma}{\gamma_o} = \frac{\gamma_{lab}-\gamma_o}{\gamma_o} = -\frac{1}{2NMc^2}\int (<n>_{Th} + \frac{1}{2})\hbar\omega\ g(\omega)d\omega \ .$$

$g(\omega)$ is obtained either experimentally (16) or using the models of lattice vibrations (16, 21-23, 53, 54). Effects of lattice-anharmonicity, temperature, pressure, the presense of the Mossbauer nucleus as impurity in the lattice and phase transitions on $\frac{\Delta\gamma}{\gamma_o}$ have been investigated experimentally as well as theoretically (16,19,21-23, 53,55-58). Using the Debye model

$$\frac{\Delta\gamma}{\gamma_o} = - \frac{1}{2Mc^2N} \{ 9Nk_B\theta_D \left[\frac{1}{8}+ (\frac{T}{\theta_D})^4 \int_0^{\theta_D/T} \frac{x^3}{e^x-1} dx \right]\}$$

which reduces to (16)

$$\frac{\Delta\gamma}{\gamma_o} = - \frac{1}{2Mc^2N} \{ 9Nk_B\theta_D \left[\frac{1}{8}+ \frac{\pi^4}{15}(\frac{T}{\theta_D})^4 \right]\} \quad \text{When } T << \theta_D \qquad (22)$$

and

$$\frac{\Delta\gamma}{\gamma_o} = - \frac{3k_BT}{2Mc^2} \left[1 + \frac{1}{20} (\frac{\theta_D}{T})^2 - \frac{1}{1680}(\frac{\theta_D}{T})^4 + \ldots \right] \text{ when } T>>\theta_D \,. \qquad (23)$$

The Experimentally observed shift,

$$\delta' = (\Delta\gamma)_{absorber} - (\Delta\gamma)_{source},$$

is generally referred to as the second order Doppler shift or the thermal shift. From Eqn. (23), it is seen that in the high temperature limit (i.e. $T >> \theta_D$), δ' is independent of θ_D. Thus in this limiting case, even if the source and absorber lattices have different θ_D, δ' depends only on the temperature difference between these lattices and thus becomes zero when the source and absorber temperatures are identical.

4.3 Quadrupole Splitting

As mentioned in the previous section, the quadrupole interaction is given by the third term in Eqn. (14):

$$H_Q = H_{el}^3 = \int_{\tau_e}\int_{\tau_n} \rho_e(\vec{r}_e)\, \rho_n(\vec{r}_n) \frac{r_n^2}{r_e^3} \frac{1}{2}(3\cos^2\theta_{en} - 1)d\tau_e d\tau_n \,.$$

In Cartesian co-ordinates this becomes

$$H_Q = \int_{\tau_e}\int_{\tau_n} \rho_n(\vec{r}_n)(\frac{3}{2}\sum_{i,j} x_{ni}x_{nj}x_{ei}x_{ej} - \frac{1}{2} r_n^2 r_e^2)\rho_e(\vec{r}_e)\frac{1}{r_e^5} d\tau_e d\tau_n \,.$$

In a shorter form, this can be written as

$$H_Q = -\frac{1}{6}\sum_{i,j} Q_{ij}(\nabla E)_{ij} \ , \tag{24}$$

where

$$Q_{ij} = \int_{\tau_n} \rho_n(\vec{r}_n)(3x_{ni}x_{nj} - \delta_{ij}\, r_n^2)\,d\tau_n \tag{25}$$

and

$$(\nabla E)_{ij} = -\int_{\tau_e} \rho_e(\vec{r}_e) \frac{\partial}{\partial x_{ei}} \frac{\partial}{\partial x_{ej}} \left(\frac{1}{r_e}\right) d\tau_e$$

$$= -\int_{\tau_e} \frac{\rho_e(\vec{r}_e)}{r_e^5} (3x_{ei}x_{ej} - \delta_{ij} r_e^2)\,d\tau_e \ . \tag{26}$$

The second rank tensors Q and ∇E are symmetric and have vanishing traces.

Since $\int \frac{\rho_e}{r_e} d\tau_e$ is the potential produced at the nucleus by all the charges surrounding it, Eqn. (26) shows that ∇E is the electric field gradient (EFG) tensor. Q as defined in Eqn. (25) is the nuclear quadrupole moment tensor. By using Wigner-Eckart theorem Q_{ij} can be written in terms of the nuclear spin $\vec{I}$ as (31,34)

$$Q_{ij} = c\left[\frac{3}{2}(I_iI_j + I_jI_i) - \delta_{ij} I^2\right] \ . \tag{27}$$

The expectation value of Q_{zz} when the nucleus is in the state $m_I = I$ is conventionally called the nuclear quadrupole moment Q(31,32,34). Thus,

$$Q = \frac{1}{e}\int (\rho_n)_{m_I=I} (3z_n^2 - r_n^2)\,d\tau_n$$

$$= \frac{1}{e} \langle II|Q_{zz}|II\rangle \ . \tag{28}$$

Here, $(\rho_n)_{m_I=I}$ is the nuclear charge density when the nucleus is in the state $|m_I=I\rangle$. Q has the dimensions of cm^2. It follows from Eqns. (27) and (28),

$$C = \frac{Q}{I(2I-1)} \ . \tag{29}$$

Denoting the electric field gradient at the origin as $V_{ij} = -(\nabla E)_{ij}$, from Eqns. (24), (27) and (29),

$$H_Q = \frac{eQ}{6I(2I-1)} \sum_{i,j} V_{ij} [\tfrac{3}{2}\{I_i I_j + I_j I_i\} - \delta_{ij} I^2]. \quad (30)$$

When the principal axes of the symmetric second rank tensor V_{ij} are along the reference axes of the coordinate system, the off diagonal elements of V_{ij} are zero. Since s electrons, which alone possess finite density at nuclear site (Fig. 4), do not contribute to EFG, the Laplace equation is satisfied by V at the nucleus which means V_{ij} is a traceless tensor. Similarly, half filled or fully filled shells do not contribute to EFG due to spherically symmetric charge distribution. Only non-s electrons in the incompletely filled shells and charges external to the ion contribute to the EFG. Thus,

$$\begin{aligned} H_Q &= \frac{eQ}{6I(2I-1)} \sum_{i=x,y,z} V_{ii}\{3I_i^2 - I^2\} \\ &= \frac{e^2qQ}{4I(2I-1)} [3I_z^2 - I^2 + \tfrac{\eta}{2}(I_+^2 + I_-^2)], \end{aligned} \quad (31)$$

where, $V_{zz} = eq$, $I_{\pm} = I_x \pm iI_y$, η, the asymmetry parameter, is given by

$$\eta = \frac{V_{xx} - V_{yy}}{V_{zz}}$$

and

$$|V_{zz}| \geqslant |V_{yy}| \geqslant |V_{xx}|,$$

which implies $0 \leqslant \eta \leqslant 1$. For an axially symmetric charge distribution around the z-axis, $V_{xx} = V_{yy}$ ($\eta = 0$). In presence of cubic symmetery, $V_{xx} = V_{yy} = V_{zz}$ and since $\sum_i V_{ii} = 0$, each of the components is identically zero.

The nuclear quadrupole moment, Q, is related to its intrinsic quadrupole moment, Q_0, by the relation (59):

$$Q = \frac{I\,(2I-1)}{(I+1)\,(2I+3)} Q_0. \quad (32)$$

Q_0 measures the deviation of nuclear charge distribution from the spherical symmetry (32,34). For a spherically symmetric nucleus, $Q_0 = 0$. For a nuclear charge distribution elongated along I, Q_0 is negative (32). It is also clear from Eqn. (32) that $Q = 0$ for $I \leqslant \frac{1}{2}$ (irrespective of the value of Q_0). For the ^{57}Fe nucleus,

Q = 0 for the ground state and Q $\approx$ +0.21 barn for the first excited state (17).

Since the EFG is a traceless tensor, the quadrupole interaction splits the nuclear energy level without shifting the centroid of the manifold of levels. This is because the trace determines the mean of the eigenvalues.

Fig. (5a) shows the splitting of ^{57}Fe nuclear levels when $\eta=0$. $6A=\frac{1}{2}e^2qQ$ is referred to as the quadrupole coupling constant. The selection rule $\Delta m_I = 0, \pm 1$ for M1 transition results in a Mossbauer spectrum consisting of two resonance lines (Fig. 5b). Similar is the situation when $\eta \neq 0$ although the eigen functions are not pure $|I_z\rangle$ states. The separation between the two lines of the observed quadrupole doublet, ΔE_Q, depends on the value of η also. For ^{57}Fe, this dependence is given by (60)

$$\Delta E_Q = \frac{1}{2}e^2qQ(1+\frac{\eta^2}{3})^{\frac{1}{2}}. \tag{33}$$

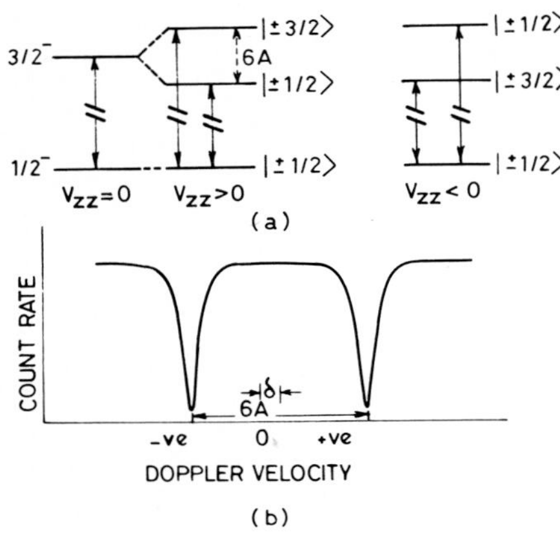

Fig. 5. (a) Nuclear level splitting due to quadrupole interaction for ^{57}Fe. (b) Mossbauer spectrum of ^{57}Fe in presence of quadrupole interaction.

4.3.1 Sign of Electric Field Gradient

Conventionally the sign of EFG is the sign of its largest component V_{zz}, (V_{zz} = eq where e is the unit positive charge), in the principal axes system. Accordingly, the sign of EFG has meaning only if the symmetry parameter η is less than 1. When $\eta = 1$, sign of EFG cannot be defined since V_{zz} is then no longer the largest component of EFG. In the case of ^{57}Fe nucleus, the sign of EFG is the same as the sign of the quadrupole coupling constant, $1/2\ e^2qQ$, since Q is a positive quantity for the 14.4 keV excited state of ^{57}Fe The theoretical evaluation of the sign and magnitude of EFG for various representative compounds is discussed later. In this section, we describe the experimental methods used for the determination of the sign of EFG. This is illustrated below using ^{57}Fe as an example. In the presence of axial symmetry, when V_{zz} is positive, $|m_I = \pm 3/2\rangle$ doublet lies higher in energy than $|m_I = \pm 1/2\rangle$ doublet, of the excited nuclear state. Fig. (5a) illustrates this. Thus the sign of EFG can be obtained from the Mossbauer spectrum (Fig. 5b) by identifying whether the higher energy line of the quadrupole doublet belongs to the $|\pm 3/2\rangle \rightleftarrows |\pm 1/2\rangle$ transition or not. The following methods have been used to identify the lines with the particular transitions. However, it may be emphasised here that the methods are not effective when $\eta > 0.6$ in which case the two excited state doublets are neither $|\pm 3/2\rangle$ nor $|\pm 1/2\rangle$ but a mixture of the two states (60).

(1) For a single crystal sample, the ratio of the intensities of the two lines are measured as a function of the angle between the direction of gamma ray propagation and a crystal axis. Theoretically, the ratio $I_{|\pm 1/2\rangle \rightleftarrows |\pm 3/2\rangle}/I_{|\pm 1/2\rangle \rightleftarrows |\pm 1/2\rangle}$ varies from 3 to 0.6 as the direction of gamma ray relative to the z axis changes from 0° to 90°, when $\eta = 0$. Here, $I_{|\pm 1/2\rangle \rightleftarrows |\pm 3/2\rangle}$ represents the intensity of the Mossbauer line corresponding to the nuclear transition $|\pm 1/2\rangle \rightleftarrows |\pm 3/2\rangle$. Thus identification of the line corresponding to $|\pm 1/2\rangle \rightleftarrows |\pm 3/2\rangle$ nuclear transition can be made by observing spectra at several orientations of the direction of gamma ray relative to the crystal axes. If the relative intensity of the line on the positive velocity side in the spectrum becomes 3 for any one of the orientations, V_{zz} is positive. This enables us to identify the direction of V_{zz} with respect to the crystalline axis also. The case when $\eta \neq 0$ is complicated as in this case:

$$\frac{I_{|\pm 1/2\rangle \rightleftarrows |\pm 3/2\rangle}}{I_{|\pm 1/2\rangle \rightleftarrows \pm 1/2\rangle}} = \frac{4\{(3+\eta^2)/3\}^{1/2} + (3\cos^2\theta - 1 + \eta\sin^2\theta\cos 2\phi)}{4\{(3+\eta^2)/3\}^{1/2} - (3\cos^2\theta - 1 + \eta\sin^2\theta\cos 2\phi)}$$

Here, θ and ϕ represent the polar coordinates of the direction of gamma ray in the principal axes system of EFG. It may be remarked here, that even in a polycrystalline absorber, the two lines of the quadrupole doublet are unequal in intensity if the f-factor is anisotropic. This effect, known as the Goldanskii-Karyagin effect (61), also complicates the determination of the sign of V_{zz} using single crystal.

(2) A preferred orientation of crystallite axes resulting from texture effects (61) can lead to unequal intensities of lines for the reason described above and in suitable circumstances enables the determination of the sign of V_{zz}.

(3) Using polarised gamma ray source, and thus inhibiting certain transitions, the two lines of the doublet in absorption spectrum can be made unequal in intensity. This procedure has been used to characterise the two lines and thus obtain the sign of V_{zz} (62).

(4) The sign of V_{zz} can also be obtained using a polycrystalline absorber and an external magnetic field. The general form of the hyperfine interaction Hamiltonian consists of two terms:

$$H_Q = \frac{eQV_{zz}}{4I(2I-1)}[\ 3I_z^2 - I(I+1)+\eta(I_x^2-I_y^2)\ ], \tag{34}$$

and

$$H_M = -g_n\beta_n H_{ext}[\ I_z\cos\theta+(I_x\cos\phi+I_y\sin\phi)\sin\theta\ ]. \tag{35}$$

Here, g_n is the nuclear g-factor, β_n is the nuclear magneton and H_{ext} is the external magnetic field oriented at polar angles θ and ϕ with respect to the principal axes of EFG.

Consider the case when $\eta=0$. Without loss of generality H_{ext} can be assumed to be in XZ plane. When $\theta=0$, the excited nuclear state splits into four and the ground state splits into two pure $|I_z\rangle$ states. When the magnetic splittings are smaller than the quadrupole splitting, a quartet and a doublet are obtained. When $\theta\neq0$, the levels $|\pm\frac{1}{2}\rangle$ mix. The admixture of $|\pm\frac{1}{2}\rangle$ with $|\pm\frac{3}{2}\rangle$ nuclear states can be neglected if the magnetic splitting is smaller than the quadrupole splitting (Fig. 6). Since the splittings depend on θ, the spectrum of a polycrystalline absorber is not a well defined quartet and a doublet (Fig. 6). When the quartet lies on the negative velocity side, V_{zz} is positive (and vice versa). The method fails when $\eta \sim 1$ in which case the quartet-doublet Mossbauer spectrum transforms into a symmetric triplet-triplet spectrum (60, 63, 64).

4.3.2 Origin of Electric Field Gradient

In solids the EFG at the nuclear site arises from three sources : (a) the charges on the neighbouring ions, (b) the valence electrons on the parent atom of the Mossbauer nucleus, and (c) the conduction

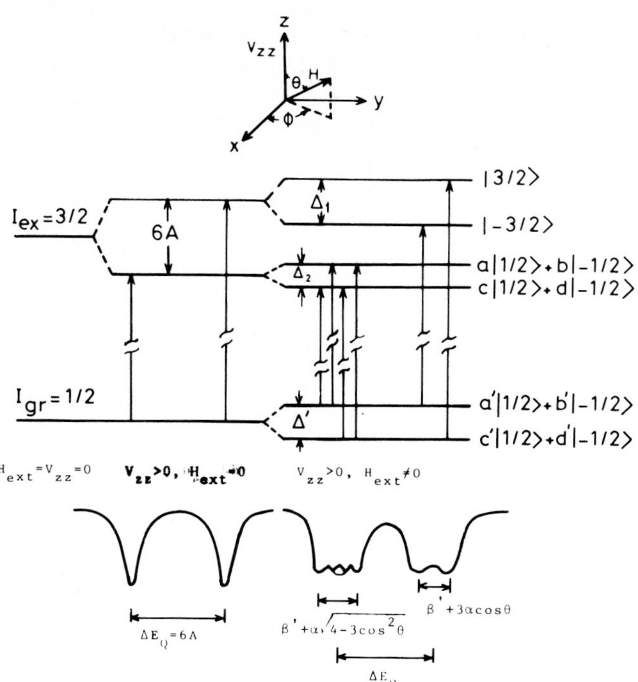

Fig. 6. Schematic representation of the splitting of nuclear levels of ^{57}*Fe by an axially symmetric electric field gradient and a magnetic field oriented at an angle* θ *with respect to* V_{zz} *when the magnetic perturbation is small in comparison to the quadrupole splitting. The mixing of pure Zeeman states* $|+1/2\rangle$ *and* $|-1/2\rangle$ *results when* $\theta \neq 0$ *as shown by the Hamiltonian given in the text.* $\Delta_1 = 3|g_{ex}|\beta_n H_{ext}\cos\theta$, $\Delta_2 = |g_{ex}|\beta_n H_{ext}(4-3\cos^2\theta)^{1/2}$, $\Delta' = g_{gr}\beta_n H_{ext}$, $\alpha = |g_{ex}|\beta_n H_{ext}$, $\beta' = g_{gr}\beta_n H_{ext}$, $\eta = 0$ *and* $\vec{H}_{ext}$ *is parallel to* γ *ray.*

electrons.

Thus the EFG can be written as

$$q = \frac{V_{zz}}{e} = q'_{lat} + q'_{val} + q'_{CE}$$

$$= (1-\gamma_\infty)q_{lat} + (1-R)q_{val} + (1-\gamma)q_{CE}$$

and

$$\eta q = \frac{V_{xx}-V_{yy}}{e} = (1-\gamma_\infty)\eta_{lat}\, q_{lat} + (1-R)\eta_{val}\, q_{val} + (1-\gamma)\eta_{CE} q_{CE} \, . \quad (36)$$

Here, (q'_{lat}, q'_{val}, q'_{CE}) and (q_{lat}, q_{val}, q_{CE}) are the EFG's produced at the nuclear and atomic sites respectively. The additional factors γ_∞, R and γ appear for the following reasons. Under the influence of

q_{lat}, etc., the electronic shells get distorted and consquently modify the EFG which appears at the nucleus. The factor γ_∞ is called the Sternheimer anti-shielding factor and is usually negative (42, 65). In case of Fe^{3+} ion, $\gamma_\infty \approx -9.14$(42,63,65,66). In case of Fe^{2+}, R $\approx$ +0.12 (67) implying that in this case the core electrons produce a shielding effect. In metallic lattices, the contribution of conduction electrons should be included. The factor $(1-\gamma)$ is nearly one (68). In metallic systems, the lattice contribution q'_{lat} may also get influenced through partial screening of the surrounding ionic charges by the conduction electrons (69-71). Also, q'_{val} need be calculated by properly ascertaining the charges localised on the ion. Physically, the effect of the conduction electrons on the EFG can be understood in the following way.

In pure metals if the conduction electrons can be described by simple plane waves, they would not contribute to EFG (51). In actual systems they contribute to the EFG because their wave functions depart significantly from plane waves, especially in the vicinity of the nucleus where they have a mixture of s and non-s characters (65,69,70,72,73). These lead to oscillations in the conduction electron density in the neighbourhood of the nucleus which is responsible for the EFG. The magnitude depends on the fraction of non-s character present in the conduction electron wave function (65,74). In some systems, q'_{CE} has a pronounced temperature dependence (74,75).

An impurity ion put in a metallic lattice and acquiring a charge different from that of the host ions, produces in the host matrix an oscillation in the conduction electron charge density (Friedel oscillations (76)) whose amplitude is a maximum at the impurity site and decreases asymptotically with distance. The EFG produced at the impurity site depends, among other things, on the charge difference between the host and the impurity ions (68).

The above approach of representing the EFG in a system as a sum of three apparently independent contributions which are correlated only through the factors γ_∞, R and γ is indeed a simplification (71,72,77). An ideal approach has been attempted by including in the Hamiltonian all possible interactions to evaluate the EFG (71,73,78,79). However, these attempts are not yet perfect (80).

It may be added that q'_{lat}, q'_{val}, and q'_{CE} are zero if the point symmetry around the Mossbauer nucleus is cubic (81,82). In certain cases, the presence of impurity ion or defect (83, 84, 85) can lower the point symmetry even if the lattice is cubic. Similarly, the presence of magnetic field (82,86) lowers the symmetry.

(a). The Lattice Contribution (q'_{lat})

If ρ_i is the ith ionic charge (assumed to be a point charge) situated at (r_i, θ_i, ϕ_i) with respect to reference axes coinciding with the principal axes of the EFG tensor and the Mossbauer nuclues is situated at the origin of the reference frame, the electrostatic Coulomb potential at the site of the Mossbauer nucleus due to the surrounding ionic charges is

$$V_{lat} = \sum_i \frac{\rho_i}{r_i} = \sum_i \frac{\rho_i}{(x_i^2 + y_i^2 + z_i^2)^{1/2}} \quad .$$

Here (x_i, y_i, z_i) are the Cartesian coordinates of ρ_i and the summation extends over all the neighbouring ions in the lattice. This yields

$$(V_{zz})_{lat} = \sum_i \rho_i \frac{3z_i^2 - r_i^2}{r_i^5}$$

$$= \sum_i \rho_i \frac{3\cos^2\theta_i - 1}{r_i^3} \quad . \qquad (37)$$

Similarly,

$$(V_{xx})_{lat} = \sum_i \rho_i \frac{3\sin^2\theta_i \cos^2\phi_i - 1}{r_i^3} \quad ,$$

$$(V_{yy})_{lat} = \sum_i \rho_i \frac{3\sin^2\theta_i \sin^2\phi_i - 1}{r_i^3} \quad .$$

This gives

$$\eta_{lat} = \frac{3}{(V_{zz})_{lat}} \sum_i \rho_i \frac{\sin^2\theta_i \cos 2\phi_i}{r_i^3} \; .. \qquad (38)$$

The summation over the different lattice points in above equation can be performed directly by including ions in a sphere of radius R centered at the ion under consideration, where R is much larger than the average lattice spacing (87). Due to the slow convergence of the terms in the summation (88), suitable summation procedures have been developed, like the summation in reciprocal lattice space (69,89) or the summation over chargeless clusters in real space (65).

The point charge model is not appropriate in many cases where it is necessary to consider the dipole moment of the ions for calculating the EFG (65,88,90). The ions at the lattice sites in general possess a dipole moment, and particularly when they are at sites without inversion point symmetry (65,91). In such a case, there is a nonzero electric field present at the ionic site, produced by the neighbouring ions, which polarises the ions and thus induces a dipole moment.

The potential due to the lattice charges is modified to

$$V_{lat} = \sum_i \left(\frac{\rho_i}{r_i} + \frac{\mu_i}{r_i^2}\right) ,$$

where μ_i is the dipole moment at the ith lattice site. The dipolar contribution is dominantly due to the anions (91) which are easily deformable. The accuracy of the lattice sum calculation can be further increased by taking into account the induced ionic quadrupole moment and the higher order multiple moments (65). In systems where the orbitals of the neighbouring ions overlap appreciably, the effect of overlap and the charge transfer should also be taken into account in the calculation (65,88,92). A detailed discussion appears elsewhere in the book (Marathe and Trautwein).

In some systems, the choice of the principal axis system of the EFG may not be obvious owing to the complexity of the crystal structure. In such a case it is necessary to derive expressions for all the components of the EFG tensor with respect to some convenient reference axes (60) and then diagonalise the resulting matrix to obtain V_{zz} and η (6,63,93).

Various methods have been developed for calculating γ_∞ such as the perturbation numerical method of Sternheimer (94), matrix method of Cohen (95), variational method of Das and Bersohn (95), unrestricted Hartree-Fock technique of Watson and Freeman (96), and linked-cluster many-body perturbation theory of Ray et al (97). The Hamiltonian for electrostatic Coluomb interaction between the electron of the ion (charge - e) and the point charge ρ_i situated at the ith lattice site at distance R_{ext} from the electron is given by

$$H_i = - \frac{e\rho_i}{R_{ext}} .$$

This can be expanded as (Eqn. 1?)

$$H_i = - \frac{e\rho_i}{R} - \frac{re\rho_i\cos\theta}{R^2} - \frac{r^2e\rho_i(3\cos^2\theta-1)}{2R^3} ------- , \quad (39)$$

where R and r are distances of ρ_i and the electron from the nucleus at the origin and θ is the angle between $\vec{R}$ and $\vec{r}$.

The first term in this expansion is the interaction obtained if the two charges are assumed to be point charges. This interaction does not distort the charge cloud and need not be considered further. The second and the third terms produce the dipole and the quadrupole polarisations of the electron charge clouds. The second term does not contribute in the first order of perturbation theory. Its contribution in the second order is much smaller than the contribution of the third term in the above expansion in the first order of perturbation (94). The contributions of the higher order terms in the above expansion is also negligibly small (94,98). Thus only the effect of the quadrupole term need be considered in the calculation of the polarisation of the electron shells on the parent atom. This perturbation term can be expressed in spherical harmonics as $c' r^2 Y_2^0(\theta,\phi)$ which shows that it mixes orbitals whose orbital quantum numbers are either equal or differ by ± 2. The mixing of electron orbitals characterised by same ℓ is called the 'radial distortion' or the 'radial excitation', whereas the mixing of the orbitals characterised by different ℓ is called the 'angular distortion' or the 'angular excitation'. This is because whereas in the former case it is the radial character of the orbitals which is getting perturbed while the angular character remains unchanged, in the later case, it is the angular character of the orbitals which gets perturbed. The radial-excitation contributions to γ_∞ are generally much larger than (and of opposite sign to) the angular-excitation contributions (94,97,99-101). Sternheimer (99) has obtained for the Fe^{3+} ion

$$\gamma_\infty = \gamma_\infty \text{ (radial excitation)} + \gamma_\infty \text{ (angular excitation)}$$

$$= \gamma_\infty (2p\rightarrow p) + \gamma_\infty (3p\rightarrow p) + \gamma_\infty (3d\rightarrow d) + \gamma_\infty \text{ (angular excitation)}$$

$$= -0.70 - 7.89 - 1.59 + 1.04 = -9.14 \quad .$$

Here, the symbol $(n\ell\rightarrow\ell')$ denotes the excitation of an electron from state $n\ell$ to ℓ'. Thus we see that the antishielding effect of γ_∞ in the case of Fe^{3+} ions arises due to the dominant contribution of the radial excitation to γ_∞. Whereas the angular excitations always give rise to a shielding effect (i.e. have a positive sign), the radial excitations may give shielding or anti-shielding effect.

(It may be mentioned here that the ns→s radial excitations do not contribute owing to the spherical symmetry of the excited orbitals). Recently, using Hartree-Fock-Slater wavefunction, γ_∞ has been calculated to be equal to -9.64 for the Fe^{3+} ion (102).

It may also be mentioned that in many solids, the free ion values of γ_∞ can be used without much inaccuracy (71). However in some systems, it may be necessary to take various solid state effects, such as overlap and covalency, change in ionic radius upon insertion in the lattice, etc., into account while calculating γ_∞ (96,103).

(b) The ionic contribution, q'_{ion}

In analogy with Eqn. (37) $(V_{zz})_{ion} \equiv eq_{ion}$ due to a single valence electron can be written as (63,65)

$$(V_{zz})_{ion} = \int -e\, \Psi_e^* \frac{3z^2-r^2}{r^5} \Psi_e d\tau_e$$

$$= \int -e\, \Psi_e^* \frac{3\cos^2\theta-1}{r^3} \Psi_e d\tau_e \quad , \qquad (40)$$

where Ψ_e is the wavefunction of the valence electron at (r,θ,ϕ) (relative to the nucleus at origin) and -e charge of the electrons. This Eqn. can be rewritten as (81):

$$(V_{zz})_{ion} = -e\left\langle \frac{3z^2-r^2}{r^5} \right\rangle$$

$$= -e\langle 3\cos^2\theta-1 \rangle \langle r^{-3} \rangle \quad . \qquad (41)$$

As will be discussed later, the angular part of $(V_{zz})_{ion}$ can be derived easily. However, the evaluation of $\langle r^{-3} \rangle$ requires knowledge of the valence electron wavefunction (63). Freeman and Watson (41,104) have calculated $\langle r^{-3} \rangle$ using unrestricted Hartree-Fock formalism for iron group ions. When there are more than one electron in the valence shell, the EFG produced at the nucleus is sum of the contributions of each of the valence electrons.

Similarly, the asymmetry parameter η_{ion}(or η_{val}) can be written as:

$$\eta_{ion} = \frac{(V_{xx})_{ion} - (V_{yy})_{ion}}{(V_{zz})_{ion}}$$

$$= - \frac{3e}{(V_{zz})_{ion}} < \frac{x^2 - y^2}{r^5} >$$

$$= - \frac{3e}{(V_{zz})_{ion}} < \sin^2\theta \cos 2\phi > < r^{-3} > \quad .$$

It is clear that a spherically symmetric charge distribution will not contribute to EFG at the nucleus. As will be shown later, the dipolar magnetic hyperfine field produced by the dipole moment of the asymmetric charge distribution of the valence electrons at the nucleus is given by (65,105)

$$H_D = 2\beta < 3\cos^2\theta - 1 > < r^{-3} > m_s \quad ,$$

where β represents the Bohr magneton and m_s is the spin magnetic quantum number of the ion. Thus, it is possible to obtain $< 3\cos^2\theta - 1 > < r^{-3} >$ from the dipolar hyperfine field. This can be used to estimate the expected value of the $(V_{zz})_{val}$. For a single electron in the valence shell (m_s = 1/2) the two are related as (105):

$$(V_{zz})_{ion} = -e\beta^{-1} H_D \ .$$

This method of obtaining $(V_{zz})_{ion}$ from H_D has been successfully used in some cases for the determination of the nuclear quadrupole moments (65).

The temperature dependence of $(V_{zz})_{lat}$ is weak and is due to the change in lattice parameters with the temperature. Thus, in the case of Fe^{3+} ion in nonmetallic host where the ionic and conduction electron contributions are negligible, $V_{zz} = (V_{zz})_{lat}$ is weakly dependent on temperature. On the other hand, $(V_{zz})_{ion}$ is highly temperature dependent. This is illustrated using the example of Fe^{2+} (81,106). The electronic configuration of Fe^{2+} ion is $3d^6$ and the ground term ^5D. The electric field gradient can be considered to be due to an electron outside the spherical core provided by $3d^5$ configuration. The extra electron can occupy any of the five d-oribtals:

$$d_0 = \frac{C}{\sqrt{3}} r^2 (3\cos^2\theta - 1) \quad ,$$

$$d_{\pm 1} = \sqrt{2}C\, r^2 \sin\theta\cos\theta\, e^{\pm i\phi},$$

$$d_{\pm 2} = \frac{C}{\sqrt{2}} r^2 \sin^2\theta\, e^{\pm 2i\phi} \quad , \qquad (42)$$

where, C is a function of r, independent of the polar angles (θ,ϕ). Alternatively, the following set of wave-functions can be used (107):

$$d_{3z^2-r^2} = \frac{C}{\sqrt{3}} (3z^2 - r^2) \quad ,$$

$$d_{x^2-y^2} = C(x^2 - y^2) \quad ,$$

$$d_{xz} = 2Cxz \quad ,$$

$$d_{yz} = 2Cyz \quad ,$$

$$d_{xy} = 2Cxy \quad . \qquad (43)$$

The $(V_{zz})_{ion}$ produced by the extra d electron in Fe^{2+} can be calculated using either the complex form of wavefunctions, Eqn. (42), or the real wave functions, Eqn. (43). Using the complex form of wavefunctions, $(V_{zz})_{ion}$ is given by (Eqn. 40)

$$(V_{zz})_{ion} = -e\int \Psi^*_{n\ell m}\, r^{-3}(3\cos^2\theta - 1)\, \Psi_{n\ell m}\, r^2 dr\, \sin\theta d\theta d\phi$$

$$= -e\langle r^{-3}\rangle \iint Y^*_{\ell m}(3\cos^2\theta - 1) Y_{\ell m} \sin\theta d\theta d\phi \ . \qquad (44)$$

Thus, EFG produced by an electron in d_0 state is (Eqn. 42)

$$(q_{ion})_{|0\rangle} = -\frac{4}{7}\langle r^{-3}\rangle \ .$$

Similarly,

$$(q_{ion})_{|\pm 1\rangle} = -\frac{2}{7}\langle r^{-3}\rangle \quad \text{and}$$

$$(q_{ion})_{|\pm 2\rangle} = \frac{4}{7}\langle r^{-3}\rangle \ . \qquad (45)$$

Same results can be obtained if q_{ion} is calculated using the real wavefunctions of Eqn. (43). In this case

$$(q_{ion})_{|3z^2-r^2>} = -\frac{4}{7} < r^{-3} >,$$

$$(q_{ion})_{|xz>} = -\frac{2}{7} < r^{-3} >,$$

$$(q_{ion})_{|yz>} = -\frac{2}{7} < r^{-3} >,$$

$$(q_{ion})_{|x^2-y^2>} = \frac{4}{7} < r^{-3} >, \quad \text{and}$$

$$(q_{ion})_{|xy>} = \frac{4}{7} < r^{-3} >. \qquad (46)$$

Similarly $(V_{xx})_{ion}$ and $(V_{yy})_{ion}$, and thus η_{ion}, can also be calculated, and the results are independent of the choice of the form of the wave functions, Eqn. (42) or (43). It may also be mentioned here that in a many electron atom, though the angular part of the one electron wavefunction is adequately described by the angular part of the hydrogen atom wavefunction, (Eqns. 42 and 43), the radial part is different (41,42) and therefore $< r^{-3} >$ is evaluated using a more realistic radial wavefunction (96,104) like the one obtained from the Hartree-Fock method. In addition, the free ion values of $< r^{-3} >$ thus obtained must be further modified when the ion is not free but is in the solid to include various solid state effects such as overlap between the wavefunctions of neighbouring ions, covalency, etc. (7,81).

In a free ion, where the orbital states with same (nl) are degenerate, the 3d electrons spend equal time in the five orbital states. Thus q_{ion} is given by

$$(q_{ion})_{free\ ion} = (q_{ion})_{|3z^2-r^2>} + (q_{ion})_{|xz>} + (q_{ion})_{|yz>}$$
$$+ (q_{ion})_{|x^2-y^2>} + (q_{ion})_{|xy>} = 0 \quad .$$

Similarly it can be shown that

$$(\eta_{ion})_{free\ ion} = 0.$$

When the ion is in a solid, the crystal field removes the orbital degeneracy of the wavefunctions, either partially or completely (81). In a cubic crystal field, the electronic ground term of the Fe^{2+} ion (^5D) gets split into two levels, a doublet (E_g) and a triplet (T_{2g}). This kind of grouping of the orbital states can be appreciated physically from Fig. 7. The $d_{3z^2-r^2}$ and $d_{x^2-y^2}$ orbitals have angular dependences in which the charge is preferentially along the coordinate axes whereas the remaining three orbitals, namely

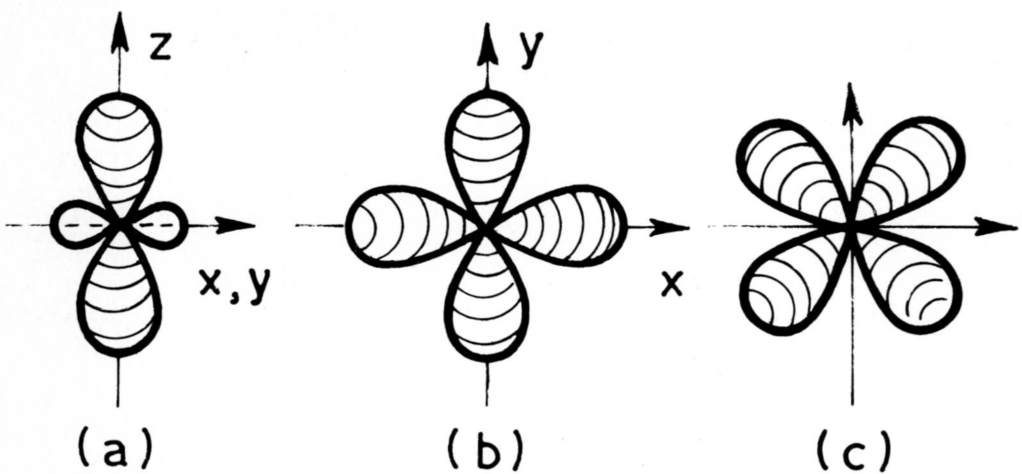

Fig. 7. Angular dependences of the probability density of (a) $d_{3z^2-r^2}$, *(b)* $d_{x^2-y^2}$, *(c)* d_{xy}, d_{xz} *or* d_{yz} *orbitals (Eqn. 43) in two dimensions. The two axes shown in the case of* d_{xy} *are* x- *and* y- *axes, and so on.*

d_{xy}, d_{xz} and d_{yz}, have angular dependences, in which the charge is preferentially located in between the coordinate axes. Thus depending upon the nearest neighbour environment the ground state of the central Fe^{2+} ion can be either $E_g(d_{3z^2-r^2},\ x^2-y^2)$ or $T_{2g}(d_{xy,xz,yz})$ (19,108). Generally, the separation between the ground state and the excited state is so large that at the temperatures concerned in experiments only the ground state is populated (81). It follows from Eqn. (46)

$$(q_{ion})_{E_g} = (q_{ion})_{|3z^2-r^2>} + (q_{ion})_{|x^2-y^2>} = 0 \quad .$$

Also, $(q_{ion})_{T_{2g}} = (q_{ion})_{|xz>} + (q_{ion})_{|yz>} + (q_{ion})_{|xy>} = 0$.

Similarly it can also be shown that,

$$(\eta_{ion})_{E_g} = (\eta_{ion})_{T_{2g}} = 0.$$

Thus $(V_{zz})_{ion}$ and η_{ion} are zero in a cubic crystal field. The ionic contribution to EFG is nonzero only when the symmetry is lower than cubic so that E_g and T_{2g} orbitals are further split and are unequally populated. In such a case, the temperature dependence of q_{ion} arises from a change in the population of the various orbital states.

The results given by Eqns. (45) and (46) can also be obtained, in a more convenient way and without the need of evaluating the angular integrals, using the Steven's operator equivalent method (81). Using this technique, we get the relation

$$<L,M|(3z^2-r^2)f(r)|L,M> = C_L <L,M|(3L_z^2-L(L+1))|L,M> \quad \text{and}$$

$$<L,M|3(x^2-y^2)f(r)|L,M> = C_L <L,M|3(L_x^2-L_y^2)|L,M> \quad , \qquad (47)$$

where

$$C_L = \frac{<L,L|(3z^2-r^2)f(r)|L,L>}{L(2L-1)} = -\frac{2}{21} <r^2 f(r)> \text{ for } L=2 \quad .$$

Here, the angular integral appearing in C_L has been evaluated using the wavefunction of $|2,2>$ state. For further details of the operator equivalent method, the article by Bhargava appearing elsewhere in the book may be referred. Thus, using the operator equivalent method, q_{ion} due to an electron in the i th orbital can be written as

$$(q_{ion})_i = -<(3z^2-r^2)r^{-5}>_i = \frac{2}{21} <r^{-3}>_i <3L_z^2-L(L+1)>_i \quad , \qquad (48)$$

where the suffix i implies that the expectation value is to be obtained using the i th orbital wavefunction.

The results contained in Eqns. (45) and (46) can be conveniently obtained using Eqn. (48). Similarly,

$$(\eta_{ion})_i = -\frac{e}{[(V_{zz})_{ion}]_i} <3(x^2-y^2)r^{-5}>_i$$

$$= \frac{e}{[(V_{zz})_{ion}]_i} \frac{2}{21} <r^{-3}>_i <3(L_x^2-L_y^2)>_i$$

$$= \frac{3}{21} \frac{e}{[(V_{zz})_{ion}]_i} <r^{-3}>_i <(L_+^2 + L_-^2)>_i \; . \qquad (49)$$

It is also clear from Eqn. (48) that the sign of the q_{ion} is determined by the angular part of the wavefunction and hence can be predicted correctly (109).

In a system, where the orbital states are split and are populated at the working temperature, each of these orbital states produce EFG at the nucleus. Thus the EFG at the nucleus can take any of the values corresponding to the ionic states. When the rate of transition between ionic levels is small compared to $(\Delta E_Q/h)$, the spectrum is expected to be the superposition of many quadrupole split patterns. Generally, this is not the case. As will be discussed later, there are various relaxation processes by which the electron spin can flip from one ionic state to another. Generally, this fluctuation frequency is much larger than the nuclear quadrupole frequency $(\Delta E_Q/h)$ and thus the nuclear spin experiences a value of the EFG which is the thermal average, given by

$$< q_{ion} >_{Th} = \frac{\sum_i (q_{ion})_i \exp(-\Delta_i/k_BT)}{\sum_i \exp(-\Delta_i/k_BT)} , \tag{50}$$

where the ith orbital state is at an energy separation Δ_i from the ground orbital state. It follows from Eqns. (48) and (50) that

$$< q_{ion} >_{Th} = \frac{2}{21} <r^{-3}> \frac{\sum_i < 3L_z^2 - L(L+1) >_i \exp(-\Delta_i/k_BT)}{\sum_i \exp(-\Delta_i/k_BT)} ,$$

where the suffix i has been dropped from $< r^{-3} >$ since the radial part of the wavefunction is same for all the orbital substates of a given (nL) manifold. Similarly,

$$(\eta_{ion})_{Th} = \frac{3}{21} \frac{< r^{-3} >}{<q_{ion}>_{Th}} \frac{\sum_i <L_+^2 + L_-^2>_i \exp(-\Delta_i/k_BT)}{\sum_i \exp(-\Delta_i/k_BT)} .$$

If the temperature is sufficiently high so that all the orbital states are equally populated and the fluctuation rate is large, then

$$< q_{ion} >_{Th} = < \eta_{ion} >_{Th} = 0 .$$

Since both q_{ion} and q_{lat} are proportional to $< r^{-3} >$, q_{ion} is generally much larger than q_{lat}. This is because the valence electrons are much nearer to the nucleus than the lattice charges. Thus the EFG observed in Fe^{3+} ($\sim$ 0.5 mm/sec.) is much smaller than the EFG observed in the Fe^{2+} ($\sim$ 3.0 mm/sec.) systems (6,63). Like quadrupole splitting, the isomer shift can also be used to distinguish between the Fe^{2+} and Fe^{3+} charge states. $\delta_{Fe^{2+}} \gtrsim 1$ mm/sec whereas $\delta_{Fe^{3+}} \lesssim 0.5$ mm/sec with respect to the 310 stainless steel absorber (6,13,19). Ions in low spin state (e.g. Fe^{II} and Fe^{III}) can also be characterised by their respective isomer shifts and quadrupole splittings (6,13,19,63). However, it may be mentioned here that in some Fe^{2+} compounds, q_{lat} has been found to be slightly larger than q_{ion} (110). This has been understood on the basis of the site symmetry, the energies of the various orbital states (110) and the temperature dependence of q_{ion} (81). It may also be mentioned that in some Fe^{3+} compounds also (111) ΔE_Q has been found to be exceptionally large, $\sim$ 1.5 mm/sec. This has been understood on the basis of the charge distribution in the lattice. Pressure and atomic volume dependences of the quadrupole interaction have also been studied and understood on the basis of the discussions given above (46). In a few Fe^{2+} complexes, ΔE_Q has been found to be anomalously large, $\sim$ 4.3 mm/sec (112). Such exceptionally large quadrupole splitting arises due to the same sign of the valence and the lattice contributions.

Thus, q_{ion} arises due to the presence of extra p- or d- or f- ... electrons outside the spherically symmetric ionic core provided by half filled or fully filled outer shell. However, for obtaining q_{ion} it is necessary to evaluate the shielding factor R also. As in the case of γ_∞, R has also been calculated using various methods such as the variational method (95,113), unrestricted Hartree-Fock method (81,104), linked cluster many body perturbation calculation (114) and the perturbation-numerical method of Sternheimer (67,115). Out of these methods, the Sternheimer's procedure is the most preferred one for calculating R and details can be found in Ref. 67. The results obtained for the case of Fe^{2+} ion are described below.

The interaction between a core electron and valence electron can be represented by Eqns. (12) and (13) with ρ_e replaced by the valence electron charge density, ρ_n replaced by the core electron charge density, and r_e replaced by the radius vector of the valence electron, $(r)_{n_e \ell_e}$. Thus the interaction Hamiltonian responsible for the polarisation of the ionic core by the valence electrons is of the same form as the third term of Eqn. (39) and therefore causes mixing of the core orbitals whose orbital quantum numbers differ by 0 or $\pm$ 2. Thus as in the case of γ_∞, in calculating R also the effects of both the radial excitations and the angular excitations must be considered. In the Sternheimer's approach (67,115), R is given by

$$R = \sum_{n\ell} [R_D(n\ell \rightarrow \ell'; n_e \ell_e) + \sum_L R_E(n\ell \rightarrow \ell'; n_e \ell_e; L)] , \qquad (51)$$

where n, ℓ are the quantum numbers associated with core electrons and n_e, ℓ_e are the quantum numbers of valence electron. R_D is the direct contribution to R due to the excitation $(n\ell \rightarrow \ell')$ of the core orbital and R_E is contribution to R due to the exchange term arising from $(n\ell \rightarrow \ell')$, Sternheimer has calculated R using Hartree-Fock wavefunctions (67). Recently Hartree-Fock-Slater wavefunctions have also been used for calculating R (67) for a Fe^{2+} ion. The results obtained are summarised below:

$$
\begin{aligned}
R &= [\,R\,(\text{angular excitation})] + [R(\text{radial excitation})] \\
&= [R_D(1s\rightarrow d) + R_E(1s\rightarrow d) + R_D(2s\rightarrow d) + R_E(2s\rightarrow d) + \\
&\quad R_D(3s\rightarrow d) + R_E(3s\rightarrow d) + R_D(2p\rightarrow f) + R_E(2p\rightarrow f) + \\
&\quad R_D(3p\rightarrow f) + R_E(3p\rightarrow f) + R_D(3d\rightarrow s) + R_E(3d\rightarrow s) + \\
&\quad R_D(3d\rightarrow g) + R_E(3d\rightarrow g)] + [R_D(2p\rightarrow p) + R_E(2p\rightarrow p) + \\
&\quad R_D(3p\rightarrow p) + R_E(3p\rightarrow p) + R_D(3d\rightarrow d) + R_E(3d\rightarrow d)] \\
&= [0.0257 + (-0.0008) + 0.0416 + (-0.0118) + 0.0184 + (-0.007) + \\
&\quad 0.0624 + (-0.0137) + 0.03 + (-0.0108) - 0.0084 + 0 + 0.0169 + 0] + \\
&\quad [-0.3027 + 0.2387 + 0.179 - 0.1773 + 0.0389 + 0] \\
&= [0.1425] + [-0.0234] \\
&= 0.1191.
\end{aligned}
$$

It is thus seen that shielding effect (i.e. the positive sign) of R is mainly due to the large net shielding effect of the angular excitations. This explains why γ_∞ causes anti-shielding whereas R causes shielding. It is also seen that, excluding the contribution of 3d→s excitation which is negligibly small $[R_D(3d\rightarrow s) + R_E(3d\rightarrow s) = -0.0084]$ owing to the spherical symmetry of the excited orbital, the angular excitations always have a shielding effect [i.e. $R_D(n\ell\rightarrow\ell\pm2) + R_E(n\ell\rightarrow\ell\pm2)$ is always positive]. On the other hand, the radial excitations have either shielding or antishielding effect [i.e. $R_D(n\ell\rightarrow\ell) + R_E(n\ell\rightarrow\ell)$ is either positive or negative]. The fact brought out by these calculations that γ_∞ is much larger than R can be physically understood as follows. Unlike lattice charges, which are always situated outside the ionic core, the valence electrons overlap the ionic core (Fig. 4). Because of this penetration, they spend some time inside the core also and therefore the net polarisation produced by them is small (42,94). The free ion values of R may get modified inside the solid owing to various effects out of which the most important one is the change in the ionic radius when the ion is bound in the lattice (67).

(c) The conduction electron contribution, q'_{CE}

q_{CE} and η_{CE} can be calculated from Eqn. (40) by replacing Ψ_e with the conduction electron wavefunction (Ψ_{CE}) which is obtained from the band structure calculations (71). The shielding factor γ can be calculated using the procedure similar to the method for calculating R and the conduction electron wave function (116). We briefly mention here some recent interesting studies of q'_{CE}. One such study is the universal correlation observed between $(q-q'_{lat})$ and q'_{lat} which, excluding few exceptions (117), is true for all the host-impurity systems in metallic hosts (72). Here, q is obtained from the measurement at the impurity site. When $(q-q'_{lat})$ is plotted against q'_{lat} for different host-impurity systems, all the data points fall on a single curve. No satisfactory theoretical explanations exists for this behaviour (72,117). However, it may be noted that in obtaining this universal correlation, q'_{lat} has been calculated assuming a point charge model and an unscreened Coulomb potential as the interaction potential between the lattice charges and the nucleus, neglecting the screening due to the conduction electrons (69,74,75). These approximations may not be valid for all the host-impurity systems considered. Further the use of the free ion value for γ_∞, in calculating q'_{lat} too may not be appropriate in all the cases. It is also to be noted that the universal correlation has been observed using mostly the source experiments (72). In these measurements it is difficult to know the exact charge state of the ions at various sites in the lattice (69). These are some of the uncertainities which come in the way of accepting the reported universal correlation as well established (71,117). It may also be realised here that the reported correlation between $(q-q'_{lat})$ and q'_{lat} cannot be treated as the correlation between q'_{CE} and q'_{lat} (72,117), since in many cases, depending upon the actual charge state of the impurity ion, the contribution of q'_{val} to q may be quite siginificant. Another interesting result regarding q in metals and dilute binary alloys (72) is the observed $T^{3/2}$ dependence of q. This result is also not properly understood (116,118), though attempts have been made to understand it on the basis of the effect of the temperature dependence of the mean square displacement of atoms on EFG when the conduction electrons are present in the lattice (119,120). Here again, as has been mentioned above, the observed temperature dependence of q should not be assumed to arise only due to the temperature dependence of q'_{CE} (116,118-120) since in many cases the temperature dependence of the q'_{val} may be quite significant. Measurements have also been done for the pressure dependence of q in metals and these results have been understood on the basis of either the shielding of the ionic potential by the conduction electrons or the scattering of the conduction electron wavefunctions by lattice vibrations (121).

4.3.3 Magnetically Induced EFG

We have seen that EFG vanishes ($q=\eta=0$) when the site symmetry is cubic. However, as has been mentioned before, a nonvanishing value of the EFG has been reported in many cubic systems also when the system either becomes magnetically ordered or is subjected to a sufficiently large external magnetic field (82,86,122-124). This EFG, which is observed only in the presence of a magnetic field, is called magnetically induced EFG, and has been observed in noncubic systems also (124).

The symmetry of a cubic system gets reduced when the system becomes magnetically ordered owing to the presence of preferred direction upon ordering, namely the direction of magnetisation. Same is the case when the system is put in an external magnetic field, $\vec{H}_{ext}$. This reduction in the symmetry makes $q'_{lat} \neq 0$. The magnitude of the induced q'_{lat} is directly proportional to the magneto-striction, $\delta\ell/\ell$, produced by the magnetic ordering or the external magnetic field (82) ($\delta\ell$ is the change in the length of the lattice (ℓ) along the direction of magnetisation or $\vec{H}_{ext}$). Since $\delta\ell/\ell$ is generally quite small, the induced q'_{lat} is small. Apart from making $q'_{lat} \neq 0$, the presence of magnetic ordering or $\vec{H}_{ext}$ can also make $q'_{val} \neq 0$ in a cubic system. This can be understood easily by considering the example of an Fe^{2+} ion (82,123). As has been discussed before $< q_{ion} >_{E_g} = < q_{ion} >_{T_{2g}} = 0$ owing to the degeneracy of the E_g and T_{2g} orbitals in a cubic crystal field. This is the situation when the spin-orbit coupling ($H_{LS}=\lambda\vec{L}.\vec{S}$) has not been taken into account. The effect of spin-orbit interaction is to split the E_g and T_{2g} states (106). Let us first consider a cubic tetrahedral environment for the Fe^{2+} ion (Fig. 8). In this case E_g is the ground state which has orbital degenency of 2 and a spin degeneracy of 5. Spin-orbit interaction has no effect on this level in the first order (106). If the magnetic field is present, it splits the level E_g into five equally spaced orbital doublets. This is because the magnetic field can remove only the spin degeneracy but not the orbital degeneracy (123). Like unperturbed E_g state, each of these doublets will produce zero EFG at the nucleus. However, the second order spin-orbit interaction removes this residual degeneracy. These levels when unequally populated give a net EFG at the nucleus (125). Let us now consider the case of a cubic octahedral environment for the Fe^{2+} ion (Fig. 8). In this case the orbital-triplet T_{2g} is the ground state. The state T_{2g} behaves as if it has an effective orbital quantum number L=1. This is called t_{2g}-p equivalence (126).

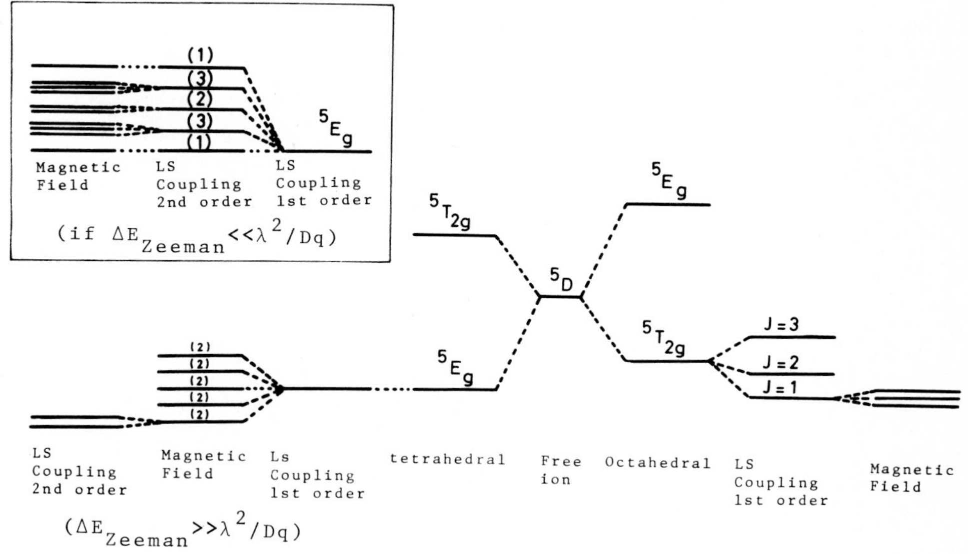

Fig. 8. Splitting of ^5D state of Fe^{2+} in presence of a magnetic field.

The spin orbit interaction splits this state into three levels characterised by an effective total angular momentum quantum number J = 1, 2 and 3 respectively. The state J = 1 lies lowest. For each of these states, q_{ion} can be calculated from Eqns. (48) - (50) if L is replaced by J and L_z is replaced by J_z. It is thus seen that $(q_{ion})_i=0$ and hence $< q_{ion} >_{Th}=0$ for any of the three states. Let us consider the case of the lowest J = 1 state only since this is the only state which is populated at the temperatures of interest (126,127). The magnetic field splits this state into three non-degenerate levels which can be characterised by the quantum number J_z = 1,2 and 3 respectively. Each of these levels produce a non-vanishing EFG i.e. $(q_{ion})_i \neq 0$ and if these levels are unequally populated, $< q_{ion} >_{Th} \neq 0$. Similarly, the magnetically induced EFG observed in rare earth compounds (122) can also be understood although spin orbit interaction is stronger than crystal field interaction in these cases, unlike in iron group ions. The presence of magnetically induced $< q_{ion} >_{Th}$ in any system can be easily found by studying the temperature dependence of EFG. $< q_{ion} >_{Th}$ depends on T as $e^{-\Delta_i/k_BT}$. Δ_i can be calculated from the value of the magnetic field and thus the temperature dependence of $< q_{ion} >_{Th}$ can be predicted. In case the Weiss molecular field is responsible for the magnetically induced EFG, the temperature dependences of $< q_{ion} >_{Th}$ and magnetisation of the system are same (10). It may be mentioned here that, in addition to producing a nonzero q'_{lat} and q'_{val}, the magnetic ordering or the presence of the H_{ext} can also make $q'_{CE} \neq 0$ in a cubic system. This happens because the magnetic ordering or H_{ext} spin polarises the conduction electrons, and thus distort the conduction electron charge cloud around the Mossbauer nucleus (82,124). However, the magnitude of the magnetically induced q'_{CE} is expected to be quite small (82,86).

4.4 Magnetic Hyperfine Splitting

When nuclei with nonzero spin $\vec{I}$ interact with a magnetic field $(\vec{H})$, the degeneracy of the levels is completely removed due to the Zeeman interaction given by

$$\mathcal{H}_m = -g_n \beta_n \ \vec{I}.\vec{H} \quad , \tag{52}$$

where β_n is the nuclear magneton and g_n is the nuclear g factor. The energy shifts are given by,

$$E_{m_I} = - g_n \beta_n H \ m_I . \tag{53}$$

As an example, consider the interaction of ^{57}Fe nucleus with $\vec{H}$ (Fig. 9). The transitions between the excited and the ground states are governed by the selection rule $\Delta m_I = 0, \pm 1$ (owing to the magnetic dipole nature of the nuclear transition). Thus, six transitions are allowed as schematically shown in Fig. 9(a). The resulting Mossbauer spectrum appears as shown in Fig. 9(b). Numbering the resonance lines as 1, 2, 3, ... from left to right, their relative intensities are:

$$I_1:I_2:I_3:I_4:I_5:I_6 = \tfrac{3}{2}(1+\cos^2\theta):2\sin^2\theta:\tfrac{1}{2}(1+\cos^2\theta):\tfrac{1}{2}(1+\cos^2\theta):$$
$$2\sin^2\theta:\tfrac{3}{2}(1+\cos^2\theta).$$

Here, θ is the angle between the directions of gamma ray and the magnetic field at the nucleus. For a polycrystalline sample, an

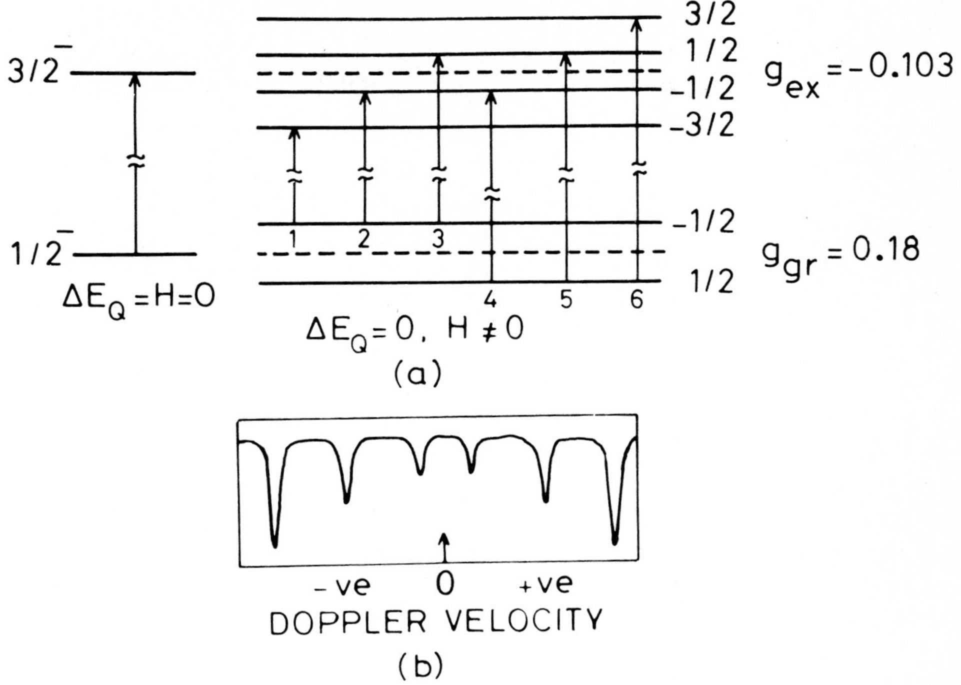

Fig. 9. (a) Zeeman splittings of nuclear levels of ^{57}*Fe and (b) the resulting Mossbauer spectrum.*

averaging over θ must be done which yields $I_1:I_2:I_3 = 3:2:1$. Deviations from this theoretical ratio of intensities (3:2:1) can occur due to the Goldanskii-Karyagin effect and the texture effect mentioned earlier. It is possible to reduce the intensity of the middle lines to zero by aligning the magnetic field at the nucleus along the direction of γ-ray propagation. Corrections to the relative intensities and to the shapes of Mossbauer resonance lines due to the finite thickness of source and absorber (128) have been discussed elsewhere in this book.

4.4.1 Origin of Magnetic Hyperfine Field

One of the most fruitful contributions of Mossbauer spectroscopy has been in the elucidation of the origin of magnetic hyperfine field in solids.

It is well known that unpaired spin of valence electrons in atoms polarises the core electrons, due to the exchange interactions. Thus there is a polarisation of the spin densities at the nucleus. This effectively produces an internal magnetic field of several kilo-Oersted in magnitude which correspond to the line splittings in the Mossbauer spectra of the order of several mm/sec. There are other important sources of the hyperfine field also (129). We discuss in the following paragraphs the nature of various interactions which lead to the magnetic hyperfine field.

(a) Fermi contact interaction:

The direct interaction of the nuclear moment with the magnetisation of the overlapping electron is called Fermi contact interaction. Only s electrons possess nonzero density at the nucleus and are thus involved in this interaction. The origin of this interaction was first explained by Fermi (130) using the relativistic formulation. A number of authors have used other approaches (130). However Ferrel (130) has given a derivation using classical approach which is briefly outlined here.

An unpaired s electron has a spherically symmetric magnetic moment distribution around the nucleus. The magnetisation M(r) thus depends only on the distance r from the origin at the centre of the nucleus. Thus, the nuclear moment is immersed in a spherically symmetric magnetic medium. The amperian current density which is equivalent to the magnetisation density M(r) is given by (131):

$$\begin{aligned}\vec{j}(\vec{r}) &= c\ \nabla\times\vec{M}(r)\\ &= c\ \mathrm{grad}|\vec{M}(r)|\times\vec{u}_M\\ &= -c\ M'(r)\ \sin\theta\,\vec{u}_\phi\end{aligned}$$

if the polar axis of the spherical coordinate system is chosen along $\vec{u}_M = \dfrac{\vec{M}(r)}{|\vec{M}(r)|}$. The grad $|\vec{M}(r)|$ is along the radial direction r, for the spherically symmetric M(r) and, $M'(r) = \dfrac{\partial|\vec{M}(r)|}{\partial r}$ (Fig. 10).

Consider the effect of the current contained in the spherical shell between r and r + dr. The magnetic moment associated with the shell can be written as (131):

$$\begin{aligned}d\vec{\mu}_M &= \frac{1}{2c}\int_\Omega \vec{r}\times\vec{j}(\vec{r})\,r^2\,dr\,d\Omega\\ &= \frac{4\pi}{3}\,r^3[-M'(r)\,dr]\,\vec{u}_M\ ,\end{aligned}$$

where $d\Omega$ is the solid angle $\sin\theta\, d\theta\, d\phi$. The current flowing through in this shell forms a surface current which is equivalent to the

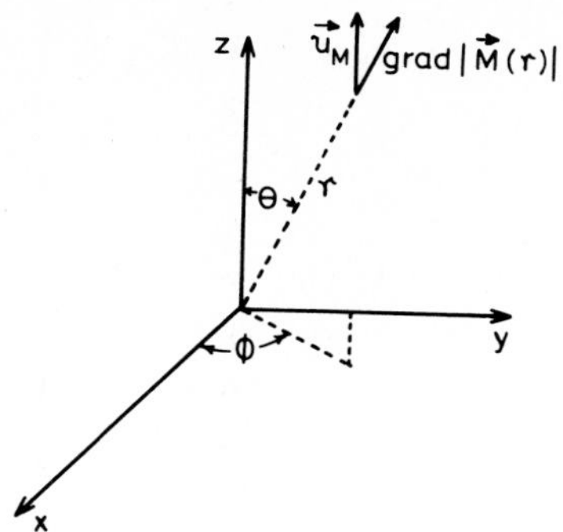

Fig. 10. Schematic representation of the magnetisation.

current which would be produced if the shell was uniformly charged and rotating about an axis parallel to the magnetisation of the shell. The magnetic field produced by this surface current at the centre of the shell is given by (33,130)

$$d\vec{B}(0) = \frac{8\pi}{3V}\, d\vec{\mu}_M,$$

where $V = \frac{4\pi}{3} r^3$ is the volume enclosed within the shell. Using these equations

$$d\vec{B}(0) = \frac{8\pi}{3}\, [-|M'(r)|\, dr]\vec{u}_M \quad .$$

Integrating over all the spherical shells, we get the magnetic field produced at the nuclear site as

$$\vec{B}(0) = \frac{8\pi}{3}\, \vec{\mu}_M \int_0^{\infty} - |M'(r)|\, dr = \frac{8\pi}{3}\, \vec{u}_M \; [\; M(0) - M(\infty) \;]$$

$$= \frac{8\pi}{3}\, \vec{u}_M\, M(0) = \frac{8\pi}{3}\, \vec{M}(0),$$

where we have used the fact that the electron magnetic moment density must vanish at the infinity, $M(\infty) = 0$. For an s-electron (33),

$$\vec{M}(0) = g\beta\vec{s}\,|\Psi(0)|^2,$$

where $\beta=-e\hbar/2mc$ is the Bohr Magneton and $|\Psi(0)|^2$ is the electron density at the nucleus. Thus,

$$\vec{B}(0) = \frac{8\pi}{3}\, g\beta\vec{s}\,|\Psi(0)|^2. \tag{54}$$

The interaction Hamiltonian for the Fermi contact interaction can be written as (Eqn. 52)

$$H = -\frac{8\pi}{3} g_n g\beta_n \beta \vec{I}.\vec{s}|\Psi(o)|^2$$

$$= A_s \vec{I}.\vec{s} \quad . \tag{55}$$

A_s is known as the magnetic hyperfine interaction constant and its value depends on the properties of the nucleus and the s electron density at the nucleus.

So far we have considered only the effect of a single s-electron. Considering the effect of all the s electrons, with spin up and spin down, the net field is given by:

$$B(o) = -\frac{8\pi}{3} g\beta s \Sigma_n \{|\Psi_{ns\uparrow}(o)|^2 - |\Psi_{ns\downarrow}(o)|^2\}. \tag{56}$$

The Fermi contact field is thus proportional to the net spin polarisation at the nucleus. By convention the negative sign has been used and has been verified experimentally (42,132).

In general the electrons involved in the Fermi contact interaction are many and can be classified as: (i) core s electrons, (ii) electrons of neighbouring ions mixed into valence shells of the parent atom, and (iii) conduction electrons which have s character close to the nucleus.

(i) Core polarisation

The effect of a net unpaired spin in the valence shell of an ion, or atom, on the core electrons is to introduce spin polarisation and thus induce a difference in the spin up and spin down electron densities at the nucleus. This could be understood in the following way.

The Slater determinant for an N electron atom or ion is

$$\Psi = \frac{1}{\sqrt{N}} \begin{vmatrix} \phi_1(\vec{x}_1) & \phi_2(\vec{x}_1) & -------- & \phi_N(\vec{x}_1) \\ \phi_1(\vec{x}_2) & \phi_2(\vec{x}_2) & -------- & \phi_N(\vec{x}_2) \\ \cdot & & & \\ \cdot & & & \\ \cdot & & & \\ \cdot & & & \\ \phi_1(\vec{x}_N) & \phi_2(\vec{x}_N) & -------- & \phi_N(\vec{x}_N) \end{vmatrix} ,$$

where $\vec{x}_i$ denotes the space and spin coordinates of the ith electron and the functions ϕ_i are the orthonormalised one-electron spin orbitals. ϕ_i is the product of spin part and the spatial part of the wave function

$$\phi_i(\vec{x}) = \psi_i(\vec{r})\chi_i(\vec{\sigma}),$$

where (38, 41),

$$\psi(\vec{r}) = R_{n\ell}(r)Y_{\ell m}(\theta,\phi)$$

and χ_i represents the spin part. The energy neglecting spin orbit and spin spin interactions can be written as (42)

$$E = \sum_{i=1}^{N} \int \phi_i^*(\vec{x}_1)\ k_{op}\ \phi_i(\vec{x}_1)d\tau_1 + \sum_{i=1}^{N} \sum_{j<i} \int\int |\phi_i(x_1)|^2 \frac{e^2}{r_{12}} |\phi_j(\vec{x}_2)|^2 d\tau_1 d\tau_2 - \sum_{i=1}^{N} \sum_{j<i} \int\int \phi_i^*(\vec{x}_1)\phi_j^*(\vec{x}_2) \frac{e^2}{r_{12}} \phi_i(\vec{x}_2)\phi_j(\vec{x}_1)d\tau_1 d\tau_2, \tag{57}$$

where $r_{12} = |\vec{r}_1-\vec{r}_2|$ is the distance between the electrons 1 and 2, k_{op} is the kinetic energy plus potential energy due to nuclear attraction, and the integrations are over the space and spin coordinates. The last term is called the exchange term. Owing to spin orthogonality, this term is nonzero only for pairs of electrons in identical spin state, $m_{s_i} = m_{s_j}$. Introducing the permutation operator P_{12}, which interchanges the coordinates of the two electrons, the energy E can be rewritten as,

$$E = \Sigma_i \int \phi_i^*(\vec{x}_1)\ k_{op}\ \phi_i(\vec{x}_1)d\tau_1 + \frac{1}{2} \Sigma_i \Sigma_j \int\int \phi_i^*(\vec{x}_1)\phi_j^*(\vec{x}_2) \frac{e^2\{1-\delta(m_{s_i},m_{s_j})P_{12}\}}{r_{12}} \phi_i(\vec{x}_1)\phi_j(\vec{x}_2)d\tau_1 d\tau_2.$$

The effective Hamiltonian of the system is given by

$$H_{eff} = k_{op} + \sum_{i\neq j} \int \phi_j(x_2) \frac{\{1-\delta(m_{s_i},m_{s_j})P_{12}\}}{r_{12}} \phi_j(\vec{x}_2)d\tau_2 \tag{58}$$

and the Hartree-Fock equation for ϕ_i is of the form

$$H_{eff}\ \phi_i(\vec{x}_1) = \varepsilon_i \phi_i(\vec{x}_1),$$

where ε_i is the one-electron energy. Eqn. (58) shows that whereas the electrons with antiparallel spins produce only Coulomb repulsion [$\delta(m_{s_i}, m_{s_j}) = 0$], the electrons with parallel spins produce both the Coulomb repulsion as well as the exchange attraction (attraction due to the negative sign of the exchange term). As a result, two electrons with parallel spins lie closer to each other than the two electrons with antiparallel spins. In a system where all the electrons, including the valence electrons, are spin paired, there is equal attraction for electrons in either spin state and thus the core electrons remain unpolarised. However if the number of electrons in the two spin states in the valence shell are unequal, it is easily seen that the core electrons with spin parallel to the net spin of the valence shell electrons experience less repulsion than the core electrons with spin antiparallel to the net spin of the valence shell electrons. The result is to make the radial parts of the wave functions of the core electrons with spin up and spin down in any shell different from each other, one of them being pushed more towards the nucleus than the other. Thus there is a net spin density of the core electrons at the nucleus leading to the Fermi contact field.

Using unrestricted Hartree-Fock technique, Watson and Freeman (41,42,133) have calculated the core polarisation contact field (hereafter denoted hy H_c) for several iron group and rare earth ions and neutral atoms. For the case of Mn^{2+} ($3d^5$), H_c is given by

$$\begin{aligned} H_c &= (H_{c\uparrow} + H_{c\downarrow})_{1s} + (H_{c\uparrow} + H_{c\downarrow})_{2s} + (H_{c\uparrow} + H_{c\downarrow})_{3s} \\ &= \{(2,502,840-2,502,870) + (2,26,670-2,28,080) + (31,210-30,470)\}\text{kOe} \\ &= -30-1410+740 \\ &= -700 \text{ kOe}. \end{aligned}$$

Here, the arrow ↑ and ↓ denote electrons with spin parallel and antiparallel to the 3d shell spin, respectively. Thus, it is seen that the contact field produced by a single s electron is very large and decreases for outer shells. However when both the 1s and 2s shells are filled, as we see here, the net effect of core polarisation is more pronounced for the 2s shell than for the 1s shell. This is because the 2s electrons have comparatively greater overlap with the 3d electrons. Though the 3s electrons overlap even more, as shown in Fig. 4, they lie partly outside the 3d shell and therefore are not so effective in producing the contact field. The sign of the fields produced by the 1s and 2s shells is negative since in their case

$|\Psi_{\downarrow}(0)|^2_{1s,2s} > |\Psi_{\uparrow}(0)|^2_{1s,2s}$. It can now be easily realised that if there is a filled 4s shell, as for instance in iron atom ($3d^6 4s^2$), a large part of it lies outside the 3d shell. The 4s electrons will then be polarised by the 3d electrons in such a way that $|\Psi_{\uparrow}(0)|^2_{4s} > |\Psi_{\downarrow}(0)|^2_{4s}$ and thus contribute a positive contact field at the nucleus. Since 3s electrons produce a positive field at the nucleus in Mn^{2+}, on the average the 3s shell behaves as if it lies "outside" the 3d shell as far as H_c is concerned.

The core polarisation field in the case of Fe^{3+} ($3d^5$) is

$$H_c = (H_c)_{1s} + (H_c)_{2s} + (H_c)_{3s}$$
$$= -50-1790+1215$$
$$= -630 \text{ kOe}$$

and for Fe^{2+} ($3d^6$)

$$H_c = (H_c)_{1s} + (H_c)_{2s} + (H_c)_{3s}$$
$$= -30-1310+790$$
$$= -550 \text{ kOe.}$$

For a neutral iron atom

$$H_c = (H_c)_{1s} + (H_c)_{2s} + (H_c)_{3s} + (H_c)_{4s}$$
$$= -16-1300+724+492$$
$$= -100 \text{ kOe.}$$

$(H_c)_{4s}$ is smaller than $(H_c)_{3s}$ owing to relatively smaller overlap between 3d and 4s wave functions which results in a smaller polarisation of the 4s shell.

Since 3d spins directly determine the core polarisation field, the interaction Hamiltonian of Eqn. (55) can be written as (134):

$$\mathcal{H} = A_c \vec{I}.\vec{S}, \tag{59}$$

where $\vec{S}$ is the spin of the valence shell. In presence of a magnetic field, the ionic spin and the nuclear spin precess independently around the field direction, say the z-axis. In this case

$$\mathcal{H} \approx A_c I_z S_z, \quad \text{if magnetic field is large.} \tag{60}$$

Only $|-S\rangle$ state is populated at very low temperatures. Thus the maximum value of the contact field is

$$(H_c)_{max} = -A'S, \tag{61}$$

where $A' = A_c/g_n\beta_n$.

In Fe^{3+} compounds, the saturation hyperfine field has been experimentally found to lie between 500 and 600 kOe. However for Fe^{2+} compounds, the saturation field has a much wider range: 220 kOe (Fe^{2+} in CoO) to 485 kOe (Fe^{2+} in Fe_3O_4). The variation in Fe^{3+} compounds is due to transferred and super-transferred hyperfine fields to be discussed later. In Fe^{2+} compounds the orbital moment of the valence shell also contributes and is responsible for the wider range of values obtained. After correcting for transferred field, orbital contribution, etc., $-A'$ has been experimentally found to be close to -190 kOe, -220 kOe and -260 kOe for the $3d^3$, $3d^5$ and $3d^8$ configurations respectively (41,135). These values agree well with the theoretical predictions (42,133).

(ii) Transferred and supertransferred hyperfine fields

In compounds of paramagnetic ions, an appreciable interaction between the ligands and the paramagnetic ion results in magnetic hyperfine field at the ligand nucleus due to electrons of the paramagnetic ion and vice versa. This is called transferred hyperfine field (41,42) and can be understood using molecular orbital theory.

Consider a system consisting of a cation and an anion. In the linear combination of atomic orbitals method (LCAO), the molecular orbital wave function, Ψ_{Mo}, is approximated as a linear combination of atomic orbitals on the ligand (ψ_1) and the paramagnetic ion (ψ_2). The ligand orbital are generally lower in energy than the cation valence orbitals. Thus (136)

$$\Psi^B_{Mo} = N_1(\psi_1+\gamma\psi_2)$$

$$\Psi^{AB}_{Mo} = N_2(\psi_2-\lambda\psi_1)$$

where N_1, N_2 are the normalisation constants and γ and λ are given by

$$\lambda = \frac{H_{12}-H_{22}S_{12}}{H_{11}-H_{22}}, \qquad \gamma = \frac{H_{12}-H_{11}S_{12}}{H_{11}-H_{22}}, \qquad \text{and} \quad \lambda = \gamma+S_{12} \tag{62}$$

Here H is the total Hamiltonian, $H_{12}=\langle\psi_2|H|\psi_1\rangle$, and $S_{12}=\langle\psi_2|\psi_1\rangle$

Ψ^{B}_{Mo} is called the bonding orbital and the state characterised by it has a lower energy than the state characterised by the antibonding orbital Ψ^{AB}_{Mo}. In a case where the ligands are $2p^6$ ions and the metal ion is from the 3d group of elements, the molecular orbitals are formed from the partially filled 3d orbitals and the completely filled 2p orbitals. Fig. (11) shows schematically the overlap of the cation and anion which lead to σ and π bonds. A molecular orbital can be defined as a σ orbital if there is no plane passing through the internuclear

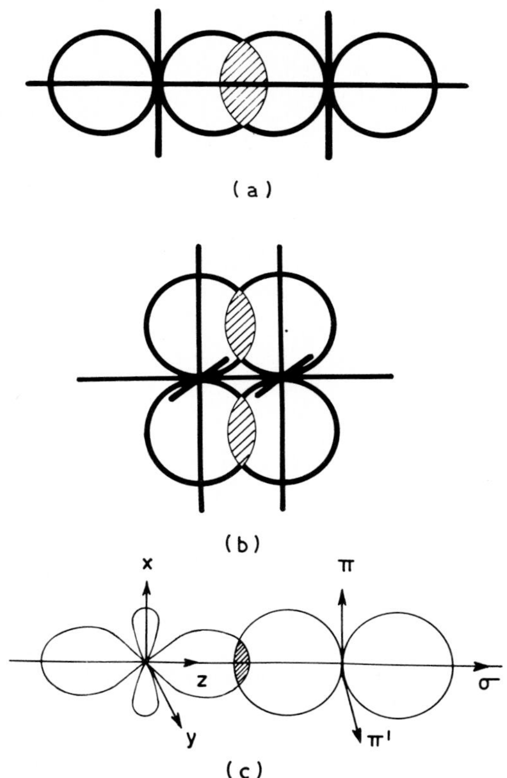

Fig. 11. Schematic diagram showing σ and π bonds. (a) σ bond between two p-orbitals (b) π bond between two p-orbitals (c) σ bond between $d_{3z^2-r^2}$ and p_z orbitals, z-axis is along the internuclear axis.

axis along which the electronic charge density is zero as in Fig. 11(a). Similarly if the internuclear axis has a nodal plane passing through it, it is called π bond (Fig. 11b). Fig. 11(c) illustrates the case of σ bonding, between $d_{3z^2-r^2}$ and p_z orbitals. The bonding molecular orbitals can be written as

$$\Psi^{B}_{Mo} = N(\psi_p+\gamma\psi_d) \quad , \tag{63}$$

where ψ_p and ψ_d are the anion p-orbital and cation d-orbital, respectively. γ gives a small admixture of the d orbital with the p orbital. The normalisation constant N is given by

$$N = [\ 1+2\gamma< \psi_p|\psi_d > + \gamma^2\ ]^{-1/2} \quad .$$

The antibonding orbital is given by,

$$\Psi^{AB}_{Mo} = N'(\psi_d-\lambda\psi_p) \quad , \tag{64}$$

where N' is given by

$$N' = [\ 1-2\lambda< \psi_p|\psi_d > + \lambda^2\ ]^{-1/2} \quad .$$

The condition of orthogonality of Ψ^{B}_{Mo} and Ψ^{AB}_{Mo} gives

$$\gamma = \frac{\lambda-<\psi_p|\psi_d>}{1-\lambda<\psi_p|\psi_d>} \quad .$$

Denoting $< \psi_p|\psi_d >$ by S_{pd} we get

$$\lambda = \frac{\gamma+S_{pd}}{1+\gamma S_{pd}} \quad .$$

This is similar to Eqn. (62) if the small term γS_{pd} is neglected. This can be rewritten as

$$\lambda = S_{pd} + \frac{\gamma(1-S_{pd}^2)}{1+\gamma S_{pd}}$$

$$= S_{pd} + \gamma' \quad ,$$

where γ' is known as the covalent mixing parameters. Thus if $S_{pd}=0$, $\gamma'=\gamma=\lambda$. These relations clarify the physical meaning of the admixture coefficients λ and γ (137). It is also seen from Eqns. (63) and (64) that if λ and γ are small, the bonding orbital consists of mostly a ligand wave function having a small admixture of the d orbital and the antibonding orbital is mainly a d orbital. As bonding orbitals are generally filled, they do not affect the magnetic properties of the system. Thus, the magnetic properties are due to the unfilled antibonding orbitals (138). Denoting the mutually perpendicular axis as x, y, z at the metal ion and as π, π', σ at the ligand ion as in Fig. 11(c), it can be seen that the only significant overlap integrals between the orbitals of the two ions are:

$$< 3z^2-r^2 | 1s >, < 3z^2-r^2 | 2s >, < 3z^2-r^2 | p_\sigma >, < xz | p_\pi > \text{ and } < yz | p_{\pi'} >.$$

As a result of these overlaps, the electrons of the metal ion produce transferred hyperfine field (THF) at the ligand nucleus and the electrons of the ligand ion produce THF at the nucleus of the metal ion. This is known as the overlap contribution to THF. Physically, the origin of THF can be understood as follows.

At the metal ion nucleus the THF arises owing to the following reasons:

(1) The change in the occupation of the two spin states of the shell changes the value of the spin $\vec{S}$ of the 3d shell. This affects the magnitude of the Fermi contact field H_c at the metal ion-nucleus (Eqns. 56 and 61).

(2) The ligand 1s, 2s and 2p electrons can get partially transferred to the empty 4s orbital of the metal ion and thus produce a contact hyperfine field at the metal ion nucleus (139). This effect is not included in the discussion above.

(3) The orbital moment thus induced in the valence shell directly contributes to the hyperfine field.

At the nucleus of the ligand ion THF appears owing to the following reasons:

(a) Consider bonding and antibonding orbitals given by Eqns. (63) and (64), and assume $\gamma=0$. Unlike ψ_p and ψ_d, they are orthogonal. Thus the overlap of the 3d electrons of the metal ion with the 1s, 2s and 2p electrons of the ligand ion, which contribute to λ, polarises the ligand ion electrons. This overlap polarisation arises as a result of the Pauli exclusion principle. Because of this, the probability of finding a spin up ligand electron in the overlap region gets decreased since the d electron also has its spin up (Eqns. 63 and 64). This increases the probability of finding a spin up ligand electron near the ligand nucleus. However the Pauli principle does not affect the probability distribution of a spin down ligand electron. This causes a net increase in the density of the spin up 1s and 2s electrons at the ligand nucleus directly resulting in a hyperfine field due to the Fermi contact interaction (137,140). The effect of the similar p-electron spin polarisation is also felt at the ligand nucleus through core polarisation of 1s and 2s shell.

(b) When $\gamma \neq 0$, the covalent mixing of the 3d electrons of the metal ion with the 1s, 2s and 2p electrons of the ligand (141,142) also causes an additional spin-unpairing of the ligand s and p electrons. The total amount of spin unpairing due to a spin up electron in a d orbital is given by the square of the sum of overlap integral and the covalency parameter. Whereas the spin-unpaired s electrons produce THF at the ligand nucleus by the contact interaction, the spin unpaired p electrons produce THF by the dipolar interaction as well as through s electron polarisation. For 1s, and 2s orbitals of the anions γ is small, and the spin unpairing is mainly due to S_{sd}.

(c) The above mentioned covalent mixing of the 3d electrons of the metal ion with the ligand orbitals contributes to THF at the ligand nucleus through the dipolar interaction of the 3d electrons with the ligand-nucleus. It may be mentioned here that orbital contribution to THF is negligibly small in the example chosen here (141,143).

(d) The d electron admixture in the p shell of the ligand also produces THF through core polarisation and enhances the contribution of 2p polarisation to THF described in (a) above.

In order to get an idea of the relative magnitudes of the various parameters involved in the THF, we now construct the molecular orbitals for the two atom cation-ligand $Fe^{3+}(3d^5)$-$O^{2-}(2p^6)$, or Fe^{3+} -F^{1-}, system (Fig. 11c). Neglecting the possible partial transfer of the ligand electrons to the empty shells of the metal ion (144), the antibonding molecular orbitals are

$$\Psi^{AB}_{3z^2-r^2} = N_\sigma [\ |3z^2-r^2\rangle - \lambda_{1s}|1s\rangle - \lambda_{2s}|2s\rangle - \lambda_\sigma|p_\sigma\rangle\ ],$$

$$\Psi^{AB}_{xz} = N_\pi\ [\ |xz\rangle - \lambda_\pi|p_\pi\rangle] \quad , \tag{65}$$

$$\Psi^{AB}_{yz} = N_{\pi'}[\ |yz\rangle - \lambda_{\pi'}|p_{\pi'}\rangle] \quad ,$$

where

$$N_\sigma = [\ 1-2\lambda_{1s}S_{1s} - 2\lambda_{2s}S_{2s} - 2\lambda_\sigma S_\sigma + \lambda^2_{1s} + \lambda^2_{2s} + \lambda^2_\sigma\ ]^{-1/2} \quad ,$$

$$N_\pi = [\ 1-2\lambda_\pi S_\pi + \lambda^2_\pi]^{-1/2} \quad ,$$

$$N_{\pi'} = [\ 1-2\lambda_{\pi'}S_{\pi'} + \lambda^2_{\pi'}\ ]^{-1/2} \quad .$$

The orbitals $\Psi^{AB}_{x^2-y^2}$ and Ψ^{AB}_{xy} do not involve in bonding and thus do not contribute to THF. Since the magnitude of the overlap integrals $S_{1s}(=\langle 3z^2-r^2|1s\rangle)$, S_{2s}, S_σ, etc. <<1 and also the covalent mixing parameters $\gamma' << 1$ we have:

$N_\sigma \approx N_\pi \approx N_{\pi'} \approx 1$.

In the case of Fe^{3+} or Mn^{2+} ion, only spin up states are occupied by the electrons and involve only spin up ligand electrons in the bonding. Thus, in the molecular orbital approach, the degree of spin unpairing in the ligand orbital is given by λ^2 (139,145). Using Eqns. (54) and (65), the Fermi contact field, produced by the s electrons, at the ligand nucleus is given by (139,145):

$$H_C = \frac{8\pi}{3}\beta N_\sigma^2 [\lambda_{2s}^2|\chi_{2s}(0)|^2 + \lambda_{1s}^2|\chi_{1s}(0)|^2 + 2\lambda_{1s}\lambda_{2s}\chi_{1s}(0)\chi_{2s}(0)], \tag{66}$$

where we have used g=2, s=1/2. $\chi_s(0)$ is the amplitude of the ligand electron wave function at the nucleus of the ligand ion. Here we have neglected the smaller terms containing $|\psi_{3z^2-r^2}(r_L)|^2$, $\psi_{3z^2-r^2}(r_L) \times \chi_{1s}(0)$ and $\psi_{3z^2-r^2}(r_L)\chi_{2s}(0)$, where $\psi_{3z^2-r^2}(r_L)$ is the metal ion wave function at the ligand nucleus which is situated at a distance r_L from the metal-ion-nucleus (143,145). Thus r_L and 0 refer to the same point but metal and ligand ion positions as centres, respectively. The dipolar field produced at the ligand nucleus by the induced magnetic moment in the p_σ ligand orbital is given by (137,143):

$$(H_D)_{p_\sigma} = \frac{\mu_{p_\sigma}}{\langle r^3\rangle_{2p}} \langle 3\cos^2\theta_{r\sigma} - 1\rangle,$$

where $\mu_{p_\sigma} = g\beta s N_\sigma^2\lambda_\sigma^2$ is the induced magnetic moment in the $2p_\sigma$ orbital due to the spin unpairing. In the expressions of H_D, r denotes the electron distance from the ligand nucleus and $\theta_{r\sigma}$ is the angle between $\vec{r}$ and the σ axis. Similarly,

$$(H_D)_{p_\pi} = \beta N_\pi^2\lambda_\pi^2 \langle r^{-3}\rangle_{2p} \langle 3\cos^2\theta_{r\pi} - 1\rangle$$

and

$$(H_D)_{p_{\pi'}} = \beta N_{\pi'}^2\lambda_{\pi'}^2 \langle r^{-3}\rangle_{2p} \langle 3\cos^2\theta_{r\pi'} - 1\rangle .$$

Finally, the dipolar field produced at the ligand nucleus by the overlapped spin density of the 3d electrons of the metal ion is given by

$$(H_D)_{3d} = \beta(2+N_\sigma^2+N_\pi^2+N_{\pi'}^2) \langle R^{-3}\rangle_{3d} \langle 3\cos^2\theta_{r\sigma} - 1\rangle,$$

where R is the distance of the 3d electrons as measured from the ligand nucleus. The transferred hyperfine field at the ligand nucleus is thus given by,

$$H_{THF} = H_c + (H_D)_{p\sigma} + (H_D)_{p\pi} + (H_D)_{p\pi'} + (H_D)_{3d} \quad .$$

Here we have considered the overlap of only one ligand ion with the metal ions. In actual practice the effect of all the neighbouring metal ions is to be considered. The same is the case when transferred hyperfine field is to be calculated at the metal ion nucleus (137, 142,143). To a good approximation, the field produced at the nucleus of the central ion is equal to the sum of the fields produced by each of the neighbouring ion. Using the procedure described above transferred hyperfine fields have been calculated at the nuclei of many anions and cations, and the agreement between the calculated and the measured values has generally been found to be satisfactory. For example, a THF of about 60 kOe has been calculated to be present at the nucleus of iodine ion, $I^-(5s^2 5p^6)$, produced by the surrounding Fe^{3+} ions, in a few iron-iodine organic complexes. This value has been found to be in agreement with the measured values using ^{129}I Mossbauer spectroscopy (146). Similarly, effects on the asymmetric charge distribution and charge density at the nuclear site due to the neighbouring ions have been theoretically estimated and have been experimentally observed (146).

The transferred hyperfine field appears at the nucleus of the diamagnetic anion due to the neighbouring cation, and vice versa, as described above. However, when two cations interact through an anion (M_1-L-M_2), the supertransferred hyperfine field appears at the nucleus of a cation M_1 due to other cation M_2 interacting through an anion L and vice versa. This is known as supertransferred hyperfine field (STHF). The magnitude of this field is very large in some cases; for example STHF $\approx$ 300 kOe at ^{121}Sb nuclei in Sb-substituted ferrites (147). However, in most of the iron compounds, STHF is much smaller (92,144,148,149). As has been discussed above, the overlap of L and M_1 ions causes spin unpairing in the orbitals of the anion. These polarised anion orbitals in turn cause core polarisation of the M_2 ion, since they overlap the M_2 ion orbitals also. Thus the electrons of the M_1 ion produce a contact hyperfine field at the site of another cation, namely the M_2 ion. Apart from the contact field, various other smaller contributions discussed above, like the dipolar field produced by the polarised non-s orbitals will also be present at the M_2 nucleus. In addition, STHF is also created at the M_2-nucleus due to a partial transfer of an electron, say a 3d electrons, of M_1 to the empty s shells (4s shell) of M_2 via an indirect process in which an electron is partially transferred from L to M_2 and another one is partially transferred from M_1 to L (139,150). This transferred electron to the s-shell of M_2 ion produces a Fermi contact field at its nucleus.

The magnitude and sign of STHF can be easily calculated using the molecular orbital approach described above (139). For simplicity, let us consider the M_1 and M_2 ions to be the $3d^5$ ions, like Mn^{2+}, Fe^{3+}, etc., and L to be a $2p^6$ ion like O^{2-}, F^-, etc., which is centrally located between M_1 and M_2. We consider only the σ bond between the $d_{3z^2-r^2}$ and $2p_z$ orbitals. The z-axis is along the inter-nuclear axis. The antibonding orbital at M_1 is not only orthogonalised to the bonding orbital predominantly at the ligand ion, but also to the core s orbitals of the M_2 ions which are also orthogonal to the bonding orbitals. The molecular orbital for core s electrons of M_2 ion can be written as

$$\Psi_{ns}(M_2) = N_{ns}\{\psi_{ns}(M_2) + S_{p,ns}[\psi_{pz}(L) + \gamma_\sigma\psi_{3z^2-r^2}(M_1)]$$
$$-S_{s,ns}[\psi_s(L) + \gamma_s\psi_{3z^2-r^2}(M_1)]\} \quad ,$$

where

$$S_{p,ns} = -\langle\psi_{p_z}(L)|\psi_{ns}(M_2)\rangle$$

$$S_{s,ns} = \langle\psi_s(L)|\psi_{ns}(M_2)\rangle \quad .$$

γ_σ and γ_s are the admixture coefficients <<1. The overlap of $\psi_{3z^2-r^2}$ on M_1 with core s orbital on M_2 ion is neglected. The terms in the [ ] brackets represent bonding orbitals predominantly formed from the ligand p_z and s orbitals, respectively. Thus, the antibonding molecular orbital at M_1 can be written as

$$\Psi_{3z^2-r^2}(M_1) = N_{3z^2}[\psi_{3z^2-r^2}(M_1) - \lambda_\sigma\psi_{pz}(L) - \lambda_s\psi_s(L)$$
$$-\sum_{n=1}^{3}\mu_{ns}\psi_{ns}(M_2) + a'\psi_{4s}(M_2)\,] \quad . \qquad (67)$$

Here, a' is the 3d → 4s charge transfer parameter from the d orbital of ion M_1 to 4s of M_2, $a' \approx 2\%$ (144,151). μ_{ns} is determined from the requirement that $\Psi_{3z^2-r^2}(M_1)$ and $\Psi_{ns}(M_2)$ are orthogonal and is given by:

$\mu_{ns} \approx \lambda_\sigma S_{p,ns} - \lambda_s S_{s,ns}$, (n=1,2,3).

λ_σ and λ_s are the cation-ligand covalent mixing parameters. Thus the STHF at the M_2 nucleus is given by (42,139):

$$H_{STHF} = \frac{8\pi}{3} g\beta S\{|\Psi_{3z^2-r^2}(M_1)|^2 \text{ at } M_2 \text{ nucleus } \}. \quad (68)$$

S is the spin of the $\Psi_{3z^2-r^2}(M_1)$ orbital. For example, $S = \frac{1}{2}$ if the ion is Fe^{3+} or Mn^{2+}. Also, since 2s and 2p orbitals of the ligand and the $\psi_{3z^2-r^2}(M_1)$ orbital of the M_1 ion have much smaller amplitudes at the nucleus of M_2 ion as compared to the amplitudes of $\psi_{ns}(M_2)$ orbitals of the M_2 ion, Eqn. (68) can be written as,

$$H_{STHF} \approx \frac{8\pi}{3} g\beta S[-\sum_{n=1}^{3} \mu_{ns}\psi_{ns}(0) + a'\psi_{4s}(0)]^2$$

$$= \sum_{n,m} H_{ns,ms} , \quad (69)$$

where the diagonal terms are given by

$$H_{ns,ns} = \frac{8\pi}{3} g\beta S \mu_{ns}^2 |\psi_{ns}(0)|^2$$

and

$$H_{4s,4s} = \frac{8\pi}{3} g\beta S a'^2 |\psi_{4s}(0)|^2 .$$

The cross terms are given by

$$H_{ns,ms} = \frac{16\pi}{3} g\beta S \mu_{ns} \mu_{ms} \psi_{ns}(0) \psi_{ms}(0)$$

and

$$H_{ns,4s} = -\frac{16\pi}{3} g\beta S a' \mu_{ns} \psi_{ns}(0) \psi_{4s}(0) .$$

Here, 0 in the bracket refers to the position of the M_2 nucleus. Using Hartree-Fock wavefunctions for the atomic orbitals and summing the contributions from all the nearest neighbour cations the value of STHF has been obtained in many cases. In case of $KMnF_3$, which is cubic, the following value of the STHF has been obtained for a central $3d^5$ ion which is octahedrally surrounded by $2p^6$ ligand ions and $3d^5$ nearest neighbour cations (139,151):

$$H_{STHF} = H_{1s,1s} + H_{2s,2s} + H_{3s,3s} + H_{4s,4s} + H_{1s,2s} + H_{1s,3s} + H_{1s,4s} + H_{2s,3s} + H_{2s,4s} + H_{3s,4s}$$

$$= (0.4 + 2.3 + 7.7 + 1.9 - 2.0 + 3.7 + 2.1 - 8.5 - 4.2 + 7.7)\ \text{kOe}$$
$$= 11.1\ \text{kOe} \quad .$$

This gives an idea about the relative contributions of the various terms in Eqn. (69) to H_{STHF}. As can be expected (150), the predominant contribution is from the 3s electrons. It may be noted that the H_{STHF} is parallel to the spin of the M_1 ion.

When the surrounding M_1 ions are diamagnetic, the STHF at the M_2 site automatically vanishes. However as long as M_2 is a paramagnetic ion, it will produce STHF at the nuclei of the diamagnetic M_1 ions. This has been experimentally verified and STHF has been observed at the site of the diamagnetic cations, like Al^{3+} (152), Ga^{3+} (153), Sn^{4+} (147,154,155), Sb^{5+} (147), Cd^{2+} (156), etc., produced by the neighbouring paramagnetic cations, like Fe^{3+}, Cr^{3+}, Ni^{2+}, etc. As in the case of the transferred hyperfine interaction, the supertransferred hyperfine interaction also affects the charge density and the electric field gradient existing at the nuclei of the M_1 and M_2 (144). This effect has also been experimentally seen in the Mossbauer spectra (92,148). For further details and the application of these theories to oxides with spinel structure the reader is referred to the article appearing elsewhere in this book.

(iii) Conduction electron contribution

An important contribution to the hyperfine field in metallic systems is due to the conduction electrons. The isomer shift can be used to assign an electronic configuration to the parent atom of the Mossbauer nucleus in the metallic system assuming the total number of electrons on the atom in the metallic system and the free atom are equal. In iron metal the electronic configuration of the iron atom is thus found to be $3d^7\ 4s^1$ (157). The 4s electrons behave as free electrons in the region away from the nucleus but as 4s orbital electrons when they are close to the nucleus. Thus, for the purpose of hyperfine interaction it is reasonable to think of an atomic configuration which includes 4s electrons too. The conduction electrons partly possess 4p character also (157). In addition, the d orbitals also broaden into bands and partly possess itinerant character (157,158). It has been found that in the case of iron metal, about 5%

of the 3d electrons are itinerant and 95% are in d bands which are sufficiently narrow that they can be considered localised (158). We call the itinerant d electrons as the d band conduction electrons.

The localised magnetic moment on iron atom in the Fe metal is 2.2β per iron atom (157). This contributes to the contact hyperfine field through the polarisation of the core electrons (H_{CP}) and conduction electrons (H_{CE}). The band electrons contribute to the magnetic hyperfine field (H_{CE}) through the following mechanisms:

(a) The localised moment on the transition metal atom, mainly due to the localised 3d electrons, polarises the 4s like electrons of the s-conduction band (158), due to exchange interaction (RKKY spin density oscillations). This in turn produces a hyperfine field (H_{CEP}) at the nucleus through the Fermi contact interaction. Apart from this moment perturbation, the charge perturbation (i.e. Fridel's charge density oscillations) also produces conduction electron spin density oscillations but its effect is an order of magnitude smaller (158).

In addition, the magnetic moment of the neighbouring atoms also cause conduction electron polarisation. Thus

$$H_{CEP} = (H_{CEP})_{self} + (H_{CEP})_{neighbours} \quad . \qquad (70)$$

(b) In many metallic systems all the magnetic moment does not arise purely from d electrons. A part of the d band possess the s character due to the interband mixing (68). Because of this a finite density of the unpaired electrons appears at the nucleus. This term will be denoted by H_{CEPM}.

It may be mentioned here that though the main contribution to H_{CEP} comes from the localised 3d electrons, the itinerant d electrons also contribute to it to some extent (157). Furthermore, interband mixing also causes the s band to possess partly d character and this also leads to RKKY type spin density oscillations (158).

Whereas H_{CEP} is always positive, i.e. the s-band electrons are polarised by the 3d electrons such that the net s-band moment is parallel to the 3d moment, H_{CEPM} is generally negative, i.e. the interband mixing adds an antiparallel component of polarisation to the s-band. Thus H_{CE} can be either positive (i.e. can have its direction parallel to the 3d moment) or negative ($H_{CE} = H_{CEP}+H_{CEPM}$).

Like H_{CE} , H_{CP} also consists of two contributions. The first contribution (H_{CPD}) arises, as discussed before, due to the polarisation of the 1s, 2s, and 3s electrons by the 3d electrons. The second contribution (H_{CPID}) is a result of indirect effects and arises as follows. Apart from polarising the 1s,2s and 3s electrons, the 3d electrons also polarise the 2p,3p and s-band electrons. These polarised electrons in turn affect the polarisation of the 1s,2s and 3s electrons, and thus give rise to a contribution, H_{CPID}. Generally H_{CPID} is 10% of H_{CPD} in metals as well as in insulators (157). Thus,

$$H_{CP}=H_{CPD} + H_{CPID} \quad . \tag{71}$$

The hyperfine field at the nucleus in a transition metal, for example iron metal, can be written as

$$\begin{aligned} H_{hf} &= H_{CP} + H_{CE} \\ &= (H_{CPD} + H_{CPID}) + (H_{CEP} + H_{CEPM}) \quad . \end{aligned} \tag{72}$$

It follows that the hyperfine field at the nucleus of a diamagnetic impurity ion diffused in the magnetically ordered metallic host is $(H_{CEP})_{neighbours} + (H_{CEPM})_{neighbours}$. However, sometimes these solute ions develop a moment in the host lattice (76). Then the other contributions appearing in Eqn. (72) also become significant. In some cases, the valence ns-electrons (say 4s electrons) of the solute impurity atom remain localised and do not go over to the conduction band. In such a case if the volume of the solute atoms is larger than the volume available upon removing a host atom, the ns-like valence electrons overlap with the host matrix and become polarised by an amount proportional to the volume misfit of the solute atom. These polarised electrons produce a hyperfine field at the solute nucleus by the Fermi contact interaction. In some cases, this contribution to H_{hf} has been found to be quite significant (158).

As a typical example, we briefly outline the theoretical evaluation of the hyperfine field in Fe metal. In iron metal, the wave function for the band electrons can be written as (157,159,160):

$$\Psi = \sum_{m=1}^{5} \lambda_m \, u_d^m(\vec{r}) + \sum_i \mu_i u_{OPW}^i \quad , \tag{73}$$

where λ_m and μ_i are the coefficients to be obtained variationally and the suffixes m and i represent the mth d orbital and the ith reciprocal lattice vector, respectively. u_d is the tight binding wave function formed from the atomic d orbitals as follows:

$$u_d^m(\vec{r}) = \sum_n e^{i\vec{k}.\vec{R}_n} \phi_d^m(\vec{r}-\vec{R}_n) \quad ,$$

where,

$$\phi_d^m(r) = \frac{P_d(r)}{r} C_2^m(\theta,\phi)$$

and

$$C_2^1 = Y_2^0, \; C_2^2 = \frac{1}{\sqrt{2}}(Y_2^2 + Y_2^{-2}), \; C_2^3 = -\frac{i}{\sqrt{2}}(Y_2^2 - Y_2^{-2}),$$

$$C_2^4 = \frac{i}{\sqrt{2}}(Y_2^1 + Y_2^{-1}), \; C_2^5 = -\frac{1}{\sqrt{2}}(Y_2^1 - Y_2^{-1}) \quad . \tag{74}$$

u_{OPW} is the OPW wave function given as:

$$u_{OPW}^{\vec{\kappa}} = e^{i(\vec{k}+\vec{\kappa}).\vec{r}} - \sum_j b_{\vec{k}+\vec{\kappa}}^j \phi_c^j \quad ,$$

where ϕ_c^j is a tight binding core-state wave function and b^j is a constant chosen to make u_{OPW} orthogonal to the core state. The wave function of Eqn. (73) represents the conduction electron wave function of the iron metal satisfactorily since the hybridisation of d-and s-like states is built into the wave function. This form of the wave function has sufficient flexibility to describe diffuse s-band states (λ=0), compact d- states (μ=0), or any hybrid of these two ($\lambda,\mu\neq 0$). Using this wave function and unrestricted Hartree-Fock formalism, H_{hf} has been calculated for iron metal after taking into account the various many body electron-electron correlation effects (145) such as the screening of the exchange interaction for the quasi-free 4s electrons etc., (157). The following result has been obtained

$$\begin{aligned}(H_{hf})_{metal} &= H_{CP} + H_{CE} \\ &= [(H_{CPD} + H_{CPID})_{1s} + (H_{CPD} + H_{CPID})_{2s} \\ &\quad + (H_{CPD} + H_{CPID})_{3s}] + H_{CE} \\ &= [(2+10) + (-296-14) + (-137+55)] + 33 \text{ kOe} \\ &= -380 + 33 \\ &= -347 \text{ kOe} \quad .\end{aligned} \tag{75}$$

This calculated value is in good agreement with the experimental saturation value of -339 kOe. Eqn. (75) gives an idea of the relative magnitudes of the various contributions. The contribution of 1s electrons is negligible while the contributions of 2s and 3s electrons are of comparable magnitude and of the same sign. The indirect polarisation has maximum effect for 3s electrons. This indirect effect on the 3s core occurs mainly through the influence of the 3p core which has been polarised by the 3d electrons. For 1s and 2s electrons, the indirect effect occurs primarily through the 2p polarisation. In total, the indirect effect is only about 10% of the direct effect in magnitude. It is also seen that the negative sign of H_{hf} arises because of the predominance of the core polarisation contribution (H_{CP}) which is about ten times larger than the conduction electron contribution. It has been estimated for iron metal that $(H_{CEP})_{neighbour} + (H_{CEPM})_{neighbour} \approx -150$ kOe (158) and thus $(H_{CEP})_{self} + (H_{CEPM})_{self} \approx 180$ kOe.

(b) Orbital moment of the valence electrons

Unquenched orbital angular momentum of the valence shell produces a hyperfine field, H_L, at the nucleus which can be calculated as follows. The orbital motion of an electron is equivalent to an electric current flowing in a loop, with the nucleus at the centre. The magnetic field produced at the centre of the loop is given by (33,131)

$$\vec{H}_L = \frac{1}{c} \int \frac{\vec{j} \times \vec{r}}{r^3} \, d\tau \quad ,$$

where $\vec{j}$ is the electronic current density in the loop and $\vec{r}$ is the radius vector of volume element $d\tau$ of the loop, assuming the origin to be at the centre of the loop. Since $\vec{j} d\tau = \vec{v} dq$ where $\vec{v}$ is the velocity of the charge dq of the volume element $d\tau$ we have

$$\vec{H}_L = -\frac{1}{c} \int \frac{\vec{r} \times \vec{v}}{r^3} \, dq \quad .$$

The quantity $\vec{r} \times \vec{v}$ is related to the orbital angular momentum of the electron by the relation $m\vec{r} \times \vec{v} = \vec{\ell}\hbar$ (m = mass of the electron). Summing over all the electrons i and using the relation $-e\langle r^{-3}\rangle = \int r^{-3}\, dq$, we have

$$\vec{H}_L = -2\beta \sum_i \vec{\ell}_i \langle r^{-3}\rangle_i \quad .$$

Since $\langle r^{-3}\rangle$ is same for all the electrons with the same quantum numbers (n,ℓ), $\vec{H}_L$ vanishes for the closed shells and half filled shell. Thus H_L is nonzero only when the shell is partially filled. In case when the LS-coupling is appropriate we have

$$\vec{H}_L = -2\beta\langle r^{-3}\rangle\vec{L} \quad . \qquad (76)$$

The interaction Hamiltonian can be written as:

$$\begin{aligned} H_L &= -\vec{\mu}_n \cdot \vec{H}_L \\ &= 2\beta g_n \beta_n \langle r^{-3}\rangle \vec{I}.\vec{L} \qquad (77) \\ &= A_L \vec{I}.\vec{L} \quad . \end{aligned}$$

In the presence of a sufficiently large external field, the off-diagonal elements of this Hamiltonian get "quenched". In such a case, H_L can be written as (129):

$$H_L = -2\beta\langle r^{-3}\rangle\overline{L_z(t)} \quad . \qquad (78)$$

The bar denotes a time average of L_z over the period of one nuclear Larmor precession. For orbital singlets, the orbital moment is quenched and L_z is zero. When the spin orbit coupling significantly mixes the nondegenerate ground state with the excited state, and thus introduces an orbital moment in the ground state, a finite value of H_L is obtained. The excited state mixing causes the g factor of the ion to deviate from spin only value of 2. In such a case (6, 19, 129):

$$H_L = -2\beta\langle r^{-3}\rangle(g-2)\overline{S_z(t)} \quad . \qquad (79)$$

The ground states of ions with d^1, d^2, d^6 and d^7 configurations have large orbital angular momentum which remain far from quenched. H_L has been estimated for Fe^{2+} $(3d^6)$ in several compounds using crystal field-theory (161). In $Fe^{3+}(3d^5)$ compounds experiments have shown $g \sim 2$ and hence $H_L \sim 0$, as expected for a half filled shell (129). H_L makes a large contribution to the hyperfine field in rare earth compounds.

(c) Spin Dipolar Interaction

The spin magnetic moment of the valence shell of the parent atom produces dipolar field, H_D, at the nuclear site which can be estimated as follows. The Hamiltonian for the dipolar interaction between the nuclear magnetic moment $\vec{\mu}_n$ and the spin magnetisation of the valence shell can be written as (131):

$$H_D = \int \left[\frac{\vec{\mu}_n \cdot \vec{M}}{r^3} - \frac{3(\vec{\mu}_n \cdot \vec{r})(\vec{M} \cdot \vec{r})}{r^5} \right] d\tau \quad . \qquad (80)$$

Here $\vec{M}(\vec{r})$ is the spin magnetisation density. The nucleus is assumed to be at the orign of the reference system. $\mathcal{H}_D$ can be rewritten as:

$$\mathcal{H}_D = -\vec{\mu}_n . \vec{H}_D \quad , \tag{81}$$

where

$$\vec{H}_D = \int [\frac{3\vec{r}(\vec{M}.\vec{r})}{r^5} - \frac{\vec{M}}{r^3}] d\tau \quad .$$

Substituting,

$$\vec{M}(\vec{r}) = g\beta \sum_i \vec{s}_i \rho_{si}$$

we get,

$$H_D = g\beta \sum_i \langle r^{-3} \rangle_i \langle 3(\vec{s}_i . \hat{r})\hat{r} - \vec{s}_i \rangle \quad ,$$

where $\hat{r} = \frac{\vec{r}}{r}$, $\langle r^{-3} \rangle_i = \int r_i^{-3} \rho_{si} d\tau$, s_i=spin of the ith electron and ρ_{si} is the electron density $|\psi(\vec{r})|^2$

In the discussion given so far, the contributions of orbital and spin moment of the valence shell are treated separately. However, if the spin orbit coupling is stronger as in rare earths, and thus the orbital and spin moments cannot be separately considered, the hyperfine field is obtained by replacing $\vec{s}_i$ by $\vec{j}_i$ in the above expression. Corrections like relativistic corrections etc. are generally absorbed into an effective value of $\langle r^{-3} \rangle$ (33).

In the cubic symmetry and in the absence of spin orbit coupling, H_D is zero. Generally, H_D is quite small for iron group ions even in non cubic systems (105). However, in rare earth systems where the orbital momentum is not quenched, H_D can be quite large. Assuming that both $\vec{S}$ and $\vec{I}$ are parallel (call it the z axis) and are making an angle θ with the radius vector $\vec{r}$, from Eqns. (80) and (81), we have

$$H_D = g\beta \langle r^{-3} \rangle \langle 3\cos^2\theta - 1 \rangle \vec{S} \quad . \tag{82}$$

The interaction Hamiltonian of Eqn. (80) can be written as:

$$\begin{aligned} \mathcal{H}_D &= -\vec{\mu}_n . \vec{H}_D = -g_n g\beta_n \beta \langle r^{-3} \rangle \langle 3\cos^2\theta - 1 \rangle \vec{I}.\vec{S} \\ &= A_D \vec{I}.\vec{S} \quad . \end{aligned} \tag{83}$$

In presence of a sufficiently strong magnetic field, $\vec{I}$ and $\vec{S}$ precess around the field direction independently. Thus the off diagonal part is negligible and is said to be "quenched", to give

$$H_D \approx A_D I_z \overline{S_z(t)} \quad , \tag{84}$$

where average is to be taken over the precession period of the nucleus as explained earlier.

A spherically symmetric charge distribution will not cause dipolar field at the nucleus; the average $<(3\cos^2\theta-1)>$ is zero in this case. Thus, when the valence shell is either half-filled (Fe^{3+}) or completely filled, dipolar contribution to the hyperfine field becomes identically equal to zero.

We have described above the various contributions to the hyperfine field. Thus, we see that in nonmetallic systems H_{hf} can be written as:

$$H_{hf} = A\overline{S_z(t)}, \tag{85}$$

when the off-diagonal part can be neglected. In absence of the quenching of the off diagonal term of the hyperfine interaction Hamiltonian, A is a tensor quantity. When the nuclear Larmor precession period, τ_L, is large in comparison to the electronic relaxation time, τ_{relax}, which is usually the case, $\overline{S_z(t)}$ can be replaced by $<S_z>_{Th}$, the thermal average of S_z, which is related to the lattice magnetisation by $M(T)=Ng\beta<S_z>_{Th}$, N being the number of magnetic ions per unit volume in the lattice. Generally A is a constant but in some cases it has been found to have a weak temperature dependence of the type (41,162)

$$A = A_o(1-\alpha T^2) \; ,$$

where α is a constant, of the order of 10^{-7} in iron compounds (47).

It may be mentioned here that apart from the hyperfine field there is one more contribution to the magnetic field at the nucleus which is called the "local field", H_{local}, and is given by (6,13,19,11,158,163)

$$H_{local} = H_{ext} - DM + \frac{4\pi}{3} M + H_{cor} \quad , \tag{86}$$

where $-DM$ is the demagnetising field depending on the shape of the magnetised domain, H_{ext} is the external magnetic field, $\frac{4\pi}{3}M$ is the Lorentz field in cubic symmetry, and H_{cor} is the correction term to be applied to $\frac{4\pi}{3}$ M for noncubic systems. The two middle terms on the right hand side in Eqn. (86) are of the order of 1 kOe (129,132,158). $\vec{H}_{cor}$ is three orders of magnitude smaller than the Lorentz field for hexagonal materials (163). It may be mentioned here that the presence of H_{ext} modifies the core polarisation, orbital and dipolar contributions also to some extent since it affects $\langle S_z \rangle_{Th}$ by changing the Boltzmann population of the spin states. The magnetic moments of the surrounding nuclei also produce dipolar field at the nuclear site, H_d. In the paramagnetic state of the solid, this field varies in direction from site to site and thus only causes broadening of the Mossbauer lines. Thus the total magnetic field present at the nuclear site, called internal field, H_{int}, can be written as:

$$\vec{H}_{int} = \vec{H}_{hf} + \vec{H}_{local} + \vec{H}_d \quad .$$

$|\psi(0)|^2$ in the contact field has the dimension of r^{-3}. Thus relativistic corrections are sometimes absorbed in the expression of H_c by rewriting it in the form

$$H_c = 2\beta(\frac{\mu_I}{I})\vec{I}.\vec{S}\langle r_c^{-3}\rangle \quad .$$

Similarly, relativistic corrections appearing in H_L and H_D are absorbed in $\langle r^{-3}\rangle$, which is different from $\langle r_c^{-3}\rangle$ appearing in the core polarisation field.

Whereas in iron group compounds the core polarisation gives the dominant contribution to the hyperfine field, in rare earth compounds it is the unquenched orbital angular momentum which gives the main contribution to the hyperfine field. In these compounds H_L is in the range 10^3-10^4 kOe, except in ions having half filled shell, like Gd^{3+} and Eu^{2+}, where the dominant contribution is from the core polarisation ($\approx$ 300 kOe). Unlike the d shells of iron group elements, the 4f shell is well shielded from the crystalline environment by the electrons in the outer orbitals 5s, 5p, etc. This physical isolation causes the 4f shell to retain much of its free ion character whatever the ion's environment. Thus while an environment can significantly change the hyperfine interactions in iron series elements the effect is very small in rare earths ions. The contact field has been estimated to be

$$H_c \approx -90(g_J-1)\vec{J} \text{ kOe},$$

for either divalent or trivalent rare earth ions.

Like in the non metallic substance, in rare earth metals also the core polarisation effects are roughly one-third of their 3d counterparts, whereas orbital and spin dipolar fields are roughly 50 per cent larger. However, the conduction electron contribution also is much more significant in rare earth metals than in transition metals. More details can be found in the articles of Taneja and Kimball, and Shenoy, elsewhere in this book.

4.5 Combined Quadrupole and Magnetic Hyperfine splitting

We discussed the Mossbauer spectrum in the presence of magnetic hyperfine interaction alone in Section 4.4. The Mossbauer spectrum of iron metal is an example which is commonly used for the calibration of the velocity scale of Mossbauer spectrometers. In such a case, the Mossbauer spectrum of ^{57}Fe is a symmetric six line spectrum with $\Delta_{16}/\Delta_{34} = 6.35$ and $\Delta_{25}/\Delta_{34} = 3.675$ (Δ_{ij} is the separation between the ith and the jth lines). Such a spectrum has been used to check the linearity of a Mossbauer spectrometer also (by comparing the experimental and standard values of the ratios Δ_{16}/Δ_{34} and Δ_{25}/Δ_{34} or by plotting the channel number vs velocity for the experimentally observed six Mossbauer lines which should be a straight line).

The nature of the Mossbauer spectrum when both the electric quadrupole and the magnetic dipole hyperfine interactions are present simultaneously in the system is discussed here. Consider the Cartesian reference frame with x, y and z axes along the principal axes of the EFG tensor ellipsoid. In general, the direction of the hyperfine field H is oriented at angles (θ,ϕ), as shown in Fig. 6. The total Hamiltonian can be written as

$$\mathcal{H} = \mathcal{H}_Q + \mathcal{H}_M \quad , \tag{87}$$

where $\mathcal{H}_Q$ and $\mathcal{H}_M$ are given by Eqns. (34) and (35), except that H_{ext} is replaced by the magnitude of the magnetic hyperfine field. The Hamiltonian can be diagonalised and the Mossbauer line positions and intensities calculated for any set of values of q, η, H, θ and ϕ. We confine our discussion to the case of ^{57}Fe only in the following. In the simple case when $\eta=0$ and $\frac{1}{2}e^2 qQ \ll g_n\beta_n H$, the eigenvalues of Eqn. (87) are given by (19,60) the analytical expression

$$E = -g_n\beta_n H m_I + (-1)^{|m_I+1/2|} \frac{e^2qQ}{4} \left(\frac{3\cos^2\theta-1}{2}\right) \quad , \tag{88}$$

where θ is the angle between the V_{zz} direction and the direction of H. This gives,

$$\Delta_{12}-\Delta_{56} = -\frac{1}{2}e^2qQ(3\cos^2\theta-1) \quad . \tag{89}$$

It is seen that even when $\frac{1}{2}e^2qQ\neq 0$ we can have $\Delta_{12}=\Delta_{56}$ if $\theta=\text{arc}\cos(\frac{1}{\sqrt{3}})$. Such situations have been observed experimentally (164). $\Delta_{12}-\Delta_{56}$ can be positive or negative (Fig. 12). In α-Fe_2O_3 at room temperature $\eta=0$, V_{zz} is positive and $\theta=90^{\circ}$. Consequently, $\Delta_{12}-\Delta_{56}$ is positive ($\approx$ 0.4 mm/sec). Fe_2O_3 is also commonly used as a standard absorber for the calibration of the velocity scale of the Mossbauer spectrometer. Since the quadrupole interaction affects the $|+m_I\rangle$ and $|-m_I\rangle$ states identically, it is obvious from Fig. (12) that H_{hf} can be obtained from the total splitting, Δ_{16}, without the knowledge of the value of $\frac{1}{2}e^2qQ$ even when both the quadrupole and the magnetic hyperfine interactions are present simultaneously in the system:

$$\Delta_{16} = (3|g_1|+g_o)\beta_n H_{hf} \quad , \tag{90}$$

when quadrupole interaction is smaller than the magnetic interaction.

5. MOSSBAUER SPECTRAL SHAPES DUE TO TIME DEPENDENT HYPERFINE INTERACTIONS

Mossbauer spectroscopy has found interesting application in the study of fluctuation processes which affect hyperfine interactions. Fig. 13 gives the approximate frequency ranges of the relaxation rates for various spin fluctuation processes and the techniques used for their study. The theoretical derivation of the analytical expression of the line shape in the presence of spin relaxation has been given elsewhere in the book (Dattagupta, Bhargava). The stochastic model of ionic spin relaxation has been found to be suitable for describing the relaxation phenomena theoretically. The case when the time dependent hyperfine Hamiltonian, $\mathcal{H}(t)$, at different times commute with each other is simple, though very useful, and is described elsewhere in the book. In this case the line shape expression is obtainable using the approach of Kubo and Anderson (166). On the other hand, when the time dependent hyperfine Hamiltonian at different times do not commute with each other, the line shape expression is obtained using the Liouville operator formalism, using an extension of Kubo and Anderson approach (167,168). This case, called the non-adiabatic case, has been dealt with by Dattagupta and by Bhargava elsewhere in this book.

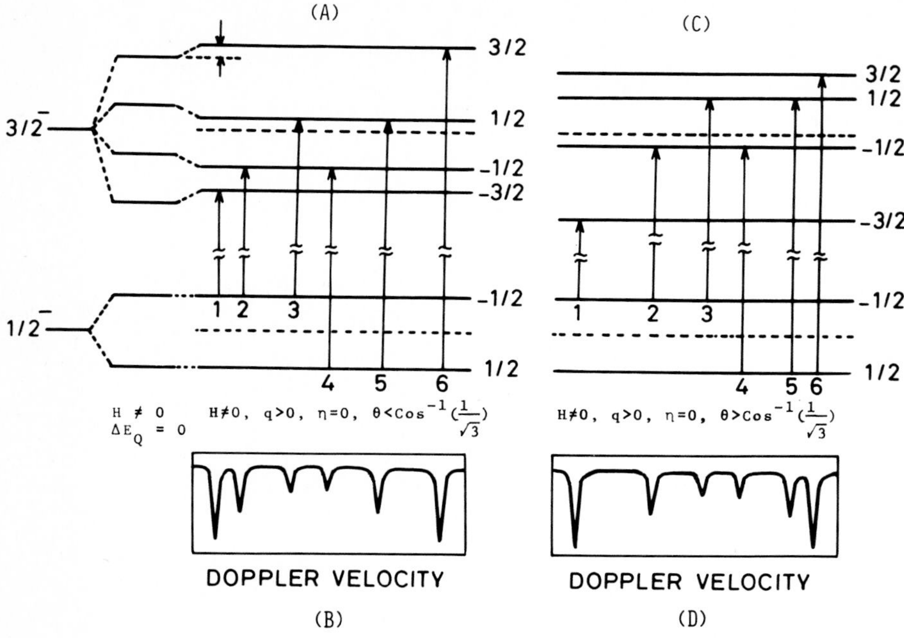

Fig. 12. Schematic representation of the nuclear level splittings and the resulting Mossbauer spectra in presence of magnetic field (H) and an axially symmetric EFG oriented at an angle θ to H, assuming $\frac{1}{2}e^2qQ \ll g_n\beta_nH$.

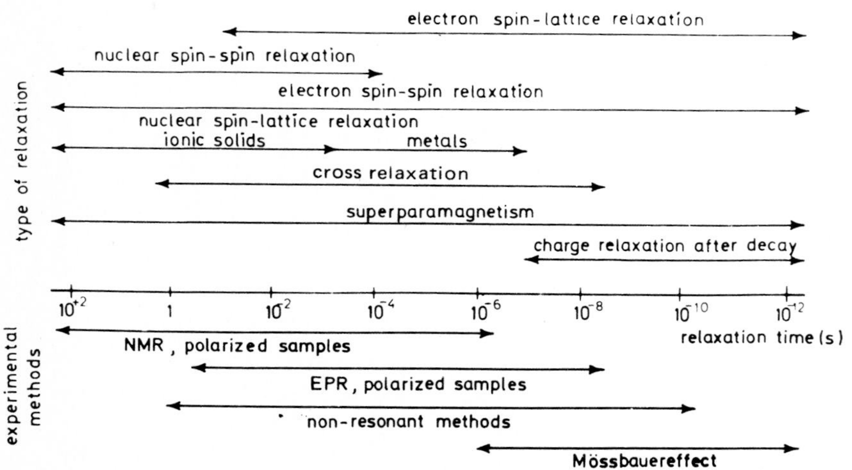

Fig. 13. Typical ranges of relaxation times of various relaxation processes and the regions in which various experimental methods can be used (taken from Ref. 165).

We shall use the line shape expression obtained for the adiabatic case to illustrate the shapes in a few interesting circumstances, and later mainly concentrate on the fluctuation processes which are responsible for the time dependence of the hyperfine interactions. The adiabatic approximation is valid when ionic levels are split by energies greater than the hyperfine interaction energies.

The expression for the line shape in the adiabatic case is (169,170),

$$I(\omega) = -\frac{2}{\Gamma}\,\mathrm{Re}\sum_{m_{gr},m_{ex}} \frac{1}{2I_{ex}+1}\,|\langle I_{gr}m_{gr}|\mathcal{H}^{+}|I_{ex}m_{ex}\rangle|^{2}\{\tilde{W}.\tilde{A}^{-1}.\tilde{1}\} \quad . (91)$$

Here Γ is the natural line width, in absence of the relaxation effects. $\mathcal{H}^{+}$ is the operator causing the nuclear transitions. $\tilde{W}$ is the row vector with elements describing the ratio of populations of successive ionic levels which are involved in the relaxation process: $\tilde{W} = \{s^5, s^4, s^3, s^2, s, 1\}$ if six levels are split by Zeeman interaction only. $\tilde{A}$ is the (nxn) matrix

$$\tilde{A} = i[\tilde{\omega} + (-\omega + i\,\frac{\Gamma}{2})\tilde{E}] + \tilde{\pi} \quad .$$

$\tilde{\omega}$ is the diagonal matrix with elements describing the n values of the line positions corresponding to a nuclear transition ($m_{ex} \rightarrow m_{gr}$) and to the n ionic states in which an ion can occur, in absence of the relaxation effects. $\tilde{\pi}$ is the (n x n) matrix with elements $\pi_{\alpha\beta}$ describing the probability of transition from the ionic state $|\alpha\rangle$ to $|\beta\rangle$ during the fluctuation process. It is in general a product of two factors, one depending on the nature of the levels involved in the transition and the other on the strength of coupling to the bath with which the energy is exchanged. $\tilde{E}$ is the (n x n) unit matrix. Thus, the effect of relaxation on any nuclear transition $m_{ex} \longrightarrow m_{gr}$ is represented by a term in the summation. The total spectral shape is obtained by summing over shapes corresponding to each nuclear transition, taking into consideration their relative intensities, as shown in the expression given above.

To illustrate the effect of spin fluctuations on spectral shape, we have computed Mossbauer spectra using Eqn. (91) for a five level relaxation process (n = 5). The ion is assumed to fluctuate among five equally spaced Zeeman levels. Fig. 14 shows the effect of a change in ionic spin fluctuation frequency on the Mossbauer line shape, when n = 5 and s = 1.0 (Fig. 14a) or s = 0.5 (Fig. 14b). Ω_e is inversely proportional to the average relaxation time of the five levels. s is

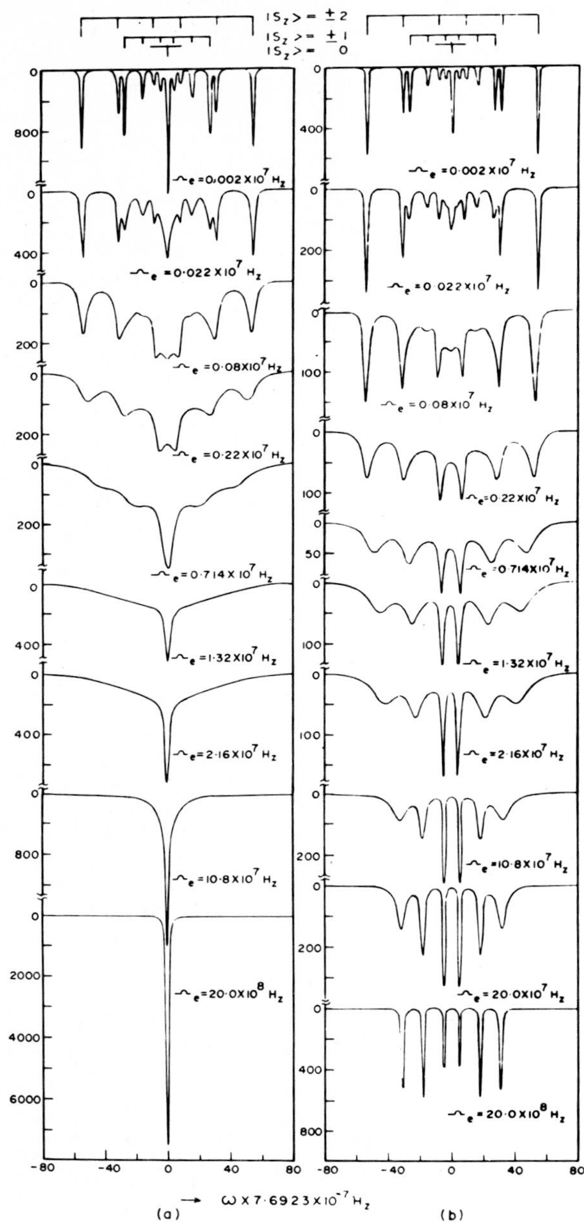

Fig. 14. Theoretical Mossbauer spectra computed using Eqn. (91) in the text when ion fluctuates among five equally spaced levels. The spectra in (a) and (b) differ in values of s parameter which is 1.0 in (a) and 0.5 in (b). When the relaxation frequency is low, spectra corresponding to the ionic Zeeman state |±2>, |±1> and |0> are well separated as shown in the figure. 1 channel≡7.69×10^{-7} Hz.

the ratio of the thermal populations of successive levels. The $n = 5$ case represents the relaxation process in certain systems such as Fe^{2+} compounds having rhombic (and sometimes also cubic and axial) crystal symmetry (106).

In the example given above, we have included only magnetic hyperfine interaction. However, when quadrupole interaction is also present, the following situations are possible:

(a) Only fluctuating EFG is present at the nucleus, magnetic field is zero. Mossbauer lineshapes have been calculated for such a case by Tjon and Blume (166).

(b) There is stationary EFG present at the nucleus alongwith the fluctuating hyperfine magnetic field. If the EFG is axially symmetric ($\eta=0$), the direction of V_{zz} can be either parallel, perpendicular, or oriented at an angle with respect to the direction of the magnetic hyperfine field $H_{hf}(t)$, which is time dependent. For the case when the EFG is parallel to $H_{hf}(t)$ and the magnetic field fluctuates along the z axis, the Hamiltonian is given by

$$\mathcal{H} = \mathcal{H}_o + P[3I_z^2-I(I+1)]+g_n\beta_nI_zhf(t) \quad , \tag{92}$$

where $P = \frac{e^2qQ}{4I(2I-1)}$ and $H_{hf}(t) = hf(t)$. Mossbauer lineshape can be calculated for such a case using Eqn. (91) but with $\tilde{\omega}$ replaced by $[\tilde{\omega} + P(3m_{ex}^2 - 15/4)\tilde{E}]$ (here we have used the fact that $I_{ex} = 3/2$ for the excited state of ^{57}Fe, the ground state of ^{57}Fe nucleus does not split due to EFG, and $P= \frac{e^2qQ}{12}$. Fig. 15 shows Mossbauer line shapes computed as a function of Ω_e using the Hamiltonian given by Eqn. (92) when $s = 1.0$ (Fig. 15a) and $s = 0.5$ (Fig. 15b). The asymmetry in the Mossbauer spectrum is a result of the presence of quadrupole interaction. When $s = 0.5$, a symmetric six line pattern is seen only when $\Omega_e/\Gamma >> 5x10^4$.

The Mossbauer lineshape for the other complicated cases when the fluctuating $\vec{H}_{hf}$ is canted at an angle $\theta \neq 0$ with respect to V_{zz}, or the EFG is not axially symmetric can be computed using the generalised stochastic theory (166).

Recently the theory of Mossbauer lineshape for the case when the hyperfine field at the nuclear site is time dependent due to the processes such as $Fe^{3+} \xrightarrow[\text{capture}]{\text{electron}} Fe^{2+}$ (171) has also been developed. These are non-stationary random processes.

6. ORIGIN OF TIME DEPENDENCE OF HYPERFINE FIELD

As has been discussed before, the time dependence of magnetic

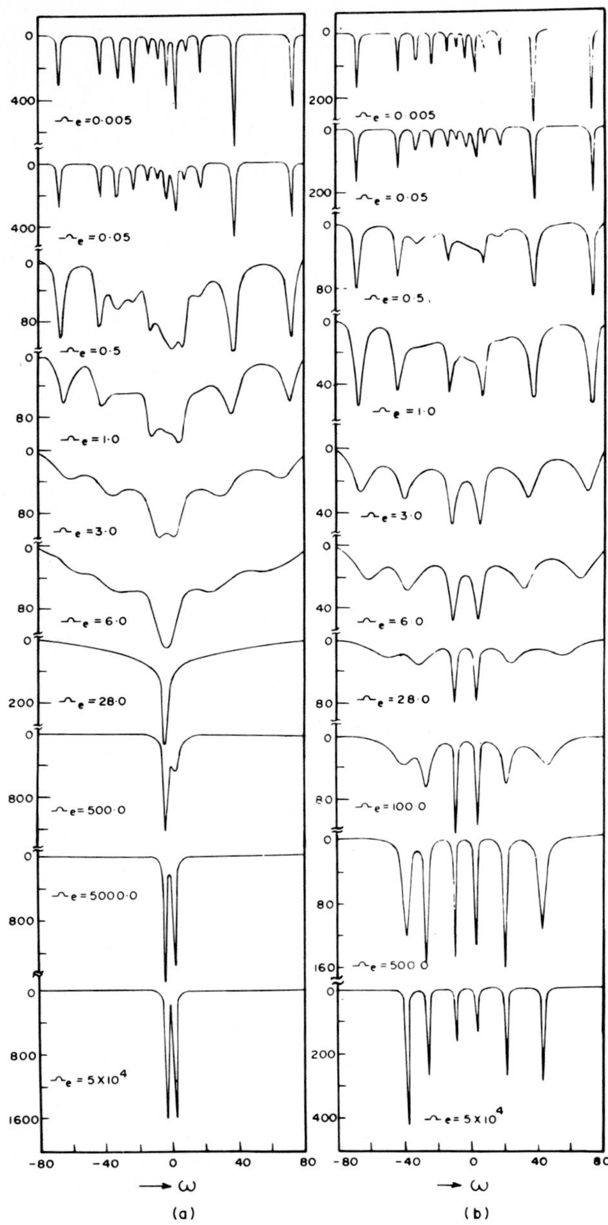

Fig. 15. Mossbauer lineshapes computed using Eqn. (91), when hyperfine field fluctuates among five values as in Fig. 14 but in presence of a static axially symmetric EFG parallel to the hyperfine magnetic field. Spectra (a) and (b) differ in the value of s-parameter which is 1.0 in (a) and 0.5 in (b), P=0.87Γ, and Ω_e and ω are expressed in units of Γ. The value of $\Gamma \approx 10^6$ Hz in the case of Fe^{57}. For -ve values of V_{zz} the asymmetry in line intensities is reversed on the velocity scale.

magnetic field at the nucleus results due to the transitions among the ionic states, known as spin fluctuations. In such a transition the ion either absorbs or emits an energy quantum equal to the energy difference between the states. The main processes responsible for the spin fluctuations can be grouped as

(1) Electronic spin relaxation, and

(2) superparamagnetic fluctuations.

6.1 Electronic Spin Relaxation

Electronic spin relaxation arises when the ionic spin exchanges energy with the neighbouring spins or lattice phonons. In this process, the ion fluctuates from a given spin state to another and stays there for a time, called electronic relaxation time, before making a transition again to some other state. Depending upon the agency with which the ion exchanges energy, electronic relaxation mechanisms in solids can be classified as follows.

6.1.1 Spin spin relaxation

In this mechanism, the ion exchanges energy with another paramagnetic ion with which it is coupled through dipolar or exchange interactions. The pairs of states of the two ions involved in the relaxation must have identical splittings so that the exchange is energy conserving. The interaction responsible for the spin spin relaxation process between the two ions with moments $\vec{\mu}_1=g_1\beta\vec{S}_1$ and $\vec{\mu}_2=g_2\beta\vec{S}_2$ can be written as

$$H_{ss} = [\frac{\vec{\mu}_1\cdot\vec{\mu}_2}{r_{12}^3} - \frac{3(\vec{\mu}_1\cdot\vec{r}_{12})(\vec{\mu}_2\cdot\vec{r}_{12})}{r_{12}^5}] + J\,\vec{S}_1\cdot\vec{S}_2$$

$$= A+B+C+D+C^*+D^* \,, \tag{93}$$

where

$$A = [J-G(3\cos^2\theta-1)]S_{1z}S_{2z},$$

$$B = [J/2 + G/4(3\cos^2\theta-1)](S_{1+}S_{2-} + S_{1-}S_{2+}) \,,$$

$$C = (-3/2)G\sin\theta\cos\theta e^{-i\phi}(S_{1+}S_{2z}+S_{1z}S_{2+}),$$

$$D = (-3/4)G\sin^2\theta e^{-2i\phi}S_{1+}S_{2+} \,,$$

and C^* and D^* are complex conjugates of C and D respectively. Here $G = g_1g_2\beta^2r_{12}^{-3}$ is the strength of the dipolar interaction,

J represents the exchange interaction, θ and ϕ are the polar angles of the vector $\vec{r}_{12}$ connecting ions 1 and 2. The operators C, D, C^* and D^* apparently do not conserve angular momentum. However, a discussion including the lattice shows that the angular momentum is concomitantly taken up by the lattice with a negligible change in rotational energy (172). This is analogous to the physical process underlying the Mossbauer effect where linear momentum is involved. It is clear that the terms C, D, C^* and D^* are important only when the splitting of the ionic levels is negligibly small (172,173). Thus the important term for the spin spin relaxation process is B of H_{ss}. Though the matrix elements of H_{ss} are temperature independant, the population of the ionic levels vary with temperature which introduces temperature dependence in Mossbauer line shape. These aspects have been discussed elsewhere in this book .

6.1.2 Spin lattice relaxation

In the spin lattice relaxation process the ionic spin exchanges energy with the lattice vibrations. In the mechanism proposed by Waller (174) for this process, the modulation of the magnetic interaction (dipolar or anisotropic exchange) between the two spins by the lattice vibrations is responsible for the energy exchange between the spins and the lattice. Since the dipolar interaction is dependent on r_{12}^{-3}, lattice vibrations can affect the dipole interaction. Thus Waller's mechanism is not only temperature dependent but also concentration dependent. The other mechanism of the spin lattice relaxation was proposed by Kronig (175) and Van Vleck (175). In this process, the lattice vibrations modulate the crystalline electric field, and thus affect the orbital motion of the electrons of the ion. This in turn acts on the spin state through spin orbit coupling. The spin-phonon interaction thus proceeds indirectly via spin-orbit and orbit-lattice interactions as the spin of the ion is not sensitive to the electric field directly. The relaxation time produced by Waller's mechanism does not compare well with the experimental measurements. The Kronig-Van Vleck mechanism is now accepted as the main cause for the spin lattice relaxation of the paramagnetic ions in magnetically dilute crystals (176), though in some cases the Waller mechanism may play an important role (177) particularly when the ion is in S state and the spin concentration is not small. In our further discussion, we will concentrate only on the Kronig-Van Vleck mechanism. The spin lattice relaxation process have been grouped into three categories. These are:

(i) Direct process,
(ii) Raman process and
(iii) Orbach process.

The direct process involves a transition of the ion between states $|A\rangle$ and $|B\rangle$ with the simultaneous emission or absorption of a phonon of energy $|E_B - E_A|$, conserving the energy. This process is also called the one phonon process. This is the dominant mode of relaxation at low temperatures. At these temperatures, only low frequency (acoustic) phonons take part in the relaxation process and thus the Debye model of lattice vibrations gives a good description of the direct process which predicts a linear dependence of relaxation rate on temperature. At higher temperatures, the two phonon processes described below dominate.

The Raman process is a two phonon process. In this process, a phonon of frequency ν_i is absorbed alongwith the simultaneous emission of another phonon of frequency ν_s when the ionic spin fluctuates between the states $|A\rangle$ and $|B\rangle$. The difference in the energies of the absorbed and emitted phonons, $h|\nu_s-\nu_i|$, is equal to the energy difference $|E_B-E_A|$. The temperature dependence of the Raman process can be obtained using the Debye model or the Einstein model. The Debye model is used when the low frequency (acoustic) phonons are involved in the process (at low temperatures) and the Einstein model is used when the high frequency (optical) phonons are involved in the relaxation process (at high temperatures). The temperature dependence of the Raman relaxation rate is quite complicated. In the limiting cases of low and high temperatures, the dependence can be summarised as follows (7,178). According to the Debye model, for Raman process $\tau_{SL}^{-1} \propto T^2$ for $T \gg \Theta_D$. For $T \ll \Theta_D$, $\tau_{SL}^{-1} \propto T^7$ for a non-Kramers ion and $\tau_{SL}^{-1} \propto T^9$ for a Kramers ion (Θ_D = Debye temperature; τ_{SL} = spin lattice relaxation time). Using Einstein model, following results are obtained. For $T \ll \Theta_E$, $\tau_{SL}^{-1} \propto \exp(-\Theta_E/T)$ and for $T > \Theta_E$, $\tau_{SL}^{-1} \propto T^2$ (Θ_E=the Einstein temperature). In presence of an external magnetic field, H_{ext}, the dependence of τ_{SL} on H_{ext} is as follows. For a direct process involving acoustic phonons, $\tau_{SL}^{-1} \propto H_{ext}^4$ for a Kramers ion and $\tau_{SL}^{-1} \propto H_{ext}^2$ for a non-Kramers ion. For a Raman process, τ_{SL} is independent of H_{ext} in both the Debye model and the Einstein model when $g\beta H_{ext} \gg D$, where D = crystal field splitting parameter. When $g\beta H_{ext} \ll D$, $\tau_{SL}^{-1} \propto H_{ext}^2$. Recently a new type of Raman process called anharmonic Raman process, has also been proposed which involves anharmonic interaction of three phonons (179).

Another two phonon spin lattice relaxation process is the Orbach process. In the Orbach relaxation process, more than two energy levels of an ion are involved and this process corresponds to an inelastic phonon scattering. A transition of the ionic spin state from $|B\rangle$ to $|C\rangle$ occurs with annihilation of a phonon of energy E_C-E_B, this is followed by the ionic transition from $|C\rangle$to $|A\rangle$ with the creation of a phonon of energy E_C-E_A ($E_C>E_B,E_A$). In the Raman process, unlike in Orbach process, $|C\rangle$ is a virtual state so that the entire phonon spectrum is available for the scattering process. The relaxation rate for an Orbach process is proportional to $\exp(-\Delta/k_BT)$, where $\Delta=E_C-E_B$, and is independent of H_{ext}.

It is worthwhile to mention here that optical phonons play an important role when the Mossbauer ion is present in the host lattice as an impurity. In such a case, if m and m_o are the masses of the impurity (Mossbauer) atoms and the atoms of the host lattice respectively, there are three possibilities. If $m<m_o$, a localised vibrational mode in a narrow frequency band appears above the Debye frequency (ω_D), in the optical branch-frequency range. If $m=m_o$, the situation is same as that of a pure lattice and acoustic modes of vibrations dominate. When $m>m_o$, there appear in the phonon spectra below ω_D the so called quasi localised modes with finite lifetime and therefore with a broadened spectrum of frequencies. Localised modes are best approximated by the Einstein model.

It may be mentioned here that the measurements of τ_{SL}, f and δ' provide complementary informations about the lattice dynamical property of the system. This is because the f-factor and the second order Doppler shift δ' are related with the motion of the Mossbauer ion whereas the spin lattice relaxation process is concerned with the relative motion of the Mossbauer ion and its ligands (180). It may also be mentioned that though neutron diffraction measurements yield at present the most reliable atomic force constants for pure (without any impurity atom) solids, Mossbauer measurements of f,τ_{SL} and δ' are rather unique for the study of impurity-host binding. Laser Raman spectroscopy and the Mossbauer effect have been useful in the study of the internal vibrational modes of a solid (181).

6.1.3 Cross relaxation

Cross relaxation, which is basically a spin spin relaxation process, involves exchange of energy between two "different" spin systems, through dipolar and exchange interactions (182,183). The two spin systems, say i and j, are said to be different when their spins (a) belong to different species, (b) flip between different pairs of levels during energy exchange, or (c) occupy different sites in the lattice. Cross relaxation occurs between two ions if the energy separations of the pairs of levels of the two ions (i and j) are equal, within the widths of the levels. The rate of this process is mainly concentration dependent. If E_α^i is the energy separation between the two levels of i and E_β^j is the energy difference between the two levels of j, the cross relaxation probability for the process in which i absorbs energy E_α^i and j loses energy E_β^j is given by (177):

$$\Omega_{ij} = \frac{2\pi}{\hbar} |\langle S_z^i, S_z^j | H_{SS} | S_z^i+1, S_z^j-1\rangle|^2 g_{\alpha\beta}(0)\, p(S_z^j) \qquad (94)$$

$p(S_z^j)$ is the probability that j is initially in the state $|S_z^j\rangle$. If the shape functions g_α^i and g_β^j are known for the lines E_α^i and E_β^j then

$$g_{\alpha\beta} = \int\int g_\alpha^i(\nu')\, g_\beta^i(\nu'')\, \delta(\nu'-\nu'')\, d\nu' d\nu'' \quad .$$

Assuming g_α^i and g_β^j to have a Gaussian shape (Fig. 16) we have

$$\Omega_{ij} = \frac{(2\pi)^{-1/2}}{h} \frac{|\langle S_z^i, S_z^j | H_{SS} | S_z^i+1, S_z^j-1\rangle|^2}{[(\Delta E_\alpha^i)^2 + (\Delta E_\beta^j)^2]^{1/2}} \exp[-\frac{(E_\alpha^i - E_\beta^j)^2}{2[\Delta E_\alpha^i)^2 + (\Delta E_\beta^j)^2]}]\, p(S_z^j), \qquad (95)$$

where $(\Delta E_\alpha^i)^2$ and $(\Delta E_\beta^j)^2$ are the second moments of the shape functions g_α^i and g_β^j respectively (172,173). The total transition probability of the Mossbauer ion i due to these relaxation transitions is obtained by adding up contributions from all pairs of levels of the neighbouring ions j which conserve the energy. Mossbauer studies have been used to study these processes. In particular, the effects of S and non-S state impurity ions from iron group (Al^{3+}, Cu^{2+}, Cr^{3+}, Co^{2+}, Mn^{2+}) and rare earths (Gd^{3+} and Dy^{3+}) group of elements on the relaxation rates of Fe^{3+} ions have been studied using host medium like Al_2O_3 and amorphous frozen aqueous solutions (182,183).

6.1.4 Dynamic Jahn-Teller effect

The removal of the degeneracy of an electronic state in a solid by lattice distortion and the consequent lowering of the point symmetry is known as the Jahn-Teller effect (184). If the ground electronic state of a paramagnetic ion is degenerate, the geometrical configuration around the ion cannot be stable, except in the following two situations:

(a) the ligand and metal ions are located on a straight line, (b) the ion of interest in the solid contains an odd number of electrons

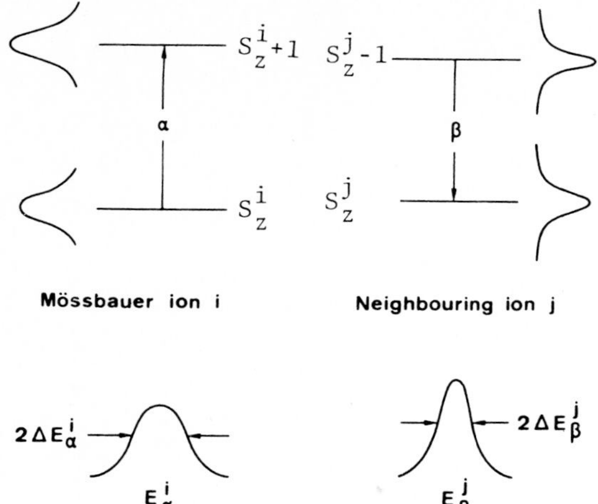

Fig. 16. Schematic representation of a cross relaxation process, in which an ion i makes a transition α, accompanied by a transition β of a neighbouring ion j, conserving energy.

and thus the electronic states must necessarily have a two fold Kramers degeneracy except in presence of an external magnetic field.

In other cases, the deformation lowers the energy of the system in one of the electronic state and therefore provides it a more stable state (185). In our further discussion, we will be mainly concerned with the Jahn-Teller effect occuring in solids.

The Hamiltonian of the system of N electrons (mass m) and N_o nuclei (mass M) can be written as

$$H = H_e+H_n+H_{en} \quad ,$$

where H_e includes kinetic energy of electrons and interelectronic coulomb repulsion, H_n includes kinetic energy of nuclei and the internuclear Coulomb interaction, and H_{en} is the Coulomb attraction between the electrons and the nuclei. The electronic wave functions for the fixed position of the nuclei, denoted by R, can be obtained from

$$\{H_e+\sum_{k>\ell}^{N_o}\sum \frac{Z_k Z_e}{R_{k\ell}} + H_{en}\}\psi_\mu(\vec{r}:\vec{R}) = U_\mu(\vec{R})\psi_\mu(\vec{r}:\vec{R}) \quad . \tag{96}$$

The second term in the bracket is the coulomb repulsion between nuclei. Thus ψ is a function of r with R appearing merely as a parameter. The familiar crystal field theory refers to the case when R represents the equilibrium positions of the nuclei, R_o. Writing the wave function of $\mathcal{H}$ as

$$\Psi = \psi(\vec{r}:\vec{R})\Phi(\vec{R}) \quad .$$

The state of nuclei is obtainable from

$$\left[\sum_{k=1}^{N_o} \frac{P_k^2}{2M_k} + U_\mu(\vec{R})\right]\Phi_{\mu\nu}(\vec{R}) = E_{\mu\nu}\Phi_{\mu\nu}(\vec{R}) \tag{97}$$

using the approximation

$$\sum_k \frac{P_k^2}{2M_k}\, \psi(\vec{r}:\vec{R})\Phi(\vec{R}) \approx \psi(\vec{r}:\vec{R}) \sum_k \frac{P_k^2}{2M_k}\, \Phi(\vec{R}) \quad ,$$

which is valid because $(\frac{m}{M})^{1/2} << 1$. The separation of the electron and the nuclear motion is known as the adiabatic approximation or the Born-Oppenheimer approximation. As will be seen below, in the presence of the Jahn-Teller distortion the separation of the nuclear and electronic parts in this way cannot, in general, be made; the nuclear motion affects electronic state.

The instantaneous potential seen by an electron in the lattice can be expanded as

$$V(R) = V(R_o) + \sum_{k'} \left(\frac{\partial V}{\partial Q_{k'}}\right)_o Q_{k'} + \sum_{k',k''} \left(\frac{\partial^2 V}{\partial Q_{k'}\partial Q_{k''}}\right)_o Q_{k'}Q_{k''} + ---, \tag{98}$$

where o denotes all nuclei at the equilibrium positions, and Q_k are the normal coordinates* (186).

In compounds of 3d group ions, the nearest environment of the paramagnetic ions is generally a tetrahedral or an octahedral coordination of ions. In these cases, it is generally adequate to include only the motions of the nearest neighbours of the central ion to obtain $Q_{k'}$ appearing in the potential experienced by an electron on the central ion. We will briefly discuss here the case of a tetrahedral complex XY_4 (T_d symmetry) which is relevant to several known experimental examples of the Jahn-Teller effect. A convenient set of Cartesian coordinates are shown in Fig. 17(a), and some of normal coordinates, which have been used in the present discussion, are

$Q_1 = \frac{1}{2}(Z_1+Z_2+Z_3+Z_4)$,

$Q_2 = \frac{1}{2}(Y_1-Y_2+Y_4-Y_3)$,

* A discussion of the normal coordinates and the solution of Eqn. (97) can be found in the article on spin lattice relaxation.

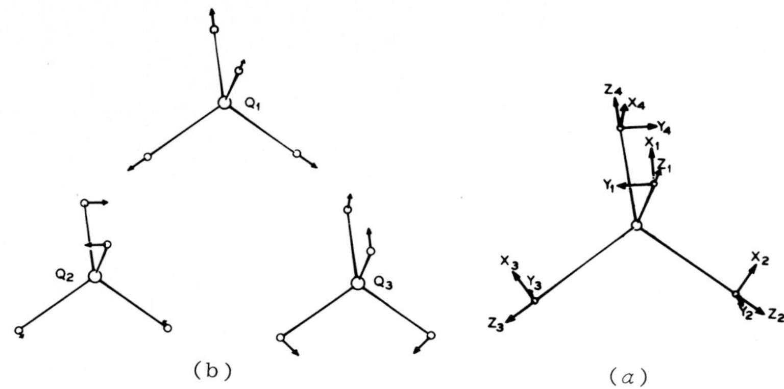

Fig. 17. (a) The Cartesian displacement vectors used in forming the normal coordinates of the tetrahedral complex XY_4. (b) normal coordinates Q_1,Q_2 and Q_3 of XY_4 complex.

$Q_3=\frac{1}{2}(X_1-X_2-X_3+X_4)$.

Q_1 transforms as the identity representation, Q_2 and Q_3 transform as $v(\equiv\varepsilon)$ and $u(\equiv\theta)$ bases of the E irreducible representation (IR). The associated nuclear motions are shown in Fig. 17(b). Other normal coordinates of the cluster can be found in several books (for example, Ref. 184). The Q_3 coordinate represents tetragonal distortion of the tetrahedron lowering the symmetry from T_d to D_{2d}. Other normal coordinates also transform as bases of irreducable representations of T_d. Thus, the normal coordinates can be labelled by $(\Gamma\gamma)$ instead of k', etc. Notice V(R) as well as each of the term in the expansion of V(R) transforms as the identity representation of T_d. In isolated tetrahedron, $Q_{k'}$'s in Eqn. (98) would represent the true normal coordinate and an independant vibrational mode. But in lattice, the molecular cluster interacts with the rest of the solid, resulting in the dependence of the energy of any of the these modes on the wave vector in the Brioullin zone, and admixture with other modes of the lattice if close in energy. The dispersion thus introduced is small if the interaction of the molecule with the rest of the lattice is weak. Nevertheless, the expansion of the potential seen by the electron using Q_k's of the cluster can be made though it may lead to other difficulties like the meaning of the momentum associated with the mode, etc. (184), if the interaction of the cluster with the rest of the lattice is not weak. Let us assume the isolation of the cluster to be complete, for simplicity. Furthermore, only those electrons are considered which occupy degenerate levels when Q_k's are zero and actively participate in the Jahn-Teller effect. They are conventionally called "active electrons". The passive electrons are the electrons in closed shells, etc. Thus in a molecular complex of a transition metal ion surrounded by anions, only the d-electrons in a degenerate level when Q's are zero are the active electrons. The Hamiltonian of the active electrons can using Eqn. (98), be written as:

$$H_a = H_e + \sum_{k'} V_{k'} Q_{k'} + \sum_{k'k''} V_{k'k''} Q_{k'} Q_{k''} + \cdots\cdots\cdots , \qquad (99)$$

where H_e is the Hamiltonian of active electrons when $\vec{Q} = 0$, $V_{k'} \equiv \left(\frac{\partial V}{\partial Q_{k'}}\right)_o$ etc. H_a, H_e and each of other terms in the expansion are invariant to symmetry operations of T_d. Consider matrix element of $\sum_{k'} V_{k'} Q_{k'}$ between eigenfunctions belonging to irreducible representation Γ of T_d, $\langle \Gamma\gamma | \sum_{k'} V_{k'} Q_{k'} | \Gamma\gamma' \rangle$. If k' represents $(\bar{\Gamma}\bar{\gamma})$ of T_d

$$\begin{aligned} \langle \Gamma\gamma | \sum_{k'} V_{k'} Q_{k'} | \Gamma\gamma' \rangle &= \sum_{\bar{\Gamma}\bar{\gamma}} \langle \Gamma\gamma | V_{\bar{\Gamma}\bar{\gamma}} | \Gamma\gamma' \rangle Q_{\bar{\Gamma}\bar{\gamma}} \\ &= \sum_{\bar{\Gamma}\bar{\gamma}} \langle \Gamma || V_{\bar{\Gamma}} || \Gamma \rangle \langle \Gamma\gamma | \Gamma\gamma' \bar{\Gamma}\bar{\gamma} \rangle Q_{\bar{\Gamma}\bar{\gamma}} \\ &\equiv \sum_{k'} h^{\Gamma}_{\gamma\gamma'}(k') Q_{k'} \quad . \end{aligned}$$

It has been assumed that $\bar{\Gamma}$ appear only once in the product representation $\Gamma\times\Gamma$. In the harmonic approximation, the nuclear Hamiltonian can be written as

$$H_n = \sum_{k'} \frac{P^2_{k'}}{2\mu_{k'}} + \frac{1}{2} \sum_{k'} K_{k'} Q^2_{k'} \qquad (100)$$

where $K_{k'} = \mu_{k'} \omega^2_{k'}$.

$\omega_{k'}$ is the frequency of the mode k'. Thus, the matrix element of the total Hamiltonian between wave-functions $|i\rangle$ and $|j\rangle$ corresponding to IR Γ are

$$H^I_{ij} = \bar{E}\delta_{ij} + \sum_{k'} \left[\left(\frac{P^2_{k'}}{2\mu_{k'}} + \frac{1}{2} \mu_{k'} \omega^2_{k'} Q^2_{k'} \right) \delta_{ij} + h_{ij}(k') Q_{k'} \right] \quad .$$

Here $\bar{E}$ includes all contributions to the electronic energy when $\vec{Q}$ is zero.

To be specific, we consider the case of an ion from the iron group with D term as the ground state when the ion is free. In the cubic crystal field, the D state splits into E and T_2 states (Fig. 8).

In cubic tetrahedral field, the E state is lower in energy. T_2 state is considerably higher in energy and need not be considered further. Since E×E = E + A_1, the matrix element $h^{\Gamma}_{\gamma\gamma}(k')$, Γ=E, is non zero if k' belongs to $E(Q_2,Q_3)$ or $A_1(Q_1)$. The later mode does not split levels and need not be considered further. Since Q_2 and Q_3 transform as v and u bases of E, we need the coefficients $\langle E\gamma|E\gamma' E\bar{\gamma}\rangle$, which are:

when $\bar{\gamma} = v$

γ' \ γ	v	u
v	0	$\frac{1}{\sqrt{2}}$
u	$\frac{1}{\sqrt{2}}$	0

and when $\bar{\gamma}$=u

γ' \ γ	v	u
v	$-\frac{1}{\sqrt{2}}$	0
u	0	$\frac{1}{\sqrt{2}}$

Thus the total Hamiltonian can be written as

$$H = [\ \bar{E}+\frac{1}{2\mu}(P_u^2+P_v^2) + \frac{1}{2}\mu\omega^2(Q_u^2+Q_v^2)\ ]\ \tilde{I}-V \begin{bmatrix} -Q_u & Q_v \\ Q_v & Q_u \end{bmatrix} \tag{101}$$

which acts on the state vector $\begin{bmatrix} a \\ b \end{bmatrix}$, where the electronic wave function is

$$\psi_e = a\psi_u+b\psi_v \quad .$$

Here, P_u and P_v are the momentum operators conjugate to Q_3 and Q_2, μ is the effective mass of the E vibrational mode of frequency ω, V is the reduced matrix element $\frac{1}{\sqrt{2}}\langle E||V_E||E\rangle$, and $\tilde{I}$ is a 2x2 unit matrix. H is called the vibronic Hamiltonian and its eigenstates the vibronic states. H is invariant to the symmetry operations of T_d. The eigenfunctions and eigenvalues of the Hamiltonian excluding the nuclear kinetic energy part are

$$E_{\pm} = \bar{E} \pm V\rho + \frac{1}{2}\mu\omega^2\rho^2 \quad , \tag{102}$$

$$\psi_+ = \psi_u\cos(\theta/2) - \psi_v\sin(\theta/2) \quad ,$$

$$\psi_- = \psi_u\sin(\theta/2) + \psi_v\cos(\theta/2) \quad , \tag{103}$$

where $Q_v(\equiv Q_\varepsilon) = \rho\sin\theta$ and $Q_u(\equiv Q_\theta) = \rho\cos\theta$.

The dependences of $E_{\pm}$ on $\pm V\rho$, $\frac{1}{2}\mu\omega^2\rho^2$, and $\pm V\rho + \frac{1}{2}\mu\omega^2\rho^2$ are separately shown in Fig. (18) when $\theta=0$. The dependence of $E_{\pm}$ on ρ and θ is shown in Fig. 19(a). Whereas E_+ increases with ρ, E_- is minimum at $\rho_0 = \frac{|V|}{\mu\omega^2}$, V is positive. $E_{\pm}$ is independant of θ. The two energy surfaces are thus cylindrically symmetric.

The energy difference (E_{JT}) between the energy at $\vec{Q}=0$ and the energy at ρ_0 is known as the Jahn-Teller energy.

$$E_{JT} = -\frac{V^2}{2\mu\omega^2}$$

Including the nuclear kinetic energy, the Schrodinger equation is

$$\left[\frac{1}{2\mu}(P_u^2+P_v^2) + (\pm V\rho + \tfrac{1}{2}\mu\omega^2\rho^2)\right]\Psi=E\Psi, \text{ or}$$

$$\left[-\frac{\hbar^2}{2\mu}\left(\frac{\partial^2}{\partial\rho^2} + \frac{1}{\rho}\frac{\partial}{\partial\rho} + \frac{1}{\rho^2}\frac{\partial^2}{\partial\theta^2}\right) + \left(\pm V\rho + \frac{1}{2}\mu\omega^2\rho^2\right)\right]\Psi=E\Psi \quad . \tag{104}$$

Let

$$\Psi = \psi_+Q_+ + \psi_-Q_- \quad . \tag{105}$$

Whereas ψ_+ and ψ_- are orthogonal to each other, Q_+ and Q_- are not so. Since the potential is independant of θ, $Q_{\pm}$ can be chosen as

$$Q_{\pm} = f_{\pm}(\rho)e^{im\theta} \quad .$$

Further, $\Psi(\theta)=\Psi(\theta+2\pi)$

and $\psi_{\pm}(\theta) = -\psi_{\pm}(\theta+2\pi)$,

show that we should have $Q_{\pm}(\theta)=-Q_{\pm}(\theta+2\pi)$.

Thus m should be an half odd integer. Substitution of Ψ in Eqn. (104) and integration over electron wave functions give

$$-\frac{\hbar^2}{2\mu}\left[\frac{\partial^2}{\partial\rho^2} + \frac{1}{\rho}\frac{\partial}{\partial\rho} - \frac{m^2+\frac{1}{4}}{\rho^2}\right]Q_{\pm} + \left(\pm V\rho + \tfrac{1}{2}\mu\omega^2\rho^2\right)Q_{\pm} - \left[\frac{\hbar^2}{2\mu\rho^2}\frac{\partial}{\partial\theta}\right]Q_{\mp} = EQ_{\pm} \quad . \tag{106}$$

The last term on the left side connects the two surfaces E_+ and E_-. Only when this term is negligible, the solutions ψ_+Q_+ or ψ_-Q_- describe E_+ or E_- surface, respectively, and the Born Oppenheimer approximation is valid.

Typically $\hbar\omega$ for the E mode is $\sim$300 cm^{-1} and $E_{JT}\sim$3000 cm^{-1}.

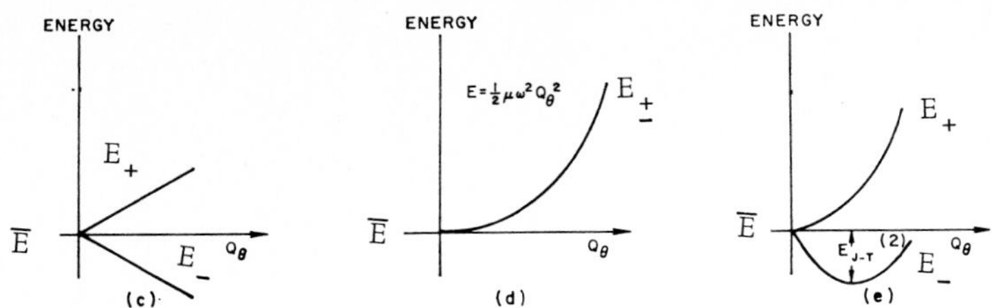

Fig. 18. The splitting of $E_{\pm}$ levels due to (a) the term linear in ρ, (b) the term quadratic in ρ^2, and (c) the combination of the linear and quadratic term, in Eqn. (102) when θ=0, assuming V >0.

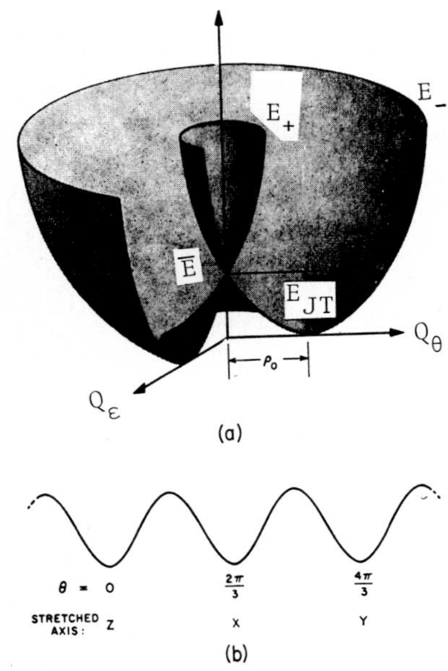

Fig. 19. (a) The dependence of $E_{\pm}$ on ρ and θ given by Eqn. (102). (b) The angular unfolded dependence of the energy E_- at $\rho = \rho_o$ when A_3 and V_2 are non zero (Ref. 185, D.P. Breen et al).

Thus

$$\frac{\hbar^2}{2\mu\rho_o^2} = \frac{\hbar^2\omega^2}{E_{JT}} \approx 8 \text{ cm}^{-1},$$

which is much smaller than $\frac{1}{2}\mu\omega^2\rho_o^2(=E_{JT})$ and $V\rho_o(=\mu\rho_o^2\omega^2=2E_{JT})$. Thus, near the minimum of the low energy surface the last term on left side in the above equation can be neglected. The term containing $\frac{\partial}{\partial\rho}$ can also be neglected near $\rho=\rho_o$. Thus

$$-\frac{\hbar^2}{2\mu}\frac{\partial^2 f_-}{\partial\rho^2} + [\ -V\rho + \tfrac{1}{2}\mu\omega^2\rho^2 + \frac{\hbar^2}{2\mu\rho^2}(m^2 + \tfrac{1}{4})\ ]f_- = Ef_- \quad . \tag{107}$$

Near the minima $\rho=\rho_o$, where the last term containing $\frac{\hbar^2}{2\mu\rho^2}$ is small, the solutions are harmonic oscillator like, with energy quantum $\hbar\omega$ but depressed in energy by E_{JT} , centred around $\rho=\rho_o$. This can be seen by rewriting $V=\mu\omega^2\rho_o$ and neglecting the small term $\frac{\hbar^2}{2\mu\rho}(m^2+\frac{1}{4})$. The lowest manifold of levels can be represented by considering the harmonic oscillator in the lowest state and adding the nuclear kinetic energy due to the circular motion:

$$E^o_m = -E_{JT} + \tfrac{1}{2}\hbar\omega + (m^2 + \tfrac{1}{4})\frac{\hbar^2}{2\mu\rho^2} \quad , \tag{108}$$

which are states centred around ρ_o. Born-Oppenheimer product functions are appropriate around ρ_o. On the other hand, in the region where the two surfaces are close to each other, at low values of ρ, the nuclear motion cannot be confined to one of them. In this case the vibronic wave function Ψ (Eqn. 105) is appropriate.

Under the influence of the nuclear kinetic energy the system therefore acquires a dynamical behaviour where the complex keeps on distorting dynamically in all possible directions of minimum energy. In other words, there is no static distortion of the complex but we have a dynamical resonance between the modes Q_v and Q_u, with all values of θ possible.

When the anharmonic term in the nuclear Hamiltonian and the second term in the expansion of V(R) (Eqn. 98), are included and symmetry considerations are used, the Hamiltonian in Eqn. (101) is replaced by

$$\mathsf{H} = \left[\bar{E} + \frac{1}{2\mu}(P_u^2+P_v^2) + \tfrac{1}{2}\mu\omega^2(Q_u^2+Q_v^2) + A_3Q_u(Q_u^2-3Q_v^2)\right]\tilde{I}$$
$$-V\begin{bmatrix}-Q_u & Q_v\\ Q_v & Q_u\end{bmatrix} + V_2\begin{bmatrix}-(Q_v^2-Q_u^2) & 2Q_uQ_v\\ 2Q_uQ_v & (Q_v^2-Q_u^2)\end{bmatrix} \quad . \tag{109}$$

The two energy surfaces, obtained when the nuclear kinetic energy part is excluded and $V_2 \ll V$ so that only diagonal matrix elements of the term containing V_2 with respect to ψ_+ and ψ_- are significant, (V>0)

$$E_\pm = \bar{E} + \tfrac{1}{2}\mu\omega^2\rho^2 + A_3\rho^3\cos 3\theta \pm V\rho \pm V_2\rho^2\cos 3\theta . \tag{110}$$

The lower energy surface shows minima when

$$\theta = 0, \frac{2\pi}{3}, \frac{4\pi}{3} \qquad \text{if} \quad (A_3\rho - V_2) < 0$$

or

$\theta=\frac{\pi}{3}, \pi, \frac{5\pi}{3}$ if $(A_3\rho - V_2) > 0$

The dependence of E_- on θ at $\rho=\rho_o$ is shown unfolded in Fig. 19(b). Thus there are three directions of minimum energy in (Q_u,Q_v) space. The three potential wells are separated by saddles. The height of these barriers depend on the anharmonicity and V_2, is typically in the range 100 to 500 cm^{-1}, much smaller than E_{JT}. As the height of these potential barriers is large in comparison to $\frac{\hbar^2}{2\mu\rho^2}$, the rotational spectrum in the absence of the barrier (Eqn. 108) is modified. The Hamiltonian excluding the nuclear kinetic energy part is also invariant to the symmetry operations of the cubic group, shows that the three directions of distortions must be physically equivalent.

When the nuclear kinetic energy is included and the non adiabatic part which causes mixing of upper and lower surfaces is neglected (appropriate in the region $\rho \sim \rho_o$ when $\hbar\omega << E_{JT}$), Eqn. (107) is replaced by

$$-\frac{\hbar^2}{2\mu}\left(\frac{\partial^2}{\partial\rho^2} + \frac{1}{\rho}\frac{\partial}{\partial\rho} - \frac{m^2+\frac{1}{4}}{\rho^2}\right)Q_{\pm} + (\pm V\rho \pm V_2\rho^2\cos 3\theta)Q_{\pm}$$

$$+(\tfrac{1}{2}\mu\omega^2\rho^2 + A_3\rho^3\cos 3\theta)Q_{\pm} = EQ_{\pm} \quad . \tag{111}$$

The vibronic wave functions corresponding to the two surfaces are the Born Oppenheiner products ψ_+Q_+ and ψ_-Q_-, respectively. The variation of the potential energy with θ is known as warping. The warping on the potential surface reduces the symmetry from axial to C_{3v} on (Q_2,Q_3) plot.

We confine our attention to the lower energy surface ψ_-Q_-. This Eqn. (111) reduces to the Eqn. (107) if the θ dependent parts, $\mathcal{H}(\theta)$, are excluded, and the term containing $\frac{\partial}{\partial\rho}$ is neglected which is a good approximation near $\rho=\rho_o$.

$$\mathcal{H}(\theta) = -\frac{\hbar^2}{2\mu\rho^2}\frac{\partial^2}{\partial\theta^2} + \rho^2\cos 3\theta(A_3\rho - V_2) \; . \tag{112}$$

The solutions of the Eqn. (107) near $\rho=\rho_o$ are harmonic oscillator states of energy quantum $\hbar\omega$, if ρ_o is large enough, and the wave functions are centred about $\rho=\rho_o$. The excited state is well separated in energy ($\gtrsim$ 300 cm^{-1}), so that we can treat $\mathcal{H}(\theta)$ as a perturbation to this ground state (indicated by subscript o). Thus the solutions of Eqn. (111) can be written as

$Q_- = f_o(\rho)\phi$, where ϕ satisfies

$$\alpha \frac{\partial^2\phi}{\partial\theta^2} + \beta\cos 3\theta\ \phi = \varepsilon\phi. \tag{113}$$

Here,

$$\alpha = \langle f_o(\rho)| - \frac{\hbar^2}{2\mu\rho^2}| f_o(\rho)\rangle, \text{ and}$$

$$\beta = \langle f_o(\rho)|\rho^2(A_3\rho - V_2)| f_o(\rho)\rangle.$$

ϕ is written as

$$\phi(\theta) = \sum_m a_m e^{im\theta}.$$

The condition $\phi(\theta) = -\phi(\theta+2\pi)$ is satisfied if m is half odd integer. The coefficients a_m's satisfy

$$-a_m(\varepsilon_n + \alpha m^2) + \frac{1}{2}\beta(a_{m-3} + a_{m+3}) = 0.$$

We obtain the sets of solutions corresponding to

$$m = \pm\frac{3}{2}, \pm\frac{9}{2}, \pm\frac{15}{2}, \text{----------} ,$$

$$m = \frac{1}{2}, -\frac{5}{2}, \frac{7}{2}, -\frac{11}{2}, \text{----------} , \text{ and}$$

$$m = -\frac{1}{2}, \frac{5}{2}, -\frac{7}{2}, \frac{11}{2}, \text{----------} .$$

The corresponding states are denoted by ϕ_A, $\phi_{A'}$, ϕ_E and $\phi_{E'}$, respectively. The vibronic states $\phi_A\psi_-$, $\phi_{A'}\psi_-$, $\phi_E\psi_-$, and $\phi_{E'}\psi_-$ correspond to A_1, A_2, E_θ and E_ε representations of the cubic group, respectively. The energy levels are shown in Fig.(20). $2|\beta|$ represents the height of the potential barrier. When $\beta=0$, $\pm m$ states are degenerate as discussed earlier (Eqn. 108). When $|\beta|/\alpha$ is small, the rapid tunneling of the complex from one potential minima to another potential minima produces the dynamic JT effect, and the splitting of the lowest excited singlet state and the ground doublet (3Γ). The tunneling rate, and consequently 3Γ, increases as β decreases. On the other hand, when $|\beta|/\alpha$ is large A_1 and E states collapse to give a triply degenerate state. These three states correspond to the three directions of distortion of minimum energy. At low temperatures, the system exists in one of the three states. As the temperature increases, the system can absorb a phonon to make a real transition to one of the excited state close to the height of the potential barrier and then decay back to one of the three ground state corresponding to different direction

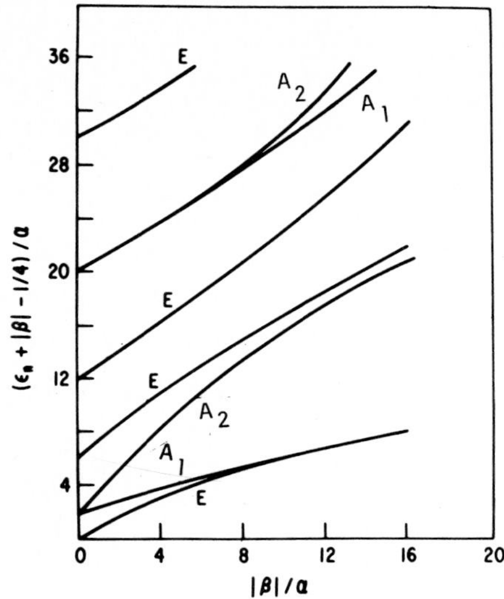

Fig. 20. The dependence of the energies of lower energy vibronic levels (ε_n) on ($|\beta|/\alpha$) in the approximation $E_{JT} >> \hbar\omega$, including the warping effects due to finite values of A_3 and V_2. It is assumed $V>0$, $\beta<0$. (Taken from Ref. 184, F.S. Ham, in S. Geschwind (Ed.), Electron Paramagnetic resonance, Plenum Press, 1972).

of distortion. The net effect is a tunneling from one direction of distortion to another. The rate of such processes is

$$\tau = \tau_o \exp(\frac{\Delta E}{k_B T}) \quad ,$$

where ΔE is the excited level above the initial ground state and τ_o depends on the life time of the excited state. This is an Orbach type of relaxation process. This has been discussed in detail by J.A. Sussmann (185).

On the other hand, when the temperature is not high enough for the above process to be important and 3Γ is small, the reorientation occurs as follows:

When 3Γ is small (<1.0 cm^{-1}), other interactions with the electronic state also affect the relaxation process and therefore need be included in the discussion. In particular, the random static strains which are unavoidable even in good lattices need be considered when 3Γ is small. The perturbation of the electronic state ψ_θ and ψ_ε can be written in general as

$$\Lambda = G_1\tilde{I} + G_2\tilde{A}_2 + G_\theta\tilde{u}_\theta + G_\varepsilon\tilde{u}_\varepsilon,$$

where

$$\tilde{\mathcal{I}} = \begin{bmatrix} 1 & 0 \\ 0 & 1 \end{bmatrix}, \quad \tilde{\mathcal{A}}_2 = \begin{bmatrix} 0 & -i \\ i & 0 \end{bmatrix}, \quad \tilde{u}_\theta = \begin{bmatrix} -1 & 0 \\ 0 & 1 \end{bmatrix}, \text{ and } \tilde{u}_\varepsilon = \begin{bmatrix} 0 & 1 \\ 1 & 0 \end{bmatrix} .$$

Here G_1, G_2, G_θ and G_ε, like V, V_2, etc., depend on the external perturbations. For example, the perturbation of ψ_θ and ψ_ε states due to static strain can be described by

$$H_S^E = V_S(e_\theta \tilde{u}_\theta + e_\varepsilon \tilde{u}_\varepsilon),$$

where

$$e_\theta = e_{zz} - \frac{1}{2}(e_{xx}+e_{yy}) \quad \text{and}$$

$$e_\varepsilon = \frac{\sqrt{3}}{2}(e_{xx}-e_{yy}). \tag{114}$$

Here, e_{ij} is a component of the strain tensor. The matrix elements of Λ within the manifold of ground state are

	Ψ_{A_1}	$\Psi_{g\theta}$	$\Psi_{g\varepsilon}$
Ψ_{A_1}	$3\Gamma+G_1$	γG_θ	γG_ε
$\Psi_{g\theta}$	γG_θ	G_1-qG_θ	qG_ε
$\Psi_{g\varepsilon}$	γG_ε	qG_ε	G_1+qG_θ

(115)

Here, $\gamma = \langle\Psi_{A_1}|\tilde{u}_\theta|\Psi_{g\theta}\rangle = \langle\Psi_{A_1}|\tilde{u}_\varepsilon|\Psi_{g\varepsilon}\rangle$,

Ψ_{A_1}, $\Psi_{g\varepsilon}$ and $\Psi_{g\theta}$ are the three lowest vibronic states described above,

$$q = -\langle\Psi_{g\theta}|u_\theta|\Psi_{g\theta}\rangle = \langle\Psi_{g\varepsilon}|u_\theta|\Psi_{g\varepsilon}\rangle$$

$$= \langle\Psi_{g\varepsilon}|u_\varepsilon|\Psi_{g\theta}\rangle,$$

and the energy of A_1 state (3Γ) relative to the ground $\Psi_{g\varepsilon}$ and $\Psi_{g\theta}$ in the absence of Λ has been included.

When 3Γ is negligible, we go over to the limit of static JT effect, no tunneling. In this case the vibronic states corresponding to the three minima in (Q_u,Q_v) plane can be written as $\Psi_i = \psi_-\Phi_i$, i=1,2,3 correspond to $\theta = 0$, $\frac{2\pi}{3}$ and $\frac{4\pi}{3}$ (or $\frac{\pi}{3}$, $\frac{3\pi}{3}$ and $\frac{5\pi}{3}$), respectively The linear combination of Ψ's which transform as A_1, E_θ and E_ε are

$$\Psi_{A_1} = \frac{1}{\sqrt{3}} (\Psi_1+\Psi_2+\Psi_3) \quad ,$$

$$\Psi_{E_\theta} = \frac{1}{\sqrt{6}} (2\Psi_1-\Psi_2-\Psi_3) \quad ,$$

$$\Psi_{E_\varepsilon} = \frac{1}{\sqrt{2}} (\Psi_2-\Psi_3) \quad . \tag{116}$$

In the limit $3\Gamma=0$, γ and q are related by (184)

$$\gamma = -q\sqrt{2} \quad .$$

Thus we obtain, using the transformation (Eqn. 116), from Eqn. (115)

	Ψ_1	Ψ_2	Ψ_3
Ψ_1	$\Gamma+G_1-2qG_\theta$	Γ	Γ
Ψ_2	Γ	$\Gamma+G_1-q(-G_\theta+\sqrt{3}G_\varepsilon)$	Γ
Ψ_3	Γ	Γ	$\Gamma+G_1-q(-G_\theta-\sqrt{3}G_\varepsilon)$

(117)

Thus the static strain in the sample can reduce the tunneling rate. When the differences between the diagonal elements are large in comparison to Γ, we obtain the static JT effect. The absence of off-diagonal terms containing G_θ and G_ε is due to the relation $\gamma=-q\sqrt{2}$ used. Such off diagonal terms are small even when $\Gamma\neq0$ but small. When the strain splitting is larger than Γ, the eigenstates to first order are given by

$$\Psi_1' = \Psi_1 - \frac{\Gamma}{\delta_{21}}\Psi_2 - \frac{\Gamma}{\delta_{31}}\Psi_3 \quad ,$$

$$\Psi_2' = \Psi_2 + \frac{\Gamma}{\delta_{21}}\Psi_1 - \frac{\Gamma}{\delta_{32}}\Psi_3 \quad ,$$

$$\Psi_3' = \Psi_3 + \frac{\Gamma}{\delta_{31}}\Psi_1 + \frac{\Gamma}{\delta_{32}}\Psi_2 \quad . \tag{118}$$

Where $\delta_{ij}=E_i-E_j$ is the energy difference between ith and jth state when $\Gamma=0$. Each of these states correspond predominantly to one of the three wells. Reorientation by one phonon or multiphonon processes can occur, involving ground states Ψ_1', Ψ_2' and Ψ_3' only, due to perturbation of the form

$$H_S^E = V_S'(e_\theta\tilde{u}_\theta+e_\varepsilon\tilde{u}_\varepsilon) \quad ,$$

where e_θ and e_ε can be expressed in terms of phonon creation and annihilation operators. This has no off-diagonal matrix elements with respect to Ψ_1, Ψ_2, Ψ_3 (Eqn. 117) and thus cannot induce reorientation transition if $\Gamma=0$ (Eqn. 118). Using the procedure discussed in detail in the chapter on spin lattice relaxation elsewhere in the book, e_θ and e_ε can be expressed* in terms of phonon creation and annihilation operators and the relaxation rates computed.

When the rate of reorientation (Ω) is smaller than the nuclear precession frequency (ω_L), Mossbauer spectra corresponding to each of the three distorted configurations is observable. On the other hand when $\Omega \geq \omega_L$, relaxation effects appear on the spectral shapes (187,188).

One of the earliest example of the Mossbauer study of JT effect is due to Tanaka et al (189). Fe^{2+} in $Fe_2Cr_2O_4$, FeV_2O_4 and $FeAl_2O_4$ occupies tetrahedral site. The EFG is significant at low temperatures when the system is tetragonally distorted. At higher temperatures the splitting disappears. The region of temperature over which this occurs is large (Fig. 21). Other examples (189) of the study of JT effect using Fe^{2+} ions have been given in the article of D.C. Price and F. Varret. The studies in ferrites and other magnetic oxides have been discussed elsewhere also in the book.

6.2 Superparamagnetic Effects

Superparamagnetic effects in the Mossbauer spectra are shown by physically small magnetic particles or by small regions, called clusters, in a solid which are more magnetic than the rest of the solid or are magnetically decoupled from the rest of the solid. In such a case, the particle or the cluster magnetisation starts fluctuating randomly among various easy directions of magnetisation due to thermal agitation (185). If V is the volume of the particle or the cluster, then the relaxation time of the particle or cluster magnetisation is given by (190):

$$\tau_F=\tau_o \exp(KV/k_BT), \tag{119}$$

where K is the anisotropy energy per unit volume. τ_o is in the range 10^{-8} to 10^{-11} sec and K is related to the anisotropy energy in various systems as follows.

---

* It is assumed that lattice vibrations are independant of the JT centre XY_4. This is questionable if the interaction of the cluster with the rest of the lattice is strong (185, F.I.B. Williams et al). The theory which takes into account the interaction when large is not well developed.

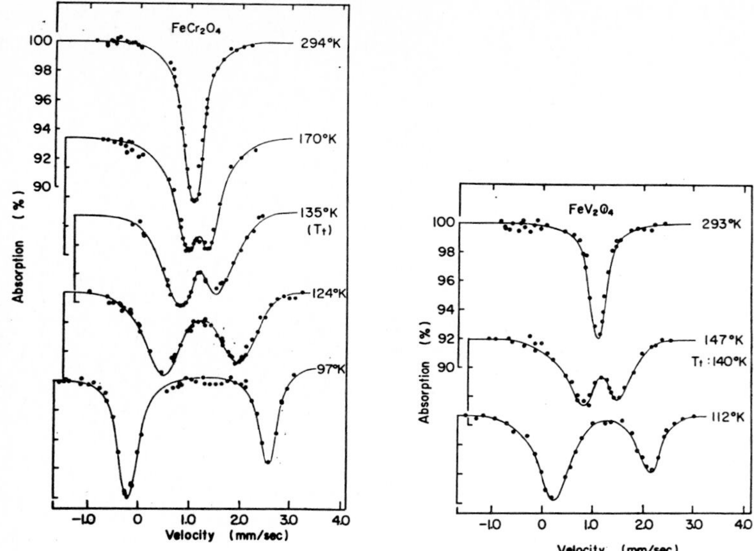

Fig. 21. (a) Mossbauer absorption spectra of $FeCr_2O_4$ at various temperatures. (b) Mossbauer absorption spectra of FeV_2O_4 (taken from Ref. 18: M. Tanaka et al).

The anisotropy energy in a system arises because of the directional dependence of the free energy of the system. Along certain directions in the lattice, called the easy directions of magnetisation the free energy is minimum. In a cubic lattice the free energy can be written as (191):

$$F = K_oV+K_1V(\alpha_1^2\alpha_2^2+\alpha_2^2\alpha_3^2+\alpha_1^2\alpha_3^2)+K_2V\alpha_1^2\alpha_2^2\alpha_3^2 + ----- ,$$

where K_1=first order anisotropy constant, K_2=second order anisotropy constant, etc., and α_1, α_2, and α_3 are the direction cosines of the cluster magnetisation $\vec{M}$ with respect to the crystal axes. The anisotropy energy is, assuming $K_2=K_3= \ldots =0$

$$E_A=F-(F)_{\alpha_1=\alpha_2=\alpha_3=0} = F-K_oV$$

$$= K_1V(\alpha_1^2\alpha_2^2 + \alpha_2^2\alpha_3^2 + \alpha_1^2\alpha_3^2).$$

In the case of an uniaxial superparamagnetic system, the anisotropy energy is given by,

$$E_A = K_1V \sin^2\theta, \qquad (120)$$

where θ is the angle between the direction of easy magnetisation and $\vec{M}$. Higher order terms are small. For an orthorhombic lattice

$$E_A=K_aV\alpha_a^2+K_bV\alpha_b^2+ -----,$$

where the suffices a and b refer to orthorhombic a and b axes and $K_a \sim 2K_b$ (191).

The anisotropy energy K is generally made up of several contributions, viz., magnetocrystalline anisotropy, shape anisotropy, stress anisotropy and surface anisotropy (190). The shape and surface anisotropies are important when the superparamagnetism is due to the physical size of the particles (192) and the stress anisotropy plays an important role only in strained lattices. The magnetocrystalline anisotropy constant is a sum of three terms (190,191,193):

$$K_{MC}=K_{MD} + K_{FS} + K_{EX} ,$$

where K_{MD}, called the magnetic dipolar anisotropy constant, arises when the magnetic dipolar interaction between the Mossbauer ion under consideration and the other ions in the lattice is anisotropic, K_{FS} arises when the interaction of the Mossbauer ion with the crystal field is anisotropic and K_{EX} is a result of the anisotropic exchange interaction. Thus, KV in Eqn. (119) is the anisotropy energy barrier in uniaxial system (E_B), existing between the two oppositely oriented directions of easy magnetisation. τ_o is generally weakly dependent on T and V (192,194).

Within the cluster (or particle) the magnetic ordering can be ferromagnetic, antiferromagnetic or ferrimagnetic (195). In the case of antiferromagnetic ordering, it is the sublattice magnetisation direction that randomly fluctuates due to thermal agitation. Also in ultrafine particles having antiferromagnetic ordering a weak moment will develop on the particle owing to surface effects. This moment fluctuate randomly because the cluster-cluster interaction is generally quite weak. The fluctuating particle (or cluster) magnetisation is responsible for the time dependence of the hyperfine field at the nuclear site. Thus, the line shape of the Mossbauer spectrum will depend on the relative magnitudes of τ_F and τ_L, (τ_L=nuclear Larmor precession time). Due to a distribution in cluster or particle size, which results in a spread in the value of τ_F, the observed Mossbauer spectrum is a sum of several Mossbauer patterns (192,194). For ^{57}Fe systems, the condition $\tau_F \sim \tau_L$ is generally achieved when $KV \sim k_BT$. Thus at room temperature particles of Fe_3O_4 with a diamater d >200A° show a well defined magnetically split Mossbauer pattern whereas when d < 100A°, a single line Mossbauer spectrum is observed (192). If the super-paramagnetic cluster (or particle) has uniaxial anisotropy, the Mossbauer lineshape can be easily calculated using Eqn. (91) and assuming n = 2 (170,192). However, if the cluster (or particle) magnetisation is fluctuating in more than two directions, the Mossbauer line shape is calculated using either the generalised stochastic theory with n>2 (167) or the perturbation approach (196).

Consider the case of an uniaxial cluster. The superparamagnetic cluster (or particle) may be regarded as a single entity with spin S' where $S' \sim 10^2-10^4$. Such an entity may be found in any one of the $(2S'+1)$ states. The variation in energy of these states is given by Eqn. (120). For the Mossbauer line shape calculation, transitions between any two of these states have to be considered. However, as shown by Eqn. (120), these states form two groups separated by an energy barrier. The energy minima correspond to states when S' is parallel or antiparallel to the symmetry axis. The two minima are separated by an energy barrier (197). Since the fluctuation rate for transitions among states which are not separated by an energy barrier (i.e. states close to the same free energy minimum) is high compared to the Larmor precession frequency, the two groups of states can be replaced by two discrete states at $\theta=0$ and $\theta=\pi$. This justifies the approximation n = 2 for the uniaxial clusters. Same explanation holds good for clusters of other symmetries also where n = 2n'. Here, n' is the number of easy directions of magnetisation. However due to the presence of large number of states near any of the energy minima among which the fluctuation is faster than τ_L^{-1}, there is a reduction in the internal field H_{int} to $H_{int} \langle\cos\theta\rangle_{av}$. The thermal average, $\langle\ \rangle_{av}$, is taken over the populated states within a well and θ is the direction corresponding to the states populated with respect to the symmetry axis. This has been experimentally observed (192).

The superparamagnetic effects have been observed experimentally in the following circumstances/systems:

(a) ultrafine particles (198)

(b) the presence of magnetic regions in the lattice surrounded by the non-magnetic ions and thus separated magnetically from the rest of the lattice (199).

(c) disappearance of the anisotropy energy and hence of the spin flip barrier (KV) when the temperature approaches T_N or T_C from below (called critical superparamagnetism) (200).

(d) small clusters of magnetic ions which are coupled, uncoupled or weakly coupled to the matrix polarisation (201)

(e) neutral precipitates of small size in a solid state reaction mixture (202)

(f) ultra thin films (203) or low dimensional magnetic lattices (204)

We may also mention here about some systems which are closely related to the superparamagnetic systems. If the superparamagnetic clusters, randomly distributed in a diamagnetic host, are small in concentration so that they interact weakly then at very low temperatures the spin gets 'frozen' in a particular orientation. In other words, the flip time becomes infinite. Such systems are called mictomagnetic systems (205). Similarly, if instead of magnetic clusters we have weakly coupled paramagnetic ions dispersed in a diamagnetic matrix, the fluctuation of the ionic spin stops at very low temperatures, i.e., the spin gets 'frozen' in a particular orientation. Such systems are called spin glasses (205) because the spins get 'frozen' in different directions. It may be noted that the above description does not give a complete picture of the mictomagnetic and spin glass systems (206). This is because, as is seen by the magnetic susceptibility behaviour of these systems, there is significant interaction present between the spins in the spin glass state (i.e. in the frozen state) and clusters in the mictomagnetic systems. Mictomagnetic systems generally have a mixture of frozen spins and frozen clusters, magnetically interacting among themselves and also with each other. Generally the mictomagnetic and spin glass systems are metallic systems, and the conduction electrons provide the spin-spin, cluster-cluster and cluster-spin couplings in these systems, through the RKKY interaction. However, recently some examples of nonmetallic spin glasses have also been found (205). Another system closely related to a superparamagnetic system is the sperimagnetic system which has a spin structure similar to that of a spin glass but does not seem to have the same type of spin-spin coupling as is found in a spin glass (207). Sperimagnetism has been found to be present in ultrafine particles of amorphous ferric gel (207). A few other important phenomena concerning the superparamagnetic ultrafine particles are non collinear spin structure at the surface, surface impurity sensitive magnetic anisotropy energy and pinning of surface spins (207).

Consider the effect of an external magnetic field on the superparamagnetic fluctuations. The presence of the field polarises the cluster and thus changes the flip time. The meaning of the cluster polarisation is as follows. For simplicity, let us consider a uniaxial cluster. The free energy of cluster is given by:

$$F = K_oV + K_1V\sin^2\theta + K_2V\sin^4\theta + ----. \tag{121}$$

We neglect the second order anisotropy constant in the following discussion.

The direction with $\theta = \frac{\pi}{2}$ is the hard axis of magnetisation and $\theta=0$ is the easy direction of magnetisation, along which the magnetisation vector can lie pointing either up (i.e. $\Theta=0^o$) or down (i.e. $\theta=\pi$) with equal probability. In the presence of an external magnetic field (H_{ext}) parallel to the symmetry axis and pointing upward, the free energy becomes:

$$F = K_oV + K_1V\sin^2\theta - \mu_{sp}H_{ext}\cos\theta. \tag{122}$$

Here, μ_{sp} is the magnetic moment of the cluster. Thus the energies of the cluster in the two states corresponding to $\theta=0$ and $\theta=\pi$ are no longer equal when $H_{ext}\neq 0$. The effect of H_{ext} on τ_F can be understood as follows. As seen from Eqn. (121), in the absence of H_{ext} the energy barrier (E_B) for the flip of the magnitisation vector from $\theta=0$ direction to $\theta=\pi$ direction and vice versa is K_1V. However, when $H_{ext}\neq 0$, the behaviour of the cluster in the two spin states differ. Denoting the barrier seen by the cluster when in the state corresponding to $\theta=0$ and π by $E_{B\uparrow}$ and $E_{B\downarrow}$ respectively, it follows from Eqn. (122)

$$E_{B\uparrow} = (F)_{\theta=\pi/2} - (F)_{\theta=0} = K_1V + |\mu_{sp}|\, H_{ext} \quad , \text{ and}$$

$$E_{B\downarrow} = K_1V - |\mu_{sp}|\, H_{ext} \quad .$$

Thus the flip times corresponding to up and down transitions respectively are given by (replacing K_1 by K)

$$\tau_{F\uparrow} = \tau_{0\uparrow} \exp \frac{KV+|\mu_{sp}|H_{ext}}{k_BT} \quad , \qquad \text{and}$$

$$\tau_{F\downarrow} = \tau_{0\downarrow} \exp \frac{KV-|\mu_{sp}|H_{ext}}{k_BT} \quad . \tag{123}$$

τ_0 is weakly dependent on E_B. The exact dependence will be discussed later. For approximate calculations, $\tau_{0\uparrow}$ can be assumed to be equal to $\tau_{0\downarrow}$. Thus,

$$\frac{\tau_{F\uparrow}}{\tau_{F\downarrow}} \approx \exp\left(\frac{2|\mu_{sp}|H_{ext}}{k_BT}\right) = s^{-1}. \tag{124}$$

An average flip time in presence of H_{ext} can be defined as follows. Let $\Omega_{F\uparrow}$ and $\Omega_{F\downarrow}$ are the flip frequencies of the cluster magnetisation when in lower and excited states, respectively. The average is defined as $\Omega = (\Omega_{F\uparrow} + \Omega_{F\downarrow})$ and $\tau_F = \frac{1}{\Omega}$. However, $\Omega_{F\uparrow} = s\Omega_{F\downarrow}$, where s is the ratio of the excited state and the ground state populations. In presence of H_{ext}, s becomes less than unity and the cluster is said to be polarised. Thus,

$$\tau_F = \frac{1}{(1+s)\Omega_{F\downarrow}} \quad .$$

In the absence of H_{ext},

$$(\Omega_{F\downarrow})_{H_{ext=0}} = (\Omega_{o\downarrow})_{H_{ext=0}} \exp(-\frac{KV}{k_BT}) \quad , \tag{125}$$

$$\frac{(\Omega_{o\downarrow})_{H_{ext=0}}}{(\Omega_{o\uparrow})_{H_{ext=0}}} = \frac{(\tau_{o\uparrow})_{H_{ext=0}}}{(\tau_{o\downarrow})_{H_{ext=0}}} = 1.$$

In the presence of H_{ext},

for $KV \gg k_BT$ and $|\mu_{sp}|H_{ext} \ll KV$ [see Eqn. 130]

$$\frac{(\Omega_{o\downarrow})_{H_{ext}}}{(\Omega_{o\uparrow})_{H_{ext}}} = \frac{(\tau_{o\uparrow})_{H_{ext}}}{(\tau_{o\downarrow})_{H_{ext}}} = \frac{2KV-|\mu_{sp}|H_{ext}}{2KV+|\mu_{sp}|H_{ext}} \quad \text{and} \tag{126}$$

$$\frac{(\Omega_{o\downarrow})_{H_{ext}}}{(\Omega_{o\downarrow})_{H_{ext=0}}} = \frac{2KV-|\mu_{sp}|H_{ext}}{2KV} = (1-\frac{|\mu_{sp}|H_{ext}}{2KV}). \tag{127}$$

Thus from Eqns. (125) - (127),

$$(\Omega_{F\downarrow})_{H_{ext}} = (\Omega_{F\downarrow})_{H_{ext=0}} (1-\frac{|\mu_{sp}|H_{ext}}{2KV})\exp(\frac{|\mu_{sp}|H_{ext}}{k_BT}) \quad , \tag{128}$$

$$(\tau_F)_{H_{ext}} \approx (\tau_F)_{H_{ext=0}} (1+\frac{|\mu_{sp}|H_{ext}}{2KV})\exp(-\frac{|\mu_{sp}|H_{ext}}{k_BT}) \quad . \tag{129}$$

This is an approximate relation valid only for $KV \gg k_BT$ and $H_{ext} \ll KV/|\mu_{sp}|$. Using the stochastic model approach of Brown (208), Aharoni and his co-workers have numerically calculated the dependence of τ_o, and thus of τ_F, on various parameters such as K, V, T, $|\mu_{sp}|$

and H_{ext} for the superparamagnetic clusters having uniaxial (209) as well as cubic (210) anisotropies. The exact dependence of τ_F on H_{ext} is quite complicated. The results obtained by Aharoni for an uniaxial superparamagnet are reproduced in Fig. (22). The dependence of τ_o on H_{ext} can be summarised as follows. For $H_{ext} = 0$ and $KV/k_BT \gtrsim 1$, τ_o is approximately given by:

$$\tau_o \approx \frac{M_s\sqrt{\pi}}{2K\gamma_o}\left(\frac{KV}{k_BT}\right)^{-1/2} ,$$

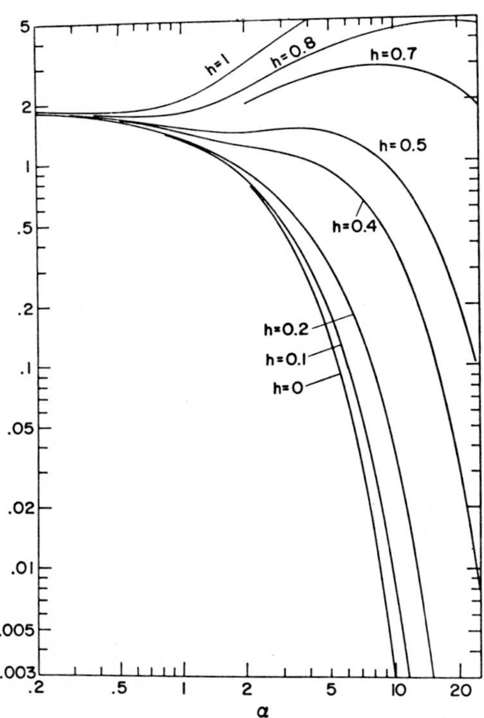

Fig. 22. The dependence of λ, which is inversely proportional to the superparamagnetic relaxation time, on $\alpha = KV/k_BT$ for various values of $h = H_{ext}\,M_s/2K$ for particles with uniaxial anisotropy in magnetic field H_{ext} at temperature T. (taken from Ref. 209).

where $\gamma_o = g\beta$ is the gyromagnetic ratio and M_s is the saturation magnetisation of the cluster [$\mu_{sp} = M_sV$]. This expression actually remains valid to a good approximation upto $H_{ext} \lesssim 0.4\ KV/|\mu_{sp}|$. When $H_{ext} \neq 0$, in the high energy barrier approximation ($KV/k_BT >> 1$) we have:

$$\tau_{o\downarrow} = \frac{M_s\sqrt{\pi}}{2K\gamma_o}\left(\frac{KV}{k_BT}\right)^{-1/2}\left[1-\left(\frac{|\mu_{sp}|H_{ext}}{2KV}\right)^2\right]\left[1-\left(\frac{|\mu_{sp}|H_{ext}}{2KV}\right)\right]^{-1} , \text{ and}$$

$$\tau_{o\uparrow} = \frac{M_s\sqrt{\pi}}{2K\gamma_o}\left(\frac{KV}{k_BT}\right)^{-1/2}\left[1-\left(\frac{|\mu_{sp}|H_{ext}}{2KV}\right)^2\right]\left[1+\left(\frac{|\mu_{sp}|H_{ext}}{2KV}\right)\right]^{-1} \quad . \qquad (130)$$

When $|\mu_{sp}|H_{ext} > 2$ KV the cluster becomes permanently polarised and superparamagnetism is destroyed. Thus for all practical purposes $|\mu_{sp}|H_{ext}/2KV < 1$. In most of the cases the term $1-(|\mu_{sp}|H_{ext}/2KV)^2$ in Eqn. (130) can be replaced by unity. Since the cluster polarisation depends on $|\mu_{sp}|$ and hence on the volume of the cluster, smaller clusters are expected to show much less influence of H_{ext} on the fluctuation frequency than the large diameter clusters. Mossbauer spectroscopy has shown this to be true unambiguously (190), Fig. 23.

An example of the temperature dependence of Mossbauer spectral shape in the presence of superparamagnetic fluctuations is given in Fig. 24. The effect of a spread in the particle size is visible in the spectrum at 78 K.

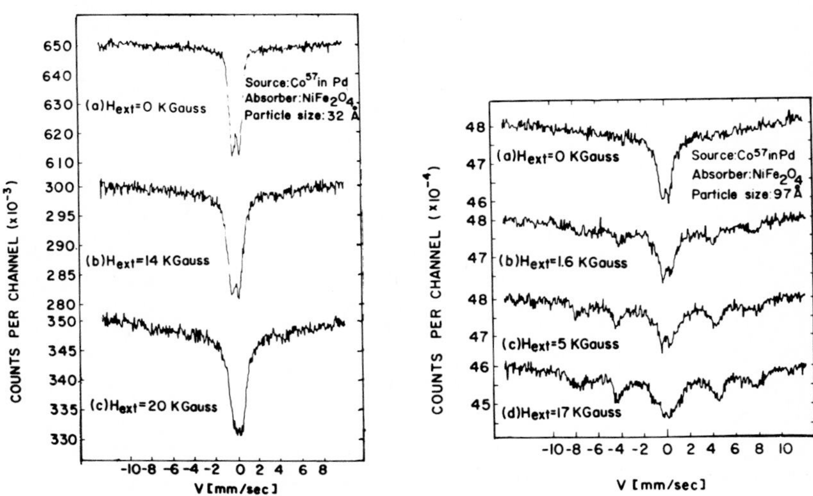

Fig. 23. Comparison of the effect of external magnetic field on particles of different sizes. (a) $NiFe_2O_4$ particles of size 32 Å at ro temperature, (b) $NiFe_2O_4$ particles of size 97 Å at room temperature (taken from Ref. 190).

6.3 Origin of Fluctuating EFG

In the above discussion, we have mainly discussed the origin of the time dependent magnetic hyperfine field. Most of the above mechanisms can also give rise to a time dependence to EFG at the nucleus obserable in Mossbauer spectra, if the ionic contribution to EFG is appreciable. In the case of the dynamic Jahn-Teller effect, the lattic contribution to EFG will also be time dependent. In Fe^{3+} compounds, where the ionic contribution to EFG is negligible and only the lattice contribution is present, time dependent EFG can arise due to either a fluctuating crystal symmetry (due to the presence of Jahn-Teller ions in the lattice) or a fluctuating vacancy in the otherwise ideal crysta lattice (17). In metallic system, the conduction electron relaxation can make the conduction electron contribution to EFG time dependent.

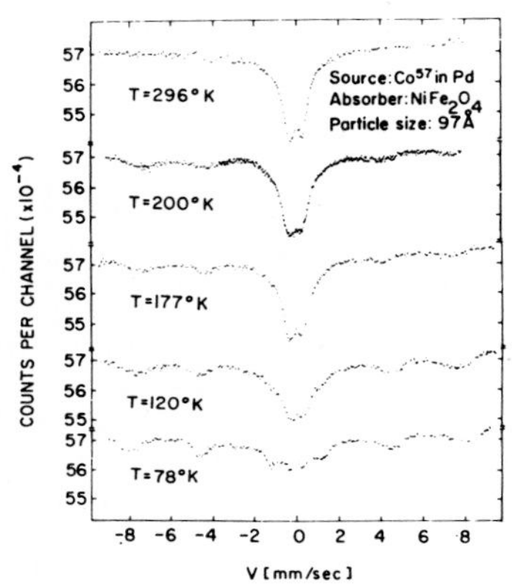

Fig. 24. Mossbauer spectra of $NiFe_2O_4$ particles of size 97 Å at various temperatures in absence of an external field (taken from Ref. 190).

7. OTHER APPLICATIONS OF MOSSBAUER SPECTROSCOPY

A few important applications of the Mossbauer spectroscopy which are not explicitly mentioned elsewhere in this book are briefly described in this section.

7.1 Magnetic Properties

7.1.1 The magnetic ordering temperature can be determined by observing the change in the gamma ray count rate at the centroid of a line of the Mossbauer spectrum above T_C or T_N when the magnetically split pattern changes to a single line (or quadrupole split) spectrum, as the tempera ture crosses T_C or T_N (Fig. 25) (64,132,211). An example of the appearance of the magnetic splitting as the temperature is lowered below T_N i shown in Fig. 26. This method has also been used to study the dependenc of the magnetic transition temperature on external pressure (Fig. 27).

7.1.2 T-dependences of lattice magnetisations

Since $H_{hf} \propto \langle S_z \rangle_{Th}$ in ionic substances, Mossbauer spectroscopy

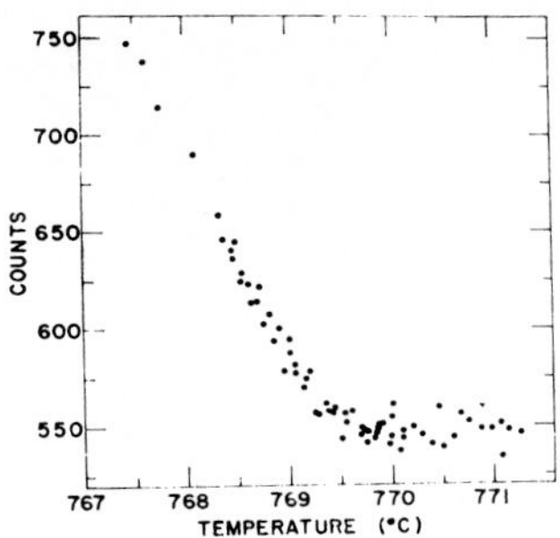

Fig. 25. The temperature dependence of the transmitted count rate in the vicinity of the transition temperature using metallic iron foil absorber. The relative velocity used correspond to the relative veloc at which maximum in the absorption spectrum occurs above T_C. The chan in the transmission is associated with the collapse of the magnetic splitting near T_C (Taken from Ref. 132, R.S. Preston et al.).

has been used to study the dependence of sublattice (or domain) magne tisation on the reduced temperature (T/T_N), impurity concentration in the lattice, external pressure, etc. (Fig. 28). These studies have provided useful information about the magnetic properties of solids (64,132,211).

7.1.3 Study of zero point spin deviation

Spin wave theories predict that whereas in the case of a ferro-magnet the magnetisation at 0° K equals the total spin value ($\langle S_z \rangle_{T=0} = S$), it is not so in the case of antiferromagnets (144,212) Assuming $H_{hf} \propto \langle S_z \rangle_{Th}$, Mossbauer measurements have been used to stud this phenomenon. For example, Mossbauer measurement on ^{193}Ir in (NH_4) $IrCl_6$ has shown that the difference $S-\langle S_z \rangle_{T=0}$ is 18%. This is in agreement with the theory (64,211).

7.1.4 Determination of critical point parameters

Near the magnetic ordering temperature, T_C or T_N, the temperatu dependence of the magnetisation can be written as:

$$M(T) = M(0).D.(1-T/T_N)^{\beta} .$$

D and β are called the critical point coefficients. Whereas, the Weis molecular field theory predicts β=1/2, the spin wave theories predict β=1/3. For a two dimensional lattice, the Ising model predicts β=1/8 and for a three-dimensional lattice it predicts β=5/16 (213). Since

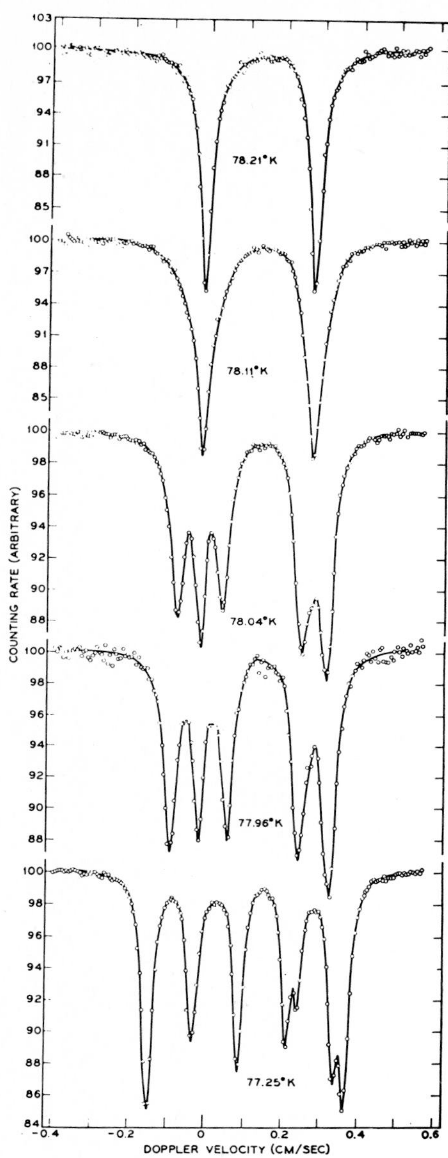

Fig. 26. Mossbauer spectra of FeF_2 in the vicinity of the Neel temperature (taken from Ref. 211, G.K. Wertheim and D.N.E. Buchanan)

$H_{hf} \propto \langle S_z \rangle_{Th}$, the above dependence gives

$$H_{hf}(T) = H_{hf}(o).D.(1-T/T_N)^{\beta} \quad .$$

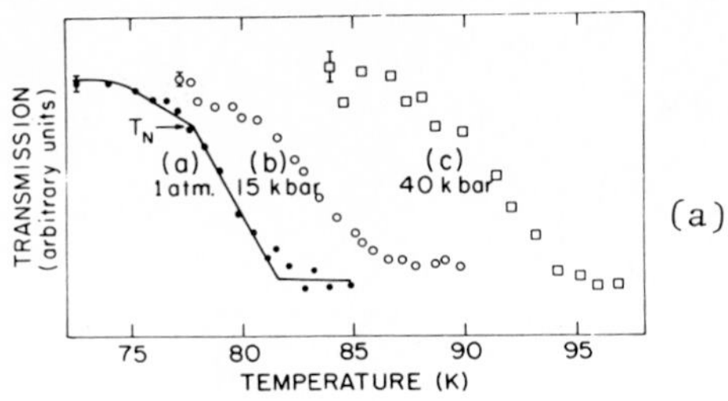

(a)

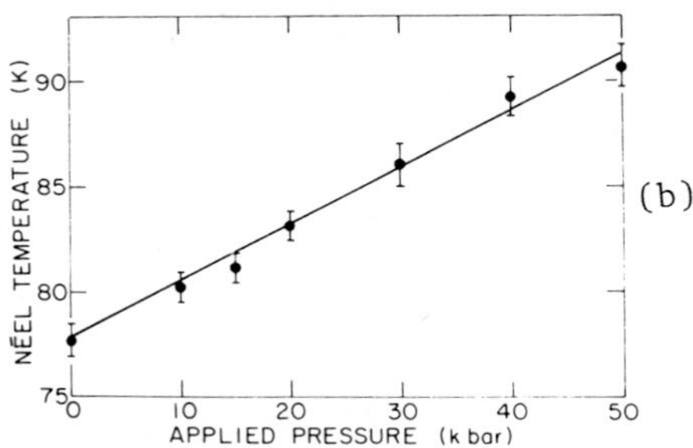

(b)

Fig. 27. (a) Temperature dependence of transmission of gamma rays through absorber FeF_2 near the magnetic transition temperature in presence of external pressure. Source is ^{57}Co in stainless steel 310 which gives one of the quadrupole split line of FeF_2 absorber near zero relative velocity. Thus the two are kept stationary for the thermal scan. (b) The dependence of T_N of FeF_2 on external pressure determined using the thermal scan method. (taken from Ref. 211, G.A. Garcia and R. Ingalls)

Thus from an experimental determination of the temperature dependence of hyperfine field close to T_N , β and D can be obtained. For example, accurate measurements of hyperfine fields in FeF_2 in the range from 77.25 to 78.21° K (T_N=78.12° K) at small intervals (Fig. 26) yield β=0.325 ± 0.005 and D=1.36 ± 0.03 (Fig. 29). Several other such studies have also been made (211).

7.1.5 Magnetic structure studies

As has been mentioned before, it is possible to find the direction of the magnetisation vector with respect to the crystallographic axes, from the analysis of the Mossbauer spectrum (64,211). Mossbauer studies in presence of an external magnetic field have also revealed relative spin arrangements in several cases. This has been discussed at length elsewhere in this book. Single crystals and polarised gamma rays have also been used for such studies (211).

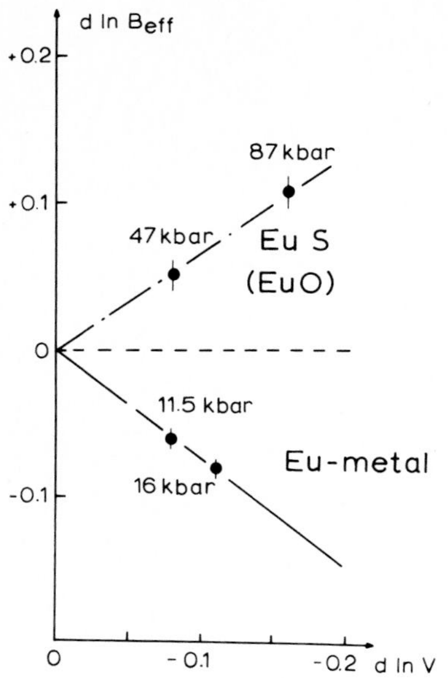

Fig. 28. The dependence of magnetic hyperfine field on the atomic volume of Eu in Eu metal and EuS. Eu is in valence state Eu^{2+} in EuS. Whereas core polarisation is largely responsible for the behaviour in EuS, the conduction electron dominates the dependence in Eu metal (taken from Ref. 211, G.M. Kalvius et al).

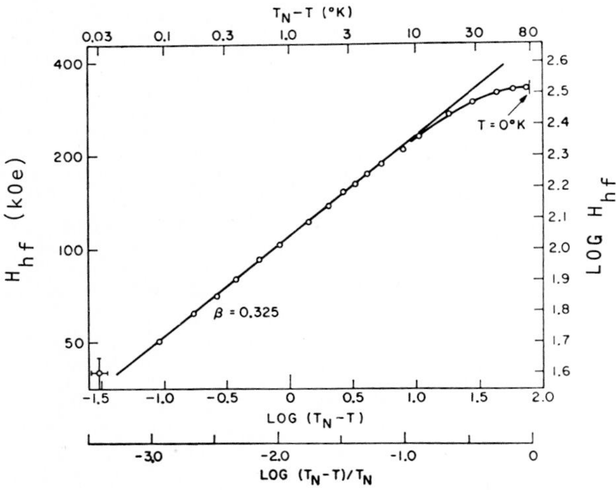

Fig. 29. The dependence of $\log_{10} H_{hf}$ on $\log_{10}(T_N - T)$ in FeF_2. (taken from Ref. 211, G.K. Wertheim and D.N.E. Buchanan)

7.1.6 Hyperfine anomaly

The magnetic hyperfine interaction Hamiltonian is $\vec{I}.\vec{A}.\vec{S}$ where $\vec{A}$ is the hyperfine interaction constant tensor. As mentioned before, $\vec{A}$ is weakly dependent on temperature in some cases. However it is generally thought to be independent of the size of the nucleus. This can be seen from Eqn. (55) where there is no term which is dependent on the nuclear size. But experimentally, in the case of the nuclei of heavy elements, $\vec{A}$ has been found to depend on the nuclear size. Thus, in these cases $\vec{A}$ is different for the excited and the ground states of the same nucleus. Similarly, $\vec{A}$ is also different for the ground nuclear states of two isotopes of the same element. When such a phenomenon occurs, the hyperfine interaction is said to be anomalous. Anomalous hyperfine interaction is experimentally measured by observing the difference between nuclear magnetic moment measured in an uniform field (H_{ext}) and the moment measured in a magnetic field of hyperfine origin (H_{hf}). The principle used in the Mossbauer study is as follows. From Mossbauer spectrum, the energy separation between the successive m-substates of the nuclear ground state, Δ_{gr} and that of the excited state, Δ_{ex} , can be obtained. This ratio is given by, assuming that the splitting of the nuclear levels is due to H_{ext} ,

$$\left(\frac{\Delta_{ex}}{\Delta_{gr}}\right)_{H_{ext}} = \frac{H_{ext} g_{ex} \beta_n I_{ex}}{H_{ext} g_{gr} \beta_n I_{gr}} = \frac{g_{ex} I_{ex}}{g_{gr} I_{gr}} \quad . \tag{131}$$

Similarly, if the nuclear levels are split by the hyperfine field, we have

$$\left(\frac{\Delta_{ex}}{\Delta_{gr}}\right)_{H_{hf}} = \frac{(H_{hf})_{ex}}{(H_{hf})_{gr}} \left(\frac{I_{ex} g_{ex}}{I_{gr} g_{gr}}\right) \quad . \tag{132}$$

From Eqns. (131) and (132), we get

$$\frac{(\Delta_{ex}/\Delta_{gr})_{H_{hf}}}{(\Delta_{ex}/\Delta_{gr})_{H_{ext}}} = \frac{(H_{hf})_{ex}}{(H_{hf})_{gr}} \quad ,$$

where $(H_{hf})_{ex}$ and $(H_{hf})_{gr}$ are the magnitudes of the hyperfine magnetic field at the nucleus when in excited and ground states respectively. As discussed earlier in this section $H_{hf}=A\langle S_z\rangle_{Th}$. Thus

$$\frac{(\Delta_{ex}/\Delta_{gr})_{H_{hf}}}{(\Delta_{ex}/\Delta_{gr})_{H_{ext}}} = \frac{A_{ex}}{A_{gr}} \quad . \tag{133}$$

In light nuclei,

$$\frac{A_{ex}}{A_{gr}} \approx 1 \quad .$$

However, in heavy nuclei, where anomalous hyperfine interaction is present, we can write

$$\frac{A_{ex}}{A_{gr}} = 1+\Delta \quad .$$

Δ is called the hyperfine anomaly and its magnitude is much smaller than unity. Thus Mossbauer measurements of $(\Delta_{ex}/\Delta_{gr})$ in H_{ext} when $H_{hf}=0$ and in H_{hf} when $H_{ext}=0$ provide value of Δ. These measurements have shown that $\Delta \approx 0$ in ^{57}Fe whereas in a heavy nuclei like ^{193}Ir , $\Delta \approx 0.0$ (211). The origin of the dependence of the hyperfine constant A on the volume of the nucleus can be understood as follows (33). In obtaining Eqn. (55) it has been assumed that $\psi(0)$ is constant over the space occupied by the nucleus. This leads to a hyperfine constant independent of the nuclear radius. However, the value of $|\psi(0)|^2$ which is maximum at the origin (at the centre of the nucleus) for an s-electron may fall by a small but significant amount over the nuclear volume, particularly for heavier nuclei whose radii are larger and for which the electronic wave functions are contracted because of the large nuclear charge. In such a case the assumption of a constant $|\psi(0)|^2$ over the nuclear volum is not satisfactory. As the nuclear radius may be significantly different for different isotopes of the same element or for the different energy states of the same nucleus, the strength of the hyperfine interaction constant A, may vary from isotope to isotope and may be very different for the nucleus in different energy states. This leads to the hyperfine anomaly. The effect is obviously relevant to the Fermi contac interaction only.

7.1.7 Giant magnetic moments

When a paramagnetic ion is diffused in a metallic nonmagnetic host it polarises the conduction electrons. If the paramagnetic ion is a Mossbauer ion like Fe, Mossbauer spectroscopy can provide valuable information about the polarisation. If the concentration of the Fe is very large, this conduction electron polarisation extends throughout the host matrix. However, if the concentration of the Mossbauer ions is very small, the conduction electron polarisation is predominant mainly in a small region around the Mossbauer ion. In this case, in some system giant magnetic moment formation occurs which arises due to the tight coupling of the Mossbauer ion with its polarised environment (211). This magnetic moment of the rigidly coupled composite Mossbauer ion-host complex is dependent on the concentration of the Fe atom (for example Fig. 30). The Mossbauer spectrum of such a system does not show

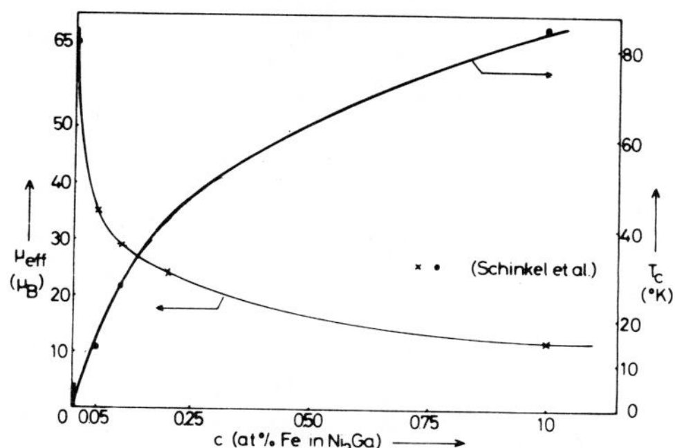

Fig. 30. The dependence of the effective magnetic moment of the host-Mossbauer ion complex and the curie temperature related to the cluster-cluster magnetic interactions on the concentration of the Fe atom in Ni_3Ga (Taken from Ref. 211, H. Maletta and R.L. Mossbauer and J.C. Schinkel, F.R. DeBoer and J. Biesterbos, Phys. Lett. 26A(1968)501).

any magnetic splitting in the absence of an external magnetic field owing to perhaps the fast superparamagnetic effect type fluctuation of the giant magnetic moment of the Mossbauer ion-host complex*. However, a magnetic splitting appears in presence of an external magnetic field which increases with it up to a saturation value (Fig. 31 gives an example). Writing the saturation value of the hyperfine field as $(H_{hf})_{Sat}$, the following relation has been found to hold good between the hyperfine field and (H_{ext}/T):

$$(H_{hf})_{H_{ext}/T} = (H_{hf})_{Sat} \cdot B_{S'}\left(\frac{\mu H_{ext}}{k_B T}\right) .$$

Here, $B_{S'}(\ )$ is the Brillouin function. μ, the magnetic moment of the Mossbauer ion-host complex, is related to its effective spin value S' by

$$\vec{\mu} = g\beta\vec{S}' .$$

---

* It may be emphasised the magnetic transition temperature shown in Fig. 30 relates to the magnetic coupling of the clusters with each other. The magnetic interactions within the cluster is much stronger.

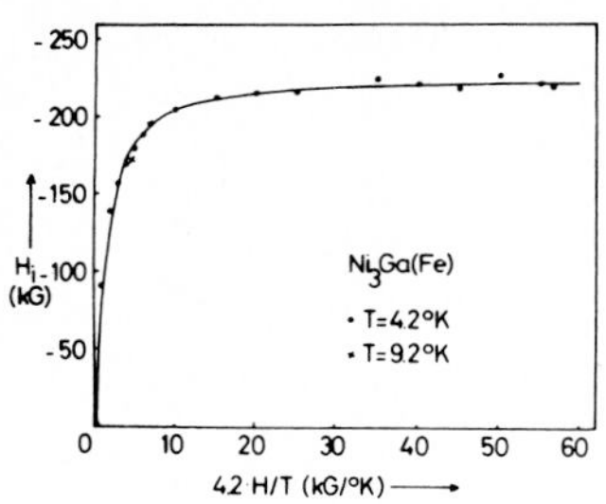

Fig. 31. The dependence of the hyperfine magnetic field at Fe in Ni_3Ga (concentration of Fe 20 ppm) on H_{ext}/T. The solid curve is the least squares fit of the experimental points to the Brillouin curve for values $(H_{hf})_{Sat}$ = -(222.6 ± 0.5)kG and S'=32.6 ± 0.8 (taken from Ref. 211, H. Maletta and R.L. Mossbauer)

However, $(H_{hf})_{Sat}$ has been found to depend only on the magnetic moment localised on the Mossbauer ion and not on μ which is the total moment of the Mossbauer ion-host complex. Taking the Fe in Ni_3Ga system as an example, μ has been found to lie between 10β and 70β. Whereas $(H_{hf})_{Sat}$ = -222 ± 0.5 kG when the concentration of Fe in Ni_3Ga is 20 ppm, μ=65.2 ± 1.5 β. As the magnitude of μ is dependent on the spatial extent of the medium over which the conduction electron polarisation around the Mossbauer ion is significant, its exact value depends not only on the nature of the host lattice but also on the concentration of the Mossbauer ion in the matrix (Fig. 30). The magnetic susceptibility technique has also been very commonly used for detecting the giant moment formation (211).

7.2 Study of Liquid Crystals

Study of the liquid crystalline state of matter is fascinating. In certain substances, the constituent molecules are highly anisotropic in shape and can be approximated to lines (rods) or ellipsoids (globes). In these cases, an additional orientational order (order in the direction of orientation of the ellipsoid axis for example), along with the positional order, is necessary for having a perfect crystalline solid state (Fig. 32a). The liquid state of matter, on the other hand, has no positional or orientational order of any kind (Fig. 32b). The liquid crystalline state is an intermediate state between these two extremes (214,215). Obviously many such states are possible. For instance, one can have a high degree of orientational order but no positional order. In this state, the molecules are all oriented approximately

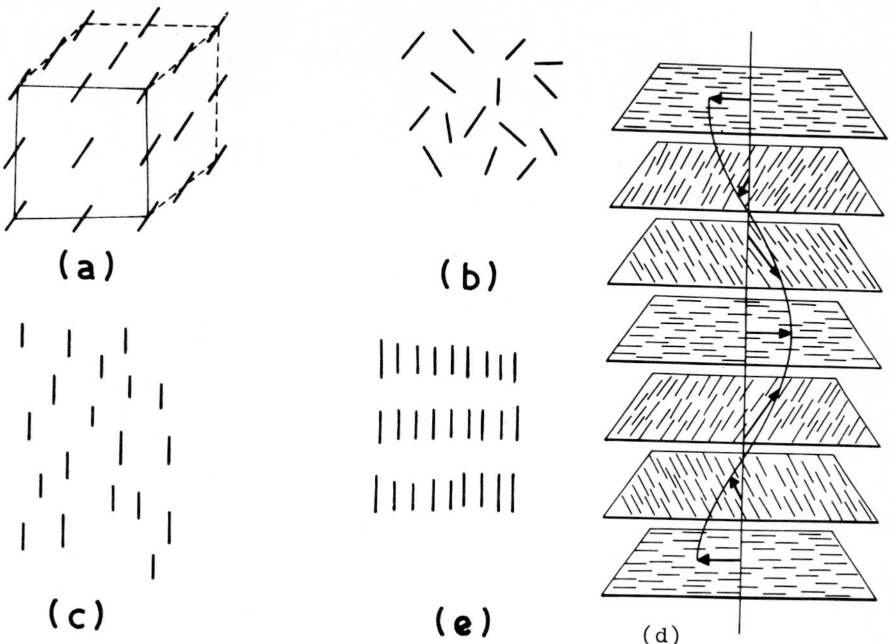

Fig. 32. Schematic representations of ordering (positional as well as orientational) (a) crystalline state, (b) liquid state, (c) nematic liquid crystalline state, (d) cholesteric liquid crystal, (e) smectie liquid crystal.

parallel to one another but their centres of gravity are randomly distributed in the medium (Fig. 32c). Such an ordered arrangement of molecules has been termed as the nematic liquid crystal. These materials show anisotropy in various physical properties as well as certain degree of fluidity. When the molecules form layered structure but the orientation of the molecules in successive planes rotates gradually through a definite angle, it is another liquid crystalline state termed as cholesteric liquid crystal (Fig. 32d). Another possible intermediate layered structure is the one in which the molecules are not only orientationally ordered but their centres of gravity are also in different planes resulting in a layered structure (Fig. 32e). Such a system can be described as layers stacked one over the other. This state is termed as smectic liquid crystal. These are the three general categories of the liquid crystals. There are many sub-categories also, like smectic A, smectic B, twisted smectic, etc. The cholesteric liquid crystal is sometimes referred to as twisted nematic and, strictly speaking, is not fundamentally different from the normal nematic phase. It may be mentioned here that a lattice having only positional order and no orientational order is called a plastic crystal.

Using Mossbauer spectroscopy solid $\xrightarrow{T_1}$ smectic $\overset{T_2}{\rightleftarrows}$ nematic phase transitions have been studied by observing discontinuities in the f-factor vs. T curve at these transitions ($T_1 < T_2$) (Fig. 33). The f-factor has been found to decrease considerably in the nematic phase. By studying the θ dependence of the f factor and of the intensity ratio of the two quadrupole doublet lines, sign of V_{zz} and the molecular and lattice contributions to the vibrational anisotropy have been determined (216). (θ is the angle between the direction of the applied electric or magnetic field and the gamma ray propagation direction). Studies have also been made of the anisotropic diffusion and the anisotropy in the coefficient of viscosity in liquid crystals. Brownian motion of particles (containing Mossbauer ion) in aligned and unaligned liquid crystals, glass transition phenomenon and order parameter in super-cooled liquid crystals etc. are some of the studies carried out in liquid crystals using the technique of Mossbauer spectroscopy (216).

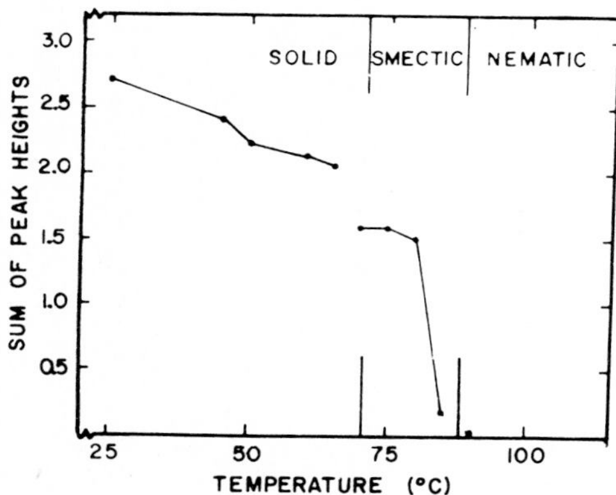

Fig. 33. The temperature dependence of the sum of the peak heights of the two lines of the quadrupole split spectrum of the unoriented 7-wt% solutions of 1,1'-diacetyl ferrocene in 4,4'-bis (heptyloxy)azoxybenzene. The plot is for the heating cycle (taken from Ref. 216, D.L. Uhrich et al, 1970).

REFERENCES

1 W.E. Lamb, Jr. Phys. Rev., 55 (1939) 190.
2 R.W. Wood, Physical Optics, Macmillan, New York, 1934.
3 P. Pringsheim, Fluorescence and Phosphorescence, Interscience New York, 1949.
4 E.R. Andrew, Nuclear Magnetic Resonance, Cambridge University Press, Cambridge, 1955.
5 W. Low, Paramagnetic Resonance in Solids, Academic, New York, 1960
6 F. Frauenfelder, The Mossbauer Effect, W.A. Benjamin Inc., New York, 1962.
7 V.I. Goldanskii and R.H. Herber (Eds.), Chemical Applications of Mossbauer Spectroscopy, Academic, New York, 1968, p. 1, 548, 249.
8 P.B. Moon, Proc. Phys. Soc., 63 (1950) 1189; ibid 64 (1951) 76.
9 R.L. Mossbauer, Z. Physik, 151 (1958) 124; Naturwissenschaften, 45 (1958) 538; Z. Naturforsch; 14(a) (1959) 211.
10 W.M. Visscher, Annals Phys., 9 (1960) 194.
11 H.J. Lipkin, Annals Phys., 9 (1960) 332.
12 A.A. Maradudin, Rev. Mod. Phys., 36 (1964) 417; K.S. Singwi and A. Sjolander, Phys. Rev., 120 (1960) 1093.
13 R.L. Mossbauer, J. Phys. (Paris) Colloq., 37 (1976) C6-5; H. Lustig, Amer. J. Phys. 29 (1961) 1; R.L. Mossbauer, Ann. Rev. Nucl. Sci. 12 (1962) 123; R.L. Mossbauer and M.J. Clauser, in A.J. Freeman and R.B. Frankel (Eds.), Hyperfine Interactions, Academic, New York, 1967, p. 497.
14 D. Raj and S.P. Puri, Phys. Lett., 29A (1969) 510; S.S. Nandwani, D. Raj and S.P. Puri, Phys. Lett., 34A (1971) 365; S.S. Nandwani and S.P. Puri, Phys. Status Solidi, 41 (1970) 199.
15 J.K. Major, in I.J. Gruverman (Ed.), Mossbauer Effect Methodology, Vol. 1, Plenum, New York, 1965, p. 89; P. Debrunner, ibid, p. 97; J.J. Spijkerman, ibid Vol. 7, 1971, p. 85.
16 R.H. Nussbaum, in I.J. Gruverman (Ed.), Mossbauer Effect Methodology, Vol. 2, Plenum, New York, 1966, p. 3; R.H. Nussbaum, D.G. Howard, W.L. Nees and C.F. Steen, Phys. Rev., 173 (1968) 653; J.A. Moyzis, Jr. G. De Pasquali and H.G. Drickamer, Phys. Rev., 172 (1968) 665; C. Zener, Phys. Rev., 49 (1936) 122; A.H. Muir, Jr., Tables and Graphs for Computing Debye-Waller Factors in Mossbauer Effect Studies, Report AI-6699, Atomics International, Canoga Park, 1962; J. Heberle, in I.J. Gruverman (Ed.), Mossbauer Effect Methodology, Vol. 7, Plenum, New York, 1971, p. 299; D.C. Gupta and N.D. Sharma, Nucl. Instrum. Methods, 143 (1977) 407.
17 J.G. Stevens and Virginia E. Stevens (Eds.), Mossbauer Effect Data Index, IFI/Plenum Data Co. (A Division of Plenum Pub. Corporation) New York, 1958, onwards.
18 G.K. Shenoy and F.E. Wagner (Eds.), Mossbauer Isomer Shifts, North Holland Pub. Co., Amsterdam, 1978.
19 G.K. Wertheim, Mossbauer Effect: Principles and Applications, Academic, New York, 1964.
20 R.D. Taylor, D.J. Erickson and T.A. Kitchens, J. Phys. (Paris) Colloq., 37 (1976) C6-35; S.S. Nandwani, D. Raj and S.P. Puri, J. Phys. C: Solid State Phys., 4 (1971) 1929; R.M. Housley and F. Hess, Phys. Rev., 146 (1966) 517; P.D. Mannheim, Phys. Rev., 165 (1968) 1011; ibid B5 (1972) 745; P.D. Mannheim and S.S. Cohen, Phys. Rev. B4 (1971) 3748.
21 D. Raj and S.P. Puri, Phys. Lett., 33A (1970) 306; ibid J. Phys. Chem. Solids, 33 (1972) 2177; P.D. Mannheim and A. Simopoulos, Phys. Rev., 165 (1968) 845; D.G. Howard and R.H. Nussbaum, Phys. Rev. B9 (1974) 794; K. Ohashi and Y.H. Ohashi, J. Phys. F: Metal Phys., 7 (1977) 1875; J.M. Grow, D.G. Howard, R.H. Nussbaum and M. Takeo, Phys. Rev. B17 (1978) 15; R.K. Puri and L.R. Gupta, Phys. Status Solidi (b), 83 (1977) 293.

22 S.S. Nandwani and S.P. Puri, Phys. Letters, 44A (1973) 459; D.C. Gupta and N.D. Sharma, J. Phys. Soc. Japan, 42 (1977) 1463; J.A. Moyzis, Jr., G. De Pasquali and H.G. Drickamer, Phys. Rev., 172 (1968) 665; A.A. Opalenko and Z. Zallam, Phys. Status Solidi (b), 99 (1980) K27; B.N. Srivastava and R.N. Tyagi, J. Phys. Chem. Solids; 31 (1970) 2154; K. Mahesh, Indian J. Pure Appl. Phys., 4 (1966) 480.

23 J.G. Dash, in I.J. Gruverman (Ed.), Mossbauer Effect Methodology, Vol. 1, Plenum, New York, 1965, p. 107; B. Kolk and B. Harwig, Nucl. Instrum. Methods, 94 (1971) 211.

24 W.H. Southwell, D.L. Decker and H.B. Vanfleet, Phys. Rev., 171 (1968) 354.

25 D.P. Johnson, Nucl. Instrum. Methods, 85 (1970) 29.

26 S. Roth and E.M. Horl, Phys. Lett., 25A (1967) 299.

27 O.I. Sumbaev, A.I. Smirnov and V.S. Zykov, Sov. Phys. JETP, 15 (1962) 82.

28 J. Ball and S.J. Lyle, Nucl. Instrum. Methods, 163 (1979) 177.

29 N. Abe and L.H. Schwartz, in I.J. Gruverman and C.W. Seidel (Eds.), Mossbauer Effect Methodology, Vol. 8, Plenum, New York, 1973, p. 249.

30 D.A. O'Conner and G. Longworth, Nucl. Instrum. Methods, 30 (1964) 290; R.L. Mossbauer, H.E. Seelbach, B. Persson, M. Bent and G. Longworth, Phys. Lett., 28A (1968) 94.

31 N.F. Ramsey, Nuclear Moments, John Wiley, New York and Chapman and Hall, London, 1953, p. 3.

32 R.D. Evans, Atomic Nucleus, McGraw Hill, New York, 1955.

33 B. Bleaney, in A.J. Freeman and B.R. Frankel (Eds.), Hyperfine Interactions, Academic, New York, 1967, p. 1.

34 M.H. Cohen and F. Reif, in F. Seitz and D. Turnbull (Eds.), Solid State Phys., 5 (1957) 321.

35 H.E. White, Introduction to Atomic Spectra, McGraw Hill, New York, 1934.

36 J.B. Goodenough, Magnetism and the Chemical Bond, Interscience, New York, 1963.

37 C.J. Ballhausen and H.B. Gray, Molecular Orbital Theory, W.A. Benjamin, New York, 1964, p. 1.

38 R.E. Watson, Phys. Rev., 118 (1960) 1036; ibid 119 (1960) 1934.

39 J.C. Slater, Quantum Theory of Molecules and Solids, Vol. 1, McGraw Hill, New York, 1963, p. 256; H.L. Schlafer and G. Gliemann, Ligand Field Theory, Wiley-Interscience, London, 1969, p. 207, 321

40 Ç.M. Moser, in A.J. Freeman and B.R. Frankel (Eds.), Hyperfine Interactions, Academic, New York, 1967, p. 95.

41 A.J. Freeman and R.E. Watson, in G.T. Rado and H. Suhl (Eds.), Magnetism, Vol. IIA, Academic, New York, 1965, p. 167; D.R. Hartree, The Calculation of Atomic Structure, John Wiley, New York, 1957.

42 R.E. Watson and A.J. Freeman, in A.J. Freeman and B.R. Frankel (Eds.), Hyperfine Interactions, Academic, New York, 1967, p. 53.

43 R. Ingalls, Phys. Rev., 155 (1967) 157; R. Ingalls, H.G. Drickamer and G. De Pasquali, Phys. Rev., 155 (1967) 165; J.P. Bocquet, Y.Y. Chu, O.C. Kistner, M.L. Perlman and G.T. Emery, Phys. Rev. Lett., 17 (1966) 809; L.R. Walker, G.K. Wertheim and V. Jaccarino, Phys. Rev. Lett., 6 (1961) 98; A.R. Bodmer, Nucl. Phys., 21 (1961) 347; G.K. Wertheim, Phys. Rev. Letters, 12 (1964) 28; G.K. Wertheim and J.H. Wernick, Phys. Rev., 123 (1961) 755; P.G. Huray, C.M. Tung, F.E. Obenshain and J.O. Thomson, J. Phys. (Paris) Colloq., 37 (1976) C6-371.

44 G.K. Shenoy and S.L. Ruby, in I.J. Gruverman (Ed.), Mossbauer Effect Methodology, Vol. 5, Plenum, New York, 1970, p. 77; B.D. Dunlap, ibid, Vol. 7, 1971, p. 123.

45 K.N. Shrivastava, Phys. Rev. B, 13 (1976) 2782; C. Bansal and K.N. Shrivastava, Chem. Phys. Lett., 64 (1979) 388.
46 G.M. Kalvius, U.F. Klein and G. Wortmann, J. Phys. (Paris) Colloq., 35 (1974) C6-139; H. Prosser, G. Wortmann, K. Syassen and W.B. Holzapfel, Z. Physik B, 24 (1976) 7; R. Yanovsky, E.R. Bauminger, I. Felner, I. Nowik and S. Ofer, Hyperfine Interact., 3 (1977) 263; D.L. Williams, J.H. Dale, W.D. Josephson and L.D. Roberts, Phys. Rev. B, 17 (1978) 1015; G. Wortmann, G. Trollmann, A. Heidemann and G.M. Kalvius. Hyperfine Interact., 4 (1978) 610; R.E. Watson and L.H. Bennett, Hyperfine Interact., 4 (1978) 806; W.T. Krakow, W.D. Josephson, P.A. Deane, D.L. Williamson and L.D. Roberts, Phys. Rev. B, 23 (1981) 499.
47 G. Kaindl, D. Salomon and G. Wortmann, in I.J. Gruvermann and C.W. Seidel (Eds.), Mossbauer Effect Methodology, Vol. 8, Plenum New York, 1973, p. 211; I.J. Zuckerman, in I.J. Gruverman (Ed.), Mossbauer Effect Methodology, Vol. 3, Plenum, New York, 1967, p. 15; M.A. Kobeissi, Lee Chow and C. Hohenemser, Hyperfine Interact., 4 (1978) 485.
48 P. Kienle, in E. Matthias and D.A. Shirley (Eds.), Hyperfine Structure and Nuclear Radiations, North Holland, Amsterdam, 1968, p. 27.
49 D.A. Shirley, Rev. Mod. Phys., 36 (1964) 339.
50 W. Kundig, Hyperfine Interact., 2 (1976) 113; H.K. Walter, H. Backe, R. Engfer, E. Kankeleit, C. Pelitjean, H. Schneucoly and W.U. Schroder, Phys. Lett., 38B (1972) 64.
51 R.W. Grant, in I.J. Gruverman (Ed.), Mossbauer Effect Methodology, Vol. 2, Plenum, New York, 1966, p. 23; Y. Hazony, ibid, Vol. 7, 1971, p. 147; A.J. Carty and H.D. Sharma, ibid, Vol. 7, 1971, p. 167; P.H. Ramy and H. Pollak, J. Appl. Phys., 36 (1965) 860 and Rev. Mod. Phys., 36 (1964) 352.
52 L. Pfeiffer, M. Eibschutz and D. Salomon, Hyperfine Interact., 4 (1978) 803; M. Eibschutz and F.J. DiSalvo, Phys. Rev. B, 15 (1977) 5181; M.J. O'Shea and M. Springford, Phys. Rev. Lett., 46(1981)1303
53 T.A. Kitchens, P.P. Craig and R.D. Taylor, in I.J. Gruverman (Ed.), Mossbauer Effect Methodology, Vol. 5, Plenum, New York, 1970, p. 123.
54 D. Raj and S.P. Puri, Phys. Status Solidi, 34 (1969) K13.
55 D. Feder and I. Nowik, J. Magn. Magn. Mater., 12 (1979) 149 and 162.
56 G.P. Gupta and K.C. Lal, Phys. Status Solidi (b), 51 (1972) 233.
57 D. Kishore, Umendra, G.P. Gupta and K.C. Lal, Phys. Status Solidi (b), 98 (1980) K51.
58 S.K. Roy, K.M. Singh and D.L. Bhattacharya, Indian J. Pure Appl. Phys., 16 (1978) 1019; A.A. Maradudin and P.A. Flinn, Phys. Rev., 126 (1962) 2059; ibid 129 (1963) 2529.
59 A. Studel, in A.J. Freeman and B.R. Frankel (Eds.), Hyperfine Interactions, Academic, New York, 1967, P. 141; J.C. Walker, ibid, p. 650.
60 R.L. Collins and J.C. Travis, in I.J. Gruverman (Ed.), Mossbauer Effect Methodology, Vol. 3, Plenum, New York, 1967, p. 123.
61 U. Gonser and H. -D Pfannes, J. Phys. (Paris) Colloq., 35 (1974) C6-113.
62 C.E. Johnson, W. Marshall and G.J. Perlow, Phys. Rev., 126 (1962) 1503; U. Gonser, in A.J. Freeman and B.R. Frankel (Eds.), Hyperfine Interactions, Academic, New York, 1967, p. 696; J.P. Stampfel and P.A. Flinn, in I.J. Gruverman (Ed.), Mossbauer Effect Methodology, Vol. 6, Plenum, New York, 1971, p. 95; R.M. Housley, ibid, Vol. 5, 1970, p. 109; U. Gonser, in E. Matthias and D.A. Shirley (Eds.), Hyperfine Structure and Nuclear Radiations, North Holland, Amsterdam, 1968, p. 343; U. Gonser, H. Sakai and W. Keune, J. Phys. (Paris) Colloq., 37 (1976) C6-709.

63 R.W. Grant, in I.J. Gruverman (Ed.), Mossbauer Effect Methodology, Vol. 2, Plenum, New York, 1966, p. 23.

64 P. Zory, Phys. Rev., 140 (1965) A1401; W. Kundig, Nucl. Instrum. Methods, 48 (1967) 219; J.K. Srivastava and R.P. Sharma, Phys. Status Solidi (b), 49 (1972) 135; J.K. Srivastava and K.G. Prasad, Phys. Status Solidi (b), 54 (1972) 755; V.A. Povitskii, A.N. Salugin, Yu.V. Baldokhin, E.F. Makarov and N.V. Elistratov, Phys. Status Solidi (b), 87 (1978) K101.

65 J.O. Artman, in I.J. Gruverman (Ed.), Mossbauer Effect Methodology Vol. 7, Plenum, New York, 1971, p. 187; C.P. Slichter, Principles of Magnetic Resonance, Harper and Row, New York, 1963, p. 160; I.F. Foulkes and B.L. Gyorffy, Phys. Rev. B, 15 (1977) 1395.

66 R.M. Sternheimer, Phys. Rev., 130 (1963) 1423.

67 R.M. Sternheimer, Phys. Rev. A, 6 (1972) 1702; S.N. Ray and T.P. Das, Phys. Rev. B, 16 (1977) 4794; K.D. Sen and P.T. Narasimhan, Phys. Rev. B, 16 (1977) 107.

68 R.E. Watson, in A.J. Freeman and B.R. Frankel (Eds.), Hyperfine Interactions, Academic, New York, 1967, p. 413; M. Belakhovsky and J. Pierre, Solid State Commun., 9 (1971) 1409; G.S. Collins, Hyperfine Interact., 4 (1978) 523; P.C. Pattnaik, M.D. Thompson and T.P. Das, Phys. Rev. B, 16 (1977) 5390.

69 M. Pomerantz and T.P. Das, Phys. Rev., 119 (1960) 70; T.P. Das and M. Pomerantz, Phys. Rev., 123 (1961) 2070; M.B. Stearns, Phys. Rev. B, 8 (1973) 4383.

70 R.E. Watson, A.C. Gossard and Y. Yafet, Phys. Rev., 140 (1965) A375 T.J. Rowland, Nuclear Magnetic Resonance in Metals, Pergamon, New York, 1961, p. 27.

71 T.P. Das, Phys. Scr., 11 (1975) 121; K.W. Lodge, J. Phys. F: Metal Phys., 6 (1976) 1989; T. Butz, Hyperfine Interact., 4 (1978) 528; E.N. Kaufmann and R. Vianden, Hyperfine Interact., 4 (1978) 532; M. Piecuch and Ch. Janot, J. Phys. (Paris) Colloq., 37 (1976) C6-359; K.D. Sen and P.T. Narasimhan, Phys. Rev. B, 16 (1977) 107.

72 R.S. Raghavan, Hyperfine Interact., 2 (1976) 29.

73 E.H. Hygh and T.P. Das, Phys. Rev., 143 (1966) 452; N.C. Mohapatra, C.M. Singal, T.P. Das and P. Jena, Phys. Rev., Lett., 29 (1972) 456

74 R.S. Raghavan and P. Raghavan, Phys. Lett., 36A (1971) 313.

75 P. Raghavan and R.S. Raghavan, Phys. Rev. Lett., 27 (1971) 724; R.S. Raghavan, E.N. Kaufmann and P. Raghavan, Phys. Rev. Lett., 34 (1975) 1280.

76 E. Daniel, in A.J. Freeman and B.R. Frankel (Eds.), Hyperfine Interactions, Academic, New York, 1967, p. 712; D. Shaltiel, ibid, p. 737; M. Shiga and Y. Nakamura, J. Phys. (Paris) Colloq., 40 (1979) C2-204.

77 F.D. Correll, Hyperfine Interact.; 4 (1978) 544.

78 P. Jena, S.D. Mohanti and T.P. Das, Phys. Rev. B, 7 (1973) 975.

79 K. Nishiyama and D. Riegel, Hyperfine Interact., 4 (1978) 490; K. Nishiyama, F. Dimmling, Th. Kornrumpf and D. Riegel, Phys. Rev. Lett., 37 (1976) 357.

80 E.N. Kaufman and R.J. Vianden, Rev. Mod. Phys., 51(1979) 161; P.C. Pattnaik, M.D. Thompson and T.P. Das, Phys. Rev. B, 16 (1977) 5390; M.D. Thompson, P.C. Pattnaik and T.P. Das, Hyperfine Interact 4 (1978) 515.

81 R. Ingalls, Phys. Rev., 133 (1964) A787.

82 T. Butz and G.M. Kalvius, Hyperfine Interact., 2 (1976) 222; G.A. Gehring, Phys. Scr., 11 (1975) 215.

83 W. Kohn and S. Vosko, Phys. Rev., 119 (1960) 912; R.P. Livi, M. Behar, I.J.R. Baumvol and F.C. Zawislak, Hyperfine Interact., 4 (1978) 540; R.P. Livi, M. Behar and F.C. Zawislak, ibid 4 (1978) 594.

84 O. Echt, E. Recknagel, G. Schatz, A. Weidinger and Th. Wichert, Hyperfine Interact., 4 (1978) 585; F. Namavar, M. Rots, R. Coussement, H. Ooms and J. Claes, ibid, p. 716.

85 T.E. Cranshaw and R.C. Mercader, J. Phys. F: Metal Phys., 6 (1976) 2129.

86 S. Ofer, P. Avivi, R. Bauminger, A. Marinov and S.G. Cohen, Phys. Rev., 120 (1960) 406; R.L. Cohen, Phys. Rev., 134 (1964) A94; B.D. Krawchuk, S.N. Ray, T.P. Das and K.J. Duff, Hyperfine Interact. 4 (1978) 352.

87 R. Bersohn, J. Chem. Phys., 29 (1958) 326; R.A. Berheim and H.S. Gutowsky, ibid 32 (1960) 1072.

88 R.G. Bernas, S.L. Segel and W.H. Jones, Jr., J. Appl. Phys., 33 (1962) 296; A. Weiss, in R. Blinc (Ed.), Magnetic Resonance and Relaxation (Proc. 14th Colloque Ampere, Ljubljana), North Holland, Amsterdam, 1967, p. 1076; N.W.G. Debye and J.J. Zuckerman, Developments in Applied Spectroscopy, Plenum, 8 (1970) 267.

89 F.W. de Wette and B.R.A. Nijboer, Physica, 24 (1958) 1105; F.W. de Wette, Phys. Rev., 123 (1961) 103; R.R. Hewitt and T.T. Taylor, Phys. Rev., 125 (1961) 524.

90 J.F. Hon, Phys. Rev., 124 (1961) 1368; G. Burns, Phys. Rev., 124 (1961) 524; B.G. Dick and T.P. Das, Phys. Rev., 127 (1962) 1053 and 1063.

91 T.T. Taylor and T.P. Das, Phys. Rev., 133 (1964) A1327; R.R. Sharma and T.P. Das, J. Chem. Phys., 41 (1964) 3581; J.O. Artman, A.H. Muir and H. Wiedersich, Phys. Rev., 173 (1968) 337; J.O. Artman, Phys. Rev., 143 (1966) 541.

92 G.A. Sawatzky and J. Hupkes, Phys. Rev. Lett., 25 (1970) 100; G.A. Sawatzky and F. van der Woude, J. Phys. (Paris) Colloq., 35 (1974) C6-47; R.R. Sharma, Phys. Rev. Lett., 26 (1971) 563; ibid 25 (1970) 1622.

93 A. Gelberg, Rev. Roum. Phys. 14 (1969) 183.

94 R. Sternheimer, Phys. Rev., 84 (1951) 244; H.M. Foley, R.M. Sternheimer and D. Tycko, Phys. Rev., 93 (1954) 734; R. Sterheimer and H.M. Foley, Phys. Rev., 92 (1953) 1460; ibid 102 (1956) 731; R.M. Sternheimer, Phys. Rev., 107 (1957) 1565; ibid 115 (1959) 1198; P.C. Pattnaik, M.D. Thompson and T.P. Das, Phys. Rev. B, 16 (1977) 5390.

95 T.P. Das and R. Bersohn, Phys. Rev., 102 (1956) 733.

96 R.E. Watson and A.J. Freeman, Phys. Rev., 131 (1963) 250; A.J. Freeman and R.E. Watson, Phys. Rev., 132 (1963) 706.

97 S.N. Ray, T. Lee, T.P. Das, R.M. Sternheimer, R.P. Gupta and S.K. Sen, Phys. Rev. A, 11 (1975) 1804.

98 K.D. Sen and P.T. Narasimhan, Phys. Rev. A, 11 (1975) 1162.

99 R.M. Sternheimer, Phys. Rev., 130 (1963) 1423; ibid 146 (1966) 140.

100 R.M. Sternheimer, M. Blume and R.F. Peierls, Phys. Rev., 173 (1968) 376.

101 G. Burns and E.G. Wikner, Phys. Rev., 121 (1961) 155.

102 K.D. Sen and P.T. Narasimhan, Phys. Rev. A, 14 (1976) 539.

103 A.C. Beri, T. Lee, T.P. Das and R.M. Sternheimer, Hyperfine Interact., 4 (1978) 509.

104 A.J. Freeman and R.E. Watson, Phys. Rev., 131 (1963) 2566.

105 F. van der Woude, Phys. Status Solidi, 17 (1966) 417; V.G. Bhide, and M.S. Multani, Phys. Rev., 139 (1965) A1983.

106 J.K. Srivastava, Phys. Status Solidi (b), 46 (1971) K93; 50 (1972) K21; W. Low and M. Weger, Phys. Rev., 118 (1960) 1119.

107 J.G. Cosgrove and R.L. Collins, J. Chem. Phys., 55 (1971) 4238; F.A. Matsen, J. Amer. Chem. Soc., 81 (1959) 2023; L.E. Orgel, An Introduction to Transition Metal Chemistry: Ligand Field Theory, Methuen and Co., London, 1960, p. 166; D.P. Johnson and R. Ingalls, Phys. Rev. B, 1 (1970) 1013.

108 C.J. Ballhausen, Introduction to Ligand Field Theory, McGraw-Hill, New York, 1962.

109 R.L. Collins, J. Chem. Phys., 42 (1965) 1072; J.R. Gabriel and S.L. Ruby, Nucl. Instrum. Methods, 36 (1965) 23.

110 M.G. Clark, G.M. Bancroft and A.J. Stone, J. Chem. Phys., 47 (1967) 4250.

111 R.W. Grant, J. Chem. Phys., 51 (1969) 1156.

112 S. Chandra, K.B. Pandeya, G.L. Sawhney and J.S. Baijal, Chem. Phys. Lett., 61 (1979) 109; E. Konig, G. Ritter, E. Lindner and I.P. Lorenz, Chem. Phys. Lett., 13 (1972) 70.

113 R. Ingalls, Phys. Rev., 128 (1962) 1155.

114 S.N. Ray and T.P. Das, Phys. Rev. B, 16 (1977) 4794.

115 R.M. Sternheimer, Phys. Rev., 164 (1967) 10; R.M. Sternheimer and R.F. Peierls, Phys. Rev. A, 3 (1971) 837.

116 N.C. Mahapatra, P.C. Pattnaik, M.D. Thompson and T.P. Das, Phys. Rev. B, 16 (1977) 3001; P.C. Pattnaik, M.D. Thompson, and T.P. Das, Phys. Rev. B, 16 (1977) 5390; M.D. Thompson, P.C. Pattnaik and T.P. Das, Hyperfine Interact., 4 (1978) 515.

117 B. Kolk, J. Phys. (Paris) Colloq., 37 (1976) C6-355; H. Ernst, E. Hagn, E. Zech and G. Eska, Phys. Rev. B, 19 (1979) 4460; M. Forker and S. Scholz, Hyperfine Interact., 9 (1981) 261.

118 R.S. Raghavan and P. Raghavan, Hyperfine Interact., 4 (1978) 535.

119 K. Nishiyama and D. Riegel, Hyperfine Interact., 4 (1978) 490.

120 P. Jena, Phys. Rev. Lett., 36 (1976) 418.

121 J.A.H. da Jornada, E.R. Fraga, R.P. Livi and F.C. Zawislak, Hyperfine Interact., 4 (1978) 589.

122 R.L. Cohen, Phys. Rev., 134 (1964) A94; S. Ofer, P. Avivi, R. Bauminger, A. Marinov and S.G. Cohen, Phys. Rev., 120 (1960) 406.

123 U. Ganiel and S. Shtrikman, Phys. Rev., 167 (1968) 258.

124 J. Gal, Z. Hadari, E.R. Bauminger, I. Nowik, S. Ofer and M. Perkal, Phys. Lett., 31A (1970) 511.

125 M. Eibschutz, S. Shtrikman and J. Tenenbaum, Phys. Lett., 24A (1967) 563.

126 G.K. Wertheim, H.J. Guggenheim, H.J. Williams and D.N.E. Buchanan, Phys. Rev., 158 (1967) 446; R. Fatehally, G.K. Shenoy, N.P. Sastry and R. Nagarajan, Phys. Lett., 25A (1967) 453; R. Fatehally, N.P. Sastry and R. Nagarajan, Phys. Status Solidi, 26 (1968) 91.

127 U. Ganiel, M. Kestigian and S. Shtrikman, Phys. Lett., 24A (1967) 577.

128 S. Margulies and J.R. Ehrman, Nucl. Instrum. Methods, 12 (1961) 131; J.M. Trooster and M.P.A. Viegers, Mossbauer Effect Ref. Data J., 1 (1978) 154.

129 W. Marshall and C.E. Johnson, J. Phys. Radium, 23 (1962) 733; F. van der Woude, Mossbauer Spectroscopy and Magnetic Properties of Iron Compounds, Ph.D. Thesis, Groningen University, Bronder-Offset, Rotterdam, 1966; F. van der Woude, Phys. Status Solidi, 17 (1966) 417; G.J. Perlow, C.E. Johnson, and W. Marshall, Phys. Rev., 140 (1965) A875.

130 E. Fermi, Z. Physik 60 (1930) 320; R.A. Ferrell, Amer. J. Phys., 28 (1960) 484; G.T. Rado, Amer. J. Phys., 30 (1962) 716; A. Abragam and M.H.L. Pryce, Proc. Roy. Soc. (London), A205 (1951) 135; W.D. Knight, in F. Seitz and D. Turnbull (Eds.), Solid State Physics, Academic, New York, 2 (1956) 93; R. Arnowitt, Phys. Rev., 92 (1953) 1002; R. Karplus and A. Klein, Phys. Rev., 85 (1952) 972 and 87 (1952) 848; H. Kopferman, Nuclear Moments (English Edition, Translated by E.E. Schncider), Academic, New York, 1958; S.M. Blinder, Theory of Atomic Hyperfine Structure, Academic, New York, 1965, p. 47.

131 J.M. Blatt and V.F. Weisskopf, Theoretical Nuclear Physics, John Wiley, New York, 1952, p. 32, 392; W.K.H. Panofsky and M. Phillips, Classical Electricity and Magnetism, Addison-Wesley, Reading (Massachusetts), 1962, p. 134, 125, 20.
132 S.S. Hanna, J. Heberle, G.J. Perlow, R.S. Preston and D.H. Vincent, Phys. Rev. Lett., 4 (1960) 513; R.S. Preston, S.S. Hanna and J. Heberle, Phys. Rev., 128 (1962) 2207; A.J.F. Boyle, D.St.P. Bunbury and C. Edwards, Phys. Rev. Lett., 5 (1960) 553.
133 E. Fermi and E. Segre, Z. Physik, 82 (1933) 729; J.C. Slater, Phys. Rev., 82 (1951) 538; R. Sternheimer, Phys. Rev., 86 (1952) 316; J.H. Wood and G.W. Pratt, Phys. Rev., 107 (1957) 995; V. Heine, Phys. Rev., 107 (1957) 1002; R.E. Watson and A.J. Freeman, Phys. Rev., 123 (1961) 2027.
134 A. Abragam and M.H.L. Pryce, Proc. Roy. Soc. (London), A205 (1951) 135; ibid A206 (1951) 164 and 173; B. Bleaney, Proc. Phys. Soc. A64 (1951) 315 and Phil. Mag. 42 (1951) 441; A. Abragam, J. Horowitz and M.H.L. Pryce, Proc. Roy. Soc., A230 (1955) 169; J.S. Griffith, The Theory of Transition Metal Ions, Cambridge Univ. Press, London, 1961.
135 H.H. Wickman and C.F. Wagner, J. Chem. Phys., 51 (1969) 435.
136 J.D. Roberts, Molecular Orbital Calculations, W.A. Benjamin, New York, 1962.
137 R. Schulman and S. Sugano, Phys. Rev., 130 (1963) 506; R. Schulman and V. Jaccarino, Phys. Rev., 108 (1957) 1219; R. Schulman and K. Knox, Phys. Rev., 119 (1960) 94; G.B. Benedek and T. Kushida, Phys. Rev., 118 (1960) 46.
138 W. Marshall, in W. Low (Ed.), Paramagnetic Resonance, Vol. 1, Academic, New York, 1963, p. 347.
139 N.L. Huang, R. Orbach and E. Simanek, Phys. Rev. Lett., 17 (1966) 134.
140 A. Mukherji and T.P. Das, Phys. Rev., 111 (1958) 1479.
141 M. Tinkham, Proc. Roy. Soc. (London) A236 (1956) 535 and 549; A.M. Clogston, J.P. Gordon, V. Jaccarino, M. Peter and L.R. Walker, Phys. Rev., 117 (1960) 1222.
142 F. Keffer, T. Oguchi, W. O'Sullivan and J. Yamashita, Phys. Rev., 115 (1959) 1553.
143 W. Marshall and R. Stuart, Phys. Rev., 123 (1961) 2048.
144 E. Simanek, N.L. Huang and R. Orbach, J. Appl. Phys., 38 (1967) 1072; J. Owen and J.H.M. Thornley, Rep. Progr. Phys., 29 (1966) 675; E. Simanek and Z. Sroubek, in S. Geschwind (Ed.), Electron Paramagnetic Resonance, Plenum, New York, 1972; S. Geschwind, in A.J. Freeman and B.R. Frankel (Eds.), Hyperfine Interactions, Academic, New York, 1967, p. 263, 258.
145 A.J. Freeman and R.E. Watson, Phys. Rev. Lett., 6 (1961) 343.
146 D. Petridis, A. Simopoulos, A. Kostikas and M. Pasternak, J. Chem. Phys., 65 (1976) 3139; J.P. Sanchez, J.M. Friedt and G.K. Shenoy, J. Phys.(Paris) Colloq., 35 (1974) C6-259.
147 B.J. Evans, in I.J. Gruverman (Ed.), Mossbauer Effect Methodology, Vol. 4, Plenum, New York, 1968, p. 139; B.J. Evans and L.J. Swartzendruber, Phys. Rev. B, 6 (1972) 223; L.J. Swartzendruber and B.J. Evans, J. Phys. (Paris) Colloq., 35 (1974) C6-265.
148 J.M.D. Coey and G.A. Sawatzky, J. Phys. C: Solid State Phys., 4 (1971) 2386; T. Shinohara, K. Sasaki and H. Watanabe, J. Phys. (Paris) Colloq., 40 (1979) C2-302.
149 F. van der Woude and G.A. Sawatzky, Phys. Rev. B, 4 (1971) 3159; Y. Miyahara and S. Iida, J. Phys. Soc. Japan, 33 (1972) 849; V.D. Doroshev, N.M. Kovtun and V.N. Seleznev, Phys. Status Solidi (a), 13 (1972) K41; M.A. Butler, M. Eibschutz and L.G. van Uitert, Phys. Lett., 37A (1971) 199; R. Valentin, H. Luft and K. Baberschke, Phys. Status Solidi (b), 48 (1971) 763.

150 J. Owen and D.R. Taylor, Phys. Rev. Lett., 16 (1966) 1164.
151 N.L. Huang, R. Orbach, E. Simanek, J. Owen and D.R. Taylor, Phys. Rev., 156 (1967) 383.
152 J. Owen and D.R. Taylor, J. Appl. Phys., 39 (1968) 791.
153 R.L. Streever and G.A. Uriano, Phys. Rev., 139 (1965) A305.
154 I. Nowik, Solid State Commun., 18 (1976) 1461; N. Bykovetz, Solid State Commun. 18 (1976) 143; M.A. Elmeguid and G. Kaindl, J. Phys. (Paris) Colloq., 40 (1979) C2-310.
155 T. Okada, H. Sekizawa and T. Yamadaya, J. Phys. (Paris) Colloq., 40 (1979) C2-299; M. Takano, Y. Takeda, M. Shimada and T. Matsuzawa, J. Phys. (Paris) Colloq., 35 (1974) C6-263.
156 H.H. Rinneberg and D.A. Shirley, Phys. Rev. Lett., 30 (1973) 1147.
157 K.J. Duff and T.P. Das, Phys. Rev. B, 3 (1971) 192 and 2294; T.P. Das, Phys. Scr., 11 (1975) 121.
158 M.B. Stearns, Phys. Rev. B, 4 (1971) 4069 and 4081; ibid B, 8 (1973) 4383; B. Kolk, Hyperfine Interact., 4 (1978) 313.
159 C.M. Singal and T.P. Das, Phys. Rev. B, 16 (1977) 5068, 5093 & 5108.
160 J. Callaway, Energy Band Theory, Academic, New York, 1964.
161 A. Okiji and J. Kanamori, J. Phys. Soc. Japan, 19 (1964) 908.
162 G.B. Benedek and J. Armstrong, J. Appl. Phys., 32 (1961) 1065; C. Kittel, in G.E. Benedek (Ed.), Magnetic Resonance at High Pressures, Interscience, New York, 1963; J.K. Srivastava and R.P. Sharma, Phys. Status Solidi (b), 49 (1972) 135.
163 B. Lax and K.J. Button, Microwave Ferrites and Ferrimagnetics, McGraw Hill, New York, 1962; L.W. McKeehan, Phys. Rev., 43 (1933) 1025.
164 J.K. Srivastava and R.P. Sharma, Phys. Status Solidi 35 (1969) 491.
165 H.H. Wickman, in E. Matthias and D.A. Shirley (Eds.), Hyperfine Structure and Nuclear Radiations, North Holland, Amsterdam, 1968, p. 930; P. Gutlich, R. Link and A. Trautwein, Mossbauer Spectroscopy and Transition Metal Chemistry, Springer-Verlag, Berlin, 1978.
166 M. Blume and J.A. Tjon, Phys. Rev., 165 (1968) 446; J.A. Tjon and M. Blume, Phys. Rev., 165 (1968) 456; M. Blume, Phys. Rev., 174 (1968) 351; Phys. Rev. Lett., 14 (1965) 96; ibid 18 (1967) 305; A.J. Dekker, in A.J. Freeman and B.R. Frankel (Eds.), Hyperfine Interactions, Academic, New York, 1967, p. 679; S. Morup and N. Thrane, Phys. Rev., B, 4 (1971) 2087.
167 M.J. Clauser and M. Blume, Phys. Rev. B, 3 (1971) 583; M.J. Clauser Phys. Rev. B, 3 (1971) 3748; S. Dattagupta and M. Blume, Phys. Rev. B, 10 (1974) 4540 and 4551, F. van der Woude and A.J. Dekker, Phys. Status Solidi, 9 (1965) 775; ibid 13 (1966) 181; Solid State Commun., 3 (1965) 319; B.C. van Zorge, F. van der Woude and W.J. Caspers, Z. Phys., 221 (1969) 113; J.K. Srivastava and R.P. Sharma, Phys. Status Solidi, 35 (1969) 491.
168 S. Banerjee and M. Blume, Phys. Rev. B, 16 (1977) 3061; S. Banerjee Phys. Rev. B, 19 (1979) 5463; S. Dattagupta, Phys. Rev. B, 12 (1975 3584; Hyperfine Interact., 4 (1978) 942; Phys. Rev. B, 16 (1977) 158; S. Dattagupta, G.K. Shenoy, B.D. Dunlop and L. Asch, Phys. Rev., B, 16 (1977) 3893.
169 A.J.F. Boyle and J.R. Gabriel, Phys. Lett., 19 (1965) 451; J.K. Srivastava, Phys. Status Solidi (b), 97 (1980) K123.
170 J.K. Srivastava, Phys. Status Solidi (b), 55 (1973) K119; S. Morup, Paramagnetic and Superparamagnetic Relaxation Phenomenon Studied by Mossbauer Spectroscopy, D.Sc. Thesis, Technical Univ. of Denmark, Lyngby, Polyteknisk Forlag, 1981; R. Nagarajan and J.K. Srivastava, Phys. Status Solidi (b), 81 (1977) 107.
171 M. Blume, J. Phys. (Paris) Colloq., 37 (1976) C6-61; S. Banerjee, J. Appl. Phys., 50 (1979) 7581.
172 N. Bloembergen, S. Shapiro, P.S. Pershan and J.O. Artman, Phys. Rev 114 (1959) 445; J.H. Van Vleck, Phys. Rev., 74 (1948) 1168; F. Keffer, Phys. Rev., 88 (1952) 686.

173 S.C. Bhargava, J.E. Knudsen and S. Morup, J. Phys. Chem. Solids, 40 (1979) 45.
174 I. Waller, Z. Physik, 79 (1932) 370.
175 R. de L. Kronig, Physica, 6 (1939) 33; J.H. Van Vleck, J. Chem. Phys., 7 (1939) 72 and Phys. Rev., 59 (1941) 730.
176 A. Abragam and B. Bleaney, Electron Paramagnetic Resonance of Transition Ions, Clarendon Press, Oxford, 1970, Chapter 10, 21.
177 W.J. Caspars, Theory of Spin Relaxation, Wiley-Interscience, New York, 1964; S.A. Al'tshuler and B.M. Kozyrev, Electron Paramagnetic Resonance, Academic, New York, 1964.
178 K.N. Shrivastava, Phys. Status Solidi (b), 51 (1972) 377; P.J. Scott and C.D. Jeffries, Phys. Rev., 127 (1962) 32; R. Orbach, Proc. Roy. Soc. (London), A264 (1961) 458 and 485; S.C. Bhargava, J.E. Knudsen and S. Morup, J. Phys. C: Solid State Phys., 12 (1979) 2879; R. Orbach and M. Blume, Phys. Rev. Lett., 8 (1962) 478; C.R. Viswanathan and G. Kaelin, Phys. Rev., 171 (1968) 992; K.W.H. Stevens, Rep. Progr. Phys., 30 (1967) 189; A.A. Manenkov and R. Orbach (Eds.), Spin Lattice Relaxation in Ionic Solids, Harper and Row, New York, 1966; R. Orbach, Proc. Phys. Soc. (London), A77 (1961) 821; C.Y. Huang, Phys. Rev., 139 (1965) A241; R. Orbach and H.J. Stapleton, in S. Geschwind (Ed.), Electron Paramagnetic Resonance, Plenum, New York, 1972, p. 121; G.C. Psaltakis and M.G. Cottom, Phys. Status Solidi (b), 103 (1981) 709.
179 R. Hernandez and M.B. Walker, Phys. Rev. B, 4 (1971) 3821.
180 J.K. Srivastava, Phys. Status Solidi (b), 97 (1980) K123; I.P. Suzdalev, A.M. Afanasev, A.S. Plachinda, V.I. Goldanskii and E.F. Makarov, Sov. Phys. - JETP, 28 (1969) 923; V.V. Svetozarov, Sov. Phys - Solid State, 12 (1970) 826; ibid 13 (1971) 1057; S.C. Bhargava, J.E. Knudsen and S. Morup, J. Phys. C: Solid State Phys., 12 (1979) 2879.
181 Y. Hazony and R.H. Herber, in I.J. Gruverman and C.W. Seidel (Eds.), Mossbauer Effect Methodology, Vol. 8, Plenum, New York, 1973, p. 107; R.H. Herber, in D. Barb and D. Tarina (Eds.), Proc. Int. Conf. on Mossbauer Spectroscopy, Vol. 2, Bucharest, 1977, Documentation Office, Central Inst. of Phys., P.O. Box 5206, Bucharest, 1977, p. 179.
182 J.K. Srivastava and B.W. Dale, Phys. Status Solidi (b), 90 (1978) 391; ibid 90 (1978) 571.
183 G. Raoult, A. Gavaix, A. Vasson and A.M. Vasson, Phys. Rev. B, 4 (1971) 3849 and J. Phys. C: Solid State Phys., 4 (1971) 3297; S. Morup and J.E. Knudsen, Chem. Phys. Lett. 40 (1976) 292; W.J.C. Grant, Phys. Rev., 134 (1964) A1554, A1565, A1574; S. Hufner and G. Weber, Phys. Lett., 13 (1964) 115; R.W. Roberts, J.H. Burgess and H.D. Tenny, Phys. Rev., 121 (1961) 997; N. Bloembergen and P.S. Pershan, in J.R. Singer (Ed.), Advances in Quantum Electronics Columbia Univ. Press, New York, 1961, p. 373; K.J. Standley and R.A. Vaughan, Electron Spin Relaxation Phenomena in Solids, Adam Hilger, London, 1969.
184 F.S. Ham, J. Phys. (Paris) Colloq., 35 (1974) C6-121; K.N. Shrivastava, Phys. Rep., 20 (1975) 137; R. Englman and D. Horn, in W. Low (Ed.), Paramagnetic Resonance, Vol. 1, Academic, New York, 1963, p. 329; M.D. Sturge, in F. Seitz, D. Turnbull and H. Ehrenreich (Eds.), Solid State Physics, Academic, New York, 20 (1967) 92; F.S. Ham, in S. Geschwind (Ed.), Electron Paramagnetic Resonance, Plenum, New York, 1972, p. 1; R. Englman, The Jahn-Teller Effect in Molecules and Crystals, Wiley-Interscience, London-New York, 1972.
185 D.P. Breen, D.C. Krupka and F.I.B. Williams, Phys. Rev., 179 (1969) 241; F.I.B. Williams, D.C. Krupka and D.P. Breen, ibid p. 255; R. Pirc, B. Zeks and P. Gosar, J. Phys. Chem. Solids, 27 (1966) 1219; F.S. Ham, Phys. Rev., 166 (1968) 307, 321; J.B. Goodenough, J. Phys. Chem. Solids, 25 (1964) 151; M.C.M. O'Brien, Proc. Roy. Soc. (London), 281A (1964) 323; J.A. Sussmann, J. Phys. Chem. Solids

28 (1967) 1643.
186 E.B. Wilson, J.C. Decius and P.C. Cross, Molecular Vibrations, McGraw Hill, New York, 1955; G. Herzberg, Infrared and Raman Spectra of Polyatomic Molecules, Van Nostrand, Princeton, New Jersey, 1945.
187 F. Hartmann- Boutron, J. Phys. (Paris), 29 (1968) 47.
188 H.M. McConnell, J. Chem. Phys., 25 (1956) 709; G.R. Liebling and H.M. McConnell, ibid 42 (1965) 3931.
189 H.R. Leider and D.N. Pipkorn, Phys. Rev., 165 (1968) 494, 500; M. Tanaka, T. Tokoro and Y. Aiyama, J. Phys. Soc. Japan, 21 (1966) 262; D.N. Pipkorn and H.R. Leider, Bull. Amer. Phys. Soc., 11 (1966) 49; F.S. Ham, J. Phys. (Paris) Colloq., 35 (1974) C6-121; J.R. Regnard, ibid, p. C6-181; J. Chappert, R.B. Frankel and N.A. Blum, Phys. Lett., 25A (1967) 149; F.S. Ham, W.M. Schwarz and M.C.M. O'Brien, Phys. Rev., 185 (1969) 548; L. Cianchi, M. Mancini and G. Spina, Lett. Nuovo Cimento, 16 (1976) 505; M.R. Spender and A.H. Morrish, Solid State Commun., 11 (1972) 1417; A.M. Van Dipen, F.K. Lotgering and J.F. Olijhoek, J. Mag. Mag. Mater, 3 (1976) 117; H. Pollak, R. Quartier, W. Bruyneel and P. Walter, J. Phys. (Paris) Colloq., 37 (1976) C6-589; G. Garcin, A. Gerard, P. Imbert and G. Jehanno, ibid 40 (1979) C2-413; V.K. Singh, R. Chandra and S. Lokanathan, Phys. Status Solidi (b) 105 (1981) K13; A. Gerard and F. Grandjean, J. Phys. C: Solid State Phys., 12 (1979) 4601.
190 D.W. Collins, J.T. Dehn and L.N. Mulay, in I.J. Gruverman (Ed.), Mossbauer Effect Methodology, Vol. 3, Plenum, New York, 1967, p. 103; J.M.D. Coey and D. Khalafalla, Phys. Status Solidi (a), 11 (1972) 229; M. Eibschutz and S. Shtrikman, J. Appl. Phys., 39 (1968) 997; W.J. Schuele, S. Shtrikman and D. Treves, ibid 36 (1965) 1010; S.Morup, J.A. Dumesic and H. Topsoe, in R.L. Cohen (Ed.), Applications of Mossbauer Spectroscopy, Vol. II, Academic, New York, 1980, p. 1.
191 J.C. Slonczewski, J. Appl. Phys., 32 (1961) 253S; H.B. Callen and E. Callen, J. Phys. Chem. Solids, 27 (1966) 1271; J.O. Artman, J.C. Murphy and S. Foner, J. Appl. Phys., 36 (1965) 986; M. Tachiki and T. Nagamiya, J. Phys. Soc. Japan, 13(1958)452; J.K. Srivastava and R.P. Sharma, Phys. Status Solidi (b), 49(1972)135; D.J. Epstein, B. Frackiewicz and R.P. Hunt, J. Appl. Phys., 32(1961)270S; R.M. Bozorth, Ferromagnetism, D. Van Nostrand, New York, 1955, p. 563.
192 T.K. McNab, R.A. Fox and A.J.F. Boyle, J. Appl. Phys., 39 (1968) 5703; J.M.D. Coey, Phys. Rev. Lett., 27 (1971) 1140; S. Morup, H. Topsoe and J. Lipka, J. Phys. (Paris) Colloq., 37 (1976) C6-287; S. Morup and H. Topsoe, Appl. Phys., 11 (1976) 63; W. Kundig, R.H. Lindquist and G. Constabaris, in R. Blinc (Ed.), Magnetic Resonance and Relaxation, Proc. 14th Colloque Ampere, Ljubljana, North Holland Amsterdam, 1966, p. 1029; D.S. Robell, J. Phys. Soc. Japan 17 (1961) 313S; G. Belozerskii and S. Simonyan, J. Phys. (Paris) Colloq., 40 (1979) C2-237.
193 S. Foner, J. Appl. Phys., 32 (1961) 63S; T. Oguchi Phys. Rev., 111 (1958) 1063.
194 W. Kundig, H. Bommel, G. Constabaris and R.H. Lindquist, Phys. Rev., 142 (1966) 327; V.N. Sharma, Phys. Lett., 19 (1965) 462.
195 I.S. Jacobs and C.P. Bean, in G.T. Rado and H. Suhl (Eds.), Magnetism, Vol. III, Academic, New York, 1963, p. 271; P.W. Selwood, Absorption and Collective Paramagnetism, Academic, New York, 1962; T. Nakamura, T. Shinjo, Y. Endoh, N. Yamamoto, M. Shiga and Y. Nakamura, Phys. Lett., 12 (1964) 178; W.J. Schuele and V.D. Deetscreek, J. Appl. Phys., 33 (1962) 1136S; J.G. Booth and J. Crangle, Proc. Phys. Soc., 79 (1962) 1278; Y. Ishikawa, J. Phys. Soc. Japan, 17 (1962) 1877; J. Appl. Phys., 35 (1964) 1054; I.P. Syzdalev, Sov. Phys. - Solid State, 12 (1970) 775.
196 W. Karas, J. Korechi and K. Krop, Acta Phys. Pol., A55 (1979) 669.

197 C.P. Bean and J.D. Livingston, J. Appl. Phys., 30 (1959) 120S; P. Roggwiller and W. Kundig, Solid State Commun., 12 (1973) 901.

198 W. Kundig, M. Kobelt, H. Appel, G. Constabaris and R.H. Lindquist, J. Phys. Chem. Solids, 30 (1969) 819; D. Schroeer, in I.J. Gruverman (Ed.), Mossbauer Effect Methodology, Vol. 5, Plenum, New York, 1970, p. 141; G. Von Eynatten and H.E. Bommel, Appl. Phys., 14 (1977) 415; J.A. Dumesic. H. Topsoe, S. Khammouma and M. Boudart, J. Catal., 37 (1975) 503; M. Hayashi, I. Tamura, Y. Fukano and S. Kanemki, Phys. Lett., 77A (1980) 332; F. Galembeck, N.F. Leite, L.C.M. Miranda, H.R. Rochenberg and H. Vargas, Phys. Status Solidi (a), 60 (1980) 63; K. Haneda and A.H. Morrish, Solid State Commun., 22 (1977) 779; S. Morup, B.S. Clausen and H. Topsoe, J. Phys. (Paris) Colloq., 40 (1979) C2-78.

199 G.A. Petitt and D.W. Forester, Phys. Rev. B, 4 (1971) 3912; J.K. Srivastava, Phys. Status Solidi (b), 55 (1973) K119; Y. Ishikawa, J. Phys. Soc. Japan, 17 (1962) 1877; J. Appl. Phys., 35 (1964) 1054; V.U.S. Rao, F.E. Huggins and G.P. Huffman, J. Appl. Phys., 50 (1979) 2408.

200 L.M. Levinson, M. Luban and S. Shtrikman, Phys. Rev., 177 (1969) 864; H.L. Wehner, G. Ritter and H.H.F. Wegner, Phys. Lett., 46A (1974) 333; J.M.D. Coey, Phys. Rev., B, 6 (1972) 3240.

201 G. Shirane, D.E. Cox, W.J. Takei and S.L. Ruby, J. Phys. Soc. Japan, 17 (1962) 1598; S.L. Ruby and G. Shirane, Phys. Rev., 123 (1961) 1239; M. Rubinstein, Solid State Commun., 8 (1970) 919; W.A. Ferrando, R. Segnan and A.I. Schindler, J. Appl. Phys., 41 (1970) 1236; Phys. Rev. B, 5 (1972) 4657; J.K. Srivastava and R.P. Sharma, J. Phys. (Paris) Colloq., 35 (1974) C6-663; R. Nagarajan, and J.K. Srivastava, Phys. Status Solidi (b), 81 (1977) 107; K. Fujimoto and M. Boudart, J. Phys. (Paris) Colloq., 40 (1979) C2-81.

202 U. Gonser, H. Wiedersich and R.W. Grant, J. Appl. Phys., 39 (1968) 1004.

203 C.E. Violet and E.L. Lee, in I.J. Gruverman (Ed.), Mossbauer Effect Methodology, Vol. 2, Plenum, New York, 1966, p. 171; K. Aggarwal and R.G. Mendiratta, Phys. Rev. B, 16 (1977) 3908; O. Massenet, Solid State Commun., 21 (1977) 337; G. Bayreuther, H. Hoffman and J. Reffle, Phys. Rev. B, 19 (1979) 1614; S. Duncan, A.H. Owens, R.J. Semper and J.C. Walker, Hyperfine Interact., 4 (1978) 886; J.L. Dormann, P. Gibart, C. Suran, J.L. Tholence and C. Sella, J. Magn. Magn. Mater., 15-18 (1980) 1121; T. Shigematsu, T. Shinjo, Y. Bando and T. Takada, ibid, p. 1367; D.G. Howard and R.H. Naussbaum, Surf. Sci., 93 (1980) L105.

204 G.D. Sultanov, N.G. Guseinov, R.M. Mirzababaev and F.A. Mirishli, Sov. Phys. -Solid State, 18 (1976) 1496; H.Th.Le Fever, F.J. Van Steenwijk and R.C. Thiel, Physica, 86-88 B+C (1977) 1269.

205 P.A. Beck, in P.A. Beck and J.T. Weber (Eds.), Magnetism in Alloys, TSM-AIME, New York, 1972, p. 211; V. Cannella and J.A. Mydosh, in Int. Conf. Magnetism, Moscow, 1973; K.H. Fischer, Phys. Rev. Lett., 34 (1975) 1438; P.A. Beck, J. Less-Common Metals, 28 (1972) 193; D.A. Smith. J. Phys. F: Metal Phys., 4 (1974) L226; F.W. Smith, Solid State Commun., 13 (1973) 1267; D. Sherrington and B.W. Southern, J. Phys. F: Metal Phys., 5 (1975) L49; P.W. Anderson, J. Appl. Phys., 49 (1978) 1599; H. Maletta and W. Felsch, J. Phys. (Paris) Colloq., 39 (1978) C6-931; A. Amamou, R. Caudron, P. Costa, J.M. Friedt, F. Gautier and B. Loegel, J. Phys. F: Metal Phys., 6 (1976) 2371; Y. Muraoka, M. Shiga and Y. Nakamura, J. Phys. (Paris) Colloq., 40 (1979) C2-213; N.H.J. Ganges, J. Katradis, A. Moukarika, A. Simopoulos, A. Kostikas and V. Papaefthimiou, J. Phys. C: Solid State Phys., 13 (1980) L357.

206 H. Maletta, J. Magn. Magn. Mater., 24 (1981) 179.

207 J.M.D. Coey and P.W. Readman, Nature, 246 (1973) 476; J.M.D. Coey, Phys. Rev. Lett., 27 (1971) 1140; M. Boudart, H. Topsoe and J.A. Dumesic, in E. Drauglis and R.I. Jeffe (Eds.), The Physical Basis for Heterogeneous Catalysis, Plenum, New York, 1975, p. 337; A.E. Berkowitz, J.A. Lahut, I.S. Jacobs, L.M. Levinson and D.W. Forester, Phys. Rev. Lett., 34 (1975) 594.

208 W.F. Brown, Jr., J. Appl. Phys., 30 (1959) 130S; J. Appl. Phys., 34 (1963) 1319; Phys. Rev., 130 (1963) 1677.

209 A. Aharoni, Phys. Rev., 135 (1964) A447; ibid 177 (1969) 793.

210 A. Aharoni, Phys. Rev., B, 7 (1973) 1103; A. Aharoni and I. Eisenstein, ibid 11 (1975) 514; I. Eisenstein and A. Aharoni, ibid 16 (1977) 1278 and 1285.

211 L. Haggstrom, T. Ericsson, R. Wappling, E. Karlsson, and K. Chandra, J. Phys. (Paris) Colloq., 35 (1974) C6-603; H.W. de Wijn, R.E. Walstedt, L.R. Walker and H.J. Guggenheim, Phys. Rev. Lett., 24 (1970) 832; F.E. Wagner, W. Potzel and T. Katila, Phys. Lett., 33A (1970) 83; G.K. Wertheim and D.N.E. Buchanan, Phys. Rev., 161 (1967) 478; K. Ono, M. Shinohara, A. Ito, N. Sakai and M. Suenaga, Phys. Rev. Lett., 24 (1970) 770; H. Pinto, G. Shachar and H. Shaked, Solid State Commun., 8 (1970) 597; S. Geller, R.W. Grant, U. Gonser, H. Wiedersich and G.P. Espinosa, Phys. Lett., 25 A (1967) 722; N. Yamamoto, J. Phys. Soc. Japan, 24 (1968) 23; G.J. Parlow, W. Henning, D. Olson and G.L. Goodman, Phys. Rev. Lett., 23 (1969) 680; H. Maletta and R.L. Mossbauer, Solid State Commun., 8 (1970) 143; J. Jach, R.J. Borg and D.Y.F. Lai, J. Appl. Phys. 42 (1971) 1611; H. Bunzel, E. Kreber and U. Gonser, J. Phys. (Paris) Colloq., 35 (1974) C6-609; H. Keller, W. Kundig and H. Arend, Physica, 86-88 B+C (1977) 683; J. Phys. (Paris) Colloq., 37 (1976) C6-629; G.P. Gupta, D.P.E. Dickson and C.E. Johnson, J. Phys. C: Solid State Phys., 11 (1978) 215; A.S. Kamzin, V.A. Bokov and G.A. Smolenskii, Sov. Phys. - JETP Lett., 27 (1978) 477; N.N. Delyagin, Yu.D. Zonnenberg, E.N. Kornienko, V.I. Krylov and V.I. Nesterov, Sov. Phys. - Solid State, 20 (1978) 148; D.C. Cook and J.D. Cashion, J. Phys. C: Solid State Phys., 12 (1979) 605; J.M. Daniels, Can. J. Phys., 57 (1979) 263; G.A. Garcia and R. Ingalls, J. Phys. Chem. Solids, 37 (1976) 211; L. Asch, I. Dezsi, T. Lohner and B. Molnar, Chem. Phys. Lett., 39 (1979) 177; D.R. Rhiger, R. Ingalls and C.M. Liu, Solid State Commun., 18 (1976) 681; U. Atzmony, M.P. Dariel and G. Dublon, Phys. Rev. B, 14 (1976) 3713; A.T. Howe and G.J. Dudley, J. Solid State Chem., 18 (1976) 149; N.J. Stone, in A. Perez and R. Coussement (Eds.), Site characterisation and Aggregation of Implanted Atoms in Materials, Plenum, New York, 1980; G.M. Kalvius, U.F. Klein and G. Wortmann, J. Phys. (Paris) Colloq., 35 (1974) C6-139.

212 M.E. Lines, J. Phys. Chem. Solids, 31 (1970) 101; L.J. de Jongh, Phys. Lett., 40A (1972) 33; G.P. Gupta, D.P.E. Dickson, C.E. Johnson and B.M. Wanklyn, J. Phys. C: Solid State Phys., 10 (1977) L459; R.E. Walstedt, H.W. de Wijn and H.J. Guggenheim, Phys. Rev. Lett., 25 (1970) 1119; D.J. Kim and B.B. Schwartz, Phys. Rev. Lett., 20 (1968) 201; 21 (1968) 1744.

213 J.S. Smart, Effective Field Theories of Magnetism, W.B. Saunders, Philadelphia, 1966; C. Kittel, Introduction to solid state physics, 3rd edn., John Wiley, New York, 1966; V. Jaccarino, in G.T. Rado and H. Suhl (Eds.), Magnetism, Vol. IIA, Academic, New York, 1965, p. 307.

214 S. Chandrasekhar, Liquid Crystals, Cambridge Univ. Press, London, 1977; G.H. Brown (Ed)., Advances in Liquid Crystals, Academic, London, Vol. I (1975), Vol. II (1976) and Vol. III (1978); G.W. Gray and P.J. Winsor (Eds.). Liquid Crystals and Plastic Crystals, Ellis Horwood, Chichester, 1974; M.J. Stephen and J.P. Straley, Rev. Mod. Phys., 46 (1974) 617.

215 H. Sackmann and D. Demus, Mol. Cryst. Liquid Cryst., 21 (1973) 239; I.G. Chistyakov, Sov. Phys. - Uspekhi, 9 (1967) 551; I. Haller, H.A. Huggins, H.R. Lilienthal and T.R. McGuire, J. Phys. Chem., 77 (1973) 950; J.D. Rowell, W.D. Phillips, L.R. Melby and M. Panar, J. Chem. Phys., 43 (1965) 3442; G.R. Luckhurst, Chem. Phys. Lett., 9 (1971) 289.

216 D.L. Uhrich, R.E. Detjen and J.M. Wilson, in I.J. Gruverman and C.W. Seidel (Eds.), Mossbauer Effect Methodology, Vol. 8, Plenum, New York, 1973, p. 175; D.L. Uhrich, J.M. Wilson and W.A. Resch, Phys. Rev. Lett., 24 (1970) 355; V.G. Bhide, M.C. Kandpal and S. Chandra, Solid State Commun., 23 (1977) 459; D.L. Uhrich and V.O. Aimiuwu, Mol. Cryst. Liquid. Cryst., 43 (1977) 295; W.J. LaPrice and D.L. Uhrich, J. Chem. Phys., 72 (1980) 678, ibid 71 (1979) 1498; V.G. Bhide and M.C. Kandpal, Phys. Rev. B, 20 (1979) 85; R.S. Preston, in G.K. Shenoy and F.E. Wagner (Eds.), Mossbauer Isomer Shifts, North Holland, Amsterdam, 1978, p. 281.

CHAPTER 2

INSTRUMENTATION FOR MÖSSBAUER SPECTROSCOPY

Geoffrey Longworth

Nuclear Physics Division, Atomic Energy Research Establishment, Harwell, England.

1. INTRODUCTION

There is still considerable interest in improvements in instrumentation for Mossbauer spectroscopy and in the development of specialised units for new applications. In the past five years about 150 references devoted specifically to instrumentation are cited in the Mossbauer Effect Data Index (1). Several excellent review articles have been published, for example those by Kalvius and Kankeleit (2) and by Cohen and Wertheim (3). In this chapter therefore we have tried to concentrate on the more recent developments in this field.

In the second section dealing with Mossbauer spectrometers, emphasis is placed on the measurement of Mossbauer spectra by velocity scanning in the time mode and on the principles which govern the choice of velocity waveform, velocity drives and associated drive circuits. The use of multiple systems based on minicomputers or microprocessors is discussed and the main methods for calibration of the velocity drive are mentioned. In Section 3 the problems encountered in data analysis are described while in Section 4 the criteria for the choice of source matrices are mentioned. In Section 5 various types of gamma ray detector for counting at high rates are described while in Section 6, detectors specifically designed to count scattered electrons, X rays or gamma rays are detailed.

The instrumentation required for measurements at low temperatures is described in Section 7. Emphasis is placed here on more specialised cryostats for matrix isolation measurements rather than on conventional cryostats and variable temperature inserts. In the final Section 8, a brief outline is given of the main types of pressure cells and furnaces required for work at high pressures or high temperatures. Apart from the references given here, the reader is referred to the invaluable Mossbauer Effect Data Index of Stevens and Stevens (1).

2. MOSSBAUER SPECTROMETERS

The majority of Mossbauer spectrometers used at the present time incorporate a velocity sweep in which the source is mounted on an electromechanical velocity drive and is swept periodically through the range of velocities of interest. The velocity spectrum is obtained by storing the counts as a function of velocity in a memory device such as a multichannel analyser or computer. The channels/addresses are normally accessed sequentially in synchronism with the velocity sweep so that the velocity spectrum may be displayed directly. This synchronisation is maintained in one of the two ways. In the first method an external clock is used to provide both the channel advance information and to synthesise the reference waveform for the velocity drive. In the second method the most significant bistable of the store address register is used to produce a square wave which may then be integrated to give a triangular velocity waveform. In this second method the data store provides the synchronisation but the correlation between each data channel and the corresponding velocity increment depends upon the linearity of the integrater.

The velocity waveforms used are typically either a triangular or sinusoidal waveform. Details of some of the recent methods for obtaining these waveforms are discussed in a later section.

In the velocity spectrometer the reference waveform is delivered to the velocity drive and a transducer is used to provide a signal proportional to the actual velocity. Comparison of this signal with the reference waveform is used to produce an error signal since in general the drive motion will deviate from that required. The suitably amplified error signal is then applied to the drive system in order to correct the velocity. The operation of such a feedback loop for a velocity drive has been discussed by Kankeleit (4) and by Cranshaw (5) and only a brief mention is given here in the section on velocity drives.

2.1 Drive waveforms

One of the main considerations which determine the choice of waveform is that the drive should be able to follow it easily. This is satisfied most easily by a sinusoidal waveform although it has some disadvantages and an attractive alternative is a sawtooth waveform with a rounded flyback (6). Both conventional sawtooth and symmetrical triangular waveforms have the disadvantage that large changes in acceleration and force are involved during the flyback which can make the moving system ring and can render the first part of the spectrum unreliable. The forces involved are smaller for the symmetrical triangle and although a data store is required with twice as many channels as in the final velocity spectrum, folding of one half of the spectrum over the other does serve to eliminate the background curvature due to changes in source detector distance.

A simple digital based triangle wave generator has been described by Taragin (7) which uses as its input the standard output pulses obtainable from a multichannel analyser (MCA). A series of up-down counters are used to count up the channel address advance pulses for the first half of the spectrum and to count them down for the second half. The output is then fed into a digital to analog converter (DAC) to produce a triangular waveform correlated with each channel of the MCA. The most significant bistable of the MCA address register is used to ensure the proper polarity to drive the up-down counters and to reset them at the beginning of each sweep.

Instead of folding the spectrum during the subsequent data processing step, folding has been done automatically by Biran et al (8) using a small computer and an address up-down counter. The triangular waveform is derived from a DAC and a scaler driven by a precision clock. For the second half of the triangular mode the waveform generator produces a 'count back' signal which counts down the addresses until it reaches the initial address when it starts the next cycle.

A rounded sawtooth waveform has been described by Cranshaw (6) in which successive linear ramps are joined by two parabolas during the flyback. This ensures that there are no discontinuities in either velocity or acceleration. This idea has been extended by Window et al (9) who used a linear ramp followed by a flyback waveform given by a simple Fourier series. The coefficients of the series were chosen such that there are no discontinuities in the first seven derivatives of the reference waveform. The linear ramp containing 1024 steps is produced by incrementing the output register of a DR 11C interface (16 bit output register, input data way and control and status register, together with a 12 bit DAC). The values of the flyback are fed in as an array during program initialisation. The same type of waveform derived in a similar fashion has been described by Graham et al (10).

Although a sinusoidal waveform is easier for the velocity drive to follow, it has several disadvantages. There is a non-linear relationship between velocity and channel number from which it is not easy to visualise what the spectrum will look like on the normal linear velocity scale. Such a scale can be produced of course during subsequent data processing. More time is spent at the velocity extremes which are usually the uninteresting parts of the spectrum. On the other hand large velocities are easily obtainable and the ease with which the waveform is followed allows the use of long drive rods between velocity drive and source. This is convenient for drives operated in vertical geometry in a cryostat where both source and absorber are cooled. The system may be driven at its mechanical resonance frequency so that demands on the drive circuitry are not excessive. It is convenient to be able to vary the frequency in order to match changes in resonance frequency due for example to changes in

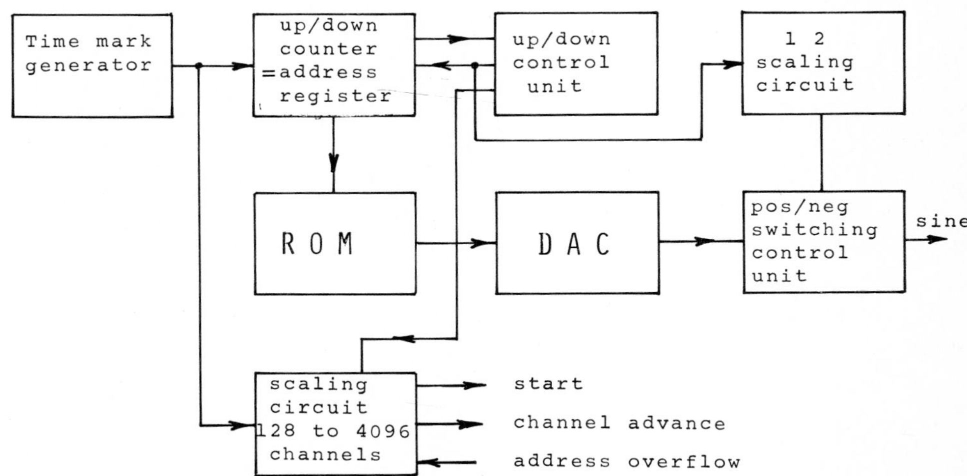

Fig. 1. *Digital sine generator (Halder and Kalvius (12)*

source loading. As with any other type of waveform it is necessary to provide a trigger pulse to start the analyser sweep, which has a fixed phase relationship with the sine wave. This has been produced at the zero velocity positions using a zero crossing detector or at the positive maximum velocity positions by using the same type of detector acting on the sine wave after integration (11). Clearly precautions must be taken to ensure high phase stability otherwise the particular velocity increment corresponding to a given channel will drift.

The alternative approach to the synchronisation of velocity and analyser sweeps has been to generate the sine wave directly from the address advance pulses using a read only memory (ROM) and DAC (12) (Figure 1). The amplitude values for the sine wave are programmed into a MOS read only memory (ROM) which contains 2^{11} addresses. Pulses from the variable time mark generator are counted in an 11 bit address register (up-down counter) running between numbers 0 and 2047. Pulse separations between 0.5 & 50 μsec in steps of 0.05 μsec may be provided.

The generator also provides the address advance pulses via a digital divider. The required sinusoidal voltage is produced by a DAC, in the form of 2048 addresses for each period of $\pi/2$. The four quarters of the sine wave are then constructed using an inverter and adder circuit. When addresses 0 and 2047 are reached in the velocity scan the address scalar is switched from up to down scaling or vice versa. The address register is also blocked for one time mark pulse at addresses 0 and 2047. In this way a dwell time of twice the normal duration is produced for these two addresses in order to synchronise the velocity exactly to 128, 256, 512, 1024 or 2048 channels of equal dwell time on the multichannel analyzer. The quality of the sine wave is given by the resolution of the programmed sine amplitudes (11 bit resolution), by the accuracy of the DAC, and by the stabilisation of the reference voltage for the DAC and operational amplifiers. If 1024 channels are used then there are 8 steps per channel in the waveform which are too small to be followed by the velocity drive. The waveform generator has been used with a double loudspeaker type drive having a resonance near 40 Hz. At this frequency velocities of up to ± 30 cm/sec are easily achieved.

2.2 Drive Systems

The velocity drives used in Mossbauer spectroscopy frequently have either a moving magnet (6) or moving coil (4) to sense the source velocity, together with a feedback control circuit. The behaviour of such a circuit has been discussed in detail by Kankeleit (4), Cranshaw (5) and by Cohen and Wertheim (3). The feedback loop is shown schematically in Figure 2. The components are a drive amplifier

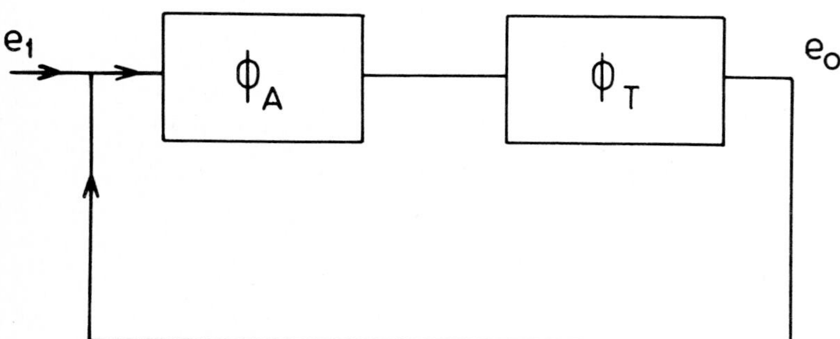

Fig. 2. *Feedback loop in Mossbauer drive system, where* ϕ_A *and* ϕ_T *are the transfer functions for the drive amplifier and velocity drive.*

with transfer function ϕ_A and velocity drive with transfer function ϕ_T. The input to the amplifier is the difference between a signal e_i, corresponding to a given reference velocity, and the output of the velocity sensor e_o. Operation of the loop ensures that $(e_i - e_o)\ \phi_A\phi_T = e_o$ so that the ratio of output to input $G=e_o/e_i=(1 + 1/\phi_A\phi_T)^{-1}$. Ideally G should be close to one so that the drive follows the required waveform closely. This implies that $\phi_A\phi_T$ should be large. To understand why this is not always possible to achieve in practice we discuss briefly the frequency response of the transfer function for the drive ϕ_T.

There are two main mechanical resonances visible in the amplitude and phase transfer function as illustrated in Figure 3 obtained with a

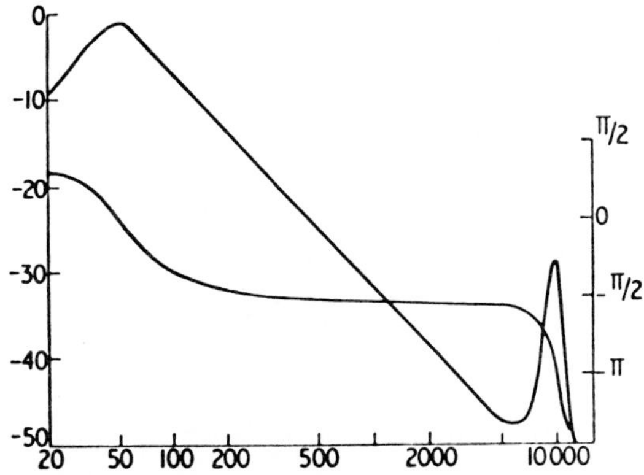

Fig. 3. Amplitude transfer function (upper figure) and phase transfer function (lower figure) as a function of frequency (in Hz) for a transducer with moving magnet velocity sensor, in units of db. (amplitude) and radians (phase).(5).

moving magnet velocity sensor. The lower resonance frequency ∿ 50 Hz comes from the mass of the whole system with the restoring force provided by the support springs holding the drive rod. The higher resonance frequency, typically about 2 kHz for moving coil pick ups(2) or somewhat higher, about 9 kHz for moving magnet pick ups, corresponds to compression of the drive rod which connects the drive coil to the velocity sensor. At this frequency the two ends of the drive rod move π out of phase with each other. In other words ϕ_T changes phase by π so that $\phi_A\phi_T$ becomes -1 and G becomes infinite. Thus the feedback now becomes positive at this frequency, sometimes called the pushrod compression resonance (3), leading to oscillation. This means that in general the drive is less able to follow waveforms with high frequency components. In practice efforts are made to make this resonance frequency as high as possible. Before discussing briefly several designs for transducers we mention a modification to the basic feedback loop. In the system as discussed (Figure 2) the error signal ($e_i - e_o$) not only corrects for deviation from linearity but also provides the whole driving waveform. In a different approach, Clark et al (13)have used an additional element with transfer function ϕ_T^{-1} which when combined with the drive transfer function ϕ_T gives a transfer function of 1 at all frequencies. Here most of the drive signal is produced outside the loop so that feedback is needed solely to correct for small velocity deviations. Hence the danger of instability is less although changes for example in loading of the drive will require adjustment of ϕ_T^{-1}.

In the following section examples are given of a velocity drive using a moving magnet and a moving coil velocity transducer, which are typical of most of the drives described in the literature.

A moving magnet design (6) is shown in Figure 4. The drive coil and

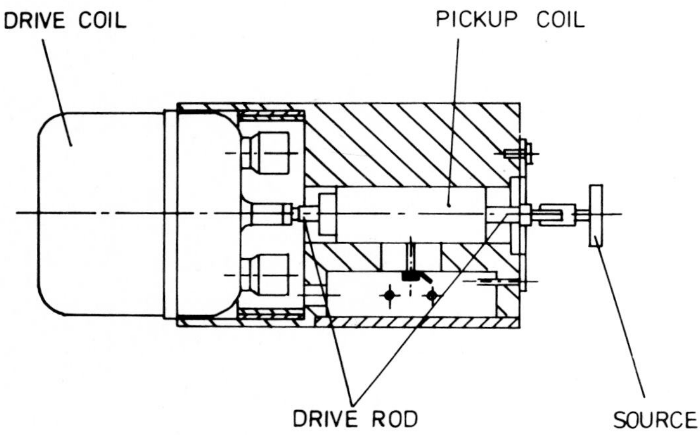

Fig. 4. Transducer with moving coil drive and moving magnet velocity sensor.

magnet assembly is a commercially obtainable unit (Ling Dynamics) used with a shaped sawtooth waveform with a period of about 13 Hz. The drive rod connecting the drive coil and a copper beryllium spider is made of cryogenic stainless steel and contains the magnet assembly which thus moves inside the pick-up coil. In order to shorten the magnet length while maintaining a low variation in sensitivity to displacement, an assembly is used consisting of two short magnets (10 mm) separated by an iron slug (6 mm) (Figure 5). The variation in sensitivity with displacement may be altered by changing the width of the air gap and it was found that for a gap of 76 μ mm the

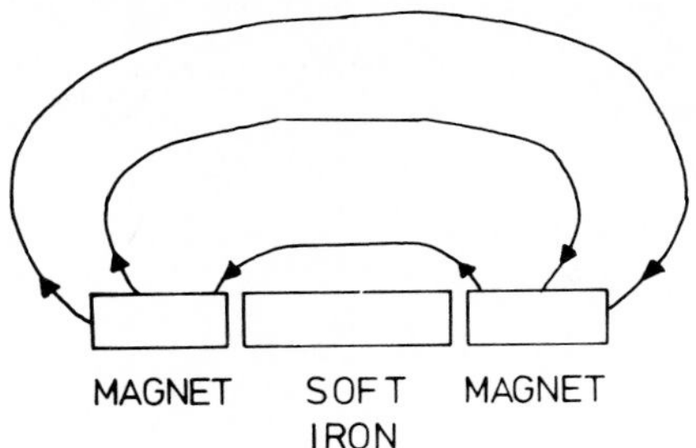

Fig. 5. Design of velocity sensor for transducer shown in Fig. 4.

sensitivity is constant to within 0.1% for displacements of up to 4.5 mm. Using such a system a differential non-linearity x of less than 0.3% may be obtained. Here x represents the difference between the greatest and smallest values of $\delta V/\delta N$ where a given velocity increment δV corresponds to a given range of channel number δN.

Most velocity drives using a moving coil pick-up are based on the original designs of Kankeleit (4). One such drive due to Wit et al (15) is shown in Figure 6. Here the drive and pick-up coils are wound on

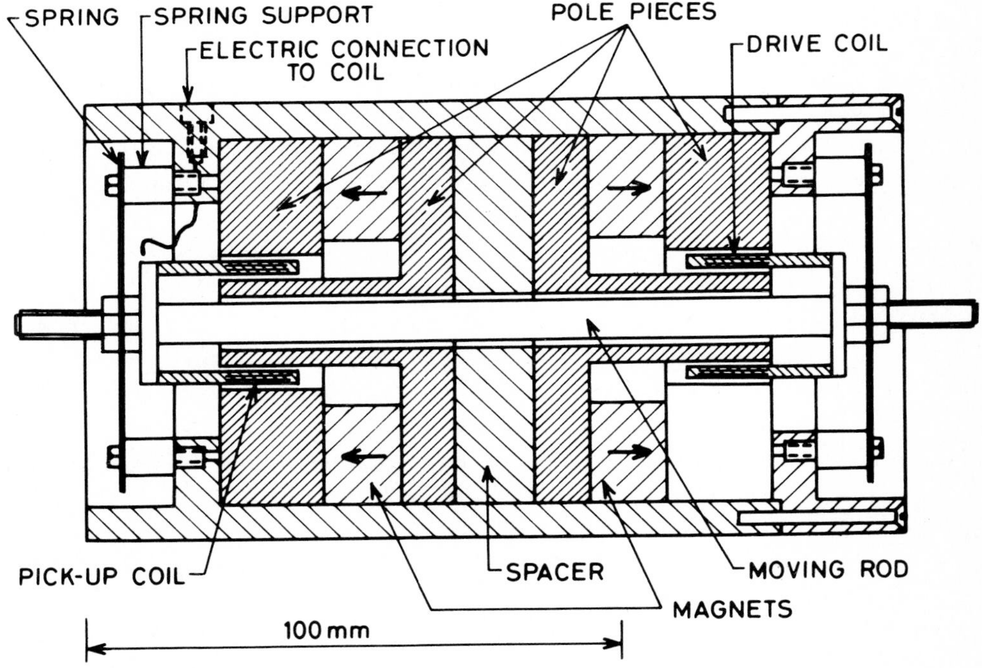

Fig. 6. Transducer with moving coil drive and velocity sensor (15).

bakelite formers glued to aluminium discs. The drive rod is made of aluminium in order to raise the resonance frequency (≃ 7 k Hz). Another way to raise this frequency would be to shorten the drive rod but this leads to increased mutual inductance coupling between the coils This in turn will produce an induced voltage in the pick-up coil when the current in the drive coil changes so that the output of the pick-up coil is no longer proportional to velocity. This coupling may be reduced with magnetic shielding. The drive is operated with a triangular waveform (3 Hz) for velocities up to 1 cm/sec and up to 100 cm/sec with a sinusoidal waveform (40 Hz). In the velocity range used for ^{57}Fe work the differential non-linearity x is 0.2%.

2.3 Data Acquisition Systems

In the past hard-wired multichannel analysers have been used to provide data storage but more recently because of their decreased cost and high reliability, mini-computers such as the PDP series made by Digital Equipment Corporation have found increasing use. There are also a few published systems (16-18) in which simple microprocessors are utilised. Such computer based spectrometers have the advantage of being more flexible in the number of addresses available and hence it is easier to design a multiple device for recording several spectra simultaneously. In addition to being a data store the computer may be used both to produce the velocity reference signal for the drive and to provide some data processing capability. Thus background correction, simple curve fitting and spectrum stripping may be performed. Since data collection is under program control, experimental variables like temperature or magnitude of applied magnetic fields may be controlled in order to record a sequence of spectra under differing conditions.

The size of the minicomputer needed is determined by the number of channels per spectrum, the number of simultaneous spectra required and by the maximum counts per channel before overflow that is required. Thus with a 12 bit word length, in order to accumulate more than 2^{12} or 4096 counts per channel without recording the overflows, a double word per channel must be used which then allows a maximum count of about 1.6×10^7 per channel. If a spectrum is to consist of 1024 channels then a typical core size of 4k words is more than adequate. In the multiple system described by Window et al (9), in which 8 spectra each with 256 channels may be accumulated, a minicomputer with 18 bit word length and 8 k words of core memory was used.

One of the problems encountered with velocity scanning spectrometers using either a computer or a multichannel analyser as a data store, is that of dead time losses. These can be considerable for high sweep frequencies when the time taken for the device to process the incoming data becomes a non-negligible part of the time spent in one channel or address.

For a sweep frequency of 50 Hz and 1024 channels, the dwell time is about 20 μsec. At the arrival of the next channel advance pulse the device must close the input to further counts, transfer the counts accumulated in the previous channel open interval from a scaler to the

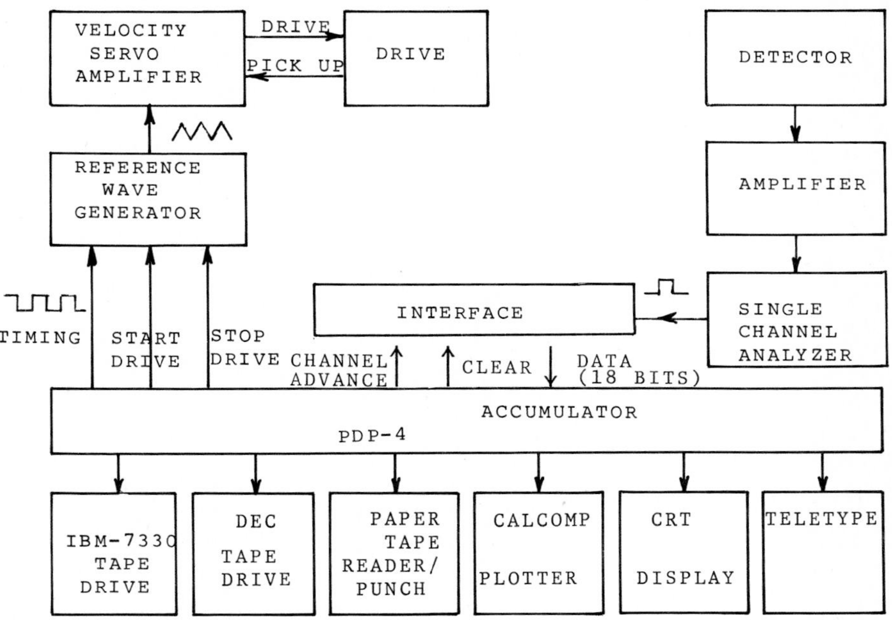

Fig. 7. Block diagram of Mossbauer spectrometer using PDP-4 computer(19)

appropriate memory location, reset this scaler, advance the channel number and then open the input again to collect further counts. The time taken for all these procedures, the memory cycle time, is of the order of 1 - 20 μsec for most multichannel analysers and hence it is not negligible in the above example.

The usual way of getting around this problem has been to use an additional scaler and buffer register before the analyser. Counts are accumulated in this scaler until the arrival of the next address advance pulse when the input is inhibited, the scaler contents are transferred in parallel to the buffer and the scaler reset, in a time of the order of 1 μsec. The scaler is then free to accept further counts while the counts in the buffer may be transferred serially to the memory relatively slowly before the arrival of the next address advance pulse. A similar arrangement is used in the computerised systems described in the literature both for single or multiple systems.

An example of a single system using a minicomputer is the one due to Wang et al (19) who used an older and slower general purpose computer PDP-4 with 8k 18 bit words of core memory. A block diagram of the system is shown in Figure 7. A commercial velocity drive is used with conventional servoamplifier. The computer is employed both to generate the pulses from which the reference waveform is derived and also the address advance pulses that determine the dwell time. Two scalers are used alternately as the buffer store at the interface. The dead time involved between successive channels is then determined by the switching time of the buffer, which is of the order of nanoseconds. The address advance pulses are obtained by using the computer memory cycle time as a clock. Since this computer is relatively slow the minimum dwell time $\simeq$ 100 μsec which for 400 channels allows a maximum frequency of 25 Hz.

The velocity reference waveform is derived from a timing pulse from the computer which sets a flip flop in the reference wave generator. The resulting square wave is made symmetrical about zero and integrated to give a triangular reference waveform.

A double word per channel is used to store the data which after using two bits for sign and overflow indication allows a maximum number of counts of $(2^{34} - 1)$ or about 2×10^{10} counts/channel. The number of channels and sweep frequency are initially supplied from the teletype punch and the counting sequence starts by the computer clearing the memory and the scalers and then issuing address advance pulses to count scalers and timing pulses to the waveform generator. A visual display of the memory contents is obtained by interrupting the data accumulation at zero drive velocity, using appropriate settings of the Accumulator Switches (ACS). Before display, the computer stops the data accumulation and the velocity drive and ceases to issue address advance commands and timing pulses. Reversing these steps allows data accumulation to continue. When an adequate spectrum has been obtained, it may be printed on the teletype or stored on magnetic (DEC) tape.

A somewhat different system has been described by Schmidt et al (20) who used a piezo-electric drive unit driven at its longitudinal resonance frequency (f_o = 16.74 kHz at room temperature) and a PDP 8/f,4k,12 bit memory computer. Two words are used per channel which can then contain up to about 1.6×10^7 counts. The data accumulation at the different velocities is obtained by measuring the time interval between a gamma ray leaving the source and a reference time occurring at the maxima of the sinusoidal waveform applied to the drive (Figure 8). This time interval defines the source velocity when the gamma ray is emitted and corresponds to an address in the computer memory. Thus in contrast to the previous system described the memory is not addressed sequentially.

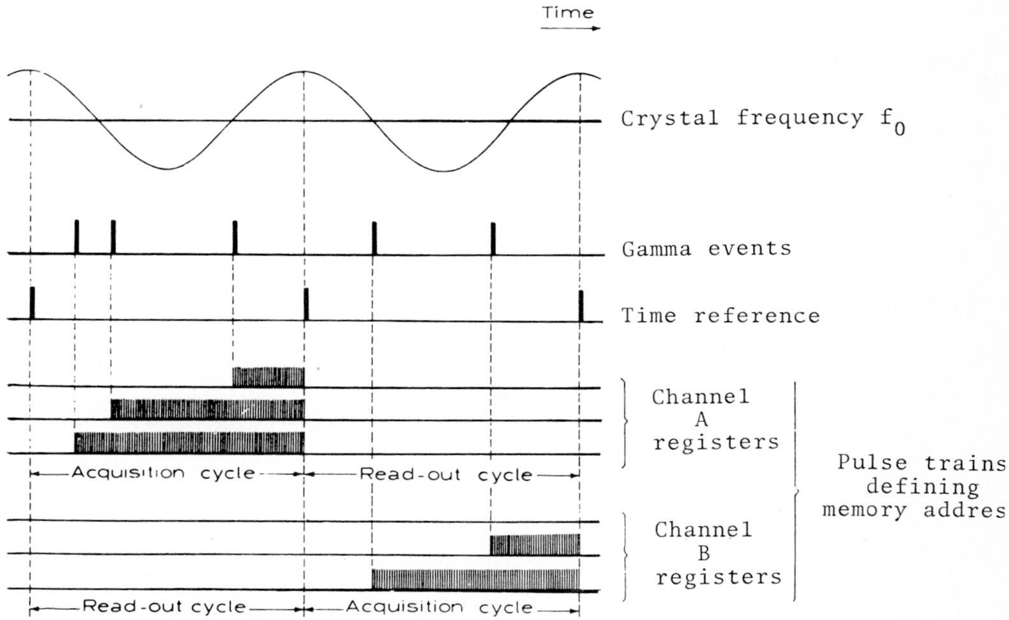

Fig. 8. Data acquisition cycle in spectrometer due to Schmidt et al (20.

The velocity drive waveform is derived from a frequency generator (Figure 9) which produces block pulses of frequency $f = nf_0$ (n = 256 or 512) which are fed to the frequency divider with a dividing ratio of n. An active filter is used to select the 1st harmonic sine wave, frequenc f_0, which after amplification is used to drive the piezo-electric crystal. A phase sensitive detector is used to stabilise the frequency generator and maintain exactly n block pulses in each crystal oscillation period. Two acquisition channels (A,B) are used alternatively in successive crystal periods in order to suppress dead time losses. The channels consist of two shift registers each connected to 3 count registers. An initial pulse from the single channel analyser in the detection system sets each shift register to the first corresponding

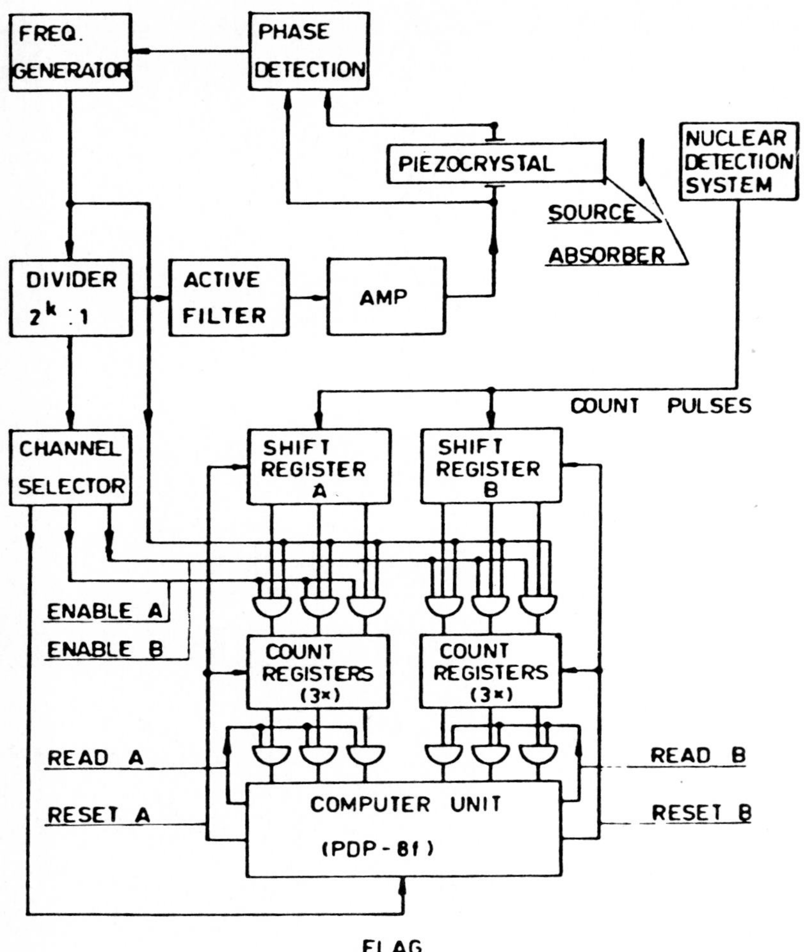

Fig. 9. Block diagram of spectrometer due to Schmidt et al(20).

count register. If a gamma ray pulse arrives during the period when channel A is open, pulses from the frequency generator, frequency *f*, are fed to the first counter of channel A. If a second gamma ray pulse is received during the same period, the shift register also opens the second counter. Since there are only three counters per channel this restricts the count rate to about 5 x 10^4 counts/sec. During the next period the channel A gates are closed, those for B opened and time intervals are measured in channel B. At the same time the computer reads in the data from each count register in channel A, resets the counter and adds one to the address corresponding to the count value. Thus while one channel is set to receive data, the other channel transfers its data to the computer memory.

An example of a multiple system has been described by Window et al (9) who used a PDP 11/10, 8k, 16 bit memory computer. This system allows the user to accumulate either 8 spectra each with 256 channels, or if required, spectra with 512 or 1024 channels by rearrangement of these groups. Both the time for which the data counters remain open and the reference waveform are here determined by a crystal clock. The analog signals for the velocity reference are derived from digital signals using 12 bit digital to analog converters. The actual waveform consists of a linear velocity sweep for time t followed by a flyback for time t/2. The shape of the flyback waveform is given by a Fourier series with the coefficients so chosen that there are no discontinuities in the first seven derivatives of the total waveform. Such a waveform is easily followed by the velocity drive. The sweep frequency was chosen to be 13 Hz. Count pulses are recorded during the linear ramp period and a selected sub-group of the memory may be displayed. During the flyback, information about the memory sub-grouping is read in, and clearing or printing out of the spectrum performed. When the system is divided into 8 sub-groups, to acquire 8 set of 256 channel spectra, the subgroups are treated in pairs, so that four lots of two are moved in turn through input buffers to the computer memory. Alternatively either 4 or 8 subgroups may be combined in blocks to give 512 or 1024 channels per spectrum.

During the linear ramp count pulses from the detection system are accumulated in eight counters (6 bits). These are accessed in turn by the interface, their contents transferred to a buffer, when they are cleared and counting resumed. This procedure takes about 2 μsec and the counter remains open for the remaining 198 μsec. During each 200 μsec time interval the clock delivers pulses every 50 μsec to initiate the above sequence in turn for each of the four sub groups. After the counter for each sub group has resumed counting the computer transfers the contents of the buffer into the current memory location, and advances to the next location.

Only one 16 bit word is used per data channel and the number of times the memory capacity 65535 is exceeded is determined during the flyback sequence and recorded in the first data channel.

Since only about 2.7k words are used for data storage and 512 words for the program, this leaves 4.8k words for further computation which may be used for instance to initiate printouts at preset times or to alter experimental parameters.

An alternative design for a multiple spectrometer allows four independent spectra to be measured with fast channel advance rates (21) Here the spectra are truly independent in that the four velocity sweeps may follow different waveforms with no phase relationships between them In one version the use of buffers for both data and address registers allows minimum dwell times of 5 μsec for one experiment or 12 μsec for four experiments with a maximum allowable count rate of 10 MHz. For dwell times less than 1 μsec the channel advance pulses are arranged to drive a 12 bit scaler in phase with the drive reference signal. A read in cycle in the computer is now commenced only when a count pulse arrives rather than at each channel advance.

Apart from the computerised systems described in the previous paragraphs there is a design due to Pettit et al (22) where the CAMAC system for interfacing a commercially available Mossbauer drive system with a PDP-9 minicomputer is used and also a spectrometer described by Graham et al (10) in which the computer (PDP 8L) is used in addition for least squares fitting and spectrum stripping.

Finally an alternative replacement for the hard wired multichannel analyser is a microprocessor. The low cost involved then means that each spectrometer can have its own data store. Such a spectrometer has been described by Player et al (17) using a Motorola MEK 6800 D1 microprocessor evaluation kit. This is used to provide a 256 channel spectrometer using 28-bit bytes for each data channel. With 768 x 8 bit bytes of random access memory (RAM) this leaves 200 bytes which is sufficient for the program store.

2.4 Absolute Velocity Measurements

The constant acceleration Mossbauer spectrometer may be calibrated either using a secondary standard, for velocities typically up to ± 10 mm/sec or by an absolute method. In the former it is straightforward to measure the six line magnetic spectrum for pure iron. The method relies upon having a very precise measurement for this material and an independent measurement of magnetic hyperfine field. It is quite sensitive since the full splitting $\simeq$ 10.7 mm/sec and it is relatively easy to measure a line position to 0.01 mm/sec leading to a sensitivity of about 0.1%. A useful check on the linearity is obtained by comparing the separations of lines 1 and 2, 2 and 3, 4 and 5, and 5 and 6.

In order to measure the velocity in an absolute method either an interferometer is used (23, 24), or a Moire fringe device (25, 26)

A simple Michelson interferometer is shown in Figure 10 where S is a laser source, M_m the moving mirror attached to the velocity drive, M_s the stationary mirror, B the beam splitter and P_A a photodiode. A bright fringe occurs for each half wavelength of displacement and the fringe frequency is proportional to velocity. The interference fringes generate pulses which are fed into the multiscaling input of a multichannel analyser. The number of counts per channel is then proportional to the modulus of the instantaneous velocity of M_m. Several problems encountered in the use of such an interferometer have been described by Cranshaw (27) and by Player et al (28) with suggested improvements.

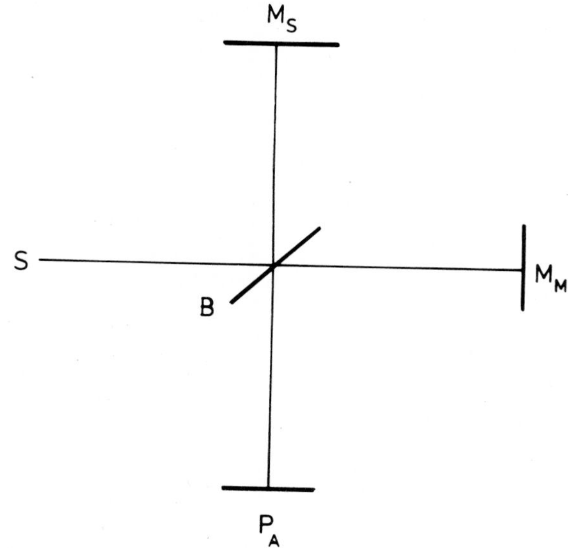

Fig. 10. Simple Michelson interferometer. Symbols defined in text.

For higher velocities Michelson interferometry becomes less useful due to the high modulation frequency and a calibration relying on Moire fringes is used. In one example (26) the light from a lamp is modulated with a frequency proportional to the velocity of a moving grating and a moire pattern is formed by imaging this grating on a stationary grating using a microscope objective. Such a device has been used up to about 80 cm/sec.

3. DATA ANALYSIS

The shape of a Mossbauer spectrum is given by the convolution of a source profile S(E, V), usually assumed to be Lorentzian in shape, with the absorber transmission ABS(E), which is an exponential function with the exponent containing a further Lorentzian function appropriate to the absorber. Many attempts have been made to perform the resulting transmission integral:

$$TRANS(V) = \int S(E,V) \times ABS(E)\, dE$$

either by numerical integration or by replacing it by the summation of an infinite series. These have usually required elaborate computation, thus restricting their use in a least squares minimisation routine, used to refine the initial values for the Mossbauer parameters. Because of this many people have used the so-called 'thin absorber' approximation in which all but the first terms of the exponential are neglected. The Mossbauer spectrum may then be constructed from the superposition of Lorentzian profiles whose positions, widths and depths are determined by the hyperfine parameters (29, 30 and 31). Although the 'thin absorber' approximation frequently is reasonable, it neglects saturation effects which are important for thicker absorbers and in such cases the spectral area may be no longer proportional to the areal density of Mossbauer atoms. The accurate determination of the relative concentrations of components in a multiphase mixture is then not possible even though the individual components are well separated. When the components overlap the situation becomes worse. For example, if a doublet with overlapping lines is fitted to the sum of two Lorentzians, the values determined for the line positions will vary with absorber thickness (32, 33).

Because of these limitations improved methods for dealing with the transmission integral have been devised. Cranshaw (34) calculates both the source profile S(E,V) and absorber transmission ABS(E) in the form of arrays and replaces the integral by a summation:

$$TRANS(V) = \sum_{E} S(E,V) \times ABS(E)$$

Both functions need be calculated only once and the value of TRANS(V) at any velocity is found from their convolution with the appropriate velocity offset. This consists merely of simple multiplications and summations.

In another technique Dibar Ure and Flinn (35) deconvolute the experimental spectrum using fast Fourier transform techniques to remove the 'blackness distortion' and obtain the absorber transmission. This generally consists of the sum of Lorentzians so that the 'thin absorber' approximation is accurate.

In a different approach, Gerdau et al (36) and Shenoy et al (37) have generated model spectra by accurate convolution of the source and absorber functions and then fitting them using the 'thin absorber' approximation. Awareness diagrams (37) are then constructed in which the values of the original input parameters are compared with their 'fitted' values, and are used to illuminate the major problems encountered in the use of the 'thin absorber' approximation.

Frequently the Mossbauer spectrum results from a combined magnetic dipole and electric quadrupole interaction. In general it is not possible to get analytical expressions for the eigenvalues of the Hamiltonian and numerical methods are used. Several programs have been written to calculate the energy levels, transition energies and intensities for particular cases (38, 39).

Finally we mention examples of a different type of fitting program used to analyse the spectra of samples in which the hyperfine parameters vary from site to site. An example is the spectrum for a disordered magnetic alloy where the distribution of hyperfine fields leads to a broad but relatively featureless six line pattern. The programs are of two types depending on whether an initial knowledge of the distribution shape is assumed. Sharon and Tsuei (40) assumed the distribution of fields P(H) to be Gaussian in an analysis of the spectra for FePdP alloys, and used an iterative procedure to refine the function parameters. The advantage of the other type of program is that no knowledge of the shape of the distribution is required (41, 42, 43). In the earliest version (41) P(H) is approximated by a trigonometrical series whose coefficients are determined in a linear least squares fit. This works well for broad distributions, and if there are also unique field values present, these may be added to the P(H) function. It is possible to get non-physical fluctuations in P(H) due for example to an ill defined background level or to inadequate counting statistics and these problems are discussed by (43) who present an improved fitting procedure.

4. SOURCES

The use of the resonances of ^{57}Fe and ^{119}Sn continues to dominate Mossbauer spectroscopy due to their convenient properties and easy availability. There are however many others which have a reasonable range of applications although they are not all available commercially. In the reviews by Kalvius et al (46) and by Stadnik (47), about twenty such resonances are listed together with details of source preparation.

In general it is necessary to incorporate the source activity into a non-magnetic, cubic lattice. Metallic matrices have the advantage that the electrical screening of the source atom charges can be made very effective so that there is little broadening due to electrostatic interactions (48). In addition one is able to avoid problems with the after-effects of nuclear decays and the handling problems are less than with the use of powdered materials. It is clearly necessary to be able to incorporate an appreciable concentration of source activity into the matrix in order to make a strong source and if possible there should be a minimum of stable isotope in the source material capable of giving broadening due to self-absorption.

5. DETECTORS FOR FAST DATA COLLECTION

Since conventional gamma ray detectors have been in use for many years and several review articles exist (2) this short section is restricted to a discussion of counting techniques for fast data collection. Measurements at high count rates are required in order to measure small absorptions and to determine lineshapes very accurately. The upper limit to the usable count rate is usually set by the detector performance rather than by the rate at which the counts can be stored. This means that the upper limit will vary from one Mossbauer transition to another depending upon the intensity and energy of any nearby gamma rays or X rays. In some cases the upper limit will be set by the available source strength.

For ^{57}Fe work most commercial proportional counters have an upper count rate limit of $\simeq$ 70,000 counts per sec in the 14.4 keV region (49). These are typically small (2" diameter), with a side window and sealed with a xenon-methane mixture.

One of the main reasons for the saturation of a proportional counter is the formation of a positive ion sheath close to the anode wire (49). This may reduce the electric field there sufficiently to inhibit gas multiplication. At high count rates the field may be reduced along the entire wire leading to a breakdown in the proportionality of the detector response.

The count rate limit has been extended by Cranshaw (50) with the use of a long, end-window counter filled with argon-methane. The advantages of argon over xenon as a stopping gas have been discussed previously (51,5). The increased wire length ($\sim$ 30 cm) necessitated by the choice of argon greatly reduces the risk of saturation. In order to avoid end effects due to changes in electric field at the end of the wire closer to the gamma ray source, the electrode is shaped in order to maintain a uniform electric field close to the wire. In this design the anode is at near earth potential and the negative high voltage is applied to a cylindrical electrode surrounding the wire. The field at one end of the wire is shaped by the use of a disc connected to the end of the cylinder, whose dimensions have been chosen so as to maintain a uniform field strength there. In this way the use of a field tube at an intermediate potential is avoided. Using such a design an energy resolution of 1.7 keV at 14 keV is obtained at a count rate in the 14 keV window of about 2 X 10^4 counts/sec. For a count rate of 10^5 counts/sec the resolution is about 3 keV so that the counter may still be used effectively. This design of counter is available commercially from Harwell Scientific Services.

In an alternative design for a high rate proportional counter Semper et al (49) use a large multiwire counter (8 cm X 15 cm X 30 cm) through which an argon-methane mixture is flowed.

Here the authors used five anode wires connected in parallel, placed one behind the other along the length of the counter in order to increase the effective wire length. The counter is used with fast amplification and discrimination. Although the energy resolution deteriorates appreciably above a count rate in the 14 keV window of about 10^5 counts per sec, it remains acceptable up to about 9 X 10^5 counts per sec. Such a count rate can be achieved with a 150 mCi ^{57}Fe source close to the counter window.

The use of tin loaded plastic scintillators has been described by (52) and (53). Plastic scintillators have decay times about two orders of magnitude less than that for NaI(Tl) but their energy resolution is very poor below 200 keV. This may be improved by loading with tin, but a plastic scintillation detector will be more effective for isotopes such as ^{119}Sn with a 'clean' pulse height spectrum rather than for ^{57}Fe whose spectrum contains other gamma rays and X rays at nearby energies to 14.4 keV. (52). For both these isotopes, however, the available source strengths are limited. Integral count rates of the order of 1 MHz have been achieved using such tin loaded scintillators to study the Mossbauer resonances in ^{170}Yb (84 keV). ^{199}Hg (158 keV) and ^{237}Np (60 keV) (53). In this work the performance of the detector is compared with those of a standard NaI(Tl) scintillator, a 'fast' NaI(Tl) scintillator and with an integrating counter technique (54, 55).

The use of a current integration technique has been described by (54, 55) for ^{197}Au(77 keV) Mossbauer studies using a NaI(Tl) scintillator. In this technique there is no energy resolution at all but the data acquisition rate can be made very high. The photomultiplier anode is connected to the inverting input of an integrating operational amplifier whose d.c. level can be adjusted by a voltage applied to the non-inverting input. The integration time constant was set to 400 μsec. The voltage output of the operational amplifier is then fed into a voltage to frequency converter and from there to a multichannel pulse height analyser operating in the multiscaling mode.

Since the d.c. output level of the operational amplifier can be adjusted independently of the detector current it is necessary to have a way of determining the true relative absorption intensity. This was done by comparing the peak intensity of the Mossbauer spectrum with the baseline curvature which may be calculated from the experimental geometry.

There is no direct information on relative absorptions with this technique but usable count rate up to and exceeding 10^8 counts per second may be achieved (53).

6. COUNTERS FOR SCATTERING GEOMETRY EXPERIMENTS

In the last ten years increasing use has been made of scattering techniques in which the decay products are detected arising from the deexcitation of the Mossbauer level in the scatterer nucleus. Due to the relatively high conversion of the 14.4 keV level in ^{57}Fe($\alpha \simeq 8$) it is more favourable to detect either K conversion electrons or the subsequent K X-rays rather than the reemitted gamma rays. The advantage of the scattering technique is firstly that spectra may be obtained for samples that are too thick for use in a transmission measurement. A spectrum produced by iron atoms in a surface layer some tens of microns thick may be measured using either scattered gamma rays or X rays. The second advantage comes with the use of conversion electrons which allows a much shallower surface layer to be studied, of the order of several tenths of a micron thick. This is due to the relatively strong absorption of 7 keV electrons in most materials.

Each type of radiation may be detected in a gas-filled proportional counter while electrostatic or magnetic analysers may be used to detect conversion electrons. The remainder of this section is concerned with a discussion of these types of detector starting with those used for electrons or X rays.

Although proportional counters have been in use in several laboratories for many years to detect conversion X rays, one of the first designs to appear in the literature was due to Fenger (56). Subsequent designs due to Swanson et al (57), Isozumi et al (58), Yagnik et al (59) and Sette Camara et al (60) differ only in details such as the anode wire structure. Figure 11 shows one such counter which has been designed in the Mossbauer group at Harwell. It has been

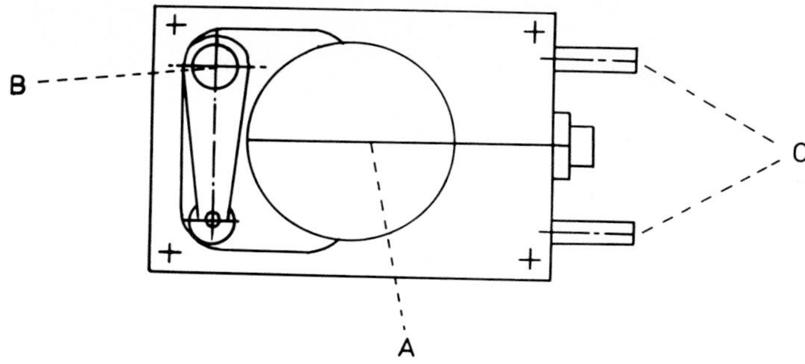

Fig. 11. Scattering counter for conversion X-rays and electrons. A-anode, B-fluorescer foil and C-gas inlet and outlet.

used to detect K conversion electrons or X rays from ^{57}Fe and the L conversion electrons from both ^{119}Sn and ^{151}Eu.

The counter body is made of brass with two copper pipes through which the gas is allowed to flow, either argon-10% methane (X rays) or helium-5% methane (electrons). A smaller percentage of methane is used in the latter case since with 10% methane a significant number of scattered X rays are stopped. There is not significantly more photoelectric absorption from gamma rays striking the counter walls using a brass counter as compared with an aluminium counter. The central anode wire is 25 μ stainless steel which is operated at a voltage of either + 800 V (electrons) or + 1000 V(X rays). The wire is sufficiently long ($\simeq$ 3 cm) so that end effects are small. Such end effects arise from the distortion of the electric field at the wire supports which can produce a local change in gas multiplication and hence an overall decrease in energy resolution. Although the resolution for counting electrons is poor it is important to achieve as good a resolution as is reasonably possible for counting X rays in order to reduce the number of background counts detected in the 6.3 keV window. The window facing the source is made of 2 mm perspex which has an electrically conducting layer of graphite on its inner surface. The back-plate of the counter is aluminium sheet which is sealed on a rubber gasket using demountable screws. A small sample may be greased to its inner surface or in order to examine a small area of a larger sample, the appropriate sized hole is cut in the back plate and the sample mounted using sealing compound. It is usual to count electrons with energies from 7 keV down to essentially zero. This therefore includes the L Auger electrons (mainly at 5 keV) which also carry Mossbauer information since they again originate from de-excitation of the Mossbauer level. In order to set a threshold to exclude noise pulses it is convenient to compare pulse height spectra (Figure 12) obtained in about 30 secs. with source on resonance with a scatterer (fluorescer) and with the source vibrating so as to destroy resonance. The fluorescer contains sufficient ^{57}Fe to give a large single line response and for an ^{57}Fe source in either rhodium or palladium, it is typically Rh Fe 20 at %. In the above design the fluorescer foil is mounted on a movable arm so that after use it can be rotated out of the beam. In this way the scatterer sample may be loaded before the threshold is set.

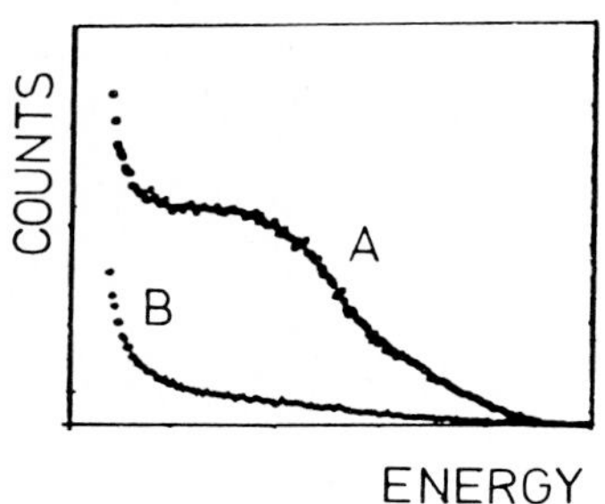

Fig. 12. Pulse height spectra for conversion electron scattering. A-on resonance, B-off resonance.

It is easily verified using a filter to reduce the incident 14.4 keV gamma rays that most of the off resonance pulse height spectrum originates with the 122 and 137 keV radiation from the ^{57}Fe source. It is due mainly to gamma rays dissipating part of their energy by creating Compton electrons at the incident window. It is relatively high because there are about 10 times as many 122 keV gamma rays as 14.4 keV gamma rays due to the highly converted 14.4 keV level. Attempts to reduce it by coupling the detector in anticoincidence with a scintillation counter placed behind the detector have been unsuccessful (56).

The counter thickness (2 cm) is chosen so that either X rays or electrons may be counted conveniently by changing the gas mixture. The absorption of the incident gamma rays in this thickness of helium is negligible and of the order of only 6% in argon, while about 60% of the scattered X rays are absorbed in argon. The energy resolution for X rays is about 2 keV at 6.3 keV.

In the case of X ray scattering most of the background comes from K X-rays produced by photoelectric absorption which clearly have the same energy as the conversion X rays. In order to reduce the absorption of the incident gamma rays in an argon filled counter, multiple anode wires have been used (57, 58). In the first design two wires are used symmetrically located on either side of the counter centreline. Since there is then a region with low electric field between the two wires operated at the same voltage, not all electrons created by the incident gamma rays will reach the wires and be counted. Although this will reduce the background, some of the scattered electrons from the sample in that region will also be lost. Isozumi et al (58) describe a three-wire-in-line structure where electrons produced in the central region are not counted at all. Their detector is in fact a double counter. The thicker chamber (2 cm) closer to the source is used to count scattered X rays which have passed through the window of the second thinner chamber (1 cm) placed directly behind the first. Using two counting systems and two memory groups of a multichannel analyser, spectra from the back scattering of electrons and X rays are obtained at the same time.

It is often of interest to measure the scattered spectrum for a sample below room temperature. This is somewhat easier to do for X rays than electrons if a proportional counter is used. The sample may be cooled to 77° K in the tail of a nitrogen cryostat and the scattered X rays which pass through a thin mylar window may be detected in the conventional counter at room temperature.

For electron scattering where the sample is inside the counter, the whole counter must be cooled. Only one group so far has published details of such a proportional counter operating at 77° K (61, 62). They in fact point out the difficulties, namely:

1. The quench gas, usually methane, has a relatively high boiling point. It is possible, however, to use pure helium.

2. The gas gain decreases with decreasing temperature and pressure so that good thermal stability of the gas filling is required.

3. Pulses are strongly reduced in size at low temperatures leading to the need for high anode voltage and large amplification and

4. The condensation of any impurities can lead to spurious pulses.

Sawicki et al (61) have described a system in which the proportional counter is held at the bottom of a tube which is filled with helium-methane to a pressure of 0.3 atm. The tube is inserted in the inner container of a liquid nitrogen cryostat. Spectra were obtained with the scatterer at 77° K. In a later version (62) both detector and source, mounted on a long drive rod on a velocity drive, are placed in the central tube of either a nitrogen or helium cryostat. Pure helium at pressures between 0.5 - 1.2 atm. is used and with such a system "electron counting can be extended down to 10 - 20° K".

Although it is possible with some difficulty to construct gas filled proportional counters for electrons which will operate at low temperatures, it is preferable to separate the cooled scatterer from the point at which the electrons are detected, which may then be at room temperatures. Such an arrangement may be realised for example by the use of an electron spectrometer between the scatterer and electron detector (63, 66). In another simpler system (67) the electrons are detected by a windowless multiplier which is evacuated but isolated from the scatterer environment in order to allow sample-surface reactions with gases to be studied. The scatterer is mounted in a small oven which allows temperatures up to 800° K to be reached. In an alternative arrangement the sample may be mounted in a cryostat for temperatures between 78° K and 400° K. The source/scatterer/detector geometry may be varied by changing the angle between the incident gamma rays and the direction of detection as well as by changing the angle of incidence by rotating the scatterer.

In order to measure the scattered spectrum at 4.2° K Massenet (65) used a longitudinal field electron analyser (Figure 13) which is similar in design to an earlier version due to Schunk et al (63). The

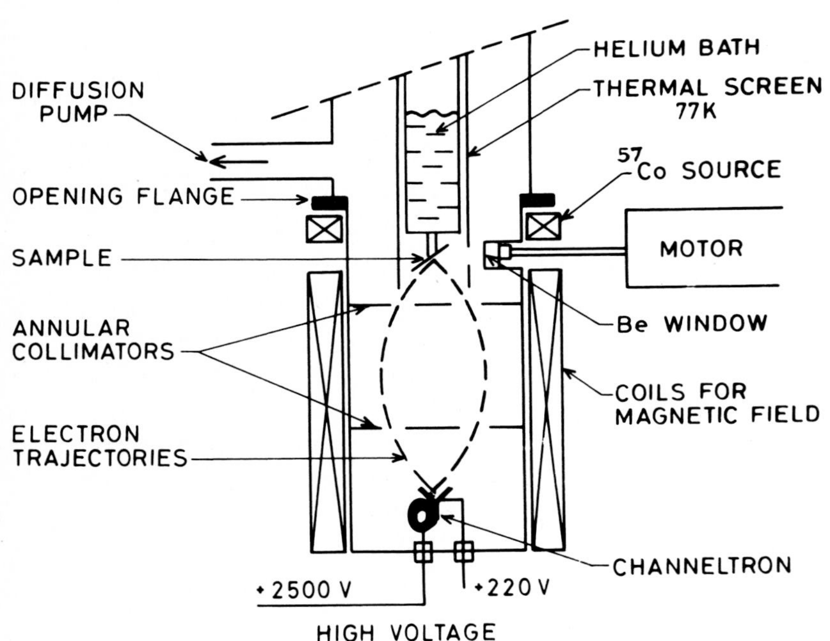

Fig. 13. Electron analyser due to Massenet 1978. (65).

scatterer is mounted on the tail of a helium cryostat and the electron trajectories in the magnetic field are selected by two annular collimators to be focussed on a tubular electron multiplier (channeltron). The energy of the electrons to be counted can be varied by changing the value of the applied magnetic field. However, the energy resolution is made intentionally poor in order to detect both the K conversion and L Auger electrons. The electron analyser and insulating space of the cryostat are maintained in vacuum using an oil diffusion pump with a liquid nitrogen baffle to reduce condensation of oil on the sample.

Baverstam et al (67, 68), Liljequist et al (69), have demonstrated that information about the scattered electron depth distribution may be obtained from scattered spectra measured for several scattered electron energies using an electron detector with good energy resolution. A brief description of this method is given in the following section. They have described an electrostatic spectrometer of the cylindrical mirror type (66) where the design has been optimised using computer calculations to produce a high luminosity consistent with good energy resolution.

Song et al (70, 71), Benczer-Koller et al (64), have used the high energy resolution of a spherical electrostatic electron spectrometer to distinguish the various S - shell electrons. A resolution of better than 0.5% is necessary, for example, in order to distinguish electrons from 3S (M_I) and 4S (N_I) configurations.

It was noted earlier that a proportional counter may be used to detect scattered electrons having energies from 7 keV down to essentially zero. In such a situation although we can say that the spectrum is produced by iron atoms in a layer about 0.2 μ thick at the surface, little can be said about the depth distribution of these atoms. It is true that an electron emitted from just below the surface will have a higher probability for escape and detection as compared to one emitted from a greater depth. Thus the overall spectrum will be weighted towards contributions from iron atoms situated close to the surface. Even if electrons over a narrow range of energies are detected in an electron spectrometer, this by itself is not sufficient to indicate the origin of these electrons since all electrons detected at a given energy will not have originated at the same depth. In the notation of Baverstam et al (68) there is a given probability $W(E, x)$ that an electron emitted at depth x will be detected at a spectrometer energy setting E. The number of counts recorded in the nth channel of a velocity spectrum at this setting E, is:

$$T(E)_n = \int_0^\infty W(E, x)\, P(x)_n\, dx,$$

where $P(x)_n$ is the probability of emission for electrons at depth x which is related to the areal density of ^{57}Fe atoms which is frequently the quantity of interest. The factor $W(E,x)$ may be obtained in a separate experiment (67).

In a scatterer $P(x)_n$ may have the form of a step function for instance when an iron metal foil has a thin oxide coating. Then $P(x)$ has a constant value within each layer and may be removed from the integral, thus:

$$T(E_j)_n = \sum_{i=1}^{I} \left| \int_{\ell i}^{\ell i+1} W(E_j, x)\, dx \right| P_{in},$$

$j = 1, 2 \ldots\ldots J.$

Here spectra are measured at several energy settings E_j. P_{in} represents the constant value of $P(x)$ over the ith layer of extent ℓi. The total number of layers is I and J is the total number of spectra measured. These equations, one for each channel n in the sets of spectra, are readily solved to determine the quantities P_{in}. In fact their solutions which are weighted sums of the measured $T(E_j)$ values, may be calculated directly on a multichannel analyser in which the memory may be divided up into several regions and in which simple algebraic manipulation of individual spectra is possible.

For the more usual experimental situation where no energy discrimination of scattered electrons is possible we can still get an idea of the relative contributions to the spectra from iron atoms at various depths. This is important for instance in the studies of the distribution of non-iron atoms which have been implanted near the surface of an iron foil (72). In general the spectral contribution due to iron atoms inside and outside the implanted layer will differ. By measuring the scattered spectra for an iron foil on which were evaporated successive layers of ^{56}Fe, the author was able to determine the probability that electrons emitted at a given depth below the surface of the iron foil would be detected in a proportional counter (Figure 14A). From this we can derive a curve which represents the relative contributions to the total spectrum from iron atoms in different layers (Figure 14B). This indicates that about 50% of the spectrum is due to iron atoms in a layer 0.1 μ thick below the surface. The form of the probability function is consistent with the usually assumed depth of 0.2 - 0.3 μ and is approximately exponential.

In this final paragraph of this section we mention several designs

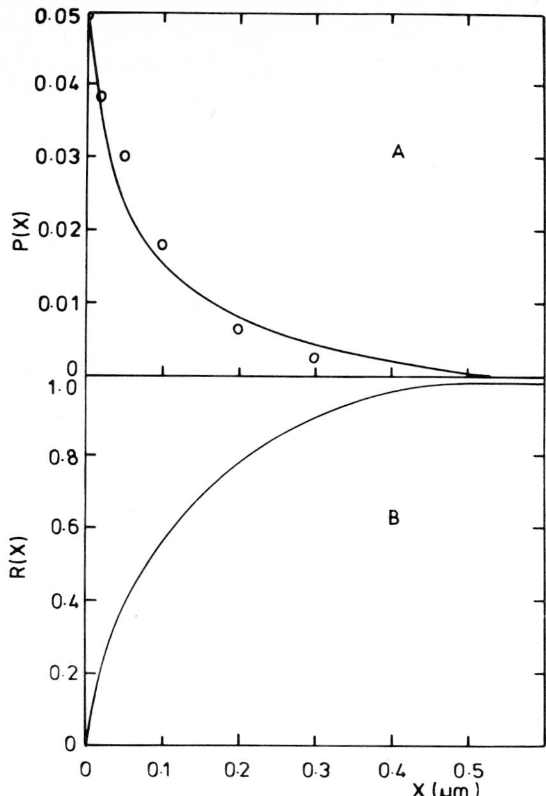

Fig. 14. Contribution to electron scattering spectrum from iron atoms at different depths. P(X)-probability of detection when emitted at depth X, R(X)-fraction of spectrum from electrons within layer 0 to X.

for counters for scattered gamma rays. Most of the early measurements on gamma ray scattering (73-75) were performed using conventional side window proportional counters ($\simeq$ 2" diameter) filled with xenon-methane. The scatterer was usually in the form of a foil and scattering angles of up to $\simeq 180^{o}$ were used. The main problem was to maintain a reasonable geometry while using sufficient shielding between source and counter to reduce the background radiation. A more favourable geometry was obtained using a toroidal counter (76-78). The basic design is shown very schematically in Figure 15. The counter is shielded from the incident gamma rays from the source S using lead or tungsten alloy shielding SH. The detector body is aluminium and the conical window (K) is either nylon or aluminium (0.003"). The anode wire, A, is supported by seven quartz rods and is in the form of an octagon (77). When the counter was filled with krypton-10% methane and sealed, the initial resolution was 1.8 keV at 14.4 keV which deteriorated during the first

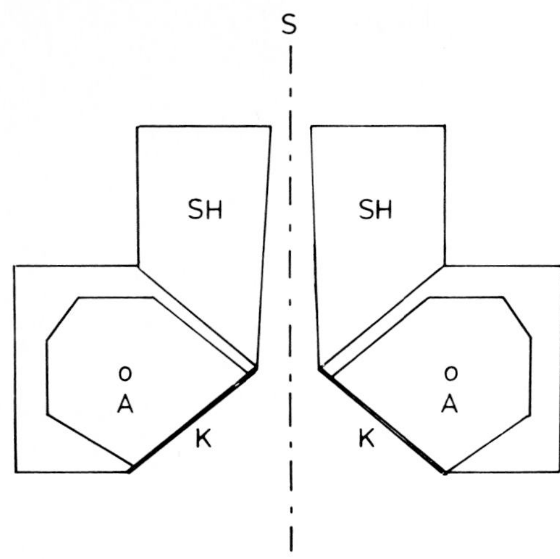

Fig. 15. Schematic drawing of scattering counter to detect X-rays or gamma rays.

few days but then remained stable for the order of a month. It was suggested that the original deterioration was due to water from the nylon window. In a later somewhat simpler version (78) the anode is supported at only four points. The departure from cylindrical geometry leads to poorer energy resolution. In order to counteract this, cylindrical grid wires were used surrounding the anode. These counters have also been used to detect iron X rays using a gas filling of argon-methane.

7. CRYOSTATS

There are several designs in the literature for liquid nitrogen or liquid helium cryostats (2, 3, 51). If only the absorber is to be cool it is convenient to use a horizontal gamma ray beam and to be able to change absorbers without breaking the vacuum seals. Examples of such cryostats (51) are shown in Figure 16. The absorber in each case is mounted on a long insulating rod and inserted in the central tube eith in air in the nitrogen cryostat or in the liquid in the helium cryosta The helium cryostat runs for 3 days on 3ℓ of liquid helium. The windows in the tail of each cryostat consist of mylar mounted using epoxy cement. These can be cycled from room temperature to 4.2° K many times without developing leaks. As an alternative to this metal cryostat Pasternak et al (79) have described the construction of an all glass helium cryostat for cooling source and absorber in a vertical geometry This has the advantages of low cost and good vacuum properties but the

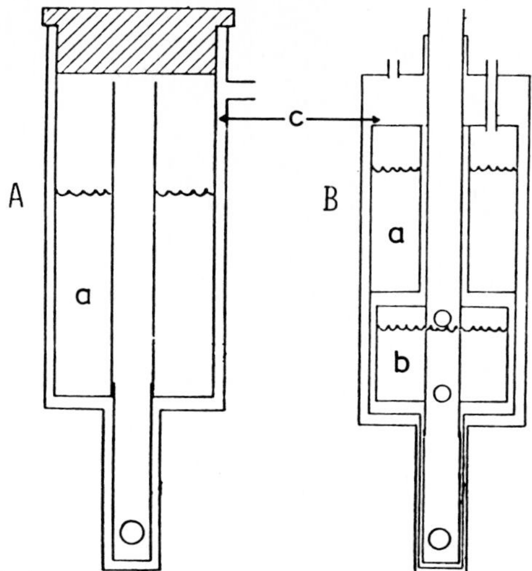

Fig. 16. Nitrogen (A) and Helium (B) cryostats where a represents liquid nitrogen, b liquid helium and c the vacuum space.

disadvantages of vulnerability and the fact that it is difficult to directly seal a window to the glass. In this design the lower part of the dewar tail is made of metal using a metal to glass seal so that a bottom window may be indium sealed.

In order to obtain temperatures between 4.2 to 90° K or 77° K and room temperature it is possible to use a small vacuum insert for the absorber (5) in which the absorber holder is heated electrically and its temperature controlled (80). The alternative approach has been to use a continuous flow cryostat (81). In this type the cryostat is mounted directly on the helium storage vessel and cooling is provided by the boil off. Helium is drawn through the system using an external flow control pump going first to a central heat exchanger carrying the absorber and then to a second heat exchanger which carries a radiation shield, before being exhausted. A heater is used for temperature regulation together with a control unit and temperatures can be maintained to better than 0.1° K. At 4.2° K the consumption of helium is 0.120 $\pm$ 0.005 ℓ/hr, while at 150°K it is 0.040 $\pm$ 0.005 ℓ/hr. The advantages of this type of cryostat are its simplicity and the rapid cool down time while the disadvantage is its vulnerability to clogging by frozen air or water vapour.

Temperatures down to 0.3° K can be obtained by pumping on ^{3}He but this is expensive and must be used in a closed cycle. Bogner et al (82) have described a continuously operating ^{3}He cryostat which allows the absorber to be mounted, while keeping its temperature below 100° K at all times. This is frequently necessary for biological samples.

For lower temperatures a ^3He - ^4He dilution refrigerator has been used. Ehnhohm et al (83) and Kalvius et al (84) describe such a cryostat which enables temperatures between 0.1 - 0.01°K to be reached with the sample mounted in the mixing chamber. Below 0.1° K there are problems connected with the thermal coupling of the sample to the cold bath and in the case of cooled sources with radioactive self-heating.

Samples which have been prepared by using a matrix isolation technique have been studied in optical, IR, UV, NMR and Mossbauer measurements. The technique consists in the incorporation of the appropriate atom or molecule at a very low concentration (< 1 pt in 1000) on a frozen gas matrix, typically a rare gas or nitrogen. Several cryostat designs (2, 85, 86, 87) have been published specifically for Mossbauer measurements on samples produced by matrix isolation in situ, and in one case (86) there is the extra provision for IR measurements.

A thin layer of isolated atoms is produced by mixing a well collimated atomic beam with a stream of rare gas and condensing the mixture on a cold substrate at a temperature between 5 - 20° K. The beam of atoms is produced using a furnace which will usually cater for evaporation temperatures of up to about 1000° C. In most designs the substrate is in close contact with liquid helium in order that it should remain at a temperature close to 4.2° K under the condition of high thermal load when the sample is co-deposited.

In the design due to Pasternak et al (87) the substrate is an aluminium foil used as the bottom window of a top loading helium cryostat. The source vibrates in the liquid helium above the substrate and an evaporation chamber is mounted under the helium cryostat which contains the furnace and radiation shield. The furnace is rotated to a point under the substrate for the deposition and then rotated to one side and replaced by a radiation shield. The detector is placed below a mylar window in the bottom of the evaporation chamber. The sample thickness is measured by measuring the attenuation of iron K X-rays.

In order to study the migration and aggregation of the atomic species it is desirable to be able to maintain the sample at temperatures up to about 30° K. If the sample has a good thermal connection with liquid helium as in the system just described it is not possible to do this by heating the sample holder electrically. Conversely if we sacrifice the good thermal contact then the temperature of the substrate may rise sufficiently on deposition that appreciable aggregation occurs. Bos et al (86) in their work on SnO molecules observed that the migration and aggregation is markedly greater at 12° K than at 6° K.

One solution has been to pump liquid helium from the cryostat through a copper heat exchanger embedded in the cold finger (2). The temperature may be varied by changing the pumping rate. A more satisfactory solution to the problem was described by Bos et al (86) (Figure 17). This cryostat has the added advantage that infra red measurements may also be made on the sample.

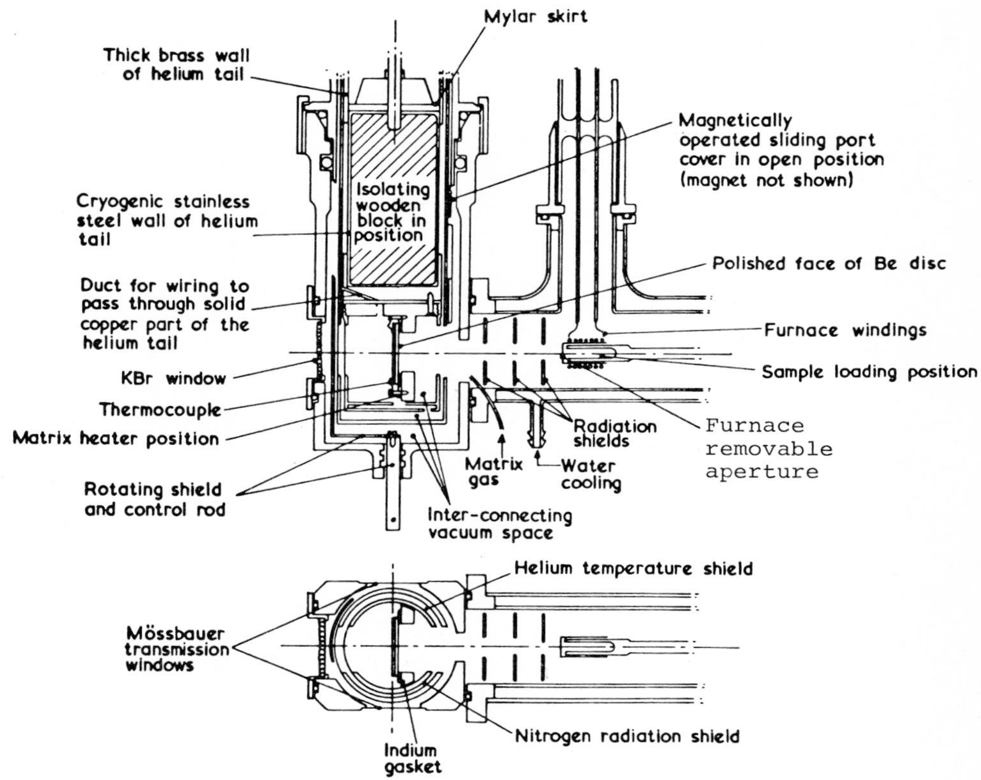

Fig. 17. Matrix isolation cryostat (Bos et al 1974).(86).

This design (Figure 17) is based on a modification of the tail of a top loading helium cryostat in which the outer vacuum-tight tail was made rotatable with one side connection to a furnace and matrix gas supply. Rotation through 180° aligned a KBr window for IR measurements while further rotation through 90° aligned two mylar windows to allow a Mossbauer transmission measurement. In addition there is a movable radiation shield attached to the liquid nitrogen tail and a rotatable shield between the outer and liquid nitrogen tails which is used to interrupt the beam from the furnace until the conditions are ready for deposition.

Two beryllium discs are mounted in the absorber holder with one, used as the substrate, being highly polished and coated with aluminium in order to make it a good IR reflector. Two thermocouples were arranged between the two discs to monitor any temperature gradients.

The matrix condensing temperature was about 6° K and when higher annealing temperatures were required a high thermal resistance was introduced between sample and liquid helium. This was achieved by inserting a closely fitting wooden plug into the tail of the dewar to provide a false bottom for the liquid helium. In addition in the original modification to the dewar, part of the inner copper tail was replaced by a cryogenic stainless steel tube. A heater on the sample holder allowed temperatures up to 30° K to be achieved. The temperature was regulated using a thermocouple activated temperature controller.

Finally we mention two specialised cryostat assemblies. The first of these is a cryostat for Mossbauer measurements on a sample following high dose neutron irradiation at 4° K with no appreciable warm up ($< 10^{\circ}$ K) between irradiation and measurement (88). Using such a system the lattice defects produced on irradiation can be studied at temperatures where their mobility is very low. The cryostat may be separated from the low temperature irradiation facility after sample transfer and it is then possible to heat the sample in order to study the annealing out of the defects.

The second design is for a helium cryostat which is coupled to an electromagnetic separator for in situ measurements on implanted sources (89). It is of interest both to implant and to measure the source at low temperature (5° K). The temperature of the source may then be varied between 5° K and 300° K while the absorber temperature remains at about 5° K.

8 HIGH PRESSURE CELLS AND FURNACES

Most cells used for measuring the effect of high pressures on the hyperfine interactions have been of the Bridgman anvil type. A Mossbauer source in the form of a foil or pellet is placed in a supporting ring of boron-lithium hydride, and compressed between the flat faces of two Bridgman anvils with the gamma rays detected in a direction at right angles to the axis of the pistons. Using such a device (90) were able to measure ^{57}Fe Mossbauer spectra at pressures up to 240 kbar. A variation of this design was used by Debrunner et al (91) to allow the pressure to be applied to an absorber. The available area for the sample in each case however is small.

There are several recent descriptions of high pressure cells which will operate at low temperatures (92-97).

Christoe et al 1969 used a Bridgman anvil technique to apply pressure at room temperature and then cooled the cell to 77° K. Pressures of up to 240 kbar could be reached. A similar type of cell was used by Williamson et al (94) which could be operated at 14° K with pressures up to 170 kbar. The sample pressure was obtained using a room temperature calibration curve of applied stress versus sample pressure and by making allowance for the effects of thermal contraction. The pressure on the sample depends critically on the alignment of the sample and gasket and it is better to measure it directly.

Schilling et al (95) measured the sample pressure using the pressure dependence of the superconducting temperature of lead (in the region 5° - 8° K). The clamp cell which they describe is also used for resistivity measurements and may be adapted for optical measurements.

A different approach to high pressure cell design for work at low temperatures has been adopted by Roberts et al (93) and by Wenzel et al (96). Here the stress is transmitted via a rod from the press to the cell in the cryostat, so that the pressure may be regulated at low temperature. Wenzel et al (96) have used this technique with a Bridgman anvil cell and a gear mechanism inside a liquid nitrogen cryostat. Pressures of up to 110 kbar could be obtained. The cell used by Roberts et al (93) for ^{197}Au measurements at 4.2^{o} K was somewhat different in that the anvils were made of B_4C which is light enough that the gamma rays can be observed along the axis of the anvil body.

The main requirements on the design of a furnace for Mossbauer measurements are high temperature stability, good temperature homogeneity and small overall size. There are many designs in the literature with temperature inhomogeneities of the order of 1° K (98-100). Coey et al (101) describe a furnace operating at temperatures up to about 500° C and situated in the bore of a 50 kOe superconducting magnet. In the design due to Kobeissi et al (102) temperatures of up to 800° C may be reached with a long term temperature stability of better than 0.02^{o} K and temperature inhomogeneity of better than 0.05^{o} K.

REFERENCES

1 J.G. Stevens and V.E. Stevens, Mossbauer Effect Data Index, Plenum Press 1975, 1976, 1977; Mossbauer Effect Reference and Data Journal, University of North Carolina, USA, 1978, 1979.

2 G.M. Kalvius and E. Kankeleit, Mossbauer Spectroscopy and its applications, IAEA, Vienna, 1972, p. 9.

3 R.L. Cohen and G.K. Wertheim, in R.V. Coleman (Ed.), Methods of Experimental Physics, Vol. 11, Solid State Physics, Academic Press, New York, 1974, p. 307.

4 E. Kankeleit, in I.J. Gruverman (Ed.), Mossbauer Effect Methodology, Volume 1, Plenum Press, New York, 1965, p. 47.

5 T.E. Cranshaw, J. Phys. E: Sci. Instrum., 7 (1974) 497.

6 T.E. Cranshaw, Nucl. Instrum. Methods, 30 (1964) 101.

7 M.T. Taragin, Nucl. Instrum. Methods, 150 (1978) 607.

8 A. Biran, A. Shoshani and P.A. Montano, Nucl. Instrum. Methods, 88 (1970) 21.

9 B. Window, B.L. Dickson, P. Routcliffe and K.K.P. Srivastava, J. Phys. E: Sci. Instrum., 7 (1974) 916.

10 M.J. Graham, D.F. Mitchell and J.R. Phillips, Nucl. Instrum. Methods, 141 (1977) 131.

11 G. Kaindl, M.R. Maier, H. Schaller and F. Wagner, Nucl. Instrum. Methods, 66 (1968) 277.

12 N. Halder and G.M. Kalvius, Nucl. Instrum. Methods, 108 (1973) 161

13 P.E. Clark, A.W. Nichol and J.S. Carlow, J. Sci. Instrum., 44 (1967) 1001.

14 T.E. Cranshaw, G. Lang and F. Placido, J. Phys. E: Sci. Instrum., 9 (1976) 10.

15 H.P. Wit, G. Hocksema, L. Niesen and H. de Waard, Nucl. Instrum. Methods, 141 (1977) 515.

16 B. Fultz and J.W. Morris, Rev. Sci. Instrum., 49 (1978) 1216.

17 M.A. Player and F.W.D. Woodhams, J. Phys. E: Sci. Instrum., 11 (1978) 191.

18 S.I. Uehara and Y. Maeda, Ann. Rep. Res. Reactor Inst. Kyoto Univ., 11 (1978) 189.

19 G.W. Wang, L.M. Chirovsky, W.P. Lee, A.J. Becker and J.L. Groves, Nucl. Instrum. Methods, 155 (1978) 273.

20 K.P. Schmidt, M. Hayse, de Raedt, G. Langouche, M. van Rossum and R. Coussement, Nucl. Instrum. Methods., 120 (1974) 287.

21 A. Forster, N. Halder, G.M. Kalvins, W. Potzel and L. Asch, J. Phys. (Paris) Colloq., 37 (1976) C6-725.

22 J.W. Pettit and R.A. Levy, Nucl. Instrum. Methods, 159 (1979) 561.

23 J.P. Biscar, W. Kundig, H. Bommel and R.S. Hargrove, Nucl. Instrum. Methods, 75 (1969) 165.

24 J.G. Cosgrove and R.L. Collins, Nucl. Instrum. Methods, 95 (1971) 269.

25 H. de Waard, Rev. Sci. Instrum., 36 (1965) 1728.

26 H.P. Wit, Rev. Sci. Instrum., 46 (1975) 927.

27 T.E. Cranshaw, J. Phys. E: Sci. Instrum., 6 (1973) 1053.

28 M.A. Player and F.W.D. Woodhams, J. Phys. E: Sci. Instrum., 9 (1976) 1148.

29 M.F. Bent, B.I. Persson and D.G. Agresti, Comput. Phys. Commun., 1 (1969) 67.

30 B.L. Chrisman and T.A. Tumolillo, Comput. Phys. Commun., 2 (1971) 322.

31 W. Wilson and L.J. Swartzendruber, Comput. Phys. Commun., 7 (1974) 151.

32 R.E. Meads, B.M. Place, F.W.D. Woodhams and R.C. Clark, Nucl. Instrum. Methods, 98 (1972) 29.

33 S.A. Wender and N. Hershkowitz, Nucl. Instrum. Methods, 98 (1972) 105.

34 T.E. Cranshaw, J. Phys. E: Sci. Instrum., 7 (1974) 122.

35 M.C. Dibar Ure and P.A. Flinn, in I.J. Gruverman (Ed.), Mossbauer Effect Methodology, Vol. 7, Plenum Press, New York, 1971. p. 245.

36 E. Gerdau, W. Rath and H. Winkler, Z. Phys., 257 (1972) 29.

37 G.K. Shenoy, J.M. Friedt, H. Maletta and S.L. Ruby, in I.J. Gruverman (Ed.), Mossbauer Effect Methodology, Vol. 9, Plenum Press, New York, 1974, p. 277.

38 J.R. Gabriel and S.L. Ruby, Nucl. Instrum. Methods, 36 (1965) 23.

39 G. Lang and B.W. Dale, Nucl. Instrum. Methods, 116 (1974) 567.

40 T.E. Sharon and C.C. Tsuei, Phys. Rev. B, 5 (1972) 1047.

41 B. Window, J. Phys. E: Sci. Instrum., 4 (1971) 401.

42 J. Hesse and A. Rubartsch, J. Phys. E: Sci. Instrum., 7 (1974) 526.

43 G. LeCaer and J.M. Dubois, J. Phys. E: Sci. Instrum., 12 (1979) 1083.

44 G.M. Kalvius, F.E. Wagner and W. Potzel, J. Phys. (Paris) Colloq., 37 (1976) C6-657.

45 Z.M. Stadnik, Mossbauer Effect Reference and Data Journal, 1 (1978) 217.

46 V.I. Goldanskii and R.H. Herber (Eds.), Chemical Applications of Mossbauer Spectroscopy, Academic Press, New York, 1968.

47 Applications of Mossbauer Effect in Chemistry and Solid State Physics, I.A.E.A., Vienna, 1966.

48 G. Longworth and B. Window, J. Phys. D: Appl. Phys., 4 (1971) 835.

49 R.J. Semper, C.R. Guarnieri and J.C. Walker, Nucl. Instrum. Methods, 129 (1975) 447.

50 T.E. Cranshaw, Commercial version of proportional counter available, Mossbauer Group, AERE Harwell, 1972.

51 G. Lang, Quart. Rev. Biophys., 3 (1970) 1.

52 J. Becker, L. Eriksson, L.C. Moberg and Z.H. Cho, Nucl. Instrum. Methods, 123 (1975) 199.

53 G.M. Kalvius, F.E. Wagner and W. Potzel, J. Phys. (Paris) Colloq., 37 (1976) C6-657.

54 M.P.A. Veigers and J.M. Trooster, Nucl. Instrum. Methods, 118 (1974) 257.

55 M.P.A. Viegers and J.M. Trooster, in G.J. Perlow (Ed.), Workshop on new directions in Mossbauer Spectroscopy, Argonne, A.I.P. Conf. Proc. Vol. 38, 1977, p. 93.

56 J. Fenger, Nucl. Instrum. Methods, 69 (1969) 268.

57 K.R. Swanson and J.J. Spijkerman, J. Appl. Phys., 41 (1970) 3155.

58 Y. Isozumo, D.I. Lee and I. Kadar, Nucl. Instrum. Methods, 120 (1974) 23.

59 C.M. Yagnik, R.A. Mazak and R.L. Collins, Nucl. Instrum. Methods, 114 (1974) 1.

60 A. Sette Camara and W. Keune, Corros. Sci. 15 (1975) 441.

61 J.A. Sawicki, B.D. Sawicka and J. Stanek, Nucl. Instrum. Methods, 138 (1976) 565.

62 J. Sawicki, J. Stanek, B. Sawicka and J. Kowalski, Proc. Int. Conf. Mossbauer Spectroscopy, Vol. 1, Bucharest, 1977, p. 15.

63 J.P. Schunk, J.M. Friedt and Y. Llabador, Rev. Phys. Appl., 10 (1975) 121.

64 N. Benczer-Koller and B. Kolk, in G.J. Perlow (Ed.), Workshop on new directions in Mossbauer spectroscopy, Argonne, A.I.P. Conf. Proc., Vol. 38, 1977, p. 107.

65 O. Massenet, Nucl. Instrum. Methods, 153 (1978) 419.

66 U. Baverstam, B. Bodlund-Ringstrom, C. Bohm, T. Ekdahl and D. Liljequist, Nucl. Instrum. Methods, 154 (1978) 401.

67 U. Baverstam, T. Ekdahl, C. Bohm, B. Ringstrom, V. Stefansson and D. Liljequist, Nucl. Instrum. Methods, 115 (1974) 373.

68 U. Baverstam, T. Ekdahl, C. Bohm, D. Liljequist and B. Ringstrom, Nucl. Instrum. Methods, 118 (1974) 313.

69 D. Liljequist, T. Ekdahl and U. Baverstam, Nucl. Instrum. Methods, 155 (1978) 529.

70 C.J. Song, J. Trooster, N. Benczer-Koller and G.M. Rothberg, Phys. Rev. Lett., 29 (1972) 1165.
71 C.J. Song, J. Trooster and N. Benczer-Koller, Phys. Rev. B, 9 (1974) 3854.
72 G. Longworth and N.E.W. Hartley, Thin Solid Films, 48 (1978) 95.
73 P.J. Black and P.B. Moon, Nature 188 (1960) 481.
74 P.J. Black, D.E. Evans and D.A. O'Connor, Proc. Roy. Soc., A270 (1962) 168.
75 P.J. Black, G. Longworth and D.A. O'Connor, Proc. Phys. Soc., 83 (1964) 925.
76 H.K. Chow, R.F. Weise and P.A. Flinn, Mossbauer spectrometry for analysis of iron compounds, US Atomic Energy Commission Report No. NSEC-4023-1, 1969.
77 B. Keisch, Nucl. Instrum. Methods, 104 (1972) 237.
78 R.A. Levy, P.A. Flinn and R.A. Hartzell, Nucl. Instrum. Methods, 131 (1975) 559.
79 M. Pasternak and S. Shamai, Cryogenics, 19 (1979) 40.
80 B. Window, J. Phys. E: Sci. Instrum., 2 (1969) 894.
81 I.R. Herbert and S.J. Campbell, Cryogenics, 16 (1976) 717.
82 L. Bogner, W. Gierisch, W. Potzel, F.J. Litterst and G.M. Kalvius, in G.J. Perlow (Ed.), Workshop on new directions in Mossbauer spectroscopy, Argonne, A.I.P. Conf. Proc., Vol. 38, 1977, p. 180.
83 G.J. Ehnholm, T.E. Katila, O.V. Lounasmaa and P. Reivari, Cryogenics, 8 (1968) 136.
84 G.M. Kalvius, T.E. Katila and O.V. Lounasmaa, in I.J. Gruverman (Ed.), Mossbauer Effect Methodology, Vol. 5, Plenum Press, New York, 1970, p. 231.
85 T.K. McNab and P.H. Barrett, in I.J. Gruverman (Ed.), Mossbauer Effect Methodology, Vol. 7, Plenum Press, New York, 1971, p. 59.
86 A. Bos, A.T. Howe, L.W. Becker and B.W. Dale, Cryogenics, 14 (1974) 47.
87 M. Psternak and S. Shamai, Nucl. Instrum. Methods, 138 (1976) 673.
88 P. Rosner, W. Vogl and G. Vogl, Nucl. Instrum. Methods, 105 (1972) 473.
89 S.A. Drentje and S.R. Reintsema, Nucl. Instrum. Methods, 133 (1976) 421.
90 D.N. Pipkorn, C.K. Edge, P. Debrunner, G. De Pasquali, H.G. Drickamer and H. Fauenfelder, Phys. Rev., 135 (1964) A1604.
91 R. Debrunner, R.W. Vaughan, A.R. Champion, J. Cohen, J. Moysis and H.G. Drickamer, Rev. Sci. Instrum., 37 (1966) 1310.
92 C.W. Christoe and H.G. Drickamer, Rev. Sci. Instrum., 40 (1969) 169.
93 L.D. Roberts, D.O. Patterson, J.O. Thomson and R.P. Levey, Phys. Rev., 179 (1969) 656.
94 D.L. Williamson, S. Bukshapan, R. Ingalls and H. Schechter, Rev. Sci. Instrum., 43 (1972) 194.
95 J.S. Schilling, U.F. Klein and W.B. Holzapfel, Rev. Sci. Instrum., 45 (1974) 1358.
96 H. Wenzel, A.K. Sherif, A.A. Opalenko and R.N. Kuz'min, Instrum. Exp. Tech., Pt. 2, 20(1) (1977) 262.
97 G.N. Stepanov and V.N. Panyushkin, Cryogenics, 17 (1977) 304.
98 T.A. Kovats and J.C. Walker, Phys. Rev., 181 (1969) 610.
99 V.S. Sundaram, V.P. Gupta and E.C. Subbarao, Rev. Sci. Instrum., 42 (1971) 1616.
100 S. Lewis and F. Flinn, Philos. Mag. 26 (1972) 977.
101 J.M.D. Coey, D.C. Price and A.H. Morrish, Rev. Sci. Instrum., 43 (1972) 54.
102 M.A. Kobeissi and C. Hohenemser, Rev. Sci. Instrum., 49 (1978) 601.

CHAPTER 3

SELECTIVE EXCITATION DOUBLE MOSSBAUER SPECTROSCOPY

Bohdan Balko[†] and Gilbert R. Hoy[††]

† National Institutes of Health, Bethesda, Maryland 20014

†† Department of Physics, Old Dominion University, Norfolk, VA 23508

1. INTRODUCTION

Although the selective excitation double Mossbauer method (SEDM) has been known for some time (1-4), it has only been applied to the study of solid state and biological problems relatively recently. The reason for the delay is that the method is rather time consuming, compared to ordinary Mossbauer spectroscopy, and the analysis is also more involved. In addition, it has taken time to demonstrate the unique utility of SEDM.

The basic reason that SEDM experiments are so interesting, is that in contrast to Mossbauer transmission experiments which measure the absorption cross section as a function of incident energy, these novel experiments measure the differential scattering cross section as a function of both the incident and scattered energies. In this way, not only the absorption energy is determined, but also the spectral energy distribution of the emitted radiation. Thus, in principle, any energetic interaction with the surroundings during the lifetime of the excited nucleus can be detected.

Let us first consider a normal Mossbauer transmission experiment as depicted in Fig. 1. As an example, a spectrum of α-Fe_2O_3 (90% enriched in ^{57}Fe) at 300K is shown in Fig. 1b. Using well known methods, the internal effective magnetic field, the electric field gradient, and the isomer (chemical) shift can be determined. One can also obtain information about the valence state of the iron ion, covalency effects, the symmetry of the iron sites, the electronic charge density at the nucleus, and the presence of inequivalent iron sites in the material. Using the relative intensities of the peaks as well as their widths, one can obtain information about the presence of inhomogeneous fields and dynamic aspects due to spin fluctuations or the motion of ions.

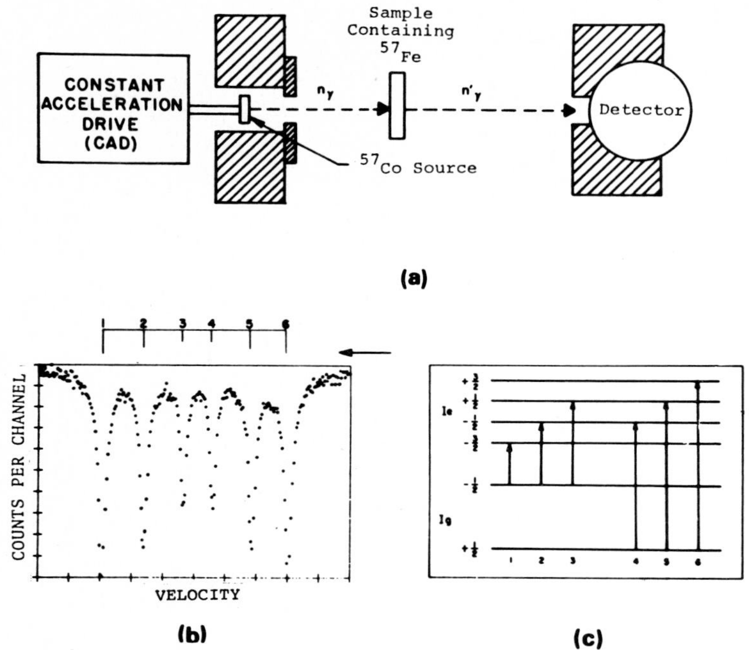

Fig. 1. (a) Mossbauer transmission apparatus; (b) transmission spectrum of α-Fe_2O_3; (c) Nuclear energy diagram of ^{57}Fe.

In contrast, the SEDM experimental set-up is shown in Fig. 2a. A "single line" source containing ^{57}Co is attached to a constant velocity drive (CVD). The CVD doppler shifts the energy of the emitted gamma radiation so that it corresponds to the desired absorption resonant energy in the sample. A constant acceleration drive (CAD) with a "single line" Mossbauer absorber is then used to energy analyze the radiation coming from the sample. Fig. 2(b) shows an SEDM spectrum obtained at 300K with an α-Fe_2O_3 (90% enriched in ^{57}Fe) sample. In this experiment, the CVD was set to excite the transition in the sample, labeled 5. Due to the magnetic dipole selection rule, the resulting SEDM spectrum shows two lines labeled 5 and 3. Notice that SEDM spectra, in general, contain fewer lines than the transmission spectra. In SEDM experiments only those energies occur which result from the decay of the selectively populated excited nuclear level.

However, if the excited nucleus can exchange energy in some fashion with its surroundings, the carefully prepared excited nuclear ensemble will be modified, and interesting and unexpected

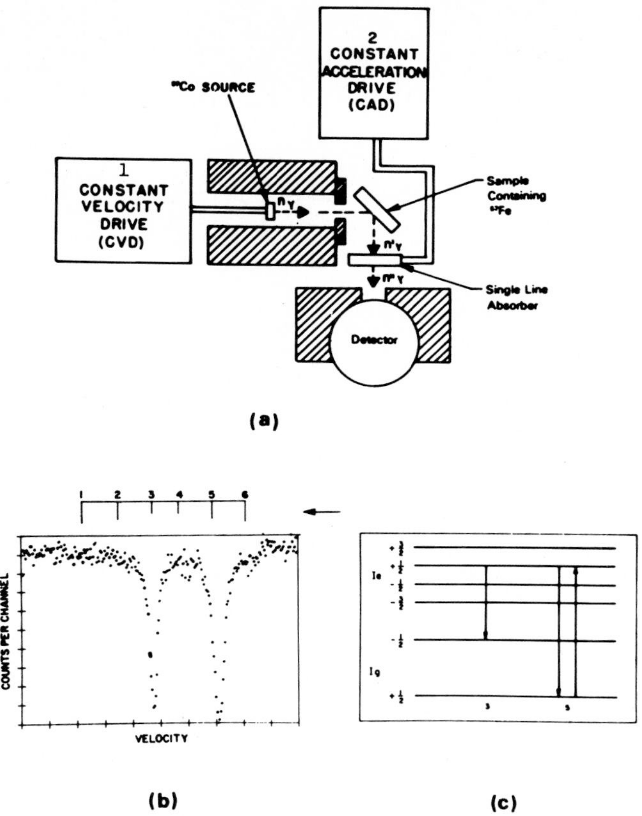

Fig. 2. (a) SEDM apparatus; (b) SEDM spectrum of α*-*Fe_2O_3*; (c) Nuclear energy diagram of* ^{57}Fe *showing the SEDM transitions.*

line shapes and/or lines may appear. One of the most intriguing aspects of the SEDM method is the study of relaxation effects (5). In ordinary Mossbauer transmission spectroscopy, relaxation processes must often be identified by spectral line broadening. However, other factors, notably inhomogeneous hyperfine fields, can also lead to a similar effect. This problem does not occur in SEDM. Thus, SEDM is a powerful, new, unambiguous method for studying dynamic processes.

To summarize, the basic idea of the technique is to have a "single line" source move at constant velocity (CVD) such that the emitted Mossbauer photons have the correct energy to excite a particular nuclear level in the material under study (called the scatterer). In this way, an excited state nuclear ensemble is carefully prepared, and the subsequently emitted nuclear, recoilless, gamma rays can be energy analyzed by using a "single line" Mossbauer absorber (called the analyzer) driven in the constant acceleration mode (CAD).

If during the lifetime of the excited nuclear state there are hyperfine interactions which perturb the prepared ensemble, the energy spectral distribution of the recoillessly emitted gamma rays will be different from that obtained if the ensemble were unperturbed. An example of such a perturbing hyperfine interaction would be when the effective magnetic field at the nucleus is flipping between "up" and "down" orientations. We will discuss these matters more fully in subsequent sections.

2. APPARATUS AND EXPERIMENTAL DETAILS

A block diagram of the SEDM apparatus is shown in Fig. 3. Most of the electronic components of the SEDM apparatus are of conventional design, and commercially available from companies which supply ordinary Mossbauer systems. The motions of the source and "analyzer" are determined by the wave forms impressed on the electronics of drive 1 and drive 2, respectively. Drive 1, the constant velocity drive (CVD), follows a rectangular wave form which can be obtained from a bistable multivibrator activated by a photocell pulser as described by Ruegg et al (6). A useful modification of this design is the addition of an RC circuit at the input to the power amplifier reducing the rise time and thus controlling the overshoot at high velocities. Since drive 1 is used to excite a particular resonance absorption transition in the scatterer by moving at a particular constant velocity, the other portions of the motion must be discriminated against by, for example, gating the multichannel analyzer off during the flyback time.

The "analyzer" is attached to drive 2. If one is interested in investigating the energy range corresponding to the total hyperfine interaction (i.e. all the 6 lines in Fig. 1(b)), the triangular waveform which produces constant acceleration over a range of velocities v to -v is applied to the drive. Triangular wave form generators are commercially available. The trapezoidal wave form is used when one wishes to study a selected section of the spectrum, such as line 2 of Fig. 1(b), in detail.

A proportional counter (Reuter-Stokes model RSG-61) filled with $XeCO_2$ at 2 atm is a good detector for iron experiments. Strong sources and enriched samples (i.e. scatterers) are very desirable.

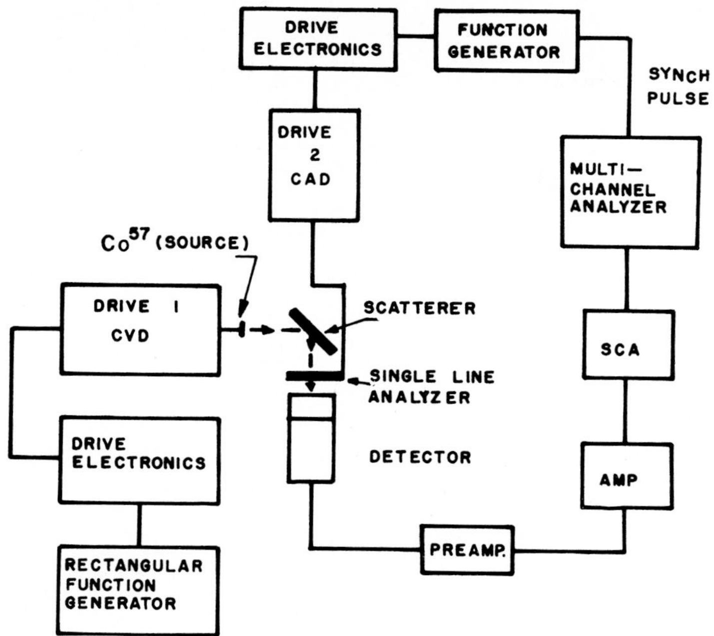

Fig. 3. Block Diagram of the SEDM Apparatus.

A 150 mCi source, usually in a palladium matrix, covering a circular area of 6 mm in diameter is a reasonable starting point. Stronger sources are now available but the price is often a limiting factor. Perhaps the use of synchrotron radiation or a new, cheaper, method of preparation of standard sources will be available in the near future to alleviate the strong source problem.

Both stainless steel and $Na_4Fe(CN)_6 \cdot 10H_2O$ are essentially "single line" absorbers which can be used as the "analyzer" component of the experimental setup. The choice of the amount of absorber material characterized by the thickness parameter $\beta_a = n_a \sigma_o f_a$, (where n_a is the number of iron nuclei per cm^2 in the absorber, σ_o is the resonant cross section, and f_a is the recoilless fraction in the absorber) is a compromise between resolution and signal-to-noise ratio. Normalized line intensities and widths (full width at half maximum) as a function of the thickness parameter β_a can be obtained from the Ruby and Hicks (7) result. An absorber of $5 \leq \beta_a \leq 15$ is a good general purpose choice. For better energy resolution absorbers with smaller values of β_a should be used. If resolution is not so important but a good signal-to-noise ratio is a problem, absorbers with larger β_a values are preferable.

In SEDM experiments proper shielding and a tight geometry are important. A good example is shown in Fig. 4. In this case the analyzer holder rod actually moves through a hole in the lead allowing the source and scatterer to be close together yet offering good shielding. All surfaces facing the beam are either made of brass or clad with 1/16" thick copper plate to absorb the lead x-ray.

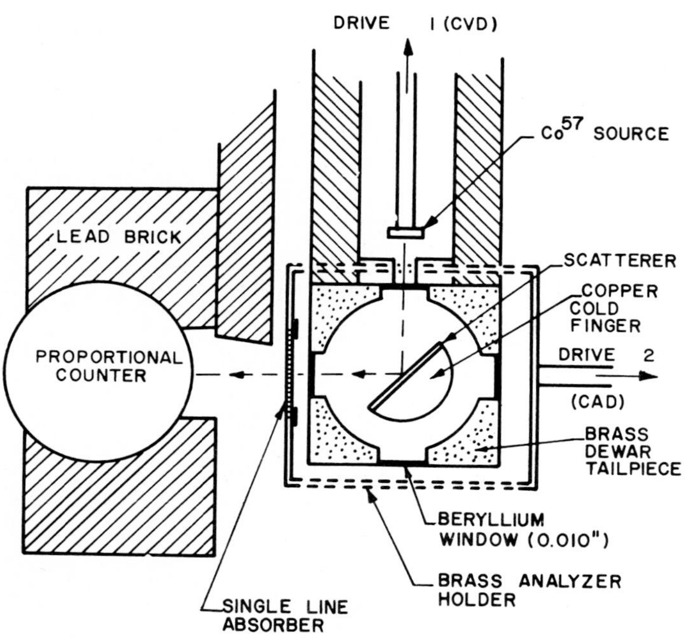

Fig. 4. SEDM geometry shielding arrangement. The horizontal sections of the analyzer holder (dashed lines) are actually 1/4" above (top part) and below (bottom part) the plane of the beams.

The general experimental procedure for conducting SEDM measurements can be summarized as follows. First, a normal Mossbauer transmission spectrum of the material to be used as the scatterer is obtained. From this spectrum one can make a correspondence between the "pick-up" voltage of the linear velocity transducer, which monitors the source velocity, and absorption peak positions. Next, the source is driven in the CVD mode over a range of velocities with the apparatus still in the transmission geometry to look in detail at the region of interest. Finally, the SEDM geometry is used with the CVD set to excite the appropriate absorption transition, and the detector and "analyzer" in the 90^0 scattering geometry.

3. SEDM LINESHAPE CALCULATION - TIME INDEPENDENT THEORY

In this section we calculate the energy spectral distribution of radiation that reaches the detector after being scattered by a "split," "thick" scatterer and transmitted through the analyzer. We restrict ourselves to a stationary environment. Effects due to time dependent hyperfine interactions and electronic relaxation will be treated in a later section.

The major effects that need to be considered are: Mossbauer processes characterized by the resonant cross section σ_0, Rayleigh events characterized by σ_R, and the electronic extinction coefficient μ_e of the scatterer which characterizes the other processes that attenuate the intensity of the radiation as it traverses the scatterer. In the present discussion we neglect coherent thickness effects (8) and multiple scattering.

Consider the situation as shown in Fig. 5(a). The intensity distribution of the source radiation, assumed to be Lorentzian, is I(E,S) where S is the Doppler energy set by the CVD. The intensity of the radiation reaching a distance x into the scatterer is,

$$I_x(E,S) = I(E,S)\exp[-\mu_T(E)\csc\alpha_1 x], \tag{1}$$

where α_1 is the angle of the incoming beam relative to the scatterer and $\mu_T(E)$ is the scatterer's total absorption coefficient. The total linear absorption coefficient $\mu_T(E)$ can be decomposed into three parts: nuclear resonant $\mu_M(E)$, coherent nonresonant μ_R (including electronic Rayleigh and nuclear Thomson scattering), and the electronic absorption coefficient μ_e which includes all incoherent, essentially energy-independent, scattering processes. Thus,

$$\mu_T(E) = \mu_M(E) + \mu_R + \mu_e . \tag{2}$$

The resonant nuclear absorption can take place at energies corresponding to the allowed nuclear transitions E_{ij}. If the nuclear energy levels are pure-m states,

$$\mu_M(E) = \sum_{i,j} \frac{nf\sigma_o(\frac{1}{2}\Gamma)^2}{(E-E_{ij})^2+(\frac{1}{2}\Gamma)^2} W_{ij}(\theta_1\phi_1), \tag{3}$$

where f is the scatterer's recoilless fraction, n is the number of resonant nuclei per unit volume, Γ is the natural linewidth, and

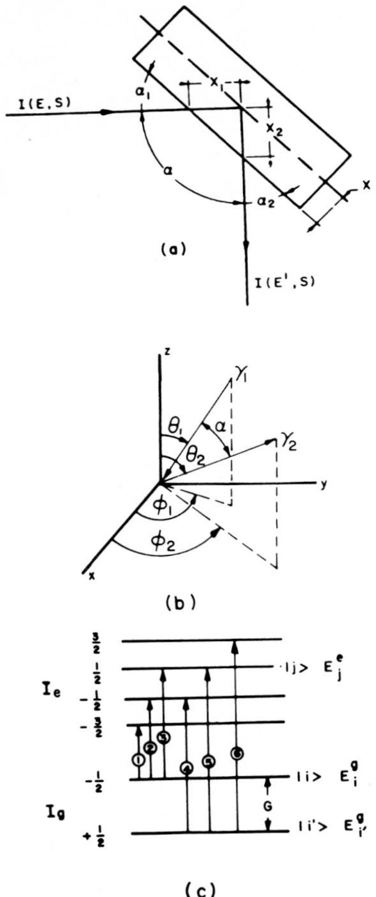

Fig. 5. (a) Scatterer orientation relative to the incident and scattered radiation directions. (b) Definition of the angles used in the scattering calculations. (c) Energy level diagram for a "split" ^{57}Fe scatterer. The allowed transitions are labeled 1, 2, 3, 4, 5, and 6.

σ_0 is the maximum resonant absorption cross section. The scattering angles are defined in Fig. 5(a). The functions $W_{ij}(\theta_1,\phi_1)$ are (9)

$$W_{ij} = |C(I_e m_j|LMI_g m_i)|^2 |\vec{X}_L^M(\theta_1,\phi_1)|^2. \tag{4}$$

Here the C's are the usual Clebsch-Gordon coefficients and the $\vec{X}_L^M$'s are vector spherical harmonics. We first consider the nuclear resonant contribution. In general this term contributes a part from recoilless absorption and recoilless emission and another part from recoilless absorption and nonrecoilless emission. However, since the scattered radiation is analyzed using the Mossbauer effect in the analyzer, we only need consider the recoilless-recoilless part. The rest contributes to the background only. The energy distribution per unit length per unit solid angle of the radiation scattered at x by resonant nuclear processes is

$$\frac{d}{dx} I_M(E',S) = \int dE I_x(E,S) nf \frac{d^2\sigma_M}{dEdE'} . \qquad (5)$$

Following Heitler, (10) and Boyle and Hall (11) we can write the differential scattering cross section as,

$$\frac{d^2\sigma_M}{dEdE'} = \sum_{i,i'} \frac{f\sigma_o(\frac{1}{2}\Gamma)^2}{(1+\alpha')} \left|\sum_j \frac{\langle i|V^+|j\rangle\langle j|V^-|i'\rangle}{[(E-E'+\Delta_{ii'})-\frac{1}{2}i\gamma][(E'-E_{ij}-\Delta_{ii'})-\frac{1}{2}i\Gamma]}\right|^2 \qquad (6)$$

where V^- and V^+ are operators responsible respectively for absorption and emission of gamma radiation by the nuclei. The other quantities are explained below. Notice that this equation includes interference effects because the sum over j is inside the squared term. In general, of course, this equation is correct. However, because at least initially, we will be dealing with nuclear energy levels that are split by an amount that is large compared to our energy resolution, we will neglect this interference effect. It should be kept in mind that the subsequent equations are not completely general. Therefore with this simplification we get,

$$\frac{d^2\sigma_M}{dEdE'} = \sum_{i,j,i'} \frac{f\sigma_o W_{ij}(\theta_1\phi_1)W_{ji'}(\theta_2\phi_2)(\frac{1}{2}\Gamma)^2}{(1+\alpha')[(E-E'+\Delta_{ii'})^2+(\frac{1}{2}\gamma)^2][(E'-E_{ij}-\Delta_{ii'})^2+(\frac{1}{2}\Gamma)^2]} , \qquad (7)$$

where α' is the internal conversion coefficient, and γ is the linewidth due to the source intensity and hence is very small; we have allowed for absorption between the nuclear levels $E_i'-E_j'$ to result in the decays $E_j'-E_i'$ and $E_j'-E_{i'}'$. $[E_{ij} = E_j'-E_i'$ and $\Delta_{ii'} = E_i'-E_{i'}'=G$ (see Fig. 5(c))]. In the above sum we must include all transitions consistent with the selection rules.

We assume that the source radiation has a Lorentzian line shape and neglect any anisotropic f dependence. Substituting Eqns. (7) and (1) into Eqn. (5) and integrating over E gives,

$$\frac{dI_M}{dx}(E',S)=\sum_{i,j,i'}\frac{I_o nf^2\sigma_o(\frac{1}{2}\Gamma_b)^2(\frac{1}{2}\Gamma)^2\exp[-\mu_T(E'-\Delta_{ii'})\csc\alpha_1 x]W_{ij}W_{ji'}}{(1+\alpha')[E'-\Delta_{ii'}-S)^2+(\frac{1}{2}\Gamma_b)^2][(E'-E_{ij}-\Delta_{ii'})^2+(\frac{1}{2}\Gamma)^2]}, \quad (8)$$

where Γ_b is the effective linewidth of the incident beam. Neglecting multiple resonant scattering, the energy distribution of the nuclear resonant scattered radiation that gets to the analyzer is

$$I_M(E',S) = \int_0^T \frac{d}{dx} I_M(E',S)\exp[-\mu_T(E')\csc\alpha_2 x]dx,$$

where T is the thickness of the scatterer. The result is

$$I_M(E',S)=\sum_{i,j,i'}\frac{nf^2\sigma_o I_o(\frac{1}{2}\Gamma_b)^2(\frac{1}{2}\Gamma)^2 W_{ij}W_{ji'}F}{(1+\alpha')[(E'-\Delta_{ii'}-S)^2+(\frac{1}{2}\Gamma_b)^2][(E'-E_{ij}-\Delta_{ii'})^2+(\frac{1}{2}\Gamma)^2]}, \quad (9)$$

where

$$F=\frac{1-\exp\{-T[\mu_T(E'-\Delta_{ii'})\csc\alpha_1+\mu_T(E')\csc\alpha_2]\}}{\mu_T(E'-\Delta_{ii'})\csc\alpha_1+\mu_T(E')\csc\alpha_2}.$$

Equation (9) is essentially the same as found in Debrunner and Morrison (12) but generalized to a "split" scatterer. The angular distribution of the resonantly scattered radiation is represented by $W_{ij}(\theta_1,\phi_1)W_{ji'}(\theta_2,\phi_2)$. If the scatterer is a powder, this factor must be averaged subject to the constraint that the scattering angle α is constant. We can denote this result by $W_{ij,ji'}(\alpha)$. These values are calculated in appendix A. For calculational purposes it is convenient to use a notation where the transitions are labeled by the lines themselves as shown in Fig. 5(c). Then Eqn. (9) can be written

$$\begin{aligned}I_M(E',S)=nA_M\Bigg(&\sum_{i=1}^{6}\frac{W_{i,i}(\alpha)}{(E'-E_i)^2+(\frac{1}{2}\Gamma)^2}\;\frac{1-\exp[-T(\csc\alpha_1+\csc\alpha_2)\mu_T(E')]}{[(E'-S)^2+(\frac{1}{2}\Gamma_b)^2](\csc\alpha_1+\csc\alpha_2)\mu_T(E')}\\ &+\sum_{i=2}^{3}\frac{W_{i,i+2}(\alpha)(1-\exp\{-T[\mu_T(E'-G)\csc\alpha_1+\mu_T(E')\csc\alpha_2]\})}{[(E'-E_i-G)^2+(\frac{1}{2}\Gamma)^2][(E'-S-G)^2+(\frac{1}{2}\Gamma_b)^2][\mu_T(E'-G)\csc\alpha_1+\mu_T(E')\csc\alpha_2]}\\ &+\sum_{i=2}^{3}\frac{W_{i+2,i}(\alpha)(1-\exp\{-T[\mu_T(E'+G)\csc\alpha_1+\mu_T(E')\csc\alpha_2]\})}{[(E'-E_{i+2}+G)^2+(\frac{1}{2}\Gamma)^2][(E'-S+G)^2+(\frac{1}{2}\Gamma_b)^2][\mu_T(E'+G)\csc\alpha_1+\mu_T(E')\csc\alpha_2]}\Bigg)\end{aligned} \quad (10)$$

where G equals the ground-state splitting, $W_{ij,ji}$, changing to the line label $W_{i,j}$, and

$$\mu_T(E') = \sum_{i=1}^{6} \frac{nf\sigma_o(\frac{1}{2}\Gamma)^2\bar{W}_i}{[(E'-E_i)^2+(\frac{1}{2}\Gamma)^2]} + \mu_R + \mu_e. \qquad (11)$$

The bar indicates an average over all angles and A_M is a constant determined by the particular experiment.

The coherent noresonant contribution to the scattered intensity is mostly Rayleigh, because at these energies the Rayleigh contribution is 10^4 times the Thomson contribution (13). The energy dependence of the Rayleigh cross section is relatively weak and so,

$$I_R(E',S) = \frac{I_o(\frac{1}{2}\Gamma_b)^2 n' f_R \sigma_R W_R(\alpha)}{(E'-S)^2+(\frac{1}{2}\Gamma_b)^2} \times \left(\frac{1-\exp[-\mu_T(E')(\csc\alpha_1+\csc\alpha_2)T]}{\mu_T(E')(\csc\alpha_1+\csc\alpha_2)}\right), \qquad (12)$$

where n' is the number of atoms per unit volume, f_R is the recoilless fraction for Rayleigh scattering ($f_R = f^2$ for $\alpha = 90^o$)[14] σ_R is the Rayleigh cross section evaluated at 14 keV (for our purposes), and

$$W(\alpha) = \frac{1+\cos^2\alpha}{\sin^3(\frac{1}{2}\alpha)}, \qquad (13)$$

The use of equation 13 has been criticized (15). However, for our purposes it contributes only a constant factor, and thus is unimportant in line shape studies. Collecting energy-dependent factors,

$$I_R(E',S) = \frac{n' A_R\{1-\exp[-\mu_T(E')(\csc\alpha_1+\csc\alpha_2)T]\}}{[(E'-S)^2+(\frac{1}{2}\Gamma_b)^2]\mu_T(E')(\csc\alpha_1+\csc\alpha_2)}, \qquad (14)$$

In general the energy distribution of the scattered radiation contains an interference term from the nuclear resonant and coherent Rayleigh-scattering processes. According to Black et al (14), at a scattering angle $\alpha=90^o$ this interference term is zero*. Because of

*Wong (16) has made a theoretical calculation which indicates that the Rayleigh and Mossbauer interference term does not vanish at a scattering angle of 90° although it does attain its minimum value. Our scattering results at a scattering angle of 90° did not show any evidence of asymmetrical line shapes. Therefore the interference

contd... next page

the unavoidable experimental angular spread about 90° there is still a possible effect due to this interference term. However, the interference term goes like $\cos\alpha$ and thus tends to average to zero for a symmetrical angular spread around $\alpha=90^{\circ}$. By integrating over the appropriate values for α, for our case, we found that the ratio of the interference term to the terms included in the above analysis is 7×10^{-6} for excitation on resonance and 3×10^{-3} for excitation ten natural line widths off resonance. As will be seen below, our experiments verify this and do not show interference effects. Thus any line-shape asymmetry is not due to this effect (17). Under these circumstances the two contributions (nuclear resonant recoilless and Rayleigh) can be calculated separately and added. The total energy dependent intensity distribution of the scattered radiation, which is subject to analysis by the analyzer, is

$$I(E',S) = I_M(E',S)+I_R(E',S),$$
$$I(E',S) = \text{Eqn. (10)} + \text{Eqn. (14)} . \qquad (15)$$

Once $I(E',S)$ is determined the effect of the analyzer can be calculated by

$$I(S,S') = \int_0^{\infty} I(E',S)e^{-\mu_a(E',S')T_a}\,dE', \qquad (16)$$

where S' is the Doppler energy of the "single-line" analyzer and

$$\mu_a = \frac{n_a f_a \sigma_o (\frac{1}{2}\Gamma)^2}{(E'-S'-E_o)^2+(\frac{1}{2}\Gamma)^2}$$
$$= \frac{\beta_a (\frac{1}{2}\Gamma)^2}{[(E'-S'-E_o)^2+(\frac{1}{2}\Gamma)^2]T_a} .$$

Equation 15 gives the SEDM spectrum for the excitation by a single line source of a scatterer in which the iron atoms have no time dependent effects. The integration over E' introduces the effect of the analyzer (single line absorber driven by the CAD) and thus

term, if it exists, is too small to be observable in our experiments, and thus has been taken to be zero in our analysis as mentioned above. Wong's result also predicts an energy dependence in the Rayleigh cross section due to the internal conversion process. In order to investigate this effect one needs to use enriched samples. However, under these conditions, the SEDM technique is not sensitive to this effect because the major deviation from energy independence occurs at the resonance energy where the resonant cross section dominates. We have made calculations which show these deviations would be less than 0.1%.

gives us a result which can be directly compared with experiment.

4. SEDM LINESHAPE CALCULATIONS AND EXPERIMENTAL RESULTS

4.1 General Characteristics of the SEDM Lineshape

In this section we will discuss experiments designed to investigate general features of SEDM spectra and to test some of the consequences of the theoretical development presented in the previous section for the time independent case. For a complete interpretation of the SEDM spectra leading to useful applications of this technique in the investigation of various physical, chemical, and biological processes we have to be able to separate out the various contributions to the lineshape. In particular we will concentrate on incoherent thickness effects and the electronic Rayleigh contribution. Relaxation effects will be dealt with in a later section.

All the experiments to be discussed were done using ^{57}Fe, and the calculations were performed assuming that the nuclear levels are pure-m states. As noted above, $\alpha=90^{\circ}$ and $\alpha_1=\alpha_2=45^{\circ}$. The E_{ij}'s and G can be determined by ordinary transmission experiments. The linear absorption coefficient, $\mu = \mu_R + \mu_e$, can be determined experimentally by measuring the attenuation of the γ-ray beam on passing through various sample thickness. For example, the values obtained were $\mu = 205$ cm^{-1} for iron and $\mu = 54$ cm^{-1} for α-Fe_2O_3.

Fig. 6 shows some SEDM results obtained with a thick enriched sample of α-Fe_2O_3. In this figure the bottom spectrum (c) was obtained in the usual transmission geometry and is presented for comparison. Notice that in Fig. 6(c) all the six lines are present and there are no relaxation effects. An SEDM spectrum obtained by exciting the first resonance is shown in Fig. 6(a). The SEDM spectrum has a single line at the excitation energy but if we excite the second line of the α-Fe_2O_3 sample two lines appear in the SEDM spectrum as shown in Fig. 6(b). This is more interesting because it shows that the two resonances are coupled by the selection rules for the transitions. Such information cannot be obtained directly from the transmission spectrum. In fact this has to be known from other experiments in order to properly analyze the transmission spectrum. It is obtained directly from the SEDM result. The SEDM spectra in general are cleaner and only lines that are coupled by selection rules or relaxation appear.

The relative strength of Rayleigh to Mossbauer processes, i.e.,

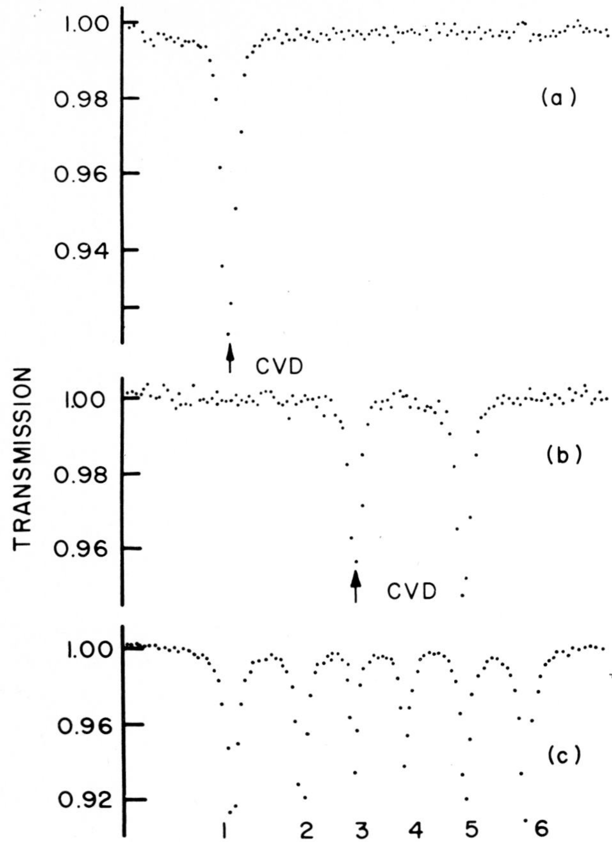

Fig. 6. (a) Experimental SEDM result for a 90%-enriched α-Fe_2O_3 scatterer at room temperature when line 1 is excited. (b) Same as (a) except line 3 is excited. (c) Full transmission spectrum for a 90%-enriched α-Fe_2O_3 absorber at room temperature.

A_R/A_M [see Eqns. (10) and (14)] must be determined for each experiment although in ideal cases it can be calculated (13). The calculated result was checked by performing the experiment shown in Fig. 6(b) and subsequent experiments presented below. In Fig. 6(b) only the peak on the left has a Rayleigh contribution. By comparing the experimental results with calculations from Eqn. (16), we found, for example, $(A_R/A_M) = 4 \times 10^{-4}$ for the 90%-enriched sample of α-Fe_2O_3. This number was further checked by exciting the same peak at different values off resonance and again comparing them with the results from Eqn. (16).

In order to get a feeling for these calculations let us consider a hypothetical scatterer in which the hyperfine field is zero, i.e., the absorption spectrum contains a single line, and the only process is the Mossbauer resonant effect, i.e., $\mu_e = \mu_R = 0$. In this calculation the CVD is set to excite the sample three natural linewidths off resonance. The results are shown in Fig. 7. There are several observations to be made. First as the thickness of the scatterer increases, i.e., as β increases, the resulting emission line shape peaks closer to the excitation energy. Notice further that all the peaks are slightly asymmetric. The reason for the location of the peak is that, due to thickness, the "effective" absorption cross section of the scatterer becomes very broad in energy and the relatively narrow incident beam determines the location. However, it is only for thick scatterers that the effective

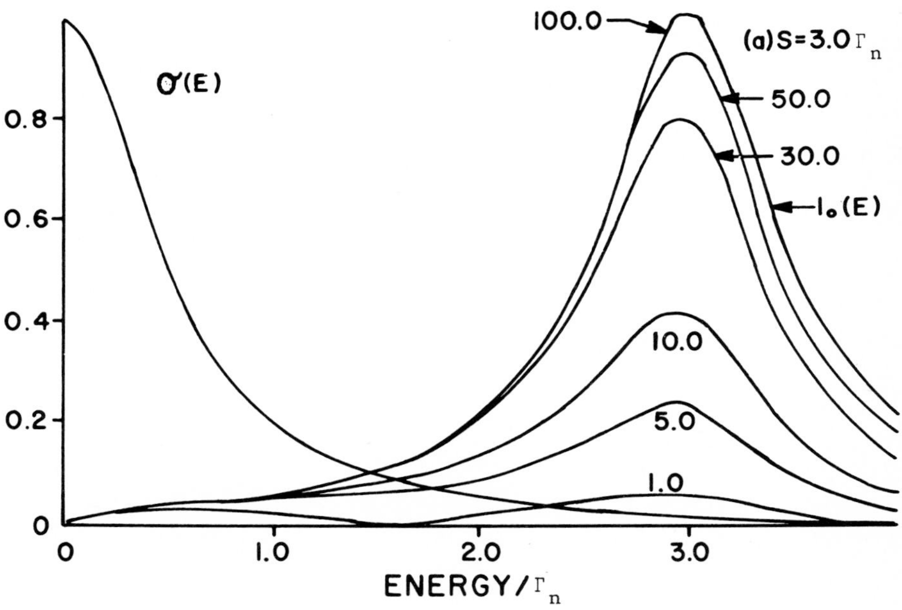

Fig. 7. Calculated emission line shapes for ideal resonant scatterers. The numbers labelling the curves indicate the thickness, β, of the hypothetical scatterers. The setting of the CVD (the Doppler shift) is 3.

absorption cross section is truly constant over the incident beam energy profile resulting in a symmetric line. On the other hand, as β decreases we approach the "single nucleus" limit as calculated classically by Moon (13) and emphasized by Boyle and Hall (11). In this limit the emission line shape has two peaks, one at the excitation energy and the other at the resonance energy. However, both peaks are extremely small compared to the results for the thicker scatters. This case is of no practical interest.

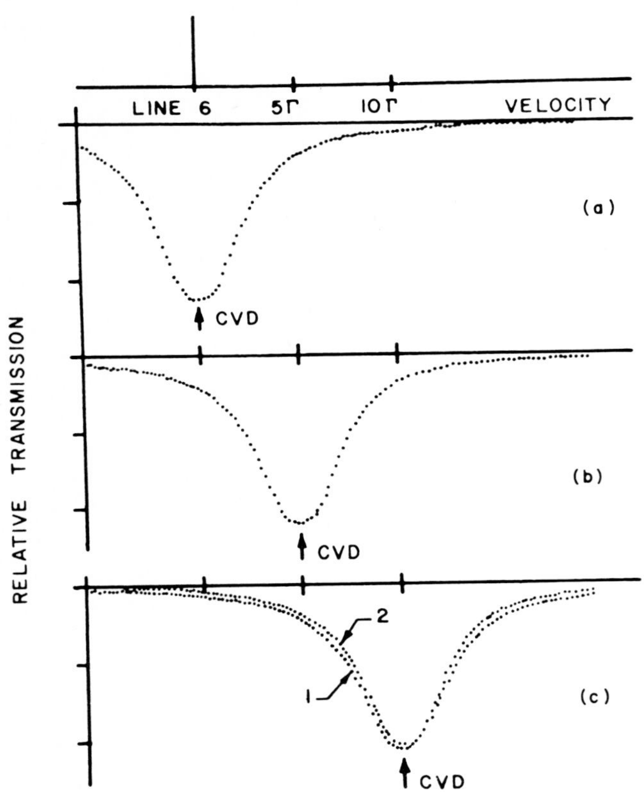

Fig. 8. Calculated SEDM results for an iron-powder scatterer when the sixth line is excited (a) on resonance, (b) 5Γ off resonance, and (c) 10Γ off resonance. In (c) curve 1 is for a 90%-enriched sample and curve 2 is for a natural sample. Notice that cases 1 and 2 give the same results for situations a and b. The vertical scales are not to be compared for a, b and c. The normalization was chosen the same so that each figure is clearly presented. In reality the percentage effect decreases as the excitation moves off resonance (see Figs. 11 and 12).

Now using the appropriate (A_R/A_M) values the calculated SEDM line shape for a real iron powder is shown in Fig. 8 when the sixth line is excited on resonance and at higher energies. In these calculations we include Rayleigh and incoherent thickness effects. The first point to notice is that in all cases the peak of the SEDM line comes at the energy of the incoming beam as set by the CVD. These calculations were made for two different types of samples: (1) a 90%-enriched iron-powder scatterer and (2) a natural iron-powder scatterer, both having a β of 286. Notice in Fig. 8 a difference between the calculated results is seen for the two different samples only for excitation 10Γ off resonance. Whereas the line shapes for (1) and (2) are very similar, the main contributions to the two lines are different. For case (1), i.e., the enriched iron powder, 95% of the peak is due to Mossbauer scattering and 5% to Rayleigh scattering. For case (2) i.e., the natural iron powder, only 25% is due to Mossbauer scattering and the rest is due to Rayleigh scattering. Notice also that, as a result of this, case (1) (see Fig. 8(c)) is slightly less symmetrical than case (2). The difference between the relative contributions of Mossbauer and Rayleigh processes is more clearly exhibited in Fig. 9. In this

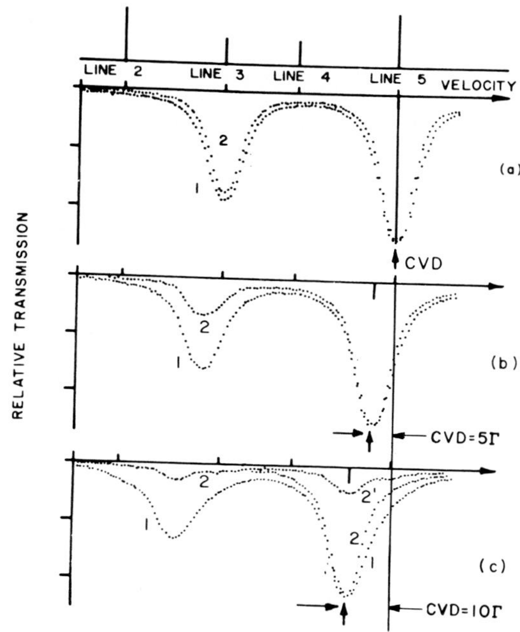

Fig. 9. Calculated SEDM results for an iron-powder scatterer when the fifth line is excited (a) on resonance, (b) 5Γ off resonance, and (c) 10Γ off resonance. The curve labeled 1 correspond to a 90%-enriched sample and curve 2 to a natural sample. Curve 2' in (c) shows the nuclear resonant contribution to the peak. The vertical scales are not to be compared for a, b, and c. The normalization shown was chosen so that the details of each spectrum could be clearly presented. In reality, the percentage effect decreases as the excitation moves off resonance.

figure we assume that the fifth line in an iron-powder scatterer is excited. In this case only the peaks on the right contain a Rayleigh contribution. Cases (1) and (2) correspond to the same type of scatterer as assumed for Fig. 8. Now, however, the SEDM spectrum consists of two peaks due to the magnetic dipole selection rule. The ratio of the two peak intensities for the enriched sample changes slightly in cases (a), (b) and (c). However, for the natural sample the change in the ratio is much larger, reflecting the increased relative importance of the Rayleigh contribution. It is clear that an experiment such as this can determine σ_R. We see that Mossbauer processes dominate Rayleigh effects for enriched scatters excited off resonance by less than 10Γ. However, for natural scatterers excited 10Γ off resonance, the SEDM peak is almost all due to Rayleigh scattering. This can be clearly seen by looking at the peak on the right-hand side of Fig. 9 for case (2). We have indicated by 2' the pure Mossbauer contribution to case (2). Notice, that 2' gives a hint of a peak at the resonance position of line 5, but these effects are completely obliterated by the large Rayleigh contribution. For experimental results using a natural iron-powder sample excited off resonance which substantiate these calculations see Fig. 9 of Reference 5.

In the above analysis we have stressed the fact that incoherent thickness effects, which arise from the absorption of the beam passing through the scatterer by the different nuclei independently, are responsible for the major characteristics of the SEDM lineshape. In particular, for a reasonably thick scatterer, when exciting a line with a β_s=50 or more off resonance, we have found that the scattered radiation is centered at the excitation energy and closely approximates the incoming beam lineshape. This was shown theoretically even while ignoring the Rayleigh contribution and considering only the purely nuclear resonant beam. We have also made calculations to study the effect of inhomogeneous broadening on the SEDM lineshape. This effect occurs when different ^{57}Fe sites are subject to slightly different hyperfine fields. Some of our calculated results are shown in Fig. 10. In these computations we have kept the total number of nuclei constant (β_s=constant) but have assumed different hyperfine fields with varying probabilities of existance. In each figure we show the scattering cross-section (dashed line) which gives the field distribution and the SEDM line shape (solid line) which results from excitations at the same position from the center of the distribution.

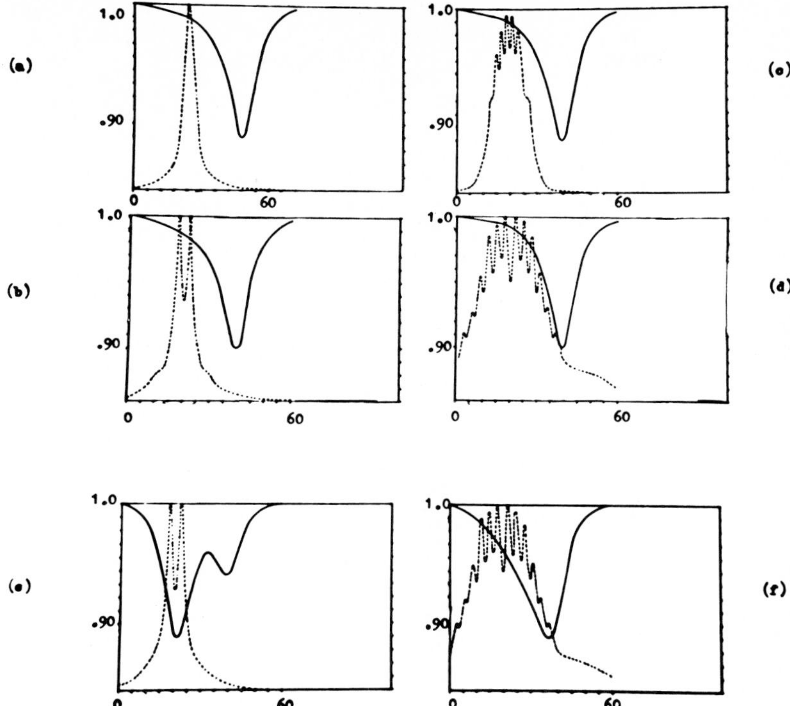

Fig. 10. SEDM line shape for various distributions of the hyperfine field. The total value of the scatterer thickness, β_s=100, is kept constant and the position of excitation from the center of the resonances remains the same for a,b,c and d. The dashed line represents the resonant cross section for a single nucleus subjected to different hyperfine environments and the solid line the calculated SEDM line shape assuming a single line absorber with β_a=14. In (e) and (f) is shown the SEDM lineshape for the same two distributions of hyperfine fields as in (b) and (d) above except in this case we assumed a thin scatterer with β_s=10.

For a thick scatterer with $\beta_s = 100$ (Figs. 10(a), (b), (c), (d)) the lineshape is not affected by the distribution and closely approximates the incoming beam. For $\beta_s = 10$ (Figs. 10(e) and (f)) this is not the case and a more complicated lineshape results. Figures 10b(10d) and 10e(10f) should be compared as they represent results for the same hyperfine field distribution but different scatters.

In order to check our calculations further we performed SEDM experiments on a 90% enriched iron powder at room temperature (298 K) and a 90%-enriched α-Fe_2O_3 powder at liquid-nitrogen temperature (77K). The CVD was set to excite the sixth line on resonance and two values off resonance for each of these samples.

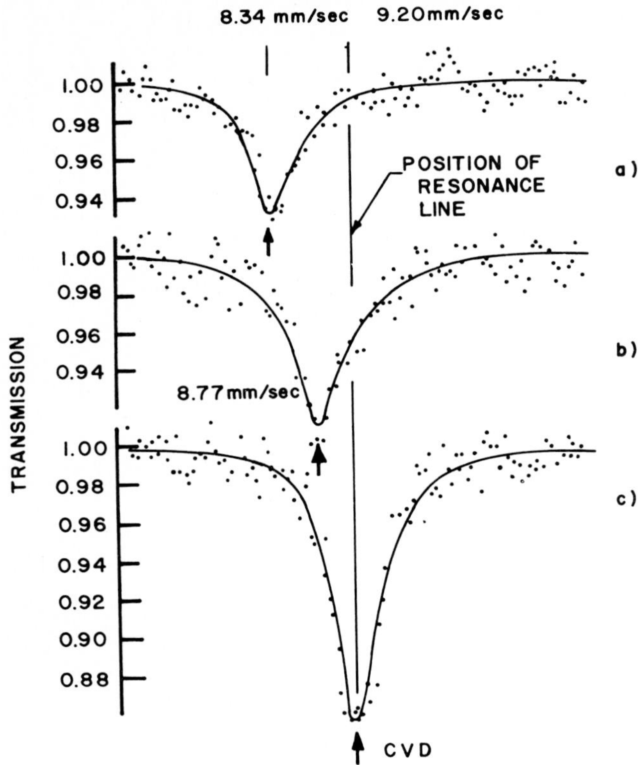

Fig. 11. Experimental SEDM spectra for an enriched α-Fe_2O_3 scatterer at 77 K. In (a) line 6 is excited 8.85Γ off resonance, (b) line 6 is excited 4.43Γ off resonance, and (c) line 6 is excited on resonance. The solid curves are calculated based on the results of Sec. 3. The asymmetry in Parts (a) and (b) is due to excitation off resonance (see Sec. 3).

The trapezoidal-function generator in each case was set to sweep the velocity range of the analyzer over a region which contains the sixth peak. Fig. 11 gives results for the enriched α-Fe_2O_3 scatterer at liquid-nitrogen temperature. The dots are the experimental results. The solid curve gives the calculated result using Eqns. (10), (14), (15) and (16) above. The calculated results involve two parameters namely, (A_R/A_M) and the over-all percentage effect for spectrum (c) in the figure. The decrease in the percentage effect owing to excitations off resonance follows from our calculations. Notice that the theory correctely predicts the change in the percent effect owing to moving off resonance, the peak position, and the total line shape. Observe in Fig. 11 that for excitations off resonance the line shape is slightly asymmetric. This is the result of the fact that (A_R/A_M) is small and so the Mossbauer resonant process is dominant (see Fig. 7). The agreement between theory and experiment is quite satisfactory.

4.2 Application of SEDM to the Measurement of the Rayleigh Scattering Cross Section

We now want to consider a method for obtaining the Rayleigh contribution (18) in some detail. All the parameters in Eqns. (9), (10), (12), and (16) are determined by the experimental conditions except for the product $\sigma_R W_R(\alpha)$ which characterizes the Rayleigh scattering contribution. The analysis of the experimental results involves the determination of the best value of $\sigma_R W_R(\alpha)$. This is accomplished by considering $\sigma_R W_R(\alpha)$ as a free parameter and comparing the resulting calculations with the experimental data.

The Mossbauer contribution is normalized by going to the thin scatterer limit and calculating the total number of scattered photons. We imagine that the hyperfine field vanishes so that all transitions occur at the same energy. We also set the Doppler shift equal to the resonance energy and evaluate Eqn. (9) on resonance. The resonant contribution has two parts: a recoilless-recoilless part, and a recoilless-nonrecoilless part. We calculate the total number of scattered photons which are resonantly absorbed. Using all these considerations and Eqn. (9) gives,

$$\frac{I_o n\sigma_o T}{1+\alpha'}[f^2+f(1-f)]\int\sum_{i,j,i'} W_{ij}W_{ji'}d\Omega = \frac{I_o nf\sigma_o T}{1+\alpha'},$$

when

$$\int\sum_{i,j,i'} W_{ij}W_{ji'}d\Omega=1.$$

This is the normalization condition we use for the Mossbauer scattering contribution.

Since we are interested in obtaining a value for the Rayleigh scattering cross section, the normalization of Eqns. (9) and (12) must be done in a consistent way. In order to do this for the Rayleigh contribution [Eqn. (12)] let $E'=S$. and consider the thin scatterer limit. Then

$$I_R=I_o n' f_R \sigma_R W_R(\alpha)T. \tag{17}$$

From Eqn. (17) we can see that the measured Rayleigh differential scattering cross section is given by

$$\frac{d\sigma_R}{d\Omega} = \sigma_R W_R(\alpha). \tag{18}$$

Moon (13) has given an expression for the differential scattering cross section

$$\frac{d\sigma_R}{d\Omega} = 8.67 \times 10^{-33} \left[\frac{ZE_o}{E}\right]^3 \frac{1+\cos^2\alpha}{2\sin^3(\alpha/2)} \text{cm}^2/\text{sr} \tag{19}$$

where Z is the atomic number of the scattering atom, E the energy of the γ-ray (14.4125 keV in our case), and E_o the self-energy of the electron (0.511 MeV).

Our approach is to obtain the best value for the product $\sigma_R W_R(\alpha)$ from Eqns. (9), (10), (12), and (16) by comparing with the experimental SEDM results at $\alpha=90^o$. The value can then be compared to that given by Eqns. (18) and (19).

In general, the Rayleigh contribution depends on the experimental configuration, because there is Rayleigh scattering from the sample holder and any other foreign material in the beam's path. For the moment, let us neglect these effects and consider an idealized case in which the only scattering occurs from the sample material.

Consider a sample material whose nuclear energy-level diagram is as shown in Fig.5(c). If one uses the SEDM technique in such a way as to excite the transition corresponding to line 2 on resonance, then radiation will be reemitted at the energies of lines 2 and 4. In this case however only the reemitted radiation at the energy of line 2 will contain a Rayleigh contribution. If one can calculate the Mossbauer contributions at the energies of lines 2 and 4, the additional contribution at line 2 is due to Rayleigh scattering. A further check can be made by exciting the transition corresponding to line 2 somewhat off resonance, because then the Rayleigh contribution would increase relative to the Mossbauer contribution. One can also excite the transition corresponding to line 4. In addition, similar experimental results can be obtained using the coupled transitions corresponding to lines 3 and 5. When such a series of measurements have been done, the results can be analyzed by using Eqns. (9), (10), (12), and (16). All the parameters are known except $\sigma_R W_R(\alpha)$ which can be determined by fitting the experimental data to the theory.

We have used two types of scatterers. One was an enriched iron powder scatterer. Because of the enrichment, it was possible to do SEDM experiments on this sample both on and off resonance. Our second sample was an ordinary steel bar 1/8 in. thick on which we previously performed single drive scattering experiments. Since the steel bar had only the naturally occurring amount of ^{57}Fe, it was practical to do SEDM experiments only by exciting on resonance.

In Figs. 12 and 13 we show our experimental results. Figure 12 shows our results for the enriched iron powder sample. In these experiments the excitation energy was set on the energy of the fifth

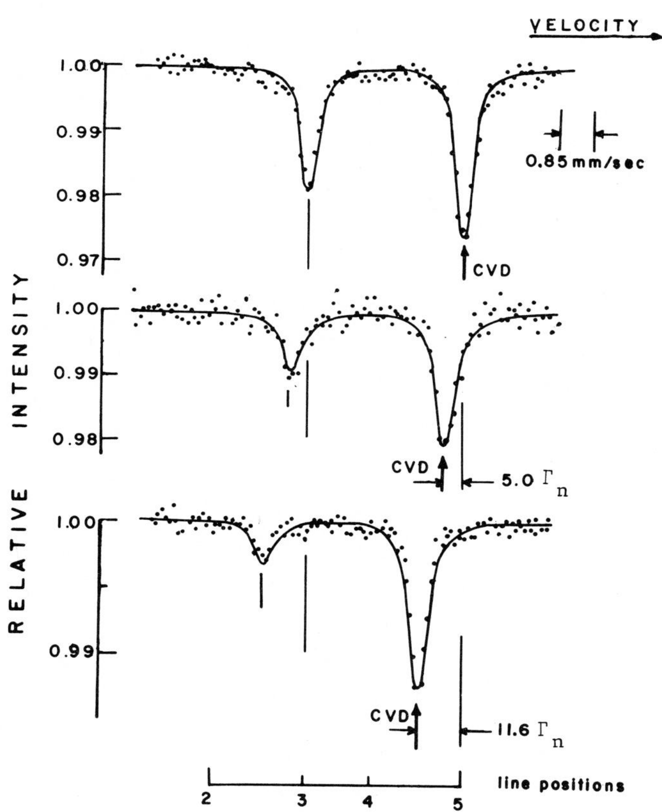

Fig. 12. Experimental SEDM spectra using a 90% enriched iron powder scatterer. The parameters for this sample are β_s*=223,* $\mu_R+\mu_e$ *= 250 cm*$^{-1}$*, and the thickness = 1.4 × 10*$^{-3}$*cm. The solid curves are theoretical fits to the experimental data using* f_R*=0.8 and* $\sigma_R W_R(\alpha)$*=10 × 10*$^{-24}$ *cm*^{3}*/sr. (*Γ_n *is the natural linewidth).*

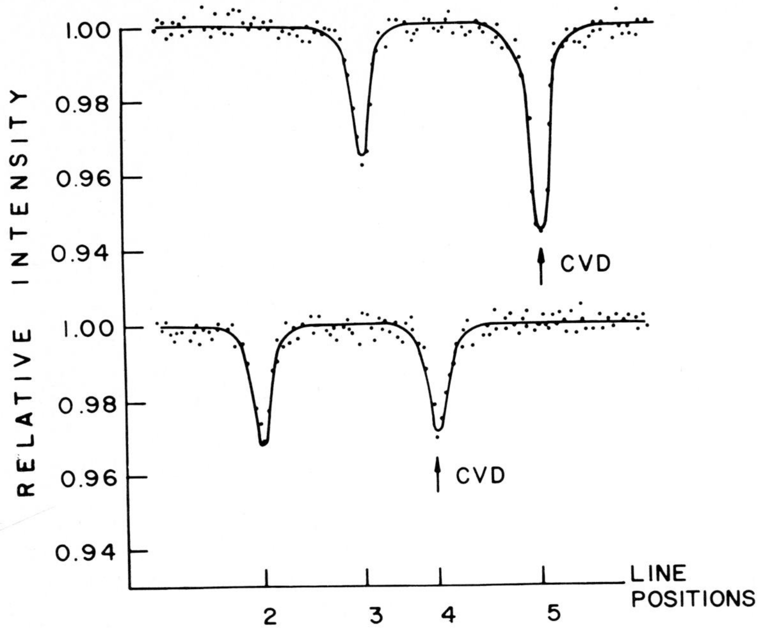

Fig. 13. Experimental SEDM spectra using a natural iron scatterer. The parameters for this sample are; Fe enrichment is 2%, β_s=700, $\mu_R+\mu_e$ =250 cm^{-1}, and the thickness is 0.2 cm. The solid curves are theoretical fits to the experimental data using f_R=0.8 and $\sigma_R W_R(\alpha)=2.5 \times 10^{-24}$ cm^2/sr.

line and at two other positions somewhat off resonance as shown in figure. Figure 13 gives our results for the steel bar scatterer. In this case the CVD was set to excite the transition corresponding to line 5, and then a second experiment was done exciting line 4. The solid curves shown in Figs. 12 and 13 are the theoretical results based on the above equations using the optimum value of $\sigma_R W_R(\alpha)$. Different results for the Rayleigh differential scattering cross section were obtained for the two samples. The value for the enriched iron powder was found to be $(d\sigma_R/d\Omega)_{90^o}$ = (10 $\pm$ 2 x 10^{-24} cm^2/sr while for the steel bar it was $(d\sigma_R/d\Omega)_{90^o}$ = (2.5 $\pm$ x 10^{-24} cm^2/sr. The measurement of the Rayleigh contribution from the steel bar experiment should represent a rather good value for the Rayleigh scattering cross section for iron. In the case of the enriched iron powder scatterer the sample had to be enclosed in a sample holder. Separate experiments showed that, despite our careful collimation, 14-keV photons were scattered into our detector from the sample holder assembly. Therefore, the results from the enriched iron powder sample do not give the Rayleigh scattering cross section for iron atoms alone.

Consider Fig. 12. It should be pointed out that when all the physical parameters needed for Eqns. (9), (10), (12), and (16) are known, there is only one free parameter used to fit all three spectra in Fig. 12. We usually take this parameter to be the percentage dip for the peak excited on resonance. In this case that would be the size of the dip on the right-hand side of the uppermost spectrum. Once this is determined, the locations and line shapes of all other peaks in Fig. 12 are determined. The agreement between theory and experiment establishes the validity of our SEDM equations.

Figure 13, in addition to the results already discussed, shows that SEDM spectra can be obtained using a natural (i.e., a nonenriched) sample. Of course strong sources are a great help in SEDM experiments due to the low counting rates from natural samples. For these experiments we used a 60-mCi source and the experiment lasted five days.

If one is doing SEDM experiments such as those shown in Figs. 12 and 13 it is tempting to find the Rayleigh contribution by calculating the resonant contribution and then considering only the ratio of the experimentally measured dip intensities. This procedure is incorrect because of thickness effects in the scatterer. Consider the calculated energy distributions depicted in Fig. 14. The solid curve in Fig. 14(a) represents the resonant absorption cross section of the ^{57}Fe nuclei in the scatterer, and the dashed line gives the energy spectrum of the incident beam. Excitation of line 5 on resonance, as shown in Fig. 14(a), leads to a scattered beam having two contributions. The nuclear resonant contribution has an energy distribution that is shown by the solid line in Fig. 14(b). The Rayleigh contribution is shown by the dashed line in Fig. 14(b). The rather odd shape of the Rayleigh contribution is due to thickness effects in the scatterer.

The final SEDM line shape is obtained by a convolution of the energy distributions shown in Fig. 14(b) with the absorption cross section of the "single line" analyzer as in Eqn. (16). Since the two contributions in Fig. 14(b) do not have the same line shape, the integrated result [Eqn. (16)], which corresponds to the actual measured count rate at a particular analyzer velocity are not proportional to the contribution's intensity at the resonance energy (line 5). Thus a correct result requires the analysis to be carried out in full, and a simple relationship between line intensities and

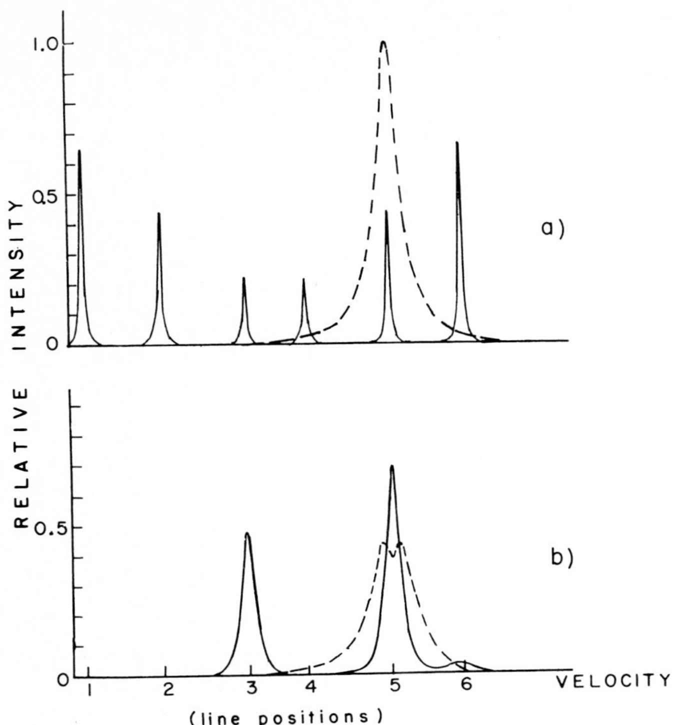

Fig. 14(a) Solid curve represents the resonant absorption cross section of the ^{57}Fe nuclei in the scatterer. The dashed line gives the energy spectrum of the incident beam which is set to excite line 5 in the scatterer. (b) The energy distribution of the resulting scattered radiation has two contributions. The solid line gives the nuclear resonant contribution, and the dashed line gives the Rayleigh contribution. Since the line shapes are not the same, no simple relationship exists between the SEDM experimentally observed line intensities and the corresponding cross sections.

cross sections does not exist.

4.3 Thickness Effects in Ordinary Scattering Experiments

In preparation for SEDM experiments both transmission and ordinary scattering experiments are usually performed to determine the position and intensity of the resonances in the sample under investigation and to optimize the experimental setup. Transmission lineshape has been extensively discussed in the literature. We will now consider the rather interesting, and somewhat unexpected, results due to thickness effects in ordinary single drive scattering experiments (19) in which the energy spectrum of the scattered radiation is not determined. Using Eqns. (5), (7), (8), and (9) we can calculate the nuclear resonant contribution to the scattered radiation. Since in ordinary scattering experiments the energy of the scattered radiation is not determined, $I_M(E',S)$ has to be integrated over the energy parameter E' to obtain the measured intensity. Thus, the number of recoillessly absorbed and recoillessly emitted events is

$$N_{MM}(S) = \int_0^\infty I_M(E',S)\,dE'. \tag{20}$$

The number of recoillessly absorbed but nonrecoillessly emitted events N_{MN} can be obtained in the following way. First we replace f^2 by $f(1-f)$ in Eqn. (9). Then we must realize that, due to the recoil, the energy distribution of the emitted photon is "smeared out" over an energy region large compared to the hyperfine splitting. This requires that the factor F in Eqn. (9) be modified since the outgoing photon is not subject to resonant absorption in this case. In this way,

$$N_{MN}(S) = \int_0^\infty dE' \sum_{i,j,i'} \frac{nf(1-f)\sigma_o I_o(\frac{1}{2}\Gamma)^2(\frac{1}{2}\Gamma_b)^2 W_{ij}W_{ji'}F_{MN}}{(1+\alpha')[(E'-\Delta_{ii'}-S)^2+(\frac{1}{2}\Gamma_b)^2][(E'-E_{ij}-\Delta_{ii'})^2+(\frac{1}{2}\Gamma)^2]}$$

where

$$F_{MN} = \frac{1-\exp\{-T[\mu_T(E'-\Delta_{ii'})\csc\alpha_1+(\mu_e+\mu_R)\csc\alpha_2]\}}{\mu_T(E'-\Delta_{ii'})\csc\alpha_1+(\mu_e+\mu_R)\csc\alpha_2}. \tag{21}$$

The coherent nonresonant contribution to the scattered intensity, which is essentially all Rayleigh scattering, can be obtained from Eqn. (12). Collecting energy-independent factors and integrating over E' we get

$$N_R(S) = \int_0^\infty \frac{n'f_R a_R(1-e^{-\mu_T(E')(\csc\alpha_1+\csc\alpha_2)T})\,dE'}{[(E'-S)^2+(\frac{1}{2}\Gamma_b)^2]\mu_T(E')(\csc\alpha_1+\csc\alpha_2)}. \tag{22}$$

The nonrecoilless Rayleigh contribution is

$$N_{NR} = \int_0^\infty \frac{dE'n'(1-f_R)a_R(1-e^{-[\mu_T(E')\csc\alpha_1+(\mu_R+\mu_e)\csc\alpha_2]T})}{[(E'-S)^2+(\frac{1}{2}\Gamma_b)^2][\mu_T(E')\csc\alpha_1+(\mu_R+\mu_e)\csc\alpha_2]}. \tag{23}$$

The total intensity distribution of the scattered radiation is the sum of four terms given by Eqns. (20), (21), (22), and (23),

$$N(S)=N_{MM}(S)+N_{MN}(S)+N_R(S)+N_{NR}(S). \tag{24}$$

The integrations over the energy parameter E' required to obtain the scattering spectrum given by Eqn. (24) can be accomplished by using standard numerical techniques.

In order to get some feeling for the various contributions in Eqn. (24) we considered the case of an enriched iron-powder scatterer. From our previous results, the ratio of the nuclear resonant to Rayleigh contribution was known for our experiments. Figure 15 shows these calculated results taking the recoilless fraction and the linear absorption coefficient to be 0.7 and 205 cm^{-1}, respectively. In addition, β was set equal to 153. The top curve in Fig. 15 gives the nuclear recoilless-recoilless contribution. Notice that the second and fifth lines are taller than the first and sixth. This is a characteristic feature for thick scatterers. On the other hand, the second curve (from the top in Fig. 15) which gives the nuclear recoilless-nonrecoilless contribution shows all peaks to be the same height. This type of saturation effect is familiar from Mossbauer transmission experiments with thick samples. The third curve, labeled R (in Fig. 15) shows the Rayleigh contribution which for our case is rather small. Notice that there are dips at the Mossbauer resonance energies. This occurs because the Rayleigh cross section is essentially constant over this energy range, but the photons are absorbed preferentially at the resonance energies on passing through the thick scatterer. The bottom curve in Fig. 15 gives the final, combined result.

The rather unusual saturation effect (i.e., the ratio of peak heights as the thickness of the scatterer increases) in scattering experiments is exhibited more graphically in Fig. 16(a). Here we have plotted the relative nuclear resonant recoilless-recoilless intensities of the first three lines for an enriched iron-power scatterer as a function of the thickness parameter β. Notice that above a certain value of β the second line becomes more intense and for large values of β the three intensities approach different limits. The reason for this particular saturation effect can be understood in the following way. The nuclear resonant scattering process involves an absorption and a subsequent emission. Note, that when line 2 (or 3) is excited, the re-emission will occur at line 2 (3) and also at line 4 (5). Now the photons emitted at line 4 will be attenuated less than those at line 2 on passing through the scatterer on the way to the detector. Therefore, more photons will get to the detector than would if all the radiation were re-emitted at the energy of line 2. The opposite is true when line 3

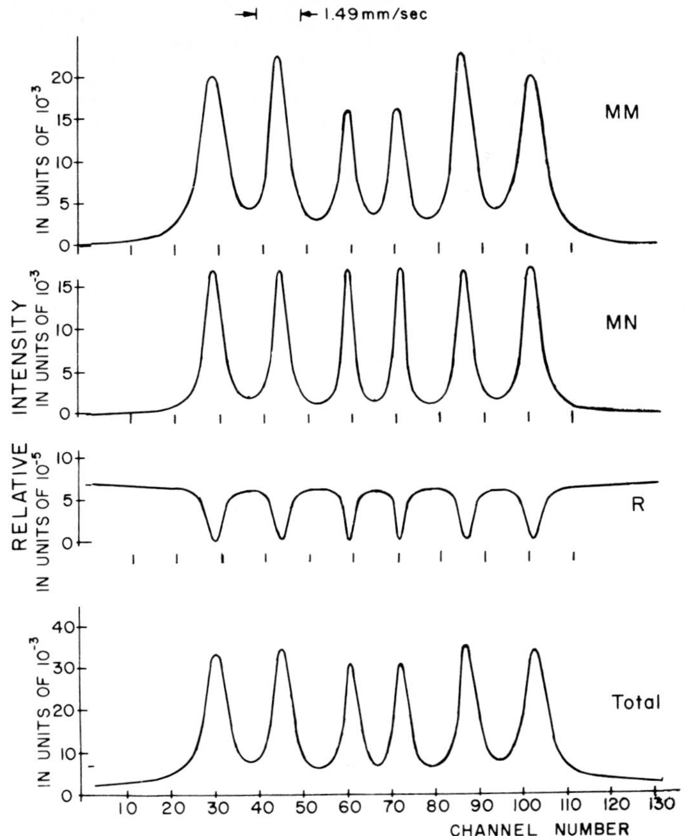

Fig. 15. Calculations showing the various contributions to the scattered Mossbauer spectrum. In these calculations the linewidth of the incident beam has been neglected. The scatterer is assumed to be a 91%-enriched iron powder having recoilless fraction of 0.7, linear-absorption coefficient 205 cm^{-1}, and thickness parameter β_s ($\beta_s = n\sigma_0 f$) of 153. The top curve gives the nuclear recoilless-nonrecoilless part. The third curve from the top gives the Rayleigh contribution and the bottom curve gives the total result.

is excited.

The total intensity distribution of the scattered radiation is given by Eqn. (24). The result depends on the value of the recoilless fraction (f) and the electronic absorption coefficient (μ_e). In Fig. 16(b), we show calculated spectra for an enriched iron-powder scatterer using two different values of μ_e, namely, 200 and 800 cm^{-1}. We see that as μ_e increases we obtain an effectively thinner scatterer and hence, narrower lines. The peak intensities are also modified slightly.

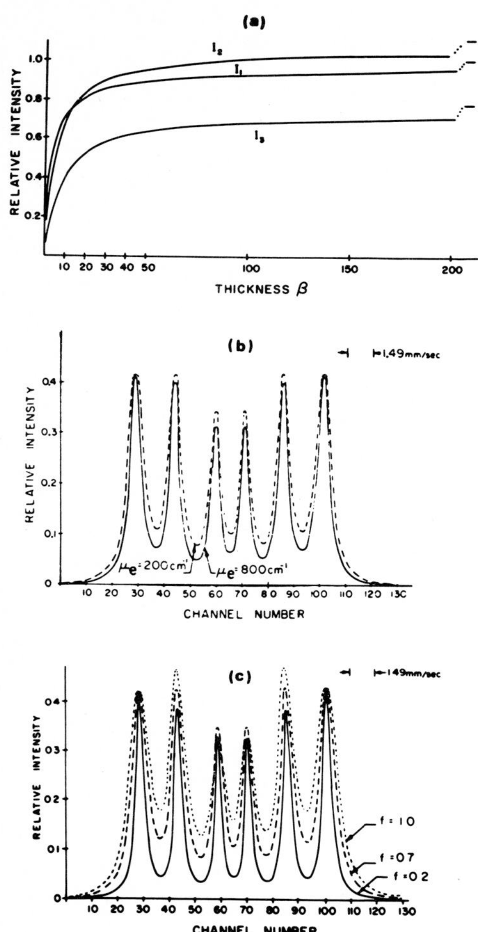

Fig. 16. (a) Calculated saturation effect is shown for iron-powder scattering. Notice the difference between such an effect in scattering as compared to ordinary transmission results. The curves are normalized so that $I_1=1$ at $\beta=\infty$. (b) Calculated results showing the effect of a changing electronic absorption coefficient on the Mossbauer scattering spectrum. We have assumed an enriched iron-power scatterer with $f=0.7$ and $\beta_s=175$. The spectra are normalized to the same intensity for line 6. (c) Same as (b) but now the electronic absorption coefficient μ_e is fixed at 200 cm^{-1} and the recoilless fraction takes on the values 0.2, 0.7, and 1.0.

In Fig. 16(c) we show the variation in the scattering line shape for different values of the recoilless fraction (f = 0.2, 0.7, and 1.0). The rather dramatic changes in the spectra indicate that such scattering experiments may be useful for measuring recoilless fractions.

5. SEDM LINESHAPE CALCULATIONS - TIME DEPENDENT THEORY

5.1 Adaptation of the emission line shape formula to SEDM

If the effective fields at the ^{57}Fe nuclei in the scatterer are time dependent, the expected SEDM result is different from the static case discussed in Sec. 3. To obtain the SEDM line shape with relaxation occurring in the scatterer, the effect of the time-dependent fields has to be included in the expression for the scattering line shape. This can be done following Blume's procedure (21). In such an approach the effective hyperfine fields are written as explicit functions of time by introducing a stochastic model. The central part of the calculation then involves the determination of the correlation function as described by Anderson (22) for the case of transmission geometry. To find the scattering line shape the appropriate stochastic average including the time-dependent fields, has to be obtained. In general, the SEDM problem is more difficult than the calculation for the transmission case. To obtain a relatively simple result we will use a less-general approach which is only valid for excitation on resonance of a completely split scatterer. In this method modifications of the well-known emission line-shape formula are needed for the SEDM case. Following Abragam (23) we write for the line shape as a function of energy ω,

$$I(\omega) = \mathrm{Re}[\vec{W}.\underline{A}^{-1}.\vec{1}], \qquad (25)$$

where $\vec{W}$ is the probability vector for the ionic states which produce effective fields at the ^{57}Fe nuclei and is time independent for an equilibrium situation, $\vec{1}$ is the unit vector, $\underline{A} = i(\underline{\omega}-\omega\underline{1}) + \underline{\pi}$, where $\underline{1}$ is the unit matrix, $\underline{\pi}$ is the probability matrix for transitions between the ionic states, and $\underline{\omega}$ is the diagonal energy matrix.

Consider the ordinary transmission situation and suppose also a two-level ionic system with both states equally probably. Then Abragam (23) shows that,

$$\underline{\pi} = \begin{bmatrix} -\Omega & \Omega \\ \Omega & -\Omega \end{bmatrix}, \qquad (26)$$

where Ω is the relaxation rate and

$$\underline{A}^{-1} = \frac{1}{\det\underline{A}} \begin{bmatrix} -i(\omega+\delta)-\Omega & -\Omega \\ -\Omega & i(-\omega+\delta)-\Omega \end{bmatrix} , \tag{27}$$

where $\pm\delta$ are the eigenvalues of $\underline{\omega}$, i.e., 2δ equals the nuclear energy difference corresponding to the two possible ionic states. The line shape for this case can be found by using Eqns. (25)-(27) and noting that $\vec{W}=(1/2,1/2)$. The result is,

$$I(\omega) = \frac{-2\Omega\delta^2}{(\delta^2-\omega^2)+4\omega^2\Omega^2} . \tag{28}$$

We now want to calculate a relaxation line-shape expression appropriate for SEDM experiments. Consider an experiment where the CVD is set to excite one of the lines in the scatterer (we are assuming here a situation where even with relaxation they can still be resolved), say the one located at $\omega=+\delta$, corresponding to the ionic spin value $S_z=-1/2$. If there were no relaxation or if $\Omega\approx 0$ we would expect the reemitted radiation to have a peak at $\omega=+\delta$ only.

Suppose we now allow Ω to increase to a value large enough so that there is a reasonable probability for the ionic spin to flip to the other orientation, $S_Z=+1/2$. There is now a finite probability for the nucleus to "see" a magnetic field in the opposite direction during the nuclear lifetime and consequently a peak at $\omega=-\delta$ should appear in the SEDM spectrum. In the analysis of this situation, it is still appropriate to consider the ionic spin-flipping process to be Markovian. It is not stationary, however, since the ionic spin probabilities P_I and P_{II} change with time. If at time t=0, the nucleus is excited and sees a field corresponding to $S_Z=+1/2$, what is the probability of its seeing the field due to $S_Z=-1/2$ at a later time? We know from Markov-process calculations (24) that,

$$W(t) = \begin{bmatrix} P_I(t) \\ P_{II}(t) \end{bmatrix} = \begin{bmatrix} \frac{1}{2}(1+e^{-2\Omega t}) \\ \frac{1}{2}(1-e^{-2\Omega t}) \end{bmatrix} , \tag{29}$$

when only one state is initially populated but the states are equally likely at equilibrium. Thus P_{II} increases, while P_I decreases with time until each equals 1/2.

To obtain the measured probabilities we need to average Eqn. (29) over the nuclear lifetime. The result is

$$\overline{W} = \begin{bmatrix} \frac{\Omega+\Gamma}{(2\Omega\ \Gamma)} \\ \\ \frac{\Omega}{(2\Omega+\Gamma)} \end{bmatrix} = \begin{bmatrix} W_1 \\ \\ W_2 \end{bmatrix} . \qquad (3o)$$

For our purposes we must generalize Eqn. (30) to include the case when the equilibrium populations of two ionic states are not equal, i.e., $(P_I/P_{II})=k$. This gives

$$\overline{W} = \begin{bmatrix} \frac{k\Omega+\Gamma}{\Gamma+(1+k)\Omega} \\ \\ \frac{\Omega}{\Gamma+(1+k)\Omega} \end{bmatrix} . \qquad (31)$$

In addition to the above, we need to modify $\underline{A}^{-1}$ [Eqn. (27)] to include the natural linewidth Γ as discussed by Abragam (25). The result for the selectively excited emission line shape including relaxation between ionic levels is,

$$I(\omega) = \frac{(a\Delta^+-\Delta^-)[\Gamma\omega+\Omega(\Delta^+-a\Delta^-)]-(1+a)\gamma(\Delta^+\Delta^-+\Gamma\gamma)}{(\Delta^+\Delta^-+\frac{1}{2}\Gamma\gamma)^2+[\Gamma\omega+\Omega(\Delta^+-a\Delta^-)]^2} , \qquad (32)$$

where

$\Delta^+=\omega+\delta$, $\Delta^-=-\omega+\delta$, $\gamma=(a+1)\Omega+(\frac{1}{2}\Gamma)$,

$a=(\Gamma/\Omega)+k$.

In Figs. 17(a) and 17(b) we show the final SEDM line shapes including the effect of the analyzer thickness for different values of the various parameters as noted on the figures. In general we note that the difference in the two peak heights depends on the ratio of the relaxation rate to nuclear lifetime (Ω/Γ). In addition the lines will be broadened according to the ratio of the relaxation rate to the energy splitting $(\Omega/2\delta)$ just as in ordinary transmission results.

The superiority of the SEDM technique over the usual approaches to the study of time-dependent effects can be demonstrated by a direct comparison of the calculated results for the two types of experiments. We consider a hypothetical case where the two peaks are 35 natural linewidths apart and calculate the line shapes for several relaxation rates. Fig. 18 shows the calculated transmission

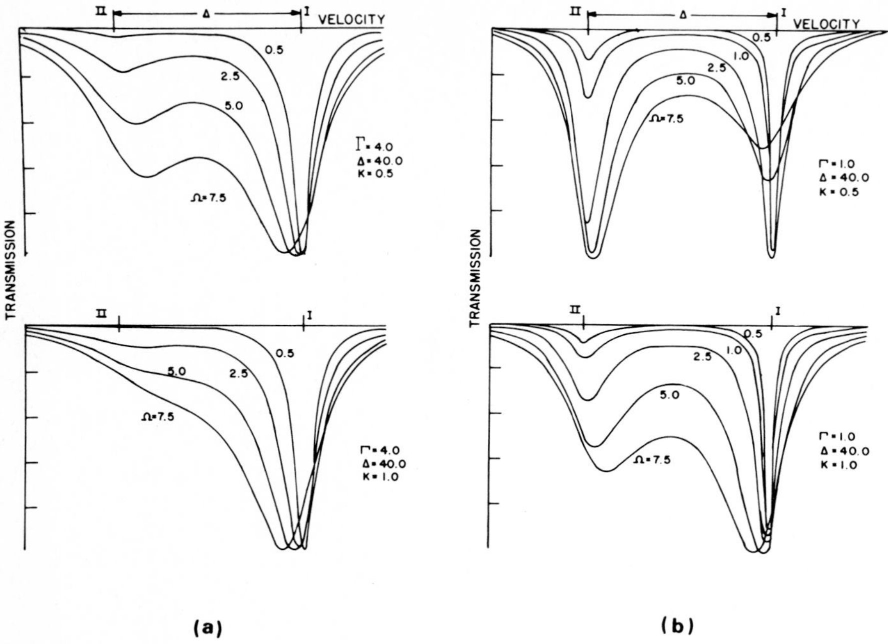

Fig. 17. Calculated SEDM relaxation spectra for different ionic population ratios $K = P_I/P_{II}$. The excitation occurs at line I and $2\delta/\Gamma=40$. Γ, δ, and Ω are expressed in energy units.

spectra and the corresponding results appropriate for the SEDM experiment. Line broadening effects due to the analyzer thickness (β=14) were included in the calculations. In Fig. 18(A) the curves labeled a and b are virtually the same as the result for no relaxation. Curves c and d show a slight broadening which under experimental conditions may be difficult to identify positively as due to relaxation, since other effects such as sample thickness, collimation, and field inhomogeneities cause a similar modification in the line shape if not accounted for properly. However, the SEDM results show the appearance of a second peak in all four cases and thus unambiguously determine the existence of relaxation in the scatterer.

5.2 Superoperator calculation of SEDM spectra

The above calculation of the SEDM spectrum for a scatterer in which the iron ion is fluctuating has limited applicability because it is based on assumptions which are not generally satisfied.

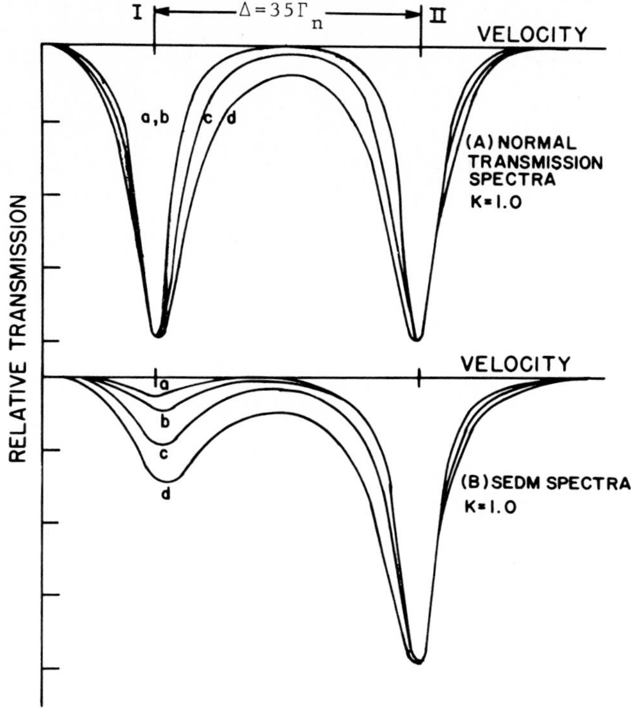

Fig. 18. Calculated relaxation spectra using (A) ordinary Mossbauer transmission geometry and (B) SEDM assuming relaxation rates in units of the natural line width Γ_n: (a) 0.25, (b) 0.5, (c) 1.24, and (d) 2.5.

This result is a good approximation only when both Ω/δ and Γ/δ are small. The above approach has the redeeming feature, however, that the spectra are easy to calculate and the major characteristics of SEDM spectra which arise from the dynamics of the iron ions in the scatterer are clearly exhibited. Thus the appearance of new lines in the SEDM spectrum as a result of energy transfer between resonances due to electronic relaxation, etc, and the broadening of lines as the relaxation rate increases are features that are revealed in this simple computation. However, the appearance in the scattered beam of a sharp elastic component at the excitation energy is not predicted by this simple theory and the difference between hot luminescence and resonance Raman is not delineated (25,26). Thus we return to a more general approach.

In this section we will discuss a more generally applicable calculation which is based on the superoperator approach. Recently three groups (27-29) presented calculations of the SEDM lineshape using this powerful technique and they should be consulted for details. Only the general approach and the application to some special cases will be presented here.

The probability for the absorption, by a Mossbauer nucleus in the initial ground state g_1, of a photon of wave vector k and frequency ω_k, out of a beam with Lorentzian lineshape and full width at half maximum Γ and the subsequent reemission of a photon with wave vector k' and frequency $\omega_{k'}$, with the nucleus ending up in the final ground state g_2 can be written as

$$W(\vec{k} \rightarrow \vec{k}\ ) = 2\mathrm{Re} \int_0^\infty dt \int_0^\infty dt_3 \int_0^\infty dt_2\ [\exp\{i(\omega_{k'}-\omega)t - \tfrac{\gamma}{2}t + i\omega_{k'}(t_3-t-t_2) - \tfrac{\Gamma}{2}(t_3-t+t_2)\}\ \overline{<V_k^+ V_{k'}^-(t_2)V_{k'}^+(t_3)V_k^-(t)>}] \qquad (33)$$

In the above expression Γ is the natural line width and

$$V_k^\pm(t) = \exp[i \int_0^t H(t')dt'][V^\pm] \exp[-i \int_0^t H(t')dt']$$

where H(t) is the Hamiltonian of the system including the time varying part and $V^\pm$ the complete interaction operator. The actual form of the Hamiltonian has been discussed in the references (30 ,31). It contains a function varying randomly with time which represents the interaction between the electron-nuclear system and the reservoir representing the rest of the system. In order to perform the computation of the expression in Eqn. (33) a stochastic average over all the electronic states, as indicated by the bar, has to be taken. In general the stochastic and quantum mechanical parts do not separate and so the average has to be taken together as shown by Blume (30). Also, if the time ordering is appropriately accounted for the above expression can be written as

$$W(\vec{k} \rightarrow \vec{k}') = I_1 + I_2 + I_3 \qquad (34)$$

where

$$I_1 = 2\mathrm{Re} \int_0^\infty dt_3 \int_0^{t_3} dt \int_0^t dt_2\ [\quad]$$

$$I_2 = 2\mathrm{Re} \int_0^\infty dt_3 \int_0^{t_3} dt_2 \int_0^{t_2} dt\ [\quad]$$

$$I_3 = 2\mathrm{Re} \int_0^\infty dt_2 \int_0^{t_2} dt_3 \int_0^{t_3} dt\ [\quad]$$

and the bracket represents the correlation function.

The superoperator approach allows one to carry out the computation of the correlation function in the presence of relaxation and to perform the time integrations at least formally (27-29). The actual computation of the SEDM spectrum is still a formidable problem as the result involves the computation of nine resonance denominators and the sum of terms which involve products of three of these taken at a time. Several authors have dealt with the problem of the formal evaluation of the above integrals. The results are usually presented as products of matrix elements of superoperators in a general form, but very little has been done to evaluate the SEDM spectra for specific cases of interest to the experimentalist and to discuss the interpretation of the individual terms in the resulting sums. The effect of incoherent thickness on the SEDM lineshape was found to be important in determining the final character of the result but again this has not been fully discussed in the literature. We have found that even though the problem of the computation of the SEDM spectra in general is quite formidable, for some specific cases of interest, as for example in the effective field approximation, the result can be presented in an algebraic closed form solution for all values of the relaxation rate. Each of the three terms in Eqn. (34) can be written as a sum of sixteen terms

$$I_1 = \sum_{p=1}^{16} I_1^p, \; I_2 = \sum_{p=1}^{16} I_2^p, \; I_3 = \sum_{p=1}^{16} I_3^p \, . \tag{35}$$

The individual terms I_1^p, I_2^p and I_3^p and are defined in appendix B and the details of the computation are presented elsewhere. Using this result we can write an expression for the radiation scattered by a thick sample in which the iron ions are undergoing time dependent fluctuations, in analogy with the stationary case (treated in Section 3), as

$$I_S(E',S) = \int_0^T dx \int dE \; W(\vec{k} \rightarrow \vec{k}') \; \exp[-\mu_T(E)\cos\alpha_1 x - \mu_T(E')\cos\alpha_2 x] \tag{36}$$

The integration over the position variable can be performed as before but the integration over the incoming beam energy E is not so trivial now and has to be, in general, performed numerically. It turns out, however, that to a good approximation the lineshape is not appreciably affected if $\mu(E)$ is replaced by $\mu(S)$, the value of the extinction coefficient at the excitation energy. This approximation obviates the difficult integration over E and saves a lot of computer time.

The results of the superoperator computation for an effective field approximation and some selected values of the relaxation rate are shown in Figs. 19 and 20. In Fig. 19 column A gives the scattering cross sections for the various relaxation rates and columns B and C give the energy distributions of the scattered radiation (results obtained from Eqn. 36) for thick ($ß_s$=600) and thin ($ß_s$=10) scatterers,

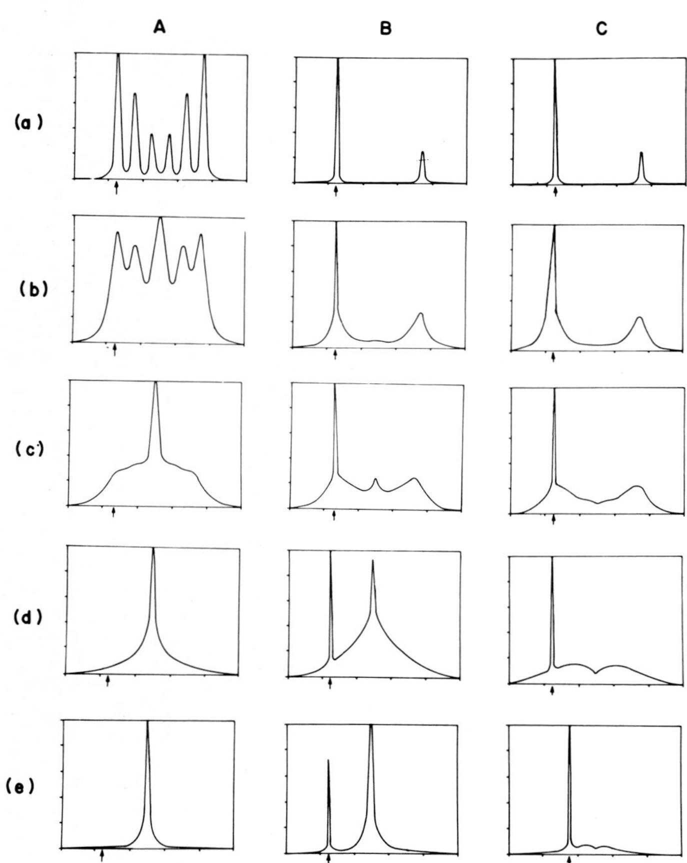

Fig. 19. Energy distribution of scattered radiation for different relaxation rates. Column A gives the scattering cross section and columns B and C the energy spectra of the scattered radiation for a thin ($ß_s$=10) and a thick ($ß_s$=600) scatterer respectively. The small vertical arrows indicate the excitation energy as determined by the CVD. The relaxation rates in units of the natural line width Γ_n are a) Ω =1, b) Ω =5, c) Ω =10, d) Ω = 25, e) Ω =100.

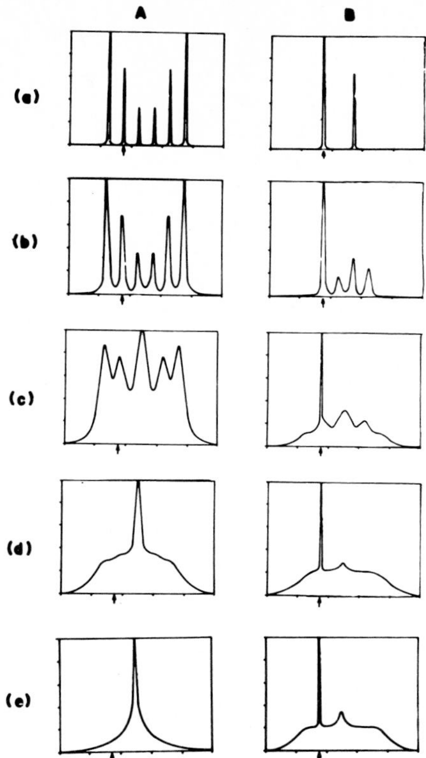

Fig. 20. Energy distribution of scattered radiation for different relaxation rates. Column A gives the scattering cross section and column B the energy spectrum of the scattered radiation for excitation as indicated by the arrows. The results are for a thick (β_s =600) scatterer. The relaxation rates are in units of Γ_n a) $\Omega=10^{-3}$, b) $\Omega=1$, c) Ω =5, d) Ω =10, e) Ω=25.

respectively. The arrows give the excitation positions as determined by the CVD. In Fig. 20 the excitation energy is different and only the results for the thick scatterer are shown. From these computations it is clear that both the relaxation rate and the thickness of the scatterer have profound effects on the SEDM lineshape. Quantitatively the super-operator calculations and the simpler calculations discussed in Section 5.1 seem to give the same results. On closer observation the superoperator computation shows a sharp peak at the excitation energy, for finite relaxation rates, which is absent in the simpler result. This peak comes from the first term (I_1) of the computation. It is a purely elastic component which is due to the resonant Raman scattering process. The other two terms (I_2 and I_3) correspond to the hot luminescence part of the scattering process which is of particular interest in the SEDM relaxation studies.

6 TIME DEPENDENT RESULTS

6.1 Application of SEDM to α-Fe_2O_3

We have used SEDM techniques to study the Morin transition in hematite. The rhombohedral crystal hematite (α-Fe_2O_3) is basically antiferromagnetic up to the Neel temperature (T_N=946 K) but also has many other interesting magnetic properties (32). In particular it has a spin-flip transition at T_M which is called the Morin transition (23). Below T_M the antiferromagnetic axis is along the [111] direction while above T_M the spins lie in the [111] basal plane. The electric field gradient (EFG) principal axis is along the [111] direction for temperatures both above and below T_M. This transition region has been studied experimentally (33-36) and also theoretically (37). Ordinary Mossbauer experiments (38) show that above T_M the characteristic six line hyperfine pattern is different from the one observed at low temperatures. This is due to the change in relative orientation of the EFG principal axis system and the direction and magnitude of the effective internal magnetic field at the ^{57}Fe nuclei (38). The difference in the Mossbauer spectra can be characterized by noting the positions of the most energetic transition at room (8.34 mm/sec) and liquid nitrogen (9.20 mm/sec) temperatures relative to a Co-Cu source. Regular Mossbauer transmission experiments near T_M show the presence of both hyperfine patterns broadened and partially resolved. Is this because both hyperfine interactions are present and the observed spectrum is simply a superposition of the two, or is each iron ion's spin flipping between the two possibilities? One could extract possible time-dependent information by applying relaxation theory to these ordinary Mossbauer spectra. Such an approach is very difficult in this case because the two hyperfine patterns differ only slightly from each other. SEDM experiments obviate this difficulty as discussed above.

In order to apply the relatively simple relaxation results of Sec. 5.1, we need to know the equilibrium populations of the two ionic states [Eqn. (31)]. For this purpose a series of ordinary transmission runs were done using a thin (β=1) sample of powdered α-Fe_2O_3. This thin sample was used, in contrast with the thick one of the same material used for SEDM, in order to obtain narrower lines and hence better energy resolution. The trapezoidal-function generator was again set to run only over the velocity range of the sixth peak. The results are shown in Fig. 21. From this figure we see that at room temperature one observes a single line which corresponds to the room-temperature hyperfine pattern. At lower temperatures a second line corresponding to a different hyperfine interaction appears and becomes increasingly more intense until at T_M = 263.5 K, the two lines are of about equal intensity. Below T_M line II increases at the expense of line I. The bottom spectrum in Fig. 21 was obtained at 77 K and shows a single narrow line at 9.20 mm/sec. A computer program was used to obtain the best

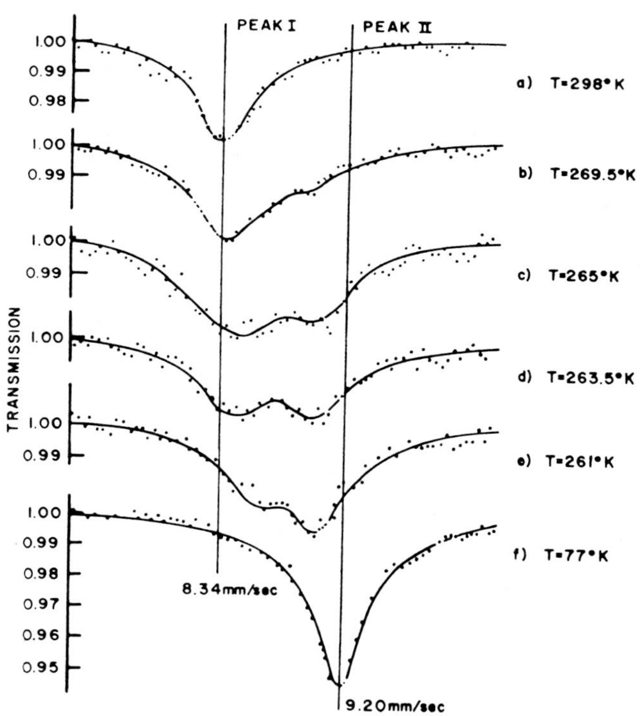

Fig. 21. Transmission spectra in the neighbourhood of the most energetic transition in α-Fe_2O_3 at different temperatures in the transition region. The sample was a thin (β=1) 90%-enriched α-Fe_2O_3.

two-line Lorentzian fit to the experimental data. From these data the equilibrium populations of the two ionic spin states P_I and P_{II} were obtained as a function of temperature. These values were used in generating the necessary theoretical SEDM relaxation spectra.

Our SEDM experiments were performed using a scatterer of 90%-enriched α-Fe_2O_3 powder, i.e., the same material as used for Fig. 21. However, the sample was made thicker to have an effective single-line thickness parameter β=350. Our first SEDM experiments were performed with the α-Fe_2O_3 scatterer at 263.5 K. The CVD was set on the energy of the sixth line at room temperature and the analyzer scanned over a velocity range in the neighborhood of the sixth peak. The results are given by the dots in Fig. 22. If these results represent simply a superposition of two static hyperfine patterns, they could be explained by using our time-independent SEDM calculations (Sec. 3). One would simply have three contributions: the Rayleigh scattering, the resonant contribution from exciting a peak at or near its resonance (8.34 mm/sec), and another resonant contribution from exciting a second peak off resonance. The calculated result according to the time-independent methods of Sec. 3 is shown in Fig. 22 by the dashed line. The dashed line

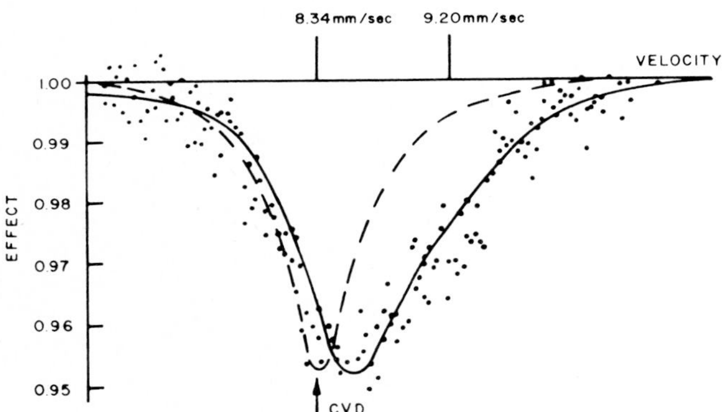

Fig. 22. An experimental SEDM spectrum of an enriched α-Fe_2O_3 *scatterer at T=263.5 K. The dashed line is computed according to Sec. 3 without time-dependent effects. The solid curve is the computed result for a relaxation time of* 1.1×10^{-7} *sec.*

can also be considered roughly as the sum of experimental spectra obtained by exciting the peak at room temperature near resonance and the peak at liquid-nitrogen temperature off resonance. The liquid-nitrogen peak was actually slightly shifted in our calculations because of the temperature dependence of the internal magnetic field. Notice that the dashed line is quite symmetric. This is to be expected for this case because the major contribution is Mossbauer resonant scattering on resonance which is symmetric. The Rayleigh contribution is also symmetric and the asymmetric term arising from resonance scattering off resonance is relatively small. However, the solid curve shows the best time-dependent calculated result. The results of experiments performed at 259.5 K and 261 K are given in reference 5. Because of the well-known hysteresis effect, which we also observed, all experiments were done by first going to liquid-nitrogen temperature and approaching each new temperature from the low-temperature side. It is worth emphasizing that the results shown in Figs. 11 and 22 were taken using the same sample and geometry, only the temperature was changed. In comparing Figs. 11 and 22 it is important to keep in mind the various contributions to the time-independent SEDM line shape.

Our results show that each iron nucleus in α-Fe_2O_3 near the Morin temperature is in a fluctuating hyperfine field. The values of the relaxation times at 263.5, 261, 259.5, and 258.5 K are respectively $(1.1 \pm 0.2) \times 10^{-7}$, $(2.3 \pm 0.4) \times 10^{-7}$, $(2.9 \pm 0.6) \times 10^{-7}$ and $(4.3 \pm 0.9) \times 10^{-7}$ sec. These results were checked by Furubayashi and Sakamoto (39) using Mossbauer magnetic diffraction spectroscopy and SEDM. They also developed a theoretical explanation for this effect.

6.2 Application of SEDM to Fe_3O_4

Although magnetite, Fe_3O_4, has been studied for quite sometime, there are still some unresolved problems. Magnetite crystallizes in a cubic inverse spinel structure above the Verwey transition (T_V=119 K). In magnetite both ferrous and ferric ions are present at the octahedral (B) sites, while only ferric ions are on the tetrahedral (A) sites. At room temperature the electrical conductivity of Fe_3O_4 is unusually high (250 Ω^{-1} cm^{-1}). The explanation of this fact is thought to be a rapid electron exchange between the ferrous and ferric ions on the octahedral (B) sites. The electrical conductivity results suggest that the electron exchange relaxation time is about 10^{-11} sec. at room temperature. However, previous Mossbauer absorption measurements (40) on magnetite at room temperature have found the electron exchange relaxation time to be 10^{-9} sec. The Mossbauer result was obtained by examining the line broadening in the hyperfine pattern associated with the octahedral (B) sites. It was the discrepancy between the electrical conductivity, and Mossbauer results that prompted our interest in magnetite (41).

The complete theory of SEDM including relaxation is complicated, nevertheless, we can make several simple observations. When there is no relaxation occurring in the scatterer, the experimental SEDM peak appears at the excitation energy with perhaps some asymmetry. However, when relaxation is present, the experimental peak is shifted toward the resonance position and the asymmetry can be quite pronounced.

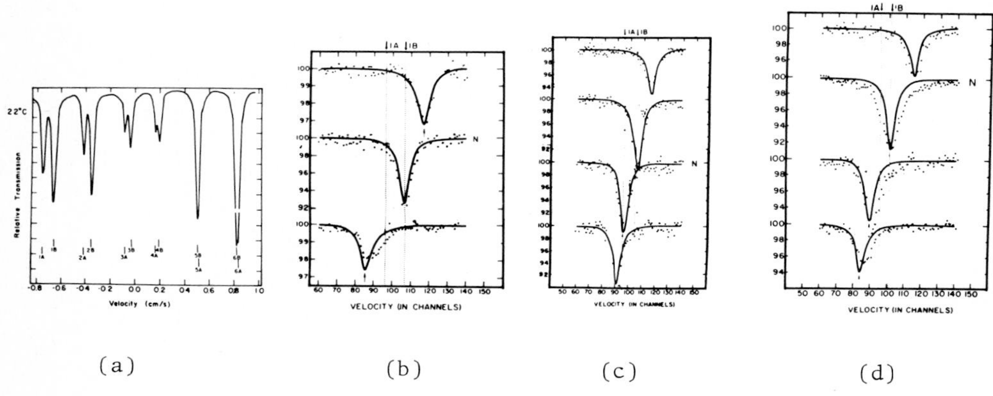

Fig. 23. (a) This figure shows a typical Mossbauer absorption spectrum using magnetite at room temperature (reference 42). The dips labeled B arise from the contribution due to the octahedral sites, and those labeled A are due to the tetrahedral sites. Our SEDM results for magnetite at room temperature (b), 125 K (c), and 112 K (d). The arrows labeled 1A and 1B and the dashed vertical lines locate the positions of those resonance (see Fig. 34). The small arrows locate the excitation energy (i.e., CVD setting) for each of the cases. The dots are the data and the solid curves give the theoretical SEDM results assuming no relaxation.

In Fig. 23(a) we show a typical Mossbauer transmission spectrum of magnetite at room temperature (42,43). The dips labeled B correspond to the hyperfine pattern due to iron nuclei on the octahedral (B) sites, while those labeled A correspond, similarly, to the tetrahedral (A) sites. We have concentrated our SEDM studies on the region in the neighborhood of the 1A and 1B dips. To be more precise, at room temperature we excited the scatterer at five different CVD settings: on the 1A resonance, on the 1B resonance, and at three other positions off these resonances. At 125 K we excited the magnetite sample at seven different positions, and at 112 K at five positions.

As mentioned above, when a non-relaxing resonance peak is

excited on or off resonance, the *resulting SEDM dip appears at the excitation energy*. We can obtain much more definitive information by looking at the actual SEDM experimental results and comparing them with non-relaxing SEDM calculations. These results are shown in Figs. 23(b),(c),(d). Consider first the room temperature results given in Fig. 23(b). We see that, except for the bottom spectrum in the figure, the data are well represented by the calculations without including relaxation effects. It is important to point out that there is only one free parameter per figure. This parameter is simply a normalization factor chosen to fit one spectrum in the figure. For the case of Fig. 23(b) the second spectrum from the top, labeled N, was selected for this purpose. The calculations for the other spectra in Fig. 23(b) have no free parameters. These calculations were made using a ratio of B to A sites equal to two and a sample thickness parameter β=336.

We see that Fig. 23(b) presents two interesting results. First, it looks as if there is no relaxation at the B sites, or the electron exchange relaxation time is so short that we obtain a "motionally narrowed" result. In the second interpretation the relaxation rate must be so rapid that the SEDM technique can not distinguish between that and the static effect. Our best estimate of the relaxation time according to the second interpretation is that it must be less than 10^{-11} sec. It certainly is not 10^{-9} sec.

Recently van Diepen et al (44) and later Haggstrom et al (45) have shown that the broadening of the Mossbauer transmission lines in α-Fe_3O_4 is due to a quadrupole and magnetic interaction and not due to relaxation which is in agreement with the first explanation of our SEDM spectra. The second result of Fig. 23(b) is even more surprising. It appears that there are relaxation effects at the A sites at room temperature. We have observed this result over a period of several months on the same sample. Others (45) have also observed indications of dynamic behavior at this site.

In Fig. 23(c) we show our results for magnetite at 125 K. The good fit of our calculated SEDM spectra to the experimental results indicates that there is no relaxation occuring at this temperature. In Fig. 23(d) we see our results at 112 K i.e,, below the Verwey transition. Previous Mossbauer measurements below the Verwey transition (43) have shown that the spectrum is composed of five components. It is of considerable interest to realize that complicated component structure does not seriously hinder SEDM relaxation studies (recall the results of Figure 10). Thus SEDM studies even below

T_V are very useful. From Fig. 23(d) we see that the SEDM experimental line shapes can not be adequately fitted using static SEDM theory. However, we do not know of a relaxation model for the B sites in the magnetite at 112 K, which we could use in order to estimate any relaxation times. We also note in Fig. 23(d) that A sites again show what appears to be relaxation effects.

In conclusion we have applied the SEDM technique to study the electron hopping time at the octahedral (B) sites in magnetite. We have found that this relaxation time is either zero or less than 10^{-11} sec at room temperature and 125 K. At 112 K the SEDM results indicate that ions on the octahedral (B) sites are behaving quite different from those at room temperature and 125 K. Relaxation may be occurring, but in the absence of a theoretical model no time estimate can be made.

6.3 Application of SEDM to ferrichrome A

Ferrichrome A is an iron containing polypeptide which has been studied in detail by transmission techniques (46) and thus serves as an appropriate sample to test the relative merits of the SEDM technique. In the transmission geometry between 1K and 300K it exhibits a Mossbauer relaxation spectrum characteristic of a paramagnetic material. At low temperatures the spectrum shows a split six line pattern. As the temperature increases the lines broaden and collapse to form a single line at 300K. This behaviour as a function of temperature was inferred by Wickman et al (46) to occur because of electronic relaxation. They assumed that the ^{57}Fe nucleus is subjected to a time dependent randomly varying effective magnetic field. The SEDM technique can show whether or not such an adiabatic spin flip model is adequate for a paramagnetic like ferrichrome A. In any case their model does explain some general features of relaxation line shape obtained in the transmission geometry. To reveal more detailed information about the relaxation mechanism the SEDM technique needs to be employed.

The results of experiments on ferrichrome A (47) at 300 K are shown in the composite Fig. 24. The top spectrum (Fig. 24(a)) was obtained in the transmission geometry using an absorber with β=22 and shows an almost complete collapse of the hyperfine pattern due to fast electronic relaxation. The arrows indicate the positions of the excitations (determined by the CVD setting) in the SEDM

(a)

(b)

(c)

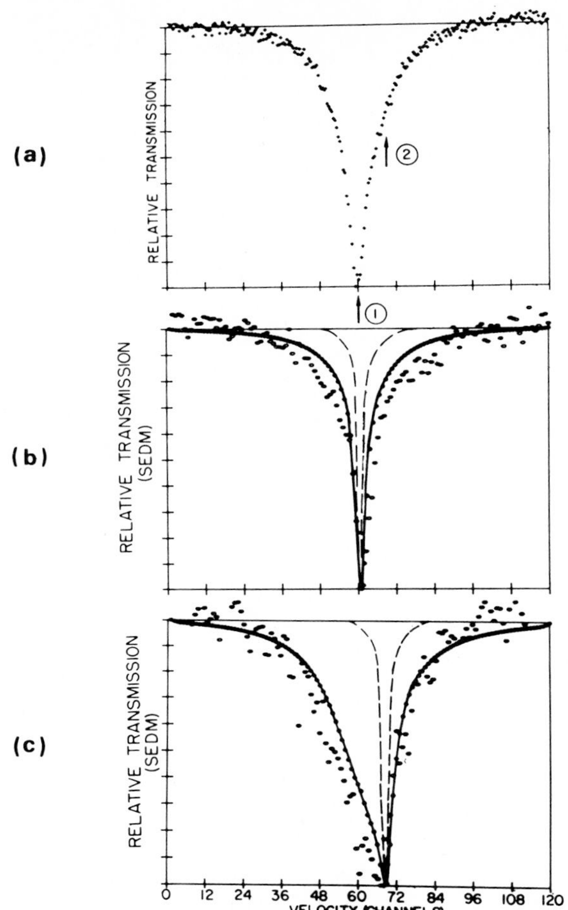

Fig. 24. This figure shows the results with ferrichrome A at 300 K [from ref 47] (1 mm/sec = 9.09 channels). Fig. 24(a) shows a transmission spectrum taken with a β=22 absorber and Fig. 24(b) and 24(c) the SEDM results with an enriched (β_s=350) scatterer. Excitations at positions 1 and 2 indicated by arrows in Fig. 24(a) lead to the SEDM results shown in Fig. 24(b) and 24(c) respectively. The dashed line represents the calculated spectrum assuming no relaxation. The solid line is a calculated SEDM spectrum assuming a relaxation time t_e=1.2 × 10^{-8} sec.

experiments for Figs. 24(b) and 24(c). The resulting SEDM lineshapes are shown in Fig. 24(b) for excitation at position 1 and in Fig. 24(c) for excitation at position 2. The dashed lines in the figure represent the expected spectra calculated assuming a static case as discussed in Section 3. Comparing this with the broader experimental

SEDM line (Fig. 24(b)) clearly indicates relaxation in the scatterer. Fig. 24(c) shows even more clearly that relaxation is occurring in the scatter. Notice the large deviation of the data from the dashed line.

The solid lines in Fig. 24(b) and 24(c) represent our calculated results based on the SEDM lineshape theory of Afanasev and Gorobchenko (27) which have been modified for thick scatterers. For this calculation the effective field-spin flip model of Wickman was used which assumes relaxation in the Kramers doublets ($S_Z = \pm 5/2$, $\pm 3/2$, $\pm 1/2$, where S is the electron spin). This mechanism reverses the effective magnetic field at the nuclear site.

In calculating the theoretical fit to these experimental data the only free parameter was the elastic Rayleigh contribution to the peak at the excitation energy. The relaxation time $t_e = 1.2 \times 10^{-9}$ sec used in these SEDM calculations was determined by finding the best fit to the transmission spectra (using the spin flip model). In general the calculated SEDM line shapes give fairly good fits to the experimental results. Deviations from the data occur in the wings of the line in Fig. 24(b) and the absence of a sharp peak at the resonance energy (Channel 61) in our calculated spectrum of Fig. 24(c).

The results obtained at 5.4 K are shown in Fig. 25. The hyperfine splitting from the $S_Z = \pm 5/2$ Kramers doublet leads to a partially resolved spectrum as can be seen from the transmission results in Fig. 25(a). In the SEDM experiments excitation of the main central peak (position 1) could be fitted well. This is shown in Fig. 25(b). However, exciting the outside peak at position 2 does not give results predicted by the spin flip theory, i.e, occurrence of a peak at 2'. An effective field flip associated with $S_Z = \pm 5/2$ flip occurring during the nuclear lifetime should give a line at the energy corresponding to position 2' in the SEDM spectrum. The calculated SEDM lineshape using the super-operator formalism and assuming this spin flip relaxation mechanism shows a peak at this position for a relaxation time of $t_e = 1.9 \times 10^{-8}$ sec. (This value of t_e was determined from a fit to the transmission spectra). The experimental SEDM results, however, clearly show a broad line in the neighborhood of the central resonance position (Channel 61) of the absorption spectrum. Notice that even in the results shown in Fig. 24(b) and 24(c) there are deviations from the SEDM spin flip model calculation.

These results show that the relaxation mechanism in ferrichrome

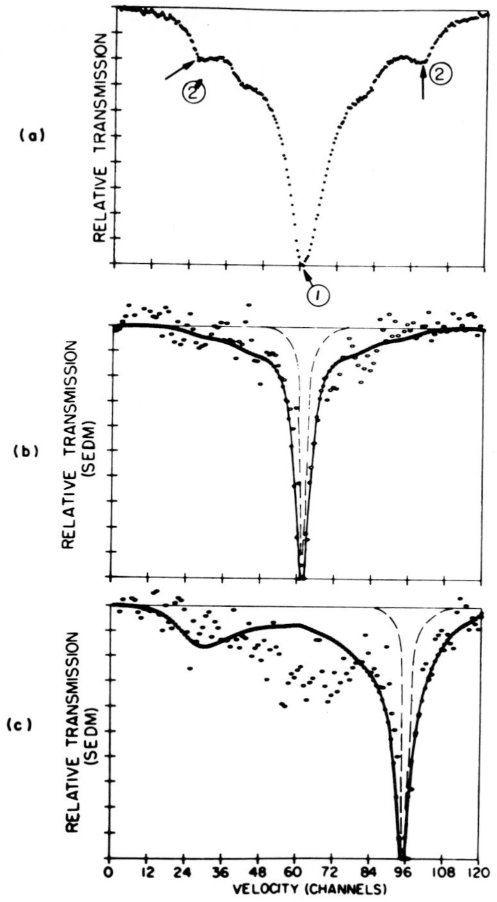

Fig. 25. This figure shows the results with ferrichrome A at 5.4 K [from ref. 47], (1 mm/sec = 9.09 channels). Fig. 25(a) shows the transmission spectrum taken with a β=22 absorber and in Fig. 25(b) and 25(c) the SEDM results with the enriched β_S=350 scatterer. Excitation positions 1 and 2 indicated by arrows in Fig. 25(a) lead to the SEDM results shown in Fig. 25(b) and 25(c) respectively. The dashed line represents the calculated spectrum assuming no relaxation. The solid line is a calculation of the SEDM line shape assuming an effective field spin flip type relaxation with t_e=1.9 x 10^{-8} sec.

A is basically different from the simple effective magnetic field spin flip model so often used. A more general model which includes other relaxation mechanisms such as transitions among the Kramers doublets or nonadiabatic terms will have to be invoked in order to explain the appearance of the broad central line in Fig. 25(b), and the deviations from the predicted result in Fig. 25(c).

Using SEDM techniques we see that new information about relaxation processes in ferrichrome A has been obtained. Further work along these lines is necessary to understand fully the electron spin dynamics in this system. Perhaps the most important conclusion we can draw from these results is that time dependent hyperfine interactions in general should be investigated using SEDM techniques. Even in investigating such a well studied and apparently simple system such as ferrichrome A this technique has advantages.

7. SUMMARY AND CONCLUSIONS

Selective excitation double Mossbauer (SEDM) experiments, although more difficult and time consuming than ordinary Mossbauer spectroscopy, can be done. These experiments can even be performed using scatterers containing the natural content of ^{57}Fe, but it is considerably easier to use enriched scatterers. The use of strong sources is also helpful.

Calculational methods have been developed (see Section 3) for SEDM that apply to a "split" scatterer, and include Rayleigh, incoherent electronic, and thickness effects. The results can be used to fit experimental SEDM data in the absence of relaxation processes. The important point in these calculations is that the scattered radiation has a peak at the excitation energy. These calculations have been verified by doing SEDM experimental runs on enriched iron powder scatterers at room temperature, and on an enriched α-Fe_2O_3 powder scatterer at liquid nitrogen temperature.

Because of the nature of the SEDM technique, the resulting spectra usually contain only a few lines. This allows the experimenter to resolve and interpret complicated experimental spectra. For example, one can look for weak transitions associated with the quadrupole mixing of the nuclear states in an effective magnetic field. Also one can indeed see if the effective field approximation is valid by considering the consequences of a full hyperfine interaction.

Perhaps the most fruitful area for SEDM research is the study of time dependent effects. We have demonstrated the advantage of the SEDM technique over regular Mossbauer spectroscopy in the study of relaxation processes (see Section 4).

SEDM time dependent studies have already produced interesting results. The first application was to the study of the nature of the Morin transition in α-Fe_2O_3. Our results show that each iron nucleus in α-Fe_2O_3 near the Morin temperature is in a fluctuating hyperfine field. We have also observed relaxation effects in α-Fe_2O_3 at room temperature. A similar investigation was conducted on magnetite (Fe_3O_4). We applied the SEDM technique to measure the electron hopping time at the octahedral (B) sites in magnetite and found that this relaxation time is either zero or less than 10^{-11} sec. at room temperature and 125K. At 112K the SEDM results for the octahedral (B) sites are quite different from those at room temperature and 125K. Relaxation may be occurring, but in the absence of a theoretical model no time estimate can be made. The most surprising result was the observation of relaxation at the A sites at all three temperatures.

It has also been recently demonstrated that the "effective field-spin flip" model (46) usually used in treating ferrichrome A is incorrect (47). The SEDM experimental spectra clearly show that the relaxation mechanism is more complicated than the originally proposed. It appears that SEDM has considerable promise for obtaining detailed information. However, much more work is needed. It is important that more researchers begin to explore this technique.

APPENDIX A.

Calculation of the Angular Dependence of the SEDM Line Intensities for a Powder Scatterer.

In Sec. 3 there are various W's which are needed in calculating the results for Eqn. (9). We will use the "line" notation as shown in Fig. 5(c).

$$W_1 = \frac{3}{16\pi}(1+\cos^2\theta),$$

$$W_2 = \frac{1}{4\pi}(1-\cos^2\theta),$$

$$W_3 = \frac{1}{16\pi}(1+\cos^2\theta),$$

$$W_4 = W_3,\ W_5=W_2,\ W_6=W_1 .$$

To find $\overline{W}$ we need to integrate over all angles.
We find,

$$\overline{W}_1 = \frac{1}{4\pi},\ \overline{W}_2 = \frac{1}{6\pi},\ \overline{W}_3 = \frac{1}{12\pi},$$

$\overline{W}_4=\overline{W}_3$, $\overline{W}_5=\overline{W}_2$, $\overline{W}_6=\overline{W}_1$.

We also need the quantities $W_{i,j}(\alpha)$. These are formed by multiplying the angular factor for the absorption process by the angular factor for the emission process and then, since we are dealing with powder samples, integrating over all angles subject to the constraint that the angle between the incoming and outgoing radiation is α. These quantities are

$$W_1W_1 = W_6W_6 = (\tfrac{3}{8\pi})^2(\tfrac{1}{4})(1+\cos^2\theta_1)(1+\cos^2\theta_2),$$

$$W_2W_2 = W_5W_5 = (\tfrac{3}{8\pi})^2(\tfrac{4}{9})(1-\cos^2\theta_1)(1-\cos^2\theta_2),$$

$$W_3W_3 = W_4W_4 = (\tfrac{3}{8\pi})^2(\tfrac{1}{36})(1+\cos^2\theta_1)(1+\cos^2\theta_2),$$

$$W_2W_4 = (\tfrac{3}{8\pi})^2(\tfrac{1}{6})(1-\cos^2\theta_1)(1+\cos^2\theta_2),$$

$$W_4W_2 = (\tfrac{3}{8\pi})^2(\tfrac{1}{9})(1+\cos^2\theta_1)(1-\cos^2\theta_2),$$

$$W_3W_5 = (\tfrac{3}{8\pi})^2(\tfrac{1}{9})(1+\cos^2\theta_1)(1-\cos^2\theta_2),$$

$$W_5W_3 = (\tfrac{3}{8\pi})^2(\tfrac{1}{9})(1-\cos^2\theta_1)(1+\cos^2\theta_2).$$

The only complication in the averaging process involves an average over $(\cos^2\theta_1\cos^2\theta_2)$ subject to the constraint. This can be done in the following way. Let $\vec{\gamma}_1$ indicate the direction of the incident photon and $\vec{\gamma}_2$ represent the direction of the outgoing photon with respect to a particular $(\vec{i}, \vec{j}, \vec{k})$ coordinate system. Then $\vec{k}\cdot\vec{\gamma}_1 = \cos\theta_1$, $\vec{k}\cdot\vec{\gamma}_2 = \cos\theta_2$, and $\vec{\gamma}_1\cdot\vec{\gamma}_2=\cos\alpha$. Now if we perform the same rotation R on $\vec{\gamma}_1$ and $\vec{\gamma}_2$, then $R\vec{\gamma}_1\cdot\vec{\gamma}_2=\cos\alpha$. The necessary average can be found by integrating $(\vec{k}\cdot R\vec{\gamma}_1)^2(\vec{k}\cdot R\vec{\gamma}_2)^2$ over all values of R. This can be done fairly easily by using group theory (48).

$$\langle\cos^2\theta_1\cos^2\theta_2\rangle = (1/8\pi)\ (\vec{k}\cdot R\vec{\gamma}_1)^2(\vec{k}\cdot R\vec{\gamma}_2)^2 \times (1-\cos\gamma)d\gamma d\Omega.$$

In this calculation the axis of rotation, R, points in the direction subtended by $d\Omega$ and the angle of rotation about this axis is γ. The above integral allows us to sum over all possible rotations R by summing over all axes and all angles of rotation.

It should be noted that $0<\gamma<\pi$. Using the rotation matrix given by Goldstein (49) and performing the above integral gives

$$\langle\cos^2\theta_1\cos^2\theta_2\rangle = \frac{1+2\cos^2\alpha}{15}$$

and

$$W_{1,1} = W_{6,6} = \left(\frac{3}{8\pi}\right)^2\left(\frac{1}{4}\right)\left(\frac{26+2\cos^2\alpha}{15}\right),$$

$$W_{2,2} = W_{5,5} = \left(\frac{3}{8\pi}\right)^2\left(\frac{4}{9}\right)\left(\frac{6+2\cos^2\alpha}{15}\right),$$

$$W_{3,3} = W_{4,4} = \frac{1}{9}\, W_{1,1},$$

$$W_{2,4} = W_{4,2} = W_{3,5} = W_{5,3} = \left(\frac{3}{8\pi}\right)^2\left(\frac{1}{9}\right)\left(\frac{14-2\cos^2\alpha}{15}\right).$$

APPENDIX B.

Closed Form Solution of the SEDM Lineshape Equation for an Effective Field Relaxation.

Although the superoperator approach gives a formal solution for the SEDM lineshape for a general hyperfine interaction and all values of the relaxation rate, the actual numerical computation of the line shape is still a formidable problem. For a system consisting of an ^{57}Fe nucleus with two ground and four excited substates and an ion with electronic spin S = 5/2 and consequently six substates the total Hamiltonian has to be represented by a 36 x 36 matrix. The resulting superoperator will be 1296 x 1296 complex matrices. There are six of these superoperator matrices which have to be inverted and multiplied in groups of three to obtain the desired lineshape. Such calculations strain the memory and speed of most presently available computers. For certain special cases, however, the problem is tractable, leading to a closed form algebraic solution. These special cases are valuable because they reveal important features of the SEDM lineshape with relaxation occurring in the scatterer and give insite into the behavior of the various terms of the solution.

In this appendix we will give the results of an SEDM lineshape calculation using the superoperator technique. We assume that the ^{57}Fe nucleus is subject to a fluctuating magnetic field. The result is valid for all values of the fluctuating rate Ω. Details of the calculations are presented elsewhere.

Consider the energy diagram of ^{57}Fe shown in Fig. 5(c) of the text. The excited nuclear substates are labeled $|j\rangle$ where j=3/2, 1/2, -1/2, -3/2 and the ground substates are labeled $|i\rangle$ where i=1/2, -1/2. In the effective field approximation we will consider the effect of the electron to be approximated by a magnetic field which can assume two orientations, up and down, which will be represented by + and -. Relaxation in the nuclear-electronic system is of the type $|j, +\rangle \rightleftarrows |j, -\rangle$ and $|i, +\rangle \rightleftarrows |i, -\rangle$ and occurs at random with a rate Ω. Since only the states with $j=\pm 1/2$ and $i=\pm 1/2$ are coupled by the electromagnetic transitions and the effective field reversal model for relaxation doesn't mix these with the $j=\pm 3/2$ states we can restrict ourselves to a four state nuclear and two states electronic system. We assume that 2δ is the splitting in the excited state and $2\delta_0$ in the ground state respectively and j_1, j_2 are the probabilities for the required transitions. This allows us to deal with processes leading to lines 2, 3, 4, and 5 of Fig. 9. By assuming $\delta_0=\delta$ and $j_1=j_2$ we can perform the calculation for lines 1 and 6. The sum of the two contributions gives us the energy distribution of the scattered radiation, $W(k \to k')$. In order to calculate the SEDM lineshape for the total sample we use equations (36), (34), (35) and $W(k \to k')$. The terms I_1^P, I_2^P, and I_3^P, for P = 1 through 16 which are needed for the computation of I_1, I_2, I_3 and $W(k \to k')$ are given in Table I. The other parameters are defined below.

ACKNOWLEDGEMENT

One of us (B.B.) would like to thank Dr. J. O'Brien for helpful discussions regarding the mathematics of the superoperator approach and Dr. B. Furubayashi for helpful comments regarding the physical interpretation of the SEDM lineshape.

TABLE I

p	A_p^2	I_1^p	I_2^p	I_3^p
1	$J_1^2\ J_1^2$	$\alpha_{12}\ \beta_{21}\ \gamma_{12}$	$(\varepsilon_{21}\alpha_{12}+\theta_1^\varepsilon\alpha_{22})\delta_{12}$	$(\varepsilon_{21}\alpha_{12}+\theta_1^\varepsilon\alpha_{22})\gamma_{11}$
2	$J_1^2\ J_2^2$	$\alpha_{11}\ \beta_{12}\ \gamma_{12}$	$(\varepsilon_{22}\alpha_{12}+\theta_2^\varepsilon\alpha_{22})\delta_{12}$	$(\varepsilon_{21}\alpha_{12}+\theta_2^\varepsilon\alpha_{22})\gamma_{21}$
3	$J_1^2\ J_2^2$	$\alpha_{12}\ \beta_{21}\ \gamma_{21}$	$(\varepsilon_{21}\alpha_{12}+\theta_1^\varepsilon\alpha_{22})\delta_{11}$	$(\varepsilon_{21}\alpha_{12}+\theta_1^\varepsilon\alpha_{22})\gamma_{11}$
4	$J_1^2\ J_2^2$	$\alpha_{11}\ \beta_{12}\ \gamma_{21}$	$(\varepsilon_{22}\alpha_{12}+\theta_2^\varepsilon\alpha_{22})\delta_{12}$	$(\varepsilon_{22}\alpha_{12}+\theta_2^\varepsilon\alpha_{22})\gamma_{22}$
5	$J_1^2\ J_2^2$	$\alpha_{21}\ \beta_{22}\ \gamma_{12}$	$(\varepsilon_{11}\alpha_{22}+\theta_1^\varepsilon\alpha_{12})\delta_{22}$	$(\varepsilon_{21}\alpha_{11}+\theta_1^\varepsilon\alpha_{21})\gamma_{12}$
6	$J_1^2\ J_2^2$	$\alpha_{22}\ \beta_{11}\ \gamma_{11}$	$(\varepsilon_{12}\alpha_{22}+\theta_2^\varepsilon\alpha_{12})\delta_{22}$	$(\varepsilon_{22}\alpha_{11}+\theta_1^\varepsilon\alpha_{21})\gamma_{21}$
7	$J_1^2\ J_2^2$	$\alpha_{21}\ \beta_{22}\ \gamma_{22}$	$(\varepsilon_{12}\alpha_{22}+\theta_2^\varepsilon\alpha_{12})\delta_{21}$	$(\varepsilon_{21}\alpha_{11}+\theta_1^\varepsilon\alpha_{21})\gamma_{11}$
8	$J_1^2\ J_1^2$	$\alpha_{22}\ \beta_{11}\ \gamma_{22}$	$(\varepsilon_{11}\alpha_{22}+\theta_1^\varepsilon\alpha_{12})\delta_{21}$	$(\varepsilon_{22}\alpha_{11}+\theta_2^\varepsilon\alpha_{21})\gamma_{22}$
9	$J_1^2\ J_2^2$	$\alpha_{21}\ \beta_{21}\ \gamma_{12}$	$(\varepsilon_{22}\alpha_{12}+\theta_2^\varepsilon\alpha_{21})\delta_{11}$	$(\varepsilon_{12}\alpha_{21}+\theta_2^\varepsilon\alpha_{11})\gamma_{11}$
10	$J_1^2\ J_2^2$	$\alpha_{22}\ \beta_{12}\ \gamma_{12}$	$(\varepsilon_{21}\alpha_{11}+\theta_1^\varepsilon\alpha_{21})\delta_{11}$	$(\varepsilon_{11}\alpha_{21}+\theta_1^\varepsilon\alpha_{11})\gamma_{22}$
11	$J_2^2\ J_2^2$	$\alpha_{21}\ \beta_{21}\ \gamma_{21}$	$(\varepsilon_{21}\alpha_{11}+\theta_1^\varepsilon\alpha_{21})\delta_{11}$	$(\varepsilon_{12}\alpha_{21}+\theta_2^\varepsilon\alpha_{11})\gamma_{12}$
12	$J_1^2\ J_2^2$	$\alpha_{22}\ \beta_{12}\ \gamma_{21}$	$(\varepsilon_{22}\alpha_{11}+\theta_2^\varepsilon\alpha_{21})\delta_{12}$	$(\varepsilon_{11}\alpha_{21}+\theta_1^\varepsilon\alpha_{11})\gamma_{21}$
13	$J_1^2\ J_2^2$	$\alpha_{22}\ \beta_{22}\ \gamma_{11}$	$(\varepsilon_{12}\alpha_{21}+\theta_2^\varepsilon\alpha_{11})\delta_{21}$	$(\varepsilon_{12}\alpha_{22}+\theta_2^\varepsilon\alpha_{12})\gamma_{11}$
14	$J_2^2\ J_2^2$	$\alpha_{11}\ \beta_{11}\ \gamma_{11}$	$(\varepsilon_{11}\alpha_{21}+\theta_1^\varepsilon\alpha_{11})\delta_{21}$	$(\varepsilon_{11}\alpha_{21}+\theta_1^\varepsilon\alpha_{12})\gamma_{22}$
15	$J_1^2\ J_2^2$	$\alpha_{12}\ \beta_{22}\ \gamma_{22}$	$(\varepsilon_{12}\alpha_{21}+\theta_1^\varepsilon\alpha_{11})\delta_{22}$	$(\varepsilon_{12}\alpha_{22}+\theta_2^\varepsilon\alpha_{12})\gamma_{12}$
16	$J_1^2\ J_2^2$	$\alpha_{11}\ \beta_{11}\ \gamma_{22}$	$(\varepsilon_{11}\alpha_{21}+\theta_1^\varepsilon\alpha_{11})\delta_{22}$	$(\varepsilon_{11}\alpha_{22}+\theta_1^\varepsilon\alpha_{12})\gamma_{21}$

The following definitions are assumed in Table I

$$\bar{\alpha}_{ij} = \alpha_{ij} + \theta_j^\alpha \qquad i = 1,2 \quad j = 1,2$$

TABLE I (contd...)

$$\bar{\gamma}_{ij} = \gamma_{ij} + \theta_j^{\gamma}$$

$$\bar{\delta}_{ij} = \delta_{ij} + \theta_j^{\delta}$$

$$\bar{\beta}_{ij} = \beta_{ij} + \theta_j^{\beta}$$

$$\bar{\varepsilon}_{ij} = \varepsilon_{ij} + \theta_j^{\varepsilon}$$

$$\alpha_{ij} = \frac{a_{kj}}{a_{ij}a_{kj}+\Omega^2} \qquad \theta_j^{\alpha} = \frac{-i\Omega}{a_{ij}a_{kj}+\Omega^2}$$

$$\gamma_{ij} = \frac{c_{kj}}{c_{ij}c_{kj}+\Omega^2} \qquad \theta_j^{\gamma} = \frac{+i\Omega}{c_{ij}c_{kj}+\Omega^2}$$

$$\delta_{ij} = \frac{d_{kj}}{d_{ij}d_{kj}+\Omega^2} \qquad \theta_j^{\delta} = \frac{-i\Omega}{d_{ij}d_{kj}+\Omega^2}$$

$$\beta_{ij} = \frac{b_{kj}}{b_{ij}b_{kj}+\Omega^2} \qquad \theta_j^{\beta} = \frac{+i\Omega}{b_{ij}b_{kj}+\Omega^2}$$

$$\varepsilon_{ij} = \frac{e_{ki}}{e_{ij}e_{kj}+\Omega^2} \qquad \theta_j^{\varepsilon} = \frac{-i\Omega}{e_{ij}e_{kj}+\Omega^2}$$

$$a_{ij} = \omega^1 - \Delta_{ij} - i(\Gamma/2 + \Omega)$$

$$c_{ij} = \omega - \Delta_{ij} + i\left(\frac{\Gamma+\Gamma_b}{2} + \Omega\right)$$

$$d_{ij} = \omega - \Delta_{ij} - i\left(\frac{\Gamma+\Gamma_b}{2} + \Omega\right)$$

$$b_{ij} = (\omega - \omega^1 - \delta_{ij}) + i\left(\frac{\Gamma_b}{2} + \Omega\right)$$

$$e_{ij} = -\delta_{ij} - i(2\Gamma+\Omega)$$

with Δ_{ij} defined as

$$\Delta_{12} = \delta - \delta_o$$

$$\Delta_{22} = (\delta-\delta_o)$$

$$\Delta_{11} = (\delta+\delta_o)$$

$$\Delta_{21} = -(\delta+\delta_o)$$

with δ_{ij} defined as

$$\delta_{11} = \delta_{21} = 0$$

$$\delta_{12} = 2\delta_o$$

$$\delta_{22} = -2\delta_o$$

REFERENCES

1 A.N. Artemev, G.V. Smirov, E.P. Stepanov, Sov. Phys. - JETP, 27 (1968) 547.

2 W. Meisel, Monatsber, Deutschen Akd. Wiss. Berlin Ger., 11 (1969) 355.

3 N.D. Heiman, J.C. Walker, L. Pfeiffer, Phys. Rev., 184 (1969) 281 and Mossbauer Effect Methodology, Vol. 6, I.J. Gruverman, (Ed.) Plenum Press, New York, 1971.

4 B. Balko, G.R. Hoy in I.J. Gruverman (Ed.), Mossbauer Effect Methodology, Vol. 9, Plenum Press, New York, 1974.

5 B. Balko, G.R. Hoy, Phys. Rev. B, 10 (1974) 36.

6 F.C. Ruegg, J.J. Spijkerman, J.J. DeVoe Jr., Rev. Sci. Instrum., 36 (1956) 356.

7 S.L. Ruby, J.M. Hicks, Rev. Sci. Instrum., 33 (1962) 27.

8 Y. Kagan, A.M. Afanasev, I.P. Perstnev, Zh. Eksp. Teor. Fiz. 54 (1968) 1530 [Sov. Phys. - JETP, 27 (1968) 819].

9 See, for example G.R. Hoy and S. Chandra, J. Chem. Phys., 47 (1967) 961.

10 W. Heitler, Quantum Theory of Radiation, Clarendon, Oxford, 1954, p. 196-204.

11 A.J.F. Boyle, H.E. Hall, Rep. Progr. Phys., 25 (1962) 441.

12 P. Debrunner, R.J. Morrison, Rev. Sci. Instrum., 36 (1965) 145.

13 P.B. Moon, Proc. Phys. Soc. Lond. A63 (1950) 80.

14 P.J. Black, G. Longworth, D.A. O'Connor, Proc. Phys. Soc. Lond., 83 (1964) 937.

15 W. Meisel, L. Keszthelyi, Hyperfine Interactions, 3 (1977) 413.

16 M.K.F. Wong, Proc. Phys. Soc. Lond., 85 (1965) 723.

17 P.J. Black, D.E. Evans, D.A. O'Conner, Proc. Roy. Soc. London, A270 (1962) 168.

18 W. Renz, H. Appel, in E. Matthias and D.A. Shirley (Eds.) Hyperfine Structure and Nuclear Radiations, North-Holland, Amsterdam, 1968, p. 370.

19 B. Balko, G.R. Hoy, Phys. Rev. B, 13 (1976) 2729.

20 B. Balko, G.R. Hoy, Phys. Rev. B, 10 (1974) 4523.

21 M. Blume, in E. Mathias and D.A. Shirley (Eds.) Hyperfine Structure and Nuclear Radiations, North- Holland, Amsterdam, 1968, p. 911.

22 P.W. Anderson, J. Phys. Soc. Jap., 9 (1954) 316.

23 A. Abragam, Nuclear Magnetic Resonance, Oxford University Press, London, 1961.

24 A. Paupolis, Probability, Random Variables and Stochastic Processes, McGraw-Hill, New York, 1965.

25 Y.R. Shen, Phys. Rev. B, 9 (1974) 622.

26 Y. Toyogawa, A. Kotani, A. Sumi, J. Phys. Soc. Japan, 42 (1977) 1495.

27 A.M. Afanasev, V.D. Gorobchenko, Sov. Phys. - JETP, 40 (1975) 1114.

28 F. Hartmann-Boutron, J. Physique 37 (1975) 537.

29 S. Banerjee, Ph.D. Dissertation, State University of New York at Stony Brook, 1977.

30 M. Blume, Phys. Rev., 174 (1968) 351.

31 M.J. Clauser, M. Blume, Phys. Rev. B, 3 (1971) 583.

32 P.J. Flanders, J.P. Remeika, Philos. Mag., 11 (1965) 1271.

33 P. Imbert, A. Gerard, C.R. Acad. Sci. (Paris), 257 (1963) 1054.

34 S.T. Lin, Phys. Rev., 116 (1959) 1447.

35 R.C. Liebermann, S.K. Banerjee, J. Appl. Phys. 41 (1970) 1414.

36 S. Foner, S.J. Williamson, J. Appl. Phys., 36 (1965) 1154.

37 J.O. Artman, J.C. Murphy, S. Foner, Phys. Rev., 138 (1965) A912.

38 F. Van der Woude, Phys. Status Solidi, 17 (1966) 417.

39 B. Furubayashi, I. Sakamoto, J. Phys. (Paris) Colloq. C2 Supplement No. 3, 40 (1979) C2-677.

40 W. Kundig, R.S. Hargrove, Solid State Commun., 7 (1969) 223.

41 B. Balko, G.R. Hoy, J. Phys. (Paris) Colloq. C6, Supplement No. 12, 37 (1976) C6-89.

42 Taken from F. Van der Woude, G.A. Swatsky, A.H. Morrish, Phys. Rev., 167 (1968) 533.

43 R.S. Hargrove, W. Kundig, Solid State Commun., 8 (1970) 303.

44 A.M. van Diepen, 2C3, Int. Conf. on Magnetism, Amsterdam, 1976.

45 L. Haggstrom, H. Annersten, T. Ericson, R. Wappling, W. Karner, S. Bjarman, J. Phys. (Paris) Colloq., Supplement No. 3, 40 (1979) C2-327.

46 H.H. Wickman, M.P. Klein, D.A. Shirley, Phys. Rev., 152 (1966) 345

47 B. Balko, E.V. Mielczarek, R.L. Berger, J. Phys. (Paris) Colloq. C2 Supplement au No. 3, 40 (1979) C2-17.

48 M. Hamermesch, Group Theory, Addison-Wesley, Reading, Mass., 1964.

49 H. Goldstein, Classical Mechanics, Addison-Wesley, Cambridge, Mass., 1953.

CHAPTER 4

THE STUDY OF METALS BY MOSSBAUER SPECTROSCOPY

T.E. Cranshaw

Nuclear Physics Division, AERE Harwell, England

1. INTRODUCTION

The contributions of Mossbauer spectroscopy to the study of metals in the last decade or so have been made over such a wide front that no short review article can possibly do justice to them all. Inevitably a selection must be made, and some important areas treated only cursorily, or not at all. In this article such areas are the study of actinide elements, the use of the Mossbauer spectrum to study lattice dynamics, the study of intermetallic compounds and particularly the rare-earth transition metal compounds, and the study of hydrogen in metals. Review articles giving information on these topics are by Brodsky (1) (actinides), Grow et al (2) (lattice dynamics) Kirchmayer and Poldy (3) (rare earth intermetallic compounds) and Wagner and Wortmann (4) (hydrogen in metals).

The space allotted to each Mossbauer nucleus in this review is roughly in proportion to the amount of published work using it, rather than to any judgment upon the importance of the results obtained. Thus, whereas ^{57}Fe appears more strongly than all the other nuclei put together, important work such as the results recently obtained with Ta, is treated in no great detail.

The layout of the article will be clear from the table of contents, although the division into the various headings obviously cannot be rigid. An interpretation of a spectrum will often involve isomer shift, quadrupole and magnetic hyperfine interaction simultaneously, and the grouping adopted here is mainly a matter of the author's, and it may be hoped the reader's, convenience.

2. INFORMATION FROM HYPERFINE INTERACTIONS

2.1 Isomer Shift

The generation of useful information from measurement of the isomer shift starts with the empirical calibration of ^{57}Fe by the observation of iron in different chemical states by Walker et al (5). This indicated that the state of iron atoms in iron metal was well represented by

$3d^7 4s$, and work since then has been concentrated on understanding the relatively small changes in isomer shift which occur when iron is placed in different metallic environments.

2.1.1 Pressure dependence

The simplest change in the natural iron environment is obtained by changing the pressure. The average value of the volume coefficient of the isomer shift of iron obtained in this way (6-8), $\delta s/\delta \ln V = 1.37$ mm sec^{-1} is mainly accounted for by scaling the 4s electron density. Shifts produced by changes in screening of 3s electrons by 3 d electrons or s-d transfer are thought to be small, and not well understood at the present time.

2.1.2 XFe alloys

The effect of dissolving iron in X Fe alloys has been discussed by Ingalls (9), and we give a summary of his work here. Figure 1 presents

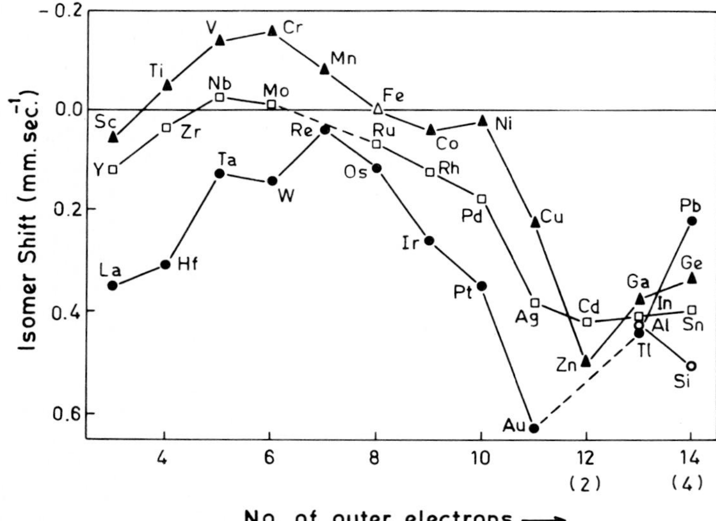

Fig. 1. Isomer shifts at ^{57}Fe in hosts of the fourth fifth and sixth periods plotted as a function of outer electron number.

the isomer shift relative to α-iron of ^{57}Fe dissolved in elements of the third, fourth and fifth series plotted as a function of number of outer electrons. The shifts peak in the middle of each transition series and then rise as the d-band becomes filled. In Figure 2, the same isomer shift data is plotted as a function of atomic volume. Bearing in mind the dependence of isomer shift on volume for pure iron, the absence of a

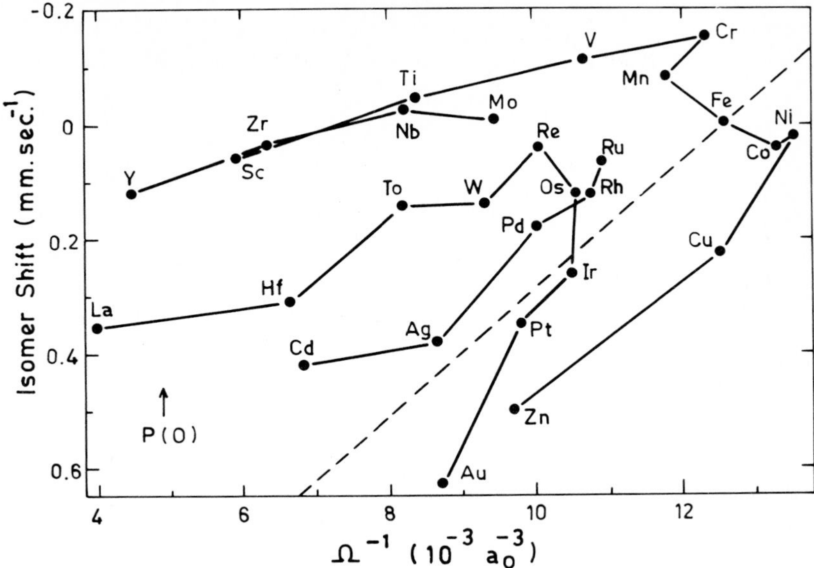

Fig. 2. The isomer shift at ^{57}Fe dissolved in hosts of the fourth fifth and sixth periods plotted as a function of host volume.

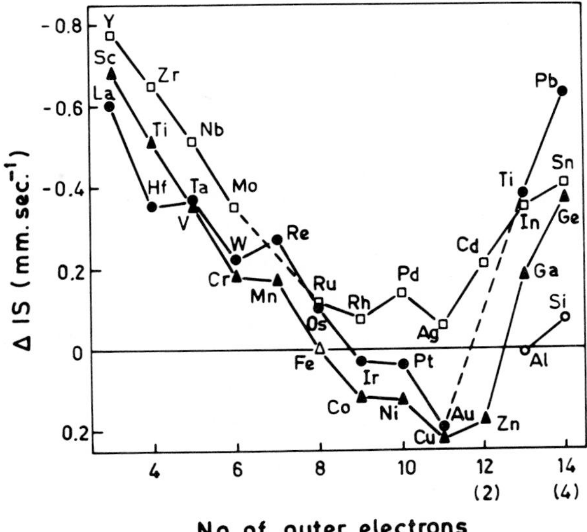

Fig. 3. Isomer shift at ^{57}Fe nuclei relative to pure iron for various hosts, corrected for volume effects.

correlation with volume is striking (10). Ingalls nevertheless suggests that the volume effect must exist, uses pressure dependence data, plotted as a dashed line in Figure 2 to correct the shift values to constant volume, and then replots the residual shifts as a function of the number of outer electrons in Figure 3. It is immediately obvious that the peak in the middle of the transition periods has disappeared, and there is now a smooth fall of isomer shift with Z until the d-band is full at Cu. After Cu there is a sharp up-turn in the shift for Fe in s-p elements.

Before considering the implications of these results, we note the correlation pointed out by van der Woude and Sawatzky (11). These authors plot the volume corrected isomer shift in M Fe alloys against the isomer shift produced at an iron nearest neighbour by M atoms in an Fe M alloy. Figures 4 and 5 show the results for transition and s-p

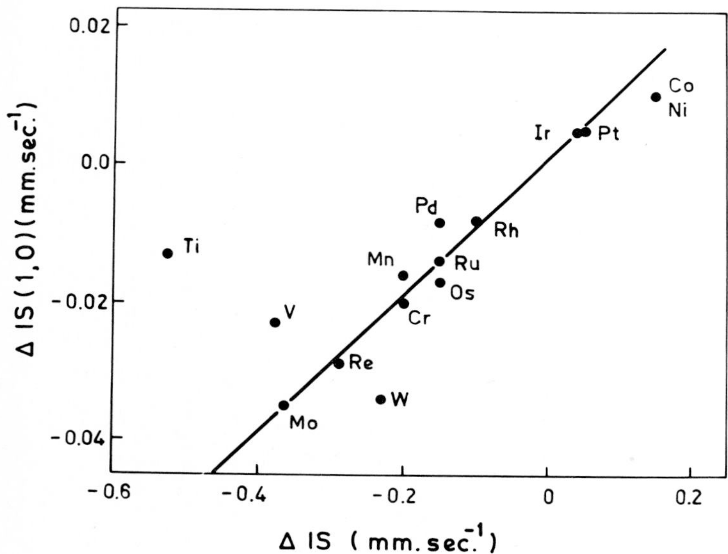

Fig. 4. Isomer shift of ^{57}Fe nuclei which are nearest neighbours to 3d, 4d and 5d impurities in pure iron plotted against the isomer shift of ^{57}Fe dissolved in the same element as host. The latter shifts are corrected for the volume effect.

elements respectively. It is noticeable that a strong correlation exists in both cases, indicating that the same mechanism is responsible for the isomer shifts. Furthermore, the slope of the solid line in Figure 4 is near 0.125, the value we would obtain if we simply scaled the effect with the number of M neighbours, strongly suggesting that the mechanism is mainly between the iron atom and its first neighbours. Comparing Figure 4 with Figure 5 further emphasizes the change in behaviour as the d-band becomes full; the slopes of the solid lines have different signs.

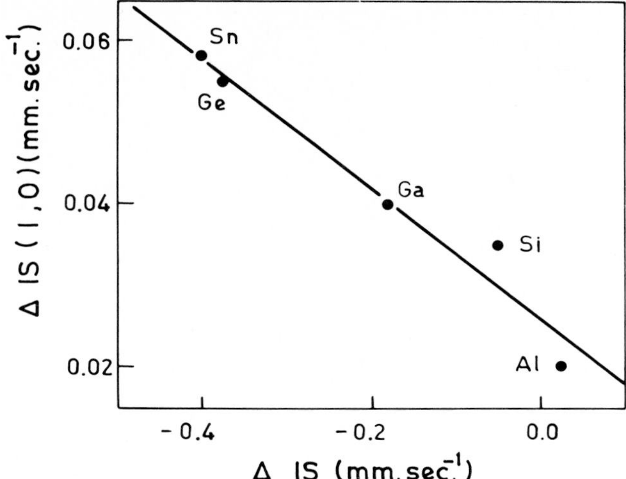

Fig. 5. Isomer shift of ^{57}Fe nuclei which are nearest neighbours to s-p impurities in pure iron plotted against the isomer shift of ^{57}Fe dissolved in the same element as host. The latter shifts are corrected for the volume effect.

The magnitude of the isomer shifts noted in Figure 1 is much smaller than would be expected if the iron atom "saw" the band structure of the host metal. It appears to be necessary to suppose that the iron atom retains approximately the electron configuration of iron metal, and the changes shown in the Figures represent a relatively small perturbation, whose exact nature is not clear. Ingalls points out that if the iron impurity is to remain electrically neutral, Figure 3 indicates a decrease of about 0.2 s-like conduction electrons in going from Sc to Zn, with the same increase in d-electrons. If the neutrality requirement is relaxed, other changes are possible.

The dependence of isomer shift on electron density of the host has been further studied in alloys, such as NiCu (12) and Cu X where X = Zn, Ga, Ge and As (13). In general, these studies are taken to illustrate a change from screening by d electrons to screening by s-p electrons. The isomer shift at the iron impurity in Cu X is a linear function of the electron density only, assuming that the impurity X contributes 2 to 5 electrons as X goes from Zn to As. The rate of decrease is faster in the alloys than in the pure metals, though part of this effect may be due to departures from randomness in the alloy (14,15).

Measurements of the ^{57}Fe and ^{197}Au isomer shift in Pd-Au alloys, and of the isomer shift of ^{57}Fe in Pd Cu alloys have been made by Longworth (16). Longworth finds that the isomer shift at ^{57}Fe is approximately constant for $o < x < .5$ and rises linearly for $.5 < x < 1$ in $Pd_{1-x}Au_x$ alloys, the discontinuity being associated with an increase in line width due to electrostatic interactions. These changes may correspond to filling of the d-band of Pd which would occur at $x = .36$ on a rigid band model, and the departure from this value presumably indicates that a detailed consideration of the mechanism of charge screening at the iron site is required. The isomer shift at ^{197}Au is non-linear, but shows no discontinuity comparable to the ^{57}Fe results, reflecting the different requirements of charge screening. Discontinuities are also observed at $x = .5$ for the isomer shift of ^{57}Fe in Pd-Cu alloys.

2.1.3 Au isomer shifts

A fruitful area for research has been the isomer shift at ^{197}Au in alloys and intermetallic compounds. In the case of ^{197}Au, $\delta R/R$ is large, so that precise measurements of electron density are possible. Studies have been made of dilute alloys of type $\underline{Au}$ X (17) and $\underline{X}$ Au (16, 19, 20)and also in complete ranges of composition (16, 18).

In the work of Roberts et al (18), Josephson et al (21), the isomer shift was found to be approximately a linear function of concentration. The values of isomer shift were interpreted in terms of a charge screening model which also accounted for the residual resistivity. It was concluded that the Au impurity represented an attractive potential to most s-conduction electrons.

A convenient basis for discussion of the Au isomer shift has been given by Cohen (22) who writes for the electron density, ρ,

$$\rho = \rho_o f(v/v_o)[1+\rho_o^{-1}\Sigma_k m_k \delta(r_k)]$$

Here ρ_o is the conduction electron density in pure gold. v/v_o is the specific volume of gold relative to the alloy, and $f(v/v_o)$ describes the effects of the change of lattice volume on the charge density. The k indicates a summation over sites, m_k the probability that an X atom occupies the kth site, and $\delta(r_k)$ is the change in electron density at the gold atom due to replacing a gold atom at r_k by an X atom.

The form of the function f is not very clear, but Cohen indicates that about 17% of the lattice volume change is seen as a volume change of the cell containing the gold atom. The electron density is then assumed to be given by $\rho\sigma(v_o/v)^{\gamma}$ with $\gamma \sim 0.86$ (23).

Constraints on the form of the scattering potential at impurity sites in gold have been deduced by Huray et al (17) who measured the isomer shifts produced by the charge perturbation round Sc, Ti, V, Cr, Mn, Fe, Co, Ni, Cu, Zn, Ga, Ge, Ag, Cd, In, Sn, Sb atoms dissolved in gold, and comparing them with the residual resistivities.

2.1.4 General remarks

Many of the points mentioned above are brought together in a paper by Watson and Bennett (24). These authors show that when volume effects are allowed for, a strong correlation exists between the isomer shifts of transition metal nuclei, (^{197}Au, ^{195}Pt, ^{193}Ir, ^{181}Ta, ^{99}Ru and ^{57}Fe) in a wide range of hosts, so that when plotted as a function of group in the periodic table, the residual isomer shifts fall on a common curve. Figure 6 shows the results of scaling and averaging the values for the six isotopes and on a suitably chosen scale, the dots show the Pauling electronegativity values. There is a striking similarity.

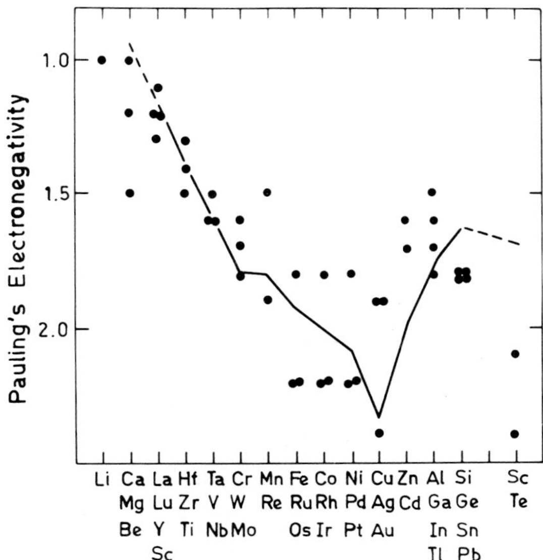

Fig. 6. Comparison of the volume corrected isomer shift with the Pauling electronegativity scale. The solid line is the result of scaling and averaging the volume corrected isomer shifts for six Mossbauer nuclei.

The change in ρ (o) can be written

$$\Delta \rho (o) = \alpha \Delta n_c + R \Delta n_d$$

where Δn_c represents the change in non-d conduction electrons, and Δn_d the screening by d-electrons. Calculations indicate that $-.5 > R > -1$, so that an increase in n_d causes a decrease in ρ (o). For approximate charge neutrality

$\Delta n_c + \Delta n_d \sim 0$

The electronegativity scale measures the total charge flow, i.e. $\Delta n_c + \Delta n_d$, whereas the isomer shift measures approximately $\Delta n_c - \Delta n_d$. The observed correlation seems to indicate that the ratio of these terms is nearly constant for transition metal impurities.

The subject of isomer shifts in metals and other systems is treated in great detail in Mossbauer Isomer Shifts, Ed. Shenoy and Wagner, (25).

2.2 Electric Field Gradients in Metals

For several nuclei Mossbauer spectroscopy provides a simple way of measuring quadrupole couplings and hence electric field gradients. In metals the main contributions to the field gradient which have to be considered arise from the lattice charges, the conduction electrons and from spin orbit coupling.

2.2.1 Efg in nominally cubic metals

In full cubic symmetry the efg vanishes, but in a ferromagnet with nominally cubic lattice, the symmetry is less than cubic because the magnetization introduces a particular axis with a direction. Thus, field gradients at ^{193}Ir in iron have been observed by Mossbauer spectroscopy by Salomon and Shirley (26), using a ^{192}Os source to achieve less than 1% concentration of impurities, and by Wagner and Potzel (27) in a 3% Ir alloy, confirming the NMR results of Aiga and Itoh (28).

Salomon and Shirley (26) analyze the apparent concentration dependence of the quadrupole coupling in Mossbauer spectra and show that the influence of near neighbour interactions is likely to be small. The value found for the coupling is $e^2qQ/2h$ à -2.05 MHz, and is attributed to spin polarization and spin orbit coupling in the Ir solute.

2.2.2 α-iron

Two papers in 1971 (29,30) reported very careful measurements of the spectrum of α-iron. Two different spectrometers were used, one a constant acceleration type, with a Michelson interferometer in the feedback loop, and one a mechanical constant velocity type. The results were in good agreement with each other, and the authors found values for the quadrupole coupling of $4\varepsilon = e^2qQ = 2.3 \pm 1.5$ and 8.8 ± 2.5 μm sec^{-1} at 298 K and 4.3 K respectively.

In a simple band model of the metal, the asymmetries of the distribution of charge and spin are both due to holes in the 3d band, so that an electric field gradient would be accompanied by a proportional dipolar magnetic field. In fact

$$H_{dip} = \mu q$$

where μ is the Bohr magneton and q the electric field gradient (31).

Substituting a value for Q of 0.2b, we find that a value of e^2qQ of 1 mm sec^{-1} would correspond to a dipolar field of 150 kOe. Thus the value of 2.3 μm sec^{-1} for e^2qQ might be expected to correspond to a dipolar magnetic field of 35 gauss. In unpublished work by the author, the change in field at ^{57}Fe nuclei in an iron single crystal when the magnetization axis lay along the [100] axis and the [111] axis was less than 6 gauss, where we would have expected about 53 gauss.

The interpretation of these results is not clear and the question was raised whether such values of the quadrupole coupling might be the result of residual strain in the sample. In an experiment using a back-scattering technique on a mild steel bar whose surface was enriched with ^{57}Fe, Mercader and Cranshaw (32) and Kjeldgaard (33), measured changes in isomer shift, quadrupole coupling and hyperfine field when the surface was strained by bending the bar.

They found a quadrupole coupling of $4\varepsilon = 2.4 \pm 0.16$ μm sec^{-1} when the surface was under a tension of about 4 kbar, i.e. near the elastic limit and $-2.08 \pm .24$ μm sec^{-1} for the same stress in compression. These couplings are of the same order of magnitude as those found by Spijkermann et al (29) and Pipkorn and Violet (8) and do not rule out the possibility that the quadrupole coupling measured by these authors in pure iron is due to residual strains.

The interpretation of the results for the strained iron are complicated by the fact that the material is polycrystalline and the strain is therefore randomly oriented with respect to the crystal axes. In a later experiment with single crystals of Fe Si alloy, Cranshaw and Mercader (34) measured the quadrupole coupling produced by tetragonal and trigonal distortions. An external magnetic field was applied to make the magnetization direction parallel or perpendicular to the strain axis.

The results are shown in Table 1. It can be seen that when the field is perpendicular to the strain axis, the measured quadrupole coupling has a value approximately $-\frac{1}{2}$ the value when the field is parallel to the strain axis, as would be expected if the principal axis of the field gradient is along the strain axis.

To discuss these results consider a cubic lattice distorted so that E_1 is the elongation along the z axis. The field gradient produced at a lattice site by unit charges on the other sites is then

TABLE 1

The measured values of isomer shift and quadrupole coupling for two directions of stress and for the magnetization parallel or perpendicular to the stress direction.

	H_{11}		$H_{\perp}$	
Direction of stress	[100]	[111]	[100]	[111]
Strain (x 10^3)	1.9	1.2	1.9	1.2
Isomer shift ($\mu m\ s^{-1}$)	0.38 ± 0.25	0.82 ± 0.09	0.74 ± 0.25	0.71 ± 0.26
$4\varepsilon(\mu m\ s^{-1})$	7.8 ± 0.72	4.89 ± 0.40	-2.88 ± 0.80	-2.80 ± 0.68
4ε/strain (x 10^{-3})	4.1 ± 0.4	4.1 ± 0.3	-1.5 ± 0.4	-2.3 ± 0.6

$$q = -3\ (E_2 - E_1)\ \Sigma\ (r^4 - 5z^4)r^{-7}$$

where the summation is taken over all sites. If the volume remains constant, we may write $E_2 - E_1 = 3/2(\delta x/x)$. The values of $q(\delta x/x)^{-1}\ a_o^3$, where a_o is the lattice spacing are then as given Table 2 for tetragonal and trigonal distortions.

TABLE 2

Calculated field gradients for trigonal and tetragonal distortions of a BCC lattice produced by unit charges on the lattice sites. The columns give the contributions from the first and second neighbour shells and the total

Distortion	1st shell	2nd shell	Σ
Tetragonal	$-16\sqrt{3}/9$	$\frac{9}{4}$	-1.08
Trigonal	$32\sqrt{3}/27$	$-\frac{3}{2}$	+0.72

If we assume unit charges at the lattice sites (which is almost certainly an overestimate because of charge screening) we can estimate the efg at the nucleus from

$$q = q(1+\gamma_\infty)$$

where γ_∞ is the Sternheimer antishielding factor, and is about 15 for iron.

By substitution in these expressions, we find values for the field gradient of the right order of magnitude, but note that according to

Table 2 the quadrupole coupling produced by charges in the first and second neighbour shell, and by the sum over all shells changes sign as the distortion is changed from the [100] axis to the [111] axis, whereas the observation shows the same sign. Moreover, no linear combination of field gradients produced by atoms in the first and second shells can be found which makes the efg distortions along the [100] and [111] axis have the same sign. It is noteworthy that the magnetostrictive constants λ_{100} and λ_{111} have different signs for iron. Thus if magnetostrictive distortion were the origin of the electric field gradient found for Ir in iron, it should have the same sign for magnetization along the [100] & [111] axes. This experiment does not appear to have been carried out.

When applied to the case of Ir in Fe, Gehring (35) and Gehring and Williams (36) find that the lattice distortion would predict an efg an order of magnitude smaller than the values observed. On the other hand, they find that spin orbit coupling can account for the observed efg's, and probably the same mechanism operates in the case of the mechanically distorted lattices.

2.2.3 Efg in non-cubic metals

In non-cubic metals, we write the electric field gradient as (37)

$$q = q_{latt}(1-\gamma_\infty) + q_{loc}(1-R_Q)$$

Here q_{latt} is the electric field gradient produced by charges at the lattice sites, and can be calculated (38,39). For example for hexagonal metals

$$q_{latt} = Z\ 0.0065-4.3584\ (c/a - 1.633)a^{-3}$$

where Z is the charge on the ion core, a and c are the lattice parameters. γ_∞ is the Sternheimer factor for charges outside the atom, q_{loc} is produced by conduction electrons close to the nucleus and R_Q the Sternheimer anti-shielding factor to allow for the distortion of the core electron distribution. Interest now centres on the behaviour of q_{loc}.

Sawicki (40) using mostly Mossbauer results of Quaim (41) on ^{57}Fe in Ge, H, Hf, Lu, Ti, Zr, Re, Zu, Cd and In observed a correlation between q and q_{latt} where

$$q = q_{exp} - q_{latt}(1-\gamma_\infty)$$

and q_{exp} is the measured value.

Further Mossbauer spectroscopic measurements of electric field gradients in hexagonal metals were made by Wortmann and Williamson (42)

(^{57}Fe in Ru, Re), Bauminger et al (43) (^{115}Gd in Gd) Wagner et al (44) (^{189}Os in Re) and Boolchand et al (45) (^{178}Hf in Hf).

The very small width of the ^{181}Ta line, and the large quadrupole moments of the nucleus in the ground and excited states permit very precise measurement of the efg in metals. Kaindl and Salomon (46) have measured the efg at ^{181}Ta in Re, Os, Hf, Ru.

Using these and other data mainly from spin precession of excited states, Raghavan et al (47,48) elaborated the correlation noticed by Sawicki (40) into a universal correlation of electronic and ionic field gradients in non-cubic metals. This correlation is not predicted by any current models of the electric field gradient in non-cubic metals and it appears that some important factor is being left out of consideration.

The temperature dependence of the electric field gradient at Fe nuclei in Be has been investigated by Janot et al (49) and Piecuch and Janot (50). They interpreted the temperature dependence as a consequence of the dependence of the local density of 3d states at the Fermi level, through the changing occupation of virtual bound states on the impurity atom (37).

2.2.4 Efg in alloys

In alloys in which atoms of different elements are distributed on a cubic space lattice, the local symmetry will normally be less than cubic so that quadrupole splittings of the Mossbauer spectrum will be expected. Preston et al (51) investigated the V-Fe system for iron concentrations between 0 and 50%. They show that over the concentration range 15-50%, the spectra are symmetrical, and that over the entire range, the isomer space shift is nearly constant. The width of the spectra rises to a maximum at a concentration of 25%, falls slightly at 36% and then rises rapidly. Preston et al point out that if the efg were produced by random numbers of V and Fe atoms in the first neighbour shell, a correlation between isomer shift and efg would be expected, leading to asymmetric spectra. Moreover the width of the spectra would have its maximum at 50%. Preston et al conclude that all the first neighbours of iron are V atoms indicating a CsCl-like ordering. The efg is then due to disorder in the second neighbour shell, which will have its maximum at 25% Fe. The quadrupole coupling produced by 1 Fe atom in the second neighbour shell is found to be 0.20 mm sec^{-1}. The rise in width above 36% Fe was attributed to the onset of magnetic order.

Large quadrupole couplings, of the order of .5 mm sec^{-1} were observed at ^{57}Fe nuclei in gold iron alloys by Ridout (52) and at ^{57}Fe used as a probe in AuX alloys, where X was Ti, V, Cr, Mn, Fe, Co, Ni, Pd, Pt and Sn, by Window (53). The magnitude of the efg correlates well with the residual resistivity, suggesting the virtual bound state model of Friedel (54) which accounts well for the magnitude of the efg. Window shows that on a Thomas Fermi model for the screening of the iron charge, about 12% of the screening charge lies outside the atomic shell because of the low density of states at the Fermi level in the gold matrix. The approach of another impurity atom causes the density of states to increase, so that the screening length in the direction of the impurity decreases. This model (55) also accounts for the reduction in charge to be screened in clusters of iron atoms, as observed in low temperature spectra. (See 4.1.2)

The quadrupole coupling at iron nuclei in atoms which are neighbours to impurity atoms can be readily observed if single crystal specimens are available. The method makes use of the fact that in a BCC lattice the first neighbours of an atom make the angle $\cos^{-1}(1/\sqrt{3})$ with the [100] direction, and the second neighbours make the same angle with the [111] direction. If an applied magnetic field is used to turn the magnetization along the [100] and [111] directions, quadrupole couplings from the 1st and 2nd neighbour shell respectively vanish. Line broadenings can be observed due to first neighbour atoms when the magnetization lies along the [111] direction, and due to second neighbour atoms when the magnetization lies along the [100] direction. The method is described in greater detail by Cranshaw (56) though it must be noted that subsequent work has shown the presence of Si impurities in the FeCr alloys used in that work, and some of the conclusions drawn there are invalid.

From the spectra, it is possible to derive q_i the quadrupole coupling, μ_i, the pseudodipole moment, ΔH_i and ΔS_i, the change in hyperfine field and isomer shift for i = 1 and 2 corresponding to first and second neighbour iron atoms of the solute atom respectively. The results for sp elements of the third and fourth period are shown in Figures 7(a) and (b), and 8(a) and (b).

As we have mentioned in 2.2.2, on a simple band model we expect a correlation between the electric field gradient and the dipolar field observed at the iron nucleus. This correlation appears to exist at the second neighbours of s-p impurities of the fourth period. Table 3 gives the values of the dipolar field calculated from the values of e^2qQ, and

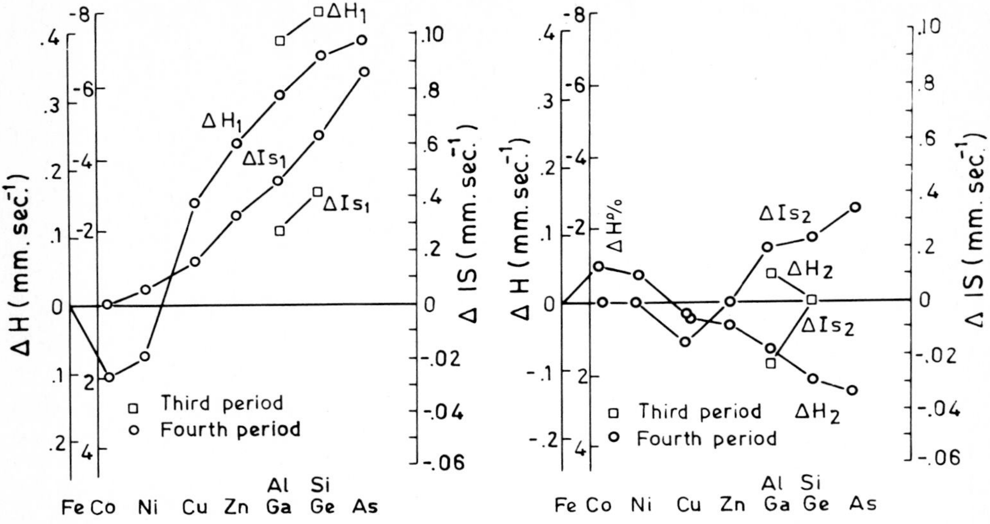

Fig. 7(a) and 7(b). The values of ΔH_1, ΔIs_1 and ΔH_2, ΔIs_2 the changes in hyperfine fields and isomer shifts at first and second neighbours of elements of the third and fourth periods.

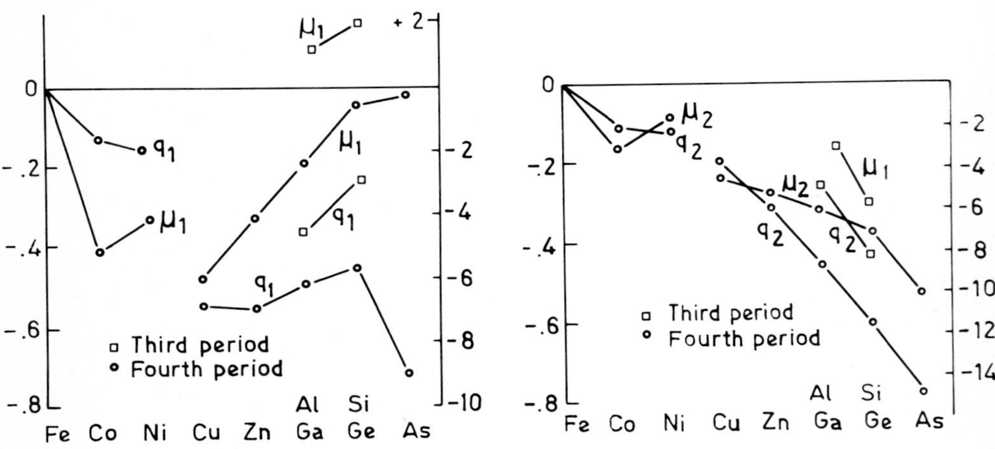

Fig. 8(a) and 8(b). The values of μ_1, q_1 and μ_2, q_2, the magnetic dipole and electric quadrupole interactions at first and second neighbours of elements of the third and fourth period. The dipolar interaction is given in terms of the equivalent dipole at the solute site, and the quadrupole interaction in velocity units.

the measured dipolar field, taking into account the defect dipolar field. The behaviour of μ_1 and q_1 seem to be much more erratic, and suggests that size mismatch may have an important influence.

TABLE 3

The dipolar field at iron atoms which are second neighbours of solute atoms, as observed, and as calculated from the observed electric field gradients using the model described in the text.

	Cu	Zu	Ga	Ge	As	Al	Si
Δ Z	3	4	5	6	7		
Δ H calc, kOe	3.0	4.7	6.8	9.0	11.6	4.4	6.6
Δ H obs, kOe	3.1	4.1	5.1	6.4	10.1	1.0	4.6

2.3 Magnetic hyperfine interaction

2.3.1 α iron

Among the first papers showing the usefulness of Mossbauer spectroscopy in the study of metals and magnetism in metals were those by Hanna et al (57), Nagle et al (58) and Preston et al (59) wherein it was shown that the hyperfine field at an iron nucleus in natural iron was -330 kOe, and that as a function of temperature its relative value was nearly the same as the reduced magnetization.

2.3.2 Fe X alloys

The studies of iron were closely followed by investigations of iron alloys (60-62). It was found that, in general, the mean hyperfine field was related to the magnetization in such a way that

$$H_{eff} \sim 120\ \mu_{Fe} + 30\ \bar{\mu}$$

where μ_{Fe} is the moment on the iron atom, and $\bar{\mu}$ the mean moment per atom. This kind of empirical expression has been found useful in a great deal of later work, though a firm theoretical basis is still lacking.

More detailed work on iron alloys showed that the spectra could be analysed into a number of components, attributed to iron atoms in different neighbour configurations (63,64). Unfortunately, the components are not resolved, and much effort has been devoted to decomposing the spectra and attributing the changes in field and isomer shift to the effects of solute atoms in different neighbour shells. It is usual to assume that the effects of atoms in the surrounding shells are simply additive, both with respect to hyperfine field and isomer shift, and often it is assumed that the alloy is a random solution.

In the paper of Wertheim et al (63) the effects of solute atoms of Ti, V, Cr, Mn, Ry, Al, Ga and Sn were studied. The hyperfine field was described by an expression of the form

$$H(m,n) = Ho(1 + an + bm)(1 + kc)$$

Here m,n are the number of first and second neighbour solute atoms, a and b are constants depending on the solute, and c is the concentratio k may be regarded as measuring a "band effect", or the unresolved effect of neighbours more distant than the second. It was found that s.p. elements produce a positive isomer shift, whereas transition elements to the left of iron produce a negative isomer shift. The magnitude of the isomer shift corresponds to a change of d-electron density not larger than 0.05 electron, whereas the change in magnetic hyperfine interaction corresponds to $\sim$ 0.2 3d electrons. "Charge contrast" is not a useful parameter since both transition elements and s,p elements cause a reduction of hyperfine field of about 6-7%.

Later work concentrated on reconciling the results from Mossbauer spectra with the results from NMR spectra where the precision and resolution are much higher. (See for example ref. 65). It became recognized that the alloys are not necessarily random, so that the intensity of a component is not an infallible guide to its coordinatior number, and that electric quadrupole and magnetic dipole fields exist a each neighbour shell which enter differently in the Mossbauer and NMR spectra (64, 66-69). At the present time, the changes in isomer shift and hyperfine fields at the first and second neighbours of a number of transition and s,p elements are reasonably well known. (34, 63, 69-72).

Although for s.p. elements at least, regularities as a function of series and period are beginning to be apparent (34, 70, 71) the interpretation of these results is still uncertain. The regularities are consistent with the interpretation in terms of charge screening processes as discussed by Friedel (54) and elaborated by Campbell and Blandin (73), Kolk (74). The impurity potential creates an oscillatory disturbance in the conduction electron polarization as described by Rudermann and Kittel (75), Kasuya (76) and Yoshida (77) which influence the value of the hyperfine field. Close to the impurity atoms there may be other effects due to overlap or size mismatch. Systematic trends and their interpretation for transition metal impurities have been discusse by Vincze and Campbell (69).

The elements Si and Al have been particularly studied because magnetization measurements and neutron diffuse scattering have indicate that these elements may make a particularly simple disturbance, namely a magnetic "hole". The saturation magnetization as a function of concentration indicates simple dilution, and the neighbouring iron atom appear to carry an unchanged moment. Over the concentration range 0 - 25%, the changes in field produced by a silicon atom at its neighbours has been shown to be remarkably constant (34, 64, 78). It is then tempting to regard the disturbance produced by such atoms as the simple inverse of the influence of the missing iron atom (64).

2.3.3 X Fe alloys

Mossbauer spectroscopy has also been very useful in the systematic measurement of "impurity" hyperfine fields in magnetic materials (79), where it has contributed about as many measurements as perturbed angular correlation methods, and in the study of Heusler alloys.

2.3.4 Heusler alloys

Heusler alloys are ferromagnetic intermetallic compounds, usually containing manganese with a magnetization of about 4 Bohr magnetons per molecule. They order in either the $L2_1$ structure, with composition X_2MnY or C1b structure with composition XMnY. Y may be any of a wide range of s.p. elements, and in particular Sn and Sb. The X site is commonly Cu, Co, Ni, Pd etc. The large separation of the Mn atoms (or other moment carrying atoms) indicates that ferromagnetism arises from indirect, conduction electron interaction rather than overlap of d-electrons.

Campbell (80) measured the hyperfine field at ^{119}Sn doped in a range of Heusler alloys, and also at ^{121}Sb in alloys when Sb was the Y element. Combining this data with results from earlier work, he was able to draw some conclusions about Heusler alloy systematics. The hyperfine field at s.p. elements in Heusler alloys follows the same trend as the same elements in other magnetic systems, becoming more positive as Z increases, and systems with similar outer electron configurations exhibit exhibit similar features.

Leiper et al (81) measured the hyperfine fields at Au and Sb sites in the Heusler alloy Au Mn Sb. The ratio H_{Au}/H_{Sb} determined experimentally could be compared with the predictions of various models proposed for the alloy, with fewer unknown parameters than are required for the prediction of the absolute values. They concluded that although the models could correctly predict the trends in hyperfine fields, they are incapable of yielding reliable specific values.

The magnetic coupling between moments in Pd_2 Mn In_{1-x} Sn_x $(0<x<1.0)$ and Pd Mn In_{1-y} Sb_y $(0.1<y<0.5)$ alloys has been investigated by Rush et al (82). These authors confirm the results of Webster and Ramadan (83) that in these series of alloys there are ferromagnetic and antiferromagnetic phases depending on the values of x and y, and the boundaries occur at the same values of electron density. Evidence is given that the proportions of magnetic and antiferromagnetic phases present are due not to gross segregation of indium or tin rich phases, but to a sensitive dependence of magnetic coupling on local electron density.

On the same samples, Rush et al show that the electron spin density changes are approximately the same as the changes in charge density (deduced from isomer shift measurements), and that the form of the changes is nearly the same function of electron density.

2.3.5 Site preferences in Heusler alloys

MacKay et al (84) have studied the results of doping ^{57}Fe and ^{57}Co into Heusler alloys CO_2 Mn X, where X = Ga, Ge, Si, Sn. They were able to show that ^{57}Fe sits preferentially on "B" sites, i.e. Mn sites, where it has 8 Co atoms in the first neighbour shell, 6 X atoms in the second neighbour shell and 12 Mn atoms in the third neighbour shell. The absence of any concentration dependence of the hyperfine field ($\sim$ - 310 KOe) is taken to indicate that the moments in the third neighbour shell have no influence on the hyperfine field. The ^{57}Fe hyperfine field at Co sites obtained by doping with ^{57}Co were found to be - 115 kOe for X = Si, Ge and Sn and - 75 kOe for X = Ga. For X = Si, Ge and Sn the moment on the Mn site is 3.6 μ_B, whereas for X = Ga, the manganese moment is 3 μ_B. The authors conclude that the changes in hyperfine field at iron sites can be understood in terms of a local contribution from core and 4s polarization and a transferred field from first neighbours.

2.3.6 The magnetism of chromium

An interesting example of the potential of Mossbauer spectroscopy is the investigation of the magnetism of chromium. Neutron diffraction results indicated the existence of spin density waves incommensurate with the metal lattice. The distribution of hyperfine fields seen on lattice sites is then given by

$$P(H) = (Ho^2 - H^2)^{-\frac{1}{2}} \qquad - Ho < H < Ho$$

Street and Window (85) prepared samples of annealed Cr containing ^{119}Sn source, and found a distribution of hyperfine fields in very good agreement with the expectation. In a later investigation, Window (86) examined unannealed powder specimens of Cr containing ^{119}Sn and also with V, Mo and Mn additions. Vanadium and molybdenum left the spin density wave relatively unperturbed, but manganese gave a distribution of H more like that of a commensurate spin wave. ^{197}Au gave a spectrum compatible with an incommensurate spin density wave, but with insufficient resolution for firm conclusions to be drawn.

2.3.7 The temperature dependence of the hyperfine field

The temperature dependence of the iron hyperfine field in pure iron was shown to be closely the same as that of the magnetization (59) and the same result was obtained for iron remote from solute atoms in alloys (11). Since the iron hyperfine field is believed to be mainly due to core polarization by the moment on the iron atom, it is natural to expect that the hyperfine field would be proportional to the expectation value of the moment, and therefore the magnetization. A sharp departure from this behaviour was observed by NMR methods for ^{55}Mn in iron by Koi et al (87). The ^{55}Mn resonance frequency was found to fall much more rapidly with temperature than the magnetization, and had fallen by 40% at T/Tc $\sim$ .6. A molecular field model of this behaviour was given by Jaccarino et al (88). According to these authors, the manganese is assumed to carry a moment which is relatively weakly coupled to the iron lattice magnetization, so that it behaves rather like a paramagnet in the molecular field of the iron. Then for the hyperfine field at temperature T

$$H_T^{Mn} = B_s \{g\beta S/kT.H_o\ \sigma_T/\sigma_o\} H_o^{Mn}$$

Here B_s is the Brillouin function for spin S, H_o is the molecular field at the manganese at T = o, σ_T is the magnetization of iron at temperature T. This may be rewritten

$$H_T^{Mn} = B_s \{\xi\ T_c/T\ \sigma_T/\sigma_o\} H_o^{Mn}$$

and families of curves for different values of s and ξ drawn. The expression fits the data of Koi et al very well for S = 3/2 and Ho = 3.7 10^6 Oe.

Another system well described by this model is that of ^{57}Fe dissolved dilutely in nickel (89), but in this case, the coupling between Fe and Ni is stronger than the coupling between Ni and Ni, and Dash et al find ξ = 2.8.

The temperature dependence of hyperfine field at iron neighbours of X solute atoms in Fe X alloys has been investigated by Cranshaw et al (90), (X =Mn), Schurer et al (91) (X = Mn and Si), van der Woude and Sawatzky (11) (X = Si, Mn, Al, Ti, V, Cr, Co, Ni), Vincze (92) (X = Mo, Ru, Rh, Pd, W, Re, Os, Ir, Pt) Vincze and Gruner (93) (X = Ti, V,

Cr, Mn, Co & Ni), Vincze and Cser (94) (X = Al, Ga), Vincze and Campbell (69) (X = Mn, Ru, Os, Ni, Pt and Pd), Vincze and Campbell (95) (X = Si, Rh).

In general the results of these investigatinns show that most impurity atoms behave "normally", i.e. the temperature dependence of the neighbours of a solute atom is very nearly the same as that of an isolated iron atom. A few solute atoms behave "anomalously", and examples are Mn, Ru, Os, Ni, Pt and Pd. In these cases, the hyperfine field at iron neighbours of a solute falls noticeably more rapidly than that at isolated iron atom.

Two distinct approaches to these observations can be discerned. In the first, (11, 90, 91), attention is concentrated on the value of hyperfine field as an indicator of <S>. Exchange couplings, four in the case of van der Woude and Sawatzky, are then introduced to account for the observed temperature dependences. This leads to the conclusion that most d-electrons are localized on the iron site, but the relatively few itinerant electrons, perhaps 10% of the whole provide 90% of the ferromagnetic coupling. In the second approach, (69, 92) attention is directed to the fact that the hyperfine field at the impurity site and the neighbour site, may have several components, and in particular, one due to core polarization and one due to conduction electron polarization. The "anomalies" are then seen as relatively small changes in one or other of these components. Campbell (96) has interpreted the ^{55}Mn in Fe temperature dependence in terms of a resonance in the local density of states at the impurity site. As the Fermi level falls with an increase in temperature, the manganese moment falls and causes the anomalies in both the manganese and iron neighbour hyperfine fields. Campbell (96) has had some success in predicting other instances of temperature anomalies.

The temperature dependence of the hyperfield field at ^{119}Sn in X Sn alloys where X is a ferromagnetic host has been studied in several works. (Jain and Cranshaw (97) (X = Co), Cranshaw (98) (X = Co), Huffman et al (99), (X = Fe, Ni), Balabanov and Delyagin (100) (X = Fe, Ni), Huffman and Dunmyre (101) (X = Co, Co Ni), Huffman and Dunmyre (X = Ni Fe), Huffman and Dunmyre (103) (X = Ni_3Fe), Price (104) (X = Fe))

Most of these alloys exhibit "anomalous" temperature dependence of the ^{119}Sn hyperfine field. Perhaps the most striking are Co Sn, where the field falls to zero, and changes sign at about 750 K, and Co_6Ni_4-(Sn) and Co_4Ni_6(Sn) where the magnitude of the hyperfine field increases to a maximum at T/Tc = .6, at which temperature the field is about 25% larger than at T = 0.

Models which have been put forward to account for these anomalies fall into three groups, finding the origin of the behaviour in (i) the orientation of a localized moment at or near the nuclear site, (101) (ii) a local dependence of magnetization different from that of the bulk (100) (iii) a strong distance dependence of the transferred components of the hyperfine field (98, 105). In the very detailed work of Price (104) on ^{119}Sn in Fe an allowance for the expansion of the iron lattice was made by using the measured pressure dependence of the hyperfine field at Sn in Fe. Price concluded that a discrepancy remained.

3. THE STUDY OF ORDERING AND SITE PREFERENCES

3.1 Detection of ordering

3.1.1 RhFe

The sensitivity of Mossbauer spectra to the immediate environment of the atom containing the Mossbauer nucleus, as seen for example in the work of Wertheim et al (63) encouraged the hope that Mossbauer spectra would provide a useful tool for the investigation of ordering in alloys. An early successful example was the work of Shirane et al (106) on RhFe alloys. In this system Cs Cl ordering occurs for Rh concentrations between 20% and 50%, leading to the existence in the ordered phase of two iron sites, Fe I on iron sites and Fe II on rhodium sites. Thus rhodium sites. Thus rhodium sites are randomly occupied by Rh and Fe II atoms. The 8 first neighbours of Fe II atoms are Rh atoms, whereas the neighbours of Fe I atoms are partly Rh atoms and partly Fe II atoms. The environment of the Fe I atoms therefore resembles that of iron atoms in the disordered alloy, whereas the environment of Fe II atoms resembles that of pure iron.

Shirane et al found that the hyperfine fields for Fe I and Fe II sites were 277 kOe and 384 kOe respectively, and the components due to the two sites are therefore readily resolved in the spectra. As may be expected, the line width obtained for the two sites varies with concentration, reflecting a distribution of hyperfine fields, and reaches a maximum at about 30% Rh. The width of the distribution was .8 mm sec^{-1} for Fe I sites and .5 mm sec^{-1} for the Fe II sites at 30% Rh concentration, indicating the importance of the near neighbour environment.

3.1.2 Fe V

Measuring the quadrupole coupling and the isomer shift instead of the magnetic coupling, Preston et al (51) (see 2.2.4) also showed the existence of Cs Cl ordering in Fe V alloys. Another early example of the use of Mossbauer spectra to reveal atomic order is the work on Fe_3Al and Fe_3Si and similar compounds. In these latter cases, the disordered material shows six broad lines, whereas in the ordered state resolved components corresponding to the A & D sites are obtained.

Attempts to extend this type of work (107-109), to alloys in which the components are not well resolved run into difficulties. Schwartz and Asano (110) have described the main difficulties as (i) accounting for the anisotropy of the hyperfine interactions and (ii) accounting for thickness distortion, and they provide an illustration with spectra of single crystal specimens of Fe 3.2% Mo. The magnetization direction could be aligned along the [100] or [111] axis in order to measure the anisotropic contribution to the hyperfine interaction (66, 68, 90, 111). Schwartz and Asano (110) conclude that from the Mossbauer spectrum one cannot, for example, obtain directly the Warren short range order parameters, but one may obtain the numbers W (i,j), the fraction of iron atoms with i nearest and j next nearest neighbour atoms and use these as a check on short range order measurements by other methods.

3.1.3 FeSi

An unequivocal indication that for Si in Fe, W(i,j) = 0 for both i and j ≠ o was obtained by Cranshaw et al (66, 90). In this particular case, a Si atom makes a large anisotropic interaction with iron atoms in its second neighbour shell, which can be readily detected in single crystal specimens, if spectra are taken with the magnetization lying along the [100] and [111] directions. It is easily seen by visual inspection of the spectra that the large second neighbour interaction is absent if i ≠ o. This type of ordering is, of course, consistent with the D03 ordering of Fe_3Si, but is here shown to exist at concentrations of less than 5%.

It was shown (112) that the intensity of the components of a spectrum of an iron 6.25% silicon single crystal could be accounted for if all the silicon atoms were confined to one FCC sublattice of the BCC iron lattice, and that some ordering of the Cu_3Au type occurred on this sublattice. Perfect order on this sublattice would imply the existence of $Fe_{15}Si$ ordering, which had been previously postulated by Papadimitriou and Genin (113) to account for a spectrum which they had obtained of an Fe 7% Si specimen. It is possible to suppose that the introduction of a Si atom into the iron lattice makes a disturbance such that other Si atoms find a repulsive potential at the first, second and fourth sites, as is required by the $Fe_{15}Si$ ordering. Cser and Naszodi (114) seek such a potential in the overlap of the charge disturbances, as measured by the isomer shift. On the other hand, Cranshaw (112) points out that the "allowed" positions of Si atoms round an iron atom subtend the tetrahedral angle, 110°, at the iron atom, and suggests that the disturbance of d electrons on the iron atom may be responsible for making the other configurations "forbidden". In any case, it is interesting to note that we have here a microscopic mechanism for bringing about the DO3 order found in several iron alloys which operates, in principle, in infinite dilution, contrasted with more usual mechanisms which depend on electron concentration or size effects.

3.2 Mainly iron-nickel alloys

3.2.1 Methods of detection of ordering

Discussing the Mossbauer spectra of Invar alloys, Window (115) pointed out that the anisotropic contributions to the hyperfine interaction could be detected even in polycrystalline samples. This is because of the different ways in which the electric quadrupole interaction and the magnetic dipole interaction shift the magnetic sub levels of the ^{57}Fe excited state. It can be seen that if the electric field gradient and the magnetic dipole interactions both act together to shift the m = 3/2 level, then they act in opposition on the 1/2 level. Hence in this case, lines 1 and 5 are broad, but 2 and 6 are narrow. If they act in opposition on the 3/2 level, lines 1 and 5 are narrow, but 2 and 6 are broad. Thus an asymmetry is introduced into the spectra.

Gros and Pauleve (116) applied these considerations to 50-50 FeNi alloys, which had been shown by Pauleve et al (117) to order in the CuAu structure with alternating planes of iron and nickel. They were able to show that the alloy consisted of two crystallites of types 1 and 2 with the quaternary axis parallel to, or perpendicular to, the magnetization. The different quadrupolar and dipolar couplings produced two resolved 6 line spectra.

The ordering of NiFe in josephinite in iron meteorites has been detected by Mossbauer spectroscopy (118). The ordering is presumed to have occurred during the very slow cooling of the meteorites (1° C in 10^{6} years).

Billard and Chamberod (119) treated the cases of ordered and disordered alloys containing 30 - 50% Ni. They found large dipolar contributions of about 6.5 kOe and quadrupole splittings of .23 mm sec^{-1}. They also showed that computation could be shortened by using Lorentz lines with a Gaussian distribution of energy with a suitably chosen width to replace the discrete spectra of the components of the alloy.

Hesse and Müller (120) used the model of Billard and Chamberod to fit the asymmetric spectra of Fe 41.3% Ni invar alloy in the range 77 K to the Curie point. Each spectrum yielded four parameters, H_o, the mean hyperfine field, H_1, the difference in field produced by the replacemen of a Ni atom by Fe in the first neighbour shell, h_2, the anisotropic dipolar effect, and w, the strength of the quadrupolar effect of such a replacement. H_o and h_2 as a function of temperature vary in a similar way to the magnetization. h_1 rises rapidly as the Curie temperature is reached, presumably because of the indirect effect of changed exchange integrals. The behaviour of w is striking, and is shown in fig. 9. Hesse and Müller point out that the thermal expansion coefficient is th sum of a lattice term and a magnetic term, and the fact that they are nearly equal and opposite gives the invar properties. These two terms may affect the quadrupolar coupling differently. Nevertheless to accoun for the large changes observed, Hesse and Müller have to appeal to "enhancement mechanisms", such as "Sternheimer anti shielding", or the steep gradient in Friedel oscillations.

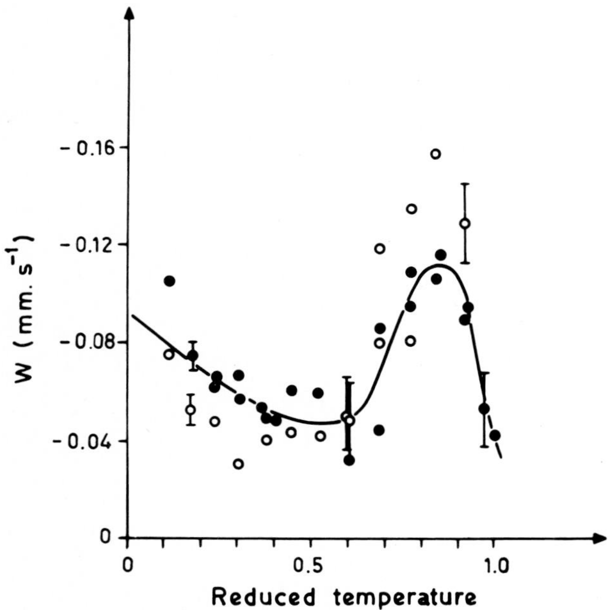

Fig. 9. The temperature dependence of W, the quadrupole interaction at ^{57}Fe nuclei in Fe 41.3 at % Ni Invar alloy.

The same method of investigation has been used by Bayreuther and Hoffman (121) in an attempt to detect the pair ordering assumed to be responsible for uniaxial magnetic anisotropy in NiFe and CoFe. Conversion electron spectra were taken of the specimens while the magnetization direction was switched periodically between the easy and hard axis. The concentration of ordered pairs is quite small and no effect was detected in the NiFe samples. Line displacements of the order of 5.10^{-5} were observed in CoFe, but in this case data for the quadrupolar and dipolar couplings are lacking, so that no quantitative comparison with the uniaxial anisotropy was possible.

The ordering of equiatomic FeCo has been studied by Montano and Seehra (122), and Eymery and Moine (123). The former authors found that the hyperfine field increases by about 3% when the material disorders. Measurements of the area of the absorption as a function of increasing temperature indicate a marked softening of the lattice at the transition temperature. Eymery and Moine produced alloys with different degrees of order by (i) quenching from temperatures below the ordering temperature, and (ii) annealing disordered material. They found asymmetric spectra at all degrees of order, and used the mean widths of the outer lines, the mean fields and the isomer shift as indicators of order. The hyperfine field was well represented by an expression of the form

$$H_{Fe} = a\mu_{Fe} + b_1\Sigma\mu_1$$

where a and b_1 are constants=70 kOe/μ_B and 8.4 kOe/μ_B respectively, μ_{Fe} is the moment on iron and μ_1 the moments on the first neighbour atoms. μ_{Co} is taken to be 1.8 μ_B and μ_{Fe} to be 3.15 μ_B. The long range order parameter η can now be deduced from measured hyperfine fields. The isomer shift is shown to be a linear function of η^2. By studying the time dependence of line width, isomer shift and hyperfine field when an initially disordered specimen is held at 480° C, the authors are able to show that the ordering proceeds homogeneously, rather than by a nucleation and growth mechanism.

3.2.2 Study of Ni_3Fe

A very beautiful example of the study of an order-disorder transition which exploits the characteristics of Mossbauer spectroscopy to the full is the work of Drijver et al (124) on Ni_3Fe. These authors took spectra of Ni_3Fe samples in an oven at temperatures between 77 K and the Curie point, 863 K. The strength of the hyperfine field and the line width were the indicators of order or disorder. Fig. 10 shows the result of hyperfine field measurements, each point corresponding to a spectrum taken during 1 day. The average field of the ordered phase decreases gradually with increasing temperature, and drops to the value of the

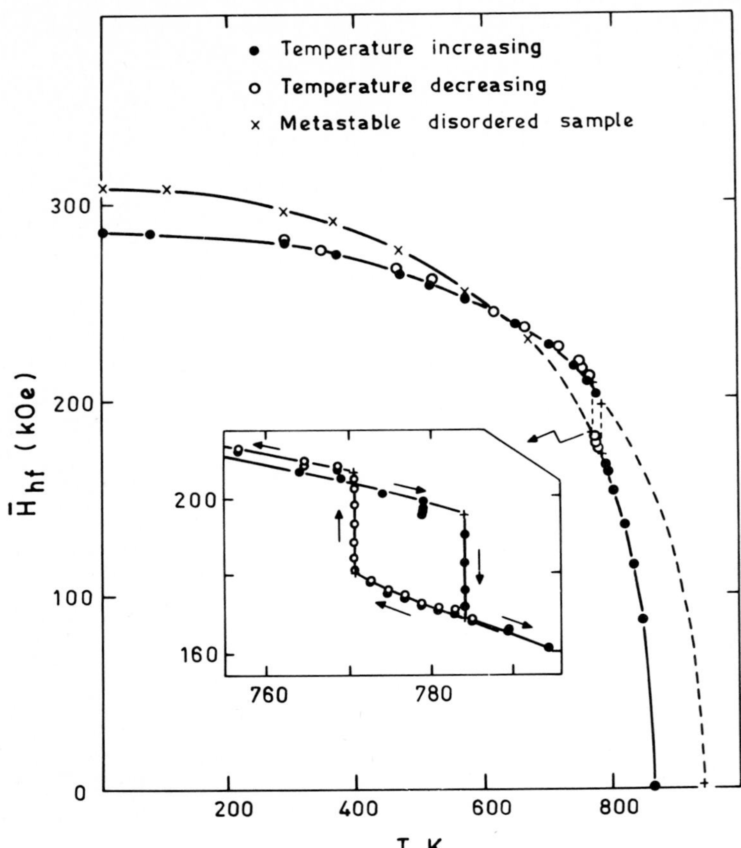

Fig. 10. Temperature dependence of the mean hyperfine field at ^{57}Fe *nuclei in* Ni_3Fe*. The inset shows the transition region in more detail. The estimated Curie temperature of the ordered phase is marked* +.

disordered phase at the transition temperature. Marked on the diagram is also the Curie point of the ordered alloy, 940 K, and the values of field found for a metastable disordered sample. On cooling from above the transition, ordering starts at a temperature about 10 K below the transition temperature for disordering. The step in hyperfine field at the transition is attributed mainly to the change in T_c.

Fig. 11 shows the time evolution of the mean field $\bar{H}_{hf}$ and the line width P at two temperatures in the transition region. Although the behaviour of $\bar{H}_{hf}$ with time could be accounted for by a nucleation and growth or homogeneous ordering models, the great width of the lines indicates a wide distribution of field strengths not consistent with homogeneous ordering. An even greater width is seen in the disordering spectra. At temperatures well outside the transition region, 758 K and

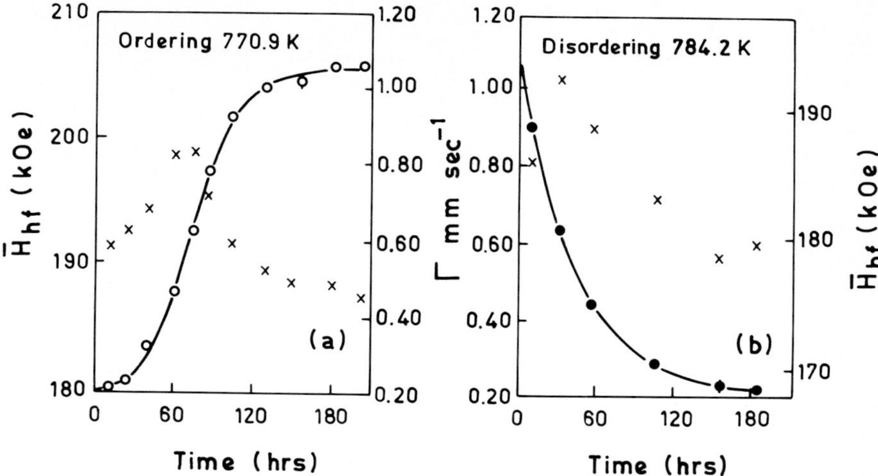

Fig. 11. The time evolution of the mean hyperfine field and the width of the field distribution for two fixed temperatures within the transition region.

793 K, no broadening is observed indicating homogeneous ordering in this case.

In an examination of the ordering of Pd_3Fe carried out by Tsurin and Men'shikov (125), spectra were taken of specimens in which the degree of order was measured by X-ray diffraction. The mean hyperfine field was shown to increase from 272 kOe to 294 kOe with increasing order and the line width fell from 1.0 to 0.8 mm sec^{-1}. In the study of ordering kinetics it was found that at 470° C a high degree of order was attained in a few hours though ordering was not complete at 60 hrs. and the partially ordered spectra could be fitted as the sum of ordered and disordered components. The authors conclude that the mechanism of ordering is predominantly heterogeneous.

3.3 Exploitation of specimens containing the source nuclei

3.3.1 FeCo

A particular property of Mossbauer spectroscopy, namely the possibility of carrying out source experiments where the source is a different chemical element from the Mossbauer isotope can be exploited to show the existence of ordering in some commercial steels containing cobalt (126). In this case, the recoil energy of the ^{57}Co decay can be assumed negligible compared with the binding energy, and the ^{57m}Fe atoms which generate the emission spectra sample the distribution of Co atoms. Belozerskii et al (126) find different field values for the emission and transmission spectra, indicating departures from a random distribution.

3.3.2 Sn intermetallic compounds

In ingenious experiments on SnSb and SnTe, Ambe and Ambe (127) make use of the recoil energy after radioactive decay to investigate the lattice positions of ^{119}Sn atoms. The ^{119}Sn atoms in their excited stat were produced by the following reactions.

$$^{119}Sb \xrightarrow{EC} {}^{119}Sn \quad (E_R << E_D) \qquad (i)$$

$$^{119m}Te \xrightarrow{EC} {}^{119}Sb \xrightarrow{EC} {}^{119}Sn \quad (E_R \approx E_D) \qquad (ii)$$

$$^{120}Sn\ (p,2n)\ {}^{119}Sb \xrightarrow{EC} {}^{119}Sn \quad (E_R >> E_D) \qquad (iii)$$

Here E_R is the recoil energy and E_D the energy required to displace an atom ($\sim$ 25 eV).

When $Sn^{119}Sb$ is used as a source, an emission line with an isomer shift 2.43 mm sec^{-1} relative to $BaSnO_3$ is found. The absorption line of ^{119}Sn in SnSb is at 2.79 mm sec^{-1}, so that the emission line at 2.43 mm sec^{-1} is attributed to ^{119}Sn on Sb sites.

When ^{119m}Te in SnTe is used as the source, two lines are observed with isomer shifts 2.24 and 3.3 mm sec^{-1}. The latter line can be attributed to ^{119}Sn on a Sn site, since the absorption line is found at 3.5 mm sec^{-1}. The former line must be attributed to ^{119}Sn on a Te site.

When ^{119}Sb is produced by the proton reaction, the emission line has an isomer shift 2.42 mm sec^{-1}, essentially the same as the source $Sn^{119}Sb$. i.e. the ^{119}Sb atoms were on Sb sites. However the emission spectrum of proton irradiated SnTe consists of two lines with shifts of 2.3 and 3.5 mm sec^{-1}. The latter line is attributed to ^{119}Sn on tin sites, and the former to ^{119}Sn on Te sites. The fact that two distinct sites are observed shows that the neighbourhood of the recoil ^{119}Sb atoms is not much disturbed.

3.4 Clustering

3.4.1 CuFe alloys

Mossbauer spectra provide a powerful means of investigating clustering and aggregation in alloys. Window et al (12) showed by the presence of quadrupole couplings and isomer shifts that iron dissolved in CuNi alloys was in a nickel rich environment, and that the nickel clusters could be partially broken up by cold work.

Very detailed work on the CuFe system has been produced by Window (128) Campbell et al (129) and Campbell and Clark (130). Window (128) prepared specimens of Cu containing 1% ^{57}Fe. In the quenched state, room temperature spectra show two components, one a single line with isomer shift .22 mm sec^{-1} with respect to α-iron, attributable to isolated iron atoms, and a doublet due to clusters of iron atoms. With ageing at 300° C for 0.6 hrs. and 1 hr. the single line is notably reduced, and the doublet becomes distorted. Window remarks that although the doublet is asymmetric, the areas under the two lines remain equal, indicating that it is made up of component doublets with different splittings and isomer shifts. In the work of Campbell et al (129) these components were found to be present with different intensities in spectra of CuFe alloys with the iron concentration going from .02 to 2%.

At 4.2 K, all specimens aged at 300°C for up to 15 hours ordered magnetically with a broad distribution of fields, which Window was able to analyse, giving a p(H) distribution. A peak at 14T could be assigned to iron with one iron neighbour, and at 20T and 21-25T to iron atoms with two and three neighbours.

After annealing at 350°C for 9 hours, very little iron is isolated from other iron atoms. As clusters grow, Window shows that even in clusters of 100 atoms, more than 80% of the atoms are surface atoms and therefore subject to large field gradients. After annealing at 500°C for 30 hours, a peak attributable to γ-iron is found.

From the spectra of cluster samples in applied fields, Window showed that in clusters of about 12 iron atoms, some iron atoms gave spectra showing only the applied field, implying that these atoms, presumably with several iron neighbours have no localized moment.

By exploiting the difference in magnetic properties of γ Fe, α Fe, and iron in solid solution, Campbell and Clark (130) showed that rapidly quenched samples (< 3 seconds) from the solid solution region, or in the as-rolled state do not contain γ Fe. γ Fe is observed in a 4.6% specimen slowly cooled (∿ 10 secs.). The iron in solution always exhibits clustering. Precipitation of γ Fe occurs during annealing at 650°C, and the γ Fe subsequently transforms to α Fe. Clark et al (131) have extended this work to the observation of samples aged at room temperature for ∿10^8 secs.They find that the fraction of isolated iron atoms decreases with ageing and the number of clusters of iron atoms increases. Assuming a a vacancy mechanism for the diffusion of iron in copper, they find a hopping rate of 7 x 10^{-10} sec^{-1}, and an activation energy Qm = 1.0 eV in good agreement with the value found by Mullen (132) for the temperature range 980 - 1300 K. Qm thus appears to be constant over the range

300 - 1300 K.

By implanting ^{57}Fe in copper, it is possible to exceed the equilibrium solubility by a large factor. Longworth and Jain (133) have studied samples prepared in this way as a function of dose, and as a function of annealing (134). In general, there could be recognized in the spectra components due to (i) isolated iron atoms, (ii) clusters of iron atoms, (iii) γ-iron precipitates. At a dose of 10^{15} atoms cm^{-2}, roughly half the iron atoms were isolated and half in clusters. At doses of 2.10^{15} to 5.10^{16}, γ-iron appears in the spectra. The number of isolated iron atoms is much less and the number of iron atoms in cluster or γ-iron much more than could be expected on a random distribution, showing the importance of the enhanced diffusion of iron during irradiation.

At the highest dose, 5.10^{16} atoms cm^{-2}, the retained dose is in fact $\sim 2.10^{16}$ atoms cm^{-2} because of sputtering of the surface. In this specimen, it is found that the number of clusters has fallen, and been presumably converted to γ-iron. However, there is an anomalously large fraction of iron with an anomalous shift in the isolated iron line. Longworth and Jain attribute this phenomenon to trapping of iron in vacancy clusters.

When implanted specimens are annealed, diffusion is found to start at temperatures of about 200°C. This is presumably due to the breakup of vacancy clusters. At higher temperatures, 300-350°C, vacancy formation occurs in copper, and the isolated iron atoms diffuse into the substrate. In implanted specimens of silver, Longworth and Jain (135) found small precipitates of superparamagnetic α-iron rather than γ-iron. The large mismatch in lattice parameters is thought to be responsible for the early conversion of clusters into α-iron precipitates.

3.4.2 Clustering in other systems

Clustering and precipitation are of course to be expected when specimens are supersaturated. Edwards et al (136) have demonstrated by Mossbauer spectroscopy of ^{119}Sn an interaction between Sn and Ni in iron at concentrations where solid solubility might be expected. Ageing at 500°C for 1000 hours produced a precipitate of Ni_3Sn_2 (see 5.1.2). ^{119}Sn spectra were taken and analysed as the superposition of 6 line spectra with different values of the hyperfine field. Values of H_{eff} less than 82 kOe, the field at Sn in pure iron, were attributed to Sn with a Ni atom in the first neighbour shell.

In a survey of Fe X Sn alloys with X = Al, Si, V, Cr, Mn, Ni, Cu, Ga, G, Mo, Ru, Rh, Pd, Os, Ir, Pt, Cranshaw (137) found the presence of a solute atom in the first neighbour shell made the hyperfine field more positive by about 30 kOe. The presence of a solute atom in the second neighbour shell appears to make the hyperfine field more negative by about the same amount. Exceptions to this rule are Co, Ni and Pd, where the change in field is about 20 kOe.

In the case of nickel and palladium, the number of Ni and Pd atoms in the first neighbour shell is much higher than the random expectation, indicating an attractive potential. In later unpublished work, Cranshaw found the potentials to be 0.28 eV and 0.1 eV in Pd and Ni respectively. In the case of s,p elements such as X = Al, Si, G, Ge a repulsive potential at both first and second neighbours was observed such that Sn X pairs are only observed in rolled specimens.

The magnitude of the field changes in these alloys is consonant with the empirical rule of Balabanov and Delyagin (100), and Delyagin and Kornienko (138) according to which the hyperfine field at Sn is given by

$$H_{eff} \sim - 140\ \mu_{nn} + 100\ \mu_{2n}$$

where μ_{nn}, μ_{2n} are the moments on the first and second neighbours of the tin atom.

Maurer et al (139) took Mossbauer spectra of ^{121}Sb in iron containing 1% Sb and 3% Ni or Cr. Quenched specimens show identical spectra, similar to the spectrum of Sb in pure iron. When the 3% Ni specimen is tempered at 500°C for 3 hours, 80% of the Sb is found to be in the form of a precipitate of $(Ni_{55-x}Fe_x)Sb_{45}$. Measurements of lattice spacing show that $x < 13$.

4. MAGNETIC IMPURITIES IN NON-MAGNETIC METALS

4.1 Magnetic properties

Certain interesting and unexpected macroscopic properties of solutions of magnetic elements in non-magnetic metals, such as the resistance minimum, and the logarithmic dependence of the resistivity on temperature have been well known since about 1950. The theoretical account of all these effects is still not complete, but in the last decade experimental methods sensitive to the state of the magnetic impurity itself or its surroundings have contributed greatly to the scope of the data. The value of Mossbauer spectroscopy in the study of these phenomena for iron impurities was recognised very early.

4.1.1 CuFe

Of all systems, most Mossbauer work has been done on CuFe. Early work was confused by the strong tendency of iron atoms to cluster, which has already been described. A significant and revealing experiment was carried out on CuFe specimens in high applied fields by Frankel et al (140).

The Mossbauer spectrum gives the value of the hyperfine field at the Fe nucleus, which may be taken as proportional to the thermal average of the moment, i.e.

$$H_{hf} = \alpha <\mu>$$

For a paramagnet,

$$<\mu> = \mu_o B_J \ (g\mu_B \ H_o/kT),$$

and therefore

$$H_{hf} = H_{sat} B_J \ (g\mu_B \ H_o/kT)$$

H_{sat} is the hyperfine field as kT → o where B_J is the Brillouin function for spin J. The plot of H_f as a function of H_o/T is shown in Fig. 12. It is clear that as T is lowered, the hyperfine field does saturate, but the magnitude of the field depends on H_o, and the saturation field increases as the applied field is increased. On the same figure is plotted the Brillouin function for J = 3/2, fitted to the data at high temperature.

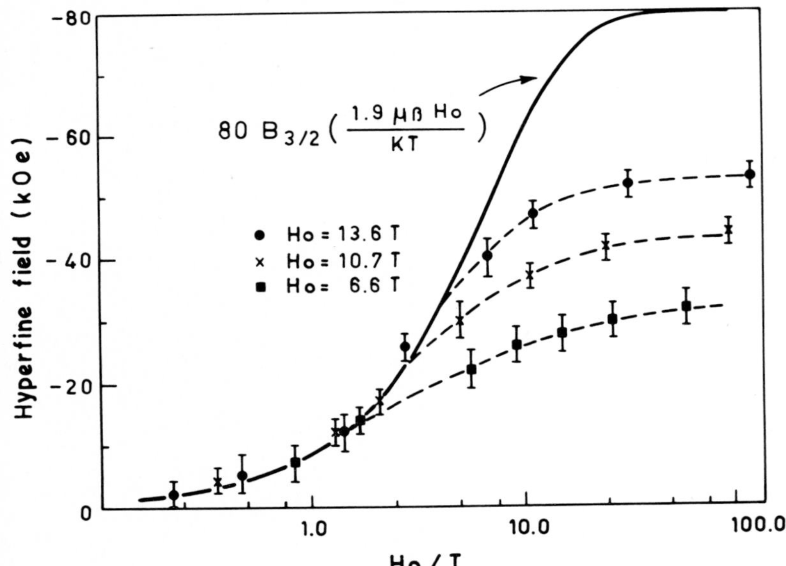

Fig. 12. H_{hf} *for* ^{57}Fe *in Cu plotted as a function of* H_o/T*, for three values of* H_o*. The solid curve is a Brillouin function for J = 3/2.*

Frankel et al introduce the idea of an effective moment μ_{eff}, which is less than μ, and is a function of H_o and T_1 i.e.

$H_{eff} = f(H_o, T)\mu^{\sigma}$

to be substituted in the above expression for $<\mu>$. By plotting H_{sat} as a function of H_o, and making a linear extrapolation, they find that the critical field where $f(H_o, T) = 1$ is about 235 kOe, well above available fields.

Later theoretical calculations (141, 142) showed that the temperature dependence of the low field susceptibility should show a Curie-Weiss behaviour, $(T+T_k)^{-1}$, where T_k is the Kondo temperature, in either the spin-fluctuation model or the spin-compensated state model. This suggests that we write for the expectation value of the spin.

$<\mu> = (\mu_o B_J g\mu_B H_o)\ K(T+T_k)$

The "local susceptibility", H/H_{app} can then be plotted as a function of $(T+T_k)^{-1}$, to obtain a straight line. The value of T_k is found to be $T_k = 27.6$ K iron in copper.

In some cases, the effect of the spin relaxation rate on the Mossbaur spectra can be observed. The spectra of CuFe alloys always show sharp spectra, indicating that relaxation is fast. In MoFe and RhFe on the other hand, broad spectra are observed, and explained as being due to a comparatively long relaxation time.

Other systems studied by Mossbauer spectroscopy are Fe in Au(143), W(144), Rh(12, 52, 144), Mo (12, 145) in which Kondo temperatures range from .4 K to 13 K.

In many of these investigations, difficulties in interpretation were introduced by interactions between the magnetic atoms, and usually attempts were made to extrapolate results to zero concentration.

Later work has concentrated on the properties of the interacting moments, in systems characterized by the terms spin-glass or mictomagnet. Let us consider the case of an element with a low Kondo temperature, i.e. we can neglect changes in the moment. Then at sufficiently low concentrations, we observe the behaviour of the single impurity. At higher concentrations, the moments interact, mainly by the RKKY interaction. When such an alloy is cooled, a temperature T_o is reached at which spins become locked in random directions. The low field susceptibility shows a marked peak, and Mossbauer spectra show a sharp transition temperature with a well defined hyperfine field. The relative internal field scales with the reduced temperature, and the relative intensities of the lines in the Mossbauer spectrum indicate a random distribution of direction of spins. This state is usually referred to as a "spin-glass".

At higher concentrations, the scaling with concentration fails because local interactions of pairs or larger clusters play a significant role. The "clusters" may be the product of metallurgical factors, such as solubility limits, or may arise simply as a result of the random distribution of solute atoms. The direction of magnetization of such clusters may vary in a super-paramagnetic fashion until the temperature is lowered to T_o, when the coupling of isolated moments by the RKKY interaction again locks the spins. This concentration regime is called "mictomagnetic".

At still higher concentrations, the "percolation limit" is approached, i.e. any moment can be regarded as coupled to any other by successive short range interactions. For the case of ferromagnetic coupling in an FCC lattice, this limit is reached at about 15%, and for antiferromagnetic coupling at about 45%.

4.1.2 AuFe

Although spin glass behaviour can be observed in CuFe, CuMnFe and other systems as described above, the best example for Mossbauer investigation is AuFe (146, 147).

The first papers described the spectra in broad outline, demonstrating that the system possessed a sharp transition temperature, and the reduced field followed a magnetization curve for S = 1 as a function of reduced temperature. The ratio of line intensities indicated a random distribution of spin directions. The line widths indicated a distribution of fields, not markedly dependent on concentration, and the critical temperature was proportional to the square root of the concentration.

More detailed work on this system was carried out by Ridout (52) and Window (55, 128). Window exploited the capacity of Mossbauer spectra to reveal quadrupole couplings and isomer shift, and showed that spectra of alloys containing 2-40% Fe at temperatures of 4.2 K and 300 K could be fitted with self consistent values of the hyperfine interactions. A Thomas-Fermi like model was used for the screening of iron in gold, and the derived parameters used for fitting the low temperature spectra. The sign of the e.f.g. was deduced to be negative. Evidence was obtained showing that the spins point along [111] crystallographic directions when the atom has two or more iron neighbours, and along directions normal to the iron-iron axis when it has one neighbour. As the size of clusters increases, the spins point along whichever [111] axis minimizes the number of Fe-Fe axes normal to it.

In a further paper, Window (148) elaborated the model proposed for Au-Fe alloys. The near neighbour exchange is assumed to have an isotropic component which aligns the spins parallel and an anisotropic part which gives preferred directions in the cluster. The clusters are coupled by an indirect further neighbour exchange. An important feature of this model is that metallurgical factors are not necessary to form the clusters, though in some cases such factors may be important.

The gold-iron alloys in the spin glass concentration regime (.49 and .76% Fe) have been examined with applied fields up to 60 kOe by Chandra and Ray (149). The effect of an applied field on the mean hyperfine field is small, indicating either that there are antiferromagnetically coupled clusters, or that the local anisotropy is large.

4.1.3 Comparison of Cu, Ag. Au matrices

The comparative behaviour of ^{57}Fe in Cu, Ag and Au has been beautifully demonstrated by Steiner et al (150). The saturation hyperfine fields extrapolated to T = 0 are shown in Fig. 13 as a function of external field. <u>Cu</u>Fe shows a nearly linear dependence of H_{sat} on H_{app}

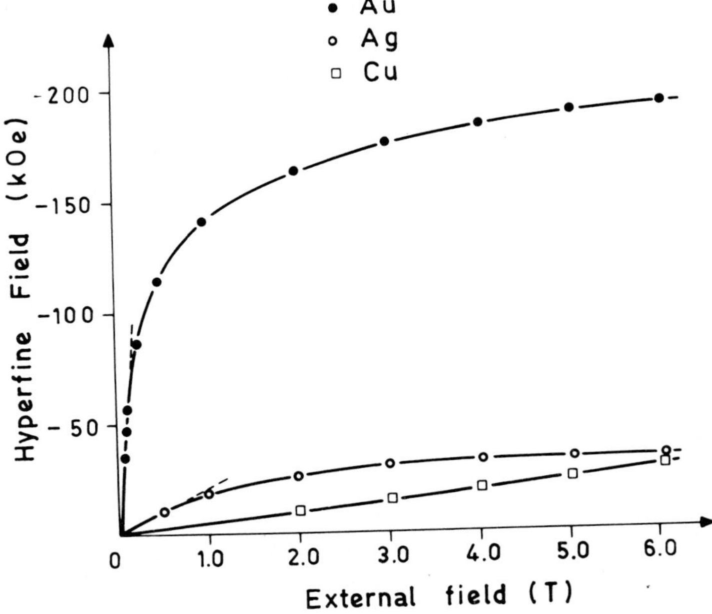

Fig. 13. Saturation hyperfine fields at nuclei of ^{57}Fe in Cu, Ag and Au for $T \rightarrow o$, as a function of external field.

up to 60 kOe because of the high Kondo temperature, ∿ 28 K. In AuFe on the contrary, the field is nearly saturated in 10 kOe. The saturation fields are also different, about 80 kOe in CuFe and 180 kOe in AuFe. In AgFe the saturation field is about 30 kOe smaller than either AuFe or CuFe, making determinations of susceptibility or Kondo temperature rather difficult. The local spin susceptibility for CuFe and AgFe was found to show Curie-Weiss behaviour at high temperatures. At low temperatures, ($T/T_k < .3$), the susceptibility shows a variation of the form

$$X(T)/X(0) = 1 - \alpha(T/T_k)^2,$$

in agreement with recent theory.

4.2 Enhanced systems

4.2.1 PdFe

A dilute magnetic system on which many Mossbauer investigations have been made is PdFe. Craig et al (151) and Maley et al (152) carried out measurements on alloys containing approximately 50 ppm Fe and external fields up to 35 kOe and 1000 ppm with fields up to 62 kOe respectively. Craig et al found no hyperfine splitting in zero applied field. The splitting increased rapidly with applied field, and reached a saturation value of -295 kOe, almost the same as the field in natural iron. A plot of reduced field against H/T could be fitted by a Brillouin curve for J = 6.5. The moment associated with the iron atom was then 12.5 μ_B. Maley et al found the best fit with values of J = 3.8 and g_{eff}= 2.9. The moment associated with one iron atom is then 11.1μ_B. Woodhams et al (153) obtained spectra of similar alloys, and were able to fit the observed magnetization with a molecular field model and J= 2.

Trousdale et al (154) observed spontaneous magnetization in alloys containing .22 to 13% Fe, and determined the dependence of the transition temperature on iron concentration, the distribution of magnetization revealed by the distribution of hyperfine fields, and the temperature dependence of the magnetization. They found that the width of the hyperfine field distribution as a function of T/T_c where T_c is the temperature corresponding to the onset of magnetization was a rapid function of concentration, being relatively broad for .2% and sharp for 7.5%. Trousdale et al interpreted this result in terms of an interaction range, σ, and found σ ∿ 6.5 Å.

4.2.2 Ni_3Ga matrix

An even stronger enhanced system similar to the above is

Ni_3Ga. Measurements of ^{57}Fe fields in this compound were made by Maletta (155), by diffusing ^{57}Co into Ni_3Ga containing 20 and 250 ppm of iron. He found that the hyperfine fields as a function of T and applied field could all be fitted with a Brillouin function, and determined the saturation field as 223 kOe in zero applied field. The magnetic moment associated with an iron atom was deduced to be 60 μ_B.

The system $Fe_xNi_{75-x}Ga_{25}$ for $x < 2\%$ has been studied by Hufner et al (156) and Steiner (150). They found two components of the hyperfine field with values extrapolated to zero concentration of 223 kOe and 200 kOe. It is reasonable to associate the 223 kOe field with iron on a Ga (A) site, where all its neighbours are Ni, and the 200 kOe field with iron on a Ni (B) site, where some of its neighbours are Ga. Further work along these lines has been reported by Drijver and van der Woude (157), who emphasize the importance of ordering in Ni_3Ga and Ni_3Al. They also point out that the A site is a site of cubic symmetry, whereas the B site is not and may show a quadrupole splitting. Moreover, the relative occupation of A and B sites is different for Co (sources) and Fe (absorbers). If there were no site preferences, n(A)/n(B) would equal 3. In the experiments, Co atoms have preference for A sites, whereas Fe atoms prefer the B sites.

4.3 Sensitivity of local moment formation to nearest neighbour environment

In the study of the formation of local moments, macroscopic measurements give information only of average moments. By taking Mossbauer spectra, one may hope to investigate distributions of local moments.

4.3.1 NbMo matrix

If the matrix is an alloy, the Anderson-Wolff band model for the appearance of a local moment would predict that the magnitude of the moment would be a continuous function of composition. For example, Fe and Co in Nb have no moment, but are magnetic in Mo. NMR experiments on Co in NbMo alloys showed two lines, corresponding to Co with and without moment. As the composition is varied, the relative intensities of the lines changes, but, in a given field, their frequency is constant. This agrees with the model of Jaccarino and Walker (158), according to which the Co has or has not a moment if it has more or less than a given number of Mo neighbours. Similar behaviour has been found in the case of ^{57}Fe in NbMo alloys by Nagasawa and Sakai (159), who deduce that Fe has a moment in the alloy when it has more than 7 Mo neighbours. However the behaviour is not simple, because Fe undergoes Kondo condensation in molybdenum (12).

4.3.2 VFe

The BCC system VFe has been studied by Shiga and Nakamura (160). The interpreted their spectra taken at 5 K by assuming a local moment on iron which is an approximately linear function of the number of iron neighbours, reaching 2.2 μ_B for 8 neighbours. In ordered alloys near 50% concentration, a marked reduction in hyperfine field was observed as the number of Fe neighbours fell.

5. METALLURGICAL APPLICATIONS

5.1 Phase analysis

When recognizable components exist in a complex Mossbauer spectrum, it is possible to deduce values for the fraction of the species in the specimen which produce these components. In many cases, the ratio of th areas under the components gives a sufficiently accurate value. If more accuracy is required, it is necessary to allow for thickness effects an differences in f, the recoilless fraction. Several methods of dealing with thickness effects have been described. (See, for example, 68, 161-165). The recoilless fraction depends upon $\langle x^2 \rangle$, the mean square value of the displacement of the Mossbauer nucleus from its mean position. Collins (166) has described a method of extrapolating the areas of components to $\langle x^2 \rangle = 0$ from spectra taken at a range of temperatures.

5.1.1 Miscibility gap and phase boundaries

A good illustration of the use of Mossbauer spectroscopy for phase analysis is the determination of the miscibility gap in the AuNi system by Howard et al (167). These authors first made calibration curves of isomer shift and Curie point against composition by quenching AuNi alloys doped with ^{57}Co from the high temperature single phase region. Alloys quenched from the two phase region gave spectra which could be analysed into two components, one gold-rich and the other nickel-rich. The composition of these phases could be determined by reference to the calibration curves, and the miscibility gap drawn out. Very good agreement with the usually accepted results of Hansen (168) were obtained.

Blasius and Gonser (169) show that in some circumstances, Mossbauer spectra can provide a very precise determination of phase boundaries, and illustrate the point with the case of the solution of iron in titanium. The solubility of iron in titanium in the temperature region

580 - 800°C is low, and in the solution α phase is in equilibrium with a phase β, Fig. 14. Then if N_α, N_β are the numbers of Fe atoms in the respective phases.

$$N_\beta/N_\alpha = C_\beta/C_\alpha \; (Co - C_\alpha)/(C_\beta - Co)$$

where C_α, C_β are the concentrations of iron atoms in the phases and Co is the average concentration, or the concentration in a high temperature, single phase region. Blasius and Gonser point out that although C_β and C_α may be widely different, N_β and N_α may be similar, so that provided that the spectra of the components α and β can be resolved, N_β/N_α can be determined very accurately. They determine the solubility of iron in titanium in the above temperature region and its temperature dependence. The solubility is of the order of 0.1%, determined with a precision ± .01%.

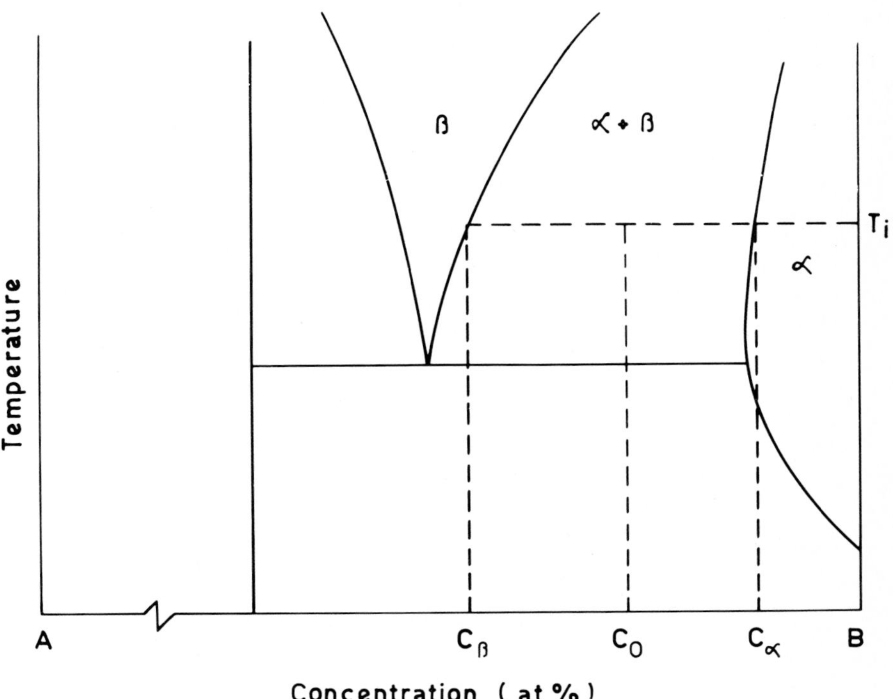

Fig. 14. Phase diagram of AB alloy

5.1.2 Clustering and precipitation

The observation of clustering in ternary alloys based on iron has been mentioned in 3.4.2. The observation of precipitates of Ni_3Sn_2

has been reported by Edwards et al (136), and of NiSb precipitates by Maurer et al (139). More work on these and similar systems will undoubtedly broaden our knowledge of their phase diagrams.

5.1.3 Grain boundary segregation

The state of tin atoms segregated at grain boundaries was investigated by Ozawa and Ishida (170, 171) who deposited ^{119m}Sn on the surface of pure iron and iron alloys prepared in a fine grained state by rapid quenching from high temperatures. After heating at $400^{o}C$, the tin was shown by autoradiography to have diffused along grain boundaries. In pure iron, and in alloys containing 3 wt % Mn and 1 wt % Ti, the Mossbauer spectrum of the tin showed a single broad line, which did not split even at 4 K. In an alloy containing 1% Ni, a small splitting was observed which the authors attribute to magnetic splitting.

The isomer shift of tin atoms in the boundary is more positive than that of the corresponding solid solution tin, and the recoilless fraction less, in conformity with the empirical rule noted by Brykhanov et al (172). Above 630 K, Ozawa and Ishida observed line broadening due to diffusion of tin atoms at the grain boundary. They suggest that grain boundary embrittlement may be caused by the weak binding of tin atoms and the anomalous state of d-electrons on neighbouring iron atoms indicated by the absence of magnetic splitting at the tin nucleus.

5.2 The determination of austenite : martensite ratios

5.2.1 Austenite:martensite ratio in steels

There is an extensive literature coverning the application of Mossbauer spectroscopy to the determination of the austenite : martensite ratio in steels. Despite early doubts raised by the complexity of the spectra of iron carbon steels, (173), much successful work has been accomplished. Using scattering geometry and 4610 steels in which the fraction of austenite produced by varying heat treatments lay between 5 and 39%, Chow et al (174) showed that the ratio of the areas under the magnetic and nonmagnetic components gave a good value for the ratio of martensite to austenite. Abe and Schwartz (175) examined powders of $Fe_{.73}Ni_{.27}$ by transmission spectroscopy, and showed that the accuracy in thin samples was comparable to that obtained with X-rays, but that corrections for thickness could be important. They also observed a surface contamination attributable to FeO.

5.2.2 Fracture investigation

In an interesting investigation of the fracture toughness of Fe 9%

Ni .1% C steel, Kim and Schwartz (176) used Mossbauer spectroscopy to determine the retained austenite. They were able to show that in the fracture surfaces the austenite present in the underformed material had been transformed into martensite by a localized mechanically induced phase transformation. By electropolishing the surface, the transformation of austenite was found to have occurred in a layer 300 μm deep. The improved fracture toughness of the steel may be attributed to the energy dissipated in the martensite phase transformation.

5.2.3 Study of lubricated sliding surfaces

Conversion electron spectroscopy was found to provide a powerful method of examining wear surfaces produced in a David Brown Two Disc machine by Cranshaw and Campany (177). In this work it was found to be possible to estimate the fractions of martensite, carbide and austenite in the wear surfaces, and in many cases to estimate the amount of dissolved carbon in the austenite component. It was shown that the first stage of breakdown of a wear surface was the production of bright polished surfaces, called "scored" surfaces consisting mainly of iron carbide. The "scuffed" surfaces characterized by gross damage and local welds between the sliding surfaces consisted of a mixture of austenite and martensite. The high percentage of carbon found in the scored provides some evidence of breakdown of the lubricant.

5.2.4 Rapid determination of austenite : martensite ratios

A method of determining the austenite : martensite ratio at great speed by utilizing different Mossbauer spectra of the components has been described by Jaggi and Rao (178). Briefly the method requires the use of two sources, one of ^{57}Co in austenite, and the other of ^{57}Co in martensite of the material in question. Then resonance counts are observed when the relative source-specimen velocity is zero, and the number of such counts is proportional to the amount of austenite and martensite respectively in the specimen. The off-resonance counts can be determined by rapid vibration, but without any necessity for an accurately controlled waveform. Jaggi and Rao point out that since only zero velocity and high velocity count rates are needed, the useful solid angle of radiation from the source can be very large, the count rates can be high and measurement times correspondingly short. Together with the simplicity of the apparatus required, this means that the method may be well suited to quality control. The principle can obviously be extended to some other measurements.

5.3 Carbon and nitrogen in solution in iron

5.3.1 Carbon steels

Austenite at temperatures in the region of 1140°C can dissolve about 8 at % of carbon. The carbon is in an octahedral interstitial site with six first neighbour iron atoms at a distance of 0.5 a, where a is the lattice parameter. The second neighbours are at a distance of 0.86 a, so that the carbon interaction is much stronger with its first neighbou atoms than any others. The spectrum of austenite, at least for low concentrations of carbon, shows a single line due to isolated iron atom and a doublet with an isomer shift of 0.06 mm sec^{-1} relative to the singlet, and a quadrupole splitting of 0.6 mm sec^{-1} due to iron atoms with one carbon neighbour. The intensity of this doublet is close to 6c where c is the carbon concentration, and can therefore be used to estimate the dissolved carbon where the concentration is unknown.

The results for carbon martensite are not so clear cut. The solubility of carbon in α-iron is only about .1 at %. When iron carbon alloys containing about 5% carbon are cooled, they partially transform martensitically to a BCT structure in which carbon atoms occupy the octahedral positions along the c axis.

The Mossbauer spectra of such alloys show mainly a six line componen similar to the spectrum of α-iron, but with additional sextets and a singlet line due to retained austenite. The attribution of the subsidiary sextets to iron sites has been the object of much work. Ino et al (179), Ron et al (180) and Gielen and Kaplow (181) analysed the spectra into three or four components (182) with field strengths 332, 342, 265 kOe or 334, 342, 330 and 265 kOe. They attributed the high field 342 kOe and low field 265 kOe to the four equatorial and two dipolar iron atoms of the six neighbours of the carbon atoms. The fields of 334 kOe were attributed to iron atoms which were 3 and 4th neighbours to a carbon atom. This attribution received convincing support from Cadeville et al (183) who took spectra of splat quenched alloys containing 1, 2, 3, 4, 5% C, and showed that the intensities of two components at 343 kOe and 269 kOe were close to 4c and 2c in all cases. They also attributed the component with a field 335 kOe to iron atoms which were third and fourth neighbours to carbon atoms. It is then possible to draw a smooth curve representing the hyperfine field at iron atoms as a function of their distance from a carbon atom (182, 184), as in Fig. 15.

The behaviour of nitrogen in austenite and martensite is generally similar to that of carbon (185).

5.3.2 Freshly quenched steels

Since 1965 it has been realized that freshly quenched austenite held

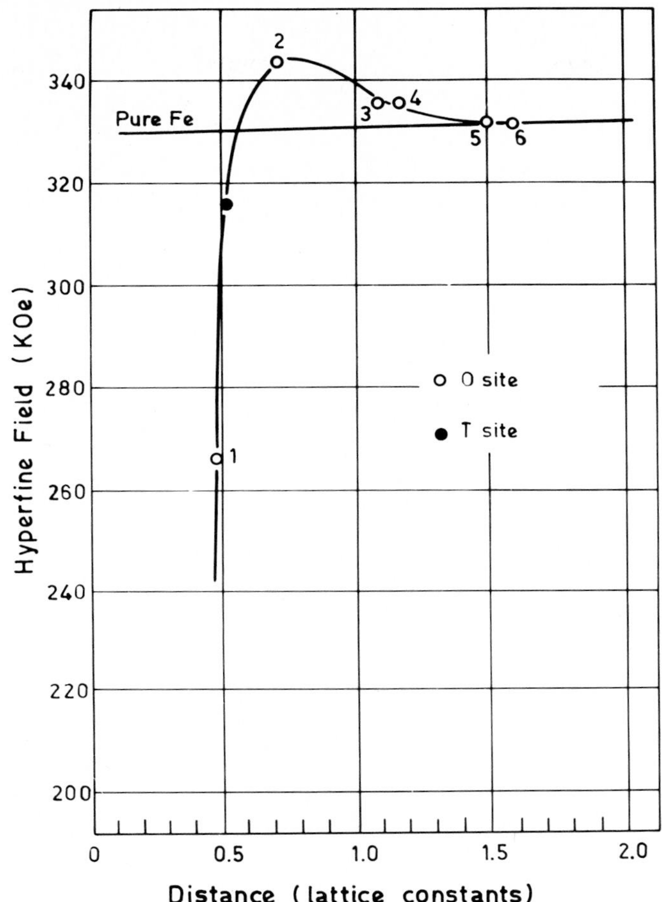

Fig. 15. Hyperfine field at ^{57}Fe nuclei in iron as a function of distance from an interstitial carbon atom. Open circles mark octahedral sites, a closed circles tetrahedral sites.

at a low temperature exists in an orthorhombic structure with "anomalously low tetragonality", i.e. c/a smaller than for the normal martensite structure. The metastable structure transforms to the normal structure at temperatures near 0°C, and has been the object of many investigations. Fujita et al (186), Fujita (184), Ino and Ito (187), Lesoille and Gielen (188), base their interpretation on the work of Lysak and Nikolin (189), and attribute new lines in the Mossbauer spectrum to neighbours of carbon in a tetrahedral interstitial site. On this interpretation, the new fields observed fit on the same graph, Fig. 15, of field against distance as deduced from the carbon in an octahedral site, and moreover during the orthorhombic to BCT transformation, Fujita finds that one half of the intensity of the octahedral components plus one quarter of the presumed tetrahedral components equals the carbon concentration, as would be expected. Fujita (190) has suggested a mechanism of the martensitic transformation which accounts for these observations. Genin and Flinn (191), Choo and Kaplow (192) and Gridnev et al (193) base their interpretations mainly on non-random distributions of the carbon atoms.

5.3.3 Nitrided materials

The Mossbauer spectra of iron nitrogen austenite, martensite and the nitrides γ^1 - Fe_4N, ε-Fe_2N_{1-x} and ζ-Fe_2N have been well characterised in number of studies (185, 194-197).

The appearance of these compounds in iron subjected to a commercial nitriding process, "tuftriding", has been studied by Longworth (198) who shows how X-ray backscatter spectra from the treated surface can be used to determine the thickness of the nitrided layer. Fig. 16 shows typical spectra. For short treatments in the salt bath (2 mins.), the spectrum consists of a singlet from nitrogen austenite, two magnetic components with fields 211 and 267 kOe from ε Fe_2N_{1-x}, and weak component at 330 kOe from the underlying iron.

Most X-rays in this technique come from a depth of about 10 μms, so that we can conclude that a 2 min. treatment makes a nitride layer of about 10 - 15 μms. As the treatment proceeds, the component due to nitrogen austenite falls as greater thicknesses of nitrides are formed.

Nitriding by gas, tuftriding and ionitriding has been investigated by Ujihira and Handa (199) who used backscatter electron spectroscopy to obtain spectra of the first .1 μm of the surface. They studied the depth distribution by grinding off the surface in about 10 μm steps. For the tuftriding process, they also find ε nitrides down to a depth of about 20 μm, and at 40 μm, they find mainly nitrogen martensite. The gas nitriding gave an uppermost surface of paramagnetic ε Fe_2N, of thickness greater than 10 μm, whereas the ion nitrided surface showed α-iron or nitrogen martensite even in the uppermost layer.

The effect of ion-implanted nitrogen has been studied by Longworth et al (200). It has been shown that the wear resistance of many steels can be increased by one to two orders of magnitude by implantation with N ions at doses in the range 10^{16} - 10^{18} ions cm^{-2} (201). In the same dose range, the conversion electron Mossbauer spectra show the appearance of nitrides of iron. The implanted nitrogen shows a high mobility, and ageing for 1 hr. at 275°C alters the distribution of nitrogen amongst different nitrides. After ageing at 500°C most of the nitrogen

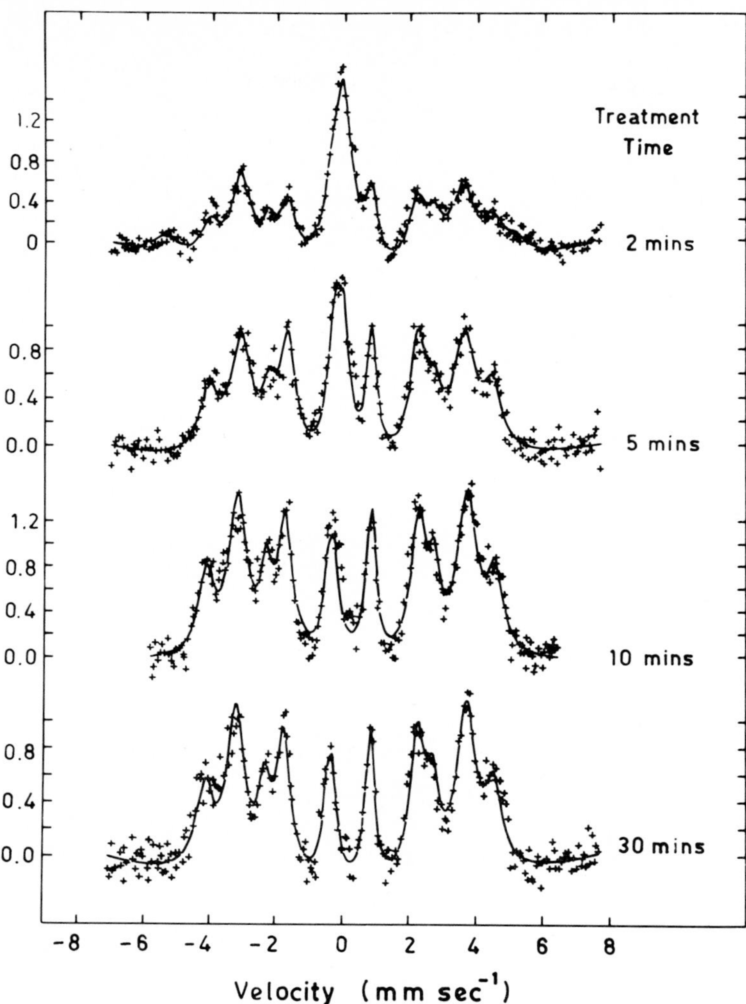

Fig. 16. X-ray backscatter spectra of nitrided surfaces as a function of treatment time in a salt bath.

had diffused away from the detection region (.1 - .3 μm). Dearnaley (202) has proposed an explanation of the surprising fact that the improvement in wear resistance survives even when 10 μm of the surface has been worn away by suggesting that nitrogen at hot spots on the surface may be carried by dislocations into the substrate. The high mobility of nitrogen indicated by these experiments is consistent with this suggestion.

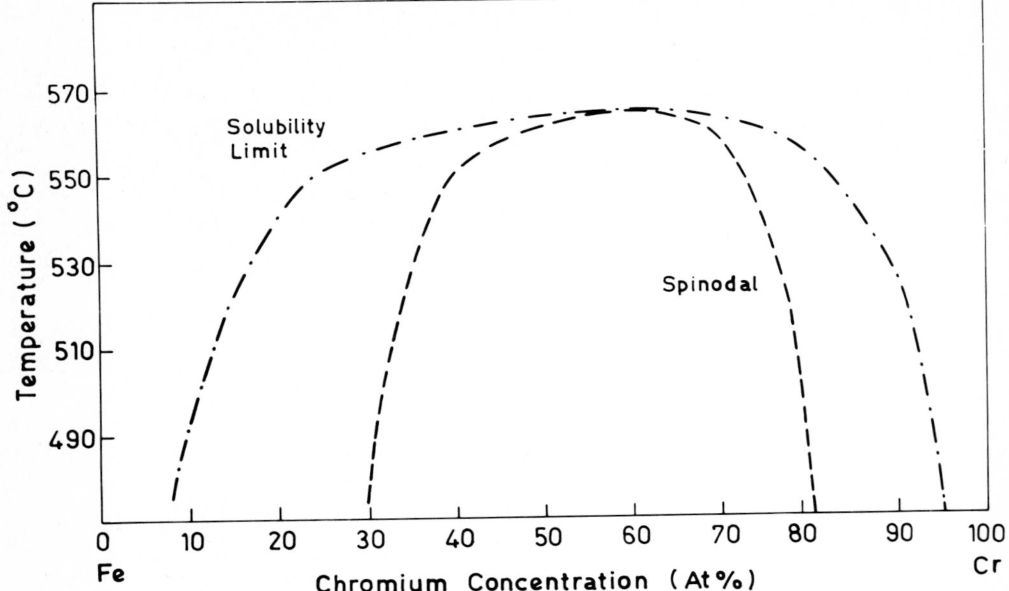

Fig. 17. Phase diagram of iron chromium system showing the solubility limit and the spinodal.

5.4 Spinodal decomposition

5.4.1 Iron chromium decomposition

Consider a phase diagram such as Fig. 17, showing the miscibility gap in the chromium iron system. The dashed line defines the spinodal region and is the locus of the points for which d^2f/dc^2 is zero where f is the free energy. Inside the spinodal region d^2f/dc^2 is negative, and outside it is positive. Consequently, inside the region, any small change in composition decreases the free energy, and an alloy quenched into this region decomposes by diffusion alone, leading to sinusoidal fluctuations of composition. Outside the region, an activation energy is required, so that decomposition can only proceed by nucleation and growth.

Schwartz (203) and Chandra and Schwartz (204) have demonstrated these effects by Mossbauer spectroscopy. A sample of Fe_4Cr_6 was aged for periods of 30, 80 and 1300 hrs, and spectra taken after each period. The first two spectra show only a broadening, corresponding to small fluctuations of composition. After 1300 hrs. a paramagnetic chromium rich component had appeared in the spectrum, but even after 2476 hrs. the spectrum had not reached complete decomposition into the two components corresponding to the ends of the miscibility gap. The gradual change in the early stages indicates that decomposition did not occur by nucleation and growth.

5.5 Amorphous alloys

Because the Mossbauer spectrum is mainly determined by the near neighbour environment of the Mossbauer nucleus, the spectrum of an amorphous alloy is not greatly different from that of a crystalline sample. Nevertheless, the interest and technological importance of amorphous iron alloys of the general type $Fe_{.8}M_{.2}$ where M is a metalloid atom has stimulated a considerable body of work.

5.5.1 Models of structure

The spectrum of an amorphous alloy consists of six broad and featureless lines, indicating either a number of discrete hyperfine fields or a continuous distribution. The procedure followed has usually been to attempt to derive a probability distribution p(H), of the field strengths, and then to relate this distribution to a model of the alloy. Tsuei et al (205) took spectra of $Fe_{80}P_{12.5}C_{7.5}$ and made a fit using about five discrete field values, and compared their relative intensities with the expectation for a random distribution of P and C atoms in the first neighbour shell. The agreement was not close, and in fact the intensities correspond more closely with the predictions of the dense random packing model, originally suggested by Gonser et al (206) for $Fe_{80}B_{20}$. However, Fujita (207) was able to fit the spectrum of $Fe_{80}P_{17}C_3$ by a model which included random second neighbour contributions as well as first, showing that it is not always necessary to assume a special lattice structure for the amorphous state. According to this view, the structure may be regarded as an aggregate of small crystallites with the regular lattice form.

5.5.2 Detailed fitting of spectra of amorphous materials

Le Caer and Dubois (208) have pointed out that spectra of amorphous alloys in the paramagnetic state show the presence of considerable quadrupole couplings (209). If this coupling can be regarded as a perturbation on the magnetic interaction, and θ is the angle between the field gradient and the magnetic field, the random distribution of θ ensures that $\langle 3 \cos^2\theta - 1 \rangle = 0$, and the mean position of each line is determined by the magnetic interaction alone. Nevertheless, each component line shape is altered by the quadrupole interaction, and Le Caer and Dubois show how this may be taken into account, in first order approximation, in the determination of p(H).

5.5.3 Temperature dependence of magnetization in amorphous materials.

The temperature dependence of the hyperfine field can be easily followed by Mossbauer spectroscopy, and has been shown (210) to fall below that for a crystalline ferromagnet, due to the absence of long range order. The Curie point for $Fe_{80}P_{12.5}C_{7.5}$ was determined to be 586 K (205), and the direction of magnetization was found to lie in the plane of the foil.

5.6 Radiation damage

The concentration of lattice defects such as vacancies or interstitial atoms produced by radiation is usually less or much less than 10^{-3}, and cannot be expected to show up directly in a Mossbauer spectrum. An effective increase in concentration can be obtained in two ways, called (i) Trapping, and (ii) Correlated damage.

5.6.1 Trapping

In the trapping method, the specimen is doped with a suitable Mossbauer nucleus (as source or absorber) and held at a temperature at which defect diffusion will take place. Then if there is an attractive potential between the Mossbauer site and the defect, trapping may occur and although the defect concentration is low, the fraction of Mossbauer nuclei with an associated defect may be easily detectable.

An example of such an experiment is found in the trapping of aluminium interstitials by ^{57}Co, (211, 212). A sample of pure Al doped with ^{57}Co was irradiated with fast neutrons at 4.6 K to produce vacancies and interstitials. The sample was annealed to increasing temperatures, and spectra taken after each anneal. Between 30 and 45 K, a new component appears in the spectrum, increases up to 180 K and disappears above 190 K. This suggests that the ^{57}Co atoms have trapped interstitia Al atoms. In later work, Vogl et al (213) studied the recoilless fraction, f, of this component as a function of temperature. They found that a remarkable reduction in f, from about .9 to .3 occurred between 11 K and 20 K. This reduction was attributed to the vibration of the ^{57}Co atom between the six equivalent sites inside the octahedral cage of the FCC lattice, as shown in Fig. 18.

5.6.2 Correlated damage

As an example of the "correlated damage" type of experiment, we cite the work of Vogl et al (214). In this work, sources of ^{197}Pt and ^{193}Os were produced in ^{196}Pt and ^{192}Os dissolved in iron by neutron capture. The recoil from the capture γ-ray is sufficient to remove the Pt or Os

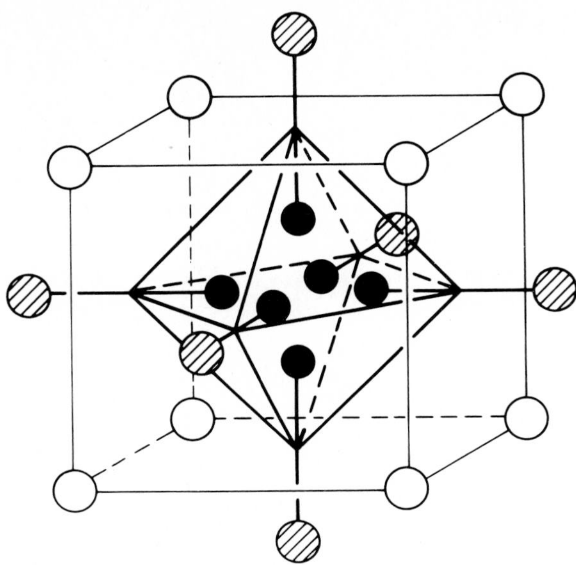

Fig. 18. Equivalent configurations for a mixed dumbbell, which can be formed by jumps of the ^{57}Co impurity atom in the octahedral cage formed by the FCC lattice. Closed circles represent ^{57}Co sites, hatched circles Al sites.

atom from its lattice site, but is insufficient to move it far. The resulting spectrum showed Pt and Os atoms in sites with a smaller hyperfine field than is found at pure solution sites.

5.7 Ion implantation

5.7.1 Implanted Fe

A survey of iron implanted in metals has been carried out by Sawicka and Sawicki (215), who find that in the spectra, iron can be recognized in the monomeric isolated state and as dimers, showing quadrupole splittings in the range .6 - .9 mm sec^{-1}. In BCC metals, the intensities of these components are usually close to the expectation for a random alloy, but in FCC metals, the dimeric state is often more strongly populated. Sawicka and Sawicki observe that the isomer shift for the dimers is in general closer to the shift for pure iron than the shift for the monomer, but otherwise follows the same trends as the shift for the monomer.

The implantation of ^{57}Fe into Cu and Ag has been studied by Longworth and Jain (133, 135) and their work is described in 3.4.1.

5.7.2 Implanted Xe

The lattice location of implanted ^{133}Xe has been extensively studied by the Groningen group, in iron (216, 217) and in the BCC metals Mo, W, Ta(218). Implanted xenon is found to be a powerful trap for vacancies, leading to the existence of at least three sites for the xenon, corresponding to substitutional sites with 0, 1, 2 or more vacancies. In addition, in order to account for the observed increase in the occupation of the other sites with annealing, or implantation temperature, another site with f $\sim$ 0 must be included, presumably a xenon atom in a vacancy cluster. When implanted in iron, the hyperfine field at the Cs nucleus falls from 276 kOe for no vacancy, to 130 kOe for one and less than 25 kOe for two or more vacancy neighbours, while the isomer shift goes from -1.1 mm sec^{-1} to -0.80 and -0.20 mm sec^{-1}.

In specimens of iron implanted with xenon below 7 K, the first recovery stage occurring below 35 K, associated with the reaction of migrating interstitials with vacancies is not observed. This is attributed to the occurrence of the stage I recovery during the heat spike of the heavy ion collisions. The time constant for the spike is estimated to be π = 1.3 10^{-11} secs. This value has been confirmed by Walker and Chien (219) in experiments on a range of rare earth nuclides with life times of the order of 10^{-10} secs.

Acknowledgements

My thanks are due to Professor C.E. Johnson, G. Longworth and R. Atkinson for their careful reading of the manuscript and their helpful comments.

REFERENCES

1 M.B. Brodsky, Rep. Progr. Phys., 41 (1978) 1547.

2 J.M. Grow, D.G. Howard, R.H. Nussbaum and M. Takeo, Phys. Rev. B, 17 (1978) 15.

3 H. Kirchmayer and C.A. Poldy, J. Magn. Magn. Mater, 8 (1978) 1.

4 F.E. Wagner and G. Wortmann, in Alefeld and Volkl (Eds.), Hydrogen in Metals. I. Basic Properties, Springer Verlag, Berlin, 1978, p. 131.

5 L.R. Walker, G.K. Wertheim and V. Jaccarino, Phys. Rev. Lett. 6 (1961) 98.

6 D.L. Williamson, S. Bukshpan and R. Ingalls, Phys. Rev. B, 6 (1972) 4194.

7 R.V. Pound, G.B. Benedek and R. Drever, Phys. Rev. Lett., 7 (1961) 405.

8 D.N. Pipkorn, C.K. Edge, P. Debrunner, G. de Pasquali, H.G. Drickamer and H. Frauenfelder, Phys. Rev., 135A (1964) 1604.

9 R. Ingalls, Solid State Commun., 14 (1974) 11.

10 S.M. Quaim, Proc. Phys. Soc., 90 (1967) 1065.

11 F. Van der Woude and G.A. Sawatzky, Phys. Rep., 12 (1974) 336.
12 B. Window, C.E. Johnson and G. Longworth, J. Phys. C: Solid State Phys., Supp. 2 S (1970) 218.
13 W.N. Cathey, J. Phys. F: Met. Phys., 8 (1978) 315.
14 B. Window, J. Phys. F: Met. Phys., 1 (1971) 533.
15 B. Window, J. Phys. Chem. Solids, 32(1971) 1059.
16 G. Longworth, J. Phys. C: Met. Phys. Suppl., 1 (1970) 581.
17 P.G. Huray, T.J. Kirthlink, F.E. Obenshain, J.O. Thomson and Cheng May Tung, Phys. Rev. B, 14 (1976) 4776.
18 L.D. Roberts, R.L. Becker, F.E. Obenshain, Phys. Rev., 137A (1965) 895.
19 F.E. Wagner, G. Wortmann, G.M. Kalvius, Phys. Lett., 42A (1973) 483.
20 T.S. Chou, M.L. Perlman and R.E. Watson, Phys. Rev. B, 14 (1976) 3248.
21 W.D. Josephson, D.W. Knoble, D.L. Williamson and L.D. Roberts, Phys. Rev. B, 14 (1976) 1480.
22 R.L. Cohen in G.K. Shenoy and F.E. Wagner (Eds.), Mossbauer Isomer Shifts, North Holland Publishing Co., 1978, p. 541.
23 T.C. Tucker, L.D. Roberts, C.W. Nestor, T.A. Carlson and F.B. Malik, Phys. Rev., 178 (1969) 998.
24 R.E. Watson and L.H. Bennett, Phys. Rev. B, 15 (1977) 5136.
25 G.K. Shenoy and F.E. Wagner (Eds.), Mossbauer Isomer Shifts, North Holland Publishing Co., 1978.
26 D. Salomon and D.A. Shirley, Phys. Rev. B, 9(1974) 29.
27 F. Wagner and W. Potzel, in G. Goldring and R. Kalish (Eds.), Hyperfine Interactions in Excited Nuclei, Gordon and Breach, New York 1971, p. 681.
28 Aiga and J. Itoh, J. Phys. Soc. Japan, 31 (1971) 1844.
29 J.J. Spijkerman, J.C. Travis, D.N. Pipkorn and C.E. Violet, Phys. Rev. Lett., 26 (1971) 323.
30 C.E. Violet and D.N. Pipkorn, J. Appl. Phys., 42 (1971) 4339.
31 W. Marshall and C.E. Johnson, J. Phys. Radium., 23 (1962) 733.
32 R.C. Mercader and T.E. Cranshaw, J. Phys. F: Met. Phys., 5 (1975) L124.
33 J. Kjeldgaard, G. Trumpy, N. Thrane and S. Morup, Proc. Int. Conf. Mossbauer Spectroscopy, Bratislava, 1975, p. 127.
34 T.E. Cranshaw and R.C. Mercader, J. Phys. F: Met. Phys., 6 (1976) 2129.
35 G.A. Gehring, Phys. Scr. 11 (1974) 215.
36 G.A. Gehring and H.C.W.L. Williams, J. Phys. F: Met. Phys., 1974, 291.
37 R.E. Watson, A.C. Gossard and Y. Yafet, Phys. Rev., 140A (1965) 375.
38 B.R.A. Nijboer and F.W. de Wette, Physica, 24 (1958) 1105.
39 T.P. Das and M. Pomerantz, Phys. Rev., 123 (1961) 2070.
40 J.A. Sawicki, Phys. Status Solidi (b) 53 (1972) K103.
41 S.M. Quaim, J. Phys., C: Solid State Phys., 2 (1969) 1434.
42 G. Wortmann and D. Williamson, Hyperfine Interactions, 1 (1975) 167.
43 E.R. Bauminger, A. Diamant, I. Felner, I. Nowick and S. Ofer, Phys. Rev. Lett., 34 (1975) 962.
44 F.E. Wagner, H. Spieler, D. Mucheida and P. Kienle, Z. Phys., 254 (1972) 112.
45 P. Boolchand, B.L. Robinson and S. Jha, Phys. Rev., 187 (1969) 475.
46 G. Kaindl and D. Salomon, Phys. Lett., 40A (1972) 179.
47 R.S. Raghavan, E.N. Kaufmann and P. Raghavan, Phys. Rev. Lett., 34 (1975) 1280.
48 P. Raghavan, E.N. Kaufmann, R.S. Raghavan, E.J. Ansaldo and R.A. Naumann, Phys. Rev. B, 13 (1976) 2835.

49 C. Janot, P. Delcroix and M. Piecuch, Phys. Rev. B, 10 (1974) 2661

50 M. Piecuch and Ch. Janot, J. Phys. (Paris) Colloq., 35 (1974) C6-291.

51 R.S. Preston, D.J. Lam, M.V. Nevitt and D.O. Van Ostenburg, Phys. Rev., 149 (1966) 440.

52 M.S. Ridout, J. Phys. C: Solid State Phys., 2 (1969) 1258.

53 B. Window, G. Longworth and C.E. Johnson, J. Phys. (Paris), 32 (1971) C1-863.

54 J. Friedel, Nuovo Cimento, 7 (1958) 287.

55 B. Window, Phil. Mag., 26 (1972) 681.

56 T.E. Cranshaw, J. Phys. F: Met. Phys., 2 (1972) 615.

57 S.S. Hanna, J. Heberle, C. Littlejohn, G.J. Perlow, R.S. Preston and D.H. Vincent, Phys. Rev. Lett., 4 (1960) 177.

58 D.E. Nagle, H. Frauenfelder, R.D. Taylor, D.R.F. Cochran and B.T. Matthias, Phys. Rev. Lett., 5 (1960) 364.

59 R.S. Preston, S.S. Hanna and J. Heberle, Phys. Rev., 128 (1962) 2207.

60 C.E. Johnson, M.S. Ridout, T.E. Cranshaw, Phys. Rev. Lett., 6 (1961) 450.

61 C.E. Johnson, M.S. Ridout, T.E. Cranshaw, Proc. Phys. Soc., 81 (1963) 1079.

62 A.J.F. Boyle and H.E. Hall, Rep. Prog. Phys., 25 (1962) 441.

63 G.K. Wertheim, V. Jaccarino, J.H. Wernick and D.N.E. Buchanan, Phys. Rev. Lett., 12 (1964) 24.

64 M.B. Stearns, Phys. Rev., 129 (1963) 1136; Ibid 147 (1966) 439; B, 4 (1971) 4069; B, 8 (1973) 4383; B, 12 (1975) 1626.

65 E.F. Mendis and L.W. Anderson, Phys. Status Solidi., 41 (1970) 375.

66 T.E. Cranshaw, C.E. Johnson and M.S. Ridout, Phys. Lett., 20 (1966) 97.

67 G. Gruner, I. Vincze and L. Cser. Solid State Commun., 10 (1972) 347.

68 T.E. Cranshaw, J. Phys. E: Sci. Instrum., 7 (1974) 1.

69 I. Vincze and I.A. Campbell, J. Phys. F: Met. Phys., 3 (1973) 647.

70 I. Vincze and A.T. Aldred, Phys. Rev. B, 9 (1974) 3845.

71 I. Vincze and A.T. Aldred, Solid State Commun., 17 (1975) 639.

72 H. Bernas and I.A. Campbell, Solid State Commun., 4 (1966) 577.

73 I.A. Campbell and A. Blandin, J. Magn. Magn. Mater., 1 (1975) 1.

74 B. Kolk, Physica 86-88B (1977) 446.

75 M.A. Rudermann and C. Kittel, Phys. Rev., 96 (1954) 99.

76 T. Kasuya, Prog. Theor. Phys., 16 (1956) 45.

77 K. Yoshida, Phys. Rev., 106 (1957) 893.

78 L. Haggstrom, L. Granas, R. Wappling and S. Devanarayanan, Phys. Scr., 7 (1973) 125.

79 G.N. Rao and A.K. Singhvi, Phys. Status Solidi (b), 84 (1977) 9.

80 C.C.M. Campbell, J. Phys. F: Met. Phys., 5 (1975) 1931.

81 W. Leiper, J.B. Rush, M.F. Thomas, C.E. Johnson, F.W.D. Woodhams, C. Blaauw and G.R. MacKay, J. Phys. F: Met. Phys., 7 (1977) 533.

82 J.D. Rush, C.E. Johnson, M.F. Thomas, D.C. Price and P.J. Webster, J. Phys. F: Met. Phys., 9 (1979) 1129.

83 P.J. Webster and M.R. Ramadan, J. Magn. Magn. Mater., 5 (1977) 51.

84 G.R. MacKay, C. Blaauw, J. Judah and W. Leiper, J. Phys. F: Met. Phys., 8 (1978) 305.

85 R. Street and B. Window, Proc. Phys. Soc., 89 (1966) 587.

86 B. Window, J. Phys. C: Solid State Phys., Vol. 3 Suppl. No. 2, Met. Phys. (1970) S 210, J. Phys. C: Solid State Phys., 3 (1970) 2156.

87 Y. Koi, A. Tsujimura and T. Hihara, J. Phys. Soc. Jap., 19 (1964) 1493.

88 V. Jaccarino, L.R. Walker and G.K. Wertheim, Phys. Rev. Lett., 13 (1964) 752.

89 J.G. Dash, B.D. Dunlop and D.G. Howard, Phys. Rev., 141 (1966) 376.

90 T.E. Cranshaw, C.E. Johnson, M.S. Ridout and G.A. Murray, Phys. Lett. 21 (1966) 481.

91 P.J. Schruer, G.A. Sawatzki and F. van der Woude, Phys. Rev. Lett., 27 (1971) 586.

92 I. Vincze, Solid State Commun., 10 (1972) 341.

93 I. Vincze and G. Gruner, Phys. Rev. Lett., 28 (1972) 178.

94 I. Vincze and L. Cser, Phys. Status Solidi (b), 50 (1972) 709.

95 I. Vincze and I.A. Campbell, Solid State Commun., 14 (1974) 795.

96 I.A. Campbell, Solid State Commun., 6 (1968) 345; J. Phys. C: Solid State Phys., 3 (1970) 2151.

97 A.P. Jain and T.E. Cranshaw, Phys. Lett., 25A (1967) 421.

98 T.E. Cranshaw, J. Appl. Phys., 40 (1969) 1481.

99 G.P. Huffman, F.G. Schwerer and G.R. Dunmyre, J. Appl. Phys., 40 (1969) 1487.

100 A.E. Balabanov and N.N. Delyagin, Sov. Phys. JETP, 30 (1970) 1054.

101 G.P. Huffman and G.R. Dumyre, J. Appl. Phys., 41 (1970) 1323.

102 G.P. Huffman and G.R. Dunmyre, J. Appl. Phys., 42 (1971) 1613.

103 G.P. Huffman and G.R. Dunmyre, AIP Conf. Proc. 5 Magn. Magn. Mater. 1971 in Graham and Rhyne (Eds.), 1972, 544.

104 D.C. Price, J. Phys. F: Met. Phys., 4 (1974) 639.

105 I. Vincze and J. Kollar, Phys. Rev. B, 6 (1972) 1066.

106 G. Shirane, C.W. Chen, P.A. Flinn and R. Nathans, Phys. Rev., 131 (1963) 183.

107 S.A. Losievskaya, Phys. Status Solidi (a), 16 (1973) 647.

108 S.A. Losievskaya and R.N. Kuzmin, Izv Akad Nauk SSSR, Metal, 3 (1972) 179.

109 I.N. Bogachev, S.D. Karakishev, V.S. Litvinov and V.V. Orchinnikov, Phys. Status Solidi (a), 24 (1974) 661.

110 L.H. Schwartz & A. Asano, J. Phys.(Paris)Colloq., 35 (1975) C6-453

111 T.E. Cranshaw, C.E. Johnson and M.S. Ridout, Proc. Int. Conf. Magnetism, Nottingham, 1964, p. 141.

112 T.E. Cranshaw, Physica 86-88B (1977) 391, Physica 86-88B (1977) 443

113 G. Papadimitriou and J.M. Genin, Phys. Status Solidi (a), 9 (1972) K19.

114 L. Cser and L.H. Naszodi, Phys. Status solidi (b), 92 (1979) K55.

115 B. Window, W.T. Oosterhuis and G. Longworth, Int. J. Magn., 6 (1974) 93.

116 Y. Gros and J. Pauleve, J. Phys. (Paris), 31 (1970) 459.

117 J. Pauleve, D. Dautreppe, J. Langier and L. Niel, J. Phys. Radium, 23 (1962) 841.

118 J. Danon, R.B. Scorzelli, I.S. Azevedo, J.F. Albertsen J.M. Knudsen, N.O. Roy-Poulsen, Y. Minai, H. Wakita, and T. Tominaga, Radiochem. Radioanal. Lett., 38 (1979) 339.

119 L. Billard and A. Chamberod, Solid State Commun., 17 (1975) 113.

120 J. Hesse and J.B. Muller, Solid State Commun., 22 (1977) 637.

121 G. Bayreuther and H. Hoffman, Physica 86-88B (1977) 297.

122 P.A. Montano and M.S. Seegra, Phys. Rev., B, 15 (1977) 2437.

123 J.P. Eymery and P. Moine, J. Phys. (Paris) Lett., 39 (1978) L23.

124 J.W. Drijver, F. van der Woude and S. Radelaar, Phys. Rev. Lett., 34 (1975) 1026.

125 V.A. Tsurin and A.Z. Men'shikov, Phys. Met. Metallogr. 45 (1979) 82.

126 G.N. Belozerskii, V.V. Dudoladov and M.I. Kazakov, Proc. Int. Conf. Bucharest, 1977, p. 383.

127 F. Ambe and S. Ambe, J. Phys. (Paris) Colloq. 37 (1976) C6-923.
128 B. Window, Phys. Rev. B, 1 (1972) 2013.
129 S.J. Campbell, P.E. Clark and P.R. Liddell, J. Phys. F: Met. Phys., 2 (1972) L114.
130 S.J. Campbell and P.E. Clark, J. Phys. F: Met. Phys., 4 (1974) 1073.
131 P.E. Clark, J.M. Cadogan, M.J. Yazxhi and S.J. Campbell, J. Phys. F: Met. Phys., 9 (1979) 379.
132 J.G. Mullen, Phys. Rev., 121 (1961) 1649.
133 G. Longworth and R. Jain, J. Phys. F: Met. Phys., 8 (1978) 351.
134 R. Jain and G. Longworth, J. Phys. F: Met. Phys., 8 (1978) 363.
135 G. Longworth and R. Jain, J. Phys. F: Met. Phys., 8 (1978) 993.
136 B.C. Edwards, B.L. Eyre and T.E. Cranshaw, Nature, 47 (1977) 269.
137 T.E. Cranshaw, J. Phys. (Paris) Colloq., 40 (1979) C2-169.
138 N.N. Delyagin and E.N. Kornienko, Sovt. Phys., JETP, 32 (1970) 832.
139 M. Maurer, M.C. Cadeville and J.P. Sanchex, Phil. Mag., 38 (1978) 739.
140 R.B. Frankel, N.A. Blum, B.B. Schwartz and Duk Joo Kim, Phys. Rev. Lett., 18 (1967) 1051.
141 N. Rivier and M.J. Zuckermann, Phys. Rev. Lett., 21 (1968) 904.
142 M.D. Daybell and W.A. Steyert, Rev. Mod. Phys., 40 (1969) 380.
143 R.D. Taylor, W.A. Steyert and D.E. Nagle, Rev. Mod. Phys., 36 (1964) 406.
144 T.A. Kitchens, W.A. Steyert and R.D. Taylor, Phys. Rev., 138A (1965) 467.
145 M.P. Maley and R.D. Taylor, Phys. Rev. B, 1 (1970) 4213.
146 R.J. Borg, R. Booth and C.E. Violet, Phys. Rev. Lett., 11 (1963) 464.
147 C.E. Violet and R.J. Borg, Phys. Rev., 149 (1966) 540.
148 B. Window, J. Magn. Magn. Mater., 1 (1975) 167.
149 A. Chandra and J. Ray, J. Phys. (Paris) Colloq., 39 (1978) C6-914.
150 P. Steiner, D. Gumprecht, W.V. Zdrojewski and Hufner, J. Phys. (Paris), 35 (1974) C6-523.
151 P.P. Craig, D.E. Nagle, W.A. Steyert and R.D. Taylor, Phys. Rev. Lett., 9 (1962) 12.
152 M.P. Maley, R.D. Taylor and J.L. Thompson, J. Appl. Phys. 38 (1967) 1249.
153 F.W.D. Woodhams, R.E. Meads and J.S. Carlow, Phys. Lett., 23 (1966) 419.
154 W.L. Trousdale, G. Longworth and T.A. Kitchens, J. Appl. Phys., 38 (1967) 922.
155 H. Maletta, Z. Phys., 250 (1972) 68.
156 S. Hufner and P. Steiner, in S.G. Cohen and M. Pasternak (Eds.), Perspectives in Mossbauer Spectroscopy, Plenum Press 1973.
157 J.W. Drijver and F. van der Woude, J. Phys. F: Met. Phys., 3 (1973) L206.
158 V. Jaccarino and L.R. Walker, Phys. Rev. Lett., 15 (1965) 255.
159 H. Nagasawa and N. Sakai, J. Phys. Soc. Jap., 27 (1969) 1150.
160 M. Shiga and Y. Nakamura, J. Phys. F: Met. Phys., 8 (1978) 177.
161 G. Lang, Nucl. Instrum. Methods, 24 (1963) 425.
162 D.W. Hafemeister and E.B. Shera, Nucl. Instrum. Methods, 41 (1966) 133.
163 M.C.D. Ure and P.A. Flinn, in I.J. Gruverman (Ed.), Mossbauer Effect Methodology, Vol. 7, 1971, p. 245.

164 T.M. Lin and R.S. Preston, in I.J. Gruverman (Ed.), Mossbauer Effect Methodology, Vol. 9, 1974, p. 205.
165 G.K. Shenoy, J.M. Friedt, H. Maletta and S.L. Ruby, in I.J. Gruverman (Ed.), Mossbauer Effect Methodology, Vol. 9, 1974, p. 277.
166 R.L. Collins, J. Phys. (Paris) Colloq., 40 (1979) C2-39.
167 E.M. Howard, C.E. Violet and R.J. Borg, Trans. AIME, 242 (1968) 1503.
168 M. Hansen and K. Anderko, Constitution of Binary Alloys, McGraw Hill Book Co., New York, 1958.
169 A. Blasius and V. Gonser, J. Phys. (Paris) Colloq., 37 (1976) C6-398.
170 T. Ozawa and Y. Ishida, J. Phys. (Paris) Colloq., 40 (1979) C2-551
171 T. Ozawa and Y. Ishida, Scr. Met., 11 (1977) 835.
172 V.A. Brykhanov, N.N. Delyagin and V.S. Shpinel, Sov. Phys. JETP, 20 (1965) 55.
173 B.W. Christ and P.M. Giles, Trans. AIME, 242 (1968) 1915.
174 H.K. Chow, R.F. Weise and P. Flinn, AEC Rept. NSEC, 4023-1, 1969.
175 N. Abe and L.H. Schwartz, Mater. Sci. Eng. 14 (1974) 239.
176 K.J. Kim and L.H. Schwartz, J. Phys. (Paris) Colloq., 37 (1976) C6-405.
177 T.E. Cranshaw and R.G. Campany, J. Phys. (Paris) Colloq., 40 (1979) C2-589.
178 N.K. Jaggi and K.R.P.M. Rao, NDT Int., 11 (1978) 281.
179 H. Ino, T. Moriya and F.E. Fujita, J. Phys. Soc. Jap., 22 (1967) 346.
180 M. Ron, A. Kidron, H. Schechter and S. Niedzwiedz, J. Appl. Phys., 38 (1967) 590.
181 P.M. Gielen and R. Kaplow, Acta Met., 15 (1967) 49.
182 T. Moriya, H. Ino, F.E. Fujita and Y. Maeda, J. Phys. Soc. Jap., 24 (1968) 60.
183 M.C. Cadeville, J.M. Friedt and C. Lerner, J. Phys. F: Met. Phys., 7 (1977) 123.
184 F.E. Fujita, in N. Doyama and S. Yoshida (Eds.), Progress in the Study of point defects, Univ. Tokyo Press 1977, 71.
185 J.M. Genin and J. Foct, Phys. Status Solidi. (a), 17 (1973) 395.
186 F.E. Fujita, C. Shiga, T. Moriya and H. Ino, J. Jap. Inst. Metals, 38 (1974) 1030.
187 H. Ino and T. Ito, J. Phys. (Paris) Colloq., 40 (1979) C2-644.
188 M. Lesoille and P.M. Gielen, Metall. Trans., 3 (1972) 2681.
189 L.L. Lysak and E.L. Nikolin, Fiz. Metal Metalloved, 30 (1970) 1189.
190 F.E. Fujita, Bull. Jap. Inst. Metals, 13 (1974) 714.
191 J.M. Genin and P.A. Flinn, Trans. Met. Soc., AIME, 242 (1968) 1419.
192 W.K. Choo and R. Kaplow, Acta Met., 21 (1973) 725.
193 V.N. Gridnev, V.G. Gavridyuk, V.V. Nemoshkalenko, Y.A. Polushkin and O.N. Razumov, Fiz. Metal Metalloved, 43 (1977) 582.
194 K.H. Eickel and W. Pitsch, Phys. Status Solidi, 39 (1970) 121.
195 T. Moriya, Y. Sumitomo, F.E. Fujita and Y. Maeda, J. Phys. Soc. Jap., 35 (1973) 1378.
196 J. Bainbridge, D.A. Channing, W.H. Whitlow and R.E. Pendlebury, J. Phys. Chem. Solids, 34 (1973) 1579.
197 N. de Cristofaro and R. Kaplow, Met. Trans., 8A (1977) 425.
198 G. Longworth, NDT International, (1977) 241 and Patent Ref. 11959 LnH.
199 Y. Ujihira and A. Handa, J. Phys. (Paris) Colloq., 40 (1979) C2-586.
200 G. Longworth and N.E.W. Hartley, Thin Solid Films, 48 (1978) 95.

201 N.E.W. Hartley, Wear, 34 (1975) 427.
202 G. Dearnaley, J.H. Freeman, R.S. Nelson and J. Stephen, Ion Implantation, North Holland, Amsterdam, 1977.
203 L.H. Schwartz, Int. J. Nondestr. Testing, 1 (1970) 353.
204 D. Chandra and L.H. Schwartz, in I.J. Gruverman (Ed.), Mossbauer Effect Methodolog-, Plenum Press, New York, 6 (1971) 79.
205 C.C. Tsuei, G. Longworth and S.C.H. Lin, Phys. Rev., 170 (1968) 60.
206 U. Gonser, M. Ghafari and H.G. Wagner, J. Magn. Magn. Mater., 8 (1978) 175.
207 F.E. Fujita, in U. Gonser (Ed.), Topics in Applied Physics, 5, Mossbauer Spectroscopy, Springer Verlag, Berlin, 1975, 201.
208 G. Le Caer and J.M. Dubois, J. Phys. E: Sci. Instrum., 12 (1979) 1083.
209 C.L. Chien, Hyperfine Interactions, 4 (1978) 869.
210 C.C. Tsuei and H. Lilienthal, Phys. Rev. B, 13 (1976) 4899.
211 W. Mansel, G. Vogl and W. Koch, Phys. Rev. Lett., 31 (1973) 359.
212 G. Vogl, W. Mansel and W. Vogl, J. Phys. F: Met. Phys., 4 (1974) 2321.
213 G. Vogl, W. Mansel and P.H. Dederichs, Phys. Rev. Lett., 36 (1976) 1497.
214 G. Vogl, A. Schaefer, W. Mansel, J. Prechtel and W. Vogl, Phys. Status Solidi (b) 59 (1973) 107.
215 B.D. Sawicka and J.A. Sawicki, J. Phys. (Paris) Colloq., 40 (1979) 576.
216 S.R. Reintsema, S.A. Drentje, P. Schurer and H. de Waard, Radiat. Eff., 24 (1975) 145.
217 S.R. Reintsema, S.A. Drentje and H. de Waard, Hyperfine Interactions, 5 (1978) 167.
218 S.R. Reintsema, E. Verbiest, J. Odeurs and H. Pattyn, J. Phys. F: Met. Phys., 9 (1979) 1511.
219 J.C. Walker and C.L. Chien, Proc. 5th Int. Conf. on Mossbauer Spectroscopy, Bratislava, Pt. 3 (1973) 560.

CHAPTER 5

THE INTENSITY TENSOR FORMULATION FOR DIPOLE TRANSITIONS (e.g. ^{57}Fe) AND ITS APPLICATION TO THE DETERMINATION OF EFG TENSOR.

R. Zimmermann

Physikalisches Institut der Universitat Erlangen-Nurnberg, Erwin-Rommel-Str. 1, 8520 Erlangen, West Germany

1. INTRODUCTION

The evaluation of the angular dependence of Mossbauer line intensities is a fundamental problem in γ-ray spectroscopy. It yields information on the hyperfine fields at the nucleus and can give insight on the electronic states of the Mossbauer ion. If measurements can be performed on single crystals the evaluation of the line intensities becomes particularly important, since it yields the necessary information for a determination of the orientation of the hyperfine fields relative to the crystallographic axes. The mathematical expressions to be used depend upon the type of multipole transition considered. In the present context we will restrict the discussion to Mossbauer radiation resulting from dipole transitions (i.e. ^{57}Fe, ^{119}Sn, ^{61}Ni nuclei). In this case the evaluation of the angular dependence of the line intensities can be considerably simplified by utilization of the intensity tensor.

The calculation of the line intensities has been extensively treated in literature [1-7]. Usually the expression for dipole radiation (electric and magnetic) is derived from a multipole expansion of the electromagnetic radiation. The angular dependence of the line intensity of the dipole term follows from that of vector spherical harmonics. The resulting formulae do not immediately show the general behaviour of the angular dependence. Thus it is convenient to return from spherical harmonics to cartesian quantities. This leads to the concept of the intensity tensor.

The aim of the present work is not so much the calculation of the line intensities but the evaluation of the line intensities. The evaluation can be, of course, carried out with any formula by using the method of trial and error. This method, however, is usually time consuming and furthermore possible solutions to experimental data may be easily overlooked. Rather it is reasonable to define quantities which contain the total information to be extracted from the angular dependence of the line intensities. The quantities should be easily obtainable from experimental data and their evaluation should be straightforward. We will see that the intensity tensor has these required properties.

The convenience of the intensity tensor formulation has been already demonstrated in the special case of single crystal ^{57}Fe quadrupole spectra [8]. By measuring spectra for different crystal orientations (relative to the γ-ray), the components of the intensity tensor were determined. From these values all possible solutions for the EFG are easily found. For a monoclinic crystal, in particular, a linear manifold of solutions can be found [8, 9]. This has not been pointed out in earlier discussions [10-12]. If this ambiguity is not recognised, it can very well lead to a wrong determination of the orientation of the EFG and the sign of the quadrupole splitting.

From the discussion given in reference [13] it is seen that the intensity tensor formulation is not restricted to ^{57}Fe quadrupolar spectra. Rather the concept of the intensity tensor can always be applied if we are concerned with nuclear dipole transitions. In the present context we will treat the problem from a general point of view. Particularly we will show that the angular dependence for polarized γ-rays can also be described by the intensity tensor. The expressions derived will be applicable not only for experiments with polarized sources and absorbers, but also for the calculation of the thickness effect in samples. Furthermore we will introduce formulae which can be applied to textured matetials.

2. THE INTENSITY TENSOR

2.1 Definition of the Intensity Tensor

We consider a Mossbauer nucleus which absorbs γ-quanta through electric or magnetic dipole transitions. To determine the angular dependence of the line intensity we specify a coordinate system O_{xyz}. This system may have any orientation relative to the crystallographic axes of the crystal. Usually it is convenient to align it with crystallographic axes. In an orthorhombic crystal, for instance, we would choose the x,y and z-axes parallel to the screw dyad axes of the crystal, i.e. parallel to the base vectors of the unit cell. With a frame of reference at hand we can specify the direction of the γ-ray for a particular measurement, either by the direction cosines of the γ-ray, which we will denote by e_x,e_y and

e_z, or by the polar angles of the direction of the γ-ray, φ and θ. Both descriptions are equivalent and connected by the equations,

$$e_x = \sin\theta\cos\phi,\ e_y = \sin\theta\sin\phi,\ e_z = \cos\theta. \quad (1)$$

A graphical display of the situation is given in Fig. 1. The line

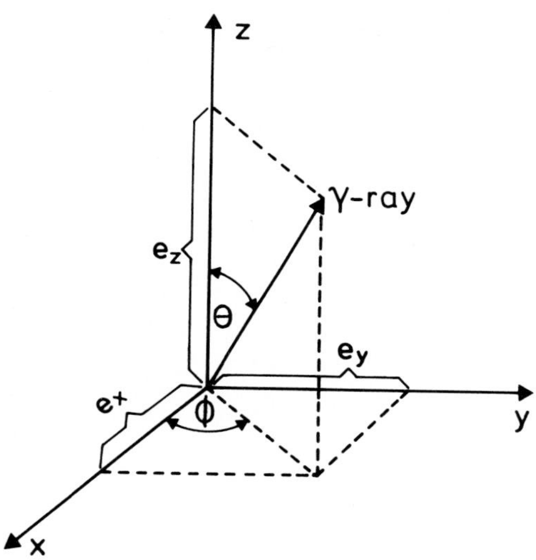

Fig. 1. The coordinate system O_{xyz} and description of the orientation of the γ-ray by direction cosines e_x, e_y, e_z and polar angles θ, φ.

intensity I depends upon the orientation of the γ-ray. In the case of an unpolarized γ-ray its angular dependence can be written in the following way [+)] [13],

$$I = \sum_{p,q = x,y,z} I_{pq}\, e_p\, e_q \quad . \quad (2)$$

The six quantities I_{pq} , p,q = x,y,z, are the components of a tensor, called intensity tensor, which is three-dimensional, real, symmetric ($I_{pq} = I_{qp}$) and of second rank. In the next section, we shall derive a formula which can be used to calculate the components of the intensity tensor from the expansion coefficients of the nuclear eigenstates. It is assumed in the derivation of the formula that we deal with a thin absorber and that the Debye Waller factor is

+) The formula can be applied also in the case of emission, if the γ-rays are detected by an unpolarized absorber.

isotropic (no Goldanskii-Karyagin effect). This assumption holds in the present context unless otherwise specified.

By Eqn. (2), the intensity I is determined up to a constant factor. This constant factor can be defined by a convention for the total intensity of the spectrum. Usually we do not worry about this constant and write the theoretical formulae for the intensity tensor in such a way that incovenient prefactors can be dropped. Sometimes, however, it is convenient to normalize the total intensity of the spectrum to 1 (e.g. for single crystal ^{57}Fe quadrupolar spectra). In this case, I is the fractional intensity of the line relative to the total intensity of the spectrum. We will use this definition whenever it is intended that the normalization to 1 is used.

Inserting Eqn. (2) into Eqn. (1), we obtain an expression for the intensity as a function of the polar angles of the γ-ray, θ and ϕ:

$$I = I_{xx}\sin^2\theta\cos^2\phi + I_{yy}\sin^2\theta\sin^2\phi + I_{zz}\cos^2\theta$$

$$+ 2I_{xy}\sin^2\theta\cos\phi\sin\phi + 2I_{xz}\cos\theta\sin\theta\cos\phi + 2I_{yz}\cos\theta\sin\theta\sin\phi.$$

Though rather lengthy, this way of writing might help to clarify what is behind Eqn. (2). If measurements are performed in one plane perpendicular to the z-axis, the intensity is given by ($\theta=\pi/2$):

$$I = I_{xx}\cos^2\phi + I_{yy}\sin^2\phi + 2I_{xy}\sin\phi\cos\phi \tag{3}$$

Hence the intensity is a sinusoidal function of the rotation angle $\phi^{\dagger}$. This function is usually graphically displayed in either of two ways, as shown in Fig. 2 and Fig. 3. Quite generally this behavior can be expressed as,

$$I = A + B\cos(2\phi - 2\phi_o) \tag{4}$$

This equation can be used to fit experimental intensities. When A, B and ϕ_o have been determined the components of the intensity tensor can be obtained from,

$$I_{xx} = A + B\cos 2\phi_o ,$$

$$I_{yy} = A - B\cos 2\phi_o , \tag{5}$$

$$I_{xy} = B\cos(\tfrac{\pi}{2} - 2\phi_o).$$

---

†Note that ϕ describes the rotation angle of the γ-ray in O_{xyz} and not the rotation angle of the crystal relative to the γ-ray (there is a difference in the sign of ϕ).

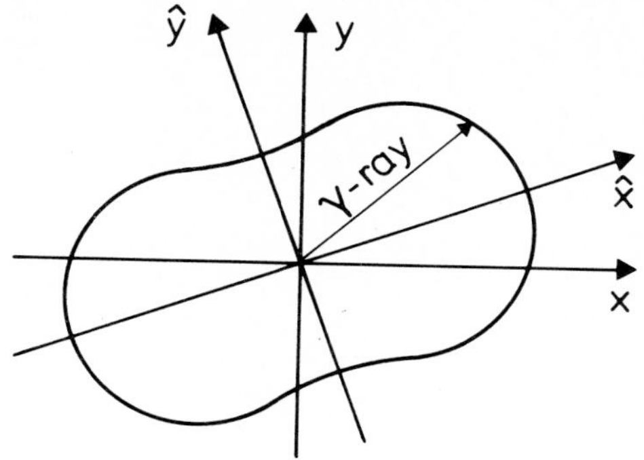

Fig. 2. Angular dependence of the line intensity in one plane. The vector drawn is to be parallel to the γ-ray. The amount of the vector gives the value of the intensity. The axes $\hat{x}$ and $\hat{y}$ correspond to the principal axes system of the two dimensional projection of the intensity tensor.

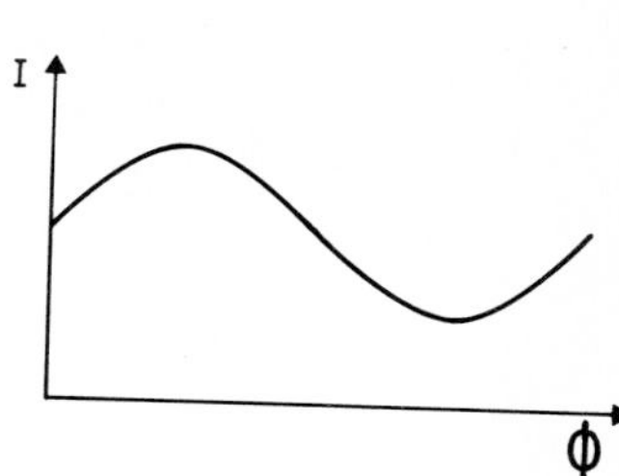

Fig.3. The intensity I as a function of the rotation angle ϕ. It has sinusoidal behavior (see Eqn. 4).

In a powder sample we have to average the intensity over the statistical orientations of the microcrystal. Carrying out the powder integration we obtain for the powder intensity I^{pow},

$$I^{pow} = \frac{1}{3}\,(I_{xx} + I_{yy} + I_{zz}). \tag{6}$$

It can be rationalized that this formula is analogous to the formula for the powder susceptibility. The powder intensity can be used to define the traceless intensity tensor,

$$T_{pq} = I_{pq} - I^{pow}\,\delta_{pq}. \tag{7}$$

This traceless tensor will play a central role in our treatment.

2.2 Experimental Determination of the Intensity Tensor

In this section, we want to discuss the determination of the intensity tensor from single crystal Mossbauer spectra. For simplicity we assume that we deal with thin single crystals which allow to neglect the thickness correction of the areas of the absorption lines. The case of thick single crystals will be discussed in Section 6. Fig. 4 shows a set of quadrupole spectra for different orientations of single crystals of Mohr salt ($(NH_4)_2Fe(SO_4)_2 \cdot 6H_2O$) [14]. The

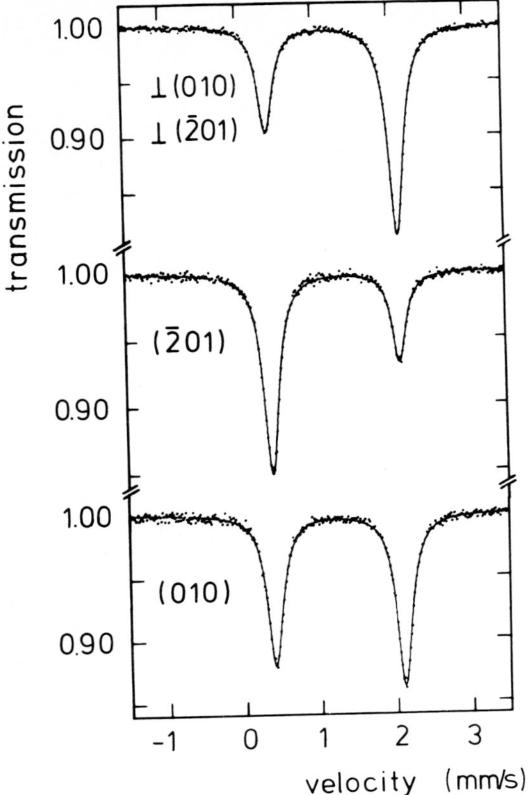

Fig. 4. Quadrupolar spectra of single crystals of Mohr salt at 300 K. Each crystal is characterised by the plane perpendicular to which the γ-ray was oriented. The planes are specified by Miller indices as indicated beside each spectrum. The plane for the uppermost spectrum was perpendicular to the (010) and (201) plane.

spectra have been fitted with Lorentzian lines. From the peak amplitude (or percentage absorption) and the line width, we can calculate the areas of the lines at high and low velocity, $A^{(2)}$ and $A^{(1)}$, respectively. The fractional intensities of the lines at high and low velocity, $I^{(2)}$ and $I^{(1)}$, are :

$$I^{(1)} = \frac{A^{(1)}}{A^{(1)} + A^{(2)}}, \quad I^{(2)} = \frac{A^{(2)}}{A^{(1)} + A^{(2)}}.$$

Due to the relation $I^{(1)} + I^{(2)} = 1$ which follows from above we can focus our attention on the angular dependence of one line of each spectrum, e.g., on the line at high velocity. For each orientation i we can determine the value for the fractional intensity $I = I^{(2)}$ and set up equation,

$$I\ (\text{Orientation } i) = I_{xx}\sin^2\theta_i\cos^2\phi_i + I_{yy}\sin^2\theta_i\sin^2\phi_i + I_{zz}\cos^2\theta_i + I_{xy}\sin^2\theta_i\sin 2\phi_i + I_{xz}\sin 2\theta_i\cos\phi_i + I_{yz}\sin 2\theta_i\ \sin\phi_i,$$

where θ_i and ϕ_i are the polar angles of the i-th orientation of the γ-ray. Combining all orientations we obtain a set of linear equations from which we can determine the intensity tensor, the number of different orientations being large enough. Since the intensity tensor is of second rank, we should need at least six orientations for its determination. In many cases, however, the intensity tensor has to satisfy some constraints (e.g. due to the symmetry group of the crystal, see Section 4) and hence we need less than six measurements. It is, of course, reasonable to make more measurements than absolutely necessary and to determine the intensity tensor by a least squares analysis. Since the set of equations is linear, the solution of the problem stays rather simple.

2.3 Calculation of the Intensity Tensor

The absorption of a γ-ray by a nucleus is accompanied by a transition of the nucleus from a ground state $|A\rangle$ to an excited state $|B\rangle$. In Mossbauer spectroscopy the eigenstates $|A\rangle$ and $|B\rangle$ belong to two different multiplets of nuclear spin j_a and j_b, respectively. The eigenstates result from diagonalizing the hyperfine field Hamiltonian in each multiplet. They can be arbitrary linear combinations of the m-quantum states $|j_a,m\rangle$, $|j_b,m'\rangle$;

$$|A\rangle = \sum_{m=-j_a}^{j_a} a_m|j_a,m\rangle\ ,\quad |B\rangle = \sum_{m'=-j_b}^{j_b} b_{m'}|j_b,m'\rangle. \tag{8}$$

In the case of a dipole transition the relative intensity I for absorption of an unpolarized γ-ray is given by (6)

$$I = |\sum_{q=-1}^{1} c_q\ \underset{\rightarrow}{\chi}^q|^2 \tag{9}$$

where $\underset{\rightarrow}{\chi}^q$ is a vector spherical harmonic for L = 1 and

$$c_q = \sum_{m=-j_a}^{j_a} a_m^* b_{m+q} \langle j_a, 1; m, q | j_b, m+q \rangle, \quad q=-1,0,1 \tag{10}$$

are the spherical components of a vector characterizing the transition. The above formula implies that thickness effects do not occur and that the recoilless fraction is isotropic. It can be applied also in the case of emission, if the γ-rays are detected by an unpolarized absorber. The Clebsch Gordan Coefficient $\langle j_a, 1; m, q | j_b, m+q \rangle$ in Eqn. (10) vanishes for $|\Delta j| = |j_a - j_b| > 1$ and hence dipole transitions can occur only for $\Delta j = 0, \pm 1$. To drive the intensity tensor we introduce the cartesian components of c_q and $\vec{\chi}^q$ (c_q and $\vec{\chi}^q$ transform contragredient):

$$\vec{\chi}^x = -\frac{1}{\sqrt{2}}(\vec{\chi}^{-1} - \vec{\chi}^1), \quad \vec{\chi}^y = \frac{i}{\sqrt{2}}(\vec{\chi}^{-1} + \vec{\chi}^1), \quad \vec{\chi}^z = \vec{\chi}; \tag{11}$$

$$c_x = -\frac{1}{\sqrt{2}}(c_{-1} - c_1), \quad c_y = -\frac{i}{\sqrt{2}}(c_{-1} + c_1), \quad c_z = c_o.$$

With these components the intensity I is expressed by a formula identical to that in Eqn. (9) just with q = x,y,z. By inserting the expressions for vector spherical harmonics [15] into the formula for the intensity and by comparing it with Eqn. (2) we obtain the components of the intensity tensor (a constant factor has been omitted):

$$I_{pq} = 2(c_x c_x^* + c_y c_y^* + c_z c_z^*)\delta_{pq} - (c_p c_q^* + c_p^* c_q), \tag{12}$$

p,q = x,y,z. Since the components c_p have vector properties, it is clear that the components I_{pq} have the properties of a tensor of the second rank. For a more a priori derivation of the intensity tensor the reader is referred to the Appendix.

Though the coefficientx c_x, c_y and c_z build up the intensity tensor in a symmetric form it is sometimes more handy to express the components I_{pq} by the coefficients c_1, c_o and c_{-1}, the result of which is given in Table. Together with the Clebsch Gordan Coefficients of Table 2, we can easily calculate the intensity tensor of any dipole transition.

If we are interested only in relative intensities we can simplify the expression for Clebsch Gordan Coefficients in Table 2 by omitting common factors. With the factors given, the total intensity of all transitions between the multiplets is equal to (4/3)($2j_b$ + 1). Thus, if we want to express the intensity I as a

TABLE 1.

The components of the intensity tensor as a function of the coefficients c_1, c_o, and c_{-1}.

$$I_{xx} = |c_1+c_{-1}|^2 + 2|c_o|^2$$

$$I_{yy} = |c_1-c_{-1}|^2 + 2|c_o|^2$$

$$I_{zz} = 2|c_1|^2 + 2|c_{-1}|^2$$

$$I_{xy} = 2|ic_1c^*_{-1}|^2$$

$$I_{xz} = \sqrt{2}|(c_1-c_{-1})c^*_o|^2$$

$$I_{yz} = \sqrt{2}|i(c_1+c_{-1})c^*_o|^2$$

TABLE 2.

Clebsch Gordan Coefficients $\langle j_a, 1; m, q|j_b, m+q\rangle$ for $j_b - j_a = o,1$ and $q = o,1$ ($q \equiv \Delta m$). The Clebsch Gordan Coefficients of the $q=-1$ transition can be obtained from the one of the $q = 1$ transition by using the formula,

$$\langle j_a,1;m,-1|j_b,m-1\rangle = (-1)^{j_b-j_a-1}\langle j_a,1;-m,1|j_b,-m+1\rangle.$$

j_b \ q	$q = o$	$q = 1$
$j_b = j_a+1$	$\sqrt{\dfrac{(j_a-m+1)(j_a+m+1)}{(2j_a+1)(j_a+1)}}$	$\sqrt{\dfrac{(j_a-m-1)(j_a-m)}{(2j_a+1)(2j_a+2)}}$
$j_b = j_a$	$\dfrac{m}{\sqrt{j_a(j_a+1)}}$	$\sqrt{\dfrac{(j_a-m-1)(j_a+m+2)}{2j_a\ (j_a+1)}}$

fraction of the total intensity we have to divide the components I_{pq} by $(4/3)(2j_b + 1)$ (or equivalently the components c_q by the root of that factor). The resulting tensor will be called normalized intensity tensor.

At the end of this section, we discuss the particular form of the intensity tensor in the case of two examples: that of magnetic hyperfine interaction and that of quadrupole interaction in ^{57}Fe. The second example, in particular, will indicate advantages of the use of the intensity tensor.

2.4 Example 1: The Intensity Tensor in the Case of Magnetic Hyperfine Interaction

In the case of magnetic hyperfine interaction we choose as usual the quantisation axis z parallel to the effective magnetic field. In this case, the eigenstates of the multiplets with spin j_a and j_b are the azimuthal quantum states $|j_a,m\rangle$ and $|j_b,m'\rangle$. Dipole transitions occur only for $q = \Delta m = 0, \pm 1$. As a result of the axial symmetry of the internal field, the intensity tensor has to be axial also, i.e. $I_{xx} = I_{yy}$ and $I_{xy} = I_{xz} = I_{yz} = 0$. Thus, only the values of the components I_{xx} and I_{zz} have to be considered in more detail. From Table 3, we can obtain their expressions in terms of Clebsch Gordan Coefficients whose values can be found from Table 2. As can

TABLE 3.

Components of the intensity tensor and powder mean value in the case of magnetic hyperfine interaction. For the components of the intensity tensor, the z-axis is to be parallel to the effective magnetic field. The Clebsch Gordan Coefficients can be obtained from Table 2.

	$q = \Delta m = \pm 1$	$q = \Delta m = 0$
$I_{xx} = I_{yy} =$	$\|\langle j_a,1;m,\pm 1\|j_b,m\pm 1\rangle\|^2$	$2\|\langle j_a,1;m,0\|j_b,m\rangle\|^2$
$I_{zz} =$	$2\,I_{xx}$	0
I^{pow}	$\frac{4}{3}\|\langle j_a,1,m\pm q\|j_b,m\pm q\rangle\|^2$	

be seen from Table 3, there are only two types of intensity tensors involved i.e. those with $I_{zz} = 2\,I_{xx}$ for $\Delta m = \pm 1$ transitions and those with $I_{zz} = 0$ for $\Delta m = 0$ transitions. Hence, we can derive

$$I = I_{xx}\,(1+\cos^2\theta) \quad \text{for } \Delta m = \pm 1 \text{ transitions}$$

and

$$I = I_{xx}\sin^2\theta \quad \text{for } \Delta m = 0 \text{ transition.}$$

These expressions are well known and presented in order to clarify the connection to the intensity tensor. Sometimes it is convenient to use a coordinate system whose z-axis is not aligned to the effective field. If h_x, h_y and h_z are the direction cosines of the effective field in O_{xyz}, the intensity tensor can be written as (p,q = x,y,z):

$$I_{pq} = \begin{cases} \frac{3}{4} I^{pow} (\delta_{pq} + h_p h_q) & \text{for } \Delta m = \pm 1 \text{ transitions} \\ \frac{3}{2} I^{pow} (\delta_{pq} - h_p h_q) & \text{for } \Delta m = 0 \text{ transitions} \end{cases}$$

2.5 Example 2: The Intensity Tensor in the Case of ^{57}Fe Quadrupolar Spectra

In the presence of an electric field gradient (EFG), the first excited state of the ^{57}Fe nucleus (j = 3/2) is split into two levels. The splitting is given by:

$$\Delta E_Q = \frac{1}{2} eQ \sqrt{V_{zz}^2 + \frac{1}{3} (V_{xx} - V_{yy})^2 + \frac{4}{3} (V_{xy}^2 + V_{xz}^2 + V_{yz}^2)}, \tag{13}$$

where e is the positive elementary charge, Q the nuclear quadrupole moment and V_{pq}(p,q = x,y,z) are the components of the EFG tensor. The nuclear ground state (j = 1/2) is not affected by the EFG. Hence the Mossbauer line is split into a doublet with quadrupole splitting ΔE_Q. In the case of a single crystal absorber the two lines will have unequal intensity, the angular dependence of which can be described by Eqn. (2). For the components of the normalized intensity tensor we find [8],

$$I_{pq} = \frac{1}{2} \delta_{pq} \pm \frac{eQ}{8|\Delta E_Q|} V_{pq}. \tag{14}$$

Here the plus (minus) sign applies for the line at higher (lower) energy. The powder mean value of each tensor is equal to 1/2. This corresponds to the fact that in a randomly oriented powder the quadrupole lines have equal line intensity. The traceless intensity tensor (see Eqn. 7) for the line at higher energy obeys the relation,

$$T_{pq} = \frac{eQ}{8|\Delta E_Q|} V_{pq} . \tag{15}$$

Thus the traceless intensity tensor is related to the EFG tensor. The anisotropy at the nucleus is directly reflected by the anisotropy of the line intensity. Eqn. (15) indicates the advantage of the use of the intensity tensor when evaluating single crystal quadrupole spectra. Assuming that the traceless intensity tensor has been determined by rotating the single crystal relative to the γ-beam[+)] we can use the following properties for its evaluation.

(1) The principal axes system of the intensity tensor is identical with the principal axes system of the EFG.

(2) The asymmetry parameter of the EFG is identical with the asymmetry parameter of the traceless intensity tensor;

$$\eta = (T_{\hat{x}\hat{x}} - T_{\hat{y}\hat{y}})/T_{\hat{z}\hat{z}} , \qquad (16)$$

where $T_{\hat{x}\hat{x}}$, $T_{\hat{y}\hat{y}}$, $T_{\hat{z}\hat{z}}$ are the components of the traceless intensity tensor in the principal axes system with $|T_{\hat{x}\hat{x}}| \leqslant |T_{\hat{y}\hat{y}}| \leqslant |T_{\hat{z}\hat{z}}|$. In the case of a non-diagonal intensity tensor, the asymmetry parameter results from the more complex equation:

$$\eta = \sqrt{3}\,\tan\left(\frac{1}{3}\arccos|I_\eta|\right), \qquad (17a)$$

where

$$I_\eta = 256\,(T_{xx}T_{yy}T_{zz} + 2T_{xy}T_{xz}T_{yz} - T_{xx}T_{yz}^2 - T_{yy}T_{xz}^2 - T_{zz}T_{xy}^2). \qquad (17b)$$

(3) The sign of the quadrupole splitting can be found from,

$$\mathrm{sign}(\Delta E_Q) = \mathrm{sign}(T_{\hat{z}\hat{z}}) = \mathrm{sign}(I_\eta). \qquad (18)$$

(4) The quantity I_Δ, defined analogously to the square of the quadrupole splitting, has a constant value,

$$I_\Delta = 16\{T_{zz}^2 + \frac{1}{3}(T_{xx} - T_{yy})^2 + \frac{4}{3}(T_{xy}^2 + T_{xz}^2 + T_{yz}^2)\} = 1. \qquad (19)$$

Hence the traceless intensity tensor is proportional to the quadrupole coupling tensor $(1/2)eQV_{pq}$ with a proportionality constant such that the quadrupole splitting is 1/4.

The above properties apply only in the case of identical lattice sites. If there is more than one equivalent lattice site per unit

---

[+)] For this determination we can use that $I_{xx} + I_{yy} + I_{zz} = 3/2$. Hence only five components of the intensity tensor have to be determined.

cell, Eqns. (16) to (19) hold only between the local intensity tensor and the local EFG tensor of each lattice site. However, what we measure is the macroscopic intensity tensor. Hence in the case of more than one equivalent lattice site, the evaluation of single crystal Mossbauer spectra has to be done with some care as will be described in Section 4.

3. GENERAL PROPERTIES OF THE INTENSITY TENSOR

In this section we discuss some properties of the intensity tensor which are valid in the case of combined electric and magnetic hyperfine interactions. These properties are concerned with the effect of symmetry on the intensity tensor, an interrelation between the components of the intensity tensor, and the use of the intensity tensor for powder samples measured in perpendicular configurations (split coil).

3.1 The effect of symmetry

One of the advantages of the utilization of intensity tensors results from the close (and obvious) relation between a symmetry at the nucleus and properties of the intensity tensor of a Mossbauer line. In the presence of an electric field gradient (EFG), for instance, the intensity tensor has to be diagonal within the principal axes system of the EFG (D_2 symmetry). Even if an effective magnetic field is parallel to one of the principal components V_{xx}, V_{yy}, V_{zz} of the EFG, the nondiagonal elements I_{xy}, I_{xz}, I_{yz} have to vanish. In both cases the principal axes system of the EFG can be found by determining that of the intensity tensor.

3.2 Interrelation between the components I_{pq}

Since the vector (c_x, c_y, c_z) is a complex vector which has, in general, six independent components, a superficial consideration will show that the intensity tensor will also have six independent components. However it can be seen from Eqn. (12) that the multiplication of $\underline{c}$ with an arbitrary phase factor does not change the components of the intensity tensor. Thus the intensity tensor has, at the most, five independent components. To write down an interrelation between the components we define,

$$I_{pq}^{(\ell)} = I^{pow}\delta_{pq} - 2\ I_{pq}. \tag{20}$$

The superfix (ℓ) has been chosen since $I_{pq}^{(\ell)}$ describes the angular dependence of the absorption of linear polarized γ-rays (see Section 5). With the above definition the interrelation results in:

$$\mathrm{Det}(I_{pq}^{(\ell)})=0. \tag{21}$$

In particular, this equation shows that one principal component of the tensor $I_{pq}^{(\ell)}$ has to be zero. In general, it can be used to calculate one component of the intensity tensor from the other five. Furthermore Eqn. (21) might be used to test whether an intensity tensor, which has been determined from experimental intensities, results indeed from a non-degenerate transition (Eqn. (21) must not hold for degenerate lines, e.g. not for ^{57}Fe quadrupole spectra).

3.3 Perpendicular Configurations

When measuring Mossbauer samples in external magnetic fields, we commonly deal with either of two kinds of configurations: one where the γ-rays are parallel to the external field (parallel configuration) and another where the γ-rays are perpendicular to the external field (perpendicular configuration in a split coil). If we want to simulate powder samples, we have to add up the spectra for different orientations of the microcrystals. In general, the orientation of a body can be described by three Euler angles α,β,γ. For the parallel configuration, a rotation of the microcrystal around the axis of the external field will not change the Mossbauer spectra. Hence the powder averaging has to be done only for two Euler angles, namely α and β. In the case of a perpendicular configuration, the rotation of a microcrystal around the magnetic field will change the intensities of the Mossbauer lines (not the positions). Thus we have to integrate over those orientations, i.e. we have also to integrate over γ. Using the intensity tensor, the latter integration is very simple. If e_p^θ and e_p^ϕ are the direction cosines of any two orthogonal directions perpendicular to the external field, the intensity found by integration is given by:

$$I = \frac{1}{2}\sum_{p,q=x,y,z} I_{pq}\,(e_p^\theta e_q^\theta + e_p^\phi e_q^\phi). \tag{22}$$

We now postulate that the unit vector $\underset{\sim}{e}$ be parallel to the external field (as it would be in the case of the parallel configuration). Since $\underset{\sim}{e}^\theta$, $\underset{\sim}{e}^\phi$ and $\underset{\sim}{e}$ are three mutual orthogonal vectors, their components satisfy the relation:

$$e_p^{\theta} e_q^{\theta} + e_p^{\phi} e_q^{\phi} + e_p e_q = \delta_{pq} \tag{23}$$

Inserting this relation into Eqn. (22), the intensity resulting from the integration over the directions perpendicular to the magnetic field is given by:

$$I = \frac{3}{2} I^{pow} - \frac{1}{2} \sum_{p,q=x,y,z} I_{pq} e_p e_q . \tag{24}$$

Hence for the perpendicular configuration, we can proceed for the powder averaging in the same way as for the parallel configuration if we use the tensor

$$I_{pq}^{(\perp)} = \frac{3}{2} I^{pow} \delta_{pq} - \frac{1}{2} I_{pq} \tag{25}$$

instead of the tensor I_{pq}.

4. SINGLE CRYSTALS WITH EQUIVALENT LATTICE SITES

4.1 General

If the Mossbauer ions sit on identical lattice sites, the experimental ratios of the line intensities correspond to the actual ratios of the line intensities of each Mossbauer ion. In this case, the evaluation of the line intensities does not contain ambiguities inherent in the symmetry of the crystal. Unfortunately, crystals with identical lattice sites are rare. Usually crystals have more than one equivalent lattice site in unit cell for the Mossbauer ions. In this case, the environment of the Mossbauer ions can differ in orientation. Hence the EFG's may be titled with respect to each other eventhough the quadrupole splitting (including sign) and the asymmetry parameter are the same. The relative orientation of the EFG's of equivalent lattice sites is not random. The EFG's of equivalent lattice sites can be transformed into each other by the symmetry transformations of the crystal. In a monoclinic crystal, for instance, the iron may sit on either of two equivalent lattice sites which are connected by a screw dyad axis.

The line intensities which result from the Mossbauer ions of a particular lattice site, say the i-th lattice site can be characterized by a local intensity tensor $I_{pq}^{(loc,i)}$. The intensity which we measure is the sum of the intensities from different lattice sites.

Thus the macroscopic intensity I can be characterized by a macroscopic intensity tensor $I_{pq}^{(m)}$ which is the mean value of the local intensity tensors

$$I_{pq}^{(m)} = \frac{1}{N} \sum_{i=1}^{N} I_{pq}^{(loc,i)}, \tag{26}$$

where N is the number of lattice sites (the division by N is necessary to keep normalized intensity tensors normalized).

The macroscopic intensity tensor has to reflect the symmetries of the crystal. In a monoclinic crystal with a screw dyad axis, for instance, one of its principal components has to be parallel to this axis. This can be seen as follows. Let the z axis be along the screw dyad axis. The two-fold symmetry (a translation can be disregarded in the present context) requires the following relations between the intensity tensor of site 1 and site 2 ($x \to -x$, $y \to -y$, $z \to z$):

$$\begin{aligned} I_{pp}^{(loc,2)} &= I_{pp}^{(loc,1)} \qquad p = x,y,z \\ I_{xy}^{(loc,2)} &= I_{xy}^{(loc,1)} \\ I_{xz}^{(loc,2)} &= -I_{xz}^{(loc,1)} \\ I_{yz}^{(loc,2)} &= -I_{yz}^{(loc,1)}. \end{aligned} \tag{27}$$

Thus the macroscopic intensity tensor of a monoclinic crystal is given by

$$\begin{aligned} I_{pp}^{(m)} &= I_{pp}^{(loc,1)} \qquad p = x,y,z \\ I_{xy}^{(m)} &= I_{xy}^{(loc,1)} \\ I_{xz}^{(m)} &= 0 \\ I_{yz}^{(m)} &= 0. \end{aligned} \tag{28}$$

The xz and yz components of the intensity tensors of the two lattice sites cancel. No matter what kind of nondiagonal elements the local intensity tensors have the macroscopic intensity tensor has one principal component (here $I_{zz}^{(m)}$) along the two-fold axis.

In an orthorhombic crystal we can have three mutual orthogonal screw dyad axes. In this case the three principal components of the macroscopic intensity tensor have to be parallel to those axes. Choosing the axes, x,y, and z along the screw dyad axes we find,

$$I_{pp}^{(m)} = I_{pp}^{(loc,1)} , \qquad p = x,y,z ,$$
$$I_{pq}^{(m)} = o , \qquad p \neq q . \tag{29}$$

The nondiagonal elements of the macroscopic intensity tensor vanish whatever may be the nondiagonal elements of the local intensity tensor $I_{pq}^{(loc,i)}$.

The cancellation of components of the local intensity tensors allows the macroscopic intensity tensor to reflect the symmetry of the crystal. This cancellation, on the other hand, is responsible for the fact that local intensity tensors cannot be determined uniquely from the macroscopic intensity tensor. In the case of ^{57}Fe quadrupolar spectra the above ambiguity has important consequences in that monoclinic or rhombic single crystals may not yield unique solutions for the local EFG and not even the sign of the quadrupole splitting may be determined in some cases. This problem is discussed in more detail in the next two examples.

4.2 Example 3: Determination of the Electric Field Gradient Tensor from Quadrupolar Spectra of Monoclinic crystals (e.g. $FeCl_2.4H_2O$)

Monoclinic crystals of $FeCl_2.4H_2O$ have a screw dyad axis along the b-axis of the unit cell. This axis is defined as the z-axis of the coordinate system O_{xyz}. Due to the twofold symmetry, the macroscopic intensity tensor has to have one principal axis along the z-axis (see Eqn. (28), $I_{xz}^{(m)} = I_{yz}^{(m)} = 0$). The other two principal axes have to lie in the a'c-plane. Neglecting thickness effects*), we can determine the components of the macroscopic intensity tensor from the ratio of areas of the two absorption lines. From literature values [10] we find that the principal axes x and y of the macroscopic intensity tensor are nearly parallel to the a' and c axes of the unit cell, respectively and that within this coordinate system the the components of the macroscopic intensity tensor for the line

*) Thickness effects are treated in Section 6.

at higher velocity are given by,

$$I_{xx}^{(m)} = 0.55,\ I_{yy}^{(m)} = 0.39,\ I_{zz}^{(m)} = 0.56,\ I_{xy}^{(m)} = 0. \tag{30}$$

From these we derive the components of the traceless macroscopic intensity tensor by subtracting 1/2 from the diagonal elements,

$$T_{xx}^{(m)} = 0.05,\ T_{yy}^{(m)} = -0.11,\ T_{zz}^{(m)} = 0.06,\ T_{xy}^{(m)} = 0. \tag{31}$$

According to Eqn. (28), those components are identical to the corresponding components of the local intensity tensors. Hence, we can write for the traceless, local intensity tensor at site 1:

$$T_{pq}^{(loc,1)} = \begin{bmatrix} T_{xx}^{(m)} & 0 & T_{xz}^{(loc,1)} \\ 0 & T_{yy}^{(m)} & T_{yz}^{(loc,1)} \\ T_{xz}^{(loc,1)} & T_{yz}^{(loc,1)} & T_{zz}^{(m)} \end{bmatrix} . \tag{32}$$

The components $T_{xz}^{(loc,1)}$ and $T_{yz}^{(loc,1)}$ are not determined by experiment. We may, however, recall that the traceless local intensity tensor must have a quadrupole splitting of 1/4 (see Eqn. (19) of Section 2.5). This condition can be written, in the present case, as:

$$(T_{xz}^{loc,1}) + (T_{yz}^{(loc,1)})^2 = \frac{3}{4}\left[\frac{1}{16} - (T_{zz}^{(m)})^2 - \frac{1}{3}(T_{xx}^{(m)} - T_{yy}^{(m)})^2\right] = L^2. \tag{33}$$

With the values of Eqn. (31), the value of L is different from zero which shows that the electric field gradients at the two lattice sites are differently oriented. As a result of Eqn. (33) the manifold of solutions for the local EFG is not two-dimensional, but one-dimensional. To describe the one-dimensional manifold we define an angle ψ by,

$$\psi = \arctan\left(T_{yz}^{(loc,1)}/T_{xz}^{(loc,1)}\right) = \arctan\left(V_{yz}/V_{xz}\right). \tag{34}$$

The traceless local intensity tensor can now be written as,

$$T_{pq}^{(loc,1)} = \begin{bmatrix} T_{xx}^{(m)} & 0 & L\cos\psi \\ 0 & T_{yy}^{(m)} & L\sin\psi \\ L\cos\psi & L\sin\psi & T_{zz}^{(m)} \end{bmatrix} . \tag{35}$$

The values of $T_{xx}^{(m)}$, $T_{yy}^{(m)}$, $T_{zz}^{(m)}$ and L are determined by experiment. The appearance of ψ shows that there are many local intensity tensors which lead to the same macroscopic intensity tensor.

The problem of finding the local EFG's is now solved, in principle, since the local EFG V_{pq} (p,q = x,y,z) is proportional to the traceless local intensity tensor:

$$eQ\, V_{pq} = 8|\Delta E_Q|\, T_{pq}^{(loc,1)}, \quad p,q = x,y,z. \tag{36}$$

It is, however, usual to diagonalize the EFG and to calculate the principal axes system, the asymmetry parameter η and the sign of the quadrupole splitting. The latter two quantities can be calculated with Eqns. (16) to (18). The orientation of the principal axes system can be described by three Euler angles α,β and γ. If $R_{\hat{q}p}$ is the p-th (p=x,y,z) component of the $\hat{q}$-th ($\hat{q}=\hat{x},\hat{y},\hat{z}$) principal axis, we may define the Euler angles by:

$$R_{\hat{q}p} = \begin{bmatrix} \cos\gamma & \sin\gamma & 0 \\ -\sin\gamma & \cos\gamma & 0 \\ 0 & 0 & 1 \end{bmatrix} \begin{bmatrix} \cos\beta & 0 & -\sin\beta \\ 0 & 1 & 0 \\ \sin\beta & 0 & \cos\beta \end{bmatrix} \begin{bmatrix} \cos\alpha & \sin\alpha & 0 \\ -\sin\alpha & \cos\alpha & 0 \\ 0 & 0 & 1 \end{bmatrix}. \tag{37}$$

This definition has the advantage that β and α are already the polar angles of the principal axis z of the EFG. Formulae for the explicit calculation of the Euler angles are found in the original publication [8] *). The results of the evaluation for $FeCl_2.4H_2O$ at 300 K are shown in Fig. 5. As can be visualized the asymmetry parameter η varies over a large region and not even the sign of the quadrupole splitting can be determined.

In Fig. 5 the values of ψ are restricted to $0<\psi<\pi/2$. The solutions for the corresponding ψ's in the other quadrants, $\pi-\psi$, $\pi+\psi$, $2\pi-\psi$ can be obtained by reflecting the local EFG's at the yz, xy, xz planes, respectively.

The ambiguity of the evaluation of zero field spectra of monoclinic crystals can be solved by additional information. If there is, for instance, a local twofold axis in the y-direction of the unit cell the value of ψ is restricted to 0 and π. Similarly the knowledge of the asymmetry parameter and the sign of the quadrupole splitting can determine the EFG up to the fundamental ambiguity between EFG's

---

*) Eqn. (21) of reference [8] contains an error and should read $\gamma = \arccos(-R_{\hat{x}z}/\sin\beta)$

which are reflected at the xz, zy or yz planes. Both quantities can be obtained from spectra of polycrystalline absorbers in applied fields, if the internal field vanishes. The latter condition is

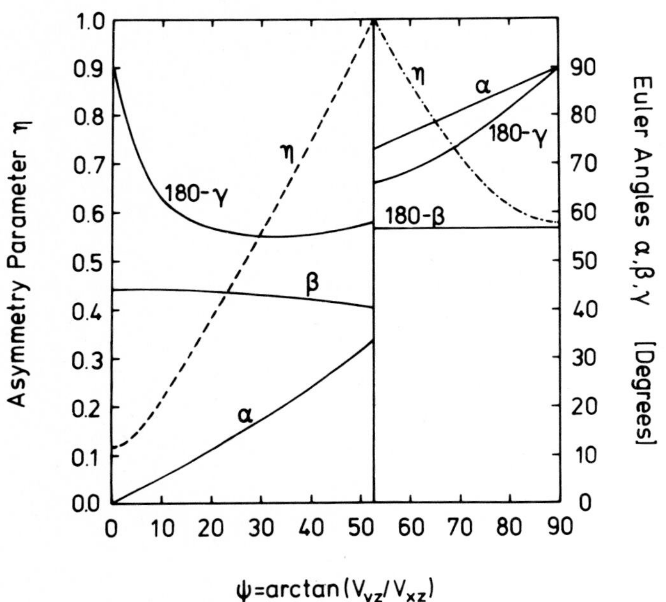

Fig. 5. The sign of the quadrupole splitting, asymmetry parameter η and Euler angles α,β,γ of the local EFG tensor in $FeCl_2.4H_2O$ as a function of the parameter ψ which describes the manifold of possible solutions. If the η line is dashed (dashed and dotted) the quadrupole splitting is positive (negative). The Euler angles β,α are identical with the polar angles θ,φ of the principal component V_{zz}^{loc}.

fulfilled for paramagnetic complexes at sufficiently high temperatures. For $FeCl_2.4H_2O$ it is unfortunately not fulfilled at room temperature. For that reason the additional information from single crystals in external fields has to be used. A method to use that information in conjunction with the above determination of the intensity tensor is described in the original publication [8]. The results were:

$$\Delta E_Q > 0 \text{ and } \eta = 0.25 \pm 0.15. \qquad (38)$$

The orientation of the EFG which belongs to these values can be found from Fig. 5.

Recently single crystal spectra of $FeCl_2.4H_2O$ have been evaluated by inclusion of the thickness effect [16]. The absolute values of the traceless macroscopic intensity tensor are slightly larger (by 0.03 in the average) according to this publication. With the new values, Fig. 5. is essentially not changed. We have abandoned a re-evaluation of the data since the changes in η appear to be much less than the error which has to be lived with then evaluating spectra with unresolved absorption lines (see Fig. 2 of Ref. 8).

4.3 Example 4. Electric Field Gradient Tensor Determination from Quadrupole Spectra of Orthorhombic Crystals (e.g. Sodium Nitroprusside)

As another example for the usefulness of the intensity tensor formulation, we choose the well known compound sodium nitroprusside [$NaFe(CN)_5NO.2H_2O$]. Single crystal measurements have been carried out by Grant et al. [17] and recently by our laboratory. From the experiment, we can derive the following components for the macroscopic traceless intensity tensor (a≡x, b≡y, c≡z):

$$T_{xx}^{(m)} = 0.12, \; T_{yy}^{(m)} = 0.01, \; T_{zz}^{(m)} = -0.13. \qquad (39)$$

This time the values for $T_{pq}^{(m)}$ have been obtained by correcting the experimental areas for the thickness effect. The method is described in Section 4. Due to the orthorhombic symmetry, the nondiagonal elements of the traceless macroscopic intensity tensor vanish (i.e. $T_{xy}^{(m)} = T_{xz}^{(m)} = T_{yz}^{(m)} = 0$.) The nondiagonal elements of the local intensity tensors are not accessible to the Mossbauer measurement. From X-ray analysis, however, [18] it is known that the iron atoms lie on mirror planes perpendicular to the c-axis (≡ z-axis) of the unit cell. In this case, one principal axis of the local EFG's has to lie parallel to the crystallographic c(≡z) axis and we can write for the local intensity tensors at the 4 lattice sites:

$$T_{xz}^{(loc,i)} = T_{yz}^{(loc,i)} = 0, \; i=1, \ldots 4. \qquad (40)$$

According to Eqn. (29), the local intensity tensor is now determined up to the component $T_{xy}^{(loc,i)}$. This component can be determined by the others, since we know that the local intensity tensor must have a "quadrupole splitting" of 1/4 (see Eqn. 19). In particular we can write:

$$T_{zz}^{(m)^2} + \frac{1}{3}(T_{xx}^{(m)} - T_{yy}^{(m)})^2 + \frac{4}{3}(T_{xy}^{(loc,i)})^2 = \frac{1}{16} .$$

Inserting the values for $T_{pp}^{(m)}$, p = x,y,z, from Eqn. (39), we thus find for sodium nitroprusside:

$$I_{xy}^{(loc,i)} = \pm 0.18.$$

The determination of the EFG is now very simple if we use the formulae for example 2 in Section 2.5. We find that the sign of the quadrupole splitting is positive and that the asymmetry parameter vanishes. Furthermore, we find that the principal axes of the EFG perpendicular to c are rotated by an angle $\delta = \pm 36^{\circ}$ around the c-axis [$\delta=(1/2)$ arctan $2I_{xy}^{loc}/(I_{xx}^{(m)} - I_{yy}^{(m)})$]. This is in agreement with the results of Grant et al. [17]. The reason we have repeated this evaluation is to show that the evaluation can be done in a straightforward manner. Due to the symmetry at the iron site, the evaluation is even unique. Otherwise a two-dimensional manifold of solutions would have to be taken into account.

5. LINE INTENSITY IN THE CASE OF POLARIZATION

In this section, we discuss the absorption intensity for polarized and partially polarized γ-rays as well as the polarization of γ-rays emitted by a source. Expressions for polarization have been given in literature [1,2], though usually for non-degenerate transitions only in which case the γ-ray is completely polarized. Here we derive expressions which also hold in the case of degenerate transitions* and in the case of single crystals with more than one equivalent lattice site. These expressions will contain macroscopic quantities (e.g. the macroscopic intensity tensor) which reflect the total information to be extracted from the angular dependence of the line intensity. This, in particular, will enable us to discuss the question in which cases polarized γ-rays can yield additional information which cannot be secured by using unpolarized γ-rays. Finally, we will make some remarks on behalf of thickness correction which, in general, has to include the effect of polarization.

---

* e.g. the intensity of each of the two quadrupolar lines of ^{57}Fe results from four transitions.

5.1 Linear Polarization

The intensity for absorption of a linearly polarized γ-ray can be obtained by use of the formula (see Appendix),

$$I = I^{pow} - 2 \sum_{p,q=x,y,z} T_{pq} n_p n_q, \tag{41}$$

where n_x, n_y, and n_z are the direction cosines of a unit vector $\vec{n}$ describing the polarization of the γ-ray. The vector $\vec{n}$ is to be parallel to the vector potential $\vec{A}$ of the electromagnetic wave in the case of electric dipole radiation and it is to be parallel to $\vec{k}\times\vec{A}$ ($\vec{k}$ = wave vector) in the case of magnetic dipole radiation. From the above equation it is seen that the absorption of a linear polarized γ-ray can be described by an intensity tensor,

$$I^{(\ell)}_{pq} = I^{pow} \delta_{pq} - 2T_{pq}, \; p,q = x,y,z, \tag{42a}$$

which determines the intensity via the equation:

$$I = \sum_{p,q=x,y,z} I^{(\ell)}_{pq} n_p n_q. \tag{42b}$$

The intensity tensor for linearly polarized γ-rays, $I^{(\ell)}_{pq}$ is again additive for degenerate transitions and it can also be applied in the case of crystals with more than one equivalent lattice site. Due to Eqn. (41), it does not contain components which could not also be found by determining the intensity tensor for unpolarized γ-rays. Hence linearly polarized γ-rays cannot yield information which could not be secured by measurements with unpolarized γ-rays. This, however, does not mean that the expense of measuring with linearly polarized γ-rays may not be worthwhile, since there is a fundamental difference between linearly polarized and unpolarized γ-rays. Linearly polarized γ-rays determine components of the intensity tensor which are perpendicular to the γ-ray. Unpolarized γ-rays determine the parallel component. Hence parallel and perpendicular components of the intensity tensor can be determined from one orientation of the crystal. This is interesting for single crystals with only one well developed face. Furthermore measurements with linearly polarized γ-rays can be important if the assumption of an isotropic recoilless fraction cannot be made [19].

5.2 Circular Polarization

The intensity for absorption of circularly polarized γ-rays can be calculated from the formula (see Appendix),

$$I = \sum_{p,q=x,y,z} I_{pq} e_p e_q \pm \sum_{p=x,y,z} I_p e_p, \tag{43}$$

where $\vec{e}$ is the unit vector parallel to the direction of the γ-ray and I_x, I_y, I_z are the components of an intensity vector defined by,

$$\begin{aligned} I_x &= 2i\,(c_y c_z^* - c_y^* c_z),\\ I_y &= 2i\,(c_z c_x^* - c_z^* c_x),\\ I_z &= 2i\,(c_x c_y^* - c_x^* c_y). \end{aligned} \tag{44}$$

The plus and minus sign in the above equation is to be used for right- and left-circulary polarized γ-rays respectively[†].

For a nondegenerate transition the intensity vector can be calculated from the components of the intensity tensor, though only up to a common sign of its components [††]. Thus circularly polarized γ-rays can indeed yield new information which cannot be secured by unpolarized or linearly polarized γ-rays. This information particularly concerns the sign of the effective magnetic field at the nucleus. In the case of vanishing effective magnetic field the intensity vector also vanishes. The spectrum for circular polarization is identical to the one for unpolarized γ-rays.

5.3 Elliptical and Partial Polarization

The polarization $\vec{\pi}$ of an elliptically polarized electromagnetic wave is of the form,

$$\vec{\pi} = a_\theta \vec{e}^{\,\theta} + a_\phi \vec{e}^{\,\theta}, \tag{45}$$

where a_θ and a_ϕ are complex coefficients normalized to 1 ($a_\theta a_\theta^* + a_\phi a_\phi^* = 1$) and $\vec{e}^{\,\theta}$, $\vec{e}^{\,\phi}$ are two orthogonal unit vectors perpendicular to the direction of the γ-ray. In anticipation of the result for the expression of the absorption intensity we define

---

[†] Note that a γ-ray is called right circulary, if its spin lies in the direction of motion. This convention is opposite to the one used in optical spectroscopy (2).

[††] This is related to the following symmetry: If the components c_p, p=x, y,z are replaced by the conjugate complex the components I_{pq} stay the same but the signs of the components I_p change.

the quantities,

$$\rho_{11} = a_\theta a_\theta^*,\ \rho_{12} = a_\theta a_\phi^*\ ,\ \rho_{21} = \rho_{12}^*\ ,\ \rho_{22} = a_\phi a_\phi^* \tag{46}$$

and

$$\sigma_{11} = \Sigma\ I_{pq}^{(\ell)}\ e_p^\theta\ e_q^\theta,$$

$$\sigma_{12} = \Sigma\ I_{pq}^{(\ell)}\ e_p^\theta\ e_q^\phi - i\ \Sigma I_p e_p,\quad \sigma_{21} = \sigma_{12}^*, \tag{47}$$

$$\sigma_{22} = \Sigma\ I_{pq}^{(\ell)}\ e_p^\phi\ e_q^\phi\ .$$

For electric dipole transitions the intensity is then given by (see Appendix),

$$I = \sigma_{11}\rho_{11} + \sigma_{12}\rho_{21} + \sigma_{21}\rho_{12} + \sigma_{22}\rho_{22}. \tag{48}$$

For magnetic dipole transitions, we have to exchange $\underset{\rightarrow}{e}^\theta$ with $\underset{\rightarrow}{e}^\phi$ and $\underset{\rightarrow}{e}^\phi$ with $-\underset{\rightarrow}{e}^\theta$ in the formula for $\sigma_{k\ell}(k,\ell = 1,2)$. The quantities $\sigma_{k\ell}$ and $\rho_{k\ell}$ can be interpreted as Hermitian 2x2 matrices which determine the intensity via the equation,

$$I\ =\ Tr\ (\sigma\rho), \tag{49}$$

(Tr(A) means the sum of diagonal elements (=trace) of a matrix A). The matrices σ and ρ have a definite physical meaning. ρ is the density matrix of the electromagnetic wave and σ is the polarization operator for the absorber. Quantities of this kind are defined in the density matrix formalism [20]. The use of this formalism is particularly convenient if the electromagnetic wave results from degenerate transitions. In this case the density matrix ρ, which enters into Eqn. (49), is obtained as the mean value of the density matrices $\rho^{(i)}$ of the individual transitions. For a N-fold degenerate transition we have,

$$\rho = \frac{1}{N}\ \Sigma\ \rho^{(i)}\ . \tag{50}$$

Hence the density matrix contains the total information on the polarization of the γ-ray and it is only that quantity (and not the individual $\rho^{(i)}$'s) which can be determined uniquely by polarization experiments.

Similar to ρ, the polarization operator of the absorber, σ is additive in the case of degenerate transitions. This is reflected by the fact that σ depends only on macroscopic quantities i.e. the macroscopic intensity tensor and the macroscopic intensity vector. Hence only these quantities can be determined from the angular dependence and the polarization dependence of the absorption intensity.

The polarization of a γ-ray emitted by a source can also be easily obtained in the density matrix formalism. The density matrix of the emitted γ-ray is given by,

$$\rho = \sigma/\Sigma \ I_{pq} \ e_p \ e_q, \tag{51}$$

where the division by the intensity had to be done in order to satisfy the condition $\rho_{11} + \rho_{22} = 1$. With the above equation the circle is closed. We can calculate the spectral intensity of source-absorber system having any polarization.

At the end of this section some examples are in order. For unpolarized γ-rays the density matrix ρ is just 1/2 times the unit matrix:

$$\rho = \begin{bmatrix} 1/2 & 0 \\ 0 & 1/2 \end{bmatrix} . \tag{52}$$

For a completely polarized γ-ray one eigenvalue of the density matrix is equal to 1, the other eigenvalue vanishes. The density matrix is diagonal in the basis of the vector which describes the polarisation of the γ-ray and the vector orthogonal to it. Some other nice properties of the density matrix are discussed in Ref. [20].

If the γ-ray is along the z-direction of O_{xyz}, the vectors $\underline{e}^{\theta}$ and $\underline{e}^{\phi}$ lie in the x,y-plane. For convenience we choose $\underline{e}^{\theta}$ along the x-axis and $\underline{e}^{\phi}$ along the y-axis. In this case the polarisation matrix has the form,

$$\sigma = \begin{bmatrix} I_{xx}^{(\ell)} & I_{xy}^{(\ell)} - i\,I_z \\ I_{xy}^{(\ell)} + i\,I_z & I_{yy}^{(\ell)} \end{bmatrix} . \tag{53}$$

Since the choice of O_{xyz} is arbitrary, we can actually always align the z-axis along the γ-ray and use the above form for the polarisation matrix. When comparing different orientations of single crystals we have to refer, in this case, the components of

$I_{pq}^{(\ell)}$ and I_p to different coordinate systems and relate the components by rotation matrices. This method may become quite tedious and hence the general form of the polarisation matrix seems more adapted to our problem of the intensity tensor determination. On the other hand the simple form of Eqn. (53) gives some insight. We see that for vanishing magnetic hyperfine interaction ($I_x = I_y = I_z = 0$), the polarisation matrix is diagonal in the principal axes system of the projection of the intensity tensor onto the plane perpendicular to the γ-ray.

5.4 The Degree of Polarization

The polarization of a Mossbauer transition has been described, in the last section, by the polarization operator σ. To set up this operator in the matrix form, we have to define the two base vectors $\underline{e}^{\theta}$ and $\underline{e}^{\phi}$ which are perpendicular to the γ-ray. When using unpolarized γ-rays, the rotation of $\underline{e}^{\theta}$ and $\underline{e}^{\phi}$ around the direction of the γ-ray cannot change the value of the area of an absorption line. Hence, the absorption intensity will depend only on the eigenvalues of the absorption matrix and not on the particular basis which belongs to the diagonal polarization matrix. For thin absorbers the absorption area of a line is proportional to the average of the eigenvalues of the polarization matrix, i.e.

$$A \propto \frac{\sigma_1 + \sigma_2}{2} , \tag{54}$$

where σ_1 and σ_2 are the eigenvalues of σ. For thick absorbers, the absorption area depends on both eigenvalues (as regards the dependence on σ). In other words, for thick absorbers the absorption area depends also on the degree of polarization. This has been pointed out by Grant et al. [17]. In this section, we want to focus our attention on the calculation of the degree of polarization. It is defined by,

$$a = \left| \frac{\sigma_1 - \sigma_2}{\sigma_1 + \sigma_2} \right| . \tag{55}$$

The degree of polarization is zero, if the absorption line does not show any polarization ($\sigma_1 = \sigma_2$). For a completely polarized absorption line (e.g. $\sigma_2 = 0$), the degree of polarization is equal to 1. the values of "a" can range only between zero and 1.

The polarisation a can also be expressed in terms of the nondiagonal polarisation matrix. We find,

$$a = \frac{\sqrt{(\sigma_{11} - \sigma_{22})^2 + 4\sigma_{12}\sigma_{21}}}{\sigma_{11} + \sigma_{22}} . \qquad (56)$$

Inserting the expression for the polarisation matrix from Eqn. (47) we find after some algebra,

$$a = \frac{\sqrt{(\sum_{p,q} T_{pq} e_p e_q)^2 - 4 \sum_{p,q} T^x_{pq} e_p e_q + (\sum I_p e_p)^2}}{I} , \qquad (57)$$

where $I = \sum I_{pq} e_p e_q$ is the intensity of the line and T^x_{pq} is the tensor of the cofactors of the matrix elements of T_{pq}, i.e.

$$T^x_{xx} = T_{yy}T_{zz} - T^2_{yz}, \quad T^x_{yy} = T_{xx}T_{zz} - T^2_{xz}, \quad T^x_{zz} = T_{xx}T_{yy} - T^2_{xy},$$
$$T^x_{xy} = T_{xz}T_{yz} - T_{zz}T_{xy}, \quad T^x_{xz} = T_{xy}T_{yz} - T_{yy}T_{xz}, \quad T^x_{yz} = T_{xy}T_{xz} - T_{xx}T_{yz}. \qquad (58)$$

Though the above formula for the degree of polarization does not appear to be simple, it has the considerable advantage that it depends on the direction cosines of the γ-ray and not on the direction cosines of $\underline{e}^\theta$ and $\underline{e}^\phi$. Hence, the above formula does not contain superflous parameters like a rotation angle of $\underline{e}^\theta$ and $\underline{e}^\phi$ around the direction of the γ-ray. Furthermore we see from Eqn. (58) that the degree of polarization depends only on the macroscopic intensity tensor and macroscopic intensity vector, and not on the corresponding local tensors.

6. THICKNESS CORRECTION

6.1 General

Up to now we have assumed that the absorbers are thin enough to be treated in the thin absorber limit. For thick absorbers, we have to take into account the fact that the intensity of the γ-ray is exponentially damped within the absorber. A general formula for the line shape of a thick absorber has been discussed by Blume and Kistner [21]. It contains the effect of dispersion and absorption. If n is the complex index of refraction of the absorber, k the wave number and z the thickness of the absorber, the unfolded line shape is given by [21],

$$L(E) = \mathrm{Tr}\,(e^{inkz}\, \rho\, e^{-in^\dagger kz}). \qquad (59)$$

Here E is the energy and ρ the density matrix of the γ-rays. The index of refraction, n, like ρ is a 2 x 2 matrix. It can be written as [21],

$$n = 1 - \frac{1}{kz}\,\frac{t\Gamma}{4}\sum_{\text{lines } \ell}\frac{\sigma_\ell}{E-E_\ell+i\Gamma/2}, \tag{60}$$

where t is the effective thickness, Γ the natural (unfolded) linewidth, σ_ℓ and E_ℓ the normalized absorption matrix and the energy of the ℓ-th line, respectively and the summation extends over all the Mossbauer absorption lines. To obtain the Mossbauer spectrum the line shape L(E) has to be folded with the source shape. Furthermore a background has to be added. Both will not be discussed here, since it has been widely treated in literature [22]. We will focus our attention on the effect of the polarisation on the thickness correction of single crystal Mossbauer spectra.

6.2 Thickness Correction of Spectra with Well Resolved Absorption Lines

For well resolved absorption lines, the angular dependence of the absorption intensity is reflected by that of the absorption areas. To drop incovenient factors we define a dimensionless area S by,

$$S = 2A/(N_\infty f_S \Gamma_A \pi), \tag{61}$$

where A is the background corrected area, N_∞ the count rate at large velocities, f_S the recoilless fraction of the source and Γ_A the natural line width of the absorber. The theoretical expression for the dimensionless area S is given by [23, 24],

$$S(tI,a) = \frac{1}{2}\{K[tI(1+a)] + [tI(1-a)]\} \tag{62}$$

where

$$K[\tau] = e^{-\tau/2}\,(I_0(\tau/2) + I_1(\tau/2)),$$

I_0 and I_1 being the zero and first order Bessel functions with imaginary argument, respectively. It may be noted that the area S depends also on the degree of polarisation, a, which has been calculated in Section 5.4.

For the thickness correction, we need an inverse function of

$s(tI,a)$. An expansion series of it has the form,§)

$$tI = S + \frac{1}{4}(1+a^2)S^2 + \frac{1}{16}(1+a^2+2a^4)S^3 + \frac{5}{384}(1+a^4+6a^6)S^4 + \ldots \quad (63)$$

For known polarisation, we can calculate the values of tI from the dimensionless areas S. This yields the fractional line intensities I whose angular dependence in turn determines the intensity tensor according to Section 2.2. Unfortunately the polarisation is not known in advance, since it is a function of the macroscopic intensity tensor and macroscopic intensity vector. Hence, the evaluation of the absorption areas has to be determined by iteration. The following iteration procedure has turned out to converge rather rapidly in the case of ^{57}Fe quadrupolar spectra. In the first step, we assume vanishing polarisation for each line. By use of Eqn. (63), we can calculate tI and fit the intensity tensor. Now we are able to calculate the values of the polarisation for each line and each orientation. With these values, we correct the thickness again via Eqn. (63) and fit the intensity tensor. We repeat this procedure until the components of the intensity tensor do not change any more within the accuracy we want to have. In the next section, we will treat an example.

6.3 Example 5: EFG Determination in Mohr Salt

Mohr salt $[(NH_4)_2Fe(SO_4)_2 \cdot 6H_2O]$ has monoclinic crystal symmetry. There are two lattice sites which can be transformed into each other by a reflection. Due to the crystal symmetry the macroscopic intensity tensor has to have one principal axis along the unit axis b. Mossbauer spectra of the compound have been shown in Fig. 4 [14]. Th results of the area determination of the absorption lines by computer fit with Lorentzian lines is given Table 4 together with the iteration procedure for the intensity tensor as described in the last section. The selfconsistent values for the principal components of the macroscopic intensity tensor for the high energy line of the

---

§) additional higher order terms are

$$+ \frac{1}{3072}(5-22a^2-47a^4-24a^6+168a^8)S^5 -$$

$$- \frac{7}{30720}(1+15a^2+35a^4+81a^6+80a^8-180a^{10})S^6 -$$

$$- \frac{1}{737280}(169+549a^2+635a^4+3087a^6+12496a^8+17640a^{10}-23760a^{12})S^7 + \ldots$$

TABLE 4

Selfconsistent determination of the intensity tensor from the angular dependence of the absorption areas obtained from thick single crystals of Mohr salt.
Line 1: Orientation of the γ-ray relative to a coordinate system O_{xyz} with z parallel to the b-axis and x parallel to the c-axis of the unit cell.
Line 2: Dimensionless areas of absorption line ℓ and absorption line r (line on the left and right hand side of the spectrum, respectively) as defined in Eqn. (61).
Line 3: First step for the thickness correction: The values of tI_ℓ and tI_r are obtained according to Eqn (63) assuming a polarization of zero.
Line 4: The effective thickness t of the absorber
Line 5: The fractional intensity of the absorption line .
Line 6: Fit of the intensity tensor. Instead of fitting a nondiagonal tensor with nonvanishing I_{xy} component we have fitted the intensity within the principal axes system O_{xyz} according to the equation,

$$I = I_{xx}\sin^2\theta\cos^2(\phi-\phi_0)+I_{yy}\sin^2\theta\sin^2(\phi-\phi_0)+I_{zz}\cos^2\theta \ .$$

The fitting procedure has taken into account that $I_{xx}+I_{yy}+I_{zz}=3/2$.
Line 7: The polarization of the absorption lines as derived from intensity tensor of line 6.
Line 8: Thickness correction of the absorption areas according to Eqn. (63) by using the polarizations from line 7.
Line 9 to 11 are similar to line 4 to 6.
Line 12: A third step in the procedure shows selfconsistency being achieved.

1	Orientation (θ,φ)		(0,0)	(90,17)	(90,-47)	(90,43)	(90,-47)
2	redced area	S_ℓ	0.435	0.309	0.541	0.347	1.02
		S_r	0.496	0.551	0.228	0.693	0.466
3	tI_ℓ for a=0		0.489	0.335	0.625	0.379	1.370
	tI_r for a=0		0.566	0.638	0.242	0.837	0.528
4	$t = tI_\ell+tI_r$		1.055	0.973	0.867	1.216	1.898
5	$I_r = (tI_r)/t$		0.537	0.656	0.279	0.688	0.278
6	Intensity tensor		$I_{\hat{x}\hat{x}}=0.695$, $I_{\hat{y}\hat{y}}=0.27$, $I_{\hat{z}\hat{z}}=0.535$ $\phi_0=34.4^\circ$				
7	Polarisation	a_ℓ	0.915	0.659	0.209	0.817	0.209
		a_r	0.794	0.346	0.540	0.374	0.541
8	tI_ℓ for $a=a_\ell$		0.542	0.346	0.629	0.403	1.384
	tI_r for $a=a_r$		0.617	0.649	0.246	0.857	0.547
9	$t = tI_\ell+tI_r$		1.159	0.995	0.875	1.260	1.931
10	$I_r=(tI_r)/t$		0.532	0.652	0.281	0.680	0.283
11	Intensity tensor		$I_{xx}=0.691$, $I_{yy}=0.273$, $I_{zz}=0.536$ $\phi_0=34.4^\circ$				
12	....						

quadrupole spectrum are at 300 K:

$$I_{xx}^{(m)} = 0.691, \; I_{yy}^{(m)} = 0.273, \; I_{zz}^{(m)} = 0.536. \qquad (64)$$

Here the z-axis is along the b-axis of the unit cell and the x-axis is rotated off the c-axis by $34.5^{\circ} \pm 1^{\circ}$ versus the a-axis. The above data characterize the angular dependence of the line intensities uniquely and should always be quoted as a direct result of the experiment. The error of the principal components of the intensity tensor is about 0.004. Note, however, that $I_{xx}^{(m)}+I_{yy}^{(m)}+I_{zz}^{(m)} = 3/2$ according to Eqn. (14). As discussed in Section 4.2, the evaluation of the macroscopic intensity tensor yields a linear manifold of possible solutions which has been plotted in Fig. 6. Fortunately the range of possible solutions is very narrow for Mohr Salt. For all solutions the sign of the quadrupole splitting is negative. The asymmetry parameter varies between 0.6 and 0.8. Each possible solution is also characterized by a certain orientation

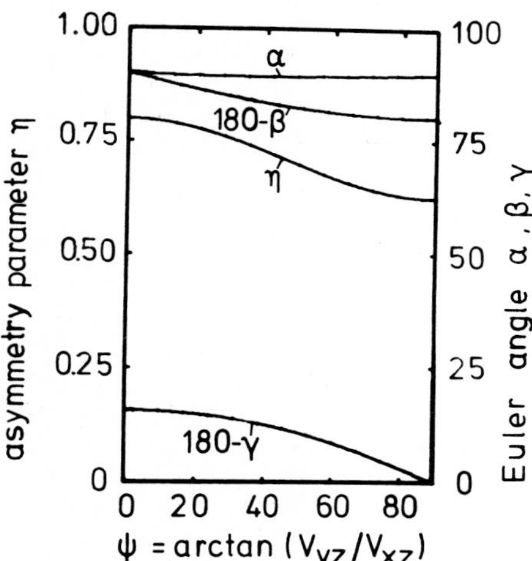

Fig. 6. The asymmetry parameter η and Euler angles α,β,γ of the local EFG tensor in Mohr salt as a function of the parameter ψ which describes the manifold of possible solutions in the case of monoclinic crystals.

of the EFG which can be related to the slightly distorted water octahedron as derived from X-ray analysis [25]. In Fig. 7, the principal axis system of the intensity tensor (=macroscopic EFG tensor)

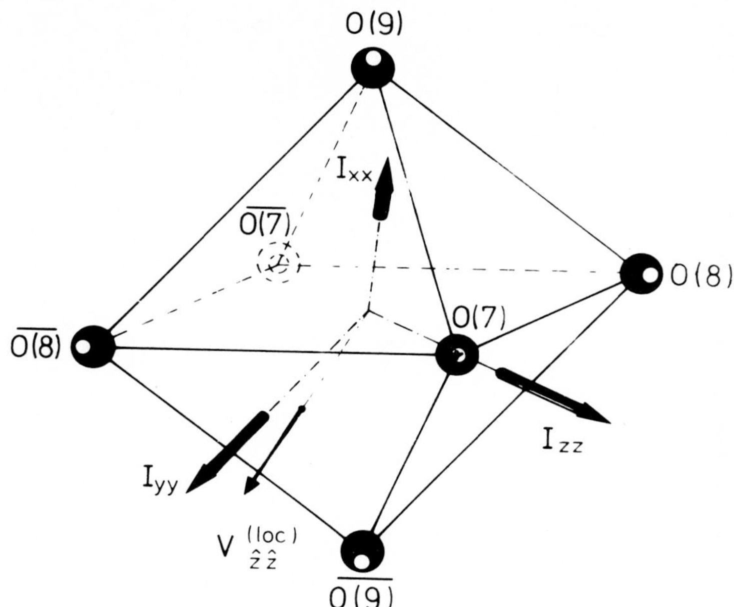

Fig. 7. Orientation of the principal axes system of the intensity tensor relative to the slightly distorted water octahedron in Mohr salt. One possible solution of the direction of the z-axis of the local EFG axis is shown.

has been plotted together with the water octahedron. Since the range of possible solutions is rather narrow according to Fig. 6, the principal axis V_{zz}^{loc} of the local EFG tensors ($|V_{xx}^{loc}|\leq|V_{yy}^{loc}|\leq|V_{zz}^{loc}|$) is very close to the y-axis of the macroscopic intensity tensor. One possible solution is plotted in Fig. 7. Hence, the z-axis of the local EFG points approximately midway between the water molecules. This result is reasonable in view of the d-electron density of the ground state having the shape of a distorted cigar (sign(ΔE_Q)<0).

The above result of a negative sign of the quadrupole splitting is rather different from what has been found earlier by Ingalls et al. [26]. They found a positive sign and an asymmetry parameter η of 0.7 at 300 K and $\eta \approx 0.1$ at 4.2 K. Our results show that the macroscopic intensity tensor at 4.2 K is nearly identical to that at 300 K which gives sign (ΔE_Q)<0 and $0.6\leq\eta\leq0.8$ at 4.2 K. The difference of both results follows clearly from our inclusion of thickness correction. This shows that thickness correction which includes the effect of the polarisation may be very important. For Mohr salt this means that it is no longer an exception to the other hexahydrates of iron(II) (e.g. $FeSiF_6 \cdot 6H_2O$, $Fe(ClO_4)_2 \cdot 6H_2O$) which feature a mainly trigonal distortion and not tetragonal distortion of the water octahedron surrounding the iron.

7. MOSSBAUER SPECTRA OF TEXTURED MATERIAL

In this section we discuss Mossbauer spectra of powder samples for which the orientation of the microcrystals is not perfectly random. Preferred orientation may occur as a result of the shape of the microcrystals (e.g. needles [27]) or as a result of preparation of the absorber (e.g. application of an external field [28]) to name just two. The distribution of the orientation of the microcrystals can be described by a distribution function $D(\phi,\theta,\gamma)$ where ϕ,θ,γ are the Euler angles describing the rotation of the crystal coordinate system O_{xyz} relative to the lab coordinate system $O_{\hat{x}\hat{y}\hat{z}}$. The Euler angles can be defined in different ways. For convenience we agree that the angles θ,ϕ represent the polar angles of the $\hat{z}$-axis in O_{xyz} and that γ corresponds to a rotation of $O_{\hat{x}\hat{y}\hat{z}}$ around $\hat{z}$. If we further agree that the z-axis is to be parallel to the direction of the (unpolarized) γ-rays we find for the powder intensity:

$$I = \frac{1}{8\pi^2} \int_0^{2\pi}\int_0^{\pi}\int_0^{2\pi} D(\phi,\theta,\gamma)\, I_{\hat{z}\hat{z}} \sin\theta d\phi d\theta d\gamma \ . \qquad (65)$$

In this formula $I_{\hat{z}\hat{z}}$ is a function of the Euler angles. If e_x, e_y, e_z are the direction cosines of the $\hat{z}$-axis in the crystal coordinate system O_{xyz}, i.e. $e_x = \sin\theta\cos\phi$, $e_y = \sin\theta\sin\phi$, $e_z = \cos\theta$ we obtain for $I_{\hat{z}\hat{z}}$,

$$I_{\hat{z}\hat{z}} = \sum_{p,q=x,y,z} I_{pq} e_p e_q . \qquad (66)$$

Inserting this result into Eqn. (65), we can express the intensity in the following form:

$$I = \sum_{p,q=x,y,z} I_{pq} D_{pq} , \qquad (67)$$

where

$$D_{pq} = \frac{1}{8\pi^2} \int_0^{2\pi}\int_0^{\pi}\int^{2\pi} D(\phi,\theta,\gamma) e_p e_q \sin\theta d\phi d\theta d\gamma . \qquad (68)$$

The components D_{pq} can be interpreted as the components of a texture tensor which is real, symmetric and of second rank. It is these components which may be determined from the Mossbauer spectra. The distribution function D is usually subject to normalization, i.e.

$$\frac{1}{8\pi^2}\int_0^{2\pi}\int_0^{\pi}\int_0^{2\pi} D(\phi,\theta,\gamma)\sin\theta d\phi d\theta d\gamma = 1. \tag{69}$$

In this case the texture tensor has a trace of 1 i.e.

$$D_{xx} + D_{yy} + D_{zz} = 1. \tag{70}$$

Furthermore the diagonal elements are nonnegative since $D(\phi,\theta,\gamma)$ is nonnegative function.

The intensity formula Eqn. (67) is of a very general nature†. It can be applied for a randomly oriented powder as well as for a single crystal. For a randomly oriented powder, the texture tensor is identical to 1/3 of the unit matrix (i.e. $D_{xx} = D_{yy} = D_{zz} = 1/3$, $D_{xy} = D_{xz} = D_{yz} = 0.$). On the other hand a single crystal which is oriented with its z-axis parallel to the γ-rays, has $D_{zz} = 1$ and all other components of the texture tensor are equal to zero. Furthermore the intensity formula is analogous to the one to be used for the powder susceptibility :

$$\chi = \sum_{p,q=x,y,z}\chi_{pq}D_{pq}. \tag{71}$$

The integral for the components D_{pq}, Eqn. (68), has three integration variables, ϕ,θ, and γ. Since the direction cosines e_p are independent of the Euler angle γ, Mossbauer spectra depend only on the average distribution function,

$$D(\phi,\theta) = \frac{1}{2\pi}\int_0^{2\pi} D(\phi,\theta,\gamma)d\gamma. \tag{72}$$

Using this distribution function the integral for D_{pq} reduces to one with two integration variables only i.e.

$$D_{pq} = \frac{1}{8\pi}\int_0^{\pi}\int_0^{2\pi} D(\phi,\theta)e_p e_q \sin\theta d\phi d\theta. \tag{73}$$

---

† Eqn. (67) can be written as the trace of the product of the intensity tensor with the texture tensor, a formula to be used in the density matrix formalism.

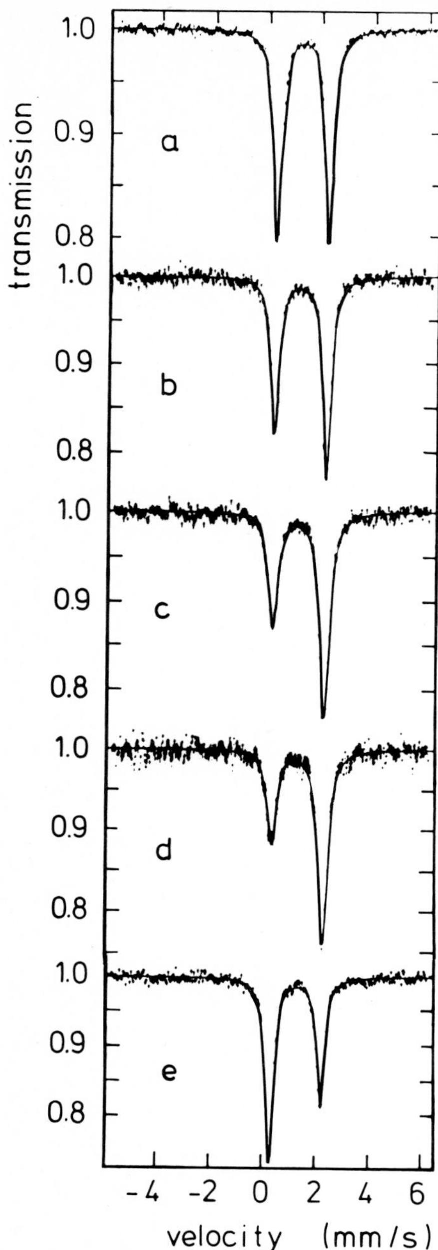

Fig. 8. Mossbauer spectra of polycrystalline $FeCO_3$ mixed with active carbon powder at 185 K after applying various longitudinal magnetic fields. (a) Original spectrum, (b) after 5 kG, (c) after 10 kG, (d) after 50 kG, (e) after 50 kG at 5 K.

This shows that the set of coefficients D_{pq} is similar to the one obtained by expanding $D(\phi,\theta)$ with respect to spherical harmonics $Y_{2m}(\theta,\phi)$.

As an example for texture problems in powder samples, we discuss $FeCO_3$ (siderite). In an effort to avoid undesireable texture effects, the material is usually ground to particle size of some microns and mixed with amorphous material (e.g. active carbon powder) to prevent direct mechanical contact of the particles. Though this method can initially yield random orientation of the microcrystals, the application of a magnetic field may very well induce texture in the material [28]. This is shown in Fig. 8. Fig. 8a corresponds to the original spectrum at 185 K and exhibits a symmetric quadrupole doublet showing no sign of texture. Fig. 8b, 8c and 8d give the zero field spectrum at 185 K after an application of 5, 10 and 50 kG, respectively for about half-an-hour before recording the Mossbauer spectrum. As can be seen, the external field induces a considerable texture in the sample.

For a quantitative description of the texture effect, we can use Eqn. (68). Since $FeCO_3$ has an axially symmetric EFG pointing along the c-axis of the microcrystals (=z-axis), the formula for the intensity can be written:

$$I = I_{xx} + (I_{zz} - I_{xx})D_{zz}. \tag{74}$$

Furthermore since we are dealing with ^{57}Fe quadrupolar spectra and a positive sign of the quadrupole splitting, we find from Eqn. (15) that $I_{xx} = 3/8$ and $I_{zz} = 6/8$ for the quadrupolar line at high energy. Thus

$$I = 3/8 + 3/8\ D_{zz}. \tag{75}$$

Evaluating the intensities of the spectra, Fig. 8a to 8d, we find D_{zz} values of 1/3, $.58 \pm .02$, $.86 \pm .01$, $.92 \pm .03$, respectively. Since $D_{zz} = 1$ corresponds to a single crystal, the application of 50 kG comes close to produce a perfectly preferred orientation with the c-axis parallel to the external field. The rotation of the microcrystals in the external field can be explained by the strong anisotropy of the susceptibility tensor χ which features $\chi_{zz} \gg \chi_{xx}, \chi_{yy}$ at 185 K [28].

$FeCO_3$ is particularly a good example for showing the effect of field induced texture, since the anisotropy of the susceptibility changes its sign at low temperatures, yielding $\chi_{xx},\chi_{yy} >> \chi_{zz}$ ($FeCO_3$ becomes antiferromagnetic below 38 K). Hence at low temperatures the application of an external field should rotate the microcrystals such that their c-axis is perpendicular to the magnetic field. This is indeed observed. Fig. 8e shows the Mossbauer spectrum at 185 K after the application of 50 kG at 5 K applied for a short period before recording the spectrum. The evaluation of the line intensity yields $D_{zz} = 0.01 \pm 0.01$ which shows that the c-axes of the microcrystals are perpendicular to the external field.

The actual distribution function cannot be derived from D_{zz}. We have rather to suggest a model for the rotation of the molecules in an external magnetic field and then we can determine the distribution function within the range of the possible distribution functions of the model. For details see Ref. [28].

8 CONCLUSION

The intensity tensor formalism has been shown to have quite a few advantages for the description of the angular dependence of line intensities. Particularly in the case of ^{57}Fe quadrupolar spectra, the application of the intensity tensor clarifies ambiguities of the problem. In this case the evaluation of the line intensities can be separated into two steps, one for the least squares analysis of the experimental data with respect to the components of the intensity tensor, and the other one, the derivation of the possible local EFG tensors from the macroscopic intensity tensor. The latter step is straight forward and does not contain any further least squares analysis.

For combined electric and magnetic hyperfine interations, the problem is more complex though it is still possible to describe the angular dependence of each line intensity by an intensity tensor. This again clarifies the angular dependence and particularly enables us to find the number of necessary parameters. Furthermore the effect of crystal symmetry can be easily discussed.

In Section 5, an intensity tensor for linearly polarized γ-rays has been derived. Since it can be uniquely expressed by the intensity tensor for unpolarized γ-rays, linearly polarized γ-rays cannot yield more information than what is obtained from unpolarized γ-rays. For some experimental situations linearly polarized γ-rays may, however, be advantageous.

Within the intensity tensor formalism the analyzing power

(density matrix) can also be derived. According to Eqn. (47), the analyzing power is written down in an arbitrary coordinate system and still stays rather simple. It can be used to include also the effect of thickness of the absorber.

In Section 6, the problem of thickness has been treated from the point of view of the intensity tensor. Since we did not wish to stay with general arguments, we have focussed our attention particularly on the thickness correction of ^{57}Fe quadrupolar spectra with rather well resolved absorption lines. Example 5 shows that thickness correction is very important even in the case of crystals as thin as 0.1 to 0.2 mm. For crystals of Mohr Salt of this thickness, the correction of thickness has yielded rather different results from those reported in earlier literature where thickness effects had not been taken into account.

Texture effects can also be discussed within the intensity tensor formalism. It differs from other treatments in that it uses Cartesian rather than polar coordinates. This is the general concept we have used. Cartesian coordinates seem more appropriate, since the EFG tensor and the hyperfine field are usually also expressed in these coordinates. If the nuclear transition has a higher character than dipole, however, spherical coordinates will be more convenient.

APPENDIX

The Hamiltonian of the interaction between the nucleus and the radiation field is given by [29]

$$\mathcal{H} = -\underline{j}.\underline{A}(\underline{r}), \tag{A1}$$

where $\underline{A}$ is the vector potential and $\underline{j}$ is proportional to the current density operator of the nucleus. For a monochromatic wave of direction $\underline{k}$ and polarization $\underline{\pi}$ the vector potential $\underline{A}$ has the form,

$$\underline{A}(\underline{r}) = \underline{A}(\underline{k},\underline{\pi})e^{i\underline{k}.\underline{r}} + \underline{A}^*(\underline{k},\pi)e^{-i\underline{k}.\underline{r}}. \tag{A2}$$

Hence the interaction Hamiltonian can be written as,

$$\mathcal{H}= \mathcal{H}_{abs} + \mathcal{H}_{em},$$

where

$$H_{abs} = \underline{j}e^{i\underline{k}.\underline{r}}\ \underline{A}(\underline{k},\underline{\pi}), \tag{A3}$$

$$H_{em} = \underline{j}e^{-i\underline{k}.\underline{r}}\ \underline{A}^{*}(\underline{k},\underline{\pi}).$$

In quantum electrodynamics $\underline{A}(\underline{k},\underline{\pi})$ and $\underline{A}^{*}(\underline{k},\underline{\pi})$ are operators for the annihilation and creation of γ-quanta, respectively. Thus H_{abs} is responsible for absorption and H_{em} is responsible for emission of γ-quanta. For multipole transitions we expand the exponential function,

$$e^{i\underline{k}.\underline{r}} = 1 + i\underline{k}.\underline{r} - \ldots. \tag{A4}$$

The first term gives rise to electric dipole transitions. Considering only this term we find for the Hamiltonian H_{abs},

$$H_{abs} = -\underline{j}.\underline{A}(\underline{k},\underline{\pi}), \tag{A5}$$

where $\underline{j}$ is independent of the direction and the polarization of the γ-ray. If $|G\rangle$ is the ground state and $|E\rangle$ the excited state the absorption intensity I is proportional to $|\langle E|\underline{j}.\underline{A}|G\rangle|^2$ i.e.

$$I \propto |\sum_{p=x,y,z} \langle E|j_p|G\rangle.A_p|^2. \tag{A6}$$

If we define $c_p = \langle E|j_p|G\rangle$, the intensity can be written as,

$$I \propto \sum_{p,q=x,y,z} c_p c_q^{*} A_p A_q^{*}. \tag{A7}$$

For emission we have to exchange $\underline{A}$ with $\underline{A}^{*}$ and $|G\rangle$ with $|E\rangle$. Since $\langle G|j_p|E\rangle = c_p^{*}$, it can be seen that the above formula holds also in the case of emission. By applying the relation between a vector operator and Clebsch Gordan Coefficients [29], we can easily prove the equivalence of the coefficients c_p defined above and those defined in Eqn. (10). Eqn. (A7) shows that a Hermitian intensity matrix $c_p c_q^{*}$ governs the angular dependence of the line intensity.

Let us now assume that we deal with linear polarization. In this case $\underline{A}$ is real, i.e. $\underline{A} = A\ \underline{n}$ where $\underline{n}$ is the unit vector along $\underline{A}$. Since $n_p n_q = n_q n_p$, the summation over p,q = x,y,z pulls out the symmetric part of the tensor $c_p c_q^{*}$. The intensity tensor for linearly polarized γ-rays is given by (we omit the factor $A^2/4$),

$$I_{pq}^{(\ell)} = 2\ (c_p c_q^{*} + c_p^{*} c_q). \tag{A8}$$

This formula corresponds to Eqn. (42a). For unpolarized γ-rays we have to calculate the mean value of the intensity for two perpendicular polarizations, e.g., $\underline{e}^{\theta}$ and $\underline{e}^{\phi}$:

$$I = \frac{1}{2} \sum_{p,q=x,y,z} I^{(\ell)}_{pq} (e^{\theta}_p e^{\theta}_q + e^{\phi}_p e^{\phi}_q). \tag{A9}$$

If $\underline{e}$ is now the unit vector parallel to the direction of the γ-ray we can use the relation for mutual orthogonal vectors:

$$e^{\theta}_p e^{\theta}_q + e^{\phi}_p e^{\phi}_q + e_p e_q = \delta_{pq}. \tag{A10}$$

Inserting this into Eqn. (A9) we obtain for the intensity tensor of unpolarized γ-beams:

$$I_{pq} = 2(c_x c^*_x + c_y c^*_y + c_z c^*_z)\delta_{pq} - (c_p c^*_q + c^*_p c_q), \tag{A11}$$

which is identical with Eqn. (12).

For right hand circularly polarized γ-rays, the components of the vector potential $\underline{A}$ are given by:

$$A_p = -\frac{A}{\sqrt{2}} (e^{\theta}_p - i e^{\phi}_p) \;, \; p = x,y,z, \tag{A12}$$

where $\underline{e}^{\theta}$, $\underline{e}^{\phi}$ and $\underline{e}$ are to form the unit vectors of a right handed coordinate system with $\underline{e}$ parallel to the direction of the γ-ray. Since

$$A_p A^*_q = \frac{A^2}{2} (e^{\theta}_p e^{\theta}_q + e^{\phi}_p e^{\phi}_q) + i\frac{A^2}{2} (e^{\phi}_p e^{\theta}_q - e^{\theta}_p e^{\phi}_q), \tag{A13}$$

the problem divides into two parts, one for the first term which has been solved in deriving the expression for an unpolarized γ-beam and the other for the second term which pulls out the antisymmetric part of the intensity matrix $c_p c^*_q$. Thus the intensity for right hand circularly polarized γ-rays is given by (up to a factor $A^2/4$),

$$I = \sum_{p,q=x,y,z} I_{pq} e_p e_q + i \sum_{p,q=x,y,z} (c_p c^*_q - c^*_p c_q)(e^{\phi}_p e^{\theta}_q - e^{\theta}_p e^{\phi}_q). \tag{A14}$$

We now use $\underline{e} = \underline{e}^{\theta} \times \underline{e}^{\phi}$, i.e.

$$e_z = e^{\theta}_x e^{\phi}_y - e^{\theta}_y e^{\phi}_x ,$$

with cyclic permutations for x,y,z. Furthermore, we define

$$I_z = i\,(c_x c_y^* - c_y c_x^*),$$

with cyclic permutations for x,y,z. Inserting both in Eqn. (A13) we find,

$$I = \sum_{p,q=x,y,z} I_{pq} e_p e_q + \sum_{p=x,y,z} I_p e_p, \qquad \text{(A15)}$$

which is identical with Eqn. (43).

For elliptical polarization the components of the vector potential $\underline{A}$ are written as,

$$A_p = A[a_\theta e_p^\theta + a_\phi e_p^\phi], \qquad \text{(A16)}$$

where the complex coefficients a_θ and a_ϕ are normalized to 1, i.e. $a_\theta^2 + a_\phi^2 = 1$. The intensity I can be begain obtained by inserting the expression for $\underline{A}$ into Eqn. (A7). By using similar algebra as for the linear and circularly polarized γ-rays we find the expression quoted in Eqn. (46) and (47).

REFERENCES

1 H.H.F. Wegener and F.E. Obenshain, Z. Phys., 163 (1961) 17.
2 H. Frauenfelder, D.E. Nagle, R.D. Taylor, D.R.F. Cochran, and W.M. Visscher, Phys. Rev., 126 (1962) 1065.
3 J.R. Gabriel and S.L. Ruby, Nucl. Instrum. Methods, 36 (1965) 23.
4 S.V. Karyagin, Sov. Phys. - Solid State, 8 (1966) 1387.
5 G.R. Hoy and S. Chandra, J. Chem. Phys., 47 (1967) 961.
6 W. Kundig, Nucl. Instrum. Methods, 48 (1967) 219.
7 A.J. Stone and W.L. Pillinger, Phys. Rev., 165 (1968) 1319.
8 R. Zimmermann, Nucl. Instrum. Methods, 128 (1975) 537.
9 E. Munck and R. Zimmermann in I.J. Gruverman and C.W. Seidel (Eds.) Mossbauer Effect Methodology, Vol. 10, Plenum Publishing Corporation, New York, 1976, p. 119.
10 P. Zory, Phys. Rev., 140 (1965) A1401.
11 J. Dannon and L. Iannarella, J. Chem. Phys., 47 (1967) 382.
12 K. Chandra and S.P. Puri, Phys. Rev., 172 (1968) 295.
13 R. Zimmermann, Chem. Phys. Lett., 34 (1975) 416.
14 R. Zimmermann and R. Doerfler, Proc. Int. Conf. Mossbauer Spectroscopy, Sept. 1979, Portoroz, Yougoslavia, Supplement to J. Phys. (Paris) Colloq., No. 1 (1980) 107.
15 S. DeBenedetti, Nuclear Interaction, Wiley, New York, 1964.
16 H. Spiering and H. Vogel, Hyperfine Int., 3 (1977) 221.
17 R.W. Grant, R.M. Housley and U. Gonser, Phys. Rev., 178 (1969) 523.
19 T.C. Gibb, Chem. Phys. Lett., 30 (1975) 137.
20 U. Fano, Rev. Mod. Phys., 29 (1957) 74.
21 M. Blume and O.C. Kistner, Phys. Rev., 171 (1968) 417.
22 See, for instance, the articles in Mossbauer Effect Methodology, Vol. 9 Ed. I.J. Gruverman, C.W. Seidel and D.K. Dieterly, Plenum Publishing Corporation, New York, 1974.

23 G.A. Bykov, Pham Zuy Hien, Sov. Phys. JETP, 16 (1963) 646.
24 R.M. Housley, R.W. Grant, and U. Gonser, Phys. Rev., 178 (1969) 514.
25 H. Montgomery, R.V. Chastain, J.J. Natt, A.M. Witkowska, and E.C. Lingafelter, Acta Cryst., 22 (1967) 775.
26 R. Ingalls, K. Ono and L. Chandler, Phys. Rev., 172 (1968) 295.
27 H.D. Pfannes and U. Gonser, Appl. Phys., 1 (1973) 93.
28 D.L. Nagy, K. Kulcsar, G. Ritter, H. Spiering, H. Vogel, and R. Zimmermann, J. Phys. Chem. Solids, 36 (1975) 759.
29 e.g. M.E. Rose, Multipole fields, John Wiley, New York, 1955.

CHAPTER 6

STATIC AND DYNAMIC CRYSTAL FIELD EFFECTS IN Fe^{2+} MOSSBAUER SPECTRA

×D.C. Price* and F. Varret†

* Dept. of Solid State Physics, Australian National University, Canberra, ACT 2600, Australia

Dept. of Physics, Royal Military College, Duntroon, ACT Australia.

† E.R.A. C.N.R.S. n° 682, Faculte´ des Sciences 72017, Le Mans, France.

1. INTRODUCTION

The Mossbauer spectrum of an ion in a crystal depends on the detailed structure of the ground state and of any low-lying excited states of the electrons of the ion. While the electrons in closed shells on the ion usually strongly influence the observed spectrum, they do so only in response to driving forces originating in the surrounding crystal lattice either directly or indirectly as a result of its influence on the electrons in the unfilled or open shell. This Chapter will deal with Mossbauer spectra of Fe^{2+} ions in insulating crystals and complexes, and therefore will be concerned with the ground and low-lying excited states of the open-shell 3d electrons of this ion and the way in which they are determined by the ion's crystalline environment.

The absolute energies and the states of the open-shell electrons of a transition metal ion depend primarily on their electrostatic (Coulomb) interactions with the nucleus and with the other electrons of the ion. Such interactions between the open-shell electrons lead to different states of the ion (i.e. different arrangements of the electrons in the open-shell orbitals) having different energies, and these different levels are known as spectral terms. In general, these terms contain degeneracies that can be lifted by other interactions to which the electrons are subject. Firstly there are the relativistic spin-orbit and spin-spin interactions involving the orbital and spin magnetic moments of the electrons on the ion, secondly, and of prime concern here, there are interactions with the other ions in the immediate vicinity of the transition metal ion, thirdly there are interactions with fields that may be applied

× Present address: CSIRO Division of Applied Physics, P.O. Box 218, Lindfield, NSW 2070, Australia.

externally to the crystal, and finally there are the higher order (non-Coulomb) hyperfine interactions with the ion's nucleus.

The first quantitative description of the way in which the electric fields produced by the ions in a crystal surrounding a transition metal ion influence its open-shell electrons was formulated by Bethe in 1929 (1), and was known as the crystal field theory. Bethe investigated, by means of symmetry concepts, how the symmetry and strength of the crystalline field affect the electronic levels of the metal ion, and this theory has since provided the basis for a comprehensive description (2-5) of the spectroscopic and magnetic properties of open-shell ions in crystals.

Because of the importance of the crystalline field to the determination of the wave functions and energy level structure of the ion, this theory has been applied to the interpretation of Fe^{2+} Mossbauer spectra for some years (6-8). With the aid of this theory Mossbauer spectroscopic determinations of the nature of the ground state wavefunction and of the single-ion anisotropy of the Fe^{2+} ion can be used to provide information to complement that obtainable by other experimental techniques, and consequently can help to elucidate particular features of the ion-crystal system being studied, to identify other interactions of the open-shell electrons that may be important in particular situations and, in favourable cases, to test certain features of the theory itself.

As the crystal field theory was originally developed, it described the effect of the crystal lattice on a transition metal ion in terms of the electrostatic field produced by point charge ions. Such a model cannot be expected to provide an adequate description of even a strongly ionic crystal, since it neglects such potentially important effects as the finite extent of the charges on the ions, the overlap and covalency of the magnetic ion's wave functions with those of the ligands and the effects of screening of the open-shell electrons by outer closed shells of the central ion. Effects such as these are important in ionic crystals (9) and become rapidly more important as the bonding between the magnetic ion and its ligands becomes stronger and more covalent, and they can only be quantitatively accounted for in a rigorous manner by the use of complex molecular orbital, or energy band, calculations (10). A hybridization of the molecular orbital method with the crystal field theory, in which the former approach is used to calculate the field strengths, is known as ligand field theory (11).

It is, however, important to realize that the symmetry of the interactions is not affected by the particular assumptions involved in calculating the strength of the crystal field. Thus, a transition metal ion at a site with a particular point symmetry in a crystal or molecule has wavefunctions with the same basic structure, and energy levels with the same degeneracies irrespective of the type of bonding it shares with its ligands and of the nature of those ligands. It has, therefore, been common practice to apply the crystal field theory to the interpretation of experimental results with the strengths of the various symmetry-defined components being regarded as empirical parameters to be determined from the data. Studies of the systematics of the parameters obtained using this approach have led to the formulation of models of the crystal (or ligand) field that relate some of the parameters to each other and provide a basis for comparing the parameters observed in different systems, and that form a link between the experimental data and the *ab initio* calculations.

One such model that is commonly used, particularly in the chemistry of transition metal complexes, is the angular overlap model (AOM) that has been developed largely by C.K. Jorgensen and coworkers (12). It is essentially an empirical molecular orbital model: it parameterizes the energy changes of open-shell orbitals due to σ- and π-bonding with particular ligands, whereas the crystal field models deal with fields (real or effective) that act on the open-shell electrons. The AOM is thus most easily visualized, and has its greatest utility, in situations where the bonding between the transition metal ion and its ligands is strong and covalent. This strong ligand field case has been discussed recently with particular reference to the interpretation of ^{57}Fe Mossbauer spectra by Gutlich et al. (10).

In this Chapter, however, consideration will only be given to cases in which the Fe^{2+} ions are in predominantly ionic environments where the crystal field is weak relative to the free ion term splittings. Another recently developed model of the crystal field, and one that is more intuitively appropriate to this situation, since it is essentially a generalization of the original point charge model, is the superposition model (SPM) proposed and investigated by Newman and coworkers (9,13). This model will be described briefly in Section 3. It may be noted that both the SPM and AOM are linear models in the sense that they assume that the effects of the individual ligands on the transition metal ion are additive.

In spite of its necessarily empirical application, the crystal field theory has been and still is a valuable tool for investigating the magnetic and spectroscopic properties of a variety of systems. Recent reviews (14,15) indicate the extent of its application to the interpretation of Fe^{2+} Mossbauer data. More recently it has been extended to include the study of vibronic (16-19) and dynamic effects (16, 20-24) in various systems, and it seems likely that the effects of the proton motion that is believed to occur in many hydrated or amined salts will also be investigated in a similar way.

In the next section, the physical basis of the crystal field, and the way it is related to the parameters determined from a $^{57}Fe^{2+}$ Mossbauer spectrum will be described. Section 3 outlines more quantitative aspects of the crystal field, with emphasis placed on features of particular concern in Mossbauer spectroscopy. Approximate methods for estimating effects of the dynamic crystal field on Mossbauer spectra are introduced, and the superposition model of the crystal field (SPM) is described briefly. Finally, Section 4 contains a brief, and somewhat selective and personalized view of the present trends and future scope for this line of research.

2. PHYSICAL BASIS FOR THE CRYSTAL FIELD INTERACTION AND ITS EFFECT ON $^{57}Fe^{2+}$ MOSSBAUER SPECTRA.

The aim of this section is to provide a more or less descriptive introduction to the effect of the crystal field on the wavefunctions of the ^5D ground term of the (high-spin) Fe^{2+} ion, and the ways in which this interaction may be manifested in the Mossbauer spectrum of the ion. The discussion is divided into three parts. In the first, the effect of a static crystal field on the orbital part of the 3d wavefunctions will be considered, since it is this part that interacts directly with the field. The dependence of the electric field gradient (EFG) at the Fe nucleus on the orbital level scheme thus produced will then be considered.

The spin degeneracy of the wavefunctions will be ignored until the second part of the discussion, when the spin-dependent intraionic interactions, and their effects on the orbital level scheme and the EFG will be introduced. The way in which the spin-orbit coupling and the crystal field combine to produce magnetic anisotropy will then be discussed and the ^{57}Fe magnetic hyperfine interaction and its crystal field-induced anisotropy introduced. Finally, magnetically-induced effects on the EFG will be mentioned.

The third part of the discussion generalizes that of the preceding two parts to include the effects of a non-static crystal field. A dynamic crystal field is expected to arise as a result of thermal vibrations of the ions in the lattice. Such motion can be classified either as harmonic (or pseudo-harmonic) vibrations of the ions in "normal" potential wells in the lattice, or as grossly anharmonic motion produced by an ion moving between different pseudo-harmonic wells, as may occur in Jahn-Teller systems or others in which more than one potential minimum is associated with a single lattice site, or in ionic diffusion through the lattice.

2.1 The Orbital Wavefunctions

2.1.1 The Free Fe^{2+} Ion Ground State (3, 4, 11, 25)

An ion is considered to be free when it is not subject to any fields, either electric or magnetic, that are generated externally to it. For example, ions in a low density gas may be considered, to a good approximation, to be free, but the closest solid state approximation to a free ion is one that is bound in an inert gas matrix. This is known as matrix isolation.

The electronic configuration of the outermost (unfilled) shell of a free Fe^{2+} ion is $3d^6$, and using Hund's rule its ground term, schematized in Fig. 1(a), can be shown to be ^5D - it has L=2, S=2.

The five one-electron 3d (ℓ=2) orbitals have angular distributions of electron density with the well-known forms shown in Figs. 2-5, and they may be expressed as linear combinations of the ℓ=2 spherical harmonics $Y_2^m(\theta,\phi)$. The free Fe^{2+} ion has one electron in each of the five 3d orbitals shown in Fig. 2, and these constitute a half-filled shell which has a spherically symmetric charge distribution. The orbital properties of the ^5D ground term of the ion are therefore similar to those of a $3d^1$ system, with one electron in a spherical potential. The lowest excited terms are a series of spin triplets $\gtrsim$ 25,000 cm^{-1} above the ground term. As long as the crystal field interaction is much weaker than the intra-ionic d-d electronic interactions, which is the case to which this discussion will be restricted, these terms will have a negligible effect on Mossbauer spectra. In this case, the crystal field can be treated as a perturbation on the free ion wavefunctions (Fig. 1(b)). (This cannot be done if the crystal field is "strong". In this case Hund's rule is not obeyed and a non-magnetic singlet ground term, referred to as "low-spin" Fe^{2+}, is observed (3, 4, 10, 11, 26) - see Fig. 1(c)).

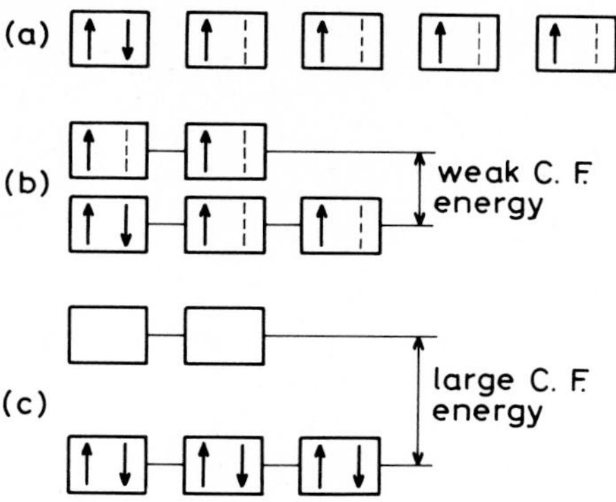

Fig. 1. Schematic representation of the one-electron 3d orbitals and of their occupation for the d^6 configuration. The boxes represent the different orbitals and the arrows represent electrons whose spin is in the direction of the arrow. (a) Free ion case. All orbitals have the same energy. (b) Weak crystal field of octahedral symmetry. The t_{2g} orbitals have lower energy than the e_g orbitals, but this is less than the repulsive energy that would be required to doubly fill all of the t_{2g} orbitals (C.F. = crystal field). (c) Strong crystal field of octahedral symmetry. Caption same as for (b), but here the crystal field energy is greater than the excess electron-electron repulsive energy, so all the t_{2g} orbitals are doubly occupied. Hund's rule is obeyed for the 3d shell as a whole in cases (a) and (b), but only within a set of orbitals (e.g. the t_{2g} orbitals) for case (c). The L, S quantum numbers for the ion are L=2, S=2 for (a) and (b) (the "high-spin" case) and L=0, S=0 ("low-spin") for (c).

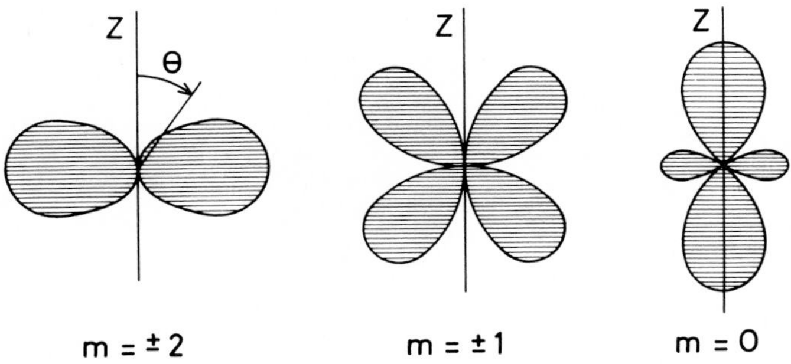

Fig. 2. The one-electron 3d orbitals as spherical harmonic functions. The shaded areas are sections of the surfaces:

$$D(\theta) = |Y_2^m(\theta,\phi)|^2 \qquad (\phi = 0 \text{ to } 2\pi).$$

These surfaces have axial symmetry about the z axis, from which θ is measured.

2.1.2 Effect of the Crystal Field on the Ground Orbital State

The crystal field may be thought of in a naive way as a non-uniform, anisotropic electric field that acts on the electrons of the paramagnetic ion. Because of its anisotropy, electrons in orbitals with different spatial distributions may have different energies. This effect may be visualized if the free-ion 3d orbitals, for which any orthogonal linear combination of the $\ell=2$ spherical harmonics may be used, are considered. Two simple and illustratively useful choices are:

(i) the spherical harmonics $Y_2^m(\theta,\phi)$ themselves. These are most appropriate when the symmetry is axial such as in a linear molecule. They are usually written in the Dirac notation as $|m\rangle$.

If the crystal field is produced by, for simplicity, a pair of charges symmetrically located on the z axis, as shown in Fig. 3, then it is clear that Y_2^0 will have the highest energy because electrons in this state will spend a considerable time close to the electrons on the ligands. $Y_2^{\pm 2}$ will lie lowest because these functions have least overlap with the ligand orbitals (see Section 2.3 below).

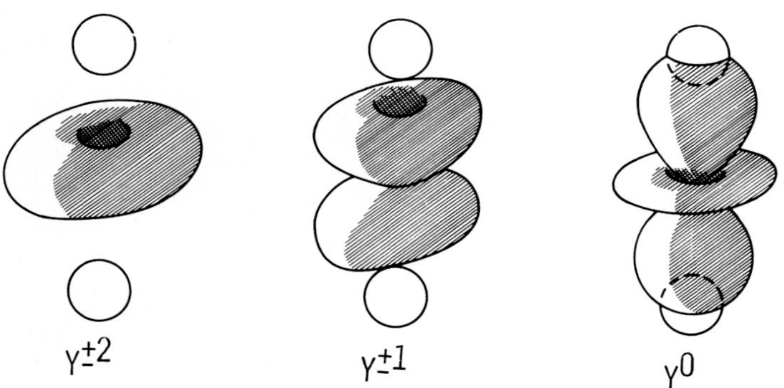

Fig. 3. The 3d orbital functions for the hypothetical case of a linear, symmetric triatomic molecule. The ligand ions are represented by the open circles. These are the same orbitals as those in Fig. 2, but represented in 3-dimensions rather than 2. It can be seen that the overlap of the 3d wavefunctions with the ligands increases from $Y_2^{\pm 2}$ *to* Y_2^0

(ii) the following real combinations, the so-called tesseral harmonics (27):

$$Y_A \equiv Z_2^0 = Y_2^0 = |0\rangle \sim 3z^2 - r^2$$

$$Y_B \equiv Z_2^2 = \frac{1}{\sqrt{2}} (Y_2^2 + Y_2^{-2}) = \frac{1}{\sqrt{2}} (|2\rangle + |-2\rangle) \sim x^2 - y^2 ,$$

$$Y_C \equiv Z_2^{-2} = \frac{-i}{\sqrt{2}} (Y_2^2 - Y_2^{-2}) = \frac{-i}{\sqrt{2}} (|2\rangle - |-2\rangle) \sim xy , \tag{1}$$

$$Y_D \equiv Z_2^1 = \frac{-1}{\sqrt{2}} (Y_2^1 - Y_2^{-1}) = \frac{-1}{\sqrt{2}} (|1\rangle - |-1\rangle) \sim xz ,$$

$$Y_E \equiv Z_2^{-1} = \frac{i}{\sqrt{2}} (Y_2^1 + Y_2^{-1}) = \frac{i}{\sqrt{2}} (|1\rangle + |-1\rangle) \sim yz.$$

The tesseral harmonics are appropriate functions to consider when the crystal field has orthorhombic (or tetragonal or cubic) symmetry, since in these cases they are eigenfunctions.

For the purposes of illustration, consider the Fe^{2+} ion to be surrounded by six anions forming a regular octahedron, as is shown in Fig. 4. It is clear from this figure that the Y_A and Y_B orbitals will have a high potential energy because their regions of highest electron density lie close to the electrons on the anions. On the other hand, the Y_C, Y_D and Y_E orbitals, which "point" between the ligands will have a lower energy. Calculations show that Y_A and Y_B have the same energy and thus form a degenerate doublet that transforms as E_g, while Y_C, Y_D and Y_E are a degenerate triplet and transform as T_{2g}. The triplet is commonly $\sim 10^4$ cm^{-1} lower in energy than the doublet.

The degeneracy of both Y_A, Y_B and Y_C, Y_D, Y_E orbitals may be removed by an appropriate distortion of the surroundings from octahedral symmetry. For instance, a tetragonal distortion that elongates the octahedron along the z axis will be seen from Fig. 4 to lower the energy of Y_A relative to that of Y_B. It also lowers Y_C relative to Y_D and Y_E (which remain degenerate) but this is more difficult to appreciate from Fig. 4. A further orthorhombic distortion will remove the degeneracy of Y_D and Y_E.

The reader may also be able to appreciate, using Fig. 4 as a guide, that if instead of six octahedrally arranged ligands there were either 8 arranged on a simple cube (as in a CsCl-type lattice) or 4 on a regular tetrahedron around the paramagnetic ion, then the Y_A and Y_B orbitals would be lower in energy than Y_C, Y_D and Y_E.

A further common situation that is worth trying to visualize is that of a trigonally distorted octahedron. In this case the

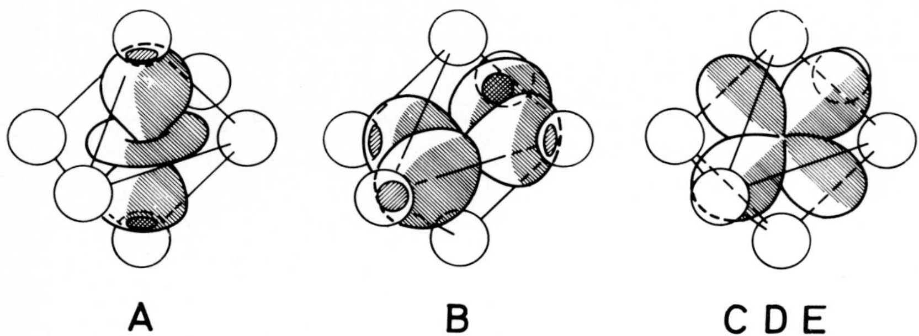

Fig. 4. The tesseral harmonic 3d wavefunctions (27), located in a regular octahedron of ligands. The orbitals labelled A-E are defined in Equations (1) of the text. The overlaps of the A and B orbitals with the ligands are shown by the small shaded areas. Only one of the C, D, E orbitals is shown (i.e. E). All have the same shape and may be obtained from the one shown by suitable rotation of the octahedron. B also has the same shape as C, D, E but is oriented differently relative to the ligands.

distortion axis (the z axis) passes through the centres of a pair of opposite triangular faces and (Fig. 5) the orbital Y_A must belong to the ground T_{2g} triplet since it overlaps very little with any of the ligands. If the octahedron is elongated along the z axis, then the iron-anion bonds are tilted closer to the Y_A orbital and consequently its energy will be raised. The behaviour of the other orbitals is less obvious pictorially and in fact different linear combinations of the Y_2^m than those in Eqn. (1) are eigenfunctions.

This simple visual picture is essentially the basis of the point-charge model for the crystal field: in it the ligands are considered to be point charges and the field they produce is calculated using simple electrostatics. Hutchings (27) has written an excellent article describing the use of the point charge model to calculate crystal fields and their effect on the energy levels of ions in crystals. Section 3 of this Chapter will require familiarity with the "operator equivalent" method introduced by Stevens and coworkers (28), and its derivation as an application of the Wigner-Eckart theorem should be understood.

Fig. 6 is a schematic diagram of the energy levels resulting from the ^5D ground term of Fe^{2+} for crystal fields arising from various simple situations, as deduced from point charge calculations.

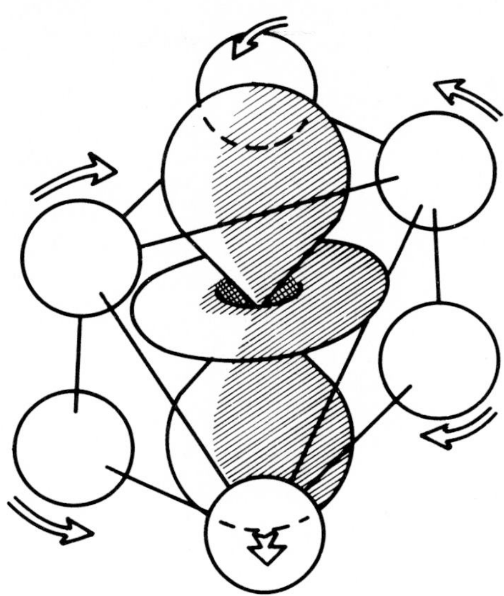

Fig. 5. A trigonally distorted octahedron, including schematically the axial $Y_A = Y_2^0$ orbital. The overlap of this orbital with the ligands is increased if the octahedron is stretched along the trigonal axis, and decreased if it is compressed.

2.1.3 The Orbital Level Scheme and the Electric Field Gradient (EFG)

The electric quadrupole interaction is the highest order multipole term in the hyperfine interaction that has been measured in Mossbauer experiments*. The interaction Hamiltonian can be expressed most neatly as:

$$H_Q = \sum_{q=-2}^{2} (-1)^q Q_2^q V_2^{-q} \ , \tag{2}$$

where Q_2^q and V_2^q are spherical tensor operators and thus transform under rotation like the spherical harmonics Y_2^q. The Q_2^q are the components of the nuclear quadrupole moment operator, and they may be expressed, using the Wigner-Eckart theorem, in terms of the nuclear spin operators. The V_2^q are the components of the electronic electric field gradient (EFG) operator. They are usually expressed in terms of their Cartesian components, the expectation values of which, evaluated at the nuclear site, are given by*:

---

* For more details, see Chapter 1.

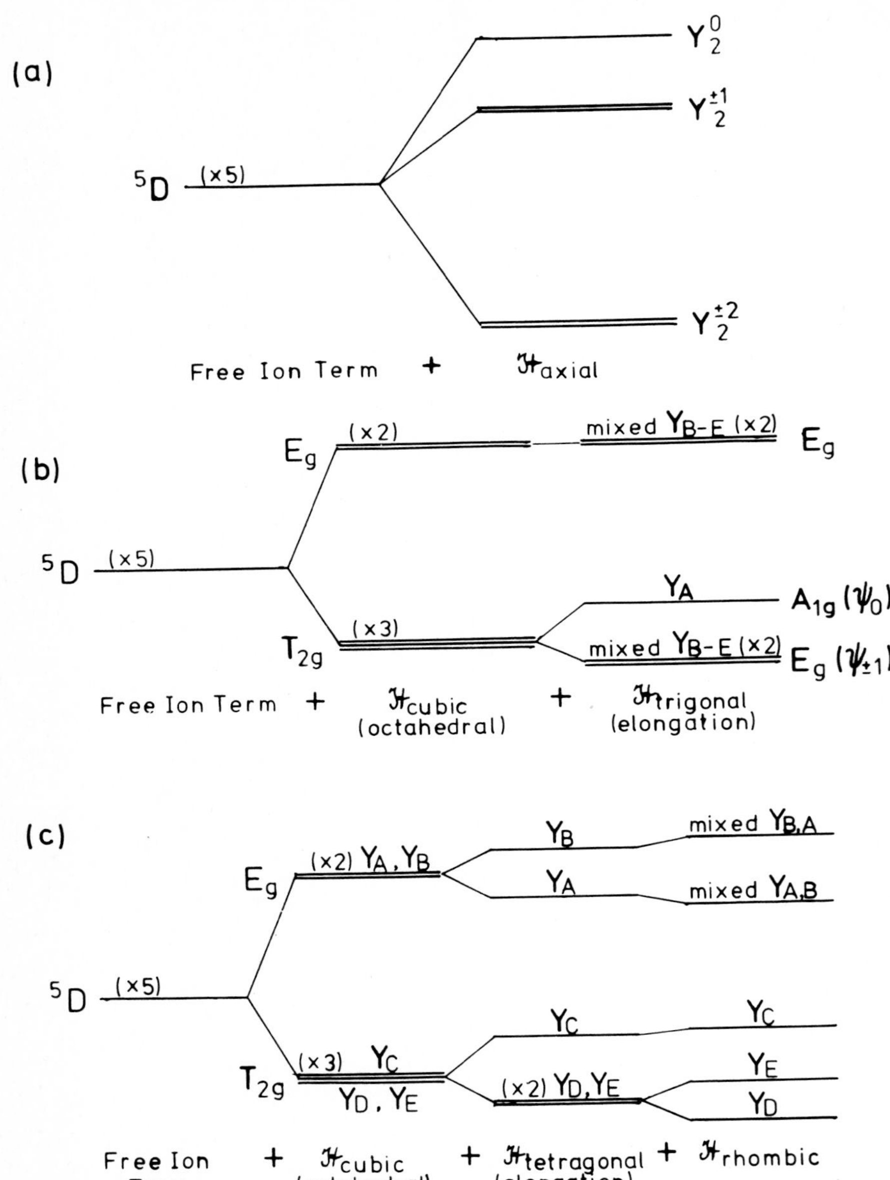

Fig. 6. Schematic diagram of the Fe^{2+} orbital levels derived from the ^5D ground term in crystal fields of (a) axial symmetry (as in Fig. 3), (b) octahedral coordination with trigonal distortion (Fig. 5) and (c) octahedral coordination with tetragonal and rhombic distortions. The orbital degeneracies (if any) are shown in brackets. For tetrahedral coordination, the dominant crystal field terms for each symmetry are of opposite sign than for octahedral coordination, so the level splittings should be reversed in each case.

$$V_{ij} = V_{ji} = \left(\frac{\partial^2 U}{\partial x_i \partial x_j}\right)_{\vec{r}=0} , \qquad (3)$$

where U(x,y,z) is the electric potential arising from the electric charges surrounding the nucleus†. It may be noted in passing that the tensor operator form of H_Q shows explicitly that there are only five independent components of the Cartesian EFG tensor (viz. V_{zz}, $V_{xx}-V_{yy}$, $V_{xy}+V_{yx}$, $V_{xz}+V_{zx}$, $V_{yz}+V_{zy}$) as follows from the validity of Poisson's, or in this case Laplace's, equation and from the symmetry of the Cartesian tensor components.

Many books and articles discuss the electric field gradient tensor in terms of the crystal field theory approach to calculations, and no more than a brief summary will be given here. An excellent account, albeit with particular reference to the point charge model, is that of Travis (29) and a good recent discussion is contained in Gutlich et al. (10). For details these articles as well as the earlier paper of Ingalls (8) may be consulted.

The symmetry of the EFG tensor at a nuclear site depends both on its own intrinsic symmetry elements as a second order symmetric Cartesian tensor (viz. inversion centre, 3 mirror planes, 3,2-fold rotation axes) as well as those of the point group of the nuclear environment. In particular, the symmetry of the EFG may be higher but may never be lower than that of the point group of the ion. Any 3-, 4- or 6-fold point group rotation axis must also be a rotation axis of the EFG tensor (in these cases $V_{xx} = V_{yy}$ so $\eta=0$ and the EFG is said to be axially symmetric). The problem of determining the complete symmetry and orientation of the EFG tensor has been considered by Zory (30) and more recently by Gibb (31). It will not be discussed further here ($\eta = \frac{V_{xx} - V_{yy}}{V_{zz}}$ is the asymmetry parameter).

The EFG may be considered to consist of two separately identifiable contributions (Ref. 10, 26, 29 for example):

(i) the"Lattice" contribution that results from all of the other charges in the lattice apart from those on the ion under consideration (32). Estimates of this contribution, based on monopole and dipole summations (33), may be made as long as the crystal structure and the dipole polarizabilities of the various ions are known. The lattice contribution is largely independent of temperature except

†The minus sign that should appear on the right-hand side of Eqn. (3) is dropped by convention.

through lattice thermal expansion effects or crystallographic changes. (ii) the "valence" contribution that results from the anisotropic charge distribution of the open-shell electrons of the ion under consideration. This contribution represents one of the major influences of the crystal field on the Mossbauer spectrum, since the crystal field determines the nature of the orbital wavefunctions and the ordering and splitting of the energy levels, and these in turn largely control the valence electron charge distribution (see, e.g., Figs. 2-6) (8).

The total EFG is therefore given by:

$$V_{ij} = (1-\gamma_{\infty})V_{ij}(\text{latt.}) + (1-R)V_{ij}(\text{val.}), \qquad (4)$$

where it is assumed, of course, that all the EFG tensors are expressed with respect to the same axes. γ_{∞} and R shielding factors that result from distortions produced in the electron density of the ion core. For Fe^{2+}, γ_{∞} has been estimated to be ~ -10 (32) and R to be ~ 0.3.

The presence of these two contributions to the EFG tensor greatly complicates interpretation of quadrupole splitting measurements, as was discussed in some detail by Travis (29). The two are obviously related, since they both result from the anisotropic charge distribution in the lattice surrounding the ion in question, but their numerical relationship is complicated. Ingalls (8) showed that within the point charge lattice model both V_{zz}(latt.) and ηV_{zz} (latt.) are proportional to second order (quadrupolar) crystal field terms so that in this case the relationship between the lattice and valence EFG contributions could be determined. However, in a more realistic situation this will not be the case because the relative magnitudes of the electrostatic, overlap and covalence contributions to V_{ij}(latt.) and to the crystal field terms will be different.

There is still no simple solution to this problem. A large number of authors have simply by-passed it by neglecting the lattice contribution on the grounds that it is generally relatively small ($\lesssim$10% of the valence contribution) but this is certainly not always the case, e.g. $MFeF_4$ (34), gillespite ($BaFeSi_4O_{10}$) (35).

The dependence of the valence electron contribution to the EFG on the particular d orbitals of the Fe^{2+} ion that are occupied can be deduced as follows. The V_{ij} components are given by:

$$V_{ij}(\text{val.}) = \int_\tau \rho(x,y,z)(3x_ix_j - r^2\delta_{ij})r^{-5}\, d\tau,$$

where $\rho(\vec{r}) = -e\psi^*(\vec{r})\psi(\vec{r})$, $\psi(\vec{r}) = R(r)Y(\theta,\phi)$ is the d orbital wavefunction and the integral is over the volume τ of the ion. This reduces to:

$$V_{ij}(\text{val.}) = \int_0^\infty -er^{-3}R^*(r)R(r)r^2dr \int_0^{2\pi} d\phi \int_0^\pi \left(\frac{3x_ix_j}{r^2} - \delta_{ij}\right)Y^*(\theta,\phi)\, Y(\theta,\phi)x\sin\theta d\theta$$

$$= -e\langle r^{-3}\rangle_{3d} \int_0^{2\pi} d\phi \int_0^\pi F(\theta,\phi)\sin\theta d\theta. \qquad (5)$$

The values of these angular integrals for the five tesseral harmonic functions Y_A ..... Y_E defined in Equations (1) are given in Table 1.

TABLE 1.
EFG components for the tesseral harmonic d orbitals.

	$Y_A = d_{z^2}$	$Y_B = d_{x^2-y^2}$	$Y_C = d_{xy}$	$Y_D = d_{xz}$	$Y_E = d_{yz}$
$V_{xx}/(e\langle r^{-3}\rangle_{3d})$	+ 2/7	- 2/7	- 2/7	- 2/7	+ 4/7
$V_{yy}/(e\langle r^{-3}\rangle_{3d})$	+ 2/7	- 2/7	- 2/7	+ 4/7	- 2/7
$V_{zz}/(e\langle r^{-3}\rangle_{3d})$	- 4/7	+ 4/7	+ 4/7	- 2/7	- 2/7
Symmetry axis	z	z	z	y	x

All these tensors are diagonal, and have axial symmetry; for $Y_{D,E}$ an axis change is required to fulfill the conventional condition $|V_{ZZ}| \geq |V_{YY}| \geq |V_{XX}|$ and yield $\eta = 0$. Crossed terms of V_{ij} between different orbitals may be found in (29).

For each of these orbitals the off-diagonal tensor components are zero, so the axes defining the wavefunctions are the EFG principal axes. Also, all give rise to an axially symmetric EFG about the axis shown in the table: for Y_D, Y_E the axes must be relabelled to satisfy the conventional condition $|V_{zz}| \geq |V_{yy}| \geq |V_{xx}|$ and make the asymmetry parameter $\eta=0$.

The valence contribution to the EFG at low temperatures

The effect of the crystal field on the EFG can be seen particularly clearly under the following conditions:

(i) the orbital ground state is sufficiently well isolate from the excited states so that there is negligible admixture of them (by, for example, the spin-orbit coupling).

(ii) the temperature is sufficiently low so that the ground state alone is occupied.

The sign and magnitude of the EFG components resulting from various crystal field symmetries when the above conditions are satisfied are given in Table 2. For cubic symmetry the contributions of the various orbitals (Y_A and Y_B or Y_C, Y_D and Y_E) cancel, resulting in a zero EFG, as expected from symmetry considerations.

It may also be noted that the magnitude of the quadrupole splitting produced by any one of these five orbitals is the same. Ganiel (42) showed that this is a general property of any singlet orbital that is a linear combination either of Y_A and Y_B or of Y_C, Y_D and Y_E (i.e. any singlet derived from the cubic E_g or T_{2g} manifolds), but it is not true in the presence of admixture between these manifolds or when the ground state is not an orbital singlet. Furthermore, Ingalls (8) showed that this quadrupole splittings, $|\Delta E_0| = \frac{2}{7} e^2Q\langle r^{-3}\rangle(1-R)$, is the maximum value that the valence contribution can have under any circumstances.

The temperature dependence of the EFG (valence)

At higher temperatures, excited orbital states become thermally populated, and if the ion relaxes between these states rapidly compared with the nuclear precession frequency (i.e. $\omega > \omega_L = \Delta E_Q/\hbar \simeq 10^8 \text{ sec}^{-1}$), as is generally true, a thermal average of the EFG associated with the various occupied states is observed. This results in a decrease of ΔE_Q with increasing temperature (8), and the rate of decrease depends on the excited state wavefunctions and on their excitation energies. As a simple example, consider the case of an Fe^{2+} ion, in a site with tetrahedral coordination that has a tetragonal compression, in Cu_2GeFeS_4 (43). The orbital states of the Fe^{2+} ion are shown in Fig. 7 and, for temperatures such that only Y_A and Y_B are significantly populated,

TABLE 2.

Calculated valence EFG components and quadrupole splitting for various crystal field symmetries under the conditions described in the text.

Site coordination	No distortion	Compressed 4-fold axis	Compressed 3-fold axis	Rhombic symmetry	Elongated 4-fold axis	Elongated 3-fold axis
Octahedral						
Ground state	$Y_C Y_D Y_E$	$Y_D Y_E$	Y_A	mixed	Y_C	mixed, doublet
V_{ZZ}	0	- 1/2	-1	$-1 \leq V_{ZZ} \leq +1$	+1	+ 1/2
η	0	0	0	$\neq 0$	0	0
$\lvert \Delta E_Q \rvert$	0	1/2	1	1	1	1/2
Examples	MgO, T>20K(20)		$FeSiF_6, 6H_2O$ (6b)	FeF_2 (36,37)	$Fe(py)_2Cl_2$ (38)	$FeCO_3$ (7)
Tetrahedral						
Ground state	$Y_A Y_B$	Y_A	mixed, doublet	$\alpha Y_A + \beta Y_B$	Y_B	mixed, doublet
V_{ZZ}	0	-1	0	$-1 \leq V_{ZZ} \leq +1$	+1	0
η		0		$\neq 0$	0	
$\lvert \Delta E_Q \rvert$	0	1	0	1	1	0
Examples	ZnS(blende) (39)	$FeCr_2O_4$ (40) (c/a<1)			FeV_2O_4 (40) (c/a>1)	
Axial						
Ground state		charges predominantly in the equatorial plane $Y_A = Y_2^0$			charges predominantly on the Z axis $Y_{B,C} = Y_2^{\pm 2}$	
V_{ZZ}		+1			-1	
$\lvert \Delta E_Q \rvert$		1			1	
Examples		$Fe_2F_5, 2H_2O$ (41)				

V_{ZZ} is expressed in units of $V_0 = \frac{4}{7}e\langle r^{-3}\rangle$

ΔE_Q is given in units of $\Delta E_0 = \frac{1}{2}\, eQV_0(1-R) \sim 4$ mm sec^{-1} for strongly ionic compounds

The quantization axis Z is the highest symmetry axis at the Fe^{2+} site.

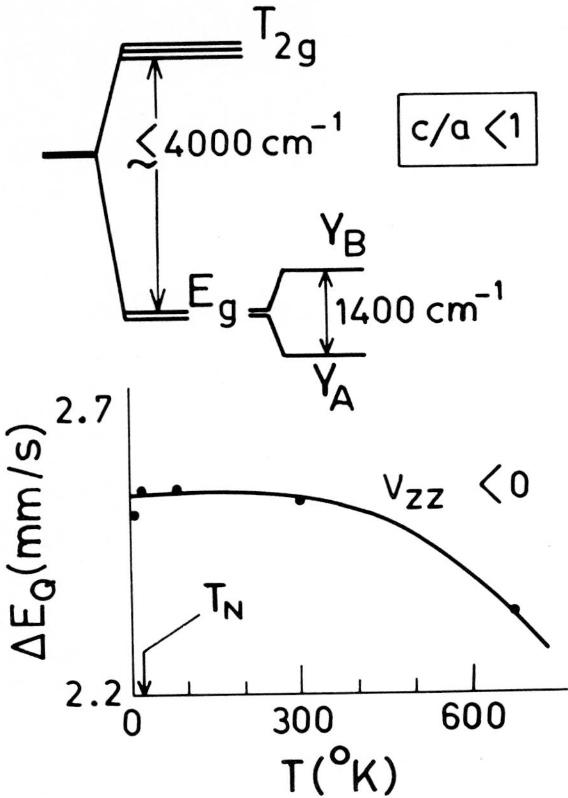

Fig. 7. Orbital level scheme (upper) and quadrupole splitting $\Delta E_Q(T)$ for Fe^{2+} in a tetrahedron compressed along a 4-fold axis. The compound is Cu_2GeFeS_4, for which X-ray data gives $c/2a \approxeq 0.99$ at room temperature. $\Delta E_Q(T)$ data is from (43).

$$\Delta E_Q^{val.}(T) = -\Delta E_0(1-\exp(-\delta/kT))/(1+\exp(-\delta/kT))$$

$$= -\Delta E_0 \tanh(-\delta/2kT), \qquad (6)$$

This relationship is shown by the solid line drawn through the ΔE_Q data in Fig. 7 for a splitting $\delta \sim 1400\ cm^{-1}$ between Y_A and Y_B. It should be emphasized that, at least at this stage, this should be regarded as a qualitative example only. Complications such as spin-orbit coupling (which however should have only a small effect here), vibronic effects, the temperature dependence of the crystal field, and the lattice EFG make detailed quantitative conclusions difficult to draw.

In the high temperature limit, the orbital states are equipopulated and their EFG contributions cancel so the lattice contribution might, in principle, be measured.

2.2 The Intra-ionic Spin Interactions

It will now be recalled that each of the orbital wavefunctions for the ^5D ground term of Fe^{2+} has five-fold spin degeneracy, and this degeneracy may be wholly or partially lifted by intra-ionic spin-dependent interactions. These interactions are (3, 4, 11):

(i) the spin-orbit coupling, which is a relativistic interaction between the orbital and spin magnetic moments of an electron. As long as its effect within a spectral term, and not between terms, is being considered, then it can be written $\mathcal{H}_{LS} = \lambda\vec{L}.\vec{S}$ where $\vec{L}$ and $\vec{S}$ are the (vector) orbital and spin operators respectively for the ion and λ is known as the spin-orbit coupling constant. Its value for the free ion can be determined from the Lande interval rule, and is -103 cm^{-1} for the free Fe^{2+} ion in the ground ^5D term.

(ii) the magnetic spin-spin interaction between electrons on the ion which, within an (L,S) term, may be written:

$$\mathcal{H}_{SS} = -\rho[(\vec{L}.\vec{S})^2 + \frac{1}{2}(\vec{L}.\vec{S}) - \frac{1}{3}L(L+1)S(S+1)]. \qquad (7)$$

This is a relatively small interaction whose existence is verified by small departures from Lande's interval rule in free-ion spectra. Pryce (44) deduced a value of +0.95 cm^{-1} for ρ by considering only the ^5D ground term of Fe^{2+}, but Trees (45) pointed out that this contains a substantial contribution from second-order spin-orbit coupling via the excited terms. However, for calculations restricted to the ^5D term of Fe^{2+}, Pryce's value would seem to be the appropriate, although it may be modified by bonding effects in a crystal, as is the spin-orbit coupling.

(iii) the magnetic hyperfine coupling, which may be written in spin-Hamiltonian form as $\mathcal{H}_{hf} = \vec{S}.[A].\vec{I}.[A]$ is the magnetic hyperfine interaction tensor, and it will be discussed more fully later in this section. $\mathcal{H}_{hf}$ has never been observed to influence the EFG at ^{57}Fe nuclei [although it has produced an observable "pseudo-quadrupole" interaction at ^{169}Tm nuclei (46, 26)], but it has been observed to mix crystal field-split states of Fe^{2+} (47).

The effect of $\mathcal{H}_{LS} + \mathcal{H}_{SS}$ on the orbital wavefunctions (with five-fold spin degeneracy) of Fe^{2+} has been extensively examined, particularly using first and second order perturbation treatments (see, Ref. 14, 15 , which contain diagrams illustrating effects on some energy level schemes). An orbital singlet, which is very commonly the Fe^{2+} ground state, is split into five levels that may be described by an S = 2 spin Hamiltonian,

$$\mathcal{H}_{spin} = DS_z^2 + \frac{1}{2}E(S_+^2 + S_-^2), \quad (8)$$

where D and E are parameters that depend on the crystal field as well as on λ and ρ.

The consequences of these spin-dependent interactions, and in particular of $\mathcal{H}_{LS}$ and $\mathcal{H}_{SS}$, for the Mossbauer and magnetic properties of the Fe^{2+} ion will now be described.

2.2.1. Influence of $\mathcal{H}_{LS} + \mathcal{H}_{SS}$ on the EFG

As well as lifting the spin degeneracy of the orbital wavefunctions, the spin-orbit and spin-spin coupling interactions mix together orbital states and consequently produce a decrease in the quadrupole splitting $|\Delta E_Q|$, similar to the thermal decrease already mentioned. This can be described by a *reduction factor* F (8) that depends on λ,ρ and on the crystal field parameters, or equivalently on crystal field splittings of specified orbital levels, δ_i. Thus, if the thermal reduction is included:

$$\Delta E_Q^{val.}(T) = \Delta E_0 \times F(\delta_i,\lambda,\rho,T) . \quad (9)$$

The effect of ρ is small and is usually neglected. ΔE_0 is, as in Eqn. (6), the *unreduced* splitting produced by the ground orbital state, values of which are listed in Tables 1 and 2.

Large effects of these interactions are observed when crystal field splittings of orbital states are smaller than or comparable with λ. For example, for Fe^{2+} in $MgCl_2$ (48) there is a small trigonal distortion from octahedral symmetry. In this case the orbital states are very strongly mixed by $\mathcal{H}_{LS}$ and the reduction factor F is ~0.5 at low temperatures. The temperature dependence of ΔE_Q is shown in Fig. 8(b). The rapid decrease of the quadrupole splitting in the temperature range 0-30° K indicates the presence of a low-lying excited state (~10 cm^{-1}) with significantly different orbital character than the ground state.

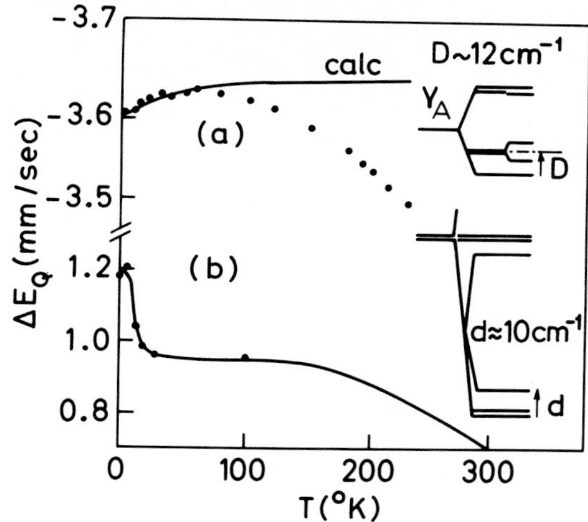

Fig. 8. Quadrupole splitting $\Delta E_Q(T)$ with solid curves derived from the energy level schemes shown, illustrating the influence of the spin interactions $H_{LS}+H_{SS}$ at low temperatures. (a) Ferrous fluosilicate (49). (b) Fe^{2+} :$MgCl_2$ (48).

Even if the ground state is a well-isolated orbital singlet, thermal population of the excited spin levels can be observed in the temperature dependence of ΔE_Q, as long as the spin-orbit coupling can introduce a significant (spin-dependent) admixture of excited orbital states. This can be seen in the results for ferrous fluosilicate ($FeSiF_6$,$6H_2O$) (49), shown in Fig. 8(a), in which the Fe^{2+} ion is in a site of octahedral coordination with a relatively large trigonal distortion (compression). (The divergence of the calculated curve from the experimental results for $T \gtrsim 70°$ K may be due to vibronic effects - see Section 3.2).

2.2.2 Ionic Magnetic Anisotropy and Magnetic Structure

At low temperature, when only its lowest spin levels are significantly populated, the ferrous ion exhibits magnetic anisotropy when the crystal field symmetry is lower than cubic. That is, at these temperatures the potential energy of a single ion in an applied magnetic field of given strength depends on the orientation of the field to the crystal axes. This angular dependence can be easily calculated by adding a Zeeman term to the Hamiltonian that acts on the ^5D ground term:

i.e. $\mathcal{H}_{total} = \mathcal{H}_{CF} + \mathcal{H}_{LS} + \mathcal{H}_{SS} + \mathcal{H}_{Zeeman}$, (10)

$$\text{where } \mathcal{H}_{Zeeman} = -\vec{M}.\vec{H}$$

$$= \mu_B(\vec{L}+2\vec{S}).\vec{H}$$

$$= \mu_B H\cos\theta(L_z + 2S_z)$$

$$+ \frac{1}{2}\mu_B H\sin\theta[(L_+ + 2S_+)e^{-i\phi} + (L_- + 2S_-)e^{-i\phi}].$$

$\vec{M}$ is the magnetic moment operator for the ion, and θ,ϕ are the polar coordinates of the magnetic field $\vec{H}$ relative to the quantization axes chosen. Note that the hyperfine interactions have been ignored in the above Hamiltonian because of their very small (generally <<kT) contribution to the energy of the ion.

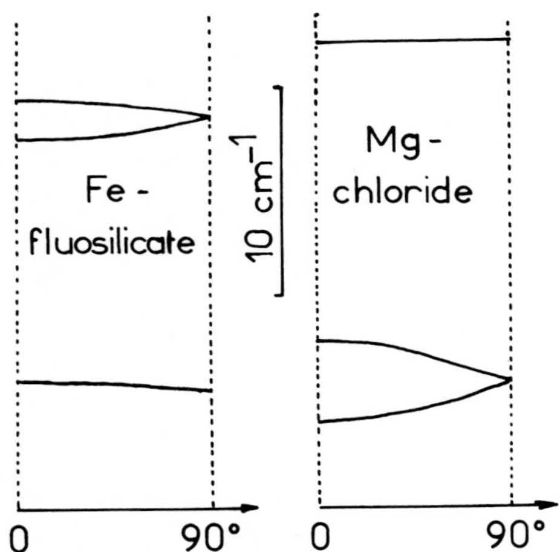

Fig. 9. The lowest three energy levels of a high-spin Fe^{2+} ion in two crystal fields of axial symmetry and a magnetic field H = 1T applied at an angle θ to the symmetry axis of the crystal field (1T=10 kOe). The two crystal fields are produced by an octahedron of ligands with: (a) trigonal compression. Calculation for ferrous fluosilicate (49, 50)
(b) trigonal elongation. $Fe^{2+}:MgCl_2$ (48).

Fig. 9. shows the dependence of the Fe^{2+} energy levels on the direction of an applied magnetic field for two typical cases where the crystal field has axial symmetry (so that the Zeeman energy depends only on θ and not on ϕ). In Fig. 9(a), for which the ground level was taken to be a singlet, the ion has minimum energy (at low temperatures) when $\theta = 90^{\circ}$, i.e. when the field lies in the XY plane, which is then known as the easy plane of magnetization. Such a situation will occur when the Fe^{2+} ion has an orbital singlet ground state with the spin-Hamiltonian (Eqn. 8) parameter $D>0$ ($E=0$ for axial symmetry).

On the other hand, Fig. 9(b) shows that the minimum energy configuration for a doublet ground state is when $\theta=0$, i.e. when the field lies along the z axis, which is then referred to as the easy axis. This situation will occur when the Fe^{2+} ion has an orbital doublet ground state in the crystal field, or when it has an orbital singlet lowest and $D<0$.

It should be remarked that the magnetic anisotropy will be temperature dependent, decreasing with increasing temperature until it becomes zero in the high temperature limit when all states are equally populated, and dependent also on the magnetic field strength. The expected dependence on field strength for a ferrous fluosilicate single crystal is given in (50).

The single-ion magnetic anisotropy may, of course, be determined by magnetic measurements (see Ref. 51 for the case of ferrous fluosilicate) or deduced from EPR or Mossbauer measurements.

Very often qualitative or semi-quantitative information about the magnetic anisotropy of an ion in a magnetically ordered crystal can be obtained from Mossbauer measurements of the relative orientation of the magnetic hyperfine field and the EFG principal axes. In a homogeneous magnetic material (i.e. one with a single type of magnetic ion) the magnetic structure is often determined by the single-ion anisotropy, but this depends on the strength and anisotropy of the inter-ionic exchange interactions. Example of definitive Mossbauer studies of materials in this category were those on $FeCl_2$ (48), $FeCO_3$ (7) and the spinels FeV_2O_4 and $GeFe_2O_4$ (40).

A considerable amount of Mossbauer work has also been done on inhomogeneous magnetic materials, and in particular in crystals with only a small amount of Fe^{2+} as a second magnetic ion. In this case there is competition between the anisotropies of the two species of magnetic ion and that of the exchange interaction in determining the magnetic structure. Mossbauer measurements are invaluable in this situation as they allow at least a semi-quantitative determination of the magnetic energy of one of the constituents. Examples of this are work on Fe^{2+} in $NiCl_2$ and $CoCl_2$ (52) and in $CoCO_3$ (53) and $MnCO_3$ (54). The earlier review of Varret (15) deals with this aspect of $^{57}Fe^{2+}$ Mossbauer studies in more detail.

Table 3 contains a summary of spin-Hamiltonian D values and magnetic anisotropies calculated (using perturbation theory) for various cases in which the crystal field has axial symmetry and where the orbital ground state is well isolated from the lowest excited state (i.e. $\delta >> \lambda$, where δ is the excitation energy of this state).

2.2.3 The Magnetic Hyperfine Interaction and its Anisotropy

In a spin-Hamiltonian formalism the magnetic dipole hyperfine interaction can be written in the form (4, 55):

$$\mathcal{H}_{hf} = \vec{S}.[A].\vec{I}, \tag{11}$$

where [A] is known as the magnetic hyperfine interaction tensor and is usually proportional to the spin-Hamiltonian electronic [g] tensor [(4), p. 650-6]. It has a complicated dependence on the spin and orbital moments of the Fe^{2+} ion, and a more explicit expression in terms of these operators will be postponed until the next section. In the great majority of cases for $^{57}Fe^{2+}$ ions, $\mathcal{H}_{hf}$ does not produce significant admixture of electronic states either because they are separated by energies much larger than hyperfine interaction energies ($\sim 10^{-2}$ cm^{-1}) or because matrix elements of $\mathcal{H}_{hf}$ connecting them are zero. Under these conditions the electronic and nuclear parts of $\mathcal{H}_{hf}$ can be evaluated separately. The effect of $\mathcal{H}_{hf}$ on the electronic states is ignored and the nuclear operator is rewritten as:

$$\mathcal{H}_{hf} = -\vec{\mu}_n.\vec{H}_{hf}$$

$$= -g_n \beta_n \vec{H}_{hf}.\vec{I}. \tag{12}$$

Here $\vec{\mu}_n = g_n \beta_n \vec{I}$ is the magnetic moment operator of the nucleus (g_n is the g-factor for the nuclear state and β_n the nuclear magneton) and $\vec{H}_{hf} = \langle \vec{S}.[A] \rangle$ is known as the magnetic hyperfine field. This is the effective field approximation.

The three major contributions to the hyperfine field are (56)*:

---

*See Chapter 1 for more details.

TABLE 3

Crystal field calculation for axial sites

Symmetry	Site coord-ination	Ground orbital wave function [5,6]	Orbital splitting $\|\delta\|$ [1]	Spin Hamilto-nian	D value [2]	Magnetic anisotropy	V_{zz}(val.) at 0° K [3]	Typical compound
Tetragonal	Octa-hedral	$\|\pm1\rangle = d_{XZ}, d_{YZ} = Y_{D,E}$	$9B_2^0$	$(-\lambda L_Z S_Z)$		easy axis Z	$-V_0/2$	
		$(\|2\rangle - \|-2\rangle)/\sqrt{2} = d_{XY} = Y_C$		DS_Z^2	$-3\rho - 4(\lambda^2/\Delta) + (\lambda^2/\delta)$	easy plane XOY	$+V_0$	$Fe(py)_2Cl_2$(38)
	Tetra-hedral	$\|0\rangle = d_{3Z^2-R^2} = Y_A$	$12B_2^0$	DS_Z^2	$3(\lambda^2/\Delta) + 3\rho$	easy plane XOY	$-V_0$	$FeCr_2O_4$(40)
		$(\|2\rangle + \|-2\rangle)\|\sqrt{2} = d_{X^2-Y^2} = Y_B$		DS_Z^2	$-3(\lambda^2/\Delta) - 3\rho$	easy axis Z	$+V_0$	FeV_2O_4(40)
Trigonal	Octa-hedral	$\|0\rangle = d_{3Z^2-R^2} = Y_A$	$9B_2^0$	DS_Z^2	$\sim 3\rho + (\lambda^2/\delta) + 2(\lambda^2/\Delta)$	easy plane XOY	$-V_0$	$FeSiF_6 \cdot 6H_2O$(66)
		$-(\|\pm1\rangle \mp \sqrt{2}\|\mp2\rangle)/\sqrt{3}$		$(-\lambda L_Z S_Z)$		easy axis Z	$+V_0/2$	$FeCO_3$(7)
	Tetra-hedral	$(\|\pm2\rangle \mp \sqrt{2}\|\mp1\rangle)\|\sqrt{3}$	0				small [4]	ZnS(Würtzite) (39)
Purely axial (Δ=0)		$\|0\rangle = d_{3Z^2-R^2} = Y_A$	$3B_2^0 = \delta''$	DS_Z^2	$3\rho + 3(\lambda^2/\delta'')$	easy plane XOY	$-V_0$	
		$\|\pm1\rangle = d_{XZ}, d_{YZ} = Y_{D,E}$		$(+\lambda L_Z S_Z)$		easy axis Z	$-V_0/2$	
		$\|\pm2\rangle = d_{XY}, d_{X^2-Y^2} = Y_{B,C}$	$12B_2^0 = 4\delta''$	$(+\lambda L_Z S_Z)$		easy axis Z	$+V_0$	$Fe_2F_5 \cdot 2H_2O$(41)

Notes:

[1] δ is the excitation energy of the lowest excited orbital state. It is assumed to be >> λ, and is given in units of the crystal field parameter B_2^0 (see Section 3.1.1, Eqn. (25)). (Fourth order contributions to δ have been neglected).

[2] Δ is the ^5E_g - $^5T_{2g}$ splitting in the cubic crystal field.

[3] V_0, the unit of V_{ZZ}, is given by $V_0 = \frac{4}{7} e \langle r^{-3} \rangle$ as in Table 2.

[4] In this case $V_{ZZ}(val) \sim 6(B_2^0/\Delta)V_0$

[5] The quantization axis OZ is the highest symmetry axis at the Fe^{2+} site. For tetragonal symmetry the axes OZ,X,Y are the cubic 4-fold axes.

[6] Wavefunctions are written in terms of the L=2 basis states $|L_Z\rangle$.

(i) the orbital field, due to the orbital motion of the 3d electrons around the nucleus, which is proportional to $\langle\vec{L}\rangle$.

(ii) the dipolar-field, produced by the dipole moment of the d electron spin distribution of the Fe^{2+} ion. It is closely related to the (valence) electric field gradient which arises from the asymmetric *charge* distribution in the d shell.

(iii) the Fermi contact term, which is proportional to $\langle\vec{S}\rangle$.

The strength and symmetry of the crystal field strongly influences the hyperfine field, and the tensor [A] may be highly anisotropic because of the anisotropy of the orbital and dipolar contributions. In particular, the dipolar contribution to [A] is proportional to the EFG tensor [V] discussed earlier (see Section 3.1). It is, however, difficult to systematize the dependence of $\vec{H}_{hf}$ on the crystal field symmetry because the different contributions have different dependences and because their signs are often such that they partially cancel each other (e.g. in $FeCl_2$ (48, 57) the net hyperfine field at $4.2^{o}K$ is $\lesssim$ 3 kOe whereas the hyperfine field observed in $FeCO_3$, in which the Fe^{2+} ion has the same point symmetry, is + 180 kOe (58)). However, the following *rules of thumb* may be applied:

(i) the orbital term is always positive when $\vec{S}$ lies in an easy direction. It may be as large as +600 kOe and thus be greater in magnitude (and opposite in sign) than the contact term.

(ii) the dipolar term has the same sign (remembering the convention) as the EFG component in the $\vec{S}$ direction and thus depends strongly on the crystal field. It may be as large as $\pm$150 kOe. in some cases.

(iii) if [A] is approximately independent of applied magnetic field strength and temperature, then widely spaced orbital levels, with only weak admixture by the spin-orbit coupling, are indicated.

The anisotropy of the magnetic hyperfine tensor can be investigated by Mossbauer measurements either on paramagnetic single crystals in applied magnetic fields (the results of an extensive study of ferrous fluosilicate are summarized in Fig. 10) or on powders in magnetic fields (59-61) (see Fig. 11).

2.2.4 Magnetically Induced Effects on the EFG

An externally applied magnetic field, or an internal molecular field, may in some cases produce a significant distortion or perturbation of the orbital wavefunctions of an ion, and thus may affect

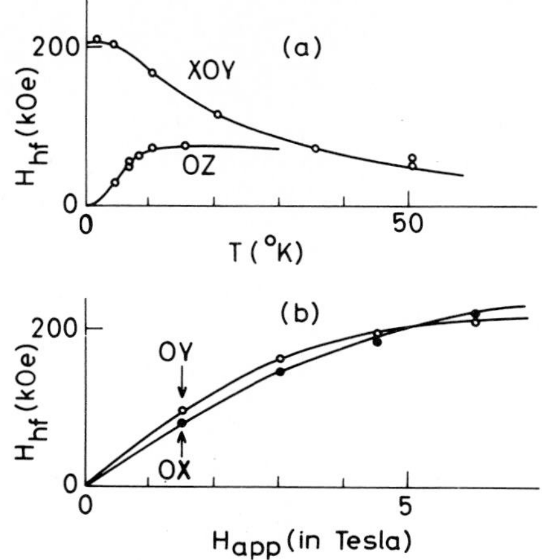

Fig. 10. Measurements of the anisotropy of the magnetic hyperfine interaction in single crystals of ferrous fluosilicate ($FeSiF_6,6H_2O$) [from (49)]. (a) as a function of temperature, with H_{app} = 5.5 T; the symbol <XOY> means that the hyperfine field in the X-Y plane has been measured as an average. (b) as a function of applied magnetic field, with T = 4.2°K. OZ is the hard magnetic axis (the trigonal of the Fe^{2+} crystal field). OY is slightly easier than OX.

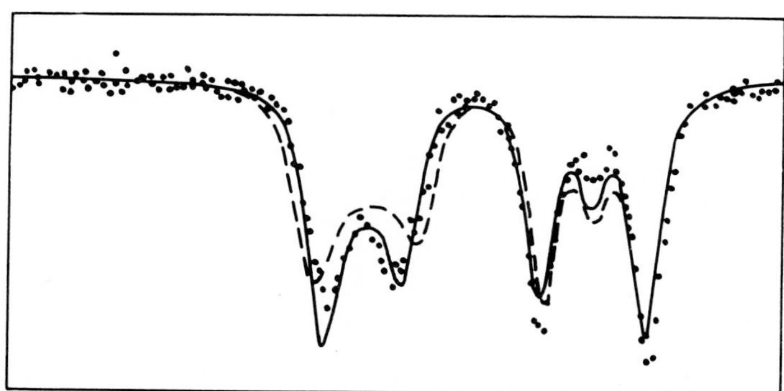

Fig. 11. Measurement of the magnetic hyperfine anisotropy from the Mossbauer spectrum of ferrous fluosilicate powder at room temperature in an applied magnetic field of 7.5 T parallel to the γ-ray beam (60). The two curves were calculated for an isotropic hyperfine tensor (dashed curve) and an anisotropic one (full curve).

the EFG. The main effect of such a field on the orbital state of the ion occurs as a result of the spin-orbit coupling - this is larger than the direct orbital Zeeman effect. Consequently, negligible effects are expected when orbital levels are widely spaced (compared with the spin-orbit coupling) and not significantly mixed. However, for cases where the crystal field splitting of orbital states is small, so that ΔE_Q is strongly reduced by spin-orbit admixture, large magnetic field effects on the EFG may be expected. Such an example (for $FeCl_2$ (48)) is shown in Fig. 12.

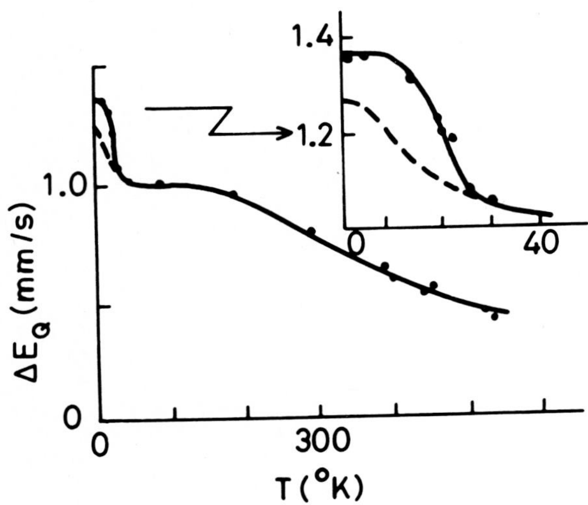

Fig. 12. The magnetic exchange (molecular field) contribution to the EFG in $FeCl_2$ (48). The solid curve is the result of a calculation that includes the molecular field contribution, which was neglected in the calculation (using the same crystal field) that yielded the dashed curve. The inset is an expansion of the low temperature (magnetically ordered) region.

Accurate measurements of magnetically induced effects resulting from the application of an external magnetic field would enable extra information on the crystal field and spin-orbit interaction strengths to be obtained, but up to date this technique has not been employed to our knowledge. It would require favourably small crystal field splittings and rather large magnetic fields.

In cubic materials, the onset of magnetic order often has associated with it the appearance of a sizeable magnetically-induced EFG, that is more or less proportional to the net magnetization (although the observation of this is often complicated by the effects of distortions from magneto-crystalline or Jahn-Teller origins) - see for instance the work on Fe^{2+}-doped in MnO, NiO (62) and $KFeF_3$ (63) for octahedral coordination and $CdCr_2S_4$ (64, 65) for tetrahedral coordination. Another feature typical of cubic, or almost cubic, sites is a decrease in the hyperfine field as the temperature is decreased at low temperatures. This decrease, examples of which can be seen in Fig. 13, occurs because the ground state is a spin singlet with a magnetic moment only by virtue of mixing by the molecular field, and the hyperfine field it produces is less than that associated with the first excited state (the magnetic properties of ground singlet systems have been described extensively - e.g. (66)).

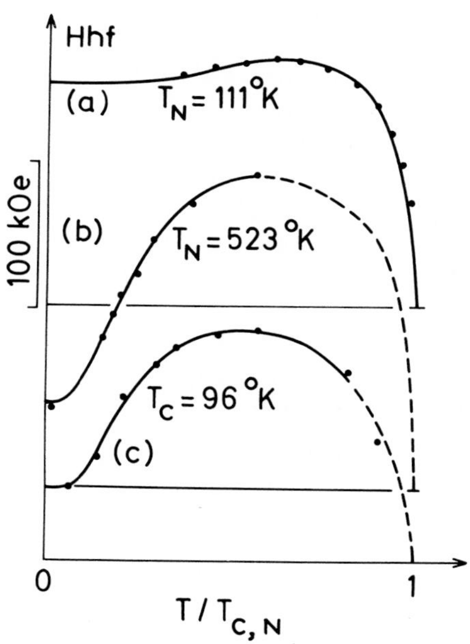

Fig. 13. Temperature dependence of the magnetic hyperfine field in cubic (or slightly distorted) Fe^{2+} sites in magnetically ordered materials. (a) $KFeF_3$ - octahedral coordination (63)
(b) NiO - octahedral coordination (62)
(c) $CdCr_2S_4$ - tetrahedral coordination (64).

Magnetically induced anisotropic EFG components have been observed and at least qualitatively accounted for in some cases (54, 62). For $FeCr_2O_4$ (40), third order perturbation calculations show that the induced value of η depends strongly on the orientation of the magnetic field in the basal plane, and that in general none of the EFG principal axes coincide with the direction of the applied field.

Table 4 contains the results of calculations of the sign of the magnetically induced contribution to the EFG for different crystal

TABLE 4.

Sign of the magnetically induced contributions to the EFG (V_{ZZ},η) for particular orientations of the magnetic field $\vec{H}_{app}$

Symmetry	Site coordination	Orbital ground state	V_{ZZ} (non-induced) [1]	$\vec{H}_{app}$ direction [2] [3]	Induced V_{ZZ} contribution	Induced $\eta = (V_{H_{app}} - V_{\perp})/V_{ZZ}$
Tetragonal	Octahedral	doublet singlet	- +	OZ,OX,OU	-, +, + +, -, -	0, +, ~0 0, -, ~0
	Tetrahedral	singlet (Y_A) singlet (Y_B)	- +	OZ,OX,OU	~0(+, -, -) ~0(+, -, -)	0, -, ~0 0, +, ~0
Trigonal	Octahedral	doublet singlet	+ -	OZ, XOY	+, - -, +	0, + 0, -
	Tetrahedral	doublet	{ - +	OZ, XOY	-, + -, +	0, - 0, +
Cubic	Octahedral	triplet	0	<100>,<111>	-, +	0, 0
	Tetrahedral	doublet	0	<100>,<111>	+, ~0 (-)	0, 0
Purely axial		singlet doublet $Y_{B,C}$	- +	OZ, XOY	-, + +, -	0, - 0, ~0 (+)

Notes: [1] See Tables 2,3
[2] OZ is the highest symmetry axis at the Fe^{2+} site
OX,OY are defined according to Fig. 15(a) (Section 3.1)
OU bisects OX,OY
XOY means any direction in the X,Y plane.
[3] For all these orientations, OZ and $\vec{H}_{app}$ are principal axes of the EFG, which would not normally be true for an arbitrary orientation of $\vec{H}_{app}$.

field symmetries. This sign depends on the orientation of the applied field $\vec{H}_{app}$. A rule of thumb may be deduced: the induced and inherent (non-induced) contributions to V_{zz} have the same sign for $\vec{H}_{app}$ parallel to Oz and opposite signs for $\vec{H}_{app}$ perpendicular to Oz. However this does not apply to states derived from the cubic E_g doublet (i.e. for tetrahedral coordination), for which Jahn-Teller effects are likely to occur and complicate the problem.

2.3 The Dynamic Crystal Field (Orbit-Lattice Interaction)

Up to this point of the discussion, the crystal field has been assumed to be equivalent to a static field acting on the paramagnetic ion, that is produced by the array of stationary charges in the surrounding lattice. But this is not the whole picture. Even at zero temperature, and more so as the temperature increases, all the ions in the crystal vibrate about their mean positions, thus varying the interatomic spacings in a time-dependent way and as a consequence modulating the crystal field. An instantaneous picture of the lattice would show distortions from the regular (time-averaged) lattice symmetry and would therefore imply an instantaneous crystal field that differs both in strength and symmetry from the average field. The rate at which this instantaneous situation changes depends on the vibration frequencies present, but as these are all less than or of the order of the Debye frequency ($\sim 10^{14}$ Hz), the electrons of any ion may be assumed to follow the displacements of the nucleus of that ion instantaneously, and the electrons of the paramagnetic ion may be expected to respond equally rapidly to the consequent changes in the crystal field, i.e. the situation is essentially adiabatic.

In order to give a qualitative idea of the effects of this time-varying field a linear triatomic molecule, such as that shown in Fig. 3, will be considered. The molecule is symmetric, with two identical ligand ions each with charge $q(<0)$ at a distance a from the central paramagnetic ion, which has net charge $-2q$ to maintain electrical neutrality. This ion has a single d electron outside the spherically symmetric core, so it could be high spin Fe^{2+}, and the crystal field that acts on it is calculated assuming the ligands may be treated as point charges. This is as shown in Fig. 14(a).

The d electron moves in a crystal field potential given by:

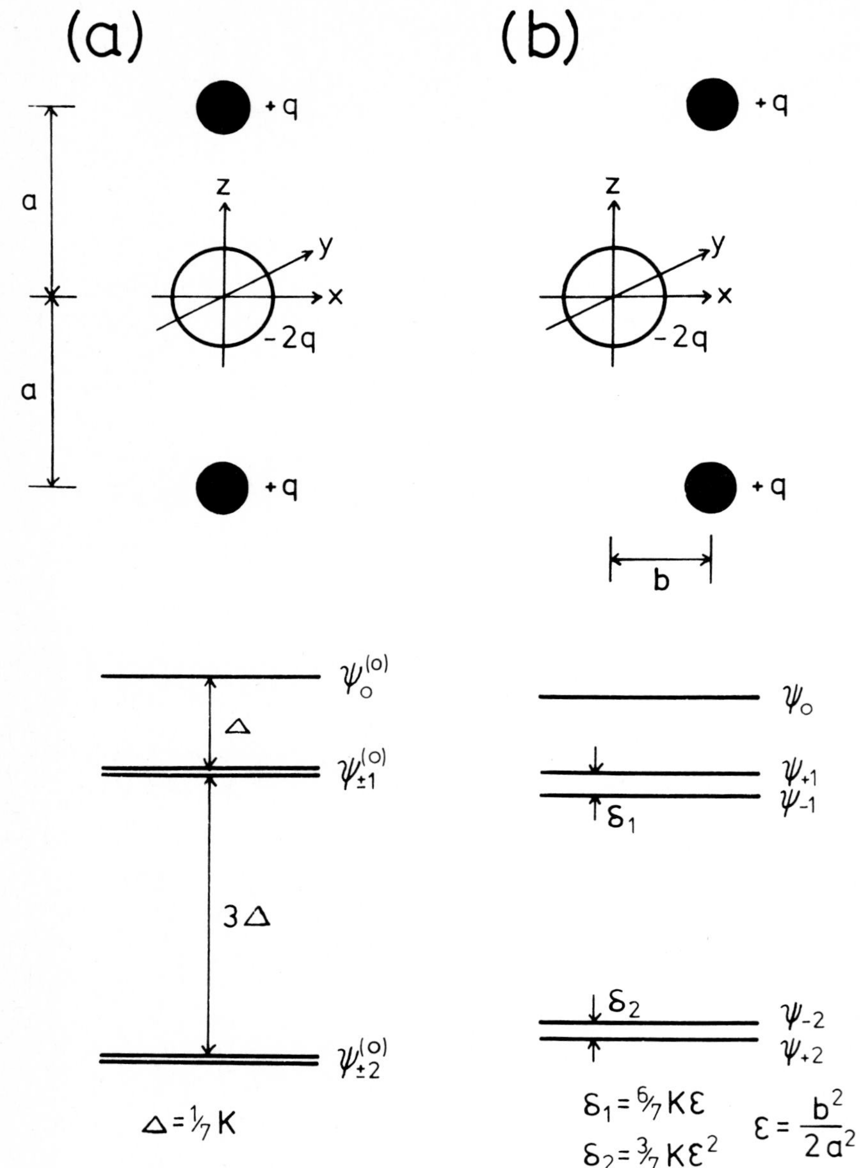

Fig. 14. Geometry (upper) and resulting energy level structure (lower) of the paramagnetic ion in a linear triatomic molecule described in the text (see also Fig. 3).
(a) the linear molecule and the wavefunctions and energy levels for a single d electron on the paramagnetic ion.
(b) the distorted molecule and the splittings produced in the energy levels by the distortion (the splittings are not drawn to scale).
All symbols are defined in the text (Section 2.3).

$$V^{o}(\vec{r}) = \frac{q}{a}[2 + \frac{1}{a^2}(3z^2 - r^2) + \frac{1}{4a^4}(35z^4 - 30r^2z^2+3r^4) + \text{higher order terms}] \quad (13)$$

$$= \frac{2q}{a}[1 + \sqrt{\frac{4\pi}{5}}\,(\frac{r^2}{a^2})Y_2^0(\theta,\phi) + \sqrt{\frac{4\pi}{9}}\,(\frac{r^4}{a^4})Y_4^0(\theta,\phi) + O(\frac{r^6}{a^6})],$$

where the series is truncated since $\vec{r} = (x,y,z) = (r,\theta,\phi)$ is assumed to be close to the nucleus of the paramagnetic ion, which is taken as the origin (i.e. $r \ll a$). The constant term represents the average ($r=0$) Coulomb interaction of the electron with the ligands. It will be dropped now and considered further later. The crystal field Hamiltonian is thus:

$$\mathcal{H}^0 = \frac{-2eq}{a}\,[\frac{r^2}{a^2}\sqrt{\frac{4\pi}{5}}\,Y_2^0 + \frac{r^4}{a^4}\sqrt{\frac{4\pi}{9}}\,Y_4^0 + \ldots\ldots]. \quad (14)$$

The d orbital eigenfunctions of this (axially symmetric) hamiltonian may be the $\ell=2$ spherical harmonics Y_2^m ($m=-2, \ldots\ldots+2$) or the tesseral harmonics, which are the real combinations defined in Equations (1). The tesseral harmonics will be used here, so the eigenfunctions are written $\psi_m^{(0)} = R(r)Z_2^m(\theta,\phi)$ where $R(r)$ is the radial wavefunction. The energies of these states are $E_m^{(0)}$, and the energy level diagram is shown Fig. 14(a). The splittings are Δ and 3Δ, where $\Delta=\frac{1}{7}(\frac{-2eq}{a})\frac{\langle r^2\rangle}{a^2}$ and $\langle r^n\rangle = \int_0^\infty r^n|R(r)|^2 r^2 dr$, if the Y_4^0 term in $\mathcal{H}^0$ is neglected.

The molecule is now distorted in such a way that the ligands are fixed and the paramagnetic ion is moved a distance b ($\ll a$) in the x direction. This geometry is shown in Fig. 14(b) and, again assuming that the ligands can be treated as point charges, the crystal field Hamiltonian becomes (including terms up to $\frac{1}{a^5}$ only):

$$\mathcal{H} = \mathcal{H}^0 + \mathcal{H}', \quad (15)$$

where
$$\mathcal{H}' = \frac{-2eq}{a}(\frac{b^2}{2a^2})\,[(-1 + \frac{3b^2}{4a^2}) + (1-\frac{3b^2}{2a^2})(\frac{r}{b})\sqrt{\frac{8\pi}{3}}\,(Y_1^{-1} - Y_1^1)$$
$$- \frac{6r^2}{a^2}\sqrt{\frac{4\pi}{5}}\,Y_2^0 + \sqrt{\frac{3}{2}}(\frac{r^2}{a^2})\sqrt{\frac{4\pi}{5}}\,(Y_2^2+Y_2^{-2}) + 2\sqrt{3}(\frac{r^3}{a^2b})\sqrt{\frac{4\pi}{7}}\,(Y_3^{-1}-Y_3^1)+\ldots].$$

The angle-independent terms are again dropped for consideration later, and the odd-order spherical harmonics (Y_1^m and Y_3^m) are ignored since they have zero matrix elements for states within a d configuration. Thus, the effective perturbation of the crystal field for the d electron may be expressed as:

$$H' = \left(\frac{-2eq}{a}\right)\left(\frac{b^2}{2a^2}\right)\left(\frac{r^2}{a^2}\right)\sqrt{\frac{4\pi}{5}}\,[-6Y_2^0 + \sqrt{\tfrac{3}{2}}\,(Y_2^2 + Y_2^{-2})]. \tag{16}$$

The wavefunctions and energy levels may now be calculated using perturbation theory. The wavefunctions, calculated to first order in the (small) expansion parameter $\varepsilon = b^2/(2a^2)$, are:

$$\psi_0^{(1)} = N^{-\frac{1}{2}}[\psi_0^{(0)} - \tfrac{\sqrt{3}}{2}\varepsilon\psi_2^{(0)}] = N^{-\frac{1}{2}}R(r)[Z_2^0 - \tfrac{\sqrt{3}}{2}\varepsilon Z_2^2],$$

$$\psi_1^{(1)} = \psi_1^{(0)} = R(r)Z_2^1,$$

$$\psi_{-1}^{(1)} = \psi_{-1}^{(0)} = (r)Z_2^{-1}, \tag{17}$$

$$\psi_{-2}^{(1)} = \psi_{-2}^{(0)} = R(r)Z_2^{-2},$$

$$\psi_2^{(1)} = N^{-\frac{1}{2}}[\psi_2^{(0)} + \tfrac{\sqrt{3}}{2}\varepsilon\psi_0^{(0)}] = N^{-\frac{1}{2}}R(r)[Z_2^2 + \tfrac{\sqrt{3}}{2}\varepsilon Z_2^0],$$

where the normalization constant is $N = 1 + \frac{3}{4}\varepsilon^2$. Thus, the perturbation induces an admixture (of order ε) of Z_2^0 and Z_2^2. To second order in the perturbation, the energies of the levels are:

$$E_0^{(2)} = K[\tfrac{2}{7} - \tfrac{12}{7}\varepsilon + \tfrac{3}{7}\varepsilon^2],$$

$$E_1^{(2)} = K[\tfrac{1}{7} - \tfrac{3}{7}\varepsilon + 0],$$

$$E_{-1}^{(2)} = K[\tfrac{1}{7} - \tfrac{9}{7}\varepsilon + 0], \tag{17b}$$

$$E_{-2}^{(2)} = K[-\tfrac{2}{7} + \tfrac{12}{7}\varepsilon + 0],$$

$$E_2^{(2)} = K[-\tfrac{2}{7} + \tfrac{12}{7}\varepsilon - \tfrac{3}{7}\varepsilon^2],$$

where $K = \left(\frac{-2eq}{a}\right)\left(\frac{\langle r^2\rangle}{a^2}\right)$. This energy level scheme is depicted in Fig. 14(b). It may be noted in passing that there is no Jahn-Teller instability of this molecule because the degenerate ground doublet is split only by a term of order $\varepsilon^2 \sim b^4$, and the elastic energy will contain a term in b^2. This can be shown if the average Coulomb energy of the molecule is considered.

When the molecule is distorted, the purely electrostatic interactions produce a force:

$$|\vec{F}| = \frac{4q^2b}{a^3}\left(1 + \frac{b^2}{a^2}\right)^{-\frac{3}{2}}$$

$$= \frac{4q^2b}{a^3}\left(1 - \frac{3b^2}{2a^2} + \ldots\ldots\right), \tag{18}$$

that acts on the central ion along the x direction in such a sense as to restore the linearity of the molecule. If only the leading term is considered, the paramagnetic ion will oscillate harmonically with total energy,

$$E_c = (n + \frac{1}{2})\hbar\omega_c ,$$

where $\omega_c = 2q(ma^3)^{-\frac{1}{2}}$, m being the mass of the ion. The potential energy for displacement b is $\int_0^b F(b')db' = 2q^2b^2/a^3$ which thus increases faster with b than the crystal field energy decreases and removes the likelihood of a static distortion.

The molecule is now considered to be oscillating in this manner at a frequency $\omega_c \sim 10^{14}$ Hz which is comparable with vibrational frequencies in solids. The crystal field Hamiltonian becomes:

$$\mathcal{H} = \mathcal{H}^0 + \mathcal{H}'(t) , \tag{19a}$$

where the time dependence of the perturbation is given by:

$$\mathcal{H}'(t) = \mathcal{H}'\sin^2\omega_c t, \tag{19b}$$

since the distortion b becomes $b_0\sin\omega_c t$. $\mathcal{H}'$ is independent of time. Because of this $\sin^2$ dependence of $\mathcal{H}'(t)$ on time, which is unusual, the numerical details of this example will not be pushed much further. In the general case, the lowest order time dependence is $\sin\omega_c t$ (and the system is susceptible to Jahn-Teller instability). However, in general terms, time dependent perturbation theory shows that two effects can result (67). Firstly the perturbation can produce a time-dependent admixture of the *stationary* states $\psi_m^{(0)}$, and secondly it can induce transitions between them. These two effects will now be briefly considered.

Time-dependent admixture of states

Because the oscillation frequency ω_c is slow compared with the rates at which electrons respond to the motion of the nucleus of the ion (the electronic relaxation rates), the wavefunctions and energy levels of the ion at any instant can be determined from the time-independent Schrodinger equation (i.e. the adiabatic approximation is assumed). Thus, the static solutions (Eqn. 7) show that the wavefunctions ψ_0 and ψ_2 will be time-dependent because the expansion parameter $\varepsilon(t)$ is now an oscillating function of time. Therefore, any static (compared with ω_c) observable calculated for the ground state ψ_2, such as the EFG, the magnetic moment or the charge density, will contain an admixture of that quantity for the state $\psi_0^{(0)}$;

i.e.

$$\langle O\rangle_2 = \langle\psi_2^{(1)}|O|\psi_2^{(1)}\rangle$$

$$= \frac{1}{N}\{\langle\psi_2^{(0)}|O|\psi_2^{(0)}\rangle + \tfrac{3}{4}\overline{\varepsilon^2(t)}\langle\psi_0^{(0)}|O|\psi_0^{(0)}\rangle\}\,, \qquad (20)$$

where N is the normalization constant and $\overline{\varepsilon^2(t)} = \frac{1}{T}\int_0^T \varepsilon^2(t)dt$ is the time average, over a long time T compared with $\frac{2\pi}{\omega_c}$, of $\varepsilon^2(t)$ and it is non-zero, even if $\varepsilon \sim b$ as it is in a more general case. This admixture is known as vibronic admixture and it can have important effects on the observables, particularly the EFG, that are measured in Fe^{2+} Mossbauer spectra. These effects will be described further later on. An important property of this admixture is to be temperature-dependent. If the temperature of the oscillator increases, as may happen if it is coupled to a thermal reservoir such as a crystal lattice, then its oscillation amplitude will increase and so therefore will the admixture coefficient ε^2.

It will also be noted that the energies of the states $E_m^{(2)}$ will also be shifted by this time-dependent perturbation - in this particular example the time average $\overline{\varepsilon(t)}$ will be non-zero as well as $\overline{\varepsilon^2(t)}$ but this will not be true in general. These energy shifts are observed as shifts in the zero-phonon optical transitions of paramagnetic ions in crystals, and have been recognized for some time (68).

Transitions between states

The simple description given above of the adiabatic behaviour of the system breaks down, however, when the energy difference between two stationary states (i.e. $E_i - E_f$) is approximately equal to a vibrational quantum $\hbar\omega_c$. In this case the admixture coefficient increases steadily with time (67) and a transition between the states ψ_i and ψ_f is induced. The transition probability per unit time is given by:

$$W \propto \rho(f)\,|\langle\psi_f|H'|\psi_i\rangle|^2 \,, \qquad (21)$$

where $\rho(f)$ is the density of final states ψ . For the general case where $H'(t) = H'\sin\omega_c t$ the constant of proportionality in this expression is $2\pi/h$ and it is known as Fermi's golden rule. In this case, the density of final states is:

$$\rho(f) = \delta(E_i^T - E_f^T),$$

where the energies E^T are total (electronic + vibrational) energies of the system.

Therefore, in the oscillating linear molecule example, the time-dependent crystal field will induce transitions between the states ψ_0 and ψ_2 as long as $\hbar\omega_c \approx E_0 - E_2$. This exchange of energy between the vibrational and electronic coordinates of the ion is known as spin-lattice relaxation. It has been studied for many years and is the subject of an immense literature. An introduction to the subject and reprints of many important papers dealing with it are contained in the book by Manenkov and Orbach (69).

In the remainder of this section two effects of the dynamic crystal field, or orbit-lattice interaction, will be discussed in rather more general terms, with particular reference to the Fe^{2+} ion, and examples of their observation in $^{57}Fe^{2+}$ Mossbauer spectroscopy given.

2.3.1 The " Quasi-molecular Cluster" Model of the Orbit-Lattice Interaction

While it was quite easy to calculate the low-order terms of the orbit-lattice interaction for the triatomic molecule considered above, it becomes rapidly more difficult as the number of ligands around the paramagnetic ion increases. Thus, for an ion in a solid it is usual to adopt an approach that utilizes the point group symmetry of the ion to reduce the number of unknowns in the dynamic interaction, just as is done for the rigid lattice in the crystal field theory. The prototype of this approach is the *quasi-molecular* model first employed by Van Vleck (70), in which it is assumed that only the nearest neighbour ligands contribute to the crystal potential acting on the paramagnetic ion. All of the vibrations of the resulting cluster may then be expressed in terms of its normal modes of vibration, which are orthogonal and which are related to the point symmetry of the cluster. This model has been discussed in detail elsewhere* in this book, may be found in the original papers of Van Vleck (70) and in the review articles of Sturge (71), Orbach and Stapleton (72) and Shrivastava (73).

Briefly, within the framework of this model, the crystalline potential acting on the open-shell electrons of the paramagnetic ion may be expanded in powders of the normal coordinates Q_k of the cluster:

$$\mathcal{H} = \mathcal{H}^0 + \sum_k (\partial \mathcal{H}^0/\partial Q_k)Q_k + \frac{1}{2}\sum_{k,\ell} (\partial^2 \mathcal{H}^0/\partial Q_k\, \partial Q_\ell)Q_k Q_\ell + \dots \quad , \tag{22}$$

where $\mathcal{H}^0$ is the static crystal field potential. The remainder of $\mathcal{H}$, which contains terms that are 1st, 2nd .... etc. order in the Q_k is the dynamic part and is referred to as the orbit-lattice interaction. This is related to the dynamic properties of the crystal by expanding the normal modes of the cluster in terms of the phonons of the crystal, and the main reason for the use of normal modes is that advantage may be taken of their orthogonality to reduce the number of terms when summations over the phonon states are made. This will be dealt with more fully in Section 3.

This model, or extensions of it involving larger clusters, has been almost universally employed in studies of vibronic effects (and in particular of the Jahn-Teller effect) and of spin-lattice relaxation. Recently, however, Stevens and coworkers (74) have developed a full lattice model of the orbit-lattice interaction which employs transformation methods to remove the linear coupling (the equivalent of the term linear in Q_k in the cluster model). While this model will undoubtedly be developed further and applied more widely in the future, the discussion in this paper will not go beyond what can be obtained from the conceptually simpler cluster model.

---

*See chapter on Spin Lattice Relaxation

2.3.2 Vibronic Admixture Effects

The classic example of vibronic admixture, and the one that has been most widely studied (see Ref. 71, 75 and references contained therein), is the Jahn-Teller effect. This describes the instability to either a static or dynamic distortion of a paramagnetic ion in a degenerate electronic orbital state due to the orbit-lattice interaction. This instability arises because non-Kramers' degeneracy can (except in linear molecules) always be lifted in first order by a distortion of the cluster and because the elastic, or quasi-elastic, energy of the cluster is a minimum for the undistorted configuration of the cluster. Thus, the electronic energy of the paramagnetic ion will decrease approximately linearly with the distortion, while the *elastic* energy will increase only quadratically, leading to a minimum energy configuration of the cluster with lower point symmetry than that of the crystal. If the vibronic interaction energy, or *Jahn-Teller energy*, is large enough then there is a static distortion of the complex. However, the overall symmetry of the crystal must be maintained, so there will, in general, be more than one equivalent distortion, and all of them produce distorted configurations of equal energy (for example, the three equivalent tetragonal distortions, along the three Cartesian axes of an octahedron). Therefore, the original purely electronic degeneracy is replaced by a more complicated, vibronic degeneracy. When the interaction energy is weaker, or the temperature higher, a cluster may either tunnel or be thermally excited from one equivalent distortion to another. The time-averaged point symmetry of each paramagnetic ion will then be that of the undistorted cluster: this is known as the dynamic Jahn-Teller effect, and it causes a reduction in the matrix elements of some orbital operators that is known as the Ham effect (75, 76).

However, only relatively few cases of significant Jahn-Teller effects involving Fe^{2+} ions have been positively identified, because the Jahn-Teller energy of these ions is generally fairly small, and because the Fe^{2+} ion occurs with an orbitally degenerate ground state fairly infrequently. Ham and coworkers (20, 77), after a detailed examination both of the theory of the dynamic Jahn-Teller effect and of the available experimental data (including the Mossbauer data of Leider and Pipkorn (21)), concluded that there was good evidence for a weak dynamic effect in the octahedral system Fe^{2+}:MgO. Following this work, Regnard, Chappert and coworkers (22, 78) subsequently identified similar but stronger effects for

Fe^{2+} in $KMgF_3$ and CaO both of which are also octahedral. Abou-Ghantous et al (74c) have associated the drastic reduction of the trigonal crystal field splitting of the $^5T_{2g}$ state of Fe^{2+} in Al_2O_3 with a second-order effect of the (dynamic) Jahn-Teller interaction, and this has been further substantiated by Bates et al (79) and Ganapol'skii (80).

There is perhaps more spectacular evidence for Jahn-Teller effects involving Fe^{2+} ions in tetrahedral sites in sulphide spinels. Spender and Morrish (81) and Van Diepen and van Stapele (82) observed a static distortion of the Fe^{2+} sites in the Mossbauer spectra of $FeCr_2S_4$ at low temperatures ($\sim 10^{o}$ K) which was associated with a cooperative Jahn-Teller transition involving pairs of Fe^{2+} ions in neighbouring A sites, while for Fe^{2+} substituted onto A sites in $CdCr_2S_4$ (64, 82a, 83), $CoCr_2S_4$ (82a) and $CdIn_2S_4$ (84), dynamic Jahn-Teller were identified. Dynamic effects have also been observed for Fe^{2+} in tetrahedral sites in compounds such as ZnS, CdTe, $MgAl_2O_4$ and others (16, 85). In particular, the very nice experiments of Garcin, Imbert et al (16) demonstrated the existence of a dynamic Jahn-Teller effect for Fe^{2+} in ZnS to verify a prediction made earlier by Ham (86), and also showed effects of slow electronic relaxation from excited states of Fe^{2+} populated out of thermal equilibrium by the radioactive decay of $^{57}Co^{2+}$.

The orbit-lattice interaction can, however, have other effects on the electronic properties of an ion than those that are generally classified as Jahn-Teller effects, although this is of course mainly a question of semantics. Even when the Jahn-Teller energy is very weak and the paramagnetic ion is in a site of sufficiently low symmetry that it has no orbital degeneracy, this interaction will mix vibronic states of the ion-lattice system in such a way as to introduce temperature-dependent reduction, analogous to the Ham effect, of electronic observables that depend on the expectation values of orbital operators. This was illustrated above for the simple case of the linear triatomic molecule. Two rather different models (17, 18) that provide a qualitative picture of the effects of vibronic coupling between the energy levels of a paramagnetic ion when the vibronic interaction is weak will be introduced in Section 3.2. The approach to this sort of calculation will be seen to differ slightly from that usually adopted in Jahn-Teller studies since it is directed primarily at determination of the electronic components of the vibronic wavefunctions rather than the electronic energy levels, and is equally applicable to orbital singlets as to degenerate states.

While a considerable amount of work has been done on the effects of vibronic coupling on EPR spectra (see, for example, the recent review of Shrivastava (73)), little consideration has been given to it in the interpretation of Mossbauer spectra. Lang (87) and subsequently Cianchi et al (88) attributed the strong temperature dependence of the ^{57}Fe quadrupole splitting in oxyhaemoglobin to the admixture induced by the rotation of a molecular ligand, and Gibb et al (89) suggested that vibronic admixture within the cubic $^{5}E_{g}$ ground states of the tetrahedrally coordinated $^{57}Fe^{2+}$ ions in tetragonally distorted sites in $(NMe_4)_2FeCl_4$ is a contributing factor to the almost linear dependence of the quadrupole splitting on temperature. The recent calculations of Bacci (17) and of Price (18), on which the later discussion in this paper will be based, indicate that vibronic effects may often make significant contributions to both the absolute value and the temperature dependence of the $^{57}Fe^{2+}$ quadrupole splitting.

2.3.3 Relaxation Effects and the Crystal Field

The term *relaxation* is used to describe a wide variety of phenomena in solid state physics in which a system makes a transition from one state to another. This discussion will be restricted to consideration of the relaxation of paramagnetic ions, and in particular high-spin Fe^{2+}, between states derived from the lowest spectroscopic term (^{5}D for Fe^{2+}) and arising from either spin-lattice (4, 69, 72, 90-93) or spin-spin processes (4, 55, 90)*.

Spin-lattice relaxation occurs as a result of interaction of the ion with the time-dependent crystal field (the orbit-lattice interaction) as was described above for the linear triatomic molecule. Spin-spin relaxation, on the other hand, is induced by magnetic interactions between ions and is not, at least in first order, due to the dynamic crystal field, but it will be discussed in this section because it is a dynamic effect. In principle, of course, the dynamic part of the crystal field may influence spin-spin relaxation both through vibronic admixture of states and through higher order relaxation processes, but, at least to our knowledge, no clear examples of this influence have yet been identified.

---

*See Chapter 1 for further details.

It will be clear that both spin-lattice and spin-spin transition rates will depend strongly in general on the strength and symmetry of the static crystal field and on any external perturbations, such as an applied magnetic field, as well as on the dynamic interactions that produce them. This is because the transition probabilities depend on the structure of the wavefunctions of the ion and on their energy spacings and occupation probabilities.

Spin-Lattice Relaxation

It was shown in Section 2.3.1 that the dynamic part of the crystal field can be expressed as a power series in the normal modes of a quasi-molecular cluster. When the ligands are vibrating, these normal coordinates become fluctuating strains of appropriate symmetry. The commonly-observed spin-lattice relaxation processes (viz. direct, Raman, second-order Raman and Orbach (4)) occur when the lowest order terms of this series induce transitions as first- and second-order perturbations.

Effects of spin-lattice relaxation have only rarely been observed in the Mossbauer spectra of Fe^{2+} ions. There are a variety of reasons for this. Firstly, as will be clear from the earlier discussion of this section, low-lying electronic states will all produce a very similar quadrupole splitting, so observation of relaxation effects in most cases can only be made if time-dependence of the magnetic hyperfine structure can be detected. Secondly, as is well known (10, 26, 55) only relaxation rates in a relatively narrow *window* can be determined from a Mossbauer spectrum. The required conditions are that the relaxation rate $\frac{1}{\tau}$ must not be much greater or much smaller than *both* the nuclear Larmor precession rate (in the effective hyperfine field approximation) and the inverse nuclear lifetime of the excited nuclear state involved in the Mossbauer transition ($\sim 10^7$ sec^{-1} for ^{57}Fe). Thirdly, in the great majority of cases, Fe^{2+} ions occur either in cubic sites in which, from symmetry considerations, both dipole and quadrupole hyperfine interactions must be identically zero, or in sites with a singlet ground state, in which the magnetic hyperfine interaction is quenched. In cases such as these, excited states may produce magnetic hyperfine structure, but in general the relaxation rates are too fast to be observable at the temperatures required to populate them sufficiently.

Nevertheless, spin-lattice relaxation effects have been observed in some instances in Fe^{2+} Mossbauer spectra, particularly in cases where the ground state is a doublet, or a pseudo-doublet. For example, when the Fe^{2+} ion is in an octahedral site with *a trigonal elongation* then the ground state is an orbital doublet, and effects of relaxation between the ground states have been observed in a number of compounds with this cation site symmetry, e.g. $FeCO_3$ (58), $Fe^{2+}:ZnCO_3$ (23), $Fe^{2+}:MgCO_3$ (94), $Fe(PyNO)_6(ClO_4)_2$ (47, 95), $Fe^{2+}:Zn(PyNO)_6(ClO_4)_2$ (47b). Price et al (23) examined in some detail the temperature dependence of the relaxation between the states of the ground doublet of $Fe^{2+}:ZnCO_3$ and verified that, in the range 10° to ~55° K, the dominant relaxation mechanism was a second-order (indirect) spin-lattice process, and because several relaxation *channels* were operative it seemed likely that both Orbach and Raman processes were important. More recently, Reiff et al (24) have observed slow electronic relaxation between the degenerate $m_s = \pm 2$ ground spin states of Fe(II) in *a distorted tetrahedral* complex. While the relaxation was temperature dependent, reaching the fast limit at ~20° K, the dominant relaxation process has not yet been identified.

Another situation in which Fe^{2+} relaxation has been observed in Mossbauer spectra, although only in the presence of an applied magnetic field, is when the ground state is a "pseudo-doublet", i.e. a magnetic doublet that has been mixed and split by a small crystal field distortion. In this situation the applied field induces a non-zero magnetic hyperfine interaction and this allows the relaxation to be observed. Zimmermann et al observed relaxation in $[FeL_4](ClO_4)_2$, where L = 1,8-naphthyridine, (96) and in $Fe(papt)_2C_6H_6$ and similar compounds (97), among the spin states of the ground orbital singlet and identified it as arising from an Orbach process. Price and Srivastava (94) also saw such effects in $Fe^{2+}:CaCO_3$ and $Fe^{2+}:CdCO_3$ although here the ground state is an orbital doublet split slightly by random strains in the crystals.

Spin-spin Relaxation

In considering spin-spin relaxation of Fe^{2+} ions it must first be appreciated that it is only important, and its effects are only observed, when spin-lattice relaxation is very slow, and as has been pointed out this is a fairly uncommon occurrence. Nevertheless, there are cases in which it is thought that spin-spin relaxation has been observed in Fe^{2+} Mossbauer spectra so it deserves some mention here.

The major interactions between two ionic magnetic moments are of exchange and dipolar origin, and explicit expressions for the interaction Hamiltonian H_{ss} have been given by many authors (4, 55, 90) (this is different from the intra-ionic spin-spin interaction mentioned earlier in Section 2.2). Relaxation may proceed in a similar manner to spin-lattice relaxation, with the analogue of the phonon bath being the ensemble of magnetic moments on the surrounding ions. Indirect spin-spin relaxation may also occur in an analogous manner to indirect spin-lattice processes, and may be important if the direct transitions are forbidden, and it is also possible that combined spin-spin and spin-lattice indirect processes may occur, but neither type of process has yet been identified, at least for Fe^{2+} ions.

Spin-spin relaxation depends strongly on the concentration of the paramagnetic ions and may be identified either through this or its temperature dependence, the latter arising from the temperature-dependent population factors. To our knowledge, all of the cases of spin-spin relaxation observed in Fe^{2+} Mossbauer spectra have been identified through the temperature-dependent line broadening it produces. Such was the case for Fe^{2+} in gillespite ($BaFeSi_4O_{10}$) (35) and in beryl (98). However, caution should be exercised in making this identification as similar line broadening may be produced by distortions or inhomogeneities in the sample. It now appears (99) that such effects might be an explanation for line broadening observed in the spectra of $FeSiF_6.6H_2O$ which had earlier been attributed to spin-spin relaxation (100). More definite assignments should rest on either concentration or applied magnetic field dependences of the relaxation rate or on direct observation that broadening is due to time-dependent effects. This may be possible with the selective excitation double Mossbauer (SEDM) technique (101).

Spin-spin relaxation was observed and identified by its concentration and magnetic field dependence for Fe^{2+} in $CaCO_3$ and $CdCO_3$ by Price (unpublished results and Ref. 94). The magnetic field dependence of the relaxation arises in an analogous way to the dependence on the crystal field, i.e. through its dependence on the electronic wavefunctions and level splittings.

3. QUANTITATIVE ASPECTS OF THE CRYSTAL FIELD AND THE CALCULATION OF $^{57}Fe^{2+}$ MOSSBAUER SPECTRA.

In this section the interactions discussed in Section 2 will be re-examined with a view to evaluate their effects numerically. The aim is not to provide a detailed formulation of the calculation of a $^{57}Fe^{2+}$ Mossbauer spectrum for any given situation, since this is neither necessary, nor is a practical proposition, but rather to give the reader an indication of the way to proceed with a calculation and of the computational tools that may be brought to bear on a particular problem. The approach to a calculation in any particular case will depend very much on the information (and its reliability) that is already available as well as that which can be gleaned from inspection of the Mossbauer results, e.g. the expected point symmetry of the Fe^{2+} ion site, values of parameters that are known from previous work, values of parameters that may be reasonably estimated either from previous work or from inspection of the Mossbauer results and, consequently, which interactions may be treated by perturbation theory, whether relaxation effects are expected and/or recognized to be present, etc.

Firstly, the calculation of a ^{57}Fe Mossbauer spectrum for a Fe^{2+} ion in a static crystal field will be discussed, although briefly because much of this material is well known. Emphasis will be placed on features of more particular concern to Mossbauer spectroscopists, such as the calculation of the hyperfine interactions. Secondly, the dynamic crystal field will be considered and methods currently available for calculation of both vibronic and dynamic effects described. Finally, a brief description of the superposition model of the crystal field (SPM) will be given, mainly from the point of view of its application to the estimation of static and dynamic crystal field parameters.

3.1 $^{57}Fe^{2+}$ Mossbauer Spectra for a Static Crystal Field

3.1.1 The Crystal Field Expansion

The most general form of the effective electric field at a point due to a surrounding lattice of ions is very complicated, but it can usually be readily simplified because:

(i) the symmetry of the crystal field at any point in the lattice depends only upon that particular point symmetry and not on the nature or strength of the interactions that it describes;

(ii) for the determination of the energies and states of open-shell electrons on an ion, interest is confined to the effect of the crystal field on those electrons, i.e. to determination of matrix elements $\langle\psi_i|\mathcal{H}_{CF}|\psi_j\rangle$ where $\psi_i(\vec{r}),\psi(\vec{r})$ are states of the open-shell electrons and $\mathcal{H}_{CF}(\vec{r})$ is the Hamiltonian describing the interaction.

While these two points may seem self-evident, they have far-reaching consequences. The first ensures that only those algebraic terms with particular symmetries can appear in the expression for the crystal field, and that the form of these can be determined purely by group theoretical arguments (102) while the second severely limits the number of terms, allowed by symmetry, that need to be considered.

A convenient general expression for the crystalline field potential is written in terms of a set of real parameters A_n^m as:

$$\mathcal{H}_{CF} = \sum_{n,m} A_n^m Z_n^m (\theta,\phi), \qquad (23)$$

where $Z_n^m(\theta,\phi)$ are the (real) tesseral harmonic functions defined by Eqn. (1) in Section 2.1. (It should be noted that there is considerable variation of notation in the literature, and consequently notation used should always be closely defined). If ψ_i,ψ_j are states belonging to the same n configuration, then $\langle\psi_i|\mathcal{H}_{CF}|\psi_j\rangle$ are only non-zero for n even and $n \leq 2\ell$, i.e. if both are 3d states then only terms in $\mathcal{H}_{CF}$ with n=2,4 need be considered (the n = 0 term produces only a constant shift of all levels).

At this stage the parameters A_n^m are regarded as empirical quantities (see Section 1), although the requirements of hermiticity and time-reversal invariance of $\mathcal{H}_{CF}$ ensure that they are real. The general expression (Eqn. 23) for the case of 3d ions thus contains 14 parameters (ignoring A_0^0) but this number can usually be greatly reduced by symmetry considerations if the quantization axes are chosen carefully. Some general rules relevant in this context are:

(a) A_n^{pm}, A_n^{-pm} (where p is an integer > 0) may be non-zero only if the point symmetry has an m-fold rotation axis. Therefore, if there is a 2-fold rotation axis parallel to the axis of quantization, Oz, or a mirror plane perpendicular to it, then the only non-zero A_n^m have m even. If there is a 3-fold axis, then m must be a multiple of 3 for $A_n^m \neq 0$.

(b) If there is a mirror plane that contains the axis of quantization, Oz, then if it is taken as the plane xOz all A_n^m terms with m<0 are zero.

The matrix elements of $\mathcal{H}_{CF}$ (Eqn. 23) may be evaluated most readily using the "operator equivalent" method based on the Wigner-Eckart theorem (102) and introduced by Stevens (28) (see also Ref. 4, 27). Using this method, matrix elements of $\mathcal{H}_{CF}$ can be written:

$$<\psi_i|\mathcal{H}_{CF}|\psi_j> = \sum_{n,m} A_n^m <\psi_i|Z_n^m(\vec{r})|\psi_j>$$

$$= \sum_{n,m} B_n^m <\psi_i|O_n^m(\vec{L})|\psi_j>. \qquad (24)$$

The $O_n^m(\vec{L})$ are equivalent operators to the $Z_n^m(\vec{r})$ and act on the (orbital) angular momentum coordinates of the wavefunctions. Tables of these operators and their matrix elements have been given by many authors, including Hutchings (27) and Abragam and Bleaney (Ref. 4, p. 862-72). The parameters B_n^m are related to the A_n^m by:

$$B_n^m = \theta_n A_n^m ,$$

where θ_n are reduced matrix elements and also are listed in (4) and (27). For the ^5D ground term of Fe^{2+}, the values are:

$$\theta_2 = <L||\alpha||L> = -\frac{2}{21} ,$$

$$\theta_4 = <L||\beta||L> = +\frac{2}{63}.$$

Therefore, as long as it is being considered only for the calculation of matrix elements within a spectral term, the crystal field Hamiltonian may be written in the form:

$$\mathcal{H}_{CF} = \sum_{n,m} B_n^m O_n^m. \qquad (25)$$

Table 5 contains expanded expressions for this effective Hamiltonian for several particular point symmetries. As will have been appreciated from comments made above, the particular form of the expansion depends on the choice of axes to which the Z_n^m(or O_n^m) functions are referred. This can be seen from the entries in Table 5, and in particular those for the cubic symmetry case. While it is, of course, obvious that the values of energy levels and other observable quantities cannot depend on the choice of axes, the simplicity of expressions and the ease with which calculations can be made can be greatly improved by a careful choice of axes. This choice is also important if comparison is to be made with crystal field parameters obtained independently: a different choice of axes will give rise to a different algebraic form for the crystal field and consequently to different parameter values. A clear example of this is in Table 5 for cubic symmetry. When a 4-fold axis is taken as the z axis, then $B_4^4 \neq 0$ and B_4^3 is zero, but when Oz is a 3-fold axis B_4^4 is zero and $B_4^3 \neq 0$. This axes dependence can be more subtle in other cases e.g. in the choice of the Ox,Oy axes when an off-axis distortion is present. Varret (15) considered this situation, and Fig. 15 shows two different axis-choices that can be made for D_2 point symmetry. Table 6 and 7 contain matrix elements for standard basis states of Fe^{2+}, $|L,L_z\rangle$ with L = 2, for two crystal field expansions (D_2 and C_{3v}) that contain all of the terms given in Table 5.

3.1.2 The Intra-ionic Spin Dependent Interactions

The intra-ionic spin-dependent interactions $\mathcal{H}_{LS}$ and $\mathcal{H}_{SS}$ were discussed in Section 2.2, and the Hamiltonians representing them were given. Because the spin-spin coupling constant ρ is small ($\lesssim 1\ cm^{-1}$), and because this interaction does not generally produce any further splitting of electronic multiplets, it is often neglected and the low energy electronic states of the Fe^{2+} ion determined from the crystal field and spin-orbit coupling interactions only. In cases where the crystal field splittings are either very large or very small compared with λ, perturbation calculations can be performed. Even in regimes in which this is not strictly valid, perturbation calculations will often yield the correct *structure* of the wavefunctions and give a physical insight into the nature of the solution and the relative importance of different interaction terms that is difficult to get from more rigorous matrix diagonalization procedures.

When adequate computer facilities are available, however, the easiest way to calculate the ionic wavefunctions accurately is to diagonalize the matrix $\mathcal{H} = \mathcal{H}_{CF} + \mathcal{H}_{LS} + \mathcal{H}_{SS}$ using the standard basis states $|L,L_z,S,S_z\rangle$. With L=2, S=2 for the ground ^5D term of Fe^{2+}, this is a 25x25 matrix, the diagonalization of which can easily be done on modern computers and some desk calculators. Alternatively,

TABLE 5

Some Equivalent Operator Representations of the Crystal Field

Point symmetry of 3d ion	Multiplicity of rotation symmetry about z-axis	x-axis	$\mathcal{H}_{CF}$
cubic	4	4	$B_4^0 O_4^0 + B_4^4 O_4^4 \quad (B_4^0 = B_4^4/5)$
cubic	4	2	$B_4^0 O_4^0 + B_4^4 O_4^4 \quad (B_4^0 = -B_4^4/5)$
cubic	3	-	$B_4^0 O_4^0 + B_4^3 O_4^3 \quad (B_4^0 = -\frac{\sqrt{2}}{40} B_4^3)$
C_{3h}	3	2	$B_2^0 O_2^0 + B_4^0 O_4^0$
C_{3v}	3	-	$B_2^0 O_2^0 + B_4^0 O_4^4 + B_4^3 O_4^3$ $= B_4^3 (O_4^3 - \frac{\sqrt{2}}{40} O_4^0)$ $+ B_2^0 O_2^0 + (B_4^0 + \frac{\sqrt{2}}{40} B_4^3) O_4^0$
D_4	4	2	$B_2^0 O_2^0 + B_4^0 O_4^0 + B_4^4 O_4^4$ $= B_4^4 (O_4^4 \pm O_4^0/5) + B_2^0 O_2^0 +$ $(B_4^0 \mp B_4^4/5) O_4^0$
D_2	2	2	$\mathcal{H}_{CF}(D_4) + B_2^2 O_2^2 + B_4^2 O_4^2$

Notes:

[1] Zox is chosen to be a mirror plane,

[2] the second forms of the expansions for C_{3v}, D_4 and D_2 symmetries are re-expressions of the first in such a way as to make the distortion (from cubic) terms axially symmetric,

[3] the $\pm$ signs in the expressions for D_4 and D_2 symmetry depend on the choice of axes (see Fig. 15 and Ref. 15). The upper signs correspond to the axes in Fig. 15(a), and the lower signs to Fig. 15(b).

TABLE 6

Matrix elements of H_{CF} for D_2 symmetry (Fe^{2+}, ^5D)

	$\|L_z = 2>$	$\|1>$	$\|0>$	$\|-1>$	$\|-2>$
$<2\|$	$-$ Dq $+6B_2^0+12B_4^0$	0	$\sqrt{6}(B_2^2+3B_4^2)$	0	$\mp$ 5 Dq
$<1\|$	0	4 Dq $-3B_2^0-48B_4^{0'}$	0	$3B_2^2-12B_4^2$	0
$<0\|$	$\sqrt{6}(B_2^2+3B_4^2)$	0	-6 Dq $-6B_2^0+72B_4^{0'}$	0	$\sqrt{6}(B_2^2+3B_4^2)$
$<-1\|$	0	$3B_2^2-12B_4^2$	0	4 Dq $-3B_2^0-48B_4^{0'}$	0
$<-2\|$	$\mp$ 5 Dq	0	$\sqrt{6}(B_2^2+3B_4^2)$	0	$-$ Dq $+6B_2^0+12B_4^{0'}$

where $10\ Dq = \mp 24\ B_4^4 = E(t_{2g}) - E(e_g)$,

$$B_4^{0'} = B_4^0 \mp B_4^4/5 \ .$$

The $\pm$ signs refer to the axis choices shown in Fig. 15a, b respectively.

TABLE 7

Matrix elements of H_{CF} *for* C_{3v} *symmetry* $(Fe^{2+}, {}^5D)$

	$\lvert L_z = 2 \rangle$	$\lvert 1 \rangle$	$\lvert 0 \rangle$	$\lvert -1 \rangle$	$\lvert -2 \rangle$
$\langle 2 \rvert$	$(2/3)$ Dq $+6B_2^0+12B_4^{0'}$	0	0	$(-10\sqrt{2}/3)$ Dq	0
$\langle 1 \rvert$	0	$(-8/3)$ Dq $-3B_2^0-48B_4^{0'}$	0	0	$(+10\sqrt{2}/3)$ Dq
$\langle 0 \rvert$	0	0	4 Dq $-6B_2^0+72B_4^{0'}$	0	0
$\langle -1 \rvert$	$(-10\sqrt{2}/3)$ Dq	0	0	$(-8/3)$ Dq $-3B_2^0-48B_4^{0'}$	0
$\langle -2 \rvert$	0	$(+10\sqrt{2}/3)$ Dq	0	0	$(2/3)$ Dq $+6B_2^0+12B_4^{0'}$

where $10\ \text{Dq} = -(9/\sqrt{2})\ B_4^3 = E(t_{2g}) - E(e_g)$

$$B_4^{0'} = B_4^0 + (\sqrt{2}/40)B_4^3.$$

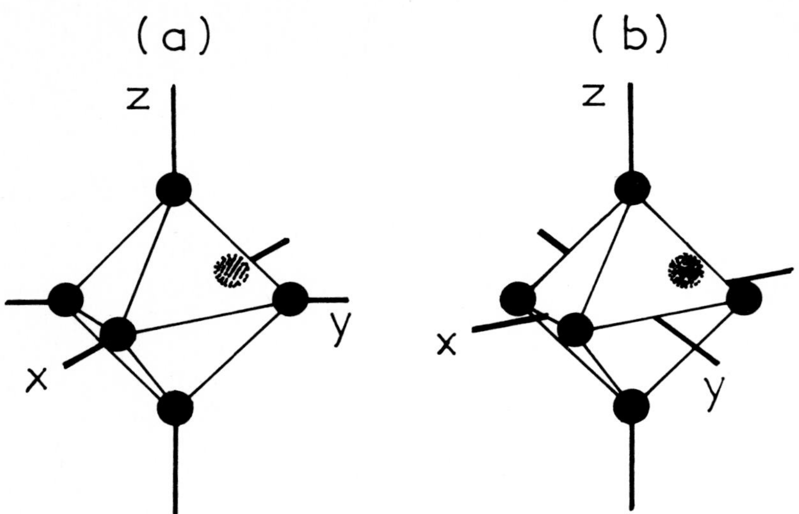

Fig. 15. Diagram of two different choices of axes that can be made for octahedral coordination. If the point symmetry at the centre of the octahedron is cubic, then Ox, Oy are 4-fold axes in (a) and 2-fold axes in (b). The effect of these different axes may be seen in Table 5. Either choice can be made for cubic (O_h) or tetragonal (D_4) symmetry, but different expressions for H_{CF} are obtained. Ingalls' (8) choice was (a), as this leaves the t_{2g} orbitals d_{xy}, d_{yz}, d_{zx} as eigenfunctions in O_h and D_4 symmetry. If the axes (b) are used, these orbitals are mixed with the e_g orbitals d_{z^2}, $d_{x^2-y^2}$. If a rhombic distortion is present (D_2 symmetry) the best choice will be determined by the nature of this distortion.

symmetry arguments can be used to factorize (or block-diagonalize) this matrix. It can be shown (103) that the 25x25 matrix for *H* can be reduced to:

two 8x8 and one 9x9 in C_{3v} symmetry,

three 6x6 and one 7x7 in D_4 symmetry,

one 13x13 and one 12x12 in D_2 symmetry.

Although there is a relatively large number of non-vanishing elements in the $\mathcal{H}$ matrix (95 in the C_{3v} case, 115 for D_2 symmetry) the matrix is simple to compute if advantage is taken of the fact that the $|L_z, S_z\rangle$ functions form a complete basis set. In this case the matrix $M(\theta \times \theta')$ associated with the product of the operators θ and θ' can be obtained as a product of the individual operator matrices i.e. $M(\theta \times \theta') = M(\theta) \times M(\theta')$. Because of this only the matrices of the

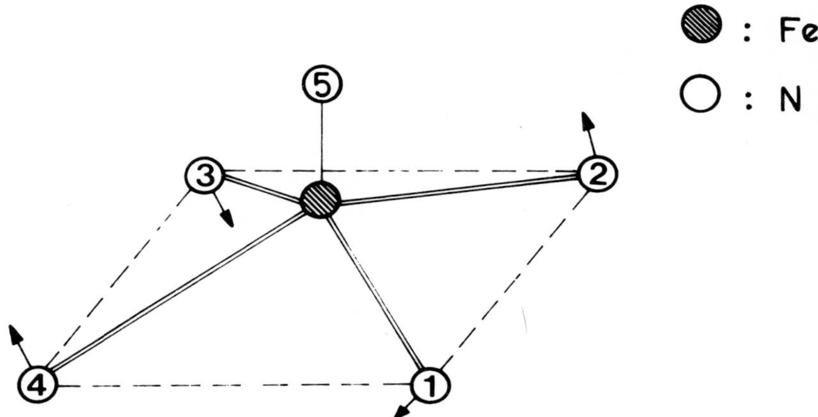

Fig. 16. Geometry of the square-pyramidal Fe(II)N_5 molecular complex in deoxyhemoglobin and deoxymyoglobin. The arrows indicate the β1 bending normal vibration mode considered by Bacci (17) in his vibronic coupling calculations.

"elementary" operators L_z, S_z, $L_\pm$, $S_\pm$ need to be specified and more complex matrices of products of these operators can be computed in the program. In order to help the reader in checking his computer program, Table 8 lists the 75 non-vanishing matrix elements of the operators $\mathcal{H}_{LS} + \mathcal{H}_{SS}$ in the standard basis.

3.1.3 Extra-Ionic Interactions

Two classes of interaction may be considered in this category, viz. those that originate within the crystal such as exchange interactions, the crystal field, which has been dealt with, and those that are applied externally such as magnetic or electric fields, or strain. These classes need not be distinguished if one is dealing with a single crystal, but external perturbations complicate the calculation of powder spectra significantly because, in general, they are not isotropic, i.e. one is imposing a directional perturbation that is determined by a laboratory axis system and which is therefore random relative to the symmetry axes of the ions in the sample. The resultant spectrum is therefore a sum (not an average) of the spectra obtained for every possible orientation of the perturbation relative to the quantization axes of the ions. If the ions have a unique quantization axis, as in any of the axially

TABLE 8

Non-vanishing matrix elements for the spin operators $H_{LS}+H_{SS}$ *in the standard basis; they are classified according to the value of* $J_z = L_z + S_z$ *(vectors of different* J_z *are not coupled together by these interactions). Note that* $\langle L_zS_z||L_z'S_z'\rangle = \langle -L_z-S_z||-L_z'-S_z'\rangle = \langle L_z'S_z'||L_zS_z\rangle = \langle S_zL_z||S_z'L_z'\rangle$. *Only values for* $J_z > 0$ *are given.*

$J_z = +4$	$\langle 22\|\|22\rangle = 4\lambda - 6\rho$
$J_z = +3$	$\langle 21\|\|21\rangle = 2\lambda + 3\rho = \langle 12\|\|12\rangle$
	$\langle 21\|\|12\rangle = 2\lambda - 9\rho = \langle 12\|\|21\rangle$
$J_z = +2$	$\langle 20\|\|20\rangle = 6\rho = \langle 02\|\|02\rangle$
	$\langle 20\|\|02\rangle = -6\rho = \langle 02\|\|20\rangle$
	$\langle 11\|\|11\rangle = \lambda-(3/2)\rho$
	$\langle 20\|\|11\rangle = \sqrt{6}\lambda - (3\sqrt{6}/2)\rho = \langle 11\|\|20\rangle = \langle 02\|\|11\rangle = \langle 11\|\|02\rangle$
$J_z = +1$	$\langle 2\text{-}1\|\|2\text{-}1\rangle = -2\lambda + 3\rho = \langle -12\|\|-12\rangle$
	$\langle 10\|\|10\rangle = -3\rho = \langle 01\|\|01\rangle$
	$\langle 10\|\|01\rangle = 3\lambda - (3/2)\rho = \langle 01\|\|10\rangle$
	$\langle 2\text{-}1\|\|10\rangle = \sqrt{6}\lambda + (3\sqrt{6}/2)\rho = \langle 10\|\|2\text{-}1\rangle = \langle -12\|\|01\rangle = \langle 01\|\|-12\rangle$
	$\langle 2\text{-}1\|\|01\rangle = -3\sqrt{6}\rho = \langle 01\|\|2\text{-}1\rangle = \langle -12\|\|10\rangle = \langle 10\|\|-12\rangle$
$J_z = 0$	$\langle 2\text{-}2\|\|2\text{-}2\rangle = -4\lambda - 6\rho = \langle -22\|\|-22\rangle$
	$\langle 1\text{-}1\|\|1\text{-}1\rangle = -\lambda-(3/2)\rho = \langle -11\|\|-11\rangle$
	$\langle 1\text{-}1\|\|-11\rangle = -9\rho = \langle -11\|\|1\text{-}1\rangle$
	$\langle 2\text{-}2\|\|1\text{-}1\rangle = 2\lambda + 9\rho = \langle 1\text{-}1\|\|2\text{-}2\rangle = \langle -22\|\|-11\rangle = \langle -11\|\|-22\rangle$
	$\langle 00\|\|00\rangle = -6\rho$
	$\langle 2\text{-}2\|\|00\rangle = -6\rho = \langle 00\|\|2\text{-}2\rangle = \langle -22\|\|00\rangle = \langle 00\|\|-22\rangle$
	$\langle 1\text{-}1\|\|00\rangle = 3\lambda+(3/2)\rho = \langle 00\|\|1\text{-}1\rangle = \langle -11\|\|00\rangle = \langle 00\|\|-11\rangle$

symmetric point symmetries, then geometric considerations dictate that this sum will be dominated by the contributions from ions with their quantization axis nearly perpendicular to the applied field.

Details of particular extra-ionic interactions will not be dealt with here as they can be found in many other publications (3, 4,5, 11). The correct treatment of exchange interactions can be very complicated although simple approximations such as the molecular field model are often used. Many references to Mossbauer work on magnetically ordered materials may be found in Varret's article (15).

Probably the most commonly-employed external perturbation is a magnetic field, and the electronic Zeeman interaction H_Z was given in Section 2.2.2 above (Eqn. 10). When calculating powder spectra it is clearly desirable to treat this interaction as a perturbation on the states determined by the crystal field and the spin-orbit coupling if this is possible. If not, the 25x25 matrix for the total Hamiltonian (now including H_Z) must now be diagonalized a large number of times to account for the angular dependence of the interaction. To compound this problem, block diagonalization of the matrix will not be possible with the applied field in an arbitrary direction.

Of course, the application of an external field has most interest when it has a large effect on the occupied states, which is just the situation in which perturbation theory may not be valid. It will often still be possible, however, to avoid repeated diagonalizations of the full 25x25 matrix by considering the effect of the field on only a relatively small number of low-lying states determined by H_{CF} and H_{LS}, i.e. by neglecting field-induced admixture between the lower states of interest and the higher excited states. In this case the repeated diagonalization may need to be done only on a relatively small matrix.

3.1.4 The Hyperfine Interactions

These interactions are obviously of particular importance to Mossbauer spectroscopists because it is the nuclear wavefunctions and energy levels that are measured in Mossbauer spectra. Information about the crystal field and other interactions is *channelled in* to the nuclei through the hyperfine interactions.

The hyperfine interactions of interest in the context of this article are the magnetic dipole and the electric quadrupole interactions and these have been discussed in some detail by, for example, Abragam (90) and, Abragam and Bleaney (4).

The electric quadrupole interaction Hamiltonian was introduced in Section 2.1.3. If the operator equivalent method is applied to the spherical tensor components (see Eqn. 2) of the nuclear quadrupole moment operator $[Q_2]$ and the EFG operator $[V_2]$ they may be written:

$$Q_2^0 = \frac{eQ}{I(2I-1)} \frac{1}{2} [3I_z^2 - I(I+1)], \quad V_2^0 = -e\langle r^{-3}\rangle\theta_2 \frac{1}{2} [3L_z^2 - L(L+1)],$$

$$Q_2^{\pm 1} = \mp \frac{eQ}{I(2I-1)} \frac{1}{2}\sqrt{\frac{3}{2}} (I_z I_\pm + I_\pm I_z), \quad V_2^{\pm 1} = \pm e\langle r^{-3}\rangle\theta_2 \frac{1}{2}\sqrt{\frac{3}{2}}(L_z L_\pm + L_\pm L_z),$$

$$Q_2^{\pm 2} = \frac{eQ}{I(2I-1)}\sqrt{\frac{3}{8}} (I_\pm)^2, \quad V_2^{\pm 2} = -e\langle r^{-3}\rangle\theta_2\sqrt{\frac{3}{8}}(L_\pm)^2. \tag{26}$$

Here -e is the charge of the electron, $\theta_2 (= -\frac{2}{21})$ is the reduced electronic matrix element introduced in Section 3.1.1 above, Q is a constant known as the quadrupole moment of the nucleus (in fact $\frac{eQ}{I(2I-1)}$ is a reduced nuclear matrix element that arises from the application of the operator equivalent method to the nuclear operators) and $\langle r^{-3}\rangle$ is the mean inverse third power of the electron distance from the nucleus, averaged over the 3d electron wavefunctions.

The quadrupole Hamiltonian (Eqn. 2), $\mathcal{H}_Q = \sum_{q=-2}^{2} (-1)^q Q_2^q V_2^{-q}$, can then be expanded in terms of the operator equivalents (Eqn. 26), converted to Cartesian form using the well-known definitions of the raising and lowering operators, or contracted to the vector form:

$$\mathcal{H}_Q = -\frac{e^2Q\langle r^{-3}\rangle}{I(2I-1)} \theta_2 . \frac{3}{2}[(\vec{L}.\vec{I})^2 + \frac{1}{2}(\vec{L}.\vec{I}) - \frac{1}{3}L(L+1)I(I+1)] , \tag{27}$$

which clearly has the same form as the intra-ionic spin-spin interaction $\mathcal{H}_{SS}$ (Eqn. 7).

The magnetic dipole hyperfine interaction was discussed in Section 2.2.3 above, although the spin-Hamiltonian formalism was used. For calculations using $|L_z, S_z\rangle$ basis states, the interaction Hamiltonian, or an equivalent operator, is most conveniently expressed in terms of $\vec{L},\vec{S}$ operators. The method of operator equivalents again allows this and yields:

$$\mathcal{H}_{hf} = -g_n\beta_n\vec{I}.\vec{H}_{hf} , \tag{28}$$

where the electronic magnetic field operator $\vec{H}_{hf}$ is given by:

$$\vec{H}_{hf} = -2\beta\langle r^{-3}\rangle\{\vec{L} - \xi[\tfrac{3}{2}\vec{L}(\vec{L}.\vec{S}) + \tfrac{3}{2}(\vec{L}.\vec{S})\vec{L} - L(L+1)\vec{S}] - \kappa\vec{S}\}.$$

The first term ($\vec{L}$) represents the orbital field, the second, in [ ] bracket, the dipolar field and the third ($\kappa\vec{S}$) the Fermi contact field. β is the Bohr magneton, $\xi = + \frac{1}{2S}\,\theta_2 = - \frac{1}{42}$ for Fe^{2+} and κ is a dimensionless positive quantity (~0.45) that arises predominantly from magnetic polarization of the Fe^{2+} ion core by the open-shell electrons and depends relatively weakly on the particular environment of the ion.

Before discussing the calculation of these interactions there are two points that are worth briefly recalling. Firstly, the expressions (27) and (28) are operator equivalents of the hyperfine interactions and as such can only be used for calculating matrix elements within a single electronic and nuclear (L,S,I) manifold. Secondly, because of different effects of screening by the ion core on different interactions, the values of the average $\langle r^{-3}\rangle$ appropriate to the different interactions are not identical, although the differences are small in practice (4, 54).

There are two ways of calculating the wavefunctions and energy levels resulting from the hyperfine interactions $\mathcal{H}_Q$ and $\mathcal{H}_{hf}$. The simplest and most commonly used involves the effective field approximation, and this will be described first. The second, in which the electrons and nucleus of the ion are treated as a single coupled quantum mechanical system will be briefly mentioned later.

The effective field approximation for hyperfine interactions was mentioned briefly in Section 2.2.3. This approximation ignores the "back interaction" of the nuclear moment on the electronic wavefunctions. It may be stated in either of two equivalent ways:

(1) In the presence of the hyperfine interactions, the wave-function for the combined quantum system of the open-shell electrons and the nucleus of the ion can be written in the product form:

$$\phi_\ell(e,n) = \psi_i(e)X_j(n) \quad ,$$

where e,n represent respectively the electron and nuclear coordinates. $X_j(n)$ can be written in terms of the basis states $|I,m_I\rangle$.

(2) Perhaps more directly, the hyperfine interactions do not couple eigenfunctions of the electronic Hamiltonian i.e. if a general hyperfine interaction Hamiltonian is written:

$$\mathcal{H}_{hf} = \sum_{q=-k}^{k} \mathcal{H}_k^q(e)(\mathcal{H}_k^q(n))^*,$$

then

$$<\psi_i(e)|\mathcal{H}_k^q(e)|\psi_{i'}(e)> = \delta_{ii'}.$$

This means that the only non-zero matrix elements are the diagonal ones, and these may be thought of as an effective field acting on the nucleus,

i.e.

$$<\phi_\ell(e,n)|\mathcal{H}_{hf}|\phi_{\ell'}(e,n)> = \sum_{q=-k}^{k} <\psi_i|\mathcal{H}_k^q(e)|\psi_i><\chi_j|\mathcal{H}_k^q(n)^*|\chi_{j'}>.$$

Since the nuclear matrix elements involved in this expression are well known (104) or can be readily calculated, the nuclear energy levels and hence the Mossbauer spectrum peak positions are determined by the expectation values of the electronic operators $\vec{H}_{hf}$ in Eqn.(28) and $[V_2]$ in Eqn.(26) or equivalently the Cartesian components of $[V_2]$:

$$V_{ij} = -e<r^{-3}>\theta_2\{\tfrac{3}{2}(L_iL_j + L_jL_i) - \delta_{ij}L(L+1)\},$$

for the eigenfunctions derived from the electronic Hamiltonian for the ion. If the electronic relaxation is very slow, then every such eigenfunction will produce a separate set of peaks in the Mossbauer spectrum, with intensities weighted by the occupation probability of the state. In the fast relaxation limit a thermal average of the individual expectation values is required.

If an external perturbation is applied to the sample, then this may interact directly with the nucleus and will appear as a real, rather than an effective, field. In the case of an external magnetic field, the nuclear Zeeman interaction:

$$\mathcal{H}_{NZ} = -g_n\beta_n\vec{I}.\vec{H}_{ext}$$

must be included in the calculation of the nuclear energy levels. Thus, the total (effective) magnetic field acting on the nucleus is obtained as the vector sum:

$$\vec{H} = <\vec{H}_{hf}> + \vec{H}_{ext}.$$

As was mentioned above, the effective field approximation is made very commonly in the interpretation of $^{57}Fe^{2+}$ Mossbauer spectra. In most cases it is a good approximation but in a few it is poor and cannot be used at all (47). Examples of the breakdown of the effective field approximation for nuclei other than ^{57}Fe include the pseudo-quadrupole effect observed in $^{169}TmCl_3.6H_2O$ (46) and the line-shift in ^{166}Er:yttrium ethyl sulphate (105). Proceeding with the calculation under these circumstances is not necessarily more difficult in principle but is certainly more time-consuming. To do the calculation rigorously one should diagonalize the Hamiltonian matrix for all of the electronic and hyperfine interactions. In this case the basis states are $|L,m_L,S,m_S,I,m_I>$ with L=2, S=2 and $I = \frac{3}{2}$ for the nuclear excited state and $I = \frac{1}{2}$ for the ^{57}Fe ground state. Thus a 100x100 and a 50x50 matrix must be diagonalized to obtain the states of the coupled electron-nucleus system derived from the 14.4 keV excited state and the ground state respectively. This computational problem may be alleviated in two ways:

(1) block diagonalization of the matrices where possible, and/or
(2) hyperfine coupling between well-separated electronic states may be ignored (except in special circumstances where it may open relaxation channels) and only relatively small matrices, with basis states $\psi_i(e)|I,m_I>$ where the ψ_i are eigenfunctions of the electronic Hamiltonian that are degenerate or nearly-degenerate, diagonalized. This was the approach adopted by Howes et al. (47).

3.1.5 Nuclear Transition Probabilities

The only remaining problem to be overcome in calculating a Mossbauer spectrum is determination of the nuclear transition probabilities. The 14.4 keV excited state of ^{57}Fe decays to the ground state by a pure magnetic dipole (M1) transition, so the transition probabilities depend on matrix elements of the form:

$$<\alpha_e I^e m_I^e | \mathcal{H}^m(\vec{k},L,m_L) | \alpha_g I^g m_I^g > \quad , \qquad (29)$$

where $\mathcal{H}^m(\vec{k},L,m_L)$ represents the magnetic interaction between the nucleus and a photon of wavevector $\vec{k}$ and angular momentum quantum numbers L,m_L. α represents the set of quantum numbers other than I,m_I required to specify the nuclear states and the indices e and g indicate the 14.4 keV excited state and the ground state of ^{57}Fe, respectively ($I^e = \frac{3}{2}$, $I^g = \frac{1}{2}$). The operator $H^m(k,L,m_L)$ can be expressed as a spherical tensor of rank L with even parity, and so for dipole (L=1) transitions it transforms as a pseudo-vector. Application of the Wigner-Eckart theorem shows that matrix elements (Eqn. 29) for different values of m_I^e, m_I^g, m_L are related by the Clebsch-Gordan coefficients $\langle I^g m_I^g L m_L | I^g L I^e m_I^e \rangle$ and these are zero unless the selection rule $m_I^e = m_I^g + m_L$ is satisfied (i.e. $m_I^e - m_I^g = m_L = 0, \pm 1$).

Explicit expressions for the transition probabilities as functions of the photon propagation direction k (relative to the nuclea quantization axis) have been given by a number of authors (104, 106) and will not be repeated here.* For the spectra of powder samples in external fields the elegant work of Lang (61) should be consulted.

3.2 Calculation of the Effects of the Dynamic Crystal Field

3.2.1 The Interaction Hamiltonian

Consider now the lowest order dynamic crystal field interaction for a quasi-molecular cluster, as was described in Section 2.3 (Eqn. 22). This interaction, which is linear in the normal modes of the cluster, was written:

$$H_{OL}^{(1)} = \sum_k \left(\frac{\partial H^0}{\partial Q_k}\right) Q_k.$$

Within the quasi-molecular cluster model the localized modes Q_k are assumed to be excited in a known way by the vibrations of the surrounding lattice, and the main problem addressed is that of describing their coupling with the open shell electrons of the central paramagnetic ion, i.e. of evaluating the effects of the interaction H_{OL} above.

Two distinct approaches to evaluating this interaction are commonly made, and they differ in the effect that the lattice mode-local mode coupling has on the local molecular vibrations. One approach is to consider the limit in which the cluster vibrations remain strongly localized and essentially monochromatic. In this limit the lattice surrounding the cluster is ignored, which is a reasonable representation of the situation in true molecular systems, in quasi-molecular solids in which the intra-molecular

---

*See Chapter 1.

bonding is much stronger than that between molecules, and for very light impurity ions in crystal lattices of relatively heavy atoms where the local impurity vibrations have a much higher frequency than the acoustic lattice vibrations.

The second approach, and that which will be of prime concern here, is to consider the local modes to be poorly localized and to have a broad frequency spectrum as a result of their resonance with the lattice vibrations. This is the approach that is most appropriate to heavy impurity ions in crystals or to calculations concerning host atoms in a lattice. The local vibrations are generally assumed to maintain the symmetry properties of the normal modes of the cluster, as is consistent with the quasi-molecular model, but to have dispersion relations characteristic of the lattice vibrations.

However, there are severe calculational difficulties with both models and each has significant advantages over the other for particular calculations. Therefore the tendency has been to choose the approach that allows the best calculation of a particular property or effect, with little regard for the physical applicability of that approach. This should not be considered as criticism but as a reflection of the computational difficulties of doing realistic calculations of vibronic effects.

The approach in which the vibrations are monochromatic, the molecular approach, has been employed extensively in calculations of the Jahn-Teller effect in solids (71, 75) as well as, of course, for its applications in molecular spectroscopy (107). Its prime advantage is that it can be applied in much the same way as static crystal field theory - group theoretical considerations can be used extensively to determine the degeneracies of vibronic states and the parameterization of the matrix elements of $\mathcal{H}_{OL}$. This allows elegant and, in some simple situations, rigorous calculations to be performed, but the complexity increases rapidly with the number of vibrations that must be considered. For the purpose of this article, however, its main disadvantage is that it does not describe the acoustic phonons that are predominantly responsible for spin-lattice relaxation.

Bacci (17) has recently used this molecular approach in a calculation of effects of vibronic admixture within the square pyramidal Fe(II) complex in deoxyhemoglobin and deoxymyoglobin (Fig. 16) on the magnetic susceptibility and the ^{57}Fe Mossbauer spectra. To reduce the complexity of the calculation he made the following assumptions:

(1) the point symmetry of the complex is C_{4v}

(2) only one local vibration, a β_1-type bending mode, was considered (see Fig. 16)

(3) only states containing 0, 1 or 2 of these vibrational quanta were included

(4) only the lowest four electronic levels of Fe(II), that lie within a thousand cm^{-1}, were considered.

While these assumptions, and those inherent in the molecular model, might appear to be rather restrictive, they allowed Bacci to justifiably conclude that vibronic effects make important contributions to the magnetic and Mossbauer properties of these compounds and to explain what had previously appeared to be a qualitative inconsistency of the crystal field in these molecules as compared with other structurally similar ones. Bacci's results are summarized* in Figs. 17 and 18, and his paper, (17), should be consulted for further details of his calculation.

When a broad frequency spectrum is introduced into the quasi-molecular vibrations calculations become even more complex because of the now effectively infinite number of modes that must be included. As a result even more explicit approximations must be made, although this is partly offset for many systems of interest because the inherent approximations of the model are somewhat less restrictive than for the localised molecular model. Orbach and Stapleton (72) have discussed this approach to the orbit-lattice coupling and have considered many of the approximations usually made in constructing the interaction Hamiltonian and in evaluating its effects.

If only the vibrational normal modes of the cluster are considered, and if they are grouped according to their symmetry properties to give normal modes of the form $Q(\Gamma,p)$ that transform as the pth subvector of the irreducible representation Γ of the point group of the cluster, then the (linear) orbit-lattice interaction Hamiltonian may be written as:

$$H_{OL}^{(1)} = \sum_{\Gamma,p} \sum_{n} a_n(\Gamma) Z_n(\Gamma,p) Q(\Gamma,p) . \qquad (30)$$

The $Z_n(\Gamma,p)$ are appropriately transforming linear combinations of the tesseral harmonic functions Z_n^m defined in Eqn.(1). For calculating

*Figs. 17 and 18 are reproduced here with the kind permission of Prof. M. Bacci.

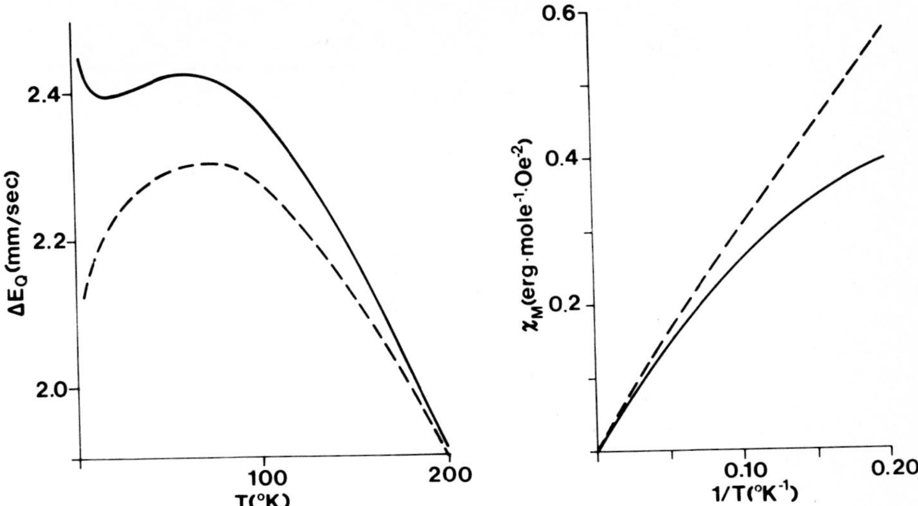

Fig. 17. Results of Bacci's vibronic coupling calculations (17). The effect of vibronic coupling on (a) the quadrupole splitting ΔE_Q and (b) the magnetic susceptibility χ_M. In each case the dashed line is obtained by neglecting vibronic coupling and the solid line by including coupling to the β_1-type bending mode (Fig. 16) with $\hbar\omega$ = 50 cm^{-1}. See (17) for other details. Parameters used were the same for both cases.

the matrix elements of $H_{OL}^{(1)}$, the same restrictions apply to the values of n that must considered as for the static crystal field (Section 3.1), and the operator equivalent method can be used to rewrite Eqn. (30) as:

$$H_{OL}^{(1)} = \sum_{\Gamma,p} \sum_n b_n(\Gamma) O_n(\Gamma,p) Q(\Gamma,p), \tag{31}$$

where the $O_n(\Gamma,p)$ are combinations of the angular momentum operators O_n^m.

Orbach and Tachiki (108) expanded* the normal coordinates of a cluster centred at $\vec{R}_j$ in terms of the phonon states of the crystal as

$$Q_j(\Gamma,p) = i \sum_{\vec{k},s} \left[\frac{\hbar}{2M\omega(\vec{k},s)}\right]^{\frac{1}{2}} [a_{\vec{k},s} \exp(i\vec{k}.\vec{R}_j) - a^{\dagger}_{\vec{k},s} \exp(-i\vec{k}.\vec{R}_j)] R(\vec{k},s,\Gamma,p). \tag{32}$$

---

*For details, see chapter on Spin Lattice Relaxation appearing elsewhere in this book.

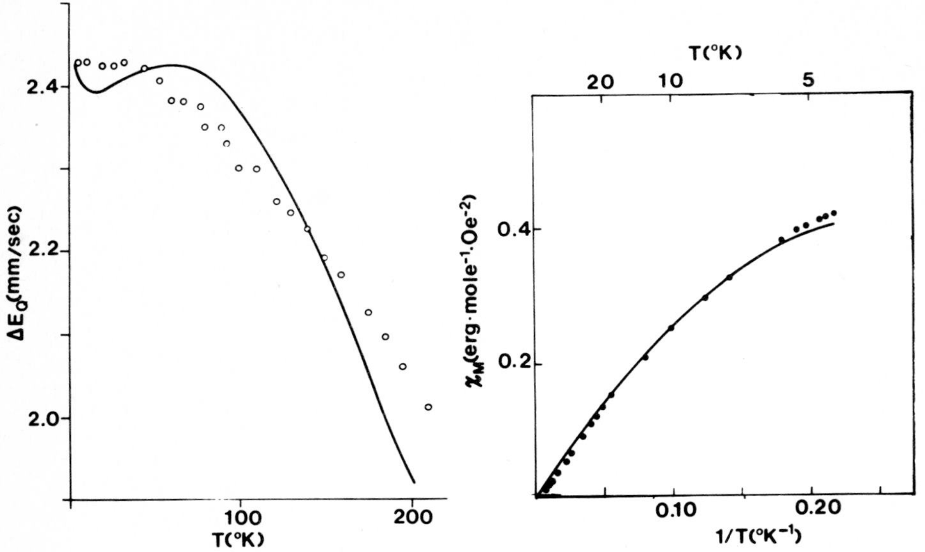

Fig. 18. Results of Bacci's vibronic coupling calculations (17) compared with experimental data for deoxyhemoglobin: (a) shows the theoretical quadrupole splitting curve compared with the experimental data of Eicher et al (127), (b) shows the calculated magnetic susceptibility compared with the experimental data of Alpert et al (128). The same parameters (17) were used to calculate both curves.

$a^{\dagger}_{\vec{k},s}$ and $a_{\vec{k},s}$ are creation and annihilation operators for a phonon of wave vector $\vec{k}$ and polarization s that has angular frequency $\omega(\vec{k},s)$. M is the mass of the crystal. The functions $R(k,s,\Gamma,p)$ are approximately transforming combinations of products of the form $e_{\mu}(\vec{k},s)\sin(k_{\nu}a)/a$ where $\vec{e}(\vec{k},s)$ is a phonon polarization vector, a is the lattice constant and μ,ν represent (Cartesian) components of the vectors (72, 108).

The relation given by Eqn. (32) forms the heart of this approach to the orbit-lattice coupling since it relates the normal modes of vibration of the cluster to the phonon states in the crystal. Let us assume that the crystal can be represented by an isotropic Debye solid. It is further assumed that the Debye cut-off wavelength is sufficiently long compared with the dimensions of the cluster that $\sin(k_{\nu}a) \simeq k_{\nu}a$: this is the long wavelength approximation, and its limitations have been discussed by Orbach and Stapleton (72). A further problem arises when the cluster is non-cubic, and that is the possibility that long wavelength phonons may not be able to excite all of the normal modes of the cluster, i.e. some of the R(Γ,p) may become zero for long wavelength phonons. This was pointed out by Curtis et al. (110), who examined the case of a D_{3h} cluster.

In the long wavelength approximation, the distortions of the cluster may be described by the lattice strain tensor $\vec{\varepsilon}$, which is a symmetric Cartesian tensor of rank 2, so $H_{OL}^{(1)}$ may be re-expressed in the form:

$$H_{OL}^{(1)} = \sum_{\Gamma,p} \sum_{n} b_n(\Gamma) O_n(\Gamma,p) \varepsilon(\Gamma,p) \quad , \tag{33}$$

where the ε(Γ,p) are those linear combinations of the strain tensor that transform as the pth subvector of the representation Γ of the point group of the cluster. Alternatively, it may be written in an analogous form to the static crystal field Hamiltonian of Equation (25) as:

$$H_{OL}^{(1)} = \sum_{n,m} b_n^m O_n^m \varepsilon_n^m . \tag{34}$$

This was the form introduced by Orbach (92) in his classic paper on spin-lattice relaxation.

3.2.2 Matrix Elements of Orbit-Lattice Hamiltonian $H_{OL}^{(1)}$

In the absence of vibronic coupling, the Hamiltonian for a crystal, or a molecular cluster, takes the form:

$$H^0 = H_{el.}(\vec{q}) + H_{nuc.}(\vec{Q}), \tag{35a}$$

where $H_{el.}(\vec{q})$ operates only on the electrons and is a function only of the electronic coordinates q. $H_{nuc.}(\vec{Q})$ is a function only of the position and momentum coordinates Q of the ion nuclei. In this approximation, the eigenstates of H can be written in the form:

$$\Phi^0(\vec{q},\vec{Q}) = \psi(\vec{q})\phi(\vec{Q}) . \tag{35b}$$

This is known as the Born-Oppenheimer approximation, and product states of this form, known as Born-Oppenheimer products, are convenient to use as basis states for calculations of vibronic effects (for a more complete discussion see, for example, Abragam and Bleaney (4; Ch. 21).

In the molecular, or localized mode, approach to the vibronic coupling problem, the vibrational wavefunctions $\phi(\vec{Q})$ are generally determined as solutions of harmonic oscillator equations for the particular mode or modes under consideration. Explicit expressions for the wavefunctions can then be used.

However, it is more difficult to do this in the non-localized, or lattice approach. In this case the nuclear Hamiltonian is written in second quantized form as:

$$H_{nuc} = \sum_{\vec{k},s} \hbar\omega(\vec{k},s) \; [a^{\dagger}_{\vec{k},s} \; a_{\vec{k},s} + \frac{1}{2}]$$

which has eigenstates that may be written in the form

$$\phi(\vec{Q}) = |N_{\alpha}(\vec{k}_1,s_1),N_{\alpha}(\vec{k}_1,s_2) \; .....N_{\alpha}(\vec{k}_i,s_j) \; .....> \; \equiv \; |\prod_{\vec{k},s} N_{\alpha}(\vec{k},s)>$$

and eigenvalues

$$E_{nuc} = \sum_{\vec{k},s} [N_{\alpha}(\vec{k},s) + \frac{1}{2}]\hbar\omega(\vec{k},s) .$$

The $N_{\alpha}(\vec{k},s)$ are the occupation numbers of the phonon states of wave-vector $\vec{k}$ and branch index s; the index α simply identifies a particular set of occupation numbers. Thus, the basis states $\phi^{0}(\vec{q},\vec{Q})$ for the vibronic calculation may be written, using Eqn. (35b):

$$\Phi^{0}_{i\alpha} = \psi_i \,|\prod_{\vec{k},s} N_{\alpha}(\vec{k},s)> \qquad (36a)$$

and they have (unperturbed) energies:

$$E^{0}_{i\alpha} = E_i + \sum_{\vec{k},s} [N_{\alpha}(\vec{k},s) + \frac{1}{2}]\hbar\omega(\vec{k},s) . \qquad (36b)$$

The most convenient electronic basis state components, ψ_i, will depend on the way in which the vibronic coupling is treated. In the following, lack of a reasonable alternative forces the use of a perturbation treatment, and in this case the ψ_i and E_i used are eigenfunctions and eigenvalues of the static crystal field H_{CF}.

Using the basis states (36a), matrix elements of the vibronic coupling $H^{(1)}_{OL}$ may be readily evaluated as follows:

$$<\Phi^0_{j\beta}|H^{(1)}_{OL}|\Phi^0_{i\alpha}> \;=\; \sum_{\Gamma,p} \sum_{n} b_n(\Gamma)<\psi_j|O_n(\Gamma,p)|\psi_i> \times$$

$$< \prod_{\vec{k},s} N_\beta(\vec{k},s)|\varepsilon(\Gamma,p)| \prod_{\vec{k},s} N_\alpha(\vec{k},s)>. \qquad (37)$$

Evaluation of the electronic matrix elements $<\psi_j|O_n|\psi_i>$ should be straightforward following the discussion of Section 3.1, while the matrix elements of the strain operator have been given in a number of places [e.g. (4, 18, 72, 92)], the only non-zero elements being:

$$<N(\vec{k},s)+1|\varepsilon(\Gamma,p)|N(\vec{k},s)> = -i\left\{\frac{\hbar[N_\alpha(\vec{k},s)+1]}{2M\omega(\vec{k},s)}\right\} R(\vec{k},s,\Gamma,p),$$

$$<N(\vec{k},s)-1|\varepsilon(\Gamma,p)|N(\vec{k},s) \;=\; i\left\{\frac{\hbar N(\vec{k},s)}{2M\omega(k,s)}\right\}R(\vec{k},s,\Gamma,p). \qquad (38)$$

3.2.3 Calculation of Vibronic Admixture Effects

In order to properly include the effects of vibronic coupling on the states of an ion in a crystal, the eigenfunctions of the Hamiltonian $H_{ion} + H_{OL} + H_{nuc}$ should be determined by matrix diagonalization. However, even if one is dealing with an isolated molecular cluster, the size of the matrix to be diagonalized rapidly becomes prohibitive - see, e.g., the work of Bacci (17) in which only one normal mode of vibration was included. Consequently, it is normally necessary to use perturbation treatments, particularly when an essentially infinite basis set such as Eqn. (36) is appropriate.

At this point, it is appropriate to note again that Stevens and coworkers (74) have developed a transformation that accounts exactly for the linear orbit-lattice coupling $H^{(1)}_{OL}$, leaving only higher order terms $H^{(2)}_{OL}$ .... to be included in a perturbation treatment. While this is undoubtedly a better approach in principle to that which will be described below, its consequences in relation to Mossbauer observables and relaxation transition rates have not yet been examined. Therefore it will not be considered further at this stage. In any case, the perturbation treatment considered here is probably more physically transparent and should provide at least a qualitative indication of the effects to be expected from the vibronic coupling.

The perturbation approach given by Price (18) will be described here, and that paper should be consulted for details. It consists of applying $\mathcal{H}_{OL}$ as a time-independent perturbation to the Born-Oppenheimer product states (Eqn. 36) that are eigenstates of $\mathcal{H}_{ion} + \mathcal{H}_{nuc}$. This is an admittedly dubious computational technique since the leading terms in $\mathcal{H}_{OL}$ are undoubtedly of the same order of magnitude, or larger, than some of the terms (e.g. $\lambda\vec{L}.\vec{S}$) in H_{ion}.

It is worthwhile considering the relationship of this approach to that which is almost universally adopted to calculate spin-lattice relaxation effects, and which will be considered briefly below. The two sets of calculations are based on an essentially identical model, with time-independent perturbation theory being used to calculate wavefunction admixture effects, while the time-dependent method is employed to account for the relaxation transitions between states. This concept of treating the interaction as being both independent of time and time-dependent was introduced in Section 2.3. It relies on the fact that the adiabatic approximation is valid in some circumstances but not in others. This is a difficulty that is normally not encountered in the localized mode approach to the vibronic interaction, since the energy of the (monochromatic) mode will generally not be nearly the same as the energy difference between two electronic states, i.e. the mode will not be involved in direct spin-lattice relaxation processes.

The perturbation approach consists simply of calculating the perturbed vibronic states to first order in time-independent perturbation theory i.e.

$$\Phi^{(1)}_{i\alpha} = A_{i\alpha}\{\Phi^{0}_{i\alpha} + \sum_{j,\beta}{}' \frac{\langle\Phi^{0}_{j\beta}|\mathcal{H}^{(1)}_{OL}|\Phi^{0}_{i\alpha}\rangle}{E^{0}_{i\alpha} - E^{0}_{j\beta}} \Phi^{0}_{j\beta}\} \tag{39}$$

and then the expectation value of an operator, such as the EFG operator [V], is calculated for these perturbed states as (18)*:

$$\langle\Phi^{(1)}_{i\alpha}|[V]|\Phi^{(1)}_{i\alpha}\rangle = |A_{i\alpha}|^{2} \{\langle\Phi^{0}_{i\alpha}|[V]|\Phi^{0}_{i\alpha}\rangle + \sum_{\substack{j\beta\\ j'}} \frac{\langle\Phi^{0}_{i\alpha}|\mathcal{H}_{OL}|\Phi^{0}_{j'\beta}\rangle\langle\Phi^{0}_{j'\beta}|[V]|\Phi^{0}_{j\beta}\rangle\langle\Phi^{0}_{j\beta}|\mathcal{H}_{OL}|\Phi^{0}_{i\alpha}\rangle}{(E^{0}_{i\alpha} - E^{0}_{j'\beta})^{*}\,(E^{0}_{i\alpha} - E^{0}_{j\beta})}\} \tag{40}$$

The symbol '', indicating complex conjugate, in the denominator of Eqn. (40) is retained since in some cases it is necessary to take into account the lifetime of the excited state by replacing, say, E_j by $E_j - i\Gamma_j$ and the energies then become complex.

where $A_{i\alpha}$ is the normalization constant.

This expression may be evaluated using the matrix elements of Eqn. (37) and Eqn. (38), and it can thus be reduced to:

$$<[V]>_i^{(1)} = \frac{<[V]>_i^0 + \sum_{jj'} <[V]>_{jj'}^0 D_{jj'}^i(T)}{1 + \sum_{jj'} D_{jj'}^i (T)} , \tag{41}$$

where:

$$<[V]>_i^0 = <\Phi_{i\alpha}^0|[V]|\Phi_{i\alpha}> = <\psi_i|[V]|\psi_i>,$$

$$<[V]>_{jj'}^0 = <\psi_j|[V]|\psi_{j'}>,$$

and

$$D_{jj'}^i(T) = (\sum_{\Gamma,p} \sum_{n,n'} b_n(\Gamma)b_{n'}(\Gamma)<\psi_i|O_n(\Gamma,p)|\psi_{j'}><\psi_j|O_{n'}(\Gamma,p)|\psi_i>) x$$

$$C_{jj'}^i(T).$$

Equation (41) shows clearly how the vibronic coupling parameter $D_{jj'}^i(T)$ modifies the observed values of the EFG tensor [V], or of any other electronic operator such as the magnetic moment (or equivalently the g values).

The factors $C_{jj'}^i(T)$ contain all of the phonon quantities, and their temperature-dependence arises from that of the phonon occupation numbers $N_\alpha(\vec{k},s)$. If the phonon spectrum is assumed to be isotropic, then they are given by:

$$C_{jj'}^i(T) = 3 \sum_{\vec{k}} [\frac{\hbar k^2}{2M\omega(k)}] \{\frac{N_\alpha(k)+1}{[E_i-E_{j'}-\hbar\omega(k)]*[E_i-E_j-\hbar\omega(k)]} +$$

$$\frac{N_\alpha(k)}{[E_i-E_{j'}+\hbar\omega(k)]*[E_i-E_j+\hbar\omega(k)]}\} . \tag{42}$$

It will be immediately apparent that for admixture between electronic states that are separated by less than the maximum phonon energy (i.e. if $E_i-E_j<h\omega_{max}$), then $C_{jj'}^i(T)$ will diverge. This will be mentioned further below, but for the time being the problem will be by-passed.

Price (18) did some model calculations of the effect of vibronic admixture between electronic states derived from the octahedral $^5T_{2g}$ level on the EFG at Fe^{2+} nuclei in sites of trigonal symmetry. Two sets of results, for trigonal splittings of opposite sign, are shown in Figs. 19 and 20. Note that, for the case where the ^5E_g orbital doublet is the crystal field ground state, (a trigonal elongation) a very crude approximation to the effect of the spin-orbit splitting of this multiplet has been made (Fig. 19). This was done both to reduce the number of states that had to be considered and to overcome the problem of divergence in $C^i_{jj'}(T)$ referred to above. It may be noted from these results that vibronic coupling can have a significant effect on the temperature dependence (as well as on the magnitude) of the EFG at low temperatures, and in this respect would appear to produce qualitatively different effects than thermal expansion of the lattice. Price (unpublished results) has shown that, largely because of this feature results such as those of Fig. 19 can provide a good qualitative explanation of the Fe^{2+} quadrupole splitting in $FeCO_3$ and dilute isomorphous compounds such as $Fe^{2+}:ZnCO_3$, $Fe^{2+}:MgCO_3$. Similarly, the curves of Fig. 20 bear a striking qualitative similarity to the Fe^{2+} quadrupole splitting in $FeSiF_6.6H_2O$, as measured by Varret and Jehanno (49) (see Fig. 8(b)), although little is known of the strength of the vibronic coupling in this compound.

It is interesting to make a comparison of the effects of the vibronic coupling at low temperatures, where it is dominated by the zero-point vibrations, as calculated using this approach (Figs. 19, 20) and using the localized mode approach (Figs. 17, 18). In both cases it is clear that there are significant effects on both the magnitude and temperature dependence of the quadrupole splitting at low temperatures, but the calculations of Bacci (17), Fig. 17 show more spectacular effects. This is probably due mainly to the fact that Bacci, by considering only a single localized mode, was able to properly include the effects of the low-lying spin-orbit levels.

In conclusion, it should be said that all that the calculations performed so far have done is to indicate in a qualitative way that vibronic coupling effects are likely to have significant effects on Fe^{2+} Mossbauer spectra and that they therefore must be considered in the interpretation of such spectra. However, quantitative evaluation of these effects is clearly difficult, particularly for the lattice mode coupling, and a transformation approach along the lines of that introduced by Stevens (74) is more likely to provide a quantitatively satisfactory way of approaching the problem.

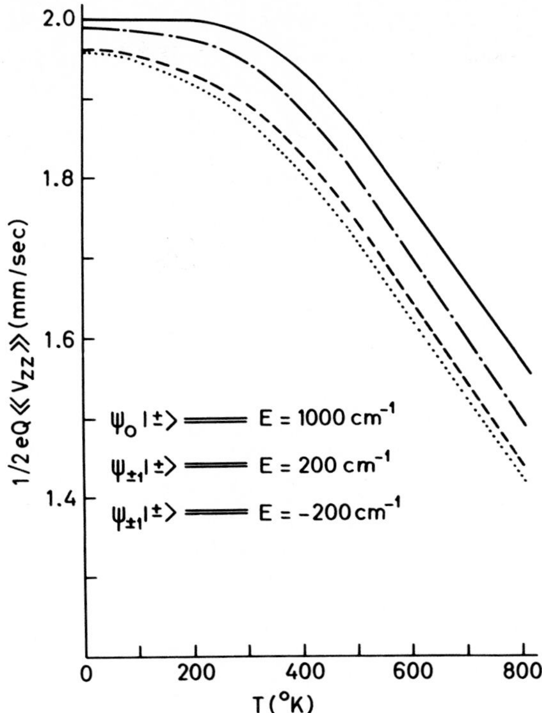

Fig. 19. Results of Price's (18) vibronic coupling calculations of the ^{57}Fe quadrupole splitting for an Fe^{2+} ion in an octahedral site with a trigonal elongation (such as in $FeCO_3$). The solid curve represents the case of no vibronic coupling (or coupling to A_{1g} vibrational modes only) while the other three curves are calculated for different values of the coupling coefficients $b_n(\Gamma)$ (Equation 33). The inset shows the wavefunctions and energy levels used in the calculation. A spin degeneracy of two (with spin states $|+\rangle$ and $|-\rangle$) was assumed, to provide a crude approximation to the effect of spin-orbit coupling on the ground orbital doublet $\psi_{\pm 1}$ (see Fig. 8(b)). Ref. (18) should be consulted for further details. Note the significant differences in the temperature dependence of the curves, particularly at low temperatures where thermal expansion effects will be less important. The symbol $\langle\langle V_{zz}\rangle\rangle$ means the thermal average of the expectation values $\langle V_{zz}\rangle$ of each wave function (18).

3.2.4 Orbit-Lattice Relaxation Effects

As has been pointed out earlier, relaxation transitions arise when H_{OL} cannot be treated as an adiabatic, or slowly time-dependent, perturbation. This occurs (in first order) when one of the frequency components of H_{OL} is comparable to the frequency difference $(E_i-E_j)/h$ of two electronic states ϕ_i,ϕ_j (67). It will be noted that this is

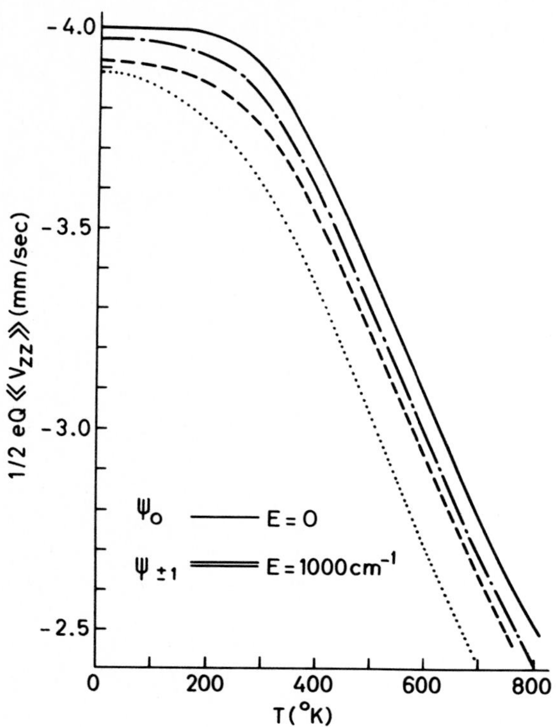

Fig. 20. Results of Price's (18) vibronic coupling calculations of ^{57}Fe quadrupole splitting for an Fe^{2+} ion in an octahedral site with a trigonal compression (such as in $FeSiF_6.6H_2O$). The solid curve represents the case of no vibronic coupling (or coupling to A_{1g} vibrational modes only), while the other three curves are calculated for different values of the coupling coefficients $b_n(\Gamma)$ (Equation 33). The inset shows the wavefunctions and energy levels used in the calculation. Ref. (18) should be consulted for further details. Note that, although; spin-orbit coupling has been neglected here, there is a qualitative similarity of these results with the experimental data of Varret and Jehanno (49) for $FeSiF_6.6H_2O$ (see Fig. 8(a)).

just the condition encountered in the previous section for which divergence occurs in the vibronic admixture coefficients.

As was pointed out in Section 2.3.3 relaxation effects have only been observed and unambiguously identified in a very few cases, and in all of these cases the effective field approximation was a good one for the hyperfine interactions produced by the electronic states between which the ion was relaxing. In this situation, the effect of the relaxation, whether of spin-spin or of spin-lattice origin, is to produce fluctuating effective fields (either magnetic dipolar and/or electric quadrupolar) on the nucleus. Using a stochastic model to describe the fluctuations, Blume and Tjon (111) calculated Mossbauer spectra for several simple but useful cases. Price et al. (23) used these results to determine Fe^{2+} relaxation rates for cases of magnetic relaxation in the presence of a non-fluctuating quadrupole interaction, and Leider and Pipkorn (21) also used this model to describe the relaxation between states of different quadrupole interactions (produced by strains) for Fe^{2+} in MgO.

If, however, electronic states for which the effective field approximation for the hyperfine interactions is not valid are involved in the relaxation, the problem of evaluating the Mossbauer spectra is much more complicated (112-118)*.

Expressions for the rates of relaxation resulting from the various first- and second-order spin-lattice processes (72, 92, 4, 23, 119) depend on the same matrix elements of the (first-order) orbit-lattice Hamiltonian that were considered in connection with the vibronic admixture effects above (e.g. Eqn. 37).

A point of interest at this stage concerns Orbach's resonance relaxation process (72,92), and the technique used for evaluating its transition rates. Consider an ion in an initial state $|i\rangle$ that undergoes a relaxation transition to the final state $|f\rangle$ via an intermediate excited state $|m\rangle$. For an Orbach process, the excitation energies of $|m\rangle$ from both $|i\rangle$ and $|f\rangle$ (i.e. Δ_{im},Δ_{mf}) are less than the maximum phonon energy $\hbar\omega_{max}$. The interesting feature is that the energy denominators in the second-order perturbation expression for the relaxation rate, which have the form $\Delta_{im} - \hbar\omega(\vec{k},s)$ will go to zero for one particular phonon frequency and will thus produce a divergence in the relaxation rate. Orbach showed that this apparent divergence could be treated by analogy with the resonance fluorescence problem in optical spectroscopy by including the finite energy width of the intermediate state that is a consequence of its finite lifetime. Therefore if this width is Γ_m, corresponding to a mean lifetime of $\hbar/\Gamma_m$, then the energy denominators take the form $(\Delta_{im} - \hbar\omega(\vec{k},s) + \frac{1}{2}\, i\Gamma_m)$, and this still produces a large resonant contribution to the relaxation rate if Γ_m is sufficiently small. Γ_m can be estimated if the modes of decay of the state $|m\rangle$ are known (see Ref. 72, 92, 23) where it was assumed that spin-lattice processes were the dominant decay mechanisms.

---

*See Chapter 1 for further details.

3.2.5 The Possibility of Resonant Vibronic Coupling

This resonant relaxation process suggests the interesting possibility of resonant effects in the vibronic admixture, as has been pointed out (19), when electronic levels are separated by less than the maximum phonon energy.

The possibility of divergence of the "phonon" factors $C^{i}_{jj'}(T)$ (Eqn. 42) was pointed out in Section 3.2.3, but was not considered further at that stage. However, it was noted in Section 3.2.4 that the condition for such a divergence (i.e. $E_i - E_j \approx \hbar\omega(\vec{k},s)$) was just the condition under which the time-independent perturbation theory might be expected to be invalid and thus render the discussion of admixture meaningless. But this is not necessarily so. If the transition rate between the states ψ_i, ψ_j induced by H_{OL} is sufficiently small (i.e. if the states have sufficiently long life-times) then they may be treated approximately as stationary states and admixture between them meaningfully discussed. Then, under these conditions, a complex energy term $\frac{1}{2}i\Gamma$ may be introduced into the energy denominators in (Eqn. 42) and the divergence removed, precisely as was done by Orbach (92).

The feature that makes such considerations useful is that this condition (i.e. large $\hbar/\Gamma$, or small Γ) is just the condition that such a resonant effect will be large, i.e. that there will be a near-divergence.

These arguments are clearly very qualitative, and so far no unambiguous evidence for such a resonant effect has been found. In order for the width Γ to be small, the electronic matrix elements $\langle\psi_i|O_n(\Gamma,p)|\psi_j\rangle$ must be very small, as must the density of phonon states of energy $\hbar\omega = |E_i - E_j|$, and only in fairly unusual circumstances will these conditions be simultaneously satisfied. One particular case, that of the Ising-like antiferromagnet $FeCO_3$, was considered in this context by Price (19) and it was found by (at best) semi-quantitative arguments that such a resonant admixture may have been observed in the Fe^{2+} Mossbauer spectra of this compound in the region of the Neel temperature.

3.3 The Superposition Model of the Crystal Field

The calculations described so far in this Section show that the shape and temperature dependence of a $^{57}Fe^{2+}$ Mossbauer spectrum depends sensitively on the symmetry and magnitude of the crystal field. While the symmetry of the static field is often known, at least approximately, from structural considerations, the magnitudes of the non-zero parameters of the crystal field often are not - their determination is frequently the reason for doing the Mossbauer experiment. However, there are often too many unknown parameters involved to allow their unambiguous determination from a set of Mossbauer data, particularly if effects of the dynamic crystal field are considered. Hazony and Ok (120) gave an excellent illustration of possible ambiguity in determining crystal field parameters from Mossbauer data in the relatively simple case of $FeCl_2$, while the work of Bacci (17) showed clearly the similarity of some static and dynamic crystal field effects. It is therefore a considerable advantage to have some independent estimates of the crystal field parameters before one attempts to interpret a Mossbauer spectrum.

A model that relates the crystal field parameters to the structural properties of the Fe^{2+} ion site may provide such estimates if independent experimental data do not exist, as well as indicating relationships between parameters (both static and dynamic). There are, of course, far more fundamental reasons for the development of models of the crystal field, but this is the reason that the superposition model (SPM) of the crystal field, that was introduced in Section 1, will be described briefly here. Further details may be found in references (9, 13, 121-123).

3.3.1 The Static Crystal Field

The fundamental assumption of the SPM is that the total crystal field can be constructed from a sum of axially symmetric contributions from surrounding ions, that depend on their distance from the paramagnetic ion. In this respect it is a generalization of the point charge electrostatic model (27) to include the large, and often dominant, contributions from overlap and covalency effects.

It is also usually assumed that only the contributions from coordinated ions (ligands) need to be included, and while this appears to be a good approximation for lanthanide crystal fields it may not be so good for iron-group ions particularly when the n=2 terms are important. However in this case a hybrid model can be used, such as was done successfully by Edgar (123), in which long-range electrostatic contributions can be calculated on the assumption that more distant ions can be treated as point charges.

If, for the moment, consideration is given only to the ligand contributions, then the crystal field parameters A_n^m (Eqn. 23) are expressed as:

$$A_n^m = \sum_i K_n^m(\theta_i,\phi_i)\bar{A}_n(R_i) \quad , \tag{43}$$

where the sum is over the ligands at $R_i = (R_i\theta_i\phi_i)$ relative to the paramagnetic ion and the $\bar{A}_n(R_i)$ are intrinsic parameters describing the axially symmetric crystal field due to a ligand at distance R_i. The "coordination factors" $K_n^m(\theta_i\phi_i)$ are functions of the angular positions of the ligands, and they are evaluated from X-ray structural data. A list of these functions has been given by Newman and Urban (13).

It is usual (though not necessary) to assume that the function $\bar{A}_n(R)$ can be approximated by a power law dependence on R, at least over a limited range of R, i.e.

$$\bar{A}_n(R) = \bar{A}_n(R_0)(R_0/R)^{t_n}. \tag{44}$$

Then $\bar{A}_n(R)$ can be specified in terms of the two parameters $\bar{A}_n=\bar{A}_n(R_0)$ and t_n.

This model provides a basis for the comparison of crystal field parameters obtained for different crystals, at least when the ligand species is the same, or for the prediction of the parameters for new systems at least to the level of accuracy required for preliminary calculation of the Mossbauer spectrum. The papers of Newman et.al. (121), Price (18) and Edgar (123) should be consulted for further details of the application of the model to 3d ion crystal fields.

3.3.2 The Dynamic Crystal Field

A particularly useful feature of the SPM is that it can be simply extended to describe the dynamic crystal field. The formalism for this, at least for the case in which only the ligands make a significant contribution (which is just the "quasi-molecular cluster" approach introduced in Section 2.3.1) has been outlined by Newman (122) and was employed to estimate the orbit-lattice coupling parameters for Fe^{2+} in $FeCO_3$ by Price (18).

Briefly, the dynamic crystal field parameters are expressed in terms of the intrinsic crystal field parameters $\bar{A}_n(R)$ and $d\bar{A}_n(R)/dR$ by "dynamic coordination factors", which are partial derivatives of the $K_n^m(\theta_i,\phi_i)$. The papers mentioned above should be consulted for further details of the formalism.

This application of the model is far from being an exact

procedure: its reliability has not been fully tested even for the more favourable case of lanthanide crystal fields (122b), and, as in the static case, it is less likely to be a good approximation for the poorly screened open-shell electrons in 3d ions. However there is relatively little data on orbit-lattice coupling coefficients for these ions and the SPM should at least provide values of sufficient accuracy to determine, for example, whether or not vibronic coupling effects are likely to be important.

4. PAST TRENDS AND POSSIBLE FURTURE DEVELOPMENTS

A very large number of investigations that have involved the evaluation of crystal field effects in the Mossbauer spectra of Fe^{2+} ions in insulating crystals have been carried out, ever since the first applications of the technique to problems in solid state chemistry and physics in the early 1960's. Many of these studies have not been aimed directly at determining the crystal field, but at investigating magnetic properties, observing phase transitions, etc. which either depend directly on the crystal field and/or rely on crystal field effects to make the necessary observations possible. Varret (15) has surveyed some of the more recent work that falls into this category. His paper indicates that this application of the Mossbauer effect has been of great importance in solid state studies, and it will undoubtedly continue to be.

There has also been a great deal of attention devoted directly to determining both the symmetry and strength of the crystal fields acting on Fe^{2+} ions, based mostly on measurements of the sign, magnitude and temperature dependence of the ^{57}Fe quadrupole splitting, either in compounds that do not order magnetically or in magnetic materials above their transition temperature. The Mossbauer technique has been particularly attractive for these studies both because of its relative simplicity as an experimental tool and because of difficulties in obtaining this sort of information for Fe^{2+} compounds from, for example, EPR, infra-red and optical absorption spectra. This line of research appeared to be realizing its ultimate potential with the determination of not only the magnitude but also the temperature dependence of crystal field parameters in FeF_2 (37). These measurements were also used to determine the quadrupole splitting ΔE_0 resulting from a single 3d electron (36) and thence to deduce 3d charge density distributions for Fe^{2+} ions in various compounds (124).

However, careful measurements on a number of compounds [e.g. $(NMe_4)FeCl_4$ (89, 125), $FeCO_3$ (126), $FeSiF_6.6H_2O$ (49)] revealed features that could not be reasonably explained on the basis of temperature-dependent static crystal field theory. Such measurements and others have led to the consideration of both dynamic and vibronic effects of the dynamic crystal field and present indications, however qualitative they may be, are that these might be of much more general relevance than had been thought. Recognition of the importance (or at least the potential importance) of these relatively small effects makes it very difficult to determine static crystal field splittings from Mossbauer spectra with anything like the accuracy hitherto thought possible. While this may be viewed pessimistically as a limitation on the usefulness of the Mossbauer technique in this particular area, it is equally true that these developments may open interesting new areas of research to the Mossbauer spectroscopist. It seems likely that more and more attention will be devoted to the observation and evaluation of relatively fine features of the spectra that nevertheless reflect important physical processes taking place in the solid. Such investigations will, of course, demand carefully planned experiments and meticulous measurement of the hyperfine interaction parameters and absorption or emission linewidths and shapes, both as functions of applied perturbations (e.g. magnetic field, stress) and of temperature. Reliable interpretation of these measurements will, to an increasing extent, require consideration of independent data, such as temperature-dependent structural data, magnetic susceptibility, anisotropy etc. Direct measurements of time-independent effects using SEDM (101) (or Mossbauer fluorescence) and/or delayed coincidence techniques will also undoubtedly be used more widely.

Therefore, emphasis in this area of research may gradually shift from using the Mossbauer effect largely as a tool to obtain crystal chemical data for use in the interpretation of other measurements, to a more fundamental role, in conjunction with other measurements, in probing the characteristics of the dynamics of paramagnetic ions in crystal as well as the effects of small static (or pseudo-static) interactions such as random strains.

ACKNOWLEDGEMENTS

We are grateful to Prof. M. Bacci for permitting us to reproduce the diagram from his paper (17a) that appear as Figs. 16, 17 and 18 here, and for supplying us with these diagrams. DCP is pleased to acknowledge the help and encouragement in this work received over a number of years from D.J. Newman, C.E. Johnson, I. Maartense and many others. FV is indebted to F. Hartmann-Boutron for initiating him to the crystal field problems.

REFERENCES

1 H. Bethe, Ann. Physik, 3 (1929) 135.
2a W. Low, Paramagnetic Resonance in Solids, Supplement 2, Solid State Physics, F. Seitz and D. Turnbull (Eds.), Academic, New York, 1960.
2b S. Sugano, Y. Tanabe and H. Kamimura, Multiplets of Transition Metal Ions in Crystals, Academic, New York, 1970.
3 J.S. Griffith, The Theory of Transition Metal Ions, C.U.P., Cambridge, 1961.
4 A. Abragam and B. Bleaney, Electron Paramagnetic Resonance, Clarendon, Oxford, 1970.
5a J.B. Goodenough, Magnetism and the Chemical Bond, Interscience, New York, 1963.
5b J. Kanamori in G.T. Rado and H. Suhl (Eds.), Magnetism, Vol. I, Academic Press, New York, 1963.
5c F.E. Mabbs and D.J. Machin, Magnetism and Transition Metal Complexes, Chapman and Hall, London, 1973.
6a A. Abragam and F. Boutron, C. Rend, Hebd. Sean. Acad. Sci., 252 (1961) 2404.
6b C.E. Johnson, W. Marshall and G.J. Perlow, Phys. Rev., 126 (1962) 1503.
7a A. Okiji and J. Kanamori, J. Phys. Soc. Japan, 19 (1964) 908.
7b K. Ono and A. Ito, J. Phys. Soc. Japan 19 (1964) 899.
8 R. Ingalls, Phys. Rev., 133A (1964) 787.
9 D.J. Newman, Adv. Phys., 20 (1971) 197.
10 P. Gutlich, R. Link and A. Trautwein, Mossbauer Spectroscopy and Transition Metal Chemistry, Springer-Verlag, Berlin, 1978.
11 C.J. Ballhausen, Introduction to Ligand Field Theory, McGraw-Hill, New York, 1962.
12a C.K. Jorgensen, R. Pappalardo and H.H. Smidtke, J. Chem. Phys., 33 (1963) 1422.
12b C.E. Schaffer and C.K. Jorgensen, Mol. Phys., 9 (1965) 401.
12c C.E. Schaffer and C.K. Jorgensen, Mat. -fys. Medd. Seist. 34, no. 13 (1965).
12d C.E. Schaffer, Structure and Bonding, 5 (1968) 68.
13 D.J. Newman and W. Urban, Adv. Phys., 24 (1975) 793.
14 F. Varret and F. Hartmann-Boutron, Ann. Phys., 3 (1968) 157.
15 F. Varret, J. Phys. (Paris) Colloq., 37 (1976) C6-437.
16a G. Garcin, P. Imbert and G. Jehanno, Solid State Commun., 21 (1977) 545.
16b P. Imbert, G. Garcin, G. Jehanno and A. Gerard in D. Barb and D. Tarina (Eds.), Proc. Int. Conf. Mossbauer Spectroscopy, Bucharest, 1977, Central Institute of Physics, Bucharest, 1977, Vol. 2, p. 123.

16c G. Garcin, A. Gerard, P. Imbert and G. Jehanno, J. Phys. (Paris) Colloq., 40 (1979) C2-413.

17a M. Bacci, J. Chem. Phys. 68 (1978) 4907.

17b M. Bacci, Chem. Phys. Letts., 48 (1977) 184.

18 D.C. Price, Aust. J. Phys. 31 (1978) 397.

19 D.C. Price, J. Phys. (Paris) Colloq., 40 (1979) C2-316.

20 F.S. Ham, Phys. Rev., 160 (1967) 328.

21a D.N. Pipkorn and H.R. Leider, Bull. Amer. Phys. Soc. 11 (1966) 49.

21b H.R. Leider and D.N. Pipkorn, Phys. Rev., 165 (1968) 494.

21c J. Chappert, R.B. Frankel, A. Misetich and N.A. Blum, Phys. Rev. 179 (1969) 578.

22 R.B. Frankel, J. Chappert, J.R. Regnard, A. Misetich and C.R. Abeldo, Phys. Rev. B, 5 (1972) 2469.

23 D.C. Price, C.E. Johnson and I. Maartense, J. Phys. C: Solid State Phys., 10 (1977) 4843.

24 W.M. Reiff, C. Nicolini and B. Dockum, J. Phys. (Paris) Colloq., 40 (1979) C2-230.

25 E.U. Condon and G.H. Shortley, Theory of Atomic Spectra, C.U.P., Cambridge, 1951.

26 N.N. Greenwood and T.C. Gibb, Mossbauer Spectroscopy, Chapman and Hall, London, 1971.

27 M.T. Hutchings, Solid State Physics, 16 (1964) 227.

28a K.W.H. Stevens, Proc. Phys. Soc., A65 (1952) 209.

28b B. Bleaney and K.W.H. Stevens, Rep. Progr. Phys., 16 (1953) 108.

29 J.C. Travis in L. May (Ed.), An Introduction to Mossbauer Spectroscopy, Plenum, New York, 1971.

30 P. Zory, Phys. Rev., 140 (1965) A1401.

31 T.C. Gibb, J. Chem. Soc. (London), Dalton Trans., 1978, p. 743.

32a A.J. Freeman and R.E. Watson, Phys. Rev., 127 (1962) 2058.

32b R. Ingalls, Phys. Rev., 128 (1962) 1155.

33 J.O. Artman, A.H. Muir and H. Wiedersich, Phys. Rev., 173 (1968) 337.

34 M. Eibschutz, G.R. Davidson and H.J. Guggenheim, Phys. Rev. B, 9 (1974) 3885.

35a M.G. Clark, M.G. Bancroft and A.J. Stone, J. Chem. Phys., 47 (1967) 4250.

35b A. Trautwein, E. Kreber, U. Gonser and F.E. Harris, J. Phys. Chem. Solids, 36 (1975) 325.

36 B.W. Dale, J. Phys. C, 4 (1971) 2705.

37 H.K. Perkins and Y. Hazony, Phys. Rev. B, 5 (1972) 7.

38a J.P. Sanchez, L. Asch and J.M. Friedt, Chem. Phys. Letts., 18 (1973) 250.

38b W.M. Reiff, R.B. Frankel, B.F. Little and G.J. Long, Chem. Phys. Letts., 28 (1974) 68.

39 A. Gerard, P. Imbert, H. Prange, F. Varret and M. Wintenberger, J. Phys. Chem. Solids, 32 (1971) 2091.

40 F. Hartmann-Boutron and P. Imbert, J. Appl. Phys., 39 (1968) 775.

41 P. Imbert, G. Jehanno, Y. Macheteau and F. Varret, J. Phys. (Paris), 37 (1976) 969.

42 U. Ganiel, Chem. Phys. Letts., 4 (1969) 87.

43 P. Imbert, F. Varret and M. Wintenberger, J. Phys. Chem. Solids, 34 (1973) 1675.

44 M.H.L. Pryce, Phys. Rev., 80 (1950) 1107.

45 R.E. Trees, Phys. Rev., 82 (1951) 683.

46a M.J. Clauser, E. Kankeleit and R.L. Mossbauer, Phys. Rev. Letts., 17 (1966) 5.

46b M.J. Clauser and R.L. Mossbauer, Phys. Rev., 178 (1969) 559.

47a B.D. Howes, D.C. Price and D. Mackey, J. Phys. (Paris) Colloq., 40 (1979) C2-286.

47b B.D. Howes, Ph.D Thesis, Australian National University, 1979 (unpublished).

48 K. Ono, A. Ito and T. Fujita, J. Phys. Soc. Japan, 19 (1964) 2119.

49 F. Varret and G. Jehanno, J. Phys. (Paris), 36 (1975) 415.

50 F. Varret, J. Phys. Chem. Solids, 37 (1976) 257.

51 D. Palumbo, Nuovo Cimento, 8 (1958) 271.

52a T. Fujita, A. Ito and K. Ono, J. Phys. Soc. Japan, 21 (1966) 1734.

52b T. Fujita, A. Ito and K. Ono, J. Phys. Soc. Japan, 27 (1969) 1143.

53 H.N. Ok, Phys. Rev., 181 (1969) 563.

54 D.C. Price, I. Maartense and A.H. Morrish, Phys. Rev. B, 9 (1974) 281.

55 H.H. Wickman and G.K. Wertheim in V. Goldanskii and R. Herber (Eds.), Chemical Applications of Mossbauer Spectroscopy, Academic Press, New York, 1968.

56 W. Marshall and C.E. Johnson, J. Phys. Radium, 23 (1962) 733.

57 D.J. Simkin, Phys. Rev., 177 (1969) 1008.

58 H.N. Ok, Phys. Rev., 185 (1969) 472.

59 H. Spiering, R. Zimmermann and G. Ritter, Phys. Status Solidi (b) 62 (1974) 124.

60 F. Varret, J. Phys. Chem. Solids, 37 (1976) 265.

61a G. Lang, J. Chem. Soc. (A), 1971, p. 3245.

61b G. Lang and B.W. Dale, Nucl. Instrum. Methods, 116 (1974) 567.

62 J.D. Siegwarth, Phys. Rev., 155 (1967) 285.

63a G.R. Davidson, M. Eibschutz and H.J. Guggenheim, Phys. Rev. B, 8 (1973) 1864.

63b A. Ito and S. Morimoto, J. Phys. Soc. Japan, 39 (1975) 884.

63c S. Morimoto and A. Ito, Jap. J. Appl. Phys., 16 (1977) 85.

64 A.M. Van Diepen and R.P. Van Stapele, Phys. Rev. B, 5 (1972) 2462.

65 H. Oudet, Thèse 3ème cycle, Paris, 1978.

66 B.R. Cooper and O. Vogt, J. Phys. (Paris) Colloq., 32 (1971) C1-958.

67 L.I. Schiff, Quantum Mechanics, 3rd edn., McGraw-Hill, New York, 1968, p. 279-92.

68a D.E. McCumber and M.D. Sturge, J. Appl. Phys., 34 (1963) 1682.

68b G.F. Imbusch, W.M. Yen, A.L. Schawlow, D.E. McCumber and M.D. Sturge, Phys. Rev., 133 (1964) A1029.

68c B. Di Bartolo and R.C. Powell, Phonons and Resonances in Solids Wiley-Interscience, New York, 1976.

69 A.A. Manenkov and R. Orbach, (Eds.), Spin-Lattice Relaxation in Ionic Solids, Harper and Row, New York, 1966.

70a J.H. Van Vleck, J. Chem. Phys., 7 (1939) 72.

70b J.H. Van Vleck, Phys. Rev., 57 (1940) 426.

71 M.D. Sturge, Solid State Physics, 20 (1967) 91.

72 R. Orbach and H.J. Stapleton, in S. Geschwind (Ed.), Electron Paramagnetic Resonance, Plenum, New York, 1972.

73 K.N. Shrivastava, Physics Reports, 20 (1975) 137.

74a K.W.H. Stevens, J. Phys. C: Solid State Phys. 2 (1969) 1934.

74b M. Abou-Ghantous, C.A. Bates, P.E. Chandler and K.W.H. Stevens, J. Phys. C: Solid State Phys., 7 (1974) 309.

74c M. Abou-Ghantous, C.A. Bates and K.W.H. Stevens, J. Phys. C: Solid State Phys., 7 (1974) 325.

75 F.S. Ham in S. Geschwind (Ed.), Electron Paramagnetic Resonance, Plenum, New York, 1972.

76 F.S. Ham, Phys. Rev., 138 (1965) A1727.

77 F.S. Ham, W.M. Schwartz and M.C.M. O'Brien, Phys. Rev., 185 (1968) 548.

78a J.R. Regnard, Solid State Commun., 12 (1973) 207.

78b T. Ray, J.R. Regnard, J.M. Laurant and A. Ribeyron, Solid State Commun., 15 (1973) 1959.

78c J.R. Regnard and J. Chappert, Solid State Commun., 15 (1974) 1539.

78d J.R. Regnard in A. Hrynkiewicz and J. Sawicki (Eds.), Proc. Int. Conf. Mossbauer Spectroscopy, Cracow, 1975, Vol. 1, p. 409.

78e T. Ray and J.R. Regnard, Phys. Rev. B, 14 (1976) 1796.
78f J.R. Regnard and T. Ray, Phys. Rev. B, 14 (1976) 1805.
79 C.A. Bates and P. Steggles, J. Phys. C, 8 (1975) 2283.
80 E.M. Ganapol'skii, Fiz. Tverd. Tela, 17 (1975) 67. [Sov. Phys. - Solid State, 17 (1975) 37.]
81a M.R. Spender and A.H. Morrish, Can. J. Phys., 50 (1972) 1125.
81b M.R. Spender and A.H. Morrish, Solid State Commun., 11 (1972) 1417
82a A.M. Van Diepen and R.P. Van Stapele, Solid State Commun., 13 (1973) 1651.
82b A.M. Van Diepen, F.K. Lotgering and J.F. Olijhoek, J. Magn. Magn. Mater., 3 (1976) 117.
83 M.R. Spender and A.H. Morrish, Can. J. Phys., 49 (1971) 2659.
84 S. Wittekoek, R.P. Van Stapele and A.W.J. Wijma, Phys. Rev. B, 7 (1973) 1667.
85 F.S. Ham and G.A. Slack, Phys. Rev. B, 4 (1971) 777.
86 F.S. Ham, J. Phys. (Paris) Colloq., 35 (1974) C6-121.
87 G. Lang, Quart. Rev. Biophys., 3 (1970) 1.
88 L. Cianchi, M. Mancini and G. Spina, Lett. Nuovo Cimento, 16 (1976) 505.
89 T.C. Gibb, N.N. Greenwood and M.D. Sastry, J. Chem. Soc. (London), Dalton Trans., 1972, p. 1947.
90 A. Abragam, The Principles of Nuclear Magnetism, Clarendon, Oxford, 1961.
91 K.W.H. Stevens, Rep. Progr. Phys., 30 (1967) 189.
92 R. Orbach, Proc. Roy. Soc. (London), A264 (1961) 458.
93 C.B.P. Finn, R. Orbach and W.P. Wolf, Proc. Phys. Soc., 77 (1961) 261.
94a D.C. Price and K.K.P. Srivastava, J. Phys. (Paris) Colloq., 37 (1976) C6-123.
94b K.K.P. Srivastava, Ph.D Thesis, Australian National University, 1976 (unpublished).
95a J.R. Sams and T.B. Tsin, J. Chem. Phys., 62 (1975) 734.
95b J.R. Sams and T.B. Tsin, Chem. Phys., 15 (1976) 209.
96 R. Zimmermann, H. Spiering and G. Ritter, Chem. Phys., 4 (1974) 133.
97a R. Zimmermann, G. Ritter, H. Spiering and D.L. Nagy, J. Phys. (Paris) Colloq., 35 (1974) C6-439.
97b R. Zimmermann, G. Ritter and H. Spiering in A. Hrynkiewicz and J. Sawicki (Eds.), Proc. Int. Conf. Mossbauer Spectroscopy, Cracow, 1975, Vol. I, p. 401.
98 D.C. Price, E.R. Vance, G. Smith, A. Edgar and B.L. Dickson, J. Phys. (Paris) Colloq., 37 (1976) C6-811.
99 F. Varret, unpublished results.
100 J. Chappert, G. Jehanno and F. Varret, J. Phys. (Paris), 38 (1977) 411.
101a B. Balko and G.R. Hoy, Phys. Rev. B, 10 (1974) 36.
101b B. Balko, E.V. Mielczarek and R.L. Berger, J. Phys. (Paris) Colloq. 40 (1979) C2-17.
102a J.W. Leech and D.J. Newman, How to Use Groups, Methuen, London, 1969.
102b V. Heine, Group Theory in Quantum Mechanics, Pergamon, New York, 1960.
102c M. Tinkham, Group Theory and Quantum Mechanics, McGraw-Hill, New York, 1964.
103 F. Varret, Thèse, Orsay, 1972.
104 W. Kundig, Nucl. Instrum. Methods, 48 (1967) 219.
105 E.R. Seidel, G. Kaindl, M.J. Clauser and R.L. Mossbauer, Phys. Lett., 25A (1967) 328.
106 J. Van Dongen Torman, R. Jagannathan and J.M. Trooster, Hyperfine Interact., 1 (1975) 135.

107 G. Herzberg, Electronic Spectra and Electronic Structure of Polyatomic Molecules, Van Nostrand, Princeton, 1966.
108 R. Orbach and M. Tachiki, Phys. Rev., 158 (1967) 524.
109 R.D. Mattuck and M.W.P. Strandberg, Phys. Rev., 119 (1966) 1204.
110 M.M. Curtis, D.J. Newman and G.E. Stedman, J. Chem. Phys., 50 (1969) 1077.
111a M. Blume and J.A. Tjon, Phys. Rev., 165 (1968) 446.
111b J.A. Tjon and M. Blume, Phys. Rev., 165 (1968) 456.
112a L.L. Hirst, J. Phys. Chem. Solids, 31 (1970) 655.
112b L.L. Hirst, J. Phys. (Paris) Colloq., 35 (1974) C6-21.
113a F. Gonzalez-Jimenez, P. Imbert and F. Hartmann-Boutron, Phys. Rev., B9 (1974) 95.
113b F. Hartmann-Boutron and D. Spanjaard, J. Phys. (Paris), 36 (1975) 307.
113c F. Hartmann-Boutron, J. Phys. (Paris), 37 (1976) 537.
113d F. Hartmann-Boutron, J. Phys. (Paris), 37 (1976) 549.
113e C. Chopin, D. Spanjaard and F. Hartmann-Boutron, J. Phys. (Paris) Colloq., 37 (1976) C6-73.
114 F. Hartmann-Boutron, Ann. Phys., 9 (1975) 285.
115a M.J. Clauser and M. Blume, Phys. Rev. B, 3 (1971) 583.
115b J. Sivardiere, M. Blume and M.J. Clauser, Hyperfine Interact. 1 (1975) 227.
116a G.K. Shenoy and B.D. Dunlap in A. Hrynkiewicz and J. Sawicki (Eds.) Proc. Int. Conf. Mossbauer Spectroscopy, Cracow, 1975, Vol. 2, p. 275.
116b G.K. Shenoy and B.D. Dunlap, Phys. Rev. B, 13 (1976) 1353.
117 D.C. Price, J. Phys. E: Sci. Instrum., 9 (1976) 336.
118 H.H.F. Wegener, in A. Hrynkiewicz and J. Sawicki (Eds.), Proc. Int. Conf. Mossbauer Spectroscopy, Cracow, 1975, Vol. 2, p. 257.
119 J.M. Baker, J. Phys. C: Solid State Phys., 4 (1971) 930.
120 Y. Hazony and H.N. Ok, Phys. Rev., 188 (1969) 591.
121 D.J. Newman, D.C. Price and W.A. Runciman, Amer. Mineral., 63 (1978) 1278.
122a D.J. Newman, Aust. J. Phys., 31 (1978) 79.
122b D.J. Newman, Aust. J. Phys., (to be published) (1980).
123 A. Edgar, J. Phys. C: Solid State Phys., 9 (1976) 4303.
124a Y. Hazony, Phys. Rev. B, 3 (1971) 711.
124b Y. Hazony in I.J. Gruverman (Ed.), Mossbauer Effect Methodology, Vol. 7, Plenum, New York, 1971, p. 147.
124c Y. Hazony, J. Phys. C: Solid State Phys., 5 (1972) 2267.
125 P.R. Edwards, C.E. Johnson and R.J.P. Williams, J. Chem. Phys., 47 (1967) 2074.
126a D.C. Price, C.E. Johnson and I. Maartense, Paper presented at the Autumn Meeting of the Chemical Society, U.K. (1973). Deposited in library of Zentralstelle fur Atomkernenergie-Dokumentation (ZAED).
126b D.L. Nagy, I. Deszi and U. Gonser, Neues Jahrb. Mineral. Monatsh. 3 (1975) 101.
127 H. Eicher, D. Bade, F. Parak, J. Chem. Phys., 64 (1976) 1446.
128 Y. Alpert and R. Banerjee, Biochim. Biophys. Acta, 405 (1975) 144.

CHAPTER 7

CALCULATION OF CHARGE DENSITY, ELECTRIC FIELD GRADIENT AND INTERNAL MAGNETIC FIELD AT THE NUCLEAR SITE USING MOLECULAR ORBITAL CLUSTER THEORY

V.R. Marathe* and A. Trautwein**

Fachbereich Angewandte Physik, Universitat des Saarlandes, 6600 Saarbrucken 11, West Germany.

1. INTRODUCTION

The mutual influence of theory and experiments upon each other has often led to fruitful results. The feedback gained by theoretical electronic structure studies, from spectroscopic measurements, helps to test the reliability of the various approximations usually involved in the calculations. On the other hand improved spectroscopic measurements may be designed on the basis of theoretical results. This interaction may finally lead to the derivation of a reliable electronic structure.

From the various types of electronic structure calculations (ligand field, molecular orbital and band structure calculations) and experimental tools (esr, Mossbauer, nmr, optical, photoelectron, susceptibility, x-ray and neutron diffraction spectroscopy) the molecular orbital (MO) and the Mossbauer method have been selected to describe the extent to which the theory can help in the interpretation of experimental data.

In Section 2 we give a short introduction to MO theory and its various approximations. The following sections describe how Mossbauer parameters are derived from the electronic structure of a compound: Section 3 deals with isomer shift, Section 4 with electric field gradient, and Section 5 with magnetic hyperfine interaction.

2. THEORETICAL ASPECTS OF MOLECULAR ORBITAL CALCULATIONS

2.1 General

The molecular orbital (MO) method is extention of the Bohr theory of electronic configuration, from atoms to molecules, where each electron is assigned a one-electron wavefunction or molecular orbital which

---

Present address

* Tata Institute of Fundamental Research, Bombay-400 004, INDIA

** Institute fur Physik, Medizinische Hochschule Lubeck, 2400 Lubeck 1, West Germany.

is a quantum mechanical analog of the atomic orbit. MO theory has its origin in the works of Hund (1), Mulliken (2), Lennard-Jones (3) and Slater (4). A rigorous mathematical framework for the MO method has been given by Roothaan (5).

If there are N nuclei and n electrons in a molecular system, the many-particle Hamiltonian operator H_{total} is given, in atomic units, as

$$H_{total} = -\sum_{k=1}^{N} \frac{1}{2M_k} \nabla_k^2 + \sum_{k>k'}^{N} \frac{Z_k Z_{k'}}{|\vec{R}_k - \vec{R}_{k'}|} - \sum_{i=1}^{n} \frac{1}{2} \nabla_i^2 - \sum_{k=1}^{N} \sum_{i=1}^{n} \frac{Z_k}{|\vec{R}_k - \vec{r}_i|} + \sum_{i>j}^{n} \frac{1}{|\vec{r}_i - \vec{r}_j|} \qquad (1)$$

where M_k, Z_k and $\vec{R}_k$ are the mass, the charge and the position vector respectively of nucleus k, and $\vec{r}_i$ is the position vector of electron i. Summations involving indices k and k' are over atomic nuclei, and those involving i and j are over electrons. The Schrodinger equation for the entire system is then

$$H_{total}\, \Psi = E\Psi, \qquad (2)$$

where Ψ is a complete wavefunction for all particles in the molecule, and E is the total energy of the system.

As the masses of the nuclei are several thousand times larger than the mass of the electrons, they move much more slowly compared to electrons. Therefore it is reasonable to assume that the electrons adjust themselves to new nuclear positions so rapidly, that at any instant the electron motion is just as it would be if the nuclei were at rest at the position they occupy at that instant. This is known as Born-Oppenheimer approximation. We can then consider only the part of the Hamiltonian which depends on the position but not on the momenta of the nuclei. This is the electronic Hamiltonian operator H_{el} given as,

$$H_{el} = \sum_{i=1}^{N} h_i + \sum_{i>j}^{n} \frac{1}{|\vec{r}_i - \vec{r}_j|}, \quad \text{where} \qquad (3)$$

$$h = -\frac{1}{2}\nabla^2 - \sum_{k=1}^{N} \frac{1}{|\vec{R}_k - \vec{r}|} \qquad (4)$$

The modified Schrodinger equation is then,

$$H_{el}\, \Psi_{el} = E_{el}\, \Psi_{el}, \qquad (5)$$

where E_{el} is the electronic energy, and Ψ_{el} is a wavefunction for all electrons in the field of fixed nuclei. The total energy of the system for a given set of nuclear positions is,

$$E = E_{el} + \sum_{k>k'}^{N} \frac{Z_k Z_{k'}}{|\vec{R}_k - \vec{R}_{k'}|} . \tag{6}$$

For an n-electron system, it is simple to associate the n electrons with an orthonormal set of n one-electron functions $\tilde{\Psi}_1, \tilde{\Psi}_2, \ldots \tilde{\Psi}_n$, called spin orbitals. Each of these functions is a product of a spatial function and a spin function,

$$\tilde{\Psi}_i(\vec{r},\vec{s}) = \Psi_i(\vec{r}) X_i(\vec{s}) . \tag{7}$$

The total n-electron wavefunction is now built up as an anti-symmetrized product of molecular spin orbitals,

$$\Psi_{el} = (n!)^{-\frac{1}{2}} \begin{vmatrix} \tilde{\Psi}_1(1) & \tilde{\Psi}_2(1) & \cdots & \tilde{\Psi}_n(1) \\ \tilde{\Psi}_1(2) & \tilde{\Psi}_2(2) & \cdots & \tilde{\Psi}_n(2) \\ \vdots & & & \vdots \\ \tilde{\Psi}_1(n) & \tilde{\Psi}_2(n) & \cdots & \tilde{\Psi}_n(n) \end{vmatrix} . \tag{8}$$

Such a determinant of spin orbitals, known as Slater determinant, is the simplest normalized wavefunction which satisfies the antisymmetry principle.

Substituting the wavefunction, Eqn. (8), into the expression for energy, Eqn. (5), we get

$$E_{el} = \int \Psi_{el}^* H_{el} \Psi_{el} \, d\tau \tag{9}$$

$$= \sum_{i=1}^{n} H_i + \frac{1}{2} \sum_{i,j=1}^{n} J_{ij} - \frac{1}{2} \sum_{i,j=1}^{n} K_{ij} , \tag{10}$$

where the orbital energies H_i, the coulomb integrals J_{ij} and the exchange integrals K_{ij} are defined by

$$H_i = \int \Psi_i^*(\vec{r}) \; h \; \Psi_i(\vec{r}) \, d\vec{r} \tag{11}$$

$$J_{ij} = \int \frac{\Psi_i^*(\vec{r}_1) \; \Psi_i(\vec{r}_1) \; \Psi_j^*(\vec{r}_2) \; \Psi_j(\vec{r}_2)}{|\vec{r}_1 - \vec{r}_2|} \, d\vec{r}_1 d\vec{r}_2 , \quad \text{and} \tag{12}$$

$$K_{ij} = \delta(m_{s_i}, m_{s_j}) \int \frac{\Psi_j^*(\vec{r}_1) \; \Psi_j(\vec{r}_1) \; \Psi_j^*(\vec{r}_2) \; \Psi_i(\vec{r}_2)}{|\vec{r}_1 - \vec{r}_2|} \, d\vec{r}_1 d\vec{r}_2 . \tag{13}$$

According to the variational principle, the best molecular orbitals are obtained by varying the one-electron functions,

$\Psi_1, \Psi_2 \ldots \Psi_n$, until the energy achieves its minimum value. The condition for the minimum of energy leads to the differential equations

$$[h_i + \sum_j (J_j - K_j)] \Psi_i = \varepsilon_i \Psi_i \quad , \tag{14}$$

where ε_i are orbital energies, and where the Coulomb operator J_i and the exchange operator K_j are defined as

$$J_j(r_1) = \int \frac{\Psi_j^*(\vec{r}_2)\ \Psi_j(\vec{r}_2)}{|\vec{r}_1 - \vec{r}_2|} d\vec{r}_2 \quad , \qquad \text{and} \tag{15}$$

$$K_j(\vec{r}_1)\ \Psi_i(\vec{r}_1) = \delta(m_{s_i}, m_{s_j}) [\int \frac{\Psi_j^*(\vec{r}_2)\ \Psi_i(\vec{r}_2)}{|\vec{r}_1 - \vec{r}_2|} d\vec{r}_2] \Psi_j(\vec{r}_1). \tag{16}$$

The set of equations defined by (14) are known as Hartree-Fock equations, and the orbitals $\Psi_1, \Psi_2 \ldots \Psi_n$, obtained from the solution of these equations, are referred to as Hartree-Fock (HF) molecular orbitals.

For molecular systems of any size, direct solution of Equation (14) is impracticable. It is convenient to approximate the HF orbitals by a linear combination of atomic orbitals (LCAO). In this approach, each molecular orbital is of the form

$$\Psi_i = \sum_\mu C_{\mu i}\ \phi_\mu \quad , \tag{17}$$

where the ϕ_μ are atomic wavefunctions, preferably real, for example Slater type orbitals (STO's), Gaussian type orbitals (GTO's), or a linear combination of STO's or GTO's.

Since the orbitals Ψ_i form an orthonormal set we require that

$$\int \Psi_i^*(\vec{r})\ \Psi_j(\vec{r}) d\vec{r} = \delta_{ij} = \sum_{\mu\nu} C_{\mu i}^* C_{\nu j} S_{\mu\nu} \quad , \tag{18}$$

where $S_{\mu\nu}$ is the overlap integral between AO's ϕ_μ and ϕ_ν

$$S_{\mu\nu} = \int \phi_\mu^*(\vec{r})\ \phi_\nu(\vec{r}) d\vec{r}. \tag{19}$$

For future applications it is useful to write down the expression for the expectation value of a spatial one-electron operator O in the LCAO approximation:

$$< O > = \sum_{i=1}^{n} n_i \int \Psi_i^*(\vec{r})\ O\ \Psi_i(\vec{r}) d\vec{r} \quad , \tag{20}$$

where n_i is the number of electrons in MO Ψ_i . Substituting relation (17) in (20) we get

$$< O > = \sum_{\mu\nu} P_{\mu\nu} O_{\mu\nu} \quad , \quad \text{with} \tag{21}$$

$$O_{\mu\nu} = \int \phi_\mu^*(\vec{r}) \; O \; \phi_\nu(\vec{r}) d\vec{r} \quad , \quad \text{and} \tag{22}$$

$$P_{\mu\nu} = \sum_{i=1}^{n} n_i \; C_{\mu i}^* \; C_{\nu i} \tag{23}$$

The matrix $P_{\mu\nu}$ is known as bond order or density matrix. Using relation (20) with operator $O = \delta(\vec{r}-\vec{r}_o)$, the density of electrons at a point $\vec{r}_o$ is given as

$$\rho(\vec{r}_o) = \sum_{\mu\nu} P_{\mu\nu} \; \phi_\mu(\vec{r}_o)\phi_\nu(\vec{r}_o). \tag{24}$$

The total number of electrons is then given as

$$n = \sum_{\mu\nu} P_{\mu\nu} S_{\mu\nu} \; . \tag{25}$$

The quantity

$$O_\mu = \sum_\nu P_{\mu\nu} \; O_{\mu\nu} \tag{26}$$

describes the contribution of orbital ϕ_μ to the expectation value of operator $\hat{O}$; then

$$<O> = \sum_\mu O_\mu \tag{27}$$

Thus the orbital charge associated with orbital ϕ_μ is given as (in a.u.)

$$q_\mu = -\sum_\nu P_{\mu\nu} \; S_{\mu\nu} \; . \tag{28}$$

This is known as Mulliken population analysis.

2.2 LCAO-MO's for Closed-Shell Systems

Substituting the MO's Ψ_i according to the LCAO approach, Eqn. (17), in the molecular orbital integrals H_i , J_{ij} and K_{ij} , the total electronic energy of the system is given as

$$E_{el} = \sum_{\mu\nu} P_{\mu\nu}H_{\mu\nu} + \frac{1}{2} \sum_{\substack{\mu,\nu \\ \lambda,\sigma}} P_{\mu\nu}P_{\lambda\sigma}[(\mu\nu|\lambda\sigma) - \frac{1}{2} (\mu\lambda|\nu\sigma)] \; , \quad \text{where} \tag{29}$$

$$H_{\mu\nu} = \int \phi_\mu^*(\vec{r}) h \; \phi_\nu(\vec{r}) d\vec{r} \quad \text{and} \tag{30}$$

$$(\mu\nu|\lambda\sigma) = \iint \frac{\phi_\mu^*(\vec{r}_1)\phi_\nu(\vec{r}_1)\phi_\lambda^*(\vec{r}_2)\phi_\sigma(\vec{r}_2)}{|\vec{r}_1 - \vec{r}_2|} \; d\vec{r}_1 d\vec{r}_2 \tag{31}$$

The coefficients $C^*_{\mu i}$ and $C_{\nu i}$ which enter $P_{\mu\nu}$ in Eqn. (23) are now optimized so as to give minimum energy. This leads to the condition

$$\sum_{\mu}(F_{\mu\nu} - \varepsilon_i S_{\mu\nu})C_{\nu i} = 0 \quad , \tag{32}$$

with the Fock matrix $F_{\mu\nu}$:

$$F_{\mu\nu} = H_{\mu\nu} + \sum_{\lambda\sigma} P_{\lambda\sigma} [(\mu\nu|\lambda\sigma) - \frac{1}{2} (\mu\lambda|\nu\sigma)]. \tag{33}$$

The set of equations represented by Eqn. (32) are known as Roothaan equations. They differ from the Hartree-Fock equations (14) in that they are algebric rather than differential equations. We can write Eqn. (32) in the matrix form

$$FC = SCE \quad , \tag{34}$$

where E is the diagonal matrix of E_i. Defining new matrices

$$F' = S^{-\frac{1}{2}}F\, S^{-\frac{1}{2}} \quad , \quad \text{and} \tag{35}$$

$$C' = S^{\frac{1}{2}}\, C \quad , \tag{36}$$

where $S^{\frac{1}{2}}$ is the square root of the overlap matrix S, Equations (34) may be transformed into

$$F'C' = C'E. \tag{37}$$

This is the form of a standard eigenvalue problem. The elements E_i of E will be roots of the determinantal equation

$$|F'_{\mu\nu} - E_i\delta_{\mu\nu}| = 0, \tag{38}$$

with the lowest roots corresponding to the occupied molecular orbitals. For each root E_i, the coefficients $C'_{\nu i}$ can be found from the linear equations

$$\sum_{\nu}(F'_{\mu\nu} - E_i\delta_{\mu\nu})C'_{\nu i} = 0, \tag{39}$$

and the coefficients $C_{\nu i}$ are then determined using

$$C = S^{-\frac{1}{2}}C'. \tag{40}$$

Since the matrix elements of the Fock matrix, $F_{\mu\nu}$, are dependent on the molecular orbitals through the elements $P_{\mu\nu}$, the Roothaan equations are solved by a series of steps:

1. An initial guess of the LCAO coefficients $C_{\nu i}$, in equation (17), is made.

2. Electrons are assigned in pairs to the molecular orbitals Ψ_i with lowest energies.
3. The density matrix $P_{\mu\nu}$ is calculated and then used to form a Fock matrix $F_{\mu\nu}$.
4. A diagonalization procedure is effected as described in Equations (34) to (40), and a new set of coefficients $C_{\nu i}$ is obtained.
5. Steps 2,3 and 4 are repeated until the coefficients $C_{\nu i}$ no longer change within a given tolerance on repeated iteration.

2.3 LCAO-MO's for Open-Shell Systems

A wavefunction for one of the spin states of an open-shell configuration with p α-electrons and q β-electrons (p>q) can be written as a Slater determinant:

$$\Psi = |\Psi_1(1)\alpha(1)\Psi_1(2)\beta(2) \;..\; \Psi_q(2q)\beta(2q)\Psi_{q+1}(2q+1)\alpha(2q+1) \;....$$

$$\Psi_p(p+q)\alpha(p+q)|. \qquad (41)$$

Wavefunctions of this type are termed restricted single determinants because the α-electrons associated with doubly occupied orbital Ψ_1, Ψ_2 ... Ψ_q are described by the same spatial functions as the β-electrons with which they are paired. However, since the total number of α-electrons differs from the total number of β-electrons, the environment of these two types of electrons is not the same. Thus their assignment to the same spatial orbital implies a restriction on the wavefunction and consequently a restriction on their spatial distribution.

In a more general procedure the p α-electrons and q β-electrons are assigned to two completely independent orthonormal sets of molecular orbitals $\Psi_1^\alpha\ \Psi_2^\alpha \ldots \Psi_p^\alpha$ and $\Psi_1^\beta\ \Psi_2^\beta \ldots \Psi_q^\beta$. The corresponding determinantal wavefunction is then

$$\Psi = |\Psi_1^\alpha(1)\alpha(1)\Psi_1^\beta(2)\beta(2) \;....\; \Psi_q^\beta(2q)\beta(2q)\Psi_{q+1}^\alpha(2q+1)\alpha(2q+1)$$

$$\Psi_{q+2}^\alpha(2q+2)\alpha(2q+2) \;....\; \Psi_p^\alpha(p+q)\alpha(p+q)|. \qquad (42)$$

Such a wavefunction is described as an unrestricted single determinant. The unrestricted wavefunction is not generally an eigenfunction of the spin operator S^2 though it is still an eigenfunction of S_z with a corresponding eigenvalue $m_s = \frac{1}{2}(p-q)$. The first selfconsistent approach to unrestricted molecular orbitals was set forth by Slater (6) and Pople and Nesbet(7).

In LCAO approximation, the two sets of molecular orbitals are written as

$$\Psi_i^{\alpha} = \sum_{\mu} C_{\mu i}^{\alpha} \phi_{\mu} \quad , \tag{43}$$

$$\Psi_i^{\beta} = \sum_{\mu} C_{\mu i}^{\beta} \phi_{\mu} \quad . \tag{44}$$

A separate density matrix may be obtained for α- and β-electrons

$$P_{\mu\nu}^{\alpha} = \sum_{i=1}^{p} C_{\mu i}^{\alpha *} C_{\nu i}^{\alpha} \quad , \qquad \text{and} \tag{45}$$

$$P_{\mu\nu}^{\beta} = \sum_{i=1}^{q} C_{\mu i}^{\beta *} C_{\nu i}^{\beta} \quad . \tag{46}$$

The full density matrix is the sum of these:

$$P_{\mu\nu} = P_{\mu\nu}^{\alpha} + P_{\mu\nu}^{\beta} \quad . \tag{47}$$

The spin density matrix $\rho_{\mu\nu}^{S}$ is defined as

$$\rho_{\mu\nu}^{S} = P_{\mu\nu}^{\alpha} - P_{\mu\nu}^{\beta} \quad . \tag{48}$$

For a closed shell system the spin density is zero, because $P_{\mu\nu}^{\alpha} = P_{\mu\nu}^{\beta}$. However for unpaired spin states $\rho_{\mu\nu}^{S}$ provides detailed information about the distribution of electron spin throughout the molecule.

When the LCAO-MO's of Eqns. (43) and (44) are used, the electronic energy of the system can be written as

$$E_{el} = \sum_{\mu\nu} H_{\mu\nu} + \frac{1}{2} \sum_{\substack{\mu\nu \\ \lambda\sigma}} \{P_{\mu\nu} P_{\lambda\sigma} (\mu\nu|\lambda\sigma) - (P_{\mu\nu}^{\alpha} P_{\lambda\sigma}^{\alpha} + P_{\mu\nu}^{\beta} P_{\lambda\sigma}^{\beta})(\mu\lambda|\nu\sigma)\} . \tag{49}$$

The coefficients $C_{\mu i}^{\alpha}$, $C_{\nu i}^{\alpha}$ and $C_{\mu i}^{\beta}$, $C_{\nu i}^{\beta}$ which enter $P_{\mu\nu}^{\alpha}$ and $P_{\mu\nu}^{\beta}$ are now independently optimized to give minimum energy. This leads to the conditions

$$\sum_{\nu} (F_{\mu\nu}^{\alpha} - \varepsilon_i^{\alpha} S_{\mu\nu}) C_{\nu i}^{\alpha} = 0, \qquad \text{and} \tag{50}$$

$$\sum_{\nu} (F_{\mu\nu}^{\beta} - \varepsilon_i^{\beta} S_{\mu\nu}) C_{\nu i}^{\beta} = 0, \tag{51}$$

where the two Fock matrices are given by

$$F_{\mu\nu}^{\alpha} = H_{\mu\nu} + \sum_{\lambda\sigma} [P_{\lambda\sigma}(\mu\nu|\lambda\sigma) - P_{\lambda\sigma}^{\alpha}(\mu\sigma|\lambda\nu)], \quad \text{and} \tag{52}$$

$$F^{\beta}_{\mu\nu} = H_{\mu\nu} + \sum_{\lambda\sigma} [P_{\lambda\sigma}(\mu\nu|\lambda\sigma) - P^{\beta}_{\lambda\sigma}(\mu\sigma|\lambda\nu)]. \quad (53)$$

Equations 50 and 51 represent a generalization of the Roothaan's equations 32 and are solved by an iterative procedure similar to that described for the closed-shell case.

In the open shell calculations described in Sections 3-5 we are within the limit of restricted Slater determinants, and we found this approximation good enough for the calculations of the hyperfine field contributions at the site of Mossbauer nuclei.

2.4 Approximate MO Calculations

Calculations in which all integrals are worked out explicitly using Roothaan's procedure, Eqns. 32, 50, 51, are termed ab initio calculations. The computation time required for such accurate calculations, especially for large systems, is prohibitively large. In many chemical or physical problems, however, qualitative or semiquantitative knowledge of the form of the molecular orbitals is sufficient to extrac the necessary information. A number of approximate MO theories have been developed within the mathematical framework of molecular orbital theory, including various simplifications within the computational procedure. Very often experimental data, on atoms and prototype molecular systems, are used to estimate values for quantities entering the calculations as parameters. These theories, known as semiempirical methods, can be classified into different categories depending on the type of approximation and parametrization involved: (1) Huckel methods (8-10), (2) Zero differential overlap methods (11-14), and (3) pseudopotential methods (15-16).

2.4.1 The simple Huckel method is generally used for π-electrons only. In this method the elements of the Fock matrix are chosen from essentially empirical conditions (8). Hoffman (9) extended this method to include also σ-electrons, which proved successful when applied to hydrocarbons. Since then a number of investigators (17-21) have introduced various refinements and have applied the method to many systems.

In the iterative extended Huckel theory (10) (IEHT), the diagonal elements of the Fock matrix take the form

$$F_{\mu\mu} = (\alpha_{\mu} + \delta\alpha_{\mu}Q_{A}), \quad (54)$$

where α_{μ} and $\delta\alpha_{\mu}$ are adjustable parameters which are strongly related to atomic ionization potentials of AO ϕ_{μ} , and Q_A is the net charge of the atom at which orbital ϕ_μ is centered. The offdiagonal matrix elements $F_{\mu\nu}$ are determined from $F_{\mu\mu}$ and $F_{\nu\nu}$ using the Wolfsberg-Helmholz approximation:

$$F_{\mu\nu} = S_{\mu\nu} K_{\mu\nu}(F_{\mu\mu} + F_{\nu\nu}), \tag{55}$$

where $K_{\mu\nu}$ is a fixed parameter, taking the value between 0.8 and 1.0.

In Cusachs approximation (17) the parameter $K_{\mu\nu}$ takes the form

$$K_{\mu\nu} = a - b\,|S_{\mu\nu}|, \tag{56}$$

where the commonly used values of a and b are 1.0 and 0.5 respectively. The iterative procedure is carried out until the solutions of the matrix Equation (34) are consistent, within a given interval, with the net atomic charge Q_A used in calculating $F_{\mu\mu}$.

2.4.2 Zero differential overlap (ZDO) methods involve the following basic approximations:

(1) Electron repulsion integrals involving the overlap distribution are assumed to be negligibly small and hence are neglected.

(2) The overlap integrals $S_{\mu\nu}$ are neglected in the normalization procedure of molecular orbitals.

(3) The one-center integrals $H_{\mu\nu}$ defined in Eqn. (30) are not neglected but are parametrized. The Roothaan equations (32) then take a form with

$$F_{\mu\mu} = H_{\mu\mu} + \sum_{\nu} P_{\nu\nu}(\mu\mu|\nu\nu) - \frac{1}{2} P_{\mu\mu}(\mu\mu|\mu\mu), \quad \text{and} \tag{57}$$

$$F_{\mu\nu} = H_{\mu\nu} - \frac{1}{2} P_{\mu\nu}(\mu\mu|\nu\nu). \tag{58}$$

There are, however, various levels of approximations in the ZDO methods which can be grouped into three different categories:

(1) Complete neglect of differential overlap (CNDO) (13,22),

(2) intermediate neglect of differential overlap (INDO) (23,24),

(3) neglect of diatomic overlap (NDDO) (13).

CNDO involves the additional approximation of making two-electron integrals depend only on the nature of the atoms A and B, to which ϕ_μ and ϕ_ν belong and not on the actual type of orbitals. The CNDO approximation does not take into consideration the differences in the interaction between two electrons with parallel or antiparallel spins, particularly if they are on the same atom. As a result, CNDO calculations are frequently unable to account for the separation of states arising from the same configuration.

In INDO approximation some account of exchange terms is taken by retaining mono-atomic differential overlap in one-center integrals. If ϕ_μ and ϕ_ν belong to the same center, the exchange terms $(\mu\nu|\mu\nu)$ are not neglected.

In NDDO approximation only the differential overlap for atomic orbitals on different atoms is neglected. The extra feature of NDDO approximation is that it retains dipole-dipole interaction since integrals of the type $(\mu\nu|\lambda\sigma)$, where ϕ_μ and ϕ_ν belong to atom A, and ϕ_λ and ϕ_σ belong to atom B, are included.

2.4.3 In pseudo potential methods, an electron is considered to be moving in the effective electric field of the atomic cores and of the electrons surrounding them. The total potential acting on an electron is given as

$$V(\vec{r}) = \sum_{k=1}^{N} V_k(\vec{r} - \vec{R}_k) \quad , \tag{59}$$

where $V_k(\vec{r}-\vec{R}_k)$ is the effective potential centered at atom k. With this approximation the multicenter electronic repulsion integrals are transformed to three-center integrals of the type $< \mu|V_k(\vec{r}-\vec{R}_k)|\nu >$. $V_k(\vec{r}-\vec{R}_k)$ is, generally, taken to be spherical. Multiple-scattering X_α methods (15) or model potential methods (16) are some of the pseudopotential methods used.

2.5 Configuration Interaction Calculations

So far we considered a Slater determinant (or a set of Slater determinants) of molecular spin orbitals as the ground state molecular wavefunction. One can generate such Slater determinants for the excited states of the molecule also. Slater determinants for the states with the same point symmetry and the same spin quantum number m_s could then mix with each other due to nonvanishing matrix elements of electronic repulsion and also due to nonorthogonality of the wavefunctions. Linear combinations of such Slater determinants would then be the appropriate wavefunction for various molecular states.

If Ψ_a, Ψ_b .... are the Slater determinents with the same symmetry and same spin quantum number m_s then the coefficients of the linear combination of these determinants to form a state (i) are given by the relation

$$\sum_b [H_{ab} - E_i \; S_{ab}]C_{bi} = 0, \tag{60}$$

where H_{ab} are the matrix elements of the electronic Hamiltonian defined in Eqn. 3, and S_{ab} refers to the overlap of many-electron space-spin states.

If various configurations are generated from a single set of orthogonal molecular spin orbitals, then they are orthogonal to each other (i.e. $S_{ab} = \delta_{ab}$), and the calculations are considerably simplified. Diagonal matrix elements of the Hamiltonian matrix are given by the total energy of the configuration

$$H_{aa} = \sum_{\mu\nu} P_{\mu\nu}[H_{\mu\nu} + \frac{1}{2}\sum_{\lambda\sigma} P_{\lambda\sigma}\{(\mu\nu|\lambda\sigma) - \frac{1}{2}(\mu\lambda|\nu\sigma)\}]. \qquad (61)$$

Offdiagonal matrix elements H_{ab} would be zero if the two configurations Ψ_a and Ψ_b differ by more than two molecular spin orbitals. If Ψ_a and Ψ_b differ because $\tilde{\Psi}_i$ is replaced by $\tilde{\Psi}_{i'}$, then

$$<\Psi_a|H_{el}|\Psi_b> = <i|h|i'> + \sum_{j\neq i}\{<ij|\frac{1}{r_{12}}|i'j> - <ij|\frac{1}{r_{12}}|ji'>\} \qquad (62)$$

If Ψ_a and Ψ_b differ because $\tilde{\Psi}_i$ is replaced by $\tilde{\Psi}_{i'}$ and $\tilde{\Psi}_j$ is replaced by $\tilde{\Psi}_{j'}$, then

$$<\Psi_a|H_{el}|\Psi_b>=<ij|\frac{1}{r_{12}}|i'j'> - <ij|\frac{1}{r_{12}}|j'i'> . \qquad (63)$$

Details of a CI program for calculations performed on a number of open-shell systems, and some of the results, which are presented in sections 3-5, are described by Trautwein and Harris (25) and also by Reschke (26).

3. INTERPRETATION OF MEASURED ISOMER SHIFTS BY CHARGE DENSITY CALCULATIONS

3.1 *General*

Electronic structure studies of systems, which are so small that *all* AO basis functions can be included in a MO calculation within reasonable computer time, directly lead to total electron densities $\rho(r)$, which are defined in Eqn. (24). Since we want to correlate isomer shifts δ with charge densities at the nuclear site ($r = 0$) of a Mossbauer isotope we are interested in $\rho(0)$ rather than $\rho(r)$:

$$\rho(0) = \sum_{ab} P_{ab}\, \phi_a(0)\phi_b(0). \qquad (64)$$

Post et al (27) have derived relativistic $\rho(0)$ values for octahedral FeF_6 and $Fe(CN)_6$ clusters with ab initio SCF MO calculations. Fig. 1 shows the linear proportionality between experimental δ and their calculated $\rho(0)$-values,

$$\Delta\delta = \alpha\ \Delta\rho(0)\ , \tag{65}$$

with a mean slope of $\alpha = -0.24 \pm 0.02\ a_o^3\ \mathrm{mm\ s^{-1}}$.

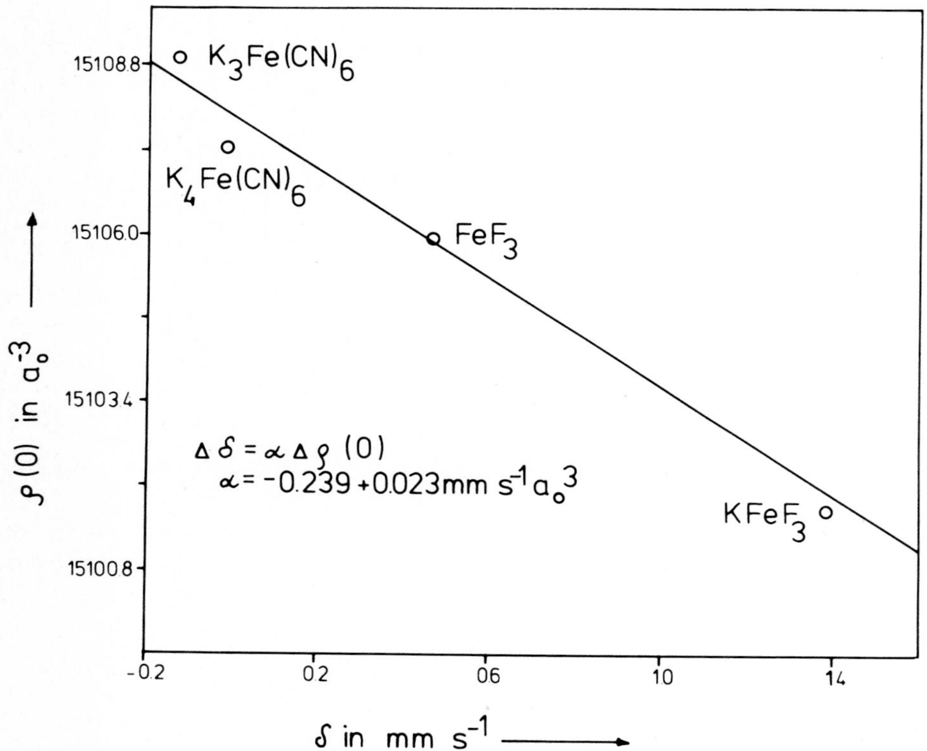

Fig. 1. Least-squares fit of calculated relativistic charge densities at the iron nucleus versus measured isomer shifts (taken from Ref. 27)

3.2 Valence Contribution to $\rho(0)$

If systems are considerably larger than FeF_6 or $Fe(CN)_6$ clusters, or if computer time limitations are severe, the MO basis set might be restricted to valence AO's only. In this case Eqn. (64) gives merely the valence contribution to $\rho(0)$:

$$\rho_{valence}(0) = \sum_{ab} P_{ab}\ \phi_a(0)\ \phi_b(0)\ . \tag{66}$$

The main contribution to $\rho_{valence}(0)$ comes from the valence-s-electrons of the Mossbauer isotope under study.

3.3 Core Contribution to $\rho(0)$

Core s-electrons of the Mossbauer isotope, which contribute considerably to $\rho(0)$, have been omitted in MO calculations with limited basis set. Their influence $\rho_{core}(0)$ upon $\rho(0)$ has to be estimated separately from the MO calculation, for example by an orthogonalization procedure. Flygare and Hafemeister (28), who were the first in applying this procedure, calculated the iodine-5s-contribution to $\rho(0)$ in iodine containing compounds. We evaluated the iron 1s, 2s, and 3s contributions to $\rho(0)$ in a series of iron containing compounds (29-32) using the relation:

$$\rho_{core}(0) = 2\sum_{i=1}^{n} |\Psi_{is}(0)|^2 , \tag{67}$$

where the core-s-orbitals Ψ_{is} are orthogonalized with respect to the MO's Ψ_j:

$$\Psi_{is}(0) = N_{is}\{\phi_{is}(0) - \sum_{j=1}^{m} \langle\Psi_j|\phi_{is}\rangle\Psi_j(0) - \sum_{k>i}^{n} \langle\Psi_{ks}|\phi_{is}\rangle\Psi_{ks}(0)\}. \tag{68}$$

The N_{is} are normalization constants, and the numbers m and n stand for the maximum number of MO's and core-s-orbitals, respectively. The ϕ_{is} are free-ion Hartree-Fock or - in the relativistic case - Dirac-Fock orbitals. One could also obtain $\Psi_{ip}(0)$ contributions to $\rho(0)$ from relativistic calculations. While these contributions are individually significant (in the order of 10 a_0^{-3}) their differences from one compound to the other are so small that they would affect $\Delta\rho(0)$ in Eqn. (65) in the third decimal place only. In order to discuss the various contributions to $\rho_{core}(0)$ we write

$$\rho_{core}(0) = \rho_{potential}(0) + \rho_{overlap}(0) , \tag{69}$$

where the *potential contribution* $\rho_{potential}(0)$ mainly represents the electron density at the Mossbauer nucleus under study would have, if it were free but with the valence electron configuration (MO-VEC) which results from the MO calculation:

$$\rho_{potential}(0) = 2\sum_{i=1}^{n} |\phi_{is}(0)|^2_{MO\text{-}VEC} . \tag{70}$$

$\phi_{is}(0)$ very much depends on the potential shielding of the Mossbauer nucleus due to a specific MO-VEC of the Mossbauer isotope within a chemical compound. Because N_{is} and the second and third term in Eqn. (68) are considerably influenced by overlap integrals between core orbitals of the Mossbauer isotope and valence orbitals of the surrounding ligands, the remaining part in Eqn. (69), $\rho_{overlap}(0)$, is termed *overlap contribution*. Collecting all contributions to $\rho(0)$ we finally have:

$$\rho(0) = \rho_{valence}(0) + \rho_{potential}(0) + \rho_{overlap}(0). \tag{71}$$

3.4 RESULTS

We take FeF_2 as the first example to illustrate the contributions coming from various parts of Eqn. (71). The computational procedure of the MO-calculation (30) for FeF_2, which was represented by a FeF_6^{4-}-cluster, was based on structural geometries (33) which varied with pressure. Fig. 2a shows the pressure dependence of the Fe 3d

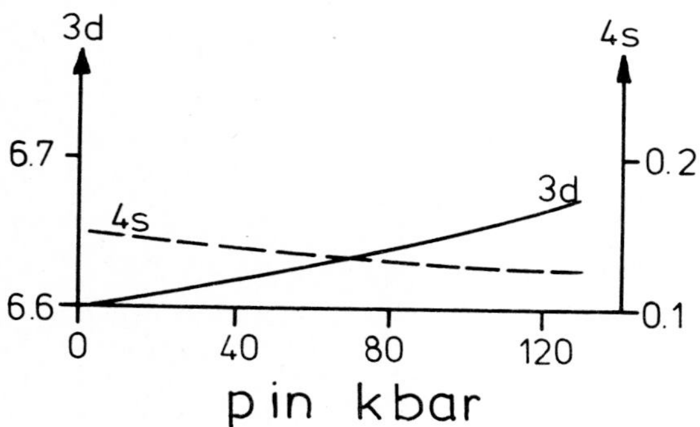

Fig. 2a. Variation of Fe 3d and Fe 4s population under the influence of pressure-dependent cluster geometries.

and Fe 4s orbital population derived from SCC-IEHT calculations, which were completed by successive inclusion of configuration interaction and spin-orbit coupling (30). Fig. 2b shows the pressure dependence of $\rho_{valence}(0)$, $\rho_{potential}(0)$, and $\rho_{overlap}(0)$. The decrease of $\rho_{valence}(0)$ with increasing pressure is due to the decrease of Fe 4s population (Fig. 2a), because the main contribution to Eqn. (66) comes from $P_{4s,4s}\ \phi_{4s}^2(0)$. The corresponding decrease of $\rho_{potential}(0)$ is due to the increase of Fe 3d population (Fig. 2a), because higher Fe 3d charge leads to stronger shielding of Fe-core

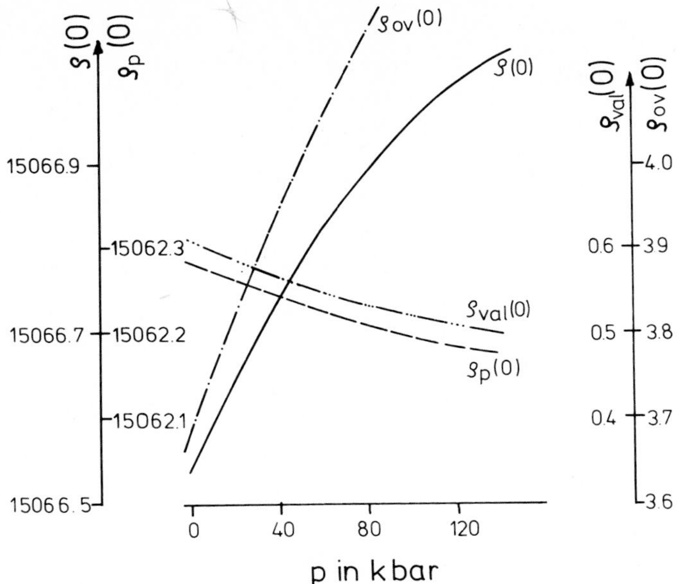

Fig. 2b. Pressure-dependent electron density $\rho(0)$ *and its contributions (in* a_0^{-3}*)* $\rho_{valence}(0)$, $\rho_{potential}(0)$, *and* $\rho_{overlap}(0)$.

electrons from the nucleus (31). By far the largest contribution to the change of $\rho(0)$ with increasing pressure, however, comes from the overlap contribution, which is due to the increased overlap integrals in Eqn. (68) between Fe-core orbitals and ligand-valence orbitals. The individual contributions from Fe 1s, Fe 2s, and Fe 3s electrons to $\rho_{overlap}(0)$ are indicated in Fig. 2c. It is characteristic for 2s electrons that $\rho^{2s}_{overlap}(0)$ decreases with increasing overlap integrals, because $\phi_{2s}(0)$ in Eqn. (68) has opposite sign compared with $\phi_{1s}(0)$ and $\phi_{3s}(0)$; (this has to do with the specific radial dependence of Hartree-Fock orbitals, zero node for 1s, one node for 2s, two nodes for 3s). Comparing the pressure-dependent $\rho(0)$ values with experimentally determined pressure-dependent isomer shifts δ, we are able to derive from Fig. 3 the isomer shift calibration constant α (see Eqn. 65); α takes the value $-0.22 \pm 0.02\ a_0^3$ mm s^{-1}, which is in reasonable agreement with the ab initio result described above.

The importance of overlap contributions in the electron charge density calculations is also reflected in a series of iron-oxygen compounds (31). From Eqn. (68) it is obvious that $\rho_{core}(0)$ depends on the choice of wavefunctions, which, in turn depend on the atomic configuration. Fig. 4 illustrates the amount of influence these

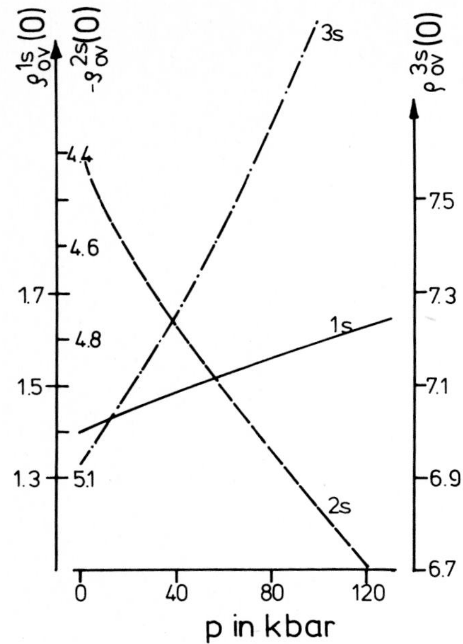

Fig. 2c. Pressure-dependent overlap contributions to ρ(0) from Fe 1s, 2s and 3s electrons.

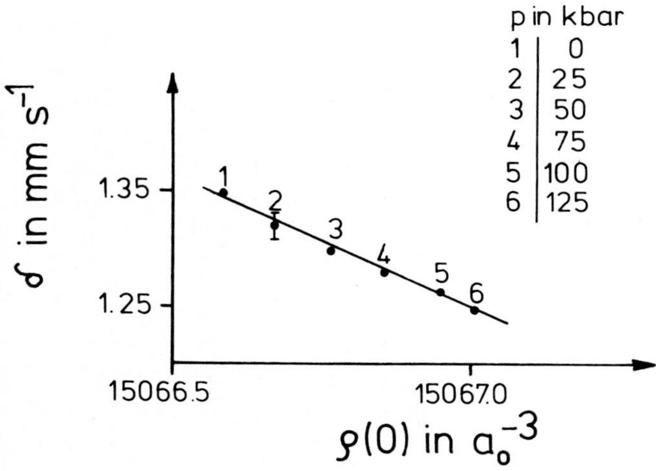

Fig. 3. Linear dependence between experimental pressure-dependent isomer shifts and calculated electron densities of FeF_2. (Experimental values at 300 K relative to metallic iron are taken from A.R. Champion et al, J. Chem. Phys., 47(1967)2583; calculated values are taken from Ref. 30).

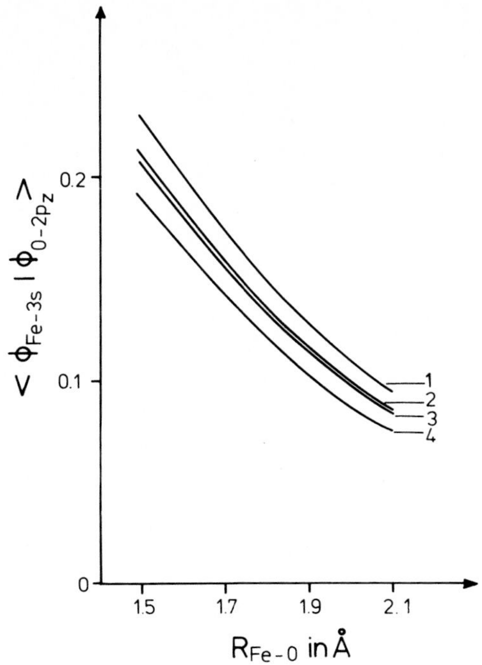

Fig. 4. Overlap integrals $<\phi_{Fe-3s}\ \phi_{0-2p}>$ depending on iron oxygen distance for various valence shell populations of iron and oxygen. Hartree-Fock wavefunctions are taken from Ref. 34, 35 and from E. Clementi and C. Roetti, Atomic Data and Nuclear Data Tables 14(1974)-3-4. (1) Fe-$3d^7$, 0-$2p^6$; (2) Fe-$3d^4$, 0-$2p^6$; (3) Fe-$3d^7$, 0-$2p^5$; (4) Fe-$3d^4$, 0-$2p^5$; (5) Fe-$3d^4$, 0-$2p^4$. (Taken from Ref. 31).

wavefunctions have upon the overlap integral $<\phi_{Fe-3s}\phi_{0-2p}>$. If we calculate overlap integrals using Clementi's HF AO's for Fe $3d^6$ configuration (34) we derive for the iron-oxygen compounds, listed in Table 1, the correlation between experimental δ and calculated $\rho(0)$ values as shown in Fig. 6, with a slope of the solid line of $\alpha = -0.165\ a_0^3\ mm\ s^{-1}$. The spread in electronic configurations in Table 1 among the various compounds under study and the $<\phi_{Fe-3s}\ \phi_{0-2p}>$ results depending on orbital populations (Fig. 4), however, indicate that it is a better procedure to calculate $\rho(0)$ by using overlap integrals which are adequate to the actual electronic configurations of Table 1. These overlap integrals are obtained, for example, from linear interpolation between curves as shown in Fig. 4. Compared to the results from Fig. 5 we now get an increased α value (31) which is $-0.195\ a_0^3\ mm\ s^{-1}$ (Fig. 6). Though the linear relation between δ and $\rho(0)$ of Eqn. (2) is well represented by most of the compounds

TABLE 1.

Iron-oxygen compounds, their relevant cluster which is used in MO calculations, $3d^m4s^n4p^0$ configuration of iron and $2s^x2p^y$ configuration of oxygen as derived from the MO cluster calculations, and experimental isomer shifts (rel. to α-Fe). Taken from Ref. 31.

NO.	Compound	Cluster	Ref. for cluster geometry	$3d^m$	$4s^n$	$4p^o$	$2s^x$	$2p^y$	δ_{exp} (mms^{-1})	Ref. for δ_{exp}
1	$Fe(ClO_4)_2 \cdot 6H_2O$	FeO_6^{10-}	a	6.272	0.159	0.437	1.851	5.674	1.41	b
2	$BiFeO_3$	FeO_6^{9-}	c	5.664	0.173	0.530	1.788	5.541	0.38 ± 0.01	d
3	bipyramidal lattice site of $BaFe_{12}O_{19}$	FeO_5^{7-}	e	5.524	0.203	0.656	1.755	5.434	0.41 ± 0.03	f
4	Fe(II) in MgO	FeO_6^{10-}	a	6.298	0.149	0.414	1.848	5.663	1.06 ± 0.01	g, h
5	α $FeSO_4$	FeO_6^{10-}	i	6.252	0.176	0.470	1.864	5.692	1.42 ± 0.02	j
5'		$Fe(SO_4)^{10-}$	i	6.066	0.132	0.322	1.479	5.230		
6	$FeTiO_3$	FeO_6^{10-}	a	6.311	0.143	0.413	1.841	5.651	1.11 ± 0.01	k, l
7	$GdFeO_3$	FeO_6^{9-}	m	5.755	0.149	0.471	1.794	5.560	0.42 ± 0.01	n
8	Fe_2TiO_5	FeO_6^{9-}	a	5.666	0.169	0.560	1.830	5.596	0.39	k
9	$FeCO_3$	FeO_6^{10-}	a	6.269	0.161	0.438	1.851	5.673	1.38 ± 0.02	j, o
9'		$Fe(CO_3)^{10-}$	a	6.096	0.122	0.307	1.458	4.960		
10	Fe(III) in MgO	FeO_6^{9-}	p	5.699	0.165	0.507	1.794	5.554	0.34	g, h
11	$SrFeO_3$	FeO_6^{8-}	a	5.419	0.189	0.605	1.745	5.440	0.01	k
11'		$FeO_6Sr_8^{8+}$	a	5.310	0.229	0.604	1.743	5.431		
12	$KFeO_2$	FeO_4^{5-}	a	5.573	0.134	0.693	1.734	5.425	0.15 ± 0.02	q, r
13	$BaFeSi_4O_{10}$	FeO_4^{6-}	s	6.179	0.196	0.507	1.815	5.533	0.756±0.002	t

14	V_2FeO_4	FeO_4^{6-}	a	6.283	0.154	0.514	1.802	5.545	1.00 ± 0.06	u
14'		$FeO_4V_{12}^{8+}$	a	6.166	0.184	0.553	1.791	5.522		
15	$BaFeO_4$	FeO_4^{2-}	a	4.293	0.162	0.782	1.518	4.813	-0.87± 0.03	v, q
15'		FeO_4^{2-}	w	5.039	0.038	0.167	1.809	4.785		
15"		$FeO_4Ba_4^{6+}$	a, x	5.036	0.043	0.167	1.810	4.810		

a. R.W.G. Wyckoff, 2nd Edn., Vo. I-III. Interscience Publishers (1965)
b. H. Spiering, D.L. Nagy and R. Zimmermann, J. Phys. C: Solid State Physics, 6(1974)231.
c. J.M. Moreaus, C. Michel, R. Gerson and W.J. James, J. Phys. Chem. Solids 32(1971)1315.
d. C. Blaau and F. van der Woude, J. Phys. C: Solid State Physics, 6(1973)1422.
e. W.D. Townes, J.H. Fung and A. Perrotta, Z. Krist., 125(1967)437.
f. A. Trautwein, E. Kreber, U. Gonser and F.E. Harris, J. Phys. Chem. Solids, 36(1975)325.
g. J.R. Regnard and J. Pelzl, Phys. Stat. Solidi (b) 56(1973)281.
h. H.R. Leider and D.N. Pipkorn, Phys. Rev., 165(1968)494.
i. D. Samaras and J. Coing Boyat, Bull. Soc. Fr. Mineral Cristallogr., 93(1970)190.
j. K. Ono and A. Ito, J. Phys. Soc. Japan, 19(1964)899.
k. G. Shirane, D.E. Cox and S.C. Ruby, Phys. Rev. 125(1962)1158.
l. R.W. Grant, R.H. Housley and S. Geller, Phys. Rev., B 5(1972)1700.
m. S. Geller and E.A. Wood, Acta Cryst., 9(1956)563.
n. M. Eibschutz, S. Shtrikman and D. Treves, Phys. Rev., 156(1967)562.
o. W. Kundig, A.B. Denison and P. Ruegsegger, Phys. Letters, 42A(1972)199.
p. E. Simanek and Z. Sroubeck, Phys. Rev., 163(1967)275.
q. W. Kerler, W. Neuwirth and E. Fluck, Z. Phys., 175(1963)200.
r. T. Ichida, T. Shinjo, Y. Bando and T. Takada, J. Phys. Soc. Japan, 29(1970)1109.
s. R.G. Burns, M.G. Clark and A.J. Stone, Inorg. Chem., 5(1966)1268.
t. M.G. Clark, G.M. Bancroft and A.J. Stone, J. Chem. Phys., 47(1967)4250.
u. U.J. Rossiter, J. Phys. Chem. Solids, 26(1965)775.
v. T. Shinjo, T. Ichida and T. Takada, J. Phys. Soc. Japan, 29(1970)111.
w. Calculated with a Fe-4p ionization potential which is consistent with atomic Hartree-Fock results of Fe-$3d^5$ and Fe-$3d^4$ 4p configuration.
x. Same as w but additionally including the next nearest Ba neighbor shell of iron in the MO cluster calculations.

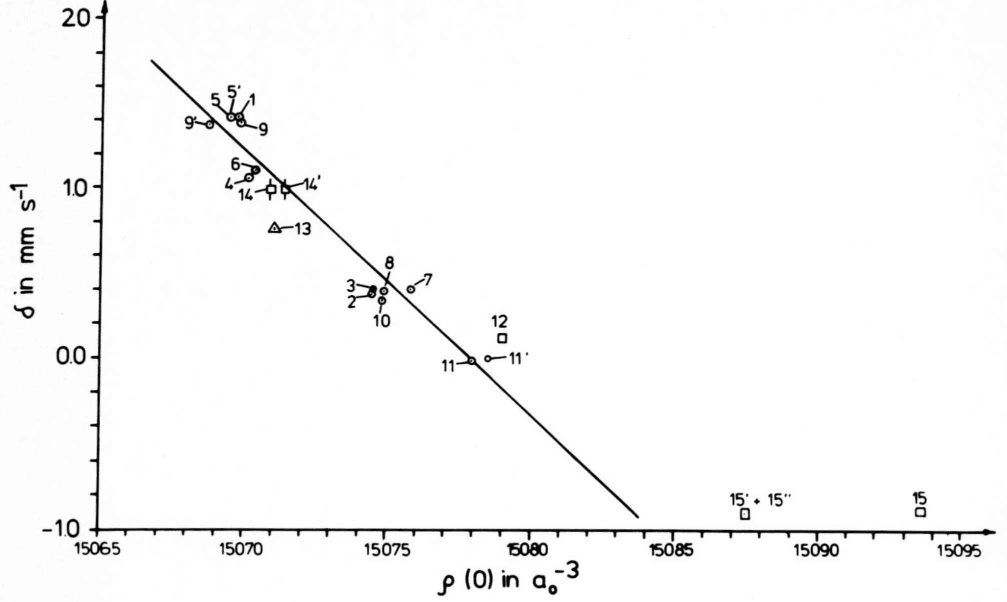

Fig. 5. Experimental isomer shifts at 300 K relative to metallic iron (Refs. see Table 1) and calculated charge densities ρ(0); slope of the solid line is $\alpha = -0.165$ mm s^{-1} a_0^3. A correction of δ due to different Debye temperatures Θ_D in the various compounds is omitted. Assuming a difference in Θ_D of 200 K (with $\Theta_D^1 = 400$ K and $\Theta_D^2 = 200$ K) between two materials leads to second second-order Doppler shift values δ_{SD} at 300K, which differ only by about 0.03 mm s^{-1} for the two compounds. O, sixfold, octahedral; □, fourfold, tetrahedral; ●, fivefold, bipyramidal; Δ, fourfold, planar. (Taken from Ref. 31).

under study, it seems that extreme cases like the Fe(VI) compound $BaFeO_4$ (no. 15 in Table 1 and Figs. 5 and 6) are beyond the scope of quantitative accuracy of our method to calculate ρ(0).

An additional interesting feature of electron charge density calculations is illustrated by the study of iron-halide compounds (32,36). The valence and potential contributions to ρ(0) depend on the model which defines the orbital populations. Calculating Fe 3d and Fe 4s populations from bond order matrix elements or from a Mulliken population analysis the $\Delta\delta$ - $\Delta\rho(0)$ correlations as indicated in Fig. 7 are derived. Note that ρ(0) is little affected by inclusion of the overlap charges in the calculations; neglecting or including them yield α values of -0.21 a_0^3 mm s^{-1} and -0.19 a_0^3 mm s^{-1}, respectively (32). The δ dependences on the ligand electronegativity are opposite for the octahedral Fe^{2+} halides compared to the tetrahedral Fe^{3+} halides (for octahedral systems $\delta_{Cl} > \delta_{Br} > \delta_I$; for tetrahedral systems

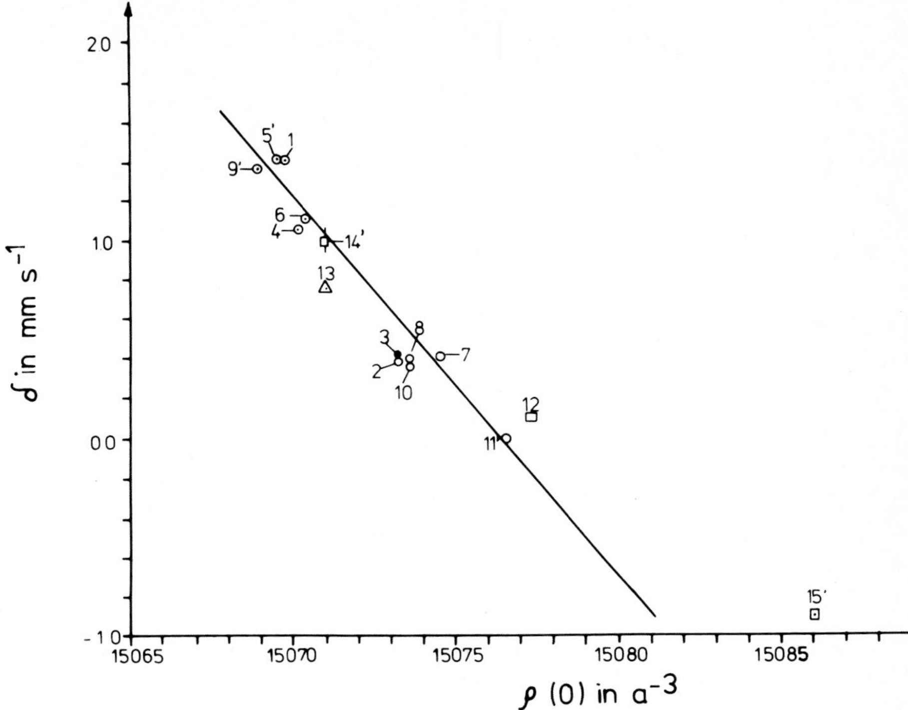

Fig. 6. Experimental isomer shift at 300 K relative to metallic iron (Refs. see Table 1) and calculated electron densities $\rho(0)$; slope of the solid line is $\alpha = -0.195$ mm s^{-1} a_0^3. (Taken from Ref. 31).

$\delta_{Cl} < \delta_{Br} \simeq \delta_I$). These reverse trends are well accounted for by the calculations and arise from different *amounts* of the several contributions to $\rho(0)$.

Completing our discussion on electron charge density calculations, which has been restricted here to iron containing compounds, we summarize the α values which are derived from different series of compounds:

from the pressure dependence of FeF_2(30) $\alpha = -0.22\ a_0^3$ mm s^{-1} ,
for iron-oxygen compounds (31) $\alpha = -0.195\ a_0^3$ mm s^{-1} ,
for iron-halides (32) $\alpha = -0.19\ a_0^3$ mm s^{-1} .

In Figure 8 we have plotted the experimentally observed isomer shifts for a number of iron-containing compounds against the shifts in the electron densities calculated by various authors using different procedures such as Relativistic Hartree-Fock (37), unrestricted Hartree-Fock (38), Band Structure (39), ab initio (27), Ms-Xα (40,41),

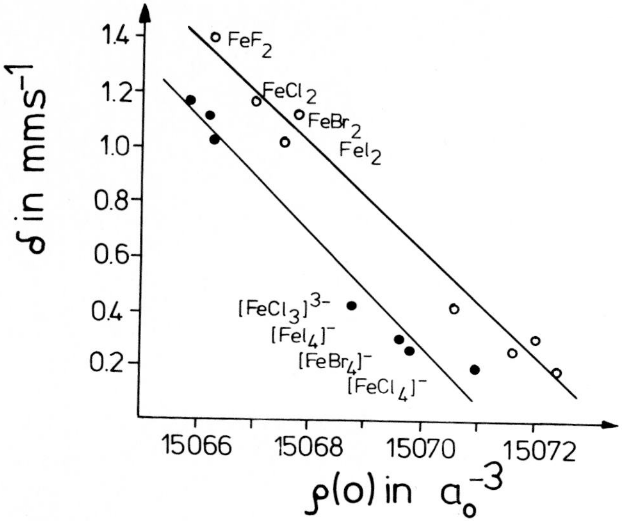

Fig. 7. Experimental isomer shifts at 300 K relative to metallic iron at 300 K versus calculated relativistic electron densities at the iron nucleus ρ(0). The isomer shift calibration constant α, which is the slope of the solid line, is -0.21 a_0^3 mm s^{-1} if overlap charge to q_{3d} and q_{4s} is neglected (full circles); with the overlap charge included (Mulliken approximation), α becomes -0.19 a_0^3 mm s^{-1} (open circles). (Taken from Ref. 32).

and Iterative Extended Hückel Technique (25,26, 29-32, 36). It is interesting to note that fairly good linear correlation is observed between the experimental δ values and calculated values of Δρ(0) with the slope α equal to -0.21 mm s^{-1} a_0^3. However this correlation is not good enough for compairing all results in one δ-ρ(0) diagram because of various approximations inherent in different MO calculations and ρ(0) estimates. We make use of our results to compare similar systems, since the calculational procedure will be least likely affected by systematic uncertainties in ρ(0) if differences Δδ and Δρ(0) between similar systems are related. A detailed review, edited by Shenoy and Wagner (37), about isomer shifts has been published recently.

4. INTERPRETATION OF MEASURED QUADRUPOLE SPLITTINGS BY ELECTRIC FIELD GRADIENT CALCULATIONS

4.1 General

Nuclei with spin quantum number I, greater than $\frac{1}{2}$, can have non-spherical charge distribution. Magnitude and shape of charge deformation are given by the nuclear quadrupole moment Q. The electric field

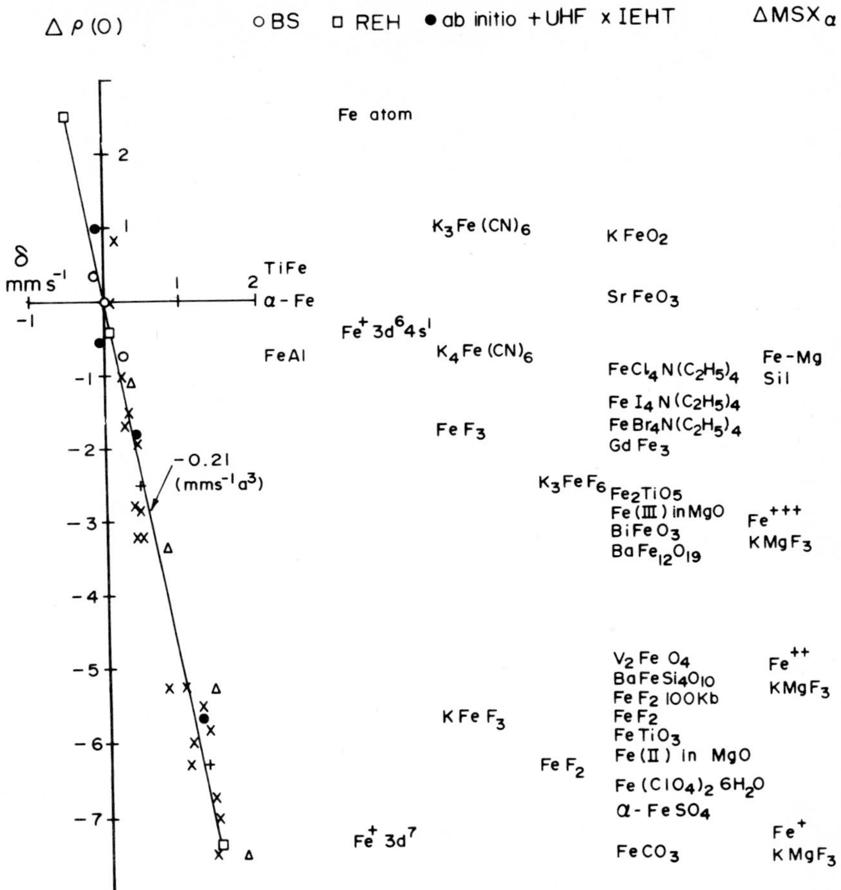

Fig. 8. Experimental isomer shifts δ at 300 K relative to metallic iron versus shifts in the electron densities, Δρ(0), calculated by various authors using Relativistic Hartree-Fock (□); unrestricted Hartree-Fock (+); Band Structure (○); ab initio (●); Iterative Extended Hückel (×); and Ms-X_α techniques (Δ).

gradient (EFG) at the nuclear site, arising due to non-spherical distribution of electronic charge, interacts with the nuclear quadrupole moment giving rise to energy splitting of the nuclear spin levels.

The EFG at the nucleus is defined as a tensor

$$V_{ab} = - \frac{\partial^2 V}{\partial a \partial b} \quad (a,b = x,y,z) \;, \tag{72}$$

where V is the electrostatic potential. It is customary to choose the coordinate system such that the field gradient tensor is diagonal, and

V_{zz} is the maximum value of the diagonal components. The Hamiltonian, describing the interaction between the nuclear quadrupole moment and the EFG is then given as

$$H_Q = \frac{eQ}{2I(2I-1)} (V_{zz}I_z^2 + V_{yy}I_y^2 + V_{xx}I_x^2)$$

$$= \frac{e^2qQ}{4I(2I-1)} [3I_z^2 - I(I+1) + \eta(I_x^2 - I_y^2)], \quad \text{where} \tag{73}$$

$$eq = V_{zz}, \quad \text{and} \tag{74}$$

$$\eta = \frac{|V_{xx} - V_{yy}|}{|V_{zz}|} . \tag{75}$$

In the case of approximate MO calculations with the valence orbitals as limited basis set, the EFG at the nuclear site of the Mossbauer atom consists of two contributions, arising from the atomic cores and the valence electrons of the molecule. The core contribution is calculated using the relation

$$V_{ab}^{core} = \sum_{k=M} Z_k [1-\gamma(|\vec{R}_k - \vec{R}_M|)]$$

$$\frac{3(a_k-a_M)(b_k-b_M)-\delta_{a_k b_k}|\vec{R}_k-\vec{R}_M|^2}{|\vec{R}_k - \vec{R}_M|^5}, \quad (a,b = x,y,z) \tag{76}$$

where $\vec{R}_M$ is the position vector of the Mossbauer atom, and $\vec{R}_k$ and Z_k are, respectively, the position vectors and the core charges of the other atoms in the molecule. The function $\gamma(r)$ is the quadrupole shielding function arising due to polarization of the core electrons of the Mossbauer atom. This function is also known as Sternheimer shielding function.

Components of the EFG due to the electrons represented by wavefunction Ψ are given by

$$V_{ab} = -e < \Psi |\hat{V}_{ab}| \Psi > \tag{77}$$

where the tensor operator $\hat{V}_{ab}$ is the sum of single electron operators $\hat{V}_{ab}(i)$ which have the form

$$\hat{V}_{ab}(i) = [1-\gamma(r_i)] \frac{3a_i b_i - r_i^2 \delta_{a_i b_i}}{r_i^5}, \quad (a_i, b_i = x_i, y_i, z_i). \tag{78}$$

Substitution of the multielectronic wavefunction Ψ by the Slater determinant of molecular orbitals Ψ_i leads to

$$V_{ab}^{el} = -e\sum_i \langle \Psi_i | \hat{V}_{ab}(i) | \Psi_i \rangle. \tag{79}$$

using Relation (21) for the expectation value of one-electron operators we get

$$V_{ab}^{el} = -e\sum_{\mu,\nu} P_{\mu\nu} \langle \phi_\mu | \hat{V}_{ab} | \phi_\nu \rangle \tag{80}$$

V_{ab}^{el} can be divided into three parts,

(1) the valence contribution which arises when both the atomic orbitals ϕ_μ and ϕ_ν belong to the Mossbauer atom,

(2) the ligand contribution which arises when both ϕ_μ and ϕ_ν do not belong to the Mossbauer atom, and

(3) the overlap contribution when only one of the orbitals ϕ_μ and ϕ_ν belong to the Mossbauer atom.

4.2 Valence Contribution to the EFG

If the atomic orbitals ϕ_μ are written as

$$\phi_\mu(r) = f_\mu(r)\, Y_{\ell_\mu m_\mu}(\Omega) \tag{81}$$

the valence contribution to the EFG is given as

$$V_{ab}^{val} = -e\sum_{\mu,\nu} P_{\mu\nu} \int_0^\infty f_\mu(r) \frac{1-\gamma(r)}{r^3} f_\nu(r) r^2 dr \int_\Omega Y_{\ell_\mu m_\mu}(\Omega) \hat{W}_{ab}(\Omega) Y_{\ell_\nu m_\nu} d\Omega \tag{82}$$

where $\hat{W}_{ab}(\Omega)$ is the angular part of the operator $\hat{V}_{ab}$. Analytical integration of the radial part gives

$$\int_0^\infty f_\mu(r) \frac{1-\gamma(r)}{r^3} f_\nu(r) r^2 dr = (1-R)\langle r^{-3} \rangle_{\mu\nu} \,. \tag{83}$$

The matrix elements of $\hat{W}_{ab}(\Omega)$ for s, p and d orbitals are given (43) in Table 2. The expectation values of $\langle r^{-3} \rangle$ depend upon the electronic

TABLE 2

Matrix elements of $\hat{W}_{ab}(\Omega)$ *for d, s, and p orbitals. Sequence of orbitals:* $(3z^2-r^2)$, xz, (x^2-y^2), yz, xy, s, p_z, p_x, p_y.

$$\langle W_{xx}\rangle = \left[\begin{array}{cccccc|ccc}
-2/7 & 0 & -2\sqrt{3}/7 & 0 & 0 & -\sqrt{5}/5 & & & \\
0 & 2/7 & 0 & 0 & 0 & 0 & & & \\
-2\sqrt{3}/7 & 0 & 2/7 & 0 & 0 & -\sqrt{3/5} & & 0 & \\
0 & 0 & 0 & -4/7 & 0 & 0 & & & \\
0 & 0 & 0 & 0 & 2/7 & 0 & & & \\
\hline
-\sqrt{5}/5 & 0 & -\sqrt{3/5} & 0 & 0 & 0 & & & \\
\hline
 & & & & & & -2/5 & 0 & 0 \\
 & & & & & & 0 & 4/5 & 0 \\
 & & & & & & 0 & 0 & -2/5
\end{array}\right]$$

$$\langle W_{yy}\rangle = \left[\begin{array}{cccccc|ccc}
-2/7 & 0 & 2\sqrt{3}/7 & 0 & 0 & -\sqrt{5}/5 & & & \\
0 & -4/7 & 0 & 0 & 0 & 0 & & & \\
2\sqrt{3}/7 & 0 & 2/7 & 0 & 0 & \sqrt{3/5} & & 0 & \\
0 & 0 & 0 & 2/7 & 0 & 0 & & & \\
0 & 0 & 0 & 0 & 2/7 & 0 & & & \\
\hline
-\sqrt{5}/5 & 0 & \sqrt{3/5} & 0 & 0 & 0 & & & \\
\hline
 & & & & & & -2/5 & 0 & 0 \\
 & & & & & & 0 & -2/5 & 0 \\
 & & & & & & 0 & 0 & 4/5
\end{array}\right]$$

$$\langle W_{zz}\rangle = \left[\begin{array}{cccccc|ccc}
4/7 & 0 & 0 & 0 & 0 & 2\sqrt{5}/5 & & & \\
0 & 2/7 & 0 & 0 & 0 & 0 & & & \\
0 & 0 & -4/7 & 0 & 0 & 0 & & 0 & \\
0 & 0 & 0 & 2/7 & 0 & 0 & & & \\
0 & 0 & 0 & 0 & -4/7 & 0 & & & \\
\hline
2\sqrt{5}/5 & 0 & 0 & 0 & 0 & 0 & & & \\
\hline
 & & & & & & 4/5 & 0 & 0 \\
 & & & & & & 0 & -2/5 & 0 \\
 & & & & & & 0 & 0 & -2/5
\end{array}\right]$$

$$\langle W_{yz} \rangle = \left[\begin{array}{cccccc|ccc}
0 & 0 & 0 & \sqrt{3}/7 & 0 & 0 & & & \\
0 & 0 & 0 & 0 & 3/7 & 0 & & & \\
0 & 0 & 0 & -3/7 & 0 & 0 & & 0 & \\
\sqrt{3}/7 & 0 & -3/7 & 0 & 0 & \sqrt{3/5} & & & \\
0 & 3/7 & 0 & 0 & 0 & 0 & & & \\
\hline
0 & 0 & 0 & \sqrt{3/5} & 0 & 0 & & & \\
\hline
 & & & & & & 0 & 0 & 3/5 \\
 & & 0 & & & & 0 & 0 & 0 \\
 & & & & & & 3/5 & 0 & 0
\end{array}\right]$$

$$\langle W_{xy} \rangle = \left[\begin{array}{cccccc|ccc}
0 & 0 & 0 & 0 & -2\sqrt{3}/7 & 0 & & & \\
0 & 0 & 0 & 3/7 & 0 & 0 & & & \\
0 & 0 & 0 & 0 & 0 & 0 & & 0 & \\
0 & 3/7 & 0 & 0 & 0 & 0 & & & \\
-2\sqrt{3}/7 & 0 & 0 & 0 & 0 & \sqrt{3/5} & & & \\
\hline
0 & 0 & 0 & 0 & \sqrt{3/5} & 0 & & & \\
\hline
 & & & & & & 0 & 0 & 0 \\
 & & 0 & & & & 0 & 0 & 3/5 \\
 & & & & & & 0 & 3/5 & 0
\end{array}\right]$$

$$\langle W_{xz} \rangle = \left[\begin{array}{cccccc|ccc}
0 & \sqrt{3}/7 & 0 & 0 & 0 & 0 & & & \\
\sqrt{3}/7 & 0 & 3/7 & 0 & 0 & \sqrt{3/5} & & & \\
0 & 3/7 & 0 & 0 & 0 & 0 & & 0 & \\
0 & 0 & 0 & 0 & 3/7 & 0 & & & \\
0 & 0 & 0 & 3/7 & 0 & 0 & & & \\
\hline
0 & \sqrt{3/5} & 0 & 0 & 0 & 0 & & & \\
\hline
 & & & & & & 0 & 3/5 & 0 \\
 & & 0 & & & & 3/5 & 0 & 0 \\
 & & & & & & 0 & 0 & 0
\end{array}\right]$$

configuration of an atom. For example $\langle r^{-3}\rangle_{3d}$ values, as derived from Hartree-Fock calculations, are 4.49 a.u. for Fe $3d^7$, 5.09 a.u. for Fe $3d^6$ and 5.73 a.u. for Fe $3d^5$ configuration, and 14.833 a.u. for I $5p^5$ and 12.727 a.u. for I $5p^6$ configuration. $(1-R)_{4p}$ for Fe is taken to be $\frac{1}{3}$ of the corresponding quantity for Fe 3d electrons (44,45).

4.3 Various Approximations in Calculating Ligand and Overlap Contributions to the EFG

There are various levels of approximation in which the ligand and the overlap contribution are calculated. We describe below five different models that have been used for calculating these contributions. In all these models, the valence contribution V_{ab}^{val} is calculated using Eqn. (82).

Model 1: Ligand and overlap contributions are combined and calculated using relation (76) in which the core charges Z_K are replaced by effective ligand charges Q_K. The Q_K are either derived from Mulliken's population analysis (28), or from adding part of the overlap charge onto the ligand charges in order to be able to describe the dipole moment of the molecule under study by effective atomic charges (10). The remaining part of the overlap charge is neglected since V_{ab}^{el} includes bond order matrix elements (Eqn. 82)) rather than effective orbital charges. This model was found to yield satisfactory results, irrespective of how the Q_K's were defined, especially for highly ionic compounds with relatively localized charges.

Model 2: In Model 1 part of the overlap charge between the Mossbauer atom and the ligands has been neglected. A correction to this is calculating the valence contribution V_{ab}^{val} using effective bond order matrix terms $P_{\mu\mu}^{eff}$, instead of $P_{\mu\mu}$. $P_{\mu\mu}^{eff}$ are calculated as

$$P_{\mu\mu}^{eff} = P_{\mu\mu} + \sum_{\nu} P_{\mu\nu} S_{\mu\nu}, \tag{84}$$

where ϕ_μ is an orbital belonging to the Mossbauer atom, while ϕ_ν is an orbital belonging to the ligand atoms. This procedure seems to overestimate the valence contribution and does not give satisfactory results especially in cases of highly covalent complexes, such as ferrocene and related compounds.

Model 3: Ligand and overlap contributions are approximated by (46)

$$V_{ab}^{overlap + ligand} = (1-\gamma_\infty)\sum_{k,\ell} \frac{(a_{k\ell}-a_M)(b_{k\ell}-b_M)-|\vec{R}_{k\ell}-\vec{R}_M|^2\delta_{ab}}{|\vec{R}_{k\ell}-\vec{R}_M|^5} Q_{k\ell}. \tag{85}$$

Summation is over all atoms k and ℓ, where at least one of them is different from the Mossbauer atom M. The charges $Q_{k\ell}$ are calculated from bond order matrix elements $P_{\mu\nu}$ and from overlap integrals $S_{\mu\nu}$ as

$$Q_{k\ell} = e(Z_k\delta_{k\ell} - \sum_{\mu\nu} P_{\mu\nu} S_{\mu\nu}). \tag{86}$$

The summation over μ includes all ϕ_μ on center k, and the summation over ν includes all ϕ_ν on center ℓ. The Cartesian coordinates $a_{k\ell}$, $b_{k\ell}$, $\vec{R}_{k\ell}$ in Eqn. (85) are chosen as if the $Q_{k\ell}$ were situated at the maximum of product $\phi_\mu\phi_\nu$. Since the so defined $Q_{k\ell}$ have distances from the Mossbauer atom generally larger than 1 Å, in Eqn. (85) $(1 - \gamma_\infty)$ was used instead of $(1-\gamma(r))$. (The Sternheimer shielding function $\gamma(r)$ reaches its saturation value γ_∞ at about 1 Å for Fe $3p^6$ and I $4d^{10}$ cores; however, one has to be cautious if iodine 5s and 5p electrons are included in the core, see section 4.4).

Model 4: The ligand contribution is calculated as in Model 1. The overlap part is derived by using a procecure (47) which transforms $\langle\phi_\mu^M|\hat{V}_{ab}|\phi_\nu^k\rangle$ into the subspace of ϕ_μ^M leading to

$$\langle\phi_\mu^M|\hat{V}_{ab}|\phi_\nu^k\rangle = \sum_{\mu'} \langle\phi_\mu^M|\hat{V}_{ab}|\phi_{\mu'}^M\rangle \langle\phi_{\mu'}^M|\phi_\nu^k\rangle. \tag{87}$$

This model overestimates the overlap contribution to V_{ab} since Q_k and $\langle\phi_{\mu'}^M|\phi_\nu^k\rangle$ both contain overlap effects. An additional approximation in this model is that in practice the AO's $\phi_{\mu'}^M$ are represented by a limited basis set only. The results obtained along this line were not always satisfactory, especially for highly covalent systems, such as ferrocene and its derivatives (48).

Model 5: The ligand and overlap contributions to the EFG are calculated using a three-dimensional numerical integration method. Each of the integrals is solved by integrating over one dimension analytically, and by integrating over the remaining two dimensions using Gauss-type integration formulae. The details of this method, which yields satisfactory accuracy with respect to the experimental results (even for ferrocene), are given by Reschke and Trautwein (48).

An important requirement for the calculation of ligand and overlap contributions to V_{ab} with the help of a numerical integration procedure is the knowledge of the Sternheimer shielding function $\gamma(r)$, which enters Eqn. (78). Due to the importance of $\gamma(r)$ we describe, in section 4.4, some methods of obtaining $\gamma(r)$, R and γ_∞.

4.4 Evaluation of Strenheimer Shielding Functions

The EFG tensor operator defined in Eqn. (78) contains the factor $(1-\gamma(r))$. This factor arises from the polarization of core electrons of the Mossbauer atom in a molecule (or solid). Core electrons are those which are excluded within the limited MO basis set. The amount of polarization depends on $\gamma(r)$ and accounts for shielding $(\gamma(r)>0)$ or antishielding $(\gamma(r)<0)$ effects in deriving the EFG. In the literature most of the work with respect to these shielding and antishielding corrections has been reported by Sternheimer and his collaborators (49). Therefore these corrections are also termed "Sternheimer procedure". Other methods used for deriving $\gamma(r)$ are variational method (50-52), many-body perturbation method (53-57), and coupled Hartree-Fock method (58-59). Most of the publications report only factors R and γ_∞ or even only γ_∞. Lauer et al (60) have recently calculated $\gamma(r)$ functions corresponding to electronic configurations which also exclude the valence orbitals, generally included in the MO calculations.

Core polarization and its effect on shielding or antishielding the Mossbauer nucleus from external charges can be visualized in two different ways. In one approach, followed by Sternheimer, one considers the interaction of the nuclear quadrupole moment with the core electrons which induces a quadrupole moment (Q_{ind}) within the electron core. The total quadrupole moment $(Q + Q_{ind})$ then interacts with the EFG which is produced by external charges (valence electrons as well as ligand charges). The total quadrupole coupling constant is then $e^2(Q + Q_{ind})q$. Within the alternative approach, one considers the interaction of the core electrons with external charges which induces an EFG in the electron core. The total EFG $(q + q_{ind})$ then interacts with the nuclear quadrupole moment. The total quadrupole coupling constant is then $e^2Q(q + q_{ind})$. Sternheimer (49b) has shown that both, Q_{ind} and q_{ind} include the same proportionality factor $(1 - R^*)$, leading to

$$e^2(Qq)_{total} = e^2(Q + Q_{ind})q = e^2Q(q + q_{ind}) = e^2Qq(1-R^*). \qquad (88)$$

The quantity R^* depends on $\gamma(r)$ by the relation

$$R^* = \frac{\langle\gamma(r)r^{-3}\rangle}{\langle r^{-3}\rangle}, \qquad (89)$$

where r is the distance of the "external" charge from the Mossbauer nucleus. In case this "external" charge is a valence electron, then the expectation values in Eqn. (89) are derived using the wavefunction of this valence electron; then R^* is denoted by R. On the other hand, if the external charge is far away from the Mossbauer atom, $\gamma(r)$ becomes constant and takes the value γ_∞; R^* then reduces to $R^*=\gamma_\infty$. For light atoms (Li, Na, F)γ_∞ is relatively small (positive or negative) compared to the large and negative γ_∞ values obtained for heavier atoms (Br, I). In a series of elements with similar electronic configuration a tendency is that γ_∞ becomes more negative with increasing atomic number (see Fig. 9). In Fig. 10 we show $\gamma(r)$ for K^+ , Sn^{2+} and Fe^{3+} . $\gamma(r)$ varies significantly for r ranging from 0 to about 3 a.u. For negative or heavy ions this range may be even wider. For the calculation of ligand (lattice) contributions to the EFG, within the point-charge model, it is, therefore, important to take the appropriate value of $\gamma(r)$ instead of γ_∞. Negative iodine (with $5p^6$ configuration) reaches its saturation value only at about 9.a.u. (60).

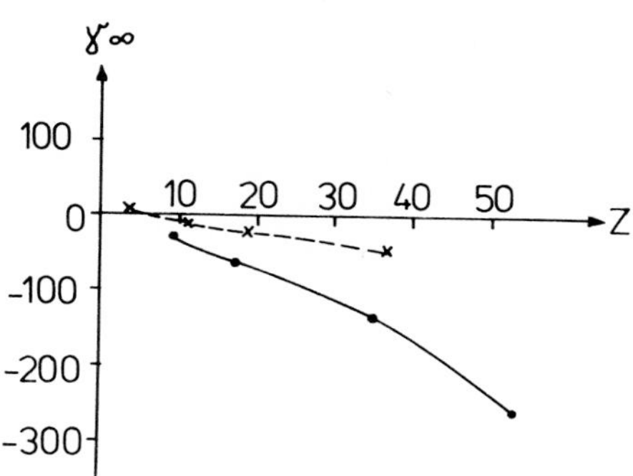

Fig. 9. Sternheimer shielding factor γ_∞ depending on the atomic number Z of elements (×) Li^o, Na^o, K^o, Rb^o, and (•) F^-, Cl^-, Br^-, I^-.

Table 3 gives the γ_∞ values for neutral and negative halides with and without the contributions due to ns, np valence orbitals. γ_∞ is significantly different when the valence orbitals are excluded from the calculations. In Fig. 11 we show $\gamma(r)$ values for Fe^{2+} and I^- with and without including the valence electrons. It is seen that the function $\gamma(r)$ saturates faster without inclusion of valence orbitals. The shielding factor R is also affected when the valence orbitals are omitted from the calculations. For example for Fe^{2+} ($3d^6$) the value of R is 0.12 (49c), however, the value 0.08 is derived (60) if the Fe 3d orbitals are omitted. It is interesting to note that many Mossbauer spectroscopists still use the old and wrong value of R = 0.32 for Fe^{2+}. Correspondingly, for I^- ($5s^2$ $5p^6$) the value of R is -0.14;

TABLE 3

Sternheimer shielding factors γ_∞ for neutral and negative halides with and without the ns, np valence orbitals.

	neutral atom		negative ion	
element	with valence orbitals	without valence orbitals	with valence orbitals	without valence orbitals
F	- 7.1	+ 0.08	- 22.3	+ 0.08
Cl	- 25.4	- 1.2	- 55.4	- 1.2
Br	- 66.0	- 6.2	-133.0	- 6.1
I	-136.0	-16.9	-254.0	-16.2

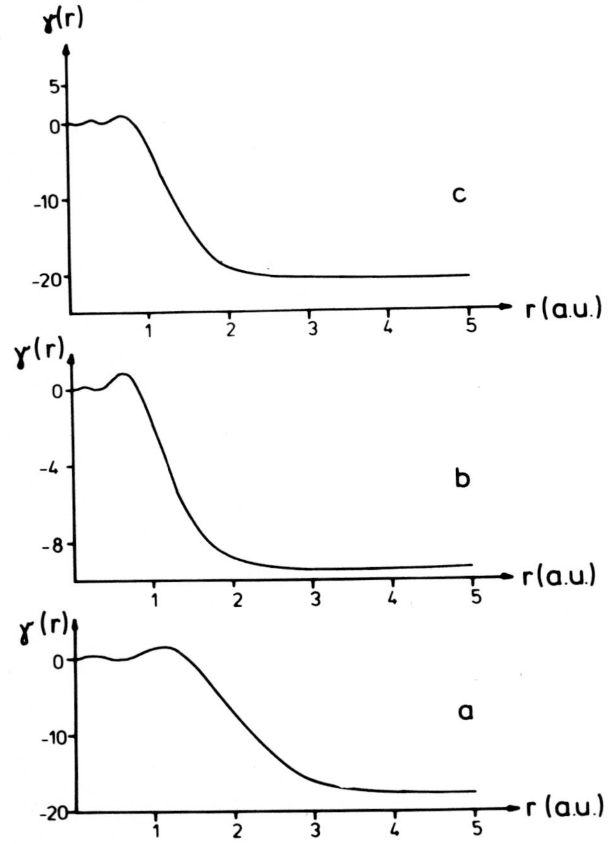

Fig. 10. Sternheimer shielding function $\gamma(r)$ for (a) K^+, (b) Fe^{3+} and Sn^{2+} (Clementi wave functions corresponding to Sn^{2+} were used for configuration [Kr] $4d^{10}5s^2$)

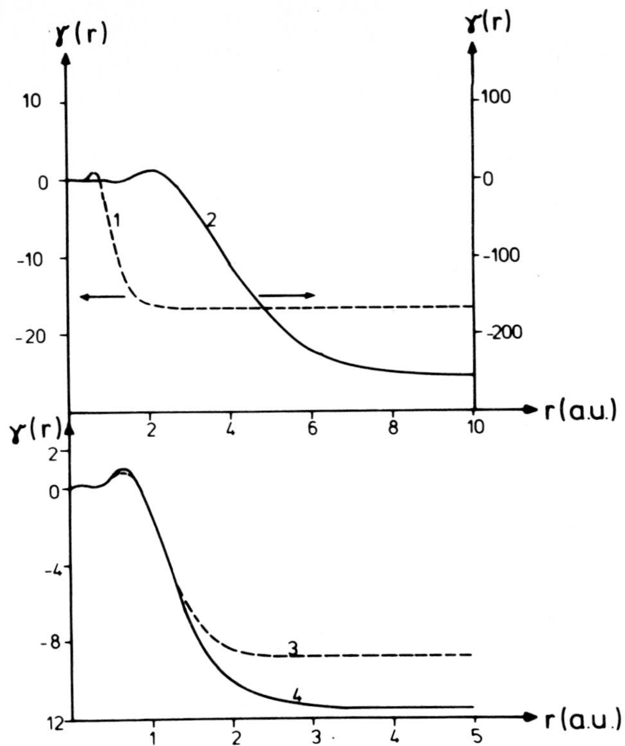

Fig. 11. Sternheimer shielding function γ(r) for I^- (1) with 5s and 5p electrons (solid line), (2) without 5s and 5p electrons (broken line), and for Fe^{2+} (3) with $3d^6$ electrons (solid line), (4) without 3d electrons (broken line).

however the value -0.18 is derived (60) if the I 5s and 5p orbitals are omitted.

4.5 Temperature Independent Quadrupole Splittings from Frozen Solutions or Powder Materials

Table 4 summarizes the experimental and calculated quadrupole splittings and asymmetry parameters corresponding to the different calculational models. Comparing experimental and theoretical results, we find that the various models lead to nearly equal results if the compound under study is highly ionic ($GdFeO_3$ and Fe_2TiO_5) or if the point symmetry of the Mossbauer atom is relatively high (the latter condition implies that several V_{ab} contributions cancel each other). When covalency becomes substantial, i.e. when the overlap charge involved is relatively large, and when the deviations from octahedral

TABLE 4

Experimental and calculated quadrupole splittings ΔE_Q^a in mms^{-1} for ferrous low-spin (S = 0) and ferric high-spin (S = $\frac{5}{2}$) compounds

Compound	Spin state	ΔE_Q exp.	Ref.	ΔE_Q calculated Model 1	Model 2	Model 3	Model 4	Model 5	M[b]
$Fe(C_5H_5)_2$	0	2.33-2.42	c-d	2.89±0.26[k]	4.03	2.40	9.70	2.63±0.24[k]	59
$Fe(C_5H_4)_2(CH_2)_3$	0	2.27±0.03	e	2.73	3.90	2.28	9.73	2.63	75
$Fe(C_5H_4)_2(CH_2)_2CO$	0	2.22±0.01	f	2.83	3.95	2.24	9.44	2.73	77
$[Fe(CN)_5H_2O]^{3-}$	0	0.80	g	0.82	0.87		0.86	0.92	55
$[Fe(CN)_5NH_3]^{3-}$	0	0.671±0.006	h	0.76	0.77		0.94	0.69	56
$GdFeO_3$	$\frac{5}{2}$	0.02±0.03	i	0.10	0.10			0.11	33
Fe_2TiO_5	$\frac{5}{2}$	0.70	j	0.65	0.65			0.78	33

a) Calculated quadrupole splittings ΔE_Q are derived from $\Delta E_Q = \frac{1}{2}e^2QV_{zz}(1 + \eta^2/3)^{\frac{1}{2}}$ with V_{zz} and $\eta = \frac{|V_{xx} - V_{yy}|}{|V_{zz}|}$ resulting from diagonalizing the calculated EFG tensor V_{pq} and with Q = 0.15 barn, (1-R) = 0.92 (60)

b) M is the number of AO basis orbitals used in the MO calculations.

c) R.A. Stukan, S.P. Gabin, A.N. Neomeganow, U.I. Goldanskii, E.F. Makarov; Teor. Eksperim. Khim., 2(1966)805.

d) U. Zahn, P. Kienle, H. Eicher, Z. Phys., 166(1962)220.

e) A.G. Nagy, I. Dezsi, M. Hillman, Institute report of the Hungarian Academy of Sciences, Budapest, KFKI-75-72, 1976.

f) M.L. Good, J. Buttone, D. Foyt, Ann. N.Y. Acad. Sci., 239(1974)193.

g) P. Gutlich, in U. Gonser (Ed.), Topics in Applied Physics, Berlin, Springer-Verlag, Vol. 5, 1975, p. 53

h) W. Kerler, W. Neuwirth, E. Fluck, P. Kuhn, B. Zimmermann, Z. Physik 173(1963)321.

i) M. Eibschutz, S. Shtrikman and D. Treves, Phys. Rev., 156(1967)562.

j) G. Shirane, D.E. Cox, S.C. Ruby, Phys. Rev., 125(1962)1158.

k) Given error corresponds to uncertainty in crystallographic data.

or tetrahedral point symmetry of the Mossbauer atom are significant, then the influence of the approximations involved in some of the models becomes critical. This is shown (48) by various ΔE_Q values for the three ferrocene derivatives $Fe(C_5H_5)_2$, $Fe(C_5H_4)_2(CH_2)_3$ and $Fe(C_5H_4)_2(CH_2)_2CO$, for which, however, the more sophisticated model 5 yields satisfactory results. To present further calculational results we summarize in Table 5 density matrix elements $P_{\mu\mu}$ corresponding to iron orbitals, atomic charges q_a of iron and its ligands, experimental and calculated ΔE_Q values (Model 5) for some of the various iron compounds which we have studied sofar. The MO interpretation of ΔE_Q values of these compounds may be summarized as follows:

1. The compounds are not characterized by iron $3d^5$ or $3d^6$ configuration but more likely by $3d^m\ 4s^n\ 4p^o$ with considerable deviation from iron charge of +3 or +2.
2. The quadrupole splittings are substantially influenced by varying population of all five iron 3d AO's even in the case of ferric and ferrous low spin compounds.
3. The quadrupole splittings are also influenced by departures of the electronic distribution from axial symmetry.
4. Asymmetric electron population in Fe 4p AO's contributes to ΔE_Q.
5. Overlap and ligand charges may have considerable effect upon ΔE_Q, like in ferrocene (46,48) and in $BaFeSiO_4$ (61-63).

For two iodine compounds (I:Fe $(\text{alkyldithiocarbamate})_2I$, and II:$N(C_2H_5)_4FeI_4$), which we discuss in further detail in section 4, we derived (64) quadrupole splittings for iodine according to Model 2, i.e. including overlap contribution implicity via the use of effective orbital charges $P_{\mu\mu}^{eff}$. The ligand contribution to the EFG was found to be negligibly small. Comparing experimental (ΔE_0^I = -14.8 to -16.6 mms^{-1}, η^I = 0.20 to 0.11[65], and ΔE_Q^{II} = -14.46 $\pm$ 0.1 mms^{-1}, η^{II} = 0 $\pm$ 0.04[66] and calculated values (ΔE_Q^I = -19 mms^{-1} η^I = 0.27 and ΔE_Q^{II} = -14 mms^{-1}, η^{II} = 0.08) we find reasonable agreement, though we believe that further improvement may be achieved by employing Model 5.

4.6 Single Crystal Results

So far we have discussed the Mossbauer results obtained from powder samples or frozen solutions only, namely ΔE_Q and η. Measurements on single crystals have certain advantages. These measurements can provide additional information, such as the sign of V_{zz} and the orientation of the principal axes system of the EFG with respect to

TABLE 5

Experimental and calculated quadrupole splittings, bond order matrix elements $P_{\mu\mu}$ and atomic charge q_a of iron and its next-nearest ligands as derived from MO calculations

Compound		$BaFeSO_4$	$Na_2Fe(CN)_5NO.H_2O$	$BaFe_{12}O_{19}$	$K_3Fe(CN)_5NH_3$	FeOCl
ΔE_Q^{exp} (mms^{-1})		0.56± 0.002[a]	1.82±0.08[b]	1.95±0.05[c]	0.65±0.03[d]	0.916± 0.001[e]
T^{exp} (K)		80	77	120	77	300
ΔE_Q^{calc} (mms^{-1})		0.35[f]	1.84[g]	1.80[f]	0.64[h]	0.83[i]
$P_{\mu\mu}$: $d_{x^2-y^2}$		1.20	1.03	1.08	0.95	1.07
$d_{3z^2-r^2}$		1.90	0.85	0.72	0.76	1.16
d_{xz}		1.01	1.51	1.05	1.79	1.08
d_{yz}		1.01	1.51	1.05	1.79	1.11
d_{xy}		1.04	1.77	1.08	1.09	1.27
s		0.20	0.11	0.19	0.14	0.19
p_z		0.25	0.20	0.25	0.20	0.23
p_x		0.15	0.22	0.20	0.23	0.24
p_y		0.15	0.22	0.20	0.23	0.23
q_a:	Fe	0.64	0.10	1.57	0	0.77
	+x	-1.70	-0.22	-1.50[l]	-0.27	-0.40[m]
	-x	-1.70	-0.22	-1.50[l]	-0.27	-0.40[m]
	+y	-1.70	-0.22	-1.50[l]	-0.27	-1.77[m]
	-y	-1.70	-0.22	-	-0.27	-1.77[m]
	+z	-	-0.14[k]	-1.60	-0.03	-1.72
	-z	-	-0.21	-1.60	-0.25	-1.72

a) Taken from (39); coordinates described in (62, 63).
b) Taken from (40); coordinates described in (61).
c) Taken from (62, 63).
d) Taken from (41); coordinates described in (71).
e) Taken from R.W. Grant, H. Wiedersich, R.M. Housley, G.P. Espinosa, J.O. Artman, Phys. Rev. B3(1971)678; coordinates described in (71)
f) Taken from (62, 63).
g) Taken from (61)
h) Taken from (61, 71)
i) Taken from (71).
k) Charge of N in NO^+ is -0.14, and charge of O is +0.473.
l) The three oxygens in the xy-plane produce D_{3h} point symmetry of the central ion.
m) The two chlorines and the two oxygens in the xy plane are situated between the x and y axes.

the crystallographic axes (67-70), which can be tested with MO calculations. For example, single crystal studies on FeOCl gave results (71) as (i) $\eta^{exp} = 0.32 \pm 0.03$, (ii) $V_{zz}^{exp} < 0$, and (iii) the direction of the Z axis of the principal axes system of the EFG tensor. The MO calculation of the EFG tensor yields (42) (i) $\eta^{calc}=0.23$, $V_{zz}^{calc} < 0$, and (iii) the calculated orientation of the principal axes system coincides with the experimentally determined principal axes system within less than 3 degrees.

4.7 Temperature Dependent Quadrupole Splitting

So far we have only considered the electric field gradient arising from valence electrons in their ground state. For the compounds in which the excited electronic states are high in energy compared to the ground state and hence remain unpopulated, the EFG tensor remains constant with respect to temperature; then the nuclear quadrupole splitting is independent over a wide range of temperature. Several high-spin ferric and low-spin ferrous compounds exhibit such a behaviour. Mossbauer nuclei in most of the diamagnetic compounds exhibit temperature independent quadrupole splitting. (In such compounds a slight variation in quadrupole splitting may arise from the change of lattice parameters with temperature).

In several ferrous high-spin compounds the quadrupole splitting was found to be considerably temperature dependent (72-76). This is mainly due to the fact that many-electron states Ψ_i (i = 1, ...) are energetically close to each other and therefore thermal energy affects the population of these states according to Boltzmann statistics. The many-electron states can mix considerably due to spin-orbit coupling, which we have neglected in our MO and CI calculations. Consideration of spin-orbit coupling will lead to new energies and eigen states which are linear-combinations of the many-electron states Ψ_i

$$e_\alpha = \sum_i C_{i\alpha} \Psi_i \quad . \tag{90}$$

An appropriate procedure for calculating the quadrupole splitting is to calculate the EFG tensor $(V_{ab})_\alpha$ for every such eigenstate e_α as

$$(V_{ab})_\alpha = \langle e_\alpha | \hat{V}_{ab} | e_\alpha \rangle \quad . \tag{91}$$

One, then, calculates the temperature average of the EFG tensor

$$\langle V_{ab} \rangle_T = \sum_\alpha (V_{ab})_\alpha \exp(-E_\alpha/kT) / \sum_\alpha \exp(-E_\alpha/kT) , \tag{92}$$

and diagonalizes $(V_{ab})_T$, which finally leads to $V_{zz}(T)$, $\eta(T)$ and the orientation of the principle axes system. A method of incorporating the spin-orbit coupling in the MO calculation of temperature dependent quadrupole splittings for Fe^{2+} compounds has been described by Zimmermann et al (77). In this method, the calculations are significantly simplified because of the assumption that the spin-orbit coupling functions $\xi(r)$ are extremely localized.

The calculated and measured nuclear quadrupole splitting $\Delta E_Q(T)$ for α-$FeSO_4$ as a function of temperature (78) are shown in Fig. 12.

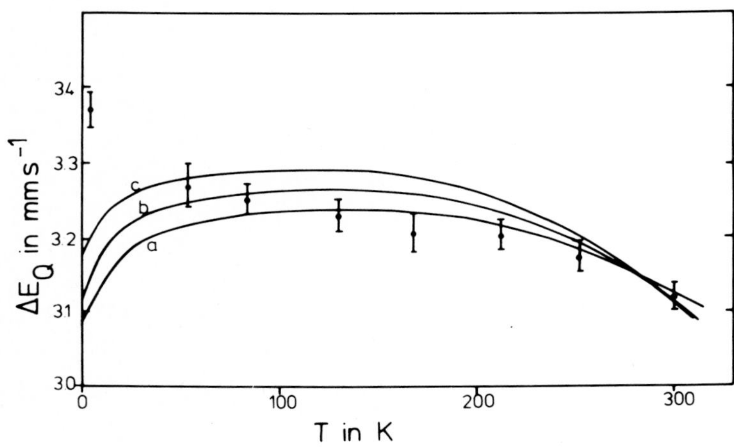

Fig. 12. Temperature dependent quadrupole splitting for α-$FeSO_4$. Experimental points are taken from Ref. 78. The three curves are calculated using the spin-orbit coupling constant as (a) -100, (b) -90 and (c) - 80 cm^{-1}

For α-$FeSO_4$, which was represented by a $[Fe(SO_4)_6]^{10-}$ cluster, with coordinates as derived from x-ray structure analysis (79), MO and subsequent spin-orbit and efg calculations were carried out. The three solid curves in Fig. 12 correspond to three different spin-orbit coupling constants. In the paramagnetic region of α-$FeSO_4$ ($T>T_N$ = 23.5 K) the calculated values of ΔE_Q are in good agreement with the experiment. Similar results have been obtained for other ferrous high-spin compounds such as deoxy-hemoglobin (80,81) and FeF_2 (30).

The difference between the calculated ΔE_Q value of 3.15 mm s^{-1} and the measured value of 3.37$\pm$0.024 mm s^{-1} for α-$FeSO_4$ at 4.2 K has been identified as magnetically induced quadrupole splitting.

4.8 Magnetically Induced Quadrupole Splitting

The presence of a magnetic field $\vec{H}$ reorients the spin states e_α according to the interaction of the electron spin $\vec{S}$ with $\vec{H}$. Depending on the magnitude of H we may get appreciable mixing between zero-field states e_α, and thus the quadrupole splitting may differ from the value in zero-field. Fig. 13 shows the field dependence of ΔE_Q at 4.2 K obtained from the calculations in which the electronic Hamiltonian includes the interaction $\beta(\vec{H}+2\vec{S})\vec{H}$ also.

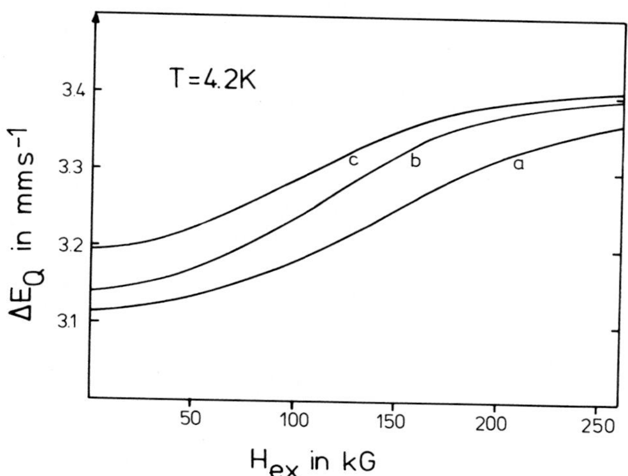

Fig. 13. Calculated values of quadrupole splitting at 4.2 K as a function of magnetic field. Curves a, b, and c correspond to spin-orbit coupling constant as given in Fig. 12.

At 4.2 K, α-$FeSO_4$ is antiferromagnetic, and it is the exchange field which interacts with the zero-field states e_α. This field is proportional to the mean value of the spin $\langle S_z\rangle_T$

$$H_{exch} = h\langle S_z\rangle_T \quad , \tag{93}$$

where h is a constant. A self-consistent solution of Eqn. (93) yields the values of $\langle S_z\rangle_T = 1.95$ and $H_{exch} = 23$ Tesla for T = 4.2 K. Using this value of H_{exch} in Fig. 13, the magnetically induced quadrupole splitting ΔE_Q at 4.2 K takes the values 3.34, 3.38 and 3.39 mm s^{-1} for three different values of the spin-orbit coupling constant. This result is now in reasonably good agreement with the experiment at 4.2 K.

5. INTERPRETATION OF MEASURED MAGNETIC HYPERFINE SPLITTINGS BY INTERNAL FIELD CALCULATIONS

5.1 General

The magnetic hyperfine splitting measured by Mossbauer spectroscopy is proportional to the effective magnetic field $\vec{H}^{eff}$ produced at a Mossbauer nucleus. $\vec{H}^{eff}$ is the vector sum of the externally applied magnetic field $\vec{H}^{ext}$ and the internal magnetic field $\vec{H}^{int}$:

$$\vec{H}^{eff} = \vec{H}^{ext} + \vec{H}^{int} \tag{94}$$

The latter consists of four parts, the contact field $\vec{H}^{C}$, the orbital field $\vec{H}^{L}$, the spin dipolar field $\vec{H}^{SD}$, and the supertransferred field $\vec{H}^{ST}$,

$$\vec{H}^{int} = \vec{H}^{C} + \vec{H}^{L} + \vec{H}^{SD} + \vec{H}^{ST} . \tag{95}$$

In the following we describe the procedure of evaluating the various contributions to $\vec{H}^{int}$ from MO calculations.

5.2 Contact Field

The contact field H^C is the contribution of the electronic spin density at the nucleus (Fermi contact term) (82):

$$H^C = \frac{16\pi}{3} \mu_B \sum_{ns} [|\phi^{\uparrow}_{ns}(0)|^2 - |\phi^{\downarrow}_{ns}(0)|^2]. \tag{96}$$

In case the spin density originates from s-electrons which have finite charge density $\rho(0)$ at the nucleus, Eqn. (96) can be written as (83):

$$H^C = \frac{16\pi}{3} \mu_B \, \rho(0) \langle S_s \rangle , \tag{97}$$

where $\langle S_s \rangle$ is the effective spin produced by these s-electrons. On the other hand, if the spin density originates from p- or d-electrons the spin-paired core-s-electrons may become spin-polarized due to exchange interaction with the p- or d-electrons; in this case Eqn. (96) can be written as

$$H^C = 2H^{core} \langle S_{p,d} \rangle + 2H^{val} \langle S_{p,d} \rangle , \tag{98}$$

where $S_{p,d}$ is the effective spin produced by the p- or d-electrons, and H^{core} and H^{val} represent the spin-polarization of core- and valence-s-electrons, respectively. Collecting all terms and using the spin direction of the system under study as z axis in the calculation, H^C takes the form

$$H^C = 2H^{core} \langle S_{p,d} \rangle_z + 2H^{val} \langle S_{p,d} \rangle_z + \frac{16\pi}{3} \mu_B \rho(0) \langle S_s \rangle_z \, . \qquad (99)$$

The core-polarization contribution to the hyperfine field at the iron nucleus (H^{core} in Eqn. 99) is approximated by interpolating the effect of the 3d orbital population (q_{3d}) between the $3d^5$ and $3d^6$ free-ion values as (83)

$$H^{core}(Fe) = -[12.6 + 1.15(q_{3d} - 5)], \text{ (in Tesla)}. \qquad (100)$$

The second term in Eqn. (100) accounts for the change of spin-polarization of the inner s-orbitals of iron (1s, 2s, and 3s) by adding Fe-3d charge to a system with $3d^5$ configuration. Several models may be considered for evaluating q_{3d}:

(i) q_{3d} is taken equal to the sum over the Fe-3d bond order matrix elements

(ii) q_{3d} includes overlap charges which are estimated from a dipole correction procedure

(iii) q_{3d} includes overlap charges estimated from a Mulliken population analysis (Eqn. 28).

It turned out that the inclusion of overlap charges, either through (ii) or (iii) leads to a better description of the actual situation than neglecting them. Table 6 illustrates the considerable differences of orbital populations according to the different models (i), (ii) and (iii) for Fe(alkyldithiocarbamate)$_2$X with X = Cl, I (abbreviated by Fe(dtc)$_2$X, the chemical and electronic structure of which is shown in Fig. 14. The S expectation values of Eqn. (99) are derived from the AO coefficients of the singly populated MO's in Fig. 14.

Similar to Eqn. (100), the dependence of the valence contribution to the hyperfine field at the iron nucleus (H^{val} in Eqn. 99) on the 3d and 4s population is described by:

$$H^{val}(Fe) = \frac{1}{2} [15.3 - 3.06(q_{3d} - 5)] q_{4s}, \text{ (in Tesla)}. \qquad (101)$$

This relation assumes that the spin polarization of the Fe4s shell is proportional to the 4s orbital charge q_{4s}.

The third term in Eqn. 99 is negligible in the case of iron compared to the first and second term, because $\rho_{4s}(0)$ is relatively small due to the partial population of the Fe4s orbital.

Turning to iodine, the expression for H^C (Eqn. 99) becomes

$$H^C(I) = 2H^{ns} \langle S_{5p} \rangle_z + \frac{16\pi}{3} \mu_B \rho_{5s}(0) \langle S_{5s} \rangle_z \, . \qquad (102)$$

TABLE 6

Diagonal bond order matrix elements $P_{\mu\mu}$, orbital charges and effective atomic charges for $Fe(dtc)_2Cl$ and $Fe(dtc)_2I$ (64).

	$Fe(dtc)_2Cl$				$Fe(dtc)_2I$		
	$P_{\mu\mu}$ a	Orbital b	Charges c		$P_{\mu\mu}$ a	Orbital b	Charges c
$3z^2-r^2$	1.05	1.13	1.13		1.14	1.18	1.18
xz	1.08	1.11	1.11		1.16	1.19	1.19
$x^2 - y^2$	1.99	1.99	1.99		1.99	1.99	1.99
yz	1.12	1.13	1.13		1.19	1.22	1.22
xy	0.80	0.93	0.93		0.79	0.92	0.93
4s	0.13	0.31	0.31		0.08	0.21	0.17
$4p_z$	0.20	0.38	0.41		0.19	0.40	0.41
$4p_x$	0.18	0.26	0.19		0.17	0.24	0.14
$4p_y$	0.19	0.41	0.45		0.18	0.40	0.43
3s	1.84	1.92	1.89	5s	1.93	1.94	1.93
$3p_z$	1.51	1.65	1.66	$5p_z$	1.34	1.54	1.55
$3p_x$	1.88	1.92	1.91	$5p_x$	1.77	1.80	1.80
$3p_y$	1.89	1.93	1.92	$5p_y$	1.81	1.83	1.83
q_{Fe}		$+0.35^b$	$+0.35^c$	q_{Fe}		$+0.25^b$	$+0.34^c$
q_{Cl}		-0.42^b	-0.38^c	q_I		-0.11^b	-0.11^c

a) Calculated according to Eqn. (23).
b) Calculated in a fashion which divides overlap charges such that the dipole moment along the bond is preserved (Ref. 21).
c) Calculated according to Mulliken population analysis (see Eqn. 28).

In this equation the first and second term of Eqn. (99) are lumped together, since the valence I5s orbital is nearly doubly occupied. The term $H^{ns}(I)$ was derived in a recent study of the tetraiodoferrate(III) salt of tetraethylammonium (32); it takes the value 24 $\pm$ 2 Tesla. For $\frac{16\pi}{3}\mu_B\rho_{5s}(0)$ we have used 59026.4 Tesla (32), including the relativistic value for $\rho_{5s}(0)$.

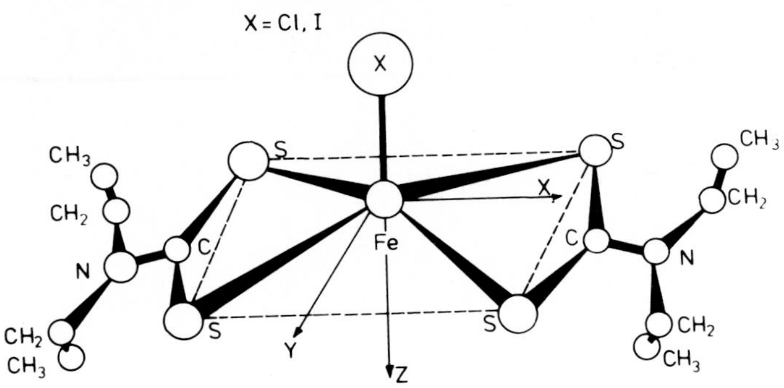

Fig. 14. $Fe(dtc)_2X$, with X = Cl, I, the cartesian coordinates of which have been taken for X = Cl from B.F. Hoskins et al, J. Chem. Soc. (A) 1668 (1970) and for X = I from P.C. Healy et al, J. Chem. Soc. (Dalton) 1369 (1972). Molecular orbitals of mainly iron character as obtained from semiempirical SCC-LCAO-IEHT calculations (selfconsistent charge-linear combination of atomic orbitals - iterative extended Huckel theory). Values $C_{\mu i}$ without and with brackets correspond to $Fe(dtc)_2Cl$ and $Fe(dtc)_2I$, respectively.

$$MO = C_{1,i}|Fe,\ 3z^2-r^2\rangle + C_{2,i}|Fe,xz\rangle + C_{3,i}|Fe,x^2-y^2\rangle + C_{4,i}|Fe,yz\rangle$$
$$C_{5,i}|Fe,xy\rangle + C_{6,i}|I,5s\rangle + C_{7,i}|I,5p_z\rangle + C_{8,i}|I,5p_x\rangle + C_{9,i}|I,$$
$$5p_y\rangle + \ldots\ldots.$$

i	μ = 1	2	3	4	5	6	7	8	9
33	+0.02 (-0.01)	+0.0 (+0.07)	-0.01 (-0.01)	+0.00 (+0.00)	+0.78 (+0.65)	-0.007 (+0.007)	-0.018 (+0.017)	-0.001 (-0.034)	-0.005 (+0.013)
32	+0.88 (+0.13)	+0.02 (-0.54)	-0.01 (-0.01)	+0.05 (-0.69)	-0.01 (-0.02)	+0.005 (+0.002)	-0.267 (-0.061)	-0.002 (-0.259)	+0.009 (-0.300)
31	-0.05 (+0.77)	-0.04 (-0.07)	-0.01 (-0.05)	+0.93 (+0.25)	+0.01 (-0.14)	-0.005 (+0.013)	+0.006 (-0.239)	+0.013 (-0.037)	+0.256 (+0.128)
30	+0.01 (-0.21)	-0.93 (-0.70)	-0.05 (+0.01)	-0.04 (+0.51)	+0.00 (+0.04)	-0.006 (-0.004)	+0.002 (+0.033)	-0.270 (-0.333)	+0.009 (-0.231)
29	+0.03 (+0.08)	-0.05 (+0.00)	+0.99 (+0.99)	-0.01 (+0.00)	+0.00 (+0.02)	+0.002 (+0.005)	-0.005 (-0.015)	-0.012 (-0.004)	-0.080 (-0.004)

5.3 Orbital Field

The orbital field H^L arises from the orbital motion of the electrons:

$$H_a^L = -2\mu_B \langle r^{-3}\rangle L_a \ , \ (a=x,y,z) \ , \qquad (1\ 3)$$

with $2\mu_B$ = 12.5283 Tesla a_0^3 , and with $\langle r^{-3}\rangle$ including potential effects as described in Section 4. The L_a are the x, y, z components of the total orbital angular momentum of the electrons of the Mossbauer isotope in the system under study. They are derived from $\langle\Psi_i|\hat{L}_a|\Psi_j\rangle$ using the many-electron MO wavefunctions Ψ_i,Ψ_j along the lines described in Section 1. In cases, where the orbital angular momentum is nearly quenched L_a may be approximated by (84)

$$L_a = (g_{ab} - 2.0023\delta_{ab})\langle S_b\rangle \ , \ (a,b = x,y,z) \qquad (104)$$

with $(g_{ab} - 2.0023\delta_{ab})$ being described in a perturbation treatment by

$$g'_{ab} = (g_{ab} - 2.0023\delta_{ab}) = 2\sum_{k=i+1}^{n} \frac{\langle i|\xi(r)\ell_a|k\rangle\langle k|\ell_b|i\rangle}{E_k - E_i} \ . \qquad (105)$$

The orbital $|i\rangle$ is the doubly occupied MO which is highest in energy, and the orbitals $|k\rangle$ are all the singly occupied MO's above. Fig. 14 illustrates the situation in $Fe(dtc)_2X$, a compound with intermediate spin $S = \frac{3}{2}$. In this specific example the doubly occupied MO is $|29\rangle$ and the singly occupied MO's are $|30\rangle$, $|31\rangle$, and $|32\rangle$. The spin-orbit coupling constants $\zeta=\langle\xi(r)\rangle$ are 525 cm^{-1} for Fe(III), 420 cm^{-1} for Fe(II), and 5000 cm^{-1} for I. To show the various contributions to L_a we discuss g'_{ab} of Eqn. (105) for $Fe(dtc)_2I$ in more detail (64):

(i) The main contribution to $|29\rangle$ comes from the $Fe(x^2-y^2)$ AO, thus one is tempted to neglect all additional AO contributions in $|29\rangle$. Under this approximation g'_{ax} values become g'_{xx} = 0.078, g'_{yy} = 0.087, and g'_{zz} = 0.

(ii) The inclusion of the $Fe(3z^2-r^2)$ AO in $|29\rangle$ leads to considerably different g'_{ax} values: g'_{xx} = 0.099, g'_{yx} = -0.015, g'_{zx} = 0, g'_{zy} = 0, g'_{yy} = 0.059, and g'_{zz} = 0.008. These values may be compared with g-factors as obtained by Ganguli et al (85): g_{xx} = 2.089, g_{yy} = 2.047, and g_{zz} = 2000. This comparison is quite satisfactory; thus we conclude that even minor AO contributions to the construction of MOs can play an important role in the evaluation of expectation values like g-factors.

(iii) Because the MOs $|29\rangle$ and $|k\rangle$ do not contain only iron contributions, it is interesting to examine the role of ligand AOs in a more general way. Since $\xi(r)$ of Eqn. (105) is an extremely localized function around the origin of the variable r we assume that any integral of the $\langle\phi_m|\xi(r)|\phi_1\rangle$ is zero, and hence a higher order correction to the g-factors is given as:

$$g''_{ab} = 2\sum_{\substack{k>29 \\ m,m'\ell,\ell'}}^{32} \frac{1}{E_k - E_{29}} C_{km}C_{km'}C_{k\ell}C_{k\ell'}\{\zeta_m\langle\phi_m|\ell_a|\phi_{m'}\rangle \langle\phi_\ell|\ell_b|\phi_{\ell'}\rangle$$

$$+ \zeta_\ell\langle\phi_\ell|\ell_a|\phi_{\ell'}\rangle \langle\phi_m|\ell_b|\phi_{m'}\rangle\}. \qquad (106)$$

The indices m and ℓ stand for "metal" and "ligand", respectively. Under this approximation we derive $g'_{xx} = 0.091$, and $g'_{yx} = -0.022$. The other factors remain the same.

(iv) All g'_{ab} contributions to $\vec{H}^{int}$ at the iodine nucleus are negligible.

$\vec{H}^L$ vanishes to zero for systems the electronic structure of which is represented by a single configuration Ψ, being energetically isolated from excited states Ψ^*, i.e. the energy differences in Eqn. (105) become large and hence g'_{ab} vanishes. This situation is reflected in ferric high-spin or ferrous low-spin systems.

5.4 Spin-Dipolar Field

The spin-dipolar field H^{SD} is the contribution to H^{int} from the magnetic moment which is associated with the spin of the electrons outside the nucleus:

$$\vec{H}^{SD} = -2\mu_B \left(\frac{3\vec{r}\ (\vec{S}.\vec{r})}{r^5} - \frac{\vec{S}}{r^3} \right). \qquad (107)$$

The x,y,z components of H^{SD} given as

$$H_a^{SD} = -2\mu_B \sum_{b=x,y,z} \frac{3ab - r^2\delta_{ab}}{r^5} S_b\ , \qquad (108)$$

may be derived using the operator equivalence method (86):

$$H_a^{SD} = -2\mu_B\xi\langle r^{-3}\rangle \sum_b [\tfrac{1}{2}(\ell_a\ell_b+\ell_b\ell_a) - \frac{\ell(\ell+1)}{3}\delta_{ab}]S_b. \qquad (109)$$

If the electron under consideration is a d or p electron, ξ takes the value $\frac{2}{7}$ or $\frac{6}{5}$, respectively. The expression $(3ab - r^2\delta_{ab})/r^5$ in Eqn. (108) also appears in EFG calculations (see Eqn. 78) and reflects the point-symmetry of the Mossbauer nucleus; cubic and tetrahedral point symmetry lead to zero EFG and to zero H^{SD}. The tetraiodoferrate cluster $[FeI_4]^-$ in $N(C_2H_5)_4FeI_4$ is an example for this situation (32) (Fig. 15): The tetrahedral point symmetry of iron in this compound yields zero EFG and $\vec{H}^{SD}$, while the lower point symmetry of iodine yields finite values for both, EFG and $\vec{H}^{SD}$.

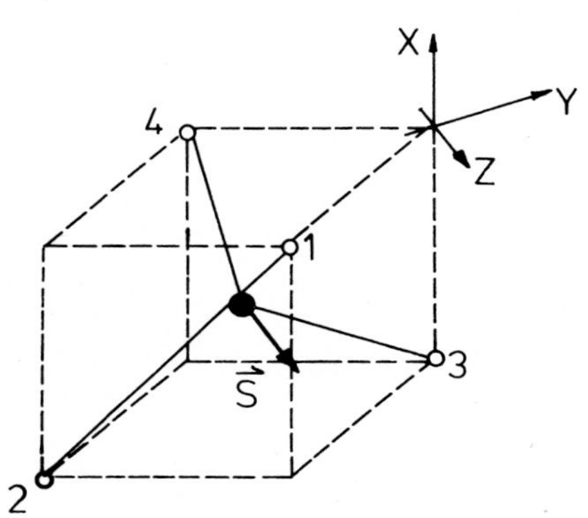

Fig. 15. System of coordinate axes used in the MO calculations. The spin is along z. The two magnetically nonequivalent iodine sites (A,B) are denoted by 1,4 and 2,3, respectively. o: iodine, :iron.

Unlike Section 4, where the EFG has been evaluated from a summation over all MO's (see Eqn. 79), H^{SD} is derived from a summation over the open shell MO's only, because they only contribute to the spin:

$$H_a^{SD} = -2\mu_B \langle r^{-3}\rangle \sum_b \sum_{\text{open shell MO's } i} C_{i\mu}C_{i\nu}\langle\phi_\mu|\hat{W}_{ab}|\phi_\nu\rangle\langle S_b^i\rangle. \qquad (110)$$

As an example we present the equations for *iodine* in $N(C_2H_5)_4FeI_4$ a compound for which the spin direction is known and taken as z axis (32):

$$H_z^{SD,val+overlap} = 2\mu_B\langle r^{-3}\rangle_{5p}(\tfrac{4}{5}\rho^S_{5p_z,5p_z} - \tfrac{2}{5}\rho^S_{5p_y,5p_y} - \tfrac{2}{5}\rho^S_{5p_x,5p_x})^{1/2} \qquad (111a)$$

$$H_y^{SD,val} = 2\mu_B\langle r^{-3}\rangle_{5p}(\tfrac{3}{5}\rho^S_{5p_y,5p_z})^{1/2} \qquad (111b)$$

$$H_x^{SD,val} = 2\mu_B \langle r^{-3}\rangle_{5p} \left(\frac{3}{5}\,\rho^S_{5p_x,5p_z}\right)^{1/2} \quad ; \tag{111c}$$

$$H_a^{SD,\ell ig} = 2\mu_B [(3R_aR_z - R^2\delta_{az})/R^5] \langle S_z^{Fe}\rangle . \tag{111d}$$

The spin densities $\rho^S_{\mu,\nu}$ (Table 7) are defined in Eqn. (48).

TABLE 7

Calculated iodine parameters (taken from Ref. 32): orbital populations and spin densities $\rho^s_{\mu\nu}$ *for* $N(C_2H_5)_4FeI_4$.

	A-site [a]	A-site [b]	B-site [a]	B-site [b]
5s	1.88 [a]		1.87 [b]	
$5p_z$	1.58		1.69	
$5p_y$	1.97		1.96	
$5p_x$	1.80		1.83	
$10^{-2}\rho^s_{5s,5s}$	0.17		0.49	
$10^{-2}\rho^s_{5p_z,5p_z}$	7.45	7.45	7.70	5.40
$10^{-2}\rho^s_{5p_y,5p_y}$	7.70	5.40	7.45	7.45
$10^{-2}\rho^s_{5p_x,5p_x}$	7.58	6.40	7.58	6.40
$10^{-2}\rho^s_{5p_y,5p_z}$	0	-	0	-
$10^{-2}\rho^s_{5p_x,5p_z}$	-0.35	-	0	-

(a) The first row in this table was derived from bond order matrix elements (Eqn. 23).

(b) The second row in this table includes overlap contributions as estimated in the Mulliken approximation (Eqn. 28).

5.5 Supertransferred Field

The supertransferred field H^{ST}, which corresponds to the interaction between magnetic ions separated by a diamagnetic ligand, depends on geometry, the number of metal-ligand-metal chains and on the magnitude of total spin. The importance of H^{ST} has been discussed by Sawatzky et al (87), who investigated the influence of nearest-neighbor cations, of bond distance, and bond angles upon H^{ST}. In some cases H^{ST} can take considerably large values as reported by Belov et al (88) and Goldanskii et al (89) for the ^{119}Sn nuclei (H^{ST}=21 Tesla in $Ca_xY_{3-x}(Fe_3)Fe_{2-x}Sn_xO_x$, and by Evans et al (90) for the ^{121}Sb nuclei (H^{ST}= 30 Tesla) in Sb-substituted ferrites. At the ^{57}Fe nuclei H^{ST} takes smaller values (up to 5 Tesla) as found by Simanek et al (91) in $LaFeO_3$ and by Sawatzky et al (92) in $CoFe_2O_4$, $MgFe_2O_4$, and $MnFe_2O_4$.

To obtain an expression for H^{ST} we consider a linear molecule as pictured in Fig. 16, where A may be a paramagnetic (S = 2) ion with spin-down, B is the cation for which we want to determine the hyperfine field H^{ST}, and O might be an oxygen between A and B. In forming MO's with four of them being singly occupied (S = 2) we shall find some of the spin-density being transferred from A to oxygen valence-orbitals $\phi_\mu : \rho^S_{\mu\mu}$ (see Eqn. 48).

This has the consequence that the overlap distortion of the B cation spin-up core-s-orbitals will be less than the spin-down

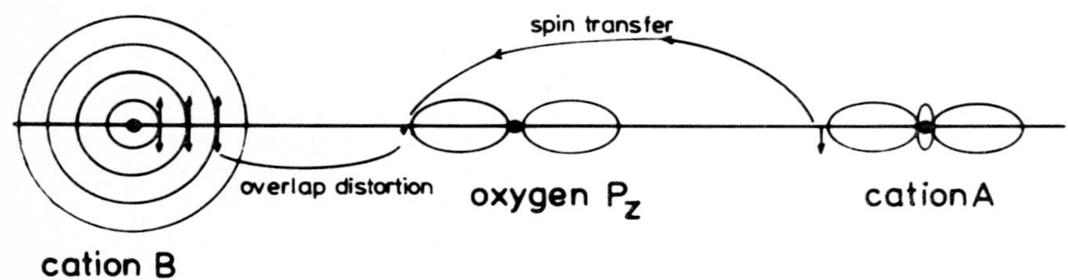

Fig. 16. Diagram of a triatomic molecule A-O-B showing the orbitals and spin transfer to be considered for calculating H^{ST}.

core-s-orbitals, causing a net spin-down density at the nucleus of site B, $\rho^{\downarrow}_B(0)$. The total overlap distortion $\rho_{overlap}(0)$ (see Eqn. 68) in Section 3 is caused by the total oxygen-2p density $P_{\mu\mu}$ (see Eqn. 23), while the net spin density $\rho^{\downarrow}_B(0)$ is caused by $\rho^S_{\mu\mu}$ only; thus it seems a reasonable approximation to derive $\rho^{\downarrow}_B(0)$ from:

$$\rho^{\downarrow}_B(0) = \rho_{overlap}(0) \frac{\rho^S_{\mu\mu}}{P_{\mu\mu}} . \qquad (112)$$

Together with the expression in Eqn. (97) we may then derive H^{ST} as:

$$H^{ST} = \frac{16\pi}{3} \mu_B \rho_{overlap}(0) \frac{\rho^S_{\mu\mu}}{P_{\mu\mu}} . \qquad (113)$$

Taking α-$FeSO_4$ as an example (77) we are concerned with only two of the six neighboring oxygens within the FeO_6 cluster belonging to a Fe-O-Fe chain through which the supertransferred process can take place. The net spin density $\rho^{\downarrow}_{Fe}(0)$ in this case is 0.022 a_0^{-3}, leading to

H^{ST} = 2.3 Tesla, which is in qualitative agreement with H^{ST} values obtained from studying orthoferrites (87, 91) and ferrimagnetic spinels (87,92). Because α-$FeSO_4$ is an antiferromagnet, H^{ST} contributes with the same sign as H^C to H^{int}, while in the case of ferromagnetic coupl-coupling H^{ST} would have opposite sign compared to H^C.

Taking $N(C_2H_5)_4FeI_4$ as an additional example we derive from Table 7 a relatively high value for the spin transfer $\rho^s_{\mu\mu}$ from iron to iodine. However, the distance from iodine to the B iron within the chain $Fe^A I_4^- \ldots Fe^B I_4^-$ is so large that its overlap contribution to $\rho_{overlap}(0)$ becomes negligible. Hence H^{ST} is also negligible for this compound; this finding is justified in view of the low ordering temperature of about 20 K (66) of this compound.

5.6 Results

From the various examples in which H^{int} has been evaluated (32,64,77,9) we select two iron-iodine compounds (32,64), $N(C_2H_5)_4FeI_4$ and $Fe[S_2CN(C_2H_5)_2]_2I$, both being paramagnetic with $S = \frac{5}{2}$ and $S = \frac{3}{2}$, respectively. These two systems have the advantage, that both nuclei, ^{57}Fe and ^{129}I, can be studied with Mossbauer spectroscopy, and that the ground state electronic configuration derived from MO calculations describes the hyperfine field $H^{int}(Fe)$ as well as $H^{int}(I)$.

In $N(C_2H_5)_4FeI_4$ we are concerned with two magnetically non-equivalent iodine sites A and B (66), which has to be reflected by the different spin-dipolar fields $S^{SD}(I_A)$ and $H^{SD}(I_B)$, because $H^C(I_A) = H^C(I_B) \neq 0$, and $H^L(I_B) = H^{ST}(I_B) = 0$ (32). The calculated components of the spin-dipolar field at the iodine nuclei in $N(C_2H_5)_4FeI_4$ are summarized in Table 8 for the two different sites A and B. The experimental difference of ± 2.3 ± 0.7 Tesla between $H^{int}(I_A)$ and $H^{int}(I_B)$ at the two sites is well accounted for by the calculated spin-dipolar contributions when overlap and ligand effects are included. However, we want to point out the lack of agreement if the overlap and ligand contribution is neglected. It is interesting to note that $H^{SD,lig}$ is of the order of 0.2 Tesla. The H_a^{SD} components can be further tested with respect to the orientation Θ_A of $H^{int}(I_A)$ and Θ_B of $H^{int}(I_B)$, respectively, relative to the EFG principal axis. The measurement yields $\Theta_A = 27^o \pm 5^o$ and $\Theta_B = 86^o \pm 5^o$. The deviation of Θ_A or Θ_B from the angle between the spin direction and the EFG principal axis is due to the x-component H_x^{SD} (Table 8). For the B site H_x^{SD} is zero, therefore, Θ_B is 90° (see Fig. 15), in agreement with the experimental value. For the A site, however, H_x^{SD} takes the value

TABLE 8

Calculated spin-dipolar fields (in Tesla) at the iodine A and B sites in $N(C_2H_5)_4FeI_4$ by including overlap contributions estimated in Mulliken approximation (taken from Ref. 32).

		z-axis	x-axis
A site	$H^{SD,val}$ + overlap	+ 1.15	- 0.19
	$H^{SD,lig}$	+ 0.24	- 0.34
B site	$H^{SD,val}$ + overlap	- 1.15	0
	$H^{SD,lig}$	- 0.24	0

-0.53 Tesla; this together with $H^{int}(I_A)$ = 10.7 $\pm$ 0.3 Tesla leads to an angle of 2.8° $\pm$ 0.1° between the spin direction (z) and $H^{int}(I_A)$. This leads finally to a calculated value Θ_A = 32.4° $\pm$ 0.1° in good agreement with the experimental one of 27° $\pm$ 5°.

The calculated internal field H^{int}(Fe) in $N(C_2H_5)_4FeI_4$,(depending on the model which is used to derive q_{3d} and q_{4s}), and the corresponding data for $N(C_2H_5)_4FeBr$ and $N(C_2H_5)_4FeCl$ are summarized (32) in Table 9 together with experimental values (66). From the electronic structure derived for $Fe[S_2CN(C_2H_5)_2]_2I$ (see Fig. 14) we evaluate the H^{int}(Fe) and H^{int}(I) contributions tabulated in Table 10 together with the H^{int}(Fe) results obtained, for $Fe[S_2SN(C_2H_5)_2]_2Cl_3$. These values show that the main contribution to H^{int} comes from H^{core} while H^{val}, H^L and H^{SD} contribute with varying degrees. Comparing the values derived from bond order matrix elements with those including overlap effects, we conclude that successful correlation between experiment and calculation is obtained if overlap contributions are included in the calculations; this conclusion holds good for internal magnetic fields as well as for isomer shifts and quadrupole splittings at the nuclei of ^{57}Fe and ^{129}I.

6. SUMMARY

We have summarised the basic principles involved in MO theory for closed and open shell configurations using the LCAO approximation. We have also considered various methods used in the calculation of electron density, EFG, and magnetic hyperfine field at the nuclear site of a Mossbauer atom.

TABLE 9

Calculated (32) and experimental (66) internal magnetic fields at the nuclear site of ^{57}Fe (in Tesla) for $FeCl_4^-$, $FeBr_4^-$, and FeI_4^- in $N(C_2H_5)_4FeX_4$ (with X = Cl, Br, I).

		$FeCl_4^-$	$FeBr_4^-$	FeI_4^-
$H^{int}(Fe)$	a	-50.78	-45.51	-48.31
	b	-45.05	-40.38	-34.46
	c	-44.40	-40.31	-36.89
	exp.	-47.0±1	-42.0±0.5	-34.4±0.5

(a) Derived from using bond order matrix elements for Fe3d and 4s charge (Eqn. 23).

(b) Includes overlap contributions as estimated from a dipole correction procedure (Ref. 21).

(c) Includes overlap contributions as estimated in the Mulliken approximation (Eqn. 28).

The electron density at the Mossbauer nucleus has contributions from the valence as well as from the core electrons. The core contribution depends on the effective configuration of the Mossbauer atom, and on the overlap of the core orbitals with the ligand orbitals.

The EFG at the nucleus arises from valence electrons in the molecule and from ligand cores. We have discussed various levels of approximation in calculating the overlap and ligand electron contributions to the EFG. The Sternheimer shielding function, $\gamma(r)$, was found to have considerable influence upon the estimation of the EFG tensor. It was essential to consider spin-orbit coupling and magnetic field effects on energetically low-lying multiplet states in understanding the temperature dependence of the EFG tensor.

The effective magnetic field at the Mossbauer nucleus was derived in terms of the externally applied magnetic field and the effective electron spin located at the Mossbauer atom. The contributions to the internal field were grouped into contact, orbital, spin-dipolar, and supertransferred fields.

Successful correlation between experimental and calculated values of isomer shifts as well as EFG and internal field is obtained if overlap contributions are included in the calculation.

Studying a large number of molecules or a series of spectroscopic observables from one compound is most important in narrowing the

TABLE 10

Calculated contributions (64) to the internal field H^{int} (in Tesla) at the nuclear site of ^{57}Fe for $Fe(dtc)_2Cl$ and at the nuclear sites of ^{57}Fe and ^{129}I for $Fe(dtc)_2I$, together with experimental results for H^{int}

	iron in $Fe(dtc)_2X$				iodine in $Fe(dtc)_2I$		
		X = Cl		X = I			
H^{1s-3s}	(a)	-25.2	(a)	-21.6	H^{1s-5s}	(c)	-6.0±0.5
	(b)	-34.7	(b)	-21.2	H^{5s}	(d)	+0.91
H^{4s}	(a)	+ 2.0	(a)	+ 0.7	H_z^{SD}		+0.05
	(b)	+ 4.73	(b)	+ 1.81	$H^{int}_{calc.}$	(e)	5.1±0.5
H_z^L		+ 5.3		+ 5.46	$H^{int}_{exp.}$		5.3-6.7
H_y^L		-		-			
H_x^L		-		- 0.91			
H_z^{SD}		-11.99		- 5.64			
H_y^{SD}		+ 0.96		- 0.19			
H_x^{SD}		- 0.13		+ 1.39			
H^{int}_{calc}	(b)	36.67	(b)	19.62			
$H^{int}_{exp.}$		32.9±1.0		17.2-22.4			

(a) Derived from using bond order matrix elements for Fe3d and 4s charge.

(b) Includes overlap contributions as estimated in the Mulliken approximation.

(c) Core polarization contribution (see Eqn. 102).

(d) Direct I5s contribution (see Eqn. 102).

(e) All H_p^L contributipns are negligible.

confidence limits of the calculated quantities. In fact, the MO calculations enable us to interpret simultaneously, all Mossbauer parameters and the results obtained from other techniques such as esr, magnetic susceptibility, nmr, optical, photo electron, x-ray and neutron diffraction spectroscopy etc.

It is, of course, always preferable to choose a method with minimum number of approximations. There is, however, good reason to derive results with less absolute accuracy, i.e. by applying methods with the increased number of approximations (like in the presently used IEHT-CI method) if the system under study is too large to work with an extended basis set, or to work out all integrals explicitly, or if the spectroscopist is interested in trends rather than in high accuracy. In future, ab initio MO and band structure calculations will probably play a more important role when large and fast computer facilities are available.

ACKNOWLEDGEMENTS

We express our thanks to the Deutsche Forschungsgemeinschaft. One of us (A.T.) wishes to convey his thanks particularly to Professor U. Gonser and Professor F.E. Harris for stimulating and fruitful discussions over many years. With respect to the discussion of specific subjects we are very much indebted to Dr. J.M. Friedt (FeX_4, X=Cl,Br,I), Dr. A. Kostikas (Fe dtc_2 X, X=Cl, I), Dipl.-Phys. S. Lauer (Sternheimer shielding), Dr. R. Reschke (numerical integration), and Dr. R. Zimmermann (spin-orbit coupling and magnetically induced EFG).

REFERENCES

1. F. Hund, Z. Physik, 40(1927) 742; ibid, 42(1927)93; 51(1928)759; 73(1931)1.
2. R.S. Mulliken, Phys. Rev., 32(1928)186; ibid, 32(1928)761. 33(1929)730; 41(1932)49.
3. J.E. Lennard-Jones, Proc. Roy. Soc. (London), A198(1949)1,14.
4. J.C. Slater, Phys. Rev., 35(1930)509; ibid, 34(1959)1293.
5. C.C.J. Roothaan, Rev. Mod. Phys., 23(1951)69.
6. J.C. Slater, Phys. Rev., 35(1930)210.
7. J.A. Pople and R.K. Nesbet, J. Chem. Phys., 22(1954)571.
8. E. Huckel, Z. Physik, 70(1931)204.
9. R. Hoffman, J. Chem. Phys., 39(1963)1397.
10. R. Rein, N. Fakuda, H. Win, G.A. Clarke and F.E. Harris, J. Chem. Phys., 45(1966)4743.
11. J.A. Pople and D.P. Santry, Mol. Phys., 7(1964)269; 9(1965)301.
12. R.G. Parr, J. Chem. Phys., 20(1952)239.
13. J.A. Pople, D.P. Santry and G.A. Segal, J. Chem. Phys., 43(1965)S129.
14. J.A. Pople, Trans. Faraday Soc., 49(1953)1373.
15. K.H. Johnson and F.C. Smith, Chem. Phys. Lett., 7(1970)541.
16. H.P. Durand and J. Barthelat, Mol. Phys., 33(1977)159; ibid, 33(1977)181; Theor. Chim. Acta, 38(1975)283.
17. L.C. Cusachs, J. Chem. Phys., 43(1965)S157.
18. L.C. Cusachs and J.W. Reynolds, J. Chem. Phys., 43(1965)S160.
19. D.G. Carrol, A.T. Armstrong and S.P. McClynn, J. Chem. Phys., 44(1966)1865.
20. C. Giessner-Prettre and A. Pullman, Theor. Chim. Acta, 9(1968)279.
21. R. Rein, G.A. Clarke and F.E. Harris, Quantum aspects of heterocyclic compounds in chemistry and biochemistry, II, Israel Academy of Sciences and Humanities, 1970.

22. J.A. Pople and G.A. Segal, J. Chem. Phys., 43(1965)S136; 47(1967)158.
23. J.A. Pople, D.L. Beveridge and P.A. Dobosh, J. Chem. Phys., 47(1967)2026.
24. S.B. Brown, M.J.S. Dewar, G.P. Ford, D.J. Nelson and H.S. Rzepa, J. Amer. Chem. Soc., 100(1978)7832 and references therein.
25. A. Trautwein and F.E. Harris, Theor. Chem. Acta, 30(1973)45.
26. R. Reschke, Dr. - Arbeit, Angewandte Physik, Universitat des Saarlandes, Saarbrucken, W. Germany, 1976, p. 13.
27. W.C. Nieuport, D. Post and P. Th. Van Duijnen, Phys. Rev. B, 17(1978)91.
28. W.H. Flygare and D.W. Hafemeister, J. Chem. Phys., 43(1965)789.
29. A. Trautwein, F.E. Harris, A.J. Freeman and J.P. Desclaux, Phys. Rev. B, 11(1975)4101.
30. R. Reschke, A. Trautwein and F.E. Harris, Phys. Rev. B, 15(1977)2708.
31. R. Reschke, A. Trautwein, and J.P. Desclaux, J. Phys. Chem. Solids, 38(1977)837.
32. J.P. Sanchez, J.M. Friedt, A. Trautwein and R. Reschke, Phys. Rev., B, 19(1979)365.
33. C.W. Christoe and H.G. Drickamer, Phys. Rev. B, 1(1970)1813.
34. E. Clementi, IBM J. Res. Develop. Suppl., 9(1965)2.
35. R.E. Watson, Phys. Rev., 111(1958)1108.
36. R. Reschke and A. Trautwein, J. Phys. (Paris) Colloq., 12(1976)C6-459.
37. H. Micklitz and F.Z. Litterst, Phys. Rev. Lett., 33(1974)480.
38. K.J. Duff, Phys. Rev. B, 9(1974)66.
39. E.V. Mielczarek and D.A. Papaconstantopoulos, Phys. Rev., B, 17(1978)4223.
40. D. Guenzburger, B. Maffeo and M.L. de Siqueira, J. Phys. Chem. Solids, 38(1977)35.
41. A.T. Kai, H. Annersten and T. Ericsson, Preprint (1979).
42. G.K. Shenoy and F.E. Wagner (Eds.), Mossbauer Isomer Shifts, North Holland, Amsterdam, 1978.
43. R. Reschke, Dr. - Arbeit, Angewandte Physik, Universitat des Saarlandes, Saarbrucken, W. Germany, 1976, pp. 63-65.
44. J.L.K.F. de Vries, C.P. Kreijzers and F. de Boer, Inorg. Chem. 11(1972)1343.
45. A. Trautwein and F.E. Harris, Phys. Rev. B, 7(1973)4755.
46. A. Trautwein, R. Reschke, I. Dezsi and F.E. Harris, J. Phys. (Paris) Colloq., 37(1976)C6-463.
47. A. Trautwein, R. Reschke, R. Zimmermann, I. Dezsi and F.E. Harris, J. de Phys. (Paris) Colloq., 35(1974)C6-235.
48. R. Reschke and A. Trautwein, Theor. Chim. Acta, 47(1978)85.
49a. R.M. Sternheimer, Phys. Rev., 80(1950)102; ibid 84(1951)244; 86(1952)316; 95(1954)736; 105(1957)158; 164(1967)10.
49b. R.M. Sternheimer, Phys. Rev., 146(1966)140.
49c. R.M. Sternheimer, Phys. Rev. A, 6(1972)1702.
49d. H.M. Foley, R.M. Sternheimer and D. Tycko, Phys. Rev., 93(1954)734.
49e. R.M. Sternheimer and H.M. Foley, Phys. Rev., 92(1953)1460; ibid, 102(1956)731.
49f. R.M. Sternheimer and R.F. Peierls, Phys. Rev. A, 3(1971)837.
50. T.P. Das and R. Bersohn, Phys. Rev., 102(1956)733.
51. E.G. Wikner and T.P. Das, Phys. Rev., 107(1957)497; ibid. 109(1958)360.
52. R. Ingalls, Phys. Rev., 128(1962)1155.
53. E.S. Chang, R.T. Pu and T.P. Das, Phys. Rev., 174(1968)16.

54. S.N. Ray, Taesul Lee and T.P. Das, Phys. Rev. A, 9(1974)93.
55. S.N. Ray, Taesul Lee, T.P. Das and R.M. Sternheimer, Phys. Rev. A, 9(1974)1108, ibid. A, 11(1975)1804.
56. M. Vajed-Samii, S.N. Ray and T.P. Das, Phys. Rev. B, 12(1975)4591.
57. S. Ahmad and D.J. Newman, J. Phys. C: Solid State Phys., 11(1978)L277.
58. J. Lahiri and A. Mukherji, Phys. Rev., 153(1967)386; ibid. 155(1967)24.
59. P.K. Mukherjee, A.P. Roy and A. Gupta, Phys. Rev. A, 17(1978)30.
60. S. Lauer, V.R. Marathe and A. Trautwein, Phys. Rev. A, 21(1980)2355
61. M.G. Clark, G.M. Bancroft and A.J. Stone, J. Chem. Phys. 47(1967)4250.
62. E. Kreber, U. Gonser, A. Trautwein and F.E. Harris, J. Chem. Phys. Solids, 36(1975)263.
63. A. Trautwein, E. Kreber, U. Gonser and F.E. Harris, J. Chem. Phys. Solids, 36(1975)325.
64. V.R. Marathe, A. Trautwein and A. Kostikas, J. Phys. (Paris) Colloq., 41(1980)C1-315.
65. A. Kostikas, D. Petridis, A. Simopoulos and M. Pasternak, Solid State Comm., 13(1973)1661.
66. J.M. Fried, D. Petridis, J.P. Sanchez, R. Reschke and A. Trautwein, Phys. Rev. B, 19(1979)360.
67. P. Zory, Phys. Rev., 140(1956)A1401.
68. A. Trautwein, Y. Maeda, U. Gonser, F. Parak and H. Formanek, Proc. 5th Int. Conf. in Mossbauer Spectroscopy, Bratislava (CSSR) Sept. 1973.
69. U. Gonser, Y. Maeda, A. Trautwein, F. Parak and H. Formanek, Z. Naturforsch., 296(1974)241.
70. R. Zimmermann, Nucl. Instrum. Methods, 128(1975)537.
71. R.W. Grant, H. Wiedersich, R.M. Housley and G.P. Espinosa, Phys. Rev. B, 3(1971)678.
72. R. Ingalls, Phys. Rev., 133(1964)A787.
73. H. Eicher and A. Trautwein, J. Chem. Phys., 50(1969)2540.
74. P.M. Champion, E. Munck, P.G. Debrunner, P.F. Hollenberg and L.P. Hager, Biochemistry, 12(1973)426.
75. P.R. Edwards, C.E. Johnson and R.J.P. Williams, J. Chem. Phys., 47(1967)2074.
76. N.N. Greenwood and T.C. Gibbs, Mossbauer Spectroscopy, Chapter 6, Chaptman and Hall, London, 1971.
77. R. Zimmermann, A. Trautwein and F.E. Harris, Phys. Rev. B, 12(1975)3902.
78. A. Wehner, Dissertation, Universitat Erlangen-Nurnberg, 1973.
79. D. Samaras and J. Coing-Boyat, Bull. Soc. Fr. Mineral Cristallogr. 93(1970)190.
80. A. Trautwein, Y. Maeda, F.E. Harris and H. Formanek, Theor. Chim. Acta, 36(1974)67.
81 A. Trautwein, R. Zimmermann and F.E. Harris, Theor. Chim. Acta, 37(1975)89.
82. A. Abragam, J. Horowitz and M.H.L. Pryce, Proc. Roy, Soc. (London), A230(1955)169.
83, R.E. Watson and A.J. Freeman, Phys. Rev., 123(1961)2027.
84. A. Abragam and M.H.L. Pryce, Proc. Roy. Soc. (London), A205(1951)135.
85. P. Ganguli, V.R. Marathe and S. Mitra, Inorg. Chem., 14(1975)970.
86. B. Bleaney, in A.J. Freeman and R.B. Frankel (Eds.), Hyperfine Interactions, Academic Press, New York, 1967.
87. G.A. Sawatzky and F. van der Woude, J. Phys. (Paris) Colloq., 35(1974)C6-47.

88. K.P. Belov and I.S. Lyubutin, Sov. Phys. -JETP Lett., 1(1965)16.
89. V.I. Goldanskii, V.A. Trukhtanov, M.N. Devisheva and V.F. Belov, Sov. Phys. -JETP Lett., 1(1965)19.
90. B.J. Evans, in I.J. Gruvermann (Ed.), Mossbauer Effect Methodology, Plenum, New York, 1967.
91. N.L. Huang, R. Orbach and E. Simanek, Phys. Rev. Lett., 17(1967)34.
92. G.A. Sawatzky, F. van der Woude and A.H. Morrish, Phys. Lett., 25A(1967)147; J. Appl. Phys., 39(1968)1204; Phys. Rev., 187(1969)747.
93. A. Trautwein and R. Zimmermann, Phys. Rev. B, 13(1976)2238.

CHAPTER 8

PARAMAGNETIC HYPERFINE STRUCTURE

K. Spartalian

Department of Physics, The University of Vermont,
Burlington, Vermont, USA

1. INTRODUCTION

The development of Mossbauer spectroscopy as a technique with a wide variety of applications has established its usefulness as a research tool with recognized influence and importance in investigative research. The Mossbauer isotope of ^{57}Fe is most extensively studied because of the fortuitous combination of experimental convenience (relatively long-lived sources, considerable natural isotopic abundance, modestly expansive instrumentation) with challenging theoretical problems. The versatility of the iron, which can exist in all paramagnetic spin states available to a transition metal ion as well as in alloys and ordered systems, makes it a prime candidate for the study of hyperfine interactions in a variety of circumstances.

The information that is most commonly available in a Mossbauer spectrum concerns the target nucleus and its environment as deduced from the strength and symmetry of the hyperfine interactions affecting the nuclear levels, the magnetic hyperfine interaction providing most of the details. The present article will be limited to the presentation of paramagnetic hyperfine structure in the Mossbauer spectra of ^{57}Fe with examples drawn from studies on materials of biological import.

The general assumption underlying the study of paramagnetic hyperfine structure is that the target iron nucleus is well isolated from other iron nuclei and can only interact with the outside world via its own electrons. The electrons themselves may interact with ligand electrons in the host material but not with electrons residing at neighboring iron nuclei. In this model the distribution of unpaired electrons gives rise to a magnetic field at the nucleus. The theoretical considerations concerning the origins of this field are discussed in section 2. Perhaps the most important use of paramagnetic hyperfine structure in Mossbauer spectra is the determination of the electronic structure of the iron ion. The link between the magnetic hyperfine interaction and models for the electronic structure are also discussed in Section 2. Certainly, the interpretation of Mossbauer spectra ought to involve modeling of the hyperfine interactions and simulation of Mossbauer spectra that match the experimental data. Section 2 discusses how the magnetic hyperfine interaction can be incorporated into the electronic model and provides general rules for the procedure and philosophy behind the set up of the spectrum simulation.

It is often necessary to design a battery of measurements for maximum information returns without redundancy. An attempt to provide the reader with guidelines for proceeding experimentally with an unknown specimen is found in Section 3. This section is intended not as a rigid laboratory protocol but rather as a series of statements about one's expectations when various experimental conditions are in effect.

Biologically relevant materials are chosen because they are most likely to exhibit paramagnetic hyperfine structure (the iron centers are usually well separated from one another in large molecules). Section 4 examines the points made in the preceding sections through the use of examples with which the author is most familiar. (Other workers in the field have significantly contributed to the knowledge presented here; their efforts are collectively acknowledged, but giving proper credit where credit is due goes beyond the scope of the present article.) The reader is referred to the review article on biological materials by Huynh and Kent appearing in this volume and to the review articles cited therein.

2. THEORETICAL CONSIDERATIONS

The philosophy behind interpreting Mossbauer spectra that exhibit paramagnetic hyperfine structure is simple: one is required to find a model on the basis of which one is able to simulate theoretical spectra that compare successfully against the experimental spectra in a variety of experimental conditions (applied magnetic fields, sample temperatures). The parameters of the theoretical model should then in principle characterize the properties of the sample and, depending on the completeness of the model, they ought to account for related magnetic data such as g-values measured by electron paramagnetic resonance (EPR) or susceptibility values.

In all situations one has to solve a Hamiltonian of the form:

$$H = H_{el} + H_{mhf} + H_{q} \tag{1}$$

H_{el} describes the electronic interactions, H_{mhf} describes the magnetic hyperfine interactions and H_q describes the quadrupole interaction. Strictly speaking, the dimensionality of the above Hamiltonian for a given electronic orbital state is (2L+1)x(2S+1)x(2I+1) where L and S are the orbital and spin angular momenta of the electronic part and I is the spin angular momentum of the nucleus (I=3/2 for the excited state, I=1/2 for the ground state). For example, in the high-spin ferric case the electronic ground state has L=0, S=5/2 so that one is faced with a 24x24 matrix for the excited nuclear state and a 12x12 matrix for the nuclear ground state. Except for cases where the spin Hamiltonian formalism is used, the electronic Hamiltonian must include at least one or two additional spin multiplets so that the size of the Hamiltonian matrix may become unmanageably large. Even though modern computers are capable of handling large matrices, it is worthwhile to reduce the dimensionality of the Hamiltonian if routine computations of Mossbauer spectra are required and especially if one uses a least-squares fitting procedure as a method for comparing theory with experiment.

The reduction is accomplished by the application of a small magnetic field ($\simeq$ 200 gauss). Fields of this magnitude dominate the Zeeman interaction of the electron spin and one may safely neglect the magnetic field of about 10 gauss created by the nucleus at the electron site. Thus, considerable simplification is introduced in the theory because the electronic and the nuclear problems can be solved separately; S and I are good quantum numbers. The nuclear spin is considered in this approximation to interact with the resultant of the applied field, $\vec{H}_{app}$, and the magnetic field created by the electron distribution. The latter is commonly called the internal field $\vec{H}_{int}$. The Hamiltonian describing the nuclear splittings for the excited state is a 4x4 matrix written as

$$H = -g_n^* \beta_n \vec{H}_{eff} \cdot \vec{I} + H_q \, , \tag{2}$$

where $\vec{H}_{eff} = \vec{H}_{int} + \vec{H}_{app}$ and g_n^* and β_n are the nuclear gyromagnetic ratio and nuclear Bohr magneton. The 2x2 ground state matrix has the same form with $H_q = 0, g_n = -1.749\ g_n^*$.

Assuming that $\vec{H}_{int}$ is known, the problem of generating a spectrum using the Hamiltonian in Equation (2) for arbitrary orientations of $\vec{H}_{app}$ has been solved and will be outlined below. In this article we will be concerned with $\vec{H}_{int}$, methods for its computation and its effects on the Mossbauer spectra.

2.1 Crystal Field Calculations

An approach that has found considerable success in the interpretatio of the Mossbauer spectra has been the crystal field calculation. One starts with a scheme for the electronic structure, namely a listing of the electronic energy levels and their associated wavefunctions. A complete listing is not necessary since $\vec{H}_{int}$ is governed by the ground state. By the same token the choice of the ground state is all-important as is the positioning in energy of all the excited states that interact with the ground state via spin-orbit coupling. There are no stringent rules as to how many states one should consider, but 4-5 judiciously chosen spin multiplets ought to suffice for most applications. Educated guesses about the electronic structure can be made on the basis of the symmetry of the molecule, the stereochemistry of the iron, x-ray crystallography, EPR, susceptibility measurements, the profile of the spectra under various experimental conditions (see below), etc.

The electronic Hamiltonian is written as

$$H_{el} = H_{cf} + H_{so} + H_z . \qquad (3)$$

H_{cf} describes the crystal-field splittings, i.e. the energy separations and possibly the admixtures among the selected spin multiplets; it is a term that basically reflects the symmetry and strength of the interaction of the iron ion with its environment. H_{so} is the spin-orbit interaction:

$$H_{so} = \zeta \sum_k \vec{\ell}_k \cdot \vec{s}_k , \qquad (4)$$

and H_z is the Zeeman interaction:

$$H_z = \beta \vec{H}_{app} \cdot \sum_k (\vec{\ell}_k + 2\vec{s}_k), \qquad (5)$$

where ζ is the spin-orbit coupling constant ($\zeta \simeq 350\ cm^{-1}$ for iron), β is the electronic Bohr magneton and the summations formally extend over all 3d electrons. The dimensionality of H_{el} is of course determined by the total number of electronic states originally selected.

Of particular interest are the zero-field splittings of the ground state because they determine the g-values as measured by EPR and the paramagnetic susceptibility at low temperature. In the absence of an applied field, the coupling of the ground state to the excited states via the spin-orbit interaction gives rise to fine structure, i.e. partial removal of the spin degeneracy of the ground state. If the spin is half-integral, the ground state splits into the appropriate number of Kramers doublets. In the case of integer spin, the spin degeneracy may be completely removed depending on the symmetry of H_{cf}.

We now turn our attention to $\vec{H}_{int}$ which for a given electronic eigenstate has three sources: (a) The orbital part, which using classical considerations can be thought of as the magnetic field produced by the electron current loop. In operator form it is written as

$$\vec{H}_{orb} = H_o \sum_k \vec{\ell}_k , \tag{6}$$

where H_o is a constant approximately equal to -61 T (1). The orbital contribution to the magnetic hyperfine field is obviously a measure of the unquenching of orbital angular momentum by the spin-orbit interaction. The orbital contribution will be significant when there is degeneracy (or near-degeneracy) in the one-electron orbital states. (b) The dipolar contribution, which can be thought of as the interaction of the electronic spin dipole with the nuclear spin dipole. It is written as

$$\vec{H}_{dip} = H_o \sum_k \{ 3(\vec{s}_k . \hat{r}_k)\hat{r}_k - \vec{s}_k \} \tag{7}$$

and is a measure of the spatial distribution of the electronic spin. The dipolar contribution is anisotropic, the anisotropy depending on the details of the angular distribution of the eigenstate. In the simple case of a single-electron in a 3d orbital the dipolar contribution has axial symmetry. In the axial direction, the dipolar contribution is twice as large and opposite in sign as the contributions in the two transverse directions. (c) The Fermi-contact contribution, which has no classical analog. The core s electrons are polarized through the exchange interaction with the 3d electrons. For some symmetries the contact interaction may also involve a small amount of 4s character present in the 3d electrons. The Fermi-contact contribution is written as

$$\vec{H}_{cont} = H_o \kappa \sum_k \vec{s}_k , \tag{8}$$

The constant κ has a value of about 0.35 which implies a Fermi-contact field of about -11 T per unpaired spin. This value for κ has found wide success in fitting experimental Mossbauer spectra (2). The contact interaction is isotropic and in the electronic ground state the contact field always opposes the applied field. A schematic depiction of the contact interaction is shown in Figure 1. The contact

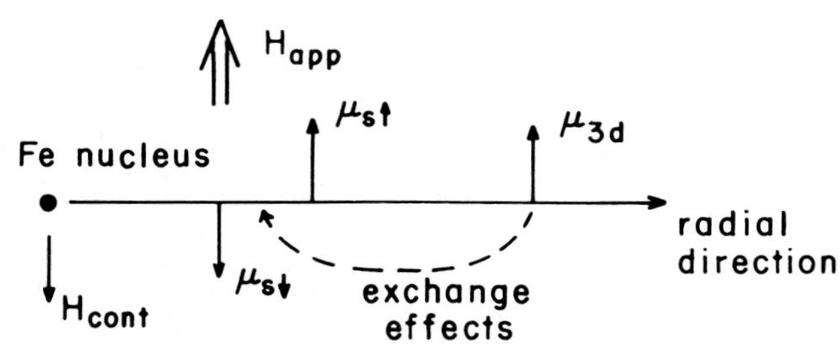

Fig. 1. Schematic representation of the Fermi contact interaction. The applied field polarizes the 3d electron moment "up". The exchange interaction between unpaired 3d electrons and the s electrons results to a net "down" field. Because the Fermi contact interaction is isotropic, the contact field opposes the applied field regardless of the orientation of the latter.

interaction is isotropic and in the electronic ground state the contact field always opposes the applied field. A schematic depiction of the contact interaction is shown in Figure 1. The contact contribution is usually dominant in high-spin (ferrous or ferric) complexes while in low-spin (ferric) complexes any of the three contributions may dominate. The contact contribution is either augmented or diminished by the dipolar contribution. Along the direction of axial symmetry of a prolate electronic orbital the dipolar contribution opposes the contact contribution in the ground state; in a direction perpendicular to the symmetry axis the opposite is true. Similar results obtain for oblate orbitals (Figure 2). Note that the summations above extend over unpaired electrons only; paired electrons do not contribute to the magnetic hyperfine field.

In writing the operators in summation form, it is assumed that the basis states are antisymmetrized products of single-electron states written as Slater determinants. The matrix elements in this case can be computed either by hand or by computer following standard procedures (3). For nearly-free ions, i.e. when the electron-electron repulsion energy is much stronger than the crystal-field energy (weak crystal field limit), it may be convenient to use the

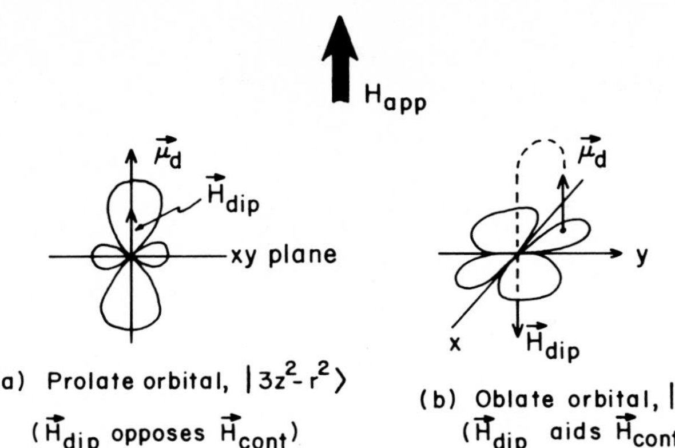

Fig. 2. Schematic representation of the dipolar field $\vec{H}_{dip}$ in the ground state of one-electron 3d orbitals. The applied field is "up" in both cases and so is the direction of the ground state dipole moment $\vec{\mu}_d$. (a) When the orbital is prolate, $\vec{H}_{dip}$ is parallel to the applied field and therefore opposes the contact field. (b) In an oblate orbital $\vec{H}_{dip}$ opposes the applied field and therefore aids $\vec{H}_{cont}$. In the transverse direction the dipolar contribution has a sign that is reversed with respect to the longitudinal direction and a magnitude half as large.

$|LM_L\ SM_S>$ representation with a "total operator" replacing the summation of single-electron operators:

$$\vec{O} = \sum_k \vec{o}_k \ , \ \vec{o}_k = \vec{s}_k \ , \ \vec{\ell}_k, \text{etc.}$$

When the crystal-field energies dominate the electron-electron repulsion effects (strong crystal field limit), one has to use a representation that suitably reflects the symmetry of the crystal field. The advantage of using the "total operators" is that in both the weak and the strong crystal field limits the computation of matrix elements is simplified through the use of the Wigner-Eckart theorem (4).

An actual calculation proceeds then as follows: The Hamiltonian in Equation (3) is written for a particular orientation of the applied field. This is diagonalized and eigenstates E_i and eigenvectors U_i are determined for the ith electronic state. The expectation of $\vec{H}_{int}$ associated with the ith state is computed in the usual manner:

$$< \vec{H}_{int} >_i = U_i^{\dagger}\vec{H}_{int}U_i .$$

This expectation value is used to find $\vec{H}_{int}$ in the appropriate relaxation limit and hence the theoretical Mossbauer spectrum through equation (2).

2.2 Spin Hamiltonians

In instances where a crystal field calculation is not practical, a convenient parameterization of the Mossbauer spectra is provided by the spin Hamiltonian formalism. A complete exposition of the subject can be found in Abragam and Bleaney (5). Spin Hamiltonians have been used to interpret Mossbauer spectra in cases where it is known that the fine structure results in a low-lying spin multiplet, well separated from the others. A spin Hamiltonian matrix then takes into account crystal field plus spin-orbit effects in the ground multiplet only; the electronic part in an applied field is written as

$$\mathcal{H}_{SH} = DS_z^2 + E\,(S_x^2 - S_y^2) + 2\,\beta\vec{H}_{app}\cdot\vec{S}\,, \tag{9}$$

where D and E are the usual spin Hamiltonian parameters. S is the effective spin assigned to the ground multiplet which has 2S+1 states. Fourth order terms in the spin matrices may be included if the spin $S \geqslant 2$, but they have found limited use in the interpretation of Mossbauer spectra (6, 7). Note that for $H_{app} = 0$, $\mathcal{H}_{SH}$ provides the zero-field splittings, i.e. simulates the coupling through the spin-orbit interaction of the ground state with the excited multiplets. The ratios of the zero-field splittings and the admixtures of the basis $|M_s\rangle$ states depend solely upon E/D while D is a factor that determines the overall energy scale. In the spin Hamiltonian approximation the jth component of the expectation value of the internal field associated with the ith eigenstate of the above matrix is related to the expectation value of the spin as follows:

$$\langle H_{int}^{(j)}\rangle_i = \sum_j A_{\ell j}\,\langle S_j\rangle_i \quad \ell = x,y,z\,. \tag{10}$$

The constants of proportionality $A_{\ell j}$ are known as the components of the magnetic hyperfine tensor (A tensor) and are variables of the calculation. In most cases the A tensor is diagonal, i.e. its principa frame coincides with the principal frame of the electric field gradient (efg).

At this point it should be emphasized that care must be exercised when using spin Hamiltonians to interpret Mossbauer spectra. Even though good fits to the data may be obtained on the basis of spin Hamiltonian theoretical models, the electronic structure implied by the spin Hamiltonian parameters, and the zero-field splittings in particular, must not be taken literally. A given Mossbauer spectrum that exhibits paramagnetic hyperfine structure is the result of a uniquely structured internal field interacting with the nuclear levels, but the set of spin Hamiltonian parameters that gives rise to this field may not be unique. Figure (3a) illustrates the parameter dependence of the transverse components of the spin expectation values associated with the ground and first excited states in the case of an axial (E=0) spin Hamiltonian with effective spin S=5/2. A similar plot with effective spin S=2 is shown in figure 3(b). The

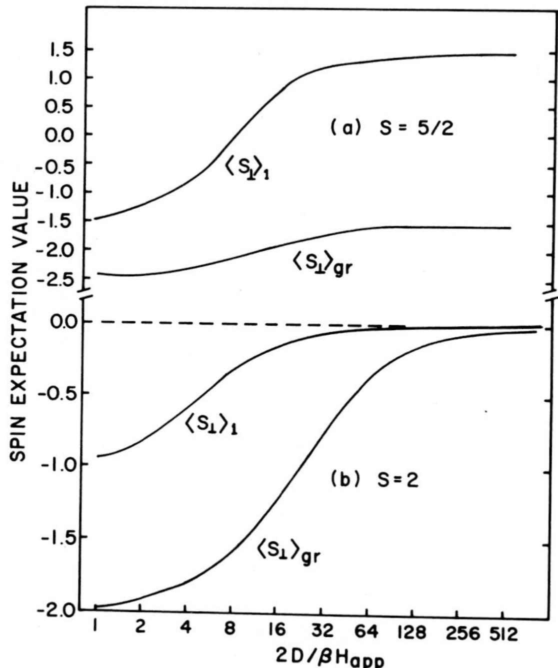

Fig. 3. Transverse spin expectation values for the ground and first excited electronic states computed with an axial spin Hamiltonian. (a) S=5/2, high-spin ferric case. (b) S=2, high-spin ferrous case. The values are given as functions of the dimensionless parameter $2D/\beta H_{app}$*. Spin saturation occurs towards the left.*

expectation values are plotted as functions of the dimensionless parameter $2D/\beta H_{app}$. It becomes apparent that the internal field as written in Equation (10) is the product of two quantities that are essentially variables of the calculation, the components of the

A-tensor and the spin expectation values. Thus, the use of the spin Hamiltonian necessitates the independent determination of the A-tensor and of D and E.

In cases where it is desirable to fit the Mossbauer spectra in order to obtain a measure of the magnitude of the internal field, it is often useful to consider a spin Hamiltonian with effective spin S=1/2. This method is applicable when the profile of the spectrum indicates that only one type of internal field acts on the nucleus. We write a spin Hamiltonian:

$$\mathcal{H}_{SH} = H_{app}^{(x)} G_{xx} S_x + H_{app}^{(y)} G_{yy} S_y + H_{app}^{(z)} G_{zz} S_z . \quad (11)$$

The G parameters are included in order to take into account that the actual electron spin is not really 1/2 and that its polarizability is a function of direction. One may be tempted to identify the G parameters as the g-values similarly defined in an EPR experiment. Such an identification would be misleading because the actual electronic states in the manifold may be strongly mixed especially in a high applied field. It would be improper to make the assumption, implicit in EPR analysis, that there exists an excited electronic state which is the time-reversed counterpart of the ground state. The 2x2 Hamiltonian matrix can be easily diagonalized and for the low energy state we obtain the spin expectation values

$$< S_x > = -\tfrac{1}{2} (G_{xx} \sin\theta \cos\phi / G),$$

$$< S_y > = -\tfrac{1}{2} (G_{yy} \sin\theta \sin\phi / G),$$

$$< S_z > = -\tfrac{1}{2} (G_{zz} \cos\theta / G) ,$$

with

$$G \equiv [(G_{xx}^2 \cos^2\phi + G_{yy}^2 \sin^2\phi) \sin^2\theta + G_{zz} \cos^2\theta]^{\frac{1}{2}}.$$

As usual the z-axis is defined by the efg and the angles θ and ϕ are the conventional spherical angles for the direction of the applied field. The negative sign indicates that the spin is antiparallel to the applied field. The magnetic hyperfine field associated with the electronic ground state,

$$\vec{H}_{int} = |\tilde{A}| . < \vec{S} > / g_n \beta_n ,$$

is taken to represent the internal field, and the computation of the Mossbauer spectrum proceeds as before. The variables in the calculatio are the values for the G_{ii} and the A_{ii}.

Note that the A_{ii} can be computed from a crystal field model (section 2.1) simply by taking the ratio of the model's prediction of the expectation values of the internal field and the spin associated with the ground electronic state:

$$A_{ii} = g_n \beta_n \langle U_{gr} | H_{int}^{(i)} | U_{gr} \rangle / \langle U_{gr} | S_i | U_{gr} \rangle .$$

In practice, when considering a crystal field model for the electronic structure, it is more convenient first to fit the spectra to a phenomenological model with S=1/2 as described in this section, and then attempt to generate the fitted parameters with the use of the crystal field model. This two-step method saves considerable computation time but cannot be justified unless, as stated earlier, only one type of internal field acts on the nucleus.

When the construction of a crystal field model is not feasible, the A-tensor can serve as a parameter for comparison purposes and as an indicator for the composition of the hyperfine field. For example, an isotropic A ($A_{xx} = A_{yy} = A_{zz}$) implies that the magnetic hyperfine field entirely originates in the contact interaction. An anisotropic A-tensor implies modification of the contact contribution by the dipolar and orbital contributions. When $\tilde{A}$ is anisotropic, the profile of the Mossbauer spectra will depend strongly on the relative strength of the components of $\tilde{A}$. As a simple illustration, assume that $G_{xx} = G_{yy} = G_{zz}$, i.e. that the electron spin is polarized along the direction of the applied field, and consider the two possibilities for the case of axial $\tilde{A}$ ($A_{xx} = A_{yy} = A_{\perp}$): (a) $A_z = 0$, $A_{\perp} \neq 0$. The probability that a site in a powder (or frozen solution) sample will have the internal field between angles θ and θ + d θ is

$$dp = \tfrac{1}{2}\sin\theta d\theta$$

The internal field has components

$$H = H_{\perp}^{(int)} = \alpha\sin\theta, \; H_z^{(int)} = 0 \; (\alpha \equiv -\tfrac{1}{2}A_{\perp}/g_n\beta_n) .$$

From this it can be easily shown that the expression

$$\frac{dp}{dH} = \frac{H}{\alpha^2 (1 - H^2/\alpha^2)^{\frac{1}{2}}}$$

gives the probability that a site has an internal field between the values H and H+dH. A plot of dP/dH as a function of H gives the probability distribution for H and is shown in Figure 4(a).

(b) $A_z \neq 0$, $A_\perp = 0$. In this case

$$H_1^{(int)} = \cos\theta,\ H_z^{(int)} = \alpha\cos\theta \quad (\alpha \equiv -\tfrac{1}{2}A_z/g_n\beta_n)$$

and similar considerations give the simple result

$$dP/dH = \text{constant}$$

which is plotted in figure 4(b).

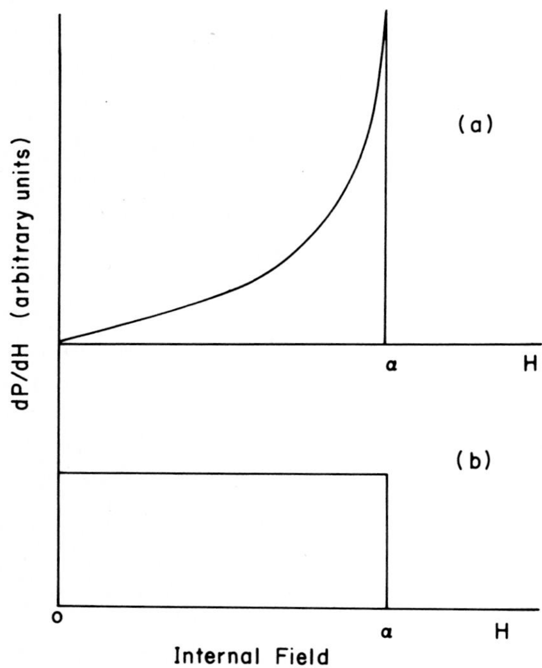

Fig. 4. Fractional probabilities for the internal field values in a polycrystalline specimen. (a) When $A_z = 0$, $A_\perp \neq 0$ the internal field is in the transverse direction and has a well defined value; the spectrum is therefore well defined. (b) When $A_z \neq 0$, $A_\perp = 0$ the internal field is in the longitudinal direction and all values are equally probable; the spectrum is therefore smeared out.

The upshot of the foregoing discussion is that in case (a) each site is favored by probability to yield a spectrum that is characteristic of nearly the same internal field value, the maximum $H^{(int)} = A/2g_n\beta_n$, and the spectra will show well-resolved lines. In case (b) value of H_{int} from zero to maximum will be equally probable, and the spectrum will be broad and featureless.

2.3 Relaxation Effects

Sharp, well-defined paramagnetic hyperfine structure is usually observed in the slow relaxation limit. This occurs when the magnetic hyperfine field at a particular instant arises from one of the states in the electronic manifold and is essentially constant over a time of order 10^{-7} seconds, the lifetime of the excited ^{57}Fe nucleus. When several electronic states are thermally accessible, the resulting Mossbauer spectrum is the sum of individual spectra, each weighted by an appropriate Boltzmann factor and each arising from the interaction of the nucleus with a particular electronic state.

For ferric salts, calculations in the slow relaxation limit are applicable at low temperatures, $T \leqslant 20$ K. This upper limit should not be taken literally, however, because if the sample is in frozen solution, solvent effects may affect the relaxation rate. For example, aqueous solutions upon freezing may precipitate the iron complex at the grain boundaries. The result is that the Mossbauer spectra may be characteristic of fast relaxation even at liquid helium temperature because of spin-spin interaction between neighboring ions even though the sample is nominally in solution. In order to guard against such effects it is advisable to use solvents that form a nice glass upon freezing.

In high-spin ferric complexes the electronic spin in the ground multiplet is relatively well-isolated from lattice vibrations since that state is formally an orbital singlet (L=0). As a result, with an appropriate solvent, it becomes possible to increase the temperature to near 20 K and still observe Mossbauer spectra exhibiting slow relaxation. With iron complexes other than high-spin ferric, slow relaxation spectra are generally observed when the question of relaxation rate becomes academic, that is when only the lowest energy state in the ground electronic multiplet contributes to the internal field, the other members of the multiplet being high in energy. Such is the case, for example, in a strong applied field at low temperature.

In the fast relaxation regime several electronic states are populated and the electrons are assumed to relax rapidly among the available states. The Mossbauer spectrum arises from the interaction of the nucleus with a single magnetic hyperfine field which is the thermal average of the fields associated with all electronic states in the manifold. The limit of fast relaxation is usually reached when the overall energy spread of the ground multiplet is of order kT. For most cases this occurs at about 30 K even in the presence of strong applied fields. When relaxation is fast, and good fits to the spectra can be obtained in the approximation that the thermal averages of the components of the internal field in a given direction are proportional to the component of the applied field in that direction (2). This statement implies that the electron spin is always polarized along the direction of the applied field and the effective field in equation (2) is written as

$$\vec{H}_{eff} = \vec{H}_{app} \cdot (\tilde{1} + \tilde{\omega}). \tag{12}$$

The symmetric tensor $\tilde{\omega}$ (sometimes referred to as the internal field coefficient) describes the effect of nuclear interaction with the induced electronic moment in an applied field. At low temperature ($\beta H_{app} / kT >> 1$), $\tilde{\omega}$ should be field dependent although its use in the slow relaxation regime is questionable because the electron spin may not necessarily be free to follow the applied field. When the internal field coefficient approximation is used to fit Mossbauer spectra, the variables of the calculation are of course the components of $\tilde{\omega}$. Note that for a powder specimen at the high end of the temperature scale the internal field, spin and paramagnetic susceptibility are proportional to each other and vary inversely with temperature. A 1/T dependence for the $\tilde{\omega}$ components is of course expected and has indeed been observed in the cases where the method has been applied (see, e.g., 8, 9).

At intermediate temperatures and fields the relaxation rate may be neither fast nor slow and the internal field changes at random from one to another of a finite number of possible values associated with the eigenstates of the electronic part of the Hamiltonian. Stochastic models (10, 11) that predict the lineshape as a function of the rates of transition of the nuclear resonant frequency have been used successfully to fit the Mossbauer spectra in a few cases (12, 13). Knowledge of the electronic structure is a prerequisite for the application of these models, and the calculations are lengthy, involved and time-consuming on the computer. The information to be obtained concerns mostly the mechanism of the relaxation process and is beyond the scope of this article.

2.4 Computation of spectra

Figure 5 shows a flowchart diagram of the basic steps that are required for the computation of the Mossbauer spectrum when the applied field is at spherical angles θ and ϕ with respect to the quantization (usually the efg) axes. For a powder or frozen solution

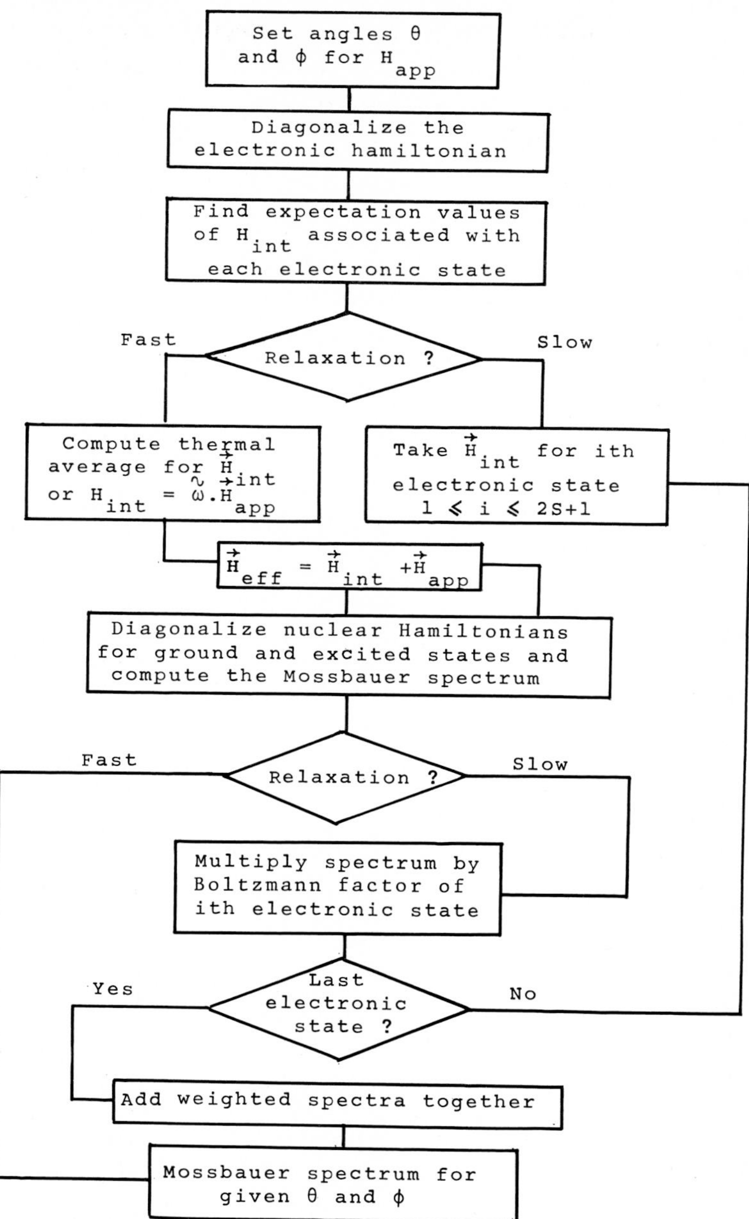

Fig. 5. Flowchart diagram of the basic steps for the computation of the Mossbauer spectrum when the applied field is at angles θ and φ with respect to the quantization axis. For a polycrystalline specimen the applied field is stepped at equal solid angle increments and the spectra from each orientation are added with equal weights.

average, spectra are calculated with the applied field stepped at equal increments of solid angle $d\Omega = d(\cos\theta)d\phi$ over an octant of the sphere and added with equal weights. When the system has axial symmetry, stepping the angle ϕ becomes redundant. For most calculations division of the octant into a 10x10 grid is sufficient. If computation time is limited, a grid as coarse as 7x7 can be used without appreciable granularity in the computed spectra.

Diagonalization of the nuclear Hamiltonian (Equation 2) yields eigenvectors as follows:

$$|\Psi_j> = \sum_{k=1}^{4} a_{jk}|m_k>, \; m_k = 3/2, \ldots, -3/2 \; ,$$

for the eigenvalue E_j^e of the excited nucleus and

$$|\Phi_i> = \sum_{k=1}^{2} b_{ik}|m_k'>, \; m_k' = \tfrac{1}{2}, -\tfrac{1}{2} \quad ,$$

for the eigenvalue E_i^g of the ground state. The coefficients a_{jk} and b_{ik} are in general complex. With the auxiliary definitions (14)

$$A = a_{ji}^* b_{i1} + (1/\sqrt{3}\,)\, a_{j2}^* \, b_{i2} \; ,$$

$$B = \sqrt{2/3}\; (a_{j2}^* b_{i1} + a_{j3}^* \, b_{i2}) \quad ,$$

$$C = a_{j4}^* b_{i2} + (1/\sqrt{3})\; a_{j3}^* b_{i1} \quad ,$$

the intensity of the line occurring at energy $E_j^e - E_i^g$ with the applied field parallel to the gamma beam is given by

$$\begin{aligned} I_{||}^{ij} &= (|A|^2 + |C|^2)\; (1 + \cos^2\theta) + 2|B|^2 \sin^2\theta \\ &+ 2 \; \mathrm{Re}\left[AC^* \sin^2\theta e^{-2i\phi}\right] + \sqrt{2}\; \mathrm{Re}\left[AB^* \sin 2\theta e^{-i\phi}\right] \\ &- \sqrt{2}\; \mathrm{Re}\left[BC^* \sin 2\theta e^{-i\phi}\right] \; . \end{aligned}$$

When the applied field is perpendicular to the gamma beam, an average has to be taken over all possible directions of observation in a plane normal to the applied field. The result is

$$I_{\perp}^{ij} = (|A|^2 + |C|^2)\ (2 + \sin^2\theta) + 2|B|^2\ (1+\cos^2\theta)$$
$$- 2\mathrm{Re}\left[AC^{*}\sin^2\theta e^{-2i\phi}\right] - \sqrt{2}\ \mathrm{Re}\ \left[AB^{*}\sin 2\theta e^{-i\phi}\right]$$
$$+ \sqrt{2}\ \mathrm{Re}\left[BC^{*}\sin 2\theta e^{-i\phi}\right] .$$

The normalization of the two intensities is arbitrary, and the angles θ and ϕ in both expressions are as always the spherical angles of the applied field with respect to the axes of quantization.

Once the positions and intensities are known, Lorentzians of suitable linewidth (Γ = 0.2-0.4 mm/s) are folded over the lines and the generation of the spectrum is complete. The Mossbauer spectrum thus simulated may be compared against the experimental data either visually or by using a least-squares fitting procedure (15).

The obvious question to be raised at this point is how one can tell from the spectra which relaxation regime is suitable for the calculation. There is no compellingly clear cut answer, but certain "rule of thumb" observations can be made. The low temperature criterion for slow relaxation is not always appropriate especially in the case of half-integral spin complexes that are not magnetically dilute. Broad, non-Lorentzian lines with long tails indicate intermediate relaxation while narrow lines are indicative of either the slow or the fast relaxation regime. If the internal field increases with decreasing temperature, at a fixed value of the applied field, relaxation is not slow. With half-integral spin complexes in a strong applied field, if it is known that the electron spin in the ground state has reached its saturation value ($2D \ll \beta H_{app}$), yet the internal field depends on the applied field, then relaxation is not slow. In the final analysis, however, the decision on the choice of the appropriate relaxation limit rests with the experimenter and is often justified *a posteriori* by the success of the calculation in fitting the data.

3. EXPERIMENTAL TECHNIQUES

A variety of experimental conditions are necessary for an unambiguous determination of the electronic structure governing magnetic hyperfine structure. This section will discuss the information that is obtained by studying the field and temperature dependence of the Mossbauer spectra and the procedures required for obtaining this information. For a more detailed account on instrumentation and sample manipulation the reader is referred to Lang (16). The discussion will be limited to either the slow or the fast relaxation limit since it is unlikely that spectra characteristic of intermediate relaxation can provide direct information on paramagnetic hyperfine structure.

Iron paramagnetic complexes can be assigned an even or odd number of outer 3d electrons. As a result the ground electronic state may have either integral spin (S=0, 1, 2) or half-integral spin (S=1/2, 3/2, 5/2). Usually, in an integral-spin complex the crystal-field symmetry is low enough so that the lowest lying member of the multiplet is a spin-singlet which, to first order, remains unaffected by an external magnetic field. With half-integral spin complexes, the lowest lying member is at least a doublet and, to first order, the separation between its members increases linearly with the applied field. Therefore, assuming slow relaxation, small applied fields distinctly sharpen the zero-field spectra by polarizing the 3d electron magnetic moment and thus "turning-on" the magnetic hyperfine interaction. Although the application of small fields to integral-spin complexes may seem pointless, it is nevertheless a handy procedure for separating out any half-integral spin components that could be present whether they be adventitious or not. High magnetic fields are necessary for observing paramagnetic hyperfine structure in integral-spin complexes and for pinpointing the zero-field splittings in half-integral spin complexes. Obviously, the experimental conditions under which Mossbauer spectra are recorded depend on the peculiarities of the sample under investigation. The description of conditions that follows is intended as a listing of routine procedures for the investigation of unknown samples and is by no means complete.

3.1 Zero-field spectra

These are useful mostly in the case of integral-spin complexes that give rise to quadrupole split patterns in zero applied field. Knowledge of the quadrupole splitting ΔE and of the isomer shift δ limits the number of variable parameters when the high-field spectra are examined. Furthermore, if the data interpretation is based on a crystal field model, the temperature dependence of the quadrupole splitting may be instrumental in at least rejecting certain choices for the electronic structure. Given an electronic state, the efg parameters may be calculated as the expectation values of a second rank tensor(17).

The zero-field spectra from half-integral spin complexes may or may not prove to be helpful. The high-spin (S=5/2) complexes tend to give rise to broad, often featureless spectra at high temperature.

At low temperature the zero-field spectra may appear to show well resolved paramagnetic hyperfine structure, but the theoretical simulation in this case is non-trivial. The experimental spectra appear to match the situation in which a small field of about 10 gauss and of random orientation acts on the electronic states, while proper simulations in exactly zero field do not seem to match the data at all (18). The apparent random field has been attributed to effects of neighboring nuclei. At high temperature, low-spin ferric complexes may exhibit quadrupole-split spectra which can sometimes be sharpened with a small applied field. The few intermediate-spin (S=3/2) examples that have been reported give rise to quadrupole-split patterns in zero field with a temperature dependent ΔE.

To summarize, it is good practice to record zero-field Mossbauer spectra from integral-spin complexes over the entire temperature range $4.2 \leqslant T \leqslant$ 200-300 K while the procedure is of limited value in the case of high and low spin ferric complexes.

3.2 Spectra in small applied fields

Small applied fields (H_{app} = 200-400 gauss) do not appreciably affect the integral spin complexes and we therefore limit the discussion to the half-integral case. It should be remembered that, although the applied field may be a few hundred gauss, the field at the nucleus is a few hundred *kilogauss* so that the applied field may be safely neglected when computing the effective field splitting of the nuclear states ($H_{eff} \simeq H_{int}$).

The Kramers degeneracy is lifted by the applied field and the magnetic moments associated with each member of a given Kramers doublet are equal and opposite when the magnetic energy ($\beta H_{app} << kT$) is much smaller than the zero-field splittings. The two magnetic hyperfine fields at the nucleus that are associated with the members of the doublet are equal in magnitude but antiparallel. As a consequence, in the slow relaxation limit, the spectra contributed by each member of the doublet will be identical. In the fast relaxation limit with $\beta H_{app} << kT$, the two members will be equally populated and the net contribution of the doublet to the thermal average of the internal field will vanish.

When relaxation is slow and the zero-field splittings are of order kT, the Mossbauer spectra in small applied field can be used to determine the zero-field splittings. As long as the contributions from at least two doublets are discernible in the spectra, measurements at two different temperatures will provide the separation between the doublets. One temperature is usually 4.2 K; temperatures below this can be obtained by pumping on the liquid helium bath, while higher temperatures can be achieved by heating the sample in an appropriate container. Thermal population or depopulation experiments as described here are of course unnecessary in the case of low-spin ferric complexes which have no zero-field splittings to speak of. The theoretical treatment of low-spin ferric complexes has been extensively presented in Lang and Oosterhuis (20).

In summary, we see that the Mossbauer spectra from half-integral spin complexes in small applied fields are useful for determining the zero-field splittings if the spectra consist of at least two contributions each weighted by an appropriate Boltzmann factor. If only one component is observed over the entire temperature range for which relaxation is slow, then the zero-field splittings cannot be determined to any reasonable degree of accuracy unless strong magnetic fields are applied to the sample.

3.3 Spectra in strong applied fields

In fields as strong as 60 kOe, the ground electronic state is well separated in energy from the others which results at low temperature, to a spectrum characteristics of a single internal field acting on the nucleus. In the case of integral-spin complexes it is most appropriate to record spectra in strong applied fields over a wide range of temperature. Strong magnetic fields mix the spin states and induce a sizable magnetic moment in the ground electronic state. The low temperature spectra will give a measure of the magnetic moment induced in the ground state, while the rate of decrease of the internal field with increasing temperature will give a measure of the spread in energy of the states in the multiplet. Usually, with integral-spin complexes at 4.2 K in a strong field the relaxation rate is irrelevant and the spectra can be treated in the slow relaxation regime; intermediate rates occur somewhere between 8 K and 20 K, while above 20 K most complexes are fast relaxing. When the dominant term in the internal field is the contact contribution which always opposes the applied field, as the temperature is raised, the internal field drops to a value that will cancel the effects of the applied field at the nucleus ($H_{eff} = 0$) in at least one principal direction if not two. The sign of the largest potential derivative (sign V_{zz}) and the value of the asymmetry parameter η can be obtained from fits to spectra in a strong field at the highest available temperature. This is a situation in which the internal field with its 1/T dependence is almost zero and therefore least likely to screen the direct Zeeman interaction of the applied field with the nuclear levels.

With half-integral spin complexes, in the low-spin (S=1/2) case, spectra in strong applied fields provide no information that cannot be supplied by measurements in small applied fields; the magnetic moment of the ground electronic state is independent of the applied field. The intermediate and high-spin ferric complexes are more interesting in strong fields because the magnetic moment of the electronic states is field-dependent; strong fields mix neighboring Kramers doublets and change the composition of the ground state (cf. Figure 3). Most, if not all, of the Mossbauer spectra from intermediate and high-spin ferric complexes have been given the spin Hamiltonian treatment. We shall, therefore, use the spin Hamiltonian formalism as a convenient parameterization of the arguments in the discussion that follows.

The magnetic energy βH_{app} is a variable of the experiment, but its size relative to the zero-field splittings, the size of 2D, determines the nature of the electronic states. When $\beta H_{app} >> 2D$, only the ground electronic state contributes to the spectrum, the spin expectation value is fixed at saturation, and the size of the A-tensor can be determined from the spectrum. The values of $\tilde{A}$ are thus fixed for use in the fits to the spectra in small applied fields. In some cases, most notably in heme proteins, the value of D is so large that the spectra cannot be made insensitive to it even for the strongest attainable field in the laboratory and that makes the separate determination of D and $\tilde{A}$ problematic. The situation can be remedied by studying the Mossbauer spectra at various field values in the strong field region ($H_{app} \geqslant 3$ T) and requiring that a single set of spin Hamiltonian parameters fit all spectra. The method leaves room for ambiguity which can, in principle, be removed by additional fits to the spectra if the contribution from the first excited electronic state can be observed at some combination of field and temperature.

4. EXAMPLES

Numerous examples in the literature illustrate the points discussed so far. For the purposes of the present work, we content ourselves with a few representative cases. The examples are arbitrarily grouped into the two general categories of integral-spin and half-integral spin complexes with three possibilities for the electron spin in each.

4.1 Integral-spin complexes

4.1.1 S=0, no unpaired electrons

A diamagnetic S=0 ground electronic state arises in $3d^6$ when all six electrons are paired in the three low-lying t_{2g} orbitals. Even though the iron in the S=0 state has no net magnetic moment, and therefore no paramagnetic hyperfine structure can be observed, the oxygenated complexes of hemoglobin and myoglobin are mentioned here not only for the sake of completeness, but also because of the controversy surrounding them. The early susceptibility measurements of Pauling and collaborators (21.23) established that oxyhemoglobin is diamagnetic at all temperatures. Oxymyoglobin, a protein whose Mossbauer spectra resemble those from oxyhemoglobin, is thought to have the same magnetic properties. However, recent susceptibility measurements by Cerdonio et al. (24, 25) provide compelling evidence that, contrary to the earlier measurements, above 90 K, there are paramagnetic contributions to the susceptibility that indicate the existence of an excited triplet state at about 150 K above the diamagnetic ground state. The signature of the excited state should, in principle, be visible in the Mossbauer spectra of oxyhemoglobin in a strong applied field. In practice, however, because the assumed triplet's separation from the ground singlet is relatively large, when the temperature is high enough to populate this excited state appreciably, its contribution to the internal field becomes vanishingly small. Nevertheless, the profiles of the Mossbauer spectra from oxyhemoglobin and oxymyoglobin in a strong applied field do depend on temperature even though the internal field seems to be zero at all temperatures: At low temperature the quadrupole splitting ΔE is about 2.2 mm/s with an asymmetry parameter $\eta = 0.2$; as the temperature is raised, ΔE decreases while $\eta \rightarrow 1$ (Lang and Spartalian, unpublished results).

Whatever the electronic structure of oxyhemoglobin and oxymyoglobin, the Mossbauer spectra in zero-field indicate relaxation effects as evidenced by the asymmetric broadening of the two-line quadrupole pattern (26). If the interpretation of Cerdonio et al. is valid, the Mossbauer spectra from these proteins are calculable according to a model whereby the electronic spin relaxes at some intermediate, temperature dependent rate between the ground singlet and the excited triplet. In contrast to this, we mention a detailed study of the oxymyoglobin Mossbauer spectra that attributes the temperature dependence of the zero-field profiles to effects arising from the second order Doppler shift, conformational changes and diffusion (27).

4.1.2 S = 1, two unpaired electrons

With two unpaired electrons the contributions to the magnetic hyperfine field depend to a large extent on the distribution of the unpaired electrons in the available orbital states. Because the spin expectation value cannot exceed the relatively small value of unity, the contact term may or may not be the dominant contribution. Ferrous tetraphenylporphyrin (FeTPP) is an intermediate-spin S = 1 complex whose Mossbauer spectra show an axial, anisotropic internal field with a value that is large in the direction transverse to the quantization axis and nearly zero in the direction along that axis (9). Furthermore, the internal field in the transverse direction points in the same direction as the applied field which indicates that, in the ground electronic state, the contact contribution is overcome by the combined dipolar and orbital contributions. A different situation is found in the case of the S = 1 complex of compound ES of cytochrome C peroxidase (28). Here the internal field is also anisotropic with a large value in the transverse direction and a small value in the longitudinal direction but, unlike the previous case, the internal field opposes the applied field. The internal field is dominated by the contact term as is indeed seen by the collapse of the spectrum at 20 K where $H_{eff} \simeq 0$.

Mossbauer spectra from both complexes were recorded in strong applied fields and fitted in the approximation of the internal field coefficients ω. The predictions of crystal field models successfully matched the temperature dependence of the ω coefficients in both compounds. In view of the differences in the internal field in the two cases, it is worthwhile to compare the two crystal field models: In both cases the two unpaired electrons have been assigned to an orbitally degenerate $|xz\rangle, |yz\rangle$ pair. The difference in the magnetic hyperfine interaction arises from the orbital states that are closest to this pair and are admixed to it by spin-orbit coupling. In the case of FeTPP a $|3z^2 - r^2\rangle$ orbital lies at 1.35ζ below the pair, while in compound ES it is an $|xy\rangle$ orbital at 5.98ζ below the pair.

4.1.3 S = 2, four unpaired electrons

A ground electronic state with S = 2 arises in $3d^6$ when four one-electron 3d orbitals are singly occupied while the fifth orbital is doubly occupied by a "spin up, spin down" pair. The four unpaired electrons in the singly occupied orbitals are expected to provide, at saturation of the spin, a contribution of about -44 T that will dominate the magnetic hyperfine field. The orbital and dipolar contributions are governed by the "spin down" electron because the five "spin up" electrons occupy a spherically symmetric subshell and do not contribute. Thus, the details of the modification of the contact contribution by the orbital and dipolar contributions should, in principle, be indicative of the orbital character of the sixth electron.

Reduced cytochrome P-450 and chloroperoxidase were the first two biological S = 2 materials whose paramagnetic hyperfine structure was extensively studied by the Mossbauer effect (29, 30). Other examples include rubredoxin (31), protocatechuate 3,4 dioxygenase (32), deoxymyoglobin and two model compounds (33, 34). In all cases Mossbauer spectra were recorded at a variety of temperatures in a variety of applied fields and fitted to a spin Hamiltonian with effective spin S = 2. Of interest here is the magnetic hyperfine tensor $\tilde{A}$, a listing of which is shown in Table I. Note that for all

TABLE I

Principal values of the magnetic hyperfine tensor $\tilde{A}$ for some biological S = 2 complexes.

	$A_x/g_n\beta_n$ (Tesla)	$A_y/g_n\beta_n$ (Tesla)	$A_z/g_n\beta_n$ (Tesla)	
Cytochrome P-450$_{cam}$	-18.0	-12.5	-15.0	Champion et al. (29)
Chloroperoxidase	-26.0	-12.0	-15.0	Champion et al. (30)
Protocatechuate 3,4 dioxygenase	-25.0	-24.0	-11.0	Zimmermann et al (32)
Rubredoxin	-20.2	- 8.4	-30.2	Schulz and Debrunner (31)
Deoxymyoglobin	- 4.9	-11.0	- 4.6	Kent et al. (34)

All entries are obtained from S = 2 spin Hamiltonian fits. In the first three complexes the principal axes of the efg do not coincide with those of $\tilde{A}$.

materials the A-tensor is anisotropic. A contact-only contribution to $\tilde{A}$ ought to yield a value $A/g_n\beta_n$ = -22 T corresponding to the usual -11 T per unpaired spin. All the entries in Table I are approximately as expected except for deoxymyoglobin. The values for the principal components of $\tilde{A}$ cannot be below the expected contact-only figure unless there is a sizable orbital contribution to the internal field. This appears not to be the case in deoxymyoglobin because significant unquenching of the orbital angular momentum is precluded by the relatively small zero-field splittings, the temperature dependence of the quadrupole splitting, and the susceptibility results. In view of these findings, Kent et al. ruled out a pure S = 2 quintet for the ground electronic state in deoxymyoglobin and, without constructing a concrete model for the electronic structure, they proposed that the ground state is probably a quintet, quantum mechanically admixed to a spin triplet. It should be noted that myoglobin does not stand by itself in exhibiting an internal field that is unusually small for a S = 2 complex; the model heme compounds studied by Kent et al. (33) and Reed et al. (35) show a similar behavior.

4.2 Half-integral spin complexes

4.2.1 One unpaired electron, S = 1/2

When the crystal field is strong in comparison with the Coulomb repulsion between electron pairs, the five electrons in $3d^5$ are distributed among three t_{2g} orbitals for a net electron spin S = 1/2. Obviously, two orbitals are doubly occupied and one is singly occupied. Crystal or ligand field theoretical treatments have been extensively applied to these complexes because of the relative simplicity of the electronic problem. The standard procedure is the "one-hole" formalism (4) in which the five-electron states in t_{2g}^5 are replaced with single-electron wave functions corresponding to the unoccupied orbital in t_{2g}. This allows easy derivation of closed form expressions for the principal components of the electric field gradient and magnetic hyperfine tensor $\tilde{A}$ (36).

In low symmetry situations, when the principal axes of a rhombic crystal field distortion do not coincide with those of the cubic field, the principal values of the electronic magnetic moment (the so-called g-tensor), the efg principal axes and the magnetic hyperfine axes will not be coincident. If, however, the cubic axes system is related to the rhombic axes by a simple Euler angle rotation, say by α about the "z-axis", the situation is simplified. The principal axes of the efg follow the rhombic axes, but the g-tensor rotates in the opposite sense, through an angle $-\alpha$ about z. The A-tensor also rotates about z but through an angle that depends on the strength of the rhombic field relative to the spin-orbit coupling constant (36).

In this simplified model, the information in the Mossbauer spectra of powder (or frozen solution) specimens of low-spin ferric complexes often suffices to determine the orientations of the g-tensor, the efg axes and the A-tensor with respect to the cubic axes despite the averaging inherent in powder spectra. The paramagnetic hyperfine structure exhibited by the low temperature spectra of the azide and cyanide complexes of myoglobin and cytochrome c peroxidase has been detailed enough to enable the determination of the crystal field parameters, namely the strength of the cubic and rhombic fields and the Euler rotation angle relating the two (37). The hyperfine parameters computed from these models correlate well with single crystal EPR experiments. It should be noted that there is a fourfold ambiguity in the dtermination of the angles relating the principal axes of the hyperfine tensors to the cubic axes. This is so because the cubic axes of the crystal field are indistinguishable and an axial rotation by an angle $+\alpha$ viewed from above the plane of rotation looks the same as a rotation by $-\alpha$ viewed from below. It is, nevertheless, true that the magnetic Mossbauer spectra of low-spin ferric powders or solutions do provide information that is unavailable by EPR on the same specimens. This does not imply that EPR measurements on single crystals of low-spin ferric complexes are rendered superfluous by Mossbauer spectroscopy; single crystal measurements are required for removing the fourfold ambiguity mentioned above.

4.2.2 Three unpaired electrons, S = 3/2

The intermediate spin S = 3/2 state arises when one of the 3d electron orbitals, usually the antibonding $|x^2 - y^2>$ lies much higher in energy than the other four orbitals as a result of a tetragonal distortion in the crystal field. In the electronic ground state, the 3d electrons fill the four low-lying one-electron orbitals according to Hund's rule of maximum spin multiplicity, which in the case of the d^5 configuration is 4 (S = 3/2). The contact contribution is expected to dominate the internal field and this is seen in the case of FeTPP(ClO_4) studied by Spartalian et al.(38). The magnetic hyperfine tensor $\tilde{A}$ is anisotropic and nearly axial, with $A_z \simeq 0$. This indicates that the contact interaction is opposed by the spin dipolar field along the z-axis which implies a prolate distribution of the electronic charge.

Similar results have been obtained from chromatium ferricytochrome c' which at neutral pH is an intermediate-spin S = 3/2 complex while at either high or low pH it is high-spin ferric (39). The electronic structure of the intermediate-spin complex has been approximated by a simple model in which a ground $^4A_2$ quartet state is quantum mechanically admixed to the $^6A_\perp$ sextet lying at about 170 cm^{-1} above it (40). This model has been able to match the EPR and susceptibility results satisfactorily. When applied to the Mossbauer spectra, however, the model is not entirely appropriate in that it predicts the observed anisotropy of $\tilde{A}$ ($A_\perp \neq 0$, $A_z \simeq 0$), but overestimates the magnitude of $\tilde{A}$. In fact, the observed value of $A_\perp/g_n\beta_n$ in ferricytochrome c' is equal to the contact-only value of -22 T. This leaves no room for the dipolar contribution which must aid the contact field in the transverse direction since it opposes the contact field in the longitudinal direction. As for the orbital contribution, there is none for the spin-orbit interaction in the Maltempo model mixes states of different spin multiplicity and, therefore, no unquenching or orbital angular momentum can take place.

The unusually small value of $A_\perp$ is not limited to chromatium ferricytochrome c'. Comparable results are obtained from FeTPP(ClO_4), Fe(OEP)ClO_4 (41) and a tetra-aza-macrocyclic iron (III) complex (42). On the other hand, the small value of $A_\perp$ is not a characteristic of all S = 3/2 complexes. The penta-coordinate dithiocarbamates studied by Wickman and Trozzolo (43) show an isotropic A-tensor which arises from a "contact-only, pure S = 3/2" contribution. The values of $A/g_n\beta_n$ for the S = 3/2 complexes mentioned above are shown in Table II for easy comparison.

TABLE II

Principal values of the magnetic hyperfine tensor $\tilde{A}$ for some intermediate spin S = 3/2 complexes.

	$A/g_n\beta_n$ (Tesla)	
Fe $[S_2CN(i - C_3H_7)_2]_2Cl$	-33.4[a]	Wickman and Trozzolo (43)
Fe $[S_2CN(CH_3)_2]_2Cl$	-33.8[a]	Wickman and Trozollo (43)
Fe (Ph_2[16]NH_4) (SPh)	-27.0[b]	Kock et al. (42)
Fe (OEP) ClO_4	-24.0[b]	Dolphin et al. (40)
Fe TPP (ClO_4). 0.5 <u>m</u>-xylene	-22.6[c]	Spartalian et al. (38)
Chromatium ferricytochrome c'	-22.0[c]	Maltempo et al. (39)

[a] A-tensor is isotropic

[b] Details for the symmetry of $\tilde{A}$ not given

[c] Only $A_\perp$ is tabulated; $A_z \simeq 0$.

4.2.3 Five unpaired spins, S = 5/2

A ground electronic state with S = 5/2 arises when each of the five 3d orbitals is singly occupied. The orbital angular momentum is nominally L = 0 for the five electrons occupy a half-filled subshell. Spin-orbit coupling will admix excited doublets to second order so that the degeneracy due to the spherical symmetry in the ground state is removed and zero field splittings are observed. The orbital and dipolar contributions to the magnetic hyperfine field will be second order corrections to the isotropic contact field. It is a good approximation to assume that the internal field is due entirely to the contact interaction. This assumption simplifies the discussion of these complexes considerably because, in small applied fields, the internal field components associated with a given electronic state are proportional to the spin expectation values which are in turn proportional to the "g-values" measured in an EPR experiment.

In zero field the electronic ground state will be split into three Kramers doublets. In all the high-spin ferric paramagnetic complexes studied by Mossbauer spectroscopy the electronic structure has been parameterized in terms of a spin Hamiltonian with effective spin S = 5/2. Calculations of Mossbauer spectra involving the S = 5/2 spin Hamiltonian have been extensively presented in the literature (see, e.g., 6, 44). Here we will explore qualitatively the effects of the spin Hamiltonian parameters upon the Mossbauer spectra. The discussion will be limited to small applied fields since the signature of the zero-field splittings fades in strong applied fields.

A study of the literature reveals that most paramagnetic high-spin ferric complexes require either an axial spin Hamiltonian with relatively large D or a "fully rhombic" spin Hamiltonian with relatively small D. The magnetic hyperfine fields associated with each Kramers doublet in the two symmetry situations show distinctive features that can best be understood by realizing the significance of the g-values associated with each doublet. For a given doublet, the g-values can be thought of as a measure of the polarizability of the electron spin by the applied field. Thus, an isotropic g-value ($g_x = g_y = g_z$) implies that the electron spin follows the applied field while an anisotropic g, say $g_\perp = 0$, $g_z \neq 0$, implies that the electron spin is largely fixed along the z-axis regardless of the orientation of the applied field. The g_i so defined are calculated by computing the quantum mechanical spin expectation value of the ith spin component with the applied field in the ith direction and multiplying the result by 4. The principal directions x,y,z are obviously referred to the principal frame in which the spin Hamiltonian is written. We define A-values by multiplying the corresponding spin values by the "contact-only" constant $A/g_n\beta_n$ = -22 T. The A-values so defined are the components of the internal field associated with a given doublet. In the slow relaxation limit, Boltzmann factors permitting, the composite Mossbauer spectrum may have as many as 24 lines. Sorting out all those lines is a laborious task, best accomplished by computer. Nevertheless, the general features of the spectra will be evident even to the half-experienced eye and it is these features that we proceed to discuss.

In a polycrystalline specimen, when the applied field is small, all three doublets give rise to sharp, well defined spectra because the magnetic moment and hence the internal field associated with each doublet are independent of the applied field. As long as there is a non-zero component of the applied field along a given direction, one obtains the full value of the magnetic hyperfine field at the nucleus.

For a spin Hamiltonian of axial symmetry the parameter E is set equal to zero and, for positive D, the g-values associated with the three Kramers doublets are (6,6,2), (0,0,6), (0,0,10) in ascending order of energy. The order is reversed for negative D. In the bottom doublet the magnetic hyperfine field is dominated by the transverse component, equal to -33 T; the contribution of the bottom doublet would be a six-line pattern with an overall spread of 10-11 mm/s. The overall spread of the spectrum is found by using the figure of 0.32 mm/s per Tesla, which for iron metal yields the calibration value of 10.66 mm/s for a field of -33 T acting on the iron nuclei. In the middle doublet the internal field is in the z-direction regardless of the orientation of the applied field and has a value of -33 T; the contribution of the middle doublet is a six-line pattern with an overall spread of 10-11 mm/s. Similarly, the top doublet has an associated internal field of -55 T along the z-axis and this gives rise to a six-line pattern with an overall spread of 17-18 mm/s. Calculated spectra of a polycrystalline (or solution) high-spin ferric complex in a situation of axial symmetry are shown in Figure 6.

For a "fully rhombic" spin Hamiltonian the parameter E is set equal to D/3 and the g-values associated with the three Kramers doublets are near the values (0,10,0), (4.3,4.3,4.3) and (0,0,10) in ascending order of energy. It is immediately obvious that the bottom and top doublets contribute six-line patterns with overall spreads of

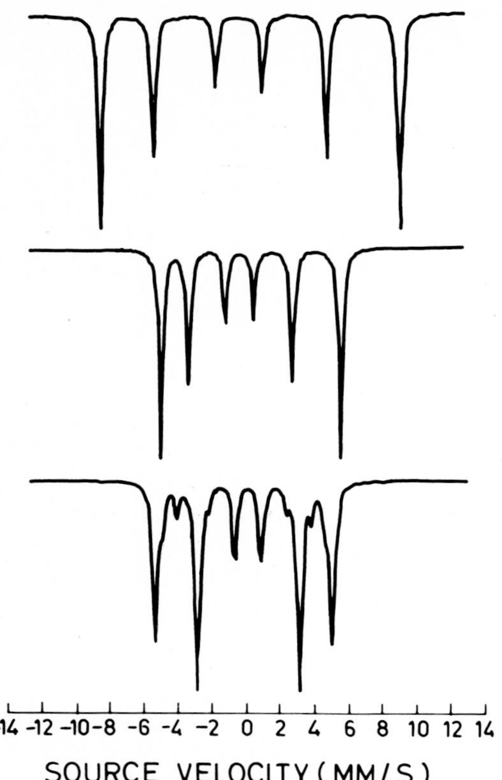

Fig. 6. Calculated spectra of a polycrystalline high-spin ferric complex with an axial spin Hamiltonian in a small applied field. The spectra are arranged in descending order of energy from top to bottom. A small perturbation by an axial efg has been added to the calculation. Some granularity is seen in the spectra even though a 10x10 grid has been used for the computation. The applied field is transverse to the γ beam.

17-18 mm/s. The direction of the internal field is fixed along the y-axis for the bottom doublet and along z for the top. Differences in the contributed spectra will be seen only when the magnetically split nuclear levels are perturbed each in their own way by the quadrupole interaction. The middle doublet shows isotropic g-values which indicates that the spin and hence the internal field follow the applied field. The value of the internal field is about -23 T and results to a six-line pattern with an overall spread of 7-8 mm/s. The axis of quantization, however, follows the applied field so that if the latter is parallel to the γ beam, the middle lines corresponding to the $\Delta m_I = 0$ nuclear transitions vanish. Calculated spectra of a polycrystalline high-spin ferric complex with a "fully rhombic" spin Hamiltonian are shown in Figure 7.

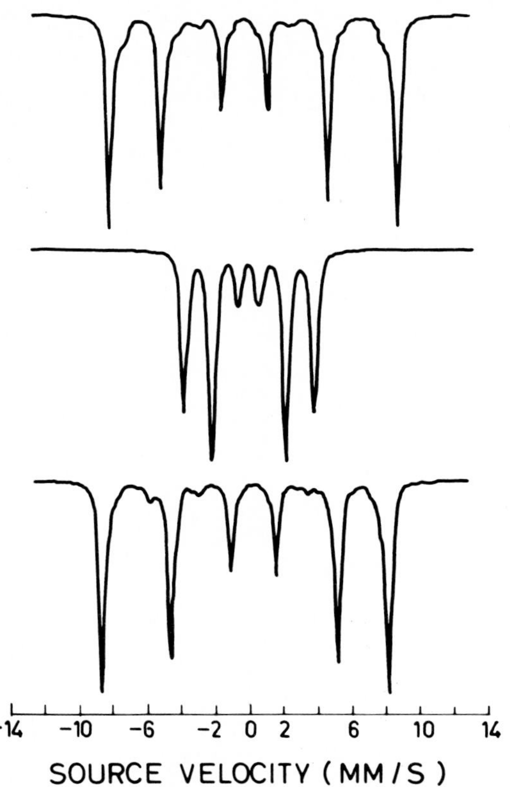

Fig. 7. Calculated spectra of a polycrystalline high-spin ferric complex with a "fully rhombic" spin Hamiltonian in a small applied field. The ordering of states is the same as in Figure 6. A small perturbation of a "fully rhombic" efg (η = 1) has been added to the calculation. The applied field is transverse to the γ beam.

It should be borne in mind that the dominant component in the observed spectrum is the contribution from the ground doublet, the others entering the picture according to their Boltzmann populations. An overall spread in the spectrum of 10-11 mm/s would indicate an axial spin Hamiltonian with positive D. On the other hand, an overall spread of 17-18 mm/s can be the result of either a "fully rhombic" situation or of an axial spin Hamiltonian with negative D. The former is usually the case, but the latter has been observed too (45). Because the middle doublet in the two situations shows markedly different values for the internal field, a distinction between the two can be made by selecting a combination of applied field and temperature values that would make this doublet observable and identifiable by depopulation experiments.

5. CONCLUSIONS

It is undoubtedly true that the examination of paramagnetic hyperfine structure in Mossbauer spectra yields information about electronic structure that is unavailable by other means. When interpreting Mossbauer spectra, one may often be faced with the difficult task of putting order to an apparently complex superposition of spectral lines. The spin Hamiltonian formulation, despite its drawbacks, is a convenient means of discussing the strength and symmetry of the interaction of the iron nucleus with its environment. Furthermore, spin Hamiltonians offer a basis of comparison for samples of the same characteristics. Nevertheless, the set of spin Hamiltonian parameters extracted from a Mossbauer spectrum is not an end in itself; one ought to take a step beyond the determination of these parameters and interpret them in the light of the underlying physical reality.

Electronic structure models that are more sophisticated than spin Hamiltonians do not abound in the literature perhaps because of the difficulties involved in generating such models. Sometimes the information obtained from Mossbauer spectra is not sufficient to distinguish one model from another without additional input from EPR and susceptibility measurements. Other times, the pieces of information available seem not to fit a complete picture at all. By far the most disturbing example of this is the magnetic hyperfine structure of deoxymyoglobin. A model for the electronic structure that accounts for the susceptibility results, the temperature dependence of the quadrupole splitting in zero field and the temperature dependence of the internal field in a strong applied field can be found if only the contact interaction constant ($H_o\kappa$) were allowed to be about half the generally accepted value of -11 T per unpaired spin. Equally disturbing is the observation that a similar situation obtains for the intermediate spin S = 3/2 complexes of chromatium ferricytochrome C' and FeTPP(ClO_4) mentioned earlier. Here, the EPR, susceptibility and Mossbauer results can be fitted using the model of Maltempo if only the value of $H_o\kappa$ were again set at about -6 T per unpaired spin.

We seek a mechanism that will affect the contact interaction only. Overlap and covalent effects which convey electron spin density from the iron to the ligands are not expected to reduce the internal field by more than about 15%. A possible mechanism could be the inclusion of considerable 4s character in the valence electron density. The 4s contribution diminishes the effects of the contact field arising from the exchange-mediated polarization of the other s electrons (1) and this is a step in the right direction. Furthermore, the spherically symmetric 4s electron density provides no valence-electron contribution to the quadrupole interaction, no contribution to susceptibility and is EPR silent. We note that the presence of 4s character in the valence electrons ought to decrease the isomer shift with respect to the value that is obtained when no such character is present (46). The isomer shift for deoxymyoglobin, at 0.92 mm/s with respect to metallic iron, is on the low side for a high-spin ferrous complex. The isomer shift for the intermediate-spin S = 3/2 complexes are typical of their kind which seems to militate against the above argument. Obviously, a concrete mathematical model is required for any credible correlation between contact interaction and isomer shift in these complexes of unusually low internal fields.

Despite the difficulties inherent in the analysis of Mossbauer spectra, the wealth of information present in paramagnetic hyperfine structure can be and has been used to determine the electronic structure at the iron site in a wide variety of complexes. The interplay of the three contributions to the magnetic hyperfine field at the nucleus often fits a unique model which can be corroborated or verified by independent magnetic measurements such as EPR and susceptibility. A thorough understanding of the nuclear hyperfine interactions coupled with the availability of modern computational facilities can promote the quest for a model for the electronic structure to an exercise that is often challenging, occasionally disappointing, but rarely boring.

REFERENCES

1 R.E. Watson and A.J. Freeman in A.J. Freeman and R.B. Frankel (Eds.), Hyperfine Interactions, Academic Press, New York, 1967, p. 53.

2 G. Lang, Quart. Rev. Biophys., 3 (1970) 1.

3 C.J. Ballhausen, Introduction to Ligand Field Theory, McGraw-Hill, New York, 1962.

4 J.S. Griffith, The Theory of Transition Metal Ions, Cambridge University Press, Cambridge, 1961.

5 A. Abragam and B. Bleaney, Electron Paramagnetic Resonance of Transition Metal Ions, Clarendon Press, Oxford, 1970.

6 W.T. Oosterhuis, Structure and Bonding, 20 (1974) 59.

7 K. Spartalian, W.T. Oosterhuis and J.B. Neilands, J. Chem. Phys. 62 (1975) 3538.

8 T.A. Kent, K. Spartalian, G. Lang and T. Yonetani Biochim. Biophys. Acta, 490 (1977) 331.
9 G. Lang, K. Spartalian, C.A. Reed and J.P. Collman J. Chem. Phys., 69 (1978) 5424.
10 M. Blume, Phys. Rev., 174 (1968) 351.
11 M.J. Clauser and M. Blume, Phys. Rev. B, 3 (1971) 583.
12 S. Winkler, C. Schulz and P.G. Debrunner, Phys. Lett., 69A (1978) 360.
13 G.R. Hoy and M.R. Corson, J. Magn. Magn. Mater., 15-18 (1980) 627.
14 G.R. Hoy and S. Chandra, J. Chem. Phys., 47 (1967) 961.
15 G. Lang and B.W. Dale, Nucl. Instrum. Methods, 116 (1974) 567.
16 G. Lang in R.L. Cohen (Ed.), Applications of Mossbauer Spectroscopy, Academic Press, 1976, p. 129.
17 H. Eicher and A. Trautwein, J. Chem. Phys., 50 (1969) 2540.
18 K. Spartalian and W.T. Oosterhuis, J. Chem. Phys., 59 (1973) 617.
19 W.T. Oosterhuis and P.J. Viccaro, Biochim. Biophys. Acta, 264 (1972) 11.
20 G. Land and W.T. Oosterhuis, J. Chem. Phys., 51 (1969) 3608.
21 L. Pauling and C.D. Coryell, Proc. Nat. Acad. Sci. U.S., 22 (1936) 210.
22 D.S. Taylor and C.D. Coryell, J. Amer. Chem. Soc., 61 (1938) 1263.
23 C.D. Coryell, L. Pauling and R.W. Dodson, J. Phys. Chem., 43 (1939) 825.
24 M. Cerdonio, A. Congiu-Castellano, F. Mogno, B. Pispisa, G.L. Romani and S. Vitale, Proc. Nat. Acad. Sci. U.S., 74 (1977) 398.
25 M. Cerdonio, A. Congiu-Castellano, L. Calabrese, S. Morante, B. Pispisa and S. Vitale, Proc. Nat. Acad. Sci. U.S., 75 (1978) 4916.
26 K. Spartalian and G. Lang, J. Phys. (Paris) Colloq., 37 (1976) C6-195.
27 H. Keller and P.G. Debrunner, Phys. Rev. Lett., 45 (1980) 68.
28 G. Lang, K. Spartalian and T. Yonetani, Biochim. Biophys. Acta, 451 (1976) 250.
29 P.M. Champion, J.D. Lipscomb, E. Munck, P.G. Debrunner and I.C. Gunsalus, Biochemistry, 14 (1975) 4151.
30 P.M. Champion, R. Chiang, E. Munck, P.G. Debrunner and L.P. Hayer, Biochemistry, 14 (1975) 4159.
31 C. Schulz and P.G. Debrunner, J. Phys. (Paris) Colloq., 37 (1976) C6-153.
32 R. Zimmermann, B.H. Huynh, E. Munck and J.D. Lipscomb, J. Chem. Phys., 69 (1978) 5463.
33 T.A. Kent, K. Spartalian, G. Lang, T. Yonetani, C.A. Reed, and J.P. Collman, Biochim. Biophys. Acta, 580 (1979) 245.
34 T.A. Kent, K. Spartalian and G. Lang, J. Chem. Phys., 71 (1979) 4899.
35 C.A. Reed, T. Mashiko, W.R. Scheidt, K. Spartalian and G. Lang, J. Amer. Chem. Soc., 102 (1980) 2302.
36 W.T. Oosterhuis and G. Lang, Phys. Rev., 178 (1969) 439.
37 D. Rhynard, K. Spartalian, G. Lang and T. Yonetani, J. Chem. Phys., 71 (1979) 3715.
38 K. Spartalian, G. Lang and C.A. Reed, J. Chem. Phys., 71 (1980) 1832.
39 M.M. Maltempo, T.H. Moss and K. Spartalian, J. Chem. Phys., 73 (1980) 2100.
40 M.M. Maltempo, Quart. Rev. Biophys., 9 (1976) 187.
41 D.H. Dolphin, J.R. Sams and T.B. Tsin, Inorg. Chem., 16 (1977) 711.
42 S. Kock, R.H. Holm and R.B. Frankel, J. Amer. Chem. Soc., 97 (1975) 6714.
43 H.H. Wickman and A.M. Trozzolo, Inorg. Chem., 7 (1968) 63.

44 G. Lang, R. Aasa, K. Garbett and R.J.P. Williams, J. Chem. Phys., 55 (1971) 4539.

45 E. Munck, R. Zimmermann, J.D. Lipscomb and L. Que, J. Phys. (Paris) Colloq., 37 (1976) C6-203.

46 L.R. Walker, G.K. Wertheim and V. Jaccarino, Phys. Rev. Lett., 6 (1961) 98.

CHAPTER 9

MOSSBAUER STUDIES OF BIOMOLECULES

B.H. Huynh and T.A. Kent

Gray Freshwater Biological Institute, University of Minnesota, Navarre, Minnesota 55392.

1. INTRODUCTION

A variety of important biological functions are catalyzed by iron containing proteins. Such functions include oxygen transport, hydrogen transport, electron transfer, nitrogen fixation, hydrogen utilization and aromatic ring cleavage. The determination of the role iron plays in these processes requires probes sensitive to the iron electronic structure. Such probes include paramagnetic susceptibility, electronic paramagnetic resonance (EPR), nuclear magnetic resonance (NMR), electron-nuclear double resonance (ENDOR) and various forms of optical spectroscopy. Each method has its own strengths and weaknesses, and many studies require the complementary use of these techniques. The potential of an additional tool for iron protein research was revealed by Mossbauer's discovery in 1958 (1) of nuclear resonance gamma ray absorption and the subsequent observation of this effect in ^{57}Fe. The first application of "Mossbauer spectroscopy" to a biomolecule was Gonser's study of hemin in 1961 (2). Since then Mossbauer spectroscopy has matured both in its technology and in its methodology and has been used fruitfully to investigate a wide range of iron proteins (3-7). In the present discussion we shall relate some recent developments in Mossbauer studies of biomolecules and attempt to convey the complementary nature of Mossbauer spectroscopy to EPR and other techniques.

Besides ^{57}Fe, other Mossbauer nuclei of biological importance are ^{40}K, ^{67}Zn, ^{127}I, and ^{129}I. However, a relatively easily detectable signal and an abundance of iron proteins have caused the vast majority of work to be done with ^{57}Fe. The choice of proteins discussed below illustrates the range of application of Mossbauer spectroscopy but, of course, also reflects the personal bias and experience of the authors. An index of publications entitled "Mossbauer Spectroscopy of Biological Materials" can be obtained from May (8).

Early applications of Mossbauer spectroscopy to biomolecules concentrated on heme proteins. The first systematic study of hemoglobin and its derivatives was reported by Lang and Marshall (4). Elegant reviews emphasizing the paramagnetic features of the Mossbauer spectra of heme proteins have been written by Debrunner (5) and Lang (6). When these studies were done the general features of the iron environment were already well known and the Mossbauer studies concentrated on explaining the observed spectra in terms of the iron electronic structure. In the process a solid foundation for interpreting paramagnetic Mossbauer spectra was laid down. The Mossbauer data also stimulated several theoretical investigations of the electronic structure of oxygenated and deoxygenated hemoglobin. Both ligand field and molecular orbital approaches have been used (See Section 3). Recently, an updated summary of Mossbauer studies of hemoproteins has been written by Munck (7) who stressed the relation of Mossbauer spectroscopy to other techniques. The electronic structure of the iron in such well known environments is of prime interest, and Mossbauer spectroscopy has and will continue to make significant contributions in this area.

When less is known about the protein of interest, Mossbauer spectroscopy may be used to answer less sophisticated, but equally important questions. A large number of iron proteins contain active centers consisting of clusters of iron atoms. It is important to know the number of irons per cluster, the number of clusters per molecule, and the oxidation states of each cluster. Mossbauer spectroscopy has become an indispensable tool for such tasks. Investigations of multi-iron systems have also shown that combining Mossbauer data with that of other techniques can provide information which other methods could not. Examples are the studies of plant-type ferredoxins and nitrogenase.

Section 2 lists a few essential experimental facts and provides a brief discussion of the hyperfine interactions. Also, an outline of the relationships between EPR and Mossbauer measurements is presented. Section 3 covers proteins containing a mono-iron center. Results obtained for desulforedoxin from *Desulfovibrio gigas* are discussed extensively. Recent developments concerning hemoglibin, low-spin ferric heme compounds, dioxygenase, and horseradish peroxidase compounds I and II are also presented in this section. In Section 4, proteins with multi-iron clusters are discussed. Nitrogenase is used as an example to demonstrate how 28 of the 30 iron atoms in that enzyme can be identified and classified into two novel types of iron clusters. The newly discovered three-iron clusters in ferredoxins from *Azotobacter vinelandii* and from *D. gigas* are discussed. Investigations of proteins containing two-iron and four-iron clusters are also included. In Section 5, we summarize the present state of the art and comment on possible trends of future developments.

2. MOSSBAUER SPECTROSCOPY

2.1 Experimental Considerations

Mossbauer spectroscopy as applied to iron proteins involves the resonant absorption of 14.4 keV γ-rays by ^{57}Fe nuclei. To ensure a good signal to noise ratio, the Mossbauer sample should have a protein concentration which provides at least a 0.5 mM ^{57}Fe concentration per inequivalent iron site. The required volume would be $\sim$ 0.5 ml with a thickness of $\sim$ 0.5 cm. Assuming typical experimental conditions, the data accumulation time for such a sample would be 12 to 24 hours. Since ^{57}Fe is 2% naturally abundant, large protein concentrations are required unless ^{57}Fe enrichment is achieved. For a given signal to noise ratio, the data accumulation time is inversely proportional to the square of the ^{57}Fe concentration. Hence, any possible enrichment of the sample in ^{57}Fe is usually worth the effort. More detailed discussions concerning the experimental techniques of biological applications of Mossbauer spectroscopy can be found in references (3) and (9).

2.2 Hyperfine Interactions

The effects of the electronic structure on the ^{57}Fe nucleus are transmitted primarily via three hyperfine interactions: the isomer shift, the electric quadrupole interaction and the magnetic hyperfine interaction. The physics of these interactions has been discussed in detail elsewhere in this volume (5, 6, 10). Here we will discuss the various terms just enough to provide a consistent notation for the sections to follow.

2.2.1 Isomer Shift

The excited ^{57}Fe nucleus is 0.1% smaller in radius than the ground state nucleus (10, 11) which causes the Mossbauer transition energy to depend on the electron density at the nucleus. This effect produces the so-called isomer shift of the Mossbauer spectrum. To allow isomer shifts to be measured independently of the nature of the source, the centroid of the room temperature absorption spectrum of metallis iron has been adopted as the zero of energy and all shifts are measured relative to it. Using this convention, we may write the isomer shift δ as

$$\delta = K_o - K\,|\Psi(0)|^2 \qquad (1)$$

where K and K_o are positive constants, and $|\Psi(0)|^2$ is the electron density at the absorber nucleus. To a good approximation the 1s and 2s electron density at the nucleus is independent of the chemical environment of the iron ion. However, the 3d electrons shield the 3s electrons and do cause a decrease in the electron density at the nucleus. The more delocalized the 3d electrons, the smaller the isomer shift. This picture is complicated by the presence of ligand electrons residing in the iron 4S orbitals. It is not always clear which of the two effects is dominant and calculàtions of δ are difficult to perform. Hence, the isomer shift is usually treated as an empirical parameter. The most valuable application of isomer shift is the characterization of the state of iron according to oxidation state, degree of covalency, and coordination number (12).

2.2.2 Electric Quadrupole Interaction

The excited (I = 3/2) state of the ^{57}Fe nucleus possess an electric quadrupole moment, and the presence of a low-symmetry electric field will tend to orient this moment. The interaction may be written as

$$H_Q = \frac{eQ}{6}\ \vec{I}.\tilde{V}.\vec{I} \quad , \qquad (2)$$

where e is the magnitude of the electron charge, Q the nuclear quadrupole moment, I the nuclear spin and $\tilde{V}$ the electric field gradient (efg) tensor. $\tilde{V}$ is real, symmetric, and traceless. In the frame in which $\tilde{V}$ is diagonal,

$$H_Q = \frac{eQ}{4} V_{\zeta\zeta}\left[I_\zeta^2 - \frac{5}{4} + \frac{\eta}{3}(I_\xi^2 - I_\sigma^2)\right],\ \eta \equiv (V_{\xi\xi} - V_{\sigma\sigma})/V_{\zeta\zeta}\quad , \qquad (3)$$

where ξ, σ, and ζ denote the principal axes of the efg and η is called the asymmetry parameter. By convention, ξ, σ, and ζ are chosen so that

$|V_{\zeta\zeta}| \geq |V_{\sigma\sigma}| \geq |V_{\xi\xi}|$, which forces $1 \geq \eta \geq 0$.

The I=3/2 quartet will be split into two doublets, while the I = 1/2 doublet will remain degenerate, resulting in two absorption peaks in the Mossbauer spectrum. Diagonalizing H_Q , we find the difference in energy of the two peaks, known as the quadrupole splitting, to be

$$\Delta E_Q = \frac{eQ\ V_{\zeta\zeta}}{2} \left(1 + \frac{\eta^2}{3}\right)^{1/2} \quad . \tag{4}$$

Sign ($V_{\zeta\zeta}$) may be measured by applying a magnetic field to a sample with randomly oriented Fe sites if the internal field due to unpaired electrons is small compared with the applied field. Computer fits to the resulting spectrum will reveal both η and sign ($V_{\zeta\zeta}$). When $\eta \simeq 0$, the absorption peak associated with the I = 3/2, $M_I = \pm 1/2$ states will be split into a triplet by the applied field. The other peak will appear as a broad, ill-defined doublet. For positive $V_{\zeta\zeta}$ the triplet appears on the low energy side. If η is large, computer fits will still determine sign ($V_{\zeta\zeta}$). When $\eta = 1$, the eigenstates of H_Q are equal mixtures of the $M_I = \pm 3/2$ and $\mp 1/2$ states and the spectrum is symmetric. In this case, sign ($V_{\zeta\zeta}$) has no meaning.

2.2.3 Magnetic Hyperfine Interaction

The ^{57}Fe nucleus possesses a magnetic moment and its energy levels will be perturbed by the local magnetic field, $\vec{H}_{eff}$. The interaction can be written as

$$H_M = - g_n\beta_n\ \vec{I}.\vec{H}_{eff} \quad , \tag{5}$$

where β_n is the nuclear magneton and g_n is the nuclear gyromagnetic ratio. g_n has the values 0.1806 ± 0.0014 and -0.1033 ± 0.0008 for the I = 1/2 and I = 3/2 states, respectively (13, 14). $\vec{H}_{eff}$ will be the sum of any applied field and the internal field created by the unpaired 3d electrons. A more explicit expression for $\vec{H}_{eff}$ is

$$\vec{H}_{eff} = \vec{H}_{app} + \vec{H}_{int}$$

$$= \vec{H}_{app} + (P/g_n\beta_n) \sum_k (\vec{\ell}_k + 3(\hat{r}_k.\vec{s}_k)\hat{r}_k - \vec{s}_k - \kappa\ \vec{s}_k) \quad , \tag{6}$$

$$P = 2\beta\ g_n\beta_n\ \langle r^{-3}\rangle$$

where the sum is over the unpaired electrons, $\vec{\ell}_k$ is the orbital angular momentum of the kth electron, $\vec{s}_k$ is the spin of the kth electron, and β is the Bohr magneton. The first term of $\vec{H}_{eff}$ is the applied field. The second is the field at the nucleus due to the orbital motion of the electrons. The third and fourth describe the field due to the dipole moment of the electrons. The last term is the Fermi contact term which accounts for any net polarization of the electronic spin at the nucleus. Such a polarization can result from the exchange interaction between the 3d electrons and the s electrons. Any admixture of s character into the 3d wave functions will also contribute to the contact field. To a good approximation, the net spin at the nucleus will be proportional to $\sum_{\kappa} \vec{s}_k$.

P and κ depend on the radial part of the wave functions and on shielding effects which are very difficult to calculate. Values have been determined empirically for many complexes. Lang and Marshall quoted values of $P/g_n\beta_n = 62$ T and $\kappa = 0.35$ for high- and low-spin ferric heme complexes (4). $P\kappa/g_n\beta_n$ is the internal field per unit spin due to the contact term. For the ferric heme complexes it would be approximately -22 T. For a variety of high-spin ferrous inorganic compounds the survey of Varret (15) shows $P\kappa/g_n\beta_n$ to be approximately -25 T.

2.3 Correlations with EPR Spectroscopy

The paramagnetic properties of an iron complex are determined by its unpaired 3d electrons. Bonding and electrostatic interactions between the iron and its ligands tend to localize the 3d electrons in space, quenching the orbital angular momentum and lifting the fivefold degeneracy of the 3d orbitals. A simple way to approximate such interactions is to assume that the effect of the ligands is to create a static electric field (the so-called crystal field) at the iron site. The crystal field approximation has been used successfully in many studies of inorganic transition metal complexes (16, 17). Calculations of the crystal field based on the assumption that a classical charge distribution can represent the ligands has met with little success. Hence the crystal field is usually treated as a free parameter and adjusted to fit experimental observations. In theory, the one-electron 3d orbitals are split and mixed by the electrostatic field. The multi-electron states are then constructed by placing the electrons in the five 3d orbitals without violating the

Pauli exclusion principle. The energy difference between terms with different numbers of unpaired electrons depends on the relative magnitudes of the Coulomb repulsion and the crystal field energies. In the weak field limit, Coulomb repulsion dominates and the ground term is that with the highest spin. In the strong field limit the ground state will be low-spin, having a maximum pairing of electron spin. Both of these cases occur in biological complexes for which the iron-ligand bonding interactions are characterized by crystal field energies ranging from 10^3 cm^{-1} to 10^4 cm^{-1}. A typical crystal field energy diagram and the pairing of electrons for some typical spin states of ferrous and ferric ions are shown in Figure 1. The eigenstates of the crystal field are spin degenerate and have zero expectation values for the orbital moment. However, the spin-orbit coupling interaction may lift any non-Kramers degeneracy and unquenched some of the orbital moment. Most iron proteins studied to date have a well isolated ground multiplet, and a convenient method of parameterizing their magnetic properties is the spin Hamiltonian formalism. A fictitious spin S is chosen and the temperature and field dependence of the electron moment is approximated by a Hamiltonian typically of the form

$$\mathcal{H}_e = D[S_z^2 - S(S+1)/3 + \frac{E}{D}(S_x^2 - S_y^2)] + \beta\vec{H}_{app}\cdot\tilde{g}_o\cdot\vec{S} \quad . \tag{7}$$

The first term approximates the zero field splittings within the multiplet caused by mixing of its states with those of other multiplets via spin-orbit coupling. If $S = 1$ or $S \geq 2$ then D and E/D cannot specify the most general splitting and mixing of the multiplet. However, only a few cases have been reported where the introduction of a more complex Hamiltonian was needed (18). The second term of $\mathcal{H}_e$ describes the interaction of the electron moment with any externally applied field. The g-"tensor" takes into account the choice of S and also any unquenched angular momentum.*

Paramagnetic complexes may be classified into two broad categories; those with an even number of unpaired electrons and those with an odd number. In the first category, the system spin

---

* In general, upon rotation the matrix g does not transform as a tensor quantity. However, we will follow convention and call it the g-tensor. For a detailed discussion of this matter, see reference 27, page 650.

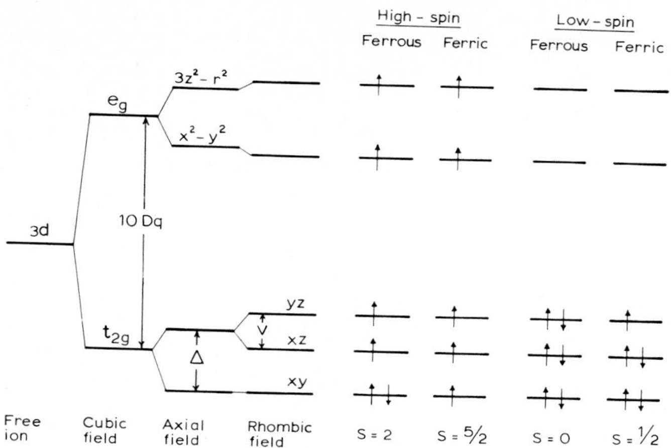

Fig. 1. Crystal field energy diagram for the 3d orbitals. Also illustrated are the pairings of the electron spins for commonly observed iron states.

will be integer and spin-orbit coupling will completely lift any spin degeneracy (except in cases of high symmetry). In most cases, no EPR signal will be observed since the splittings of the spin states are larger than the energy of the incident photons (typically, $h\nu \simeq 0.3\ cm^{-1}$)**. However, according to Kramers' theorem (19) half-integer spin states will always be at least doubly degenerate in the absence of a magnetic field, and EPR transitions can be induced within each Kramers doublet. An EPR spectrum is usually interpreted by treating each doublet as a spin 1/2 doublet, a valid approach if βH_{app} is small compared with the zero-field splittings. The spin Hamiltonian of Equation (7) then reduces to

$$H_e' = \beta \vec{H}_{app} \cdot \tilde{g} \cdot \vec{S}' , \qquad S' = 1/2 \tag{8}$$

The principal components of the effective g-tensor are readily determined from the EPR spectrum and from the resonance requirement

$$h\nu = \beta |\vec{H}_{app} \cdot g| , \tag{9}$$

where ν is the operating frequency.

** Numerous EPR data are available for the triplet state (S=1) of organic molecules and of Ni^{2+} complexes.

Assuming that $\tilde{g}_o = 2$ and βH_{app} is small compared with the zero field splittings, we may calculate the principal components of the effective g-tensor for each doublet of an S = 3/2 or S = 5/2 system from the relation

$$g_i = 4 \langle S_i \rangle_k \quad , \quad i = x,y,z \quad , \tag{10}$$

where $\langle S_i \rangle_k$ is the expectation value of $\vec{S}$ for the kth state when $\vec{H}_{app}$ is along $\hat{x}$, $\hat{y}$ or $\hat{z}$ as defined in Equation (7). The calculated g-values and the zero field splittings of the S = 5/2 and S = 3/2 multiplets are illustrated as a function of E/D in Figures 2 and 3. The applications of EPR to biological compounds are well documented (20) with a recent review by Palmer (21) being very useful.

The correlation between Mossbauer and EPR spectroscopies arises from the hyperfine interactions between the electrons and the nucleus. A spin Hamiltonian which describes the excited nucleus may be written as

$$\mathcal{H}_n = - g_n \beta_n \vec{H}_{app} . \vec{I} + \vec{S} . \vec{A} . \vec{I} + \mathcal{H}_Q \, , \qquad I = 3/2 \quad . \tag{11}$$

The first term is the nuclear Zeeman interaction, the second term approximates the magnetic hyperfine interaction of Equations (5) and (6), and $\mathcal{H}_Q$ is the electric quadrupole interaction. A Hamiltonian for nuclear ground states can be written by modifying Equation (11) by setting I = 1/2, $\mathcal{H}_Q = 0$ and multiplying A by the ratio of the g_n values (-1.748).

The magnetic hyperfine interaction is smaller than 10^{-2} cm^{-1}. If the eigenstates of $\mathcal{H}_e$ are singlets separated by much larger energies, then the electron and nuclear spins will be effectively decoupled. In other words, the expectation value of $\vec{S}$ may be substituted for the operator $\vec{S}$ in $\mathcal{H}_n$. For an integer spin complex, this situation exists even in the absence of an applied field and the expectation value $\langle \vec{S} \rangle$ equals zero. The resulting Mossbauer spectrum will be a simple quadrupole pair. For a Kramers systems in zero applied field, the electron and nuclear Hamiltonian must be solved simultaneously (22). Fortunately, a field of a few hundred gauss is sufficient to decouple $\vec{S}$ and $\vec{I}$, again separating the electron and nuclear problems.

A factor having a major impact on the Mossbauer spectrum is the electronic relaxation rate. The electrons are in thermal

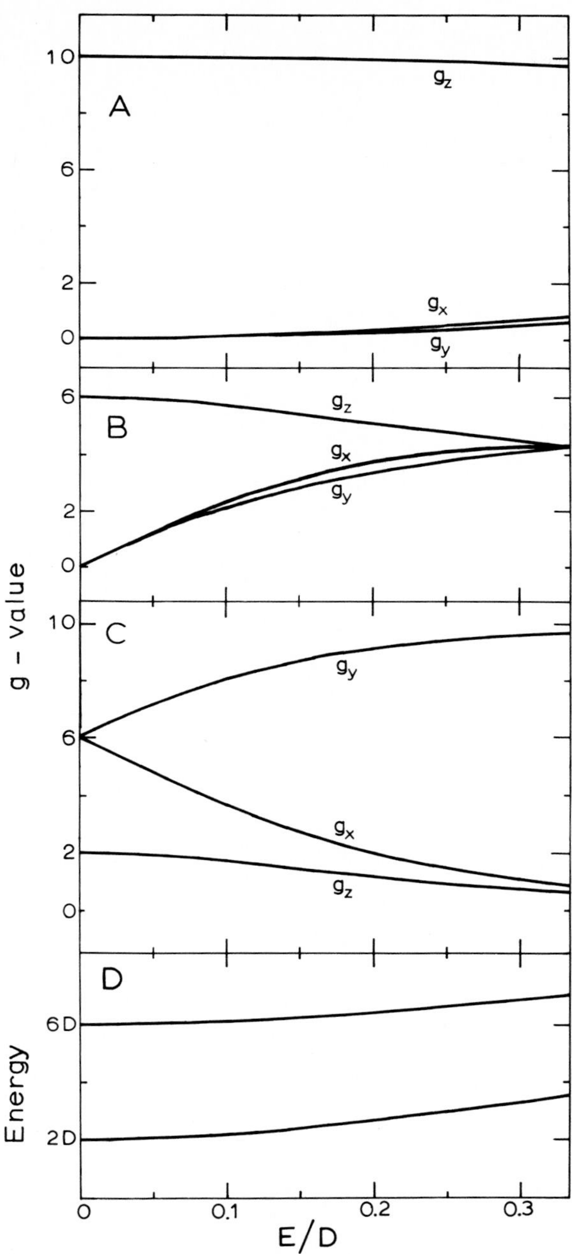

Figure 2. Effective g-values as a function of E/D with D>0 for (A) the upper most, (B) the middle and (C) the ground doublets of the S=5/2 sextet. The energies of the excited doublets relative to the ground doublet are shown in (D). The curves were calculated using Equations (7) and (10).

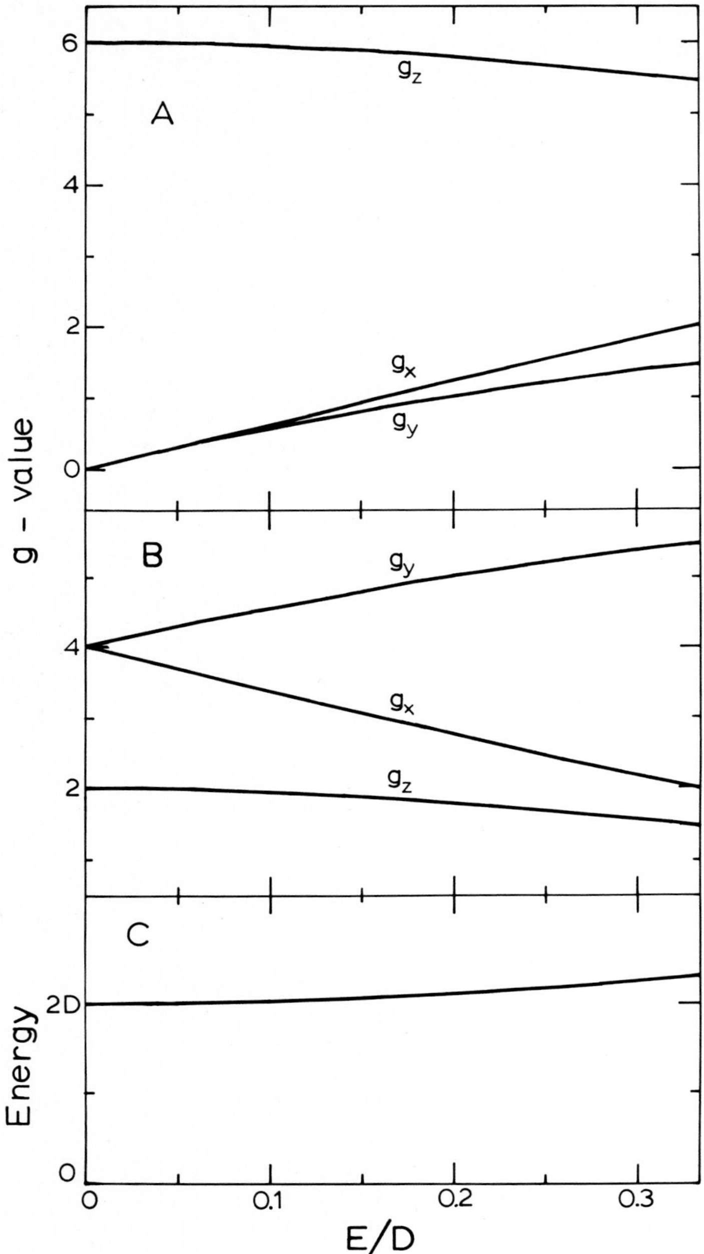

Fig. 3. Effective g-values as a function of E/D with D > 0 for (A) the upper and (B) the lower doublets of the S = 3/2 quartet. The energy of the upper doublet relative to the lower doublet is shown in (C). The curves were calculated using the spin Hamiltonian of Equation (7).

equilibrium with lattice vibrations and the populations of the various states are given by appropriate Boltzman factors. The relaxation rate between states may vary greatly with temperature and from complex to complex. If the relaxation rate is fast compared with the nuclear precession frequency ($\sim 10^7$ Hz) then the nucleus will sense only the thermal average of the electronic states. The Mossbauer spectrum would be simulated by substituting the thermal average of $\vec{S}$ into H_n. In the slow relaxation limit the nucleus senses each electronic states, and the Mossbauer spectrum is a superposition of the spectra produced by all populated states weighted by the appropriate Boltzman factors. The Mossbauer spectra of a slow relaxing Kramers system and their relation to the g-values obtained from EPR experiments are discussed below. Slow relaxation is common for biological complexes at 4.2 K.

The effective field of Eqn. (6) may be approximated as

$$\vec{H}_{eff} = \vec{H}_{app} + \vec{H}_{int} = \vec{H}_{app} - \tilde{A}.\langle\vec{S}\rangle_i/g_n\beta_n \quad , \qquad (12)$$

where the subscript i refers to the ith electronic state. Assuming that $g_o = 2$, $\tilde{A}$ is isotropic and $\vec{H}_{app}$ is small, the effective field produced by a Kramers doublet is given by

$$\vec{H}_{eff} \simeq \vec{H}_{int} = \frac{A}{4g_h}(h_x g_x^2\,\hat{x} + h_y g_y^2\,\hat{y} + h_z g_z^2\,\hat{z}) \quad , \qquad (13)$$

where $g_h = [h_x^2\, g_x^2 + h_y^2 + h_z^2\, g_z^2]^{1/2}$ and h_x, h_y and h_z are the direction cosines of $\vec{H}_{app}$. The intensities of the Mossbauer absorption peaks depend on the angle θ between $\vec{H}_{eff}$ and the γ-beam. If $\tilde{g}$ is isotropic, $\vec{H}_{eff}$ will be along $\vec{H}_{app}$ and will be the same at all the iron sites of the sample. In this case a significant difference would be observed between spectra recorded with $\vec{H}_{app}$ parallel and with $\vec{H}_{app}$ perpendicular to the γ-beam. When the g_i's are different but comparable to each other, the direction of the internal field can still be effectively oriented by the external field. Field-direction-dependent spectra will also be observed. On the other hand, if the induced moment is extremely anisotropic, i.e. two of the g_i's are close to zero, Equation (13) indicates that $\vec{H}_{eff}$ is fixed relative to the molecule. For a powder sample, θ will be averaged over all possible values and the Mossbauer spectrum will be independent of the direction of the applied field. This situation identifies an "EPR silent" Kramers doublet. For a powder sample Aasa and Vanngard (23) have shown that the EPR absorption intensity at a field corresponding to one of the g_i depends on the other two principal components as expressed below.

$$I_i \propto \frac{(g_j^2 + g_k^2)}{[(1 - g_j^2/g_i^2)(1-g_k^2/g_i^2)]^{\frac{1}{2}}} \quad . \tag{14}$$

If two of the g_i's are close to zero then the EPR signal will be difficult or impossible to observe. Such situations occur for the $|\pm 5/2\rangle$ or $|\pm 3/2\rangle$ doublet of an axially symmetric (E/D = 0) S = 5/2 or S = 3/2 system.

To summarize, electronic systems with slow relaxation rates may be differentiated and characterized by using EPR and Mossbauer spectroscopies. For a system with an even number of electrons, EPR is generally not applicable. In the absence of an applied field the corresponding Mossbauer spectrum is a quadrupole doublet. A paramagnetic system ($S \neq 0$) can be distinguished from a diamagnetic system by the application of an external magnetic field. For a system with an odd number of electrons, the Mossbauer spectrum is magnetically broadened even in the absence of an applied field. In the presence of a small magnetic field two different types of Kramers doublets can be distinguished. One, an "EPR silent" doublet with two g-values close to zero, will have a field-direction-independent Mossbauer spectrum. The other type, an EPR active doublet with at least two g-values significantly different from zero, will have a field-direction-dependent Mossbauer spectrum.

A strength of Mossbauer spectroscopy is that it detects all the iron present (including impurities) regardless of their oxidation state. Each spectral component can be accurately determined to better than a few percent. Its major weakness is that the data accumulation time is long and, in most cases, isotopic substitution is required. EPR spectroscopy requires no isotopic substitution and allows for fast data accumulation. Reasonable spectra can be obtained from samples with protein concentration in the order of 10 μM. However, conventional EPR spectroscopy is limited to half-integer spin systems. Since there is a strong correlation between EPR and Mossbauer data, the EPR active center can be unambiguously assigned to the corresponding Mossbauer spectral component. The number of iron atoms per active EPR center can be determined by correlating the relative absorption intensities of the Mossbauer spectral components, the iron content of the sample, and the spin density found from the EPR data. This method is extremely useful in identifying and characterizing iron centers in a multi-iron protein (see Section 4).

3. PROTEINS WITH MONO-IRON CENTERS

3.1 Low-spin Ferric

In biological compounds most low-spin ferric ions are found in heme proteins. The heme group consists of an iron atom and a porphyrin. The iron atom is situated at the center of the porphyrin which consists of four pyrrole rings bridged by CH groups. The iron atom is coordinated to the four pyrole nitrogens in a roughly planar configuration. The fifth and sixth ligands coordinate to the iron atom approximately along the heme normal.

As illustrated in Figure 1, the low-spin state results from the cubic field splitting being larger than the electron pairing energy. The low-spin ferric state, which has a $(t_{2g})^5$ configuration, can be described as a filled t_{2g} shell plus a single electron hole occupying one of the t_{2g} orbitals. A theory based on this assumption was introduced by Griffith (24), and has been successfully applied to the analysis of EPR data of low-spin ferric complexes (25,26). The crystal field is assumed to have rhombic symmetry and, hence, the theoretical model has only three parameters. Two are Δ and V, the axial and rhombic crystal field parameters, respectively. The eigenstates of the crystal field are the three Kramers doublets $|xy,\pm\rangle$, $|xz,\pm\rangle$ and $|yz,\pm\rangle$. Their energies may be expressed in terms of Δ and V. An isotropic orbital reduction factor, k, is introduced to account for the effects of bonds involving the 3d orbitals. The three doublets are mixed by the spin-orbit interaction, $-\lambda\vec{L}.\vec{S}$. At low temperatures the EPR and Mossbauer spectra result from the lowest doublet. This doublet can be expressed as

$$|\alpha\rangle = a|xy,+\rangle + b|yz,-\rangle + ic|xz,-\rangle \quad , \tag{15}$$

$$|\beta\rangle = a|xy,-\rangle - b|yz,+\rangle + ic|xz,+\rangle \quad ,$$

where $a^2 + b^2 + c^2 = 1$, $i^2 = -1$ and, for example, $|xy, +\rangle$ is the state with the electron hole occupying the xy orbital and its spin expectation value along positive $\hat{z}$. The coefficients a, b, and c are real and depend only on Δ/λ and V/λ. The corresponding principal g-values can be written in terms of a, b and c as

$$g_x = 2(a^2 - b^2 + c^2 + 2kac) \quad ,$$

$$g_y = 2(a^2 + b^2 - c^2 + 2kab) \quad , \tag{16}$$

$$g_z = 2(a^2 - b^2 - c^2 - 2kbc) \quad .$$

Consequently, the coefficients and the orbital reduction factor k can be determined from the g-values obtained from EPR measurements. The ligand field parameters Δ/λ and V/λ can then be computed from the values of a, b and c.

Since the planar coordination of the iron is restricted by the heme group, a comparison of reported ligand field parameters for different low-spin ferric heme compounds should provide information pertinent to the axial ligands. Blumberg and Peisach (25) have classified a large number of low-spin ferric heme proteins by analyzing EPR data. They have shown that low-spin ferric heme compounds can be grouped into different regions of a crystal field diagram which is a plot of V/λ versus Δ/λ.

However, conventional EPR spectroscopy measures only the magnitudes of the g_i and not their signs. Hence, it does not completely determine Δ/λ and V/λ. Abragam and Bleaney (27) have discussed this problem at length and have shown that the signs of two components of the effective g-tensor are changed simultaneously by either a spatial rotation of the molecular coordinates or by a different choice of the basis states. However, the sign of the product $(g_x g_y g_z)$ is invariant and reflects the magnitude of the unquenched orbital moment. Therefore, for the classification of low-spin ferric heme compounds, it is important to know the sign of $(g_x g_y g_z)$. The sign may be determined by EPR spectroscopy if circularly polarized radiation is used. In the following we discuss, within the framework of Griffith's theory, how the sign of $(g_x g_y g_z)$ can also be determined from Mossbauer measurements.

Oosterhuis and Lang (28) have extended the Griffith model to include the magnetic hyperfine interaction. The matrix A can be computed using the wavefunctions of the ground Kramers doublet given in equation (15).

$$A_x = P\,\{4ac-(1+\kappa)(a^2-b^2+c^2) - \tfrac{3}{7}(b^2-3a^2-3c^2) - \tfrac{6}{7}b(a+c)\} \quad ,$$

$$A_y = P\,\{4ab-(1+\kappa)(a^2+b^2-c^2) - \tfrac{3}{7}(c^2-3a^2-3b^2) - \tfrac{6}{7}c(a+b)\} \quad , \qquad (17)$$

$$A_z = P\,\{-4bc-(1+\kappa)(a^2-b^2-c^2) + \tfrac{3}{7}(a^2-3b^2-3c^2) + \tfrac{6}{7}a(b+c)\} \quad .$$

The parameter κ is introduced into the formula to account for the Fermi contact interaction. In principle, the contribution of the valence 3d electrons to the efg tensor can be calculated from the three Kramers doublets. However, lattice charges, covalency effects, and the σ-electrons donated to the e_g and 4p orbitals from the ligands can all contribute significantly to the efg tensor. It is therefore advisable to treat the efg as an empirical quantity.

Similar to the effective g-tensor, the signs of the components of the A-tensor are changed by permuting coordinate axes or by relabelling the basis states. However, the product (g_iA_i) is invariant. Since the internal field is a function of (g_iA_i), the sign information is contained in the paramagnetic Mossbauer spectrum. In the following, EPR and Mossbauer data of cytochrome c_2 from *Rhodospirillum rubrum (29)* are presented as an example of how the sign of $(g_xg_yg_z)$ may be determined.

Cytochrome c_2 consists of a single polypeptide chain and a heme group. In a c-type cytochrome, the heme group is covalently bound to the polypeptide chain through the sulfur atoms of two cysteine residues. The axial ligands of the iron atom are a histidine nitrogen and a methionine sulfur (30). In the oxidized state the iron ion is low-spin ferric. An EPR spectrum of ferricytochrome c_2 is shown in Figure 4. The resonances at $g_z = 3.13$, $g_y = 2.11$ and $g_x = 1.23$ are attributed to the ferricytochrome c_2. The sharp axial type signal at $g = 6.0$ and $g = 2.0$ indicate the presence of a minority species. The concentration of the minority species is estimated to be less than 5% of that of the ferricytochrome c_2.

The Mossbauer spectrum of ferricytochrome c_2 at 200 K is a quadrupole doublet (the electronic spin relaxation rate is fast). A least squares fit of two Lorentzians to the data yielded $\Delta E_Q = (2.20 \pm .02)$mm/s and $\delta = (0.25 \pm .02)$ mm/s. No significant temperature dependence of ΔE_Q was observed between 4.2 K and 200 K. The Mossbauer spectrum of ferricytochrome c_2 recorded at 4.2 K in parallel applied field is displayed in Figure 5B. Spectra taken at 8 K and 12 K are identical to that in Figure 5, indicating a slow electronic relaxation rate at 4.2 K. Since g_z is sizably larger than g_x and g_y, the internal field is predominantly oriented along the

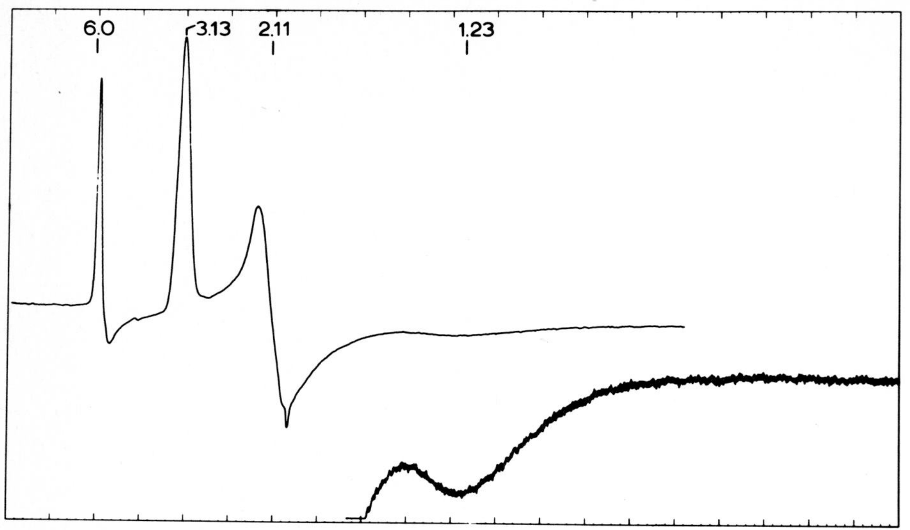

Fig. 4. EPR spectrum of ^{57}Fe-enriched ferricytochrome c_2. Conditions: T = 13 k, frequency 9.046 GHz, microwave power 10mW, modulation amplitude 1 mT, sweep rate 0.25 T/min, time constant 6s; receiver gain 320 and 3200 in the low and high field regions, respectively. The ordinate is an arbitrary linear function of the first derivative of the microwave absorption with respect to the magnetic field. The features at g = 5.95 and g = 2 correspond to a minor (2-3%) contaminant. (From ref. 29).

z axis (see Equation 13). The spectra are dominated by the magnetic hyperfine component A_z and the efg component V_{zz}. This allows a fairly precise determination of A_z and the asymmetry parameter η. The total splitting of the spectrum increases with increasing external field indicating that $(g_z A_z) > 0$. The remaining unknown parameters were obtained by comparing experimental data with theoretical spectra simulated by means of equations (8) and (11). The solid curve in Figure 5B is a simulated spectrum generated from the parameters listed in Table I. The theoretical fit can be improved if a Gaussian distribution of the crystal field parameters Δ and V is used. Dwivedi et al. (31) have demonstrated that a Gaussian distribution of Δ and V, as obtained from fitting the EPR line shapes, yields theoretical spectra which fit the Mossbauer data of cytochrome c-551 much better than those generated from discrete values of Δ and V.

We can now use the measured g-values and the theoretical model described above to calculate the hyperfine parameters. Using $\kappa = 0.35$

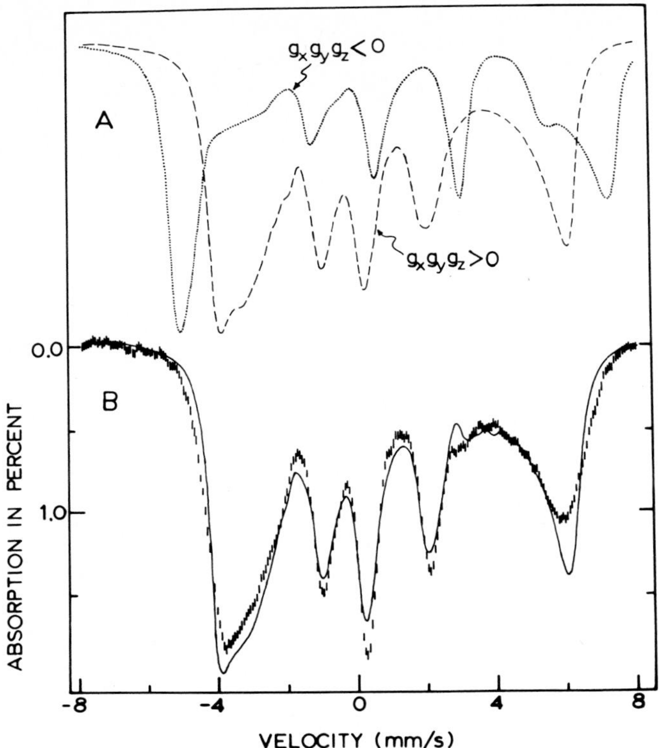

Fig. 5. Mossbauer spectrum of ^{57}Fe-enriched ferricytochrome c_2 recorded at 4.2 K with an applied field of 0.06 T parallel to the γ-beam (B). The solid curves in (A) and (B) are S = 1/2 simulations as described in the text. (Adapted from ref. 29)

and $P/g_n\beta_n$ = -62.6 T, the model yields, for $(g_xg_yg_z) > 0$, magnetic hyperfine values in good agreement with the experimental values. On the other hand, for $(g_xg_yg_z) < 0$, the model yields different hyperfine parameters (see Table I). To give the reader a feeling for the sensitivity of the Mossbauer spectra to the sign of $(g_xg_yg_z)$, we have plotted in Figure 5A the two simulated spectra corresponding to the two solutions listed in Table I.

The two solutions reflect two physically distinct situations. When the crystal field splittings are much larger than λ, the unquenched orbital angular momentum is small and the product $(g_xg_yg_z)$ is positive. On the other hand, when the product $(g_xg_yg_z)$ is negative the unquenched orbital moment is large and the crystal field splittings are comparable to λ. The Mossbauer spectra of such a low-spin ferric complex will have two distinguishing

TABLE 1.

Ligand field and hyperfine parameters for ferricytochrome c_2

	Experiment	Theory $g_xg_yg_z > 0$	Theory $g_xg_yg_z < 0$
$\Delta/k\lambda$		3.08	0.61
$V/k\lambda$		1.33	0.26
k		0.97	1.18
g_x	1.23 (4)[a]	+ 1.23	- 1.23
g_y	2.11 (3)	+ 2.11	+ 2.11
g_z	3.13 (2)	+ 3.13	+ 3.13
$A_x/g_n\beta_n$ (T)	-35 (5)	-35.9	-50.4
$A_y/g_n\beta_n$ (T)	25 (7)	+19.0	+57.0
$A_z/g_n\beta_n$ (T)	62.5 (15)	+62.5	+77.9
g_xA_x	< 0	< 0	> 0
g_yA_y	> 0	> 0	> 0
g_zA_z	> 0	> 0	> 0
ΔE_Q (mm/s)[b]	+ 2.26 (2)		
δ (mm/s)[b]	0.31 (2)		
η	-1.5 (1)		

a. The numbers in parentheses give the uncertainties in units of the least significant digit. The uncertainties in the hyperfine parameters were estimated by visually comparing the data with theoretical Mossbauer spectra computed from equations (8) and (11).

b. Values at 4.2 K

features. Due to the comparable contributions of the three t_{2g} orbitals into the ground state wave function, a small quadrupole splitting is expected. Also, the large unquenched orbital moment will produce large magnetic hyperfine coupling constants. The overall magnetic splitting of the Mossbauer spectrum is therefore expected to be large, as illustrated in Figure 5A.

For biological compounds, it seems to be a general situation that Δ and V are much larger than λ. Using published Mossbauer data (32, 33, 34), Huynh et al. (29) found that $(g_xg_yg_z)$ is positive for the following low-spin ferric heme proteins: hemoglobin azide, cytochrome P-450 from *Pseudomonas putida*, chloroperoxidase from *Caldariomyces fumago*, cytochrome c from *Torula utilis* and cytochrome b_5 from calf liver. In their analysis of the EPR data of low-spin ferric heme compounds, Blumberg and Peisach (25) assumed that $(g_x g_y g_z)$ is positive. The Mossbauer data support their assumption. However, metmyoglobin hydroxide might be an exception. It has been brought to our attention* that linear field effect data on metmyoglobin hydroxide (35) are better accounted for if V and Δ are assumed to be smaller than λ. If this is true, it would suggest a negative $(g_x g_y g_z)$. The Mossbauer data of $K_3Fe(CN)_6$ (28) indicate that low-spin ferric compounds with negative $(g_x g_y g_z)$ do exist.

In the above discussion we have assumed rhombic or higher symmetry. Data analysis is relatively simple because the g-tensor and the A-tensor share a common principal axis system. In reality, the symmetry can be lower than rhombic. To account for a more general situation, Oosterhuis and Lang (28) rotated the principal axes system of the rhombic field with respect to that of the cubic field. More parameters, the Euler angles, are introduced into the theory. Recently, this crystal field model has been used to study the symmetry of the iron environment in the azide and cyanide complexes of myoglobin and cytochrome c peroxidase (36, 37). Within the theoretical model the relative orientation of the principal axis systems of the g-tensor, the A-tensor, and the cubic field potential can be obtained by fitting the Mossbauer data. For myoglobin azide and myoglobin cyanide, the orientation of the g-tensor with respect to the molecular heme axes has been determined by single crystal EPR studies (38, 39). In the case of myoglobin azide, the axes of the cubic field appear to coincide with the molecular heme axes. For myoglobin cyanide, the result is inconclusive. Rhynard et al. (36) have pointed out that the coincidence found in myoglobin azide is probably not generally true in heme complexes. Correlation of crystal field axes with molecular axes can only be achieved by single crystal experiments.

3.2 High-Spin Ferric

The electronic configuration of a high-spin ferric ion is $3d^5$ with the five unpaired electrons each occupying a different 3d orbital. (See Figure 1). The system ground state is ^6S (or $^6A_1$ in

---

*Dr. J. Peisach (Private communication)

cubic field symmetry). The electronic charge distribution is nearly isotropic resulting in a small quadrupole splitting (< 1.0 mm/s, Ref. 40) and a small dipole contribution to the hyperfine coupling tensor A. Since the ground multiplet is ^6S, the orbital contribution is also small. Therefore, the A- and g_o-tensors are fairly isotropic. For biological high-spin ferric compounds the components of $\tilde{A}$ are within 15% of their average value. The isotropic A and small quadrupole interaction causes the magnetic Mossbauer spectrum to be a well resolved six-line pattern. Data analysis is further facilitated by the existence of EPR data. In the past, hemoproteins have attracted much attention in review articles. Here we will report Mossbauer and EPR investigations on some non-heme, high-spin ferric proteins.

3.2.1 Oxidized Rubredoxin

Rubredoxin (Rd) is an iron-sulfur protein. Its metal center consists of a single iron atom coordinated to four cysteine sulfurs in an approximately tetrahedral fashion (41). In the oxidized state the iron is high-spin ferric. The EPR spectra of rubredoxins from *Pseudomonas oleovarans* (42), *Chloropseudomonas ethylica, Clostridium pasteurianum* (43), *Desulfovibrio gigas* (44) and *Peptostreptococcus elsdenii* (45) all exhibit resonances near g = 4.3 and g = 9.4. These resonances can be explained by an S = 5/2 spin Hamiltonian with E/D = 0.28 (Ref. 42, also see Figure 2). The signal at g = 9.4 is assigned to the ground doublet while the characteristic derivative-type signal at g = 4.3 results from the first excited doublet.

Extensive Mossbauer investigations on Rd from *Clostridium pasteurianum* have been carried out (46, 47). The Mossbauer data can be explained reasonably well by an S = 5/2 spin Hamiltonian with E/D = 0.23. The zero field splittings were found to be small. In the temperature range from 4.2 K to 20 K, all three Kramers doublets are appreciably populated and the electronic relaxation rate is slow. In weak applied field the observed Mossbauer spectrum is a super-position of three subspectra, each subspectrum resulting from one of the Kramer's doublets. This situation provides the simplest case for extracting a set of spin Hamiltonian parameters from Mossbauer data. With E/D = 0.23 the effective g-values are very anisotropic for the ground and highest doublets; (g_x = 1.65, g_y = 9.27, g_z = 1.02) and (g_x = 0.38, g_y = 0.30,

g_z = 9.85), respectively. The Mossbauer subspectra resulting from the lowest and highest doublets are therefore sensitive to the hyperfine parameters along the y and z directions, respectively, allowing accurate determination of A_y, A_z, V_{yy}, and V_{zz}. The remaining unknown, A_x, can then be estimated by fitting the total spectrum using the spin Hamiltonian formalism. Temperature dependent studies have revealed that the energy separation of the three Kramers doublets can not be approximated by a quadratic S = 5/2 spin Hamiltonian. The energies have to be treated as independent parameters. A set of fine and hyperfine parameters for Rd as determined by Schulz and Debrunner (46) is listed in Table 2. X-ray

TABLE 2.

Spin Hamiltonian parameters for oxidized and reduced Rd from Clostridium pasteurianum and Dx from Desulfovibrio gigas.

	Oxidized		Reduced	
	Rd[a]	Dx[b]	Ra[a]	Dx[b]
D (cm^{-1})	1.9 (3)[c]	2.2 (3)	7.6	- 6 (1)
E/D	0.23 (2)	0.080 (5)	0.28	0.19 (5)
g_{ox}	2.0	2.0	2.11	2.08
g_{oy}	2.0	2.0	2.19	2.02
g_{oz}	2.0	2.0	2.00	2.20
$A_x/g_n\beta_n$ (T)	-16.5 (11)	-15.4 (10)	-20.1	-20 (3)
$A_y/g_n\beta_n$ (T)	-15.9 (3)	-15.4 (5)	- 8.3	-20 (3)
$A_z/g_n\beta_n$ (T)	-16.9 (3)	-15.4 (10)	-30.1	- 6.7 (3)
ΔE_Q (mm/s)	- 0.50 (5)	- 0.75 (5)	- 3.25 (1)	3.55 (2)
δ (mm/s)	0.32 (2)	0.25 (6)	0.70 (2)	0.70 (2)
η	0.2 (1)	0.6 (3)	0.65 (10)	0.35 (5)
β (degrees)[d]		90		10

(a) Ref. 46

(b) Ref. 53

(c) Values in parentheses give uncertainties in units of the least significant digit.

(d) β is the polar angle between the z-axes of the principal axis frames of the zero field splitting D-tensor and the efg-tensor.

crystallographic measurements indicate that the geometries of some synthetic model complexes (48) are very similar to the active center of Rd. The set of values listed, therefore, probably represent parameters for ferric ions with non-constrained tetrahedrally coordinated sulfurs.

Certain characteristic features of Rd are worth emphasizing. The average A-value for Rd gives a saturation field ($H_{sat} = 5\ A/2\ g_n\beta_n$) of 41 T which is about 25% smaller than that found for typical octahedral oxygen coordination (55 T, Ref. 49) and is about 20% smaller than that found for high-spin ferric heme proteins (about 500 kOe, Ref. 7). The small isomer shift (~ 0.3 mm/s) is also characteristic of tetrahedral sulfur coordination. The isomer shift found for ferric high-spin hemoproteins are typically in the range from 0.4 mm/s to 0.5 mm/s (6,7). The characteristic reduction of A and δ in Rd may result from extensive delocalization of the 3d orbitals. The relatively small zero field splitting of Rd, however, may not be characteristic of tetrahedral sulfur environments. Most ferric high-spin non-heme proteins also have small zero-field splittings. The unusually large D (6 - 15 cm^{-1}) found for high-spin ferric hemoproteins (7) might be more characteristic for its own species. Even among heme proteins, cytochrome P-450 has a rather small D of ~ 4 cm^{-1} (50).

3.2.2 Oxidized Desulforedoxin

Desulforedoxin (Dx) from *Desulfovibrio gigas*, a sulfate reducing bacterium, is a relatively small protein of molecular weight 7,900 and contains two identical subunits and two iron atoms (51). The amino acid sequence is known (51, 52) and each subunit has four cysteine residues. Two cysteines are arranged as in rubredoxin, i.e. Cys-x-y-Cys, while the other two are adjacent to each other (x and y represent residues other than cysteine). No X-ray crystallographic data have been reported. Studies on the sulfhydryl content of apo- and holo-protein suggest that all four cysteines are coordinated to the iron (53). EPR investigations indicate that the iron in native Dx is high-spin ferric (44, 53). Native Dx shows a nearly axial EPR signal (E/D ~ 0.08) which differs significantly from the nearly rhombic signal of Rd (E/D ~ 0.28). Optical data of Dx resemble those of Rd but are not identical. If it is true that the four cysteines are coordinated to the iron, the spectroscopic differences between desulforedoxin and rubredoxin might reflect the stereochemical constraints imposed by the unusual cysteine sequence.

An EPR spectrum of oxidized Dx taken at 10 K is shown in Figure 6. The observed g-values can be explained by an S = 5/2 spin Hamiltonian with E/D = 0.08 and g = 2.0. The theoretical energy levels and effective g-values for such a system are also shown in Figure 6. The observed resonances at g = 4.1, 7.7 and 1.8 are attributed to the ground doublet and the resonance at g = 5.7 to the first excited doublet. Resonances for the highest doublet are not observable (see Section 2.3). The weak resonances at g = 8.9 and 2.0 are assigned to a minority species (~ 10% of the total EPR absorption). From the relative intensities of the resonances at g = 7.7 and 5.7, the zero field splitting parameter D, was estimated to be about 2 cm^{-1} (53).

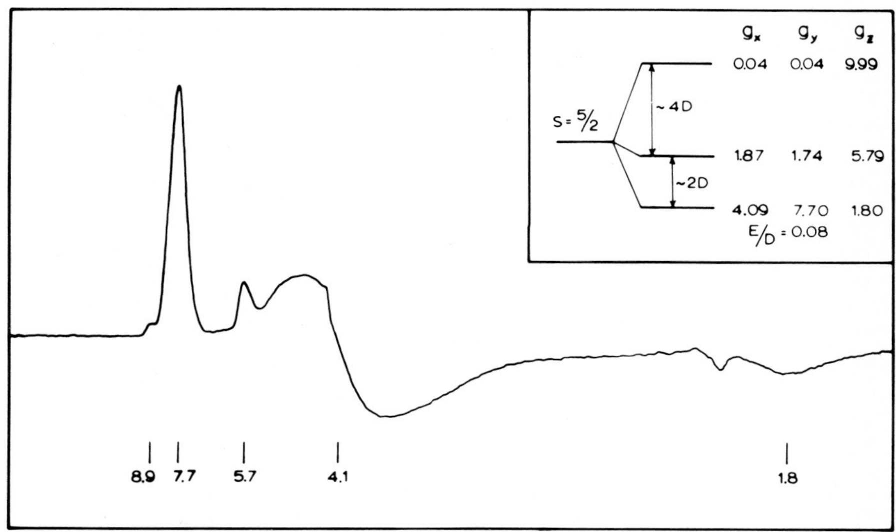

Fig. 6. EPR spectrum of desulforedoxin at 4.8 K, microwave power 0.2 MW, frequency 9.24 GHz, modulation amplitude 10 G, protein concentration 1.1 mM. The inset gives the energy level scheme and theoretical g-values for g_o = 2.0 and E/D = 0.08. (From ref. 53).

A Mossbauer spectrum of oxidized Dx recorded in our laboratory consists of two major subcomponents; a magnetic spectrum corresponding to the EPR signal and a quadrupole pair with ΔE_Q = (0.75 ± 0.05) mm/s and δ = (0.25 ± 0.06) mm/s. The spectrum was taken at 4.2 K with a parallel applied field of 0.6 kOe. The absorption of the quadrupole pair amounts to about 40% of the total absorption which rules out associating this absorption to the impurity observed in the EPR spectrum. The following observations suggest that the two spectral components both result from desulforedoxin chromophores, some relaxing slow and some fast: (1) The quadrupole pair has ΔE_Q typical of high-spin ferric ions and an isomer shift identical to that of the magnetic component. (2) Samples with different ratios of the two spectral components yield the same optical spectrum. (3) Mossbauer spectra recorded with a strong applied field ($H_{app} \geq 10$ kOe) have only one spectral component. (4) Reduced Dx shows only one sharp quadrupole pair typical of high-spin ferrous ion.

Subsequently, we have found that the magnetic component and the EPR intensity can be increased by increasing salt concentration in the sample. At high salt concentration, the fast relaxing component disappears in the Mossbauer spectrum and the EPR intensity reaches its maximum. This observation indicated that only at sufficiently high salt concentration was it possible to correctly quantitate the EPR signal in terms of spin concentration. The sample which yielded the EPR spectrum in Figure 6 had a spin concentration of 0.93 spin/Fe. (The numbers are corrected for the $g = 8.9$ impurity.). The Mossbauer spectra of that sample are shown in Figures 7 and 8. Since the salt concentration usually effects the solubility of proteins, the observed rapid relaxing component might arise from increased spin-spin or spin-lattice interaction in aggregated material. Also, since each molecule of Dx contains two iron atoms, one cannot dismiss the possibility of intramolecular spin-spin interaction between the two. It is conceivable that the structure of the protein is perturbed at high salt concentrations, decoupling the spins.

Figure 7 shows a Mossbauer spectrum of Dx recorded at 4.2 K with a magnetic field of 0.06 T applied parallel to the observed γ-radiation. Unlike that of Rd, the spectrum is not a well resolved six-line pattern. This observation can be readily understood in terms of the EPR measurements. Two of the effective g-values of the ground doublet of Dx are comparable ($g_x = 4.1$, $g_y = 7.7$, g_z is small). The induced internal field and, hence, the nuclear spin are effectively oriented in the xy-plane. Different efg components throughout the xy-plane are sampled by the nuclear spin, broadening the absorption lines. With the value of ΔE_Q known from the above-mentioned fast relaxing component, the only unknowns in the spin Hamiltonian are the hyperfine coupling tensor A and the asymmetry parameter η. If we assume an isotropic A, its value can be determined directly from the total splitting of the spectrum in Figure 7.

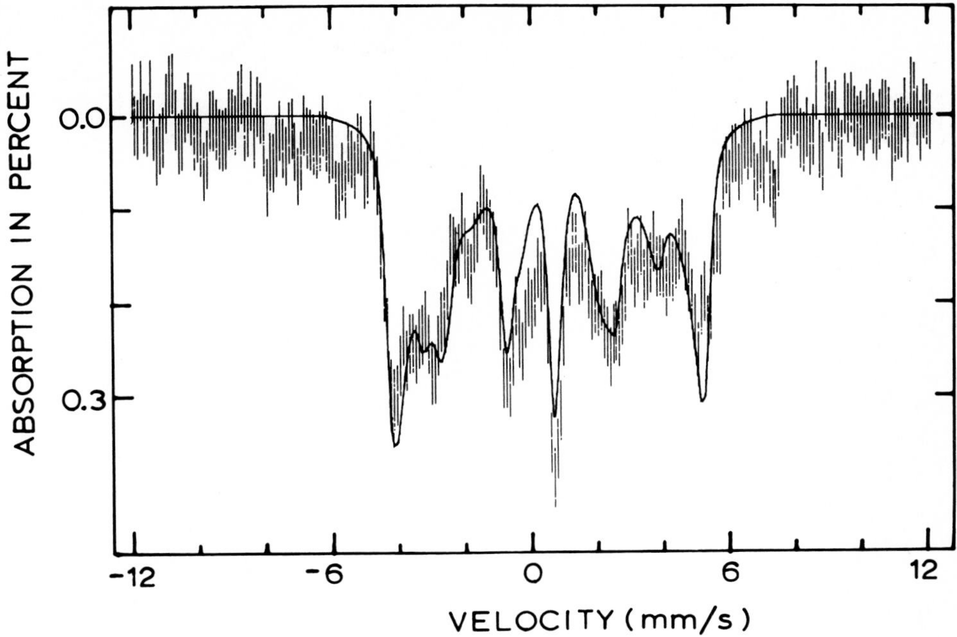

Fig. 7. Mossbauer spectrum of ferric desulforedoxin recorded at 4.2 K with an applied field of 0.06 T parallel to the γ-beam. The solid curve is a S = 5/2 spin Hamiltonian simulation as described in the text. (From ref. 53)

In applied fields such that $g \beta H \simeq D$, the electronic Zeeman interaction mixes the zero-field eigenstates significantly. Hence, the high field Mossbauer data are sensitive to the zero field splittings of the sextet. A spectrum of oxidized Dx in an applied field of 4.5 T is shown in Figure 8. Numerous simulations have shown D to be $(2.2 \pm 0.3)\ cm^{-1}$ which is in good agreement with the EPR measurements.

The spin Hamiltonian parameters found for Dx are listed in Table 2. Like Rd, the saturation field and the isomer shift of Dx are characteristic of a tetrahedral sulfur environment. Unlike Rd, the zero-field parameter E/D of Dx is small, indicating different environments at the two active sites. Care should be exercised when interpreting EPR data with regard to the molecular symmetry. Emptage et al. (54) have pointed out that as long as only one excited quartet state is effectively mixed into the ground sextet by spin-orbit coupling, the EPR data will show axial symmetry (E/D = 0) regardless of the crystal field symmetry.

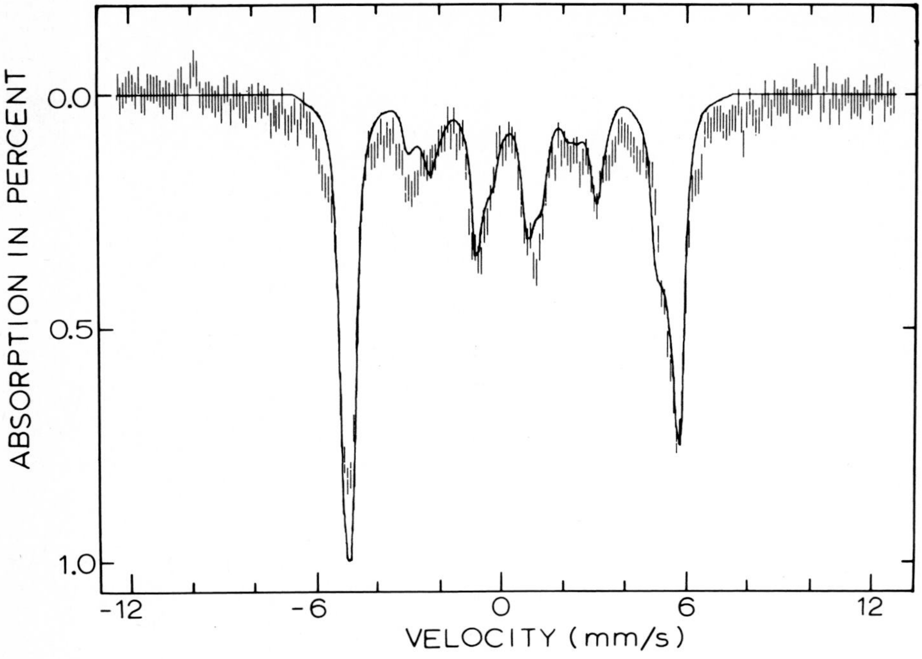

Fig. 8. Mossbauer spectrum of ferric desulforedoxin recorded at 4.2 K with a parallel field of 4.5 T. The solid curve is a S=5/2 spin Hamiltonian simulation as described in the text. (From Ref. 53)

Besides extracting spin Hamiltonian parameters, other useful information can be obtained from Mossbauer data. For example, the extinction coefficients of the chromophore in Dx were first determined to be 7.0 x 10^3 /M/cm and 1.2 x 10^4 /M/cm at 507 nm and 370 nm respectively (51). Such measurements are difficult since the presence of any apoprotein or iron impurities can lead to erroneous results. By combining optical spectroscopy, iron content determination and Mossbauer spectroscopy, corrected extinction coefficients were determined for Dx (4.5 x 10^3 /M/cm and 7.8 x 10^3 /M/cm at 507 nm and 370 nm, respectively, Ref. 53). Mossbauer spectroscopy was used to determine the ratio of chromophore iron to total iron in the sample. Also, the Mossbauer spectrum revealed an iron component with rapidly relaxing electronic spin which was not detected by EPR. Correct spin quantitation may only be obtained either by taking the rapidly relaxing spins into account or by changing the protein environment until all the spins become detectable by EPR. There are numerous examples of EPR quantitations in the literature which have yielded spin densities less than was expected on the basis of other measurements. Fast electronic spin relaxation could be the cause of some of the problems

3.2.3 Intradiol Dioxygenase

Protocatechuate 3,4-dioxygenase from *Pseudomonas aeruginosa* is an enzyme which catalyzes the cleavage of protocatechuate, a six-membered carbon ring with a COO^- group at position 1 and OH groups at positions 3 and 4. This protein has a molecular weight of 70,000 and consists of eight identical subunits. Each subunit contains one iron atom and four polypeptide chains organized in an $\alpha_2\beta_2$ structure (55, 56). The EPR spectrum of the native enzyme exhibits resonances at g = 9.5 and 4.2, typical of high-spin ferric compounds. Interpreting the EPR data with the spin Hamiltonian of equation (7) yields E/D = 0.29. Temperature studies of the EPR spectra reveal a zero-field splitting parameter D = 1.5 cm^{-1} (57). Based on these observations, it was erroneously suggested that the iron ligand environment in 3,4-dioxygenase was similar to that of ferric rubredoxin. A Mossbauer investigation (58a), however, revealed a saturation field of 525 kOe which is atypical of tetrahedral sulfur ligation.

A Mossbauer spectrum of 3,4-dioxygenase (58a) recorded at 1.5 K in a parallel field of 0.06 T is shown in Figure 9. This spectrum can be readily understood in terms of the EPR data. With

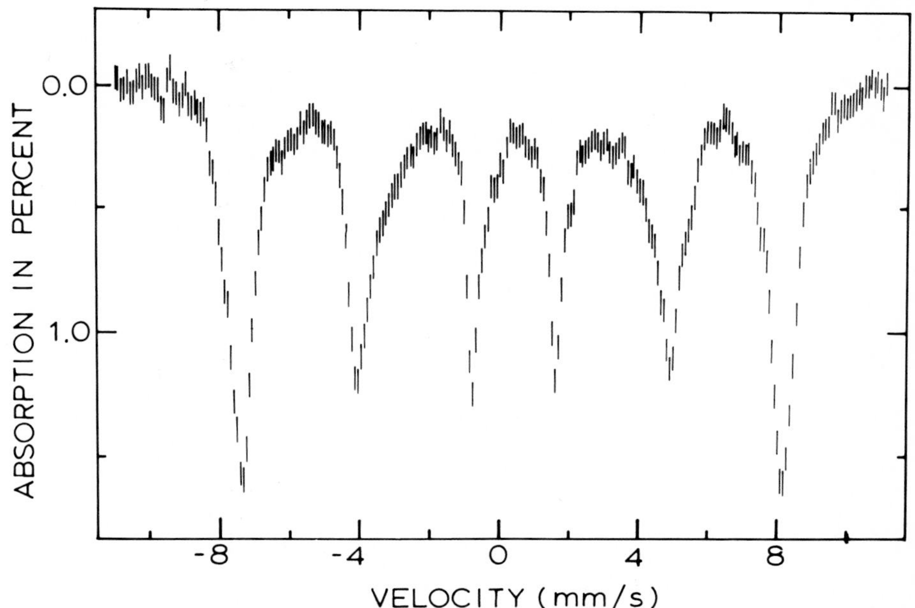

Fig. 9. Mossbauer spectrum of ^{57}Fe-enriched protocatechuate 3,4-dioxygenase recorded at 1.5 K with a 0.06 T field applied parallel to the γ-beam (adapted from Ref. 58a).

E/D = 0.29, the g-values of the ground doublet are very anisotropic. The orientation of the effective field is heavily weighted along the g = 9.5 component resulting in the well resolved six-line pattern. The hyperfine coupling constant A can be determined from the splitting of the spectrum. The obtained saturation field (525 kOe) is about 20% larger than that of ferric rubredoxin. Compared with saturation fields of other complexes, the observed value indicates a more ionic, oxygen-nitrogen ligation. Recent resonance Raman studies have suggested one or two tyrosine ligands (58b).

Much interest in dioxygenase has been aroused by the observation of a long-lived oxygenated intermediate complex using the slow substrate 3,4-dihydroxyphenylpropionate (Ref. 59, half life ~ 30 s). Mossbauer and EPR investigations of this intermediate have been performed by Que et al. (58a). They found that the EPR spectrum recorded at 12 K shows prominent resonances at g = 6.7 and 5.3. However, Mossbauer spectra recorded at 4.2 K exhibit no field direction dependence indicating that the Mossbauer spectrum results from a doublet with extremely anisotropic effective g-values. Consequently, the EPR and Mossbauer spectra do not result from the same electronic doublet. These data were explained by assuming a negative zero-field splitting parameter. With D negative and E/D < 1/3, the ground doublet is predominantly the $S_z = \pm 5/2$ states and has very anisotropic g-tensor ($g_x \simeq g_y \simeq 0$, $g_z \simeq 10$). The EPR resonance is vanishingly small for this doublet and the corresponding Mossbauer spectrum is independent of field direction. The observed EPR resonances are attributed to the highest excited doublet. Temperature studies of the EPR spectra confirmed this interpretation.

Presently, among all observed high-spin ferric proteins, a negative D is found only for the ternary-complex of 3,4-dioxygenase. Even though X-ray crystallographic measurements have not been reported and the ligation of the iron is not known, the EPR and Mossbauer data clearly established that 3,4-dioxygenase belongs to a unique class of iron protein which can be distinguished from heme proteins and from iron-sulfur proteins.

3.3 High-spin Ferrous

The electronic configuration of a high-spin ferrous ion is illustrated schematically in Figure 1. The ground term of the free ion is ^5D and the net electronic spin is S = 2. The crystal field will split the ^5D term into five quintets. Unlike a Kramers system, the spin degeneracy of the quintets will be completely lifted by spin-orbit coupling except in cases of unusually high symmetry. For biological high-spin ferrous complexes, the zero-field splittings within the ground quintet are typically $\gtrsim 1\ cm^{-1}$ and prevent conventional EPR studies. In the absence of an applied field, $\langle\vec{S}\rangle = 0$ and the zero-field Mossbauer spectrum will be a simple quadrupole pair. Observed isomer shifts (0.7 mm/s - 1.5 mm/s) are significantly larger than those of ferric ions or low-spin ferrous ions. Typical quadrupole splittings for high-spin ferrous biological complexes range from 2 mm/s to 4 mm/s. Since the ground state originates from a ^5D term, there exists dipolar interactions between the nuclear and the electronic spins. The hyperfine coupling tensor A is therefore anisotropic. Spin-orbit interactions unquench the orbital moment which tends to enhance the anisotropy of A. Typical ratios of the smallest to the largest components of A are about 0.5 or less. The lack of EPR data and the anisotropy of the magnetic hyperfine interaction do complicate the analysis of high-spin ferrous spectra. However, important information can be extracted if spectra are recorded over a wide range of temperatures and applied fields.

According to their Mossbauer spectra, biological high-spin ferrous compounds may be separated into two general categories. In the first category, the compounds have a well isolated ground quintet. The observed paramagnetic properties can be described reasonably well by an S = 2 spin Hamiltonian. The first excited multiplet may be nearly $10^3\ cm^{-1}$ above the ground quintet resulting in a temperature independent quadrupole splitting. Application of a strong magnetic field induces well resolved hyperfine absorption lines. The reduced forms of desulforedoxin (53), rubredoxin (46), 3,4-dioxygenase (60) and P-450 (61) belong to this category. The second category includes compounds which have strong temperature dependent ΔE_Q. The first excited multiplet may lie within a few hundred cm^{-1} of the ground quintet. Attempts to describe such a system with a spin Hamiltonian formalism usually result in unreasonable parameter values. Presently, no theory has satisfactorily explained all the observed phenomena. Deoxymyoglobin, deoxyhemoglobin, reduced cytochrome c' and reduced horseradish peroxidase (62) belong to this second category. In the following, reduced desulforedoxin and deoxymyoglobin are used as examples of these two types of high-spin ferrous complexes.

3.3.1 Reduced Desulforedoxin

Figure 10 shows a Mossbauer spectrum of reduced Dx recorded at 4.2 K in zero applied field. The weak absorption at about 0 mm/s

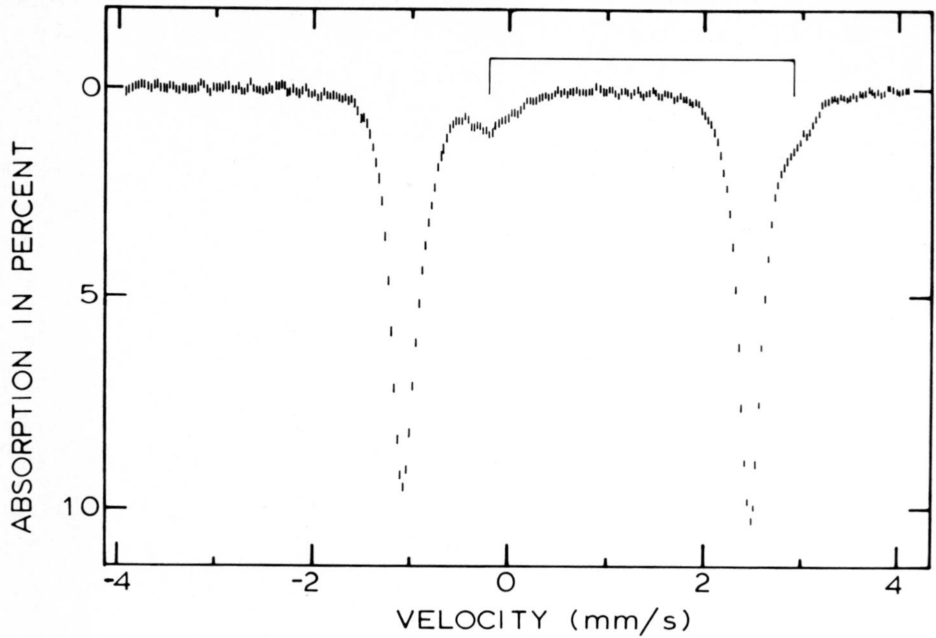

Fig. 10. Mossbauer spectrum of dithionite-reduced desulforedoxin recorded in zero applied field at 4.2 K. The ferrous minority species indicated by the bracket quantitates to approximately 10% of the total iron. (From Ref. 53)

and 3 mm/s (indicated by the bracket) is the reduced minority species corresponding to the material which produced the EPR signal at g = 8.9 in the oxidized sample. (See Section 3.2). The Mossbauer parameters for reduced Dx are $\Delta E_Q = 3.55$ mm/s and $\delta = 0.70$ mm/s. The quadrupole splitting is essentially temperature independent with $\Delta E_Q = 3.50$ mm/s at 200 K. The isomer shift is typical for high-spin ferrous ions with tetrahedral sulfur coordination. Figure 11A shows a spectrum of reduced Dx recorded at 200 K in a field of 5.5 T applied parallel to the γ-beam. For comparison, a spectrum of reduced rubredoxin from *D. gigas* recorded at the same field and temperature is shown in Figure 11B. At this temperature H_{int} is small and $H_{eff} \simeq H_{app}$. The reversed doublet and triplet absorption pattern shows clearly that $\Delta E_Q > 0$, $\eta < 0.5$ for Dx and $\Delta E_Q < 0$, $\eta < 0.5$ for Rd. This observation suggests that the ground orbital of reduced Dx is oblate (d_{xy}-like) while reduced Rd is prolate (d_{z^2}-like).

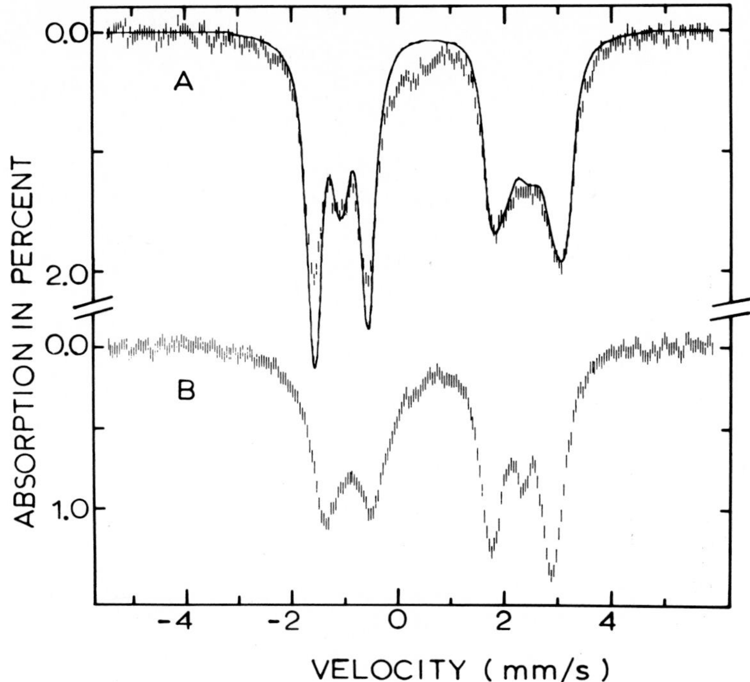

Fig. 11. Mossbauer spectra of reduced desulforedoxin (A) and D. gigas rubredoxin (B) recorded at 200 K with a parallel field of 5.5 T. Note that the triplet and doublet patterns are reversed. The solid curve is a S = 2 spin Hamiltonian simulation using the parameters listed in Table II. Fast relaxation was assumed. (From Ref. 53).

Figure 12 shows four spectra of reduced Dx taken at 4.2 K in various parallel fields. The magnetic splitting almost reaches its maximum value for H_{app} = 5 kOe indicating that the electronic moment is saturated at least in one direction and that the electronic relaxation rate is slow. The sharp absorption lines indicate that $\vec{H}_{eff}$ is constant with respect to the molecule, evidence of an extremely anisotropic induced moment. Within an S = 2 spin Hamiltonian approximation such an anisotropic situation occurs for $D < 0$ and $E/D < 1/3$. The two lowest spin states are predominantly $|S_z = \pm 2\rangle$ and their zero-field energy separation Δ is given by

$$\Delta = 2|D| \left(\sqrt{1 + 3 (E/D)^2} - 1\right) \quad (18)$$

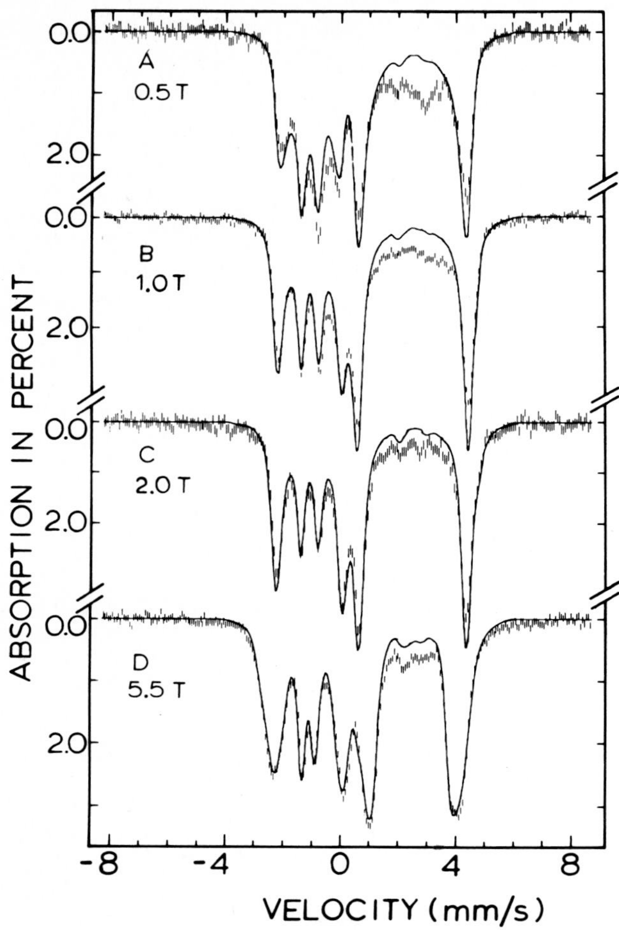

Fig. 12. Mossbauer spectra of reduced desulforedoxin recorded at 4.2 K in parallel fields as indicated. The solid curves are S = 2 spin Hamiltonian simulations generated from the parameters listed in Table 2. The deviations between theory and experiment around +3 mm/s are most likely due to the impurity iron. (From Ref. 53).

In this case the saturation behavior of the induced moment is governed chiefly by Δ (63). From spectra recorded over a range of applied field we determined that $\Delta = 0.6\ cm^{-1}$. For $H_{app} = 0.5$ T the induced electron moment and the effective field are predominantly oriented along the z-axis. The sharp absorption lines of Figure 12A allow the angle between $\vec{H}_{eff}$ and the direction of the largest efg component, $\hat{\zeta}$, to be determined (64). We found $\hat{\zeta}$ to be $10^\circ \pm 3^\circ$ away from H_{eff}. The spectra are very sensitive to A_z and are best fitted with $A_z/g_n\beta_n$ = 6.7 T. To determine parameters A_x and A_y we used the high temperature, high field data shown in Figure 11A. Since ΔE_Q is positive and η less than 0.5 the magnetic splitting of the low energy (low velocity) absorption band is controlled by A_x and A_y. Assuming $A_x = A_y$, Moura et al. (53) found that -23 T $< A_x/g_n\beta_n < -17$ T. With this restriction in A_x and A_y, D can now be determined from the spectrum shown in figure 12D. An applied field of 5.5 T is strong enough to induce a significant electronic moment in the xy plane resulting in the broadening of the spectrum. The magnitude of this broadening is sensitive to the value of D and a range for D can be inferred from the spectrum. We found $-7\ cm^{-1} < D < -4\ cm^{-1}$. The value of E/D was then determined from Equation 18.

From the above considerations, the spin Hamiltonian parameters have been determined within a rather narrow range. More importantly, we have gained some insight into the physics of the system. The final refinements of the parameter values require computer stimulations. The parameters which yielded the best fit simultaneously to all the spectra are listed in Table 2. The solid lines in Figures 11 and 12 were generated from that parameter set. Considering that the sample was about 10% impure, the fits to the data are very satisfactory. For comparison, the parameters for reduced Rd from *C. pasteurianum* are also listed in Table 2.

The anisotropy in $\tilde{A}$ can be conveniently understood in terms of a pertubation expression (65) which approximates equation (6).

$$\vec{H}_{int} = \tilde{A}.\vec{S}/g_n\beta_n \quad ,$$

$$A_{ij} = [A^c + (g_{oi}-2)A^L]\,\delta_{ij} + A^d\ell_{ij} \quad , i,j=x,y,z \quad , \tag{19}$$

where A^c represents the isotropic Fermi contact contribution, $(g_{oi}-2)A^L$ the orbital contribution, and $A^d\ell_{ij}$ the dipole contribution. The theoretical values of $A^c/g_n\beta_n$ = -27.5 T, $A^L/g_n\beta_n$ = 56.5 T and $A^d/g_n\beta_n$ = 4.0 T have been quoted for free ferrous ions (65, 66). The orbital contribution results from the orbital moment unquenched by spin-orbit coupling. Within the ^5D term, the theoretical minimum value for g_{oi} is 2.0. Hence, the orbital term opposes the Fermi contact term. Using a second order perturbation approach, the g-factor may be written as

$$g_{ox} = g_{oz} - 2(D-E)/\lambda \quad ,$$
$$g_{oy} = g_{oz} - 2(D+E)/\lambda \quad , \tag{20}$$

where λ is the spin-orbit coupling constant and has the value of $-103\ cm^{-1}$ for the free ferrous ion. The dipole term ℓ_{ij} is proportional to the valence efg

$$V_{ij}^{val} = \frac{2}{7} e^2 \langle r^{-3} \rangle \ell_{ij} \tag{21}$$

For desulforedoxin, the parameter D is negative and $E/D < 1/3$. From equation (20), we notice that the largest orbital contribution to $\tilde{A}$ is along the z-direction. The quadrupole splitting of Dx is positive and $\eta < 0.5$. The efg component along the z-direction is the largest and is positive. Consequently, A_z is reduced relative to A^c by both the orbital and dipole contributions. Along the x and y directions, the efg component is negative. The dipole contribution almost cancels the orbital contribution resulting in a magnitude of A_x and A_y comparable to A^c. For Rd, the situation is reversed with the parameter D being positive. The maximum g component is g_y. The efg component V_{yy} is positive. Consequently, A_y is the smallest component in $\tilde{A}$ and the other two components are comparable to A^c. (See table 2.)

3.3.2 Deoxymyoglobin

Myoglobin and hemoglobin are the oxygen-carrying proteins in vertebrates. Hemoglobin consists of four polypeptide chains in a $\alpha_2\beta_2$ configuration, each subunit containing one heme group. Myoglobin consists of a single polypeptide chain and one heme group. The first successful protein X-ray structural analysis was performed on myoglobin (67). In deoxymyoglobin (Mb) and deoxyhemoglobin (Hb) the iron atom is five coordinated. Four of the ligands are the pyrrole nitrogens of the porphyrin ring. The fifth ligand is the nitrogen of the proximal histidine. The sixth position is vacant and is the oxygen binding site.

Recently, Mossbauer investigations have been performed on deoxymyoglobin and deoxyhemoglobin in applied fields up to 6 T (68, 69, 70). The Mossbauer spectrum of Mb taken at 4.2 K in the absence of an applied field is a quadrupole pair. The quadrupole splitting, $\Delta E_Q = 2.22$ mm/s, and the isomer shift, $\delta = 0.92$ mm/s, are typical of high-spin ferrous ions. Compared with rubredoxin, the isomer shift of Mb reflects a more ionic nitrogen and/or oxygen ligation. The quadrupole splitting decreases from 2.22 mm/s at 4.2 K to 1.8 mm/s at 195 K, suggesting low-lying excited multiplets. Deoxyhemoglobin shows a similar temperature dependent behavior in ΔE_Q. Shown in Figure 13 are the Mossbauer spectra of Mb recorded at different temperatures in an applied field of 60 kOe. The doublet and triplet patterns of the spectra taken at high temperatures demonstrate clearly that ΔE_Q is negative. The solid lines in Figure 13 are least-squares fits to the data based on a phenomenological expression of the internal field which may be written as

$$(H_{int})_p = \omega_p (H_{app})_p \quad , \quad p = \xi, \sigma, \zeta \quad , \tag{22}$$

where ω_p are functions of both H_{app} and T. The ω-tensor was found to have axial symmetry with its largest component along the largest efg component (68). The observed efg tensor and the anisotropy of the internal field both suggest that myoglobin has a somewhat distorted prolate ground orbital (d_{z^2}-like). By combining these results with the single crystal measurements of Gonser et al. (71), Kent et al. (68) were able to find the relative orientation of the efg tensor and the heme group. The heme normal was found to coincide with none of the efg principal axes.

Figure 14 shows a Mossbauer spectrum of Mb recorded at 4.2 K in an applied field of 6 T. The spectrum is broad and unresolved. The solid line in Figure 14 is an S = 2 spin Hamiltonian simulation with the g-tensor, the A-tensor and the efg-tensor all having a common principal axes system (xyz). The parameters used (70) are $\Delta E_Q = -2.28$ mm/s, $\eta = 0.7$, $A_x/g_n\beta_n = -4.9$ T, $A_y/g_n\beta_n = -4.7$ T, $A_z/g_n\beta_n = -10.9$ T, $D = -2.4\ cm^{-1}$ and $E/D = 1$. Using a coordinate system x'y'z' in which the z'-axis coincides with the y-axis, we may write $D' = +4.8\ cm^{-1}$ and $E'/D' = 0$ in agreement with susceptibility studies (72). However, the magnetic hyperfine coupling constants are only about half as large as expected for typical high-spin ferrous ions and the spin Hamiltonian parameters quoted above do not explain the observed high temperature spectra. Also, far-infrared magnetic resonance has been used to probe the low-lying electronic states of Hb and Mb (73). Within the energy range of zero to 15 cm^{-1} only one excited state at $\sim 3.5\ cm^{-1}$ was detected. Using the spin Hamiltonian parameters $D = 5.3\ cm^{-1}$ and $E/D = 0.17$ obtained from fitting the susceptibility data, one would expect excited spin states at 3.0 cm^{-1} and 8.4 cm^{-1}. Hence, while the spin Hamiltonian approach is flexible enough to fit the susceptibility data, it does not provide an adequate model for the electronic structure of Hb or Mb. Furthermore, Kent et al. (70) have exhausted all possible

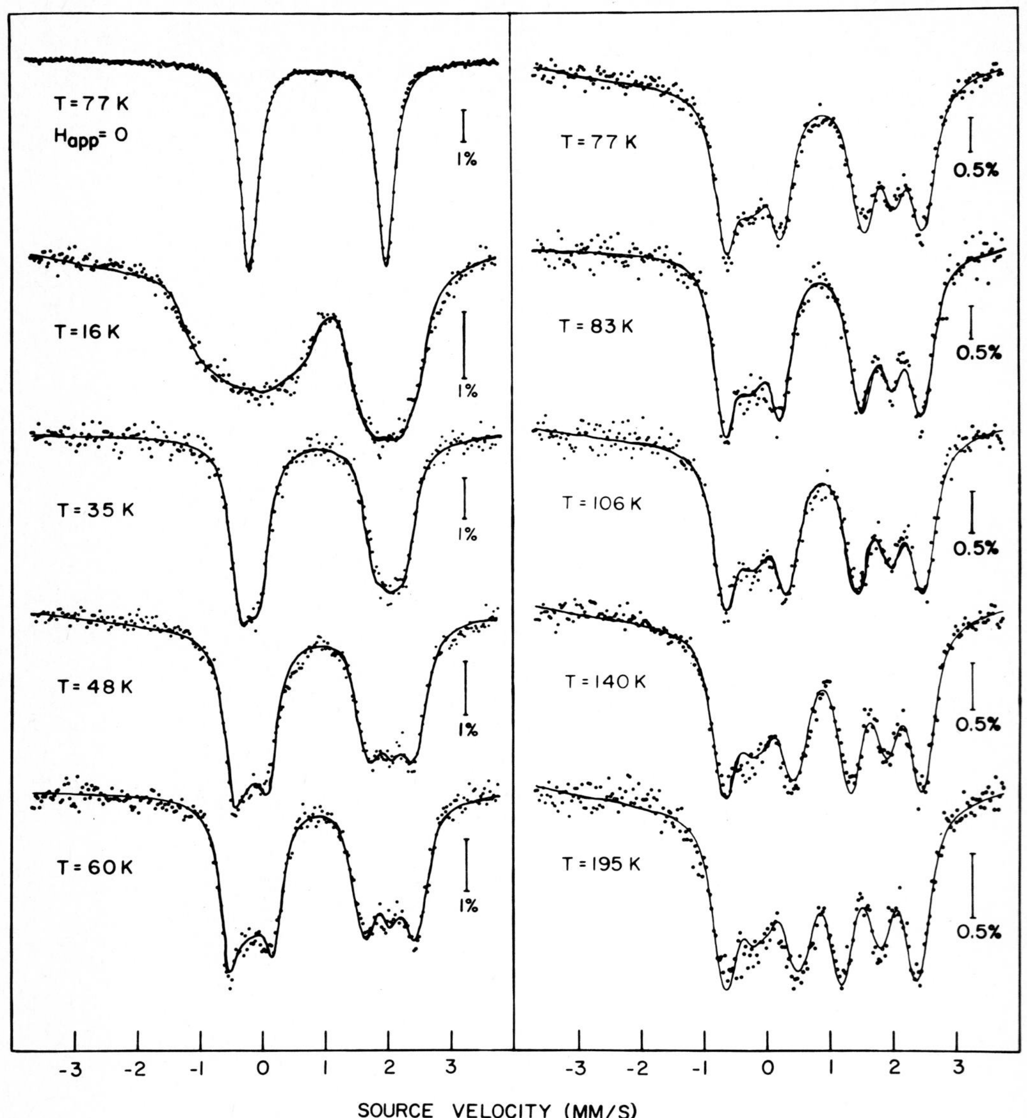

Fig. 13. Mossbauer spectra of ^{57}Fe-enriched myoglobin recorded in a transverse magnetic field of 6 T at temperatures as indicated. The top spectrum on the left was recorded in zero field at 77 K. The solid curves are calculations as described in the text. (From Ref. 68)

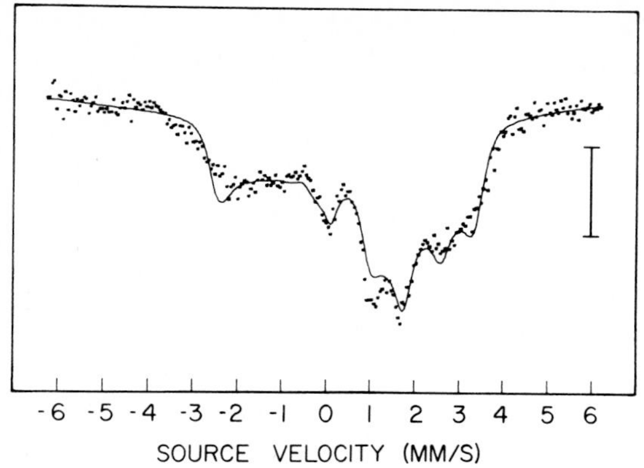

Fig. 14. Mossbauer spectrum of ^{57}Fe-enriched myoglobin recorded in a transverse applied field of 6 T and 4.2 K. The solid curve is a S = 2 spin Hamiltonian simulation generated from parameters given in the text. The vertical bar indicates 1% absorption. (From Ref. 70).

combinations of energy schemes within the ^{5}D term in trying to match the 4.2 K spin Hamiltonian parameters; the high-field, high-temperature data; and the magnetic susceptibility data. They concluded that no solution could be found within the ^{5}D term. One of the major difficulties in constructing an acceptable energy scheme from a ^{5}D term is that the small magnetic hyperfine constant suggests a large orbital contribution while the susceptibility data imply little unquenched orbital moment.

In an earlier attempt to explain the temperature dependence of ΔE_Q in Hb, Eicher and Trautwein (74) introduced a crystal field model which took into account all 210 states of the $3d^6$ rhombic configuration. They assumed that the iron environment had symmetry and that the heme normal was a two-fold symmetry axis. They found one solution which yielded the temperature dependence of ΔE_Q. The resulting ground quintet is $^5B_2$ (in D_{4h} representation) which yields a positive ΔE_Q, in disagreement with later experiments. Due to the effect of a low-lying ^{3}E state, the resulting energy scheme yields a magnetic susceptibility far below the experimental value (72). Huynh et al. (75) used the same method to search for an energy scheme that yields the temperature dependence of both the quadrupole splitting and magnetic susceptibility. They found a solution which has one of the ^{5}E states as the ground state. The first excited state is $^5B_2$ and is only 50 cm^{-1} above the ground state. Their solution also yields a positive ΔE_Q. Both calculations were done prior to the high-field studies and do not yield an energy scheme that is compatible with those of more recent data. However, since this theory takes into account all 210 $3d^6$ states, it is worth further consideration especially in light of the high-field data. The attractiveness of this model is that it contains only five parameters: four crystal field parameters and one covalency factor. An obvious advantage is that the inclusion of lower spin ($S < 2$) configurations might allow enough orbital mixing to reduce the internal field and yet not increase the magnetic susceptibility to an unacceptable value. Since the experimental data suggest a ground orbital of d_{z^2} symmetry, a solution may perhaps to be found in the parameter space with negative tetragonal crystal field. Proof of the existence of such a solution awaits actual calculations.

Presently, there exist two major difficulties in applying molecula orbital (MO) calculation to the high-spin ferrous heme. First, the heme is a large, complex molecule and, second, it is an open shell system. For large molecules, MO calculations starting from first principles appear to be unrealistic. A semi-empirical approach would have t introduce many ill-defined parameters which would render any results doubtful. In contrast to a closed shell system ($S = 0$), the wave functions of an open shell system cannot be described as a single determinant of doubly occupied MOs. Thus, many theoretical difficulties arise. Discussions on the theoretical treatments of open shell system can be found in standard text books (76). Nonetheless, Seno et al. (77) have recently used a semi-empirical MO calculation to study the low-lying multiplets of the heme group in Hb and Mb. Their calculated energy level scheme is similar to that reported by Huynh et al. (75). As discussed above, such an energy scheme is incompatible with the high field Mossbauer data. Moreover, the MOs used by Seno et al. for constructing the open shell configurations were obtained from a closed shell SCF-MO calculation. The system state energies were estimated through a limited configuration interaction calculation. The estimated energies have an uncertainty of a few eV while the energy separations of the low-lying multiplets are 10^{-2} - 10^{-1} eV. Thus, any agreement between experiment and such a theoretical approach can only be considered fortuitous.

3.4 Low-spin Ferrous

To a first approximation the electronic configuration of a low-spin ferrous ion is $(t_{2g})^6$ in a cubic field representation. (See Fig. 1

Each t_{2g} orbital is occupied by two spin-paired electrons yielding a diamagnetic (S = 0) complex. The charge distribution of the valence electrons has cubic symmetry producing a zero efg at the nucleus. An asymmetric covalent charge distribution can be induced through an asymmetric covalent bonding to the ligands. Nevertheless, small quadrupole splittings are expected for low-spin ferrous compounds, To date, hemoproteins are the only biomolecules known to contain the low-spin ferrous ion. In the absence of an applied field, the Mossbauer spectrum is a quadrupole pair with observed quadrupole splittings ranging from 0.36 mm/s for carboxyhemoglobin (4) to 1.17 mm/s for ferrocytochrome c (34). The signs of $V_{\zeta\zeta}$ of these two complexes are positive. Observed isomeric shifts are between 0.25 - 0.45 mm/s at 4.2 K.

Among low-spin ferrous compounds, oxygenated hemoglobin (HbO_2) is of particular interest. Because of its biological significance, the nature of the iron-oxygen bond in HbO_2 has been subjected to extensive investigations (78-80). Much controversy is focused on whether the iron-oxygen bond on HbO_2 should be described as Fe^{2+}-O_2 (78) or Fe^{3+}-O_2^- (79). Early magnetic susceptibility measurements (81, 82) indicated that HbO_2 was diamagnetic at room temperature. However, the observed ΔE_Q is unusually large for low-spin ferrous compounds. At 1.2 K the quadrupole splitting of HbO_2 is 2.24 mm/s and decreases monotonically to 1.89 mm/s at 195 K (4). A Mossbauer spectrum of HbO_2 recorded at 4.2 K in a transverse field of 62 kOe (7) indicates that the system is diamagnetic and the largest efg component is negative. This large and negative ΔE_Q may indicate that the iron electronic state is more suitably described as a $(t_{2g})^5$ configuration rather than $(t_{2g})^6$. This observation has been used as an argument supporting a Fe^{3+} - O_2^- bond in HbO_2. Recent theoretical studies on HbO_2 (83) suggest that the arguments for Fe^{2+}-O_2 or Fe^{3+}-O_2^- are largely semantic. According to the theoretical model, bonding of O_2 to the heme involves a significant charge rearrangement. There is substantial σ-electron donation to the iron atom from the oxygen molecule and this charge donation is almost exactly balanced by the π-electron back donation from the iron. The large and netative ΔE_Q is primarily caused by the asymmetric π-back donations from the electrons occupying the d_{xz} and d_{yz} orbitals and by the σ-donation to the d_{z^2} orbital. Other experimental data (79, 84) that were thought to support a Fe^{3+}-O_2^- model were also explained by these theoretical calculations. The essential element in interpreting the experimental data was shown to be not the net charge associated with each atom but the nature of the bonding and the attendant orbital populations.

Another interesting feature of the Mossbauer data of HbO_2 is its temperature dependent ΔE_Q. For deoxyhemoglobin the temperature dependent ΔE_Q can be explained by assuming low lying excited states. A similar assumption for HbO_2 would inevitably result in detectable paramagnetism at room temperature. The early magnetic susceptibility measurements (81, 82) do not support this assumption. As an alternative explanation, Spartalian et al. (85) have postulated conformational excitations involving different oxygen orientations. Although X-ray diffraction data of oxygenated model complexes support this idea of the oxygen molecule revolving about the Fe-O bond, it is not clear that such a situation can occur in the protein. In fact, X-ray diffraction data of MbO_2 (86) show only one orientation for the Fe-O-O bonds. Recent magnetic susceptibility data (87) disagree with the earlier measurements and indicate the existence of paramagnetic excited levels which are thermally accessible at 300 K. One way to resolve the issue would be to perform a high-field (> 50 kOe), high-temperature (T > 150 K), Mossbauer investigation and determine whether H_{eff} equals H_{app}. The experimental uncertainty of H_{eff} would have to be less than 0.1 T for the results to be meaningful which is close to the limit practically obtainable at present.

3.5 Other States

The sections above pertain to the four oxidation and spin states most commonly found in iron proteins. However, other states do occur. For example, during the enzymatic reaction of horseradish peroxidase with peroxide, two intermediates can be isolated (88). They are called compounds I and II which are two and one oxidizing equivalents, respectively, above the native enzyme. Magnetic susceptibility measurements (89) suggest a spin S = 1 assignment for compound II, and a S = 1 iron plus a S = 1/2 radical assignment for compound I. Also, when cytochrome c peroxidase reacts with hydroperoxide, an intermediate called compounds ES can be isolated (90). Similar to compound I, compound ES is two oxidizing equivalent away from the native ferric enzyme and is thought to contain both a S = 1 Fe and a S = 1/2 free radical (91). In the absence of an applied field, the Mossbauer spectra (92-94) of all three compounds recorded at 4.2 K show a quadrupole doublet indicating that the iron has integer spin. The ΔE_Q values are 1.25 mm/s, 1.61 mm/s and 1.55 mm/s for compounds I, II and ES, respectively. Their isomer shifts (< 0.1 mm/s) rule out a Fe(II) assignment and are consistent with a Fe(IV) ion. Hence, the Mossbauer data support a $S = 1$ Fe(IV) assignment for all three compounds. The additional oxidizing equivalent in compounds I and ES therefore must be localized away from the iron center. Schulz et al. (92) have shown that the Mossbauer and EPR data of compound I can be interpreted as a $S = 1$ Fe(IV) ion weakly coupled to a $S = 1/2$ radical.

Similar to that of a high-spin ferrous ion, the hyperfine coupling tensor A of an $S = 1$ Fe(IV) ion is anisotropic. The experimental anisotropy in A has been interpreted with a crystal field model for a d^4, $S = 1$ complex (95). However, the observed ΔE_Q appears to be too small in comparison with the theory. This is not surprising since the electrons occupying bonding 3d orbitals as well as the iron ligands can substantially affect the efg but can not contribute to the internal field or the paramagnetic susceptibility.

Another intermediate spin complex has been observed when any of several ferrous proteins react with nitric oxide. The product yields an almost axial EPR signal characteristic of a $S = 3/2$ spin system (96, 97). The observation of such an EPR signal when NO is passed over native protocatechuate 4,5-dioxygenase in the absence of reducing agents is, therefore, evidence of a ferrous site in the active form of this extradiol dioxygenase (98). Preliminary Mossbauer data indicates that its active site is indeed ferrous in contrast with intradiol dioxygenases which have ferric sites. Another extradiol dioxygenase, metapyrocatechase from *Pseudomonas arvilla*, has also been reported to contain ferrous ion (99).

4. PROTEINS WITH MULTI-IRON CLUSTERS

4.1 Two-iron Ferredoxins

Two-iron ferredoxins, also known as plant-type ferroxins* contain a cluster consisting of two iron atoms and two labile sulfurs. The cluster is denoted by [2Fe-2S]. The biological roles and physical properties of 2Fe ferredoxins have been extensively studied, and their primary function appears to be electron transfer. The structure of the [2Fe-2S] cluster was first deduced from magnetic resonance spectroscopic data (100) and from the studies of synthetic model complexes (101). Only recently has an X-ray diffraction analysis

*Two-iron ferredoxin was once thought to be unique to plants. Isolation of such proteins from bacteria has shown that this is not the case.

of a ferredoxin from *Spirulina plastensis* been reported (102). At a resolution of 2.5 A the structure of the active center appears to be similar to its synthetic analog. The cluster geometry may be described as two tetrahedral sulfur coordinated irons having two labile sulfurs as common ligands. The other four ligands are cysteine sulfurs which link the cluster to the polypeptide chain.

Two-iron ferredoxins can be stabilized in two oxidation states with midpoint potentials of different proteins ranging from -250 mV to -420 mV. In the reduced state a nearly axial EPR signal at $g_z = 2.05$ and $g_i = 1.94$ is observed, an indication of a spin S = 1/2 system (103). EPR spectra of proteins isotopically enriched with either Fe^{57} or S^{33} exhibit hyperfine broadenings which suggests both the irons and the labile sulfurs are part of the paramagnetic center (103 - 106). ENDOR studies (107) of the Fe^{57}-enriched ferredoxin have accurately determined the principal values of the magnetic hyperfine tensor for each of the iron atoms. Mossbauer spectra (108, 109) of the reduced cluster recorded at high temperatures show two, well-resolved, quadrupole pairs of approximately equal intensity. One exhibits parameters ($\Delta E_Q \simeq 3$ mm/s and $\delta \simeq 0.7$ mm/s) typical of high-spin ferrous ions while the other has parameters ($\Delta E_Q \simeq 0.7$ mm/s and $\delta \simeq 0.3$ mm/s) typical of high-spin ferric ions. The isomer shifts of both strongly suggest tetrahedral sulfur coordinateion. (See Section 3). High field spectra reveal that the ferrous ion has a positive A-tensor while that of the ferric ion is negative. The above observations are nicely explained by the antiferromagnetic coupling of a high-spin ferric ion with a high-spin ferrous ion to form an S = 1/2 system (110, 111). If the experimental values of $\tilde{g}$ and $\tilde{A}$ for oxidized and reduced rubredoxin are used for the intrinsic values of the ferric and ferrous sites, respectively, then such a theoretical model yields the g-values and the magnetic hyperfine tensors observed for the coupled system.

The Mossbauer spectrum of oxidized two-iron ferredoxin recorded at 4.2 K and zero magnetic field shows a single quadrupole pair, indicating that the two irons of the oxidized cluster are equivalent (108, 109). The Mossbauer parameters for proteins isolated from different species are the same within narrow ranges: $\Delta E_Q = 0.6$ mm/s to 0.7 mm/s, and $\delta = 0.25$ mm/s to hedral sulfur coordination. High field studies indicate that the oxidized protein is diamagnetic. Again, these observations are consistent with the theoretical model of two antiferromagnetically coupled spins. In this case, both irons are high-spin ferric and the two S = 5/2 spins couple to form a S = 0 ground state. Magnetic susceptibility measurements of oxidized spinach ferredoxin reveals that the protein is diamagnetic at low temperatures (< 77 K) (112) and becomes paramagnetic at higher temperatures (113). These observations strongly support an antiferromagnetically coupled spin system. If the exchange coupling is expressed as $-2J\,\vec{S}_1 \cdot \vec{S}_2$, $S_1=S_2=5/2$, the data yield $J=-182\ cm^{-1}$. Recent magnetic susceptibility data obtained for the two-iron ferredoxin from blue-green alga *Spirulina maxima* (114) are consistent with that of spinach ferredoxin.

4.2 Four-iron and Eight-iron Proteins

Another major class of iron-sulfur proteins consists of the high potential iron protein (HiPIP) plus the four-iron and the eight-iron ferredoxins. All these proteins contain a cluster which is composed of four iron atoms bridged by four labile sulfurs to form a distorted cubane structure (denoted by [4Fe-4S]). An eight-iron ferredoxin contains two clusters of this type. Like two-iron ferredoxin, four-iron ferredoxin can be stabilized in two oxidation states with a midpoint potential of about -400 mV. When reduced, four-iron ferredoxins exhibit characteristic EPR signals at $g_z \simeq 2.07$, $g_\perp \simeq 1.94$. The oxidized protein shows a very weak isotropic signal at $g \simeq 2.02$ (115, 116). HiPIP also operates in two oxidation states, but with a midpoint potential at about +350 mV. Unlike ferredoxin, reduced HiPIP shows no EPR resonance. The oxidized form, however, exhibits an EPR signal characteristic of an S = 1/2 system with the averaged g-value larger than 2.0 (117).

X-ray diffraction results have been reported for reduced HiPIP from *Chromatium vinosum* and for the oxidized ferredoxin from *Peptococcus aerogenes* (118, 119). Within the experimental uncertainties, the structures of the active centers in these two proteins are indistinguishable. To explain these observations Carter et al. (120) proposed a three-state model for the four-iron clusters; namely, the $[4Fe\text{-}4S]^{3+}$, $[4Fe\text{-}4S]^{2+}$ and $[4Fe\text{-}4S]^{1+}$ states (121). The (3+) and (2+) states are two and one oxidizing equivalents, respectively, above the (1+) state. Ferredoxin operates between the (1+) and (2+) states, while HiPIP operates between the (2+) and (3+) states. This hypothesis is also supported by studies of synthetic analogs (122) and by the observations of a Fd-like EPR signal for super-reduced HiPIP (123) and a HiPIP-like EPR signal for super-oxidized ferredoxin (124).

Four-iron proteins have also been investigated extensively by Mossbauer spectroscopy (125, 126). Unlike two-iron ferredoxins, the four-iron proteins do not exhibit Mossbauer spectra having well-resolved quadrupole pairs of ferric or ferrous character. The spectra consist of unresolved quadrupole pairs with broad, non-Lorentzian line shapes. Attempts have been made to decompose the spectra of *Chromatium* HiPIP and *Bacillus* Fd. The averaged Mossbauer parameters are listed in Table 3. It is obvious that the irons in reduced HiPIP and in oxidized Fd exhibit similar parameters which also support Carter's three-state model. Of particular interest is the *averaged* isomer shift which increases by about 0.1 mm/s per each added equivalent of electrons. Taking into consideration that the isomer shift of rubredoxin increases by about 0.4 mm/s upon reduction, the observed values for the [4Fe-4S] clusters suggest that each added electron is distributed among the four iron atoms. High-field Mossbauer studies indicate that reduced HiPIP and oxidized Fd are diamagnetic while oxidized HiPIP and reduced Fd are paramagnetic, in agreement with the EPR data. The spin-coupling within the [4Fe-4S] cluster is reflected in the positive and negative magnetic hyperfine constants observed for both oxidized HiPIP and reduced Fd. Detailed information about EPR and Mossbauer studies on the [4Fe-4S] proteins may be found in the review article by Cammack et al. (127).

4.3 Proteins Containing Three-Iron Clusters

In the early 1970's, Shethna (128) and Yoch and Arnon (129) isolated an iron-sulfur protein from *Azotobacter vinelandii* having 7-8 iron atoms and a comparable amount of labile sulfur. Redox titrations and EPR investigations (130,131) revealed that this protein contained two redox centers with midpoint potentials at +340 mV and -420 mV. As isolated, the protein is in the "semi-reduced state" with the low-potential center in its oxidized form and the high-potential center in its reduced form. A nearly isotropic EPR signal at g = 2.01 is observed and is associated with the oxidized low-potential center (131). A one-electron reduction fully reduces the protein and silences the EPR signal. The high-potential center in the semi-reduced proteins is EPR silent but may be detected by EPR after the oxidation of the protein with ferricyanide. This center also exhibits an isotropic g = 2.01 EPR signal. Among the three familiar types of iron-sulfur clusters only the [4Fe-4S] cluster in HiPIP has an EPR signal when oxidized. It was, therefore, erroneously assumed that both centers were [4Fe-4S] clusters

TABLE 3.

Average Mossbauer parameters for Chromatium HiPIP and Bacillus stearothermophilus ferredoxin

	HiPIP[a]				Ferredoxin[b]			
	Oxidized		Reduced		Oxidized		Reduced	
Temperature (K)	ΔE_Q (mm/s)	δ (mm/s)	ΔE_Q (mm/s)	δ (mm/s)	ΔE_Q (mm/s)	δ (mm/s)	ΔE_Q (mm/s)	δ (mm/s)
4.2	0.96	0.35	1.14	0.43	1.12	0.43	1.61	0.54
77	0.83	0.33	1.11	0.42	0.97	0.43	1.52	0.54
195	0.73	0.28	0.97	0.38	0.79	0.38	1.34	0.49

(a) Ref. 126a

(b) Ref. 125

operating between the (+3) and (+2) states, even though the redox potential of a HiPIP center is generally about 700 mV above that of the low-potential center. Recently, a combined EPR and Mossbauer investigation (132) demonstrated that the high-potential center in the A. *vinelandii* ferredoxin is indeed a HiPIP type [4Fe-4S] cluster. However, the low-potential center proved to be a novel cluster containing three iron atoms. These results were supported by preliminary analysis of X-ray diffraction data (133). Our group in Minnesota has since found the three-iron cluster in a variety of iron-sulfur proteins, including beef heart aconitase, ferredoxin II (Fd II) from *Desulfovibrio gigas*, an iron-sulfur protein from *Thermus thermophilus*, ferredoxin from *Methanosarcina barkerii* and glutamate synthase from *A. vinelandii*. The Mossbauer and EPR data of Fd II (134) are used below to illustrate the physical properties of the three-iron cluster.

Two oligomeric forms of ferredoxins have been isolated from *D. gigas*, the tetrameric Fd II and the trimeric Fd I. Both forms have the same monomeric subunit which consists of 57 amino acids of known sequence (135). Fd II has been shown to transfer electrons between cytochrome c_3 and the sulfite reductase system while Fd I serves as an electron carrier in the phosphoroclastic reaction (136). As isolated, Fd II exhibits a fairly isotropic EPR signal at about $g = 2.01$. A typical EPR spectrum of Fd II taken at 8 K is shown in Figure 15. At about 16 K the EPR spectrum starts to broaden indicating the increase of the spin relaxation rate at high temperatures. The spectrum shown in Figure 15 can be simulated with Gaussian lineshapes of widths 1.5, 3.5 and 8.0 mT at $g = 2.02$, 2.00 and 1.97, respectively. Spin quantitation of the EPR spectra taken at 6, 8 and 12 K gave (0.93 ± 0.12) spins/3 irons. Amino acid analyses and iron quantitation yielded an average value of 2.97 irons per monomer (134). No iron impurities were detected by either EPR or Mossbauer spectroscopy.

Figure 16 shows a Mossbauer spectrum of oxidized Fd II taken at 1.5 K in a parallel field of 0.06 T. The spectrum shows three distinct iron environments with different magnetic hyperfine coupling constants. Attempts have been made to decompose the spectrum into three subcomponents using a spin Hamiltonian with effective spin $S = 1/2$. The theoretical spectrum of such a decomposition is plotted over the experimental data in Figure 16. Also shown are the theoretical simulations of the three subspectra. The obtained magnetic

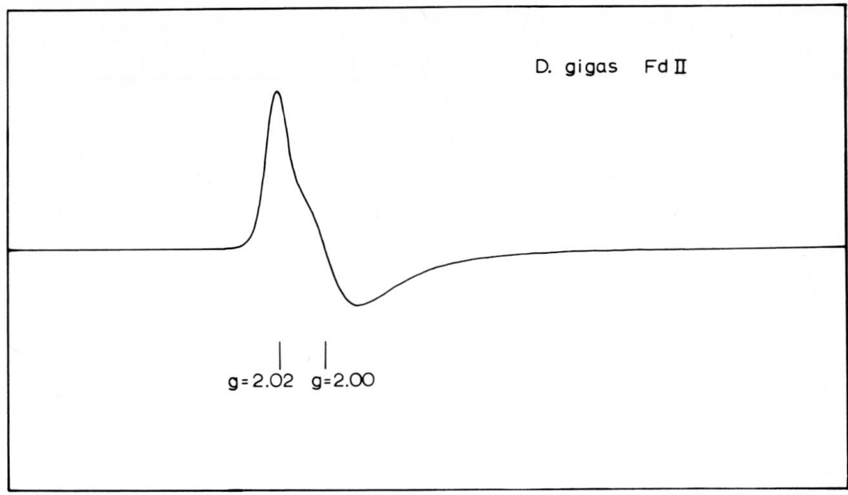

Fig. 15. EPR spectrum of oxidized Ferredoxin II from Desulfovibrio gigas. Conditions: T = 8 K, frequency 9.23 GHz, microwave power 10 μW, modulation amplitude 0.4 mT, modulation frequency 100 kHz.

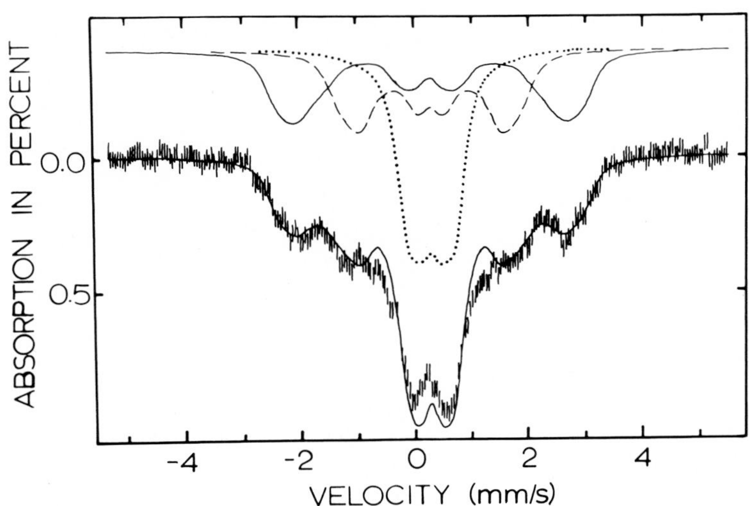

Fig. 16. Mossbauer spectrum of oxidized Fd II from D. gigas recorded at 1.5 K in a field of 0.06 T applied parallel to the γ-beam. The solid line plotted over the data is a superposition of the three spectra above. The three were computed from Equation 8 and 11 with the parameters quoted in the text. (From Ref. 134).

hyperfine coupling constants* are A_x = -27 MHz, A_y = A_z = -44 MHz for site 1 (solid line), A_x = +29 MHz, A_y = A_z = +16 MHz for site 2 (dashed line) and $|A_x| = |A_y| = |A_z|$ = 3.5 MHz for site 3 (dotted line). Due to the poor resolution of the spectrum the values quoted are tentative. At higher temperature (> 20 K), the Mossbauer spectrum collapses into a single quadrupole pair (Figure 17). The solid line in Figure 17 is a least-squares fit to the spectrum assuming two Lorentzian lines. The fit yields 0.28 mm/s full width at half maximum for both absorption lines indicating that all three subsites yield the same spectrum at high temperature. The quadrupole splitting, ΔE_Q = (0.54 $\pm$ 0.03) mm/s, and isomer shift, δ = (0.27 $\pm$ 0.03) mm/s, suggest high-spin ferric ions (S = 5/2) with tetrahedral coordination of predominantly sulfur atoms. However, these parameters do not rule out the possibility that a site might have one oxygenic or nitrogenous ligand since no model complexes are available for comparison.

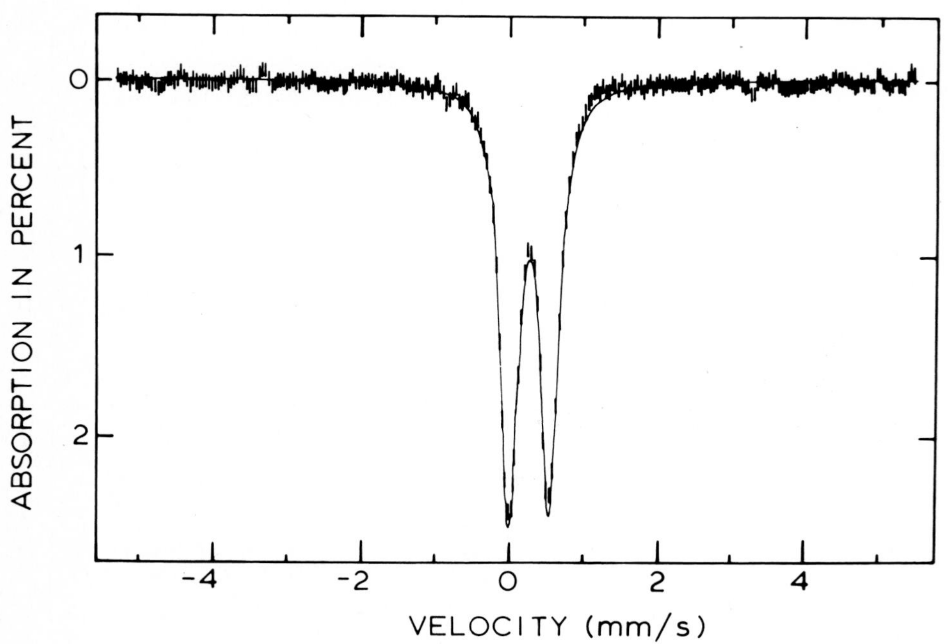

Fig. 17. Zero field Mossbauer spectrum of oxidized Fd II from D. gigas recorded at 77 K. The solid line is the result of fitting two Lorentzian peaks to the data. (Adapted from Ref. 134).

* The quoted A-values are for the nuclear ground states; an effective field of 0.73 T will split the I=1/2 states by $g_n \beta_n H_{eff}$ = 1 MHz.

At low temperature (< 4.2 K), the field direction dependence of sites 1 and 2 demonstrate clearly that they are associated with the EPR signal. However, it is difficult to observe the field direction dependence of site 3 because of its small hyperfine coupling constant. On the other hand, its ΔE_Q and δ are of typical high-spin ferric ions. Consequently, site 3 must be a structural component of a spin-coupled cluster or else an EPR signal of a high-spin ferric ion should be observed. Furthermore, if site 3 were not part of the EPR-active cluster, another half-integer electronic spin has to be postulated since sites 1 and 2 are S = 5/2 irons and cannot couple to form the observed half-integer system spin. Also, the observed hyperfine fields may be explained in terms of a spin-coupling scheme involving three high-spin ferric ions (see below). The differences between the observed A's and that of rubredoxin (~ 55 MHz) are so large that it is unreasonable to postulate that covalency effects could cause these differences.

Upon reduction, Fd II exhibits no EPR signal. Figure 18A shows a Mossbauer spectrum of reduced Fd II recorded at 4.2 K in the absence of a magnetic field. Two well-resolved quadrupole doublets, designated doublets I and II, are observed. The solid line in Figure 18A is the result of fitting two doublets to the data with the assumption that the intensities of the high and low energy lines of each doublet are the same. Most interestingly, the two species represented by doublets I and II are found to be present in the ratio of 2:1. The fit yielded a ratio of (2.01 ± 0.03). The obtained Mossbauer parameters are ΔE_Q = (1.47 ± 0.03)mm/s and δ = (0.46 ± 0.02) mm/s for doublet I, and ΔE_Q = (0.47 ± 0.02) mm/s and δ = (0.30 ± 0.02) mm/s for doublet II. These values suggest the iron associated with doublet II is high-spin ferric in character, similar to the irons in oxidized Fd II. The parameters of doublet I suggest a formal iron oxidation state between Fe^{2+} and Fe^{3+}, i.e. upon reduction the additional electron is shared by two of the iron atoms. The absence of magnetic broadening of the zero field spectrum suggests that reduced Fd II has integer spin.

Figure 18B shows a Mossbauer spectrum of reduced Fd II taken at 4.2 K in a parallel applied field of 0.06 T. The fact that a field of a few hundred gauss can elicit such substantial broadening of both doublets leads to the following conclusions: (1) The irons associated with doublets I and II share a common electronic system. (2) The system spin, S, is larger than zero. (3) The lowest two electronic spin singlets are closely spaced in energy. Their energy

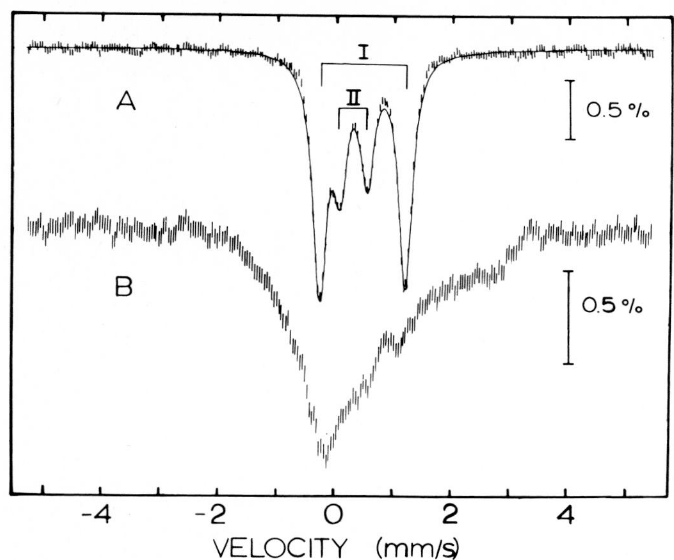

Fig. 18. Mossbauer spectra of reduced Fd II from D. gigas recorded at 4.2 K in (A) zero applied field and (B) a 0.06 T field applied parallel to the γ-beam. The solid line in (A) is a least squares fit to two symmetric quadrupole pairs to the data.

separation can be inferred from the spectrum shown in Figure 18B and was found to be approximately 0.35 cm^{-1}. (4) The electronic spin relaxation rate is slow compared with the nuclear precession frequency. High field studies reveal that the iron nucleus associated with doublet II experiences a positive internal magnetic field, proving that the site is a member of a spin-coupled structure. In strong applied fields, the two equivalent iron sites of doublet I remain indistinguishable and experience a negative internal magnetic field.

The discovery of the 3Fe clusters illustrates the particular strengths of Mossbauer spectroscopy in analyzing multi-iron structures. Hence, we will summarize the data which led to that discovery. The EPR data of oxidized Fd II revealed the presence of only one type of paramagnetic center, and spin quantitation yielded 0.9 spins/3 iron atoms. The Mossbauer spectra of oxidized Fd II recorded at 4.2 K contained three spectral components having intensity ratios 1:1:1. High temperature spectra suggested that all three sites were high-spin ferric ions. Analysis of these data showed that the three iron sites all belonged to the EPR active center. The Mossbauer spectra of reduced Fd II revealed two quadrupole pairs which were present in the ratio 2:1. The behavior of the spectra in response to applied magnetic field suggested strongly that all three iron sites share a common electronic spin. The system spin S was found to be integer and larger than zero. Also, spin coupling was indicated by the observation of positive internal fields in both oxidized and reduced Fd II. Taken together, the data above strongly suggest that the active center of Fd II is a covalently linked structure containing three iron atoms.

A spin-coupling model (137) has been proposed for the oxidized 3Fe cluster. As implied by the Mossbauer data, the model assumes that the oxidized 3Fe cluster consists of three high-spin ferric ions with their spins coupled to form a system spin S = 1/2. The magnetic properties of the system may be described by the Hamiltonian

$$\mathcal{H} = \sum_{i=1}^{3} (g_{oi}\beta\vec{H}_{app}\cdot\vec{S}_i + a_i\vec{S}_i\cdot\vec{I}_i) + \mathcal{H}_c \quad , \tag{23}$$

where $S_i = 5/2$ and where i designates the iron sites. To a very good approximation the electronic Zeeman interaction and the magnetic hyperfine interaction are isotropic for high-spin ferric ions, i.e. g_{oi} and a_i are scalars with $g_{oi} = 2.0$. We can ignore zero field splitting terms in Eq. (23) because we will consider only ground multiplets with S = 1/2. $\mathcal{H}_c$ describes the exchange coupling among the three iron atoms and was assumed to be

$$\mathcal{H}_c = J_{12}\vec{S}_1\cdot\vec{S}_2 + J_{13}\vec{S}_1\cdot\vec{S}_3 + J_{23}\vec{S}_2\cdot\vec{S}_3 \quad . \tag{24}$$

For simplicity the exchange couplings among the three sites were taken to be isotropic.

Since $\vec{S}^2$ commutes with $\hat{\mathcal{H}}$, S is a good quantum number. For $S_1 = S_2 = S_3 = 5/2$ only two configurations with S = 1/2 occur in the coupled system. These may be obtained by coupling S_2 and S_3 to an intermediate spin S' = 2 or 3, and then coupling S' with S_1 to obtain S = 1/2. These two configurations can be expressed in a ket notation: $|S_2S_3(S')S_1;\ SM\rangle$. Since $\hat{\mathcal{H}}_c$ can only mix states with the same total spin, the most general state with S = 1/2, M = +1/2 is a linear combination of these two kets:

$$|+\rangle = \sqrt{1-\alpha^2}\ |S_2S_3(2)S_1;\tfrac{1}{2}\ \tfrac{1}{2}\rangle + \alpha|S_2S_3(3)S_1;\tfrac{1}{2}\ \tfrac{1}{2}\rangle \quad , \tag{25}$$

where α is a mixing parameter, $-1 \leq \alpha \leq 1$. (For a more detailed discussion on the spin-coupling of multi-spin systems the reader is referred to reference 138). In order to express the observed g-values in terms of the g-factors of the uncoupled ions, the Zeeman term was evaluated with $\vec{H}_{app}$ along the z axis.

$$< + |g\beta H_{app} S_z| +> \; = \; <+ |\sum_{i=1}^{3} g_i \beta H_{app} S_{iz}| + > \tag{26}$$

which yields g = 2.0 since $S_z = S_{1z} + S_{2z} + S_{3z}$. Thus, the observation of an isotropic EPR signal at g = 2.01 reflects the fact that the constituent iron atoms have isotropic g_{oi}=2. The observed magnetic hyperfine coupling constants, A_i, are related to the uncoupled magnetic constants, a_i, by

$$< + |A_i S_z| +> = < +|a_i S_{iz}| + > \tag{27}$$

or

$$A_i = 2a_i <+|S_{iz}|+> \quad . \tag{28}$$

Assuming that $a_1 = a_2 = a_3 = -20$ MHz (a reasonable value taken from several tetrahedral sulfur coordinated high-spin ferric complexes) Kent et al. (137) found that with $\alpha^2 = 0.01$ equation (28) yields $A_1 = -45$ MHz, $A_2 = +20$ MHz, and $A_3 = +6$ MHz. This result is in excellent agreement with the observed values for the 3Fe cluster in *A. vinelandii* ferredoxin [$A_1 = -(41 \pm 3)$ MHz, $A_2 = +(18 \pm 3)$ MHz and $|A_3| = (5 \pm 3)$ MHz] and is in reasonable agreement with those obtained for Fd II [$A_1 = -(38 \pm 7)$ MHz, $A_2 = +(20 \pm 7)$ MHz and $|A_3| = (4 \pm 3)$ MHz]. The quoted uncertainties are our best current estimates. The large uncertainties quoted for Fd II arise from the broad absorption features observed at low temperature (see Figure 16). Both heterogeneities and magnetic anisotropies could be at the root of this broadening. Since the mixing parameter α is a function of J_{ij}, the value found for α^2 places restrictions on the J_{ij}. Using $\alpha^2 = 0.01$ and the criterion that $|+>$ should be the ground state, Kent et al. found that $J_{23}>J_{12}>J_{13}>0$; $1>J_{13}/J_{23}>0.5$; $1>J_{12}/J_{23}>0.6$; and $0.8\ J_{12} \geq J_{13}$.

Two important conclusions may be drawn from this simple spin-coupling model. First, the three iron sites are intrinsically rubredoxin like, i.e. high-spin ferric ions with distorted tetrahedral sulfur coordination. The differences in the observed magnetic constants reflect the relative orientation of the coupled spins. Second, the exchange coupling constants among the three iron sites are comparable to within a factor of two. This finding indicates that the 3Fe cluster is a fairly symmetric, covalently linked structure, and should not be considered as a [2Fe - 2S] cluster weakly attached to a third iron atom.

4.4 Nitrogenase

The nitrogenase enzyme complex catalyzes the reduction of molecular nitrogen to ammonia, allowing atmospheric nitrogen to enter the biosphere. This enzyme complex consists of two proteins designated the iron protein and the molybdenum-iron (MoFe) protein (139, 140). The iron protein (also known as component II or nitrogenase reductase) has a molecular weight of approximately 60,000 and comprises two polypeptide chains and one (4Fe-4S) cluster. The MoFe protein (also known as component I or, simply, nitrogenase) contains approximately 30 iron atoms, comparable amounts of acid-labile sulfur, and approximately 2 Mo atoms. It consists of four subunits and one or two cofactors and has a molecular weight on the order of 220,000. In this section we will discuss how 28 of the 30 iron atoms in MoFe protein can be partitioned and characterized into two novel types of redox centers. At present, only EPR and Mossbauer spectroscopy together can provide such a detailed analysis of the redox centers of the MoFe protein.

4.4.1 Native MoFe Protein

The MoFe protein is very sensitive to oxygen, and stringent anaerobic conditions are required during the purification procedure. The protein as isolated shows EPR resonances at $g = 4.3$, 3.7 and 2.0 (141). These signals result from the $S_z = \pm 1/2$ doublet of an $S = 3/2$ system. The observed g-values can be reproduced very well using an $S = 3/2$ spin Hamiltonian with $E/D = 0.055$ (142). Temperature studies of the EPR signal indicate that the zero field parameter D is approximately $+ 5.5\ cm^{-1}$. Spin quantitation of the EPR signal yielded about 1 spin/Mo atom.

Figure 19A shows a spectrum of the native MoFe protein from *Clostridium pasteurianum* recorded at 4.2 K in a parallel applied field of 0.06 T. The spectrum is a superposition of four components, three quadrupole pairs plus a paramagnetic spectrum. The three quadrupole pairs are designated as spectral components D, Fe^{2+} and S. The paramagnetic spectrum is termed component M^N. Since the

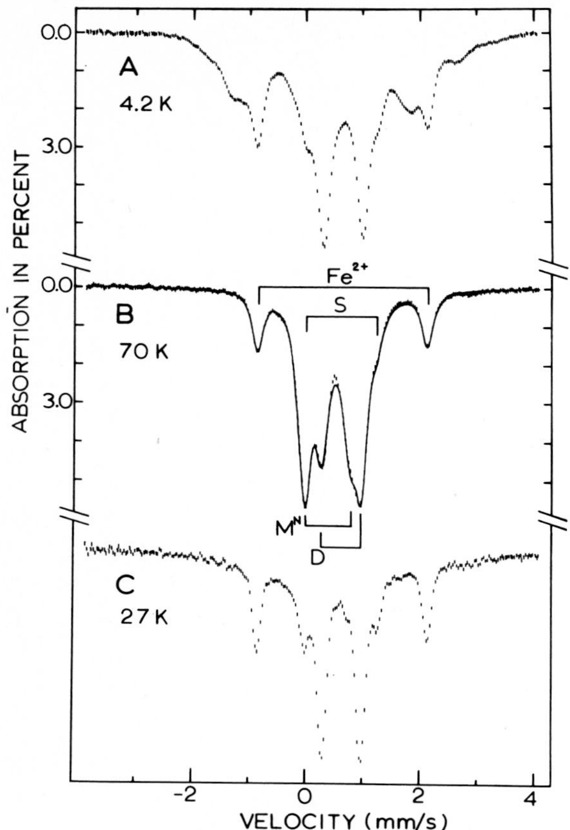

Fig. 19. Mossbauer spectra of ^{57}Fe-enriched native MoFe protein of nitrogenase from C. pasteurianum W5 recorded in a parallel field of 0.06 T at (A) 4.2 K and (B) 70 K. The solid line in (B) is the result of fitting four quadrupole pairs to the data with constraints as described in the text. The deconvoluted spectrum in (C) is also described in the text. (From Ref. 149).

EPR spectrum of the Mossbauer sample showed no other resonance except the characteristic S = 3/2 signals, component M^N must be associated with the S = 3/2 center. We will show in the following that the field direction dependence of spectrum M^N is in agreement with the observed g-values. In addition, EPR and Mossbauer investigations of the cofactor materials have firmly established that the S = 3/2 center resides in the cofactor (143). Therefore, the S = 3/2 center is designated as the cofactor M center.

At 70 K the spin relaxation rate of the cofactor M center is fast, and component M^N collapses into a quadrupole doublet as indicated in Figure 19B. In order to obtain reliable results from fitting such a complex spectrum, it is essential to reduce the number of adjustable parameters. Therefore, the following procedure was used to estimate the absorption line positions prior to the fitting. A spectrum recorded at 27 K, a temperature at which the relaxation rate of the M center is in the intermediate region. As a result the spectral component M^N is broad and featureless. Then a Fourier deconvolution method (144) was used to remove the linewidth contribution of the Co^{57} source. The resulting spectrum is shown in Figure 19C. The quadrupole pairs of components D, Fe^{2+} and S are clearly discernable and their line positions are very well defined. The line positions of the collapsed M^N quadrupole pair at 70 K are estimated from a difference spectrum of Figures 19A and 19B. During these data manipulations, a limit on the line-width of each absorption line was also acquired. With all the positions predetermined and a crude limit for each line-width, the least-squares fit of the 70 K data could be performed with little ambiguity. The solid line shown in Figure 19B is the result of such a fit. The Mossbauer parameters and the absorption percentage of each component is listed in Table 4. A similar method was used to analyze the Mossbauer spectra of the MoFe protein from *Azotobacter vinelandii* (145). The results are also listed in Table 4. The published data of the MoFe protein from *Klebsiella pneumoniae* shows similar results (146). Of particular interest is the parameters obtained for component Fe^{2+}. Its quadrupole splitting (3 mm/s) and isomer shift (0.65 - 0.70 mm/s) are typical of tetrahedral-sulfur-coordinated, high-spin ferrous ions. However, Mossbauer data taken in strong applied fields indicate that component Fe^{2+} is diamagnetic (S = 0). Therefore, component Fe^{2+} must represent an iron site which resides in a spin-coupled cluster. In addition high field studies show that component D is also diamagnetic.

Iron quantitation (147, 148) and Mossbauer data (see below) have shown that each MoFe protein molecule contains a total of (30 ± 2) iron atoms. The relative absorption areas listed in Table 4 indicate that each MoFe protein molecule contains approximately 12 iron atoms associated with component D, 4 with component Fe^{2+}, 2 with component S, and 12 with component M^N. Quantitation of the EPR signal indicates that there are two M centers per molecule. It follows then that each M center contains six iron atoms.

In order to study the spectrum of M^N in detail, the contributions of components D, Fe^{2+} and S were removed from the raw data using the

TABLE 4.

Mossbauer parameters for MoFe proteins from Clostridium pasteurianum and from Azotobacter vinelandii

Spectral Component	*C. pasteurianum*[a]			*A. vinelandii*[b]		
	ΔE_Q (mm/s)	δ (mm/s)	Relative absorption area (%)	ΔE_Q (mm/s)	δ (mm/s)	Total absorption (%)
Fe^{2+}	3.00 (2)[c]	0.64 (2)	12.8 (5)	3.02 (2)	0.69 (2)	13.0 (5)
D	0.70 (2)	0.64 (2)	39 (2)	0.81 (2)	0.64 (2)	42 (2)
S	1.37 (5)	0.64 (5)	~6	1.37 (4)	0.64 (4)	~6
M^N	0.8	0.41	42 (2)	0.74 (3)	0.40 (3)	40 (2)
M^{ox}	0.7 - 1.1	0.35		0.8 -1.05	0.37	
M^R	0.9 - 1.2	0.46		0.95 - 1.3	0.36	

(a) Ref. 149

(b) References 142, 145 and 151

(c) Values in parentheses give uncertainties in units of the least significant digit. If no uncertainties are quoted, the parameters are average values of unresolved subcomponents.

$S \geq 3/2$, $D \leq 0$, and $E/D \simeq 0$. In such an extremely anisotropic system the Mossbauer spectrum is only sensitive to quantities along the z-axis, and spectral simulations are simplified tremendously. Consequently, spectral decomposition of the M1 spectrum is possible. For detailed analysis, the readers are referred elsewhere (149,151). Although the decomposition might not be unique, several important conslusions can be readily drawn: (1) The M1 spectra can be decomposed into eight subspectra, suggesting two sets of slightly inequivalent P clusters in the oxidized form. (2) Each subspectrum amounts to about 6.7% of the total absorption indicating that the MoFe protein contains n x 15 irons per molecule. From all the spectroscopic and chemical evidence, it was concluded that n = 2. At high temperature the M1 spectrum collapses into several quadrupole pairs. The absorption intensity of the weakest resolved pair also amounts to 6.7% of the total absorption. (3) Of the eight subcomponents, two have a positive hyperfine coupling constant A indicating the spin-coupled nature of the M1 spectrum.

4.4.3 The MoFe Protein Under Fixing Condition

The $S = 3/2$ EPR signal of the MoFe protein decreases to about 10% of its maximum intensity under nitrogen fixing conditions, i.e. when MgATP, a proper reductant, and the Fe protein are added to the MoFe protein under N_2 atmosphere. Figure 22 shows a Mossbauer spectrum of *C. pasteurianum* MoFe protein prepared under fixing conditions. The spectrum was recorded at 4.2 K. Components Fe^{2+} and D retain their native oxidation state. The magnetic M^N spectrum is transformed into a quadrupole pair (M^R), indicating that a one electron reduction process has transformed the $S = 3/2$ center into an integer spin system. High-field studies indicate that the reduced M center is paramagnetic, $S > 0$. The Mossbauer parameters for M^R are listed in Table 4.

4.4.4 Nitrogenase Summary

Two novel types of redox centers in the MoFe protein have been discovered, identified and characterized. The MoFe protein contains two M centers and four P clusters. The M center consists, most probably, of six spin-coupled iron atoms. It resides in the cofactor and can be stabilized in three oxidation states, $M^{OX} \underset{\leftarrow}{\overset{e^-}{\rightarrow}} M^N \underset{\leftarrow}{\overset{e^-}{\rightarrow}} M^R$. The native state M^N has a spin $S = 3/2$ and its ground doublet is EPR active. A one electron oxidation process transforms M^N into

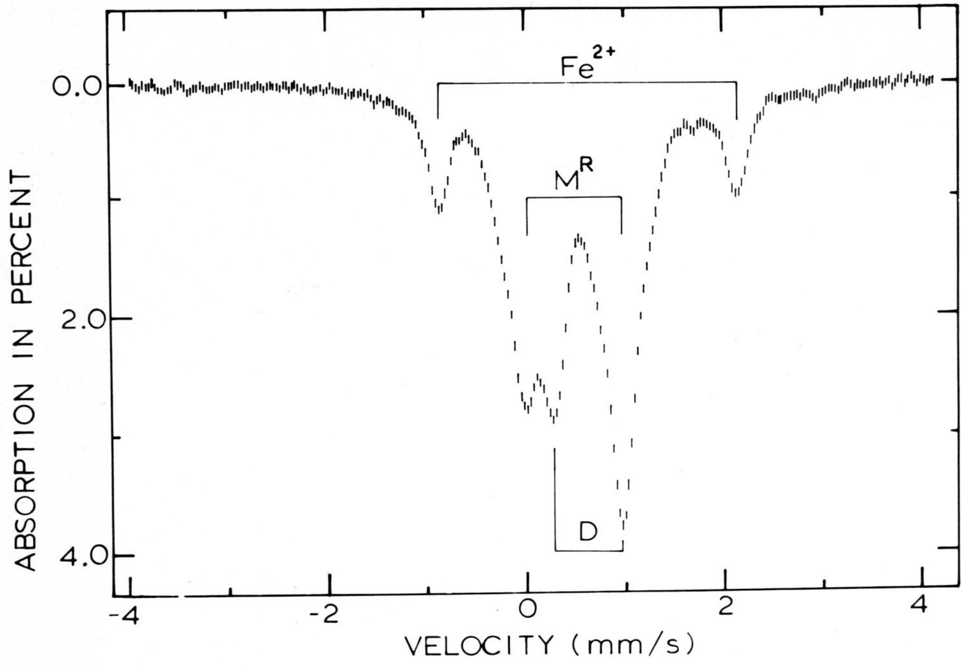

Fig. 22. Mossbauer spectrum of C. pasteurium MoFe protein in the presence of Fe-protein, MgATP, N_2 and dithionite. The spectrum was recorded at 4.2 K in a field of 0.06 T applied parallel to the γ-beam. (From Ref. 149).

state M^{ox} which is a diamagnetic (S = 0) state. The reduced state M^R is attained under nitrogen fixing condition and has an integer spin S > 0. Recent extended X-ray absorption fine structure (EXAFS) studies (152,153) strongly suggest that Mo is part of the spin-coupled M center. The P cluster contains four spin-coupled iron atoms, most probably bridged by labile sulfur. Two oxidation states, P^{ox} and P^N, have been characterized. The native state P^N is diamagnetic. The oxidized state P^{ox} has a half integer spin $S \geq 3/2$, and is characterized by a negative zero field parameter D. Recent cluster extrusion experiments (154) confirm the existence of four [4Fe-4S] clusters in the MoFe protein.

Little information has been acquired for component S because the spectrum of MoFe protein is complex and component S has relatively weak absorption (~ 6% of the total iron absorption). However, component S is belived not to be an impurity but an essential part of the MoFe protein for the following reasons. Component S is present in equal amounts in the MoFe proteins from *A. vinelandii* and *C. pasteurianum*. The *K. pneumoniae* MoFe protein also appears to contain comparable amounts of component S. For all the proteins studied, component S has practically identical Mossbauer parameters (ΔE_Q = 1.37 mm/s, $\delta \simeq$ 0.65 mm/s). These values are atypical of commonly observed copro-iron. Also, no EPR signal attributable to component S has been observed despite the fact that samples have been studied over a wide range of redox conditions.

5 SUMMARY

A major contribution of Mossbauer spectroscopy to the study of biological molecules has been the determination of the iron electronic states in many proteins. In particular, Mossbauer spectroscopy is indispensable in the study of low-spin ferrous complexes since diamagnetic compounds are impossible to detect by susceptibility or magnetic resonance techniques. The paramagnetic Mossbauer spectra of ferric ions, along with their complementary EPR data, are fairly well understood. The observed hyperfine parameters are sensitive to the iron-ligand environment and are useful for the classification of such complexes. For high-spin ferrous complexes two situations have arisen. One occurs when the ground quintet is well separated from excited multiplets, a situation for which the spin Hamiltonian approximation is valid. Mossbauer spectra recorded over wide ranges of temperatures and applied fields are required for the determination of the fine and hyperfine parameters. Presently, the extraction of structural information from such parameters is difficult at best. The other case occurs when there are low-lying, excited multiplets. The spin Hamiltonian approach is no longer adequate and the data interpretation becomes more complex. Ligand field calculations including all $3d^6$ configurations may be suitable but require further investigation. Molecular orbital calculations have many drawbacks and should not be applied to such complex systems at the present time. Obviously, more detailed theoretical and experimental studies of high-spin ferrous systems are needed. In general, biological molecules with mono-iron centers provide the simplest case for studying the electronic structure of iron proteins. Mossbauer spectroscopy will undoubtedly continue to contribute to this research area.

Applications of Mossbauer spectroscopy to the study of multi-iron clusters is still in its infancy. However, significant progress has been made including the discovery of several novel iron clusters.

These early successes should stimulate further investigations of multi-cluster proteins and, hopefully, also stimulate progress in both experimental and theoretical methods. Many such complex molecules contain spectroscopically unresolved iron centers, and meaningful interpretation of their spectra is very difficult (see, for example, Ref. 132). Selective isotopic enrichment would greatly simplify this problem and research in this area should be pursued. We have demonstrated in the earlier sections that Mossbauer spectroscopy used in conjunction with other techniques can provide structural information beyond the characterization or iron electronic states. Such combination of techniques will most likely be necessary for future studies of complex biomolecules. For example, by using electrochemistry and Mossbauer spectroscopy one could determine the redox potential of a particular iron center in a multicluster protein. Using stopped-flow/rapid-quench techniques one may be able to arrest transient states during enzymatic cycles. Investigation of these transient states with Mossbauer spectroscopy could provide information concerning the electron transfer rate and sequence among the iron centers. Moreover, with selective isotopic enrichment one would be able to investigate inter- and intra- molecular electron transfer. Such studies would generate information pertinent to the understanding of important biological processes.

Several important topics have, of necessity, been neglected in the present discussion. For instance, iron storage and transport compounds have been the subject of several Mossbauer studies (18, 155, 156, 157, 158). Iron transport compounds, such as enterobactin, have been found to have a high affinity for ferric ions and a relatively weak affinity for ferrous ions. Mossbauer data have firmly established that at high or neutral pH the chelated irons are high-spin ferric. More recent Mossbauer and EPR investigations of enterobactin and its synthetic analogues reveal a delicate Fe^{2+}/Fe^{3+} balance which depends upon solvent effects and pH (G.B. Wong, V.L. Pecoraro, T.A. Kent, and B.H. Huynh, unpublished results). Since there are pH and potential gradients across biological membranes, the observed pH and solvent dependence may play an important role in the function of these compounds. The mechanisms for the translocation of iron through membranes and for the release of iron intracellulary are not fully understood. Mossbauer spectroscopy can be used to elucidate these processes. Another subject not discussed here is the effect of intermediate relaxation rates on Mossbauer spectra. Interested readers are referred elsewhere (159,160). Mossbauer emission spectroscopy has been applied to heme complexes (7,161) and may offer some unique advantages for protein work, Also, Mossbauer spectroscopy has been applied to the study of whole organisms. Examples are the studies of magnetotatic bacteria (162) and frozen *Escherichia coli* cells (163).

The preceeding sections clearly illustrate the utility of Mossbauer studies of iron proteins. Information to be gained ranges from the relative abundance of inequivalent irons in a given sample to the details of the iron electronic structure. Iron proteins are being isolated at an ever accelerating rate, and Mossbauer spectroscopy will play a key role in characterizing those proteins and help to determine their functions and mechanisms.

ACKNOWLEDGEMENTS

We would like to express our sincere appreciation to our collaborators, Drs. E. Munck, I. Moura, J.J.G. Moura, M. Emptage, J.D. Lipscomb, A.V. Xavier, W.H. Orme-Johnson, G. Lang, K. Spartalian, and T. Yonetani for stimulating discussions, experimental assistance and moral support. We thank Ms. D. Bischke for her preparation of the many revisions of this paper. This work was supported by USA NIH grant GM-22701 and NIH grant PCM-8005610.

REFERENCES

1. R.L. Mossbauer, Z. Physik, 151 (1958) 124.
2. U. Gonser, in D.M.J. Compton and A.H. Schoen (Eds.), Proc. 2nd Int. Conf. on Mossbauer Effect, John Wiley, New York, 1962, p. 281.
3. E. Munck, in S. Fleischer and L. Packer (Eds.), Methods in Enzymology, Vol. 54, Academic Press, New York, 1978, p. 346.
4. G. Lang and W. Marshall, Proc. Phys. Soc., 87 (1966) 3.
5. P.G. Debrunner, in D.W. Urry (Ed.), Spectroscopic Approaches to Biomolecular Conformation, American Medical Association Press, Chicago, 1969, p. 209.
6. G. Lang, Quart. Rev. Biophys., 3 (1970) 1.
7. E. Munck, The Porphyrins, 4 (1979) 379.
8. L. May, An Introduction to Mossbauer Spectroscopy, New York, Plenum, 1971.
9. G. Lang, in R.L. Cohen (Ed.), Applications of Mossbauer Spectroscopy, Vol. 1, Academic Press, New York, 1976, p. 129.
10. V.I. Goldanskii and E.F. Makarov, in V.I. Goldanskii and R.H. Herbert (Eds.), Chemical Applications of Mossbauer Spectroscopy, Academic Press, New York, 1968, p. 1.
11. R. Walker, G.K. Wertheim and V. Jaccarino, Phys. Rev. Lett., 6 (1961) 98.
12. N.N. Greenwood and T.C. Gibb, Mossbauer Spectroscopy, Chapman and Hall, London, 1971, p. 91.
13. R.S. Preston, S.S. Hanna and J. Heberle, Phys. Rev., 128 (1962) 2207.

14. C.E. Violet and D.N. Pipkorn, J. Appl. Phys., 42 (1971) 4339.
15. F. Varret, J. Phys. (Paris) Colloq., 37 (1976) C6-437.
16. J.S. Griffith, The Theory of Transition Metal Ions, Cambridge University Press, London, 1971.
17. C.J. Ballhausen, Introduction to Ligand Field Theory, McGraw-Hill, New York, 1962.
18. K. Spartalian, W.T. Oosterhuis and J.B. Neilands, J. Chem. Phys., 62 (1975) 3538.
19. K. Kramers, Akad. van Wetenschappen, Amsterdam, 33 (1930) 959. See also ref. 18, p. 205.
20. H.M. Swartz, J.R. Bolton and D.C. Borg, Biological Applications of Electron Spin Resonance, Wiley-Interscience, New York, 1972.
21. G. Palmer, in D.W. Darnall and R.G. Wilkins (Eds.), Advances in Inorganic Biochemistry, Vol. II, Elsevier, North Holland, 1980, p. 153.
22. G. Lang and W.T. Oosterhuis, J. Chem. Phys., 51 (1969) 3608.
23. R. Aasa and T. Vanngard, J. Magn. Resonance 19 (1975) 308.
24. J.S. Griffith, Mol. Phys., 21 (1971) 135.
25. W.E. Blumberg and J. Peisach, Advan. Chem. Ser., 100 (1971) 271.
26. C.P.S. Taylor, Biochim. Biophys. Acta, 491 (1977) 137.
27. A. Abragam and B. Bleaney, Electron Paramagnetic Resonance of Paramagnetic Ions, Clarendon, Oxford, 1970. pp. 138, 653.
28. W.T. Oosterhuis and G. Lang, Phys. Rev., 178 (1969) 439.
29. B.H. Huynh, M.H. Emptage and E. Munck, Biochim. Biophys. Acta, 534 (1978) 295.
30. F.R. Salemme, S.T. Freer, N.H. Xuong, R.A. Alden and J. Kraut, J. Biol. Chem., 248 (1973) 3910.
31. A. Dwivedi, W.A. Toscano, Jr. and P.G. Debrunner, Biochim. Biophys. Acta, 576 (1979) 502.
32. M. Sharrock, P.G. Debrunner, C. Schulz, J.D. Lipscomb, V. Marshall and I.C. Gunsalus, Biochim. Biophys. Acta, 420 (1976) 8
33. P.M. Champion, E. Munck, P.G. Debrunner, P.F. Hollenberg and L.P. Hager, Biochemistry, 12 (1973) 426.
34. G. Lang, D. Herbert and Y. Yonetani, J. Chem. Phys., 49 (1968) 944.
35. W.B. Mims and J. Peisach, J. Chem. Phys., 64 (1976) 1074.
36. D. Rhynard, G. Lang, K. Spartalian and T. Yonetani, J. Chem. Phys., 71 (1979) 3715.
37. T. Harami, J. Chem. Phys., 71 (1979) 1309.
38. G.A. Helcke, D.J.E. Ingram and E.F. Slade, Proc. Roy. Soc., London, Ser. B, 169 (1968) 275.
39. H. Hori, Biochim. Biophys. Acta, 251 (1971) 227.
40. E. Munck and P.M. Champion, J. Phys. (Paris) Colloq., 35 (1974) C6 - 33.
41. (a) K.D. Watenpaugh, L.C. Sieker and L.H. Jensen, J. Mol. Biol., 131 (1979) 509.
(b) D. Watenpaugh, L.C. Sieker and L.H. Jensen, ibid, 138 (1980) 615.
42. J. Peisach, W.E. Blumberg, E.T. Lode and M.J. Coon, J. Biol. Chem., 246 (1971) 5877.
43. K.K. Rao, M.C.W. Evans, R. Cammack, D.O. Hall, C.L. Thompson, P.J. Jackson and C.E. Johnson, Biochem. J., 129 (1972) 1063.
44. I. Moura, A.V. Xavier, R. Cammack, M. Bruschi and J. LeGall, Biochim. Biophys. Acta, 533 (1978) 156.
45. N.M. Atherton, K. Garbett, R.D. Gillard, R. Mason, S.J. Mayhew, J.L. Peel and L.E. Stangroom, Nature, 212 (1966) 590.
46. C. Schulz and P.G. Debrunner, J. Phys. (Paris) Colloq., 37 (1976) C6-153.
47. P.G. Debrunner, E. Munck, L. Que and C.E. Schulz, in W. Lovenberg, (Ed.), Iron-Sulfur Proteins, Vol. III, Academic Press, New York, 1977, p. 381.

48. R.W. Lane, J.A. Ibers, R.B. Frankel, G.C. Papaefthymiou and R.H. Holm, J. Amer. Chem. Soc., 99 (1977) 84.
49. W.T. Oosterhuis, Struct. Bonding, 20 (1974) 59.
50. R. Tsai, C.A. Yu, I.C. Gunsalus, J. Peisach, W. Blumberg, W.H. Orme-Johnson and H. Beinert, Proc. Nat. Acad. Sci. USA, 66 (1970) 1157.
51. M. Bruschi, I. Moura, A.V. Xavier, J. Legall and L.C. Sieker, Biochem. Biophys. Res. Commun., 90 (1979) 596.
52. I. Moura, M. Bruschi, J. LeGall, J.J.G. Moura and A.V. Xavier, Biochim. Biophys. Res. Commun., 75 (1977) 1037.
53. I. Moura, B.H. Huynh, R.P. Hausinger, J. LeGall, A.V. Xavier and E. Munck, J. Biol. Chem., 255 (1980) 2493.
54. M.H. Emptage, R. Zimmermann, L. Que, Jr., E. Munck, W.D. Hamilton and W.H. Orme-Johnson., Biochim. Biophys. Acta, 495 (1977) 12.
55. H. Fujisawa, M. Uyeda, T. Kojima, M. Nozaki and O. Hayaishi, J. Biol. Chem., 247 (1972) 4414.
56. J. Lipscomb, J. Howard, T. Lorsbach and J. Wood, Fed. Proc., 35 (1976) 1536.
57. W.E. Blumberg and J. Peisach, Ann. N.Y. Acad. Sci., 222 (1973) 539.
58. (a) L. Que, Jr., J.D. Lipscomb, R. Zimmermann, E. Munck, N.R. Orme-Johnson and W.H. Orme-Johnson, Biochim. Biophys. Acta, 452 (1976) 320.
(b) L. Que, Jr., and R.H. Heistand, II, J. Amer. Chem. Soc., 101 (1979) 2219.
59. H. Fujisawa, K. Hiromi, M. Uyeda, S. Okuno, M. Nozaki and O. Hayaishi, J. Biol. J. Biol. Chem., 247 (1972) 4422.
60. R. Zimmermann, B.H. Huynh, E. Munck and J.D. Lipscomb, J. Chem. Phys., 69 (1978) 5463.
61. P.M. Champion, J.D. Lipscomb, E. Munck, P. Debrunner and I.C. Gunsalus, Biochemistry, 14 (1975) 4151.
62. P.M. Champion, R. Chiang, E. Munck and L.P. Hager, Biochemistry, 14 (1975) 4159.
63. R. Zimmermann, G. Ritter, H. Spiering and D.L. Nagy, J. Phys. (Paris) Colloq., 35 (1974) C6-439.
64. J. Van Dongen Torman, R. Jagannathan and J.M. Trooster, Hyperfine Interact., 1 (1975) 135.
65. R. Zimmermann, H. Spiering and G. Ritter, Chem. Phys., 4 (1974) 133
66. A.J. Freeman and R.E. Watson, Phys. Rev., 131 (1963) 2566.
67. J.C. Kendrew, R.E. Dickerson, B.E. Strandberg, R.G. Hart, D.R. Davies, D.C. Phillips and V.C. Shore, Nature, 185 (1960) 422.
68. T.A. Kent, K. Spartalian, G. Lang and T. Yonetani, Biochim. Biophys. Acta, 490 (1977) 331.
69. T.A. Kent, K. Spartalian, G. Lang, T. Yonetani, C.A. Reed and J.P. Collman, Biochim. Biophys. Acta, 580 (1979) 245.
70. T.A. Kent, K. Spartalian and G. Lang, J. Chem. Phys., 71 (1979) 4899.
71. U. Gonser, Y. Maeda, A. Trautwein, F. Parak and H. Formanek, Z. Naturforsch, 29b (1974) 241.
72. (a) N. Nakano, J. Otsuka and A. Tasaki, Biochim. Biophys. Acta, 236 (1971) 222.
(b) N. Nakano, J. Otsuka and A. Tasaki, ibid, 278 (1972) 355.
73. P.M. Champion and A.J. Sieves, J. Chem. Phys., 72 (1980) 1569.
74. (a) H. Eicher and A. Trautwein, J. Chem. Phys., 50 (1969) 2540.
(b) H. Eicher and A. Trautwein, ibid, 52 (1970) 932.
75. B.H. Huynh, G.C. Papaefthymiou, C.S. Yen, J.L. Groves and C.S. Wu, J. Chem. Phys., 61 (1974) 3750.
76. M.J.S. Dewar, The Molecular Orbital Theory of Organic Chemistry, McGraw-Hill, New York, 1969.
77. (a) Y. Seno, N. Kameda and J. Otsuka, J. Chem. Phys., 72 (1980) 6048.
(b) Y. Seno, N. Kameda and J. Otsuka, ibid, 72 (1980) 6059.

78. L. Pauling, Nature, 203 (1964) 182; Stanford Med. Bull., 6 (1948) 215.
79. J.J. Weiss, Nature, 202 (1964) 83.
80. (a) W.A. Goddard, III, and B.D. Olafson, Proc. Nat. Acad. Sci. USA, 72 (1975) 2335.
(b) D.B. Olafson and W.A. Goddard, III, ibid, 74 (1977) 1315.
81. L. Pauling and C. Coryell, Proc. Nat. Acad. Sci. U.S.A., 22 (1936) 210.
82. D.S. Taylor and C.D. Coryell, J. Amer. Chem. Soc., 60 (1938) 1177.
83. (a) D.A. Case, B.H. Huynh and M. Karplus, J. Amer. Chem. Soc. 101 (1979) 4433.
(b) B.H. Huynh, D.A. Case and M. Karplus, ibid, 99 (1977) 6103.
84. (a) J.C. Maxwell, J.A. Volpe, C.H. Varlow and W.S. Caughey, Biochem. Biophys. Res. Commun., 58 (1974) 166.
(b) C.H. Barlow, J.C. Maxwell, W.J. Wallace and W.S. Caughey, ibid, 55 (1973) 91.
85. K. Spartalian, G. Lang, J.P. Collman, R.R. Gagne and C.A. Reed, J. Chem., Phys., 63 (1975) 5375.
86. S.E.V. Phillips, Nature, 273 (1978) 247.
87. (a) M. Cergonio, A. Congiu-Castellano, L. Calabrese, S. Morante, B. Pispisa and S. Vitale, Proc. Nat. Acad. Sci. USA, 75 (1978) 4916
(b) M. Cerdonio, A. Congin-Castellano, F. Mogno, B. Pispisa, G.L. Romani and S. Vitale, ibid, 74 (1977) 398.
88. H.B. Dunford and J.S. Stillman, Coord. Chem. Rev., 19 (1976) 187.
89. H. Theorell and A. Ehrenberg, Arch. Biochem. Biophys., 41 (1952) 41 (1952) 442.
90. (a) T. Yonetani, J. Biol. Chem., 240 (1965) 4509.
(b) T. Yonetani, ibid, 241 (1966) 2562.
91. T. Iizuka, M. Kotani and T. Yonetani, Biochim. Biophys. Acta, 167 (1968) 257.
92. C.E. Schulz, P.W. Devaney, H. Winkler, P.G. Debrunner, N. Doan, R. Chiang, R. Rutter and L.P. Hager, FEBS Letters, 103 (1979) 102.
93. C. Schulz, R. Chiang and P.G. Debrunner, J. Phys. (Paris) Colloq., 40 (1979) C2-534.
94. G. Lang, K. Spartalian and T. Yonetani, Biochim. Biophys. Acta, 451 (1976) 250.
95. W.T. Oosterhuis and G. Lang, J. Chem. Phys., 58 (1973) 4757.
96. P.R. Rich, J.C. Salerno, J.S. Leigh and W.D. Bonner, Jr., FEBS Letters, 93 (1978) 323.
97. J.R. Galpin, G.A. Veldink, J.F.G. Vliegenthart and J. Boldingh, Biochim. Biophys. Acta, 536 (1978) 356.
98. J.D. Lipscomb, B.H. Huynh and E. Munck, Fed. Proc., 38 (1979) 731.
99. Y. Tatsuno, Y. Saeki, M. Nozaki, S. Otsuka and Y. Maeda, FEBS Letters 112 (1980) 83.
100. R.H. Sands and W.R. Dunham, Quart. Rev. Biophys., 7 (1975) 443.
101. (a) J.A. Ibers and R.H. Holm, Science, 209 (1980) 223.
(b) R.H. Holm and J.A. Ibers, in W. Lovenberg (Ed.), Iron-Sulfur Proteins, Vol. III, Academic Press, New York, 1977, p. 206.
102. T. Tsukihara, K. Fukuyama, H. Tahara, Y. Katsube, Y. Matsuura, N. Tanaka, M. Kakudo, K. Wada and H. Matsubara, J. Biochem. (Tokyo), 84 (1978) 1645.
103. W.H. Orme-Johnson and R.H. Sands, in W. Lovenberg (Ed.), Iron-Sulfur Proteins, Vol. II, Academic Press, New York, 1973, p. 195.
104. J.C.M. Tsibris, R.L. Tsai, I.C. Gunsalus, W.H. Orme-Johnson, R.E. Hansen and H. Beinert, Proc. Nat. Acad. Sci. USA, 59 (1968) 959.
105. D.V. Der Vartanian, W.H. Orme-Johnson, R.E. Hansen, H. Beinert, R.L. Tsai, J.C.M. Tsibris, R.C. Bartholomaus and I.C. Gunsalus, Biochem. Biophys. Res. Commun., 26 (1967) 569.

106. W.H. Orme-Johnson, R.E. Hansen, H. Beinert, J.C.M. Tsibris, R.C. Bartholomaus and I.C. Gunsalus, Proc. Nat. Acad. Sci. USA, 60 (1968) 368.
107. J. Fritz, R. Anderson, J. Fee, G. Palmer, R.H. Sands, J.C.M. Tsibris, Tsibris, I.C. Gunsalus, W.H. Orme-Johnson and H. Beinert, Biochim. Biophys. Acta, 253 (1971) 110.
108. W.R. Dunham, A.J. Bearden, I.T. Sulmeen, G. Palmer, R.H. Sands, W.H. Orme-Johnson and H. Beinert, Biochim. Biophys. Acta, 253 (1971) 134.
109. E. Munck, P.G. Debrunner, J.C.M. Tsibris and I.C. Gunsalus, Biochemistry, 11 (1972) 855.
110. J.F. Gibson, D.O. Hall, J.H.M. Thornley and F.R. Whatley, Proc. Nat. Acad. Sci. USA, 56 (1966) 987.
111. J.H.M. Thornley, J.F. Gibson, F.R. Whatley and D.O. Hall, Biochem. Biophys. Res. Commun., 24 (1966) 877.
112. T.H. Moss, D. Petering and G. Palmer, J. Biol. Chem. 244 (1969) 2275.
113. G. Palmer, W.R. Dunham, J.A. Fee, R.H. Sands, T. Iizuka and T. Yonetani, Biochim. Biophys. Acta, 245 (1971) 201.
114. L. Petersson, R. Cammack and K.K. Rao, Biochimica et Biophysica Acta, 622 (1980) 18.
115. G. Palmer, R.H. Sands and L.E. Mortenson, Biochem. Biophys. Res. Commun., 23 (1966) 357.
116. R.N. Mullinger, R. Cammack, K.K. Rao, D.O. Hall, D.P.E. Dickson, C.E. Johnson, J.D. Rush and A. Simopoulos, Biochem. J., 151 (1975) 75.
117. G. Palmer, H. Brintzinger, R.W. Estabrook and R.H. Sands, in A. Ehrenberg, B.G. Malmstrom and Vanngard (Eds.), Magnetic Resonance in Biological Systems, Pergamon, Oxford, 1967, p. 159.
118. C.W. Carter, Jr., S.T. Freer, Ng. H. Xuong, R.A. Alden and J. Kraut, Cold Spring Harbor Symp. Quant. Biol., 36 (1971) 381.
119. E.T. Adman, L.C. Sieker and L.H. Jensen, J. Biol. Chem., 248 (1973) 3987.
120. C.W. Carter, Jr., J. Kraut, S.T. Freer, R.A. Alden, L.C. Sieker, E.T. Adman and L.H. Jensen, Proc. Nat. Acad. Sci. USA, 69 (1972) 3526.
121. Nomenclature Committee of the International Union of Biochemistry (NC-IUB) "Nomenclature of Iron-Sulfur Proteins" Biochim. Biophys. Acta, 549 (1979) 101.
122. E.J. Laskowski, R.B. Frankel, W.O. Gillum. G.C. Papaefthymiou, J. Renaud, J.A. Ibers and R.H. Holm, J. Amer. Chem. Soc., 100 (1978) 5322.
123. R. Cammack, Biochem. Biophys. Res. Commun., 54 (1973) 548.
124. W.V. Sweeney, A.J. Bearden and J.C. Rabinowitz, Biochem. Biophys. Res. Commun., 59 (1974) 188.
125. P. Middeton, D.P.E. Dickson, C.E. Johnson and J.D. Rush, Eur. J. Biochem., 88 (1978) 135.
126. (a) P. Middleton, D.P.E. Dickson, C.E. Johnson and J.D. Rush, Eur. J. Biochem., 104 (1980) 289.
(b) D.P.E. Dickson, C.E. Johnson, R. Cammack, M.C.W. Evans, D.O. Hall and K.K. Rao, Biochem. J., 139 (1974) 105.
127. R. Cammack, D.P.E. Dickson and C.E. Johnson, in W. Lovenberg (Ed.), Iron-Sulfur Proteins, Vol. III, Academic Press, New York, 1977, p. 283.
128. Y.I. Shethna, Biochim. Biophys. Acta, 205 (1970) 58.
129. D.C. Yoch and D.I. Arnon, J. Biol. Chem., 247 (1972) 4514.
130. D.C. Yoch and R.P. Carithers, J. Bacteriol., 136 (1978) 822.
131. W.V. Sweeney and J.C. Rabinowitz, J. Biol. Chem., 250 (1975) 7842.
132. M.H. Emptage, T.A. Kent, B.H. Huynh, J. Rawlings, W.H. Orme-Johnson and E. Munck, J. Biol. Chem., 255 (1980) 1793.

133. C.D. Stout, D. Ghosh, V. Pattabhi and A.H. Robbins, J. Biol. Chem. 255 (1980) 1797.
134. B.H. Huynh, J.J.G. Moura, I. Moura, T.A. Kent, J. LeGall, A.V. Xavier and E. Munck, J. Biol. Chem., 255 (1980) 3242.
135. (a) J. Travis, J.D. Newman, J. LeGall and H.D. Peck, Jr., Biochem. Biophys. Res. Commun., 45 (1971) 452.
(b) M. Bruschi, ibid, 91 (1979) 623.
136. J.J.G. Moura, A.V. Xavier, E.E. Hatchikian and J. Legall, FEBS Lett., 89 (1978) 177.
137. T.A. Kent, B.H. Huynh and E. Munck, to be published in Proc. Nat. Acad. Sci. USA
138. J.S. Griffith, Struct. Bonding, 10 (1972) 87.
139. W.H. Orme-Johnson and L.C. Davis, in W. Lovenberg (Ed.), Iron-Sulfur Proteins, Vol. III, Academic Press, New York, 1977, p. 16.
140. L.E. Mortenson, Methods. Enzymol., 24 (1972) 446.
141. L.C. Davis, V.K. Shah, W.J. Brill and W.H. Orme-Johnson, Biochim. Biophys. Acta, 256 (1972) 512.
142. E. Munck, H. Rhodes, W.H. Orme-Johnson, L.C. Davis, W.J. Brill and V.K. Shah, Biochim. Biophys. Acta, 400 (1975) 32.
143. J. Rawlings, V.K. Shah, J.R. Chisnell, W.J. Brill, R. Zimmermann, E. Munck and W.H. Orme-Johnson, J. Biol. Chem., 253 (1978) 1001.
144. M. Celia Dibar-Ure and P.A. Flinn, in I.J. Gruverman (Ed.), Mossbauer Effect Methodology, Vol. 7, Plenum Press, New York, p. 245.
145. B.H. Huynh, E. Munck and W.H. Orme-Johnson, Biochim. Biophys. Acta 576 (1979) 192.
146. B.E. Smith and G. Lang, Biochem. J. (London), 137 (1974) 169.
147. V.K. Shah and W.J. Brill, Biochim. Biophys. Acta, 305 (1973) 445.
148. J. Postgate, in R. Ondarza, D. Edmundson and T.P. Singer (Eds.), Mechanism of Oxidative Enzymes, Plenum Press, New York, p. 173.
149. B.H. Huynh, M.T. Henzl, J.A. Christner, R. Zimmermann, W.H. Orme-Johnson and E. Munck, Biochim. Biophys. Acta, 623 (1980) 124.
150. B.H. Huynh, E. Munck and W.H. Orme-Johnson, J. Phys. (Paris) Colloq., 40 (1979) C2-526.
151. R. Zimmermann, E. Munck, W.J. Brill, V.K. Shah, M.T. Henzl, J. Rawlings and W.H. Orme-Johnson, Biochim. Biophys. Acta, 537 (1978) 185.
152. T.E. Wolff, J.M. Berg, C. Warrick, K.O. Hodgson, R.H. Holm and R.B. Frankel, J. Amer. Chem. Soc., 100 (1978) 4630.
153. T.E. Wolff, J.M. Berg, K.O. Hodgson, R.B. Frankel and R.H. Holm, J. Amer. Chem. Soc., 101 (1979) 4140.
154. D.M. Kurtz, Jr., R.S. McMillan, B.K. Burgess, L.E. Mortenson and R.H. Holm., Proc. Nat. Acad. Sci. USA, 76 (1979) 4986.
155. J.L. Bock and G. Lang, Biochim. Biophys. Acta, 264 (1972) 245.
156. K. Spartalian, W.T. Oosterhuis and N. Smarra, Biochim. Biophys. Acta, 399 (1975) 203.
157. W.T. Oosterhuis and K. Spartalian, in R.L. Cohen (Ed.), Applications of Mossbauer Spectroscopy, Academic Press, New York, 1976, p. 141.
158. R.C. Hider, J. Silver, J.B. Neilands, I.E.G. Morrison and L.V.C. Rees, FEBS Lett., 102 (1979) 325.
159. H.H. Wickman, M.P. Klein and D.A. Shirley, Phys. Rev., 152 (1966) 345.
160. G.R. Hoy and M.R. Corson, J. Magn. Magn. Mater., 15-18 (1980) 627.
161. N. Ikeda, J. Akashi, R. Seki, Radiochem. Radioanal. Lett., 42 (1980) 235.
162. R.B. Frankel, R.P. Blakemore and R.S. Wolfe, Science, 203 (1979) 1355.
163. E.R. Bauminger, S.G. Cohen, D.P.E. Dickson, A. Levy, S. Ofer and J. Yariv, Biochim. Biophys. Acta, 623 (1980) 237.

CHAPTER 10

MAGNETIC INTERACTIONS IN SUPERCONDUCTORS STUDIED BY MOSSBAUER SPECTROSCOPY

G.K. Shenoy

Argonne National Laboratory, Argonne, Illinois 60439, USA

1. INTRODUCTION

The interaction between magnetism and superconductivity has now been studied over two decades. Experimentally this field of research has generated many fundamental concepts, questions and paradoxes. There are materials in which the superconductivity is destroyed by the addition of a small amount of 3d or 4f magnetic atoms (1,2). In contrast, 4f and 3d atoms are part of the lattice in $ErRh_4B_4$ and $Lu_2Fe_3Si_5$ which are superconductors with T_c of 8.6 K and 6.0 K, respectively (3,4). A large number of alloys demonstrate the coexistence of magnetism and superconductivity (5), while in compounds like $ErRh_4B_4$ and $Ho_{1.2}Mo_6S_8$ the superconductivity is destroyed by the onset of magnetic order (6,7). In many systems there is a decline in the critical field value by the introduction of magnetic atoms (1,2). On the other hand, in some Chevrel phase ternary compounds like $SnMo_6S_8$, replacing Sn^{2+} by magnetic Eu^{2+} ions enhances the critical field (8). Understanding of the diverse behavior of different materials demands microscopic knowledge of many fundamental quantities related to the electro-magnetic interaction in superconductors.

Superconductivity results from a condensation of Cooper-pairs of electrons with opposite spin and momenta at the Fermi surface. The presence of magnetic spins polarizes the conduction electrons, which in a superconductor will produce depairing of the Cooper pairs. The ability of a magnetic atom to destroy the superconductivity is hence governed both by the size of the local magnetic moment and by the strength of its exchange coupling (j) to the conduction electron spins.

The spatial extent of the magnetic correlation, ξ_M, compared to the coherence length of a Cooper pair, ξ_s, would strongly determine the phenomenon of coexistence of magnetism and superconductivity.

Thus for example, if $\xi_M << \xi_s$, a coexistence of these two orders will be possible.

Usually the external magnetic fields destroy superconductivity. However, if the magnetic atoms could generate an internal field which would oppose the external field, then we can realize high-field superconductors. This mechanism proposed by Jaccarino and Peter (9) seem to be applicable in $Sn_{1-x}Eu_xMo_6S_8$. A microscopic description of the phenomenon would require knowledge of direction of the spin polarization produced by Eu atoms at the site of superconducting electrons.

It is clear from the above that microscopic experiments giving information on the exchange interaction, local magnetic order and the magnetic correlation, and spatial distribution of spin polarization in the lattice are needed at least to partly understand the influence of the magnetic interactions in superconductors. There is no single technique capable of providing all this information. However, the Mossbauer effect in conjuction with other microscopic tools (ESR,NMR and neutron scattering) has played a major role in evaluating the behavior of many of the magnetic superconductors.

The present chapter is an attempt to present some of the significant contributions in this area with an emphasis on the Mossbauer studies. No attempt is made to detail each and every study reported in the literature. A brief breakdown of remaining sections is as follows.

In Section 2 we relate the information gained from the spin relaxation studies in the normal state of superconductors to the depression in T_c due to pair breaking effects. The complexity of the relaxation phenomenon in the superconducting state is then presented.

The coexistence of magnetic order and superconducting order is discussed in Section 3. The contribution of Mossbauer experiments to this subject is considerable. Along with other techniques it is clearly established that superconductivity coexists only with spin-glass type short range magnetism and antiferromagnetism.

The Mossbauer effect studies have contributed a great deal in obtaining the magnitude of local moments in many alloys and ternary compounds. How the value of these moments determine the superconducting behavior of some of these systems is not completely understood. Section 4 will present the experimental results on a number of studies on local moment determination.

There is a dilemma of missing magnetic moment when Mossbauer and neutron results are compared in the case of reentrant superconductors (5). This is presented in Section 5. In Section 6, we review the question of enhancement of critical fields and the Jaccarino-Peter mechanism.

2. SPIN-RELAXATION IN SUPERCONDUCTORS

2.1 Background:

Substitution of magnetic impurities in a superconducting matrix is known to weaken the superconducting behavior. The impurities of 3d transition series influence the superconductivity more drastically than the 4f rare-earth impurities. This is due to the highly localized nature of 4f electrons. The magnetic atoms cause spin flip scattering of singlet Cooper pairs through the exchange interaction of the local moment with the conduction electron moment. The theory developed by Abrikosov and Gor'kov (10) to describe the influence of magnetic atoms assumes such an exchange interaction to be

$$H_I = -2j\,\bar{S}.\bar{\sigma} \quad , \tag{1}$$

where s is the paramagnetic impurity spin and σ is the conduction electron spin. The exchange interaction parameter is denoted by j. The inverse life time characterizing this interacting system will now determine the superconducting behavior of the electrons and the spin-relaxation rate of the magnetic atom. Within the first Born-Approximation, the depression in T_c for low impurity concentration x is given by

$$\Delta T_c = -\frac{\pi}{4}\tau_s^{-1} \tag{2}$$

where

$$\tau_s^{-1} = x\,j^2 N(E_F)\;S(S+1)/\hbar \tag{3}$$

$N(E_F)$ is the density of states at the Fermi level (for one spin direction).

This discussion will now be extended to obtain an expression for the temperature dependence of the relaxation-rate of a paramagnetic spin in a superconductor. If i and j are the initial and final state of the electron spin, and k and ℓ that of the magnetic spin, the relaxation rate is given by (11)

$$W_{j\ell,ik} = \frac{2\pi}{\hbar} \left| j,\ell|H_I|i,k \right|^2 \delta(E_j + E_\ell - E_i - E_k) \quad . \tag{4}$$

Considering the contributions from all the electrons

$$W = \sum_{k,\ell} W_{j\ell,ik} \; f(E_i) \; [1 - f(E_j)] \quad , \tag{5}$$

where f(E) is the Fermi distribution. In the *normal state of the superconductor* this leads to

$$W = \frac{2\pi}{\hbar} \; j^2 \; N(E_F)^2 \; g^2 \; kT \quad , \tag{6}$$

where g is the g-factor of the impurity spin. This is the Korringa relation which shows that the spin-relaxation rate is linearly dependent on the temperature. Such measurements permit one to deduce the product $|jN(E_F)|$ which also determines ΔT_c.

As one enters the *superconducting state* of the material, a gap in the energy states of the electrons will appear. This influences, the product f(E)[1-f(E)] and in turn dramatically alters the relaxation rate. We will return to this later on.

Because of the strong coupling of 3d spins to the conduction electrons, very few Mossbauer investigations of their relaxation rate have been carried out. For the rare-earth atoms the 4f moment is highly localized. In this case the orbital angular momentum is not quenched and we have to replace $\vec{S}$ by the projection of the total angular momentum $\vec{J}$. This leads to

$$\tau_s^{-1} = xN(E_F)j^2 \; (g_J-1)^2 \; J(J+1)/h \tag{7}$$

and

$$W = \frac{2\pi}{\hbar} \; (g_J-1)^2 \; j^2 \; N(E_F)^2 \; kT \quad , \tag{8}$$

where g_J is the Lande g factor.

Furthermore, crystal fields can introduce a different ground state Γ_i with a g-factor by $g(\Gamma_i)$. This will also influence the analysis of τ_s^{-1} and W. If the crystal fields are large, given (g_J-1) will have to be replaced by $(g_J-1)\; g(\Gamma_i)/g_J$ in the above expressions. However, if there are crystal field states at intermediate energies a detailed analysis is needed in evaluating both τ_s^{-1} and W. Even if there are states at considerably higher values compared to T_c, exchange induced transitions (from Van Vleck off-diagonal terms) may contribute to τ_s^{-1}.

2.2 Relaxation in the Normal State of Thorium:

Thorium metal is a superconductor with T_c = 1.4 K. Addition of rare-earth impurities strongly reduces T_c. For example, approximately 5 at % of Er in Th makes it a normal metal down to 0.2° K (12). The spin relaxation-rate for Er and Dy impurities in Th metal have been investigated from Mossbauer spectroscopy.

In the cubic crystalline electric field of Th metal the free-ion ground state of $Er^{3+}(4I_{15/2})$ and $Dy^{3+}(^6H_{15/2})$ splits into three quartets $[\Gamma_8^{(1)}, \Gamma_8^{(2)}, \Gamma_8^{(3)}]$ and two doublets (Γ_6 and Γ_7) (13). In Th both rare-earth atoms have Γ_7 as the ground state, which is a magnetically isotropic Kramers doublet (14,15). In the absence of relaxation the paramagnetic hyperfine structure is described by the Hamiltonian (13)

$$H_{hf} = Ag(\Gamma_i)\ \bar{I}.\bar{S}/g_J \quad . \tag{9}$$

Here S is an effective spin 1/2 and A is the hyperfine coupling constant. For the $0^+ \rightarrow 2^+$ transition with 80.6 keV energy in ^{166}Er and 86.8 keV energy in ^{160}Dy, the Mossbauer spectrum produced by the above hyperfine Hamiltonian consists of two resonances located at $Ag(\Gamma_7)/g_J$ and $-3\ Ag(\Gamma_7)/g_J$ with relative intensities of 3 and 2, respectively. Spin relaxation brings in a time dependence through S(t) in H_{hf} which modifies the lineshape of the Mossbauer spectrum. A detailed analysis of the lineshape allows one to obtain the relaxation rate W (16,17).

Such measurements have been performed using Mossbauer spectrocopy scopy of Er and Dy in Th metal. In Fig. 1 we present the data of W vs. T for the Dy in Th system (15). The linear relation between W and T above T_c permits us to obtain the values of j. In Table 1 the values of W/T derived from the Mossbauer studies along with the values for j are given.

It is of great interest to compare these values of j with those obtained from the measurement of ΔT_c. The j values obtained using Eqns. (2) and (7) along with $g(\Gamma_7)$ values are also in Table 1. The two sets of j values obtained from Mossbauer relaxation measurements

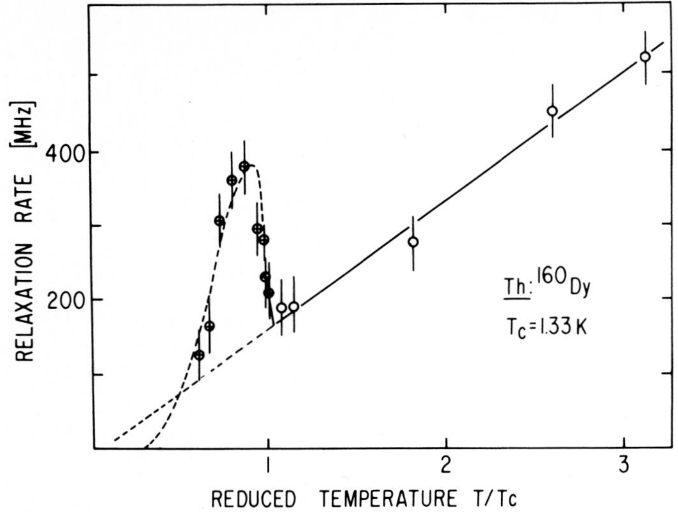

Fig. 1. Relaxation rate of Dy^{3+} spin in Th metal as a function of reduced temperature. Above T_c the relaxation rate is linear with temperature (Ref. 19).

TABLE 1

Comparison of the exchange integral deduced from the Mossbauer studies j(ME) with that from ΔT_c measurement using Abrikosov-Gor'kov theory, j(AG) for Er and Dy impurities in Thorium.

Resonance Isotope	W/T MHz/°K	$g(\Gamma_7)$	$N(E_F)$ states/ev atom spin	j(ME) (eV)	$\Delta T_c/x$ (°K/a%)	j(AG) (eV)	Ref.
^{166}Er	85±13	6.84±0.05	0.55	0.025±0.002	0.21	0.023	14
^{160}Dy	122±10	7.60±0.05	0.55	0.022±0.002	0.20	0.024	15

and the T_c depression measurements agree very well.

It must be pointed out that relaxation studies can only be done on very dilute samples and thus one can justify comparison only with the ΔT_c measurements done in the low concentration range. If the magnetic impurity concentration is larger spin-spin interactions can complicate the analysis of relaxation patterns. The inferences drawn from the ESR linewidth measurements (18) for Er and Dy impurity in thorium metal agree very well with the Mossbauer effect studies.

2.3 Relaxation in the Superconducting State of Thorium:

The measurement and interpretation of the relaxation rate for ^{160}Dy in the superconducting state of Th has been given by Wagner et al. (19). The analysis of the data cannot be carried out using conventional relaxation lineshapes are discussed in the last section.

In the normal state the spin flip energy is transferred to the conduction electron bath which has an energy of the order of E_F. This is large compared to the hyperfine energies involved in the problem, viz. $\pm 5Ag(\Gamma_7)/2g_J$. Hence the white noise approximation is valid. In the superconducting state the electrons are paired and the relaxation is through creation and annihilation of the quasi-particles. Near T_c the gap is narrow and the energy involved in the transition between the states below and above the gap is comparable to the hyperfine energies. Thus the white noise approximation is not valid (20). This introduces a hyperfine frequency dependence in the relaxation matrix and a complete lineshape calculation becomes rather involved (19,21).

In Fig. 1 we present the relaxation rate deduced by Wagner et al (19) in the superconducting state of Th metal. We will present a qualitative description of this behavior. Below T_c the number of normal electrons drastically reduces, thus hindering any spin relaxation. In addition, the opening of the energy gap modifies the value of $N(E_F)$. Since there are no states available in the gap, they will be piled up on either side of the gap. When the gap is small, the pile up of states within the interval will increase the relaxation rate rather abruptly as observed. On lowering the temperature the gap becomes wider than kT and the rate will decrease. Since the relaxation rate is proportional to f(1-f) in Eqn. (5), it will peak about the Fermi energy. On approaching T = 0, the density of states in the superconducting state is lower than in the normal state and this results in a lower rate of relaxation. The functional variation of f(1-f) between T_c and 0° K makes the relaxation rate to reduce exponentially with temperature. This is indeed seen in Fig. 1.

It is appropriate to note that the Mossbauer effect studies discussed here do not need external magnetic fields. This is important in type-1 superconductors like Th where external fields will modify the superconducting behavior. Other resonance methods (ESR, NMR) needing external fields cannot always be applied in such situations.

2.4 Relaxation in Ternary Superconductors:

Recently a group of ternary compounds known as Chevrel phases have been extensively investigated for their superconducting properties (22). A typical material in this class in $SnMo_6S_8$ which has T_c = 10.4 °K. What is peculiar about this compound is that replacing Sn^{2+} ions by magnetic $Eu^{2+}(S_{7/2})$ ions does not decrease T_c over a wide composition range (8). Figure 2 shows the variation

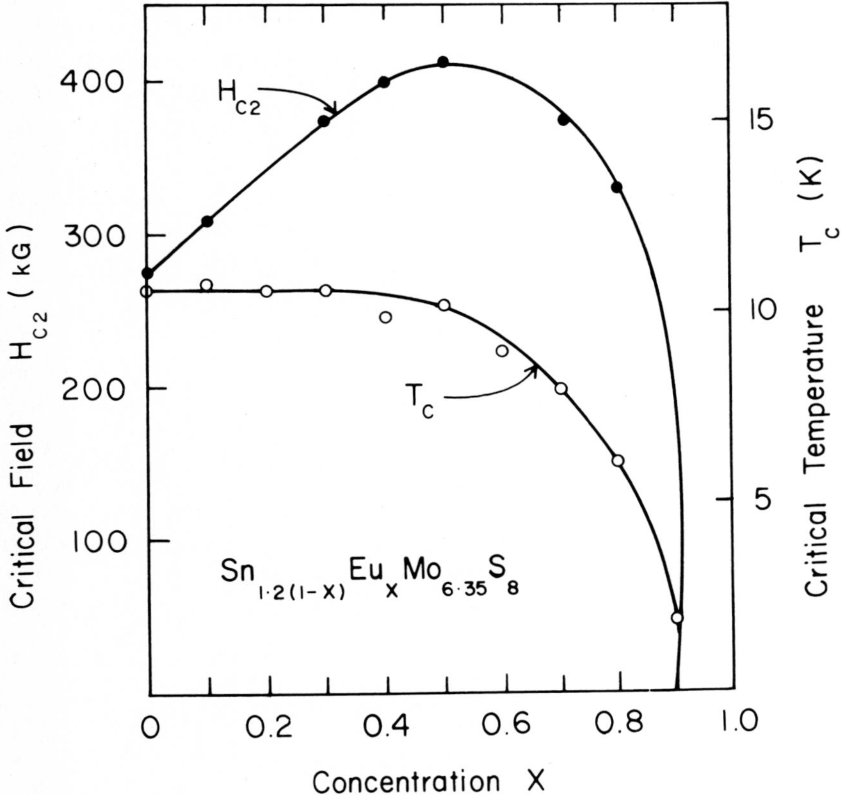

Fig. 2. Concentration dependence of the transition temperature T_c (open circles) critical field H_{c2}(solid circles) in $Sn_{1.2(1-x)}Eu_xMo_{6.35}S_8$ (Ref. 8).

of T_c in $Sn_{(1-x)}Eu_xMo_6S_8$ as a function of x. Mossbauer and magnetization measurements verify that Eu is divalent, and hence has a large magnetic moment (of $7\mu_B$) (23). However, for $x \leq 0.5$, T_c remains almost unaltered. Within the framework of Abrikosov-Gorkov theory, this is qualitatively understood by assuming either j or $N(E_F)$ or both to be small in this system such that Eqn. (7) predicts small ΔT_c.

As was demonstrated earlier, the Mossbauer effect can be used to deduce the value of $|jN(E_F)|$ through the measurement of relaxation rate, W. When W is comparable to the Larmor precession rates, the analysis of the line-shape changes yields the value W. The Chevrel phase compound $Sn_{0.75}Eu_{0.25}Mo_6S_8$ has been studied (24,25) for relaxation between 1.5 and 300° K using the 21.6 keV Mossbauer resonance in ^{151}Eu. The hyperfine field at the Eu nucleus is such that the relaxation rate is large compared to the Lamor precession rates involved. This produces only a broadening of the Mossbauer line over the investigated temperature range. The excess linewidth $\Delta\Gamma$ can, however, be related to the relaxation rate (24,25) using the perturbation procedure of Bradford and Marshall (26). Describing the fluctuating Hamiltonian by

$$H(t) = A\ \bar{I}.\bar{S}(t) \qquad (10)$$

where A is the ground-state hyperfine constant for ^{151}Eu, we obtain

$$W = \frac{S(S+1)}{\hbar^2}\frac{A^2}{4}\ (35-68\ R_g + 45\ R_g^2)/\Delta\Gamma \quad , \qquad (11)$$

where R_g is the ratio of the nuclear excited state g-factor to that of the ground-state. A general expression for this problem applicable to other Mossbauer transitions as well has been published recently (27).

From the measurement of the variation of the excess linewidth with temperature, and using Eqn. (11), the relaxation rate has been obtained (25) (Fig. 3). The relaxation rate is linearly dependent on temperature as expected. The non-zero intercept is presumably due to spin-spin relaxation. From the slope of Fig. 3, the value of $|jN(E_F)|$ can be found. In Table 2 we have compared this value with that in other ternary Chevrel phase compounds. In Eu doped $PbMo_6S_8$ compounds on the other hand considerably larger values of $|jN(E_F)|$ have been obtained (30). The value for Eu in $LaAl_2$ is an order of magnitude larger (29) than that found in ternaries.

The small value of the coupling $|jN(E_F)|$ in the Chevrel phase compounds at least partially provides an answer for the weak suppression of T_c in the presence of magnetic atoms shown in Fig. 2. Detailed band structure calculations of Freeman and Jarlborg (32) show that the smallness of $jN(E_F)$ in Chevrel phase compounds is due to small values of the projected density of states $N(E_F)$ at the magnetic atom.

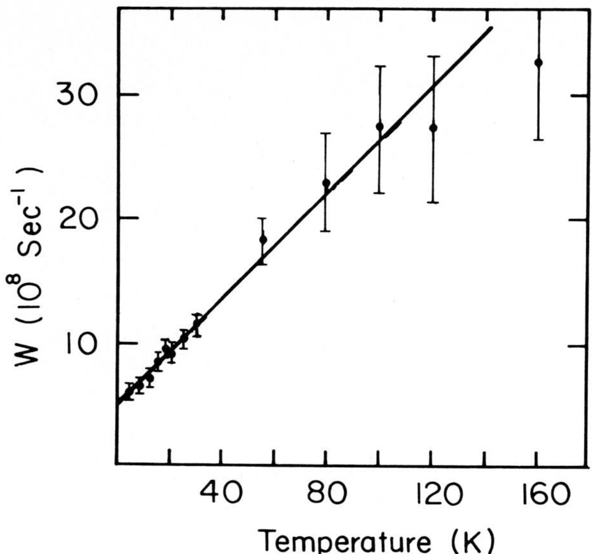

Fig. 3. Temperature dependence of the electronic relaxation rate of Eu spin in $Eu_{0.25}Sn_{0.75}Mo_6S_8$ (Ref. 25).

TABLE 2.

Values of the product of exchange parameter and density of states obtained from resonance methods.

	$jN(E_F)$ per atom spin	References
$Eu_{0.25}Sn_{0.75}Mo_6S_8$	0.003	25
$GdMo_6Se_8$	0.0047	28
1 % Gd in $SnMo_6S_8$	0.0061	29
0.91% Gd in $PbMo_6S_8$	0.0050	29
Eu in $LaAl_2$	0.03	31

3. INTERACTION BETWEEN MAGNETIC ORDER AND SUPERCONDUCTIVITY:

Coexistence of superconductivity and ferromagnetism was proposed by Matthias et al (33) through the investigation of Laves phase pseudo-binary compounds of the type $Ce_{1-x}R_xRu_2$ where R represents a magnetic rare-earth. Careful a.c. susceptibility measurements established the phase diagram of such systems (34). In Fig. 4 we present the phase diagram for $Ce_{1-x}Gd_xRu_2$ system where the extrapolation of T_c and T_m reveals the coexistence region. From an extension

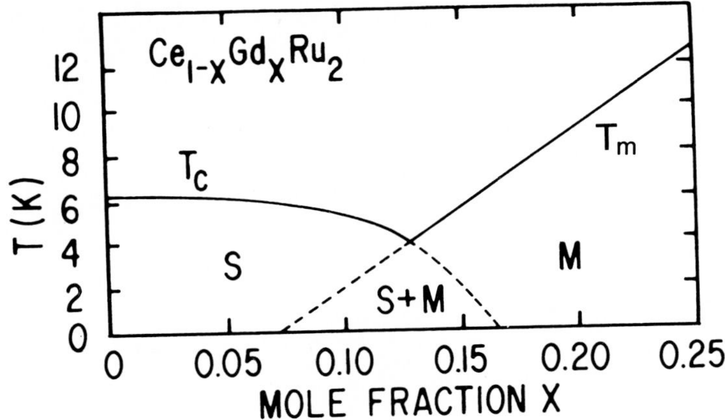

Fig. 4. The magnetic-superconducting phase diagram for $Ce_{1-x}Gd_xRu_2$ (Ref. 34).

of the work of Abrikosov and Gor'kov, a coexistence of these phenomena is theoretically predicted by Gor'kov and Rusinov (35).

The first microscopic study of the phenomena came from the Mossbauer effect studies of Taylor et al (36,37). They introduced tracer amounts of ^{57}Fe in the $Ce_{1-x}Gd_xRu_2$, in which Fe atoms substituted for the Ru atoms. The measurements conclusively presented the evidence for the magnetic order around a fraction of the ^{57}Fe atoms consistent with the phase-diagram. We present in Fig. 5 the measured hyperfine magnetic field at the Fe nucleus as a function of temperature. The value of T_m deduced from these microscopic measurements agreed with that obtained from the susceptibility measurement used to establish the phase diagram. This suggests that the conduction electrons responsible for the RKKY coupling of the magnetic moments on Gd atoms do not determine the superconducting behavior of these alloys. However, it must be recognized that the Mossbauer effect represents only the local magnetic order and not the long-range magnetic order (usually established through neutron diffraction studies). One possible explanation of the coexistence phenomenon in these alloys assumes two sublattices exhibiting the two types of orders. If the spatial extent (or the correlation length)

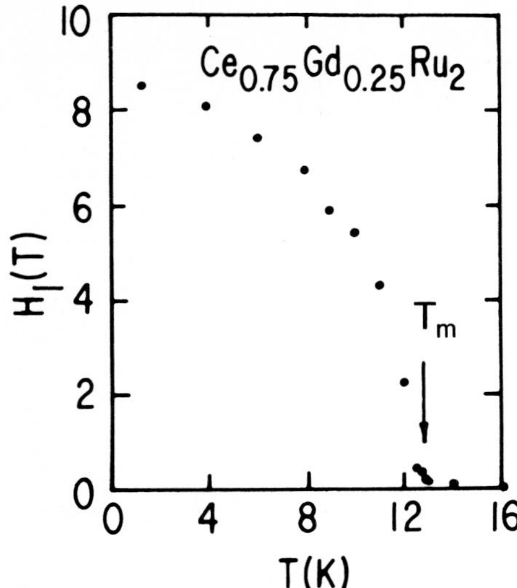

Fig. 5. Temperature dependence of hyperfine magnetic field H_i (in Tesla) at ^{57}Fe nuclei in superconducting spin glass $Ce_{0.75}Gd_{0.25}Ru_2$. T_c is the magnetic ordering temperature from Fig. 4 (Ref. 36).

of the ferromagnetic order, ξ_M, is small compared to the superconducting coherence length, ξ_s, then we expect a coexistence.

The most conclusive evidence in this regard has come from the inelastic neutron studies of Roth et al (38) and Lynn et al (39). The studies find the maximum value of ξ_M to be about 15 Å in $Ce_{0.8}Tb_{0.2}Ru_2$ and about 80 Å in $Ce_{0.73}Ho_{0.27}Ru_2$. In view of the facts, the presence of spontaneous magnetic ordering evidenced from Mossbauer studies suggests these systems to be spin-glasses with no long-range magnetic order. Aptly, Roth (5) has called these superconducting spin glasses.

The ^{57}Fe substituted investigations have now been carried out on other $Ce_{1-x}R_xRu_2$ alloys with almost similar results (37,40).

The evidence for the spin-glass behavior also comes from the ^{155}Gd Mossbauer study of $Ce_{1-x}Gd_xRu_2$ compounds (41). In this case, however, the temperature dependence of the hyperfine field of the ^{155}Gd nucleus does not follow a simple S = 7/2 Brillouin function except for x = 0.5 sample (Fig. 6). A modified molecular field model with a temperature dependent Curie-Weiss temperature has been used to describe much of this data (for X < 0.5). The analysis shows a slow onset of magnetic order which implies a nearly critical

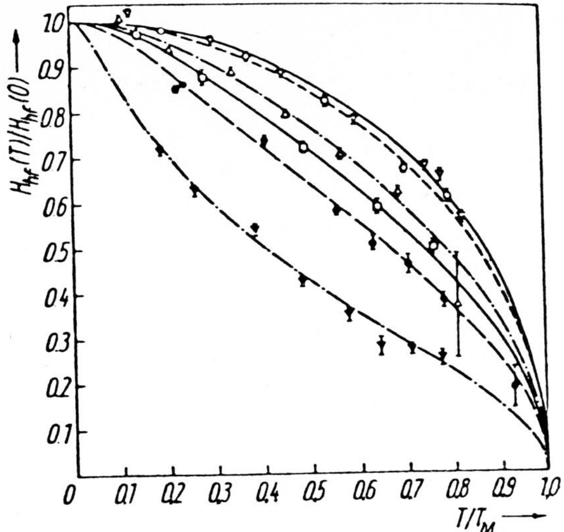

Fig. 6. Temperature dependence of the reduced hyperfine fields at ^{155}Gd *nuclei in* $Gd_xCe_{1-x}Ru_2$ *for various x values. Curves from top to bottom are for x = 0.5, 0.35, 0.24, 0.18, 0.135 and 0.11 (Ref. 41).*

behavior of these magnetic alloys over a wide temperature range. This behavior is consistent with the Edwards-Anderson model (42) of spin-glasses.

The coexistence phenomenon in $La_{1-x}Eu_x$ alloys has been studied by Steiner et al. (43) using ^{151}Eu Mossbauer measurements. The Mossbauer effect has primarily aided in detecting spontaneous magnetic ordering of this system and Fig. 7 summarizes their results. For higher Eu concentrations in La($x \geq 0.015$), the magnetic order is a result of the freezing in two or more ferromagnetically coupled Eu spins. In the lower concentration range ($x < 0.015$) Eu atoms are well separated and individual spins are "frozen in" randomly. The ordering temperature in this concentration range is proportional to square-root of the concentration, x. This square-root behavior is typical for spin-glasses (42,44,45) and again leads us to conclude that the phenomenon of coexistence depicted in Fig. 7 is due to spatial extent of the two interactions.

The spin-glass behavior is not observed in ternary compounds (like $ErRh_4B_4$ and $HoMo_6S_8$). They also do not exhibit coexistence phenomenon. With the onset of ferromagnetism the superconductivity vanishes. This will be discussed in detail in the next section. Here, it should however be pointed out that the pseudo-ternaries behave in many respects like the pseudo-binary compounds. Bolz et al. (46) have studied the magnetic ordering in the Chevrel phase compounds

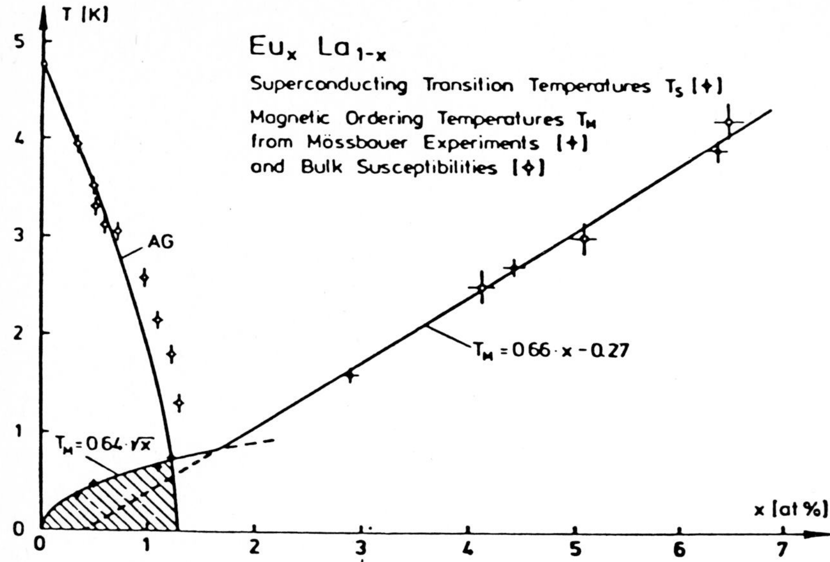

Fig. 7. T_c *and* T_M *for* Eu_xLa_{1-x}*. AG is the Abrikosov-Gor'kov curve. The shaded area represents the region where the compounds show superconducting spin-glass behavior (Ref. 43).*

of the type $Sn_{1-x}Eu_xMo_6S_8$ using the Mossbauer effect in ^{151}Eu. The sample studied by them with T_c of 8° K and T_m below 0.5^o K showed the presence of magnetically ordered regions separated by paramagnetic Eu atoms. Since the weak magnetic dipolar interaction is primarily responsible for their ordered magnetism, it is expected that some of the Eu atoms situated in magnetically ordered clusters have randomly oriented spin directions. This order hence appears to be of spin-glass type.

In all the alloys that we have discussed here there is a statistical distribution of magnetic atoms in the structure. Over certain concentration ranges the magnetic interaction freezes in the spin-glass behavior. These typically have short magnetic correlation lengths compared to the superconducting coherence length and thus the two phenomena can coexist. The Mossbauer effect has played a major role in unraveling many mysteries of the coexistence phenomenon on a microscopic scale.

We have casually used the word spin-glass without questioning the influence of superconducting forces on its formation. The spin-glass order has been referred to as "crypto ferromagnetism" by Anderson and Suhl (47) in a discussion of the coexistence phenomenon. They considered the modification of the RKKY coupling between the local moments by the presence of the superconducting interactions. This results in a peak in the conduction electron susceptibility at non-zero value of the wavevector (rather than at zero wavevector as in ferromagnets). This produces the crypto-ferromagnetic state. Physically this is perhaps identical to the spin-glass state.

The suppression of long-range magnetic correlation in above discussed systems is discussed differently by Blount and Varma (48). They consider the interaction between the electromagnetic field and the superconducting state. As the value of ξ_M increases, the superconducting currents can cancel any macroscopic magnetic field in the sample. This will then require more energy to establish a long-range order than a short-range order.

4. LOCAL MOMENTS IN SUPERCONDUCTORS

As was pointed out earlier the problem of statistical fluctuation in magnetic atom distribution is absent in ternary compounds. The magnetic atoms have a well defined site in the lattice. In many of the ternary compounds one observes either superconducting or magnetic order, or both. If the order is ferromagnetic as in $ErRh_4B_4$ and $Ho_{1.2}Mo_6S_8$, the superconductivity is destroyed when the ferromagnetic order condenses (6,7). However, antiferromagnetic order can coexist with superconductivity. Typical examples are $R_{1.2}Mo_6S_8$ (R = Gd, Tb, Dy) (49-51) and $SmRh_4B_4$ (52). These observations are consistent with the discussion of the last section. In a lattice containing regularly placed magnetic atoms, in the ferromagnetic state the order will be a long-range one. Then ξ_M being infinity, the coherence of the Cooper pairs is destroyed. On the other hand, in an antiferromagnet, the average magnetization (or the correlation) is rather small compared to ξ_s for the two order to coexist.

It was pointed out earlier that in the absence of crystalline electric fields, ΔT_c is governed by the factor $G = (g-1)^2 J(J+1)$, the deGennes factor [see Eqn. (7)]. It is also true that within the mean field approximation, the magnetic ordering temperature, T_m, in conductors is proportional to G. In Fig. 8 a plot of T_m and $-\Delta T_c$ vs deGennes factor for the heavy rare-earth based Mo sulphides is presented (53). The slope of T_m vs G allows one to obtain (54) the value of j to be 0.055 eV. The value can also be estimated from the slope of ΔT_c vs G plot using Eqn. (2) and is found to be 0.025 eV. We would normally expect these two numbers to be equal, although generally they agree with those obtained from the relaxation studies (Table 2).

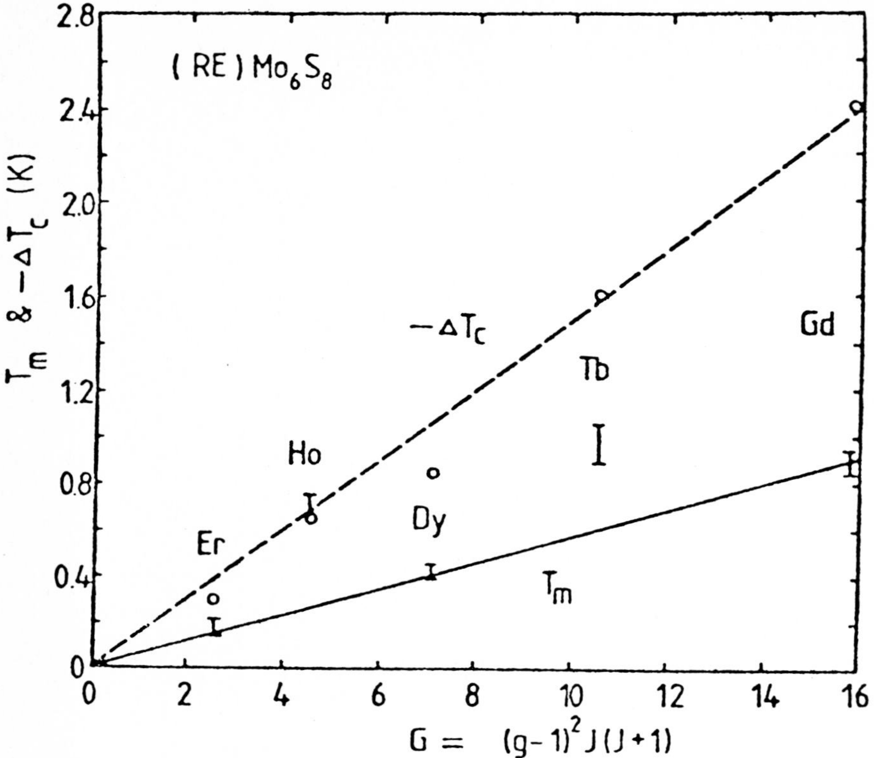

Fig. 8. Plot of T_m and - T_c vs deGennes factor, G, for heavy rare-earth Mo sulphides (Ref. 53).

There are many reasons for the j values to be different when derived from T_m and T_c measurements. First we notice that all the experimental values in Fig. 8 do not obey the linear relation. This is due to the crystal field effects which will modify the magnetic character of the ground state and excited states. This will influence both the T_c and T_m expressions as discussed in Section 2 (55). In the above picture we have assumed an isotropic RKKY picture. It is likely that j is anisotropic and that the electrons responsible for the magnetism and superconductivity are not the same. This will alter the $N(E_F)$ values which determine T_m and T_c. Furthermore, the dipolar coupling may play a major role in determining T_m. What is important at present time is to understand the rare-earth crystal fields and deduce the magnetic moment (or g-factors) for the ground state in order to carry out a more complete evaluation of the problem. The needed data can be obtained using the Mossbauer effect of the rare-earths, although there is no systematic work as yet in this direction. $ErMo_6S_8$, $DyMo_6S_8$ and $YbMo_6S_8$ have been investigated using Mossbauer spectroscopy (56-58).

In another series of ternary compounds RRh_4B_4 there is again very little resemblance between experimentally measured and predicted variations in T_c and T_m with G (59). There are however measurements of the rare-earth ground state moment available in compounds with R = Dy, Er and Tm (60). This permits us to modify the expressions for T_c and T_m by replacing g_J with $(g_{J-1})\ g(\Gamma_i)/g_J$. This also does not provide a unique value for the product $|N(E_F)j|$ expected from the above discussion. Neglect of the crystal field excited states and the anisotropy of $g(\Gamma_i)$ may partly account for this situation.

There are only a few Fe-based compounds which are superconducting since the strong depairing of Cooper pairs is unavoidable in the presence of an Fe moment. The question naturally arises regarding the origin of superconductivity in those few compounds. In table 3 we have listed the Fe-based superconductors.

TABLE 3.
Superconducting transition temperatures of iron based compounds

Compound	T_c(K)	T_m(K)	Reference
Th_7Fe_3	1.86	-	61
U_6Fe	3.9	-	62
Zr_2Fe	0.17	-	63
$Sc_2Fe_3Si_5$	4.6	-	64,65
$Y_2Fe_3Si_5$	2.4-2.0	-	64
$Lu_2Fe_3Si_5$	6.1-5.8	-	64
$Dy_2Fe_3Si_5$	-	4.0	66,67
$Tm_2Fe_3Si_5$	1.7	1.1	68
$TlFe_3Te_3$	4.0	-	69

Only under special circumstances can one avoid depairing interactions. A weak exchange coupling can keep the superconducting electrons from depairing, as in some of the rare-earth based Chevrel compounds. It is also possible that Fe carries no magnetic moment.

To achieve this situation one should have unusual crystal structure and electronic structure.

The Mossbauer effect is an excellent microscopic tool for measuring the local magnetization. The value of the hyperfine magnetic field at the Fe nucleus in the ordered state of the material is roughly proportional to the moment on the atom and permits us a sensitivity of the order of 0.03 μ_B in the measurement.

In the absence of a magnetic order, one can obtain the same information by applying external magnetic fields, H_{ext}. The field experienced by the nucleus is then $H_{ext} + H_i$ where H_i is the induced field. In a situation with no magnetic moment on the atom H_i will be zero. This is true in diamagnets and Pauli paramagnets. On the other hand, H_i will be non-zero if the atom carries a moment or the neighbouring paramagnetic atoms induce a spin polarization.

Some of the iron-based compounds from Table 3 have been investigated using the ^{57}Fe Mossbauer effect to evaluate some of the above ideas. External magnetic measurements (65,70) on $Sc_2Fe_3Si_5$ and Th_7Fe_3 show $H_i = 0$ in both these systems. Hence the absence of Fe magnetism allows the superconductivity to exist.

A complex coordination for Fe atom is known to produce a low-spin zero moment situation. When Fe is coordinated by 12 or 13 atoms, the Fe goes into a low-spin state with a filled band or filled d-orbital configuration. This is the case in $Sc_2Fe_3Si_2$ and other superconducting compounds in this class. Since Fe has simple coordination in Th_7Fe_3, the absence of d-moment is associated with a low density of the projected d-states of the Fe site. The distinct mechanisms responsible for quenching the d-moment in these two compounds is reflected in the values of their isomer shifts and support the above picture (70).

The origins of superconductivity can be traced now to 3d-electrons in $Sc_2Fe_3Si_5$ and 6d electrons in Th_7Fe_3. In $Dy_2Fe_3Si_5$ there is a moment of 7 μ_B on Dy atoms (66). In external field studies it is found that H_i is not zero due to a spin polarization at Fe originating from the Dy moments. The coupling between the Dy moment and the conduction electrons is anti-ferromagnetic (see Fig. 9) (71) and is adequate to destroy the superconductivity of the Fe 3d electrons. The other criterion for realizing superconductivity is the requirement that the atoms donating the superconducting electrons in the lattice form clusters (72). This, for example, is true in Chevrel phase compounds in which Mo atoms are clustered, and the Rh atoms are clustered in

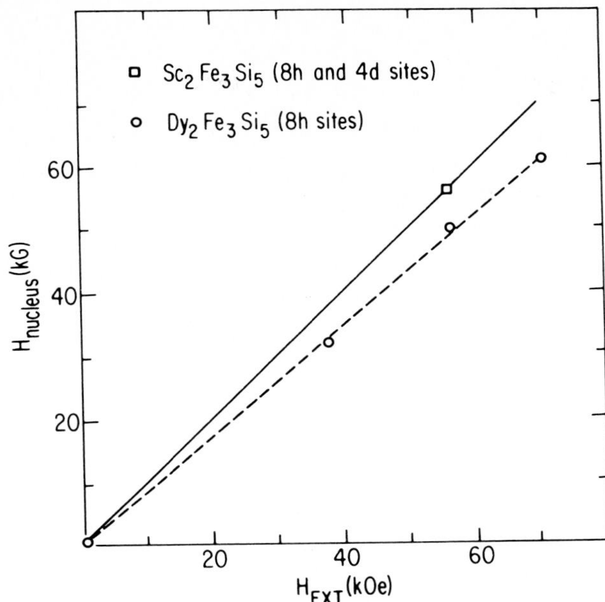

Fig. 9. Measured hyperfine field at ^{57}Fe nuclei in $Sc_2Fe_3Si_5$ and $Dy_2Fe_3Si_5$ at 4.2 K as a function of external magnetic field. The Dy moments induce a negative spin polarization at Fe site in $Dy_2Fe_3Si_5$ (Ref. 60).

RRh_4B_4 compounds. In $R_2Fe_3Si_5$, Fe atoms occupying the 8h site satisfy this criterion. In Th_7Fe_3, the distance between near neighbor Th atoms is even smaller than in Th metal, providing the superconductivity of the 6d electrons.

It is now known (68) that $Tm_2Fe_3Si_5$ is a superconductor below 1.7 K while all the other rare-earth based iron-silicides of this class show magnetic order. Preliminary studies using ^{169}Tm Mossbauer resonance show that Tm has a singlet ground state with no magnetic moment (73). With no moment on either Fe or Tm, the compound is a superconductor.

5. MISSING MAGNETIC MOMENT IN REENTRANT SUPERCONDUCTORS

The ternary compound $ErRh_4B_4$ becomes superconducting at T_c = 8.6 K and ferromagnetic at T_m = 0.9 K. This reentrant superconducting behavior is also observed (74) in $ErRh_{1.1}Sn_{3.6}$ in which T_c = 1.36 K and T_m = 0.46 K. Detailed investigation of Mossbauer resonance using the 80.6 keV transition in ^{166}Er has been carried out in both these compounds down to temperatures well below T_m. This permits one to obtain the value of the hyperfine field ar the Er nucleus in the ferromagnetic state. The hyperfine field is made up of a core polarization term, an orbital current term and a conduction electron polarization term. Of these the orbital term accounts for more than 98% of the hyperfine field. This term is proportional to $\langle J_z \rangle_T$ and hence to the magnetic moment. A linear relationship between the measured field and the measured moment on the atom in a large number of rare-earth compounds has been established (24).

In Table 4 the values of the measured hyperfine field and the deduced moments from the Mossbauer investigations are given (75,76).

TABLE 4

Magnetic hyperfine field, H_n, the values of T_c and T_m, and the magnetic moments from Mossbauer and neutron diffraction studies in Er-based ternary superconductors.

	T_c K	T_m K	H_n kOe	μ(ME) μ_B	μ(N) μ_B	Ref.
$ErRh_4B_4$	8.6	0.96	7670	8.3	5.6	75
$ErRh_{1.1}Sn_{3.6}$	1.36	0.46	7030	7.6	-	76

In both of these compounds the value of the ordered moment is lower than the free-ion moment of Er^{3+} (9.0 μ_B). This indicates the presence of crystal-field interactions which reduce the moment below the free-ion value.

It is curious that the value of the moment on Er atom in $ErRh_4B_4$ measured from neutron diffraction (77) is approximately 30% lower than the value measured from the Mossbauer studies (Table 4). In $ErRh_{1.1}Sn_{3.6}$ although neutron studies establish the ferromagnetic order, lack of true long-range order even at 0.07 K does not permit them to obtain a value for the moment (74). This should be contrasted with a unique moment measured in Mossbauer experiments.

The discrepancy in the moment values measured from the two techniques in $ErRh_4B_4$ are outside the experimental errors. Single crystal neutron diffraction studies even have not resolved the difference (78). One of the following reasons can account for the missing moment:

(1) In neutron diffraction, only that part of the moment showing a long-range order is measurable. In the Mossbauer effect, since one measures single-ion correlation, because of the slow spin relaxation rate in $ErRh_4B_4$, one obtains the full value of the moment. This would imply 30% of the moment is disordered and only a component shows long-range ferromagnetism.

(2) The material may contain ferromagnetic domains separated by large domain walls or a disordered magnetic lattice. In this case neutron intensities would not provide the proper average of ordered moments. The Mossbauer effect, cannot present distinct spectra for the atoms in the domains and the domain walls (or disordered spins).

More work is needed to resolve the missing moment anomaly in $ErRh_4B_4$. The phenomenon responsible for this behavior is even more pronounced in $ErRh_{1.1}Sn_{3.6}$ where the long-range correlation is completely absent. What is clear is that the ferromagnetic order in both these reentrant materials is complex even well below T_m. This is perhaps due to a continued interaction between superconductivity and magnetic order.

6. ENHANCEMENT OF THE CRITICAL FIELD

In Fig. 2, we show the variation of the upper critical field with concentration of magnetic Eu atoms in the Chevrel phase compound $Sn_{1.2(1-x)}Eu_xMo_6S_8$ measured by Fischer et al (8). As mentioned in the introduction, such an enhancement was suggested by Jaccarino and Peter (9) as a possibility if the paramagnetic atom induced a spin polarization with a sign that would compensate the applied magnetic field.

In order to obtain information on the sign of the spin polarization at Eu and Mo site, Mossbauer effect and NMR experiments were done on a $Sn_{0.5}Eu_{0.5}Mo_6S_8$ compound (23). The ^{151}Eu Mossbauer effect measurements indicated the spin polarization to be positive (parallel to the applied field) at the Eu site. The ^{95}Mo NMR measurement showed negative spin polarization at the Mo site. Since it is presumed that the superconductivity arises from the Mo 4d-electrons, the negative spin polarization will provide the compensation of the external field. This is a direct microscopic conformation of the Jaccarino-Peter mechanism describing the enhancement of the critical field.

The enhancement of the critical field is an important aspect of the technological use for the ternary superconductors. In Gd doped $PbMo_6S_8$, critical fields as high as 700 kG have been realized.

7. CONCLUSIONS:

The study of the interaction between magnetism and superconductivity is a rich subject. The use of many different available techniques is imperative in their understanding. The Mossbauer effect along with other microscopic methods has played a pivotal role in this regard. In this chapter we have made an attempt to provide the reader with some important applications of the Mossbauer effect to this field, and pointed to some outstanding problems where a great many other Mossbauer experiments could contribute.

The discovery of novel materials like the ternary superconductors has added much to the field of magnetic superconductors. This, however, should only be considered the beginning of a new and potentially important field. The Mossbauer results on a few systems have already helped in clarifying many thoughts in the field.

ACKNOWLEDGEMENTS

I would like to thank various collaborators in this field of research and in particular Dr. B.D. Dunlap for critically reading this manuscript.

REFERENCES

1. M.B. Maple, in H. Suhl (Ed.), Magnetism, Vol. V, Academic Press, New York, 1973, p. 289.
2. K.H. Bennemann and J.W. Garland, in J.I. Budnick and M.P. Kawatra (Eds.), Dynamical Aspects of Critical Phenomena, Gordon and Breach New York, 1972, p. 512.
3. B.T. Matthias, E. Corenzwit, J.M. Vandenberg and H.E. Barz, Proc. Nat. Acad. Sci. USA, 74 (1977) 1334.
4. H.F. Braun, in G.K. Shenoy, B.D. Dunlap and F.Y. Fradin (Eds.), Ternary Superconductors, North-Holland Publishing Co., Amsterdam, 1980, p. 225.
5. See for example, S. Roth, Appl. Phys., 15 (1978) 1.
6. W.A. Fertig, D.C. Johnston, L.E. Delong, R.W. McCallum, M.B. Maple and B.T. Matthias, Phys. Rev. Lett., 38 (1977) 987.
7. M. Ishikawa and Ø. Fischer, Solid State Commun., 23 (1977) 37.
8. Ø. Fischer, M. Decroux, S. Roth, R. Chevrel and M. Sergent, J. Phys. C : Solid State Phys., 8 (1975) L474.
9. V. Jaccarino and M. Peter, Phys. Rev. Lett., 9 (1962) 290.
10. H.A. Abrikosov and L.P. Gor'kov, Sov. Phys. - JETP, 12 (1961) 898.
11. C.P. Slichter, Principles of Magnetic Resonance, Harper and Row, New York, 1963, p. 121.
12. R.P. Guertin, Thesis, University of Rochester (unpublished) 1967.

13. G.K. Shenoy, J. Stohr and G.M. Kalvius, Solid State Commun., 13 (1973) 909.
14. J. Stohr, W. Wagner and G.K. Shenoy, Phys. Lett., 47A (1974) 177.
15. W. Wagner, J. Phys. (Paris) Colloq., 37 (1976) C6-133.
16. F. Gonzalez-Jimenez, P. Imbert and F. Hartmann-Boutron, Phys. Rev. B, 9 (1974) 95.
17. G.K. Shenoy and B.D. Dunlap, Phys. Rev. B, 13 (1976) 1353.
18. D. Davidov, R. Orbach, C. Rettori, D. Shaltiel, L.J. Tao and B. Ricks, Phys. Rev. B, 5 (1972) 1711.
19. W. Wagner, G.M. Kalvius and V.D. Gorobchenko, J. Phys. (Paris) Colloq., 41 (1980) C1-243.
20. S. Dattagupta, G.K. Shenoy, B.D. Dunlap and L. Asch, Phys. Rev. B, 16 (1977) 3893.
21. F. Hartmann-Boutron, J. Phys. (Paris), 41 (1980) 1289; W. Wagner, F.J. Litterst, V.D. Gorobchenko, A.M. Afanasev and G.M. Kalvius, J. Phys. C : Solid State Phys. (in print).
22. Ø. Fischer, Appl. Phys., 16 (1978) 1.
23. F.Y. Fradin, G.K. Shenoy, B.D. Dunlap, A.T. Aldred and C.W. Kimball Phys. Rev. Lett., 28 (1977) 719.
24. G.K. Shenoy, B.D. Dunlap, F.Y. Fradin, C.W. Kimball, W. Potzel, F. Probst and G.M. Kalvius, J. Appl. Phys., 50 (1979) 1872.
25. B.D. Dunlap, G.K. Shenoy, F.Y. Fradin, C.D. Barnet and C.W. Kimball, J. Magn. Magn. Mater., 13 (1979) 319.
26. E. Bradford and W. Marshall, Proc. Phys. Soc. (London), 87 (1966) 731.
27. F. Hartmann-Boutron, J. Phys. (Paris) Colloq., 41 (1980) C1-223.
28. S. Oseroff, R. Calvo, D.C. Johnston, M.B. Maple, R.W. McCallum and R.N. Shelton, Solid State Commun., 27 (1978) 201.
29. R. Odermatt, M. Hardiman and J. van Meijel, Solid State Commun., 32 (1979) 1227.
30. T.S. Radhakrishnan, M.P. Janawadkar, S.H. Deware, H.G. Deware, R.G. Pillay, R. Janaki, A.M. Umarji and G.V. Subba Rao, in G.K. Shenoy, B.D. Dunlap and F.Y. Fradin (Eds.), Ternary Superconductors, North-Holland, New York, 1980, p. 91.
31. G. Koopman, U. Engel, K. Baberschke an S. Huffner, Solid State Commun., 11 (1972) 1197.
32. A.J. Freeman and T. Jarlborg, in G.K. Shenoy, B.D. Dunlap and F.Y. Fradin (Eds.), Ternary Superconductors, North-Holland, New York, 1980, p. 59.
33. B.T. Matthias, H. Suhl and E. Corenzwit, Phys. Rev. Lett., 1 (1958) 449.
34. M. Wilhelm and B. Hillenbrand, J. Phys. Chem. Solids, 31 (1970) 559; Physica, 55 (1971) 608; Z. Naturforsch., 26a (1971) 141; Phys. Lett., 31A (1970) 448.
35. L.P. Gor'kov and A.J. Rusinov, Sov. Phys. - JETP, 19 (1964) 922.
36. R.D. Taylor, W.R. Decker, D.J. Erickson, A.L. Giorgi, B.T. Matthias, C.E. Olsen and E.G. Szklarz, in W.J. O'Sullivan, K.D. Timmerhaus and E.F. Hammel (Eds.), Low Temperature Physics, Vol. 2, Plenum Press, New York, p. 605.
37. D.J. Erickson, C.E. Olsen and R.D. Taylor, in I.J. Gruverman and C.W. Seidel (Eds.), Mossbauer Effect Methodology, Vol. 8, Plenum Press, New York, 1973, p. 73.
38. S. Roth, K. Ibel and W. Just, J. Phys. C : Solid State Phys., 6 (1973) 3465.
39. J.W. Lynn, D.E. Moncton, L. Passel and W. Thomlinson, Phys. Rev. B, 21 (1980) 70.
40. J.O. Willis, D.J. Erickson, G.E. Olsen and R.D. Taylor, Phys. Rev. B, 21 (1980) 79.
41. K. Reubenbauer, J. Fink, H. Schmidt, G. Czjzek and K. Tomala, Phys. Status Solidi (b), 84 (1977) 611.
42. S.F. Edwards and P.W. Anderson, J. Phys. F : Met. Phys. 5 (1975) 965

43. P. Steiner, D. Gumprecht and S. Hufner, Phys. Rev. Lett., 30 (1973) 1132; P. Steiner and G. Gmprecht, Solid State Commun., 22 (1977) 501.
44. D. Sherrington & B.W. Southern, J. Phys. F:Met. Phys. 5 (1975) L49.
45. K.H. Fischer, Phys. Rev. Lett., 34 (1975) 1438.
46. J. Bolz, G. Crecelius, H. Maletta and F. Pobell, J. Low Temp. Phys., 28 (1977) 61.
47. P.W. Anderson and H. Suhl, Phys. Rev., 116 (1959) 898.
48. E.I. Blount and C.M. Varma, Phys. Rev. Lett., 42 (1979) 1079.
49. G. Shirane, W. Thomlinson, M. Ishikawa and Ø. Fischer, Solid State Commun., 31 (1979) 1773.
50. W. Thomlinson, G. Shirane, D.E. Moncton, M. Ishikawa and Ø. Fischer, J. Appl. Phys., 50 (1978) 1981.
51. D.E. Moncton, G. Shirane, W. Thomlinson, M. Ishikawa and Ø. Fischer, Phys. Rev. Lett., 41 (1978) 1133.
52. H.R. Ott, W. Odoni, H.C. Hamaker and M.B. Maple, Phys. Lett., 75A (1980) 243.
53. M. Ishikawa, in G.K. Shenoy, B.D. Dunlap and F.Y. Fradin (Eds.), Ternary Superconductors, North-Holland, New York, 1980, p. 43.
54. P.G. deGennes, C.R. Acad. Sci. (Paris), 247 (1958) 1836.
55. P. Fulde and I. Peschel, Adv. Phys., 21 (1972) 1.
56. G.K. Shenoy, B.D. Dunlap and F.Y. Fradin (unpublished).
57. G.K. Shenoy, D. Hinks, P.J. Viccaro, F. Probst (to be published).
58. P. Bonville, J.A. Hodges, P. Imbert, G. Jehanno, R. Chevrel and M. Sergent, Rev. Phys. Appl., 15 (1980) 1134.
59. B.D. Dunlap and G.K. Shenoy, Proc. Int. Conf. on Hyperfine Interactions, Berlin, 1980, Hyperfine Interactions (in press).
60. G.K. Shenoy, P.J. Viccaro, D. Niarchos, J.D. Cashion, B.D. Dunlap and F.Y. Fradin in G.K. Shenoy, B.D. Dunlap and F.Y. Fradin (Eds.), Ternary Superconductors, North-Holland, New York, 1980, p. 163.
61. B.T. Matthias, V.B. Compton and E. Corenzwit, J. Phys. Chem. Solids, 19 (1961) 130.
62. B.S. Chandrasekhar and J.K. Hulm, J. Phys. Chem. Solids, 7 (1958) 259.
63. E.E. Havinga, H. Damsma and J.M. Kanis, J. Less-Common Metals, 27 (1972) 281.
64. H.F. Braun, Phys. Lett., 75A (1980) 386.
65. J.D. Cashion, G.K. Shenoy, D. Niarchos, P.J. Viccaro and C.M. Falco, Phys. Lett., 79A (1980) 454.
66. J.D. Cashion, G.K. Shenoy, D. Niarchos, P.J. Viccaro, A.T. Aldred C.M. Falco, J. Appl. Phys., 52 (1981) 2180.
67. H.F. Braun, F. Acker and C.U. Segre, Bull. Amer. Phys. Soc., 25 (1980) 232.
68. C.U. Segre and H.F. Braun (private communication).
69. W. Honle, H.G. von Schnering, A. Lipka and K. Yvon, J. Less-Common Metals, 71 (1980) 135.
70. J.D. Cashion, G.K. Shenoy, D. Niarchos, P.J. Viccaro and C.M. Falco in E.N. Kaufmann and G.K. Shenoy (Eds.), Nuclear and Electron Resonance Spectroscopies as Applied to Materials Science, North-Holland, New York, 1981.
71. G.K. Shenoy, P.J. Viccaro, J.D. Cashion, D. Niarchos, B.D. Dunlap, F. Probst and J.P. Remeika in G.K. Shenoy, B.D. Dunlap and F.Y. Fradin (Eds.), Ternary Superconductors, North-Holland, New York, 1980, p. 233.
72. J.M. Vandenberg & B.T. Matthias, Science, 198 (1977) 194.
73. G.K. Shenoy, D. Niarchos, C.V. Segre & H.F. Braun (to be published)
74. J.P. Remeika, G.P. Espinosa, A.S. Cooper, H. Barz, J.M. Rowell, D.B. McWhan, J.M. Vandenberg, D.E. Moncton, Z. Fisk, L.D. Woolf, H.C. Hamaker, M.B. Maple, G. Shirane and W. Thomlinson, Solid State Commun., 34 (1980) 923.

75. G.K. Shenoy, B.D. Dunlap, F.Y. Fradin, S.K. Sinha, C.W. Kimball, W. Potzel, F. Probst and G.M. Kalvius, Phys. Rev. B, 21 (1980) 3886
76. G.K. Shenoy, F. Probst, J.D. Cashion, P.J. Viccaro, D. Niarchos, B.D. Dunlap and J.P. Remeika, Solid State Commun., 37 (1980) 53.
77. D.E. Moncton, D.B. McWhan, J. Eckert, G. Shirane and W. Thomlinson, Phys. Rev. Lett., 39 (1977) 1164.
78. S.K. Sinha, H.A. Mook, D.G. Hinks and G.W. Crabtree, Bull. Amer. Phys. Soc., 26 (1981) 277.

CHAPTER 11

STOCHASTIC THEORY OF RELAXATION EFFECTS ON MOSSBAUER LINESHAPE

S. Dattagupta*

Reactor Research Centre, Kalpakkam 603102, India

Hahn-Meitner-Institut fur Kernforschung Berlin GmbH
D-1000 Berlin 39, West Germany†

1. INTRODUCTION

The study of Mossbauer effect in solids and viscous liquids involves in general a many-body system of which the resonating nucleus forms a constituent. The properties, static as well as dynamic, of the various kinds of interactions in the system affect the resonance emission or absorption of γ-rays by the nucleus. An examination of the Mossbauer lineshape can therefore reveal important knowledge of such properties.

The theoretical study of the lineshape is a complicated one involving, as in other problems of condensed matter physics, many degrees of freedom. The latter include that of the nucleus itself plus the degrees of freedom of its environment consisting of other nuclei, electrons, phonons, etc. For an approximate evaluation of the Mossbauer lineshape, one may, in principle, start with the complete Hamiltonian for the entire system and attempt to develop suitable perturbation theories, as are common in other branches of many-body physics. Such *ab initio* formalism is however difficult to carry out successfully in many cases. An alternative approach, which has been found to be rather useful in recent years in dealing especially with relaxation effects, is to recognise that the fluctuations in the surroundings of nucleus produce explicitly time-dependent interactions for the latter. In this the so-called stochastic theory one constructs a randomly time-dependent Hamiltonian, modelled in such a way as to incorporate the fluctuations in the surroundings that are *relevant* for the probe nucleus. With this approximate form of the Hamiltonian and suitable assumptions regarding the nature of the fluctuations, one tries to give an exact solution for the lineshape.

---

† Address during the summer of 1979.

* Present Address: School of Physics, University of Hyderabad, Central University P.O., Hyderabad-500 134, India.

The aim of the present article is to describe the stochastic theory of lineshape and its applications. Our objective is to stress not so much on the mathematical formalism but on the underlying physical concepts. In the stochastic approach, a few terms in the total Hamiltonian, which describe what we shall call the *subsystem*, are treated exactly. The rest of the terms are lumped into what is fashionably known as the *heat bath*. The role of the latter is to drive thermal fluctuations into the subsystem. The heat bath, for most purposes, can be treated entirely classically. The spirit behind the stochastic theory of Mossbauer lineshape is therefore not dissimilar to the concepts extant in irreversible statistical mechanics. Also, the very idea of replacing the surroundings of a "tagged" particle (in this case, the nucleus) by a fluctuating heat bath is quite akin to the widely studied theory of the Brownian motion.

The stochastic theory of Mossbauer lineshape has evolved in the following sequence. Blume (1), in first applying the ideas of Anderson and Kubo (2), developed in connection with magnetic resonance studies, considered the case in which the characteristic frequencies of the radiating system undergo random modulations due to interactions with the surroundings. In this theory, the nucleus itself constitutes the subsystem and the interactions in the subsystem, like those in the heat bath, are also treated classically. The need to describe the nuclear interactions quantum mechanically was later recognized by Blume and Tjon, who had performed a quantum generalisation of the Anderson-Kubo model (3,4). A more complicated situation may arise, however, in paramagnetic systems. Here the surroundings do not interact directly with the nucleus but do so with the ion that contains the nucleus. The latter feels the presence of the surroundings via its magnetic hyperfine interaction with the ionic spin. In such cases, the subsystem consists of the coupled nucleus-ion system, which has to be treated exactly and quantum mechanically, while the interaction between the ionic spin and its surroundings may be taken into account by classical fluctuating fields. A stochastic theory of this type was presented by Clauser and Blume (5) which was later extended by this author (6) to cover situations in which the rate of fluctuations in the heat bath may be comparable to the hyperfine frequencies.

In presenting lineshape expressions in this article, we deviate from the historical sequence in which the stochastic theory has developed. Instead we give a comprehensive treatment which includes the essential results of all the earlier models. Each model is then discussed in terms of a physical example which helps illustrate the usefulness of the study of Mossbauer lineshape in various problems of applied science.

As mentioned earlier, there exist also numerous *ab initio* approaches in which the lineshape is calculated within perturbation theory approximations (7-15). While the stochastic treatments have the distinct advantage of not relying on perturbation-theory arguments and the ability to describe in a natural way the problems related to diffusion (see Sections 5, 8, 9 and 10), the *ab initio* calculations, on the other hand, have the merit of treating the interaction between the ion and its surroundings on a microscopic basis. Thus the two approaches are complementary to each other and this aspect can be readily seen in the brief description in Section 7 of Mossbauer spectra in paramagnets.

Until now in our discussion of time-dependent effects on Mossbauer lineshape, we have regarded the Mossbauer nucleus to be stationary; relaxation is caused by its surroundings. The stochastic model discussed above can, however, be generalized to include the case in which the nucleus itself is in a state of motion during the emission or absorption of a γ-ray (16). This motion may occur, due to thermal agitation, in the form of solid-like jump diffusion or liquid-like continuous diffusion or a combination of both, and generally leads to a broadening of the lines. On the other hand, each diffusive step may take the nucleus to a new environment which can give rise to relaxation effects. This interplay of relaxation and diffusion effects on the Mossbauer lineshape is the subject of Sections 8-10.

In Section 11, we conclude with a brief summary of the scope of the article.

2. EXPRESSION FOR THE LINESHAPE

We follow here the derivation of Blume and Tjon (3). The probability of emission of a photon with wave vector $\vec{k}$ and frequency ω by the system, which in the process of emitting the photon makes a transition from the initial state $|i\rangle$ to its final state $|f\rangle$, is given by the Wigner-Weisskopf expression:

$$F_{if}(\omega,\vec{k}) = \frac{|\langle f|A|i\rangle|^2}{(\omega + E_f - E_i)^2 + \frac{1}{4}\Gamma^2} , \tag{1}$$

where A is the interaction between the nucleus and the electromagnetic field of the photon. The states $|i\rangle$ and $|f\rangle$ represent, in general, eigenstates of the entire many-body system including nuclear spin quantum numbers, electronic quantum numbers, etc. and, E_i and E_f are the respective eigenvalues of the total Hamiltonian H. The quantity Γ, which is the inverse of the natural lifetime of the excited state $|i\rangle$, is assumed to be independent of the particular sublevel of the excited state. (We work here in units of $\hbar=1$).

Following a procedure due to Van Hove (17), Eqn. (1) can be cast into the form:

$$F_{if}(\omega,k) = \frac{2}{\Gamma} \mathrm{Re} \int_0^{\infty} dt e^{i\omega t - \frac{\Gamma}{2}t} \langle i|A^{\dagger}|f\rangle\langle f|e^{i\mathsf{H}t} Ae^{-i\mathsf{H}t}|i\rangle, \tag{2}$$

where the symbol "Re" indicates that only the real part of the expression is to be considered. In the lineshape problem, the interactions with the surroundings cause perturbations in both the excited and ground states of the nucleus. It is therefore natural to deal with the Liouville operator associated with H (and denoted by $\mathsf{H}^{\times}$) whose eigenvalues give the possible transition frequencies of the system (18). In terms of the Liouville operator, the time-development of A in the Heisenberg picture can be expressed as:

$$A(t) = e^{i\mathsf{H}t} Ae^{-i\mathsf{H}t} = (e^{i\mathsf{H}^{\times}t} A). \tag{3}$$

For our purposes, we need just the following few properties of the Liouville operator associated with any ordinary operator B:

$$\langle f|(e^{iB^{\times}t} A)|i\rangle = \sum_{i'f'} (fi|e^{iB^{\times}t}|f'i')\langle f'|A|i'\rangle, \tag{4}$$

where

$$(fi|e^{iB^{\times}t}|f'i') = \langle f|e^{iBt}|f'\rangle\langle i'|e^{-iBt}|i\rangle. \tag{5}$$

The matrix elements of a Liouville operator are thus given in terms of four indices (with "states" denoted by round brackets). Also,

$$(fi|B^{\times}|f'i') = \langle f|B|f'\rangle\, \delta_{ii'} - \langle i'|B|i\rangle\, \delta_{ff'}. \tag{6}$$

We shall introduce further

$$U(t) = e^{i\mathsf{H}^{\times}t}\,, \tag{7}$$

where U(t) is the time-development operator for the entire system. Next we note that the experimentally observed emission probability is simply obtained by averaging Eqn. (2) over all possible initial states $|i\rangle$ and summing over all final states $|f\rangle$ of the emitter. Together with Eqn. (7) we then have:

$$F(\omega,\vec{k}) = \sum_{if} p_i W_{if}(\omega,\vec{k})$$
$$= \frac{2}{\Gamma} \mathrm{Re} \int_0^\infty dt\, e^{i\omega t - \frac{\Gamma}{2}t} \langle A^\dagger (U(t)A)\rangle, \tag{8}$$

where p_i is the probability that the initial state $|i\rangle$ occurs, and the average is defined by:

$$\langle O\rangle = \sum_i p_i \langle i|O|i\rangle = \mathrm{Tr}(\rho O), \tag{9}$$

ρ being the density matrix for the entire system.

In the theory of lineshape, it is assumed that the coupling between the subsystem and the heat bath is much weaker than the thermal energy $k_B T$ (k_B=Boltzmann's constant). In that case the statistical average indicated by $\langle ...\rangle$ in Eqn. (8) factors:

$$\langle A^\dagger (U(t)A)\rangle \approx \mathrm{Tr}_o\{\rho_o\, A^\dagger (U(t))_{av} A\}, \tag{10}$$

where Tr_o indicates a trace over the variables of the subsystem alone described by the density matrix ρ_o, and $(...)_{av}$ denotes the average over the heat bath properties.

3. STOCHASTIC THEORY

We adopt a physical picture in which the subsystem is assumed to be subject to fluctuations due to its coupling with the surroundings which make the Hamiltonian jump between different forms (5,6). The latter can be written as:

$$\mathcal{H}(t) = \mathcal{H}_o + \sum_{j=1}^{n} V_j f_j(t), \tag{11}$$

where $\mathcal{H}_o$ is the static part of the Hamiltonian, V_j are operators belonging to the subsystem, and f_j are random functions of time which can be chosen to take different values corresponding to the various forms of the Hamiltonian. The meaning of these terms will be made clear in terms of the physical examples to be discussed in the subsequent sections. For the present it is important, however, to note that in general,

$$[\, H_o, V_j \,] \neq 0, \; [\, V_j, V_{j'} \,] \neq 0, \quad j \neq j' \quad . \tag{12}$$

The model can be summarized as follows. At random instants of time the subsystem is subject to s pulses which instantaneously change the form of the Hamiltonian [See Eqn. (16) below]. The pulses are assumed to be randomly distributed with a Poisson distribution (19), and in between pulses, the subsystem is assumed to be unperturbed. In the linear vector space spanned by the values of f_j's, the time development operator [cf., Eqn. 7] can be viewed as a matrix which is constructed as:

$$U(t) = e^{iH^{\times}t_1}\, \mathcal{J}\, e^{iH^{\times}(t_2-t_1)}\, \mathcal{J} \ldots \mathcal{J}\, e^{iH^{\times}(t-t_s)} \quad , \tag{13}$$

where $\mathcal{J}$ is a transition probability-matrix,

$$H^{\times} = H^{\times} + \sum_j V_j^{\times} F_j , \tag{14}$$

F_j being a diagonal matrix in the vector space referred to in the above, whose elements are the possible values of f_j:

$$(a|F_j|b) = \delta_{aj}\delta_{ab} . \tag{15}$$

From Eqn. (13) to (15), the matrix elements of U(t) can be written as:

$$(a|U(t)|b) = \sum_{cde\ldots} e^{i(H_o^{\times}+V_a^{\times})t_1}\, \mathcal{J}_{ac}\, e^{i(H_o^{\times}+V_c^{\times})(t_2-t_1)}\, \mathcal{J}_{cd}$$

$$\times\, e^{i(H_o^{\times}+V_d^{\times})(t_3-t_2)} \ldots \mathcal{J}_{eb}\, e^{i(H_o^{\times}+V_b^{\times})(t-t_s)} \quad . \tag{16}$$

Equation (16) has the following interpretation. At t = 0, the subsystem "sees" a Hamiltonian $H_o^{\times}+V_a$ in the stochastic state |a). The quantum-mechanical state of the subsystem then evolves in time until t_1 with the appropriate time-development operator $\exp[i(H_o^{\times}+V_a^{\times})t_1]$. At t_1, a pulse hits the subsystem which has a probability $\mathcal{J}_{ac}$ of throwing it (instantaneously) into a stochastic state |c) governed by the Hamiltonian $H_o^{\times}+V_c$. The state then develops in time under $H_o^{\times}+V_c$ until t_2 at which point another pulse makes it jump into a stochastic state |d), and so on. Since all possible intermediate stochastic states have to be considered, we sum over the variables c, d, e..... Note that the state $|e)$ is the one in which the system is found immediately prior to the pulse at t_s.

The average we seek in the lineshape expression cf., Eqn. 10 is obtained by summing over the final stochastic states $|b)$, averaging over the initial states $|a)$ and averaging over the type and location (in time) of the pulses. Noting that for a Poisson system, the probability of the occurrence of a pulse during the interval t and t+dt is given by $\nu e^{-\nu t}$ dt, where ν^{-1} is the mean time between pulses (19), we have from Eqn. (13),

$$(U(t))_{av} = \sum_{s=0}^{\infty} \int_0^t dt_s \int_0^{t_s} dt_{s-1} \ldots \int_0^{t_2} dt \sum_{1ab} p_a (a| e^{-(\nu - iH^{\times})t_1} (\nu J)$$

$$\times\ e^{-(\nu - iH^{\times})(t_2 - t_1)} (\nu J) \ldots (\nu J)\ e^{-(\nu - iH^{\times})(t - t_s)} |b), \tag{17}$$

where p_a is the *a priori* probability of the occurrence of the initial stochastic state $|a)$.

Using the convolution theorem, the Laplace transform of Eqn. (17) can be written as:

$$(\tilde{U}(p))_{av} = \sum_{ab} p_a (a| \{\sum_{s=0}^{\infty} \frac{1}{p+\nu-iH^{\times}} (\nu J \frac{1}{p+\nu-iH^{\times}})^s\} |b)$$

$$= \sum_{ab} p_a (a| (p - iH^{\times} - i\sum_j V_j^{\times} F_j - W)^{-1} |b), \tag{18}$$

using Eqn. (14) and introducing

$$W = \nu(J-1). \tag{19}$$

The Mossbauer lineshape can finally be expressed as [cf., Eqn. 8 and 10]:

$$F(p,\vec{k}) = \mathrm{Re}\ \mathrm{Tr}_o \{\rho_o A^{\dagger} [(\tilde{U}(p))_{av} A]\}, \tag{20}$$

where $(\tilde{U}(p))_{av}$ is given by Eqn. (18) and the Laplace transform variable is defined by,

$$p = -i\omega + \frac{\Gamma}{2}. \tag{21}$$

The central result of the lineshape theory, as given by Eqn. (18) above, can also be obtained via a different derivation due to Blume (4).

3.1 Meaning of the Transition Matrix W

In order to give a physical interpretation to the matrix W, we wish to investigate the structure of $(a|P(t)|b)$ which is defined as the conditional probability that the stochastic state is $|b)$ at time t, given that the state was $|a)$ at $t = 0$. In terms of the model presented in Section 3, it is easy to see that the P-matrix is given by:

$$P(t) = \sum_{s=0}^{\infty} \int_0^t dt_s \ldots \int_0^{t_2} dt_1\ e^{-\nu t_1} (\nu J) e^{-\nu(t_2 - t_1)} \ldots (\nu J) e^{-\nu(t-t_s)}$$

$$= \exp(Wt). \tag{22}$$

where W is defined in Eqn. (19).

Equation (22) is the solution of the Chapman-Kolmogorov equation which defines a stationary Markov process (20):

$$(a|P(t)|b) = \sum_c (a|P(J)|c)(c|P(t-J)|b). \tag{23}$$

Since Eqn. (22) yields

$$W = \left(\frac{dp}{dt}\right)_{t=0} , \tag{24}$$

$(a|W|b)$ defines the probability per unit time of an instantaneous jump of the stochastic state from $|a)$ to $|b)$.

The conservation of total probability leads to

$$\sum_b (a|W|b) = 0. \tag{25}$$

Furthermore, for most applications, the heat bath is assumed to be in thermal equilibrium which implies detailed balance of transitions:

$$p_a(a|W|b) = p_b(b|W|a) , \tag{26}$$

where p_a is defined following Eqn. (17) (p_a, for example, may be the Boltzmann factor).

It is evident from the present discussion that the stochastic model of Section 3 is based on the tacit assumption that the fluctuations in the heat bath are governed by a stationary Markov process. Indeed this assumption is the major input in most stochastic theories of Mossbauer line shape treated to date.

4. THE CLASSICAL ANDERSON-KUBO-BLUME MODEL

Consider the problem in which an ^{57}Fe Mossbauer nucleus (with an excited state of spin $I_1 = 3/2$ and a ground state of spin $I_0 = 1/2$) finds itself in a static electric field gradient along the Z axis and a fluctuating magnetic field which jumps at random between the values $\pm h$, also along the Z axis (3). The Hamiltonian can be written in the form of Eqn. (11) with

$$n=2,\ \mathcal{H}_0=Q(3I_z^2 - I^2),\ V_1=-V_2=g\mu hI_z,\ f_1=\tfrac{1}{2}(1+f),\ f_2=\tfrac{1}{2}(1-f), \tag{27}$$

where f(t) is a random function of time, which takes on the values ± 1; Q is the quadrupole moment of the nucleus, g its g-factor and μ is the nuclear magneton. This model describes in an approximate way the nucleus in the presence of relaxing electronic spins.

We note that in the present case, the nucleus itself is taken to constitute the subsystem. Also the two different forms of the Hamiltonian commute with each other. In particular, if the Z axis is chosen as the axis of quantisation, the two forms of the Hamiltonian remain diagonal at all times. This implies that the Liouville operators $\mathcal{H}_0^{\times}$ and $V_j^{\times}$ can simply be replaced by the corresponding eigenvalues which are just the characteristic frequencies of transitions between the various nuclear levels. The solution for the lineshape as given in Eqn. (18) is therefore completely equivalent to the random frequency modulation theory of Anderson and Kubo, introduced by them in magnetic resonance studies (2) and adapted to the Mossbauer case by Blume (1).

To illustrate how the lineshape theory of Section 3 can be applied to the present situation, we note that associated with two forms of the Hamiltonian, the stochastic states can be represented by $|1)$ and $|-1)$, and in the vector space spanned by these states, F_j's are 2 x 2 matrices [cf., Eqn. (15) and (27)]:

$$F_1 = \begin{bmatrix} 1 & 0 \\ 0 & 0 \end{bmatrix}, \qquad F_2 = \begin{bmatrix} 0 & 0 \\ 0 & 1 \end{bmatrix}. \tag{28}$$

Also, if the magnetic field is assumed to jump between the values +h and -h with equal probability, then the transition matrix can be written as:

$$W = \begin{bmatrix} -\lambda & \lambda \\ \lambda & -\lambda \end{bmatrix}, \qquad (29)$$

where λ is the rate of jump. Notice that the diagonal elements of the W-matrix are constructed from the condition given by Eqn. (25). Further,

$$p_a = \frac{1}{2}, \quad a = 1,2, \qquad (30)$$

so that the detailed balance relation (26) is trivially satisfied.

Assuming that the sublevels in the excited state of the nucleus are equally populated thermally, the Mossbauer lineshape can be expressed, by writing out the trace in Eqn. (20), as:

$$F(p,\vec{k}) = (2I_1+1)^{-1} \mathrm{Re} \sum_{M_1,M_0} <M_1|A^{\dagger}|M_0><M_0|[(\tilde{U}(p))_{av}\ A]|M_1>, \qquad (31)$$

where M_1 and M_0 are the angular momentum quantum numbers in the excited and ground states of the nucleus, respectively. Since $(\tilde{U}(p))_{av}$ is a Liouville operator in the nuclear angular momentum space [cf., Eqn. (18)], we can write Eqn. (31), using Eqn(4), as:

$$F(p,\vec{k}) = (2I_1+1)^{-1} \times \mathrm{Re} \sum_{M_1,M_0} |<M_1|A^{\dagger}|M_0>|^2\ (M_0M_1|(\tilde{U}(p))_{av}|M_0M_1), \qquad (32)$$

where use has been made of the fact that $(U(p))_{av}$ is diagonal [cf., Eqn. (27)].

Next we substitute Eqn.(18 and 27) into Eqn. (32), employ the property of Liouville operators [cf., Eqn. (6)]and make use of the fact that the quadrupolar interaction is zero in the ground state of ^{57}Fe. We thus obtain:

$$F(p) = \mathrm{Re} \sum_{M_1M_0} |<M_1|A^{\dagger}|M_0>|^2$$

$$\times \frac{1}{2} \sum_{ab} (a|[p+i\alpha(M_1)-i\beta(M_0,M_1)(F_1,F_2)-W]^{-1}|b), \qquad (33)$$

where we have dropped the index k and the factor $(2I_1+1)$. In Eqn. (33),

$$\alpha(M_1) = Q(3M_1^2 - \frac{15}{4}),$$

$$\beta(M_0,M_1) = \mu(g_0M_0 - g_1M_1)h, \qquad (34)$$

g_o and g_1 being the respective g-factors in the ground and excited states of the nucleus.

The matrix elements $\langle M_1|A^{\dagger}|M_o\rangle$ are essentially proportional to Clebsch-Gordan coefficients which determine the intensity and polarisation of the individual lines (21). The lineshape is then obtained from Eqn. (33) by inverting a 2x2 matrix, having inserted Eqn. (28) and (29). The result can be written in a simple analytic form:

$$F(p) = \mathrm{Re}\sum_{M_1 M_o} |\langle M_1|A^{\dagger}|M_o\rangle|^2 [p+i\alpha + \frac{\beta^2}{(p+i\alpha+2\lambda)}]^{-1} , \qquad (35)$$

where the dependence of α and β on the angular momentum indices has been suppressed, for the sake of brevity.

Equation (35) can be discussed in physical terms in the two extreme cases of very slow ($\lambda \simeq o$) and very rapid ($\lambda >> \beta$) relaxations. For $\lambda \simeq o$,

$$F(p) \simeq \mathrm{Re}\sum_{M_1 M_o} |\langle M_1|A^{\dagger}|M_o\rangle|^2 \left(\frac{1/2}{p+i(\alpha+\beta)} + \frac{1/2}{p+i(\alpha-\beta)}\right). \qquad (36)$$

In this case the static spectrum is obtained, and if the magnetic interaction is much stronger than the quadrupolar interaction (i.e., $h >> Q$), the full six-line Zeeman pattern for ^{57}Fe is seen (see Fig. 1).

On the other hand, when λ is very large, Eqn. (35) yields:

$$F(p) \simeq \mathrm{Re}\sum_{M_1 M_o} |\langle M_1|A^{\dagger}|M_o\rangle|^2 (p+i\alpha)^{-1}. \qquad (37)$$

In this rapid regime of relaxation, the nucleus "sees" a zero time-averaged magnetic field (Fig. 1). The magnetic splitting then collapses with the emergence of a symmetric quadrupole doublet for $\lambda/\Gamma > 5\times 10^3$ (the emitter is assumed to be a powder).

The behaviour of the lineshape in the intermediate regimes of relaxation is demonstrated by the plots in Fig. 1, following Reference (3). Since the β-values are different for different transitions [cf., Eqn. 34], the collapse of the magnetic structure, as determined by the factor $\beta^2/(p+i\alpha+2\lambda)$ in Eqn. (35), proceeds at different rates. This explains the asymmetry of the pattern for intermediate values of λ.

5. NONSECULAR EFFECTS

Theories such as the one described in Section 4 in which one

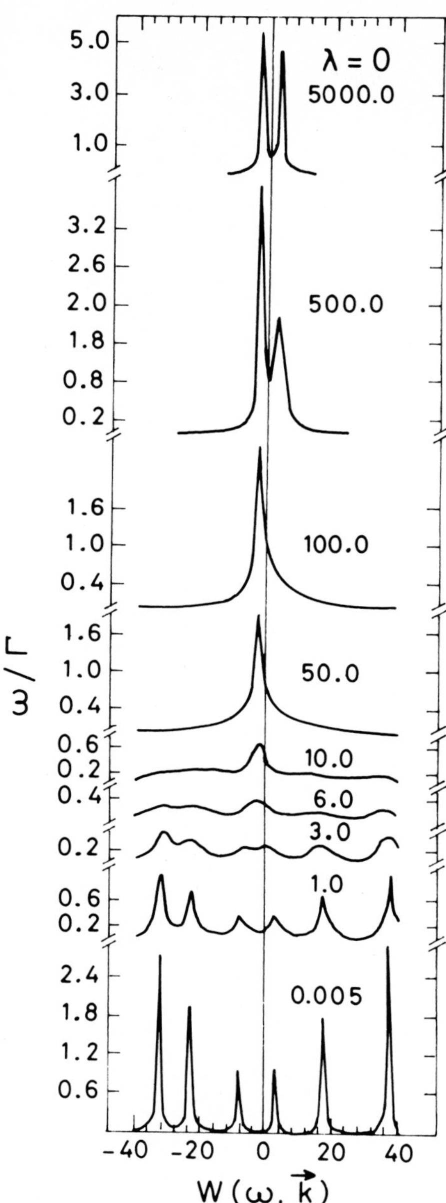

Fig. 1. Mossbauer lineshapes from reference (3) with the Hamiltonian of Eqn. (27), for different values of the relaxation rate λ. The values of the quadrupole coupling constant Q and the effective magnetic field h are taken as Q = 0.8736 and h = 143.67, in units of Γ, the natural linewidth. Also $I_1=3/2$, $I_0=1/2$, $g_1=0.102$ and $g_0= -0.806$, as for ^{57}Fe.

deals with simultaneously diagonalisable forms of the Hamiltonian are known as secular theories. Similar stochastic models restricted to secular effects in Mossbauer lineshape have been employed by van der Woude and Dekker (22), Wegener (23), Wickman, Klein and Shirley (24), and Boyle and Gabriel (25). In many examples of Mossbauer studies, however, the different Hamiltonian forms which the nucleus experiences do not commute with one another. Such nonsecular effects were treated within stochastic theory by Blume (4), and several examples were analysed (3, 21, 26). The Blume solution is given by Eqn. (18) where the subsystem again is just the nucleus, but unlike in the previous example, the restrictions given by Eqn. (12) apply now. We illustrate how nonsecular effects come into play by means of a metallurgical application in Section 5.1.

5.1 Diffusion of Interstitials in Austenite

The Mossbauer lineshape of ^{57}Fe in austenitic steel is influenced by the presence of interstitials (C or N) in the following manner (26). When the interstitial is at a site which is a nearest neighbour to the ^{57}Fe nucleus, an electric field gradient is created which causes a quadrupole splitting. In addition to this, there is also an isomer shift (27). However, the isomer shift when the interstitial is a nearest neighbour to the γ-emitting nucleus is different from the isomer shift when it is not a nearest neighbour. The quadrupole splitting and the isomer shift are the only two effects to be considered because in the austenitic phase there is no magnetic hyperfine splitting of the Mossbauer spectra.

With the increase of temperature, the interstitials start to diffuse from site to site. As an interstitial jumps among the six nearest neighbour positions of an ^{57}Fe nucleus (since the concentrations of both interstitials and radioactive ^{57}Fe atoms are rather low, the probability of the presence of more than one interstitial or more than one radioactive atom in one unit cell can be neglected), the electric field gradient at the nucleus changes direction abruptly between $\pm x$, $\pm y$ and $\pm z$ axes (see Fig. 2). The Hamiltonian can be constructed as:

$$\mathcal{H}(t) = \frac{1}{4}(1-f)(4-f^2)\mathcal{H}_1 + \frac{1}{3}f(4-f^2)[\mathcal{H}_0+Q(3I_z^2-I^2)]$$
$$+ \frac{1}{24}f(f+2)(f^2-1)[\mathcal{H}_0+Q(3I_x^2-I^2)]+ \frac{1}{24}f(f-2)(f^2-1)[\mathcal{H}_0+Q(3I_y^2-I^2)], \quad (38)$$

where $\mathcal{H}_0$ and $\mathcal{H}_1$ are the interactions which produce the isomer shift, when the interstitial is a nearest neighbour and when it is not a nearest neighbour, respectively. The random function f(t) is now chosen to jump between the values 0, 1, 2 and -2 (cf., Section 4).

A slight rearrangement enables us to cast Eqn. (38) exactly into the form given by Eqn. (11) in which:

$$n = 3,\ V_1 = \mathcal{H}_1-\mathcal{H}_0,\ f_1 = \frac{1}{4}(1-f)(4-f^2),$$
$$V_2 = Q(3I_z^2-I^2),\ f_2= \frac{1}{24}f[8(4-f^2)-f(f^2-1)], \quad (39)$$
$$V_3 = Q(I_+^2 + I_-^2),\ f_3 = \frac{1}{8}f(f^2-1),$$

where I_+ and I_- are raising and lowering angular momentum operators (28). Although $\mathcal{H}_0^\times$ and $\mathcal{H}_1^\times$ are c-numbers, V_2 and V_3 are operators which do not commute with each other. In particular, V_3 can cause transitions between the eigenstates of V_2 (nonsecular effects).

The Mossbauer lineshape is again given by Eqn. (18) and (20), but now, since the quadrupolar interaction is zero in the ground state, $(\tilde{U}(p))_{av}$ is just an ordinary operator (not a Liouville operator)

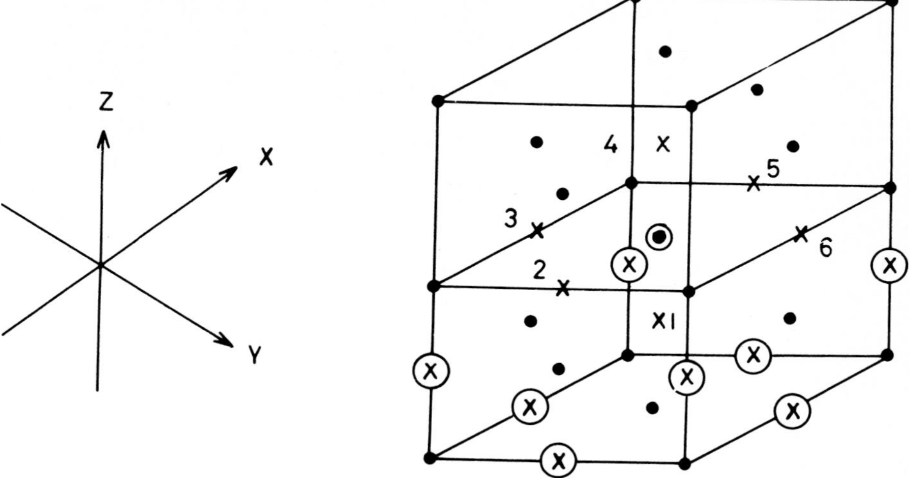

Fig. 2: The f.c.c. structure of austenite ●=Fe atoms; ⊙=radioactive Mossbauer atom; X=interstitial sites which are nearest neighbour to the Mossbauer atom; and ⊗=interstitial sites which are not nearest neighbour to the Mossbauer atom. Here by "radioactive Mossbauer atom" we mean an ^{57}Fe atom which has absorbed 14.4 keV gamma rays (incident on it from the ^{57}Co Mossbauer source) and is in the excited state. All other Fe atoms, shown by ●, are the ^{56}Fe atoms and thus are not Mossbauer atoms.

which acts only on the excited state of the nucleus. We have:

$$F(p) = \mathrm{Re} \sum_{M_0 M_1 M_1'} \langle M_1|A^\dagger|M_0\rangle\langle M_0|A|M_1'\rangle$$

$$\times\langle M_1|\{\Sigma_{ab} p_a (a|[p+i\delta_0+i(\delta_1-\delta_0)F_1+iV_2F_2+iV_3F_3-W]^{-1}|b)\}|M_1'\rangle, \qquad (40)$$

where δ_0 and δ_1 are the isomer shifts associated with H_0 and H_1, respectively.

The F-matrices can be written as (cf., Eqn. (39)):

$$F_1 = \begin{bmatrix} 1 & 0 & 0 & 0 \\ 0 & 0 & 0 & 0 \\ 0 & 0 & 0 & 0 \\ 0 & 0 & 0 & 0 \end{bmatrix}, \quad F_2 = \begin{bmatrix} 0 & 0 & 0 & 0 \\ 0 & 1 & 0 & 0 \\ 0 & 0 & -\frac{1}{2} & 0 \\ 0 & 0 & 0 & -\frac{1}{2} \end{bmatrix}, \quad F_3 = \begin{bmatrix} 0 & 0 & 0 & 0 \\ 0 & 0 & 0 & 0 \\ 0 & 0 & \frac{3}{4} & 0 \\ 0 & 0 & 0 & -\frac{3}{4} \end{bmatrix} \qquad (41)$$

while the transition matrix has the form:

$$W = \begin{bmatrix} -3\lambda'' & \lambda'' & \lambda'' & \lambda'' \\ \lambda' & -(2\lambda+\lambda') & \lambda & \lambda \\ \lambda' & \lambda & -(2\lambda+\lambda') & \lambda \\ \lambda' & \lambda & \lambda & -(2\lambda+\lambda') \end{bmatrix} \qquad (42)$$

where λ is the probability per unit time that the interstitial jumps from one nearest neighbour site to another, λ' is the probability per unit time of jumps from a nearest neighbour to a non-nearest neighbour site, and λ'' that for the reverse hops. As stated earlier, the diagonal elements of W are constructed using the probability conservation condition given by Eqn. (25).

5.1.1 Relations between λ, λ' and λ''

The rate constants λ, λ' and λ'' are not independent as can be seen through the following considerations. Since austenite has a face-central cubic (f.c.c.) structure (Fig. 2), if the interstitial is at one of the six nearest neighbour positions, say that corresponding to f = 1, (labelled 1 in Fig. 2), it can jump to 12 sites (assuming jumps over nearest neighbour distance only). Of these 12 sites, only four (labelled 2, 3, 5 and 6 in Fig. 2) are nearest neighbours to the Mossbauer nucleus (indicated by the encircled dot in Fig. 2) which produces an electric field gradient along x, -x, y or -y direction. However, field gradients along x and -x cause the same quadrupolar interaction. Thus:

$$\lambda = \frac{2}{12}\nu = \frac{1}{6}\nu, \qquad (43)$$

where ν^{-1} is the mean time between successive jumps. Similarly

$$\lambda' = \frac{8}{12}\nu = 4\lambda. \qquad (44)$$

To find λ'', it is necessary to realise that the stochastic state for which f = 0 represents *any* non-nearest neighbour site. But a jump from *any* non-nearest neighbour site does not necessarily take the interstitial to a nearest (say that marked 1 in Fig. 2) neighbour site. This can happen only if the interstitial came from one of the eight neighbouring sites (encircled crosses in Fig. 2) each of which has an occupational probability equal to X, the concentration of interstitials. Therefore,

$$\lambda'' = \chi \times 8 \times \frac{2}{12} \nu = 8\chi\lambda \tag{45}$$

The *a priori* probabilities p_a can be determined from the detailed balance requirement given by Eqn. (26) and the relation,

$$\sum_{a=1}^{4} p_a = 1, \tag{46}$$

so that:

$$p_1 = \frac{\lambda'}{\lambda'+3\lambda''}, \quad p_2 = p_3 = p_4 = \frac{\lambda''}{\lambda'+3\lambda''} . \tag{47}$$

The p's can also be expressed in terms of X using Eqn. (44) and (45).

5.1.2 Analysis of the Lineshape

In this case since $[V_2, V_3] \neq 0$, the stochastic and quantum aspects of the problem do not separate in contrast to the example of Section 4. The solution, as given by Eqn. (40), involves the inversion of a 16x16 matrix corresponding to the fact that there are four possible stochastic variables and four angular momentum indices for the nucleus in its excited state. It is possible, however, to obtain an analytic solution in this case [for details see Reference 26], which can be written as:

$$F(p) = \mathrm{Re}\ \frac{1}{\lambda'+3\lambda''}\ \frac{\lambda'+[\lambda'(2\lambda''-\lambda)+\lambda''(p+i\delta_1+3\lambda'')]g}{(p+i\delta_1+3\lambda'')(1-\lambda g) - \lambda'\lambda'' g}, \tag{48}$$

where

$$g = 3(p+i\delta_o+\lambda'+3\lambda)[(p+i\delta_o+\lambda'+3\lambda)^2 + 9Q^2]^{-1}. \tag{49}$$

In order to gain some feeling for the results embodied in Eqn. (48) and (49), we discuss below a few special cases (which may not, however, correspond to any real situation in austenite).

(1) $\lambda' = 0$ in Eqn. (42): This, from Eqn. (47), implies,

$$p_1 = 0, \ p_2 = p_3 = p_4 = \frac{1}{3}, \tag{50}$$

i.e. the interstitial is always restricted to only nearest neighbour positions of the nucleus. In this case we have to treat merely the effect of fluctuating electric field gradients along $\pm x$, $\pm y$ and $\pm z$. The situation corresponds also to a different physical problem considered by Tjon and Blume (3). In the latter, such fluctuations are attributed to dynamical Jahn-Teller distortions in the vicinity of the nucleus. From Eqn. (48) and (49), we now obtain:

$$F(p) = \text{Re}\ \frac{(p+i\delta_o+3\lambda)}{(p+i\delta_o)(p+i\delta_o+3\lambda)+9Q^2}, \qquad (51)$$

which is identical to the Tjon-Blume result (if we set $\delta_o=0$).
(2) $\lambda'' = 0$ in Eqn. (42): From Eqn. (47),

$$p_1 = 1,\ p_2 = p_3 = p_4 = 0. \qquad (52)$$

This means that the interstitial is *always* at a non-nearest neighbour site and can never occupy any nearest neighbour site. From Eqn. (48) and (49):

$$F(p) = \text{Re}\ \frac{1}{p+i\delta_1}, \qquad (53)$$

which, as expected, yields just a single isomer shifted line.
(3) $Q = 0$: From Eqn. (48),

$$F(p) = \text{Re}\ (\lambda'+3\lambda'')^{-1}[\lambda'(p+i\delta_o)+3\lambda''(p+i\delta_1)+(\lambda'+3\lambda'')^2]\times$$

$$\times[(p+i\delta_1)(p+i\delta_o)+\lambda'(p+i\delta_1)+3\lambda''(p+i\delta_o)]^{-1}. \qquad (54)$$

The lineshape in this case corresponds to a situation in which the diffusion of the interstitial from nearest neighbour to non nearest neighbour positions causes fluctuating isomer shifts at the nucleus. Such phenomena have been observed for diffusion of hydrogen in Tantalum metal (28a).

5.1.3 Computed Spectra

Returning now to the result given by Eqn. (48) and (49) for austenite, the Mossbauer emission lineshape $F(\omega)$ is plotted as a function of the frequency ω in Figs. 3a-3e for different values of λ. The value of carbon concentration is taken as X=0.04. If the interstitials are stationary, λ is zero and the spectrum, as expected, is a superposition of a quadrupole-split doublet and an isomer-shifted line (Fig. 3a) (29). As the interstitials start moving with the increase of λ, the lines start to broaden because of relaxation effects. When the motion is rapid (λ is large), the doublet collapses

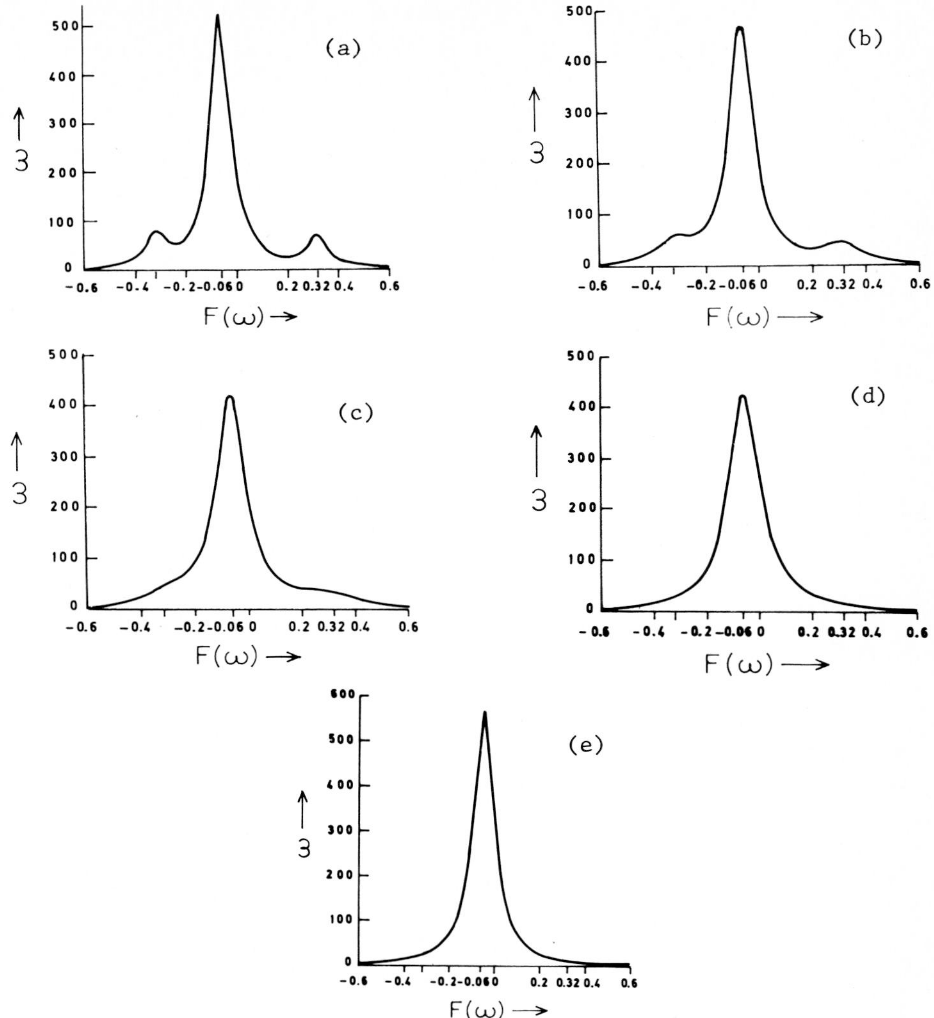

Fig. 3a. Mossbauer spectrum for $\lambda=0$ which corresponds to a very low temperature. The two outer peaks are the quadrupole split lines while the inner peak is the isomer-shifted unsplit line due to those Mossbauer atoms which do not have an interstitial atom as a nearest neighbour.

Fig. 3b. Mossbauer spectrum for $\lambda=0.005$ mm/sec. The lines are broadened because of quadrupolar relaxations.

Fig. 3c. The spectrum for $\lambda=0.015$ mm/sec. The left member of the quadrupole doublet has almost disappeared while the right one still has a shoulder.

Fig. 3d. The spectrum for $\lambda=0.1$ mm/sec. The quadrupole doublet has collapsed.

Fig. 3e. The completely motionally-narrowed spectrum for $\lambda=0.3$ mm/sec. This corresponds to a very high temperature.

In Figs. 3a-3e, the ordinates are in arbitrary units whereas the abscissae are in mm/sec, $3Q=0.32$ mm/sec following reference (30), and $\Gamma=0.1$ mm/sec.

(Fig. 3d) because the nucleus cannot follow such a fast-changing electric field gradient (30). (Recall a similar effect for a relaxing magnetic field in Section 4). Finally, when λ is very large, the spectrum degenerates into a narrow line (Fig. 3e) which has a width equal to the natural width Γ, a phenomenon known as "the motional narrowing" in nuclear magnetic resonance (31). As in the previous case of Section 4, the spectrum is asymmetric due to the presence of the isomer shift (Fig. 3b). As λ increases, the left member of the quadrupole doublet collapses first (Fig. 3c) since it is located closer to the centre of gravity of the spectrum. This is typical of an asymmetric relaxation pattern (1, 32).

From a fitting of the theoretical curves with the experimental spectra, the concentration and the diffusion coefficient of the interstitials in austenite can be evaluated (26).

6. PARAMAGNETIC RELAXATION EFFECTS

In the model treated in Section 5, the effect of the environment on the nucleus is replaced by classical fluctuating fields albeit the nucleus itself is treated quantum mechanically. This so-called effective field approximation is valid when fluctuating electric field gradients due to diffusive motion of interstitials or vacancies or Jahn-Teller distortions in the surroundings of the nucleus are considered, as discussed in Section 5. The approximation works well too in the study of paramagnetic relaxation effects when the off-diagonal terms in the hyperfine interaction are quenched, say in a strong applied magnetic field. In describing paramagnetic relaxation effects in Kramer's degenerate ions, however, the effective field approximation breaks down. Here the entire coupled nucleus-electron system has to be treated exactly and quantum mechanically. Thus the subsystem now has to be expanded in order to accommodate the electronic degrees of freedom as well. In this case the heat-bath (describing fluctuations due to spin-phonon interactions, coupling of the ionic spin to the conduction electrons, ionic spin-spin interactions of the dipole or exchange nature, etc.) acts directly on the electrons. The nucleus feels the effect of the fluctuations only indirectly through its magnetic hyperfine coupling with the electrons.

The paramagnetic relaxation effects can be treated again by using the result given by Eqn. (18) except now the terms in the Hamiltonian are to be interpreted differently. The static part H_0

is to describe the magnetic hyperfine interaction which, for example, in an isotropic case, is given by:

$$\mathcal{H}_o = a\,\vec{I}.\vec{S} \quad , \tag{55}$$

where a is the hyperfine constant and, $\vec{I}$ and $\vec{S}$ are the effective nuclear and electronic spins respectively. For modelling the coupling terms, we may imagine a situation where the electronic spin experiences an effective magnetic field $\vec{H}$, that changes its magnitude and direction at random between $\pm x$, $\pm y$, and $\pm z$ axes. Thus, one may have:

$$V_1=H_xS_x,\ V_2=-H_xS_x,\ V_3=H_yS_y,\ V_4=-H_yS_y,\ V_5=H_zS_z,\ V_6=-H_zS_z. \tag{56}$$

The solution for the lineshape (cf., Eqn. (18)) involves the inversion (or diagonalisation) of a matrix whose dimension is $(2I_1+1).(2I_0+1)\ (2S+1)^2 n$, where 1 and 0 refer to the excited and ground states of the nucleus as before and n is the number of different forms of the Hamiltonian [six, in the example of Eqn. (56)]. Clearly, the computational labour is increased since the dimension of the matrix is larger than $(2I_1+1)\ (2I_0+1)n$, the one considered in Section 5. However, as we shall show next, by making a special assumption about the matrix $\mathcal{J}$, the stochastic variables can be completely eliminated, however large n may be, thus considerably reducing the (computational) complexity of the problem.

6.1 $\mathcal{J}$ Matrix in the RPA

We assume that the probability of transition of the Hamiltonian from one form to another does not depend on the initial stochastic state from which the Hamiltonian makes a jump. This means that the new state to which the Hamiltonian jumps is completely uncorrelated to the old state, and so we may refer to this as a random-phase approximation (RPA). Admittedly, the RPA is a crude description of the dynamics of the system. However, it is to be noted that the RPA is applied only to the dynamics of the system exterior to the coupled nucleus-electron system, and the latter, of course, is treated exactly. Since the Mossbauer nucleus is only an indirect "observer" of the dynamics of the surrounding spin system, it is not expected to be overly sensitive to the detailed aspects of the relaxation mechanisms.

The RPA assumes:

$$\mathcal{J}_{ab} = p_b \quad , \tag{57}$$

which is independent of a, where p_b is the occupational probability of the state $|b)$. This form of the $\mathcal{J}$ matrix obviously satisfies the detailed balance requirement (Eqn. 26; see also Eqn. (19)). Following References (6) and (21), the solution for the RPA model can be written:

$$(\tilde{U}(p))_{av} = \frac{(\tilde{U}^o(p+\nu))_{av}}{1-(\tilde{U}^o(p+\nu))_{av}} \quad , \tag{58}$$

where

$$(\tilde{U}^o(p+\nu))_{av} = \sum_{j=1}^{n} p_j[(p+\nu)-iH_o^{\times} - iV_j^{\times}]^{-1} \quad . \tag{59}$$

The Liouville operator $(\tilde{U}(p))_{av}$ is thus given entirely in terms of $(\tilde{U}^o(p+\nu))_{av}$ which has a dimension of $(2I_1+1)(2I_o+1)(2S+1)^2$. This is quite manageable on a computer in most cases.

It should be emphasised that in cases where the RPA is not expected to be valid, one can still go back and use Eqn. (18) for the lineshape. This of course requires a reasonable model for the $\mathcal{J}$- or the W-matrix, depending on the physical situation at hand. The RPA, however, in addition to giving mathematically simpler expressions, also leads to several existing lineshape theory results in a quite straightforward manner. We discuss this below.

6.1.1 Very Fast Relaxation : $\nu \gg V_j^{\times}$, $\nu \gg H_o^{\times}$

We now have (6):

$$(\tilde{U}(p))_{av} \approx (p-iH_o^{\times} - R)^{-1} \quad , \tag{60}$$

where

$$R \approx -\sum_j p_j \frac{(V_j^{\times})^2}{\nu} \tag{61}$$

Thus we find that in this limiting case, the hyperfine interaction has no influence on the relaxation processes, and the relaxation matrix R is independent of frequency ω [cf. Eqn. 61]. This is sometimes referred to as the white-noise case (33).

The first attempt in the stochastic theory of Mossbauer lineshape to go beyond the effective field approximation and treat the coupled nucleus-ion system properly, is due to Clauser and Blume (5). Their theory, however, is restricted to the white-noise approximation (WNA).

To demonstrate the connection between the results given by Eqn. (60) and (58), and those of the Clauser-Blume model, we specialize to the case of effective electronic spin $S = \frac{1}{2}$, assume isotropic relaxation, i.e.

$$H_x = H_y = H_z = H, \tag{62}$$

in Eqn. (56), and

$$p_j = \frac{1}{6}, \tag{63}$$

the infinite-temperature case. We then have from Eqn. (61) and (56),

$$R = - \frac{H^2}{3\nu} [(S_x^{\times})^2 + (S_y^{\times})^2 + (S_z^{\times})^2] . \tag{64}$$

Using the matrix elements of $S = \frac{1}{2}$ Liouville operators (34), the matrix R can be written as (the rows and columns are labelled by $\frac{1}{2}\frac{1}{2}$, $-\frac{1}{2} - \frac{1}{2}$, $\frac{1}{2} - \frac{1}{2}$, $-\frac{1}{2}\frac{1}{2}$, respectively):

$$R = \frac{H^2}{3\nu} \begin{bmatrix} -1 & 1 & 0 & 0 \\ 1 & -1 & 0 & 0 \\ 0 & 0 & -2 & 0 \\ 0 & 0 & 0 & -2 \end{bmatrix} . \tag{65}$$

This is exactly the matrix given by Clauser and Blume in Table II of their paper (5), if we make the identification

$$\frac{H^2}{\nu} = 2\lambda (\mathrm{Sin}^2 \frac{h}{2})_{av}, \tag{66}$$

where in the Clauser-Blume notation, h is the (dimensionless) measure of the strength of the pulses that "hit" the electronic system and λ^{-1} is the mean time between successive pulses. A comment is in order here. Even in this regime of relaxation, the ratio of the effective relaxation rate (= H^2/ν) and the hyperfine frequency can cover a wide range of values, and a variety of relaxation behaviour can be discussed within the WNA (35).

The Clauser-Blume model based on the WNA has been used to analyse the effects of off-diagonal hyperfine interaction on Mossbauer spectra in ferric hemin (32) and Γ_8 quartet systems (36), among others.

6.1.2 Intermediate Case of Relaxation: $\nu >> V_j^{\times}$, $\nu \sim \mathcal{H}_o^{\times}$

This case corresponds to an exchange or dipolar-coupled system or a spin-phonon system in which the relaxation rate is much larger than the instantaneous strength of the effective feld at the ion ($\nu >> V_j^{\times}$). However, ν can be comparable to the hyperfine coupling as in rare-earth and actinide salts. Using a result given in Reference (21), Eqn. (18) can be approximately written as:

$$(\tilde{U}(p))_{av} \approx [p - i\mathcal{H}_o^{\times} + \sum_j p_j V_j^{\times} (p+\nu-i\mathcal{H}_o^{\times})^{-1} V_j^{\times}]^{-1} \quad , \tag{67}$$

or we may express:

$$(\tilde{U}(p))_{av} = (p - i\mathcal{H}_o^{\times} + R(p))^{-1} \quad , \tag{68}$$

where

$$R(p) = \sum_j p_j V_j^{\times} (p+\nu - i\mathcal{H}_o^{\times})^{-1} V_j^{\times} \quad . \tag{69}$$

Note that unlike the WNA case compare Eqns.[(60) with (68)] the relaxation matrix R(p) has an explicit dependence on the static part $\mathcal{H}_o$ of the Hamiltonian. This frequency dependence (or the breakdown of the WNA) can be extremely important in some systems, as is exemplified below.

We discuss here the Mossbauer relaxation spectra of $Cs_2NaYbCl_6$ measured using the 84.6 keV, $0^+ \rightarrow 2^+$ transition in ^{170}Yb (37). The Yb^{3+} ion is in a cubic crystal field with the ground state being a well-isolated Kramer's doublet. In the presence of a small applied field h_{ext}, the unperturbed Hamiltonian is:

$$\mathcal{H}_o = a\,\vec{I}.\vec{S} - g\mu_B S_z h_{ext}, \tag{70}$$

where g is the gyromagnetic ratio for the Γ_6 state, μ_B is the Bohr magneton and the effective spin $S = \frac{1}{2}$. Relaxation in this system is caused by dipolar interaction between ionic moments and the effect of the latter on the central ion that contains the Mossbauer nucleus may be modelled after Eqn. (56). Furthermore, the presence of cubic symmetry and high temperatures (compared to the dipolar interaction) permit us to use Eqns. (62) and (63).

Spectra at 4.2° K in external fields of 0,350 and 450 G and with a = -699 MHz (obtained from prior ENDOR studies) are shown in Fig. 4 [see also Reference 38]. The dashed lines are results given by the WNA Eqns. [(60) and (70)] while the solid lines show fits using the relaxation matrix of Eqn. (69) together with Eqns. (62) and (63). The latter, quite clearly, gives a much better description of the data than the WNA. The reason is obvious. Estimates show that the spin-spin relaxation rate in this system is of the same order as the hyperfine frequency (38). The remaining discrepancies between the data and the theory may be attributed to our assumption of the isotropic nature of the relaxation mechanism (cf., Eqn. 62), and the use of the perturbation theory result of Eqn. (67). A more improved analysis of the data with the help of Eqns. (58) and (59) has not been attempted yet.

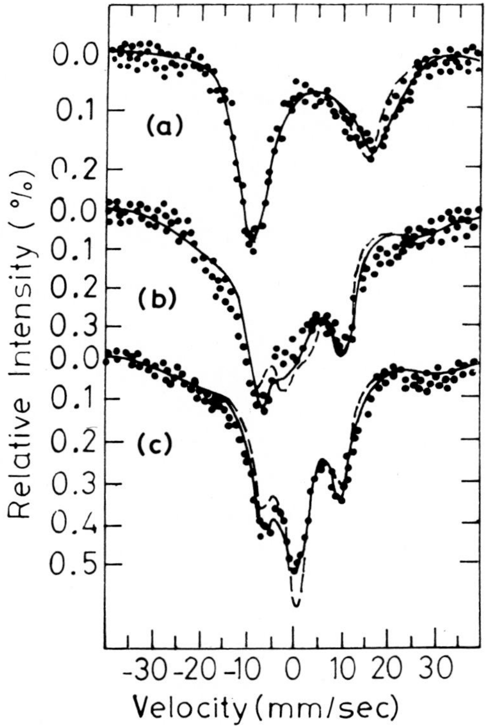

Fig. 4. Experimental spectra of $Cs_2NaYbCl_6$ measured at 4.2° K in external magnetic fields of (a) 0G, (b) 350 G and (c) 450 G [reference (37)]. Shown also are the least-squares fits to the data using theories described in the text.

7. Ab INITIO APPROACHES

In the stochastic theory of lineshape, a system with many degrees of freedom is viewed through the eyes of a smaller subsystem. The idea is of course motivated by an actual experimental situation wherein the subsystem acts as a probe. The effect of various interactions in the system felt by the subsystem is modelled in terms of a Hamiltonian for the latter alone which has random properties. There exists, at the back of our mind, an "invisible" heat bath whose role is perceived to cause fluctuations in the subsystem. In order to demonstrate how a heat bath can be brought into the picture in a more visible form, one has to make a first-principles analysis of the lineshape problem based on techniques of many body physics.

As mentioned in the introduction, numerous ab initio lineshape calculations exist in the literature, and all these are based on perturbation theories of one kind or another (7-15). The foremost among these are the diagrammatic treatments of Afanas'ev and Kagan (7). Then followed the studies by Gabriel et al (9), Schwegler (11), and Afanes'ev and Gorobschenko (14). These are based on the projection operator formalism of Zwanzig, developed in connection with the theories of irreversible statistical mechanics (39). A different attack to the problem was initiated by Hirst (10) and expanded and analysed in depth by Hartmann-Boutron and her coworkers (12,13). This method uses the equation of motion of the density matrix as elaborated by Abragam following the pioneering work of Bloch, Wangsness and Redfield (31). Here we shall follow the treatment given by Dattagupta et al (38) which utilizes a resolvent operator method such as used by Fano in pressure broadening problems of atomic spectroscopy (40). This approach has been demonstrated to yield results which are equivalent to those obtained by the projection operator and the diagrammatic treatments (41). Recently, Hartmann-Boutron has made a thorough comparison between the resolvent and the density matrix methods and has established their equivalence too (15). We shall see that in the resolvent method the results are derived with a minimum of difficulty with regard to limits of integration, time ordering of operators, decoupling approximations, etc.

Our aim is to calculate [cf. Eqns. 8 and 9]:

$$F(p) = \mathrm{Re}\ \mathrm{Tr}\{\rho A^{\dagger}(\tilde{U}(p)A)\}, \qquad (71)$$

where ρ is the density matrix of the entire manybody system:

$$\rho = \frac{\exp(-\beta \mathsf{H})}{\mathrm{Tr}\{\exp(-\beta \mathsf{H})\}} , \qquad \beta = (k_B T)^{-1} , \tag{72}$$

and U(p) is the resolvent [cf. Eqn. 7]:

$$\tilde{U}(p) = (p - i\mathsf{H}^{\times})^{-1} . \tag{73}$$

The basic idea of the theory is to split the Hamiltonian into a subsystem part H_o , a heat bath part H_b and a part V that describes a *weak* interaction between the two:

$$\mathsf{H} = \mathsf{H}_o + V + \mathsf{H}_b . \tag{74}$$

The heat bath has no *direct* effect on the subsystem which means

$$[\mathsf{H}_o, \mathsf{H}_b] = 0, \quad [A, \mathsf{H}_b] = 0. \tag{75}$$

We seek a solution in which V is treated perturbatively:

$$\tilde{U}(p) = \tilde{U}^o(p) + \tilde{U}^o(p)\, \tilde{G}(p)\, \tilde{U}^o(p), \tag{76}$$

where

$$\tilde{U}^o(p) = (p - i\mathsf{H}_o^{\times} - i\mathsf{H}_b^{\times})^{-1} . \tag{77}$$

and

$$\tilde{G}(p) = (iV^{\times}) \sum_{s=0}^{\infty} (iU^o(p)V^{\times})^s . \tag{78}$$

The assumption of weak coupling ($V \ll k_B T$) and the fact that $[\mathsf{H}_o, \mathsf{H}_b] = 0$, imply that the density matrix is factorisable:

$$\rho \approx \rho_o \otimes \rho_b . \tag{79}$$

Note that the same assumption given by Eqn. (79) is used in stochastic theories too [cf. discussion preceding, Eqn. 10].

The lineshape, after regrouping of terms, can be expressed as (38):

$$F(p) = \mathrm{Re}\, \mathrm{Tr}_o\{\rho_o A^{\dagger}[(p - i\mathsf{H}_o^{\times} + R_c(p))^{-1} A]\} , \tag{80}$$

where

$$R_c(p) = \sum_{s=0}^{\infty}(-1)^{s+1}(\tilde{G}(p))_{av}[(p-i\mathcal{H}_O^{\times})^{-1}(\tilde{G}(p))_{av}]^s \quad , \tag{81}$$

$$(\tilde{G}(p))_{av} = \sum_{i_b,i'_b} \langle i_b|\rho_b|i_b\rangle\ (i_b i_b|\tilde{G}(p)|i'_b i'_b). \tag{82}$$

The subscript c in Eqn. (81) stands for "connected" and is attributed to the fact that R_c has the character of a "linked diagram" expansion in manybody physics (40). Note further that Eqn. (82) lends a more detailed meaning, in terms of the bath variables, to the stochastic average considered earlier [cf., Eqn. (10)].

Equation (80) is formally identical with several other results mentioned in the beginning of this section, except that the assumptions used here are minimal, primarily in the form of the Hamiltonian, Eqn. (74), and in the factorisation of the density matrix, Eqn. (79). In addition, there is an implicit assumption that the series of Eqns. (78) and (81) converge. This relies on the physical assumption that the bath has a very large number of degrees of freedom. As a result, the eigenvalues of $\mathcal{H}_b$ will cover a large range of values. Therefore, the quantity $(V^{\times}\tilde{U}^{O}(p)\ V^{\times})_{av}$ will be effectively small (40). Given that, however, $R_c(p)$ can be calculated to any degree of approximation desired. Upto the second order in $V^{\times}$, we have (38):

$$R_c(p) \approx \sum_{i_b i'_b} \langle i_b|\rho_b|i_b\rangle(i_b i_b|V^{\times}(p-i\mathcal{H}_O^{\times} - i\mathcal{H}_b^{\times})^{-1}\ V^{\times}|i'_b i'_b). \tag{83}$$

From this point it is only a few more steps before we can exhibit the equivalence between $R_c(p)$ and the relaxation matrix R(p) in Eqn. (69) derived earlier from the stochastic formalism. What we ought to do is to write,

$$V = S_k\ H_k, \quad (k = x,y,z), \tag{84}$$

where H_k is a bath operator, express the matrix elements of Eqn. (83) in terms of correlation functions of H_k by making prior use of the properties of Liouville operators listed in Section 2 (see Reference (38) for details), and assume for the correlation function:

$$\langle H_k(t)H_\ell(O)\rangle_b = \langle H_k^2\rangle \exp(-\nu t)\quad \delta_{k\ell}\ , \tag{85}$$

where ν^{-1} is the correlation time. In the stochastic theory language,

Eqn. (85) follows from a stationary Gaussian-Markov process (42).

Equation (85) provides a basic connotation to the parameter ν which enters into the stochastic model as the mean frequency of pulses (see Section 3). In a true *ab initio* calculation, the correlation function, instead of being modelled as in Eqn. (85), is to be calculated from first principles. Needless to say, such a calculation which requires a detailed knowledge of all the interactions present in the bath, is a complicated problem and may not always be tractable. The stochastic model, on the other hand, sidetracks an elaborate study of the variables pertaining to the bath and yields expressions for the lineshape in terms of parameters which are directly accessible in the laboratory.

8. EFFECT OF NUCLEAR MOTION ON RELAXATION SPECTRA

In the problems discussed so far it has been tacitly assumed that the Mossbauer nucleus is rigidly fixed to a site; relaxation is caused by fluctuations in its surroundings. However, the effect on the relaxation spectra when the nucleus itself is in a state of motion can also be physically interesting and relevant in some experimental cases (16).

The situation that we want to discuss here occurs when the atom carrying the Mossbauer nucleus, due to thermally activated diffusion, jumps into a new environment. A new environment may mean a new Hamiltonian for the nucleus (see Section 9 and 10) which produces relaxation effects of the sort discussed earlier. On the other hand, the diffusive motion itself is known to cause a broadening of the lines (43). Qualitatively, what goes on is this. Each time the nucleus makes a jump, the phase of the emitted or absorbed γ-ray changes abruptly. At the same time, however, the nucleus may find itself in a different energy state so that the frequency of the radiation also changes at random (as in the example of Section 4). We are dealing here with a situation in which both phase and frequency modulations are involved simultaneously.

The Mossbauer lineshape expression in the present case is a straightforward extension of Eqn. (8), in which the transition operator A has to be generalized to $Ae^{i\vec{k}.\vec{R}}$ where $\vec{R}$ is the position operator of the center of mass of the nucleus and $\vec{k}$ the wave vector of the γ-ray. Thus,

$$F(p,\vec{k}) = \mathrm{Re}\int_0^\infty dt\ e^{-pt}\langle (A^\dagger e^{-i\vec{k}.\vec{R}})[U(t)(A\ e^{i\vec{k}.\vec{R}})]\rangle. \quad (86)$$

We may rewrite Eqn. (86) as:

$$F(p,\vec{k}) = \mathrm{Re}\int_0^\infty dt\ e^{-pt}\int d\vec{r}\ e^{i\vec{k}.\vec{r}}\ \Phi(\vec{r},t), \tag{87}$$

where, by definition,

$$\Phi(\vec{r},t) = (2\pi)^{-3}\int d\vec{k}'\ e^{-i\vec{k}'.\vec{r}} \langle (A^\dagger e^{-i\vec{k}'.\vec{R}})[U(t)(A\ e^{i\vec{k}'.\vec{R}})]\rangle . \tag{88}$$

In the usual treatment of diffusion broadening of Mossbauer lines, the internal quantum states of the nucleus are assumed to be not influenced by the heat bath (43). This implies:

$$\Phi(\vec{r},t) = \mathrm{Tr}_o(\rho_o A^\dagger A)\ G_s(\vec{r},t), \tag{89}$$

where

$$G_s(\vec{r},t) = (2\pi)^{-3}\int d\vec{k}'\ e^{-i\vec{k}'.\vec{R}}\ \mathrm{Tr}_b(\rho_b\ e^{-i\vec{k}'.\vec{R}}\ e^{i\vec{k}'.\vec{R}(t)}) , \tag{90}$$

using notations as before and introducing,

$$\vec{R}(t) = (U(t)\ \vec{R}(0)). \tag{91}$$

The quantity $G_s(\vec{r},t)$ is the well known Van Hove self-correlation function which is widely studied in neutron (44) and light scattering experiments (45). Here we shall treat the motion of the atom classically in which case G_s can be interpreted as the probability of finding the atom at $\vec{r}$ at time t given that it was at the origin at t = 0 (17). Further, it is assumed that the contributions of the diffusive and vibrational motion to G_s can be separated (46) so that the latter may be ignored for the present discussion.

In deriving the lineshape we adopt the same model as discussed in Section 3 except that the effect of the bath pulses is now viewed as to make the atom jump instantaneously to a *new* site. There it "sees" a different Hamiltonian (see also Sections 9 and 10). In the spirit of the model described earlier, we may write Eqn. (88) as:

$$\Phi(\vec{r},t) = \mathrm{Tr}_o\{\rho_o A^\dagger[(U(\vec{r},t))_{av}A]\}, \tag{92}$$

where $U(\vec{r},t)$ determines now both the time as well as the position development of the nucleus under the influence of the heat bath. In the purely diffusive case, $(U(\vec{r},t))_{av}$ reduces to $G_s(\vec{r},t)$

[cf., Eqn. 89].

In analogy with Eqn. (13) we may introduce the matrix

$$U(\vec{r},t) = e^{i\mathcal{H}^{\times}t_1} g(\vec{0},\vec{r}_1)\mathcal{J}\, e^{i\mathcal{H}^{\times}(t_2-t_1)}\, g(\vec{r}_1,\vec{r}_2)\mathcal{J}\, e^{i\mathcal{H}^{\times}(t_3-t_2)}$$
$$\times \ldots g(\vec{r}_{s-1}, \vec{r}_s = \vec{r})\mathcal{J}\, e^{i\mathcal{H}^{\times}(t-t_s)} \quad , \tag{93}$$

where $\mathcal{H}^{\times}$ and $\mathcal{J}$ have the same meaning as before, while $g(\vec{r}_1,\vec{r}_2)$ is the probability that a jump takes the nucleus from $\vec{r}_1$ to $\vec{r}_2$ (47). Since the stochastic process is assumed to be stationary, g and $\mathcal{J}$ do not depent on the time at which the jump occurs.

The average of $U(\vec{r},t)$ can be written as in Eqn. (17) with an additional (S-1)-fold integral over the position coordinates $\vec{r}_1$, $\vec{r}_2$, ... $\vec{r}_{s-1}$ (16). Assuming translational invariance:

$$g(\vec{r}_1,\vec{r}_2) = g(\vec{r}_2 - \vec{r}_1), \tag{94}$$

and using the convolution theorem, the Fourier (with respect to space) - Laplace (with respect to time) transform of $(U(\vec{r},t))_{av}$ can be written in analogy to Eqn. (18) as:

$$(\tilde{U}(p,\vec{k}))_{av} = \sum_{ab} p_a(a|(p-i\mathcal{H}_0^{\times}-i\sum_j V_j^{\times}F_j + \nu-\nu g(\vec{k})\mathcal{J})^{-1}|b) \quad , \tag{95}$$

where

$$g(\vec{k}) = \int d\vec{r}\, e^{i\vec{k}.\vec{r}}\, g(\vec{r}). \tag{96}$$

Subsitution of Eqn. (95) into Eqn. (20) yields the lineshape expression.

We note that if the nucleus is rigidly fixed and relaxations are viewed to be caused by fluctuations in the surroundings, $g(\vec{r}) = \delta(r)$, i.e. $g(\vec{k}) = 1.$, and Eqn. (95) reduces to our earlier result given by Eqn. (18). On the other hand, if the quantum states of the nucleus are assumed to be unaffected by the motion of the nucleus, we may ignore $\mathcal{H}_0$ and V_j in Eqn. (95). It can be shown then (16)

$$(\tilde{U}(p,\vec{k}))_{av} = (p+\nu(1-g(\vec{k}))^{-1} \tag{97}$$

Recalling Eqn. (21), Eqn. (97) predicts a Mossbauer line centered around $\omega = 0$ and broadened by an amount $[\Gamma+2\nu(1-g(\vec{k}))]$, where ν^{-1} is the mean time between successive jumps (43).

Before concluding this section we should point out that the above treatment assumes a completely uncorrelated jump model. This is valid only if the atom carrying the nucleus diffuses in the lattice like an ordinary interstitial. For diffusion via the vacancy mechanism, however, correlation effects are quite important (48, 49). The study of the effects of relaxation in that case is much more complicated than the one presented above. Secondly, it has been assumed that the jumps are instantaneous. On the other hand, it is possible, especially in viscous liquids, that in between two residential sites, the atom diffuses like a Brownian particle (50). One then has to deal with two different times, the mean residence time and the mean diffusion time. The effect of relaxation in this case can also be studied by suitably extending the treatment given above (51).

In the next two sections, applications of the ideas developed here to two different examples are discussed.

9. MOLECULAR MOTIONS IN LIQUIDS

We consider a situation in a viscous liquid in which an ^{57}Fe atom is located at the center of a molecule and the charge distribution of the surrounding atoms in the molecule produces an axial electric field gradient (EFG) at the nucleus. At random instants of time, the molecule containing the resonating nucleus is assumed to be subject to collisions with the other molecules. Each such collision makes the center of the molecule undergo a translational Brownian motion and at the same time tilts the axis of the EFG with respect to the nucleus. Our objective is to study the simultaneous effects of the translational motion of the nucleus and the rotational motion of the EFG direction on the Mossbauer lineshape using the model introduced in Section 8 (52).

The rotational motion of the molecules in the liquid is assumed to be described by an isotropic diffusion equation:

$$\frac{\partial \Psi}{\partial t} = d\, \nabla_{\Omega}^{2}\, \Psi, \qquad (98)$$

where $\Psi(\phi,\theta;t) = \Psi(\Omega;t)$ is the probability that the EFG has the orientation Ω with respect to the nucleus at time t, and where d is known as the rotational diffusion constant (31).

In the present problem, the Euler angles Ω_0, Ω_1, etc., which define the possible orientations of the EFG replace the stochastic variables a, b, etc. introduced earlier. Further the summation over a, b in Eqn. (95), for example, has to be replaced by integrals over Ω's, since the stochastic variables can now assume continuous values. Recalling the fact that ^{57}Fe has no quadrupole moment in the ground state (see the comments preceding Eqn. (40)), Eqn. (95), for the present problem, can be expressed as:

$$(\tilde{U}(p,\vec{k}))_{av}=\int\int d\Omega_o d\Omega p(\Omega_o)(\Omega_o|[p-i\sum_{m=-2}^{2}V_{2m}\underset{\sim}{Y}_{2m}+\nu-\nu g(\vec{k})\mathcal{J}]^{-1}|\Omega), \tag{99}$$

where V_{2m}'s are operators describing the quadrupolar interaction in the excited state of the nucleus:

$$V_{20} = Q(\tfrac{4}{5}\pi)^{\frac{1}{2}}(3I_z^2 - I^2),$$

$$V_{2\pm1}=\mp Q(\tfrac{6}{5}\pi)^{\frac{1}{2}}(I_{\mp}I_z + I_z I_{\mp}), \tag{100}$$

$$V_{2\pm2}= Q(\tfrac{6}{5}\pi)^{\frac{1}{2}}(I_{\mp})^2,$$

$\underset{\sim}{Y}_{2m}$ is a diagonal matrix [analogue of F_j defined by Eqn. 15]:

$$(\Omega_o|\underset{\sim}{Y}_{2m}|\Omega) = Y_{2m}(\Omega_o)\,\delta(\Omega_o - \Omega), \tag{101}$$

the Y's being the spherical harmonics, and $p(\Omega_o)$ denotes the *a priori* probability of the occurrence of the orientation Ω_o ($p(\Omega_o)$ is the analogue of p_a discussed previously).

The elements of the $\mathcal{J}$ matrix are to be obtained from $(\Omega_o|P(t)|\Omega)$ which defines the conditional probability that the EFG has the orientation Ω at time t, *given* that it has the orientation Ω_o at time zero (cf. Section 3.1). From Eqns. (22) and (19), the Laplace transform of the P-matrix can be written as:

$$\tilde{P}(p) = [p + \nu(1-\mathcal{J})]^{-1} \quad . \tag{102}$$

On the other hand, in the rotational diffusion model, $(\Omega_o|P(t)|\Omega)$ is the solution of Eqn. (98) (cf., (31)):

$$(\Omega_o|P(t)|\Omega) = \sum_{\ell m} Y^*_{\ell m}(\Omega_o)Y_{\ell m}(\Omega)\, e^{-d\ell(\ell+1)t} \quad , \tag{103}$$

with the initial condition:

$$(\Omega_o|P(0)|\Omega) = \delta(\Omega_o - \Omega). \tag{104}$$

Comparing Eqns. (102) and (103),

$$(\Omega_o|[p+\nu(1-\mathcal{J})]^{-1}|\Omega) = \sum_{\ell m} \frac{Y^*_{\ell m}(\Omega_o)\, Y_{\ell m}(\Omega)}{p+ d\ell(\ell+1)} . \tag{105}$$

Equation (99) can be shown to yield a closed analytic expression [for details, see Reference 52]:

$$(\tilde{U}(p,\vec{k}))_{av} = \left[p+\nu(1-g(\vec{k})) + \frac{9Q^2}{p+\nu(1-g(\vec{k}))+6dg(\vec{k})}\right], \tag{106}$$

which determines the lineshape.

The discussion of Eqn. (106) becomes clearer if a relationship can be established between ν and d. This can be achieved via the following phenomenological considerations.

Based on a random walk model, we define a translational diffusion coefficient

$$D = \frac{1}{6}\,\nu\langle r^2\rangle, \tag{107}$$

where $\langle r^2\rangle$ is the mean-square jump distance for the atom carrying the nucleus. On the other hand, if we regard the molecule (with the nucleus at the center) as a sphere of radius a whose center undergoes a translational Brownian motion while the sphere itself performs a rotational Brownian motion, and assume the validity of the Einstein-Stokes model (31), then

$$D = \frac{k_B T}{6\pi a\eta}\ , \qquad d = \frac{k_B T}{6\pi a^3 \eta}\ , \tag{108}$$

where η is the viscosity of the liquid.

Combining Eqns. (107) and (108),

$$6d = \frac{6D}{a^2} = \frac{\nu\langle r^2\rangle}{a^2} . \tag{109}$$

Equation (106) then yields:

$$(\tilde{U}(p,\vec{k}))_{av} = \left[p+\nu(1-g(\vec{k})) + \frac{9Q^2}{p+\nu(1-g(\vec{k}))+\nu g(\vec{k})\langle r^2\rangle/a^2}\right]^{-1} \tag{110}$$

If the impact with the surrounding molecules is so strong as to make the center of the molecule jump a distance of the order of the molecular radius (a is typically 2 - 3 Å), $<r^2> \simeq a^2$. In that case:

$$(\tilde{U}(p,\vec{k}))_{av} \approx [p+\nu(1-g(\vec{k})) + \frac{9Q^2}{p+\nu}]^{-1}, \tag{111}$$

a result that can be obtained independently in the so-called strong collision approximation (16).

On the other hand, when the jump lengths are much shorter than the wavelength of the γ-ray, we have $|\vec{k}.\vec{r}| << 1$. For ^{57}Fe ($k \approx 7.3$ Å^{-1}), this yields a mean jump distance <<0.137 Å. In that case, we have, from Eqn. (96),

$$g(\vec{k}) \approx 1 - \frac{1}{6} k^2 <r^2>, \tag{112}$$

so that from Eqn. (107),

$$\nu(1-g(\vec{k})) \approx Dk^2. \tag{113}$$

Equation (110) now reduces to:

$$(\tilde{U}(p,\vec{k}))_{av} = [p+Dk^2 + \frac{9Q^2}{p+Dk^2(1-<r^2>/a^2) + \frac{6D}{a^2}}]^{-1} . \tag{114}$$

But in this case, $<r^2> << a^2$ and $k^2 >> 6/a^2$ (since a is 2-3 Å). Therefore, Eqn. (114) equals approximately,

$$(\tilde{U}(p,\vec{k}))_{av} \simeq [p+Dk^2 + \frac{9Q^2}{p + Dk^2}]^{-1}, \tag{115}$$

which predicts just the diffusion broadening of two quadrupole split lines without any relaxation effect. This conclusion is also supported by physical intuition, because in this case the directional change in the EFG due to collisions (measured by $(<r^2>/a^2)^{1/2}$) is so small that the relaxation effects can be neglected.

Equation (110) has been used by Flinn et al to analyse the effect of relaxation and diffusion on the Mossbauer spectra in supercooled phosphoric acid and in the viscous organic liquid butyl phithalate (53). These systems seem to conform to the situation described by Eqns. (113) to (115). Hence, rotational diffusion and consequent quadrupolar relaxation could not be seen.

10. ELECTRON-IRRADIATION STUDY IN Al-Co

We discuss here the Mossbauer data obtained between 4.2° and 30° K in Al containing Co impurities and single (i.e. isolated) Al self-interstitials produced by prior irradiation at 100° K with 2.8 MeV electrons (part of the Co impurities was in the form of radioactive ^{57}Co Mossbauer isotope which decays into ^{57}Fe) (54). The interstitials are highly mobile unless trapped by Co impurities. The results were interpreted by Vogl et al on the basis of a model in which the Co atom is constrained to occupy one of the six dumbbell configurations (with an interstitial Al as partner) inside a "cage" (Fig. 5) (55, 56). Within the life time of the Mossbauer state the

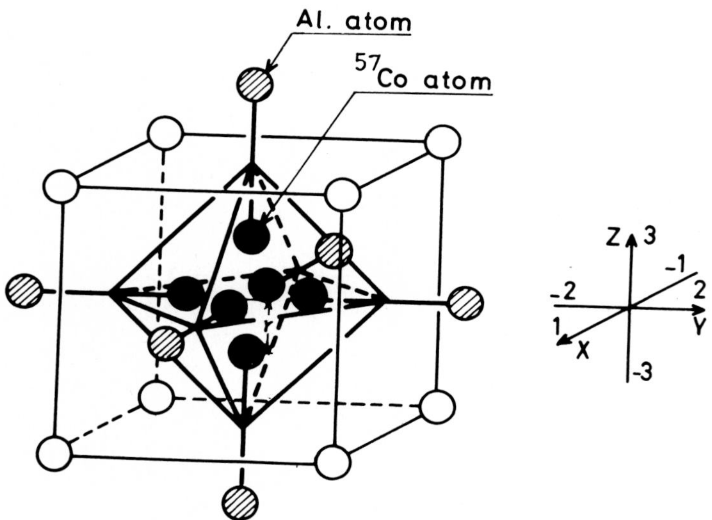

Fig. 5. Equivalent configurations for a mixed dumbbell, which can be formed by jumps of the ^{57}Co-impurity-atom in the octahedral "cage" in the FCC lattice of Al. The inset on the right shows the variables that are associated with the six positions of the impurity. Here O represents a normal Al atom and ⊘ represents an Al atom which forms a dumbbell partner with a Co atom.

Co atom is likely to have an appreciable probability of jumping to a neighbouring dumbbell position via a thermally activated process.

The jump diffusion of the Co atom inside the cage is expected to lead to a broadening of the type discussed in Section 8. On the other hand, as the Co atom jumps between the six positions, the electric field gradient at the nucleus (caused by axial distortion of the surrounding charge distribution) jumps abruptly between

$\pm x$, $\pm y$ and $\pm z$ axes (see Fig. 5). The time fluctuation of the quadrupolar interaction is thus expected to lead to relaxation effects similar to the ones treated in Section 5.

The present problem, however, differs in an important respect from that considered in Section 8. In the latter, the diffusion takes place in an infinite lattice whereas the diffusion is now constrained to occur in a finite region. This means that within the observation time, the nucleus has a finite probability of returning to its initial position. This phenomenon, which is well-known in the quasielastic neutron scattering study (44), has an important effect on the Mossbauer lineshape too as indicated below.

The problem has to be formulated in a slightly different manner than what has been described in Section 8 in order to account for the finite (six, in the present case) number of possible sites. We may start from Eqn. (86), factor out the averages over the variables of the subsystem (the nucleus, in this case) and the bath as in Eqn. (10), and write for the lineshape:

$$F(p,\vec{k}) = Tr_o\{\rho_o\ A^{\dagger}(e^{-i\vec{k}.\vec{R}}\ \tilde{U}(p)\ e^{i\vec{k}.\vec{R}})_{av}\ A\}. \tag{116}$$

Note that the operators $e^{\pm i\vec{k}.\vec{R}}$ depend on the bath variable R and are therefore included within $(....)_{av}$.

Following the theory developed in Section 3, we have:

$$F(p,\vec{k}) = Tr_o\{\rho_o\ A^{\dagger}[\Sigma_{ab}\ p_a e^{-i\vec{k}.\vec{R}_a}(a|(p-i\sum_{j=1}^{6} V_j^{\times}F_j-W)^{-1}|b)e^{i\vec{k}.\vec{R}_b}]A\}, \tag{117}$$

where, for the present application, H_o equals zero and the V_j's are the different forms of the quadrupolar interaction, e.g. (see Fig. 5)

$$V_{\pm 1} = Q(3I_x^2 - I^2),\ V_{\pm 2} = Q(3I_y^2 - I^2),\ V_{\pm 3} = Q(3I_z^2 - I^2). \tag{118}$$

Assuming nearest neighbour jumps only, the transition matrix W can be constructed as (cf. Eqns. (29) and (42)):

$$W = \begin{bmatrix} -4\nu & 0 & \nu & \nu & \nu & \nu \\ 0 & -4\nu & \nu & \nu & \nu & \nu \\ \nu & \nu & -4\nu & 0 & \nu & \nu \\ \nu & \nu & 0 & -4\nu & \nu & \nu \\ \nu & \nu & \nu & \nu & -4\nu & 0 \\ \nu & \nu & \nu & \nu & 0 & -4\nu \end{bmatrix} \tag{119}$$

where the rows and columns are labelled by a=1, -1, 2, -2, 3 and -3, respectively, and $\nu^{-1} = \tau$ is the mean time between successive jumps. Also, since the six sites inside the cage are equivalent,

$$p_a = \frac{1}{6}. \tag{120}$$

The evaluation of the lineshape involves now the inversion of a 24x24 matrix corresponding to the fact that there are six possible values of a and four angular momentum indices (recall that the quadrupolar interaction is zero in the ground state of ^{57}Fe). Exploiting again the symmetry of the quadrupolar interaction for the $I_1 = 3/2$ excited state of the nucleus (as has been done also in Sections 5 and 9), and the form of the W matrix, it is possible to obtain the following analytic expression for the lineshape in polycrystals (56):

$$F(p,k) = \mathrm{Re}\{\left[(g(p+6\nu))^{-1}-6\nu\right]^{-1} \times \frac{1}{6}\left(1+\frac{\sin 2kr}{2kr}+\frac{4\sin\sqrt{2}kr}{\sqrt{2}kr}\right)$$

$$-\left(\frac{\alpha}{p+6\nu}\right)^2 \times \frac{1}{6}\left(1+\frac{\sin 2kr}{2kr}-\frac{2\sin\sqrt{2}kr}{\sqrt{2}\,kr}\right) + g(p+4\nu)\,\frac{1}{2}\left(1-\frac{\sin 2kr}{2kr}\right)$$

$$+g(p+6\nu) \times \frac{1}{3}\left(1+\frac{\sin 2kr}{2kr}-\frac{2\sin\sqrt{2}kr}{\sqrt{2}\,kr}\right)\left[1+\frac{1}{2}\left(\frac{\alpha}{p+6\nu}\right)^2\right]\} \tag{121}$$

where r is the distance between the cage-center and any one of the six positions of the Co-atom (Fig. 5),

$$g(p) \equiv \frac{p}{p^2+\alpha^2}, \tag{122}$$

and

$$\alpha^2 = 9Q^2.$$

We note that if the quadrupolar interaction is neglected ($\alpha^2=0$), Eqn. (121) reduces to:

$$F(p,k) = \mathrm{Re}\{\frac{1}{p}\,\frac{1}{6}\left(1+\frac{\sin 2kr}{2kr}+\frac{4\sin\sqrt{2}kr}{\sqrt{2}\,kr}\right)$$

$$+\frac{1}{p+4\nu}\,\frac{1}{2}\left(1-\frac{\sin 2kr}{2kr}\right)+\frac{1}{p+6\nu}\,\frac{1}{3}\left(1+\frac{\sin 2kr}{2kr}-\frac{2\sin\sqrt{2}kr}{\sqrt{2}\,kr}\right). \tag{123}$$

In addition, if $\nu \gg \Gamma/2$ (high temperatures), Eqn. (123) yields (recall Eqn. (21)):

$$F(p,k) \approx \mathrm{Re}\,\frac{1}{p}\,\frac{1}{6}\,\left(1+\frac{\sin 2kr}{2kr}+\frac{4\sin\sqrt{2}kr}{\sqrt{2}\,kr}\right), \tag{124}$$

which gives a single line centered around $\omega = 0$, *unbroadened* by diffusion, and with the intensity decreased by the k-dependent factor. The occurrence of a similar "elastic" component, a consequence of restricted motion, is well-known in quasielastic neutron scattering (44). It has also been discussed for light scattering from atoms in a finite box (57).

Using the experimental values of r, Γ and α (54, 55), we plot the lineshape given by Eqn. (121) for different values of ν in Fig. 6. For ^{57}Fe in Al, the natural linewidth is so much greater than the quadrupole splitting that the latter is unresolved even at very low values of ν. The relaxation effect is expected to be much more important in systems for which the quadrupole splitting is larger than that in Al-Co system.

The temperature-dependence of the jump rate ν is usually given by:

$$\nu = \nu_0 \exp(-E/k_B T). \tag{125}$$

The values of ν_0 and the activation energy E can be deduced from a fitting of the temperature-dependent experimental spectra with the theoretical plots of Fig. 6.

11. CONCLUSIONS

We have presented in this article a survey of the stochastic models that have been used to date to analyse relaxation effects in Mossbauer spectra. We have endeavoured to underscore the physical concepts behind the mathematical formalism by means of several illustrative examples. These examples are discussed within the framework of a single stochastic model. The latter can be used, by an appropriate interpretation of the different terms of the lineshape expression, in a variety of physical situations. It is found that the stochastic models are well suited for describing problems involving diffusion or Brownian motion for which an *ab initio* formulation is not practical, if not impossible. Even in cases in which such *ab initio* treatments are feasible, the results are shown here to be derivable in the stochastic theory via a simpler mathematical route.

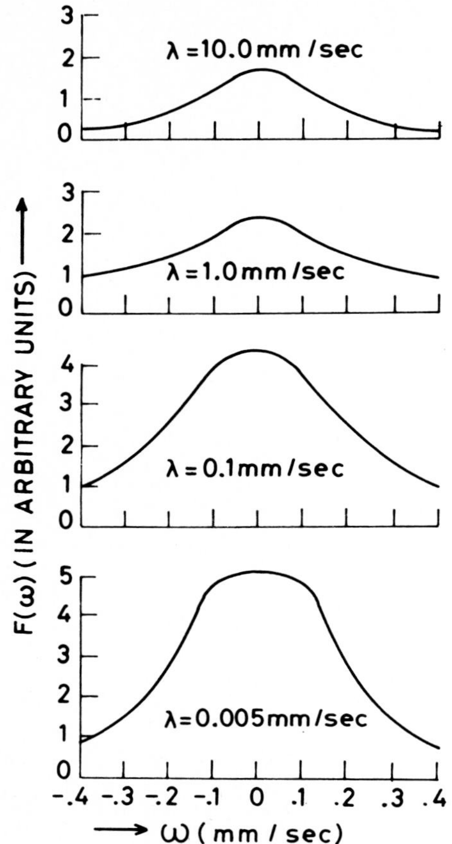

Fig. 6. Intensity $F(\omega)$ versus frequency ω for various values of λ; ($\lambda=6\nu$, ν being the jump rate). We have used $\Gamma=0.285$ mm/sec and $2\alpha = 0.18$ mm/sec (Reference 54).

Two distinct kinds of relaxation phenomena are considered; one, in which the nucleus is fixed but finds itself in fluctuating surroundings (Sections 2-7) and the other, in which the nucleus is mobile and jumps from one environment to another (Sections 8-10). We have discussed cases for which the stochastic process may be regarded as stationary. The relaxation effects in nonstationary situations are not treated here (58). In addition, we have restricted our discussion to emission or absorption experiments in which the polarisation of the γ-ray is not measured. The effect of relaxation on the polarisation (58,59) or on the resonantly scattered radiation (60,61) is outside the scope of this article.

The usage of Liouville operators, which is motivated by

practical requirements of lineshape problems (and not any fascination for mathematical elegance), enables us to give reasonably compact expressions for the lineshape. Throughout the article, we have stressed on the matrix matrix method rather than handling rate equations or master equations for the probability. The matrix method is suitably geared for taking into consideration quantum nature of the interactions (non-secular effects). In many cases the matrices are of tractable dimensions and can be easily handled on a computer. Frequently, analytic solutions are possible too, as has been exemplified here. We have also used the Hamiltonian method (i.e. the Heisenberg picture) in contrast to the density matrix method (the Schrodinger picture). Although the two approaches are completely equivalent, we find the former conceptually simpler (62).

The major input into the stochastic theory is the transition matrix W which is a by-product of the assumption that the fluctuations in the heat bath are governed by a stationary Markov process. The matrix W can be of finite (Sections 4, 5, 6 and 10) or infinite (Section 9) dimension depending on the nature of the problem. We have attempted to explain the construction of W (and the F-matrices) with the aid of physical examples.

Finally, it should be mentioned that the examples chosen here are merely typical and not exhaustive. It is nevertheless hoped that these examples will serve to explain to both experimentalists and theorists the basic ideas underlying the stochastic theory of Mossbauer lineshape.

ACKNOWLEDGEMENTS

This manuscript was cast into its final form when the author was a guest scientist at the Hahn-Meitner-Institut, Berlin. The author is grateful to G. Vogl and the members of the "Hyperfine Interactions" group of the Institute for their warm hospitality. Much of the work presented here was stimulated during the author's association with M. Blume to whom he is deeply indebted.

REFERENCES

1. M. Blume, Phys. Rev. Lett. 14(1965)96 and 18(1967)305.
2. P.W. Anderson, J. Phys. Soc. Japan 9(1954)316; R. Kubo, ibid 9(1954)935.
3. M. Blume and J.A. Tjon, Phys. Rev. 165(1968)446; J.A. Tjon and M. Blume, ibid 165(1968)456.
4. M. Blume, Phys. Rev. 174(1968)351.
5. M.J. Clauser and M. Blume, Phys. Rev. B 3(1971)583.

6. S. Dattagupta, Phys. Rev. B, 16 (1977) 158.
7. A.M. Afanas'ev and Yu. Kagan, Zh. Eksperim i Teor. Fiz. 45 (1963) 1660 [English transl. Sov. Phys. - JETP 18 (1964) 1139]; Yu. Kagan and A.M. Afanas'ev, ibid 47 (1964) 1108 [English transl. JETP 20 (1965) 743].
8. E. Bradford and W. Marshall, Proc. Phys. Soc.(London), 87(1966)731.
9. H. Gabriel, J. Bosse and K. Rander, Phys. Status Solidi, 27 (1968) 301.
10. L.L. Hirst, J. Phys. Chem. Solids, 31 (1970) 655.
11. H. Schwegler, Phys. Status Solidi, 41 (1970) 353.
12. F. Gonzalez-Jimenez, P. Imbert and F. Hartmann-Boutron, Phys. Rev. B, 9 (1974) 95; F. Hartmann-Boutron, Ann. Phys. (Paris), 9 (1975) 285.
13. C. Chopin, F. Hartmann-Boutron and D. Spanjaard, J. Phys. (Paris), Colloq., 35 (1974) C6-433.
14. A.M. Afanas'ev and V.D. Gorobschenko, Phys. Status Solidi, 73 (1976) 73.
15. F. Hartmann-Boutron, J. Phys. (Paris), 40 (1979) 57.
16. S. Dattagupta, Phys. Rev. B, 12 (1975) 47.
17. L. Van Hove, Phys. Rev., 95 (1954) 249.
18. For the properties of Liouville operators, see the appendix of the paper by M. Blume, reference (4).
19. R.L. Stratonovich, Topics in the Theory of Random Noise, Gordon and Breach, New York, 1963, Vol. I, p. 153.
20. See for instance, N.G. Van Kampen, Phys. Reports, 24 (1976) 171.
21. S. Dattagupta and M. Blume, Phys. Rev. B, 10 (1974) 4540.
22. F. van der Woude and A.J. Dekker, Phys. Status Solidi, 9 (1965) 775.
23. H. Wegener, Z. Physik, 186 (1965) 498.
24. H.H. Wickman, M.P. Klein and D.A. Shirley, Phys. Rev., 152 (1966) 345.
25. A.J.F. Boyle and J.R. Gabriel, Phys. Lett., 19 (1965) 451.
26. S. Dattagupta, Phil. Mag., 33 (1976) 59.
27. See for example, G.K. Wertheim, Mossbauer Effect: Principles and Applications, Academic, New York, 1964.
28. A. Messiah, Quantum Mechanics, North Holland, Amsterdam, 1965, Vol. II, Chap. VIII.
28a. A. Heidemann, G. Kaindl, D. Salomon, H. Wipf and G. Wortmann, Phys. Rev. Lett., 36 (1976) 213.
29. P.A. Flinn in C.A. Ziegler (Ed.), Applications of Low Energy X- and γ-rays, Gordon and Breach, New York, 1972, p. 123.
30. S.J. Lewis and P.A. Flinn, Phys. Status Solidi, 26 (1968) K 51.
31. A. Abragam, The Theory of Nuclear Magnetism, Oxford University Press, London, 1961, Chap. X.
32. S. Dattagupta, Phys. Rev. B, 12 (1975) 3584.
33. G.K. Shenoy, B.D. Dunlap, S. Dattagupta, and L. Asch, Phys. Rev. Lett., 37 (1976) 539.
34. See for instance, S. Dattagupta and M. Blume, Phys. Rev. B, 10 (1974) 4551.
35. For a more detailed discussion of this point, see reference (15). Also for comments on the validity or the breakdown of the WNA in different physical cases, see reference (33).
36. J. Sivardiere and M. Blume, Hyperfine Interact., 1 (1975) 283.
37. G.K. Shenoy, R. Poinsot, L. Asch, J.M. Friedt and B.D. Dunlap, Phys. Lett., 49A (1974) 429; also G.K. Shenoy, L. Asch, J.M. Friedt, and B.D. Dunlap, J. Phys. (Paris) Colloq., 35 (1974) C6-425, and B.D. Dunlap, G.K. Shenoy, S. Dattagupta, and L. Asch, Physica, 86-88B (1977) 1267.
38. S. Dattagupta, G.K. Shenoy, B.D. Dunlap and L. Asch, Phys. Rev. B, 16 (1977) 3893.

39. R. Zwanzig, Physica, 30 (1964) 1109, and J. Chem. Phys., 33 (1960) 1338.
40. U. Fano, Phys. Rev., 131 (1963) 259.
41. S. Dattagupta, Ph.D. Thesis, St. John's University, 1974 (unpublished).
42. M.C. Wang and G.E. Uhlenbeck, Rev. Mod. Phys., 17 (1945) 323.
43. K.S. Singwi and A. Sjolander, Phys. Rev., 120 (1960) 1093.
44. T. Springer, Quasielastic Neutron Scattering for the Investigation of Diffusive Motions in Solids and Liquids Springer Tracts in Modern Physics, 64 (1972) 55.
45. B.J. Berne and R. Pecora, Dynamic Light Scattering, John Wiley, New York, 1976.
46. For conditions under which this can be achieved, see C.T. Chudley and R.J. Elliot, Proc. Phys. Soc. London, 77 (1961) 353.
47. The theory is equally applicable when the two probabilities g and J are not independent, as assumed here. In that case we have to replace $g(\vec{r}_1,\vec{r}_2)$ by a more general matrix $T(\vec{r}_1,\vec{r}_2)$.
48. M.C. Dibar-Ure and P.A. Flinn, Phys. Rev. B, 15 (1977) 1261.
49. O. Bender and K. Schroeder, Phys. Rev. B, 19 (1979) 3399.
50. K.S. Singwi and A. Sjolander, Phys. Rev., 119 (1960) 863.
51. S. Dattagupta, unpublished.
52. S. Dattagupta, Phys. Rev. B, 14 (1976) 1329.
53. P.A. Flinn, B.J. Zabransky and S.L. Ruby, J. Phys. (Paris) Colloq., 37 (1976) C6-739; also, S.L. Ruby, B.J. Zabransky and P.A. Flinn, ibid, 37 (1976) C6-745.
54. G. Vogl and W. Mansel, Proc. Int. Conf. Fundamental Aspects of Radiation Damage in Metals, Gatlinburg, 1975, p. 349.
55. G. Vogl, W. Mansel and P.H. Dederichs, Phys. Rev. Lett., 36 (1976) 1497.
56. S. Dattagupta, Solid State Commun., 24 (1977) 19.
57. R.H. Dicke, Phys. Rev., 89 (1953) 472.
58. M. Blume, J. Phys. (Paris) Colloq., 37 (1976) C6-61.
59. S. Banerjee and M. Blume, Phys. Rev. B, 16 (1977) 3061.
60. F. Hartmann-Boutron, J. Phys. (Paris) Colloq. 37 (1976) C6-71.
61. S. Banerjee, Phys. Rev. B, 19 (1979) 5463.
62. For a discussion, see S. Dattagupta and M. Blume, Phys. Rev., A, 14 (1976) 480.

CHAPTER 12

SPIN LATTICE RELAXATION

S.C. Bhargava

Nuclear Physics Division, Bhabha Atomic Research Centre,
Bombay-400 085, India.

1. INTRODUCTION

The spin lattice relaxation has been the subject of several review articles (1-4). It has been generally presumed in these descriptions that the experimental determination of the electronic spin lattice relaxation times (τ_{SL}) can be done with the electron paramagnetic resonance and the discussion is accordingly directed to this method. Recent advances in Mossbauer Spectroscopy (5-10) have revealed the excellent potential of this technique for the study of the relaxation processes, viz., spin-spin and spin-lattice relaxation, superparamagnetic fluctuations, and other fluctuation processes, if the frequency with which the hyperfine field fluctuates is close to the Larmor precession frequency of the nuclear spin (ω_L). In this method, experimental conditions like the temperature of the solid under investigation and the strength of the external magnetic field can be controlled very conveniently. The line widths and the shape of the Mossbauer spectrum are not only sensitive to the relaxation times of the ionic levels but also to the relaxation mode affecting the shape (spin-spin, spin-lattice, superparamagnetic relaxation, etc.) and thus clearly distinguishes various processes of relaxation if spectra can be obtained under a few specifically different conditions.

Even though the Mossbauer spectra showing relaxation effects are observable in conventional manner easily, the determination of the relaxation frequencies of the different relaxation processes from the spectral shape requires systematic computational approach. This aspect is dealt with in sufficient detail in Secs. 5 and 6. Theoretically, the Mossbauer line shapes in presence of spin relaxations are simulated using the widely accepted stochastic model of ionic spin relaxation. The theory using the Liouville operator formalism has been discussed elsewhere in the book. Nevertheless, it has been felt necessary to discuss in greater detail those aspects of the theory which are helpful in the actual simulation of the spectral shapes under a variety of experimental conditions.

There are three aspects which must necessarily be considered in the study of the spin lattice relaxation process using Mossbauer spectroscopy and are discussed in Secs. 2-5. Firstly, the eigenstates and the eigenvalues of the manifold of electronic states populated under the experimental conditions must be known. Secondly, the normal modes of the lattice vibrations involved in the spin lattice relaxation should be identified and described using appropriate model (Einstein or Debye model) if the interpretation of the observed behaviour of the relaxation frequences is to be done adequately. Experimental methods, like the neutron diffraction, Raman scattering, provide the required information. It is adequate to consider the long wavelength phonons in the harmonic approximation, when the investigations of τ_{SL} cover low temperatures only. But if the investigations cover higher temperatures as is easily possible in Mossbauer spectroscopy (as the method is sensitive upto $\tau_{SL} \sim 10^{-10}$ s), it is only appropriate to examine the implications of relaxing these assumptions. Finally, the spin phonon interactions must be considered. These consist of terms which are product of functions of operators which cause transitions between ionic states and functions of ionic displacements from the equilibrium positions due to lattice vibrations and can create or destroy phonons connected with the displacements as we shall see in detail later on. We shall be interested in details of this aspect as it would help in relating the observed relaxation frequencies with the underlying mode of lattice vibrations.

In the last section briefly describing the earlier experimental investigations, the great potential of these studies which has not yet been adequately used is emphasized. This has been primarily due to the tendency to use approximate methods of determining relaxation frequencies, with emphasis on trends rather than accurate analysis. Indeed, it is reasonable to expect that Mossbauer spectroscopy will contribute significantly to our understanding of the dynamical properties under a wide variety of conditions.

2. ELECTRONIC STATES OF IONS

It is essential to know the splittings as well as the eigenfunctions of the electronic levels of the ion which are thermally populated at the temperatures concerned not only for the computation of the paramagnetic hyperfine spectrum, but also for the simulation of the Mossbauer line shape in presence of the spin relaxation effects. We start with a brief mention of the electronic states in free ions and proceed to include the interactions to which it is subjected in the solid. Alternatively, one employs group theoretical method to derive the splitting of the ground levels using symmetry considerations in terms of a few parameters. This approach is particularly suitable for ions with half filled d or f shells (S-state ions). The perturbation theory approach for these S-state ions has been described elsewhere by R.R. Sharma in this book.

2.1 Electronics States of Free Ions

The spin independent part of the electronic Hamiltonian can be written as

$$H = \sum_i \left(\frac{\vec{p}_i^{\,2}}{2m} - \frac{Ze^2}{r_i}\right) + \sum_{i>j} \frac{e^2}{|\vec{r}_i - \vec{r}_j|} \quad . \tag{1}$$

Here, the nucleus is at the origin and the summation in the last term include each pair of the electrons only once. In central field or independent particle approximation, all the electrons are assumed to move independently in identical potential. We neglect the fact that interelectron repulsion between any pair of electrons depends on the states occupied by the two electrons. Thus the Hamiltonian is approximated as

$$H_c = \sum_i \left(\frac{\vec{p}_i^{\,2}}{2m} + V(r_i)\right) \quad . \tag{2}$$

An eigenfunction of H_c is given by the Slater determinant (or a linear combination of Slater determinants) formed from the eigenstates occupied by the Z electrons χ_1 , - - - - , χ_p , ϕ_1 , - - - - , ϕ_n . Here χ and ϕ denote states belonging to closed and open shells, respectively. These states, characterised by n, ℓ^2, ℓ_z and s_z, are

$$\phi_{n\ell m} = P_{n\ell}(r)\; Y_\ell^m(\theta,\phi)\; \chi_{m_s}(\vec{S}) \quad , \tag{3}$$

and are denoted as $|n\ell\ m\ m_s\rangle$. A change in V(r) changes $P_{n\ell}(r)$ only, which can be determined using the self consistent field method (11).

A configuration is defined when the number of electrons in each level is specified. A configuration in which all lowest energy levels are occupied defines ground state configuration. We can write

$$H = H_c + H', \text{ where } H' = \sum_{i>j} \frac{e^2}{|\vec{r}_i - \vec{r}_j|} - \sum_i \left(\frac{Ze^2}{r_i} + V(r_i)\right) \quad .$$

Different configurations are separated by large energy spacing. Thus, one can neglect matrix element of H' or any other small perturbation between eigenstate Ψ and Ψ' representing different configurations and consider the effect of perturbations within the subspace of ground configuration. H' partly removes the degeneracy of states corresponding to a configuration. However, H (like H_c) is invariant to overall rotation, rotation of spin variables alone and to rotation of orbital variables alone and thus commute with $\vec{J}$, $\vec{L}$ and $\vec{S}$. Thus

$$<\gamma LSM_L M_S | H' | \gamma' L' S' M_{L'} M_{S'}> = \delta_{LL'} \delta_{SS'} \delta_{M_L M_{L'}} \delta_{M_S M_{S'}} H'^{LS}_{\gamma\gamma'} \quad ,$$

where γ distinguish states with same $\vec{L},\vec{S}$. H' causes splitting of states with different $(\vec{L},\vec{S})$, known as terms. According to Hund's rule, levels of maximum spin multiplicity lie lowest. Thus for the case of $3d^5$ configuration (Fe^{3+} and Mn^{2+}) the ground and a few excited terms, denoted as ^{2L+1}S, are as shown in Fig. 1.

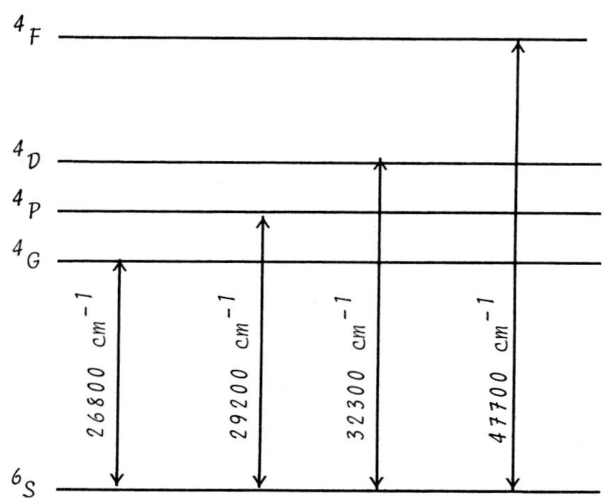

Fig. 1. Energy levels of the free Mn^{2+} ion-$3d^5$ configuration (12).

Another important interaction to be included is the spin orbit interaction which in a spherically symmetric potential V(r) is given by

$$H^i_{SO} = \frac{\hbar^2}{2m^2c^2} \vec{\ell}_i . \vec{s}_i \left(\frac{1}{r_i} \frac{dV(r_i)}{dr_i}\right) \quad . \tag{4}$$

It commutes with $\vec{j}_i$ but not with $\vec{\ell}_i$ and $\vec{s}_i$. If $H' >> H_{SO}$ as is true for lighter atoms, $Z < 50$, the perturbation H_{SO} is applied to the eigenstates of $H_c + H'$. When the separation in energy of terms is comparatively large we can apply the perturbation within the subspace of a term. Using Wigner Eckart theorem, the effect of $\sum_i g(r_i)\vec{\ell}_i.\vec{s}_i$ within the subspace of $(\vec{L},\vec{S})$ can be obtained by replacing it with $A\,\vec{L}.\vec{S}$, where A is a constant characteristic of $(\vec{L},\vec{S})$. The leads to splitting of levels into states characterised by $\vec{J} = \vec{L} + \vec{S}$, $\vec{J}$ taking values from $|L-S|$ to $L+S$.

If $H_{SO} >> H'$ as is true for heavy ions, $Z > 80$, the eigenstates of $H_C + H_{SO}$ are obtained in the first place. The states characterised by $j = \ell \pm 1/2$ and denoted by $|n\ \ell\ j\ m_j>$ are occupied by the electrons. H' gives splitting of the degenerate levels occupied by the electrons, to give states characterised by $\vec{J}$. In the following discussion it is assumed that $H' > H_{SO}$. Finally, we mention the spin dipole-dipole interaction between electrons. This interaction is small ($<1\ cm^{-1}$)

2.2 Electronic States of Ions in Solid

In a solid, the ions interact with the neighbouring ions. The interaction of the ion with the crystal electric field due to the neighbouring ions is described in the next section using a simple approach, known as the crystal field theory. In this approach, the overlap of the electrons on the paramagnetic ion with the neighbouring ions is neglected and each of the ions in the neighbourhood is assigned a charge located at its centre. The assignment of a centre to a well bound group like $(NO_3)^-$ or $(SO_4)^{--}$ presents difficulty. Further, if a dipole moment is assigned to such groups and other ions, orientation of dipoles must also be specified to include its effect.

The relative strength of the perturbations, H', H_{SO} , and H_{cr} decide the order in which the effect of these perturbations on the eigenstates of H_C (configurations) should be considered. If some of them are comparable in strength, they need be simultaneously included. Ions of the iron group generally exist as octahedral complexes in solids. The interaction of the ion with the cubic component of the crystal potential (H_{cubic}) is stronger ($\sim 10^4\ cm^{-1}$) than other interactions including H_{SO} ($\sim 10^3\ cm^{-1}$). Thus for these ions $H' > H_{cubic} > H_{SO}$. On the other hand, for ions of the rare earth group $H' > H_{SO} > H_{cr}$.

2.2.1 Crystal Field

Let the centre of the paramagnetic ion be at the origin. The potential at $P(\vec{r})$ due to the neighbouring charges q_i at $\vec{R}_i$ (fig. 2) is

$$V(\vec{r}) = \sum_i \frac{q_i}{r_i} = \sum_i \frac{q_i}{(R_i^2 + r^2 - 2R_i r \operatorname{Cos} W_i)^{1/2}}$$

$$= \sum_i \sum_{\ell=0}^{\infty} q_i R_i^{-\ell-1}\, r^{\ell}\, P_{\ell}(\operatorname{Cos} W_i)$$

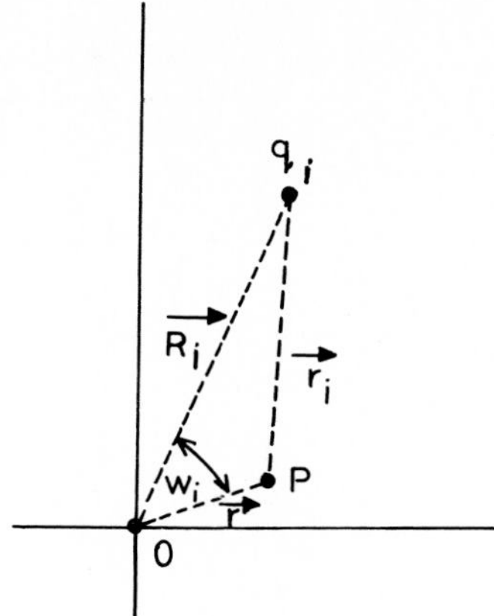

Fig. 2. Relative positions of P and the charges q_i. The centre of the paramagnetic ion is at O.

$$= \sum_i \sum_{\ell=0}^{\infty} \sum_{m_\ell=-\ell}^{+\ell} q_i R_\ell^{-\ell-1} \; r^\ell C_\ell^m(\theta,\phi) C_\ell^{m^*}(\theta_i,\phi_i) \qquad ,$$

where (r,θ,ϕ) and (R_i,θ_i,ϕ_i) refer to spherical polar coordinates of point P and the centre of the charge q_i, respectively, and W_i is the angle between $\vec{R}_i$ and $\vec{r}$. It has been assumed that $r/R_i < 1$, i.e., there is no overlap of the ligand orbitals with the electrons of the paramagnetic ions. We have used the relation

$$P_\ell(\mathrm{Cos}W_i) = \sum_{m=-\ell}^{+\ell} C_\ell^m(\theta,\phi) C_\ell^{m^*}(\theta_i,\phi_i) \; .$$

The Hamiltonian of the interaction of an electron on the paramagnetic ion with the crystal field:

$$\mathcal{H}_{cr} = \sum_{\ell=0}^{\infty} \sum_{m=-\ell}^{+\ell} q_{\ell m} \; r^\ell C_\ell^m(\theta,\phi)$$

$$= \sum_{\ell=0}^{\infty} \sum_{m=-\ell}^{+\ell} B_{\ell m} \; r^\ell \; Y_\ell^m(\theta,\phi) \qquad , \qquad (5)$$

where $q_{\ell m} = \sum_i eq_i \; R_i^{-\ell-1} \; C_\ell^{m^*}(\theta_i,\phi_i)$

and $B_{\ell m} = \left(\frac{4\pi}{2\ell+1}\right)^{1/2} q_{\ell m} \; .$

In Appendix 1, the equations relating C_ℓ^m with θ and ϕ have been given alongwith other useful relations. We need matrix element of $\mathcal{H}_{cr}$ between Slater determinant (or a linear combination of Slater determinants corresponding to the specific value of $\vec{L},\vec{S}$) representing the ground state of the paramagnetic ion. The Slater determinants are formed using orthonormal spin orbitals corresponding to closed shells (χ) and unfilled shells (ϕ), d or f shell. The matrix element $<\Psi|H_{cr}|\Psi>$ of H_{cr}, which is a one electron operators, is sum of one electron matrix elements of the form $<\Psi_a|H_{cr}|\Psi_a>$ where Ψ_a represents the spin orbitals forming the Slater determinants. The contribution of the closed shell electrons is $\sum_a <\chi_a|H_{cr}|\chi_a>$. If the term in H_{cr} with $\ell=0$, which merely shifts the levels, is not included, this contribution is zero. Closed shell electrons do not contribute to $<\Psi|H_{cr}|\Psi'>$ also. This can be seen using the orthonormal property of the spin orbitals and the fact that the two determinants Ψ and Ψ' have identical compositions as far as the spin orbitals belonging to the closed shells are concerned. Thus the closed shells can be left out of further consideration. Matrix elements between unfilled spin orbitals (Eqn. 3) is

$$<\phi_a|H_{cr}|\phi_b> = \sum_{\ell'}\sum_{m'} <r^{\ell'}>q_{\ell'm'} \ <Y_{\ell}^{m}(\theta,\phi)|C_{\ell'}^{m'}(\theta,\phi)|Y_{\ell}^{m''}(\theta,\phi)> \times \delta(m_s,m_s') \ ,$$

$$\text{where } <r^{\ell'}> = \int_0^{\infty} r^{\ell'} |P_{n\ell}(r)|^2 \ r^2 dr \ . \tag{6}$$

Consider the matrix element

$$C = \int_0^{\pi}\int_0^{2\pi} Y_{\ell}^{m*}(\theta,\phi) C_{\ell'}^{m'} \ Y_{\ell''}^{m''}(\theta,\phi) \operatorname{Sin}\theta d\theta d\phi \ . \tag{7}$$

Integration over ϕ shows that $C=0$ unless $m = m' + m''$. Values of the integral have been listed in Ref. (13). The integration shows that C is non zero when $\ell+\ell' + \ell'' =$ even, and $|\ell - \ell''| < \ell' < \ell + \ell''$. Thus, values of ℓ' to be included in H_{cr} are restricted, depending on ℓ and ℓ''. Further simplification is obtained by using symmetry considerations.

1. If H_{cr} possesses axial symmetry around Z axis, it should be independent of ϕ, $m=0$.
2. If the symmetry is tetragonal around Z-axis, H_{cr} should be invariant to a change in ϕ by $\pi/2$: $e^{im\phi} = e^{im(\phi+\pi/2)}$. Thus, $m = \pm 4$ or an integral multiple of it including zero.
3. If the arrangement of neighbouring ions is invariant to inversion, H_{cr} should have even parity.

Thus, using the notation

$$U_{\ell}^{o} = q_{\ell o} \ r^{\ell} \ C_{\ell}^{o}(\theta,\phi)$$

$$U_{\ell}^{m} = q_{\ell m} \ r^{\ell} \ C_{\ell}^{m}(\theta,\phi) + q_{\ell -m} \ r^{\ell} \ C_{\ell}^{-m}(\theta,\phi) ,$$

the simplified expressions of $\mathcal{H}_{cr}$ for $\ell=\ell''$ in Eqn. 7 can be written as:

$$\mathcal{H}_{tetragonal} = U_2^o + U_4^o + U_4^4 + U_6^o + U_6^4 \tag{8a}$$

$$\mathcal{H}_{axial} = U_2^o + U_4^o + U_6^o \tag{8b}$$

$$\mathcal{H}_{trigonal} = U_2^o + U_4^o + U_4^3 + U_6^o + U_6^3 + U_6^6 \tag{8c}$$

$$\mathcal{H}_{hexagonal} = U_2^o + U_4^o + U_6^o + U_6^6 \quad . \tag{8d}$$

As an example, we compute $q_{\ell m}$ for the case of an ion surrounded by six charges (q) at ($\pm$R, o, o), (o, $\pm$R, o) and (o, o, $\pm$R) which produce the crystal field. It will be seen in sections to follow, the term with $\ell=2$ is not allowed. Symmetry considerations also show m=o, $\pm$4. Summing over all the six charges, using the values of C_ℓ^m given in Appendix 1, give $q_{4o} = \frac{7eq}{2R^5}$, and $q_{44} = [\frac{35}{8}]^{1/2} \frac{eq}{R^5}$.

$$\mathcal{H}_{cubic} = r^4[q_{4o}\, C_4^o(\theta,\phi) + q_{44}(C_4^4(\theta,\phi) + C_4^{-4}(\theta,\phi))] + \text{terms containing } C_\ell^m \text{ with higher powers of } \ell. \tag{9}$$

2.2.2 Splitting of Electronic State derived from Symmetry Considerations:

Consider the group G of symmetry operations R which leave the system invariant. The Hamiltonian is invariant under transformations R, $R\mathcal{H}R^{-1} = \mathcal{H}$. Thus if $\mathcal{H}\Psi = E\Psi$, we have $R\mathcal{H}\Psi = \mathcal{H}R\Psi = ER\Psi$, implying that if Ψ is an eigenfunction of $\mathcal{H}$, $R\Psi$ is also an eigenfunction of $\mathcal{H}$, corresponding to E. Thus, if level E_i is g fold degenerate and $\Psi_{i\ell}(\ell=1, ---, g)$ are the corresponding orthonormal eigenfunctions, $R\,\Psi_{i\ell}$ will in general be a linear combination of $\Psi_{im}(m=1, ---, g)$:

$$R\Psi_{i\ell} = \sum_m D^i_{m\ell}(R)\Psi_{im} \quad .$$

The matrix $D^i(R)$ forms the representation of R which in this case is irreducible and $\Psi_{i\ell}$'s the bases of the representation. Thus, to any given energy level corresponds an irreducible representation of the group with dimensionally equal to the degeneracy of the level. Furthermore, the eigenstates corresponding to the level form bases of the irreducible representation.

Consider an electron in the central field of the effective nuclear charge. The wave functions of the $2(2\ell+1)$ fold degenerate eigenstate (including spin degeneracy) are given in Eqn. (3). The crystal field leave the spin part unaffected which can thus be left out of consideration for the present purpose. When the ion is free the electronic Hamiltonian H_c is invariant to rotation by any angle about any axis. This group of symmetry operations is known as the continuous rotation group. Consider rotation about Z axis by an angle α :

$$R_\alpha \phi_{n\ell m}(\vec{r}) = \sum_{m'=-\ell}^{+\ell} D^\ell_{m'm}(R_\alpha)\phi_{n\ell m'} \quad .$$

Using $R_\alpha Y^m_\ell(\theta,\phi) = e^{im\alpha}\ Y^m_\ell(\theta,\phi)$,

$$D^\ell(R_\alpha) = \begin{bmatrix} e^{i\ell\alpha} & 0 & 0 & \cdots\cdots & 0 \\ 0 & e^{i(\ell-1)\alpha} & 0 & \cdots\cdots & 0 \\ 0 & 0 & e^{i(\ell-2)\alpha} & \cdots\cdots & 0 \\ \cdot & \cdot & \cdot & \cdots\cdots & \cdot \\ \cdot & \cdot & \cdot & \cdots\cdots & \cdot \\ \cdot & \cdot & \cdot & \cdots\cdots & \cdot \\ 0 & 0 & 0 & \cdots\cdots & e^{-i\ell\alpha} \end{bmatrix}$$

forms $(2\ell+1)$ dimensional irreducible representation of R_α with $\phi_{n\ell m}$ $(m = -\ell, - - -, +\ell)$ as bases. The character of the representation

$$\chi^\ell(R_\alpha) = \sum_{m=\ell}^{-\ell} e^{im\alpha} = \frac{\mathrm{Sin}(\ell+1/2)\alpha}{\mathrm{Sin}\alpha/2} \quad . \tag{10}$$

Rotation about any other axis by the angle α belong to the same class, would give same character. When this atom is not free but is in solid surrounded by six point charges giving octahedral symmetry, only a small number of the symmetry operations of the continuous rotation group still leave the system invariant i.e., 'O' group is a subgroup of the continuous rotation group. If $\phi_{n\ell m}$, $m=-\ell, - - -, \ell$, are used as bases to obtain the representation of this subgroup also, the character table obtained is as given in the Table 1. The representations are, however, reducible and can be reduced into irreducible representations (IR) of the 'O' group using the conventional method. The character table of the 'O' group is given in Appendix 2.

The representation matrix corresponding to any symmetry operation

TABLE 1

Character table of 'O' group obtained using $\phi_{n\ell m}$ as bases and Eqn. (10).

ℓ	E	$3C_4^2(= 2C_2)$	$6C_4$	C_3	$6C_2$
0	1	1	1	1	1
1	3	-1	1	0	-1
2	5	1	-1	-1	1
3	7	-1	-1	1	-1

R of the subgroup obtained using ϕ_{n2m} as bases can be reduced to the block factored form

$$\begin{bmatrix} D^E & o \\ o & D^{T_2} \end{bmatrix} .$$

D^E and D^{T_2} are IR matrices of the 'O' group. This implies that the electronic state of the free ion with five fold degeneracy (excluding spin degeneracy) split when the symmetry is lowered into a doubly degenerate level and a triply degenerate level corresponding to E and T_2 IR of the 'O' group respectively. The splitting for other value of ℓ are given in Table 2.

TABLE 2.

Splitting of the reducible representations obtained using $\phi_{n\ell m}$, $m=-\ell$, - - -, $+\ell$, as bases into the IR of 'O' group.

ℓ	Irreducible representations of 'O' group
0	A_1
1	T_1
2	$E + T_2$
3	$A_2 + T_1 + T_2$
4	$A_1 + E + T_1 + T_2$
5	$E + 2T_1 + T_2$
6	$A_1 + A_2 + E + T_1 + 2T_2$

Correspondingly, the group of bases $\phi_{n\ell m}$, $m=-\ell$, - - -, $+\ell$ split. Thus the five fold degenerate functions ϕ_{32m} split into two groups of degenerate states, $(\phi_{320}, \frac{1}{\sqrt{2}}(\phi_{322} + \phi_{32-2}))$ and $(\phi_{321}, \phi_{32-1}, \frac{1}{\sqrt{2}}(\phi_{322} - \phi_{32-2}))$.

The bases of the second set provide an alternative set of real functions functions, transforming as yz, xz, and xy, respectively. These real states are denoted as *(U, V)* and (ξ, η, ζ), respectively.

It may be noted, the character of the identity operation (E) gives the dimensionality of the representation and hence the degeneracy of the corresponding level, if the representation is irreducible. It can be verified, for any R the sum of characters of the IR into which the reducible representation splits is equal to the character of the reducible representation, i.e.,

$$\chi^{\ell}(R) = \sum_{k} C(k\ell)\chi^{k}(R) \quad ,$$

where $\chi^{k}(R)$ denotes the character of the kth IR of the 'O' group and $C(k\ell)$ represents the number of times it appears in the reducible representation.

O is a subgroup of O_h (Appendix 2). In O_h symmetry, the inherent property of bases ($\phi_{n\ell m}$) under the operation of inversion is indicated by attaching subscript g to A_1, T_2, E , etc., when the bases functions are invariant to inversion (when ℓ=0,2,4, --), and u when it changes sign under inversion (when ℓ=1,3, --).

It is interesting to consider the splitting in tetragonal field. Out of the 24 operations of the 'O' group, only 8 symmetry operations leave the system invariant in presence of the D_4 symmetry. Thus D_4 group is not only a subgroup of the continuous rotation group but also a subgroup of 'O' group. The character table of D_4 symmetry group is given in the Appendix 2. The states corresponding to E and T_2 split as shown below when the symmetry is lowered to D_4 (Table 3).

TABLE 3

Character table obtained when bases functions of E and T_2 irredicuble representations of 'O' group are used to represent D_4.

Bases of IR of 'O' group	E	$2C_4$	C_4^2	$2C_2$	$2C_2'$	Splitting into IR of D_4 group
U, V	2	0	2	0	2	$A_1 + B_1$
ξ, η, ζ	3	-1	-1	+1	-1	$B_2 + E$

The level splitting as the symmetry is lowered is schematically shown in Fig. 3. So far only a single electron in various states, ℓ=0,1,2 ---, has been considered. These considerations similarly apply to various terms, L=0,1,2---, obtained when there are more than one electron in the

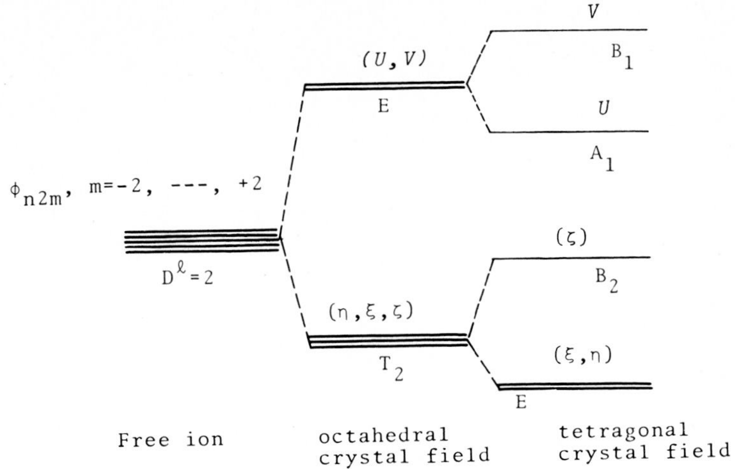

Fig. 3. Splitting of the d orbitals of the free ion when the symmetry is lowered to octahedral and tetragonal symmetries.

free atom. This can be easily seen from a consideration of the eigenfunctions of these terms (14) which also have orbital degeneracies given by (2L+1), and similar dependence on ϕ, $\exp(iM\phi)$, $M=-L,---, +L$.

2.2.3 Form of Ligand Field Potential deduced from Symmetry Considerations

Ligand field potential is invariant to any symmetry operation of the group G which leaves the system invariant. Thus, it has same transformation property as the base function of the identity representation of G. Thus, the cubic potential should have the same transformation properties as base function of A_{1g} representation of the O_h group. The base function of A_{1g} representation can be written as

$$\sum_{\ell,m} A_{\ell m} C_\ell^m(\theta,\phi) \quad , \tag{11}$$

where C_ℓ^m are the bases of irreducible representation D^ℓ of the continuous rotation group. However, not all values of ℓ are allowed. It is clear from Table 2 that the reducible representations obtained with C_ℓ^m, ℓ = 0, 4 and 6 as bases split into IR of O_h group which include A_{1g} when the symmetry is O_h. Thus base of A_{1g} representations, and consequently the expression of cubic field potential, is function of form Eqn. (11) with ℓ = 0, 4, 6, etc. This clarifies why the term with ℓ = 2 was excluded earlier in the expression for H_{cubic}. Similarly, th identity representation of D_{4h} is obtained when E_g representation of 'O_h' group splits when the symmetry is lowered to D_{4h}. The E_g represen tations of O_h group are obtained (Table 2) when the reducible representation corresponding to ℓ = 2, 4 and 6 are reduced. Thus the potential of D_{4h} is of form Eqn. 11 containing terms with ℓ = 2, 4 and 6. The U base of E_g IR of O_h provide basis of the identity representation of D_{4h}.

2.2.4 Wigner-Eckart theorem and the Operator Equivalent Method :

Consider the group of symmetry operations O_h. The irreducible tensor operator $X(\Gamma)$, is denfined as follows:

The components, $X_\gamma(\Gamma)$, of the irreducible tensor operator (ITO) transform in the same way as the bases, $\phi(\Gamma\gamma)$, of the IR Γ under symmetry operations of the group. Thus,

$$RX_\gamma(\Gamma)R^{-1} = \Sigma_{\gamma'} X_{\gamma'}(\Gamma)\, D^\Gamma_{\gamma'\gamma}(R) ,$$

where R is a symmetry operation of the group. The crystal field potential can be considered as such ITO.* The matrix element of the ITO, $\langle\alpha\Gamma\gamma|X_{\bar{\gamma}}(\bar{\Gamma})|\alpha'\Gamma'\gamma'\rangle$, is zero unless Γ representation is contained in the product representation $\bar{\Gamma}X\Gamma'$ so that the integrand transforms as the totally symmetric representation (A_{1g}) of the group (O_h). Thus, the value of the matrix element of an operator transforming as the bases of an IR $\bar{\Gamma}$ between bases belonging to the representations Γ and Γ' (13) can be shown to be

$$\langle\alpha\Gamma\gamma|X_{\bar{\gamma}}(\bar{\Gamma})|\alpha'\Gamma'\gamma'\rangle = \frac{1}{[D(\Gamma)]^{1/2}} \langle\alpha\Gamma||X(\bar{\Gamma})||\alpha'\Gamma'\rangle \langle\Gamma\gamma|\Gamma'\gamma'\bar{\Gamma}\bar{\gamma}\rangle \qquad (12)$$

This is known as Wigner Eckart Theorem. $\langle\alpha\Gamma||X(\bar{\Gamma})||\alpha'\Gamma'\rangle$ is known as the reduced matrix element which is independent of γ, γ' and $\bar{\gamma}$. It is assumed that $\bar{\Gamma}$ appears only once in $\Gamma X\Gamma'$. $\langle\Gamma\gamma|\Gamma'\gamma'\bar{\Gamma}\bar{\gamma}\rangle$ is known as Clebsch Gorden Coefficient. $D(\Gamma)$ denotes the dimension of IR Γ Similar considerations apply when Γ,Γ', and $\bar{\Gamma}$ are IR of any other group G.

The special case when Γ,Γ' and $\bar{\Gamma}$ are the representations of the continuous rotation group is of great utility to us and is discussed further below. The ITO in this case is also called spherical tensor

---

* For example, cubic potential and tetragonal potentials transform as base of A_{1g} and U base of E_g IR of the 'O_h' group, respectively.

$$V(A_{1g}) = (x^4 + y^4 + z^4 - 3/5r^4) + 6^{th} \text{ order cubic harmonic}$$

$$V_U(E_g) = (3z^2 - r^2) + 4^{th} \text{ order harmonic and soon.}$$

operator. The equivalent definition of irreducible spherical tensor operator (components T_k^q , q=k, - - -, +k) can be shown to be

$$[J_\pm , T_k^q] = \sqrt{k(k+1)-q(q\pm 1)} \; T_k^{q\pm 1} , \tag{13}$$

$$[J_Z, T_k^q] = q \; T_k^q .$$

The (2ℓ+1) spherical harmonics Y_ℓ^m considered as operators are the standard components of ITO of order ℓ. When ℓ=0, it is a scaler. When ℓ=1, Y_1^1, Y_1^0 and Y_1^{-1}, which are proportional to (x+iy), Z, and (x-iy), respectively, represent the standard components of ITO of order unity. It may be noted, similar combinations of components of any vector $\vec{V}$(e.g. $V_x + iV_y$, V_z, $V_x - iV_y$) also have same transformation properties and are also components of ITO. The Wigner Eckart theorem in the case when the group G is a continuous rotation group reads

$$\langle\tau JM| T_k^q |\tau' J' M'\rangle = \frac{1}{(2J+1)^{1/2}} \langle\tau J|| T_k ||\tau' J'\rangle \langle JM|J'M'kq\rangle \tag{14}$$

Consider the case when τ=τ' and J=J'. Similar matrix element appeared earlier in Eqn. (6). The matrix element of the second order ITO $S_2^0 = 3J_z^2 - J(J+1)$ is given by

$$\langle JM|S_2^0|JM'\rangle = [3M^2-J(J+1)]\delta_{MM'} = \frac{1}{(2J+1)^{1/2}} \langle J||S_2||J\rangle \langle JM|JM'20\rangle .$$

The matrix element of any other second order tensor operator T_2^0 is

$$\langle JM|T_2^0|JM'\rangle = \frac{1}{(2J+1)^{1/2}} \langle J||T_2||J\rangle \langle JM|JM'20\rangle .$$ Thus,

$$\frac{\langle JM|T_2^0|JM'\rangle}{\langle JM|S_2^0|JM'\rangle} = \frac{\langle J||T_2||J\rangle}{\langle J||S_2||J\rangle} = \beta(J,2),$$ which shows

$$\langle JM|T_2^0|JM'\rangle = \beta\langle JM|S_2^0|JM'\rangle = \beta[3M^2 - J(J+1)]\delta_{MM'} .$$

This shows, if β is known, which depends on J and k, we can find all matrix elements of T_k^q , q = -k, ---, +k, using the matrix elements of S_k^q , which can be found easily as is shown above for the case when S_k^q is S_2^0. The matrix element of $\mathcal{H}_{cr}$ (Eqn. 5, Appendix 3) between states |JM> and |JM'> can be obtained using

$$\langle JM|T_{\ell m}|JM'\rangle = \langle r^\ell\rangle \beta(J,\ell)\langle JM|O_{\ell m}|JM'\rangle \quad \text{where} \tag{15}$$

$$\beta(J,\ell) = \frac{\langle J||T_\ell||J\rangle}{\langle J||O_\ell||J\rangle} .$$

The values of β have been evaluated in several references (Ref. 5, p. 670). The operators $O_{\ell m}$ required for evaluation of matrix elements of $\mathcal{H}_{cr}$ are given in Appendix 3.

2.3 Spin Hamiltonian

In this section, the spin dependent interactions are considered. The spin Hamiltonian method for iron group ions in non S-ground state when free using perturbation theory was developed by Pryce (16) and Abragam and Pryce (17), and is briefly discussed below.

The ground term of the free ion for various electron configurations are:

Configuration	d^1	d^2	d^3	d^4	d^5	d^6	d^7	d^8	d^9
ground term	^2D	^3F	^4F	^5D	^6S	^5D	^4F	^3F	^2D

The splitting of the D and F terms due to the cubic component of the crystal field is generally larger than the effect of the spin dependant perturbations and consequently perturbation method can be used to include the effect of these smaller interactions. The splitting of the D and F terms by the cubic field can give a singlet (Fig. 4), a doublet, or a triplet ground state, excluding the spin degeneracy. While considering the effect of smaller perturbations which act on the spin part of the wavefunction also, it is sufficient to include the excited states derived from the same term. Using the operator equivalent method, restricting to manifold of states belonging to $(\vec{L},\vec{S})$ ground term only, these smaller perturbations can be expressed as

$$\mathcal{H}_{so} = \lambda\vec{L}.\vec{S}, \tag{16}$$

$$\mathcal{H}_{ss} = -P\Sigma_{p,q}\{\frac{L_pL_q + L_qL_p}{2} - \frac{1}{3}L(L+1)\delta_{pq}\}S_pS_q \quad , \tag{17}$$

$$\mathcal{H}_{z} = \beta\vec{H}.(\vec{L}+2\vec{S}), \tag{18}$$

and low symmetry crystalline field present in addition to the cubic field, denoted by $\mathcal{H}_{LOW}$. The hyperfine interactions are much smaller and need not be included in the present discussions. The order of splitting due to various interactions are:

$\mathcal{H}_{cubic} \approx 10^4\ cm^{-1}$, $\mathcal{H}_{so} \approx 10^2\ cm^{-1}$, $\mathcal{H}_{LOW} \approx 10^2\ cm^{-1}$

$\mathcal{H}_{z} \approx 1\ cm^{-1}$, $\mathcal{H}_{ss} \approx 1\ cm^{-1}$ and $\mathcal{H}_{hyperfine} \approx 10^{-2}\ cm^{-1}$

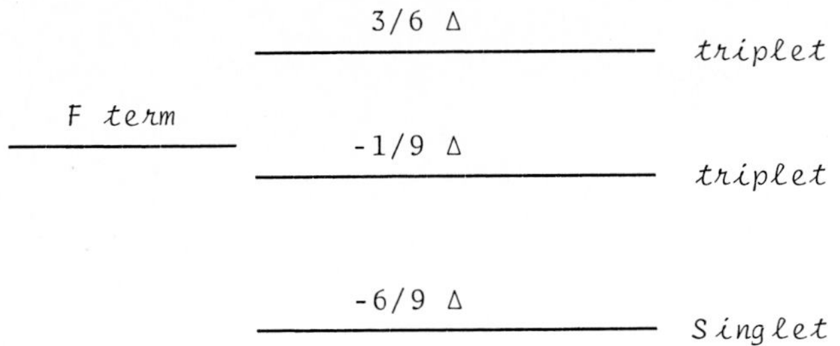

Fig. 4. Splitting of ground state of d^3 and d^8 configuration.

The derivation of the spin Hamiltonian using the perturbation theory for the case when the ground state in presence of the (cubic) crystalline field is an orbital singlet is given below. The ground state of ions of configurations d^1, d^4, d^6, and d^9 is the D term, which splits into T_{2g} and E_g levels in cubic field. The method described below is also applicable to the case when the splitting of the doubly degenerate E_g level by low symmetry fields give an orbitally non degenerate ground state, as will be seen later. The perturbation theory method to obtain the splitting of ground S term (e.g. Fe^{3+} ion) is discussed elsewhere in the book by R.R. Sharma.

Let the unperturbed Hamiltonian $\mathcal{H}_o$ has eigenstates and eigenvalues denoted by ϕ_n and ε_n^0. The subscript i, j, ... will be used for states belonging to the ground manifold and μ,γ,--- for states belonging to the manifold of excited state (Fig. 5) which is assumed to be well separated from the ground manifold (Appendix 2 of Ref. 14). The states Ψ are the eigenstates of the complete Hamiltonian $\mathcal{H}_o + \mathcal{H}_1$. The eigenvalues of $\mathcal{H}_o + \mathcal{H}_1$ are denoted by ε. Expand Ψ: $\Psi = \sum_j a_j\phi_j + \sum_\mu a_\mu\phi_\mu$, in terms of the complete set ϕ_n. It has been shown using perturbation theory that the eigenvalues of the ground manifold corresponding to $(\mathcal{H}_o + \mathcal{H}_1)$ can be obtained using effective Hamiltonian ($\mathcal{H}_{eff}$), known as the spin Hamiltonian, which operates on Ψ_{eff} instead of Ψ, i.e. $\mathcal{H}_{eff}\Psi_{eff} = \varepsilon\Psi_{eff}$,

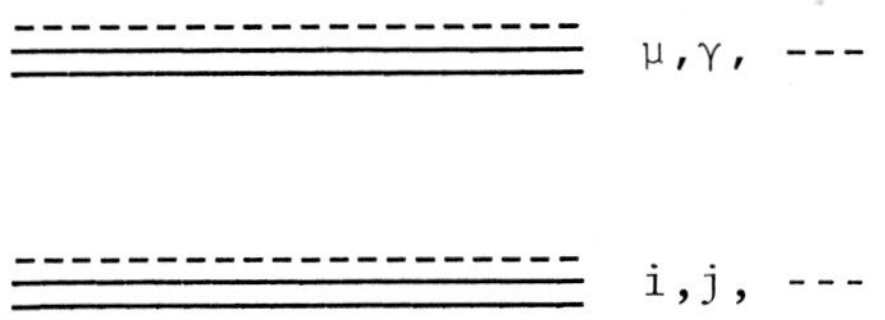

Fig. 5. The Eigenvalues of H_o are assumed to form two well separated groups.

where

$$H_{eff} = H_o + H_1 + \sum_{\mu} \frac{H_1|\mu\rangle\langle\mu|H_1}{\varepsilon-\varepsilon_{\mu}^0} + \frac{H_1|\mu\rangle\langle\mu|H_1|\gamma\rangle\langle\gamma|H_1}{(\varepsilon-\varepsilon_{\mu}^0)(\varepsilon-\varepsilon_{\gamma}^0)} + \cdots , \tag{19}$$

$$\Psi_{eff} = \sum_j a_j \phi_j \quad . \tag{20}$$

To second order in H_1, ε can be replaced by an average value for the set of ground levels j, k, etc., if the separation in energy from the manifold of excited levels μ, γ, --- is large. In the present case

$$H_o = H_c + H' + H_{cubic} \tag{21a}$$

$$H_1 = H_{so} + H_z + H_{ss} + H_{LOW} \tag{21b}$$

$$\langle OSM_S|H_{eff}|OSM_S'\rangle = \varepsilon_0\delta_{M_SM_S'} + \langle OSM_S|H_1|OSM_S'\rangle + \sum_{\mu M_S''} \frac{\langle OSM_S|H_1|\mu SM_S''\rangle\langle\mu SM_S''|H_1|OSM_S'\rangle}{(\varepsilon^0 - \varepsilon_{\mu}^0)} \tag{22}$$

neglecting higher order terms. It may be noted, spin multiplicity for the excited states and the ground states are same, as they are derived from the same term $(\vec{L},\vec{S}).H_{eff}$ acting on ground states give same eigen-values as $H_o + H_1$ operating on $\Psi = \sum_j a_j\phi_j + \sum_{\mu} a_{\mu}\phi_{\mu}$. For simplicity H_{ss} and H_{LOW} are not included in the following derivation of H_{eff}. The third term on the right side of Eqn. (22) contains the matrix element

$$\langle OSM_S|\lambda\vec{L}.\vec{S} + \beta\vec{H}.(\vec{L}+2\vec{S})|\mu SM_S''\rangle = \sum_{m=x,y,z} \sum_{\gamma, M_S'''} \langle OSM_S|\lambda S_m + \beta H_m|\gamma SM_S'''\rangle \times \langle\gamma SM_S'''|L_m|\mu SM_S''\rangle ,$$

which is non zero if $|o\rangle = |\gamma\rangle$ and $M_S''' = M_S''$. Thus, the third term on the right of Eqn. (22) can be rewritten as

$$-\sum_{m,n} \Lambda_{mn} \langle OSM_S|\lambda S_m + \beta H_m|OSM_S''\rangle \langle OSM_S''|\lambda S_n + \beta H_n|OSM_S'\rangle , \quad \text{where}$$

$$\Lambda_{mn}^{M_S''} = \sum_{\mu} \frac{\langle O|L_m|\mu\rangle\langle\mu|L_n|o\rangle}{\varepsilon_{\mu}^0 - \varepsilon^0} \quad .$$

Eqn. (22) can be rewritten as

$$\langle OSM_S|H_{eff}|OSM_S'\rangle = \varepsilon_0\delta_{M_SM_S'} + \langle OSM_S|2\beta\vec{H}.\vec{S}|OSM_S'\rangle - \sum_{m,n} \Lambda_{mn} \langle OSM_S|\lambda^2 S_m S_n + 2\lambda\beta S_m H_n + \beta^2 H_m H_n|OSM_S'\rangle \quad .$$

Thus, $\mathcal{H}_{eff}$ can be written as

$$\mathcal{H}_{eff} = \varepsilon_0 + \sum_{m,n} [2\beta H_m S_m \delta_{mn} - \Lambda_{mn}(\lambda^2 S_m S_n + 2\lambda\beta S_m H_n + \beta^2 H_m H_n)]$$

$$= \varepsilon_0 + \beta\vec{H}.g.\vec{S} + \vec{S}.D.\vec{S} - \beta^2\vec{H}.\Lambda.\vec{H}, \qquad (23)$$

where $g_{mn} = 2\delta_{mn} - 2\lambda\Lambda_{mn}$, $D_{mn} = -\Lambda_{mn}\lambda^2$.

The integration over spatial variables have been done. Thus $\mathcal{H}_{eff}$ consists of spin operators only and is known as the Spin Hamiltonian. This must be diagonalised within the subspace of (2S+1) ground states $|OSM_S\rangle$ to obtain the eigenvalues and thus the splitting of the manifold of the ground states.

Choosing the coordinate axes to coincide with the principal axes of the g tensor, and excluding ε_0 and $-\beta^2\ \vec{H}.\Lambda.\vec{H}$, which do not contribute to the splitting of the (2S+1) levels,

$$\mathcal{H}_{SP} = \beta(g_{xx}H_xS_x + g_{yy}H_yS_y + g_{zz}H_zS_z) + D(3S_z^2 - S(S+1)) + \frac{1}{2}E(S_+^2 + S_-^2),$$

where $D = \frac{D_z}{2}$ and $E = \frac{1}{2}(D_x - D_y)$. (24)

When the low symmetry components of the crystal field which are comparable to the strength of the spin orbit coupling under consideration removes the orbital degeneracy of the ground state in the cubic field to yield an orbitally nondegenerate ground state $|o\rangle$ and states $|i\rangle$, the method described above can still be used if the matrix elements of $L_m (m=x,y,z)$, $\langle o|L_m|i\rangle$, are zero for all $|i\rangle$. The theory is inadequate if the ground term of the free ion is a S-state.

2.4. Spin Hamiltonian Derived Using Symmetry Considerations:

As discussed earlier, the spin Hamiltonian is an effective Hamiltonian which acting within the subspace of the manifold of the ground levels gives eigenvalues of the total Hamiltonian $\mathcal{H}_0+\mathcal{H}_1$. We get no information about the actual eigenstates of $\mathcal{H}_0+\mathcal{H}_1$, using this method. In this section, the symmetry considerations are used to derive the spin Hamiltonian for the orbitally singlet state, which is the ground state of the spin independent part of the Hamiltonian including crystal field interactions. Thus ions with S-term as ground state when free are also included, unlike in the perturbation method described above.

First, consider the spin orbit interaction. When the spin orbit coupling is included in the Hamiltonian, the wave functions of the ground manifold mix with the excited states. Let the new (2S+1) wave functions of the ground manifold be denoted by Ψ_i and the matrix of the Hamiltonian including spin orbit coupling take partially diagonal form:

	Ψ_1	Ψ_2	- - - -	Ψ_{2S+1}	.	.	.	.
Ψ_1	x	x	- - - -	x				
Ψ_2	x	x	- - - -	x				
.	.	.						
.	.	.						
Ψ_{2S+1}	x	x	- - - -	x				
.					x	x	x	x
.					x	x	x	x
.					.	.	.	.
.					.	.	.	.

Instead of finding out Ψ_i's and the matrix, to calculate the eigenvalues of the ground manifold when the spin orbit interaction is included, we reproduce the hermitian submatrix of dimension (2S+1) with $\Theta(SM)$, M=-S, --- , +S, as bases and Hamiltonian a linear combination of indepe-ndant spin operators but involving parameters. Thus a 3x3 matrix can be formed with $\Theta(1M)$, $M=\pm 1$, o, as bases and a linear combination of independant spin operators, $1, S_z, S_z^2, S_x, S_y$, $(S_xS_z + S_zS_x)$, $(S_yS_z + S_zS_y)$, $(S_x^2 - S_y^2)$ and $(S_xS_y + S_yS_x)$. The linear combination is called the spin Hamiltonian.

$$H_{SP} = Y1 + \frac{1}{2}(X-Z)S_Z + \frac{1}{2}(X+Z-2Y)S_Z^2 + \frac{1}{\sqrt{2}}(C+e)S_X - \frac{1}{\sqrt{2}}(d+f)S_y - \frac{1}{\sqrt{2}}(C-e) \times$$

$$(S_xS_z + S_zS_x) + \frac{1}{\sqrt{2}}(f-d)(S_yS_z + S_zS_y) + a(S_x^2 - S_y^2) + b(S_xS_y + S_yS_x).$$

The matrix of H_{SP} with bases $\Theta(1M)$ is

	$\Theta(1,1)$	$\Theta(1,0)$	$\Theta(1,-1)$
$\Theta(1,1)$	X	e+id	a-ib
$\Theta(1,0)$	e-id	Y	C+if
$\Theta(1,-1)$	a+ib	C-if	Z

We next use symmetry considerations and time reversal invariance condition to reduce the number of parameters appearing in H_{SP}. The condition of time reversal invariance leads to zero value of the coeff-icients of terms proportional to the product $S_x^pS_y^qS_z^r$ with p+q+r equal to an odd integer. Thus,

$$H_{SP} = \Upsilon 1+(x-y)S_z^2+\sqrt{2}C(S_xS_z+S_zS_x)+\sqrt{2}f(S_yS_z+S_zS_y)+a(S_x^2-S_y^2)+b(S_xS_y+S_yS_x) .$$

The transformation properties of the unperturbed wave functions of the ground state ${}^{2S+1}A$, Ψ_i and $\Theta(S,M)$ are same under symmetry operations which leave the system invariant. This can be seen as follows. As we saw earlier, cubic field splits the term ${}^{2S+1}L$ into states which form basis of IR of cubic point group. When spin orbit interaction is included in the Hamiltonian, we must apply identical symmetry operations in spin space (R_s) and orbital space (R^o) simultaneously so that H_{so} remains invariant. Such a symmetry operation is denoted by $R^{os}=R^o\times R^s$. In case of rotation in spin space, the rotation by 2π is not an identity operation if the spin is half integral (13). Instead, the identity operation is a rotation by 4π. We assume, the period of rotation in orbital space is also 4π and thus replace cubic rotation group representation with cubic double group representation (Appendix 2) which is appropriate for rotations in the spin space also.

The operation R^{os} transforms wave function as

$$R^{os}\ \Psi(\alpha S\Gamma M_s\gamma) = \sum_{M_s'\gamma'} \Psi(\alpha S\Gamma M_s'\gamma')\ D^S_{M_s'M_s}(R^S)D_{\gamma'\gamma}(R^o) .$$

To obtain the IR of the cubic double group contained in the product representation $D^\Gamma\times D^S$, we proceed as follows:

Just as we reduce representation with $\phi_{n\ell m}$, $m=-\ell, \ldots, +\ell$, as bases into IRs of O group, it is needed to reduce representation with $|SM_s>$, $M_s=-S, \ldots\ldots, +S$, as bases (bases of the continuous rotation group in spin space) into IRs of cubic double group. For integral values of S, the reduction in the two cases are identical (Table 2). For half integr integral spin S, the reduction is given in Table 4 below:

TABLE 4

Decomposition of reducible representation obtained with $|SM_s>$, $M_s=-S, \ldots, +S$, as bases, into IR of the cubic double group.

S	Irreducible representation
1/2	E_1
3/2	G
5/2	E_2 + G
7/2	E_1 + E_2 + G

Next, the IRs of the cubic double group contained in the product representation $D^S x D^\Gamma$ are obtained using Table 5. The product representation $(E_2$ + G)xA is E_2 + G. Thus IR obtained from the product representations

$D^{\Gamma} x D^{S}$ are same* as the IR obtained from D^{S}. ^{6}A state splits into a double degenerate and a four degenerate states. R^{SO} leave H_{SO} invariant. Thus H_{SO} transforms as the bases of the identity representation of the double group and can give non zero matrix elements between states corresponding to same IR obtained from ^{6}A and other, excited, states with R^{OS}. Thus the transformation properties of the unperturbed state and the perturbed states are identical to the transformation properties of the spin part of the wave function.

TABLE 5

Decomposition of the product representation $\Gamma x \Gamma'$ *into IR of the cubic double group.*

Γ \ Γ'	A_1	A_2	E	T_1	T_2	E_1	E_2	G
E_1	E_1	E_2	G	E_1+G	E_2+G	A_1+T_1	A_2+T_2	$E+T_1+T_2$
E_2	E_2	E_1	G	E_2+G	E_1+G	A_2+T_2	A_1+T_1	$E+T_1+T_2$
G	G	G	E_1+E_2+G	E_1+E_2+2G	E_1+E_2+2G	$E+T_1+T_2$	$E+T_1+T_2$	A_1+A_2+E+ $2T_1+2T_2$

The effect of low symmetry fields can be likewise included. If the double group correspond to the group of symmetry operations which leave the system invariant, V_{LOW} also transforms as identify representation of this double group and thus gives matrix elements which are non zero if it is diagonal in the IR of the double group arising from ^{6}A and other terms of the free ion. Also, the orbital as well as spin part of the wave function of ^{2S+1}A term is unaffected by changes in crystal potential.

As the transformation properties of $\Theta(SM)$ and Ψ_i are identical, the effective Hamiltonian should also be invariant to symmetry operations which leave the system invariant as the real Hamiltonian is. Thus if the symmetry is axial with Z as the symmetry axis, the coefficients C, f, a and b are zero. The remaining term can be suitably rearranged to give

$$H_{SP} = D[3S_z^2 - S(S+1)]$$

* This is not so if ground state in the cubic field is a non S state which thus necessitates a different and slightly more complicated approach for obtaining the spin Hamiltonian (13).

In general, one can construct (2S+1) dimensional hermitian matrix with bases Θ(SM) and a linear combination of Irreducible spin tensor operators which are independent and possess well known transformation properties under symmetry operations. When cubic symmetry is present, it is better to work with combinations of irreducible spin tensor operators S_k^q which have same transformation properties as bases of A_{1g} IR of the cubic group. Thus the spin Hamiltonian for ^6S state of Fe^{3+} or Mn^{2+} in presence of cubic symmetry should contain spin operators S_k^q with k=o or ± 4.

In case the Z axis is along the tetragonal axis, $H_{SP}^{cubic} = aS_{40} + bS_{44}$, where

$$S_{40} = 35S_z^4 - 30S(S+1)S_z^2 + 25S_z^2 - 6S(S+1) + 3S^2(S+1)^2 ,$$

$$S_{44} = \frac{1}{2}[S_+^4 + S_-^4] .$$

S_{40} and S_{44} are obtained from $O_{\ell m}$ in Appendix 3 by replacing J_x, J_y, J_z and J with S_x, S_y, S_z and S, respectively. If trigonal axis is chosen as the Z axis, terms proportional to S_{43} and S_{40} only are allowed where

$$S_{43} = \frac{1}{4}[S_z(S_+^3 + S_-^3) + (S_+^3 + S_-^3)S_z]$$

Inclusion of Zeeman Interaction. Within the subspace of a term ^{2S+1}L, the Zeeman interaction can be written as $\beta\vec{H}.(\vec{L}+2\vec{S})$. The Zeeman interaction and consequently the Hamiltonian including the Zeeman interaction is time invariant if time reversal operator is applied simultaneously to both, the electron system as well as to the source of magnetic field. Similarly, Zeeman term is invariant to any rotations of the electron system and the magnetic field by the same angle simultaneously, as $\vec{H}, \vec{L}$ and $\vec{S}$ transform like the axial vector. We again form the desired hermitian submatrix with bases Θ(SM) and an effective Hamiltonian consisting of a linear combination of Irreducible spin tensor operators but with coefficients of these terms depending on the external magnetic field also.

$$H_{SP} = \sum_{k,q} \alpha_{kq}(\vec{H}) S_k^q , \quad \text{where} \tag{25}$$

$$\alpha_{kq}(\vec{H}) = \sum_n \sum_{\substack{\alpha,\beta,\gamma \\ \alpha+\beta+\gamma=n}} A^n_{\alpha\beta\gamma}(kq) H_x^\alpha H_y^\beta H_z^\gamma .$$

The terms corresponding to higher values of n are smaller and will not be considered further. Thus

$$\alpha_{kq}(\vec{H}) \sim A^{o}(kq) + a_1(kq)H_x + a_2(kq)H_y + a_3(kq)H_z \quad .$$

Field independant terms were discussed earlier. To satisfy the time reversal invariance condition, field dependent terms with odd integral values of k alone are nonzero. Consider the terms with k=1,

$$\alpha'_{11}(\vec{H})S_1^1 + \alpha'_{10}(\vec{H})S_1^o + \alpha'_{1-1}(\vec{H})S_1^{-1} \quad , \quad \text{where} \quad \alpha'_{kq}(\vec{H}) = \alpha_{kq}(\vec{H}) - A^{o}(kq).$$

It involves nine independant coefficients and can be rewritten in the familiar form $\beta(\vec{H}.g.\vec{S})$. Similarly, contributions from terms with k=3 and 5 in Eqn. (25) can be written.

In the presence of cubic symmetry $\beta(\vec{H}.g.\vec{S})$ reduces to the familiar form $\beta g(\vec{H}.\vec{S})$. Similarly, in the presence of cubic symmetry, the terms with k=3 contribute

$$g_1\beta\,[H_xS_x^3 + H_yS_y^3 + H_zS_z^3]$$

and the terms corresponding to k=5 contribute

$$g_2\beta\,[S_x^5H_x+S_y^5H_y+S_z^5H_z] + g_3\beta\,[S_x(S_y^4+S_z^4)H_x+S_y(S_z^4+S_x^4)H_y+S_z(S_x^4+S_y^4)H_z] \; .$$

It is instructive to compare these expressions with the field independant expression of cubic field

$$H_{cubic} = C_4(x^4+y^4+z^4-\tfrac{3}{5}r^4)+D_6\{(x^6+y^6+z^6)+\tfrac{15}{4}[(x^4+z^4)y^2+(y^4+x^4)z^2 + (z^4+y^4)x^2]-\tfrac{15}{14}r^6\}$$

However, g_1, g_2, and g_3 are small.

3. LATTICE VIBRATIONS

The theory of lattice vibrations can be found in several excellent reviews and books. In this section, we introduce various relations which have been used in sections to follow, closely following references 18 and 19.

The equilibrium position of kth atom in ℓth primitive cell can be written as

$$\vec{X}(\ell k) = \vec{X}(\ell) + \vec{X}(k)$$

where $\vec{X}(\ell)$ gives the location of the primitive cell in the crystal. k takes r values if there are r atoms in the primitive cell. At any instant t, the displacement of the kth atom from the equilibrium position is denoted by $\vec{u}(\ell k)$. The Hamiltonian and the equations of motions in the harmonic approximation are

$$H = \frac{1}{2}\sum_{\ell,k,\alpha} m_k \dot{u}_\alpha^2(\ell k)+\phi_0+\sum_{\substack{\alpha,\beta,\ell,\\ k,\ell',k'}} \frac{\partial^2\phi}{\partial u_\alpha(\ell k)\partial u_\beta(\ell' k')}\Big|_0 u_\alpha(\ell k)u_\beta(\ell' k'),$$

$$p_\alpha(\ell k) = m_k\ddot{u}_\alpha(\ell k) = -\sum_{\ell',k',\beta} \frac{\partial^2\phi}{\partial u_\alpha(\ell k)\partial u_\beta(\ell' k')}\Big|_0 u_\beta(\ell' k'),\quad \alpha,\beta=x,y,z\ .$$

Here, m_k is the mass of the kth atom. In a periodic lattice, the wavelike solutions are appropriate.

$$u_\alpha(\ell k) = \frac{1}{\sqrt{m_k}} U_\alpha(k|\vec{q})\exp[i(\vec{q}.\vec{X}(\ell)-\omega(q)t)] \quad . \tag{26}$$

$\vec{q}$ is the wave vector, and $(m_k)^{-1/2}$ is added for convinience. Substitution of these solutions in the 3r equations (26) corresponding to the 3r values of k and α give 3r×3r secular determinant which provide 3r real values of $\omega^2(\vec{q})$, corresponding to the wave vector $\vec{q}$, denoted by $\omega_j^2(\vec{q})$, j=1, --,3r. For any $\omega_j^2(\vec{q})$, the set of equations provide the 3r components of the corresponding eigenvector, $U_\alpha(k|\vec{q}j)$. For normalisation U_α is expressed as

$$U_\alpha(k|\vec{q}j) = A(\vec{q}j)e_\alpha(k|\vec{q}j) \quad ,$$

where e_α is a dimensionless quantity satisfying Orthonormality and closure conditions :

$$\sum_{\alpha,k} e_\alpha^*(k|\vec{q}j)e_\alpha(k|\vec{q}j') = \delta_{jj'} \ , \quad \sum_j e_\alpha^*(k|\vec{q}j)e_\beta(k'|\vec{q}j) = \delta_{\alpha\beta}\delta_{kk'} \ . \tag{27}$$

The general solution is thus

$$u_\alpha(\ell k)=\frac{1}{\sqrt{Nm_k}}\sum_{\vec{q},j} e_\alpha(k|\vec{q}j)Q(\vec{q}j)\exp(i\vec{q}.\vec{X}(\ell)),\quad \text{where } Q(\vec{q}j)=A(\vec{q}j)e^{-i\omega_j(\vec{q})t} \ ,$$

N is the number of primitive cells, $\sqrt{N}$ is added for convinience.

The components of $\vec{e}(k|\vec{q}j)$, commonly known as the polarisation vector, determine the pattern of displacement of the atoms in a primitive cell corresponding to a particular mode of vibration $(\vec{q},j)$. The 3r values of $\omega_j^2(\vec{q})$ as a function of $\vec{q}$, along any direction in the Brillouin Zone constitute the 3r phonon branches.

Substituting the expression of $u_\alpha(\ell k)$ in the Hamiltonian give

$$H = \frac{1}{2}\sum_{\vec{q}j} [\dot{Q}^*(\vec{q},j)\dot{Q}(\vec{q},j)+\omega_j^2(\vec{q})Q^*(\vec{q},j)Q(\vec{q},j)]. \tag{28}$$

The Hamiltonian splits into a sum of terms one for each independant mode $(\vec{q},j)$. $Q(\vec{q}j)$, j=1, --, 3r, are called normal coordinates, describes an independant modes of vibration of the crystal. Define $\dot{a}^*_{\vec{q}j}$ and $a_{\vec{q}j}$ by

$$Q(\vec{q},j) = [\frac{\hbar}{2\omega_j(\vec{q})}]^{1/2}(a^*_{-\vec{q}j} + a_{\vec{q}j})$$

$$\dot{Q}(\vec{q},j) = i[\frac{\hbar\omega_j(\vec{q})}{2}]^{1/2}(a^*_{-\vec{q}j} - a_{\vec{q}j}) \quad .$$

The condition $Q(-\vec{q},j) = Q^*(\vec{q},j)$, necessary for the displacements to be real, is satisfied. The Hamiltonian takes the form

$$\mathcal{H} = \sum_{\vec{q},j} \hbar\omega_j(\vec{q})[a^*_{\vec{q}j} \; a_{\vec{q}j} + \frac{1}{2}] = \sum_{\vec{q},j} \mathcal{H}(\vec{q},j) \quad . \tag{29}$$

An eigenstates of $\mathcal{H}$ is a product of the eigenstates of the harmonic oscillator Hamiltonians $\mathcal{H}(\vec{q},j)$ corresponding to the independant modes $(\vec{q},j)$, j=1, --, 3r.

The eigenstate of the oscillator corresponding to the eigenvalue $\hbar\omega_j(\vec{q})(n_{\vec{q}j}+\frac{1}{2})$ is denoted by $|n_j(\vec{q})>$. $n_{\vec{q}j}$ takes values 0,1,2, ---.

$$a^*_{\vec{q}j} \; a_{\vec{q}j} |n_j(\vec{q})> = n_{\vec{q}j} |n_j(\vec{q})>. \tag{30}$$

Further, using commutation relations satisfied by $a^*_{\vec{q}j}$ and $a_{\vec{q}j}$, it can be shown

$$a^*_{\vec{q}j} |n_j(\vec{q})> = (n_{\vec{q}j}+1)^{1/2}|n_j(\vec{q})+1>$$

$$a_{\vec{q}j} |n_j(q)> = n^{1/2}_{\vec{q}j}|n_j(\vec{q})-1> \; . \tag{31}$$

$a^*_{\vec{q}j}$ and $a_{\vec{q}j}$ are known as the creation and annihilation operators as they create and destroy phonon in state corresponding to $(\vec{q},j)$, respectively. For lattice in thermal equilibrium

$$<n_{\vec{q}j}>_{Th} = [\exp(\hbar\omega_j(\vec{q})/kT)-1]^{-1} \tag{32}$$

4. SPIN LATTICE RELAXATION

4.1. Temperature and Magnetic Field Dependences of the Spin Lattice Relaxation Frequencies

The Hamiltonian can be considered as a sum of three parts

$$H_T = H_{ion} + H_{lattice} + H'_{SL} \quad . \tag{33}$$

The ionic levels in the solid in the absence of vibrations of atoms or ions in the lattice are eigenstates of H_{ion}. $H_{lattice}$ describes the lattice vibration and H'_{SL} is the interaction between the ion and the lattice which is responsible for the transitions between states of the paramagnetic ion accompanied by the creation or annihilation of phonons, conserving energy.

In general, the interaction of the paramagnetic ion with other ions at $\vec{R}_i$ in the lattice can be expressed as (20)

$$H_{int} = \sum_s (\sum_i B^s(\vec{R}_i))O^s \quad . \tag{34}$$

Here, O^s is the operator which causes the transitions between the ionic states. $B^s(\vec{R}_i)$ is dependent on the relative coordinate of the ith ion with respect to the magnetic ion under consideration. For example, consider the interaction of the paramagnetic ion with the crystal field, Eqn. 5. Here, ℓ,m refers to s and $\sum_i B^s(\vec{R}_i)$ is q_{lm}. The periodic changes in the interatomic distances due to the lattice waves which thus modulates H_{int} through $B^s(\vec{R}_i)$, is responsible for the spin lattice relaxation. The Taylor series expansion of $B^s(\vec{R}^o_i+\vec{u}_i)$ in powers of the displacements (Fig. 6) gives

$$H_{int} = \sum_{i,s} B^s(\vec{R}^o_i)O^s + \sum_{\substack{i,s\\ \alpha,\beta=x,y,z}} \frac{\partial B^s}{\partial R_{i\alpha}}\Big|_0 \frac{\partial u_{i\alpha}}{\partial\beta} R^o_{i\beta}O^s + \sum_{\substack{i,s,\\ \alpha,\beta,\gamma,\delta=x,y,z}} \frac{\partial B^s}{\partial R_{i\alpha}\partial R_{i\beta}}\Big|_0 \frac{\partial u_{i\alpha}}{\partial\gamma}\frac{\partial u_{i\beta}}{\partial\delta} R^o_{i\gamma}R^o_{i\delta}O^s + \text{----}$$

Here, the relation $(u_{i\alpha}-u_\alpha) = \sum_\beta \frac{\partial u_{i\alpha}}{\partial\beta} R^o_{i\beta}$ has been used, which is valid when the change in displacement is small over the distance $R^o_{i\beta}$, i.e., for long wavelength elastic waves. The interaction H'_{SL} is given by,

$$H'_{SL} = H_{int} - \sum_{i,s} B^s(\vec{R}^o_i)O^s \quad ,$$ which can be rewritten as

$$H_{SL} = \sum_{\alpha,\beta} G^1_{\alpha\beta}e_{\alpha\beta} + \sum_{\alpha,\beta,\gamma,\delta} G^2_{\alpha\beta\gamma\delta}\, e_{\alpha\beta}e_{\gamma\delta} + \text{---} \quad , \tag{35}$$

where all derivatives $\frac{\partial u_{i\alpha}}{\partial\beta}$ have been replaced by the strain tensor components,

$$e_{\alpha\beta} = \frac{1}{2}\left(\frac{\partial u_{i\alpha}}{\partial\beta} + \frac{\partial u_{i\beta}}{\partial\alpha}\right) \; , \quad G^1_{\alpha\beta} = \sum_{i,s}\frac{1}{2}\left(\frac{\partial B^s}{\partial R_{i\alpha}}\Big|_0 R^o_{i\beta} + \frac{\partial B^s}{\partial R_{i\beta}}\Big|_0 R^o_{i\alpha}\right)O^s \; ,$$ and

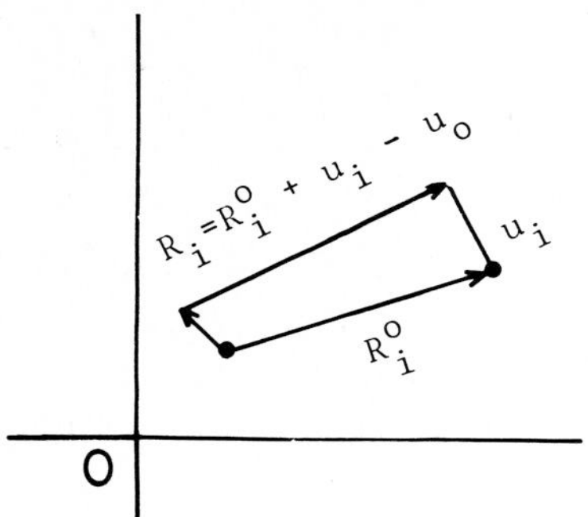

Fig. 6. The position of the neighbouring ion relative to the paramagnetic ion, R_i. The equilibrium positions are marked by '•'

$$G^2_{\alpha\beta\gamma\delta} = \sum_{i,s} \frac{1}{4}\Big(\frac{\partial^2 B^s}{\partial R_{i\alpha}\partial R_{i\beta}}\Big|_0 R^o_{i\gamma}R^o_{i\delta} + \frac{\partial^2 B^s}{\partial R_{i\alpha}\partial R_{i\delta}}\Big|_0 R^o_{i\gamma}R^o_{i\beta} + \frac{\partial^2 B^s}{\partial R_{i\beta}\partial R_{i\gamma}}\Big|_0 R^o_{i\alpha}R^o_{i\delta} + \frac{\partial^2 B^s}{\partial R_{i\gamma}\partial R_{i\delta}}\Big|_0 R^o_{i\alpha}R^o_{i\beta}\Big)O^s \ .$$

Using Eqns. (26)-(28)

$$\frac{\partial u_\alpha}{\partial \beta} = \sum_{\vec{q},j} i\Big[\frac{\hbar}{2Nm_k\omega_j(\vec{q})}\Big]^{1/2}\big[q_\beta e_\alpha(k|\vec{q}j)a_{\vec{q}j}\exp(i\vec{q}.\vec{X}(\ell)) - q_\beta e^*_\alpha(k|\vec{q}j)a^*_{\vec{q}j}\exp(-i\vec{q}.\vec{X}(\ell))\big] \ . \qquad (36)$$

When there is only one atom per primitive cell e_α is real, giving

$$e_{\alpha\beta} = \frac{i}{2}\Big[\frac{\hbar}{2Nm}\Big]^{1/2}\sum_{\vec{q},j}[\omega_j(\vec{q})]^{-1/2}\big[a_{\vec{q}j}\exp(i\vec{q}.\vec{X}(\ell)) - a^*_{\vec{q}j}\exp(-i\vec{q}.\vec{X}(\ell))\big](e_\alpha q_\beta + e_\beta q_\alpha). \qquad (37)$$

The matrix element of $e_{\alpha\beta}$ between lattice states $|n\rangle$ and $|n'\rangle$, which differ by one phonon more in mode $(\vec{q},j)$ in $|n'\rangle$

$$\langle n|e_{\alpha\beta}|n'\rangle = \langle \ldots, n_j(\vec{q}) \ldots |e_{\alpha\beta}| \ldots, n_j(\vec{q})+1, \ldots\rangle$$
$$= \frac{i}{2}\Big[\frac{\hbar}{2Nm\omega_j(\vec{q})}\Big]^{1/2}\vec{q}\exp(i\vec{q}.\vec{X}(\ell))\Big(e_\alpha\frac{q_\beta}{\vec{q}} + e_\beta\frac{q_\alpha}{\vec{q}}\Big)\sqrt{n_j(\vec{q})+1}. \qquad (38)$$

4.1.1 Direct Process

In the direct process of spin-lattice relaxation, the transition of the ionic state from $|m\rangle$ to $|m'\rangle$ is accompanied by creation or annihilation of a phonon such that the energy is conserved. Clearly the term linear in H_{SL} in the relation for H'_{SL} can give such a process. The transition probability

$$\Omega_D^{m,n\to m',n'} = \frac{2\pi}{\hbar}|\langle m,n|\sum_{\alpha,\beta} G^1_{\alpha\beta} e_{\alpha\beta}|m',n'\rangle|^2 \delta(E_m+E_n - E_{m'}-E_{n'}) \, , \qquad (39)$$

where $|n\rangle$ and $|n'\rangle$ represent the initial and final lattice states, respectively. The transition probability $\Omega_D^{m\to m'}$ is obtained by averaging over the initial states $|n\rangle$, taking into account the thermal population, and summing over the final states, satisfying the energy conservation. Thus

$$\Omega_D^{m\to m'} = \frac{2\pi}{\hbar} \sum_{q,j} |\langle m,n|\sum_{\alpha,\beta} G^1_{\alpha\beta} e_{\alpha\beta}|m',n'\rangle|^2_{Th} \delta(E_m-E_{m'} + E_n-E_{n'}) \qquad (40)$$

The subscript Th denotes the thermal averaging, and the summation is over final state. Replacing summation with an integral and using Debye approximation for the phonon spectrum, which is satisfactory at low temperatures where mainly acoustic modes of low energy are appreciably populated, the temperature dependence is obtained. Thus,

$$\Omega_D^{m\to m'} = \frac{2\pi}{\hbar} \sum_j \frac{V}{8\pi^3} \int \sum_{\alpha,\beta} |\langle m|G^1_{\alpha\beta}|m'\rangle \langle n|e_{\alpha\beta}|n'\rangle|^2_{Th} \delta(E_m-E_{m'} + E_n - E_{n'}) q^2 dq\Omega$$

V is the volume of the N cells in the crystal, $d\Omega$ is the solid angle. The substitution of $\langle n|e_{\alpha\beta}|n'\rangle$ (Eqn. 38) gives

$$\Omega_D^{m\to m'} \propto \frac{(\hbar\omega)^2 \exp\frac{(\hbar\omega)}{kT}}{\exp(\frac{\hbar\omega}{kT})-1} \qquad (41)$$

where $E_m-E_{m'}=\hbar\omega$. At temperature such that $kT \gg \hbar\omega$, which is of the order of a few cm^{-1}, $\Omega_D^{m\to m'} \propto \omega^2 T$. On the other hand, when $kT \ll \hbar\omega$, $\Omega_D^{m\to m'} \propto \omega^3$.

4.1.2 Two Phonon Processes

The process of relaxation in which the transition of the ionic state from $|m\rangle$ to $|m'\rangle$ is accompanied by the creation or annihilation of two phonons or creation of a phonon and annihilation of another phonon such that the energy is conserved is known as the Raman process. The energy conservation in the two cases can be expressed as

$E_{m'}-E_m = \hbar\omega_j(\vec{q}) + \hbar\omega_{j'}(\vec{q}')$ and $E_{m'}-E_m = \hbar\omega_j(\vec{q}) - \hbar\omega_{j'}(\vec{q}')$, respectively, Evidently, the range of energy in the phonon spectrum is highly restricted in the first of the two possibilities and thus this process is normally insignificant compared to the second mode in which the entire phonon spectrum is used.

The second term in the expansion of H'_{SL} (Eqn. 35) gives two phonon process due to the matrix element

$$\Omega_{R1}^{m\to m'} = \frac{2\pi}{\hbar}\left[\frac{\hbar}{2Nm_k}\right]^2\left(\frac{V}{8\pi^3}\right)^2 \sum_j \int \left|\sum_{\alpha,\beta,\gamma,\delta} \langle m|G^2_{\alpha\beta\gamma\delta}|m'\rangle \frac{1}{2}\left(e_\alpha \frac{q_\beta}{\vec{q}} + e_\beta \frac{q_\alpha}{\vec{q}}\right) \times \right.$$

$$\left.\frac{1}{2}\left(e_\gamma \frac{q'_\delta}{\vec{q}'} + e_\delta \frac{q'_\gamma}{\vec{q}'}\right)\right|^2_{Th} \frac{\exp\left(\frac{\hbar\omega_j(q)}{kT}\right)}{\left[\exp\left(\frac{\hbar\omega_j(q)}{kT}\right)-1\right]\left[\exp\left(\frac{\hbar\omega_{j'}(q')}{kT}\right)-1\right]} \times$$

$$\delta(E_m - E_{m'} - \hbar\omega_{j'}(q') + \hbar\omega_j(q)) \frac{q^4 q'^4}{\omega_j(q)\omega_{j'}(q')} dq dq' d\Omega d\Omega' \qquad (42)$$

This gives a temperature dependence,

$$\Omega_{R1}^{m\to m'} \propto \left(\frac{\hbar}{16\pi^4\rho^2\nu^{10}}\right) \int_0^{\omega_D = \frac{k\Theta_D}{\hbar}} \frac{\omega^6 \exp\left(\frac{\hbar\omega}{kT}\right)}{\left[\exp\left(\frac{\hbar\omega}{kT}\right)-1\right]^2} d\omega \; . \qquad (43)$$

The difference between $\hbar\omega(q)$ and $\hbar\omega(q')$ is negligible in comparison to the thermal energy, Θ_D is the Debye Cut off frequency, ρ is the crystal density, and ν is the velocity of longitudinal as well as transurse sound wave.

If Einstein's model is used instead of the Debye model which is quite appropriate for optical phonons with little dispersion in their $\vec{q}$ dependence,

$$\Omega_{R1}^{m\to m'} \propto \frac{\exp\left(\frac{\Theta_E}{T}\right)}{\left[\exp\left(\frac{\Theta_E}{T}\right)-1\right]^2}, \text{ where } k\Theta_E = \hbar\omega_E \qquad (44)$$

Thus in Debye Approximation* ,

---

* The integral $I_n = \int_0^{k\Theta/\hbar} \frac{\omega^n \exp\left(\frac{\hbar\omega}{kT}\right)}{\left[\exp\left(\frac{\hbar\omega}{kT}\right)-1\right]^2} d\omega$ is

$= \int_0^\infty \omega^n \exp\left(-\frac{\hbar\omega}{kT}\right) d\omega \sim n!\left(\frac{kT}{\hbar}\right)^{n+1}$ when $T/\Theta \ll 1$

$= (n-1)^{-1}(k/\Theta)^{n+1}\,\Theta^{n-1}T^2$ when $T \gg \Theta$.

$$\Omega_{R1}^{m\to m'} \propto T^7 \text{ for } T << \frac{\hbar\omega_D}{k}$$

$$\propto T^2 \text{ for } T >> \frac{\hbar\omega_D}{k} \quad .$$

In Einstein approximation,

$$\Omega_{R1}^{m\to m'} \propto T^2 \text{ for } T >> \frac{\hbar\omega_D}{k} \quad .$$

The leading term in H'_{SL} also contribute to the Raman process of relaxation but in second order of perturbation theory. The matrix element for this process is

$$<m,n|H|m',n'> = \sum_{m'',n''} \frac{<m,n|H|m'',n''> \; <m'',n''|H|m',n'>}{E_m + E_n - E_{m''} - E_{n''}}$$

$$= \sum_{\alpha,\beta,\gamma,\delta} \frac{<m\; G^1_{\alpha\beta}|m''><m''|G^1_{\gamma\delta}|m'><n_j(\vec{q}),n_{j'}(\vec{q}\,')|e_{\alpha\beta}|n_j(\vec{q})+1,n_{j'}(\vec{q}\,')>}{E_m - E_{m''} - \hbar\omega_j(\vec{q})} \times$$

$$<n_j(\vec{q})+1,n_{j'}(\vec{q}\,')|e_{\gamma\delta}|n_j(\vec{q})+1,n_{j'}(\vec{q}\,')-1>$$

$$+ \sum_{\alpha,\beta,\gamma,\delta} \frac{<m|G^1_{\alpha\beta}|m''><m''|G^1_{\gamma\delta}|m'><n_j(\vec{q}),n_{j'}(\vec{q}\,')|e_{\alpha\beta}|n_j(\vec{q}),n_{j'}(\vec{q}\,')-1>}{E_m - E_{m''} + \hbar\omega_{j'}(\vec{q}\,')}$$

$$<n_j(\vec{q}),n_{j'}(\vec{q}\,')-1|e_{\gamma\delta}|n_j(\vec{q})+1,n_{j'}(\vec{q}\,')-1> \quad . \tag{45}$$

The two intermediate states of the lattice have been schematically shown in Fig. 7. The lattice state is represented by the occupation numbers of lattice modes in which a phonon has been either created or annihilated. The transition probability can be calculated using procedure similar to the one used for Eqn. (41). In Debye approximation, this gives a temperature dependence:

$$\Omega_{R2}^{m\to m'} \propto \left(\frac{h}{8\pi^4\rho^2\nu^{10}} \right) \int_0^{\omega_D} \frac{\omega^6 \exp(\frac{\hbar\omega}{kT})}{[\exp(\frac{\hbar\omega}{kT})-1]^2} \; \frac{(E_{m''}-E_m)^2 + (\hbar\omega)^2}{[(E_{m''} - E_m)^2 - (\hbar\omega)^2]^2} \, d\omega$$

when $\hbar\omega_D < (E_{m''} - E_m)$, the range of integration does not include the frequency when $[(E_{m''} - E_m) - \hbar\omega] \to 0$ as ω_D represents the highest frequency of phonons available. In this case, assuming $\hbar\omega << E_{m''}-E_m$

$$\Omega_{R2}^{m\to m'} \propto \int_0^{\omega_D} \frac{\omega^6 \exp(\frac{\hbar\omega}{kT})}{[\exp(\frac{\hbar\omega}{kT})-1]^2} \, d\omega \tag{46}$$

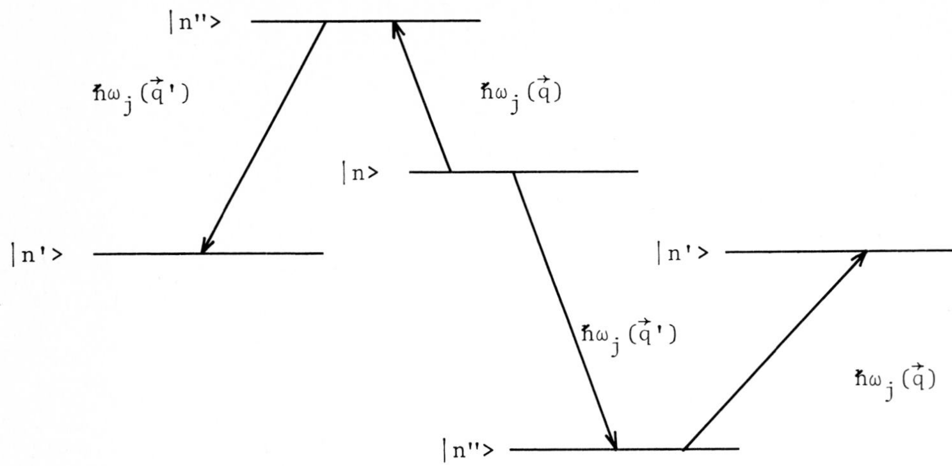

Fig. 7. Schematic representation of the Raman relaxation process. |n''> can be higher or lower in energy than |n>. The two terms in the matrix element for the second order Raman relaxation process, Eqn. 45, represents the two situations, respectively.

which has the same temperature dependence as $\Omega_{R1}^{m\to m'}$. On the other hand, if $E_m - E_{m''} < \hbar\omega_D$, the range of integration includes the frequency at which the integrand diverges. Even if the excited level is not appreciably populated thermally, the resonance nature of the process makes it very effective. For the evaluation of the transition probability in this case the method outlined by Heitler (22) for treating the problem of resonance fluorescence is used (23,24). The finite width of the level $|m''\rangle$ due to its finite lifetime is taken into consideration. Thus, $[(E_{m''}-E_m)^2 - \hbar^2\omega^2]^2$ in the denominator of the integrand is replaced with $[(E_{m''}-E_m)+\hbar\omega]^2|(E_{m''}-E_m-\hbar\omega + i\frac{\Gamma}{2})|^2$ where Γ is the width of the intermediate level $|m''\rangle$, which can be calculated from the probabilities per unit time of transition $|m''\rangle \rightarrow |m\rangle$ and $|m'\rangle$. If the maxima in the integrand at $E_m-E_{m''}=\hbar\omega$ is sharp, it is possible to split the integral into two parts.

1. *The Non Resonant Part,* which excludes the range of frequency around $E_m-E_{m''}$. The behaviour when the range of integration is such that $\hbar\omega \ll E_{m''}-E_m$ was described earlier. On the other hand, in the range of integration such that $E_m-E_{m''} \ll \hbar\omega$, the integral can be obtained using approximation $\hbar\omega \pm (E_{m''}-E_m) \approx \hbar\omega$. This gives the temperature dependence

$$\Omega_{R2}^{m\to m'} \propto \int_0^{\omega_D} \frac{\omega^4 \exp(\frac{\hbar\omega}{kT})}{[\exp(\frac{\hbar\omega}{kT})-1]^2}\, d\omega \ . \qquad (47)$$

The lower range of the integral has been made zero as it would not significantly change the value of the integral. This is proportional to T^5, when $T \ll \Theta_D$, and proportional to T^2 when $T \gg \Theta_D$.

2. *The Resonant Part* (20,23), can also be easily found by replacing $[(\hbar\omega-(E_{m''}-E_m))^2 + \frac{\Gamma^2}{4}]^{-1}$ with $\frac{2\pi}{\Gamma}\,\delta(\hbar\omega-(E_{m''}-E_m))$, which gives

$$\Omega_{RR2}^{m\to m'} \propto \exp(-\frac{E_{m''}-E_m}{kT}) \quad . \tag{48}$$

This becomes small if $kT \gg E_{m''}-E_m$.

4.2 The Waller Process

The spin-spin interaction between two paramagnetic ions, dipolar as well as exchange interaction, depends on the interionic separation (R). Lattice vibrations modulate this interaction through periodic variation in the interatomic separation and thus can cause transitions between ionic states. This mode may become important (25) when the other mechanism (Van-Vleck mechanism) is less important (4), due to the quenching of orbital moment. Thus, this process need be considered for the S-state ions (Fe^{3+}, Mn^{2+} etc.). Accordingly, we assume that the paramagnetic ions possess spin moments only.

The general form of the interaction of the paramagnetic ion with other ion is given as

$$H_{int} = \sum_s B^s(\vec{R})O^s \quad .$$

The terms in the spin spin interaction which change the spin state of the paramagnetic ion, without causing a change in the state of the interacting ion at $\vec{R}$ is of the form

$$B^z S_x S_z' + B^{yz} S_y S_z' \quad . \tag{49}$$

Similarly, the part of the spin spin interaction which change the spin state of both the interacting ions is of the form

$$B^{xx} S_x S_x' + B^{yy} S_y S_y' + B^{xy} S_x S_y' + B^{yx} S_y S_x' \quad . \tag{50}$$

It can be easily seen that if H_{int} is the dipolar interaction between the spins of the two ions,

$$H_{ss} = g^2\beta^2[\frac{\vec{S}.\vec{S}'}{R^3} - \frac{3(\vec{S}.\vec{R})(\vec{S}'.\vec{R})}{R^5}] \; , \; B^{xz} = 3g^2\beta^2\, R^{-5}\, R_x R_z, \quad \text{and so on}$$

Thus, $G^1_{\alpha\beta}$ and $G^2_{\alpha\beta\gamma\delta}$ for the two processes (Eqn. 35) can be computed. Averaging the matrix elements $<m|G^1_{\alpha\beta}|m'>$ & $<m|G^2_{\alpha\beta\gamma\delta}|m'>$ over m and m' and the directions of $\vec{R}$, it can be shown (24) that the probability of the process in which the state of both the interacting ions change is greater than the process in which the state of only one of the two ion changes. Furthermore, the average transition probability of the change i in the spin state of the paramagnetic ion is proportional to $Z(g^4\beta^4/R^6)S^2(S+1)^2$ for both the processes in Direct spin lattice relaxation mode as well as in Raman process due to the second term in H'_{SL}.

The isotropic exchange interaction between the ions commute with Zeeman energy and thus does not contribute to the spin lattice relaxation. The anisotropic exchange interaction

$$H_{int} = J(R)[\vec{S}.\vec{S}' - \frac{3}{R^2}(\vec{S}.\vec{R})(\vec{S}'.\vec{R})]$$

is similar in form to the dipole dipole interaction and its contribution to the relaxation can be similarly found.

4.3 Van Vleck Mechanism

The crystal field interaction of the ion, H_{cr}, is dependent on the inter-nuclear separations and provides the most important mechanism (26) of spin lattice relaxation, unless the ground term of the free ion is an S-state and thus excited states (terms) are very far in energy from it when the Waller's mechanism (25) also become relatively important. It was seen earlier (Eqns. 5 and 15)

$$H_{cr} = \sum_{k,q\geq 0} \beta\, q_{kq}<r^k>O_{kq} = \sum_{k,q} V_{kq} \quad , \tag{51}$$

where the constants β appear as a result of the use of the operator equivalent method. Thus the superscript s in $\sum_s B^sO^s$ refer to (k,q) and B^s and O^s denote $\beta <r^k>q_{kq}$ and O_{kq}, respectively. This gives,

$$G^1_{\alpha\beta} = \frac{1}{2}\sum_{k,q} \beta <r^k>(\frac{\partial q_{kq}}{\partial R_\alpha}|_0 R^o_\beta + \frac{\partial q_{kq}}{\partial R_\beta}|_0 R^o_\alpha)O_{kq} \quad .$$

It was seen earlier (Section 2.2.1) that $q_{kq} \propto \frac{1}{R^{k+1}}$. Thus, Orbach used the approximation

$$G^1_{\alpha\beta} = \sum_{k,q} \beta <r^k>q_{kq}O_{kq} = \sum_{k,q} V_{kq} \quad .$$

Similar approximation is used for G^2 ,

$$G^2_{\alpha\beta\gamma\delta} = \sum_{k,q\geq 0} V_{kq} \quad .$$

The matrix elements $<M|G^1_{\alpha\beta}|M'>$ and $<M|G^2_{\alpha\beta\gamma\delta}|M'>$ are non zero and easy to calculate when the levels M and M' correspond to non Kramers states and the formulation described earlier is directly applicable. To see the order of magnitude of $G^1_{\alpha\beta}$ and $G^2_{\alpha\beta\gamma\delta}$, consider terms with $k = 2$. G^1 and G^2 are $\sim \frac{e^2<r^2>}{R^3}$ in this case, the relaxation probability $\sim (\frac{e^2<r^2>}{R^3})^2$. The strength of Waller mechanism, on the other hand, is $\sim (\frac{g^2\beta^2}{R^3})^2$. Thus in this case the ratio

$$(\frac{e^2<r^2>}{R^3} \ / \ \frac{g^2\beta^2}{R^3})^2 \sim 10^8 \quad ,$$

showing the Waller's mechanism is negligible*. Further, the contribution to relaxation probability decreases with increasing R. On the other hand, if $|M>$ and $|M'>$ are the two component states of a Kramers doublet

$$<M|H_{cr}|M'> = 0$$

in absence of an external magnetic field. In presence of an external field $\vec{H}$, the admixture of the wave functions of the state $|c>$ and $|c'>$ of an excited Kramers doublet with the wave function of the ground states give

$$|M_1> = |M> - \frac{<M|g\beta\vec{H}.(\vec{L}+2\vec{S})|c>}{E_c - E_M}|c>,$$

$$|M_1'> = |M'> - \frac{<M'|g\beta\vec{H}.(\vec{L}+2\vec{S})|c'>}{E_{c'} - E_{M'}}|c'> \quad .$$

From the properties of the Kramers doublet, if $<M|\vec{L}+2\vec{S}|c>\neq 0$, $<M|\vec{L}+2\vec{S}|c'> = 0$. The matrix element of ΣV_{kq} between $|M_1>$ and $|M_1'>$ is

* $<r^2>$ and $<r^4>$ for 3d electrons of some iron group ions, in atomic unit, are given below (27)

Ion	term	$<r^2>$	$<r^4>$
Mn^{2+}	^6S	1.528	5.325
Fe^{3+}	^6S	1.141	2.765
Fe^{2+}	^5D	1.391	4.530
Co^{2+}	^4F	1.262	3.861

nonzero, contributes an extra factor $(\frac{Hg\beta}{E_c-E_M})^2$ to the relaxation transition probabilities $\Omega_D^{m\to m'}$ and $\Omega_{R1}^{m\to m'}$ in comparison to the case when the states $|M>$ and $|M'>$ are of non Kramers type. When $|M>$, $|M'>$, $|c>$ and $|c'>$ represent states of rare earth ion corresponding to the manifold of (J^2, J_z), the interaction with the external magnetic field $g\beta\vec{H}.\vec{J}$ need be substituted for $g\beta\vec{H}.(\vec{L}+2\vec{S})$ in the above equation.

The presence of $(\frac{Hg\beta}{E_c-E_M})^2$ reduces the strength of Van Vleck's process in comparison to the case when $|M>$ and $|M'>$ are non Kramers type. Thus if the splitting due to $\vec{H}$, $g\beta H$, $\sim 1\ cm^{-1}$ and $(E_c-E_m)\sim 10^4\ cm^{-1}$, the extent of reduction is sufficient to make the Waller's process important. Waller's process is further characterised by its strong dependence on the concentration of the paramagnetic ions.

4.4 Spin Phonon Coupling and Internal Modes

The alternate approach to deal with the spin lattice relaxation process which is suitable to ions of the iron group, is described in this section.

In general, the iron group ions in solids exist as well as bound cluster of atoms or dipoles, e.g., $[Fe(H_2O)_6]^{3+}$ complex with Fe^{3+} ion at the centre of the six water dipoles forming an octahedral environment. The binding forces within the cluster are far stronger than with other atoms and other clusters in the solid. It is thus reasonable to start with a consideration of modulation of the crystal field at the central paramagnetic ion due to vibrational modes of this cluster. The frequencies of these modes are affected when the cluster is in solid, giving rise to $\vec{q}$-dependence, but the change is small due to the weak binding with other atoms in the solid. The lattice modes of the solid thus derived from the internal modes of the cluster are far more influential in modulating the crystal field at the paramagnetic ion because not only the moving atoms are closer but also lead to large changes in the distances of the central ion from other ions. Other modes in which the cluster moves as a whole, contribute little to the relaxation process of the central ion.

The vibrational modes of a cluster of n atoms can be obtained using procedure outlined below for a cluster with octahedral coordination of ligands. The paramagnetic ion at the centre is surrounded by the six charges at $(\pm R, 0, 0)$, $(0, \pm R, 0)$ and $(0, 0, \pm R)$. The total degrees of freedom for the motion of the seven atoms is 21. To each atom is assigned Cartesian displacement vectors, (X_o,Y_o,Z_o), ... , (X_6,Y_6,Z_6), as shown in Fig. 8(a). Let us examine the transformation properties of the 21 displacement vectors. An identity operation leaves all the displacement vectors unchanged. This operation can be represented by a 21 x 21 unit matrix of trace 21. A two fold rotation about the z axis transforms the displacements as shown in Fig. 8(b). The transformation can be represented in matrix form as:

	$X_oY_oZ_o$	$X_1Y_1Z_1$	$X_2Y_2Z_2$	$X_3Y_3Z_3$	$X_4Y_4Z_4$	$X_5Y_5Z_5$	$X_6Y_6Z_6$
X_o	-1 o o						
Y_o	o -1 o						
Z_o	o o 1						
X_1					-1 o o		
Y_1					o -1 o		
Z_1					o o 1		
X_2						-1 o o	
Y_2						o -1 o	
Z_2						o o 1	
X_3				-1 o o			
Y_3				o -1 o			
Z_3				o o 1			
X_4		-1 o o					
Y_4		o -1 o					
Z_4		o o 1					
X_5			-1 o o				
Y_5			o -1 o				
Z_5			o o 1				
X_6							-1 o o
Y_6							o -1 o
Z_6							o o 1

The trace of this representation matrix is (-3). Similarly, the traces of the representation matrices corresponding to other symmetry operations can be obtained to yield the character table:

Representation	E	$8C_3$	$6C_2$	$6C_4$	$3C_2(=C_4^2)$	i	$6S_4$	$8S_6$	$3\sigma_h$	$6\sigma_v$
Γ	21	o	-1	3	-3	-3	-1	o	5	3

This reducible representation can be split into irreducible representa-

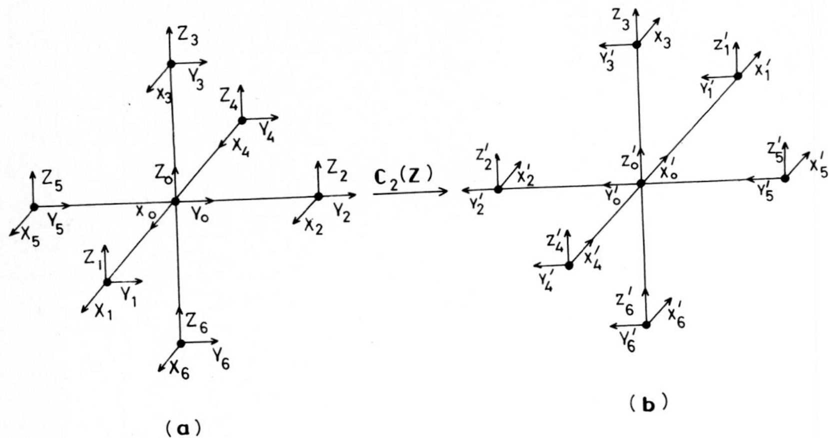

Fig. 8. The paramagnetic ion at (0,0,0) surrounded by the six ions at (±R, 0,0), (0,±R,0) and (0,0,±R). The rotation by 180° around Z-axis changes the cartesian displacement vectors as shown in this figure.

tions of the 'O_h' group using the conventional method:

$$= A_{1g} + E_g + T_{1g} + 3T_{1u} + T_{2g} + T_{2u} \quad .$$

The subscript g denotes a representation which is symmetric with respect to inversion and u denotes one which is asymmetric to inversion. T_{1g} and one T_{1u} representations correspond to the pure rotation and the translation of the cluster, respectively, and are not of interest.

The bases vectors of the irreducible representations, which are 21 in number, are the linear combinations of the 21 displacement vectors, and are called the normal coordinates (Q). They represent independent vibrational modes of the cluster. The normal coordinates corresponding to the symmetric representations are (Fig. 9).

$$Q_1 = \frac{1}{\sqrt{6}}(X_1 - X_4 + Y_2 - Y_5 + Z_3 - Z_6),$$

$$Q_2 = \frac{1}{2}(X_1 - X_4 - Y_2 + Y_5),$$

$$Q_3 = \frac{1}{\sqrt{3}}(\frac{1}{2}(X_1 - X_4 + Y_2 - Y_5) - Z_3 + Z_6),$$

$$Q_4 = \frac{1}{2}(Y_1 - Y_4 + X_2 - X_5),$$

$$Q_5 = \frac{1}{2}(Z_1 - Z_4 + X_3 - X_6),$$

$$Q_6 = \frac{1}{2}(Z_2 - Z_5 + Y_3 - Y_6).$$

It is helpful to note, xy, xz, and yz transform under symmetry

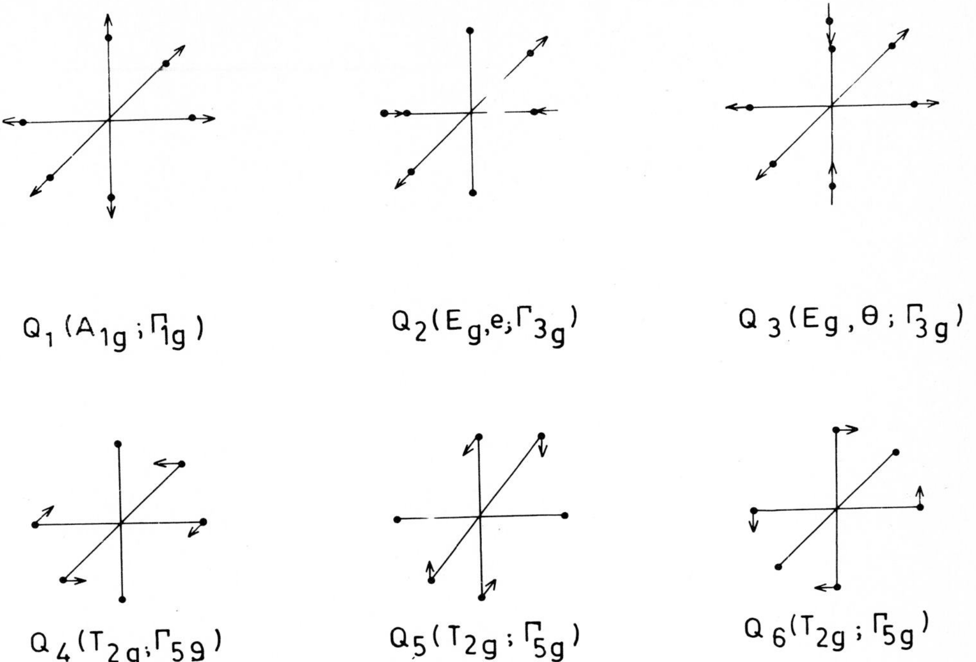

Fig. 9. The symmetrical normal modes of vibration of the octahedral complex shown in Fig. 8.

operations of the O_h group as Q_4, Q_5 and Q_6 normal coordinates respectively. Similarly $3z^2-x^2-y^2$ and x^2-y^2 correspond to Q_3 and Q_2 respectively. The expressions for other normal coordinates can be found in Ref. (2). Expanding the crystal field potential in series of the normal coordinates,

$$H_{cr} = H^o_{cr} + \sum_{n,m}\sum_{f}\left(\frac{\partial V_{nm}}{\partial Q_f}\right)_o Q_f + \frac{1}{2}\sum_{n,m}\sum_{f,f'}\left(\frac{\partial^2 V_{nm}}{\partial Q_f \partial Q_{f'}}\right)_o Q_f Q_{f'} + \text{--} \quad (52)$$

$$= H^o_{cr} + \sum_{f} V_f\, Q_f + \frac{1}{2}\sum_{f,f'} V_{ff'}\, Q_f Q_{f'} + \text{-----}\,,$$

where V_f and $V_{ff'}$ contain parts which operate on the ionic states causing transitions between them as well as the parts which depend on coordinates of the neighbouring charges. While considering the transitions among the states $|L,S,M_L,M_s>$ of a term $(\vec{L},\vec{S})$, the operator equivalent method can be used to express the operator part in V's in terms of L_x, L_y and L_z. For the case of the six charges e' at the octahedral positions $(\pm R,o,o)$, $(o,\pm R,o)$, and $(o,o,\pm R)$, V_1, V_2 ...., V_6 were calculated by Van Vleck (21) and are given below in terms of the coordinates x,y,z of the d electron of the central ion and R.

$$V_2 = \Sigma(A(x^2-y^2) + B(x^4-y^4) + \ldots\ldots)$$

$$V_3 = \Sigma(A(x^2+y^2 - 2z^2) + B(x^4-y^4-2z^4) + \ldots\ldots)/\sqrt{3}$$

$$V_4 = \Sigma(Cxy + E(x^3y + yx^3) + \ldots\ldots)$$

$$V_5 = \Sigma(Cxz + E(x^3z + xz^3) + \ldots\ldots)$$

$$V_6 = \Sigma(Cyz + E(y^3z + yz^3) + \ldots\ldots) \quad \text{with}$$

$$A = \frac{1}{4}\, ee'\, (18R^{-4} - 75R^{-6}\, r^2), \quad B = 175\, ee'/8R^6 \quad ,$$

$$C = ee'(-6R^{-4} + 15R^{-6}\, r^2) \quad , \quad E = -35\, ee'/2R^6$$

Here, the summation is over all the d-electrons on the central ion. Restricting to terms upto second order,

$$V_2 = \Sigma F(Y_2^{+2} + Y_2^{-2}), \quad V_3 = -\Sigma\sqrt{6}FY_2^o \ , \quad V_4 = \Sigma\frac{2i}{3}F(Y_2^{+2} - Y_2^{-2}),$$

$$V_5 = \Sigma\frac{2}{3}F(Y_2^{+1} - Y_2^{-1}), \quad V_6 = \Sigma\frac{2i}{3}\, F(Y_2^{+1} + Y_2^{-1}), \quad \text{where} \quad F = 9(\frac{2\pi}{15})^{1/2}\, \frac{ee'r^2}{R^4},$$

When the cluster is in a solid, the vibrational modes of the cluster become the normal modes of the lattice. Not only the energy of these modes acquire $\vec{q}$ dependence but also admixture from other modes which leave the transformation properties of the normal coordinates unchanged. If the primitive cell contains only the cluster under consideration, the number of modes j corresponding to $\vec{q}$ is same as that of the isolated cluster. If, however, there are other atoms or molecular clusters in the primitive cell, the number of modes j for any $\vec{q}$ in the solid is larger.

The displacement vectors $X_o, Y_o, \ldots\ldots, X_6, Y_6, Z_6$ in the solid can be expressed in terms of the creation an annihilation operators of the phonons using the expression for u_α given earlier (Eqn. 26). Thus

$$u_\alpha(\vec{X}(\ell)+\vec{a}) - u_\alpha(\vec{X}(\ell)) = i\sum_{\vec{q},j} [\frac{\hbar}{2Nm\omega_j(\vec{q})}]^{1/2} e_\alpha \times$$

$$[a_{\vec{q}j} \exp(i\vec{q}.\vec{X}(\ell)) - a^*_{\vec{q}j} \exp(-i\vec{q}.\vec{X}(\ell))] \mathrm{Sin}\vec{q}.\vec{a} \qquad (53)$$

assuming $\mathrm{Cos}\vec{q}.\vec{a}\sim 1$, which is true when the wavelength of the lattice mode is large compared to $\vec{a}$, and one atom per primitive cell, for simplicity. Thus,

$$X_1 = \sum_{\vec{q},j} i\left[\frac{\hbar}{2Nm\omega_j(\vec{q})}\right]^{1/2} e_x[a_{\vec{q}j}\exp(i\vec{q}.\vec{X}(\ell))-a^*_{\vec{q}j}\exp(-i\vec{q}.\vec{X}(\ell))]\mathrm{Sin}(q_xR) \quad ,$$

and so on. Thus,

$$Q_M = i\sum_{\vec{q},j}\left[\frac{\hbar}{2Nm\omega_j(\vec{q})}\right]^{1/2} A_M[a_{\vec{q}j}\exp(i\vec{q}.\vec{X}(\ell))-a^*_{\vec{q}j}\exp(-i\vec{q}.\vec{X}(\ell))] \quad , \quad (54)$$

assuming the central paramagnetic ion is at $X(\ell)$. Here,

$$A_2 = e_x\mathrm{Sin}q_xR - e_y\mathrm{Sin}q_yR \;, \quad A_3 = \frac{1}{\sqrt{3}}(e_x\mathrm{Sin}(q_xR)+e_y\mathrm{Sin}(q_yR)-2e_z\mathrm{Sin}(q_zR)),$$

$$A_4 = e_y\mathrm{Sin}(q_xR) + e_x\mathrm{Sin}(q_yR), \quad A_5 = e_z\mathrm{Sin}(q_xR) + e_x\mathrm{Sin}(q_zR),$$

$$A_6 = e_z\mathrm{Sin}(q_yR) + e_y\mathrm{Sin}(q_zR),$$

Instead of using Q_1, ..., Q_6 as normal coordinate of the vibrational modes, it is possible to use the linear combinations

$$Q(E_g,e) = \frac{1}{2}Q_2 \;, \quad Q(E_g,\Theta) = \frac{1}{2}Q_3 \;, \quad Q(T_{2g},1) = -\frac{i}{2\sqrt{2}}[Q_6 + iQ_5] \;,$$

$$Q(T_{2g},o) = \frac{i}{2}Q_4 \;, \quad Q(T_{2g},-1) = \frac{i}{2\sqrt{2}}(Q_6 - iQ_5)$$

Expanding $\mathcal{H}_{cr}$ in series in powers of Q thus defined, the first term in the expansion of $(\mathcal{H}_{cr} - \mathcal{H}^o_{cr})$ can be written as

$$(\mathcal{H}_{cr} - \mathcal{H}^o_{cr}) = \sum_{\substack{m=\varepsilon,\Theta\\ \ell=2,4,\ldots}} V(E_g,\ell)C(E_g\ell,m)Q(E_g,m) + \sum_{\substack{m=\pm1,o\\ \ell=2,4,\ldots}} V(T_{2g},\ell)$$
$$C(T_{2g}\ell,m)Q(T_{2g},-m)(-1)^m + ----- \quad (55)$$

where C's are the functions of the spherical harmonics. We evaluate a typical term in this sum:

$$V(T_{2g},2)C(T_{2g}2,1)Q(T_{2g},1) = -\sum_i\frac{\partial V_{cr}}{\partial Q_i}\frac{\partial Q_i}{\partial Q(T_{2g},1)}Q(T_{2g},1)$$
$$= -2\sqrt{2}(V_5-iV_6)Q(T_{2g},-1)$$
$$= \frac{8\sqrt{2}}{3}\Sigma FY_2^{-1}\,Q(T_{2g},-1) \quad .$$

Thus, $C(T_{2g}2,1) = Y_2^{-1}$. Similarly

$$V(T_{2g},2)C(T_{2g}2,-1)Q(T_{2g},+1) = -\sum_i\frac{\partial V_{cr}}{\partial Q_i}\frac{\partial Q_i}{\partial Q(T_{2g},+1)}Q(T_{2g},+1)$$
$$= -2\sqrt{2}[V_5+iV_6]Q(T_{2g},+1)$$

$$= -\frac{8\sqrt{2}}{3}\Sigma FY_2^{+1}\ Q(T_{2g},+1) \quad .$$

Thus, $C(T_{2g}2,-1)=-Y_2^1$, and so on. Here, we have denoted $\sum_{\ell,m} V_{\ell m}$ by V_{cr}. Similarly, other terms in Eqn. 55 can be obtained (28). It is seen that $V(T_{2g},2)$, $V(E_g,2) \propto R^{-3}$ and $V(T_{2g},4)$ and $V(E_g,4) \propto R^{-5}$. Thus, the corresponding relaxation probabilities falls off as R^{-6} and R^{-10}, respectively. It is thus sufficient to include only the nearest neighbours as is done in the cluster consideration. The contribution of the next nearest neighbour should be negligibly small.

In the long wavelength approximation, the relative displacement of atoms is small. Thus we can write

$$e_{xx} = \frac{X_1}{R} = -\frac{X_4}{R}, \quad e_{yy} = \frac{Y_2}{R} = -\frac{Y_5}{R}, \quad e_{zz} = \frac{Z_3}{R} = -\frac{Z_6}{R},$$

$$e_{xy} = \frac{1}{2R}(Y_1+X_2) = -\frac{1}{2R}(Y_4+X_5) \quad , \quad e_{yz} = \frac{1}{2R}(Y_3+Z_2) = -\frac{1}{2R}(Y_6+Z_5) \quad ,$$

$$e_{xz} = \frac{1}{2R}(X_3+Z_1) = -\frac{1}{2R}(X_6+Z_4).$$

Thus, Q's can be reexpressed as

$$\varepsilon(A_{1g}) = \frac{Q(A_{1g})}{R} = \frac{1}{\sqrt{6}}(e_{xx}+e_{yy}+e_{zz}), \quad \varepsilon(E_g,e) = \frac{Q(E_g,e)}{R} = \frac{1}{2}(e_{xx}-e_{yy}) \ ,$$

$$\varepsilon(E_g,\Theta) = \frac{1}{\sqrt{3}}\ [\tfrac{1}{2}(e_{xx}+e_{yy})-e_{zz}], \quad \varepsilon(T_{2g},1) = -\frac{i}{\sqrt{2}}(e_{yz}+ie_{xz}) \quad ,$$

$$\varepsilon(T_{2g},1) = -\frac{i}{\sqrt{2}}(e_{yz}+ie_{xz}), \quad \varepsilon(T_{2g},0) = ie_{xy}, \quad \varepsilon(T_{2g},-1) = \frac{i}{\sqrt{2}}(e_{yz}-ie_{xz}).$$

Accordingly, $(H_{cr} - H^o_{cr})$ can be reexpressed in a series in ε(Eqns. 63).

4.5 Dynamical Spin Hamiltonian

4.5.1 Perturbation Theory Method

We had discussed the spin Hamiltonian method of including the effect of the admixture of excited states into the ground state on the splitting of the spin independant Hamiltonian states. Similarly, the effect of the admixture of the excited states into the ground states by small perturbations on the transition probabilities between states of the spin independant Hamiltonian by V_f and $V_{ff'}$ is included by using an effective Hamiltonian known as the dynamical spin Hamiltonian. In this section, the method of Mattuck and Strandberg (28) is briefly outlined. The method is not applicable to ions, which when free, have S-term as the ground state. It is assumed,

$H_c + H' > H_{cr} > H_{SO}$.

Thus, the method is applicable to iron group ions only. The Hamiltonian is split as

$$H = H_A + P + H_{lattice} + H'_{SL} \tag{56}$$

where $H_A = H_c + H' + 2\beta\vec{H}.\vec{S} + H^o_{cr}$,

$P = \lambda\vec{L}.\vec{S} + 2\beta\vec{L}.\vec{H}$,

$H_o = H_A + P$, and

$H'_{SL} = \sum_f V_f Q_f + \sum_{f,f'} V_{ff'} Q_f Q_{f'} + \cdots$.

It is further assumed here, the levels of the ground state manifold among which the relaxation transitions occur correspond to a orbital non degenerate state of H_A. Thus the number of these levels is (2S+1), where $\vec{S}$ is the spin of the term to which the levels belong. As discussed in see 2.3.1, the eigenstates of H_A are divided into two groups. One group of states denoted by ϕ_j, ϕ_k, ϕ_ℓ etc. are the low energy states, known as ground states. They are well separated from excited states denoted by ϕ_μ, ϕ_γ, etc., (Fig. 5). The matrix element of V_f and $V_{ff'}$ between any two states, $\langle OM_s|V_f|OM'_s\rangle$ is zero, since V_f and $V_{ff'}$ do not act on spin parts of the wavefunction. Thus, V_f or $V_{ff'}$ does not cause relaxation transitions between the unperturbed ground states of H_A. If, however, the admixture of the excited states ϕ_μ, ϕ_γ, etc., with non zero orbital moment, into the ground states due to P is considered, the states $|0,M_s\rangle$ and $|0,M'_s\rangle$ change to Ψ_s and Ψ'_s ,

$\langle\Psi_s|V_f|\Psi'_s\rangle \neq 0$ and $\langle\Psi_s|V_{ff'}|\Psi'_s\rangle \neq 0$.

The solutions of the time independent part of the Hamiltonian (H_A+P), Ψ_r and $\Psi_{r'}$, can be expressed as linear combination of wavefunctions of H_A.

$$\Psi_r = \sum_j a^r_j \phi_j + \sum_\mu a^r_\mu \phi_\mu \quad , \quad \Psi_{r'} = \sum_j a^{r'}_j \phi_j + \sum_\mu a^{r'}_\mu \phi_\mu \quad . \tag{57}$$

The corresponding energies are ε^r and $\varepsilon^{r'}$ which are close to the unperturbed energies ε^o_j and ε^o_ℓ , respectively, i.e. ε^r and $\varepsilon^{r'}$ tend to ε^o_j and ε^o_ℓ , respectively, as $P \rightarrow 0$.

It has been shown (14), using procedure similar to the ons used to derive the effective Hamiltonian described earlier, that to calculate the matrix element $\langle\Psi_r|V_f|\Psi_{r'}\rangle$ the wave functions Ψ_r and $\Psi_{r'}$ can be replaced by the corresponding eigenstate of the spin Hamiltonian

$\sum_j a_j^r \phi_j$ and $\sum_j a_j^{r'} \phi_j$, respectively, if V_f(or $V_{ff'}$) is replaced by

$$H'_{eff} = V_f + \sum_\mu \left(\frac{V_f|\mu><\mu|P}{(\varepsilon^{r'} - \varepsilon_\mu^o)} + \frac{P|\mu><\mu|V_f}{(\varepsilon^r - \varepsilon_\mu^o)} \right) +$$

$$\sum_{\mu,\gamma} \frac{V_f|\mu><\mu|P|\gamma><\gamma|P+P|\mu><\mu|V_f|\gamma><\gamma|P+P|\mu><\mu|P|\gamma><\gamma|V_f}{(\varepsilon - \varepsilon_\mu^o)(\varepsilon - \varepsilon_\gamma^o)} + \ldots. \quad (58)$$

where ε is the average value of the energy for the ground set of levels.

As in the derivation of the spin Hamiltonian, the integration over the orbital variables are carried out to obtain Hamiltonian consisting of spin operators only, known as the Dynamical spin Hamiltonian (28,29).

4.6 Dynamical Spin Hamiltonian; Symmetry Considerations

The derivation of the spin Hamiltonian using symmetry considerations was discussed earlier for the case when the levels of the ground manifold belong to orbitally nondegenerate state. The case when the ground manifold belong to orbitally degenerate state is more complicated and is discussed in Ref. (13) in detail. The coefficients in the spin Hamiltonian D,E,....., etc., depend on the internuclear separation though they appear as parameters in the formulation. Thus, in the spin Hamiltonian, the quantities which correspond to B^s and O^s in $\sum_s B^s O^s$ (Eqn. 34) are the parameters D,E, etc., and the Irreducible spin tensor operators, respectively. Similarly, dynamical spin Hamiltonian can be rewritten. Thus, for a three fold degenerate ground manifold, S=1, the part which is linear in displacements can be written as $H^1_{DSH} = \vec{S}.D.\vec{S}$, where $D_{\alpha\beta} = G_{\alpha\beta\gamma\delta} e_{\gamma\delta}$ is a second rank tensor. D should be symmetric, $D_{\alpha\beta} = D_{\beta\alpha}$, so that terms linear in spin operators do not appear, to satisfy the condition of the time reversal invariance. Thus,

$$H^1_{DSH} = \frac{1}{2} \sum_{\alpha,\beta,\gamma,\delta} (S_\alpha S_\beta + S_\beta S_\alpha) G_{\alpha\beta\gamma\delta} e_{\gamma\delta}. \quad (59)$$

The fourth rank tensor G is at best a parameter here, just as D,E etc. are parameters in the spin Hamiltonian. Nevertheless, the number of components of this tensor which are nonzero is generally very small and can be found using symmetry considerations. *The transformation properties of G are similar to that of the elastic constant tensor.* Voigt's notation is used to express the tensor components: xx≡1, yy≡2, zz≡3, yz≡4, xz≡5, xy≡6.

Consider the case of cubic symmetry. Symmetry considerations yield only three nonzero components of the tensor G:

$$G_{11} = G_{22} = G_{33} = G_1 , \quad G_{44} = G_{55} = G_{66} = G_3 ,$$

$$G_{12} = G_{13} = G_{31} = G_{21} = G_{32} = G_{23} = G_2 \quad . \tag{60}$$

Furthermore, we can put $Tr[H^1_{DSH}] = \sum_{S_z} <S_z|D_{xx}S_x^2+D_{yy}S_y^2+D_{zz}S_z^2|S_z>=o$, as we are not interested in terms giving uniform shift of the levels. Thus,

$(D_{11} + D_{22} + D_{33})S(S+1) = o$, which yield,

$$G_{1k}+G_{2k}+G_{3k}=o, \; k=1, \; \ldots, \; 6, \tag{61}$$

as e_k is arbitrary. Thus Eqn. 60 gives $G_1+2G2=o$. Using these informations, for the case of cubic symmetry

$$H^1_{DSH} = \tfrac{3}{2}G_{11}(S_x^2e_1+S_y^2e_2+S_z^2e_3)+G_{44}[(S_xS_y+S_yS_x)e_6+(S_xS_z+S_zS_x)e_5 + (S_yS_z+S_zS_y)e_4] \quad . \tag{62}$$

Alternatively, this can be expressed as

$$H^1_{DSH} = G_{11}[\tfrac{1}{2}(3S_z^2-S(S+1))\varepsilon(E_g,\Theta)+\tfrac{\sqrt{3}}{2}\,\tfrac{1}{2}(S_+^2+S_-^2)\varepsilon(E_g,\varepsilon)]+$$

$$G_{44}[-\sqrt{\tfrac{2}{3}}\,\tfrac{1}{2}(S_zS_+ + S_+S_z)\varepsilon(T_{2g},1)+\tfrac{1}{\sqrt{3}}\,\tfrac{1}{2}(S_+^2 - S_-^2)\varepsilon(T_{2g},o)$$

$$-\sqrt{\tfrac{2}{3}}\,\tfrac{1}{2}(S_zS_- + S_-S_z)\varepsilon(T_{2g},-1)] \quad .$$

This expression is same as the part with $\ell=2$ in

$$(H_{cr}-H^o_{cr}) = \sum_{\substack{m=\Theta,\varepsilon \\ \ell=2,4,\ldots}} V(E_g,\ell)C(E_g\ell,m)\varepsilon(E_g,m) + \sum_{\substack{m=o,\pm 1 \\ \ell=2,4}} (-1)^m \, V(T_{2g},\ell) \, C(T_{2g}\ell,m)\varepsilon(T_{2g},-m) \tag{63}$$

but with $V(E_g,\ell)$ and $V(T_{2g},\ell)$ replaced by the parameters G_{11} and G_{44}, respectively. C's represent Irreducible Spin tensor operators discussed earlier. Notice, V is independent m, as shown here explicitly.

For S>1, terms containing higher powers of p+q+r in the product $S_x^pS_y^qS_z^r$ must be included. An additional term which satisfy the condition of time reversal invariance is

$$\frac{1}{4}\sum_{\alpha,\beta,\gamma,\delta,\xi,\eta}(S_\alpha S_\beta + S_\beta S_\alpha)\;(S_\gamma S_\delta + S_\delta S_\gamma)G_{\alpha\beta\gamma\delta\xi\eta}e_{\xi\eta} \quad . \tag{64}$$

Again, using symmetry considerations corresponding to cubic point symmetry, this expression can be reduced to the part in Eqn. (63) corresponding to $\ell=4$, with V's as parameters. For Raman process, the term quadratic in the strain tensor components is of the form

$$H' = \frac{1}{2} \sum_{\alpha,\beta,\gamma,\delta,\xi,\eta} (S_\alpha S_\beta + S_\beta S_\gamma) G_{\alpha\beta\gamma\delta\xi\eta} e_{\gamma\delta} e_{\xi\eta} . \tag{65}$$

It has been remarked by Koloskova (30), contribution made by the terms containing higher than the second power in spin variables is small.

Similarly, the term in the spin Hamiltonian $\vec{S}.g.\vec{H}$ in which the dependence on the interatomic distances is carried by the g tensor leads to a term in the dynamical spin Hamiltonian linear in strain tensor component:

$$\frac{1}{2} \sum_{\alpha,\beta,\gamma,\delta} (S_\alpha H_\beta + H_\alpha S_\beta) G'_{\alpha\beta\gamma\delta} e_{\gamma\delta} . \tag{66}$$

The source of the field H should have same transformations applied as S, for the Zeeman term to remain invariant. Thus G' also transforms like the Elastic Constant tensor. To derive the form of this term in cubic symmetry, we use the relations,

$$G'_{11} = G'_{22} = G'_{33} , \quad G'_{44} = G'_{55} = G'_{66} ,$$

$$G'_{12} = G'_{13} = G'_{23} = G'_{21} = G'_{31} = G'_{32} , \text{ which gives,}$$

$$S_x H_x [G'_{11} e_1 + G'_{12}(e_2+e_3)] + S_y H_y [G'_{22} e_2 + G'_{23}(e_1+e_3)] + S_z H_z [G'_{33} e_3 + G'_{31}(e_1+e_2)]$$

$$+ G'_{44} [(S_x H_y + S_y H_x) e_6 + (S_x H_z + S_z H_x) e_5 + (S_z H_y + S_y H_z) e_4] .$$

This can be reexpressed as,

$$\vec{H}.\vec{S}. \tfrac{1}{3}(G'_{11}+2G'_{12})(e_1+e_2+e_3) + (\vec{H}.\vec{S}-3H_z S_z). \tfrac{1}{3}(G'_{11}-G'_{12})(\tfrac{1}{2}(e_1+e_2)-e_3)$$

$$+(H_x S_x - H_y S_y)(G'_{11} - G'_{12}). \tfrac{1}{2}(e_1 - e_2) + G'_{44}[(H_y S_z + H_z S_y) e_4$$

$$+(H_x S_z + S_x H_z) e_5 + (H_x S_y + S_x H_y) e_6] . \tag{67}$$

Notice, each term is a product of two factors with identical transformation properties and a constant. For example, first term is $(H_x S_x + H_y S_y + H_z S_z). \frac{1}{3}(G'_{11}+2G'_{12})(e_{xx} + e_{yy} + e_{zz})$, and so on.

We shall return to the discussion of the dynamical spin Hamiltonian in the section dealing with the experimental results.

5. MOSSBAUER SPECTRAL SHAPES IN THE PRESENCE OF SPIN RELAXATION EFFECTS :

The theoretical treatments of the effects of spin relaxation on Mossbauer line shapes have been successful in describing the shapes observed experimentally. The stochastic model (5-7, 31, 32) is widely accepted and has provided useful informations about the dynamical part of the interactions in solid. The method using the Liouville operator formalism provides a general expression for the Mossbauer line shape, applicable to various situations. A simpler method is, however, adequate to obtain the expression of the line shape when the magnetic hyperfine interaction can be assumed to be of the form $AI_zS_z(t)$ and the electric field gradient is axially symmetric with Z as the symmetry axis, or zero (31). The significance of the form of the Hamiltonian can be realised from the consideration of a nuclear transition from the excited state $|I_1m_1\rangle$ to the ground state $|I_om_o\rangle$. The effect of the time dependence of $S_z(t)$ merely changes the energy of the nuclear substates without causing any transition between the substances of the nuclear excited state ($|I_1m_1\rangle \rightarrow |I_1m_1'\rangle$) or the ground states ($|I_om_o\rangle \longrightarrow |I_om_o'\rangle$). Thus, even though the energy of the nuclear transition $|I_om_o\rangle \rightleftarrows |I_1m_1\rangle$ changes due to the time dependence of $S_z(t)$ (the relaxation process), the pair of sublevels remain unchanged. Consequently, the effect of the relaxation process on each nuclear transition can be treated separately. This simple situation, when the Hamiltonians at different times commute with each other, occurs frequently and is known as the adiabatic case. A simple approach, following the Kubo and Anderson's method (33), can be used to obtain the expression of the line shape in this case and is described elsewhere in this book.

On the other hand, when the off diagonal terms of the hyperfine interaction,

$$H_{MHI} = A_xI_xS_x + A_yI_yS_y + A_zI_zS_z \quad ,$$

are significant, or when the EFG is not axially symmetric, or symmetric but not along Z direction, etc., so that the Hamiltonians at different times do not commute with each other, the line shape expression is obtained using Liouville operator formalism (5-7). This is described by Dattagupta elsewhere in this book. In this case, not only the time dependence of the hyperfine interaction due to fluctuation processes in the bath with which the ion interacts must be taken into account, but also the quantum mechanical transitions between the substates of the excited nuclear level ($|I_1m_1\rangle \rightarrow |I_1m_1'\rangle$) or the ground level ($|I_om_o\rangle \rightarrow |I_om_o'\rangle$) need be taken into consideration.

The discussion in the following paragraphs aims to emphasize, or rather reemphasize, those aspects of the theory which are useful in the simulation of the line shapes showing relaxation effects theoretically.

5.1 Stochastic Model of Ion Spin Relaxation

The probability of emission of a photon of wave vector $\vec{k}$, frequency ω, is given by (31)

$$P(\vec{k},\omega) = \frac{2}{\Gamma}\int_0^\infty d\tau\, e^{i\tau(\omega+i\frac{\Gamma}{2})} \sum_{\lambda,\alpha} \rho(\lambda)\langle\lambda|U^-|\alpha\rangle\langle\alpha| e^{i\frac{\mathcal{H}}{\hbar}t} U^+ e^{-i\frac{\mathcal{H}}{\hbar}t}|\lambda\rangle \quad . \quad (68)$$

$|\lambda\rangle$ and $|\alpha\rangle$ are the eigenstates of the system Hamiltonian $\mathcal{H}$ which includes the nucleus and the entire solid but excludes the interaction U responsible for the transition between the nuclear ground and the excited state. The nucleus is in the excited state in $|\lambda\rangle$ and in the ground state in $|\alpha\rangle$. U^+ de-excites the nucleus, $U^- = (U^+)^\dagger$. Γ is the natural width of any of the excited nuclear level, $|I_1 m_1\rangle$. $\rho(\alpha)$ is the thermal population of $|\alpha\rangle$. When $\mathcal{H}$ is time dependent, the term $\langle\alpha| e^{i\frac{\mathcal{H}}{\hbar}t} U^+ e^{-i\frac{\mathcal{H}}{\hbar}t}|\lambda\rangle$ in the above expression must be replaced with

$$\left(\langle\alpha|\exp_-\left(\frac{i}{\hbar}\int_0^t \mathcal{H}(t')dt'\right) U^+ \exp_+\left(-\frac{i}{\hbar}\int_0^t \mathcal{H}(t')dt'\right)|\lambda\rangle\right)_{av} \quad .$$

Here $(\ )_{av}$ denotes an average over the fluctuation paths which the system can choose during the period 0-t. We shall come back to a detailed discussion of it later on. The subscript - and + on the exponentials indicate negative and positive time ordering, respectively. To clarify, the matrix element can be explicity written as

$$\left(\langle\alpha|\exp\left(\frac{i}{\hbar}\int_0^{t_1} \mathcal{H}(t')dt'\right) \text{---} \exp\left(\frac{i}{\hbar}\int_{t_{n-1}}^{t} \mathcal{H}(t')dt'\right)|U^+|\exp\left(-\frac{i}{\hbar}\int_{t_{n-1}}^{t} \mathcal{H}(t')dt'\right) \text{---} \exp\left(-\frac{i}{\hbar}\int_0^{t_1} \mathcal{H}(t')dt'\right)|\lambda\rangle\right)_{av} \quad .$$

$\mathcal{H}(t)$ includes the hyperfine interaction, the interactions of the electrons with each other and with other ions in the solid, the static and the dynamical parts, and other interactions in the solid. The evaluation of this is the central problem of the relaxation theories. It is a many body problem and an exact solution is not to be expected. Two approaches have been used to evaluate this. We follow the stochastic model here which has been experimentally found to be very satisfactory. The alternative approach using the perturbation theory provides the expression of line shape in limiting conditions only and can be found in References (34-36).

For clarity, consider the specific case of the Fe^{3+} ion, in high spin state. In presence of an external field greater than a few hundred Gauss along Z axis, the magnetic hyperfine interaction can be approximated as

$$A_x I_x S_x + A_y I_y S_y + A_z I_z S_z \sim A_z I_z S_z \quad ,$$

since the electronic and the nuclear spins precess differently around the external field, thus *quenching* the x and y parts of the interaction. This can be written as $g_N \beta_N I_z H_{hf}$, where $H_{hf} = A_z S_z / g_N \beta_N$. g_N is the g factor of the nuclear state concerned, and β_N is the nuclear magneton. The effect of the interaction of the paramagnetic ion with other ions in the solid or the lattice vibration is to induce transitions between the Zeeman state of the central ion. Without going into the details of these interactions, it is assumed that the spin state of the ion changes at random times. The change occurs instantaneously in comparison to the period for which the ion stays in a spin state, on the average, and is assumed to be a stationary Markov process (31). Thus, the actual interactions responsible for the relaxation are replaced by the changes occurring in the spin state of the ion instantaneously at random times. They are referred to as stochastic fluctuations. The system Hamiltonian can be replaced by

$$\mathcal{H}(t) = \mathcal{H}_o' + H_{hf}(t)\; I_z g_N \beta_N \quad , \tag{69}$$

where $H_{hf}(t)$ is the time dependent internal field at the nucleus along Z-axis, the Quadrupole interaction is assumed zero for simplicity, and

$$\mathcal{H}_o' | I_o m_o > = E_o | I_o m_o > \; . \tag{70}$$

Here, E_o is the energy of the ground nuclear state, in absence of magnetic field and the Quadrupole interactions. It is important to note the absence of terms in $\mathcal{H}$ like $A_x I_x S_x$, $A_y I_y S_y$, etc. which can cause transitions between the nuclear substates of the ground or the excited manifold of states.

With these assumptions, the problem finally reduces to the evaluation of $[i \int_0^t \delta S_z(t')dt']$, where $S_z(t')$ changes at random instants of time due to the stochastic fluctuations and $\hbar\delta = A_o m_o - A_1 m_1$. This can be done using the procedure of Kubo and Anderson (33). The details can be found elsewhere in this book.

On the other hand, if the terms like $\frac{A}{2}(I_+ S_- + I_- S_+)$, $I_x^2 - \frac{1}{3}I(I+1)$, etc., are not negligible, it is not only necessary to consider the stochastic fluctuations, but also consider the quantum mechanical transitions among the substates of the excited and ground nuclear levels due to these terms. Thus, we must split the period from 0 to t into time intervals separated by instants at which the stochastic pulses appear (6,32). In each interval thus obtained, the time development occurs quantum mechanically. We outline the procedure followed and assumptions made in this case. We need to evaluate

$$(\langle I_o m_o|\exp_-[\tfrac{i}{\hbar}\int_0^t H(t')dt']|I_o m_o'\rangle\langle I_1 m_1'|\exp_+[-\tfrac{i}{\hbar}\int_0^t H(t')dt']|I_1 m_1\rangle)_{av}\,.$$

The evaluation of this expression is made difficult by the presence of two time ordered series, in general. The stochastic average is easily performed when the random fluctuations are arranged in a single time ordered sequence as in adiabatic case mentioned above. The evaluation when two time ordered series are present, is done using Liouville operator formalism. Before discussing this procedure we describe briefly the properties of the Liouville operators (31).

The Liouville operator A^x corresponding to any operator A when acts on a quantum mechanical operator B gives

$$A^x B = AB - BA = [A,B] \quad . \tag{71}$$

Thus, whereas an ordinary operator, like A,B, etc., acting on an state vector in general gives a different state vector, the Liouville operator acting on an ordinary operator in general gives a different operator. Consider

$$\begin{aligned}\langle\mu|A^x B|\gamma\rangle &= \sum_{\mu'}\langle\mu|A|\mu'\rangle\langle\mu'|B|\gamma\rangle - \sum_{\gamma'}\langle\mu|B|\gamma'\rangle\langle\gamma'|A|\gamma\rangle \\ &= \sum_{\mu',\gamma'}(\mu\gamma|A^x|\mu'\gamma')\langle\mu'|B|\gamma'\rangle \quad , \quad \text{where}\end{aligned} \tag{72}$$

$(\mu\gamma|A^x|\mu'\gamma') = \{\langle\mu|A|\mu'\rangle\,\delta_{\gamma\gamma'} - \langle\gamma'|A|\gamma\rangle\delta_{\mu\mu'}\}$ defines the matrix element of A^x in terms of the matrix element of A.

Let $e^{\lambda A}\,B\,e^{-\lambda A} = F(\lambda)B$. To find $F(\lambda)$, we differentiate

$$\frac{d}{d\lambda}[F(\lambda)B] = e^{\lambda A}[A,B]e^{-\lambda A} = A^x e^{\lambda A}Be^{-\lambda A} = A^x F(\lambda)B \quad .$$

This has the solution $F(\lambda)B = \exp(\lambda A^x)B$. Thus, $F(\lambda) = e^{\lambda A^x}$. When $\lambda=1$, $A = \frac{i}{\hbar}Ht$, the above relation gives

$$B(t) = e^{i\frac{H}{\hbar}t}\,B\,e^{-i\frac{H}{\hbar}t} = e^{\frac{i}{\hbar}H^x t}\,B \quad .$$

When H is time dependent,

$$B(t) = \exp_-[\tfrac{i}{\hbar}\int_0^t H(t')dt']\,B\exp_+[-\tfrac{i}{\hbar}\int_0^t H(t')dt'] = \exp_-[\tfrac{i}{\hbar}\int_0^t H^x(t')dt']B\,. \tag{73}$$

Finally, let us see the eigenvalues and eigenoperators of $\mathcal{H}^x$. Let $|\mu\rangle$ and $|\gamma\rangle$ be the eigenvectors of $\mathcal{H}$ corresponding to the eigenvalue E_μ and E_γ , respectively.

$$\mathcal{H}^x|\mu\rangle\langle\gamma| = \mathcal{H}|\mu\rangle\langle\gamma| - |\mu\rangle\langle\gamma|\mathcal{H} = (E_\mu - E_\gamma)|\mu\rangle\langle\gamma| \ . \qquad (74)$$

Thus, $|\mu\rangle\langle\gamma|$ is an eigenoperator of $\mathcal{H}^x$ and $(E_\mu - E_\gamma)$ is the corresponding eigenvalue. The eigenvalues of $\mathcal{H}^x$ are more meaningful in spectroscopic studies than the eigenvalues E_μ or E_γ of $\mathcal{H}$ with arbitrary zero of energy. Thus, the Liouville operator of $\mathcal{H}$ is more useful than $\mathcal{H}$ itself. Symbolically, this is expressed as

$$\mathcal{H}^x|\mu\gamma) = (E_\mu - E_\gamma)|\mu\gamma) \quad .$$

The above relations can be generalised for nonhermitian operator A by replacing Eqn. (71) with $A^xB = A^\dagger B - BA$. This yields,

$$e^{iA^\dagger}Be^{-iA} = e^{iA^x}B \quad ,$$

$$\exp_-[\tfrac{i}{\hbar}\int_0^t \mathcal{H}^\dagger(t')dt']U^+\exp_+[-\tfrac{i}{\hbar}\mathcal{H}(t')dt'] = (\exp_-[\tfrac{i}{\hbar}\int_0^t \mathcal{H}^x(t')dt'])U^+ \ . \qquad (75)$$

Let us write the system Hamiltonian in the form

$$\mathcal{H} = \mathcal{H}_o + \sum_i V_i \ \delta(t-t_i) \qquad (76)$$

including the part which give quantum mechanical transitions between substates of nuclear ground or excited state in $\mathcal{H}_o$. The subsystem described by $\mathcal{H}_o$ interacts with the rest of the solid, known as bath, at random instants of time t_i's only*. These pulses at t_i , known as the stochastic pulses, can cause transitions between different states of the subsystem (ionic spin state in the example to follow) under consideration.

When the subsystem is the paramagnetic ion, the interaction of the electron spin of the paramagnetic ion with other ions and the lattice is replaced by stochastic pulses which act at instants t_i only and change the spin state of the central ion. For example, the spin spin interaction of the central ion with a neighbouring ion

$$\mathcal{H}_{ss} = \vec{S}_1 \cdot \frac{g_1 g_2 \beta^2}{r_{12}^3}[\vec{S}_2 - \frac{3\vec{r}_{12}(\vec{S}_2\cdot\vec{r}_{12})}{r_{12}^2}] \qquad (77)$$

* A different form of the Hamiltonian appropriate when the stochastic fluctuation in the bath changes the form of the hyperfine interaction Hamiltonian is considered by DattaGupta elsewhere in this book. This occurs, for example, when vacancy diffusion changes the principal axes of the EFG tensor.

is thus replaced by the model interaction $\sum_i \vec{S}_1 \cdot \vec{h}_i \, \delta(t - t_i)$. The time dependence of the wave function governed by the above model Hamiltonian, $\mathcal{H}$, (Eqn. 76), is

$$\psi(t) = e^{-\frac{i}{\hbar}\mathcal{H}_o(t-t_n)} e^{-\frac{i}{\hbar}V_n} e^{-\frac{i}{\hbar}\mathcal{H}_o(t_n-t_{n-1})} \cdots e^{-\frac{i}{\hbar}V_1} e^{-\frac{i}{\hbar}\mathcal{H}_o(t_1)} \psi(0). \quad (78)$$

It has been assumed that there are exactly n stochastic pulses V_1, V_2, ... V_n in the time interval o-t which act at times t_1, ..., t_n, respectively. n is a random variable. The probability of occurrence of n pulses in time t is related to the fluctuation frequency as we shall see below. $\mathcal{H}_o$ can cause transition among the substates of a nuclear state.

The operator V changes the state of the subsystem. In the example given above V changes the spin state of the ion. Taking V_i to be hermitian implies that the transition induced by V_i between any two states $|a\rangle$ and $|b\rangle$ of the subsystem are equally probable in either direction. This can be true only at infinite temperature so that all the levels of the system are equally populated. At finite temperatures, the ratio of the transition probabilities is

$$p_{a\rightarrow b} : p_{b\rightarrow a} = \exp\left(\frac{E_a - E_b}{kT}\right)$$

Thus, in the model in which we divide the system into the subsystem and the bath and replace the interaction of the subsystem with the bath by $\Sigma V_i \delta(t-t_i)$, we must allow the model Hamiltonian to be nonhermitian, to keep the thermal equilibrium in the populations of levels of the subsystem. Accordingly, the time dependence of an operator is given by Eqn. 75 when $\mathcal{H}$ is time dependent (Eqn. 76). Thus, we can write

$$\begin{aligned}(U^+(t))_{av} &= (\exp_-[\tfrac{i}{\hbar}\textstyle\int_0^t \mathcal{H}^x(t')dt'])_{av} U^+ \\ &= (\exp(\tfrac{i}{\hbar}\mathcal{H}_o^x t_1)\exp(\tfrac{i}{\hbar}V_1^x)\exp(\tfrac{i}{\hbar}\mathcal{H}_o^x(t_2-t_1)) \ldots \exp(\tfrac{i}{\hbar}V_n^x) \\ &\quad \exp(\tfrac{i}{\hbar}\mathcal{H}_o^x(t-t_n)))_{av} U^+ . \end{aligned} \quad (79)$$

The average $(U^+(t))_{av}$ includes (1) an average over the instants of time at which the stochastic pulses appear (they appear at t_1, t_2, ... t_n in the Eqn. (79)), (2) an average over the type of pulses (over $\vec{h}_i$ in the example of spin spin interaction mentioned above).

Assuming that these two types of averages are independant of each other and that the types of successive pulses ($\vec{h}_i$ and $\vec{h}_{i\pm 1}$ in the example given above) are uncorrelated;

$$(U^{+}(t))_{av} = (u_o(t_1)\tau_{av}u_o(t_2-t_1)\tau_{av} \cdot\cdot \tau_{av}\, u_o(t-t_n))_{av}U^{+} \quad , \qquad (80)$$

where, $u_o(t) = \exp(\frac{i}{\hbar}H_o^x t)$, $\tau_{av} = (\exp(\frac{i}{\hbar}V_i^x))_{av}$.

The average over the types of pulses is included in τ_{av} . The other type of average over the instants t_i is denoted by $(\ )_{av}$ on the R.H.S. of the above equation. This can be written as (6)

$$(U^{+}(t))_{av} = \sum_{n=0}^{\infty} P_n(t) \int_0^t dt_n \int_0^{t_n} dt_{n-1} \cdots \int_0^{t_2} dt_1 W(t_1,t_2, --, t_n; t) \times$$

$$u_o(t_1)\tau_{av} \cdots \tau_{av}u_o(t-t_n)U^{+} \ .$$

Here, $P_n(t)$ is the probability that n stochastic pulses appear in time t. Assuming that the stochastic pulses appear randomly, $P_n(t)$ is given by the Poission's distribution

$$P_n(t) = \frac{(\lambda t)^n}{n!} e^{-\lambda t} \quad ,$$

where λ is the mean frequency of the pulses. $W(t_1, --, t_n; t)dt_1 dt_2 ---$ dt_n is the probability that the stochastic pulses appear on the sub-system at t_1 in dt_1 , t_2 in dt_2 , --, t_n in dt_n . Since pulses appear randomly, $W(t_1 , --, t_n; t) = n!t^{-n}$. Thus

$$(U^{+}(t))_{av} = \sum_{n=0}^{\infty} \int_0^t dt_n \cdots \int_0^{t_2} dt_1\, e^{(\frac{i}{\hbar}H_o^x-\lambda)t_1}(\lambda\tau_{av})e^{(\frac{i}{\hbar}H_o^x-\lambda)(t_2-t_1)} \cdots$$

$$e^{(\frac{i}{\hbar}H_o^x-\lambda)(t-t_n)}U^{+} \ .$$ This gives

$$\begin{aligned} F(p) &= \langle U^{-}\int_0^{\infty} dt\, e^{-pt}(\exp_{-}(\frac{i}{\hbar}\int_0^t H_o^x(t')dt'))_{av}U^{+}\rangle \\ &= \langle U^{-}(p-W-\frac{i}{\hbar}H_o^x)^{-1}\, U^{+}\rangle \\ &= \sum_{\lambda,\lambda',\alpha,\alpha'} \rho(\lambda)\langle\lambda|U^{-}|\alpha\rangle(\alpha\lambda|(p-W-\frac{i}{\hbar}H_o^x)^{-1}|\alpha'\lambda')\langle\alpha'|U^{+}|\lambda'\rangle \ . \end{aligned} \qquad (81)$$

Here, $W=\lambda(\tau_{av}-1)$. The line shape is given by

$$P(\overline{k},\omega) = \frac{2}{\Gamma}\, \mathrm{Re}\, F(p), \quad \text{where } p = -i\omega+\frac{\Gamma}{2} \qquad (82)$$

Thus, the line shape is dependent on the matrix elements $(\alpha\lambda|(p-W-iH_o^x)^{-1}|\alpha'\lambda')$. If the ionic spin is S, there are $(2I_1+1)(2S+1)$ states $|\lambda\rangle$ of the subsystem in which the nucleus is in the excited state and $(2I_o+1)(2S+1)$ states $|\alpha\rangle$ in which the nucleus is in the ground state. Thus, the matrix formed by $(\alpha\lambda|\quad|\alpha'\lambda')$ is of dimension

$(2S+1)^2(2I_1+1)(2I_o+1)$. The matrix element of H_o^x, which is hermitian, are

$$(\mu\gamma|H_o^x|\mu'\gamma') = \delta_{\gamma\gamma'}\langle\mu|H_o|\mu'\rangle - \delta_{\mu\mu'}\langle\gamma|H_o|\gamma'\rangle \quad . \tag{83}$$

Furthermore, the matrix element of W is obtained using

$$\langle\mu|e^{\frac{i}{\hbar}V^x}U^+|\gamma\rangle = \langle\mu|e^{\frac{i}{\hbar}V^\dagger}U^+ e^{-\frac{i}{\hbar}V}|\gamma\rangle \quad \text{or}$$

$$(\mu\gamma|e^{\frac{i}{\hbar}V^x}|\mu'\gamma') = \langle\mu|e^{\frac{i}{\hbar}V^\dagger}|\mu'\rangle\langle\gamma'|e^{-\frac{i}{\hbar}V}|\gamma\rangle \ . \tag{84}$$

5.2 Relaxation Matrices of Typical Stochastic Model Hamiltonians

A few examples have been provided by Clauser and Blume (6) to clarify the mathematical procedure involved in the simulation of the Mossbauer spectrum in presence of the relaxation effects. We discuss a few important details from an example which is useful. Let $\frac{V}{\hbar} = \vec{h}.\vec{S}$ (Eqn. 76).

$$e^{-i\vec{h}.\vec{S}} = \text{Cos}|\vec{h}.\vec{S}| - i\frac{\vec{h}.\vec{S}}{|\vec{h}.\vec{S}|}\text{Sin}|\vec{h}.\vec{S}| \tag{85}$$

$\vec{h}$ varies in magnitude and direction. When $|\vec{h}.\vec{S}|=0$ or $\pi, e^{-iV/\hbar}=1$. In this case the operator does not change the state of the system. The effectiveness of the stochastic pulse is maximum when $|\vec{h}.\vec{S}|=\pi/2$. Thus the transition probability depends on $|\vec{h}.\vec{S}|$ as well as $(\vec{h}.\vec{S})/|\vec{h}.\vec{S}|$.

We consider the case when

$|\vec{h}.\vec{S}|=\pi/2$, $h = \pi$ and $h_z = 0$.

$$e^{-i\vec{h}.\vec{S}} = -i\frac{h_xS_x + h_yS_y}{|\vec{h}.\vec{S}|} = -i(S_-e^{i\theta} + S_+e^{-i\theta}) \quad ,$$

where $\text{Sin}\theta = h_y/h$, $\qquad \text{Cos}\theta = h_x/x$.

We assume S=1/2 and denote $|\pm 1/2\rangle$ states by $|\pm\rangle$, respectively. It follows from Eqn. (84) :

$$(++\,|e^{-iV^x}|\,--) = (ie^{i\theta})^*(ie^{i\theta}) = 1 \qquad .$$

Similarly, other matrix elements can be calculated. The operator V cannot cause transition between nuclear states. Thus, $\langle\mu'|e^{-iV/\hbar}|\mu\rangle$ is nonzero only if the nuclear state is same in $|\mu'\rangle$ and $|\mu\rangle$. When the field of bath on the subsystem changes, the perturbation on the subsystem due to the bath changes. The total effect of the field from the bath, over the interval the field remains constant, is replaced by random instantaneous pulse $\vec{h}$ which is equally effective.

A stochastic pulse at an instant t_i causes transition between spin states with a phase change of Θ which depend on h_x and h_y in the example under consideration. Thus, to obtain average over the type of pulses which can appear, τ_{av}, we must average over the values Θ can take. Thus

$$(+- |e^{-iV^x/\hbar}|-+)= e^{-2i\theta} \quad .$$

has an average value zero, and so on. This enable us to compute the matrix elements $(\mu\gamma|W|\mu'\gamma')$ where $W=\lambda(\tau_{av}-1)$.

Further, let us assume $H_o = GS_z + AI_zS_z$, as is the case when the splitting of ionic levels produced by an external magnetic field is large in comparison to the crystal field interactions and off diagonal terms of the magnetic hyperfine interactions. The matrix element $\langle\mu|H_o|\mu'\rangle$ is finite only if nucleus is in the same state in $|\mu\rangle$ & $|\mu'\rangle$. The matrix elements $(\mu\gamma|H_o^x|\mu'\gamma')$ can be found using Eqn. (83). Thus the matrix elements given in Table 6 are obtained

TABLE 6

Matrix formed by $(\mu(m_1), \gamma(m_o)|p-W-iH_o^x|\mu'(m_1), \gamma'(m_o))$ *when* $\tau=i(e^{-i\theta}S_+ + e^{i\theta}S_-)$, $H_o=GS_z + AI_zS_z$, and $S=1/2$.

$\mu\gamma$ \ $\mu'\gamma'$	++	--	+-	-+
++	$\lambda+p+i(\beta_o-\beta_1)$	$-\lambda$	0	0
--	$-\lambda$	$\lambda+p+i(\beta_o-\beta_1)$	0	0
+-	0	0	$\lambda+p-iG-i(\beta_o+\beta_1)$	0
-+	0	0	0	$\lambda+p+iG+i(\beta_1-\beta_o)$

Here, $\mu(m_1)$ indicates that the nucleus is in $|m_1\rangle$ state when the subsystem is in $|\mu\rangle$ state, $\beta_o=\frac{1}{2}A_om_o$, & $\beta_1=\frac{1}{2}A_1m_1$. Similarly matrices can be written for other pairs of nuclear levels (m_o',m_1'). The matrices for different pairs of m_o and m_1 levels differ in the values of β_o and β_1 only.

5.3 Simulation of Line Shape in Presence of Relaxation Effects

The line shape can be obtained by calculating $F(p)$ at each of the frequencies (ω) which are contained in the experimental spectrum. This is however time consuming since the matrix $(\mu\gamma|| \; ||\mu'\gamma')$ must be inverted at each of the values of ω. Clauser (5) has pointed out that it is necessary to invert the matrix only once to simulate the line shape completely. Besides, providing great economy in the computer time required, and thus enabling larger dimension matrices to be handled, the procedure provides much better insight into the effect of relaxation on the Mossbauer lines by providing broadening and shift of each of the component line due to the relaxation effect. To appreciate the procedure first let us consider a simple case. It is known that the eigenfunction of an operator and its inverse are same. Consider a Hermitian operator A. If $|i\rangle$ and λ_i are the eigenfunction and eigenvalue of A, it follows that λ_i^{-1} and $(\lambda_i-\omega)^{-1}$ are the eigenvalues of A^{-1} and $(A-\omega)^{-1}$, where ω is a number. Thus,

$$\langle a|(A-\omega)^{-1}|b\rangle = \Sigma \frac{\langle a|i\rangle\langle i|b\rangle}{\lambda_i-\omega} \quad . \tag{86}$$

Thus, the matrix element $\langle a|(A-\omega)^{-1}|b\rangle$ corresponding to any ω can be obtained by using the simple Eqn. 86 if the λ_i's and $|i\rangle$'s of A are known This need be done only once as A is independant of ω. Similar simplication can be used when A is not an ordinary operator but the superoperator $(-H_o^x - i\frac{\Gamma}{2} + iW)$, denoted by P, as shown below. Some complication nevertheless appears because the operator to which P correspond is not hermitian.

Consider the eigenvalue equation $PB_\alpha = p_\alpha B_\alpha$, where B_α is the eigenvector and p_α the corresponding eigenvalue of P. This gives

$$\sum_{k,\ell} (ij|P|k\ell)\ \langle k|B_\alpha|\ell\rangle = p_\alpha\langle i|B_\alpha|j\rangle \ . \tag{87}$$

Let the hermitian conjugate of P has an eigenvector denoted by C_β and the corresponding eigenvalue be p_β^*.

$$P^\dagger C_\beta = p_\beta^* C_\beta$$

$$\sum_{i,j} (k\ell|P^\dagger|ij)\ \langle i|C_\beta|j\rangle = p_\beta^*\langle k|C_\beta|\ell\rangle \ , \quad \text{where}$$

$$(k\ell|P^\dagger|ij) = \langle k|P^\dagger|i\rangle\ \delta_{\ell j} - \langle \ell|P^\dagger|j\rangle\delta_{ki} \quad .$$

$P^\dagger$ is the superoperator corresponding to the operator $P^\dagger$. This yields

$$\sum_{i,j} \langle i|C_\beta|j\rangle^*\ (ij|P|k\ell) = p_\beta\langle k|C_\beta|\ell\rangle^* \ . \tag{88}$$

Multiply Eqn. (87) by $\langle i|C_\beta|j\rangle^*$ and sum over i and j, and Eqn. (88) by $\langle k|B_\alpha|\ell\rangle$ and sum over k and ℓ; and substract the two equations thus obtained :

$$\sum_{k,\ell} p_\beta\langle k|C_\beta|\ell\rangle^*\ \langle k|B_\alpha|\ell\rangle = \sum_{i,j} p_\alpha\langle i|B_\alpha|j\rangle\ \langle i|C_\beta|j\rangle^* \ . \tag{89}$$

Thus, either $\beta=\alpha$ or $\sum\limits_{k,\ell} \langle k|C_\beta|\ell\rangle^* \langle k|B_\alpha|\ell\rangle = \sum\limits_{\ell} \langle \ell|C_\beta^\dagger B_\alpha|\ell\rangle = 0$, $\beta\neq\alpha$.

Since the eigenvalue equation do not determine the normalisation, we may choose

$$\sum_\ell \langle \ell|C_\beta^\dagger B_\alpha|\ell\rangle = \delta_{\alpha\beta} \text{ which is denoted as } (C_\beta|B_\alpha) = \delta_{\alpha\beta} \quad , \tag{90}$$

This yields $\sum\limits_\alpha |B_\alpha)(C_\alpha| = 1.$ (91)

We make use of the fact that P, $P-\omega$ and $(P-\omega)^{-1}$ have same eigenvectors. Let the eigenvalue corresponding to B_α be denoted by p_α, $p_\alpha-\omega$ & $(p_\alpha-\omega)^{-1}$, respectively.

Using Eqn. (91), it can be easily shown

$$(P-\omega)^{-1} = \sum_\alpha \frac{|B_\alpha)(C_\alpha|}{p_\alpha-\omega} . \tag{92}$$

The spectral shape is obtained using Eqn. 81 and 82, and the notation expressed in Eqn. 90.

$$\begin{aligned} F(p) &= i\mathrm{Tr}[\rho U^-(\omega-P)^{-1}U^+] \\ &= i(\rho U^+|(\omega-P)^{-1}|U^+) \\ &= \sum_\alpha i \frac{(\rho U^+|B_\alpha)(C_\alpha|U^+)}{\omega-p_\alpha} \\ &= \frac{i\sum\limits_{\alpha,\lambda} \rho(\lambda)\langle\lambda|U^-|\alpha\rangle\langle\alpha|B_\alpha|\lambda\rangle \sum\limits_{\alpha'\lambda'} \langle\lambda'|C_\alpha^\dagger|\alpha'\rangle\langle\alpha'|U^+|\lambda'\rangle}{(\omega-p_\alpha)} \end{aligned} \tag{93}$$

If $p_\alpha=\omega_R+i\omega_I$ and $(\rho U^+|B_\alpha)(C_\alpha|U^+) = a_R+ib_I$, where the subscript R and I denote the real and the imaginary parts, the spectral shape is

$$\frac{2}{\Gamma}\ \mathrm{Re}\ F(p) = \frac{2}{\Gamma}\ \mathrm{Re}\ \frac{i\sum\limits_\alpha (a_R+ib_I)}{\omega-\omega_R-i\omega_I} = \frac{2}{\Gamma}\sum_\alpha \frac{b_I(\omega_R-\omega)-a_R\omega_I}{(\omega-\omega_R)^2+\omega_I^2} \quad . \tag{94}$$

As an example we consider a case with known result. We assume that $W=0$ so that $P=-H_o^x-i\frac{\Gamma}{2}$. Since Γ is a number, the eigenvector of P and H_o^x are same. P is Hermitian in this example, B_α and C_α are identical. Let $B_\alpha \equiv B_{ij} = |i\rangle\langle j| \equiv |ij)$. It can be easily shown using the properties of the Liouville operators given earlier:

$P|ij)=(\omega_{ij}-i\,\frac{\Gamma}{2})$ ij), where $\omega_{ij}=E_j-E_i$ and

$H_o|i\rangle = E_i|i\rangle$, $H_o|j\rangle = E_j|j\rangle$.

Thus, the spectral shape is

$$\frac{2}{\Gamma}\sum \frac{a_{ij}\Gamma/2}{(\omega-\omega_{ij})^2+(\frac{\Gamma}{2})^2} \text{ , where } (\rho U^+|ij)(ij|U^+)=a_{ij} \text{ , or}$$

$$\rho(j) \langle j|U^-|i\rangle \langle i|U^+|j\rangle = \alpha_{ij} = \rho(j)|\langle j|U^-|i\rangle|^2 .$$

To conclude this section, it may be remarked that the methodology of the simulation of Mossbauer line shape using the stochastic model is well known. The economy in computer time obtained using the method shown by Clauser opens up the possibility of taking up least squares analysis to fit the experimental spectra with the theory.

6. EXPERIMENTAL RESULTS

Relaxation effects in Mossbauer spectra have been found in several investigations. These investigations cover a wide variety of substances and can be grouped, for the present discussion, as :

1. Fe^{3+} in alums and similar substances
2. Iron in biological molecules
3. Compounds of rare earth elements
4. Compounds containing Fe^{2+}
5. Other substances : frozen solutions, ion exchange resins etc.

In the following, we summarise the results obtained in these substances except the relaxation of Fe^{2+} ions and the rare earth elements which have been included in other chapters of the book.

6.1 Alums

The chemical formula is M^{1+} M^{3+} $(RO_4)_2$ $12H_2O$. Here, M^{1+} represents the monovalent ion (K^{1+}, Rb^+, Cs^+, Tl^+, NH_4^+, or $CH_3NH_2^+$), M^{3+} represents the trivalent ion (Al^{3+}, Fe^{3+}, Cr^{3+}, V^{3+}, In^{3+}, Ga^{3+}, etc.) and R can be S, Se or Te. Each unit cell (Fig. 10) contains four formula units. They have been further classified as α, β or γ alums depending on the size of the monovalent atom which effects the closeness and consequently the relative orientation of the neighbouring groups H_2O and $(SO_4)^{2-}$. The monovalent group is small in γ alums and the large in β alums.

In an investigation by Campbell and Bendetti (40,41), $NH_4(Al,Fe)(SO_4)_2.12H_2O$, $NH_4(Fe,Al)(SO_4)_2.12D_2O$ and $K(Fe,Al)(SO_4)_2.12D_2O$, with various concentrations of iron, were investigated at 4,20,77 and 300 K.

Zero field spectra of hydrated and deutrated samples of the compositions with Fe less than 9% at 4.2, 20, 77 and 300 K showed the presence of spin lattice relaxation effects. At higher concentrations of Fe, the spin-spin relaxation causes the disappearance of a well defined magnetic splitting even at 20° K. Effect of external transverse magnetic fields of 35,200, and 1200 Gauss on the spectral shape of the spectrum of $NH_4(Fe,Al)(SO_4).12D_2O$ with 0.72% Fe was also investigated at 20 K. The analysis made is insufficient to reveal the behaviour of the relaxation time. It is, however, noteworthy that even at 300 K relaxation broadening exist in the spectra of all samples, in agreement with the results obtained in other alums, discussed below in more detail.

In the investigation by Bhargava et al (8) on $NH_4(Fe,Al)(SO_4)_2 \cdot 12H_2O$, Fe:Al 2-3%, spectral shapes were investigated in the temperature range from 5.2 K to 293 K in zero field as well as in transverse field of 12.3 kG. The effect of external fields upto 12.3 kG on the spectral shapes was also investigated at temperatures 82, 157 and 224 K. Mossbauer spectra obtained in presence of the field of 12.3 kG were quantitatively analysed to obtain the temperature dependences of the relaxation frequencies. The transition probabilities between various ionic levels due to the spin lattice relaxation processes can be expressed in terms of the parameters of the dynamical spin Hamiltonian

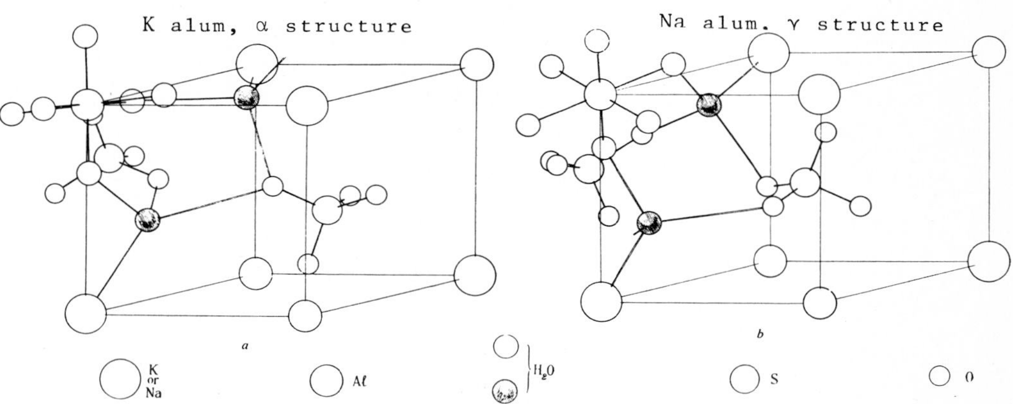

Fig. 10. Comparison of the structure of α and γ alums, showing arrangement of octahedral around Al at (1/2,0,1/2) and the arrangement of bonds to other water molecules (H. Lipkin, Proc. Royal Soc. London A151 (1935)347).

(H_{DSH}) as described by Suzdalev et al (10). The static and dynamical spin Hamiltonians are

$$H_{SP} = 2\beta\vec{H}.\vec{S} + D(S_z^2 - \tfrac{1}{3}S(S+1)) + \lambda(S_x^2 - S_y^2)$$

$$H_{DSH}^i = d^i(S_z^2 - \tfrac{1}{3}S(S+1)) + e^i(S_x^2 - S_y^2) + f^i(S_xS_y + S_yS_x) + g^i(S_xS_z + S_zS_x)$$
$$+ h^i(S_yS_z + S_zS_y) \quad . \tag{95}$$

Here, the axes have been so chosen to diagonalise the static part of the symmetric D tensor. The constants in H_{DSH}^i for i = 1 depend linearly on strain tensor components, for i = 2 depend quadratically on the strain tensor components, and so on. $D \ll 2\mu_\beta H$ in the whole temperature range investigated, even when H = 5.7 kG, for which $2\mu_\beta H \sim 0.5\ cm^{-1}$ and $\lambda=0$ (42). It is thus assumed, the ionic levels are pure Zeeman states in the presence of an external field H>6 kG, and the dynamical spin Hamiltonian gives rise to the transitions between the Zeeman states $|M\rangle$. Thus, the relaxation is longitudinal. The Raman process due to H^2_{DSH} contributes to the transition probabilities:

$$\Omega^{\mu\rightarrow\gamma}_{R1} = \frac{2\pi}{\hbar} \sum_{q,r} (|\langle\gamma;\ n_q-1,\ n_r+1|H^2_{DSH}|\mu;\ n_q, n_r\rangle|^2)_{Th}\,\delta(\hbar\omega_q-\hbar\omega_r-\Delta), \qquad (96)$$

where $\Delta = E_\gamma - E_\mu$. Denoting

$$\Omega^1_{SL} = \frac{4\pi}{\hbar} \sum_{q,r} (|\langle n_q-1,\ n_r+1|g^2+ih^2|n_q, n_r\rangle|^2)_{Th} \ ,$$

$$\Omega^2_{SL} = \frac{4\pi}{\hbar} \sum_{q,r} (|\langle n_q-1,\ n_r+1|e^2+if^2|n_q, n_r\rangle|^2)_{Th} \ , \qquad (97)$$

the transition probabilities obtained are as shown in Fig. 11. The contribution of the direct process due to H^1_{DSH} are also given by same quantities but with Ω^1_{SL} and Ω^2_{SL} replaced by

$$\Omega^{1D}_{SL} = \frac{4\pi}{\hbar} \sum_q (|\langle n_q+1|g^1+ih^1|n_q\rangle|^2)_{Th}\ \delta(h\omega_q+\Delta) \quad \text{and}$$

$$\Omega^{2D}_{SL} = \frac{4\pi}{\hbar} \sum_q (|\langle n_q+1|e^1+if^1|n_q\rangle|^2)_{Th}\ \delta(h\omega_q+\Delta) \quad , \text{ respectively.}$$

Raman process due to H^1_{DSH} in second order of perturbation is neglected completely. It is also assumed $(\Omega^1_{SL}:\Omega^2_{SL})$ is independent of temperature and the external magnetic field. Theoretical line shapes simulated using the Stochastic model of ionic spin relaxation fitted the experimementally observed line shapes remarkably well. Parameters in the theoretical expression for the line shape, viz., $H_{int}(S_z = \pm 5/2)$, line widths Γ and relative intensities corresponding to the six nuclear transitions, and Ω_{SS} are obtained from the spectrum at 5° K in a field of 400 Gauss. Thus, only $(\Omega^1_{SL}/\Omega^2_{SL})=1.5\ \Omega_{SL}$ was treated as free parameter. Thus temperature and magnetic field dependences of Ω_{SL} thus obtained, and typical line shapes have been reproduced in Figs. 12 and 13. The temperature dependence of the relaxation time could not be fitted either using Debye model or using Einstein model alone (Fig. 13(a)). Whereas experimentally $\Omega_{SL} \propto T^{0.5}$ in the range of temperature between 150 and 295 K, all theoretical models predict T^2 dependence *at sufficiently high temperatures*. Experiments could not be continued at higher temperatures as the alum starts loosing water rapidly. It is more appropriate to use the Einstein model as phonons in optical modes are expected to be mainly

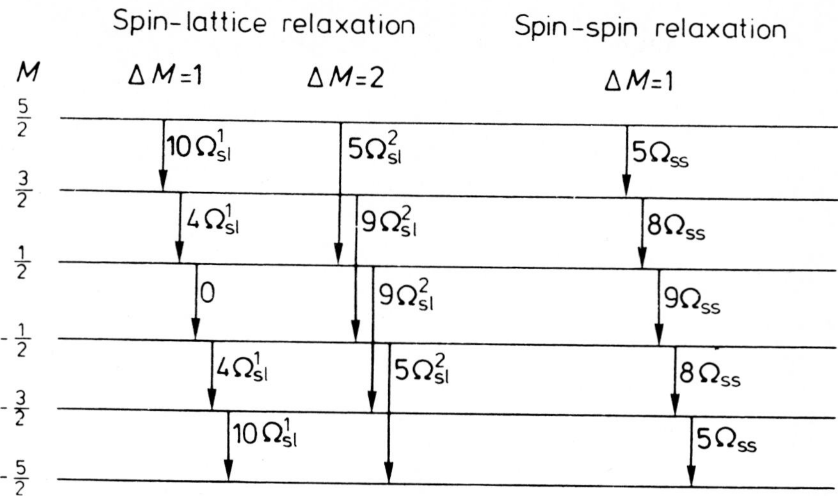

Fig. 11. Transition probability rates between the six electronic Zeeman states of Fe^{3+} *ion due to spin-lattice and spin-spin relaxation processes processes.*

responsible for the spin relaxation in the range of temperature investigated. In this work, the Einstein temperature characterising the E_g mode of the $[Fe(H_2O)_6]^{3+}$ complex in the alum was used to simulate the theoretical temperature dependence of Ω_{SL}. This gave the fit as shown with the broken line in Fig. 13(b). Yet another discrepency was found between the dependence of the relaxation frequencies on H_{ext} and the theory (Fig. (Fig. 13(c)). Whereas Raman process involving pure Zeeman levels should be field independent, experimentally Ω_{SL} is found to be field dependent, more so at 158 K.

EPR measurements on Cr^{3+} in Ammonium aluminium alum showed (43) the width of the satellite lines ($\pm$ 3/2$\rightleftarrows$$\pm$1/2) just above the structural phase transition temperature 60^0 K is unusually large and decreased when the temperature was increased. Although this behaviour is similar to the observation in other systems like $SrTiO_3$, it differs in the order of magnitude of the distortion and the temperature range over which the motional narrowing is effective. The line broadening was attributed to fluctuations of the crystal field associated with the phase transitions. Near T_c , the fluctuations are quasistatic and the linewidths of the EPR lines is determined by quasistatic spread in the crystal field parameters. As the temperature increases the correlation time of the fluctuations τ_c decreases with a resulting narrowing of the satellite

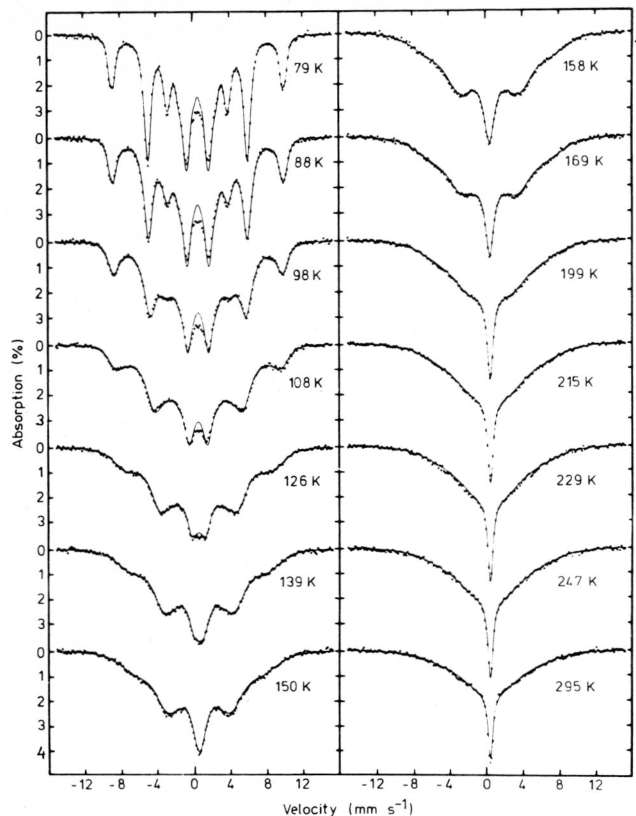

Fig. 12. Mossbauer spectra of Fe^{57} in $NH_4Al(SO_4)_2.12H_2O$ (Fe:Al=2 at %) in presence of a transverse field of 12.3 kG. The solid curves show theoretical line shapes simulated using the stochastic model of relaxation (adiabatic case). Taken from Ref. 8.

lines, until at high temperatures the fluctuations become too fast to influence the line shape. In a semiquantitative way, the spectral dens density $D(\omega)$ of fluctuations of the crystal field parameters is given

$$D(\omega) = \frac{2\sigma^2\tau_c}{(1+\omega^2\tau_c^2)} \quad ,$$

where we use for the correlation time the dielectric relaxation time in the NH_4Al alum above T_c ,

$$\tau_c = \tau_o \exp(T_o/T) \quad ,$$

with $\tau_o \sim 6.6 \times 10^{-15}$ s, $T_o \sim 1370$ K and σ is the rms value of the

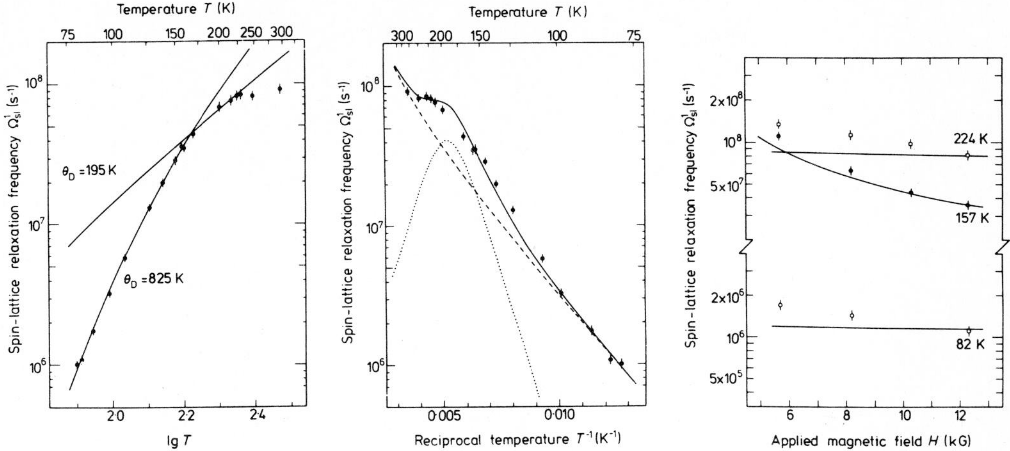

Fig. 13. The dependences of the spin lattice relaxation rate Ω_{SL} in $NH_4Al(Fe)$ alum on temperature and transverse external magnetic field. (a) The temperature dependence in presence of external field of 12.3 kG. The two solid curves correspond to the Debye model. (b) The temperature dependence in presence of external field of 12.3 kG. The broken curve correspond to the Einstein model (Eqn. 100) with $C=2.8\times10^8\ s^{-1}$ and $\Theta_E=450$ K. The dotted curve corresponds to the contribution from the critical spin lattice relaxation process (Eqn. 98) with parameters described in the text. The sum of these two contribution gives the temperature dependence shown by the solid line. (c) The dependence on external field at 82, 157 and 224 K. (taken from Ref. 8).

fluctuations of the crystal field parameter in question.

The anomalous temperature dependence of Ω_{SL} is thus explained by postulating that the direct spin lattice relaxation is induced by the critical crystal field fluctuations. The probability that the fluctuations give rise to the direct process is proportional to the spectral density $D(\omega)$ of the fluctuations at $\omega = \Delta/\hbar$

$$\Omega^D_{SL} = \frac{A\tau_c}{1+(\tau_c\Delta/\hbar)^2} \quad . \tag{98}$$

A similar expression hold good for $\Delta M=2$ transitions. When the contribution, shown by the dotted curve in Fig. 13(b) with $A = 1.8 \times 10^{19}\ S^{-1}$, and $T_o = 1260$ K is added, the good agreement shown by the full line is obtained.

Although it has been possible to simulate the experimentally observed temperature dependence of Ω_{SL} , a few improvements appear possible. 1. The coefficients f^1, g^1 and h^1 depend on internal modes transforming as bases of T_{2g} (the three coefficient transform as Q_4 , Q_5 and Q_6 respectively) whereas d^1 and e^1 transform as bases of E_g

representation (θ) and (ε), respectively. The phonons relating to these coefficient are related to different optic branches. In case of a phase transition, these coefficients may not show similar temperature dependences as has been assumed in the analysis. The argument is valid if contribution of lattice modes from other branches connected with motion of other groups like $(NH_4)^{+1}$, $(SO_4)^{-2}$ are included. The Einstein temperatures appropriate to different optic branches are different. In the fitting in this work only one value connected with the E_g mode has been assumed appropriate to describe the temperature dependence of all coefficients.

2. In presence of a field of 12.3 kG, it may be necessary to include terms in dynamical spin Hamiltonian which depend linearly on spin components, like, $(S_xH_z + S_zH_x)e_5$, $(S_yH_z + S_zH_y)e_4$ etc.

3. This kind of behaviour of τ_{SL} at higher temperatures has been found in almost all alums. As we shall see later, even in pure $NH_4Fe(SO_4)_2 \cdot 12H_2O$ the lines are not fully narrowed even at 300 K.

4. At higher temperatures, the presence of short wave length phonons must be taken into consideration.

In the investigation by Morup and Thrane (44) on $NH_4Fe(SO_4)_2 \cdot 12H_2O$ in the temperature range from 85 to 250 K, using magnetic fields upto 5 kG, similar behaviour of τ_{SL} at high temperature has been found.

Using a number of spectra in the temperature range 85-250 K in presence of the field of 5 kG, τ vs. T was obtained which is reproduced in Fig. (14). As Ω_{SS} is temperature independent, the observed dependence on temperature is due to τ_{SL} . The behaviour is similar to the experimental observation described above.

In the investigation by Morup and Thrane (45) on $(NH_4)_3FeF_6$ which is tetragonal below 263 K and cubic above this temperature similar behaviour of relaxation time has been found. The relaxation time does not change appreciably in the temperature interval from 264 to 348 K. In the ferric alum, Fe-Fe distance is 8.65 A^o, whereas in $(NH_4)_3FeF_6$, this separation is 6.3 A^o. It is not clear if the temperature dependence of τ and $(NH_4)_3FeF_6$ in the range from 264 to 348 K is due to the dominance of spin-spin relaxation over the spin lattice relaxation or due to the behaviour of τ_{SL} similar to the one described above in the alums.

The investigations of Dezsi et al (46) on M^{1+} $Fe(SO_4)_2 \cdot 12H_2O$ ($M^+ = NH_4^+$, K^+ , Rb^+, Cs^+ or Tl^+) are interesting in view of the observation of the temperature dependence of τ_{SL} of the Fe^{3+} ions in $NH_4Fe(SO_4)_2 \cdot 12H_2O$ by Morup and Thrane described earlier and because the spin spin relaxation times do not significantly vary in these substances as the Fe-Fe distances are almost same. It is interesting to check if

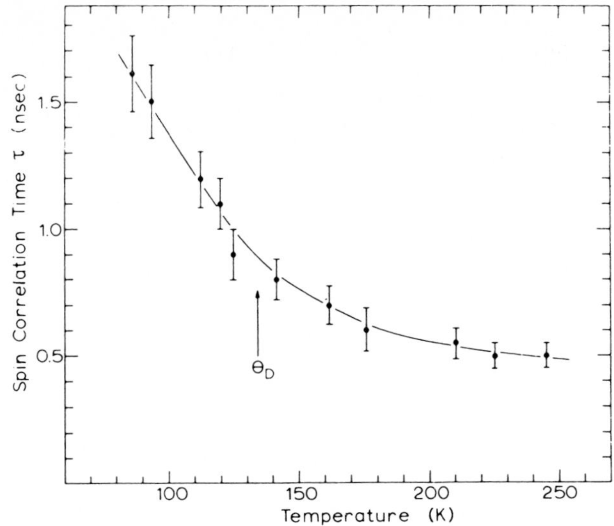

Fig. 14. The temperature dependence of the spin lattice relaxation time of Fe^{+3} ion in ferric alum in presence of a transverse field of 5 kG (taken from Ref. 44). Notice the behaviour is similar to the dependence of Ω^{1}_{SL} in Fig. 15.

the complete narrowing of the absorption line occurs at 300 K or above. These measurements were made in zero field. The linewidths are not strongly dependent on temperature and the lines are broad even at 300 K in all the alums investigated. The spectrum of Cs alum is reproduced in Fig. 15. This alum belong to the β group, unlike other alums investigated. Thus, the anomalous behaviour of τ_{SL} found in ferric and aluminium alums may be present in these alums too.

In the investigations by Barb et al (47) on $K_3[Fe(C_2O_4)_3].3H_2O$ and $(NH_3)_3[Fe(C_2O_4)_3].3H_2O$ similar features in relaxation line broadening have been observed. The measurements were made in the absence of an external field, in the range of temperature from 77 to 300 K. The absorption lines of both the absorbers are broad and the broadening is weakly dependent on temperature.

6.2 Spin Relaxation in Biological Molecules

Iron occurs in several biological molecules. When in Fe^{3+} state, the the spin lattice relaxation time of the ferric ion becomes large at low temperatures and thus paramagnetic hyperfine structure is observable at the liquid helium temperature. At higher temperatures, the spectral shape show effects of spin relaxation which are visible in some cases

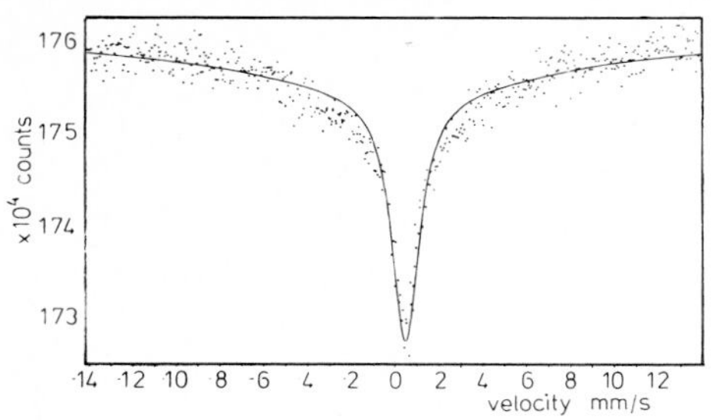

Fig. 15. Mossbauer spectrum of $CsFe(SO_4)_2.12H_2O$ (β-alum) at 87 K (taken from Ref. 46).

at as high temperatures as 250° K. However, these shapes have not been subjected to quantitative analysis, to derive the temperature dependence of τ_{SL}. On the contrary, it has been generally attempted in these earlier studies to suppress the relaxation effects even if it requires experimental temperatures and other conditions which are too far from the conditions in which these biological molecules actually function. The aim has been to characterise the state of iron with the help of the parameters of the static part of spin Hamiltonian, which necessitates lowering of the temperature. The characterisation of biological properties using the parameters of the dynamical spin Hamiltonian has the obvious advantage that experimental conditions can be close to the conditions in which these molecules operate. It must be emphasized, the spin lattice relaxation times are critically dependent on the near neighbour environment. Thus, choosing a medium for these molecules which alters the near neighbourhood would hardly provide meaningful results, except in the case when these molecules actually function in this medium Here, the features of the spectra showing relaxation effects obtained in the earlier studies on biological substances are qualitatively summaris summarised.

Spin relaxation effects in Mossbauer spectra of crystalline ferrice chlorohemin have been observed (48,49) in the temperature range from 4.2 to 300 K. The explanation by Blume (50) assumes the presence of spin spin relaxation effects only. Paramagnetic hyperfine splitting was observable at 4.2 K when Fe-Fe separation was increased using tetrahydrofuran (THF) and dimethyl formamide (DMF) solvents (51).

One of the earliest investigations in which the spin relaxation effect in Mossbauer spectra was observed is due to Wickmann et al (52) on ferrichrome A, which is naturally occurring (in fungi) iron transport protein. The structural formula is reproduced in Fig. 16(a). Iron exists

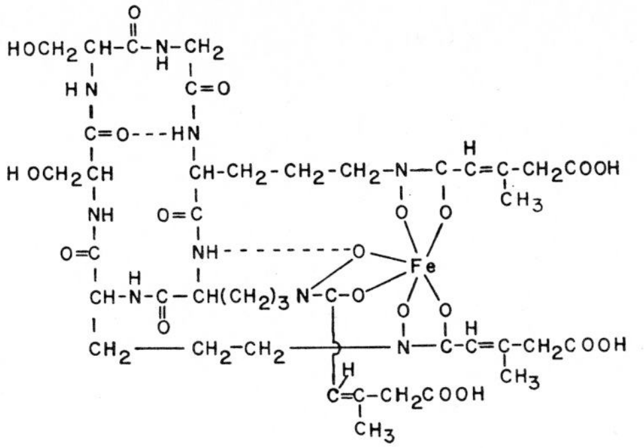

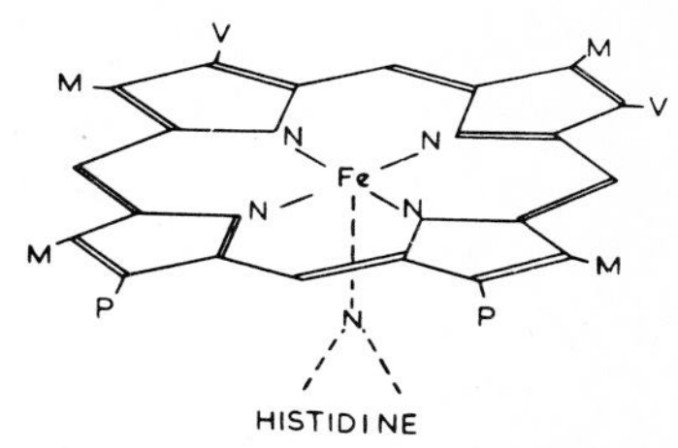

M,V, & P represents methyl, Vinyl and proprionic acid side chains

Fig. 16. (a) The structural formula of Ferrichrome A (52). (b) The structure of the Haem group (53).

in high spin Fe^{3+} state. The zero field spectra obtained in the temperature range 0.98 to 300° K showed the presence of spin relaxation effects. The application of a transverse magnetic field of 18 kG increased the magnetic splitting and narrowed the component lines considerably. The spin relaxation is appreciable is shown by the effect of magnetic dilution at 4.2 K which narrows the lines considerably. The spin lattice relaxation, however, dominantly effects the line shape at least at temperatures above 10° K. The shape of the zero field spectra were compared with theoretically simultaed shapes using two level relaxation formula, even though six levels participates in the relaxation process.

6.2.1 Relaxation Effects in Haem Proteins

In haem proteins, iron exist in a small unit known as haem (Fig. 16(b)), which is a combination of iron and protoporphyrin IX, is very stable, and can exist outside protein too. The iron is connected to the protein globin through the imidazole group of histidine. The nature of the ligand at the sixth position can vary and governs the state of iron in the haem group. Iron exist in low spin ferric state with CN^- at the sixth position, high spin ferric state with H_2O at the sixth position and so on. Ferrous and Ferric hemoglobins are represented by Hb and Hi respectively.

Haemoglobin: Iron in the ferrous state does not exhibit magnetic hyperfine splitting in absence of an external field except in HbNO (53). The g value of HbNO (isotropic with magnitude 2.03) indicates that the orbital moment is highly quenched giving larger value of τ_{SL}, like τ_{SL} of Fe^{3+} ion in high spin state. Though the magnetic splitting in zero field spectrum is clearly visible even at 195 K, no definite structure appears even at 1.2 K.

In HiF, iron exist as ferric ion in high spin state. Although the zero field spectra show magnetic splitting even at 195 K, the spectral shape in absence of an external field is not well defined even at 1.2 K. The shape gets well defined in presence of a small external field of 0.5 kG at 4.2 K (Fig. 17). This field decouples electron and nuclear spin precessions as well as splits the Kramers doublets, though the splitting is small in comparison to the splitting between adjacent doublets (2D = +14 cm^{-1}). The splitting of the lowest doublets $|\pm 1/2>$ which alone is singificantly populated at 4.2 K, when a small magnetic field (H>100 Gauss) is applied, changes the relaxation time sufficiently to cause a remarkable change in the shape of the spectrum. In acid methaemoglobin (PH6) in which H_2O is thought to be in the sixth coordination position, iron appears to be in high spin state (S=5/2) but the quadrupole splitting is large (∿2.0 mm/sec), showing deviation from spherical symmetry of the charge distribution of the Fe^{3+} ion. At 1.2 K, the paramagnetic hyperfine splitting in the zero field spectrum is well defined but the relaxation time decreases more rapidly with the increase in temperature than in HiF, consistent with the observation of the large quadrupole splitting (53).

Iron in HiCN exist as low spin ferric ion and shows a symmetric doublet at 195 K, an asymmetric doublet at 77 K and well resolved paramagnetic hyperfine splitting at 4.2 and 1.2 K, in zero field (53). Iron in HiN_3, Haemoglobin azide, also exist as ferric iron in low spin state. The relaxation broadening is visible at the liquid nitrogen temperature. The line shapes are, however, comparatively less well defined even at low temperature (4.2 and 1.2 K), in zero external field (53).

Myoglobin: In comparison to haemoglobin which possesses four haem groups per molecule, myoglobin molecule contains one haem group only. The Mossbauer spectra of myoglobin containing iron in ferric form show spin relaxation effects (54,55). Spectra of acid met myoglobin in 0.1M phosphate buffer PH6 and in 0.5M phosphate buffer PH6, which are not well defined in zero field, are well defined in the presence of a small field (H>100 Gauss) which decouples the electronic and the nucleus spins and splits the Kramers doublet. At 195° K, the spectrum is an asymmetric doublet due to relaxation effects, corresponds to larger values of the spin lattice relaxation frequencies. Similar features are observed in the spectra of myoglobin fluoride which contain iron in high spin Fe^{3+} state. The relaxation effects are not significantly different when myoglobin containing protohaem is replaced with myoglobin containing mesohaem prosthetic group. Paramagnetic hyperfine splitting at the

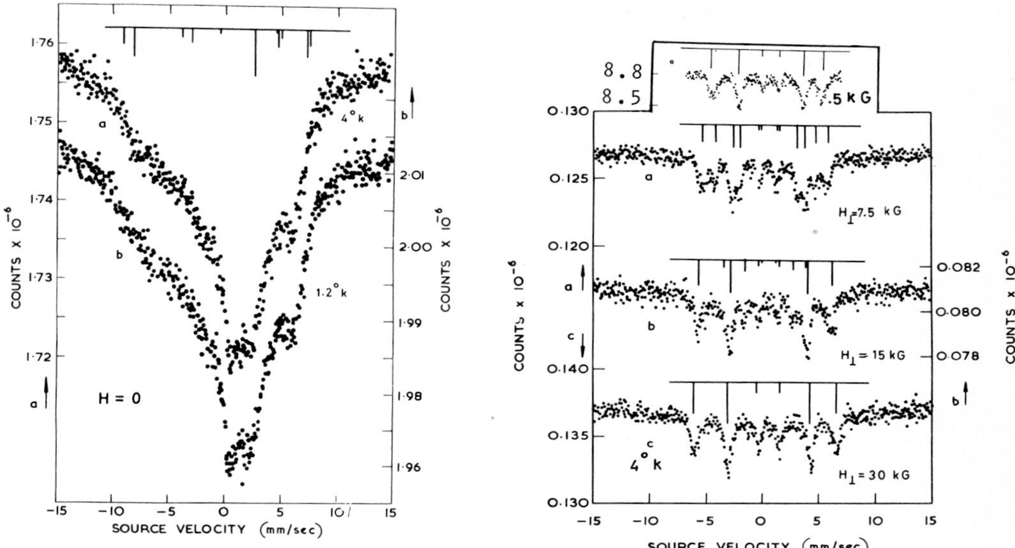

Fig. 17. Effect of the external magnetic field on spectral shape of Haemoglobin fluoride. The axial crystal field (2D=14 cm^{-1}) splits the ^6S state. AT low temperatures, only $|\pm 1/2\rangle$ doublet is significantly populated. The relaxation time between these two states increase sharply as the external field splits the doublet (taken from Ref. 39).

helium temperatures is also observed in the spectra of myoglobin azide (MbN_3) containing iron in low spin ferric state, showing large value of the spin lattice relaxation frequencies at the low temperature.

Cytochrome C: Spectra of lyophilized ferric cytochrome C & frozen solution of ferric cytochrome C in PH6 0.15M phosphate buffer were obtained by Lang et al (56). The ferric samples exhibit the general properties of a low spin ferric material. Spectra at 4.2 K and 1.8 K showed poorly resolved magnetic hyperfine splitting corresponding to a field of the order of 500 kG. The application of a small external field affects the spectral considerably giving better defined line shapes. Magnetic splitting is visible at 77 K but is absent at 195 K.

Protohaem and Mesohaem Cytochrome C peroxidases and their fluorides: The spectra of Cytochrome C peroxidase fluoride (57) are similar to those of Haemoglobin fluoride (Fig. 17). Iron is in high spin ferric state in both cases. The zero field spectra at 4.2 K of meso and proto cytochrome C peroxidase fluorides are magnetically split but poorly defined. The application of a field of 100 Gauss changes the spectral shapes remarkably and makes it sharply defined. The magnetic splitting is present at 77 K too but is absent in spectrum at 195 K, showing

τ_{SL} is small at this temperature. Spectra of proto cytochrome C and meso cytochrome C peroxidase in neutral 0.2M phosphate buffer showed components corresponding to low and high spin ferric species. The shapes change remarkably when a field of 500 Gauss is applied at 4.2 K. The spectral shape at 77 K is dominated by the effects of spin lattice relaxation. At 195 K, the relaxation rate becomes fast enough to eliminate magnetic splitting almost completely.

6.2.2 Relaxation Effects in Iron-Sulpher Proteins

Rebredoxins (Rd) is the simplest of all iron sulpher proteins. The active centre is a strongly distorted tetrahedron $Fe\text{-}S_4$ (Fig. 18(a)) with one of the Fe-S bond unusually short (2.5 A^o). Spin relaxation effects have been observed in Mossbauer spectra of oxidised Rd (iron in high spin Fe^{3+} state) from Cl. Paşteurianum (58-60) and the green photosynthetic bacterium chloropseudomonas ethylica (58). The spectra of the two are not significantly different. The spectrum at 4.2 K is a well resolved sextat. The relaxation effects are observable at higher temperatures. The spectrum at the 77 K show low absorption intensity due to the relaxation broadening, though the magnetic splitting is present. *Desulforedoxin (Dx):* from Desulfovibrio gigas (and other iron sulpher proteins) is discussed extensively in the article by Huynh and Kent elsewhere in this book. Mossbauer spectrum of oxidised Dx at 4.2 K in a field of 600 Gauss show well defined magnetic splitting due to the low value of the spin relaxation frequencies.

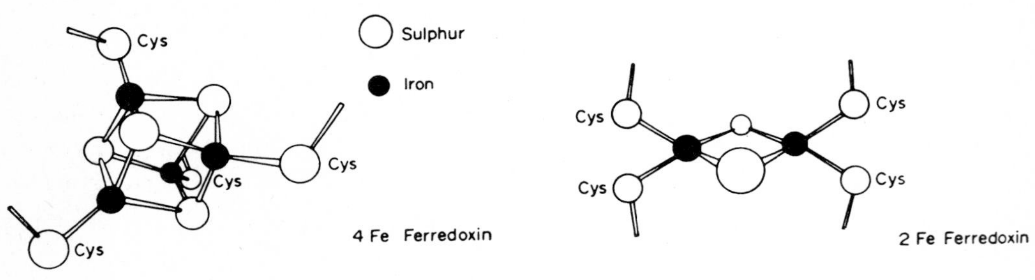

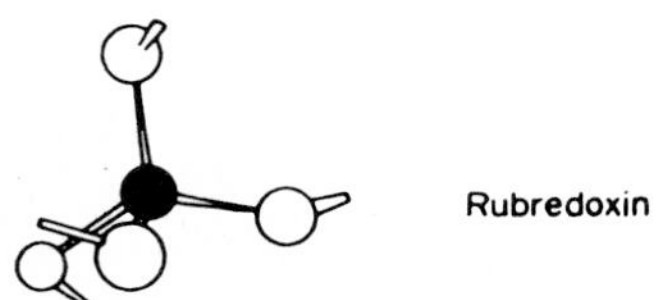

Fig. 18. Structure of Fe-S centres (58-64).

Putidaredoxin: This enzyme belongs to the class of iron sulpher protein, and contains two iron atoms per molecule. The investigation by Frauenfelder et al (61) showed that the Mossbauer spectra of the reduced putidaredoxin (which contains an iron ion in high spin Fe^{2+} state anti-ferromagnetically coupled to the other iron ion in high spin Fe^{3+} state, Fig. 18(b)) at low temperatures show magnetic splittings. The spectral shape changes little upto 80 K. The relaxation time even at 250 K is long enough to effect the spectral shape.

Plant type Ferrodoxin: The enzyme is a 2Fe protein similar to Putidaredoxin and andrenodoxin, the enzyme found in animals. At low temperatures, the spectrum of the reduced protein in a moderate magnetic field show well defined paramagnetic hyperfine splitting (62). In the reduced form the two iron ions are in high spin Fe^{2+} and Fe^{3+} states, respectively, coupled antiferromagnetically.

Iron sulpher proteins with active centres containing four iron ions and four labile sulpher atoms also exhibit magnetic splitting at low temperatures under suitable conditions (63,64), but are far more complicated for relaxation studies.

6.2.3 Iron Transport Proteins

They have high affinity for ferric iron (65,66). Ferrichrome A was discussed earlier. Deferoxamine was investigated by Bock and Lang (67) at several temperatures and also in the presence of an external field. Experiments were done with frozen solution of the complex in 0.1M citrate buffer. It was found that even at low temperatures, well resolved magnetic split spectra of Fe^{3+} in Deferoxamine were obtained only with magnetically dilute samples (4μ mole Fe in 0.8 cc). Magnetic splitting persisted at 77 K and 195 K, but is absent at 233 K.

Myobactin P: Mossbauer spectra of the quick frozen solution in methanol at 4.2 K, in zero field and in a transverse field of 1.3 kG, show magnetic splitting (68). The spectra show relaxation effects. Spin spin relaxation cannot be made negligible due to limited solubility in methanol. Zero field spectrum of the myobactin is very similar to the spectrum of ferrichrome A at 4.2 K.

Enterobactin: Spectra of frozen solution, prepared by mixing methanol solution of $FeCl_3$ and iron free entrobactin in 1 ml methanol in molar ratio 1:2 and then brought to PH 7.0 with sodium carbonate Soln, in zero field upto 77 K show large magnetic splittings (69). The Application of an external field sharpens the absorption lines. Spectra in field of 1.3 kG obtained upto 19.5 K show spin lattice relaxation effects. Spectra are very well defined in comparison to the spectra of transferrins and myobactin P.

Transferrins: They have high affinity for Fe^{3+}, molecular weight of about 80000 and two iron binding sites per molecules, are found in the milk of mammals (Lactotransferrins), in the eggs of white birds (conalbumins or ovotransferrins), and in the blood sera of vertebrates (70). Mossbauer spectra of iron in human and rabbit transferrins (iron in high spin ferric state) show well resolved magnetic splitting at 4.2 and 78 K, in zero field as well as in field of 470 Gauss (71,72). At higher temperatures, the magnetic hyperfine structure begins to smear out. However, it does not collapse before the reduction in the recoilless fraction due to higher temperature makes the observation of spectrum difficult. Conalbumin from hen's egg white was studied by Aisen et al (73). The spectrum at 4.2 K in field of 550 Gauss show well resolved magnetic splittings.

6.3 Other Substances

The temperature dependence of the spin lattice relaxation time of Fe^{3+} ion in $[Fe(H_2O)_6]^{3+}$ complex in amorphous medium was investigated by Knudsen and Morup (74). For this purpose, frozen solutions of (1) 0.025 M $Fe(NO_3)_3$, 0.5 M HNO_3 and 50% (by volume) glycerol and (2) 0.25 M $Fe(NO_3)_3$, and 7.0 M HNO_3 were used. Spectral shapes obtained in the temperature range from 80 to 165 K in the presence of an external field of 12.4 kG were analysed using the stochastic model for longitudinal relaxation. As the Zeeman splitting is large compared to the crystal field splitting, the theoretical procedure described earlier (Section 6.1) which includes $\Delta M = \pm 1$ and $\Delta M = \pm 2$ transition between the Zeeman levels was used. The temperature dependence of Ω_{SL} thus obtained is shown in Fig. 19. The investigation shows the large dependence of Ω_{SL} of the Fe^{3+} ions on the medium in which the complex $[Fe(H_2O)_6]^{3+}$ exist.

Suzdalev et al (10) investigated a number of ion exchange resins containing Fe^{3+} ions. Ion exchange resins with functional group SO_3^- (Sulforesins) have structural formula shown in Fig. 20. The Fe^{3+} ions are introduced by the ion exchange of H^+ from an acid aqueous solution of $FeCl_3$. The sorption of polar absorbent such as water, and consequent increase in volume, leading to larger separation of the magnetic ions, results in magnetically split Mossbauer spectra in the temperature range upto 250 K. The spectrum shape becomes better defined as the number of water molecule increases from one to six per iron ion in sulforesins. The spectrum of the sample with adsorbed water and low concentration of Fe^{3+} (1.6%) were analysed in which Fe^{3+}- Fe^{3+} separation is large. The Hamiltonian describing the interaction of the Fe^{3+} ion with crystal

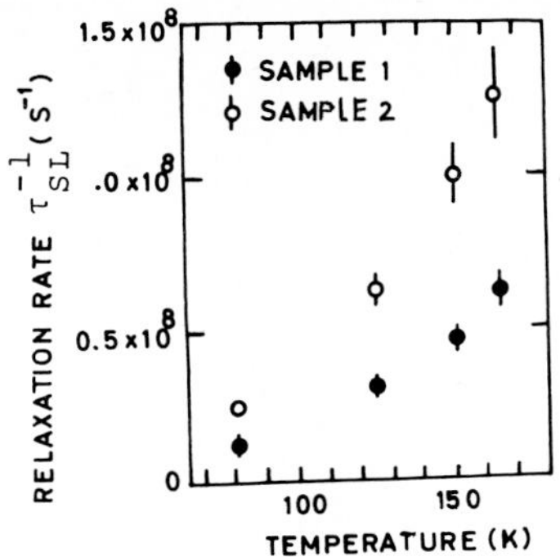

Fig. 19. The temperature dependence of the spin lattice relaxation rate of Fe^{3+} *in the frozen solutions (1) and (2) in the presence of a transverse field of 12.4 kG (taken from Ref. 74)*

field through spin orbit interaction is split into the static and the dynamical part as described in Section 6.1. H_{SP} splits the ^6S state of the Fe^{3+} ion into there Kramers doublets. In the range of temperature investigated (90-250 K) the contribution of Raman process due to the term quadratic in the displacements of the neighbouring ions in the first order of perturbation theory alone is significant. The assumption $\lambda=0$ simplifies the analysis and reduces the number of parameters required to describe the relaxation processes. In this case, the various transitions are as shown in Fig. 21. At $T >> \Delta$, the population of all the levels can be considered equal. The transition probabilities P_{ij} were calculated earlier in Section 6.1, and can be expressed using two parameters only Ω^1_{SL} and Ω^2_{SL}. When the relaxation frequency is low such that it only broadens the lines without changing their positions, the broadening of the spectral lines corresponding to the ith ionic state can be expressed as

$$\Delta\Gamma_i = \sum_k P_{ik} \quad .$$

Here, P_{ik} is the probability of transition from ionic states $|i\rangle$ to the state $|k\rangle$. Thus, the broadening of the absorption lines arising from the three doublets are

$$\Delta\Gamma_{\pm 5/2} = 40\Omega^2_{SL} + 80\Omega^1_{SL}$$

$$\Delta\Gamma_{\pm 3/2} = 72\Omega^2_{SL} + 112\Omega^1_{SL}$$

$$\Delta\Gamma_{\pm 1/2} = 112\Omega^2_{SL} + 32\Omega^1_{SL} \quad . \tag{99}$$

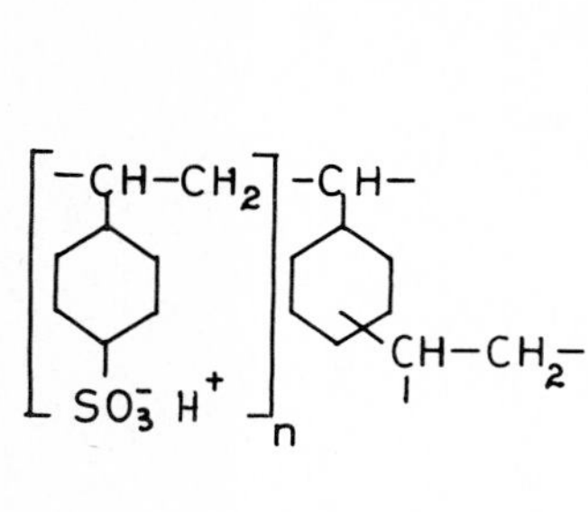

Fig. 20. Structural formula of sulforesin.

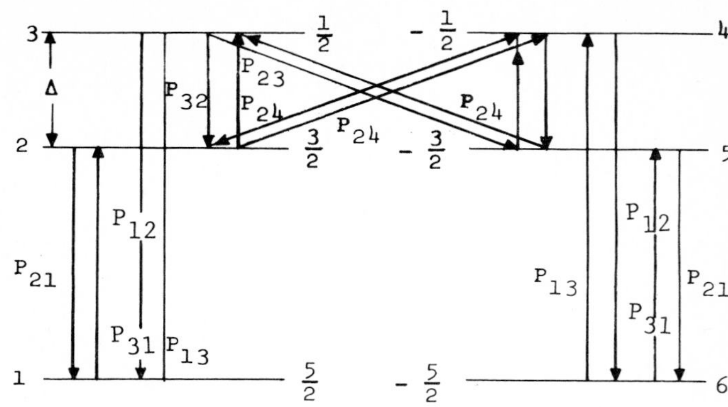

Fig. 21. Transition probability rates due to the spin lattice first order Raman relaxation as well as direct processes. $P_{12}=80\Omega_{SL}$, $P_{13}=40\Omega_{SL}$, $P_{23}=32\Omega_{SL}$. $P_{12}=P_{21}$, $P_{13}=P_{31}$, $P_{23}=P_{32}$, $T>>\Delta$ (taken from Ref. 10).

The contribution of spin spin relaxation is assumed negligible. The temperature dependence of the parameters Ω^1_{SL} and Ω^2_{SL} for the first order Raman process can be expressed as:

1. Using the Einstein model

$$\Omega^{1,2}_{SL} = C_{1,2}\,\frac{\exp(\theta_E/T)}{[\exp(\theta_E/T)-1]^2} \tag{100}$$

2. Using the Debye model

$$\Omega^{1,2}_{SL} = d_{1,2}\int_0^{\theta_D/T}\frac{x^7e^x dx}{(e^x-1)^2} \tag{101}$$

where θ_E and θ_D are Einstein and Debye temperatures, respectively, and $C_{1,2}$ and $d_{1,2}$ are constants. Using the resolved outer line of the component spectrum corresponding to $|S_z = \pm 5/2>$, the temperature dependence of $\Delta\Gamma_{\pm 5/2}$ could be determined upto 150° K. Thus,

$$\frac{1}{\tau} = \frac{1}{\tau_o} + \frac{1}{\tau_{SL}} \quad \text{and} \quad \tau^{-1}_{SL} = 40\Omega^2_{SL} + 80\Omega^1_{SL}$$

where τ^{-1} is the temperature independent broadening due to source other then the spin lattice relaxation. Using Einstein model, best agreement is obtained with $\theta_E \sim 325$ K and C=3.56. Using Debye model, the agreement is best when $\theta_D=400$ K and d=0.65. The hyperfine splitting appears following the adsorption of the water in sulforesin, is accompanied by a decrease in the recoilless fraction. Other resins with similar structures but the functional group SO_3^- replaced by PO_3^- and CoO^- do not show paramagnetic hyperfine splitting following the adsorption of water and the consquent increase in Fe-Fe separation or any appreciable reduction in the recoilless fraction as the temperature rises to the room temperature. The increase in τ_{SL} in sulforesin is thus concluded to be connected with the fact that water weakens the coupling between Fe^{3+} and the SO_3^- in the resin.

Mossbauer investigation of 2.26% of $K_3Fe(CN)_6$ in $K_2CO(CN)_6$, which provide the diamagnetic host, was made by Oosterhuis and Lang (75). At 20° K magnetically split spectrum was observed which was successfully fitted using the stochastic model of relaxation by Shenoy and Dunlop (76). On the other hand EPR measurements (77) in the temperature range 1.5 to 4.5 K showed $\tau_{SL}^{-1} = 5.4T + 0.0054T^9$ which gives $\tau_{SL}^{-1} = 2765$ MHz at 20° K. This frequency is sufficiently large to collapse any magnetic splitting in the Mossbauer spectrum. The reason for this disagreement is not clear.

ACKNOWLEDGEMENTS

It is a pleasure to thank Dr. K.V. Bhagwat for a critical reading of the draft of this article.

APPENDIX 1

$$P_\ell^m(\cos\theta) = \sin^m\theta \frac{d^m}{d^m(\cos\theta)} P_\ell(\cos\theta) , \qquad m \geq 0 .$$

$$Y_\ell^m(\theta,\phi) = (-)^m \left[\frac{2\ell+1}{4\pi} \frac{(\ell-m)!}{(\ell+m)!}\right]^{1/2} P_\ell^m(\cos\theta)\exp(im\phi) , \qquad m \geq 0 .$$

$$Y_\ell^{-m}(\theta,\phi) = (-)^m Y_\ell^{m*}(\theta,\phi) .$$

APPENDIX 2

The 24 symmetry operations of the 'O' group form five classes:

1. The Identity operations (E) .
2. The rotation by ($\pm\pi/2$) about x, y, and z axes (C_4) .
3. The rotation by π about x, y, and z axes (C_4^2) .
4. The rotation by $2\pi/3$ and $4\pi/3$ about (111), $(1\bar{1}1)$, $(\bar{1}11)$, and $(11\bar{1})$ axes (C_3) .
5. The rotation by π about (011), (101), (110), $(0\bar{1}1)$, $(\bar{1}01)$, and $(\bar{1}10)$ axes (C_2) .

When the Symmetry is lowered to tetragonal, D_4 , only eight of the 24 symmetry operations of the 'O' group still leave the system invariant. The character table of D_4 group is given below.

TABLE 2.1

The Character table of the O group.

Irreducible representation	E	$6c_4$	$3c_4^2$	$8c_3$	$6c_2$
A_1	1	1	1	1	1
A_2	1	-1	1	1	-1
E	2	0	2	-1	0
T_1	3	1	-1	0	-1
T_2	3	-1	-1	0	1

TABLE 2.2

The Character table of the D_4 group.

Irreducible representation	E	$2C_4$ $C_4(z),C_4^3(z)$	$C_4^2(z)$	$2C_2$ $C_2(110),C_2(\bar{1}10)$	$2C_2'$ $C_2(x),C_2(y)$
A_1	1	1	1	1	1
A_2	1	1	1	-1	-1
B_1	1	-1	1	-1	1
B_2	1	-1	1	1	-1
E	2	0	-2	0	0

The double group is obtained by regarding the rotation by 4π about an axis as the identity operation and the rotation by 2π as one of the symmetry operations of the double group. It can be verified that the symmetry operations of the double group corresponding to the 'O' group described above form eight classes. The character table is given below. In this table $RC_4(z)$ denotes a rotation of $2\pi+\pi/4$ about the z axis, and so on.

The O_h group is the product of O and the inversion group (elements: inversion operator i and $i^2=E$). The character table of the O_h group is given below.

APPENDIX 3

When $B_{\ell m}$ is real

$$H_{cr} = \Sigma\left(\sum_{\ell=0}^{\infty} B_{\ell 0} r^{\ell} Y_{\ell}^{0} + \sum_{\ell=0}^{\infty}\sum_{m>0}^{\ell} B_{\ell m} r^{\ell}[Y_{\ell}^{m} + Y_{\ell}^{m*}]\right)$$

$$= \left(\sum_{\ell=0}^{\infty} B_{\ell 0} T_{\ell 0} + \sum_{\ell=0}^{\infty}\sum_{m>0}^{\ell} B_{\ell m} T_{\ell m}\right) \quad .$$

TABLE 2.3

The Character table of the cubic double group.

IR	E	R	$3C_4, 3C_4^3R$	$3C_4^3, 3C_4R$	$3C_4^2, 3C_4^2R$	$4C_3, 4C_3^2R$	$4C_3^2, 4C_3R$	$6C_2, 6C_2R$
A_1	1	1	1	1	1	1	1	1
A_2	1	1	-1	-1	1	1	1	-1
E	2	2	0	0	2	-1	-1	0
T_1	3	3	1	1	-1	0	0	-1
T_2	3	3	-1	-1	-1	0	0	1
E_1	2	-2	$\sqrt{2}$	$-\sqrt{2}$	0	1	-1	0
E_2	2	-2	$-\sqrt{2}$	$\sqrt{2}$	0	1	-1	0
G	4	-4	0	0	0	-1	1	0

TABLE 2.4

The Character table of the O_h group.

O_h	E	$6C_4$	$3C_4^2$	$8C_3$	$6C_2$	i	$6S_4(C_4xi)$	$3\sigma_h(C_2xi)$	$8S_6(C_3xi)$	$6\sigma_d(C_2xi)$
A_{1g}	1	1	1	1	1	1	1	1	1	1
A_{2g}	1	-1	1	1	-1	1	-1	1	1	-1
E_g	2	0	2	-1	0	2	0	2	-1	0
T_{1g}	3	1	-1	0	-1	3	1	-1	0	-1
T_{2g}	3	-1	-1	0	1	3	-1	-1	0	1
A_{1u}	1	1	1	1	1	-1	-1	-1	-1	-1
A_{2u}	1	-1	1	1	-1	-1	1	-1	-1	1
E_u	2	0	2	-1	0	-2	0	-2	1	0
T_{1u}	3	1	-1	0	-1	-3	-1	1	0	1
T_{2u}	3	-1	-1	0	1	-3	1	1	0	-1

Here, the summation Σ outside the bracket in $\mathcal{H}_{cr}$ denotes inclusion of all the valence electrons of the central ion.

In Equation (15), the matrix elements of $T_{\ell m}$ between states $|JM\rangle$ are considered. However, if the matrix elements of $T_{\ell m}$, $m = -\ell, --, \ell$, between spin orbitals $\phi_{n\ell m}$ are desired, Eqn. (15) is replaced by

$$\langle \ell'm'|T_{\ell m}|\ell'm''\rangle = \langle r^{\ell}\rangle \frac{\langle \ell'||T_{\ell}||\ell'\rangle}{\langle \ell'||O_{\ell}||\ell'\rangle} \langle \ell'm'|O_{\ell}^{m}|\ell'm''\rangle$$

Where O_{ℓ}^{m} are obtained from the expressions given below by replacing J_x, J_y, J_z, and J with ℓ_x, ℓ_y, ℓ_z, and ℓ, respectively. Similarly the matrix element of $T_{\ell m}$ between states $|LM\rangle$ can be obtained.

The form of the operators corresponding to $T_{\ell m}$ and suitable when matrix elements between states $|JM\rangle$ are desired are given in Table 3.1 below:

TABLE 3.1
Equivalent operators of the polynomials $T_{\ell m} = \Sigma\ r^{\ell}[Y_{\ell}^{m} + Y_{\ell}^{m*}]$

$T_{\ell m}$	Corresponding operator $O_{\ell m}$
$T_{20} = \Sigma(3z^2-r^2)$	$\alpha\langle r^2\rangle[3J_z^2-J(J+1)]$
$T_{22} = \Sigma(x^2-y^2)$	$\alpha\langle r^2\rangle\frac{1}{2}[J_+^2+J_-^2]$
$T_{40} = \Sigma(35z^4-30r^2z^2+3r^4)$	$\beta'\langle r^4\rangle[35J_z^4-30J(J+1)J_z^2+25J_z^2-6J(J+1)+$ $+3J^2(J+1)^2]$
$T_{42} = \Sigma(7z^2-r^2)(x^2-y^2)$	$\beta'\langle r^4\rangle\frac{1}{2}[(7J_z^2-J(J+1)-5)(J_+^2+J_-^2)$ $+(J_+^2+J_-^2)(7J_z^2-J(J+1)-5)]$
$T_{43} = \Sigma xz(x^2-3y^2)$	$\beta'\langle r^4\rangle\frac{1}{2}[J_z(J_+^3+J_-^3)+(J_+^3+J_-^3)J_z]$
$T_{44} = \Sigma(x^4-6x^2y^2+y^4)$	$\beta'\langle r^4\rangle\frac{1}{2}[J_+^4+J_-^4]$

REFERENCES

1. R. Orbach and H.J. Stapleton, in S. Geschwind (Ed.), Electron Paramagnetic Resonance, Plenum Press, New York London, 1972, p.121
2. K.N. Shrivastava, Phys. Rep. 20 (1975) 137.
3. E.B. Tucker, in W. Mason (Ed.), Physical Acoustics, Vol. 4A, Academic Press, New York, 1966, p. 47.
4. S.A. Al'tshuler, B.I. Kochelaev, and A.M. Leushin, USPEKHI Fiz. Nauk, 75 (1961) 459 [English Transl : Sov. Phys. USPEKHI, 4 (1962) 880.
5. M.J. Clauser, Phys. Rev. B, 3 (1971) 3748.
6. M.J. Clauser and M. Blume, Phys. Rev. B, 3 (1971) 583.
7. G.K. Shenoy and B.D. Dunlop, Phys. Rev. B, 13 (1976) 1353, 3709.
8. S.C. Bhargava, J.E. Knudsen and S. Morup, J. Phys. C : Solid State Phys., 12 (1979) 2879.
9. D.W. Forester and W.A. Ferrando, Phys. Rev. B, 13 (1976) 3991.
10. I.P. Suzdalev, A.M. Afanas'ev, A.S. Plachinda, V.I. Goldanskii, and E.F. Makarov, Sov. Phys. - JETP, 28 (1969) 923.
11. A. Messiah, Quantum Mechanics, North Holland, Amsterdam, 1961.
12. R.R. Sharma, T.P. Das, and R. Orbach, Phys. Rev., 149 (1966) 257.
13. S. Sugano, Y. Tanabe, and H. Kamimura, Multiplets of Transition - Metal Ions in Crystals, Academic Press, New York, 1970.
14. H.J. Zeiger and G.W. Pratt, Magnetic Interactions in Solids, Clarendon Press, Oxford, 1973.
15. A. Abragam and B. Bleaney, Electron Paramagnetic Resonance of Transition Ions, Clarendon Press, Oxford, 1970.
16. H.M.L. Pryce, Proc. Phys. Soc., A63 (1950) 25.
17. A. Abragam and H.M.L. Pryce, Proc. Roy. Soc., A205 (1951) 135.
18. A.A. Maradudin, E.W. Montroll, and G.H. Weiss, in F. Seitz and D. Turnball (Eds.), Solid State Phys., Suppl. 3, Academic Press, New York, 1963.

19. G. Venkataraman, L.A. Feldkamp, and V.C. Sahni, Dynamics of Perfect Crystals, MIT Press, Cambridge, 1975.
20. S.A. Al'tshuler and B.M. Kozyrev, Electron Paramagnetic Resonance Academic Press, New York and London, 1964.
21. J.H. Van Vleck, Phys. Rev., 57 (1940) 426, J. Chem. Phys., 7 (1939) 72.
22. W. Heitler, The Quantum Theory of Radiation, Clarendon Press, Oxford, 1957.
23. R. Orbach, Proc. Roy. Soc. (London), A264 (1961) 458.
24. C.P.B. Finn, R. Orbach and W.P. Wolf, Proc. Phys. Soc. (London), 77 (1961) 261.
25. I. Waller, Z. Physik, 79 (1932) 370.
26. R. de L. Kronig, Physica, 6 (1939) 33.
27. Michel-Calendini and Kibler, Theor. Chim. Acta, 10 (1968) 37.
28. R. Orbach and M. Tachiki, Phys. Rev., 158 (1967) 524.
29. R.D. Mattuck and N.W.P. Strandberg, Phys. Rev., 119 (1960) 1204.
30. N.G. Koloskova, Fiz. Tverd. Tela 5 (1963) 61[English Transl. : Sov. Phys. - Solid State, 5 (1963) 40]
31. M. Blume, Phys. Rev. B, 3 (1971) 3748; Phys. Rev. Lett., 14 (1965) 96; Phys. Rev., 174 (1968) 351.
32. J.A. Tjon and M. Blume, Phys. Rev., 165 (1968) 456, M. Blume and J.A. Tjon, Phys. Rev., 165 (1968) 446.
33. R. Kubo, J. Phys. Soc. Japan, 9 (1954) 935, P.W. Anderson, J. Phys. Soc. Japan, 9 (1954) 316.
34. H. Wegener, Z. Phys., 186 (1965)498.
35. E. Bradford and W. Marshall, Proc. Phys. Soc., 87 (1966) 731.
36. A.M. Afanasev and Yu. Kagan, Sovt. Phys. JETP, 18 (1964) 1139.
37. C.E. Johnson, Phys. Lett., 21 (1966) 491.
38. F.E. Obeushain, L.D. Roberts, C.E. Coleman, and D.W. Forester, Phys. Rev. Lett., 14 (1965) 365.
39. G. Lang and W. Marshall, Proc. Phys. Soc., 87 (1966) 3.
40. L. Campbell and S. DeBenedetti, Phys. Lett., 20 (1966) 102.
41. L.E. Campbell and S. DeBendetti, Phys. Rev., 167 (1968) 556.
42. J. Ubbink, J.A. Poulis and C.J. Gorter, Physica 17 (1951) 213.
43. R. Chicault and R. Buisson, J. Phys. (Paris), 38 (1977) 795.
44. S. Morup and N. Thrane, Phys. Rev. B, 4 (1971) 2087.
45. S. Morup and N. Thrane, Phys. Rev. B, 8 (1973) 1020.
46. I. Dezsi, T. Lohner, D.L. Nagy and A.M. Afanasev, J. Phys. (Paris) Colloq., 35 (1974) C6-449.
47. D. Barb, L. Diamandescu and D. Tarabasanu, J. Phys. (Paris) Colloq., 37 (1976) C6-113; Proc. Int. Conf. on Mossbauer Spectroscopy, Bucharest, Romania, 1977, p. 219.
48. A.J. Bearden, T.H. Moss, W.S. Caughey, and C.A. Beaudreau, Proc. Nat. Acad. Sci., 53 (1965) 1246.
49. R.M. Housley and H. De Waard, Phys. Lett., 21 (1966) 90.
50. M. Blume, Phys. Rev. Lett., 18 (1967) 305.
51. G. Lang, T. Asakura, T. Yonetani, Phys. Rev. Lett., 24 (1970) 981.
52. H.H. Wickman, M.P. Klein and D.A. Shirley, J. Chem. Phys., 42 (1965) 2113; Phys. Rev., 152 (1966) 345.
53. G. Lang, Quart. Rev. Biophys., 3 (1970) 1
54. G. Lang, T. Asakura and T. Yonetani, Biochim. Biophys. Acta., 214 (1970) 381.
55. T. Harami, J. Chem. Phys., 71 (1979) 1309.
56. G. Lang, D. Herbert and T. Yonetani, J. Chem. Phys., 49 (1968) 944.
57. G. Lang, T. Asakura, and T. Yonetani, J. Phys. C : Solid State Phys., 2 (1969) 2246.
58. K.K. Rao, M.C.W. Evans, R. Cammack, D.O. Hall, C.L. Thompson, P.J. Jackson, and C.E. Johnson, Biochem. J., 129 (1972) 1063.
59. C. Schulz and P.G. Debreunner, J. Phys. (Paris) Colloq., 37 (1976) C6-153.

60. W.D. Phillips, M. Poe, J.F. Weiher, C.C. McDonald, and W. Lovenberg, Nature, 227 (1970) 574.
61. H. Frauenfelder, I.C. Gunsalus, E. Munck, MossBauer Spectroscopy and its Applications, IAEA, Vienna, 1972, p. 231.
62. W.R. Dunham, A.J. Bearden, I.T. Salmeen, G. Palmev, R.H. Sands, W.H. Orme-Johnson and H. Beinert, Biochim. Biophys. Acta, 253 (1971) 134.
63. R.N. Mullinger, R. Cammack, K.K. Rao, D.P.E. Dickson, C.E. Johnson, J.D. Rush, A. Simopoulos, Biochem. J., 151 (1975) 75; D.O. Hall, K.K. Rao, R. Cammack, Sci. Prog. (London) 62 (1975) 285.
64. D.P.E. Dickson, C.E. Johnson, P. Middleton, J.D. Rush, R. Cammack, D.O. Hill, R.N. Mullinger, and K.K. Rao, J. Phys. (Paris) Colloq., 37 (1976) C6-171.
65. K. Spartalian, W.T. Oosterhuis, and B. Window, in I.J. Gruverman, (Ed.), Mossbauer Effect Methodology, Vol. 8, Plenum Press, New York, 1973, p. 137.
66. W.T. Oosterhuis and K. Spartalian, in S.G. Cohen (Ed.), Applications of Mossbauer Spectroscopy, Vol., Academic Press, New York, London, 1976, p. 141.
67. J. Bock and G. Land, Biochimi. Biophys. Acta, 264 (1972) 245.
68. W.T. Oosterhuis and K. Spartalian, J. Phys. (Paris) Colloq., 35 (1974) C6-347.
69. K. Spartalian, W.T. Oosterhuis, and J.B. Nielands, J. Chem. Phys., 62 (1975) 3538.
70. J.B. Nielands, Struct. Bonding, 1 (1966) 59.
71. C.P. Tsang, A.J.F. Boyle, and E.H. Morgan; Biochim. Biophys. Acta, 328 (1973) 84.
72. K. Spartalian and W.T. Oosterhuis, J. Chem. Phys., 59 (1973) 617.
73. P. Aisen, G. Lang, and R.C. Woodworth, J. Biol. Chem., 248 (1973) 649.
74. J.E. Knudsen and S. Morup, Int. Conf. on Mossbauer Spectroscopy, Bucharest, Romania, 1977, p. 205.
75. W.T. Oosterhuis and G. Lang, Phys. Rev., 178 (1969) 439.
76. G.K. Shenoy and B.D. Dunlap, Phys. Rev. B, 13 (1976) 3709.
77. T. Bray, G.C. Brown, Jr., and A. Kiel, Phys. Rev., 127 (1962) 730.

CHAPTER 13

THEORY OF ZERO-FIELD SPLITTING, SPIN-LATTICE COUPLING CONSTANTS AND NUCLEAR QUADRUPOLE INTERACTIONS OF S-STATE IONS IN SOLIDS

R.R. Sharma

Department of Physics, University of Illinois, Chicago, Ill. 60680.

1. INTRODUCTION

The advent of various resonance methods such as nuclear magnetic resonance, the electron paramagnetic resonance and the Mossbauer effect has offered practical and important means of investigating the microscopic properties of ions in solids.

Bethe (1) discussed as early as in 1929 the splitting of electron energy levels of ions (and atoms) by crystal fields using group theoretical techniques. Since then the crystal or ligand field theory has been widely adopted by Van Vleck (2-8) and Schlapp and Penney (9-10) and it pervades much of the current literature (11-29) on impurity ions in solids and transition metal ion complexes.

Since its discovery (30) the applications of the EPR (and also of the Mossbauer effect and the NMR) have spread in different branches of physical and biological sciences. The basic ideas and theoretical methods for the empirical analysis of the EPR spectra have been described in detail by Abragram and Bleaney (31) and Low (32). Basically, no difficulty arises in the study of the EPR spectra of non S-state ions where the total angular momentum quantum number is different from zero. However, in case of S-state ions (Mn^{2+} and Fe^{3+} having $3d^5(^6S)$ and Eu^{2+} and Gd^{3+} having $4f^7(^8S)$ configurations) the interpretation is tedious and complex. A review of the theoretical analysis of S-state ions has recently been given by Newman and Urban (33). The empirical spin-Hamiltonian Parameters for Gd^{3+} have been fully collected by Buckmaster and Shing (34). Many experimental results exist in the literature also for Mn^{2+}, Fe^{3+} and Eu^{2+}.

The important applications of the resonance techniques have been in determining the site symmetries, orientation of symmetry axes, the co-ordination number and axes, the co-ordination number and oxidation stats (35), ionicities (or valencies), magnetization and magnetic moments, electronic charge and spin-distributions (36-40), electron configurations, covalency effects, localized moments, crystal field parameters and splitting of crystal field states of ions in crystals. They are also useful in the ordering mechanisms of spins and, more recently, in investigating metal-semimetal (41), magnetic and structural phase transitions (42-47), amorphous materials (48), super-ionic conductors (49-52), intermetallic compounds (53,54), small particles; surface effects and linear systems (55-57), biomolecules (56-66), geological materials (67-68), and art and pottery (69-71).

Though vast experimental data have been collected for the S-state ions in various crystalline environments from resonance experiments (28, 30-34, 72-77), the relevant theory has not been fully developed. As for ^{6}S-state ions Van Vleck and Penny (78) in their paper had pointed out the possibility of explaining the ground state splitting (known as zero field splitting) in non-cubic crystalline fields. However, they had not made any explicit analysis of the conditions under which one can obtain such a splitting. Subsequently, Pryce (79), Watanabe (80), Blume and Orbach (81), Orbach, Das and Sharma (82), proposed various mechanisms. Kondo (83,84) has considered the covalency effects on the zero-field splitting of ^{6}S-state ions and arrived at the conclusion that such effects are indeed important. Chakravarty (85) studied Pryce mechanism in greater detail making use of the Slater type d-orbitals to facilitate analytic solutions. Leushin (86-88) has also made an attempt to understand the zero-field splitting by examining relatively the Blue-Orbach and Pryce mechanisms. A parametric approach has been proposed by Mcfarlene (89) by relating the optical data for the excited state to the zero-field splitting parameters in a trigonal environment. A most complete study up to date for the ^{6}S-state zero-field splitting has been made by Sharma, Das and Orbach (90-92) by taking the case of Mn^{2+} in different crystals. They made accurate quantitative evaluation of the zero-field splittings arising from various mechanisms, including overlap effects due to nearest neighbor ligands and inferred that the Blume-Orbach mechanism is the most dominant process. They explained the sign and also the magnitude of the splitting reasonably well, considering the first-principles nature of calculations. For Fe^{3+} in distorted MgO the calculated results by Sharma (93) following the theory in Ref. 90-92 were very satisfactory particularly when the local distortion corrections were employed as pointed out by Borg and Ray (94).

A useful discussion of the spin-lattice coupling constants from various mechanisms and overlap effects is given in Ref. 93. Several authors (95-104) have estimated the zero-field splitting and spin-lattice coefficients of Mn^{2+} and Fe^{3+} in different crystals using the known mechanisms, especially, the Blume-Orbach mechanism which is found to provide both the sign and correct order of magnitude of the parameters.

The situation for the ^8S-state ions from the rare earth group is, on the other hand, quite opposite. Wybourne (105, 106) considered eight different mechanisms (including the relativistic mechanism proposed by him) for Gd^{3+} in lathanum ethyl sulfate crystal and obtained the net zero-field splitting which is opposite in sign and twice in magnitude compared to the experimental value. Newman (107) has also proposed two specific mechanisms based on the charge conjugation invariant components of the crystal field. The relativistic mechanism of Wybourne (106) has been studied for Mn^{2+} by Van Heuvelen (108) and by Hagston and Lowther (109). Encouraged by the reasonable agreement obtained by Sharma et al (90-92) for Mn^{2+}, Calvo et al (110), Schlottman (111) and others investigated the Blume-Orbach mechanism for Gd^{3+}spin-lattice relaxation in various crystalline environments.

Recently, Gill (112) has reviewed the literature on the establishment of thermal equilibrium in paramagnetic crystals. Reference may be made to the article by Orbach and Stapleton (113) on spin-lattice relaxation in the book by Geschwind (114). Other references are Marrenkov and Orbach (115), Poole and Farach (116), Standley and Vaughan (117), Tucker and Rampton (118), Veastelle and Curtis (119), Tucker (120) and by Stevens (121).

For the theoretical discussions of the nuclear quadrupole interactions attention is drawn to the review articles by Cohen and Reif (122), Das and Hahn (123) and Artman (28) and to the book by Lucken (124). Readers who are interested in determination of the experimental parameters from nuclear quadrupole resonance spectroscopy may consult the paper by Collins and Travis (125). Among the papers which have helped in recent developments in the theory of nuclear quadrupole interactions in ionic crystals the noteworthy are the ones by Bersohn (126), Burns (127), Brun and Hafner (128), Taylor and Das (129), Sharma and Das (130), Grant et al (131), Artman and Murphy (132), Raymond (133), Hafner and Raymond (134), Taylor (135), Sawatzky and Hupkes (136), Sharma and co-workers (26, 27, 137-139) and others.

Section 2 describes briefly the Spin-Hamiltonian and the various parameters involved in it, covering also a short description of the dynamic Spin-lattice Hamiltonian in terms of the spin-lattice coupling constants. Details of the theory of zero-field splitting, spin-lattice coupling constants and nuclear quadrupole interactions of the relevant experimental data for the S-state ions in various crystalline environments follow in Sections 3, 4 and 5. Only the second-order spin-lattice coupling constants will be considered owing to the lack of theoretical and experimental results for fourth-order spin-lattice coupling constants. Concluding remarks form Sec. 6 which also indicates the directions for future work.

2. SPIN-HAMILTONIAN

Extensive knowledge has already been gathered on the properties of magnetic ions in a wide variety of lattices - see for example review articles by Orton (74), Bleany and Stevens (72), Bowers and Owen (73), Griffith (11), Low (32) and others. Along with the experimental works, many theoretical studies were also undertaken which gave rise to the Spin-Hamiltonian formalism. Van Vleck (140) appears to be the first who introduced the concept of the Spin-Hamiltonian. It was rederived in detail by Pryce (141) and developed by Abragam and Pryce (142) and Bleaney and Stevens (72). Though the paramagnetism arises in part due to orbital angular momentum, it is customary to describe the state of the system by means of the Spin-Hamiltonian. The Spin-Hamiltonian is of great value since not only the experimental information (especially low lying states of magnetic ions) can be summed up succinctly and comprehensively but also the theorist finds it very acceptable.

The Spin-Hamiltonian contains only the components of the effective or fictitious spin S whose degeneracy (2S + 1) is equal to the number of lowest lying levels. In general, the fictitious spin and the real spin are different although in case of S-state ions they are the same.

The form of the Spin-Hamiltonian is determined by the surroundings. The magnitude of the terms in the Hamiltonian are, however, indicative of the strength of the different types of interactions involved. The Spin-Hamiltonian with the most commonly occurring terms can be written as follows (142):

$$\begin{aligned} H_S = {} & D(3S_z^2 - S(S+1)) + E(S_x^2 - S_y^2) + \frac{e^2Qq}{4I(2I-1)}\{(3I_z^2 - I(I+1)) + \eta(I_x^2 - I_y^2)\} \\ & + \vec{S}.\tilde{A}.\vec{I} + \vec{S}.\tilde{g}.\vec{H} + \vec{I}.\tilde{\lambda}.\vec{H} \\ & + a\{S_x^4 + S_y^4 + S_z^4 - \tfrac{1}{5}S(S+1)(3S^2 + 3S - 1)\} \\ & + F(35S_1^4 - 30S(S+1)S_1^2 + 25S_1^2 - 6S(S+1) + 3S^2(S+1)^2) \end{aligned} \qquad (1)$$

The first and second terms are usually called the zero-field splitting terms and appear whenever there is a departure from cubic symmetry. The second term occurs only when there is lack of axial symmetry as in the case of orthorhombic symmetry. Thus E is zero whenever the symmetry around the paramagnetic ion site is tetragonal or trigonal. In any case both the first and second terms are present in the Spin-Hamiltonisn only when the total spin S is greater than 1/2. The third and fourth terms in H_S represent the energy of the interaction of the nuclear quadrupole moment with the field gradient at the site of the nucleus with spin I. Q is the quadrupole moment of the nucleus; eq is the electric field gradient V_{zz} at the nucleus (V being the potential at the nucleus) and η is the asymmetry parameter ($\eta = \frac{V_{xx} - V_{yy}}{V_{zz}}$ with the convention that $|V_{xx}|<|V_{yy}|<|V_{zz}|$). This interaction is nonzero if the nuclear spin is greater than 1/2 and there is departure from cubic symmetry. The fifth term shows the hyperfine interaction between the nuclear spin I and electron spin S. The sixth and seventh terms are the Zeeman interaction energies of the electronic spin S and nuclear spin I respectively. Here $\vec{H}$ stands for the applied magnetic field and $\tilde{g}$ is in general a tensor which reduces to the free ion g-factor when the crystalline field is removed. The eighth and nineth terms are called cubic terms and are quartic in the components of the total spin S.

Equation 1 represents the static Spin-Hamiltonian. In cases of spin-lattice relaxation process or uniform stress measurements one requires the "dynamic-Spin-Hamiltonian" which can be written as (140-143)

$$H_S^{(d)} = \vec{S}.\ \tilde{d}.\ \vec{S} \tag{2}$$

where $\vec{d}$ is the second-rank tensor given by

$$\tilde{d} = \hat{G}.\ \tilde{\varepsilon} \equiv G_{\alpha\beta\gamma\delta}\ \varepsilon_{\gamma\delta}$$

with the convention of the repeated indices to be summed over. $\tilde{\varepsilon}$ is the strain tensor whose components are

$$\varepsilon_{\gamma\delta} = \frac{1}{2}\left| \frac{\partial\mu_\gamma}{\partial\chi_\delta} + \frac{\partial\mu_\delta}{\partial\chi_\gamma} \right| \tag{3}$$

where $\vec{\mu}(\vec{r})$ denotes the displacement of the lattice at point $\vec{r}$. $G_{\alpha\beta\gamma\delta}$ are termed as strain coefficients.

Invoking time-reversal invariance of the system and neglecting the uniform shift of the S manifold, Blume and Orbach (81, 113, 144, 145) have reduced (2) to the following simplified form, for a cubic point symmetry at the magnetic ion,

$$
\begin{aligned}
H_S^{(d)} = G_{11} & \{\tfrac{1}{2}[3S^2 - S(S+1)]\ \varepsilon(\Gamma_{3g},\theta) \\
& + \tfrac{\sqrt{3}}{4}[S_+^2 - S_-^2]\ \varepsilon(\Gamma_{3g},e)\} \\
& + G_{44}\ \{-\tfrac{\sqrt{2}}{\sqrt{3}}.\tfrac{1}{2}(S_zS_+ + S_+S_z)\ \varepsilon(\Gamma_{5g},1) \\
& + \tfrac{\sqrt{3}}{6}[S_+^2 - S_-^2]\varepsilon(\Gamma_{5g},0) \\
& -\tfrac{\sqrt{2}}{\sqrt{3}}\ \tfrac{1}{2}(S_zS_- + S_-S_z)\ \varepsilon(\Gamma_{5g},-1)\}
\end{aligned} \qquad (4)
$$

where

$$
\begin{aligned}
\varepsilon(\Gamma_{3g},\theta) &= \tfrac{1}{2}(2\varepsilon_{zz} - \varepsilon_{xx} - \varepsilon_{yy}) \\
\varepsilon(\Gamma_{3g},e) &= \tfrac{1}{2}\sqrt{3}\ (\varepsilon_{xx} - \varepsilon_{yy}) \\
\varepsilon(\Gamma_{5g},1) &= -i\tfrac{\sqrt{3}}{\sqrt{2}}\ (\varepsilon_{yz} + i\varepsilon_{xz}) \\
\varepsilon(\Gamma_{5g},0) &= i\sqrt{3}\ \varepsilon_{xy} \\
\varepsilon(\Gamma_{5g},-1) &= i\tfrac{\sqrt{3}}{2}(\varepsilon_{yz} - i\varepsilon_{xz})
\end{aligned} \qquad (5)
$$

In deriving the above results the operator equivalent methods (146) have been used and the long wavelength limits (147) (uniform distortion) have been imposed. The frequencies involved are usually low. So the long wavelength limit turns out to be a good approximation especially for the one phonon relaxation. For two phonon relaxation phenomenon the situation may be quite different as it is evident from the investigations of Feldman et al (148) and others (146).

It is customary to write the spin-Hamiltonian of a noncubic crystal in the form

$$H_S^{(d)} = D(3S_z^2 - S(S + 1)) + E(S_x^2 - S_y^2) \quad (6)$$

which is analogous to the first two terms in Eq. (1). For stresses P along three different directions, Feher (149) has expressed D and E in terms of the stress coefficients C_{ij} (in Voigt notation) as follows:

(i) P || [001]

$$D = \frac{1}{2} C_{11} P; \; E = 0$$

(ii) P || [110]

$$D = -\frac{1}{4} C_{11} P \text{ and } E = \frac{1}{2} C_{44} P$$

(iii) P || [111]

$$D = \frac{1}{3} C_{44} P \text{ and } E = 0$$

It is clear from Eq. (4) that the number of independent elements of G which are nonzero are only two, namely, G_{11} and G_{44} in case of the cubic symmetry. In general, the number of independent components are different in different site symmetries. As for, example, there are ten independent components (150) of the strain coupling constant tensor G for a 3d-magnetic ion occupying an aluminium site in Al_2O_3.

Though for the iron group ions only the second order coefficients, G_{11} and G_{44} (which are also denoted by $G_{3g}^{(2)}$ and $G_{5g}^{(2)}$, respectively) contribute, there are second order as well as the fourth order coefficients (110, 111, 151), $G_{3g}^{(4)}$ and $G_{5g}^{(4)}$ which contribute to the dynamic spin-Hamiltonian for a 4f metal ion in a cubic symmetry. For a tegragonal distortion the spin-Hamiltonian assumes the form

$$H_S^{(d)} = G_{3g}^{(2)} O_2^0 + G_{3g}^{(4)} (O_4^0 - 7 O_4^4) \varepsilon_{3g,\theta}$$

whereas, for a trigonal distortion

$$H_S^{(d)} = \{G_{5d}^{(2)} \frac{1}{2i} (S_+^2 - S_-^2) + G_{5g}^{(4)} \frac{1}{4i} [(7S_z^2 - S(S+1)-5)(S_+^2 - S_-^2)$$

$$+ (S_+^2 - S_-^2)(7S_z^2 - S(S+1)-5)]\}\varepsilon_{5g,\zeta}$$

where O_n^m are the Steven's spin operators (152) and

$$\varepsilon_{3g,\theta} = \frac{1}{4}(2\varepsilon_{zz} - \varepsilon_{xx} - \varepsilon_{yy})$$

and

$$\varepsilon_{5g,\zeta} = \varepsilon_{xy}$$

The strain coefficients G_{ij} and the stress coefficients C_{ij} are also termed as spin-lattice coupling constants.

3. THEORY OF ZERO-FIELD SPLITTING AND SECOND ORDER SPIN-LATTICE COEFFICIENTS

The theory of zero-field splitting can be applied also for the calculations of the second-order spin-lattice coupling constants. In the following we shall give details of the theory and describe mechanisms which are responsibile for the splittings and the coupling constants.

3.1 Half-filled Shell Iron-group Ions

3.1.1 The Hamiltonian:

The Hamiltonian of a paramagnetic ion in a surrounding (rhombic) crystalline field is given by

$$H = -\sum_i \frac{\hbar^2}{2m} \nabla_i^2 - \sum_i \frac{Ze^2}{r_{iN}} + \sum_{i<j} \frac{e^2}{r_{ij}} + V_C$$

$$+ H_{SO} + H_{SS} + V_2^0 + V_2^2 + V_4^0 + V_4^4 \qquad (7)$$

The first term represents the kinetic energy of the electrons, the second term the Coulomb interaction between the electrons and the nucleus of the paramagnetic ion with nuclear charge Ze and third term the Coulomb interaction between the electrons. The fourth term is the cubic field,

$$V_C = B_4^0 \sum_i r_i^4 [Y_4^0(i) + \sqrt{\frac{5}{14}} (Y_4^4(i) + Y_4^{-4}(i)] \qquad (8)$$

For a $3d^5$ ion the first three terms of Eq. (7) give rise to the ground state ^6S and the excited states ^4P, ^4F, ^4G and ^4D and the doublets and other terms (153). The ground state and the quartets for Mn^{2+} are shown in Fig. 1. The difference in energies between the quartets is of the same order of magnitude as the strength of the cubic field and, therefore, the Ritz variational principle is usually adopted to obtain the effect of the cubic field on the quartets. Fig. 1. also depicts the cubic field $^4\Gamma_4$ states which are derived by admixing the free ion quartets by means of the cubic crystalline field.

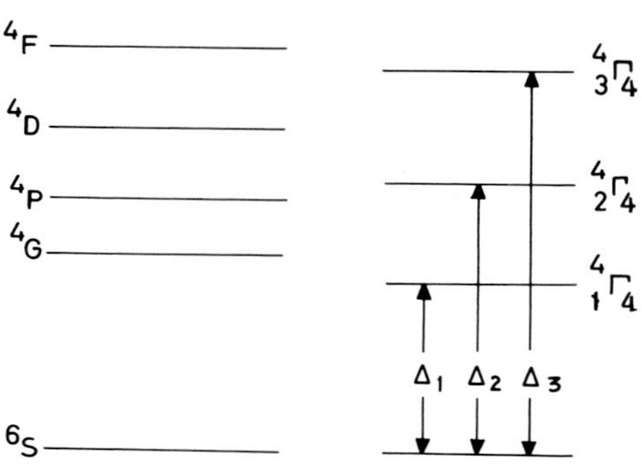

FIG: 1

Fig. 1. Schematic representation of level splittings of a $3d^5(^6S)$ ion. The atomic energy levels are shown on the left whereas the levels appropriate to a cubic field are shown on the right.

The other terms in the Hamiltonian (7) are small and are treated by the quantum mechanical perturbation theory. H_{SO} is the spin-orbit interaction,

$$H_{SO} = \sum_i \zeta(r_i)\, \vec{\ell}_i \cdot \vec{s}_i \tag{9}$$

where $\zeta(r_i)$ is the spin-orbit coupling constant and $\vec{\ell}_i$ and $\vec{s}_i$ are the the angular and spin momentum vectors of the ith electron. H_{SS} is the spin-spin interaction between the electrons:

$$H_{SS} = g^2\beta^2 \sum_{i<j} \left[\frac{(\vec{s}_i \cdot \vec{s}_j)}{r_{ij}^3} - \frac{3(\vec{s}_i \cdot \vec{r}_{ij})(\vec{s}_j \cdot \vec{r}_{ij})}{r_{ij}^5} \right] \tag{10}$$

where r_{ij} is the distance between the electrons i and j, g is the g-factor and β is the Bohr magneton.

V_2^0 and V_2^2 define the axial and nonaxial crystal potential terms:

$$V_2^0 = -B_2^0 \sum_i r_i^2 Y_2^0(i) \tag{11}$$

$$V_2^2 = -B_2^2 \sum_i r_i^2 (Y_2^2(i) + Y_2^{-2}(i)) \tag{12}$$

whereas V_4^0 and V_4^4 are the unbalanced part of the cubic field,

$$V_4^0 = -(B_4^0)' \sum_i r_i^4 Y_4^0(i) \tag{13}$$

and

$$V_4^2 = -B_4^2 \sum_i r_i^4 [Y_4^2(i) + Y_4^{-2}(i)] \tag{14}$$

B_ℓ^m's in the above equations are the crystal field parameters associated with the location of the paramagnetic ion, generated by the surrounding charge distribution. Depending on the symmetry of the location of the ion, some of the crystal potential terms in Eq. (7) may be absent or some new terms may appear. (14, 20, 32, 154). As for example a V_4^3 type term will be present in a trigonal symmetry (155, 156) as in case of $CdCl_2$: Mn^{2+}. The higher order crystal potential terms (32) of the type V_6^m are also needed if the paramagnetic ion involved is an f-state ion.

The zero-field splitting occurs because of the complicated interactions between the electrons and the crystal fields in high orders. The mechanisms known for the splitting of ^6S and ^8S ground states are as follows.

3.1.2 Mechanisms for Zero-field Splitting and Second Order Spin-lattice Coupling Constants

As mentioned in Sec. 1 the origin of the zero-field splitting was first discussed qualitatively by Van Vleck and Penney (78) who proposed admixtures of excited orbitally asymmetric states into the ground state via the spin-orbit interaction and perturbations due to axial crystalline electric fields. However, they did not reveal explicit conditions under which the zero-field splitting may occur.

Since the diagonal matrix elements of an electric operator in a half-filled shell (S-state ions) vanish one requires to consider the effects of high orders of crystalline fields and spin-orbit coupling for the splitting.

(a) Spin-spin (Pryce) Mechanism:

This mechanism, first proposed by Pryce (79) and further extended by Chakravarty (85) and Sharma, Das and Orbach (90), considers the distortion of the ground state wavefunction of the ion by the electronic dipole-dipole interaction which allows a nonvanishing value for the matrix element of the axial (or rhombic) crystalline fields. The Pryce mechanism is expressed as (for $3d^5(^6S)$ ions)

$$D_P \;\alpha\; \frac{\langle {}^6S(3d^5)\,|V_{ax}|\,{}^6D(3d^4(4s)\rangle\,\langle {}^6D(3d^4(4s)\,|H_{SS}|\,{}^6S(3d^5)\rangle}{E({}^6S(3d^5)) - E({}^6D(3d^4 4s))} \qquad (15)$$

which involves the matrix elements of the spin-spin interaction H_{ss} and the axial potential V_{ax} between the states $^6S(3d^5)$ and $^6D(3d^4 4s)$. This mechanism has been shown pictorially in Fig. 2.

Pryce's paper represents the first meaningful quantitative treatment of the axial field splitting of S-state ions. For estimation of the splitting he extracted the strength of the spin-spin interaction from free ion excited state splittings. This method has been criticized by Blume and Watson (157) and by Leushin (86). As it is clear from Eq. (15) Pryce considered only the s-like admixtures. It has been shown in Ref. 90 - using the Sternheimer's perturbation method (158-162) that, besides s-like admixtures, there are also d- and g- like admixtures via spin-spin interaction which give rise to the zero-field splitting.

It has been observed that the d and g-excitations (particularly the g) can be equally important compared with the s-excitations and that the d-excitations yield opposite sign to the zero-field splitting. The complete spin-spin contributions to the zero-field splitting parameter D and E (as defined by Eq. 1) have been derived to be (92),

$$D_{SS} = D_{SS}(d\rightarrow S) + D_{SS}(d\rightarrow d) + D_{SS}(d\rightarrow g) \qquad (16)$$

with

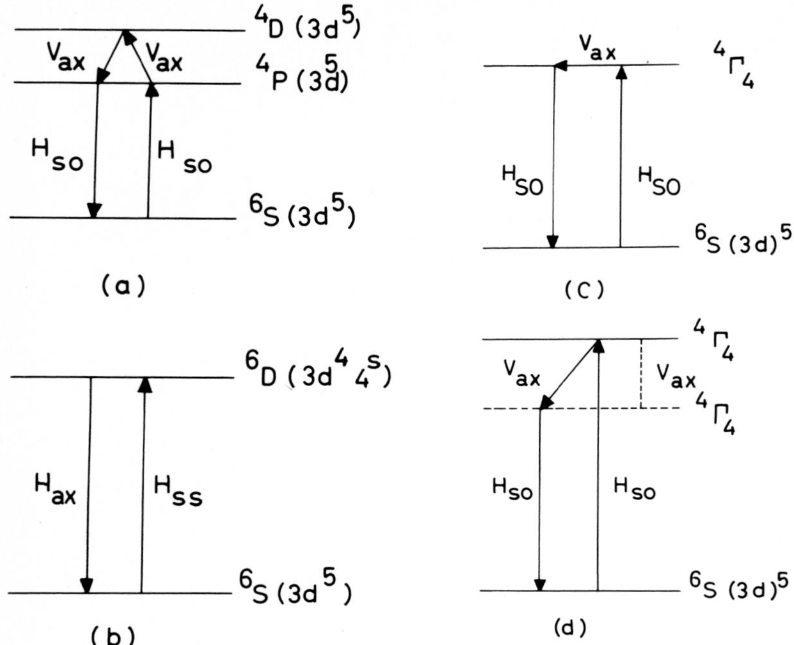

Fig. 2. Pictorial representations of various mechanisms responsible for the zero-field splitting of a $3d^5(^6S)$ ion. H_{so} and H_{ss} indicate the involvement of the spin-orbit and spin-spin interactions while V_{ax} shows the axial crystalline field acting on the ion. (a) Watanabe mechanism (b) Pryce mechanism (c) Blume-Orbach mechanism and (d) Orbach-Das-Sharma mechanism.

$$D_{ss}(d\to s) = -(g^2\beta^2 B_2^0/20\sqrt{5}a_0^3)\ \{3.57771h^{0,3}_{d\to s} - 5.36656g^{2,5}_{d\to s}\} \tag{17}$$

$$D_{ss}(d\to d) = -(g^2\beta^2 B_2^0/20\sqrt{5}a_0^3)\{-1.27775g^{0,3}_{d\to d} + 5.11101h^{0,3}_{d\to d} -2.19043(g^{2,5}_{d\to d} + h^{2,5}_{d\to d})\} \tag{18}$$

$$D_{ss}(d\to g) = -(g^2\beta^2 B_2^0/20\sqrt{5}a_0^3)\ \{ 9.19982h^{0,3}_{d\to g} - 0.10952g^{2,5}_{d\to g} -5.47608h^{2,5}_{d\to g} - 19.16629h^{4,7}_{d\to g}\} \tag{19}$$

and

$$E_{ss} = E_{ss}(d\to s) + E_{ss}(d\to d) + E_{ss}(d\to g) \tag{20}$$

with

$$E_{ss}(d\to s) = -(g^2\beta^2 B_2^2/40\sqrt{5}a_0^3)\ \{8.76356h_{d\to s}^{0,3} - 13.1453g_{d\to s}^{2,5}\}\ , \tag{21}$$

$$E_{ss}(d\to d) = -(g^2\beta^2 B_2^2/40\sqrt{5}a_0^3)\ \{-3.12984g_{d\to d}^{0,3} + 12.51942h_{d\to d}^{0,3} - 5.36544(g_{d\to d}^{2,5} + h_{d\to d}^{2,5})\} \tag{22}$$

$$E_{ss}(d\to g) = -(g^2\beta^2 B_2^2/40\sqrt{5}a_0^3)\ \{22.5349h_{d\to g}^{0,3} - 0.26827g_{d\to g}^{2,5} - 13.4136h_{d\to g}^{2,5} - 46.9476h_{d\to g}^{4,7}\}\ . \tag{23}$$

where a_0 is the Bohr radius.

The quantities $g_{d\to\ell}^{n,m}$ and $h_{d\to\ell}^{n,m}$ denote the integrals

$$g_{d\to\ell}^{n,m} = \int_0^\infty dr_1\ \frac{\{u_d^0(1)\}^2}{r_1^m} \int_0^{r_1} r_2^n u_d^0(2)u_{d\to\ell}^1(2)dr_2 \tag{24}$$

and

$$h_{d\to\ell}^{n,m} = \int_0^\infty dr_2\ \frac{u_d^0(2)}{r_2^m}\ u_{d\to\ell}^1(2) \int_0^{r_2} r_1^n\{u_d^0(1)\}^2\ dr_1 \tag{25}$$

In the above expressions, u_d^0 is r times the radial 3d wave function of the paramagnetic ion. The functions $u_{d\to\ell}^1$ (ℓ=s, d, g) are the perturbations on u_d^0 due to the axial and rhombic potentials V_2^0 and V_2^2 and are the solutions of the differential equations

$$\{-\frac{d^2}{dr^2} - \frac{6}{r^2} + \frac{1}{u_d^0}\frac{d^2u_d^0}{dr^2}\}\ u_{d\to s}^1 = r^2u_d^0\ , \tag{26}$$

$$\left\{ -\frac{d^2}{dr^2} + \frac{1}{u_d^0}\frac{d^2u_d^0}{dr^2} \right\} u_{d\to d}^1 = r^2u_d^0 - \langle u_d^0|r^2|u_d^0\rangle , \tag{27}$$

$$\left\{ -\frac{d^2}{dr^2} + \frac{14}{r^2} + \frac{1}{u_d^0}\frac{d^2u_d^0}{dr^2} \right\} u_{d\to g}^1 = r^2u_d^0 . \tag{28}$$

(b) Watanabe Mechanism:

Watanabe (80) suggested the admixture of the excited $|^4P\rangle$ state into the ground $|^6S\rangle$ state by the spin-orbit coupling H_{so}. The nonzero matrix element of the axial field was then obtained between the excited $|^4P\rangle$ state admixed into the ground level and the excited $|^4D\rangle$ state. Accordingly,

$$D_W \propto \frac{\langle ^6S|H_{so}|^4P\rangle \langle ^4P|V_{ax}|^4D\rangle \langle ^4D|V_{ax}|^4P\rangle \langle ^4P|H_{so}|^6S\rangle}{[E(^6S) - E(^4P)]^2 [E(^6S) - E(^4D)]} \tag{29}$$

where all the term values are constructed only from the 3d configuration. See also Fig. 2a for the pictorial representation of this mechanism. Watanabe found

$$D_W = -\frac{1}{70}\frac{\zeta^2}{\Delta_{DS}}\frac{\langle r^2\rangle^2}{(\Delta_{PS})^2}(B_2^0)^2, \tag{30}$$

where Δ_{DS}, Δ_{PS} are the energy differences $E(^4D) - E(^6S)$, $E(^4P) - E(^6S)$, respectively.

Watanabe did not take into account the presence of the cubic field. If the cubic field is considered, his expression will alter since ^4P state will no longer be "pure" but will incorporate the admixtures of ^4F and ^4G levels. In the presence of the cubic field the Watanabe contribution becomes (90)

$$D_{WC} = -\frac{1}{70}\frac{\zeta^2}{\Delta_{DS}}\langle r^2\rangle^2(B_2^0)^2|P_{\alpha\alpha} + \frac{4}{7}P_{\alpha\beta}|^2 , \tag{31}$$

where

$$p_{\alpha\alpha} = \sum_{i=1}^{3} \alpha_i^2/\Delta_i \quad , \tag{32}$$

$$p_{\alpha\beta} = \sum_{i=1}^{3} \alpha_i\beta_i/\Delta_i \quad , \tag{33}$$

where α_i, β_i (and γ_i) are the admixing parameters and Δ_i (see also Fig. 1) are the eigenvalues obtained by diagonalizing the $^4\Gamma_4$ matrix in the cubic field (81,90).

(c) Blume-Orbach Mechanism:

Blume-Orbach (81) mechanism (Fig. 2c) considers the admixture of the excited quartet states $|^4\Gamma_4\rangle$ appropriate to the cubic field into the ground ^6S state via spin-orbit interaction and the first order matrix elements of the axial and rhombic crystalline fields. Thus,

$$D_{BO} \propto \frac{\langle ^6S|H_{so}|^4\Gamma_4\rangle \langle ^4\Gamma_4|V_{ax}|^4\Gamma_4\rangle \langle ^4\Gamma_4|H_{so}|^6S\rangle}{[E(^4\Gamma_4) - E(^6S)]^2} \tag{34}$$

Not only that the $\ell = 2$ terms (Eqns. 11 and 12) but also the $\ell = 4$ terms (Eqns. 13 and 14) contribute to D and E. These extra terms do not contribute to D_W because of the triangular rule which makes the matrix elements between D and P states to vanish. One finds the contributions to D and E from Blume-Orbach mechanism (90) as

$$D_{BO} = (B_4^0)' \ (\sqrt{5}/36)\langle r^4\rangle \ [\zeta^2 p_{\alpha\gamma}(2p_{\alpha\alpha}-p_{\alpha\beta})] \tag{35}$$

and

$$E_{BO} = -(B_4^2) \ (\sqrt{2}/6) \ \langle r^4\rangle \ [\zeta^2 p_{\alpha\gamma}(2p_{\alpha\alpha}-p_{\alpha\beta})] \tag{36}$$

where $\langle r^4\rangle$ is the expectation value of r^4 for a 3d electron of the paramagnetic ion and ζ is the spin-orbit coupling constant for the electron of this ion. $(B_4^0)'$ and B_4^2 are the unbalanced axial noncubic and the fourth-order rhombic crystal fields, respectively. The quantites $p_{\alpha\alpha}$, $p_{\alpha\beta}$, and $p_{\alpha\gamma}$ are defined by Eqs. 32 and 33, and

$$p_{\alpha\gamma} = \sum_{i=1}^{3} \frac{\alpha_i \gamma_i}{\Delta_i} , \tag{37}$$

It is interesting to observe that

$$D_{BO} / E_{BO} = - \frac{1}{6} \sqrt{(5/2)} \; (B_4^0)' \; / \; B_4^2 \qquad ,$$

independent of the strength of the cubic field and the probability of the admixtures of the excited quartet states.

In case of trigonal symmetry, not only $(B_4^0)'$ but also the axial field B_2^0 yields contribution. Denoting the contributions from these fields to D by the symbols BO-1 and BO-2, respectively, one obtains (99),

$$D_{BO-1} = \frac{\sqrt{5}}{14} \; (B_2^0) \; \langle r^2 \rangle \zeta^2 \; p_{\alpha\beta} p_{\alpha\gamma} \tag{38}$$

and

$$D_{BO-2} = - \frac{\sqrt{5}}{189} \; (B_4^0)' \; \langle r^4 \rangle \zeta^2 \; p_{\alpha\gamma} (7p_{\alpha\alpha} + 4p_{\alpha\beta}) \tag{39}$$

for the trigonal symmetry.

(d) Orbach-Das-Sharma Mechanism:

Orbach, Das and Sharma (82) pointed out an alternate mechanism (see Fig. 2d) which involves configuration interaction via the combined effects of spin-orbit interaction and axial crystalline field. If ${}^4\Gamma_4{}'$ denotes the quartet states perturbed by the first order axial field (denoting the perturbation by "prime") then

$$D_{ODS} \propto \frac{\langle {}^6S | H_{so} | {}^4\Gamma_4{}' \rangle \langle {}^4\Gamma_4{}' | V_{ax} | {}^4\Gamma_4' \rangle \langle {}^4\Gamma_4{}' | H_{so} | {}^6S \rangle}{[E({}^6S) - E({}^4\Gamma_4{}')]^2} \tag{40}$$

Explicitly,

$$D_{ODS} = (B_2^0)^2 \ (\sqrt{5} \ / \ 192\pi) \ [\zeta^2 p_{\alpha\gamma}(2p_{\alpha\alpha} - p_{\alpha\beta})] \ (M_2 - 4M_1 + 3M_0), \tag{41}$$

where

$$M_m = (8\pi/5) \sum_{\ell=0,2,4} a_{m\ell} \langle u_d^0 | r^2 | u_{d\to\ell}^{(1)} \rangle \tag{42}$$

with

$$a = \begin{bmatrix} a_{20} & a_{22} & a_{24} \\ a_{10} & a_{12} & a_{14} \\ a_{00} & a_{02} & a_{02} \end{bmatrix} = \begin{bmatrix} 0 & 4/49 & 3/49 \\ 0 & 1/49 & 6/49 \\ 1/5 & 4/49 & 36/245 \end{bmatrix} \tag{43}$$

and $u_{d\to\ell}^{(1)}$ being solutions of the differential Equations 26-28.

(e) Covalency Mechanism:

In fact, this mechanism is frequently referred to as a model which accounts for only part of the zero-field splitting. Kondo in 1960 suggested that the anisotropic covalent admixtures (due to overlap and charge transfer effects) of the ligand orbitals in the 3d-orbitals of the paramagnetic ion can contribute to D and E. He derived approximate expressions for D and E and fitted the experimental data of Mn^{2+} in MnF_2 and strained MgO in terms of the covalency parameters. A complete treatment of the covalency model has been given in Ref. 91 using the spin-spin and spin-orbit interactions and accounting for "local", "nonlocal" and "distant" effects. The notations adopted were similar to those used by Shulman and Sugano (163), Watson and Freeman (164) and Simanek and Sroubek (40, 165). For the details of derivations the reader is referred to Ref. 91. For the sake of clarity and completeness the final expressions arising from the overlap and charge-transfer covalency effects have been listed in Appendix A.

(f) Wybourne's Relativistic Effect (RE(Wyb)):

The relativistic effect was noted first of all by Wybourne (105, 106) in the context of the zero-field splitting of Gd^{3+}. Following Wybourne other authors (108, 109, 166-171) have estimated the relativistic contribution to the zero-field splitting of Mn^{2+} and Fe^{3+}. This mechanism will be discussed more in detail in Sections 3.2 and 4.

3.2 The Rare-Earth Ions:

The Hamiltonian of a rare earth $(4f^7)$ ion embedded in a crystal is analogous to Equation 7. As mentioned in Section 3.1 it contains also the higher order crystal potential terms (32). Also for rare-earth ions the spin-orbit interaction associated with 4f electrons is greater than the strength of the cubic crystal field which really demands attention while applying the perturbation treatments. In the following we discuss the splitting mechanisms appropriate to the rare-earth $4f^7(^8S)$ ions. For the sake of convenience the splitting mechanisms will be classified into four groups depending on the interactions involved, as follows:

3.2.1 Crystal Field (CF) Mechanisms:

(a) Hutchison-Judd-Pope Mechanism: This is a fourth order mechanism (172) which is linear in the axial crystal field and cubic in the spin-orbit interaction.

Symbolically,

$$D_{HJP} \propto \langle ^8S_{7/2}|H_{SO}|^6P_{7/2}\rangle \langle ^6P_{7/2}|H_{SO}|^6D_{7/2}\rangle \langle ^6D_{7/2}|V_{ax}|^6P_{7/2}\rangle \langle ^6P_{7/2}|H_{SO}|^8S_{7/2}\rangle$$

(b) Judd Mechanism: This is a fourth order mechanism (172, 173) quadratic in both the axial crystal field strength and the spin-orbit interaction

$$D_J \propto \langle ^8S_{7/2}|H_{SO}|^6P_{7/2}\rangle \langle ^6P_{7/2}|V_{ax}|^6L_{J''}\rangle \langle ^6L_{J''}|V_{ax}|^6P_{7/2}\rangle \langle ^6P_{7/2}|H_{SO}|^8S_{7/2}\rangle$$

(c) Rajnak - Wybourne's Mechanism: This is a crystal field configuration mixing mechanism represented as (105, 174)

$$D_{RW} \propto -\Sigma_X \frac{< f^n SLJM|V_{ax}|X> <X|V_{ax}|f^7 SL'M'J'M'>}{\Delta E_X}$$

where X stands for the excited configurations and E_X is the corresponding positive excitation energies. This effect has been shown by Rajnak and Wybourne to produce only a uniform shift of the levels without any zero-order splitting. Thus this mechanism can be safely neglected.

(d) Fifth-order configuration interaction mechanism:- This has been described by Wybourne (105) as

$$D_{CIW} \propto < f^7\ {}^8S|V_E|f^5\ {}^6x\ 5d^2\ {}^8S > < f^5\ {}^6x\ 5d^2\ {}^8S|H_{SO}|f^5\ {}^6x'5d^2\ {}^{2s+1}P >$$

$$< f^5\ {}^6x'5d^2\ {}^{2s+1}P|V_{ax}|f^5\ {}^6x''5d^2\ {}^{2s+1}P > < f^5\ {}^6x''5d^2\ {}^{2s+1}P$$

$$|H_{SO}|f^5\ {}^6x5d^2\ {}^8S > < f^5\ {}^6x5d^2\ {}^8S|V_E|f^7\ {}^8S >$$

where V_E is the Coulomb potential; x, x' and x" are excited state configurations. This mechanism does not contribute to the splitting larger than a few percent of the observed splitting.

3.2.2 Relativistic Effect of Wybourne (RE(Wyb)):

As mentioned in Section 3.1 the importance of this effect was pointed out first of all by Wybourne (105, 106). If the relativistic wave functions are used the crystal field matrix elements between the states of differing spin no longer vanish. Consequently, the second order contribution to D due to this mechanism turns out to be

$$D_{RE(Wyb)} \propto <{}^8SM|H_{SO}|{}^6P_{7/2}M> <{}^6P_{7/2}M|V_{ax}|{}^8S_{7/2}M>$$

where it is assumed that the wavefunctions to be used are the relativistic ones.

It is important to understand the relativistic effect in non-relativistic terms. Andriessen et al (175) have shown that the relativistic mechanism of Wybourne contains two mechanisms, one proposed by Lulek (176) which is due to the isotropic contribution of the potential gradient part of the spin-orbit operator arising from the electrostatic potential and the other, due to the anisotropic part as discussed by Chatterjee at al (177). It has been shown (176, 177) that in mixing the excited $n\ell$ states into the ground $n\ell$ states via spin-orbit interaction, the dominant contribution to the relativistic effect is not due to the anisotropy of the electrostatic potential. Also the effect of the overlap and charge transfer covalency on the Wybourne's relativistic model has been considered by Chatterjee at al (177).

3.2.3 Spin-Spin-Correlation Effect:

It contains two mechanisms, one proposed by Wybourne (105) and the other due to Pryce (79, 105).

(a) Wybourne's spin-spin correlation mechanism: This is a third order mechanism prepresented by Wybourne (105)

$$D_{SSW} \propto \langle {}^8S_{7/2}|H_{ss}|{}^6D_{7/2} \rangle \langle {}^6D_{7/2}|V_C|{}^6P_{7/2} \rangle \langle {}^6P_{7/2}|H_{so}|{}^8S_{7/2} \rangle$$

(b) Pryce Mechanism: This mechanism has already been described in Section 3.1 in context with the iron group ion. For a rare earth ion one writes (79, 105)

$$D_P \propto \langle 4f^7\ {}^8S_{7/2}\ M'|H_{ss}|4f^6\ ({}^7F)\ 6p\ {}^8D_{7/2}M \rangle \langle 4f^6\ ({}^7F)\ 6p\ {}^8D_{7/2}M\ |V_{ax}|4f^7\ {}^8S_{7/2}M' \rangle$$

It should be pointed out that the above expression gives only a part of f→p type of excitation. Analogous to d→s, d→d and d→g excitations as pointed out for the iron group ions we would expect here also f→f and f→h type of excitations.

3.2.4 Correlation Crystal Field Mechanism:

This concerns the modification of the Coulomb interaction between the open-shell electrons because of the crystalline environment. This was first described by Newman and Bishton (178) in general terms and later evaluated by Wybourne (105) to estimate its contribution for the zero-field splitting of Gd^{3+}. Symbolically it is represented as

$$D_{ccF} \propto \left(- \frac{2}{\Delta E_{av}}\right) \sum_{X} < n\ell\ {}^{N}\psi | V_{ax} | X > < X | V_{E} | \ell\ {}^{N}\psi' >$$

where V_E is the Coulomb potential and X represents excited states and ΔE_{av} is the positive average energy of excitation.

Newman (107) has pointed out that there is apparently another mechanism which could produce splitting by giving rise to charge conjugation invariant component to the crystal field. It can be shown (33), however, that this mechanism is contained in the CCF mechanism because it includes the Coulomb admixtures of the nf states into the 4f states.

4. INTERPRETATION OF ZERO-FIELD SPLITTING AND SPIN-LATTICE COUPLING CONSTANTS

4.1 Iron Group $3d^5(^6S)$ Ions

As mentioned in Sec. 1 the initial attempts to explain the zero-field splittings were made by Van Vleck and Penney (78). Pryce (79), Watanabe (80), Blume and Orbach (81), Chakravarty (85), Leushin (86-88), McFarlene (89), Orbach, Das and Sharma (82), and others. The most complete calculations to date have been performed by Sharma, Das and Orbach (90-92) for Mn^{2+} in MnF_2, ZnF_2 and stressed MgO and by Sharma (93) for Fe^{3+} in stressed MgO. They considered point-multiple as well as the covalency contributions and studied all important mechanisms in the non-relativistic frame-work. The lattice sums for the calculation of crystalline fields were evaluated using convergent methods (179, 180). The relevant overlap and other two-center integrals were computed accurately by means of the convenient α-function technique (181-184) utilizing the available SCF atomic Hartree-Fock Orbitals. In order to examine the relative importance, the various contributions to the parameters D and E have been summarized in Table 1 for $ZnF_2:Mn^{2+}$ along with the experimental results(185). For the evaluation of spin-spin contributions use was made of the Eqs. 16-28 and the expressions of Appendix A. The perusal of Table 1 reveals that, in view of the ab initio procedure adopted, the calculated results predict correct signs and yield reasonable order of magnitudes of D and E parameters. Though all processes seem

TABLE 1

Compilation of various theoretical contributions to D and E in the Spin-Hamiltonian $H_s = D\,(3S_z^2 - S(S+1)) + E(S_x^2 - S_y^2)$, for Mn^{2+} in ZnF_2. For comparison the experimental values of D and E have also been given. D and E are all expressed in units of 10^{-4} cm^{-1}.

Contribution	Process	D	E
	Point - Multipole Model		
Spin - Orbit	ODS	+ 1.59	
Spin - Spin*	d→s	+ 4.53	+ 1.85
	d→d	- 0.44	- 0.18
	d→g	+ 2.25	+ 1.01
	Total	+ 6.35	+ 2.68
Spin - Orbit*	Watanabe	- 1.01	
	WC	- 1.31	
	ODS + WC	+ 0.28	
	BO	+27.51	-99.25
Total (point-charge)		+34.13	-96.57
	Overlap - Model		
Spin - Spin**		+ 7.87	+54.06
Spin - Orbit**		- 5.29	- 5.33
Total (Overlap)		+ 2.58	+48.73
Experiment †		+10.50	-113.50

* Refs. 90, 92

** Refs. 91, 92

† Refs. 185

to be important it is noteworthy that a single mechanism, namely, the BO-mechanism gives correct sign and order of magnitude of both the parameters D and E and that by combining the contributions from the remaining point-multipole mechanisms the results change only slightly. The overlap contribution, however, does not appear to improve the results. It may be due partly to our ignorance of the local geometry (which depends on the local elastic strain constants) of Mn^{2+} in ZnF_2 matrix and partly to the neglect of the charge transfer covalency.

In the process of calculations (155) of the zero-field splitting parameter D for $CdCl_2:Mn^{2+}$, it has been found that the expressions for BO, ODS and WC mechanisms obtained in Refs. 90-92 for the tetragonal symmetry are modified in case of trigonal symmetry as a result of the appearance of different combinations of ^4F and ^4G states which admix into the ground state. Also, in the trigonal symmetry, as mentioned earlier, not only the fourth-degree crystal fields give rise to the BO mechanism, as in case of tetragonal symmetry, but also the second-degree crystal fields; the contributions from the second- and fourth-degree crystal fields to BO mechanism are denoted by D_{BO-1} and D_{BO-2}, respectively. For comparison purposes we tabulate the contributions from various mechanisms for Mn^{2+} and Fe^{3+} in tetragonal and rhombic symmetries (90-93) in Table 2 and for Mn^{2+} in trigonal symmetry (155) in Table 3. The contribution from the relavistic effect of Wybourne (105, 106) as determined by Dreyboudt and Silver (186) has also been listed in Table 2.

TABLE 2.

List of the theoretical contributions to the Spin-Hamiltonian parameters D and E arising from various processes for both Mn^{2+} and Fe^{3+} in axial and rhombic symmetries. D and E are in units of cm^{-1} and the crystal fields B_n^m are in units of $e^2/2a_o^{n+1}$ where a_o is the Bohr radius.

	Mn^{2+}			Fe^{3+}		
Process	D	E	Ref.	D	E	Ref.
Spin-Spin	$-0.073B_2^0$	$-0.093B_2^2$	90,92	$-0.16\ B_2^0$	$-0.045\ B_2^2$	93
Blume - Orbach	$4.34\ (B_4^0)'$	$-16.47B_4^2$	90,92	$4.7\ (B_4^0)'$	$-17.7\ B_4^2$	93
Watanabe (with cubic Field)	$-1.74(B_2^0)^2$	-	90,92	$-1.2\ (B_2^0)^2$	-	93
Orbach - Das Sharma	$2.10\ (B_2^0)^2$	-	90,92	$0.53\ (B_2^0)^2$	-	93
Relativistic Effect	$-1.54\ B_2^0$	$-1.88\ B_2^2$	186	-	-	

TABLE 3.

Tabulation of the theoretical results for the zero-field splitting parameters D from various mechanisms for Mn^{2+} in trigonal symmetry. D is in units of cm^{-1} and the crystal fields B_n^m in units of $e^2/2a_o^{n+1}$, a_o being the Bohr radius. The values are from Ref. 99.

Process	D
BO - 1	$-0.042\ B_2^0$
BO - 2	$-2.19\ (B_4^0)'$
Orbach - Das-Sharma	$-1.10\ (B_2^0)^2$
Spin-Spin	$-0.073\ B_2^0$
Watanabe (WC)	$-1.72\ (B_2^0)^2$

We also compile the values of D and E calculated by various researchers for different systems in Table 4 which also lists the processes taken into account for the calculations and the relavant experimental results. In the following we discuss briefly the results of Table 4. In case of $MgO:Mn^{2+}$ the calculations (92) include the mechanisms from the point multipole model as well as the overlap model. Though the calculations are very satisfactory with respect to the experimental data (149), considering the first-principles calculations, it has been pointed out (94) that the local distortion effects improve the results appreciably. Similar conclusions can be derived for the D and E calculations (93) in $MgO:Fe^{3+}$. In case of vacancy-Mn^{2+} pair spectra III_1 and III_2 in NaCl and KCl the calculated results (187) from the point-charge mechanisms and the overlap process are of the correct sign, they fall short by an order of magnitude relative to the experimental data by Walkins (188). The main reasons for this are (i) our ignorance of the structure near the vacancy-Mn^{2+} pair and (ii) neglect of the correct contributions from the electronic multipoles in the lattice generated by the defect.

For the derivation of expressions for D for various mechanisms (particularly, BO-1, BO-2, WC and ODS) appropriate to trigonal symmetry one may refer to Ref. 155. These expressions when applied (155) to the case of $CdCl_2:Mn^{2+}$ yield $D = 4.92 \times 10^{-4}\ cm^{-1}$ which compares favorably with the single crystal experimental value (189) ($< +5 \times 10^{-4}\ cm^{-1}$). Similar calculations (190) for $CdBr_2$ and $MgCl_2$

Table IV: Compilation of the Calculated and experimental values of the zero-field splitting parameters D and E in the Spin-Hamiltonian $H_s = D(3S_z^2 - S(S + 1)) + E(S_x^2 - S_y^2)$ *for various systems in units of* 10^{-4} cm^{-1}.

System	Processes	Calculated D	Calculated E	Calculated Ref.	Experimental D	Experimental E	Experimental Ref.
$MnF_2:Mn^{2+}$	SS + BO + WC + ODS ($10\ Dq=10{,}000\ cm^{-1}$)	16.01	-92.98	(90-92)	11.5	-121.5	(185)
	SS + BO + WC + ODS ($10\ Dq= 7800\ cm^{-1}$)	12.6	-	(190)			
$ZnF_2:Mn^{2+}$	SS + BO + WC + ODS ($10\ Dq=10{,}000\ cm^{-1}$)	34.13	-96.57	(90-92)	10.5	-113.5	(185)
	SS + BO + WC + ODS ($10\ Dq= 7800\ cm^{-1}$)	25.0	-	(190)			
	SS + BO + WC + ODS + Overlap	36.71	-47.84	(190)	10.5	-113.5	
$MgO:Mn^{2+}$ (stressed)	SS + BO + WC + ODS	(a) 0.62	(b) - 0.42	(92)	(a) 2.09	(b) -0.62	(149)
$MgO:Mn^{2+}$ (stressed)	SS + BO + WC + ODS + Overlap	(a) 0.58	(b) - 0.34	(92)			
$MgO:Fe^{3+}$ (stressed)	SS + BO + WC + ODS	(a) 0.62	(b) - 0.90	(93)	(a) 7.65	(b) -1.62	(149)
$MgO:Fe^{3+}$	SS + BO + WC + ODS + Overlap	(a) 0.57	(b) - 0.84	(93)			
$NaCl:Mn^{2+}$ (+ Vacancy) Spectra III_1	SS + BO + WC + ODS + Overlap	-5.23	2.21	(187)	-45.0	40.7	(188)
$NaCl:Mn^{2+}$ (+ Vacancy) Spectra III_2	SS + BO + WC + ODS + Overlap	3.67	-	(187)	43.67	-	(188)
$KCl:Mn^{2+}$ (+ Vacancy) Spectra III_1	SS + BO + WC + ODS + Overlap	-2.47	3.05	(187)	-66.0	41.0	(188)
$KCl:Mn^{2+}$ (+ Vacancy) Spectra III_2	SS + BO + WC + ODS + Overlap	5.06	-	(187)	58.0	-	(188)

(a) corresponds to stress along (001)
(b) corresponds to stress along (110)

$CdCl_2:Mn^{2+}$	SS + (BO-1) + (BO-2) + WC + ODS	+4.92		(155, 156)	<(+5)		(189)
$CdBr_2:Mn^{2+}$	SS + (BO-1) + (BO-2) + WC + ODS	3.6		(190)	+67.3		(189)
$MgCl_2:Mn^{2+}$	SS + (BO-1) + (BO-2) + WC + ODS	-7.5		(190)	-43.8		(189)
$Ca(OH)_2:Mn^{2+}$ (T=0K)	SS + ODS + WC + BO	-26.8		(97)	-7.2		(97)
$Ca(OH)_2:Mn^{2+}$ (T=300 K)	SS + ODS + WC + BO	-25.1		(97)	-2.5		(97)
α-$LiIO_3:Fe^{3+}$	SS + (BO-1) + (BO-2) + RE (Wyb.) + WC + HJP + ODS	+95.0		(171)	+879.8		(171)
$YAlG:Mn^{2+}$ (Tetrahedral site) Dq=900 cm^{-1}	SS + BO + WC + ODS	+168.4		(98)	-159.0		(98)
$YAlG:Mn^{2+}$ (Octahedral site) Dq=900 cm^{-1}	SS + BO + WC + ODS	10.5		(98)	-134.6		(98)
$YAlG:Mn^{2+}$ (Dodecahedral site) Dq=900 cm^{-1}	SS + BO + WC + ODS	73.3		(98)	-97.5 or 122.1		(98)
$YAlG:Mn^{2+}$ (Dodecahedral site) Dq=900 cm^{-1}	BO + SS		+109.7	(98)		-146.6 or +72.7	(98)
$Cd_2V_2O_7:Mn^{2+}$	SS + BO	-162.7	+18.3	(101)	-130.3	+36.3	(101)
$Zn_2P_2O_7:Mn^{2+}$	SS + BO	-270.7	+35.7	(101)	-320	+3.3	(101)
$Mg_2P_2O_7:Mn^{2+}$	SS + BO	-296	+33.0	(101)	-337.7	+1.7	(101)
Mn^{2+} in							
$La_2Mg_3(NO_3)_{12}:24H_2O$	RE (Wyb)+ Lulek + SS +BO + SS (Wyb.) + JHP +WC						
Site-I		-63.5		(168)	-62.9		(191)
Site-II		+30.6		(169)	+ 7.0		(191)
$K_2ZnF_4:Mn^{2+}$	Overlap("local" only)	12.0		(192)	+12.0		(192)
$K_2MgF_4:Mn^{2+}$	Overlap("local" only)	20.8		(193)	+35.7		(194)

$AlPO_4:Fe^{3+}$	BO	General Agreement	(104)	(104)
Quartz:Fe^{3+}	BO	General Agreement		
RbCl:Mn^{2+}	Phenominological		(195)	

predict correct signs for D, though value smaller than the experimental values (189).

The theory of Refs. 90 and 92 has been applied in the point-charge model by Holuj, Quick and Rosen (97) for studying the origin of the parameter D for Mn^{2+} in $Ca(OH)_2$. They considered BO, WC, ODS and SS contributions and obtained correct sign for D. The magnitudes of D at T = 0 and 300 K were, however, about four and ten times less than the respective experimental values (97). They did not evaluate the overlap effects which may come out to be important. It must be mentioned that the $Ca(OH)_2$ is a difficult system to handle theoretically due to the fact that the net charge on the OH-group is very uncertain to determine correctly the crystal fields.

Karthe (171) has faced a similar situation in interpreting D for α-$LiIO_3:Fe^{3+}$. He evaluated contributions from mechanisms not only from BO-1, BO-2, WC, SS and ODS but also from the relativistic effect of Wybourne (RE(Wyb.)) and Hutchison-Judd-Pope (HJP) mechanisms. The D value came out to be of right sign but an order of magnitude too small relative to the experimental value (171). The RE(Wyb.), BO-1 and HJP contributions were predominant, BO-1 almost cancelling HJP contribution. The largest contribution which is from RE(Wyb.), was also about five times smaller than the experimental value. Karthe did not estimate the overlap contributions.

The values of D for Mn^{2+} at the three sites (tetrahedral, octahedral and dodecahedral) in yttrium aluminium garnet have been obtained by Hodges et al. (98) by evaluating the contributions from SS, BO, WC and ODS under point-monopole approximation. For tetrahedral and dodecahedral sites the expressions of Refs. 90 and 92 were adopted whereas for the octahedral site the expressions of Leushin (86) and Andriessen et al. (169) were used. The agreement of the calculated values with the experiments were found to be poor. The obvious reasons for it are: (i) the lack of knowledge of the local distortions due to the mismatch in the size of the impurity ion Mn^{2+} and the lattice site it substitutes, (ii) neglect of the overlap and charge transfer covalency effects and (iii) the omission of the electronic multipoles on the ions for the calculation of crystal fields.

The SS and BO expressions of Refs. 90 and 92 have also been used by Stager (101) to calculate D and E parameters in $CdV_2O_7:Mn^{2+}$ $Zn_2P_2O_7:Mn^{2+}$ and $MgP_2O_7:Mn^{2+}$ obtaining good agreement for D but not for E. Chatterjee and Van Ormondt (168) calculated D values for Mn^{2+} at site I and Site II in $La_2Mg_3(NO_3)_{12}.24H_2O$ by evaluating not only the BO, SS, and WC mechanisms but also RE(Wyb.) mechanisms. Their calculations yield good agreement with the experiments (191) for site I but not for site II. The main uncertainty in these calculations center on the assumption of unjustified charges on oxygen, hydrogen and nitrogen species. They also neglected the overlap effects.

Following the treatment of Sharma, Das and Orbach (91) the "local" part of the overlap contribution to D has been estimated by Folan (192) in $K_2ZnF_4:Mn^{2+}$ and by Narayana (193) in $K_2MgF_4:Mn^{2+}$. They obtained good agreement with the experiments (192, 194). The agreement appears to be fortuitous since they have not added the remaining overlap contributions (91) and failed to include the contributions from the other important mechanisms.

The D and E parameters for Mn^{2+} in $AlPO_4$ and quartz have been estimated by Lang et al. (104) from the BO mechanism following closely the procedure discussed by Sharma et al. (90, 92) to examine the dependence of the parameters on the atomic positions. Phenomenological analysis of the zero-field parameters for Mn^{2+} at axial sites of the alkali chlorides and fluorides are also available (195) in terms of the superposition model of Newman and co-workers (196) which involves a power law variation of the effect of the crystal fields with distance of the ions (usually first and second nearest neighbors) in the lattice. For trigonal ligand field, Coulomb repulsion and spin-orbit matrices concerning d^5 configuration the reference may be made to the computation by Hemple (197). In semiempirical - extended Huckel - model, the calculations of the zero-field splitting parameters for hemin carried out by Das and co-workers (60) may be mentioned. They conclude that the spin-orbit interaction dominates the spin-spin interaction. Their calculations account for the experimental zero-field splitting but suffer from the uncertainty in estimations of the energy demoninators.

Since the BO- mechanism is emportant for many systems and it gives nonvanishing result in a given cubic crystalline field, it seems justifiable to study the variation of D with the strength of the cubic field. This has been done in Ref. 190 considering five cases of Mn^{2+} in MnF_2, ZnF_2, $CdCl_2$, $CdBr_2$ and $MgCl_2$. The usual belief that the zero-field splitting increases with the strength of the cubic field, has been found to be incorrect. An interesting result that the splitting can be zero even in the presence of an axial field, has also been pointed out. Furthermore, the splitting can change sign, in certain circumstances, as a result of the variation in the strength of the cubic field.

As for the spin-lattice coupling constants for $3d^5(^6S)$ ions the calculations have been made in MgO and CaO for both Mn^{2+} and Fe^{3+}. The calculated values of G_{11}(i.e. $G^{(2)}_{\Gamma 3g}$) for MgO:Mn^{2+} considering SS, WC, ODS and BO contributions (92) are smaller than the experimental values (148) by about a factor 3 whereas G_{44}(i.e. $G^{(2)}_{\Gamma 5g}$) is only a factor 1.4 smaller. The inclusion of RE(Wyb.) contributions (167) on the other hand spoils the good agreement with G_{44} and makes G_{11} about a factor 1.7 larger than the experimental values. Similar situation exists for MgO:Fe^{3+} if one compares the calculations of Sharma (93) and of Schlottmann and Pessaggi (167) with the experimental values (148). The calculations for CaO:Mn^{2+} and CaO:Fe^{3+} carried out by Schlottmann and Pessaggi (167) taking into account BO and RE(Wyb.) contributions yields G_{11} about a factor four and G_{44} about an order of magnitude larger than the corresponding experimental values (198).

Uniaxial stress method has been used also to determine experimentally (96) a complete set of nine spin-lattice coefficients for Fe^{3+} in TiO_2. However, no theoretical calculations have yet been reported for their interpretation.

4.2 Rare-Earth Group $4f^7(^8S)$ Ions

Various mechanisms proposed for the zero-field splitting of ^8S state ions have already been described in Sec. 3. Detailed calculations by Wybourne (105) resulted in the wrong sign for the splitting of Gd^{3+} in a lanthanum ethyl sulfate crystal. Also calculations by Lacroix (199) yielded result too small in magnitude in comparison with the observed value. Role of orbit-lattice interaction has also been investigated for the splitting by Huang (200) and Menne (201). Huang's calculation (200) yielded the phonon induced splitting for Gd^{3+} in CaF_2 of right sign, but a factor 40 smaller than the experimental value. Menne (201) concluded that the effective orbit-lattice interaction, without configuration mixing, should be discarded as a possible mechanism for the splitting.

Having known the importance of the BO mechanism for the iron group ^6S state ions from Ref. 90-92, Calvo et al. (110) calculated the contribution of the BO mechanism to the second order spin-lattice coefficients for Gd^{3+} in CaF_2. Similar calculations have been made by Schlottmann (111) for Gd^{3+} in CaO, ThO_2 and CeO_2. It has been concluded (111) that the BO-mechanism is not important for explaining the splittings in case of Gd^{3+} as against $3d^5(^6S)$ ions. This is not surprising since the spin-orbit coupling constant ζ for the rare-earth ions is much greater than the strength of the crystaline fields (10 Dq) with the result that the BO- mechanism as it is formulated for $3d^5$ ions (where $\zeta < 10Dq$) is not suitable for the rare-earth ions. In fact, even after including the contributions from the RE(Wyb.) and HJP mechanisms the values for G_{11} and G_{44} were found to be greater than the experimental value and G_{44} with the wrong sign. Also, for $ThO_2:Gd^{3+}$ and $GeO_2:Gd^{3+}$ both G_{11} and G_{44} can not be fitted by considering BO, RE(Wyb.) and HJP mechanisms. The better agreement for the rare-earth ions is conceivable only when one evaluates correctly those mechanisms which take J as a good quantum number and appropriately incorporates the fact that the spin-orbit coupling constant is larger than the strength of the crystalline fields.

Though the second- (and also the fourth-) order spin-lattice coupling constants have been measured (202) for $BaF_2:Eu^{2+}$, no calculations have yet been available for their theoretical explanations.

Newman and co-workers (196) have also made attempts to explain the splitting of Gd^{3+} and Eu^{2+} in terms of the superposition model which is more of a phenomenological approach. It must be remarked that a relativistic configuration interaction in conjunction with the appropriate orders of crystal fields and other interactions (spin-spin and spin-orbit) using many-body techniques as adopted by Andriessen et al. (203) for investigating the hyperfine interactions in free Gd^{3+} should be examined in explaining the zero-field splitting for $4f^7(^8S)$ state ions.

4.3 Temperature Variation of the Zero-field Splitting and Spin-lattice Coupling Constants

The temperature variation of an observable in solids is usually divided into two parts: implicit, that due to thermal expansion of the lattice and explicit, that due to the lattice vibrations. The implicit temperature dependence of the parameter D has been estimated for $CdCl_2:Mn^{2+}$ in Ref. 156 by evaluating the contributions of BO-1, BO-2, SS, ODS and WC mechanisms in point-multipole approximation. In this system it is possible to explain the variation of D with temperature only for temperatures below 300°K. These results await corrections due to the explicit variation of D. The temperature variation of D has also been studied in $Ca(OH)_2:Mn^{2+}$ in the range 80°-300° K by Holuj et al. (97). The phonon modulation of the crystal fields was invoked by them to interpret D as a function of T. They find that the lattice vibrations contributed about 80% of the observed results. Also, the implicit temperature variation of D employing the theory of Sharma et al. (90,92) in the point-charge and point-dipole approximation was estimated and the contribution was determined to be significant.

The temperature dependence of the ground state splitting of Mn^{2+} with nearest cation vacancy in alkali halides has been interpreted by Pfister et al. (204) by assuming a resonant vibration of frequency ω coupling to the center. The resonant vibration modulates the rhombic crystal field which produces the zero-field splitting of Mn^{2+} in the framework of the relativistic treatment of van Heuvelen (108). The functional variation was determined to be (204)

$$D(T) = D(0) + D_1 \coth\left(\frac{\hbar\omega}{2kT}\right).$$

The variation of D with temperature has also been discussed by Milsch and Windsch (205) for Mn^{2+} in guanidinium aluminium sulfate hexahydrate $(C(NH_2)_3Al(SO_4)_2.6H_2O)$, by Serway (166) for Mn^{2+} in $CaCO_3$ in terms of thermal lattice expansion and local lattice vibration effects. Besides the local vibration and the lattice phonons, Bill (206,207) considered also the anharmonic effects to interpret the parameter D for an ^8S state ion in an axial field.

5. NUCLEAR QUADRUPOLE INTERACTION

It has widely been recognized that the study of nuclear quadrupole interactions offers a great potential for investigating the electronic structure of atoms, ions and complexes in various environments. From Eq. 1 it is clear that the nuclear quadrupole interaction (122-124,209) involves the product of nuclear quadrupole moment Q and the electric field gradient q. Only the product Qq is determined through experiments by studying the nuclear levels and therefore accurate evaluation of q or Q is very crucial. If q is known precisely, in conjunction with the experimental data it is possible to obtain the sign and magnitude of Q. This is an important information since the nuclear models are not yet sufficiently accurate to yield correct value of Q. Furthermore, the precise and accurate evaluation of q will be very helpful in studying the electronic charge distributions surrounding the nucleus.

Iron nucleus has been the subject of extensive study. For this nucleus, Eq. 1 yields the quadrupole splitting

$$\Delta E_Q = \frac{1}{2} e^2 Qq(1 + \frac{\eta^2}{3})^{\frac{1}{2}} \quad (44)$$

for transitions between spin $\frac{1}{2}$ ground state and spin 3/2 excited state of ^{57m}Fe. The asymmetry parameter η is zero if the nucleus is located at the site with axial symmetry.

For the evaluation of the electric field gradient (EFG) one needs to know the accurate charge distribution in the crystal. Since the wave functions for the electrons in the crystal are not known accurately, one faces a problem of how to obtain EFG as correctly as possible. In ionic solids the initial attempts to evaluate EFG have centered on visualizing anions and cations in the solids as array of appropriate point charges. However, since the ions are not really point charges obvious improvement over this point-charge model has been accomplished by assuming multipoles on the lattice ions. In addition the electrons surrounding the nucleus with the quadrupole moment are distorted by the presence of the external charges or charge distribution, the effective EFG experienced by the nucleus differs from the bare-value by a shielding factor (known as the Sternheimer shielding factor). Many researchers (162, 209-213) have calculated the shielding factors for various atoms and ions. The shielding factors for systems with atomic number A $\geq$ 16 are, in general, negative. As an example, the shielding factor, γ_∞, for Fe^{3+} is -9.18. A discussion of the shielding factor is also given in the chapter by Marathe and Trautwein appearing in this book.

Significant progress has been made in recent years by the theories which incorporate the overlap distortions due to the surrounding ions and also the polarization distortions of the electron charge distribution surrounding the nucleus in question. Following the suggestion of Ref. 130, Taylor (135), Sawalzky and Hupkes (136) and, Sharma (137) have shown the importance of the overlap effects, particularly, of the overlap distortions of the p-electrons of the central ion. The p-electrons being closer to the nucleus are found to contribute dominantly to the EFG due to the overlap distortion.

Taylor, Sawatzky and Hupkes consider only a part of the overlap contribution whereas Sharma's theory (26, 137-139) is more general and complete and takes into account all contributions from the surrounding ligand charge distributions as well as the contribution from the rest of the lattice. Later developments, particularly by Sharma and Sharma (27), include also the polarization distortions of the ligands of the ions in question together with the complete overlap distortion. In order to be systematic we classify the theoretical results as arising from various contributions and discuss them as follows:

5.1 Purely Point-multipole Contributions:

In the simplest approximation, the (ionic) crystal is viewed as an arrangement of point charges (the point-charge approximation). The next stage of approximation is achieved by incorporating higher multipoles e.g. dipoles, quadrupoles etc. besides the monopoles on the location of the ions. The EFG in these approximations is calculated by straightforward summation of the contributions from the individual ions. However, the convergence is usually a problem associated with this summation since the net charge inside a sphere is effectively nonzero and the contributions from such a sphere is subjected to what is known as the deBoer surface effect. This surface effect can be eliminated and the convergence achieved either by forming groups of ions into clusters of net zero electric charge and using direct lattice summation method as described by Wood (180) or by transforming the direct lattice sums into appropriate reciprocal lattice sums as done by Nijboer and deWette (179). These methods (26, 28, 90, 93) and other procedures (214, 215) for evaluation of convergent lattice sums have been used widely in the literature.

Because of the complexity of the problem usually the monopoles and dipoles are assumed on the lattice and higher multipoles are neglected. In dipole approximation the selection of the dipolar polarizabilities of the ionic species in the solids has remained always a problem. Possible values of the dipolar polarizabilities of ions in crystals have been deduced by Tessman et al (216), Bolton et al. (217) and Boswara (218). By distinguishing between several dipolar polarizabilities in BeO Taylor and Das (129) have deduced the polarizability α of O^{2-} equal to 2.19 $Å^3$ which compares with Pauling's value (219) of 3.88 $Å^3$. An estimation of the quadrupolar and higher polarizabilities of ions is a rather difficult task (220,221)

The method of evaluation of dipole moments in the lattice self-consistently has been laid down by Taylor and Das (129), Sharma and Das (130) and Artman and Murphy (132) and has been used and extended subsequently by others. The method (in the linear approximation) essentially consists of solving self-consistently the equations of the form:

$$\vec{P} = \alpha(K\vec{P} + \vec{E}) \tag{45}$$

which represents that for every ionic site where a dipole $\vec{P}$ is present, the dipole moment is the polarizability α (of the ion) times the net electric field at the site where the electric field is generated by the monopoles as well as the dipoles. The electric field due to the monopoles $\vec{E}$ and due to the dipoles denoted by $K\vec{P}$ are calculated by lattice summation methods as discussed above.

Monopoles and dipoles are then used to evaluate the EFGs by resorting again to the lattice summation methods. If the unshielded EFG is expressed by q', the shielded q is obtained by means of the Sternheimer antishielding factor $(1 - \gamma_\infty)$ as

$$q = (1 - \gamma_\infty)\, q' \tag{46}$$

in the monopole or multipole approximation.

There are articles (28, 77) which review, though partially, the data on nuclear quadrupole interactions. Several examples of the evaluation of EFG in ionic crystals may be cited (126-134, 209, 222-227). Hudson and Whitfield (223) have studied the sensitivity of the crystal parameter u on EFG in spinels $ZnFe_2O_4$ and $CdFe_2O_4$. Recent calculations on these systems have been carried out by Evans et al (224). They found that the dipolar contribution to the EFG is substantial. Evans et al find best fit to the nuclear quadrupole splitting of Fe by assuming $\alpha\ (O^{2-})$ to be 0.8 $Å^3$. In FeOCl grant et al (131) find that the Mossbauer data can be explained provided that one assumes the quadrupole moment of Fe^{57m}, Q = 0.33b and $\alpha(Cl^-) = \alpha(O^{2-}) = 1Å^3$. From the point multipole calculations in α-Fe_2O_3, Artman et al (222) deduces $Q(Fe^{57m}) = 0.28$b. Recent calculations by Taft (226) include dipolar contribution to EFG in α-$NaFeO_2$, $CuFeO_2$ and $AgFeO_2$. He deduces the O^{2-} polarizability ranging from 1.0 to 1.4 $Å^3$ by assuming fractional effective charges on Fe in these systems.

If one couples the calculated EFGs with the corresponding experimental quadrupole coupling constants in various ferric systems, one obtains (127, 130, 209, 222) $Q(Fe^{57m})$ values which range from 0.1 to 0.59b heading to a consensus of $\sim$ 0.3b. On the other hand, the Q values obtained from the ferrous compounds (228) are consistently close to 0.18b. This well-known ferrous-ferric anomaly has been resolved by means of a theory (138) which incorporates the contributions from the electronic charge distribution in the immediate neighborhood of Fe nucleus and from the multipoles on the remaining lattice. In the following we consider the development of the theories which include the electronic contributions.

5.2 Theories Based on the Electronic Contributions and the Point Multipoles on the Lattice:

A free S-state ion is a spherical symmetric system and this and this has led one to believe that the electronic contribution to EFG is zero even when the ion is located in the crystal. Clearly, this concept is wrong, In fact, it was pointed out in Ref. 130 that the covalency (overlap as well as charge transfer) distortions could give rise to important contributions to EFG's in crystals. This indeed was found to be correct by Taylor (135) by evaluating the "local" contributions (229) to the EFG due to the overlap distortion of the electrons surrounding the central ion by the ligands. Similar conclusions were derived by Sawatzky and Hupkes (136) who also used the "local" approximation. Sharma's (137,138) theory (hereafter referred to as S-theory) on the other hand is more complete as it takes into account all electronic contributions such as "local", "nonlocal" and "distant" as well as contributions from the point-multipoles on the rest of the lattice, incorporating adequately the shielding effects.

The S-theory has been applied very systematically to various systems to assess its validity. By considering a molecular complex consisting of the central ion with the nuclear quadrupole moment and its ligands embedded in a crystal, it has been able (i) to derive the correct value (137) of the nuclear quadrupole moment of Al^{27} considering Al_2O_3, (ii) to resolve (138) the ferrous-ferric anomaly of Fe^{57m} in cases of Fe_2O_3 and $Al_2O_3:Fe^{3+}$ and, as a result, establishing the quadrupole moment $Q(Fe^{57m})$ to be 0.18 $\pm$ 0.02 b and (iii) to explain (139) consistently the then existing experimental quadrupole coupling constants of Fe^{3+} associated with two inequivalent sites in yttrium iron garnet employing the deduced value $Q(Fe^{57m}) = 0.18 \pm 0.02$ b (IV) to estimate (26) the nuclear quadrupole coupling constants in several rare-earth iron garnets and (VI) in explaining the nuclear quadrupole splitting of Fe^{57m} in hemin (66), hemoglobin hydroxoxide and hemoglobin cyanide (230).

According to the S-theory, the contribution to the EFG is divided into two parts: (i) the metal-ligand complex consisting of a central ion, where the electric field gradient is to be calculated and its nearest neighbors (ligands) and (ii) the remaining crystal. The EFG is produced by the electron charge density and nuclear charges on the metal ligand complex and the multipoles on the rest of the crystal. The electronic charge density of the complex is calculated in the Hartree-Fock approximation using the many-electron wavefunction as a Slater determinant formed by one-electron molecular orbitals Ψ_i. The electron contribution to EFG is given by

$$q_{el} = e\sum_i \varepsilon_i < \Psi_i | (3 \cos^2\theta_i -1)/r_i^3 | \Psi_i > \qquad (47)$$

where e represents the charge of an electron (including sign) and ε_i the occupancy number of the molecular orbitals Ψ_i. The orbitals Ψ_i are the bonding and antibonding orbitals which are linear combinations of the metal ion orbitals ψ_α^0 and the ligand orbitals $X_{g\beta}$ where α and β specify the quantum numbers of the relevant orbitals. The location of a ligand is designated by the symbol g. If only the overlap effects are included the molecular orbitals may be written as

$$\Psi_i \equiv \begin{cases} \psi_\alpha^{ab} = N_\alpha(\psi_\alpha^o - \sum_{g\beta} S_{g\alpha\beta} X_{g\beta}) \\ \psi_{g\beta}^{b} = X_{g\beta} \end{cases} \qquad (48)$$

where ab and b have been used to show the antibonding and bonding orbitals. $S_{g\beta\alpha}$ are the overlap integrals

$$S_{g\beta\alpha} = < X_{g\beta} | \psi_\alpha^0 > \qquad (49)$$

The factor N_α is the normalization constant

$$|N_\alpha|^2 = (1 - \sum_{g\beta} |S_{g\beta\alpha}|^2)^{-1} \tag{50}$$

The effective EFG, after incorporating the shielding effects, is

$$q = (1-R)\,(q'_\ell + q'_{n\ell}) + (1-\gamma_\infty)\,(q'_d + q'_{db} + q'_n + q'_{RL}) \tag{51}$$

where, using $(3\cos^2\theta - 1) = 2\surd(\frac{4\pi}{5})\, Y_2^0$,

$$q'_\ell = 2e\sqrt{(\frac{4\pi}{5})} \sum_\alpha \varepsilon_\alpha |N_\alpha|^2 < \psi_\alpha^0 | \frac{Y_2^0}{r^3} | \psi_\alpha^0 > \tag{52}$$

$$q'_{n\ell} = -2e\sqrt{(\frac{4\pi}{5})} \sum_{\alpha g\beta} \varepsilon_\alpha \{ N_\alpha < \psi_\alpha^0 | \frac{Y_2^0}{r^3} | \chi_{g\beta} > S_{g\beta\alpha}$$

$$+ N_\alpha^* < \chi_{g\beta} | \frac{Y_2^0}{r^3} | \psi_\alpha^0 > S_{g\beta\alpha}^* \} , \tag{53}$$

$$q'_d = 2e\sqrt{(\frac{4\pi}{5})} \sum_{\alpha g\beta g''\beta''} \varepsilon_\alpha |N_\alpha|^2 S_{g\beta\alpha}^* S_{g''\beta''\alpha}$$

$$< \chi_{g\beta} | \frac{Y_2^0}{r^3} | \chi_{g''\beta''} > , \tag{54}$$

$$q'_{db} = 2e\sqrt{(\frac{4\pi}{5})} \sum_{g\beta} \varepsilon_{g\beta} < \chi_{g\beta} | \frac{Y_2^0}{r^3} | \chi_{g\beta} > , \tag{55}$$

and

$$q'_n = \sum_g \frac{\zeta_{ng} |e| (3\cos^2\theta_g - 1)}{a_g^3} \tag{56}$$

The factors $1 - R$ and $1 - \gamma_\infty$ are the appropriate shielding factors. The components of the EFG without shielding effects have been shown by "primes". The subscripts ℓ, $n\ell$ and d represent the "local", "nonlocal", and "distant" parts of the field gradient due to the antibonding orbitals. The "local," "nonlocal," and "distant" terms have been categorized according as the ligand orbitals are involved not at all, once or twice. The subscript db signifies the field gradient from the (distant) bonding orbitals. ε_{α} and $\varepsilon_{g\beta}$ take into consideration the number of electrons occupying the orbitals ψ_{α}^{0} and $X_{g\beta}$, respectively. q_n' in Eqn. 56 gives the field gradient due to the nuclear charges of the ligands. $\zeta_{ng}|e|$ is the total charge of the ligand nucleus reduced by the amount appropriate to the total charge of those electrons on the ligand orbitals which have not been accounted for in forming the bonding orbitals. q_{RL}' in Eq. 51 is the contribution from the multipoles in the crystal external to the molecular complex under consideration.

It must be emphasized that for the nonlocal terms the factor $1 - R$ instead of $1 - \gamma_{\infty}$ has been introduced due to the fact that the electron density $\psi_{\alpha}^{0}{*}X_{g\beta}$ involved in Eq. 53 is dominant in the local region, i.e., close to the cation site.

Only a short account of the S-theory has been presented as above. For further details the reference should be made to Refs. 26 and 137-139.

The S-theory is quite general which involved, as it is clear from the above treatment, the construction of the bonding as well as anti-bonding orbitals. Though, in principle, the molecular orbitals involve both the overlap and charge transfer covalencies (91), only the overlap covalency has been taken into account simply because the correct estimates of the charge-transfer covalencies are not available and because the overlap matrix elements (as well as other required multicenter integrals) can be evaluated correctly by means of the α-function technique (181-184). In case one desires to incorporate the charge transfer covalency effects, one merely modifies the molecular orbitals in the manner given by Shulman and Sugano (163) and formulated by Sharma et al (91).

The Taylor-Sawatzky-Hupkes (TSH) theory, though incomplete, contains important part of the contributions from the S-theory. It should, however, be pointed out that Sengupta et al (231) have applied the TSH theory to analyze the Mossbauer data associated with Fe^{57m} in FeOCl. They were unable to fit the experimental results with reasonable values of the parameters. It would be interesting to employ the S-theory in FeOCl and test its validity in this system. It might be necessary to consider the charge transfer covalency to explain the experimental data in this case.

The nuclear quadrupole interactions in iron-fluoride compounds (K_3FeF_6, Na_3FeF_6 and $(NH_4)_3FeF_6$) have been investigated in both point-charge and covalent models by Christoe and Drickamer (232). These investigations show that the covalency effects are appreciable. They require to consider small changes in local symmetry with pressure to interpret their experimental data (232). The S-theory has been applied by Flygen et al (233) to calculate the EFGs at Al^{3+} in octahedral and tetrahedral sites in rare-earth aluminium garnets MeAlG (Me = Lu, Yb, Y and Gd) taking the overlaps of aluminium and oxygen wavefunctions into consideration. The calculated results for the tetrahedral-sites are in good agreement with the experiments (234-236) whereas for the octahedral sites they are an order of magnitude too large. The main cause for this is the neglect of the contributions to EFG coming from the induced dipoles on the oxygen ions and the distortion of the ligand oxygen ion wavefunctions due to the local electric fields, as will be discussed later.

The calculations based on the overlapping of the filled electron shells have also been used by Seryshev et al (237) to interpret the nuclear quadrupole interactions of Na^{23} nuclei in $NaNO_3$. However, they estimated only the "local" contribution to the field gradient which yielded the net theoretical value only 8-15 percent of the experimental values (237).

In cases where the electric fields are present at the ligand sites, the electronic charge distribution of the ligands are greatly distorted because of the polarization effect. Sharma and Sharma (27) have modified the S-theory by explicitly incorporating the polarizations of the ligand wave-functions. Because of the lack of space the details of this theory will not be presented here. The reader is recommended to consult Ref. 27 and 66 for details. The ligand-polarization effects have been found to be important not only when the ligands are chlorine ions (27) but also for other ions such as nitrogen as in case of ferric hemin (66) where about 29% of the experimental value of the nuclear quadrupole splitting of Fe^{57m} is due to the ligand polarization effects. In case of FeOCl and in rare-earth iron garnets (239) it seems that the modified S-theory with ligand polarization effects could lead to the correct results since both the O^{2-} and Cl^- ions are very polarizable and they are located at points without center of inversion symmetry.

There have also been calculations (240,241) available based on the molecular orbitals of the complexes ($FeCl_4$ and $NaCl_4$) without accounting for the shielding effects. It should be emphasized that if a complex is embedded in a crystal, the shielding effects of the electrons on the complex on the field gradient due to the external charges in the crystal become operative. In that case one faces a problem of finding the shielding factors of the complex instead of the ions, which is a more difficult task. It may be noted that in Ref. 240 and 241 the contributions of the type q'_{RL} which are known to change the results substantially because of the shielding effects have been omitted.

The relativistic effects on the molecular orbitals may be important in certain cases especially for $Gd^{3+}(^8S)$ state ions. Such effects have not yet been estimated for the EFG calculations. Also, the many body effects will be interesting to be examined on the EFG calculations. The theory of temperature and pressure dependence of the nuclear quadrupole interactions (232, 237, 239) of S-state ions in solids has not yet developed and demands special attention.

6. CONCLUDING REMARKS

Various investigations reveal that the zero-field splitting parameters and the spin-lattice coupling constants associated with the iron group ^6S state ions in the crystals can be explained reasonably well in the nonrelativistic frame-work provided that one calculates the crystal fields from convergent lattice sums and includes the contribution from the BO- mechanism which predominates in most cases studied. Better agreement is possible by including the local distortion effects in the crystals surrounding the impurity ion. It is suggestive that the relativistic effects and all important mechanisms should be studied in systems on a more thorough basis without resorting to approximations such as fractional charge approximations on the lattice sites. Moreover, the charge transfer covalency effects demand attention which have consistently been neglected. For the rare-earth S-state ions the good agreement of the zero-field splitting with the experiment has not yet been possible because of the complexities of the problem which require accurate evaluation of the expressions of many mechanisms which are of complicated nature. Also, in rare-earth S-state ions no overlap and charge-transfer covalency effects have yet been investigated. Such effects combined with correct evaluations of all point-multipole mechanisms are expected to give clues to the improvements which are needed to bring the calculated results close to the experiments. The lattice distortion and polarization effects are important in these cases also.

As for the nuclear quadrupole interactions Fe^{57m} is the only nucleus which has been studied extensively. The existing theory predicts correctly the sign and magnitude of the quadrupole splitting of Fe^{57m} in many cases. However, there are many improvements which are needed. Most important ones are connected with (i) more precise estimates of the Sternheimer antishielding factors (213). (ii) determination of the relaxation and distortion of the lattice surrounding the impurity (iii) precise determination of the polarization distortion of the ligands because of electric field of the crystal (27) and (iv) estimation of the relativistic and correlation effects on the wavefunctions. All these points are expected to influence the calculated quadrupole splitting results. Besides the quadrupole coupling constants, the asymmetry parameters, if nonzero by symmetry, must also be evaluated to test the validity of the theories. Other S-state nuclei such as Mn^{2+}, Gd^{3+}, Eu^{2+} have not yet been investigated theoretically. Similar calculations on these nuclei will be very interesting to find out how these species interact in various environments. When the theories and calculational techniques are sufficiently advanced, future attempts may be directed towards eliminating the use of the antishielding factors and the polarizabilities for the evaluation of quadrupole splitting parameters by appropriately modifying the theories and calculations. This, of course, will require the knowledge of correct wavefunctions for the whole crystal containing the impurity which seems almost impossible at present. The calculational time will also increase simultaneously since the multi-electron wavefunctions involved will require evaluation of many multi-center integrals.

APPENDIX A

In the following we give the explicit expressions for D and E arising from covalency effects on a 3d (^6S) ion.

The local contribution to D from spin-spin interaction is given by

$$D^{\ell}_{ss} = D^{\ell}_{ss}(s) + D^{\ell}_{ss}(\sigma) + D^{\ell}_{ss}(\pi). \tag{A1}$$

where

$$D^{\ell}_{ss}(s) = (6D_0/7)\ \{[S^2_s(O_{ax}) - \gamma^2_s(O_{ax})] - [S^2_s(O_{eq}) - \gamma^2_s(O_{eq})]\}$$
$$[\ f^{0,3}_{d,d} - (8/7)f^{2,5}_{d,d}\ ] \quad . \tag{A2}$$

The above expression gives the contributions arising from overlap and charge-transfer effects of 2s wavefunctions of the ligand ions. Here

$$D_0 = -g^2\beta^2/20a^3_0, \tag{A3}$$

and $S_s(O_{ax})$ and $\gamma_s(O_{ax})$ are the overlap and charge-transfer covalency parameters for the ligand ion designated by O_{ax} (see Fig. (A.1)). $S_s(O_{eq})$ and $\gamma_s(O_{eq})$ carry similar meanings. The symbol $f_{d,d}{}^{n,m}$ represents the integrals

$$f_{d,d}{}^{n,m} = \iint u^0_d(1)\ u^0_d(2)\ (r<^n/r>^m)u^0_d(1)u^0_d(2)dr_1dr_2. \tag{A4}$$

The expressions for $D^{\ell}_{ss}(\sigma)$ and $D^{\ell}_{ss}(\pi)$, the contributions arising from $2p_\sigma$ and $2p_\pi$ orbitals of the ligand ions, can be obtained from (A2) on replacing s by σ and π, respectively. The overlap parameters $S_i(O_j)$ ($i = s, \sigma, \pi$; j = eq, ax) are the matrix elements

$$S_i(O_j) = <\ u^0_d|\alpha_2(O_jLM|a_jr)\ > \tag{A5}$$

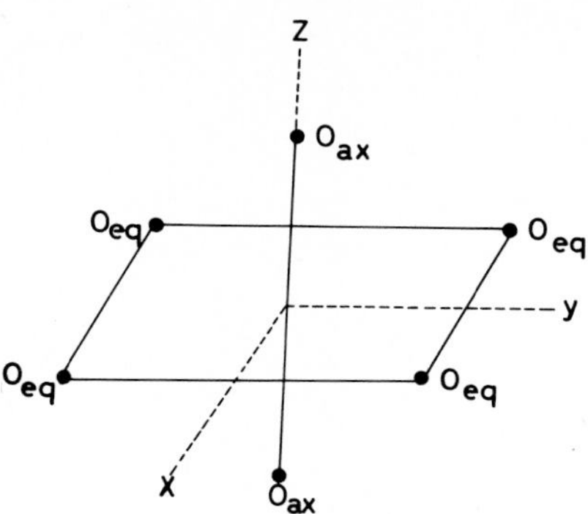

Fig. A.1: Locations of the ligand ions surrounding a $3d^5(^6S)$ ion. O_{eq} and O_{ax} represent the ligands at the equatorial and axial positions, respectively.

with s corresponding to L = 0, M = 0, σ to L = 1, M = 0 and π to L = 1, M = ± 1, a_j being the distance between the central ion and the ligand ion designated by O_j. The symbol $\alpha_\ell(O_jLM) \mid a_jr)$ denotes the α functions (183).

The local contribution to E via spin-spin interaction has been shown to vanish exactly, so that

$$E^{\ell}_{ss} = 0. \tag{A6}$$

As regards the nonlocal contribution (92) to D from spin-spin interaction, we write

$$D^{n\ell}_{ss} = D^{n\ell}_{ss}(s) + D^{n\ell}_{ss}(\sigma) + D^{n\ell}_{ss}(\pi) \qquad \text{with} \tag{A7}$$

$$\begin{aligned} D^{n\ell}_{ss}(s) = -\ 4D_0[([s_8(O_{eq}) + \gamma_8(O_{eq})]\ \{ -1.7889h^{0,3}_{d,0}(O_{eq}s) \\ + 2.6833g^{2,5}_{d,0}(O_{eq}s) + 0.2857g^{0,3}_{d,2}(O_{eq}s) \\ - 1.1429h^{0,3}_{d,2}(O_{eq}s) + 0.4898[g^{2,5}_{d,2}(O_{eq}s) + h^{2,5}_{d,2}(O_{eq}s)] \\ - 1.5333h^{0,3}_{d,4}(O_{eq}s) + 0.0183g^{2,5}_{d,4}(O_{eq}s) + 0.9127h^{2,5}_{d,4} \\ (O_{eq}s) + 3.1944h^{4,7}_{d,4}(O_{eq}s)\ \}\ ) - (eq\to ax)] \end{aligned} \tag{A8}$$

where (eq → ax) represents all the terms contained in the preceding bracket ( ) with the replacement of eq by ax in the integrals $S_i(O_{eq})$, $g^{n,m}_{d,\ell}(O_{eq}i)$, and $h^{n,m}_{d,\ell}(O_{eq}i)$. The integrals $g^{n,m}_{d,\ell}(O_j i)$ and $h^{n,m}_{d,\ell}(O_j i)$ are defined by

$$g^{n,m}_{d,\ell}(O_j i) = \int_0^\infty \frac{[u^0_d(1)]^2}{r_1^m} dr_1 \int_0^{r_1} u^0_d(2) r_2^n \, \alpha_\ell(O_j LM \mid a_j r_2) dr_2 \qquad (A9)$$

and

$$h^{n,m}_{d,\ell}(O_j i) = \int_0^\infty dr_2 \frac{u^0_d(2)}{r_2^m} \alpha_\ell(O_j LM \mid a_j r_2) \int_0^{r_2} [u^0_d(1)]^2 \, r_1^n \, dr_1 \qquad (A10)$$

with

$i = s(L=0,\ M=0),\quad \sigma(L=1,\ M=0),$

$\pi(L=1,\ M=\pm 1);\ j=eq,\ ax.$

The expression for $D^{n\ell}_{ss}(\sigma)$ can be obtained simply by replacing s by σ in (A8). As regards $D^{n\ell}_{ss}(\pi)$ we have

$$\begin{aligned} D^{n\ell}_{ss}(\pi) = -4D_0[([S_\pi(O_{eq}) + \gamma_\pi(O_{eq})]\ \{0.2857g^{0,3}_{d,2}(O_{eq}\pi) \\ -1.1429h^{0,3}_{d,2}(O_{eq}\pi) + 0.4898[g^{2,5}_{d,2}(O_{eq}\pi) + h^{2,5}_{d,2}(O_{eq}\pi)] \\ -2.7994h^{0,3}_{d,4}(O_{eq}\pi) + 0.0333g^{2,5}_{d,4}(O_{eq}\pi) + 1.6663h^{2,5}_{d,4} \\ (O_{eq}\pi) + 5.8321h^{4,7}_{d,4}(O_{eq}\pi)\}) - (eq \rightarrow ax)]. \end{aligned} \qquad (A11)$$

where (eq → ax) represents the terms contained in the preceding bracket ( ) with the replacement of eq by ax.

It can be shown that the distant contribution via spin-spin interaction gives rise to terms of order higher than two in overlap and charge-transfer covalency parameters. Thus, up to second order in overlap and charge transfer, the distant contribution D_{ss}^{d} is zero.

The nonlocal contribution to E from spin-spin mechanism is

$$E_{ss}^{n\ell} = E_{ss}^{n\ell}(s) + E_{ss}^{n\ell}(\sigma) + E_{ss}^{n\ell}(\pi) , \tag{A12}$$

where

$$\begin{aligned} E_{ss}^{n\ell}(s) = & -4D_0\cos 2p[S_s(O_{eq}) + \gamma_s(O_{eq})]\ \{5.3666h_{d,0}^{0,3}(O_{eq}s) \\ & -8.0498g_{d,0}^{2,5}(O_{eq}s) - 0.8571g_{d,2}^{0,3}(O_{eq}s) + 3.4286h_{d,2}^{0,3}(O_{eq}s) \\ & -1.4694[g_{d,2}^{2,5}(O_{eq}s) + h_{d,2}^{2,5}(O_{eq}s)] + 4.5999h_{d,4}^{0,3}(O_{eq}s) \\ & -0.0548g_{d,4}^{2,5}(O_{eq}s) - 2.7380h_{d,4}^{2,5}(O_{eq}s) - 9.5831h_{d,4}^{4,7}(O_{eq}s)\}. \end{aligned} \tag{A13}$$

For obtaining $E_{ss}^{n\ell}(\sigma)$ we simply replace s by σ in the above expression. For $E_{ss}^{n\ell}(\pi)$ we have

$$\begin{aligned} E_{ss}^{n\ell}(\pi) = & -4D_0\cos 2p[S_\pi(O_{eq}) + \gamma_\pi(O_{eq})]\ \{-0.8571g_{d,2}^{0,3}(O_{eq}\pi) \\ & +3.4286h_{d,2}^{0,3}(O_{eq}\pi) - 1.4694[g_{d,2}^{2,5}(O_{eq}\pi) + h_{d,2}^{2,5}(O_{eq}\pi) \\ & +8.3982h_{d,4}^{0,3}(O_{eq}\pi) - 0.1000g_{d,4}^{2,5}(O_{eq}\pi) - 4.9990h_{d,4}^{2,5}(O_{eq}\pi)] \\ & -17.496h_{d,4}^{4,7}(O_{eq}\pi)\} \end{aligned} \tag{A14}$$

It can also be seen that, up to second order in overlap and charge transfer, the distant contribution E_{ss}^{d} vanishes.

Now we list the local, nonlocal, and distant contributions to D and E arising from spin-orbit interaction. The local contribution to D is

$$D_{so}^{\ell} = D_{so}^{\ell}(s) + D_{so}^{\ell}(\sigma) + D_{so}^{\ell}(\pi) , \tag{A15}$$

where

$$D_{so}^{\ell}(s) = [(\zeta_{d,d})^2/10\Delta] \{[S_s^2(O_{eq}) - \gamma_s^2(O_{eq})]$$

$$-[S_s^2(O_{ax}) - \gamma_s^2(O_{ax})]\} \quad (A16)$$

where $\zeta_{d,d}$ is the spin-orbit coupling constant for the electron on the paramagnetic ion and Δ is the average energy denominator. The expressions for $D_{so}^{\ell}(\sigma)$ and $D_{so}^{\ell}(\pi)$ can be obtained from (A16) simply by replacing s by σ and π, respectively.

The nonlocal contributions to D via spin-orbit mechanism

$$D_{so}^{n\ell} = D_{so}^{n\ell}(s) + D_{so}^{n\ell}(\sigma) + D_{so}^{n\ell}(\pi) \quad (A17)$$

with

$$D_{so}^{n\ell}(s) = -(\zeta_{d,d}/5\Delta) \{\zeta_{d,2}(O_{eq}s) [S_s(O_{eq}) + \gamma_s(O_{eq})]$$

$$- \zeta_{d,2}(O_{ax}s) [S_s(O_{ax}) + \gamma_s(O_{ax})]\}, \quad (A18)$$

where

$$\zeta_{d,\ell}(O_j i) = \int u_d^0 \zeta(r) \alpha_\ell(O_j LM \mid a_j r)\, dr \quad (A19)$$

with

$$\zeta(r) = (e^2\hbar^2/4m^2c^2a_0^3) r^{-1}\, dV(r)/dr\ . \quad (A20)$$

V(r) in (A20) is the potential seen by the electron. The potential V(r) and the distance r are in units of $e^2/2a_0$ and a_0, respectively. The quantities $D_{so}^{n\ell}(\sigma)$ and $D_{so}^{n\ell}(\pi)$ in (A17) can be obtained by putting σ and π for s in (A18), respectively. The distant contribution to D from the spin-orbit interaction is

$$D_{so}^{d} = -(2/15\Delta)\zeta_{d,d}\zeta_{p,p}\{[\lambda_\pi^2(O_{eq}) - \lambda_\pi^2(O_{ax})]$$

$$- \sqrt{3}[\lambda_\sigma(O_{eq})\lambda_\pi(O_{eq}) - \lambda_\sigma(O_{ax})\lambda_\pi(O_{ax})]\} \quad (A21)$$

with

$$\lambda_i(O_j) = S_i(O_j) + \gamma_i(O_j)$$

for i=s, σ, π, j=eq, ax. The parameter

$$\zeta_{p,p} = \langle u_{2p}^0 | \zeta(r) | u_{2p}^0 \rangle \quad \text{(A22)}$$

is the spin-orbit coupling constant for the 2p electron of the ligand ion .

The local contribution to E from spin-orbit mechanism exactly vanishes:

$$E_{so}^{\ell} = 0. \quad \text{(A23)}$$

The nonlocal contribution to E from spin-orbit interaction is

$$E_{so}^{n\ell} = E_{so}^{n\ell}(s) + E_{so}^{n\ell}(\sigma) + E_{so}^{n\ell}(\pi) , \quad \text{(A24)}$$

where

$$E_{so}^{n\ell}(s) = \frac{3}{5}(\zeta_{d,d}/\Delta) \ (\cos 2p)\zeta_{d,2}(O_{eq}s)$$
$$[S_s(O_{eq}) + \gamma_s(O_{eq})] \quad \text{(A25)}$$

and $E_{so}^{n\ell}(\sigma)$ and $E_{so}^{n\ell}(\pi)$ are obtained from (A25) by substituting σ and π for s, respectively.

Lastly, the distant contribution to E via spin-orbit mechanism is

$$E_{so}^{d} = (2\zeta_{d,d}/5\Delta)\zeta_{p,p}(\cos 2p) \ \{[S_\pi(O_{eq}) + \gamma_\pi(O_{eq})]^2$$
$$-\sqrt{3}[S_\pi(O_{eq}) + \gamma_\pi(O_{eq})] \ [S_\sigma(O_{eq}) + \gamma_\sigma(O_{eq})]\} \quad \text{(A26)}$$

REFERENCES

1. H.A. Bethe, Ann. Physik, 3(1929) 133.
2. J.H. VanVleck, The Theory of Electric and Magnetic Susceptibilities, Oxford, 1932.
3. J.H. VanVleck, J. Chem. Phys., 3(1935) 803.
4. J.H. VanVleck, J. Chem. Phys., 3(1935) 807.
5. J.H. VanVleck, Phys. Rev. 41(1932) 208.
6. J.H. VanVleck and W.G. Penney, Phil. Mag. 17(1934) 961.

7. J.H. VanVleck, J. Chem. Phys., 7 (1939) 61.
8. J.H. VanVleck, J. Chem. Phys., 41 (1937) 67.
9. W.G. Penney and R. Schlapp, Phys. Rev., 41 (1932) 194.
10. R. Schlapp and W.G. Penney, Phys. Rev., 42 (1932) 666.
11. J.H. Griffith, The Theory of Transition Metal Ions, Cambridge University Press, 1961.
12. J. Ferquson, in S.J. Lippard (Ed.), Progr. Inorg. Chem., Vol. 12, Inter Science, 1970, p. 159.
13. C.J. Ballhausen, Introduction to Ligand Field Theory, McGraw Hill, New York, 1962.
14. D.S. McClure, Solid State Phys., 9 (1959) 399.
15. W.A. Runciman, Rept. Progr. Phys. 21 (1958) 30.
16. J.L. Prother, Atomic Energy Levels in Crystals, National Bureau of Standards, Monograph, 19, 1961.
17. T.M. Dunn, in J. Lewis and R.G. Wilkins (Eds.), Modern Co-ordination Chemistry, Interscience, New York, 1960, p. 299.
18. T.M. Dunn, D.S. McClure, and R.G. Pearson, Some Aspects of Crystal Field Theory, Harper and Row, New York, 1965.
19. M.T. Hutchings, Solid State Phys. 16 (1964) 227.
20. J.N. Murrell, S.F.A. Kettle and J.M. Tedder, Valence Theory, Wiley, London, 1965.
21. H.L. Schlafer and G. Gliemann, Basic Principles of Ligand Field Theory, Wiley - Interscience, New York, 1969.
22. B.N. Figgis, Introduction to Ligand Fields, Interscience, New York, 1967.
23. G. Berthier, Adv. Quantum Chem. 8 (1974) 183.
24. H. Olive and S. Olive, Co-ordination and Catalysis, Verlag Chemie. Weinheim, New York, 1977.
25. G. de Brouckere, Adv. Chem. Phys., 37 (1978) 203.
26. R.R. Sharma, Phys. Rev. B, 6 (1972) 4310.
27. A.K. Sharma and R.R. Sharma, Phys. Rev. B, 6 (1974) 3792.
28. J.O. Artman, "Electric Field Gradient Calculations", in I.J. Gruverman (Ed.), Mossbauer Effect Methodology, Vol. 7, Plenum Press, New York, 1971.
29. R.R. Sharma, Phys. Rev. A, 18 (1978) 726; R.R. Sharma and R. Kolman, J. Chem. Phys. 68 (1978) 2516.
30. E.J. Zavoisky, J. Phys. (USSR), 9 (1945) 211.
31. A. Abragram and B. Bleaney, Electron Paramagnetic Resonance of Transition Metal Ions, Clarendon Press, Oxford, 1970.
32. W. Low, Paramagnetic Resonance, Solid State Physics, Suppl. 2, Acad. Press, New York, 1960.
33. D.J. Newman and W. Urban, Adv. Phys., 24 (1975) 793.
34. H.A. Buckmaster and Y.H. Shing, Phys. Status Solidi A12 (1972) 325.
35. J.R. Akridge, B. Srour, C. Meyer, Y. Gros and J.H. Kennedy, J. Solid State Chem., 25 (1978) 169.
36. R.R. Sharma and A.K. Sharma, Phys. Rev. Lett., 29 (1972) 122; see also the references cited therein.
37. W.C. Nieuwpoort, D. Post and P.Th.van Duijnen, Phys. Rev. B, 17 (1978) 91; K.J. Duff, Phys. Rev. B, 9 (1974) 66; V.P. Romanov et al., Soviet. J. Low Temp. Phys. 3 (1978) 715.
38. A.J. Freeman and R.E. Watson in G.T. Rado and H. Suhl (Eds.), Magnetism, Vol. 2A, Academic Press, 1965, p. 167.
39. A.J. Freeman and D.E. Ellis, J. Phys. (Paris) Colloq., 35 (1974) C6-3.
40. E. Simanek and Z. Sroubek, in S. Geshwind (Ed.), Electron Paramagnetic Resonance, Plenum, New York, 1972, p. 535.
41. Y. Takeda, S. Naka, M. Takano, T. Shinjo, T. Takada and M. Shimada, Mater. Res. Bull., 13 (1978) 61.
42. R. Lang, C. Calvo and W.R. Datars, Can. J. Phys., 55 (1977) 1613.
43. J. Pebler, D. Reinen, K. Schmidt and F. Steffens, J. Solid State Chem., 25 (1978) 107.

44. G.J. Sultanov, R.M. Mirzababayev, I.H. Ismailzade and N.G. Hyseynov, Ferroelectrics, 18 (1978) 227.
45. Y. Nishihara, S. Ogawa and S. Wake, J. Phys. C : Solid State Phys., 11 (1978) 1935.
46. G. Asti, M. Carbucicchio, A. Deriu, E. Lucchini and G. Sloccari, J. Magn. Magn. Mater., 8 (1978) 65.
47. K. Binder, Solid State Commun., 24 (1977) 401.
48. O. Horie, Y. Syono, Y. Nakagawa, A. Ito, K. Okamura and S. Yajima, Solid State Commun., 25 (1978) 423.
49. C. Evora and V. Jaccarino, Phys. Rev. Lett., 39 (1977) 1554.
50. R.C. Barklie, D.O'Donnell and A. Murtagh, J. Phys. C : Solid State Phys., 10 (1977) 4815.
51. R.C. Barklie and K.O'Dnnell, J. Phys. C : Solid State Phys., 10 (1977) 4127.
52. J. Antoine, D. Vivien, J. Thery, R. Collongues and J. Livage, J. Solid State Chem., 21 (1977) 349.
53. F.J.van Steenwijk, H.Th. Lefever and R.C. Thiel, Physica, 92B (1977) 52.
54. J.W. Kim and J.S. Karra, Phys. Rev., B, 15 (1977) 2538.
55. K. Hameda and A.H. Morrish, Phys. Lett., 64A (1977) 259; Surf. Sci., 77 (1978) 584.
56. R.R. Sharma and A.M. Stoneham, J. Chem. Soc., Faraday Trans. II, 72 (1976) 913; R.R. Sharma, Bull. Amer. Phys. Soc., 21 (1976) 305; ibid. 21 (1976) 1314; see also A.J. Tench, Surf. Sci., 25 (1971) 625.
57. Y. Tazuke, J. Phys. Soc. Jap., 42 (1977) 1617; R.J. Birgenau and G. Shirane, Phys. Today, 1978, p. 32.
58. L. May, in L. May (Ed.), An Introduction to Mossbauer Spectroscopy, Plenum, New York, 1971, p. 180; see also, L. May, Index of Publications in Mossbauer Spectroscopy in Biological Materials, The Catholic University of America, Washington D.C., 1978.
59. V.G. Bhide, Mossbauer and its applications, McGraw-Hill, New York, 1975.
60. P.S. Han, T.P. Das and M.F. Rettig, J. Chem. Phys., 56 (1972) 3862.
61. M. Weissbluth, Struct. Bonding, 2 (1967) 1.
62. U. Gonser and R.W. Grant, Biophys. J, 5 (1965) 823; also in I.J. Gruverman (Ed.), Mossbauer Effect Methodology, Plenum, New York, 1965, p. 24.
63. C.E. Johnson, in U. Gonser (Ed.), Topics in Applied Physics, Vol. 5, Springer Verlag, New York, 1975, p. 139.
64. A. Trautwein, in J.D. Dunitz, P. Hemmerich, R.H. Holm, J.A. Abers, C.K. Jorgensen, J.B. Neilands, D. Reinen and R.J.P. Williams, (Eds.), Structure and Bonding, Vol. 20, Springer Verlag, New York, 1974, p. 101.
65. G. Lang, Quart. Rev. Biophys., 3 (1970) 1; G. Lang and W. Marshall, in R. Chance (Ed.), Hemes and Hemoproteins, Academic, New York, 1966.
66. P. Moutscs, J.G. Adams III and R.R. Sharma, J. Chem. Phys., 60 (1974) 1447; R.R. Sharma and P. Mastoris, J. Phys. (Paris) Colloq., 35 (1974) C6-359; R.R. Sharma and P. Moutsos, Phys. Rev. B, 11 (1975) 1840.
67. P.E. Russell and P.A. Montana, J. Appl. Phys., 49 (1978) 1573; V.G. Hill, C. Weir, R.L. Collins, D. Hoch, D. Radcliffe and C. Wynter, Hyperfine Interact., 4 (1978) 444; D. Olivier, J.C. Vedrine and H. Pezerat, J. Solid State Chem., 20 (1977) 267.
68. S.S. Hafner in U. Gonser (Ed.), Topics in Applied Physics, Springer-Verlag, New York, 1975, p. 167.
69. B. Keisch in R.L. Cohen (Ed.), Application of Mossbauer Spectroscopy, Vol. 1, Academic Press, 1976.

70. A. Kostikas, A. Simopoulos and N.H. Gangas, in R.L. Cohen (Ed.), Application of Mossbauer Spectroscopy, Vol. 1, Academic Press, 1976
71. A.K. Singh, B.K. Jain and K. Chandra, J. Phys. D: Appl. Phys., 11 (1978) 55.
72. B. Bleaney and K.W.H. Stevens, Rep. Progr. Phys., 16 (1953) 108.
73. K.D. Bowers and J. Owen, Rep. Progr. Phys., 18 (1955) 304.
74. J.W. Orton, Rep. Progr. Phys., 22 (1959) 204.
75. W. Low and E.L. Offenbacher in F. Seitz and D. Turnbull (Eds.), Solid State Phys., Vol. 17, Academic Press, 1965, p. 135.
76. H.A. Buckmaster and D.B. Delay, Magn. Res. Review 3 (1974) 127. R.A. Vaughan, ibid. 4 (1975) 25; H.A. Buckmaster and D.B. Delay, ibid. 4 (1976) 63; H.A. Buckmaster, ibid. 2 (1973) 273; R.T. Schumacher, ibid. 1 (1972) 123; J.O. Artman, ibid. 1 (1972) 169; H.J. Stapleton, ibid. 1 (1972) 65.
77. T.A. Scott, Mag. Res. Review, 2 (1973) 69; J.G. Stevens, ibid. 2 (1973) 97; T.A. Scott, ibid. 2 (1973)221; 1 (1972) 103; D.A. Shirley, ibid. 1 (1972) 143.
78. J.H. VanVleck and W.G. Penney, Phil. Mag., 17 (1934) 961.
79. M.H.L. Pryce, Phys. Rev., 80 (1950) 1107.
80. H. Watanabe, Progr. Theor. Phys. (Kyoto), 18 (1957) 405.
81. M. Blume and R. Orbach, Phys. Rev., 127 (1962) 1587.
82. R. Orbach, T.P. Das and R.R. Sharma, Proc. Int. Conf. on Magnetism, Nottingham, 1964, p. 330.
83. J. Kondo, Progr. Theor. Phys. (Kyoto), 23 (1960) 106.
84. J. Kondo, Progr. Theor. Phys. (Kyoto), 28 (1962) 1026.
85. A.S. Chakravarty, J. Chem. Phys., 39 (1963) 1004.
86. A.M. Leushin, Sov. Phys.-Solid State, 5 (1964) 1711.
87. A.M. Leushin, Sov. Phys.-Solid State, 5 (1963) 440.
88. A.M. Leushin, Sov. Phys.-Solid State, 4 (1962) 1148.
89. R.M. Macfarlane, J. Chem. Phys., 39 (1963) 3118; ibid. 42 (1965) 442L.
90. R.R. Sharma, T.P. Das and R. Orbach, Phys. Rev., 149 (1966) 257. see also Ref. (190) for the improved value of D for Mn^{2+} in MnF_2 and ZnF_2 using correct value of the parameter 10Dq.
91. R.R. Sharma, T.P. Das and R. Orbach, Phys. Rev., 155 (1967) 338.
92. R.R. Sharma, T.P. Das and R. Orbach, Phys. Rev., 171 (1968) 378.
93. R.R. Sharma, Phys. Rev., 176 (1968) 467.
94. M. Borg and D.K. Ray, Phys. Rev. B, 1 (1970) 4144.
95. B. Milsch and W. Windsch, Phys. Status Solidi (b), 89 (1978) 241.
96. J. Szumowski, J. Phys. C: Solid State Phys., 10 (1977) 3641.
97. F. Holuj, S.M. Quick and M. Rosen, Phys. Rev. B, 6 (1972) 3169.
98. J.A. Hodges, J.L. Dormann and H. Makram, Phys. Status Solidi, 35 (1969) 53.
99. R.R. Sharma, Phys. Rev. B, 3 (1971) 76.
100. R.R. Sharma, J. Appl. Phys., 42 (1971) 1572.
101 C.V. Stager, Can. J. Phys., 46 (1968) 807; to remove the misunderstanding created in this paper see also Sharma (Ref. 102) for the correct interpretation of the crystalline fields for the Spin-Hamiltonian.
102 R.R. Sharma, Can. J. Phys., 47 (1969) 1185.
103 G.D. Watkins, Phys. Rev., 113 (1959) 79.
104 R. Lang, C. Calvo and W.R. Datars, Can. J. Phys., 55 (1977) 1613. see also U. Krauss and G. Lehmann, Z. Naturforsch. Teil A, 30 (1975) 28 for $AIPO_4:Fe^{3+}$ measurements and L.M. Matarrese, J.S. Wells, and R.L. Peterson, J. Chem. Phys., 50 (1969) 2350 for Fe^{3+} in quartz.
105 B.G. Wybourne, Phys. Rev., 148 (1966) 317.
106 B.G. Wybourne, J. Chem. Phys., 43 (1965) 4506.
107 D.J. Newman, Chem. Phys. Lett., 6 (1970) 288.
108 A. Van Heuvelen, J. Chem. Phys., 46 (1967) 4903.

109 W.E. Hagston and J.E. Lowther, J. Phys. Chem. Solids, 34 (1973) 1773.

110 R. Calvo, M.C.G. Passeggi and M. Tovar, Phys. Rev. B, 4 (1971) 2876; for corrections see, ibid. B, 5 (1972) 4651.

111 P. Schlottmann, Phys. Status Solidi (b), 51 (1972) 913.

112 J.C. Gill, Rep. Progr. Phys., 38 (1975) 91; see also the references cited therein.

113. R. Orbach and H.J. Stapleton in S. Gaschwind (Ed.), Electron Paramagnetic Resonance, Plenum Press, New York, 1972.

114 S. Geschwind (Ed.), Electron Paramagnetic Resonance, Plenum Press, New York, 1972.

115 A.A. Manenkov and R. Orbach (Eds.), Spin-lattice Relaxation in Ionic Solids, Harper and Row, New York, 1966.

116 C.P. Poole and H.A. Farach, Relaxation in Magnetic Resonance, Academic Press, New York, 1971.

117 K.J. Standley and R.A. Vaughan, Electron Spin Relaxation Phenomenon in Solids, Adam-Hilger, London, 1969.

118 J.W. Tucker and V.W. Rampton, Microwave Ultrasonics in Solid State Physics, North-Holland, Amsterdam, 1972.

119 J.C. Verstelle and D.A. Curtis, in H.J.P. Wign (Ed.), Handbuch der Physik 18/1, Springer-Verlag, 1968, p. 1.

120 E.B. Tucker, in W.P. Mason (Ed.), Physical Acoustics, Vol. IV A, Academic Press, 1966, p. 47.

121 K.W.H. Stevens, Rep. Progr. Phys., 30 (1967) 189.

122 M.H. Cohen and F. Reif, Solid State Phys., 5 (1957) 321.

123 T.P. Das and E.L. Hahn, Solid State Phys. Suppl., 1 (1958) 1.

124 E.A.C. Lucken, Nuclear Quadrupole Coupling Constants, Academic Press, New York, 1969. Unfortunately, there are some typographical errors in this book which must be corrected.

125 R.L. Collins and J.C. Travis in I.J. Gruverman (Ed.), Mossbauer Effect Methodology, Vol. 3, Plenum Press, New York, 1967, p. 123.

126 R. Bersohn, J. Chem. Phys., 29 (1958) 326.

127 G. Burns, Phys. Rev., 124 (1961) 524; see also ibid. 123 (1961) 1634.

128 E. Brun and S. Hafner, Z. Krist. 117 (1962) 63.

129 T.T. Taylor and T.P. Das, Phys. Rev., 133 (1964) A1327.

130 R.R. Sharma and T.P. Das, J. Chem. Phys., 41 (1964) 3581.

131 R.W. Grant, H. Wiedersich, R.M. Housley, G.H. Espinosa and J.O. Artman, Phys. Rev. B, 3 (1971) 678.

132 J.O. Artman and J.C. Murphy, Phys. Rev., 135 (1964) A1622.

133 M. Raymond, Phys. Rev. B, 3 (1971) 3692.

134 S. Hafner and M. Raymonds, J. Chem. Phys., 49 (1968) 3570.

135 D.R. Taylor, J. Chem. Phys., 48 (1968) 536.

136 G.A. Sawatzky and J. Hupkes, Phys. Rev. Lett., 25 (1970) 100.

137 R.R. Sharma, Phys. Rev. Lett., 25 (1970) 1622.

138 R.R. Sharma, Phys. Rev. Lett., 26 (1971) 563.

139 R.R. Sharma and B.N. Teng, Phys. Rev. Lett., 27 (1971) 679.

140 J.H. Van Vleck, J. Chem. Phys., 7 (1939) 72; Phys. Rev., 57 (1940) 426.

141 M.H.L. Pryce, Proc. Phys. Soc. (London), A63 (1950) 25.

142 A. Abragam and M.H.L. Pryce, Proc. Roy. Soc., A205 (1951) 135; ibid, A206 (1951) 164; A206 (1951) 173.

143 R.D. Mattuck and N.W.P. Strandberg, Phys. Rev. Lett., 3 (1959) 369.

144 K.N. Shrivastava, Phys. Rep., 20C (1975) 137.

145 R. Buisson and M. Borg, Phys. Rev. B, 1 (1970) 3577; M. Borg, R. Buisson and C. Jacolin, Phys. Rev. B, 1 (1970) 1917.

146 P.G. Klemens, Phys. Rev., 125 (1962) 1795; D.L. Mills, Phys. Rev., 146 (1966) 336; A.A. Maradudin, Solid State Phys., 19 (1966) 1.

147 A.L. Schawlow, A.H. Piksis and S. Sugano, Phys. Rev., 122 (1961) 1469.
148 D.W. Feldman, J.G. Castle Jr and J. Murphy, Phys. Rev., 138 (1965) 1208.
149 G. Watkins and Elsa Feher, Bull. Amer. Phys. Soc., 7 (1962) 29; Elsa Rosenvasser Feher, Phys. Rev., 136 (1964) A145.
150 P.L. Donoho, Phys. Rev., 133 (1964) A1080; R.B. Hemphill, P.L. Donoho and E.D. McDonald, Phys. Rev., 146 (1966) 329.
151 R. Calvo, R.A. Isaacson and Z. Sroubek, Phys. Rev., 177 (1969) 484.
152 R. Orbach, Proc. Roy. Soc. (London), 264 (1961) 458.
153 Charlotte E. Moore, Atomic Energy Levels, National Bureau of Standards Circular No. 467, U.S. Government Publishing and Printing Office, Washington, D.C., 1949.
154 Mandel Sachs, Solid State Theory, McGraw Hill, New York, 1963, Chapter 4.
155 R.R. Sharma, Phys. Rev. B, 3 (1971) 76.
156 R.R. Sharma, Phys. Rev. B, 2 (1970) 3316.
157 M. Blume and R.E. Watson, Phys. Rev., 139 (1965) A1209.
158 R.M. Sternheimer and H. Foley, Phys. Rev., 102 (1965) 961.
159 T.P. Das and R. Bersohn, Phys. Rev., 102 (1956) 733.
160 E.G. Wikner and T.P. Das, Phys. Rev., 109 (1958) 360.
161 A. Dalgarno, Proc. Roy. Soc. (London), A251 (1959) 282.
162 P.G. Khubchandani, R.R. Sharma and T.P. Das, Phys. Rev., 126 (1962) 594.
163 R.G. Shulman and S. Sugano, Phys. Rev., 130 (1963) 506; K. Knox, R.G. Shulman and S. Sugano, ibid. 130 (1963) 512; S. Sugano and R.G. Shulman, ibid 130 (1963) 517.
164 R.E. Watson and A.J. Freeman, Phys. Rev., 134 (1964) A1526.
165 E. Simanek and Z. Sroubek, Phys. Status Solidi, 4 (1964) 251.
166 R.A. Serway, Phys. Rev. B, 3 (1971) 608.
167 P. Schlottmann and M.V.G. Passeggi, Phys. Status Solidi (b), 52 (1972) K107.
168 R. Chatterjee and D. van Ormondt, Phys. Lett., A, 33 (1970) 147.
169 J. Andriessen, G. de Jong and D. van Ormondt, Phys. Lett., 26A (1968) 617.
170 The relativistic contribution to the crystal field splitting parameters has been important for Mn^{2+} in NaCl. See W. Dreybrodt and D. Silver, Phys. Status Solidi, 34 (1969) 559.
171 W. Karthe, Phys. Status Solidi (b), 81 (1977) 323.
172 C.A. Hutchison, B.R. Judd and D.F.D. Pope, Proc. Phys. Soc. (London), B, 70 (1957) 514.
173 B.R. Judd, Thesis, Univ. of Oxford, 1955 (unpublished).
174 K. Rajnak and B.G. Wybourne, J. Chem. Phys., 41 (1964) 565.
175 J. Andriessen, R. Chatterjee and D. Ormondt, J. Phys. C : Solid State Phys., 6 (1973) L288; ibid. 7 (1974) L339.
176 T. Lulek, Acta. Phys. Pol. A, 36 (1969) 551; ibid. 40 (1971) 797; see also T. Lulek, Phys. Status Solidi, 39 (1970) K105.
177 R. Chatterjee, D.J. Newman and C.D. Taylor, J. Phys. C : Solid State Phys., 6 (1973) 706; see also H.A. Buckmaster, R. Chatterjee and Y.H. Shing, Can. J. Phys., 50 (1972) 991.
178 D.J. Newman and S.S. Bishton, Chem. Phys. Lett., 2 (1968) 616.
179 B.R.A. Nijboer and F.W. Dewette, Physica, 23 (1957) 309; F.W. DeWette, Physica, 25 (1959) 1225; see also F.E. Harris and H.J. Monkhrost, Phys. Rev. B, 2 (1970) 4400 for Fourier representation method.
180 For a description of direct lattice summation method using neutral groups of ions see Ref. (93) and R.H. Wood, J. Chem. Phys., 32 (1960) 1690. See also Refs. 28 and 56 and J.V. Calara and J.D. Miller, J. Chem. Phys., 65 (1976) 843.
181 P.O. Lowdin, Adv. Phys., 5 (1956) 1.

182 R.R. Sharma, J. Math. Phys., 9 (1968) 505; R.R. Sharma, AERE - Harwell Report No. - T.P. 601, 1975; K.J. Duff. Int. J. Quantum Chem., 5 (1971) 111; A.N. Jette, Int. J. Quantum Chem., 7 (1973) 131; R.R. Sharma, Int. J. Quantum Chem., 10 (1976) 1075; M. Viccaro, Int. J. Quantum Chem., 10 (1976) 1081.

183 R.R. Sharma, Phys. Rev. A, 13 (1976) 517. This reference derives different forms of general analytical expressions for the expansion of a general Slater orbital from one center onto the other, very useful for the analytical evaluation of multi-center integrals. Here, one also finds a general and closed expression for the two-center overlap integrals between two general Slater orbitals. For some simplifications see also H.J. Silverstone and P.K. Moats, Phys. Rev. A, 16 (1977) and M.A. Rashid (to appear).

184 R.R. Sharma and B. Zohuri, J. Comput. Phys., 25 (1977) 199.

185 M. Tinkham, Proc. Roy. Soc. (London), A 236 (1956) 535.

186 W. Dreybodt and D. Silver, Phys. Status Solidi, 34 (1969) 559.

187 C. Hofer and R.R. Sharma, Phys. Rev. B, 3 (1971) 696.

188 G.D. Watkins, Phys. Rev., 113 (1959) 79.

189 H.G. Hoeve and D.O. Van Ostenberg, Phys. Rev., 167 (1968) 245. For experimental data at 20K in $CdCl_2:Mn^{2+}$ refer also to T.P.P. Hall, W. Hayes and F.I.B. Williams, Proc. Phys. Soc. (London), 78 (1961) 883. The value of D for the powdered sample of $CdCl_2:Mn^{2+}$ have been observed to be $\sim +1 \times 10^{-4}$ cm^{-1} by H. Koga, K. Horari and O. Matumura, J. Phys. Soc. Japan, 15 (1960) 1340. A justification for this value of D has also been presented in Ref. (155).

190 R.R. Sharma, J. Appl. Phys., 42 (1971) 1572.

191 D. Van Ormondt, T. Thalhammer, J. Holland and B.M.M. Brandt, Proc. of the XIV Colloque Ampere, North Holland, 1967, p. 272.

192 V.J. Folan, Phys. Rev. B, 7 (1973) 2771.

193 P.A. Narayana, Phys. Rev. B, 10 (1974) 2676.

194 A.H.M. Schrama, P.I.J. Wouters and H.W. de Wijn, Phys. Rev. B, 2 (1970) 1235; see also S. Oqawa, J. Phys. Soc. Japan, 15 (1960) 1475.

195 J. Rubio and W.K. Cory, J. Chem. Phys., 69 (1978) 4792.

196 D.J. Newman, Adv. Phys., 21 (1971) 197; D.J. Newman and W. Urban, J. Phys. C : Solid State Phys., 5 (1972) 3101; G.E. Stedman and D.J. Newman, J. Phys., C: Solid State Phys., 7 (1974) 2347; ibid. 8 (1975) 1070; D.J. Newman and E. Siegel, ibid. 9 (1976) 4285.

197 J.P. Hempel, J. Chem. Phys., 64 (1976) 4307.

198 R. Calvo, Z. Sroubek, R.S. Rubins and P. Zimmermann, Phys. Lett., 27A (1968) 143.

198a S. Oseroff and R. Calvo, Phys. Lett., 32A (1970) 393.

199 R. Lacroix, Archs. Sci. Fasc. Spec. Ampere 11 (1958) 141.

200 C-Y. Huang, Phys. Rev., 159 (1967) 683.

201 T.J. Menne, J. Phys. Chem. Solids, 28 (1967) 1629.

202 N. Guskos, J. Kuriata and T. Rewaj, Phys. Status Solidi (b), 80 (1977) K25; J.W. Hopson and A.N. Nolle, Bull. Amer. Phys. Soc., 13 (1968) 885.

203 J. Andriessen, D. van Ormondt, S.N. Ray and T.P. Das, J. Phys. B : At. Mol. Phys., 11 (1978) 2601.

204 G. Pfister, W. Dreybrodt and W. Assmus, Phys. Status Solidi, 36 (1969) 351.

205 B. Milsch and W. Windsch, Phys. Status Solidi (b), 89 (1978) 241.

206 H. Bill, Phys. Status Solidi (b), 89 (1978) K49.

207 See also D. Nicollin and H. Bill, J. Phys. C : Solid State Phys., 11 (1978) 4803 for experimental contribution to the study of S-state ions in ionic crystals.

208 C.P. Slichter, Principles of Magnetic Resonance; Harper and Row, New York, 1977.

209 R.M. Sternheimer, Phys. Rev., 130 (1963) 1423.
210 T.P. Das and R. Bersohn, Phys. Rev., 102 (1956) 733.
211 R.E. Watson and A.J. Freeman, Phys. Rev., 131 (1963) 250.
212 S.N. Ray, T. Lee and T.P. Das, Phys. Rev. A, 9 (1974) 93.
213 See also M. Vajed - Samii, S.N. Ray and T.P. Das, Phys. Rev. B, 12 (1975) 4591.
214 R.R. Hewitt and T.T. Taylor, Phys. Rev., 125 (1962) 524.
215 T.T. Taylor, Phys. Rev., 127 (1962) 120.
216 J.R. Tessman, A.H. Kahn, and W. Schockley, Phys. Rev., 92 (1953) 890.
217 H.C. Bolton, W. Fawcett and I.D.C. Gurney, Proc. Phys. Soc. (London), 80 (1962) 199.
218 I.M. Boswara, Phys. Rev., B, 1 (1970) 1698.
219 L. Pauling, Proc. Roy. Soc. (London), A, 114 (1927) 191.
220 An estimate of the quadrupole polarizability of O^{2-} equal to 7.0 A^{05} has been obtained by G. Burns and E.G. Wikner, Phys. Rev., 121 (1961) 155.
221 The quadrupolar polarizability in Al_2O_3 has been discussed in M. Raymond and S.S. Hafner, Phys. Rev., B, 1 (1970) 979.
222 J.O. Artman, A.H. Muire, Jr. and H. Wiedersich, Phys. Rev., 173 (1968) 337; see also the references cited therein.
223 A. Hudson and H.J. Whitfield, Mol. Phys., 12 (1967) 165.
224 B.J. Evans, S.S. Hafner and H.P. Weber, J. Chem. Phys., 55 (1971) 5282.
225 D.J. Bellafoire and M.E. Caspari,Hyperfine Interact. 3 (1977) 173. This deals with $^{154}Gd^{3+}$ in rare-earth iron garnets. The nearest neighbor point ion model yields too low a value of the EFG at Gd-Site.
226 C.A. Taft, J. Phys. C: Solid State Phys., 10 (1977) L369; see also the references cited therein.
227 M. Rosenberg, S. Mandache, H. Niculescu-Majewska, G. Filotti and V. Gomolka, J. Appl. Phys., 41 (1970) 1114.
228 A.J. Freeman and R.E. Watson, Phys. Rev., 131 (1963) 2566; C.E. Johnson, Proc. Phys. Soc. London, 92 (1967) 748; M. Weissbluth and J.E. Malling, J. Chem. Phys., 47 (1967) 4166; F.S. Ham, Phys. Rev., 160 (1967) 328; R. Ingalls, Phys. Rev., 188 (1969) 1045. These references deduce $Q(Fe^{57m})$ = 0.18, 0.18, 0.17, 0.17 and 0.21 $\pm$ 0.03 b, respectively.
229 The "local", "nonlocal" and "distant" terms have been defined according as in the expectation value of the EFG operator the ligand orbitals appear not at all, once or twice, respectively.
230 R.R. Gupta and R.R. Sharma, Bull. Amer. Phys. Soc., 24 (1979) 319.
231 D. Sengupta, J.O. Artman and G.A. Sawatzky, Phys. Rev. B, 4 (1971) 1484.
232 C.W. Christoe and H.G. Drickamer, Phys. Rev. B, 1 (1970) 1813.
233 M. Ya. Flyagin, A.E. Nikoforov and A.N. Men, Chem. Phys. Lett., 21 (1973) 549.
234 D.T. Edmonds and A.J. Lindop, J. Appl. Phys., 39 (1968) 1008.
235 T. Tsang and S. Chose, Phys. Status Solidi (b), 48 (1971) K117.
236 V.H. Schmidt and E.D. Jones, Phys. Rev. B, 1 (1970) 1978. In this reference the point-charge contribution has also been obtained by assuming the charge on oxygen as 1.2e.
237 S.A. Seryshev, I.S. Vinogradova and V.M. Buznik, Sov. Phys.-Solid State, 16 (1974) 565.
238 H.K. Perkins and Y. Hazony, Phys. Rev. B, 5 (1972) 7.
239 R.M. Housley and R.W. Grant, Phys. Rev. Lett., 29 (1972) 203.
240 D.E. Ellis, F.W. Averill, J. Chem. Phys., 60 (1974) 2856.
241 D.R. Taylor, Phys. Rev. Lett., 29 (1972) 1086.
242 A. Rosen and D.E. Ellis, Chem. Phys. Lett., 27 (1974) 595.

CHAPTER 14

RADIO FREQUENCY, ACOUSTIC, MICROWAVE AND OPTICAL PERTURBATIONS OF MOSSBAUER SPECTRA

J.K. Srivastava

Tata Institute of Fundamental Research, Bombay-400005, India

1. INTRODUCTION

As has been discussed in various other chapters in this book, one of the important applications of Mossbauer spectroscopy is the study of electron spin dynamics in solids through its spectral line-shape. Such a study is of fundamental importance since the electron spin fluctuations are caused by the mutual energy exchange between the electronic spin (either individual, as in electronic relaxation effects, or in isolated clusters, as in superparamagnetic effects) and some other agency such as any other neighbouring spin, phonons, conduction electrons, magnons, etc. As this energy exchange goes on all the time in all the systems, spin fluctuation effects are always present. Depending upon the relative magnitudes of the characteristic time associated with the spin fluctuation and that involved in certain physical phenomena, these spin fluctuation effects manifest themselves in various physical properties. For example, they are responsible for the random time dependence of the hyperfine fields at the nuclear site (1), lowering of the critical temperature in superconductors doped with magnetic impurities (2), resistivity minima in Kondo systems (3), critical attenuation of ultrasound near magnetic transition temperature (3) and many such physical phenomena. Whereas generally in magnetically ordered systems and concentrated paramagnetic systems, these spin fluctuations are very fast ($\sim 10^{12}$ Hz at $\sim 10^{\circ}$ K) (4), in systems like spin glasses and mictomagnets, ionic spin fluctuations are frozen (5). As can be visualised, if there is a resonant field (microwave, acoustic or optical) present causing transitions among the ionic levels, then depending upon the intensity of the resonant field the rate of spin fluctuations and also the population in various ionic levels can be altered to any desired extent. This in turn will influence various physical phenomena which are governed by the rate of spin fluctuations and the population of ionic levels. Similarly a resonant radio frequency (rf) radiation, causing transitions among the nuclear Zeeman states, can influence various physical phenomena which are dependent on the nuclear relaxation time and the population of nuclear Zeeman states. Such studies have been made using a variety of experimental techniques (6) including Mossbauer effect (7). In this chapter, we review the Mossbauer studies. Apart from the effect of resonant fields, the influence of nonresonant fields on Mossbauer spectra has also been studied (7). As has been discussed in other chapters, the Mossbauer technique senses the ionic spin fluctuations through the random time dependence of the hyperfine fields at the nuclear site. Such a time dependence gives rise to anomalous Mossbauer lineshapes which are characterised by the intense inner lines as compared to the broad outer peaks (8). The effects of various resonant and nonresonant fields on the anomalous and normal Mossbauer spectra are discussed in this chapter. The topics covered can be grouped in the following categories:

(1) Effect of rf radiation on Mossbauer spectrum:
   (a) collapse of the Mossbauer spectrum by rf radiation;
   (b) production of sidebands in the Mossbauer spectrum by rf field;
   (c) Mossbauer-NMR double resonance;
   (d) effect of rf field on Mossbauer spectrum when the magnitude of the field is comparable to that of the internal magnetic field present at the nucleus.

(2) Effect of resonant and nonresonant acoustic vibrations on Mossbauer spectrum:
   (a) ultrasonic field induced Mossbauer sidebands;
   (b) effect of coherent ultrasonic excitations on Mossbauer spectrum;
   (c) Mossbauer-ultrasonic experiments with extra resonant filter;
   (d) Mossbauer-NAR double resonance and gamma acoustic resonance.

(3) Effect of microwave radiation on Mossbauer spectrum (Mossbauer-ESR double resonance and related experiments).

(4) Effect of optical radiation on Mossbauer spectrum:
   (a) double gamma and optical resonance;
   (b) Mossbauer-optical double resonance;
   (c) Microwave-optical-Mossbauer triple resonance.

2. EFFECT OF RF RADIATION ON MOSSBAUER SPECTRUM

2.1 The rf Collapse Effect

The collapse of a magnetically split Mossbauer pattern to a single line (or quadrupole doublet) spectrum by the externally applied radio frequency (rf) field is called the rf collapse effect (7,9). So far, the rf collapse experiments have been performed in the transmission geometry where the single line source, mounted on the drive, is moved and the stationary absorber is subjected to the rf field (7,9-16). In majority of the experiments, the rf field has been applied in the plane of the absorber (10). The rf collapse effect is essentially similar to the superparamagnetic effect described in Chapter 1. In certain magnetic materials, like permalloy, invar etc., called soft magnetic materials, the external magnetic field required for magnetic saturation, i.e. required for magnetising the material completely in the direction of the external magnetic field, $\vec{H}_{sat}$, is quite small being only a few Oe (7,11,12). Magnetic fields of this magnitude are available from the commonly existing rf generators. Thus when a soft magnetic material is placed in a rf field, $\vec{H}_{rf}$, with $H_{rf} > H_{sat}$, the material gets magnetised with the magnetisation ($\vec{M}$) direction fluctuating alongwith the direction of $\vec{H}_{rf}$ at the applied radio frequency, ν_{rf}. This, as in the superparamagnetic relaxation case, gives rise to a hyperfine field, $\vec{H}_{hf}$, at the nuclear site which is fluctuating in direction by 180° at a frequency equal to ν_{rf}. This is because the direction and magnitude of $\vec{H}_{hf}$ is dependent on the direction and magnitude of $\vec{M}$ when the dominant contribution to $\vec{H}_{hf}$ is from the core polarisation as in Fe-compounds (see Chapter 1). Thus as happens in the superparamagnetic relaxation case, when ν_{rf} exceeds the nuclear Larmor precession frequency, ν_L , the H_{hf} at the nucleus averages out to zero within the nuclear Larmor precession time and consequently the magnetically split Mossbauer pattern collapses to a single line (or quadrupole-split) Mossbauer spectrum. This explains the mechanism of the rf collapse effect. However, as may be obvious, the condition $\nu_{rf} > \nu_L$ can cause the rf collapse only if the magnetisation, $\vec{M}$, follows the oscillations of the applied rf field. In other words for collapse to occur, the switching time (τ_s) , i.e. the time required to reverse the magnetisation direction in the material when the direction of the magnetising field gets changed, should be comparable to or shorter than the period of the applied rf field. τ_s depends both on the nature of the material and on the magnitude of the driving field (7,11-13). It has been found that in soft magnetic materials, the condition $\tau_s \sim 1/\nu_{rf}$ is generally satisfied alongwith the condition $H_{rf} > H_{sat}$ (7,11-13). In these systems $\nu_L \sim 20$ MHz and the values $H_{rf} \sim 10$ Oe and $\nu_{rf} \sim 40$ MHz are generally sufficient for obtaining the rf collapse when the thickness of the sample t $\sim$10-20 μm (12). For thicker metallic samples, skin depth problem (14) and for thinner samples surface anisotropy problem, which results in a large H_{sat} (15,16), makes it necessary to have somewhat larger value of H_{rf} for obtaining the rf collapse (12,14-16). These results show that $\tau_s \sim 10$ nsec in these systems when rf collapse occurs (7,11-13) indicating that the uniform rotation of magnetisation is the dominant mechanism of magnetisation reversal in these materials in the region of rf collapse (7,11-13). Thus the magnetisation reversal does not proceed via domain wall motion (for which $\tau_s \sim 500$ nsec (7,11-13)) as was thought by some earlier workers (17).

The rf collapse studies as a function of H_{rf} show that when $H_{rf} < H_{sat}$, and thus $\tau_s > 1/\nu_{rf}$, the magnetically split Mossbauer spectrum does not collapse to a single line but collapses to a broad (triangular shape) spectrum with unresolved or poorly resolved hyperfine structure lines (11,12). The effect of ν_{rf} on Mossbauer spectrum is similar (7,9,11,12) (Fig. 1). As seen in Fig. 1 in the collapsed spectrum (i.e. when $\nu_{rf} \geq 46$ MHz), alongwith the collapsed central single line, collapsed rf induced sidebands also appear. We will discuss these sidebands in the next section. These sidebands appear on both the sides of the main Mossbauer spectrum and are separated from it by $\pm\, n\, \nu_{rf} (n=1,2,3,...)$. They show that the material under investigation is magnetically ordered (7,9-13). At small ν_{rf} (Fig. 1), like the main spectrum, these sidebands are also not collapsed. When $\nu_{rf} \sim \nu_L$ (Fig. 1e), the spectrum looks washed out because both the main spectrum as well as the sidebands are considerably broad and also because many sidebands located at $\pm\, n\, \nu_{rf}$ with respect to the central pattern are present in the velocity range in which the spectrum has been recorded. As will be discussed later, the sidebands can be quenched and thus the collapse effect and the sideband formation can be observed separately (12,18). It may be mentioned that in the metallic systems higher the value of ν_{rf}, larger is the rf field, H_{rf}, required for obtaining the rf collapse. This is due to the skin depth effect (12).

The rf collapse studies have also been carried out in presence of a d.c. (static) external magnetic field, $\vec{H}_{ext}$, applied parallel to the direction of $\vec{H}_{rf}$. In presence of $\vec{H}_{ext}$, the magnetising field, $\vec{H}_M$, is given by $\vec{H}_M = \vec{H}_{ext} + \vec{H}_{rf}$ and thus when $\vec{H}_{ext} > \vec{H}_{rf}$, $\vec{H}_M$, and hence $\vec{H}_{hf}$, always points in the same direction during the nuclear Larmor precession period and the hyperfine structure of the collapsed Mossbauer spectrum is restored back (7,9).

Studies have also been made in a few amorphous materials (13,19) and it has been found that apart from producing the collapse and

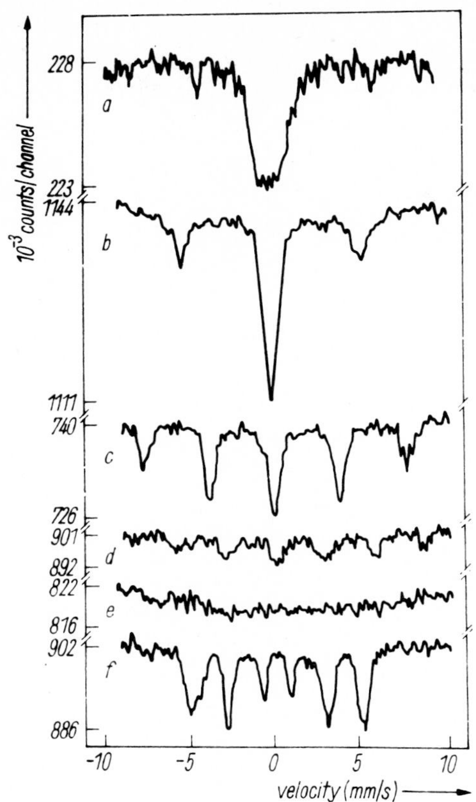

Fig. 1. Mossbauer spectra obtained for 12 μm permalloy foil as a function of ν_{rf}. (a) to (e) at rf field frequency of 85, 64, 46, 33.5 and 21 MHz respectively; (f) without rf field. $\nu_L \approx$ 21 MHz (12).

sideband effects, the rf field also causes crystallisation of the amorphous substance at a temperature much below its crystallisation temperature. This happens probably because the magnetostrictively induced vibrations within the sample initiate the crystallisation process at temperatures much lower than the normal crystallisation temperature (13,19). The rf induced crystallisation has been found to depend on the rf field intensity and also on the time for which the rf field is applied. The amorphous substances with zero magnetostriction have been found to show no evidence of rf field induced crystallisation (19).

Recently, careful measurements by Kotlicki (19) have shown that in metallic samples, like permalloy and invar, the rf heating due to eddy currents is quite significant and must be carefully taken into consideration while studying the collapse effect. The problem of rf heating is also important for nonmetallic samples like the planar ferrite $Ba_2Zn_2Fe_{12}O_{22}$ where the rf collapse effect has been observed (19). However in any case, the observed collapse effect cannot be completely attributed to the heating effect because of the presence of rf sidebands (next section). Also it is difficult to understand the effect of $\vec{H}_{ext}$, applied parallel to $\vec{H}_{rf}$, on the rf collapse if the collapse is attributed to the rf heating effect (7,9,10).

2.2 The rf Induced Sidebands

Most of the rf induced sideband experiments have been performed in the transmission geometry where the rf field has been applied to the stationary absorber (7,15,16,18-28) though a few experiments have also been carried out in the scattering geometry (29) or when the rf field has been applied to the source (30). We mainly discuss here the transmission geometry experiments when the stationary absorber is subjected to the rf field. Fig. 2 is a result of such an experiment showing the rf sidebands produced in the Mossbauer spectra of a thin iron foil when subjected to a rf field of 7.5 Oe peak amplitude, applied in the plane of the foil (31). The top spectrum of this figure (ν_{rf} = 138.28 MHz) clearly shows the repetition of the central six line pattern symmetrically on both the positive and the negative velocity sides. These are the rf induced sidebands (commonly called the rf sidebands) and, as has been mentioned in the earlier section, are separated from the central Mossbauer pattern by $\pm n\nu_{rf}$ (7,29,31). This can be seen in Fig. (1b-1d) also where the rf sidebands are present when the central Mossbauer pattern is a single line. Two types of explanation have been proposed for the origin of the rf sidebands. According to the first explanation, unlike the rf collapse effect, which has a magnetic origin, the rf sidebands have magneto-acoustic origin (7,12,25,27,29,31) and arise due to the magnetostriction effect (7,25,29,31,32). As a consequence of this effect, the applied rf field generates acoustic vibrations within the material (say absorber) which make the Mossbauer nuclei vibrate at the applied radio frequency. This causes frequency modulation of the incoming Mossbauer gamma rays due to Doppler effect (7,33) resulting in the formation of sideband absorption lines at $\nu_0 \pm n\nu_{rf}$, where ν_0 is the position of the parent absorption line (7,31,33,34). A large amount of experimental evidence has been collected by now in support of the above mentioned explanation (7,15, 16, 18-31). For example, the rf sideband effects are observed only in ferromagnetic or ferrimagnetic substances but not in paramagnetic,

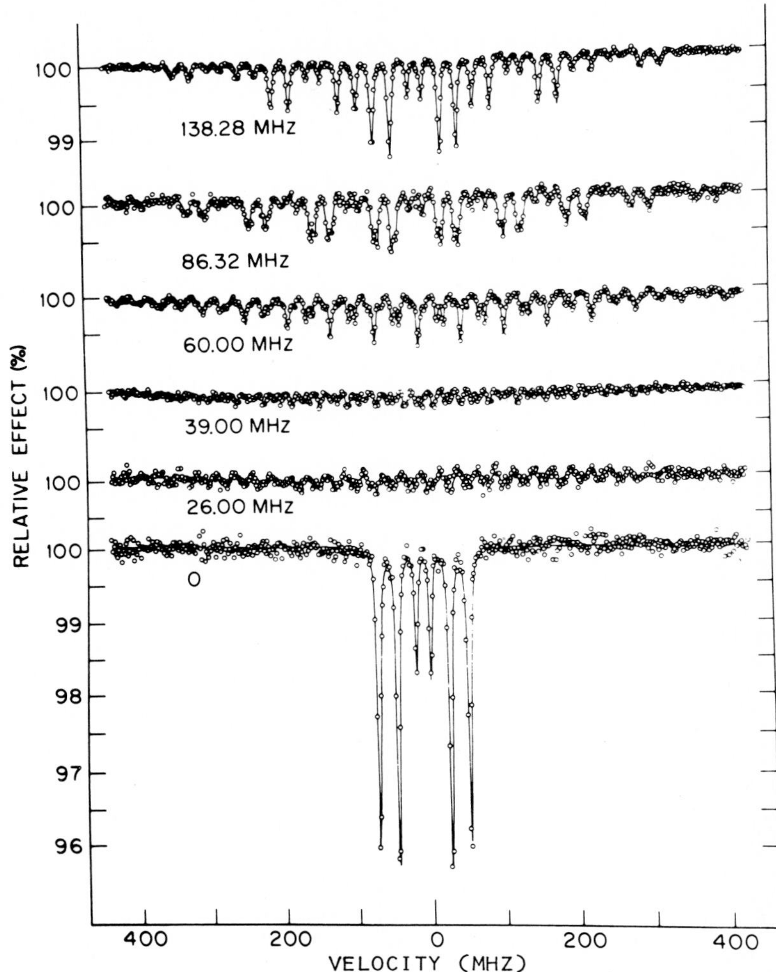

Fig. 2. Mossbauer spectra obtained for an hydrogen annealed 8 μm iron foil at different ν_{rf}. H_{rf} = *7.5 Oe (31).*

diamagnetic or antiferromagnetic materials (7,26,27,29,31). Thus a foil of 310 stainless steel (S.S.) has no rf sideband effect at room temperature where it is paramagnetic (27). However when its surfaces are electroplated with a ferromagnetic material like Ni, large rf sideband effects are observed in this system (27). In this case the acoustic vibrations are generated in the Ni layers and are transmitted to the S.S. lattice. Electroplating the 310 S.S. with diamagnetic Ag or Cu shows no sideband effect when the electroplated foil is subjected to the rf field (27,35). These results indicate that the rf induced sideband effect is basically an acoustic effect and the role of ferromagnetism is only to generate the acoustic vibrations in the material via magnetostriction. Thus phenomena like domain wall motion are not responsible for the appearance of the rf sidebands in the Mossbauer spectra. Measurements made in presence of a static external magnetic field have given a further confirmation of this conclusion (7,31). $\vec{H}_{ext}$ applied in the plane of the foil in a direction perpendicular to the direction of $\vec{H}_{rf}$, has been found to have almost no effect on the sidebands even when $H_{ext} > H_{sat}$, a situation when there are almost no domain walls present in the sample. When $\vec{H}_{ext}$ is parallel to $\vec{H}_{rf}$, the sideband effect decreases rapidly with increasing $\vec{H}_{ext}$ and finally gets quenched. This happens because in this case the net magnitude of the field causing fluctuation of $\vec{M}$ progressively decreases as $\vec{H}_{ext}$ increases and becomes larger than $\vec{H}_{rf}$. It may be mentioned here that both the metallic (like iron, invar, permalloy etc.) as well as nonmetallic (like $FeBo_3$, ferrites, α-Fe_2O_3 above Morin temperature etc.) substances have been found to show the rf sideband effect (7,12,21,25,29,31). This immediately shows that the mechanisms involving eddy current (36) need not be responsible for the generation of rf induced sidebands in the Mossbauer spectra. Measurements carried out on small particles of α-Fe_2O_3 (25) and mosaic of tiny squares of thin natural iron foil (31) have also revealed the acoustic nature of the sideband effect. The modulation index, m, which is a measure of the intensity of the sidebands (7,21,27,28,31), has been found to decrease with decreasing particle diameter, d, and finally when d becomes smaller than $\lambda_a/2$, the sidebands get quenched; (λ_a , the wavelength of the acoustic vibration, is equal to v/ν_{rf} where v is the velocity of the acoustic waves in the material and $m = x_o/\bar{\lambda}$, x_o being the amplitude of the induced acoustic vibration of the Mossbauer nucleus and λ the wavelength of the Mossbauer gamma rays). This result is easily understood since the acoustic vibrations get damped when $d < \lambda_a/2$. Thus large rf sideband effects are observed only in foils or single crystals but not in powders (29). The fact that the rf collapse and the rf sideband effects have different origins is also demonstrated by the observation that the sidebands can be quenched by coating the surface of the absorber foil by damping substances like silicon grease, adhesive "scotch" tape etc. without affecting the rf collapse effect (Fig. 3)

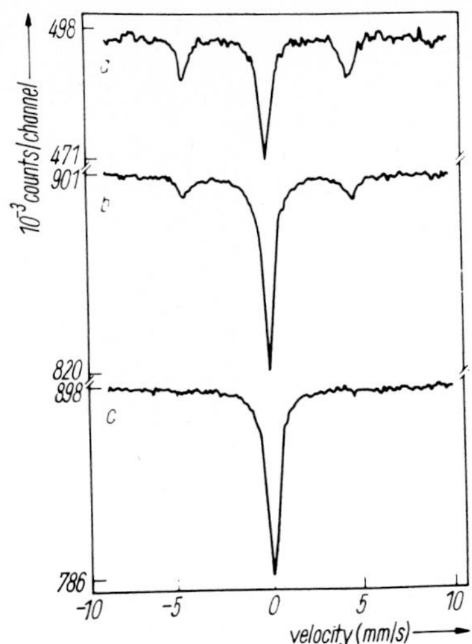

Fig. 3. Mossbauer spectra obtained for 18 μm invar foil for (a) clean sample, (b) sample covered with silicon grease and (c) sample covered with adhesive tape. ν_{rf}=53 MHz and H_{rf}=5.5 Oe. $\nu_L \approx 21$ MHz (12).

(12,18). Silicon grease does not affect the magnitude of H_{rf} inside the sample but it increases the acoustic energy loss of the sample owing to its high acoustic resistivity as compared to air resulting in the damping of the acoustic vibrations (12,18,21). It is also evident that since the stiffness of the sample does not change with the coating, the rf sideband effect is not connected with the bulk vibration of the absorber foil but with the internal vibrations induced within the foil (21,34).

It may be mentioned here that in the beginning, two problems were encountered with the magnetostriction model. The first problem arose when it was realised that in the metallic samples even when the foil thickness was many times larger than the skin depth, rf field affected the entire volume of the sample (and not only the surface atoms) while producing the rf induced effects (7,31). At present it is believed that it happens due to a strong frequency dependence of the magnetic permeability, μ, which decreases from its static (i.e. no rf field present) value of >> 6000 for a 12 μm thick invar foil to perhaps ∿1 when a rf field of 50 MHz is applied to foil (14); (the skin depth $\delta = \frac{1}{2\pi}\sqrt{(\frac{\rho}{\mu\nu_{rf}})}$, ρ = specific resistance). The second problem arises when the magnitude of x_o, and hence of m, is calculated by assuming that for the sideband production, only magnetostrictive changes in the foil thickness have importance since it is only the nuclear Doppler motion in this direction that causes the required frequency modulation of the Mossbauer gamma rays (7,20,30). The magnitude of m calculated in this way is found to be too small. To overcome this difficulty, a modified magnetostriction model has been proposed (7,31) where it is assumed that the rf field induced magnetostrictive vibrations are initially produced with large amplitude in the plane of the absorber foil i.e. perpendicular to the direction of the gamma rays. These acoustic vibrations then propagate within the foil and get scattered due to their interaction with the grain boundaries, surfaces and crystal dislocations. As a result the vibrational amplitude may develop a large component along the gamma ray axis leading to large m. Experimental support to this modified model has been obtained by performing rf sideband measurements in invar samples of two types (15,16). The sample of the first type was an invar foil of 12μm thickness whereas the second type sample was a sandwitch sample consisting of 5 layers of 2.5 μm thick invar foils separated from each other by thin mica layers. There was remarkable increase in the rf sideband effect in the sandwitch samples. This can be easily understood on the basis of the modified magnetostriction model. Since the sandwitch sample contains more surfaces which play an important role in the scattering process, the scattering of acoustic waves would be more effective, and consequently the sideband effect will be more pronounced, in sandwitch sample than in bulk sample of the same total thickness. Another support to the modified model comes from the fact that the rf sideband intensity has been found to increase when the absorber is tilted towards the gamma ray propagation direction (31). It may be mentioned here that experiments have shown that a high acoustic Q is also responsible for obtaining large m in the samples studied (24,28). The modified magnetostriction model predicts (7): $m^2 \propto (1/\nu_{rf})^r$ where $r \sim 3.5$. This fact has been brought out experimentally to some extent (7,25,31) though the exact dependence of the rf sideband intensity on m is complicated (7,27,31). This also rules out the eddy current model (36) which predicts an increase in the sideband intensity with ν_{rf} (35). Chmielowski and Kotlicki (36) have indicated that the magnetostrictively excited antisymmetrical mode of foil vibrations may be responsible for the large x_o in the direction perpendicular to the rf magnetic field.

A systematic study has been made of the dependence of the rf sideband effect on H_{rf}, thickness of the foil, anisotropy of magnetostriction, crystallographic order-disorder transition etc. in several ferromagnetic metallic systems like iron, invar (62.98% Fe-37% Ni-0.02%

Mn-61.98% Fe-38% Ni-0.02% C, etc.), permalloy (49.7% Fe-50% Ni-0.3% Mn-58% Fe-42% Ni, etc.), vicalloy (Fe-Co-V), Ni_3Fe etc. (21,26). The modulation index m, i.e. the intensity of the sidebands, has been found to increase with increasing H_{rf} and with decreasing foil thickness (21,26). This is because with increasing foil thickness, the skin depth effect becomes significant and as H_{rf} inside the foil decreases, x_o also decreases resulting in a lower value of m. Mossbauer spectra show that with the increasing sideband intensity, the intensity of the central Mossbauer pattern decreases and the higher order sidebands (i.e. sidebands corresponding to higher values of n) also become observable (21). This in some cases leads to a smearing out of the Mossbauer spectrum (26) (see the earlier section also). In vicalloy (52 at. % Co, 8 at. % V and 40 at. % Fe), it has been found that the sidebands are more pronounced when H_{rf} is applied along the direction of rolling, a direction in which the static magnetostriction is the largest (26). Studies made in the alloy Fe-28 at. % Ni, which contains both the ferromagnetic α-phase and the nonmagnetic γ-phase, have shown that the Mossbauer pattern belonging to the nonmagnetic phase also gets perturbed in the rf field indicating that the acoustic vibrations, which are generated in the ferromagnetic phase, spread all over the sample (26). Measurements made in single crystals of Fe-4 at. % Si alloy show a remarkable increase in the sideband effect when $\vec{H}_{rf}$ is applied along the easy axis of magnetisation (Fig. 4) (26). This is because relatively large amplitude vibrations can be induced when $\vec{H}_{rf}$ is applied along the easy magnetisation direction. It has been found in the Ni_3Fe system that the rf sideband effects are more pronounced in the crystallographically ordered structure, which has a large static magnetostriction, than in disordered structure (26).

The rf sideband studies have been made in several amorphous substances (like $Fe_{80}B_{20}$, $Fe_{40}Ni_{40}P_{14}B_6$, $Fe_{80}P_{16}C_3B_1$, $Fe_{75}P_{15}C_{10}$, $Fe_{40}Ni_{40}B_{20}$, $Fe_5Co_{75}B_{15}Si_5$ etc.) and the magnetostriction model has been found to be satisfactory (13,19,22-24). The sidebands have also been observed in the ^{57}Fe implanted iron foils and it has been concluded that the surface atoms have larger x_o than the atoms situated inside (25). It may be mentioned here that the early rf-Mossbauer experiments of Perlow (30) and Matthias (30), showing a general smearing out of Mossbauer pattern or a change in the transmitted gamma ray counts at zero velocity when the rf field is applied to the source or absorber, can also be understood on the basis of the sideband effects (7,29,31). It may be mentioned here that apart from the rf induced sidebands which are a consequence of magnetostriction,

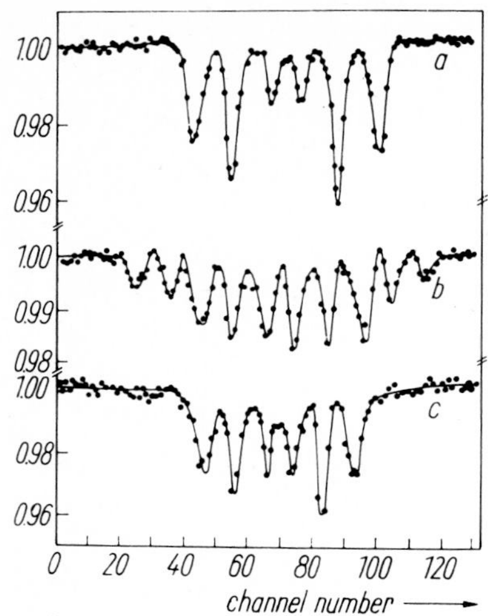

Fig. 4. Mossbauer spectra of a Fe - 4 at. % Si single crystal in rf field of different orientations relative to the easy axis of magnetisation [100]. (a) $H_{rf}=0$; (b) $H_{rf}=5$ Oe, $\vec{H}_{rf}$ || [100]; (c) $H_{rf}=5$ Oe $\vec{H}_{rf}$ ⊥ [100] (26).

sidebands have also been observed in the Mossbauer spectra when the externally generated vibrations are fed to the absorber by a piezo-electric transducer fastened to it (7,34) (described later on in detai or owing to a direct time dependent dipolar interaction of a quite intense rf magnetic field with the magnetic moments of the excited and ground nuclear states (37).

The relaxation time, τ, of the rf induced acoustic vibrations, after the rf field is switched off, has also been measured in some systems (26,28,38). These measurements have shown a high value of acoustic Q (Q ∿ 580) for thin iron foils which is independent of ν_{rf} and foil thickness in the 10-40 MHz frequency range and 5-25 μm thickness range ($Q = \pi\nu_{rf}\tau$). They also showed an inverse dependence of τ on ν_{rf} (τ = 11.4 μsec at 16 MHz) and an independence of Q on foil purity, sample annealing processes and H_{ext} indicating that neither the domain wall motion nor the scattering mechanisms from impurities are the main damping mechanisms (28). The observation that τ is sensitive to the foil mounting technique, decreasing with increased bonding of the foil to the substrate, indicates that perhaps the surface loss is an important loss mechanism causing the damping of the rf induced acoustic vibrations (28). Measurements have also shown that τ is much smaller in the disordered Ni_3Fe alloy as compared to its value in the ordered alloy and that the acoustic vibrations are not excited in the alloys with zero magnetostriction (26).

According to the second explanation, proposed by Olariu et al. (36), the rf sidebands arise due to the annihilation and creation of rf photons (*and not the acoustic phonons*) by the gamma ray field. Thus it is not the acoustic Doppler effect, but the shift in the transition energy of the excited nuclear level emitting gamma photon, due to the addition or substraction of the energy of the rf photon, which is responsible for the generation of the rf sidebands. The shortcomings (36) of the magnetostriction model like the assumption of large x_o due to scattering of acoustic vibrations in the lattice, incapability to explain the appearance of the sidebands at $\pm n\nu_{rf}$ (and not at $\pm 2n\nu_{rf}$) when $H_{ext} = 0$* and the observed asymmetry in the intensity of the sidebands in cases where the sidebands were excited on one of the Mossbauer lines of the magnetically split spectrum, isolated by the selective excitation of the $|\frac{1}{2}, \frac{1}{2}> \rightarrow |3/2, 3/2>$ nuclear transition of ^{57}Fe (next section), etc. are satisfactorily removed in the model of Olariu et al. (36). The model however requires the presence of a very intense rf field at the nuclear site ($H_{rf} \sim 10$ kOe for the M1 nuclear transition) for the multiphoton generation of the gamma ray sidebands. Olariu et al. assume that the phenomenon which is responsible for the rf collapse effect, namely the fluctuation of $\vec{H}_{hf}$ at the applied rf field frequency in ferro- or ferrimagnetic systems, is also responsible for providing a $H_{rf} \gtrsim 10$ kOe at nucleus since in such a case $\vec{H}_{hf}$ essentially acts like $\vec{H}_{rf}$ due to the fluctuations. This

* The factor of 2 appears because for each cycle of the rf field, the magnetostrictive change in the characteristic lattice dimension l, occurs twice. In the presence of H_{ext}, however, the situation is different. In this case, the magnetisation induced in the sample is given by,

$$\vec{M}(t) = \vec{M}_{static} + \vec{M}_{rf}\, e^{2\pi i \nu_{rf} t},$$

which gives an elongation Δl as (7,36):

$$\Delta l = \frac{3}{2}\Lambda\, l \left(\frac{2M_{static}\, M_{rf}}{M_{sat}^2}\, e^{2\pi i \nu_{rf} t} + \frac{M_{rf}^2}{M_{sat}^2}\, e^{2\pi i . 2\nu_{rf} t}\right)$$

(Λ=magnetostrictive constant, M_{sat}=saturation magnetisation). Thus when $M_{static}=0$, the magnetostrictive oscillations are generated at the frequency $2\nu_{rf}$. However when $M_{static} >> M_{rf}$, the harmonic term is negligible or very small.

explanation, however, has one difficulty. Since the fluctuation of $\vec{H}_{hf}$ occurs only when $H_{rf} > H_{sat}$, the multiphoton model perhaps cannot explain the fact that the rf sidebands have been observed in many systems even when $H_{rf} < H_{sat}$ and there is no collapse effect. Also, phenomena like quenching of the sidebands by surface coating, generation of sidebands in Ni plated S.S. etc. too cannot be understood easily on the basis of the multiphoton sideband model.

It may be mentioned here that the idea of multiphoton generation of sidebands is essentially contained in the gamma magnetic resonance proposal of Mitin (36). According to Mitin, the gamma magnetic resonance is a two quantum process involving the absorption of a gamma quantum and the simultaneous emission or absorption of a rf photon by the Mossbauer nucleus. Mitin has shown that this two quantum process also allows the forbidden $\Delta m_I = \pm 2$ nuclear transitions for the ^{57}Fe case and that the resonant frequency of the two quantum transition differs from that of the single quantum transition (i.e. normal Mossbauer transition) by $\pm \nu_{rf}$; (Δm_I is the change in the magnetic quantum number m_I due to nuclear transition).

2.3 Mossbauer-NMR Double Resonance

In this section we describe two types of experiments which have been carried out to observe the Mossbauer-NMR double resonance (7,39-42). In these experiments, the NMR transition either produces new Mossbauer lines in the spectrum or changes the symmetry of the Mossbauer pattern in the m^{o} K range by altering the populations of the nuclear Zeeman states. A third variety of the Mossbauer-NMR experiments is described in the next section where the NMR produced distortion of the original Mossbauer line has also been studied (43,44). Such a distortion is expected to be observed when $H_{rf} \sim H_{int}$ (45-50) where H_{int} is the internal magnetic field present at the nuclear site; ($\vec{H}_{int} = \vec{H}_{ext} + \vec{H}_{hf}$, neglecting the small demagnetisation correction etc. (see Chapter 1)).

In the experiments of the first type (39-41), the source, ^{57}Co diffused in iron crystal, is cooled in the m^{o} K region and the single line absorber is kept at the room temperature. The Doppler motion is given to the absorber and the Mossbauer spectra are recorded in the transmission geometry. At such a low temperature, the nuclear Zeeman states are unequally populated and consequently the observed Mossbauer spectrum is asymmetric about the zero velocity position i.e. the corresponding positive and negative velocity lines of the spectrum are not equally intense (40,41). When the nuclear magnetic resonance (NMR) transitions are induced in the Zeeman states of ^{57}Co, which has a zero field resonance frequency of 295.5 MHz, the difference in the populations of the Zeeman states is decreased; (an intense NMR field can make the populations equal). This in turn results in a decrease in the asymmetry of the Mossbauer spectrum. Thus by observing the change in the count rate, when the absorber velocity is fixed at a particular Mossbauer line (generally the lowest energy line of the ^{57}Fe Mossbauer spectrum where the initial effect on intensity due to the nuclear polarization is maximum), as a function of ν_{rf}, the Mossbauer-NMR double resonance is detected. The change in the count rate is maximum when $h\nu_{rf}$ equals the energy difference between the successive Zeeman states i.e. when the resonance occurs; (h = Planck's constant). The effect of H_{ext} and H_{rf} has also been studied on the count rate at resonance. The intensity of the observed NMR signal (i.e. the percentage change observed in the count rate) has been found to increase with increasing H_{rf} or decreasing H_{ext}. This happens because the effective rf field at the nucleus in a ferromagnetic material is given by (40), $(H_{rf})_{eff} = H_{rf}\ [\ 1 + (H_{hf})/(H_{ext}+H_A)\ ]$ when $H_{ext} > H_{sat}$; (H_A = effective anisotropy field acting on the electronic spins). Thus increasing H_{rf} or decreasing H_{ext}, increases $(H_{rf})_{eff}$ which in turn decreases the population difference among the nuclear Zeeman states resulting in a large change in the observed count rate. Apart from studying the count rate at a constant velocity, the complete Mossbauer spectrum has also been recorded and the effect of NMR transitions on it has been seen (40)

In the Mossbauer-NMR experiments of the second type (7,42), two Mossbauer drives are used and the experiment is performed in a combined scattering-transmission geometry (51). The first drive operates at constant velocity and a single line ^{57}Co source is mounted on it. The gamma rays emitted from the source fall on an iron scatterer (in the powder form to eliminate any rf sideband effect) and are scattered at 90°. The Mossbauer spectrum of the scattered gamma rays is recorded in the usual transmission geometry using a single line absorber mounted on the second drive operating in the constant acceleration mode. The principle of the experiment is as follows. A definite amount of Doppler velocity is given to the source and thus only one of the Mossbauer transitions, say the highest energy transition ($|\frac{1}{2}, \frac{1}{2}\rangle \rightarrow |\frac{3}{2}, \frac{3}{2}\rangle$), is excited in the scatterer (Fig. 5). The first excited states of the scatterer nuclei thus become polarised and since the nuclear spin relaxation time is greater than the lifetime ($\sim 10^{-7}$ sec) in the first excited state, they remain polarised till decay. The transmission

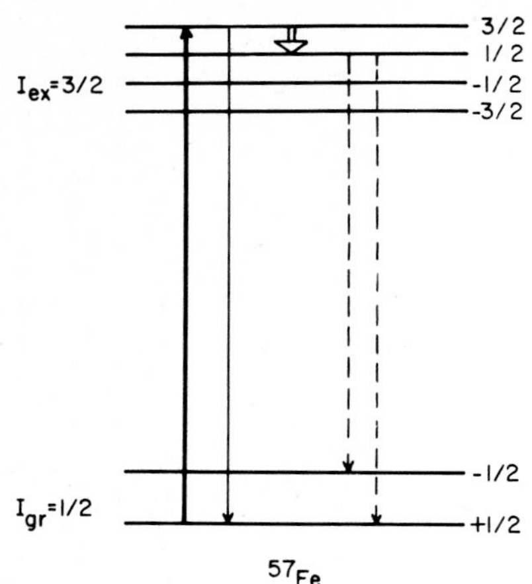

Fig. 5. Figure showing the selectively excited |1/2, 1/2> → |3/2, 3/2> transition of ^{57}Fe (thick line), normally occurring emission transition (thin line) and NMR induced emission transitions (dashed lines). The double arrow indicates the NMR transitions between |3/2, 3/2> ⇄ |3/2, 1/2> Zeeman states of the excited nuclear level.

geometry Mossbauer spectrum, recorded using the second drive, consequently contains only one Mossbauer line situated at a velocity corresponding to the Mossbauer transition $|3/2, 3/2\rangle \rightarrow |\frac{1}{2}, \frac{1}{2}\rangle$. However if the NMR transitions are induced in the excited nuclear states (having zero field resonance frequency, ν_{res} = 26 MHz) in the scatterer, as shown in Fig. 5 apart from the original $|3/2, 3/2\rangle \rightarrow |1/2, 1/2\rangle$ Mossbauer emissive transition, two additional Mossbauer transitions, namely $|\frac{3}{2}, \frac{1}{2}\rangle \rightarrow |\frac{1}{2}, \frac{1}{2}\rangle$ and $|\frac{3}{2}, \frac{1}{2}\rangle \rightarrow |\frac{1}{2}, -\frac{1}{2}\rangle$, also take place. Thus two extra resonance lines appear in the transmission Mossbauer spectrum. The appearance of these additional lines indicates that the Mossbauer-NMR double resonance has occurred in the scatterer (7,42). It has been seen that these extra lines appear only when $\nu_{rf} = \nu_{res}$. Experiments have been performed both when $H_{ext} = 0$ and when $H_{ext} \neq 0$, and it has been found that the intensities of the NMR induced Mossbauer lines increase with increasing H_{ext} for perhaps $H_{ext} \lesssim H_{sat}$ (42). The explanation for this behaviour lies in the fact that it is necessary to magnetise the scatterer to saturation along $\vec{H}_{ext}$, which is at right angles to $\vec{H}_{rf}$ in the plane of the scatterer, in order that all the rf field intensity is utilised in causing the NMR transitions (50). The intensities of the extra lines have also been found to increase with increasing H_{rf} . An explanation for this effect has been given earlier in this section. A modification of the above method has also been used for detecting the Mossbauer-NMR double resonance (7). In this modified method, the second drive is also operated at a constant velocity and its velocity is fixed at the tip of the original, i.e. $|\frac{3}{2}, \frac{3}{2}> \rightarrow |\frac{1}{2}, \frac{1}{2}>$, Mossbauer line. The transmitted count rate is then measured as a function of ν_{rf} and when ν_{rf} equals ν_{res} , a peak is observed in the count rate vs. ν_{rf} curve. This happens because of the appearance of the additional two lines, at two other velocities (to which the second drive is not tuned), when $\nu_{rf} = \nu_{res}$. The behaviour of this Mossbauer-NMR peak as a function of H_{ext} and H_{rf} shows all the essential features of the NMR induced Mossbauer lines discussed above (7). It may be realised that if there was no ferromagnetic enhancement of H_{rf} in ferromagnetic materials (i.e. $(H_{rf})_{eff} >> H_{rf}$ not satisfied), it would have not been possible to induce NMR transitions within the short nuclear lifetime ($\sim 10^{-7}$ sec) of 14.4 keV state of ^{57}Fe (7,41). A simple calculation shows that an rf field of about 10 kOe is needed for obtaining a measurable NMR transition probability in the 14.4 keV state (7) whereas the laboratory produced H_{rf} is generally $\sim$ 100 Oe (7,42). It may also be noted that the conventional NMR detection methods (52) cannot be used for studying the excited state NMR because of the extremely low concentration of the excited nuclei in any sample. This brings out the importance of the Mossbauer-NMR double resonance method for studying the excited state nuclear relaxation time (41).

2.4 Mossbauer-NMR Double Resonance when $H_{rf} \sim H_{int}$

The effect of NMR transitions on the shape of the Mossbauer spectrum has been theoretically investigated by many workers (45-50). These theories predict an appreciable distortion of each line of the Mossbauer spectrum when $\nu_{rf} = \nu_{res}$, where ν_{res}, the resonance frequency for NMR transition, is equal to Δ/h, Δ being the energy difference between the successive nuclear Zeeman states; (Δ_{ex} = excited state Zeeman splitting, Δ_{gr} = ground state Zeeman splitting). The effect of resonant rf field is quite appreciable whether the NMR transitions take place in the excited nuclear state or in the ground nuclear state or in both (45-47). Not only a change in ν_{rf} , but also a change in H_{rf} has an appreciable effect on the shape of each Mossbauer line. Theories (45-50) predict that when $\nu_{rf} = \nu_{res}$, as H_{rf} increases each of the Mossbauer lines in the spectrum becomes progressively broader and finally splits into (2I+1) components when $H_{rf} > \frac{2\Gamma}{g_n\beta_n}$ (45,49,50); here I is the nuclear spin (I_{ex} = excited state spin, I_{gr} = ground state spin), Γ is the levelwidth of the excited nuclear state, β_n is the nuclear magneton and g_n = nuclear g-factor (g_{ex} = excited state g-factor and g_{gr} = ground state g-factor). The separation between the successive components of the split Mossbauer line is directly proportional to H_{rf} . When the NMR transitions simultaneously take place in both the ground and the excited states, a situation which is possible if $g_{ex} = g_{gr}$, each Mossbauer line is expected to split into $(2I_{ex}+1).(2I_{gr}+1)$ components (46,47). Fig. 6 (46) shows the dependence of the spectral lineshape of the Mossbauer spectrum on the frequency of the applied rf field for the ^{57}Fe case when $H_{rf}/H_{int} \sim 0.1$ and $\vec{H}_{rf}$ is perpendicular to $\vec{H}_{int}$ in the plane of the sample which itself is at right angles to the gamma ray direction ($\vec{k}$). It is clearly seen that when $\nu_{rf} = (\nu_{res})_{ex}$ (i.e. 26 MHz) or $(\nu_{res})_{gr}$ (i.e. 45.4 MHz) the original Mossbauer lines, shown by the dashed line curves, split into symmetrically intense components. All the split components are not resolved owing to their close separation as compared to Γ, the natural linewidth of the lines (Γ = 1.6 MHz). As the deviation from resonance increases, the split components become asymmetrically intense and finally the original unsplit Mossbauer line is restored back when $|\nu_{rf} - \nu_{res}| > \frac{2\Gamma}{\hbar}$ (45,47). Physically, this resonant distortion of Mossbauer lineshape has been understood as follows (7). The condition $g_n\beta_nH_{rf} > 2\Gamma$ is achieved when $H_{rf} \sim H_{int}$ (43-50); (this is because for observing the split Mossbauer spectrum, we should have $g_n\beta_nH_{int} > \Gamma$ (49,50)). In such a case, the nuclear moment begins to precess about H_{rf} when $\nu_{rf} = \nu_{res}$ (7) causing a broadening and then splitting of the Mossbauer lines, the magnitudes of distortions being proportional to H_{rf} (43).

These theoretical predictions have been verified, to a great extent, only recently (43,44). As can be easily visualised, the main problem in verifying these theories is to obtain a large value of H_{rf} at the nuclear site. A simple calculation shows that for obtaining the condition $g_n\beta_nH_{rf} \sim \Gamma$ for the case of ^{57}Fe, H_{rf} should be greater than 10 kOe (44). Theoretically, there are three possible ways to overcome this problem. The first possibility is to use the ferromagnetic samples and utilise the fact that in these cases $(H_{rf})_{eff} >> H_{rf}$, where H_{rf} is the amplitude of the rf field applied to the sample and $(H_{rf})_{eff}$ is the amplitude of the rf field present at the nuclear site. However the ferromagnetic samples give rise to magnetostrictively induced rf

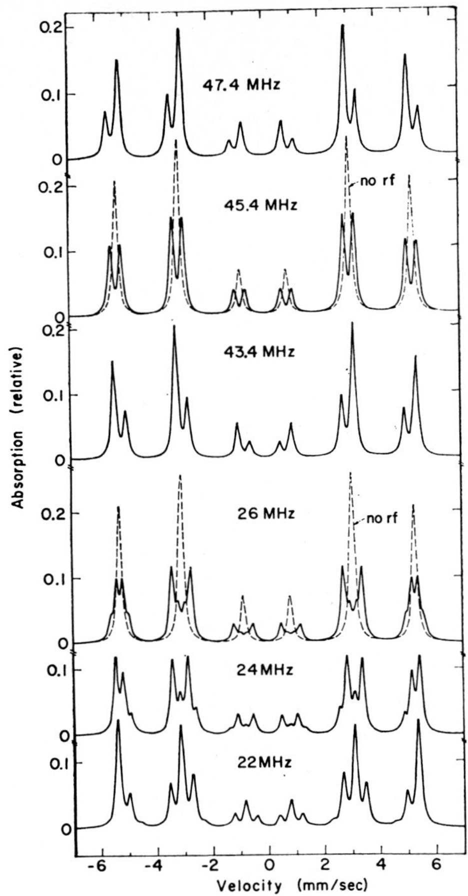

Fig. 6. Mossbauer spectra of ^{57}Fe in a source with magnetic splitting (H_{int}=330 kOe) exposed to rf field of different frequencies versus a single line absorber; $\vec{H}_{int} \perp \vec{k}$ and $H_{rf}/H_{int}=0.1$ (46).

sidebands which will mask any NMR produced lineshape distortion (7, 30). The sidebands can be quenched by the use of powder samples, but in that case inconveniently large H_{rf} is required due to the presence of the large demagnetising fields (7). Another possibility is to use paramagnetic samples where there are no rf sidebands. Fortunately, in certain circumstances even in paramagnetic samples the condition $(H_{rf})_{eff} \gg H_{rf}$ can be obtained (44). One such situation arises in a magnetically dilute paramagnetic system like [$Al(NO_3)_3 \cdot 9H_2O$] : xFe^{3+}, where $x \leq 0.5$ mol. % , when $H_{ext} \geq 100$ Oe and $T \leq 77^{o}$ K (44). In this case the observed Mossbauer spectrum is a superposition of three Mossbauer patterns arising respectively from the $|\pm \frac{5}{2}\rangle$ ($H_{hf} \sim 550$ kOe), $|\pm \frac{3}{2}\rangle$ ($H_{hf} \sim 330$ kOe) and $|\pm \frac{1}{2}\rangle$ ($H_{hf} \sim 110$ kOe) Fe^{3+}-Kramers doublets (53). The exact value of the hyperfine field produced by the $|\pm \frac{1}{2}\rangle$ doublet, $(H_{hf})_{\pm 1/2}$, in case of $[A\ell(NO_3)_3.9H_2O]:Fe^{3+}$ system is 255 kOe (44) and since hyperfine coupling tensor $\vec{A}$ is fully isotropic for $|\pm 1/2\rangle$ (54), $(\vec{H}_{hf})_{\pm\frac{1}{2}}$ is always parallel to $\vec{H}_{ext}$ (44). This makes $(H_{rf})_{eff} \sim (\frac{H_{int}}{H_{ext}})H_{rf}$ (44) and thus when $H_{ext} = 100$ Oe and $H_{rf} = 10$ Oe, $g_n\beta_n(H_{rf})_{eff} \sim \Gamma_{exp}$ (where $\Gamma_{exp} \sim 7\Gamma$) (44). Thus it should be possible to observe the NMR produced distortion of the Mossbauer lines belonging to the $|\pm \frac{1}{2}\rangle$ Kramers doublet in the $[Al(NO_3)_3 \cdot 9H_2O]:Fe^{3+}$ system. Mossbauer measurements carried out in this system unambiguously reveal such an effect (44). A broadening of the Mossbauer lines belonging to the $|\pm \frac{1}{2}\rangle$ Kramers doublet is clearly seen when the system used as absorber is subjected to the resonant rf field ($\nu_{res} = 19$ MHz for the excited nuclear state, $\vec{H}_{rf} \perp \vec{H}_{ext}$). At the same time, the Mossbauer patterns belonging to $|\pm \frac{5}{2}\rangle$ and $|\pm \frac{3}{2}\rangle$ doublets remain unaffected by the rf field. We now come to the third possibility of achieving the condition $g_n\beta_n(H_{rf})_{eff} > 2\Gamma$. This possibility requires the use of diamagnetic samples. In this case, though $(H_{rf})_{eff} = H_{rf}$ and $H_{int} = H_{ext}$, it is possible to achieve the desired condition for the excited nuclear state if both the magnetic moment (μ_{ex}) and the lifetime (τ) of the excited state are large. One such case is ^{181}Ta nucleus where the NMR transitions have shown a remarkable effect on the Mossbauer spectrum (43). The first excited state (6.25 keV) of ^{181}Ta has a half life of 6.8 μsec and μ_{ex} of 5.3 β_n; ($\nu_{res} = 3$ MHz when $H_{ext} = 3.4$ kOe). The source ^{181}W in tungsten matrix is subjected to the resonant ($\nu_{rf} = 3$ MHz) as well as the nonresonant ($\nu_{rf} = 4$ MHz) rf fields; ($H_{rf} = 150$, 300 and 360 Oe). The rf field is applied in a direction perpendicular to both the $\vec{H}_{ext}$ ($H_{ext} = 3.4$ kOe) and the gamma ray direction, and the Mossbauer spectrum is recorded in the usual transmission geometry using a tantalum absorber (43). These measurements have shown that the rf field affects the Mossbauer spectrum only when $\nu_{rf} = \nu_{res}$. Also the rf field has no effect on the spectrum when $H_{ext}=0$ i.e. when there is no Zeeman splitting of the nuclear levels by H_{ext}. In other words, the observed effect of rf field arises only due to the NMR transitions. At $\nu_{rf} = \nu_{res}$, when H_{rf} is increased from 150 Oe to 360 Oe, the shape of the Mossbauer spectrum changes drastically, the hyperfine structure of the Mossbauer spectrum getting washed away due to the broadening and splitting of the Mossbauer lines (43). The change in the shape of the spectrum has been found to be consistent with the theoretical predictions (45,46).

It may be mentioned here that if the NMR transitions are not taking place among all the Zeeman states of the excited nuclear level but are occurring only between a selected pair of the Zeeman states, the theories (45,47-50) predict that only those Mossbauer lines of the spectrum which originate from the initial and final states of this selected pair get perturbed by the resonant rf field. Such a selective NMR excitation is possible either due to the unequal spacing of the Zeeman states arising for example owing to the simultaneous presence of the magnetic hyperfine interaction and the quadrupole interaction or due to the fact that the NMR transitions have been selectively excited by an experiment of the type described in the earlier section where the Mossbauer-NMR double resonance has been observed by a combined scattering-transmission geometry Mossbauer experiment (Heiman et al. (42)). By extending the earlier calculations (45,46) to include the off-diagonal matrix elements in the expression for the Mossbauer transition probability, the theories (47-50) are able to account for the two extra lines observed by Heiman et al. (42). However, like the earlier calculations (45,46), they (49,50) also predict an appreciable broadening and splitting of the original (i.e. $|\frac{3}{2}, \frac{3}{2}\rangle \rightarrow |\frac{1}{2}, \frac{1}{2}\rangle$) Mossbauer line when $\nu_{rf} = \nu_{res}$ and $g_n\beta_n(H_{rf})_{eff} > 2\Gamma$. The fact that Heiman et al. (42) did not observe such a distortion of the original line has been explained by showing that in their case the condition $g_n\beta_n(H_{rf})_{eff} > 2\Gamma$ was not satisfied (49,50). The intensity ratios of the original Mossbauer line and the two extra lines are a function of $(H_{rf})_{eff}$ at the scatterer nuclei (49,50). From this it is concluded that in the experiment of Heiman et al. (42), $(H_{rf})_{eff} \sim 8$ kOe whereas the condition $g_n\beta_n(H_{rf})_{eff} > 2\Gamma$ requires $(H_{rf})_{eff} \sim 17$ kOe (49,50).

3. EFFECT OF RESONANT AND NONRESONANT ACOUSTIC VIBRATIONS ON MOSSBAUER SPECTRUM

3.1 Ultrasonic Field Induced Mossbauer Sidebands

Ruby and Bolef (34) were the first to observe the effect of ultrasonic vibrations on the Mossbauer spectrum. They subjected the single line source (^{57}Co in 321 S.S.) to an ultrasonic field of 20 MHz frequency. Experimentally, this is achieved by bonding the source, with the help of an acoustic glue like perspex cement, durofix adhesive, glycerine, distilled water, araldite epoxy etc., to a piezoelectric crystal like quartz (acting as the ultrasonic transducer) and applying a rf voltage to the crystal. The ultrasonic vibrations are generated in the piezoelectric crystal and transmitted to the glued source (5,55-57). Ruby and Bolef (34) mounted the entire source-ultrasonic transducer assembly on a Mossbauer drive and recorded the Mossbauer spectrum in a conventional transmission geometry using a stationary single line absorber (321 S.S.). The spectrum showed in addition to the central absorption dip, satellite lines (symmetrical sidebands) separated from the central dip by $\pm n \nu_{rf}$, where ν_{rf} is the frequency of the rf voltage applied to the piezoelectric crystal, being equal to the frequency of the ultrasonic vibrations, ν_{us}, excited in the source and $n = 1,2,3....$. The sideband intensity decreased with increasing n. Further, the intensities of the sidebands were found to depend on the voltage, V, applied to the piezoelectric crystal i.e. on the amplitude of the ultrasonic vibrations excited in the source. The sidebands became less intense with decreasing V and finally vanished when V = 0. Since the time of Ruby and Bolef's experiment (34), many Mossbauer-ultrasonic studies have been reported (7,55-74). The experiments have been performed in transmission geometry and the ultrasonically induced Mossbauer sidebands have been observed whether the ultrasonic vibrations are excited in the source (7,34,73,74) or in the absorber (58,59,61,63,66) or in both (58). Experiments have also been carried out by coating both the surfaces of the piezoelectric crystal with the absorber material for obtaining enhanced Mossbauer absorption (59,60).

We have already discussed in the earlier section, the origin of the rf sidebands. It has been stated there* that it is the acoustic (i.e. ultrasonic) vibrations of the Mossbauer nuclei in the lattice which are responsible for the production of the sidebands. These vibrations cause frequency modulation of the incoming or outgoing Mossbauer gamma rays via Doppler effect and thus the gamma ray spectrum which originally contained only the carrier frequency ν_o, splits up into lines of frequencies ν_o, $\nu_o \pm n\nu_{us}$ (67). The same explanation holds good here also; the only difference is that whereas in the case of the rf induced sidebands the acoustic vibrations are generated within the source (or absorber) lattice directly as a result of magnetostriction, in the present case the acoustic vibrations are generated externally

* Here we refer to the magnetostriction model of the rf sidebands.

and then transmitted to the source (or absorber) lattice. It may be recalled at this point that even in the absence of any magnetostrictively induced or ultrasonically induced acoustic vibrations, in the normal case also the frequency modulation of the incoming or outgoing gamma rays does occur owing to the lattice vibrations (7,33). For the case of an Einstein solid, where all the lattice oscillators have a unique frequency ν_E , it amounts to a splitting of the original gamma ray line of width Γ and frequency ν_o, into many lines, each of width Γ, of frequencies ν_o, $\nu_o \pm \nu_E$, $\nu_o \pm 2\nu_E$, ... respectively. In the case of a Debye solid, where the lattice oscillators have a frequency distribution, only the central unshifted line has the natural linewidth Γ and all the satellites are broadened, overlap each other giving rise to a continuum. In the quantum mechanical language, the nth sidebands correspond to the annihilation and creation of n phonons when the phonon field interacts with the gamma radiation field. However, since Mossbauer effect is a zero phonon process (see Chapter 1), the central unshifted line of the frequency modulated spectrum, having natural width Γ, can be identified with the Mossbauer line and its intensity directly gives the Lamb-Mossbauer factor, f. Mossbauer effect is thus insensitive to the wings of the frequency modulated gamma ray spectrum. Consequencly, the interaction of the thermal phonons with the gamma radiation field cannot be studied in a normal Mossbauer experiment. It is this restriction which motivated Ruby and Bolef (34) to generate phonons acoustically and study their interaction with the gamma rays. It is obvious now that the intensity of the ultrasonically induced, or magnetostrictively induced, sidebands can be calculated either classically, on the basis of the frequency modulation theory, (7,31,33,34,59,67-69) or quantum mechanically from the view point of the annihilation or creation of acoustic phonons by the gamma ray field (7,31,69-71). Both the approaches give the same result. In the following paragraph we briefly sketch the classical approach for deriving the dependence of the acoustic (ultrasonically or magnetostrictively induced) sidebands on the frequency, amplitude, wavevector and degree of coherence of the acoustic vibrations in the lattice.

The amplitude of the unmodulated gamma radiation field is given by:

$$\vec{E}(t) = \vec{E}_o \, e^{i\omega_o t} \, e^{-\Gamma t/2\hbar} \quad , \qquad (1)$$

where $\nu_o(=\omega_o/2\pi)$ is the gamma ray frequency and the normalisation condition is, $|\vec{E}_o|^2 = 1$. If due to any reason the frequency of the gamma ray is time dependent, Eqn. (1) gets modified as:

$$\vec{E}(t) = \vec{E}_o \exp[\, i \int_o^t \omega(t')dt'\,] e^{-\Gamma t/2\hbar} \quad . \tag{2}$$

We now consider the case when the gamma ray of frequency ν_o is incident on an absorber whose nuclei are undergoing acoustic vibrations along the gamma ray propagation direction (X direction) with amplitude (i.e. maximum displacement) x_o and frequency ν_{us} $(=\omega_{us}/2\pi)$. The motion of the vibrating nucleus is described by:

$$x(t') = x_o \sin \omega_{us} t' \quad . \tag{3}$$

Because of this vibration the frequency of the incoming gamma ray is Doppler shifted to,

$$\omega(t') = \omega_o\{\, 1 + \frac{v(t')}{c}\,\} \quad , \tag{4}$$

where $v(t') = dx(t')/dt'$. From Eqns. (2)-(4), we have the amplitude of the Doppler shifted gamma ray as:

$$\vec{E}(t) = \vec{E}_o \exp(i\omega_o t + im \sin \omega_{us} t) e^{-\Gamma t/2\hbar} \quad , \tag{5}$$

where the modulation index $m = \frac{x_o}{\lambda\!\!\!^{-}}$, $\hbar=h/2\pi$, h = Planck's constant and $\lambda\!\!\!^{-} = \lambda/2\pi$. General expression for m is, $m = \vec{k}.\vec{x}_o$ where $\vec{k}$ is the wave-vector of the gamma ray and thus when $\vec{k}$ and $\vec{x}_o$ are not parallel, a $\cos\theta$ term appears in the expression of m. Since,

$$\exp(im \sin\theta) = \sum_{n=-\infty}^{+\infty} J_n(m) \exp(in\theta),$$

where $J_n(m)$ is the nth unmodified Bessel function of the first kind with argument m, Eqn. (5) can be rewritten as:

$$\vec{E}(t) = \vec{E}_o \sum_{n=-\infty}^{+\infty} J_n(x_o/\lambda\!\!\!^{-}) \exp\{[\, i(\omega_o + n\omega_{us}) - \Gamma/2\hbar]\, t\,\} \, . \tag{6}$$

This equation shows that as far as the absorber nuclei are concerned, the incident gamma rays are frequency modulated, which is a result of the acoustic vibrations in the absorber, and the frequency modulated spectrum contains in addition to the original line at the carrier frequency ν_o, lines of frequencies ν_o, $\nu_o \pm \nu_{us}$, $\nu_o \pm 2\nu_{us}$, ... These additional lines are the sidebands (satellites). The frequency spectrum of the modulated wave train of Eqn. (6) is given by (33),

$$I(\omega) = |g(\omega)|^2 ,$$

where $g(\omega)$, the Fourier transform of $E(t)$, is:

$$g(\omega) = \frac{1}{(2\pi)^{1/2}} \int_{-\infty}^{+\infty} \exp(-i\omega t)E(t)dt \quad .$$

From these equations we obtain,

$$I(\omega) = \frac{\Gamma}{2\pi\hbar} \sum_{n=-\infty}^{+\infty} \frac{J_n^2(m)}{\{[\ \omega-(\omega_o+n\omega_{us})\ ]^2+(\Gamma/2\hbar)^2\}} \quad .$$

The intensity of the nth sideband, W_n , is thus given by,

$$W_n \propto J_n^2(m) \quad . \tag{7}$$

Eqn. (7) shows that W_n is an oscillatory function of both n and m. For instance, the intensity of the central line, W_o , approaches zero when m approaches the values 2.4, 5.5, 8.6, ... Similarly for large x_o , the nth sideband is an oscillatory function of n (73). In this derivation we have assumed that all of the absorber nuclei are vibrating with the same amplitude, x_o. This is called coherent regime and arises when $\Delta\nu_{us} << \Gamma$, where $\Delta\nu_{us}$ is the bandwidth of the ultrasonic vibrations excited in the absorber. This condition is expected to be realised in practice if the thickness of the absorber foil, $t_{abs} << \lambda_{us}$; (λ_{us} , the wavelength of the acoustic wave excited in the absorber, is equal to v_{us}/ν_{us} where $v_{us} \sim 5 \times 10^5$ cm/sec (7) is the velocity of sound in the lattice). This is because in such a case one can consider the ultrasonic transducer as simply vibrating the entire foil with a sinusoidal velocity i.e. all the Mossbauer nuclei can be considered to be vibrating with the same amplitude and phase (27,34,61,63,68,69,71,72).

We now consider the second case when $\Delta\nu_{us} >> \Gamma$. This is called the case of incoherent excitation and arises when there is a distribution in x_o i.e. when all the Mossbauer nuclei are not vibrating with the same amplitude. If $P(x_o)$ is the distribution function, then from Eqn. (7),

$$W_n \propto \int_0^{\infty} P(x_o)\ J_n^2(x_o/\lambda\!\!\!^-)\ dx_o \quad . \tag{8}$$

(Eqn. 7 corresponds to the case when $P(x_o)$ is a Dirac delta function). Abragam (69) has shown that the quantum mechanical and classical approaches give identical results for this second case if $P(x_o)$ is assumed to be a Rayleigh distribution function. Such a distribution corresponds to the case when

$$x(t') = \alpha \sin \omega_{us} t' + \beta \cos \omega_{us} t' ,$$

where α and β both have a Gaussian distribution. The resulting amplitude is, $x_o = (\alpha^2 + \beta^2)^{1/2}$. Such a situation may occur in practice due to various reasons such as reflections of the acoustic waves at the irregular surfaces of the absorber resulting in the formation of a complex pattern of standing and travelling waves, etc. (59,68). If $(x_o)_{max}$ is the value of x_o at which $P(x_o)$ is maximum, then the distribution functions for α and β are:

$$P(u) = \frac{1}{(x_o)_{max} \cdot (2\pi)^{1/2}} \exp[-\frac{u^2}{2(x_o)^2_{max}}]; \; u = \alpha, \beta ,$$

and Rayleigh distribution function is:

$$P(x_o) = \frac{x_o}{(x_o)^2_{max}} \exp[-\frac{x_o^2}{2(x_o)^2_{max}}] \; . \qquad (9)$$

From Eqns. (8) and (9):

$$W_n \propto \exp[-\frac{(x_o)^2_{max}}{\lambda\!\!\!^{-2}}] . I_n[\frac{(x_o)^2_{max}}{\lambda\!\!\!^{-2}}] \quad , \qquad (10)$$

where $I_n(m^2)$ are the modified Bessel functions of the first kind. Unlike Eqn. (7), Eqn. (10) predicts the absence of any oscillation in W_n with either n or m. The central peak (i.e. zeroth order sideband, n=0) decreases monotonically with increasing m and the satellite intensities first increase and then decrease to zero, nowhere exceeding the central line intensity.

It has been shown by Mishory and Bolef (7) that the coherent regime case (Eqn. 7) corresponds to the situation when the relaxation time of the acoustic phonons excited in the absorber (or source) is much longer as compared to the nuclear lifetime, τ_N. In this case the acoustic phonons are unable to interact with the thermal phonons within the nuclear lifetime. The incoherent phonon excitation case (Eqn. 10) corresponds to the situation when the acoustic phonon relaxation time is very short. In this case, the relaxation processes broaden the ultrasonic bandwidth, $\Delta\nu_{us}$.

Most of the Mossbauer-ultrasonic experiments have been found to belong to the category of incoherent excitation (7,34,59). However, recently a few examples have been found where the sideband formation is due to the coherent excitation of the ultrasonic vibrations (61,63). We will discuss this case in the next section. It may be mentioned here that in the case of the magnetostrictively induced sidebands though the sideband intensities show somewhat better agreement with Eqn. (10), they do not follow either Eqn. (10) or Eqn. (7) fully and the situation is somewhat complex (7,21,31,35,36).

The Mossbauer-ultrasonic experiments have many possible applications (7). One of the obvious applications is the use of the sidebands for absolute velocity calibration of the Mossbauer spectrometers (59). The positions of the sidebands depend only on one parameter, viz. the frequency of the ultrasonic vibrations and this provides a convenient tool for directly calibrating the Mossbauer spectrometer (34,59). Sidebands have also been used for the construction of a variable frequency ultrasonic Mossbauer drive (7,73,74). To understand its working principle, we consider the case when the Mossbauer source is a single line source and the absorber has a magnetically split spectrum. In ultrasonic Mossbauer drive, both the source and the absorber are held at zero velocity. The ultrasonic vibrations are fed to the source and the driving frequency is swept slowly from zero MHz. The driving voltage is so chosen that only the first order sidebands have appreciable intensity. With the increasing frequency, both the first order sidebands slowly move away from the central unshifted line and whenever these sidebands overlap any absorption line of the absorber, the transmitted gamma ray count rate is decreased. In this way, the complete Mossbauer spectrum is traced out.

It has been observed that when large ultrasonic power is fed to the source, the self absorption in the source decreases (7). It has been proposed that this effect can be used to modulate the Q factor of a gamma ray laser (72). Physically the effect arises because at high acoustic powers, the local vibrations of the nuclei cause a detuning of the gamma ray resonance and the shift in the gamma ray frequency, $\Delta\nu_\gamma = |\nu_{res} - \nu_\gamma|$, is of the order of or greater than Γ; (for self absorption, i.e. for resonance, $\Delta\nu_\gamma < \Gamma$). The effect is pronounced only for thick sources since in the case of thin sources, the vibration of the sample as a whole occurs which cannot affect the self absorption process. This is because a detuning of the gamma ray resonance cannot be caused if the emitting and the absorbing nuclei in the source material are moving with zero relative velocity. Once the self absorption in the source is suppressed, its f-factor can be measured very accurately (7).

In the field of Mossbauer-ultrasonic study several theoretical predictions exist which are yet to verified experimentally (68, 70, 75, 76). One of them is the effect on the Mossbauer spectrum of ultrasonic vibrations which have an amplitude which is decaying during the nuclear lifetime (68). Calculations show that in such a case the central line becomes narrower than Γ. In all the above discussions, it has been assumed that $\nu_{us} > \Gamma$ and, the sidebands are clearly separated from the central line and from each other. An interesting situation arises when this condition is not satisfied (75). In such a case calculations show that the spectral line splits into two broad peaks whose separation and widths depend on the amplitude of the ultrasonic vibrations excited in the sample. Calculations have also been done for the sideband intensities in the case of partial coherence i.e. when $\Delta\nu_{us} \sim \Gamma$ (70). This situation has been found to be more applicable to the case of magnetostrictively induced sidebands (27). The effect of anharmonic ultrasonic vibrations on the Mossbauer spectrum has also been theoretically investigated (76) and it has been shown that in the case of coherent excitation, where in the normal case the sideband intensities are an oscillatory function of n, with increasing anharmonicity the intensities of the central line and weak satellites increase whereas those of strong satellites decrease. Such anharmonicity is expected to exist when the amplitude of the acoustic vibrations produced in the sample is very large.

It may be mentioned here that the Mossbauer-ultrasonic experiments have been carried out both in metallic (7,34,59-61,63,74) as well as in nonmetallic (58,66,73) systems. Most of the experiments have been performed when the sample (source or absorber) subjected to the ultrasonic vibration originally contains a single line Mossbauer spectrum. Recently an experiment has been reported (66) where the sodium nitroprusside absorber, which shows a quadrupole doublet Mossbauer spectrum, has been subjected to the ultrasonic field. The quadrupole doublet has been found to become asymmetric in the presence of the ultrasonic vibrations. This has been explained on the basis that the mean square vibrational displacement of the Mossbauer nuclei is expected to be more in the direction of the ultrasonic field. Thus at high acoustic power the vibrations of the nuclei in the sample may become significantly anisotropic. Surprisingly, no Mossbauer-ultrasonic experiment has been performed so far in a magnetically ordered sample. Many phenomena like magnon-phonon interaction (57), possibility of acoustic ferromagnetic resonance (77), etc. can be investigated by a Mossbauer-ultrasonic experiment in a magnetic system. Another important application of Mossbauer-ultrasonic experiments, unexplored so far, perhaps lies in the study of the electron-phonon interaction in semiconductors (57).

3.2 Effect of Coherent Ultrasonic Excitations on Mossbauer Spectrum

In the case of coherent excitations, the intensity of the central line (n=0) and that of the sidebands ($n = \pm 1, \pm 2, \pm 3, \ldots$) is given by Eqn. (7). Only recently this expression has been experimentally verified and the situation of coherent excitation realised in practice (61,63). In accordance with Eqn. (7), the experiments show oscillations in W_n with m. Particularly at $m \sim 2.4$, the central line intensity gets reduced to the order of the experimental error. Such a situation can possibly be used for separating the elastic Rayleigh scattering and the nuclear resonant scattering. Theoretical investigations have been carried out to study this possibility when coherent ultrasonic vibrations are excited in a Mossbauer scatterer (78). It has been shown that when $\vec{k} \perp \vec{x}_o$ and $\vec{x}_o$ is sufficiently large, it is only the Rayleigh scattering which contributes to the scattering cross-section.

The coherent excitation Mossbauer-ultrasonic experiments (61,63) have shown that the homogeneity of the ultrasonic vibrational amplitude excited in the piezoelectric transducer and its homogeneous transmission to the Mossbauer sample are essential for observing the coherent excitation. Because of this, the thickness of the transmitting medium (i.e. acoustic glue), t_{gl}, plays an important role and it should be equal to $n\lambda_{us}/2$ ($n = 1, 2, 3, \ldots$) for a maximum transmission of the ultrasonic power to the sample. The results depend substantially on the accuracy with which the condition $t_{gl} = n\lambda_{us}/2$ is satisfied in the entire area of contact of the sample and the glue (61). In addition, the surfaces of the transducer and the sample should not be uneven (roughness <1 micron) and there should not be any air island trapped in the acoustic glue (63). In other words, the acoustic glue should be perfectly homogeneous. If these conditions are not satisfied, then the amplitude of acoustic vibrations will vary from point to point in the sample even though the ultrasonic wave may be coherent. In such a case, Eqn. (7) gets modified to:

$$W_n \propto \int J_n^2 \, [\vec{k}.\vec{x}_o(r)] \, dV, \tag{11}$$

i.e. one has to take volume average of $J_n^2(m)$ for obtaining W_n. Eqn. (11)

shows that in this case, like the incoherent case, there is no oscillation in W_n with n or m (62). The thickness $t_{gl} = n\lambda_{us}/4$ has also been tried and has been found to be more suitable for a homogeneous transmission of the ultrasonic vibrations (63).

3.3 Mossbauer-Ultrasonic Experiments with Extra Resonant Filter

Asher et al. (60) were the first to carry out such an experiment where an extra Mossbauer absorber, called resonant filter, was introduced in between the analysing absorber (i.e. the original absorber) and the source. In the experiment of Asher et al. both the source and the filter were held stationary with respect to each other and were mounted and simultaneously moved on the same Mossbauer drive. The analysing absorber was kept stationary. Whereas both the source and the analysing absorber were unsplit having a single line Mossbauer spectrum, two types of resonant filters were used. The first one was a single line Mossbauer absorber (i.e. unsplit filter) and the second one was a magnetically split Mossbauer absorber (i.e. split filter). Ultrasonic vibrations were excited only in the resonant filter. In case of the split filter experiment, ν_{us} was chosen to be equal to the separation between the frequencies of the source single line and one of the inner lines of the magnetically split spectrum of the resonant filter. In both the cases, viz. split and unsplit filter experiments, the observed Mossbauer spectrum contained a single line in the centre and sidebands on both the sides situated symmetrically at $\nu_o \pm n\nu_{us}$; ν_o=frequency of the central line and n = 1, 2, 3, ... However though the positions of the sidebands were symmetric, their intensities were not (60,64). This intensity-asymmetry was found to decrease with increasing m. No effect was observed when a 'dummy' filter (with no Mossbauer nucleus in it) was used in place of the resonant filter. The production of the sidebands can be physically understood by realising that in the presence of ultrasonic vibrations, the resonant filter acts like a frequency modulated source when the gamma ray absorption and reemission takes place. In the case of the split filter experiment, the central line arises as a result of the overlap of the central single line of the source and one of the first order (i.e. one phonon) sidebands of the resonant filter. Thus in this case, the corresponding (i.e. +n and -n) sidebands in the Mossbauer spectrum do not correspond to the annihilation and creation of the equal number of phonons. This explains the asymmetry observed in the corresponding sideband intensities in the case of the split filter experiment. The intensity-asymmetry observed in the case of the unsplit filter experiment is entirely due to a different reason (64,71,79,80). It has been understood on the basis of the phase difference existing between the acoustic vibrations in the source and in the resonant filter (64,71,79, 80). It has been shown that the intensities of the corresponding positive and negative sidebands (for odd n) are unequal when the source and the resonant filter are ultrasonically modulated in antiphase. This asymmetry is expected to be more pronounced for smaller m (64).

Resonant filters have been used by Perlow and his co-workers (65) to observe the quantum beats of Mossbauer gamma rays. Their results are summarised in Figs. 7 and 8. Fig. (7a) shows the conventional transmission geometry Mossbauer spectrum recorded by Perlow and co-workers (65) with a stationary single line source and a single line absorber in motion on a Mossbauer drive. Fig. (7b) shows the spectrum obtained when

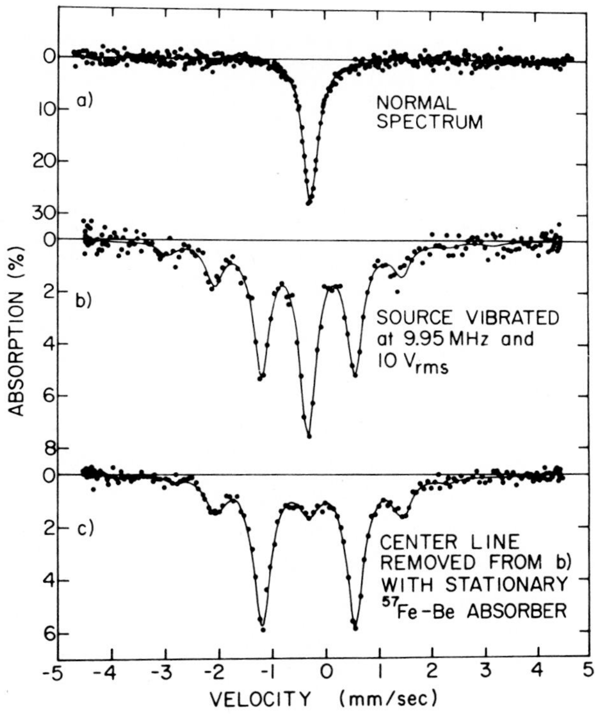

Fig. 7. Velocity spectra (a) unmodulated, (b) frequency modulated at 9.95 MHz and (c) frequency modulated and filtered (65).

the ultrasonic vibrations were excited in the source. When a single line resonant filter, stationary and without any ultrasonic vibrations, was introduced in between the ultrasonically vibrating source and the analysing absorber, the spectrum of Fig. (7c) was obtained. As seen there, the central line has almost disappeared because of the resonant absorption in the filter. Finally, the analysing absorber was removed and the time spectrum of gamma radiation emerging out of the filter was recorded in a fast timing circuit. Fig. 8 shows the time spectrum obtained. The beat formation is clearly seen; in this case the beat

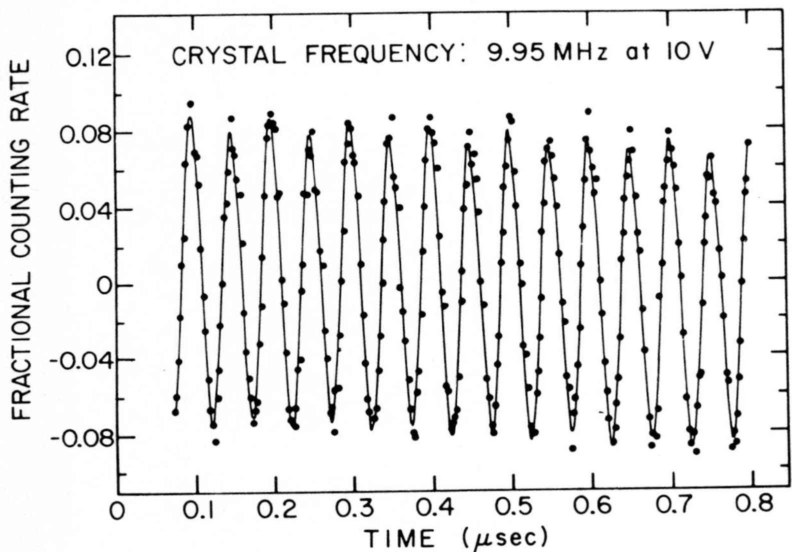

Fig. 8. Time spectrum with radiation of Fig. 7(c) (65).

cycle is equal to $2\nu_{us}$ i.e. the effect of other harmonics is much smaller. Physically, the beat formation is a result of the interference between the different frequency components of the frequency modulated gamma wave. The interference exactly cancels when all the frequency components are present i.e. when there is no resonant filter in the gamma ray path (65). This can be understood as follows. If the nuclear excited state is formed at the time $t=t_o$, Eqn. (5) can be rewritten as:

$$E(t,t_o) = e^{-\Gamma(t-t_o)/2\hbar} \exp(i\omega_o t + im \sin\omega_{us} t) \quad \text{for} \quad t \geq t_o \qquad (12)$$

$$= 0 \qquad \text{for} \quad t < t_o \qquad (13)$$

The frequency spectrum of Eqn. (12) is given by

$$I(\omega) = \int_{t_o} dt_o \; |g(\omega)|^2$$

and the results are the same as given before for the $t_o = 0$ case. The time spectrum of Eqn. (12) is given by,

$$I(t) = \int_{t_o} |E(t,t_o)|^2 \, dt_o$$

and using the conditions of Eqns. (12) and (13), it can be shown that I(t) is actually independent of t in this case. However, this is not the case when the resonant filter is present and the central line is completely absorbed (Fig. 7c). In such a case, Eqn. (12) gets modified to:

$$E'(t,t_o) = e^{-\Gamma(t-t_o)/2\hbar}\{e^{i(\omega_o t + m\sin\omega_{us} t)} - J_o(m)e^{i\omega_o t}\}. \qquad (14)$$

The corresponding time spectrum is,

$$I'(t) = \int_{t_o} |E'(t,t_o)|^2 \, dt_o$$

$$= 1 - J_o^2(m) - 4J_o(m) \sum_{n=1}^{\infty} J_{2n}(m)\cos 2n\omega_{us} t \quad . \qquad (15)$$

Thus I'(t) is time dependent and contains only even harmonics, predominantly the second harmonic (n=1) as observed experimentally (Fig. 8)

3.4 Mossbauer-NAR Double Resonance and Gamma Acoustic Resonance

Recently Gabrielyan et al. (81) have theoretically considered the effect of resonant acoustic field on the Mossbauer spectrum. Taking the example of ^{57}Fe, they consider the case when there is an axially symmetric electric field gradient present which splits the excited nuclear state into two levels, $|\frac{3}{2}, \pm\frac{3}{2}>$ (called level 3) and $|\frac{3}{2}, \pm\frac{1}{2}>$ (called level 2) with energies E_3 and E_2 respectively; $E_3 > E_2$ and $\nu_{23}=(E_3-E_2)/h$ is the frequency difference between the levels 3 and 2. Similarly, calling the ground state, $|\frac{1}{2}, \pm\frac{1}{2}>$, as the level 1, let ν_{12} and ν_{13} be the frequency differences between the levels 1 and 2, and levels 1 and 3 respectively. Gabrielyan et al. (81) consider the case when the frequency of the incident gamma ray, ν_γ , is equal to ν_{12} and the frequency of acoustic vibrations excited in the system, ν_{us} , is equal to ν_{23}. In such a case two possibilities arise:

(a) A gamma photon of frequency ν_{12} is absorbed inducing the transition $1 \rightarrow 2$ and then the nucleus absorbs an acoustic quantum of frequency ν_{23}. This two step process, causing the transition $1 \rightarrow 2 \rightarrow 3$, is the Mossbauer-NAR double resonance. The $2 \rightarrow 3$ is the nuclear acoustic resonance (NAR) transition.

(b) The nucleus simultaneously absorbs a gamma photon (frequency ν_{12}) and an acoustic phonon (frequency ν_{23}) and makes a direct transition from level 1 to level 3. This transition $1 \rightarrow 3$ is a two quantum transition and is called gamma acoustic resonance.

The transition probabilities $W_{1 \rightarrow 3}$ and $W_{1 \rightarrow 2 \rightarrow 3}$ have been calculated as a function of m, θ and φ when the gamma ray direction makes an angle θ with V_{zz} and an angle φ with the direction of ultrasonic excitation. It has been found that for $m \lesssim 1$, $\phi = \frac{\pi}{2}$ and any arbitrary value of θ, $W_{1 \rightarrow 2 \rightarrow 3} >> W_{1 \rightarrow 3}$ whereas for $m \gtrsim 1$, $\theta=0$ and $\phi=0$, $W_{1 \rightarrow 3} >> W_{1 \rightarrow 2 \rightarrow 3}$. The sidebands have significant intensity only for $m \gtrsim 0.5$.

Experimentally, the transitions $W_{1 \rightarrow 3}$ and $W_{1 \rightarrow 2 \rightarrow 3}$ can be detected in a manner similar to the detection of the Mossbauer-NMR double resonance by Heiman et al. (42) (Section 2.3). A single line source mounted on a constant velocity drive can provide gamma rays of frequency ν_{12}. These gamma rays may be allowed to fall on a scatterer tilted at 45° with respect to the gamma ray beam and the spectrum of the scattered radiation at 90° can be analysed by an analysing absorber mounted on a constant acceleration drive. The resonant ultrasonic vibrations are to be excited in the scatterer, which should have only an axially symmetric electric field gradient at its nucleus. If the acoustic vibrations are able to populate the level 3 in the scatterer by inducing either $1 \rightarrow 3$ or $1 \rightarrow 2 \rightarrow 3$ transitions, in the spectrum of the scattered radiation new Mossbauer line will be seen when $\nu_{us} = \nu_{23}$.

Another interesting experiment not yet performed is the Mossbauer-APR double resonance; (APR = acoustic paramagnetic resonance). This experiment in theory and practice will be similar to the Mossbauer-ESR double resonance experiment described in the next section. However unlike the Mossbauer-ESR case, in the Mossbauer-APR experiment $\Delta S_z = \pm 2$ transitions are also allowed. For these studies ($\nu_{us} \sim$ GHz) the piezoelectric transducer like CdS, ZnO, ZnS etc. can be vacuum deposited directly on the absorber or source (7,57). The fact that APR has been observed in compounds containing iron (57,82) makes this proposed experiment quite attractive (83).

4. EFFECT OF MICROWAVE RADIATION ON MOSSBAUER SPECTRUM

In this section we describe an area of Mossbauer research which is yet to be explored experimentally (7). A few theoretical investigations exist (84-86) and an experiment has also been attempted (87) but the effect of microwave radiation, resonant or nonresonant, on the Mossbauer spectrum is yet to be seen in an unambiguous way (7). Discussions given in earlier sections show that the microwave magnetic field may not produce any sideband effect due to its very high frequency (GHz range) as the sideband effect is found to decrease with increasing ν_{rf}. However the microwave field, $\vec{H}_{micro}$, can cause a collapse of the Mossbauer spectrum in a ferromagnetic substance, in the absence of a $H_{ext}>H_{micro}$, in the same way as the rf magnetic field does, provided $H_{micro}>H_{sat}$. Any other effect of a nonresonant microwave field on the Mossbauer spectrum is difficult to predict at this stage. However the effect of a resonant microwave radiation can be predicted somewhat easily. As has been mentioned in Section 1, the resonant microwave radiation, causing electron spin resonance (ESR) transitions among the ionic Zeeman states, affects both the spin population of these states and the rate of spin fluctuation among them. The effect of population change on Mossbauer spectrum is not appreciable (84,87,88) and the main effect is that of the change in the spin fluctuation rate (86,88). It is this latter effect of the ESR radiation which actually makes the Mossbauer-ESR double resonance quite interesting. In (86) a systematic theoretical investigation has been carried out, including both the effect of the population change as well as the change in the fluctuation rate, to study the effect of ESR transitions on Mossbauer spectrum for systems which are either paramagnetic or magnetically ordered. We briefly sketch here the theoretical steps involved in the calculations and discuss the results obtained.

As discussed in Chapter 1 and also at other places in this book, the central problem in calculating the Mossbauer lineshape in presence of spin fluctuation effects is to evaluate the correlation function,

$$G(t) = (\sum_{\lambda} p_{\lambda} < \lambda|\mathcal{H}^{(-)}\ \mathcal{H}^{(+)}(t)|\lambda>)_{av} \tag{16}$$

which appears in the lineshape expression as:

$$I(\omega) = \frac{2}{\Gamma}\ \mathrm{Re} \int_0^{\infty} dt\ \exp(-i\omega t - \frac{1}{2}\Gamma t) . G(t) \ , \tag{17}$$

where "Re" means the "real part". In Eqn. (16), $\mathcal{H}^{(+)}$ is the Hamiltonian causing gamma ray transition from $|\lambda> \rightarrow |\alpha>$, p_{λ} is the probability for the state $|\lambda\rangle$ to occur, $\mathbb{H}^{(-)} = \mathbb{H}^{(+)\dagger}$, $\mathbb{H}^{(+)}(t) = U(t)\, \mathbb{H}^{(+)}\, U^{\dagger}(t)$, $U(t) = \exp[-i \int_0^t \mathbb{H}(t')dt']$, $\mathbb{H}(t)$ is the time dependent hyperfine interaction Hamiltonian and $(\ )_{av}$ denotes the average over the stochastic degrees of freedom in the Hamiltonian. In the Kubo-Anderson-Blume approach (Chapter 1), the random fluctuation of the ionic spin gives rise to a random time dependent hyperfine field at the nuclear site which can be written as,

$$H_{hf}(t) = hf(t),$$

and thus

$$\mathbb{H}(t) = \mathbb{H}_o + P[3I_z^2 - I(I+1)] + g_n\beta_n I_z hf(t), \tag{18}$$

assuming that the off diagonal elements of the hyperfine interaction Hamiltonian are quenched due to the presence of a small magnetic field ($\sim$ 100 Oe for the Fe^{3+} case) at the ionic site (8,86). The Hamiltonian has been written for the case when there is an axially symmetric electric field gradient present at the nucleus in a direction parallel to the fluctuating hyperfine field. We consider here the case of ^{57}Fe nucleus; $P = \frac{1}{12} e^2qQ$. Finally, the correlation function can be written as,

$$G(t) = \frac{1}{2I_{ex}+1} \sum_{m_{ex},m_{gr}} |\langle I_{gr}m_{gr}|\mathbb{H}^{(+)}|I_{ex}m_{ex}\rangle|^2 . \phi(t), \tag{19}$$

where

$$\phi(t) = \{ \exp[i\int_0^t \omega(t')dt'] \}_{av} \quad \text{and} \tag{20}$$

$$\omega(t') = i(g_{ex}m_{ex} - g_{gr}m_{gr})\beta_n h\int_0^t f(t')dt'$$

$$= i\delta\int_0^t f(t')dt' \quad . \tag{21}$$

(In the ^{57}Fe case, $g_{ex} \equiv |g_{ex}|$ owing to the negative sign of g_{ex} for the 14.4 keV state). In the Kubo-Anderson-Blume approach, $f(t)$ is assumed to be a stationary Markoff function which gives:

$$I(\omega) = -\frac{2}{\Gamma} \text{Re} \sum_{m_{ex},m_{gr}} \frac{1}{2I_{ex}+1} . |\langle I_{gr}m_{gr}|\mathbb{H}^{(+)}|I_{ex}m_{ex}\rangle|^2 . \{\vec{W}.\vec{A}^{-1}.\vec{1}\}, \tag{22}$$

where $\vec{1}$ denotes a unit column vector, $\vec{W}$ is a row vector whose elements give the spin distribution over different S_z states and $\vec{A}^{-1}$ is the inverse of the matrix $\overleftrightarrow{A}$ which contains the description of the relaxation process in the form

$$\overleftrightarrow{A} = i\{ \overleftrightarrow{\omega} + [i\Gamma/2 + P(3m_{ex}^2 - 15/4) - \omega]\overleftrightarrow{E} \} + \overleftrightarrow{\pi} \quad . \tag{23}$$

Various symbols are defined in Chapter 1. Taking the example of Fe^{3+} ion and equally spaced Zeeman states,

$$\vec{W} = \frac{1}{Z} [s^5 \; s^4 \; s^3 \; s^2 \; s \; 1] \quad , \tag{24}$$

where $Z = \Sigma_{y=0}^{5} s^y$ is the partition function and $s = \exp(-\frac{\Delta}{k_B T})$, Δ being the separation between the successive Zeeman states. In the paramagnetic case and when $g\beta H_{ext} << k_B T$, $\Delta \sim 0$ and $s \sim 1$. In a magnetically ordered system, $\Delta > 0$ and $s < 1$. In the case of unequally spaced levels, the powers of s are to be calculated properly. The elements π_{ij} of the matrix $\overleftrightarrow{\pi}$ for the transitions $|i\rangle \rightarrow |j\rangle$, in which the ionic spin emits an energy quantum Δ, are obtained by multiplying the quantum mechanical transition probability, $|\langle S_j|S_{\pm}|S_i\rangle|^2$ or in the case of unequally spaced levels $|\langle S_j|S_{\pm}|S_i\rangle|^2 . |\langle S_i|S_{\mp}|S_j\rangle|^2$ (89), by a parameter Ω_e called the spin flip frequency. Here we assume a spin spin relaxation case with selection rule $\Delta S_z = \pm 1$; this is actually the case for an Fe^{3+} ion (8). For absorptive transitions, $|j\rangle \rightarrow |i\rangle$, π_{ji} are obtained by replacing Ω_e by Ω_a where $\Omega_a = s\Omega_e$. The diagonal elements π_{ii} are given by $\Sigma_j \pi_{ij} = 0$ which is a result of the stationary Markoff nature of f(t). An average relaxation time or 'spin correlation time' is defined as

$$\tau = \frac{1}{q(\Omega_a + \Omega_e)} = \frac{1}{q(1 + s)\Omega_e}$$

where the denominator is the sum of all the transition probabilities divided by the total number of transitions. For the Fe^{3+} case, $q = 7$. In the presence of ESR transitions, between $|i\rangle$ and $|j\rangle$, the transition probability π_{ij} gets enhanced and the spin distribution over the different Zeeman states gets altered. In such a case, the elements of matrix $\overleftrightarrow{\pi}$ are obtained by replacing Ω_e by $\Omega_e + \Omega_o$ and Ω_a by $\Omega_a + \Omega_o$, where the magnitude of Ω_o depends on the applied microwave power. The elements of $\vec{W}$ are calculated by writing the proper rate equations (90) in the presence of ESR transitions. For example, when the microwave radiation is inducing ESR transitions among all the successive Zeeman states of the Fe^{3+} ion, Eqn. (24) gets modified to:

$$\vec{W} = \frac{1}{\Sigma_{y=0}^{5} p^y} [p^5 \; p^4 \; p^3 \; p^2 \; p \; 1] \quad ,$$

where

$$p = \left(\frac{s\Omega_e + \Omega_o}{\Omega_e + \Omega_o} \right) .$$

The effect of selective ESR transitions on Mossbauer relaxation spectra are shown in Fig. 9 when the six Zeeman states of the Fe^{3+} ion are involved in the relaxation process (n=6). The lineshapes are given for ESR transitions

(a) $|-5/2\rangle \rightleftarrows |-3/2\rangle$, (b) $|-3/2\rangle \rightleftarrows |-1/2\rangle$, (c) $|-1/2\rangle \rightleftarrows |1/2\rangle$,

(d) $|1/2\rangle \rightleftarrows |3/2\rangle$, (e) $|3/2\rangle \rightleftarrows |5/2\rangle$, (f) $|-3/2\rangle \rightleftarrows |-1/2\rangle$ and

$|1/2\rangle \rightleftarrows |3/2\rangle$ simultaneously, and (g) among all the levels simultaneously when (A) $s = 1$, $\Omega_e/\Gamma = 3.0$, $P/\Gamma = 2.0$, $\Omega_o/\Gamma = 28.0$ (Fig. 9A) and (B) $s = 0.5$, $\Omega_e/\Gamma = 6.0$, $P/\Gamma = 2.0$ and $\Omega_o/\Gamma = 28.0$ (Fig. 9B). These selective transitions, viz. (a) to (f), are possible only when the S_z

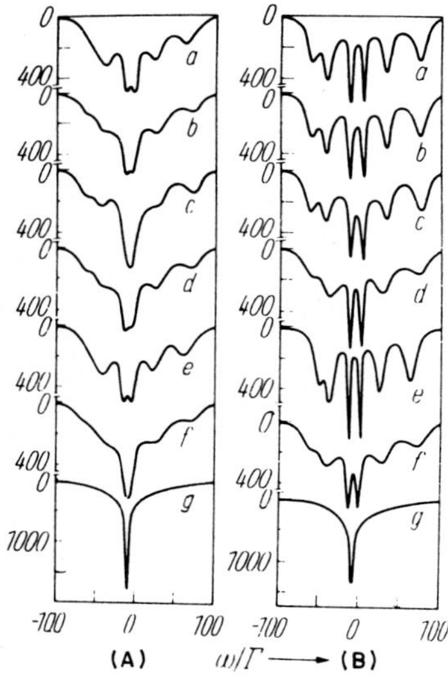

Fig. 9. Effect of selective ESR transitions on Mossbauer relaxation spectra. (A) n = 6, s = 1 and (B) n = 6, s = 0.5.

states are unequally spaced and therefore give some idea about the crystal field splitting of the ionic ground state. We have considered here the case when $\Omega_e \sim \nu_L$, where ν_L , the nuclear Larmor precession frequency, is $\sim\Gamma$. The effect of ESR transitions, particularly $|1/2\rangle \rightleftarrows |-1/2\rangle$, on the Mossbauer spectrum is much more pronounced for the case when $\Omega_e \ll \nu_L$ and $s \sim 1$ (88). In this case a paramagnetic hyperfine structure is found to be present in the Mossbauer spectrum when Ω_o = 0 (91). The situation of Fig. (9B), viz. s < 1, corresponds to the case of a magnetically ordered system. In this case the Zeeman states are mainly split by the Weiss molecular field and the use of $\vec{H}_{ext}$, which is at right angles to $\vec{H}_{micro}$, is only to magnetise the sample in its direction. The fact that the ESR transitions have been observed in magnetically ordered systems by Tinkham and his co-workers (92) makes this kind of experiments very attractive. It has been pointed out (86) that the Mossbauer-ESR double resonance experiment can also be used to distinguish between the electronic relaxation and the superparamagnetic effects. It is difficult to distinguish between these effects as they give rise to similar Mossbauer lineshapes. As a result there has been some doubt about attributing the observed anomalous Mossbauer lineshapes to either of these effects in systems like mixed ferrites in magnetically ordered state (86,93). Recently the mechanism of domain wall oscillation has also been invoked to explain the anomalous Mossbauer lineshapes observed in mixed ferrite and like systems (94). A Mossbauer-ESR experiment can provide a conclusive test for the validity of this mechanism also. The central doublet of the anomalous spectrum, observed at room temperature (94), is expected to remain undisturbed by the ESR radiation if it arises due to the domain wall oscillations (88). Fig. 10 shows the Mossbauer lineshapes, in presence of ESR transitions, calculated for a system where the anomalous lineshape is due to (a) the electric relaxation effects (Fig. 10a) and (b) the superparamagnetic effects (Fig. 10b). In the latter case, the effect of ESR transitions on the lineshape mainly arises due to the change in the elements of matrix $\overleftrightarrow{\omega}$ in Eqn. (23) (86). For Fig. (10a), which belongs to a system where the electronic relaxation effects are present, it has been assumed that n = 2, s = 1, P = 0.0, Ω_e/Γ =10.0 and for Fig. (10b), which is for a superparamagnetic system, n = 2, s = 0.5, P = 0.0, Ω_e/Γ = 5 x 10^4, n' = 2, s' = 1 and Ω'_e/Γ = 10.0 (86). Here n' is the number of easy directions of magnetisation of the superparamagnetic cluster, s' is the probability of the cluster magnetisation to be along a particular easy direction and Ω'_e is the flip frequency for the cluster magnetisation; n' = 2 means that the cluster has an uniaxial anisotropy (Chapter 1). The difference between the two sets of spectra is quite evident. This difference exists in other cases also (for example n = 6, s < 1). A general conclusion is that in all the cases, there is almost no change in the lineshape in a superparamagnetic system till Ω_o/Γ increases to $\sim 10^4$. A change can thus occur only near the ESR line saturation. It may be mentioned here that the combined Mossbauer-ESR experiment is an improvement over the ESR technique for

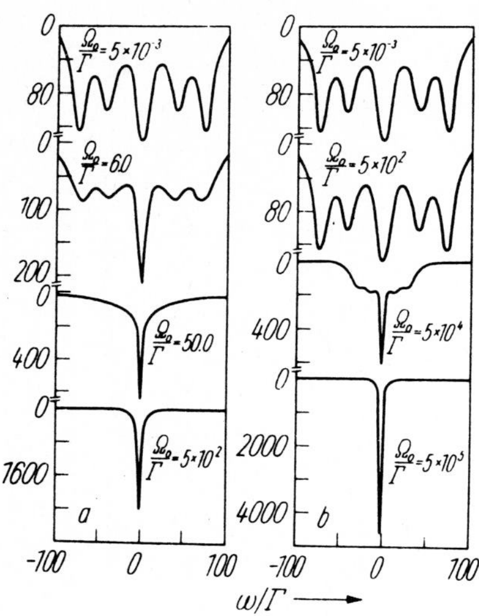

Fig. 10. Dependence of Mossbauer lineshape on the applied microwave power for a system where (a) electronic relaxation effects are present (n=2, s=1) and (b) superparamagnetic effects are present (n'=2, s'=1).

measuring the electronic relaxation time, since the conventionally used saturation recovery and spin-echo techniques in ESR experiments cannot measure the relaxation times as short as the Mossbauer-ESR experiment can do due to certain limitations such as the time response of the recording system (95).

There are many experiments similar to the Mossbauer-ESR double resonance which look promising. In magnetically ordered systems, it is necessary to work near the magnetic transition temperature (T_C or T_N) to bring the ESR frequency in the microwave (GHz) range by decreasing the Weiss molecular field. This makes it necessary to have a very accurate temperature control because of the Brillouin function dependence of the Weiss field on temperature. This problem can be avoided by using infrared radiation instead of microwave radiation and by working at temperatures much below T_C(or T_N) where the Weiss field is quite large and is weakly temperature dependent. Tinkham and his co-workers (92) have observed the ESR transition in magnetically ordered substances in this way only by using the infrared radiation. This approach is similar in concept to the laser magnetic resonance (LMR) (96) and we can call such Mossbauer-ESR experiments as Mossbauer-LMR experiments. Another advantage of using a laser source, which can be a tunable dye laser, is the availability of large radiation flux which may be required in certain studies. In these experiments, instead of infrared radiation, microwave radiations of two different frequencies can also be simultaneously used provided the sum of the two microwave frequencies is equal to the desired ESR frequency (97). It is also possible to use infrared radiation and microwave radiation simultaneously to induce resonant transitions when the ESR frequency is quite large at low temperatures (98). Similarly studying Mossbauer lineshape in presence of ferromagnetic, ferrimagnetic or antiferromagnetic resonance, or electron-nuclear double resonance (ENDOR) can also be quite interesting (99,100). Other studies of this type are combination of Mossbauer experiment with electron-electron double resonance (ELDOR), a Mossbauer-ELDOR triple resonance experiment, for the study of cross relaxation phenomenon (101). Many of these combined experiments have the advantage of being capable of measuring very short relaxation times in the same way as a Mossbauer-ESR experiment is.

5. EFFECT OF OPTICAL RADIATION ON MOSSBAUER SPECTRUM

In this section, like earlier sections, we discuss only those studies where the perturbing radiation (optical) field is presented in the source or absorber during the Mossbauer measurement. We shall not thus consider here the experiments like Mossbauer studies of photochromism where the optical radiation of a particular wavelength changes the charge state of the Mossbauer ion and cases in which the initial charge state can be restored by irradiation with the light of another wavelength (102) or optical detection of Mossbauer resonance where the resonant absorption of Mossbauer gamma rays enhances the optical fluorescence in certain systems (103). One of the experiments which falls in the scope of this section, is the suggestion of gamma-optical double resonance (104). In this proposed experiment, the Mossbauer gamma rays are passed through a gas containing the Mossbauer nuclei. A resonant laser beam, saturating one of the optical transitions of the Mossbauer atoms, is also simultaneously passed through the gas along the gamma ray direction (parallel or antiparallel). A third laser beam photoionises the optically excited atoms which are then removed from the gas by a small d.c. electric field. Thus a lack of atoms, absorbing the gamma radiation incident from the Mossbauer source, arises when the laser frequency coincides with the resonance frequency of the optical transition in the gas where significant thermal motion of the atoms is present. As a result a dip in the gamma ray absorption is observed when the transmitted counts are recorded as a function of the laser frequency. The advantage of this method is that it enables one to observe gamma ray resonance in the neighbourhood of Mossbauer transition in the range

$$\Gamma_a/\nu_a < (\frac{\nu_{nucl}}{\nu_\gamma} - 1) < \frac{(\delta\omega_{Dopp})_{opt}}{\nu_a}$$

which is not possible by the existing method; Γ_a is the levelwidth of the excited atomic state, ν_a and $(\delta\omega_{Dopp})_{opt}$ respectively are the frequency and Doppler width of the atomic transition, ν_{nucl} is the frequency of the nuclear transition and ν_γ is the gamma ray frequency.

Another important experiment is the microwave-optical-Mossbauer triple resonance (MOMTR) proposed recently (99). This is essentially a Mossbauer resonance study in the presence of microwave-optical double resonance (MODR) (101, 105), i.e. in presence of the ESR transitions among the Zeeman states of the optically excited ionic level. The experiment basically aims at studying the electronic relaxation phenomenon for the excited ionic state though, as has been pointed out (99), similar to the combined Mossbauer-ESR experiment, this experiment can also unambiguously distinguish between the electronic relaxation and the superparamagnetic effects. Many examples of optical excitation in Fe^{2+}/Fe^{3+}-compounds are known (106) and we briefly sketch here the principle of the MOMTR experiment by considering the example of Fe^{3+} ions diffused in a trigonally distorted lattice like ZnS(107) or $LiAl_5O_8$ (108). Such systems have been studied in detail by the conventional Mossbauer and ESR techniques and the necessary data for them exist (53,91,109). Several studies have been made for the optical absorption (reflectance) and emission in axially distorted Fe^{3+}-compounds (106-108,110) and these studies show that when the Fe^{3+} (and also the isoelectronic Mn^{2+}) ion has an octahedral co-ordination in the host lattice, it gives rise to a red luminescence at about 6800 A° whereas when it has a tetrahedral co-ordination, a green luminescence occurs at about 5000 A° (108) owing to the transitions between the ground state and the first excited state of the ion. Fig. 11 shows the energy level diagram of Fe^{3+} ion in a trigonally distorted lattice. Various transition energies marked in the figure are approximately true for a system where the Fe^{3+} ion has a tetrahedral co-ordination (107). Under the trigonal distortion of the crystal field, the first excited state, ^4G, of the Fe^{3+} ion splits into five levels whereas the ground state remains unperturbed. The observed green luminescence at ∿5000 A° and red luminescence at ∿6800 A° have been recognised to be arising from the $^4A_2 \rightleftarrows {}^6A_1$ electric dipole transition (107, 108).

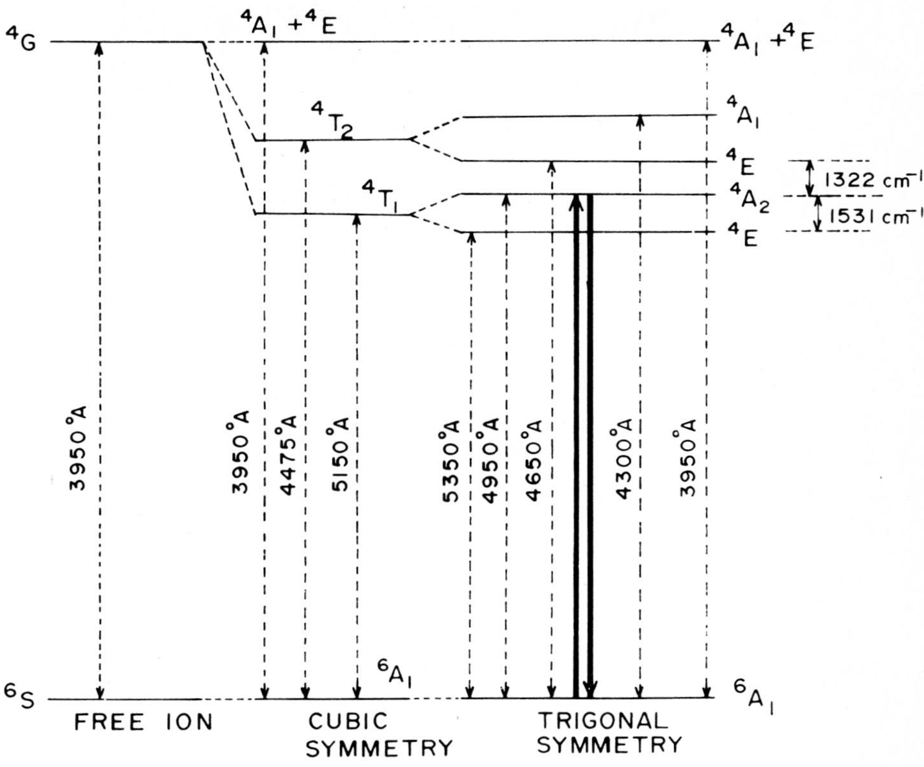

Fig. 11. Approximate positions of crystal field split energy levels of Fe^{3+} ion in a tetrahedrally coordinated environment.

This optical transition is quite suitable for the MOMTR experiment and we consider only this transition in this discussion. Fig. 12 shows the splitting of the crystal field states $^4A_2$ and $^6A_1$ due to the presence of the spin-orbit coupling and magnetic field (either external or internal i.e. Weiss molecular field). For the magnetically ordered case, we assume the presence of a small polarising external magnetic field which provides the quantisation direction for the ionic spin whether in the excited state ($^4A_2$) or in the ground state ($^6A_1$). Since both the $^4A_2$ and the $^6A_1$ states are orbital singlets, first order spin orbit coupling has no effect on them and the magnetic field splits them into four and six levels (S_z-states) respectively. If the magnetic field is small, these levels may be unequally spaced owing to the effect of second order spin orbit coupling whose effect on the splitting of these levels is ~ 1 cm^{-1}. Owing to their singlet nature, $^4A_2$ and $^6A_1$ states do

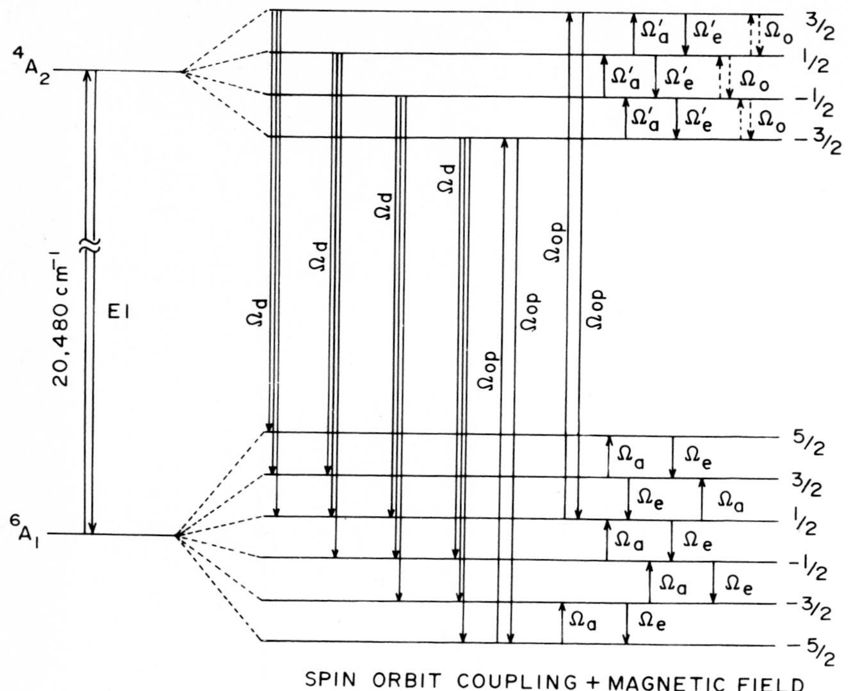

Fig. 12. Splittings of $^4A_2$ and $^6A_1$ states. Various transitions shown here are described in the text.

not contribute to the electric field gradient at the nucleus which, thus, arises only from the lattice ions. Assuming that $\vec{H}_{ext}$ is along the symmetry axis [111], we have a situation where the Hamiltonian of Eqn. (18) is valid and the Mossbauer lineshape is given by Eqns. (22) and (23).

As shown in Fig. (12), the S_z-states of $^4A_2$ are coupled with the S_z-states of $^6A_1$ through optical transitions (Ω_{op}). Various kinds of transitions taking place among the different S_z-states are shown in Fig. 12. The nucleus senses the effect of all these shake-ups through the random time dependence of the hyperfine field. In Fig. 12, Ω_a' and Ω_e' are the same quantities for the excited state which are denoted by Ω_a and Ω_e respectively for the ground state. Ω_{op} is the spin flip frequency corresponding to the spin fluctuations arising due to the stimulated optical transitions and is proportional to the input optical power. The quantity Ω_d is the decay frequency arising from the natural decay of the excited state and is proportional to $1/\tau$ where τ, the lifetime of the $^4A_2$ state, is $\sim 10^{-8}$ sec. Any multiphonon relaxation effect (111), inducing $^4A_2 \rightleftarrows {^6A_1}$ transition, has been neglected owing to the orbital singlet nature of the $^4A_2$ and $^6A_1$ states. $\vec{W}$ is calculat by writing the proper rate equations (90) for all the S_z-states and by solving the simultaneous equations so obtained numerically on a computer. Elements of $\overleftrightarrow{\pi}$ are calculated as before by incorporating Ω_a, Ω_e, Ω_{op}, Ω_d, Ω_o, Ω'_a and Ω'_e properly into the calculations and taki care of various selection rules (8,99). The results are shown in Figs. (13) and (14), where P = 0.0. The effects of Ω_o and Ω_{op} on Mossbauer lineshape are very clearly seen. Thus first without inducin

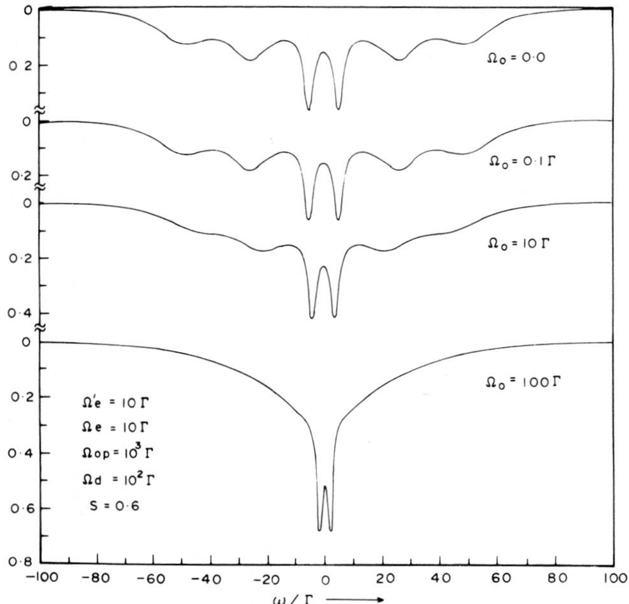

Fig. 13. Effect of excited state ESR on Mossbauer spectrum.

the resonant optical and microwave transitions, Ω_e can be obtained from the shape of the Mossbauer spectrum. Once Ω_e is known, optical and microwave frequencies can be tuned to resonance and, as is evident from Figs. (13) and (14), Ω'_e can be estimated since Ω_o and Ω_{op} are known quantities. This gives the electronic relaxation time of the excited state. It may be mentioned here that like other combined experiments, MOMTR is an improvement over the MODR technique for measuring the electronic relaxation time of the excited state. MOMTR experiments can also be done when the ionic ground state or excited state or both are no longer orbital singlets and also when the ESR transitions are taking place simultaneously among the S_z-states of both the ground

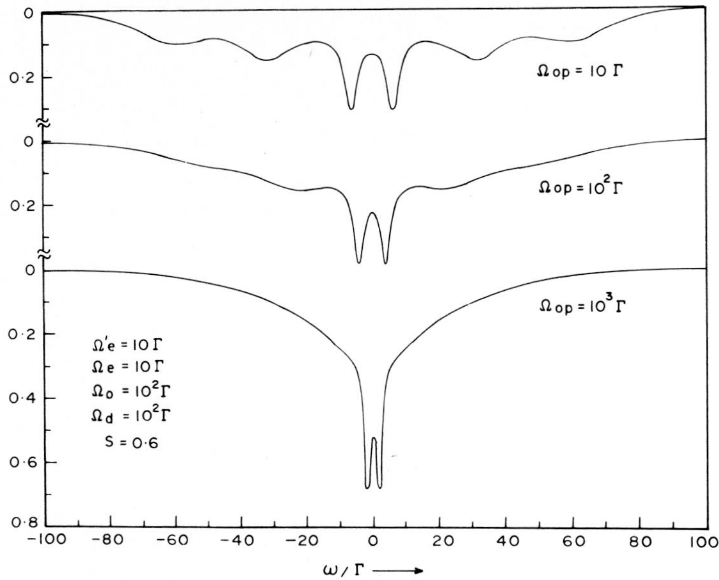

Fig. 14. Effect of optical power on Mossbauer spectrum for a given microwave power causing ESR transitions in excited ionic state.

as well as the excited ionic states (88). When the excited state of the ion is not an orbital singlet, phenomena like Jahn-Teller relaxation (Chapter 1) of the excited state can be studied by the MOMTR technique. The case of $\Omega_o = 0$ in Fig. 13 represents the situation of Mossbauer-optical double resonance (84). It has been found that in a MOMTR experiment (i.e. $\Omega_o \neq 0$), the Mossbauer lineshape is more sensitive to Ω_e or Ω_e' than it is in a Mossbauer-optical double resonance experiment (i.e. when $\Omega_o = 0$) (8,88). The rare earth systems, for which enough optical resonance studies exist (108,112-114), are also good candidates for the MOMTR experiments. Fig. 15 shows the energy level diagram of Gd^{3+} ion in the hexagonal $LaCl_3$ lattice (114). The observed optical transitions, shown by the full line arrows, are situated in the ultraviolet region. Any of these transitions can be conveniently used for observing the triple resonance.

There are many more studies which can be performed by a combined Mossbauer-optical experiment. These include studies like laser induced nuclear orientation, recoil induced gamma ray sidebands (laser compensation of gamma ray recoil energy loss), etc. (104,115).

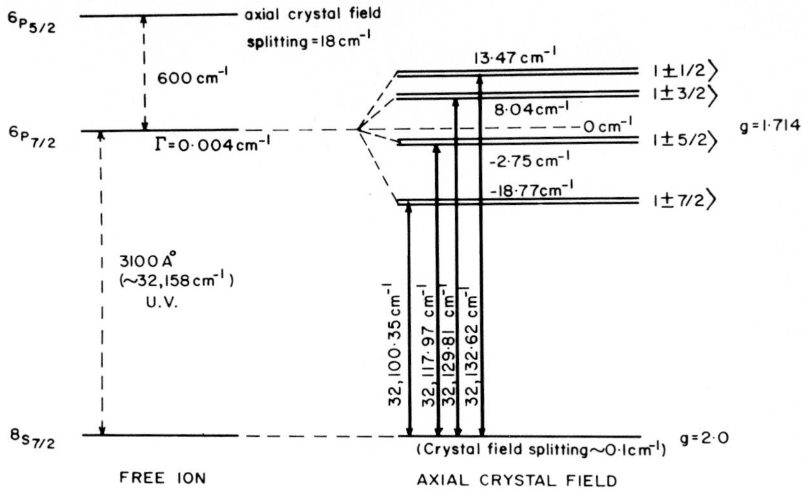

Fig. 15. Energy level diagram of a Gd^{3+} ion in hexagonal $LaCl_3$ lattice.

6. SUMMARY

In conclusion, the radiation perturbation of Mossbauer resonance is a rich field in which at present more number of theoretical predictions exist than the number of experiments actually performed. Also though in some cases the earlier predictions have been experimentally verified, in many cases where the experiments have been done an adequate theoretical understanding of the phenomenon is still lacking (as for example is seen in Section 2.2). Moreover, in some cases even the experimental situation does not seem to be very clear (see Section 2.1, (19,7,9,10)). It appears that in the coming years, more and more combined experiments will be carried out in the microwave region (ultrasonic field or electromagnetic radiation field perturbation) which is still unexplored. Due to the availability of the high power tunable laser radiation, the combined Mossbauer-optical experiments offer another attractive unexplored area having scope for discovering many new phenomena.

ACKNOWLEDGEMENT

The author is thankful to Dr. V.R. Marathe for critically going through the manuscript.

REFERENCES

1. For example, see Chapter 1 of the this book.
2. For example, see the Chapter of G.K. Shenoy in this book.
3. J. Kondo, Progr. Theor. Phys., 32 (1964) 37; R. Orbach, in C.D. Graham, Jr., G.H. Lander and J.J. Rhyne (Eds.), A.I.P. Conf. Proc. No. 24: Mag. and Mag. Materials - 1974, A.I.P., New York, 1975, p. 3; B. Golding, Phys. Rev. Letters 20 (1968) 5; G.E. Laramore and L.P. Kadanoff, Phys. Rev., 187 (1969) 619; R. Hasegawa, Phys. Lett., 38A (1972) 5; I. Riess and A. Ron, Phys. Rev. B, 4 (1971) 4099.
4. F. van der Woude and A.J. Dekker, Phys. Status Solidi, 9 (1965) 775; S.C. Bhargava and N. Zeman, Phys. Rev. B, 21 (1980) 1717; J.K. Srivastava and B.W. Dale, Phys. Status Solidi (b), 90 (1978) 391.
5. J.A. Mydosh, in C.D. Graham, Jr., G.H. Lander and J.J. Rhyne (Eds.), A.I.P. Conf. Proc. No. 24: Magnetism and Magnetic Materials - 1974, A.I.P., New York, 1975, p. 131; J. Villain, Z. Physik B, 33 (1979) 31; W. Kinzel, ibid, 46 (1982) 59; Also see the Chapter of T.E. Cranshaw in this book.
6. J.K. Srivastava and K.G. Prasad, Phys. Status Solidi (b), 56 (1973) K117; J.K. Srivastava and R. Nagarajan, Proc. Nucl. Phys. and Solid State Phys. Symp. (India) 19C (1976) 347; U.H. Kopvillem and L.N. Shakhmuratova, Phys. Lett., 42A (1972) 63; D. Spanjaard and F. Hartmann - Boutron, J. Phys. (Paris), 33 (1972) 565; H. Kempter and E. Klein, Z. Physik, A281 (1977) 341; J. Bosse and H. Gabriel, Phys. Rev. B, 5 (1972) 4269; P.M. Chaikin, Solid State Commun., 12 (1973) 35; C. Rettori, D. Davidov, P. Chaikin and R. Orbach, Phys. Rev. Lett., 30 (1973) 437; U. Engel, K. Baberschke, G. Koopmann, S. Hufner and M. Wilhelm, Solid State Commun., 12 (1973) 977; D. Davidov, A. Chelkowski, C. Rettori, R. Orbach and M.B. Maple, Phys. Rev. B, 7 (1973) 1029; W. Meisel and L. Keszthelyi, Hyperfine Interact., 3 (1977) 413; E. Matthias, D.A. Shirely, M.P. Klein and N. Edelstein, Phys. Rev. Lett., 16 (1966) 974; E. Matthias, in E. Matthias and D.A. Shirley (Eds.), Hyperfine Structure and Nuclear Radiations, North-Holland, Amsterdam, 1968, p. 815; D.A. Shirley, ibid, p. 843; W.J. Ince, Phys. Rev., 184 (1969) 574.
7. L. Pfeiffer, J. Phys. (Paris) Colloq., 35 (1974) C1-67; J. Mishory and D.I. Bolef, in I.J. Gruverman (Ed.), Mossbauer Effect Methodology, Vol. 4, Plenum, New York, 1968, p. 13; N.D. Heiman, J.C. Walker and L. Pfeiffer, ibid, Vol. 6, 1971, p. 123; L. Pfeiffer, ibid, Vol. 7, 1971, p. 263.
8. J.K. Srivastava, Phys. Status Solidi (b), 50 (1972) K21; 46 (1971) K93; 55 (1973) K119; 97 (1980) K123; R. Nagarajan and J.K. Srivastava, ibid, 81 (1977) 107; J.K. Srivastava and R.P. Sharma, J. Phys. (Paris) Colloq., 35 (1974) C6-663.
9. L. Pfeiffer, J. Appl. Phys., 42 (1971) 1725.
10. M. Kopcewicz, Phys. Status Solidi (a), 60 (1980) 43.
11. M. Kopcewicz, Solid State Commun., 19 (1976) 719.
12. M. Kopcewicz, Phys. Status Solidi (a), 46 (1978) 675.
13. M. Kopcewicz, U. Gonser and H.G. Wagner, Appl. Phys., 23 (1980) 1.
14. G. Karczewski, M. Kopcewicz and A. Kotlicki, J. Phys. (Paris) Colloq., 41 (1980) C1-217.
15. M. Kopcewicz and G. Karczewski, J. Phys. (Paris) Colloq., 41 (1980) C1-215.
16. M. Kopcewicz, J. Phys. Chem. Solids, 42 (1981) 77.
17. G.J. Perlow, Phys. Rev., 172 (1968) 319; Yu. V. Baldokhin, E.F. Makarov, A.V. Mitin and V.A. Povicki, Proc. 5th Int. Conf. on Mossbauer Spectroscopy, Bratislava, 1973, Czechoslovak Atomic Energy Commission, Praha, 1975, p. 609.

18. M. Kopcewicz and A. Kotlicki, J. Phys. Chem. Solids, 41 (1980) 631.
19. M. Kopcewicz, U. Gonser and H.G. Wagner, Paper presented at the Int. Conf. on the Applications of the Mossbauer Effect, Jaipur (India), December 14-18, 1981, Abstract No. Fri-P_V-RE-2; A. Kotlicki, Hyperfine Interact., 10 (1981) 1167; G. Albanese, G. Asti and S. Rinaldi, Nuovo Cimento, 6B (1971) 153.
20. A. Asti, G. Albanese and C. Bucci, Phys. Rev., 184 (1969) 260.
21. M. Kopcewicz, A. Kotlicki and M. Szefer, Phys. Status Solidi (b), 72 (1975) 701.
22. C.L. Chien, Phys. Rev. B, 18 (1978) 1003.
23. R. Hasegawa and C.L. Chien, Solid State Commun., 18 (1976) 913.
24. C.L. Chien, R. Hasegawa and J.C. Walker, AIP Conf. Proc., 24 (1975) 127.
25. L. Pfeiffer, AIP Conf. Proc., 5 (1971) 796; M. Chmielowski, A. Kotlicki and A. Wojtasiewicz, Hyperfine Interact., 10 (1981) 1133.
26. Yu. V. Baldokhin, V.A. Marakov, E.F. Makarov and V.A. Povitskii, Phys. Status Solidi (a), 27 (1975) 265.
27. C.L. Chien and J.C. Walker, Phys. Rev. B, 13 (1976) 1876.
28. N. Heiman, R.K. Hester and S.P. Weeks, Phys. Rev. B, 8 (1973) 3145.
29. N.D. Heiman, L. Pfeiffer and J.C. Walker, Phys. Rev. Lett., 21 (1968) 93.
30. G.J. Perlow, in H. Frauenfelder and H. Lustig (Eds.), Allerton House Conf., Univ. of Illinois Rept. (AFOSRTN 0-98), 1960 (Unpublished); E. Matthias, in E. Matthias and D.A. Shirley (Eds.), Hyperfine Structure and Nuclear Radiations, North-Holland, Amsterdam, 1968, p. 815; G.J. Perlow, Phys. Rev., 172 (1968) 319.
31. L. Pfeiffer, N.D. Heiman and J.C. Walker, Phys. Rev. B, 6 (1972) 74.
32. R.M. Bozorth, Ferromagnetism, D. Van Nostrand, New York, 1955.
33. H. Frauenfelder, The Mossbauer Effect, W.A. Benjamin, New York, 1962, pp. 17-20, 65-69, and the papers reprinted there; G.K. Wertheim, Mossbauer Effect: Principles and Applications, Academic, New York, 1964, p. 108; F.D. Murnaghan, Introduction to Applied Mathematics, John Wiley, New York, 1948, p. 337.
34. S.L. Ruby and D.I. Bolef, Phys. Rev. Lett., 5 (1960) 5.
35. C.L. Chien and J.C. Walker, in A.Z. Hrynkiewicz and J.A. Sawicki (Eds.), Proc. Int. Conf. Mossbauer Spectroscopy, Vol. 1, Cracow (Poland) 1975, Wykonano w. Powielarni Akademii Gorniczo - Hutniczej im. S. Staszica, Cracow, 1975, p. 47.
36. I.A. Dubovtsev, P.S. Zyryanov and N.P. Fillippova, Sov. Phys. - JETP, 38 (1974) 509; M. Chmielowski and A. Kotlicki, Hyperfine Interact., 10 (1981) 1129; S. Olariu, I. Popescu and C.B. Collins, Phys. Rev. C, 23 (1981) 1007; A.V. Mitin, Sov. Phys. - JETP, 25 (1967) 1062; Sov. Phys. - Doklady, 15 (1971) 827.
37. P.J. West and E. Matthias, Z. Physik A, 288 (1978) 369.
38. L. Pfeiffer and C.P. Lichtenwalner, Rev. Sci. Instrum., 44 (1973) 1500.
39. G. Cain, Phys. Lett., 38A (1972) 279.
40. G.J. Cain, J.A. Barclay and J.D. Cashion, J. Low Temp. Phys., 19 (1975) 513; S.V. Vonsovskii, Magnetism, Vol. II, Keter Pub. House, Jerusalem (John Wiley, New York), 1974, p. 1203; J.M. Winter, J. Phys. Radium, 23 (1962) 556.
41. R. Laurenz, E. Klein and W.D. Brewer, Z. Physik, 270 (1974) 233.
42. N.D. Heiman, J.C. Walker and L. Pfeiffer, Phys. Rev., 184 (1969) 281.
43. V.K. Voitovetskii, S.M. Cheremisin and S.B. Sazonov, Phys. Lett., 83A (1981) 81; V.K. Voitovetskii, S.M. Cheremisin, S.B. Sazonov and I.V. Kurchatov, J. Phys. (Paris) Colloq., 41 (1980) C1-467.
44. S.S. Yakimov, A.R. Mkrtchyan, V.N. Zarubin, K.V. Serbinov and V.V. Sergeev, JETP Lett., 26 (1977) 13.

45. M.N. Hack and M. Hammermesh, Nuovo Cimento, 19 (1961) 546.
46. H. Gabriel, Phys. Rev., 184 (1969) 359.
47. B. Krishnamurthy and K.P. Sinha, Phys. Status Solidi (b), 55 (1973) 427.
48. B. Krishnamurthy and K.P. Sinha, J. Magn. Resonance, 17 (1975) 189.
49. Sh. Sh. Bashkirov and E.K. Sadykov, Sov. Phys. - Solid State, 20 (1978) 1988.
50. Sh. Sh. Bashkirov, A.L. Belyanin and E.K. Sadykov, Phys. Status Solidi (b), 93 (1979) 437.
51. A.N. Artemev, G.V. Smirnov and E.P. Stepanov, Sov. Phys. - JETP, 27 (1968) 547.
52. A.M. Portis and R.H. Lindquist, in G.T. Rado and H. Suhl (Eds.), Magnetism Vol. IIA, Academic, New York, 1965, p. 357.
53. L. Epstein and A. Wachtel, Appl. Phys. Lett., 10 (1967) 246; J.K. Srivastava and K.G. Prasad, Phys. Lett., 40A (1972) 37; G.K. Wertheim and J.P. Remeika, in L.V. Gerven (Ed.), Nuclear Magnetic Resonance and Relaxation in Solids (Proc. 13th Colloque Ampere), North-Holland, Amsterdam, 1965, p. 1.
54. H.H. Wickman and G.K. Wertheim, in V.I. Goldanskii and R.H. Herber (Eds.), Chemical Applications of Mossbauer Spectroscopy, Academic, New York, 1968, p. 548.
55. A.R. Arakelyan, G.A. Arutyunyan, R.G. Gabrielyan, L.A. Kocharyan, A.R. Mkrtchyan and G.N. Nadzharyan, Sov. Phys. - Acoustics, 24 (1978) 461.
56. L.T. Tsankov, Nucl. Instrum. Methods, 160 (1979) 195.
57. J.W. Tucker and V.W. Rampton, Microwave Ultrasonics in Solid State Physics, North-Holland, Amsterdam, 1972.
58. V.A. Burov, V.A. Krasilnikov and O. Yu. Sukharevskaya, Sov. Phys. - JETP, 16 (1963) 837; H.E. Bommel, in D.M. Compton and A.H. Schoen (Eds.), Proc. Second Int. Conf. on Mossbauer Effect, Saclay, France (1961), John Wiley, New York, 1962, p. 229.
59. T.E. Cranshaw and P. Reivari, Proc. Phys. Soc. (London), 90 (1967) 1059.
60. J. Asher, T.E. Cranshaw and D.A. O'Connor, J. Phys. A: Math., Nucl. Gen., 7 (1974) 410.
61. A.R. Mkrtchyan, A.F. Arakelyan, G.A. Arutyunyan and L.A. Kocharyan, JETP Lett., 26 (1977) 449.
62. T.M. Aivazyan, I.M. Aivazyan, A.R. Mkrtchyan and L.A. Kocharyan, Phys. Status Solidi (b), 64 (1974) 757.
63. A.R. Mkrtchyan, G.A. Arutyunyan, A.R. Arakelyan and R.G. Gabrielyan Phys. Status Solidi (b), 92 (1979) 23.
64. J.D. Cashion and P.E. Clark, J. Phys. (Paris) Colloq., 40 (1979) C2-44.
65. G.J. Perlow, Phys. Rev. Lett., 40 (1978) 896; G.J. Perlow, J.E. Monahan and W. Potzel, J. Phys. (Paris) Colloq., 41 (1980) C1-85.
66. A.R. Mkrtchyan, L.A. Kocharyan, A.R. Arakelyan and L.G. Karamyan, Sov. Phys. - Solid State, 19 (1977) 1067.
67. S. Goldman, Frequency Analysis, Modulation and Noise, McGraw Hill, New York, 1948.
68. G. Kornfeld, Phys. Rev., 177 (1969) 494.
69. A. Abragam, L'Effect Mossbauer, Gordon and Breach, New York, 1964, pp. 22-24; A. Abragam, Compt. Rend., 250 (1960) 4334; A. Abragam, in C. DeWitt, B. Dreyfus and P. de Gennes (Eds.), Low Temperature Physics, Gordon and Breach, New York, 1962, p. 481.
70. E.K. Sadykov, Sov. Phys. - Solid State, 19 (1977) 963; E.F. Makarov and A.V. Mitin, Sov. Phys. - Uspekhi, 19 (1976) 741.
71. M.A. Andreeva, R.N. Kuzmin and S.F. Oparina, Phys. Status Solidi (b), 71 (1975) K143.
72. A.V. Mitin, Sov. Phys. - Solid State, 20 (1978) 941.

73. G.J. Perlow, W. Potzel, R.M. Kash and H. de Waard, J. Phys. (Paris) Colloq., 35 (1974) C6-197.
74. D.I. Bolef and J. Mishory, Appl. Phys. Lett., 11 (1967) 321.
75. A.R. Buev, Sov. Phys. - Solid State, 18 (1976) 1046.
76. T.M. Aivazyan, Yu. M. Aivazyan and A.V. Khachoyan, Sov. Phys. - Solid State, 19 (1977) 1887.
77. P.A. Fedders, I. Wu, J.G. Miller and D.I. Bolef, Phys. Rev. Lett., 32 (1974) 1443.
78. R.G. Gabrielyan and A.R. Mkrtchyan, Phys. Status Solidi (b), 99 (1980) K75.
79. L. Tsankov and Ts. Bonchev, J. Phys. (Paris) Colloq., 41 (1980) C1-475.
80. A.R. Buev, Sov. Phys. - Solid State 21 (1979) 1239.
81. R.G. Gabrielyan, L.A. Kocharyan, A.R. Mkrtchyan and S.S. Tumanyan, Phys. Status Solidi (b), 73 (1976) 681; K.N. Shrivastava and K.W.H. Stevens, J. Phys. C: Solid State Phys., 3 (1970) L64.
82. J.L. Lewiner, P.H.E. Meijer and J.K. Wigmore, Phys. Rev., 185 (1969) 546.
83. J.K. Srivastava, details to be published.
84. C.P. Poole, Jr. and H.A. Farach, J. Magn. Resonance, 1 (1969) 551.
85. I.B. Bersuker and S.A. Borshch, Phys. Status Solidi (b), 49 (1972) K71.
86. J.K. Srivastava, Phys. Status Solidi (b), 55 (1973) K119.
87. J.A. Lock and J.F. Reichert, J. Magn. Resonance, 7 (1972) 74.
88. J.K. Srivastava, unpublished results.
89. M. Blume, Phys. Rev. Lett., 14 (1965) 96; 18 (1967) 305.
90. E.R. Andrew, Nuclear Magnetic Resonance, Cambridge Univ. Press, Cambridge, 1955, p. 213.
91. J.K. Srivastava and B.W. Dale, Phys. Status Solidi (b), 90 (1978) 391; 90 (1978) 571; T. Buch, B. Clerjaud, B. Lambert and P. Kovacs, Phys. Rev. B, 7 (1973) 184; R. Parrot, C. Blanchard and D. Boulanger, Phys. Lett., 34A (1971) 109.
92. M. Tinkham, Phys. Rev., 124 (1961) 311; J. Appl. Phys., 33 (1962) 1248; A.J. Sievers III and M. Tinkham, Phys. Rev., 124 (1961) 321; M. Tinkham, Proc. Int. Conf. on Magnetism and Crystallography., Kyoto, Japan, 1961; B. Lax and K.J. Button, Microwave Ferrites and Ferrimagnetics, McGraw Hill, New York, 1962, pp. 287-291.
93. S.C. Bhargava and N. Zeman, Phys. Rev. B, 21 (1980) 1717; T.M. Uen and P.K. Tseng, ibid, 25 (1982) 1848.
94. C.M. Srivastava, S.N. Shringi and R.G. Srivastava, Phys. Rev., B, 14 (1976) 2041; C.M. Srivastava, S.N. Shringi and A.S. Bommanavar, J. Phys. (Paris) Colloq., 38 (1977) C1-43; C.M. Srivastava, S.N. Shringi and S.M. Joglekar, paper presented at the Int. Conf. on the Applications of Mossbauer Effect, Jaipur (India), December 14-18 (1981), Abstract No. Fri-P_V-RE-3.
95. J.S. Hyde, in L. Kevan and R.N. Schwartz (Eds.), Time Domain Electron Spin Resonance, John Wiley, New York, 1979, p. 1.
96. K.M. Evenson, H.P. Broida, J.S. Wells, R.J. Mahler and M. Mizushima, Phys. Rev. Lett., 21 (1968) 1038; P.B. Davies, D.K. Russell, D.R. Smith and B.A. Thrush, in M.A. West (Ed.), Lasers in Chemistry, Proc. Conf. held at Royal Institution, London, 31 May- 2 June, 1977, Elsevier, Amsterdam, 1977, p. 97.
97. T. Oka, Can. J. Phys., 47 (1969) 2343; T. Oka and T. Shimizu, Phys. Rev. A, 2 (1970) 587.
98. T. Oka and T. Shimizu, Appl. Phys. Lett., 19 (1971) 88.
99. J.K. Srivastava, Proc. Nuclear Physics and Solid State Physics Symp., Calcutta (India), Dec. 22-26, 1975, Vol. 18C, p. 470.
100. J.K. Srivastava, details to be published elsewhere.

101. M.M. Dorio and J.H. Freed (Eds.), Multiple Electron Resonance Spectroscopy, Plenum, New York, 1979; L. Kevan and L.D. Kispert, Electron Spin Double Resonance Spectroscopy, John Wiley, New York, 1976; J.J. Krebs and R.K. Jeck, Phys. Rev. B, 5 (1972) 3499.
102. B.W. Faughnan and Z.J. Kiss, Phys. Rev. Lett., 21 (1968) 1331; R. Duncan, Jr., B. Faughnan and Z. Kiss, Appl. Optics, 9 (1970) 2236; M.S. Multani, TIFR Solid State Phys. Gp. Rept. (1972); paper presented at Indo-Soviet Conf. on Solid State Materials, Bangalore (India) February 1972.
103. C.P. Lichtenwalner, H.J. Guggenheim and L. Pfeiffer, Phys. Lett., 56A (1976) 117.
104. V.S. Letokhov, Phys. Lett., 43A (1973) 179.
105. S.M. Kulpa and J. Nemarich, Phys. Lett., 50A (1975) 461; L.L. Chase, Phys. Rev. B, 2 (1970) 2308; R. Solarz, D.H. Levy, K. Abe and R.F. Curl, J. Chem. Phys. 60 (1974) 1158; R. Solarz and D.H. Levy, ibid, p. 842; T. Tanaka, R.W. Field and D.O. Harris, ibid, 61 (1974) 3401.
106. U. Durr and R. Weber, Phys. Status Solidi (b), 60 (1973) 733; G.A. Prinz, D.W. Forester and J.L. Lewis, Phys. Rev. B, 8 (1973) 2155; F.S. Ham and G.A. Slack, ibid, B, 4 (1971) 777; K. Moorjani and N. McAvoy, ibid, 132 (1963) 504; W. Low and M. Weger, ibid, 118 (1960) 1130; M. Dvir and W. Low, Phys. Rev., 119 (1960) 1587; P.L. Richards, W.S. Caughey, H. Eberspaecher, G. Feher and M. Malley, J. Chem. Soc., 47 (1967) 1187; G.C. Brackett, P.L. Richards and W.S. Caughey, J. Chem. Phys., 54 (1971) 4383; G. Feher and P.L. Richards, in A. Ehrenberg, B.G. Malmstrom and T. Vanngard (Eds.), Magnetic Resonance in Biological Systems, Pergamon, Oxford, 1967, p. 141; H.M. Crosswhite and H.W. Moos (Eds.), Optical Properties of Ions in Crystals, Interscience, New York, 1967.
107. D. Curie, Compt. Rend., 258 (1964) 3269; G. Gergely, J. Phys. Radium, 17 (1956) 679; D. Curie and J.S. Prener, in M. Aven and J.S. Prener (Eds.), Physics and Chemistry of II - IV Compounds, North-Holland, Amsterdam, 1967, Chapter 9; P.M. Jaffe and E. Banks, J. Electrochem. Soc., 111 (1964) 52; D.T. Palumbo and J.J. Brown, Jr., ibid, 118 (1971) 1159.
108. P.M. Jaffe, J. Electrochem. Soc., 115 (1968) 1203; N.T. Melamed, F. de S. Barros, P.J. Viccaro and J.O. Artman, Phys. Rev. B, 5 (1972) 3377; N.T. Melamed, P.J. Viccaro, J.O. Artman and F. de S. Barros, J. Luminescence, 1, 2 (1970) 348; in F. Williams (Ed.), Proc. Int. Conf. on Luminescence, North-Holland, Amsterdam, 1970, p. 348.
109. P.J. Viccaro, F. de S. Barros and W.T. Oosterhuis, Phys. Rev. B, 5 (1972) 4257; G. Lang, R. Aasa, K. Garbett and R.J.P. Williams, J. Chem. Phys., 55 (1971) 4539.
110. D.S. McClure, J. Chem. Phys., 36 (1962) 2757; N.S. Hush and R.J.M. Hobbs, in F.A. Cotton (Ed.), Progr. Inorg. Chem., Vol. 10, Interscience, New York, 1968, p. 259; J. Ferguson, in S.J. Lippard (Ed.), ibid, Vol. 12, 1970, p. 159; D.L. Wood and J.P. Remeika, J. Appl. Phys., 38 (1967) 1038; D.S. McClure, Solid State Phys., (Eds. F. Seitz and D. Turnbull, Academic, New York) 9 (1959) 400; F.A. Hummel and J.F. Sarver, J. Electrochem. Soc., 111 (1964) 252; D.T. Palumbo and J.J. Brown, Jr., ibid, 117 (1970) 1184; W. Lehmann, ibid, 127 (1980) 503.
111. W.D. Partlow and H.W. Moos, Phys. Rev., 157 (1967) 252.
112. M.J. Weber, Phys. Rev. B, 4 (1971) 3153; G.H. Dieke and L.A. Hall, J. Chem. Phys., 27 (1957) 465; C.C. Lagos, J. Electrochem. Soc., 117 (1970) 1189; L. Ozawa and P.M. Jaffe, ibid, p. 1297; J. Th. W. de Hair and W.L. Konijnendijk, ibid, 127 (1980) 161; B.G. Wybourne, Spectroscopic Properties of Rare Earths, Interscience, New York, 1965.

113. H. Yamamoto and T. Kano, J. Electrochem. Soc., 126 (1979) 305; D.J. Robbins, B. Cockayne, J.L. Glasper and B. Lent, ibid, 126 (1979) 1221; G. Szigeti (Ed.) Proc. Int. Conf. on Luminescence, Vol. 2, Akademiai Kiado, Budapest, 1966.
114. G.H. Dieke, in H.M. Crosswhite and H. Crosswhite (Eds.), Spectra and Energy Levels of Rare Earth Ions in Crystals, Interscience (John Wiley), New York, 1968, p. 249, 140.
115. D.E. Murnick and M.S. Feld, Ann. Rev. Nucl. Part. Sci., 29 (1979) 411; C.B. Collins, S. Olariu, M. Petrascu and I. Popescu, Phys. Rev. Lett., 42 (1979) 1397; W. Demtroder, Phys. Repts. (Phys. Lett., C) 7 (1973) 223.

CHAPTER 15

MOSSBAUER SPECTROSCOPY OF RARE EARTHS AND THEIR INTERMETALLIC COMPOUNDS

S.P. Taneja[+†] and C.W. Kimball[*§]

+ Department of Physics, Punjab Agricultural University, LUDHIANA - 141004, India

* Department of Physics, Northern Illinois University, De Kalb, IL 60115, USA

1. INTRODUCTION

Mossbauer spectroscopy has developed into a valuable tool for studying properties of rare earth materials. The Mossbauer measurements determine the coupling between the nucleus and its surrounding electrons via the hyperfine interactions. The 4f electrons of the rare earth ion are strongly localized and well screened. These electrons interact with the crystalline electric field (CEF) produced by ions which surround its position in the lattice. Spin orbit coupling dominates this interaction, and, in most rare earths, the total angular momentum J remains a good quantum number and is responsible for the interesting variety of magnetic properties exhibited by rare earths and their compounds.

In this review we do not give an exhaustive survey but rather an account of recent Mossbauer studies, both experimental and theoretical. Electronic structure and magnetic properties of rare earths and their intermetallic compounds will be discussed so as to update previous reviews on this topic (1-5). A brief account of hyperfine interaction theory, relaxation phenomena, magnetocrystalline anisotropy and magnetostrictive effects in rare earths is presented. These concepts are used in studies of rare earth ions and their surroundings in solids. Because of space limitations only selected topics of current interest are treated.

---

† Present address: Simon Fraser University, Burnaby, B.C. Canada
Work supported by Punjab State Plan Scheme 37 and University Grants Commission (India)

§ Work supported by a National Science Foundation Grant (DMR 7809773)

Mossbauer results for intermetallic compounds such as RFe_2 (Laves phase), R-iron ternaries, RFe_3, R_2Fe_{17} and R_6Fe_{23} (R = rare earth), have been discussed for the iron effect only. Attention has been focused on the recent Mossbauer studies of rare earth metals and alloys, rare earth-amorphous and rare earth-chalcogenide compounds. Rare earth hydrogen-absorbing intermetallics have been excluded as those Mossbauer studies have recently been reviewed by Cohen (6). For rare earth superconducting compounds the reader is referred to an article by Shenoy elsewhere in this book.

2. REVIEW OF THE THEORY

2.1 Hyperfine Interactions in Rare Earths

In this section we present a brief review of the relativistic and non-relativistic theory of hyperfine interactions in the rare earths, namely, the electric monopole interaction, the magnetic dipole interaction and the electric quadrupole interactions. For in-depth studies the reader is referred to the general reviews (7-12) of these topics.[£]

2.1.1 Electric Monopole Interaction (Isomer Shift)

The isomer shift is due to the interaction of the nuclear charge and the electron charge density at the nucleus and is proportional to the change in the nuclear charge radius between the ground and excited states, that is,

$$\delta = \frac{2\pi}{3} Ze^2 \Delta\langle r^2\rangle \Delta|\psi(0)|^2,$$

where $\Delta\langle r^2\rangle = \langle r^2\rangle_{ex} - \langle r^2\rangle_{gd}$ and $\Delta|\psi(0)|^2$ is the difference in the total relativistic electron density at the nucleus for two absorbers of different chemical surroundings. In rare earths, different contributions to the total electronic density at the nucleus are:

1. The 4s electrons of the inner shells which are constant irrespective of the surroundings;
2. The shielding of the 5s electrons by the 4f electrons, which increases the density if a 4f electron is removed. This term which changes with the 4f electron configuration is the dominant one in rare earth isomer shift.

£ Also see chapter 1.

3. Variable electron density due to the 6s electrons, which are valence electrons and are completely removed in ideal ionic compounds. The 6s contribution is present in significant amounts in metallic and intermetallic compounds where the free atom's 6s electrons go into the conduction band and have some density at the rare earth nucleus.

A relativistic treatment of the electron wave function shows that the total density of s electrons is increased by an order of magnitude in the heavy elements when compared to the value for non-relativistic electrons. If non-relativistic calculations of the charge density are used, a correction factor S(Z) has to be included in Eqn. 1. Thus

$$\delta = \frac{2\pi}{3} Ze^2 \; S(Z) \; \Delta|\psi_{nr}(0)|^2 \; \Delta\langle r^2\rangle \qquad (2)$$

where $|\psi_{nr}(0)|^2$ is the non-relativistic density of the s electrons at the nucleus. In the relativistic limit, the $p_{\frac{1}{2}}$ electrons also contribute to $|\psi(0)|^2$, the total density at the nucleus. Fairly reliable absolute values are available for both $\Delta\langle r^2\rangle$ and $|\psi(0)|^2$. The quantitative evaluation of isomer shifts is thus most advanced for the rare earths.

2.1.2 Magnetic Dipole Interaction

The magnetic hyperfine interaction is described by the Hamiltonian

$$H_m = A_x I_x J_x + A_y I_y J_y + A_z I_z J_z \qquad (3)$$

where the A_i are components of the magnetic hyperfine tensor in principal coordinates, I_i are components of the nuclear angular momentum and the J_i of the electronic angular momentum. In many cases, especially in the presence of magnetic ordering, only one term of this expression is significant. In that case, the Hamiltonian can be equivalently written as

$$H_m = -g_I \mu_n I_z H_{eff} = -\mu \, H_{eff} \qquad (4)$$

where μ is the nuclear magnetic moment, g_I the nuclear gyromagnetic ratio, μ_n the nuclear magneton and H_{eff} is an effective magnetic field at the nucleus. This "effective field" approach is often applicable in Mossbauer spectroscopy of rare earth magnetic materials.

For rare earth ions the effective field ranges up to 10 MOe. The effective field at the rare earth nucleus arises from several sources:

$$\bar{H}_{eff.} = \bar{H}_{orb.} + \bar{H}_{core} + \bar{H}_{cond.} + \bar{H}_{res.} \quad (5)$$

These different contributions are:

1. The orbital hyperfine field, $\bar{H}_{orb.}$* produced by the open shell electrons is of the order of 8 MOe. In the non-relativistic theory, the orbital contribution is given by

$$\bar{H}^{nr}_{orb} = 2\mu_B \sum_i [\bar{\ell}_i - \bar{s}_i + 3\bar{r}_o (\bar{r}_o \cdot \bar{s}_i)] \langle r^{-3} \rangle \quad (6)$$

where $\bar{\ell}_i$ and $\bar{s}_i$ are the orbital and spin angular momenta of the ith electron, μ_B is Bohr magneton and $\bar{r}_o$ is the unit vector connecting the nucleus and the ith electron; the sum runs over all of the electrons in unfilled shells. In almost all of the rare earth ions J is a good quantum number. In terms of total angular momentum, $\bar{H}_{orb.}$ becomes proportional to the ionic magnetic moment $g_j\mu_B J$. In certain cases, e.g., Sm^{3+} and Eu^{3+} where there is a mixing of various J states into the ground state (J is not a good quantum number), the proportionality does not hold.

2. Core polarization field $\bar{H}_{core}$ results from the unpaired spin density at the nucleus and is produced by the exchange interaction between the spins of 4f electrons and the spins of inner electrons. The interaction deforms the radial distribution of electrons with spin up differently from those with spin down. This contribution is of the order of -100 kOe.

3. The field $\bar{H}_{cond.}$ arises from the spin polarization of the conduction electrons. In magnetically ordered rare earth intermetallic compounds and metals, the conduction electron contribution is of the order of 100 kOe.

4. The orbital contribution from 5d electrons is very small. Externally applied fields, demagnetizing fields and Lorentz fields also make small contributions to H_{res}.

The magnetic field contribution from $\bar{H}_{orb.}$ is generally much larger than the other contributions. For heavy rare earths the

---

* $\bar{H}_{orb.}$ as used here includes contributions both from orbital angular momentum and from spin-dipolar interactions.

non-relativistic approach to $\bar{H}_{orb.}$ will be inaccurate as one should consider 4f electrons to be relativistic. The hyperfine Hamiltonian, which encompasses the relativistic effects, is given (8) by

$$\bar{H}^{rel}_{orb} = 2\mu_B \sum_i [\bar{\ell}_i \langle r_\ell^{-3}\rangle + \{3\bar{r}_o(\bar{r}_o \cdot \bar{s}_i) - \bar{s}_i\} \langle r_{sd}^{-3}\rangle + \bar{s}_i \langle r_s^{-3}\rangle] \qquad (7)$$

The first two terms are the usual non-relativistic results. The third term, proportional to electronic spin, has the appearance of a contact interaction or a core polarization and is purely relativistic in origin. In the non-relativistic limit,

$$\langle r_\ell^{-3}\rangle = \langle r_{sd}^{-3}\rangle = \langle r^{-3}\rangle \text{ and } \langle r_s^{-3}\rangle = 0,$$

and Eqn. 7 becomes Eqn. 6. Values for the radial integrals of Eqn. 7, obtained from relativistic Hartree-Fock self-consistent field calculations, are listed in Table 1 for all of the rare earth ions.

TABLE 1

Radial parameters and free-ion hyperfine fields for lanthanide ions (8)

		$\langle r_{orb}^{-3}\rangle$ (a.u.)	$\langle r_{sd}^{-3}\rangle$ (a.u.)	$\langle r_s^{-3}\rangle$ (a.u.)	H_{orb} (kOe)	H_{sd} (kOe)	H_s (kOe)	H_{hf} (kOe)
Ce^{3+}	$4f^1$	4.338	4.566	-0.108	1551	326	5	1882
Pr^{3+}	$4f^2$	4.909	5.179	-0.127	2948	299	13	3260
Nd^{3+}	$4f^3$	5.498	5.815	-0.149	3939	139	23	4101
Pm^{3+}	$4f^4$	6.109	6.481	-0.176	4280	-138	35	4177
Sm^{3+}	$4f^5$	6.743	7.174	-0.203	3615	-444	45	3216
Eu^{2+}	$4f^7$	6.776	7.320	-0.259	0	0	-113	-113
Eu^{3+}	$4f^6$	7.403	7.904	-0.236	0	0	0	0
Gd^{3+}	$4f^7$	8.099	8.673	-0.270	0	0	-118	-118
Tb^{3+}	$4f^8$	8.826	9.481	-0.308	3312	158	-116	3354
Dy^{3+}	$4f^9$	9.585	10.333	-0.352	5995	431	-110	6316
Ho^{3+}	$4f^{10}$	10.374	11.221	-0.398	7787	187	-100	7874
Er^{3+}	$4f^{11}$	11.197	12.156	-0.451	8404	-217	-85	8102
Tm^{3+}	$4f^{12}$	12.054	13.134	-0.508	7540	-548	-64	6928
Yb^{3+}	$4f^{13}$	12.946	14.161	-0.572	4859	-591	-36	4232

For a Hund's rule state specified by L, S and J, the sums of Eqn. 7 may be evaluated to give

$$\overline{H}^{rel}_{orb} = a[(2-g_J)\langle r_\ell^{-3}\rangle + \alpha sd\langle r_{sd}^{-3}\rangle + (g_J-1)\langle r_s^{-3}\rangle]J$$

$$= H_\ell + H_{sd} + H_s \qquad (8)$$

where g_J is the gyromagnetic ratio, $\ell = 3$ for f electrons, and $a = 0.1251 \times 10^6$ Oe/a.u. Table 1 also lists the results of evaluating Eqn. 8 for various lanthanide ions (One should realize that relativity plays an important role here, not only in changing the form of $\overline{H}_{orb}$, but also in causing a gross change in the values of $\langle r^{-3}\rangle$). The term H_s in Eqn. 8, for a state in which J is a good quantum number, becomes approximately

$$H_s = 125\ (g_J-1)\ \langle r_s^{-3}\rangle\ \text{kOe} \qquad (9)$$

This form is generally assumed for core polarization; for example, in the lathanides one frequently takes

$$H_{core} = -90\ (g_J-1)J\ \text{kOe} \qquad (10)$$

It must be emphasized that H_s arises strictly from the open shell electrons and occurs because electrons with spin projection $+\frac{1}{2}$ and $-\frac{1}{2}$ in the open shell have different radial distributions. Actual core polarization, discussed above, has a rather different physical origin, but because of the similarity between Eqns. 9 and 10, the two contributions are experimentally indistinguishable. Thus, the experimental "core polarization" hyperfine fields can be written as

$$H^{exp}_{core} = H_s + H_{core} \qquad (11)$$

Thus any measured values of H^{exp}_{core} must take the relativistic term into account. This will be of special importance in rare earth ions with a half-filled shell; such as Eu^{2+}, for which the entire hyperfine field will be due to the terms in Eqn. 11. From the results in Table 1 it can be seen that there is a noticeable contribution to the hyperfine field, though small, due to relativistic effects.

2.1.3 Electric Quadrupole Interaction

The electric quadrupole interaction is due to the interaction of the nuclear quadrupole moment Q with the electric field gradient (EFG) at the nucleus. The Hamiltonian can be written as

$$H_q = \frac{e^2qQ}{4I(2I-1)} [(3\hat{I}_z^2 - \hat{I}^2) + \eta/2 (\hat{I}_+^2 + \hat{I}_-^2)] \tag{12}$$

where I and Q are the spin and quadrupole moment of the nucleus in the ground or the excited state, $\hat{I}$, $\hat{I}_z$, $\hat{I}_+$, $\hat{I}_-$ are angular momentum operators, q is the principal component of the electric field gradient (EFG) and η is the asymmetry parameter. For the evaluation of Mossbauer spectra, the eigenvalues and eigenfunctions of the above Hamiltonian for the excited and ground state of the nucleus are computed, and from these the hyperfine line positions and intensities are obtained.

In rare earths the EFG at the nucleus is the sum of two terms

$$q = (1-\gamma_\infty)\ q^{(lat.)} + (1-R)q^{\langle 4f \rangle}{}_T \tag{13}$$

The first term is the EFG arising from the non-spherical distribution of the ionic charges of the lattice. Since $q^{(lat.)}$ orginates from the charges outside the atom, the electronic shells will be polarized. The effect on the EFG at the nucleus is taken into account by γ_∞, the Sternheimer antishielding factor. The enhancement can be substantial, as $\gamma_\infty = -80$ in the rare earths. In the second term $q^{\langle 4f \rangle}{}_T$ represents the contribution to the electric field gradient produced by the 4f electrons; the shielding of $q^{\langle 4f \rangle}$ by inner electron shells is represented by a Sternheimer shielding factor R whose value is 0.1 to 0.4 in rare earths.

It may be pointed out that all three hyperfine interactions are usually present simultaneously and a complete analysis of the Mossbauer spectrum yields their individual values.

2.2 Relaxation Phenomena, Magnetic Anisotropy and Magnetostriction

In Mossbauer spectroscopy one does not observe the instantaneous value of the magnetic hyperfine field, but the average over a time of the order of the precession period of the nuclear moment. That is, relaxation occurs between the various ionic states which are populated at the temperature of measurements. In Eqn. 13, $q^{\langle 4f \rangle}{}_T$ is the average over the several electronic states occupied at a given sample temperature. In quadrupole interactions relaxation is always fast and only the mean $q^{<4f>}{}_T$ can be measured. In isomer shift investigations relaxation processes are of no importance (because the occupation of different states belonging to the same configuration, e.g., crystalline levels of the ground state multiplet, has a negligible influence on the change in electron density $(\Delta|\psi(0)|^2)$; the problem can be dealt with statistically as explained in Sec. 2.1.1. Spin fluctuation of either Mossbauer or other atoms determine, in part, the properties of the measured magnetic hyperfine spectra. In many rare earth compounds the relaxation between the 4f crystal field substates can be slow compared to the precession time of the nucleus in the hyperfine field. In that case the 4f electrons may stay in one state sufficiently long to allow one to observe a hyperfine splitting corresponding to that of the populated substate. This effect allows the determination of hyperfine interactions in the paramagnetic state.

Magnetocrystalline anisotropy may be present due to dipolar interactions between magnetic moments, anisotropic exchange or crystal field effects. In compounds of rare earths and 3d transition metals the anisotropy has contributions from both the rare earth sublattice and the 3d lattice. The dominant contribution to the anisotropy in these compounds is due to the effect of crystalline electric fields on the 4f-electron wave functions. The crystal field-induced anisotropy will be appreciable only at or below temperatures comparable in magnitude to the crystal field splitting. Thus, the crystal field Hamiltonian can be written as

$$H_c = \sum_{k=0}^{\infty} \sum_{q=0}^{k} A_k^q \sum_i f_{kq} (\bar{r}_i) \tag{14}$$

where the i summation is over all of the 4f electrons and the A_k^q are the lattice coefficients; f_{kq} can be expressed in terms of Stevens operator equivalents through

$$f_{kq} (\vec{r}) = \theta_k \langle r^k \rangle O_k^q \tag{15}$$

Here O_k^q are operators in terms of J_x, J_y and J_z; θ_k's are Stevens constants characteristic of a given rare earth element. Equation 14 can be written as

$$H_c = \sum_{k=0}^{\infty} \sum_{q=0} B_k^q O_k^q \tag{16}$$

where $B_k^q = A_k^q \langle r^k \rangle \theta_k$ are the crystal field parameters. For crystal fields of cubic symmetry, Eqn. 16 reduces to

$$H_c = B_4^0[O_4^0 + 5\ O_4^4] + B_6^0[O_6^0 - 21\ O_6^6] \tag{17}$$

The eigenfunctions and energies corresponding to the crystal-field-split states of the ground J multiplet can be calculated by diagonalization of H_c. In order to describe the magnetocrystalline anisotropy in terms of crystal fields one has to take into account the fact that the eigenfunctions, which determine the magnitude of the spin and the orbital moment of a given crystal-field state, also have a given symmetry. The exchange energy (4) determined by the Hamiltonian

$$H_{ex} = 2\mu_B\ \bar{H}_{ex} \cdot \bar{s} \tag{18}$$

where $\bar{H}_{ex}$ represents the exchange field exerted on the R ions by the surrounding R and M ions, will therefore be anisotropic depending on the properties of the particular crystal field state as well as on the direction of $\bar{H}_{ex}$; (M = another cation, say Fe ion for example). This necessitates the inclusion of H_{ex} in the diagonalization procedure, leading to the total perturbing Hamiltonian,

$$H_t = H_c + H_{ex} \tag{19}$$

The diagonalization is performed for many different directions $\bar{n}$ of H_{ex} with the quantities $A_k \langle r^k \rangle$ and H_{ex} as parameters. The resulting energy values E_i are used to calculate the partition function and the Helmholtz free energy

$$\bar{F}\ (\bar{n}, T) = -k_B T\ \ell n\ Z(\bar{n}, T) \tag{20}$$

The direction of easy magnetization at a given temperature is the direction of $\bar{n}$. The bulk magnetocrystalline free energy E $(\bar{n}, T)$ of a cubic system can be expanded into a power series of the direction cosines α of the direction of magnetization $\bar{n}$ and cube edges (4):

$$E(\bar{n}, T) = K_o + K_1(\alpha_1^2\alpha_2^2 + \alpha_2^2\alpha_3^2 + \alpha_1^2\alpha_3^2) + K_2\alpha_1^2\alpha_2^2\alpha_3^2 + ---- \tag{21}$$

The anisotropy constants K_i can be obtained by replacement of $E(\bar{n}, T)$ by $F(\bar{n}, T)$ of Eqn. 20 and then both sides of the equation are calculated for a number of different directions of $\bar{n}$, after which the quantities K_i can be derived.

Extremely large magnetostrictive strains at low temperatures are present in some heavy rare earth metals. The origin of this phenomenon is thought to be the large spin-orbit interaction between the electron spins and the spatially anisotropic 4f charge cloud. The cubic rare-earth iron Laves phase compounds also exhibit large magnetostriction at temperatures much higher than for pure rare earth metals (13). The magnetostriction remains large at room temperature because of the large rare earth-iron exchange interaction which aligns the rare earth spins even at high temperatures. The magnetostrictive strains are proportional to the magnetic field strength above the Neel temperature; below this temperature they become highly sensitive to the magnetic field strength because of changes in the spin structure. In rare earth metals and rare earth iron Laves phase compounds there is a large internal magnetic field present and the variation of the hyperfine parameters with temperature can be correlated with the magnetostrictive strains.

3. SURVEY AND DISCUSSION OF EXPERIMENTAL AND THEORETICAL MOSSBAUER RESULTS

3.1. Rare Earth Metals and Ions

The electronic structure of the rare earths, important for Mossbauer studies, has been described in detail by Bauminger et al (11)*. The 4f valence electrons are well localized inside the atom. The CEF interaction does not break the spin-orbit coupling of the 4f electrons and J remains a good quantum number. Most rare earth ions are trivalent with an electronic structure described by a pure $4f^n$ configuration. Some of the rare earth ions also form a divalent or tetravalent state which is an admixture with the 4f. The electronic structure of rare earth metals can be described as that of a trivalent ion with a $4f^n$ configuration plus a (5d-6s) conduction band. The rare earth metals La, Pr and Nd have hexagonal structures; Sm and Eu are rhombohedral and cubic, respectively. The heavy rare earths, i.e., Gd to Tm, all have h.c.p. structure; Yb is cubic.

---

* A large body of definitive theoretical work has been treated by A.J. Freeman and his collaborators. Detailed references for specific applications are given in Ref. 1-6, 11.

3.1.1 Mossbauer Results on Rare Earth Metals

Mossbauer measurements of the 22.5 keV levels of $^{149}Sm^{3+}$ in samarium metal have been reported at room temperature (14). The isomer shift was observed to be 0.25(4) mm/sec which is typical in the $4f^5$ configuration of the trivalent samarium 4f shell.

Mossbauer studies of the 21.6 keV level of ^{151}Eu at 4.2° K in europium metal were made by Klein et al (15) at high pressures. The pressure dependence of isomer shift yielded $d\delta/dP = 4.8(5) \times 10^{-2}$ mm/sec/kbar. The magnetic hyperfine splitting is reduced with increasing pressure ($\frac{dB}{dP}$ = 0.133(8) T/kbar). The major contribution to the volume dependence of isomer shift arises from a congruent compression of the 6s electrons; charge transfer effects were negligible. The observed decrease of hyperfine splitting under pressure was attributed to an increase in the conduction electron polarization by the 4f electrons.

Bauminger et al (16) performed recoilless absorption studies of the 86 keV transition of ^{155}Gd in single-crystal and powder absorbers of Gd metal at 4.2°K. In the single crystal the hyperfine field acting on Gd nuclei is 373(5) kOe and the quadrupole interaction parameter $e^2qQ/4$ = 26.4(7) MHz with θ = 28(2)° (θ is the angle between the direction of magnetization and c-axis). The angle θ varies 52° to 75° in different powder samples. The hyperfine field in the Gd-single crystal is ∿ 13% larger than in Gd-metal powder. This difference has been attributed to the effects of the difference in the angle θ.

Weiss and Langhoff (17) observed that the Mossbauer line of the 58 keV transition in ^{159}Tb can be used for the analysis of phonon spectra. A principal application of these studies is the investigation of fine structures in phonon spectra.

Using the 25.6 keV transition of ^{161}Dy in Dy metal, the temperature dependence of the hyperfine field has been reported by Loh et al (18-20). The Mossbauer spectra near the Neel temperature (18) do not show the relaxation effects previously observed in dysprosium metal (19-20). From the plot of reduced effective hyperfine field vs $(1-T/T_N)$ the value of the critical exponent was found to be β = 0.335(10) and the critical temperature T_N = 180.35° K. Bowden et al (21) used the connection between the Mossbauer effect in ^{161}Dy and magnetic anisotropy in dysprosium metal to probe the limits of the Callen and Callen (22) theory of magnetic anisotropy. The difference between the Mossbauer data and the theoretical calculations with a modified C and C model were partially resolved.

Reese and Barnes (23) studied the magnetic structure of holmium metal using the 80.6 keV Mossbauer transition in ^{166}Er. Erbium was introduced as a dilute impurity ($\approx$ 10 ppm) in a holmium single-crystal and measurements were made between 4.2-50° K. The internal magnetic field at 4.2° K is 7.50(15) MOe and, with increasing temperature, decreases more rapidly than predicted by a simple molecular field-free ion approximation. From the molecular field model the exchange energy is of the order of the Neel temperature (133° K), and the CEF interaction is of the order of 25° K.

Hyperfine interactions of 80.6 keV transitions of ^{166}Er in Er metal have been reported (24-25) between 4.2-80° K to investigate the magnetic structures of erbium metal. The analysis of Mossbauer data at 4.2° K yielded a value of 26(3)$^{\circ}$ for the semi-apex angle of a ferromagnetic spiral or cone structure. In the high temperature phase (52 < T < 80° K) the antiferromagnetic alignment of the magnetic moments is along the c-axis. The magnitude of the effective field extrapolated to T = 0 is 7.75(15) MOe, which agrees with previous results (26).

The Mossbauer spectrum of Tm metal at 5° K has been taken by Wit and Niesen (27) using an implanted source of ^{169}Er in Al. The ratio of the excited and ground state nuclear magnetic moments is μ_{ex}/μ_{gd} = -2.223(13) and the quadrupole splitting e^2qQ/h = 1933(12) MHz = 28.52(17) cm/sec. These values are in agreement with those reported previously (28) but are more accurate. The Mossbauer experiments on ^{170}Tm (Yb) in Tm metal at 4.2° K show magnetic interaction (29) with H_{eff} = 91(20) kOe and isomer shift δ = -0.082 mm/sec. The measured quadrupole interaction is almost the same at 4.2° K ($T < T_N$) and at 60° K ($T > T_N$). Its value is $eV_{zz}Q/8$ = -0.62(4) mm/sec which gives V_{zz} = 0.66(4).10^{18} V. cm^{-2}; this compares well with V_{zz} = 0.52 x 10^{18} V. cm^{-2} reported earlier (30) from Mossbauer measurements on ^{169}Tm in Tm metal.

Mossbauer effect study of the isomer shift and magnetic behavior of ^{170}Yb in Yb metal has been reported (29,31) at 4.2° K. The isomer shift is -0.095(20) mm/sec (relative to $TmA\ell_2$); the quadrupole interaction between Yb metal and Tm metal is thought to arise because Yb is closer to the ideal hexagonal compact phase than Tm. The similarity between ^{170}Yb isomer shifts in Yb metal and Yb in Tm metal suggests the existence of a fractional valence of the same order in both cases, assuming that the conduction electron contributions are similar (2.2 in Yb metal). The effective field at the site of ^{170}Yb

impurity in thulium metal is smaller than the Yb^{3+} effective field in magnetic matrices by at least one order of magnitude. It was concluded that the Yb impurity is non-magnetic in Tm metal and H_{eff} arises from the s-band conduction electron polarization by the Tm^{3+} ionic magnetization at low temperature, through the s-f exchange interaction.

3.1.2 Rare Earth Ions

Our review of the Mossbauer results on rare earth ions will cover the literature on this topic which appeared after review by Bauminger et al (11).

Lanthanum has two resonances, 166 keV and 10 keV γ-transitions in ^{139}La and ^{137}La, respectively. The Mossbauer effect of ^{137}La has been reported using $^{137}CeO_2$ as a source (32) with La_2O_3 and $LaCrO_3$ as absorbers. From the Mossbauer analysis, the ratio of quadrupole moments for the ground state and excited state was found to be Q(7/2)/Q(5/2) = 0.98(2). Using Q(7/2) = 0.26(8)b (33) the value of the excited state quadrupole moment is 0.26(8)b.

Many rare earth atoms in conducting systems (particularly Ce, Sm, Eu, Tm and Yb) show fluctuating valence configurations. Inter-configuration fluctuations can be seen as a function of temperature if valence configurations are energetically close with their separation of order kT. A detailed picture involves the promotion of an electron from the conduction band at the Fermi level to the localized $4f^{n-1}$ level which forms the new configuration $4f^n$. Since the Mossbauer effect can easily identify different valence states through the isomer shift measurements, it is a valuable tool in describing the valence fluctuation phenomena (12). Mossbauer measurements on the 22.5 keV level of ^{149}Sm and 122 keV level of ^{152}Sm have been reported for different samarium compounds (14,34-37). The mixed valence state of Sm, in SmS under pressure and in $Sm_{1-x}R_xS$ (R = rare earth), for different samarium concentrations will be discussed later. Mossbauer data on some other samarium compounds have been listed in Table 2. It was observed by Mossbauer measurements (34, 35) and XPS studies (38) that Sm in SmB_6 exists in the mixed valent state. From Mossbauer studies of ^{152}Sm (122 keV) in SmS, $Sm_{.3}Pr_{.7}S$, SmB_6 and $SmBe_{13}$ it was concluded that ^{152}Sm, in comparison to ^{149}Sm (25 keV), is more sensitive to charge fluctuations. The observed width and position of the absorption line in $Sm_{.3}Pr_{.7}S$ and SmB_6 show that the mixed valence state, even at 4.2° K, is a

TABLE 2

Isomer shifts and line widths of ^{149}*Sm in various absorbers (34)*

	δ mm/sec	Γ mm/sec
Sm^{3+}		
SmF_3	0.01(2) 0.00(6)	2.15(10) --
$SmCl_3$	-0.01(4)	2.66(16)
Sm_2O_3(m)	0.04(2) 0.01(6)	2.70(12) --
Sm_2S_3	0.03(4) 0.03(6)	3.66(20) --
Sm_2Se_3	-0.02(3)	2.56(12)
Sm_2Te_3	-0.10(6)	3.60(20)
SmN	0.07(2)	2.45(10)
SmP	0.04(2)	2.24(10)
SmAs	0.06(4)	2.58(16)
SmSb	0.00(4)	2.42(16)
Sm^{2+}		
SmF_2	0.90(4)	--
$SmCl_2$	-0.7	--
SmSe	-0.71(3) -0.68(8)	2.15(12) --
SmTe	-0.72(4) -0.60(8)	-3.04(16)
$Sm^{2+/3+}$		
SmB_6	-0.33(2) -0.40(4)	1.15(8)
Sm_3S_4	-0.19(4)	

fluctuating charge state (see ref. 37 for detailed discussion).

Europium has four Mossbauer isotopes, two of which possess high natural isotopic abundance with high sensitivity for isomer shifts. The preferred one is ^{151}Eu with a 21.6 keV γ-transition formed from the decay of ^{151}Sm whose half-life is 87 years. The second is the 103 keV transition of ^{153}Eu. Europium occurs in two valence states in metallic systems. Europium Mossbauer spectroscopy is well suited to study the valence state of Eu in these systems, due to the large difference in isomer shift and magnetic behavior between Eu^{2+} and Eu^{3+}.

A large number of ^{151}Eu Mossbauer and magnetic measurements have been reported in europium compounds at a wide range of temperatures (39-53). Hyperfine parameters in various divalent and trivalent europium compounds are listed in Table 3. Buschow and co-workers (42-45,49,52,53) observed that $EuZn_5$, $EuMg_5$, Eu_2Mg_{17} and Eu_6Cd_5 order antiferromagnetically, while $EuZn_2$ and $EuCd_2$ show metamagnetic behavior.

The ground state configurations of Eu^{3+} and Eu^{2+} are $4f^6$ and $4f^7$, respectively. These two states are generally well separated in energy. An intermediate valence is observed in compounds where the energies corresponding to these two configurations do not differ too much. Due to interconfigurational fluctuations between the two states, these compounds are often referred to as fluctuating valence compounds (54). Mossbauer studies of ^{151}Eu in $EuSi_2Cu_2$, $EuRh_2$, $EuA_{2-x}B_x$, $Eu_{1-x}R_xA_{2-y}B_y$ (R = rare earth, A = Rh, Ni, Ir and B = Pt, Aℓ) have been reported by Kalvius et al and Bauminger et al (55-63) at different temperatures, pressures, and concentrations. It was shown that Mossbauer data could be analyzed to yield interconfigurational excitation energies, charge fluctuation rates and the dependence of excitation energies on the local environment (e.g., nearest neighbours), temperature and pressure. From the experimental observations, the following general conclusions were drawn:

(i) In these systems the interconfigurational excitation energy depends strongly on the local environment and, particularly, on the nature of the first nearest neighbours.

(ii) The average excitation energy for a given first nearest neighbour shell composition is temperature dependent.

(iii) The isomer shift measurements yield the dependence of the excitation energies on composition (x,y), environment, temperature and pressure.

(iv) The extreme limits of the observed isomer shifts indicate that the total width is not extremely small relative to excitation energy and that the ionic states are not pure 4f states.

Recent Mossbauer measurements (64-66) with $EuNi_xZn_{5-x}$, $EuNi_xCu_{5-x}$, $EuFe_4Aℓ_8$, $EuCu_4Aℓ_8$ and $EuMn_4Aℓ_8$ show Eu in a mixed valence state. In the first two systems, mixed valence is interpreted in terms of fast fluctuations between the $4f^6$ and $4f^7$ configurations.

In $EuCu_4Al_8$ the Eu is divalent; in $EuMn_4Al_8$ and $EuFe_4Al_8$ the Eu is in a mixed valence state. The dominant valence of the Eu is 3+ in $EuFe_4Al_8$ and the Eu sublattice does not order independently of the iron sublattice. In $EuMn_4Al_8$, although Eu is in a mixed valence state, the Eu sublattice does order by itself and Mn sublattice does not order (65,66).

The recoilless absorption measurements of 103 keV γ-ray of ^{153}Eu have been reported in $Eu_2Ti_2O_7$ and by implanting ^{153}Eu nuclei in SmS, SmSi semiconductors, anhydrous rare earth trichlorides, alkaline-earth fluorides, Sc_xY_{1-x} and metallic iron (67-73) at low temperatures. The results obtained in a single crystal of $Eu_2Ti_2O_7$ showed the presence of a large Goldanskii-Karyagin effect (67) consistent with previous observations in $Eu_2Ti_2O_7$ powder samples (74). Isomer shift data for SmS and SmSi were analyzed to determine the change in electronic structure with Eu doping (68,69). The results showed that, for SmS in the semiconducting phase, the conduction band is primarily of the 5d character with the d electrons localized on trivalent doping ions.

There are seven Mossbauer resonances in various Gd isotopes. The most commonly used is the 86.5 keV γ-transition in ^{155}Gd. Two other resonances are the 105 keV and the 60 keV transitions in ^{155}Gd; the first is a poor choice as compared to the 86.5 keV transition of the same isotope and the latter has not been used because of very broad linewidth. Very little work has been reported on the other transitions such as 123 keV in ^{154}Gd and 89 keV in ^{156}Gd. The 64 keV transition is ^{157}Gd has much higher resolution but the difficulty in its preparation restricts its use for Mossbauer measurements (11).

The recoilless absorption spectrum of the 123 keV γ-ray of ^{154}Gd in $Gd_2Ti_2O_7$ has been measured at 4.1° K by Bauminger et al (67). The analysis yields the quadrupole interaction parameter $e^2qQ(123\ keV) = (928 \pm 8)$ MHz which leads to $^{154}Q(123) = -1.85 \pm 0.20$ b.

Systematic hyperfine interaction studies have been reported for the 86.5 keV transition of ^{155}Gd in various Gd absorbers (50, 65,66,75-84). The Mossbauer results have been summarized in Table 4. Because of the low recoilless fraction of the γ-ray, the source was kept at 4.2° K. The isomer shift of the GdPd alloy is the most negative of all the gadolinium compounds. This is attributed to a very large conduction electron density, rather than any decrease in the f electron density (75). In GdM_4Al_8 (M = Cr, Mn, Fe and Al)

TABLE 3.

Hyperfine parameters of 21.6 keV γ-ray of ^{151}Eu in various absorbers

Compounds	Temperature (^oK)	T_c, T_N (^oK, Moss.)	δ (mm/sec)	e^2qQ (mm/sec)	$\|H_{eff}(0)\|$** (kOe)	Ref.
(1) Divalent Compounds						
$EuCl_2$	--	--	-13.4(1)*	-13.53(29)	--	46
EuSi	4.2	63(5)	-10.8(2)*	- 2.0	210(10) (at 4.2^{o}K)	41
EuGe	4.2	38(1)	-11.3(3)*	- 2.4	231(8) (at 4.2^{o}K)	41
EuSn	4.2	30.5(5)	-12.0(1)*	± 0.0	215(3) (at 4.2^{o}K)	41
$EuPt_2$	4.2	100(Mag.)	- 8.55(5)	0.005	33(1)	52
	RT		- 8.53(5)	--	--	52
$EuPt_{2.5}$	RT	67(Mag.)	- 8.40(5)	--	--	52
			1.5(1)			
$EuZn_2$	--	29.3(3)	- 9.1(1)	--	238(3)	42
$EuMg_2$	4.2	30	- 9.02(4)	--	161(3)	49
$EuCd_2$	4.2	36(1)	-10.0(1)	0.0(1)	243(3) (at 4.2^{o}K)	45
$EuSi_2$	4.2	<77	-10.8(2)*	3.2	320(10) (at 4.2^{o}K)	41
$EuGe_2$	4.2	<77	-12.0(3)*	3.2	256(10) (at 4.2^{o}K)	41
$EuCu_5$	8.0	57	- 8.2(1)	12.0(1)	269(5)	43,44
$EuZn_5$	4.2	10.6(3)	-10.0(1)	<.4	292(3)	42,44
$EuMg_5$	--	7.8	-10.2(1)	--	196(3)	49
$EuAg_5$	--	18	-10.1(1)	--	245(3)	44
$EuAu_5$	--	14	-10.4(1)	--	274(3)	44
$EuCd_6$	--	2.25(5)	-10.8(1)	--	240(3) (at .83°K)	45
$EuCd_{11}$	--	2.75(5)	-12.2(1)	1(1)	258(3) (at .83°K)	45
$EuZn_{13}$	4.2	3.19(3)	-11.75(1)	<.4	283(2)	42

TABLE 3. (continued)

Compounds	Temperature (^oK)	T_c, T_N (^oK, Moss)	δ (mm/sec)	e^2qQ (mm/sec)	$\|H_{eff}(0)\|$** (kOe)	Ref.
Eu_6Cd_5	4.2	75(3)	- 8.9(1)	--	150(2) (at 4.2^{o}K)	45
		20(3)				
Eu_2Mg_{17}	--	7.9	-11.1(1)		215(5) (at .86°K)	49
			-11.2(1)			
$EuTiO_3$	--	5.5(2)	-13.40(5)*	--	325(7)	40
Eu_2TiO_4	--	7.8(2)	-12.8(1)*	-10.8(1)	305(3)	40
$Eu_3Ti_2O_7$	2.3	8(3)	-12.6(2)*	- 8.6(5)	--	40
			-13.8(2)*	very small	--	40
EuB_6 (powder)		12.5(5)			308 (at 4.2^{o}K)	48
(s. crystal)	4.2	6.3(3)	-12.7(1)		262 (at 4.2^{o}K)	48
(2) <u>Trivalent Compounds</u>						
Eu_2O_3 (cubic)	--	--	+ 0.88(1)	- 5.20		
"	--	--	+ 1.02(1)	- 4.76	--	39
Eu_2O_3 (monoclinic)	--	--	+ 1.09(1)	- 4.00	--	39
$EuPt_5$	RT	--	- 2.0(5)	--	--	
			+ 1.7(1)	--	--	52

* Isomer shift in these compounds is relative to Sm_2O_3, while in all others it is relative to EuF_3/SmF_3.

** This is the absolute value of H_{eff} extrapolated to 0° K.

TABLE 4.

Hyperfine interaction parameters of 86.5 keV γ-ray of ^{155}Gd in various absorbers.

Compounds	Temperature (^{o}K)	T_c, T_N (^{o}K)	δ (mm/sec)	V_{zz}** (10^{17} V cm^{-2})	H_{eff} (kOe)	θ* (degrees)	Ref.
GdCu	4.2	--	-0.375(8)	--	--	--	76†
GdAg	4.2	--	-0.307(8)	--	--	--	76
GdZn	4.2	--	-0.108(10)	--	--	--	76
$Gd_{.018}Pr_{.982}$	1.2	10	--	2.5(1) (at $10^{o}K$)	197(4)	90	50
$Gd_{.1}Pd_{.9}$	4.2	4.8(3)	0.06(2)	--	100(20)	--	75††
$GdMn_2$	4.2	--	0.107(2)	--	-101(1)	--	77††
$GdFe_2$	4.2	--	0.067(1)	--	434.8(6)	--	77
$GdCo_2$	4.2	--	0.043(2)	--	+ 35(3)	--	77
$GdNi_2$	4.2	--	0.008(2)	--	-120(5)	--	77
$GdFe_3$ (cubic) (hex)	4.2	--	0.067(2)	- 1.52(9)	399(1)	81(2)	
			0.262(7)	8.31(6)	224(2)	90(2)	77
$GdCo_3$ (cubic) (hex)	4.2	--	0.070(2)	- 1.60(7)	117(4)	0(5)	77
			0.222(3)	8.58(4)	47(3)	32(5)	
$GdNi_3$ (cubuc) (hex)	4.2	--	0.062(2)	- 0.74(3)	- 80(3)	0(2)	77
			0.208(4)	10.59(3)	-138(2)	22(2)	
$GdPd_3$	4.2	7.5	0.06(2)	--	145(10)	--	75
$GdNi_5$	1.15	--	0.27(1)	10.3(6)	-252(5)	0	79
$GdCo_5$	4.2	--	0.24(1)	9.7(1)	± 50(10)	0	79
Gd_2Ni_{17}	4.2	--	0.24(1)	7.5(2)	-270(20)		
			0.28(1)	5.1(2)	-115(20)	0	79
Gd_2Co_{17}	4.2	--	0.24(1)	5.2(2)	± 70(10)	90	79
Gd_2Fe_{17}	4.2	--	0.26(1)	4.4(1)	+210(10)	90	79

TABLE 4. (continued)

Compounds	Temperature (^oK)	T_c, T_N (^oK)	δ (mm/sec)	Quadrupole Interaction (MHz)	H_{eff} (kOe)	θ* (degrees)	Ref.
Gd_6Mn_{23}	4.2	--	0.140(5)	+10.3(2)	+210(10)	60(2)	77
$GdVO_4$	4.2	2.5	0.66(2)	4.59(9)	--	--	80
$GdCr_2Si_2$	4.1	--	--	-237(1)	-244(2)	0(5)	84
$GdMn_2Si_2$	4.1	--	--	-214(3)	-243(3)	90(5)	84
$GdFe_2Si_2$	4.1	9.0(2)	--	-305(3)	-283(6)	90(5)	84
$GdCo_2Si_2$	4.1	46.4(5)	--	-169(3)	-289(10)	78(1)	84
$GdNi_2Si_2$	4.1	11.6(5)	--	- 50(3)	-266(5)	70(5)	84
$GdCu_2Si_2$	4.1	--	--	87(3)	-256(2)	90(10)	84
$GdRh_2Si_2$	4.1	--	--	-238(3)	-315(2)	90(2)	84
$GdPd_2Si_2$	4.1	17.7(2)	--	- 18(4)	-266(1)	60(3)	84
$GdPt_2Si_2$	4.1	7.2(2)	--	167(8)	-276(3)	70(5)	84
$GdAu_2Si_2$	4.1	--	--	- 45(5)	-261(2)	60(4)	84
$GdCr_4A\ell_8$	4.1	8(2)	0.44(5)	301(4)	172(4)	80(7)	66
	77		0.40(5)	295(4)	--	--	
$GdMn_4A\ell_8$	4.1	28(2)	0.43(5)	217(4)	102(5)	46(6)	66
	77		0.39(5)	218(6)	--	--	
$GdFe_4A\ell_8$	4.1	11	0.4(1)	-238(5)	188(5)	44(4)	65,66
	77		0.4(1)	-228(5)	24(20)	--	
$GdCu_4A\ell_8$	4.1	32(1)	0.4(1)	- 83	166(4)	60(6)	66
	77		0.4(1)	- 83(1)	--	--	
$MnGd_2S_4$	4.2	--	0.44(1)	165	--	--	81

* θ is the angle between the direction of the hyperfine field and the local symmetry axis.

** V_{zz} is the electric field gradient tensor.

† Isomer shift is relative to $La_{0.95}Sm_{0.05}A\ell_2$ source.

†† Isomer shift relative to ^{155}EuPd and in all others is relative to $SmPd_3$.

the quadrupole and magnetic hyperfine interactions are of comparable size at 4.2° K and the character of the Mossbauer spectra depends on the angle θ. This behavior is attributed to the fact that the Gd^{3+} is an S state ion with small single ion anisotropy and the small crystalline changes easily tilt the Gd magnetization. Further, the change of sign on the electric field gradient (Table 4) from positive for M = Cr and Mn, to negative for Fe and Cu may arise either from slightly different positions of the ions in the unit cell or from different charge states of various M ions (66). From Mossbauer studies of GdM_2Si_2 compounds (M = Fe, Co, Mn, Cr, Cu, Rh, Pd, Pt and Au) Gd is found to order antiferromagnetically but the M ion is non-magnetic (84). Cook and Cashion (82) observed that the magnetic ordering direction in $GdAlO_3$ is along the orthorhombic a-axis rather than the b-axis.

Recoilless absorption of the 105 keV γ-ray of ^{155}Gd in $Gd_2M_2O_7$ (M = Ti, Sn, Ru, Ir and Cr) has been reported (67). Isomer shift in all of these compounds is 0.57(2) mm/sec relative to $SmPd_3$. A Goldanskii-Karyagin effect was observed in all compounds which decreased in magnitude with increasing mass of M ions and was attributed to the change in character of external d electrons.

There are four Mossbauer transitions in ^{161}Dy but the 25.6 keV resonance is most convenient for studies over a wide range of temperature (11). Some hyperfine interaction studies have also been reported using the 86.8 keV ($2^+ \rightarrow 0^+$) γ-resonance of ^{160}Dy.

Mossbauer measurements with the 25.6 keV transition in ^{161}Dy have been reported for different materials at different temperatures (66,85-90) and are listed in Table 5. One notices that the hyperfine interaction values are close to the free ion values in $DyFe_4Al_8$, $DyMn_4Al_8$ and $DyCr_4Al_8$. It was therefore concluded that the magnetic moment on each Dy ion at 4.1°K is 10 μ_B. These systems order antiferromagnetically (66). The low temperature Mossbauer spectrum of both ionic sites (C_2,C_{3i}) in Dy_2O_3 indicates a slow relaxation frequency. The magnetic splitting of $Dy_2O_3(C_{3i})$ and $DyPO_4$ is the same; thus the ground state in these systems is almost a pure $|\pm 15/2\rangle$ Kramer's doublet (86,87).

The magnetic hyperfine field and quadrupole interaction have also been reported in Dy-Si, Dy-Ga, Dy-Sb, Dy-Sc, and Dy-Mg alloys (91-95). In dilute alloys of Dy-Si and Dy-Ga no variation in magnetic hyperfine field was observed; the ground state was inferred to be a $|\pm 15/2\rangle$ doublet in these systems. The constant magnitude of the quadrupole interaction was attributed to the predominant contribution

TABLE 5.

Hyperfine interaction parameters of 25.6 keV γ-ray of ^{161}Dy in different materials. I.S. is relative to GdF_3 source. (a) *and* (c) *stand for amorphous and crystalline respectively.*

Materials	Temperature (^oK)	T_c, T_N (^oK)	δ (mm/sec)	$g_N\mu_N H_{eff}/h$ (MHz)	e^2qQ (MHz)	Ref.
Dy^{3+} ion				-840	2560	66
$DyCr_4Al_8$	4.1	16(2)	1.3(1)	-835(30)	2360(120)	66
$DyMn_4Al_8$	4.1	19(2)	1.5(1)	-835(30)	2485(120)	66
$DyFe_4Al_8$	4.1	25	1.5(1)	-845(30)	2545(120)	66
$DyCu_4Al_8$	4.1	17(1)	1.5(1)	-740(30)	1880(120)	66
$DyPO_4(D_{2d})$	2.5	3.39	--	-831(4)	--	86
		--	--		--	
$Dy_2O_3(C_2)$	2.5	--	--	-755	--	87
(C_{3i})		--	--	-830		
$DyCo_3$	(a) 4.2	∼ 900	--	+870	2069	247
	(c) 4.2	450	--	+883	2483	85
$DyFe_3$	(a) 4.2	350	--	+890	2173	247
	(c) 4.2	600	--	+904	2648	247
$DyNi_3$	(a) 4.2	47	--	+840	2160	247
	(c) 4.2	69	--	+850	2586	247
				H_{hf}(kOe)		
$Dy(OH)_3$	1.7	3.5	--	5500(100)	2360(60)	89
DyIG	above T_N	70.0(5)	--	--	$\gtrsim$1600	90

of the 4f electrons (91). Studies in Dy-Sc show two different ordering systems (95) (i) with concentration $\leq$ 25 at % Dy alloys exhibit no static long range magnetic order but show a slow electronic-spin-relaxation mechanism, and (ii) with concentration $\geq$ 35 at % Dy alloys show a temperature dependence of the hyperfine field that indicates long range magnetic order.

Hyperfine studies have been reported (96-101) on ^{161}Dy (25.6 keV) substitutionally implanted in iron, nickel and aluminum. The Mossbauer spectrum at 5° K in ^{161}Dy<u>Fe</u> shows H_{eff} = 6.20(3) MOe and $e^2qQ/4$ = 2.76(6) cm/sec. ^{161}Dy<u>Ni</u> shows two components, one with H_{hf} = -5.85(3) MOe and $e^2qQ/4$ = 2.78(6) cm/sec, the second with $|H_{hf}|$ = 4.64(3) MOe and $e^2qQ/4$ = 1.66(6) cm/sec (99) (these values have been extrapolated to 0° K). In almost all cases about half of the rare earth impurities are in substitutional sites. The temperature dependence of the hyperfine field in these systems indicates fast relaxation behavior of the electronic moment at the substitutional sites. For the iron host the exchange interaction dominates the cubic crystalline electric field, but for nickel both interactions are of comparable magnitude, leading to a decrease in the hyperfine interaction.

As stated above, the 86.8 keV Mossbauer resonance of ^{160}Dy has been employed to study the paramagnetic hyperfine structure of Dy impurities in the hexagonal metal Zr and in the cubic metal Au (102-104). From the hyperfine splitting in Zr, the CEF ground state is inferred to be an almost pure Γ_7 Kramer's doublet with $g_z >> g_x$ and in Au it is a Γ_8 (1) quartet slightly perturbed by the contribution from the first excited state Γ_8 (2).

Erbium has four even isotopes, i.e., ^{164}Er, ^{166}Er, ^{168}Er, ^{170}Er, all having a ($2^+ \rightarrow 0^+$) rotational E2-transition. The 80.6 keV transition in ^{166}Er has been used most frequently for Mossbauer studies. The odd isotope is ^{167}Er which is very rarely used for Mossbauer studies due to its broad line width (11).

The hyperfine coupling and the 4f-conduction electron exchange interaction of localized Er moments in Au, Ag, Th, Zr and Y dilute alloys have been investigated by ^{166}Er Mossbauer spectroscopy (105-110). The hyperfine parameters are listed in Table 6a. The values of A are particularly large for Th and Y alloys. In the Th alloys a larger contribution is attributed to the d-s hybridized bands than to the s-type bands in noble metals (107), and in Y alloys to the higher conduction electron density of Y metal states (108).

The temperature dependence of the hyperfine interaction of

TABLE 6a.

Hyperfine interaction and exchange coupling constant for ^{166}Er and ^{167}Er in various hosts.

	A^{166}(Moss) (MHz)	A^{167}/g_J(Moss.) (G)	A^{167}/g_J(EPR) (G)	$\|J_{sf}\|$(Moss.) (eV)	Ref.
Au:Er	247(3)	76.0(8)	75.5(5)	0.07(1)	107
Ag:Er	247(4)	76(1)	76(1)	0.10(1)	107
Th:Er	255.5(4.0)	78.7(1.0)	75(2)	0.025(2)	107
	--	75.5(1.0)	--	--	110
Zr:Er	250	--	--	0.035	109
Y:Er	260(3)	--	--	0.046	108

TABLE 6b.

Mossbauer parameters of ^{166}Er and ^{169}Tm in various absorbers.†

	Temperature (^oK)	T_c, T_N (^oK)	$\|H_{hf}\|$ (MOe)	$g_N\mu_N H_{eff}/h$ (MHz)	e^2qQ (MHz)	Ref.
Er (free ion)	--	--	--	1840	1100	66
$ErCr_4Al_8$	4.1	14(2)	--	1790(50)	690(66)	66
$ErMn_4Al_8$	4.1	15(2)	--	1735(50)	570(60)	66
$ErFe_4Al_8$	4.1	25	--	1610(50)	560(60)	66
$ErCu_4Al_8$	4.1	6(1)	--	1350(150)	490(60)	66
ErAu	<10	--	8.3(2)	--	--	113
ErFe	4.2	--	7.68(13)*	--	+748(24)*	112
ErNi (ns)	--	--	7.01(14)	--	+520(80)	112
(s)	--	--	4.76(9)	--	+296(28)	112
TmFe	--	--	6.25(4)	--	339(13)	99
TmNi (s)	--	--	1.43(2)	--	70(6)	99
(ns)	--	--	6.27(3)	--	463(7)	99

* Values onwards are extrapolated to 0^oK. s = substitutional, ns = non-substitutional

† Isomer shifts are very small in all these absorbers.

^{166}Er impurities implanted in iron and nickel (97-99,111-113) behaves in the same way as described above for Dy implanted in iron and nickel. The Mossbauer parameters for $ErM_4A\ell_4$ (M = Cr, Mn, Fe, Cu) are listed in Table 6b. The magnetic hyperfine interactions are again, as in the Dy system, close to that expected for the free ion, showing that the magnetic moment of each Er ion is close to 9 μ_B. These systems order antiferromagnetically (66).

The isotope ^{169}Tm has an 8.42 keV Mossbauer resonance. Hyperfine studies of Tm impurities implanted in iron, nickel and aluminium and some thulium intermetallics ($TmFe_2$, Tm_2Fe_{17}, $TmCo_3$, $TmZn_2$, $Tm_3A\ell_5O_{12}$, $Tm_3Ga_5O_{12}$, $TmMo_6S_8$) have been reported using the above resonance (97-99,114-119). Mossbauer parameters are listed in Table 6b. In $TmFe_2$ the saturation electric quadrupole interaction was measured to be 523(10) MHz. No appreciable Mossbauer absorption was observed for $TmNi_2$, $TmNi_3$ and $TmCo_3$ at low temperatures (115). In $TmZn_2$, the hyperfine field in the magnetically ordered state (T_N = 4-5° K) agrees well with the value for the free Tm^{3+} ion. An asymmetrically broadened quadrupole doublet in the paramagnetic temperature region is attributed to a relatively fast spin-lattice relaxation between low lying crystal field states (116).

Ytterbium has four stable even isotopes, ^{170}Yb, ^{172}Yb, ^{174}Yb, and ^{176}Yb, all having rotational ($2^+ \rightarrow 0^+$) transitions near 80 keV. The odd isotope ^{171}Yb has two useful Mossbauer levels at 67 keV and 76 keV, both belonging to a ground state rotational band. Another odd isotope ^{173}Yb with 78.7 keV transition has a very broad resonance line width and is not very useful for Mossbauer measurements. The best choices are the 84.2 keV transition in ^{170}Yb and the 67 keV transition in ^{171}Yb (11). Hyperfine interactions and electronic relaxation of ^{170}Yb in different compounds have been studied by Mossbauer measurements (29,31,65-66,118,120-139). Mossbauer parameters for different absorbers have been summarized in Table 7. An extensive review on the Mossbauer spectroscopy of ^{170}Yb has been published by Nowik and Bauminger (31). We shall therefore present only the salient features of the data published thereafter. Isomer shift data for YbB_4, YbB_{12}, and $YbA\ell_2$ indicate that Yb has an intermediate valence state in these compounds (29). In $YbA\ell_3$, Yb has a valency close to 3 (136). Mossbauer and susceptibility measurements show that the Yb ion is divalent in $YbMn_4A\ell_8$ and trivalent in $YbFe_4A\ell_8$. However, in $YbCr_4A\ell_8$ and to some extent in $YbCu_4A\ell_8$, it is in a mixed valent state (66).

TABLE 7.

Hyperfine interaction parameters and spin relaxation times for ^{170}Yb in different absorbers. I.S. is relative to $TmA\ell_2$

Compounds	Temp. (^oK)	Ground State	I.S. (mm/sec)	$g_n\mu_nH_{eff}$ (MHz)	Hyperfine Inter. Constant A (mm/sec)	e^2qQ (MHz)	Spin Relaxation Time at 4.2^oK(ns)	Ref.
$YbSo_4$	--	--	-0.42(6)	--	--	520(15)	--	31
$CaF_2(Yb^{2+})$	4.2	--	-0.274(10)	--	--	0	--	29
YbB_6	4.2	--	-0.263(8)	--	--	0	--	29
$YbA\ell_2$	4.2	--	-0.063(10)	--	--	0	--	29
YbB_4	4.2	--	+0.020(10)	--	--	<163	--	29
YbB_{12}	4.2	--	0.038(10)	--	--	0	--	29
$YbA\ell_3$	4.2	--	0.034(10)	--	--	0	--	29
	4.2	--	0.082(12)	--	--	--	--	136
$YbCr_4A\ell_8$	4.1	--	-0.2	--	--	385(10)	--	66
$YbMn_4A\ell_8$	4.1	--	-0.2	--	--	-340(10)	--	66
$YbFe_4A\ell_8$	4.1	--	--	--	--	2280(30)	--	65,66
$YbCu_4A\ell_8$	4.1	--	+0.1	--	--	210(10)	--	66
Yb:Au	4.2	Γ_7	--	--	13.3	--	1.11(4)	120
$CaF_2(Yb^{3+})$	4.2	Γ_7	0.08(5)	--	13.18(7)	--	1.2	126
YbES	4.2	--	0.13(7)	--	A11=13.3(2)	904(16)	--	127
$Cs_2YaYbC\ell_6$	4.2	Γ_6	--	--	-10.9(1)	--	4.7(5)	121,122
$Rb_2NaYbC\ell_6$	4.2	Γ_6	--	--	-10.8(3)	--	0.98(5)	123
$YbAuNi_4$	4.2	Γ_6	--	--	-10.3(4)	--	0.96(5)	130
$Yb:TmBe_{13}$	4.2	Γ_6	--	--	- 8.5(9)	--	1.07(6)	130
$Yb:TmFe_2$		$\sim\|7/2\rangle$	--	1130(30)	--	2400(250)	0.70(15)	129
$Yb:TmA\ell O_3$	4.2	$\|\pm 7/2\rangle$	--	--	27.1(2)	2200	--	135
$YbFeO_3$	4.1	--	0	270(3)	--	1660(60)	--	128
$YbCrO_3$	1.3	--	--	~25	--	2060(30)	--	31
$YbVO_3$	4.1	--	--	~188	--	2020(60)	--	31

Hyperfine interactions and relaxation effects have been investigated in $YbCl_3 \cdot 6H_2O$ and $Cs_2NaYbCl_6$ by means of the 66.7 keV Mossbauer resonance in ^{171}Yb (140,141). At low temperatures in $YbCl_3 \cdot 6H_2O$ a magnetic hyperfine field of 2900(30) kOe was observed with negligible quadrupole interaction. The measured spin-spin relaxation time was 6 ns.

3.2 Rare Earth Iron Compounds

Rare earths form an enormous number of crystalline and amorphous compounds with 3d transition metals. The interest in these compounds derives from the intrinsic properties of both components; i.e., from the high magnetic moment per atom, the strong single ion magnetocrystalline anisotropy of the rare earth and from the high magnetic coupling strength of the moment in the 3d transition metal. All of these compounds (particularly those with iron) exhibit large magnetostriction as well as relatively high magnetic saturation and magnetic ordering temperatures. Such properties make these compounds potentially useful materials for various technical applications; e.g., high power magnetostrictive transducers and micropositioning devices of various kinds. Magnetic anisotropy and magnetostrictive properties have been correlated with Mossbauer parameters and extensive studies done over the past few years. In the following sections we review the results deduced from the Mossbauer studies of the rare earth-iron compounds.

3.2.1 RFe_2 (Laves Phase) Compounds

The RFe_2 (Laves phase) compounds possess large magnetic anisotropies and magnetostriction as well as high magnetic ordering temperatures (Table 8). In the magnetostrictive materials, spin fluctuation in the magnetic critical region influence the phonon distribution and cause sharp changes in the ultrasonic attenuation coefficients and the velocity of sound (142). Other physical quantities which are sensitive to the phonon distribution should show similar phenomena. Two such quantities, measurable by Mossbauer techniques, are the Mossbauer recoil free fraction (f) and the second order Doppler shift (δ). In crystals with weak magnon-phonon coupling the temperature dependence of f(T) and δ(T) is a smooth, monotonically decreasing function of temperature. In the presence of strong magnon-phonon coupling, spin fluctuations are expected near the critical temperature which may affect both f(T) and δ(T).

TABLE 8.

Curie temperatures T_c(^oK) for crystalline (c) and amorphous (a) RFe_2 compounds.

Compounds		$CeFe_2$	$SmFe_2$	$GdFe_2$	$TbFe_2$	$DyFe_2$	$HoFe_2$	$ErFe_2$	$TmFe_2$	YFe_2	Ref.
(i) <u>Magnetic measurements</u>											
	(c)	235	688	796	697	635	597	574	566	535	4,148
	(a)	--	--	500	388	299	200	105	--	55	240,242
(ii) <u>Mossbauer measurements</u>											
	(c)					633(1)	596(1)	576(1)	563(1)	533(1)	148
	(c)				693(3)	633(1)	593(1)				162

The RFe_2 cubic Laves compounds with large magnetostrictive coupling constants (143,144) have been intensively investigated. The RFe_2 compounds with the exception of $CeFe_2$ are ferrimagnets and crystallize in the cubic Laves phase $MgCu_2$ type structure. The rare earth ions are situated on a diamond sub-lattice and the iron ions on a corner-sharing network of regular tetrahedra. The site symmetry of the iron ions is $\bar{3}m$ with each being crystallographically equivalent and the three-fold axis in the [111] direction. The point symmetry of the rare earth site is cubic ($\bar{4}$ 3m). The Fe-Fe exchange is much stronger than the R-Fe exchange.

Mossbauer studies on ^{57}Fe have revealed that, even though all the RFe_2 compounds have an identical crystallographic structure, various types of spectra (145) are observed. The appearance of the different spectra was interpreted in terms of different directions of the magnetization axis relative to the crystallographic axes of the unit cell on the basis of a simple crystal field model. It was found that the direction of magnetization is determined by the sign of parameter B_4 in Eqn. 17. With the direction of magnetization $\vec{n}$ along [111], three out of four Fe atoms have an angle θ between the principal quadrupolar field axis and a magnetic field direction 70° 32', while for the fourth atom θ is zero. Thus, two magnetically inequivalent iron sites exist, giving rise to a Mossbauer spectrum below T_c which is a superposition of two six line patterns with an intensity ratio of 3:1. This situation prevails in $TbFe_2$, $ErFe_2$, $TmFe_2$ and YFe_2 compounds. When the direction of easy magnetization $\vec{n}$ is along [100], all iron atoms are equivalent and θ is the same for all (54° 44'). A simple six line spectrum is obtained below T_c, as is seen for $HoFe_2$, $DyFe_2$, and at low temperature for $PrFe_2$ and $YbFe_2$. With $\vec{n}$ parallel to the third major cubic axis [110], two magnetically inequivalent iron sites with a population ratio 1:1 are present. This is observed in $NdFe_2$ and $SmFe_2$ at low temperatures. This angular dependence is very well depicted in Fig. 1.

Mossbauer studies on ^{57}Fe in polycrystalline RFe_2 (R = Tb, Dy, Ho, Er, Tm and Y) in the temperature range 4.2-900° K (145-153) and for R = Ce, Pr, Nd, Sm, Gd, Yb and Lu in the temperature range 4.2-300° K (154-163) will be discussed. In recent Mossbauer studies Shechter et al (150) and Bukshpan et al (151,152) have confirmed previously reported results by Kimball et al (146,147). Fig. 2 shows the temperature dependence of the Fe magnetic hyperfine field in RFe_2 (R = Tb, Dy, Ho) Laves phases compounds as reported by different

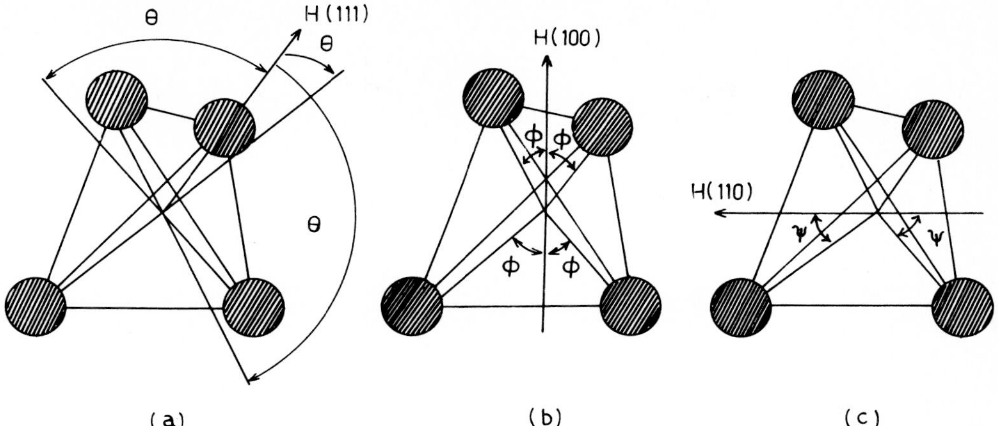

Fig. 1. The location of the iron atoms in the $MgCu_2$ structure relative to the [111],[110] and [100] axial direction.

authors (146,147,150,151). For $ErFe_2$ a similar behavior of hyperfine fields was observed for each value of θ (146,147). In addition the direction of magnetization in $ErFe_2$ was found to rotate from [111] to[110] near 490° K, which is the compensation point. The effective hyperfine fields H_{eff} measured on Fe nuclei in RFe_2 are listed in Table 9 (this is an updated Table A2 of Ref. 4).

Kimball et al (146,147) have found that the isomer shift, the quadrupolar coupling and the Debye-Waller factor undergo rather abrupt changes when the temperature is increased through the magnetic ordering temperature T_c in $HoFe_2$, $ErFe_2$ and $TbFe_2$. The latter two anomalies were attributed to the disappearance of the spontaneous magnetostriction above T_c, whereas the isomer shift change was ascribed to a decrease in the s-electron density at the Fe nuclei due to a shift in the Fermi level near T_c. X-ray measurements (164) showed that at 300° K the cubic unit cell of $TbFe_2$ is appreciably distorted along one of the [111] axes. This distortion has been detected by Mossbauer measurements (146,147) through the sensitivity of the spectrum to the value of θ. Bowden (149) also drew attention to the fact that the ^{57}Fe nuclear spin moments may not, in general, point exactly along the direction of magnetization, but are tipped away from these directions through the action of local dipolar fields.

Shechter et al (150) and Bukshpan et al (151) have shown that an increase in the quadrupole splitting near T_c is due to an anisotropic softening of the iron vibrational modes. The occurrence of a

TABLE 9.

Hyperfine fields H_{eff} measured on Fe nuclei in various RFe_2 compounds. (1) and (11) refer to two non-equivalent Fe sites when the easy direction of magnetization is not [100]. Asterisk indicates easy direction changes at higher temperature.

Compounds	Easy Direction	(1)	H_{eff} (kOe)	(11)	T (^oK)	Ref.
$CeFe_2$	[100]		156		80	4
	[100]*		156		80	4
	[100]		165		4.2	4
$PrFe_2$	[100]		190		4	161
$NdFe_2$	[110]	190(H_{av})			4.2	161
	[100]		177		300	161
$SmFe_2$	[110]	216		190	80	4
	[110]*	217		198	4.2	4
	[110]*	214		190	80	4
$GdFe_2$	Complex	238		224	80	4
	[111]	240		255	4.2	4
	[100]		225.6		80	4
	[110]	238		225	78	4
$TbFe_2$	[111]	226		197	80	4
	[111]	235		206	4.2	4
$DyFe_2$	[100]		228		80	4
$HoFe_2$	[100]		221		80	4
	[100]		228		20	4
$ErFe_2$	[111]	228		204	80	4
$TmFe_2$	[111]	216		202	80	4
$YbFe_2$	[100]*		206		5	158
$LuFe_2$	[110]	207		203	80	4
	[uuw]		214.6		80	4
YFe_2	[111]	221		221	80	4
	[111]	215		208	1.7	4
	[111]	220		210	4.2	4

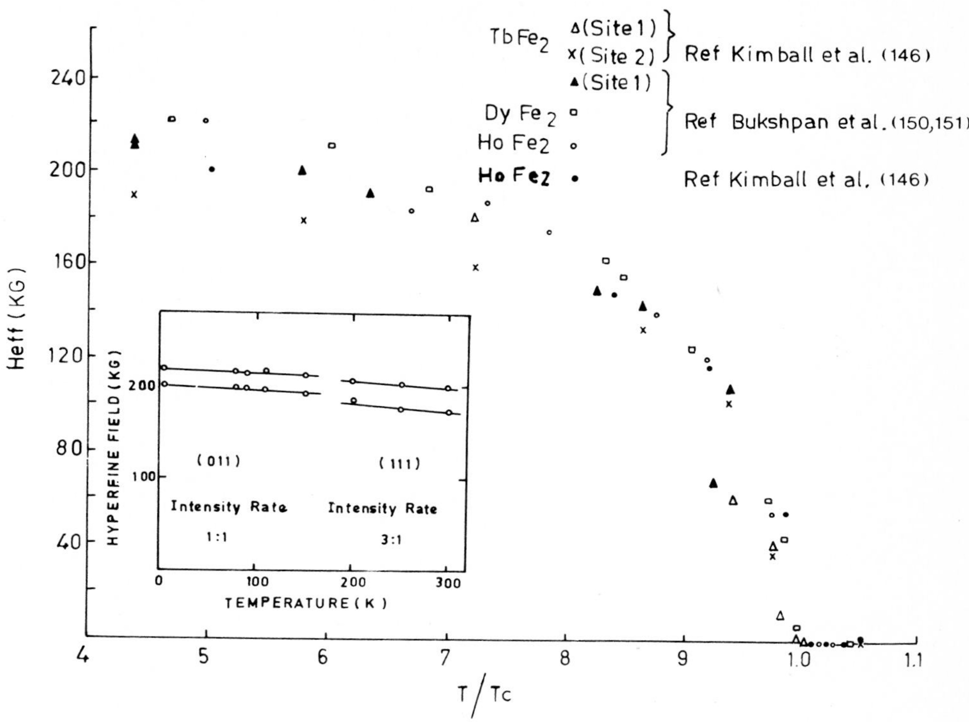

Fig. 2. Temperature dependence of the hyperfine fields in $TbFe_2$, $DyFe_2$ *and* $HoFe_2$. *The insert shows the hyperfine fields in* $SmFe_2$ *as a function of temperature (Van Diepen et al (154)).*

softening in the phonon spectrum near T_c may reduce the Mossbauer recoil free fraction in the critical region. The temperature dependence of 'f' suggests a critical reduction by about 40% in $DyFe_2$ and 30% in $TbFe_2$. The slopes of center shift vs. temperature change abruptly in the critical region. No such anomalies were observed in YFe_2 indicating that the observed phenomena are to be associated with a large magnetoelastic coupling present in $DyFe_2$ and $TbFe_2$ but absent in YFe_2 (13). Recently Bukshpan and Nowik (165) have shown that the change in Mossbauer recoil free fraction, isomer shift and electric field gradient at the iron site near the magnetic ordering temperature in the magnetostrictive materials can be explained in terms of a mean square vibrational amplitude $\langle\langle u^2\rangle\rangle$

which changes sharply and in an anisotropic manner near T_c.

The magnetocrystalline anisotropy in RFe_2 (R = Tb, Dy, Ho, Er and Tm) Laves compounds is mainly due to the interaction between the 4f electrons of the rare earth ions and crystalline field. Other contributions come from the magnetic anisotropy of the iron sub-lattice and from the possible anisotropy of Fe-rare earth exchange interactions.

Mossbauer studies (154,155) on $SmFe_2$ revealed that a peculiar behaviour with spin reorientation takes place over a wide temperature range. At temperatures below 80° K the easy direction of magnetization of the samarium sublattice is directed along the [110] direction, while at room temperature, it is along the [111] direction. This transition takes place around 175° K. The occurrence of spin reorientation as a function of temperature was carefully studied (154,155), and detailed calculations, based on the single ion crystal field model, were carried out (154) using the Hamiltonian (Eqn. 19) with an additional spin-orbit coupling term $\lambda\bar{L}.\bar{S}$. Fig. 3 shows the regions of easy direction of magnetization as a function of the crystal field parameters for four temperatures and an exchange field of $\mu_B H_{ex}/k$ = 130° K. The experimentally observed easy direction of magnetization can be accounted for if $A_4\langle r^4\rangle$ and $A_6\langle r^6\rangle$ are approximately of the same magnitude with $A_4\langle r^4\rangle$ positive and $A_6\langle r^6\rangle$ negative. It was also shown that the sixth order crystal field parameter plays a significant role in determining the behaviour of $SmFe_2$. The insert in Fig. 2 shows the temperature dependence of the hyperfine field in $SmFe_2$.

The magnetic anisotropy in $CeFe_2$ is entirely due to that of the iron sublattice (156). The temperature dependence of the Mossbauer spectra indicated the presence of a single iron site, which corresponds to the [100] easy axis of the magnetization up to 150° K. Above 230° K the spectrum is a doublet. Between 150-230° K the spectra show complex behaviour with a spin reorientation similar to that of $SmFe_2$.

In $GdFe_2$, Gd^{3+} is an S-state-ion and does not interact with the crystal field. On the basis of their experimental results, Atzomony and Dariel (156) observed that the magnetic anisotropy in this compound is due to the Gd-Fe exchange interaction which is so large that the Fe-Fe interaction becomes negligible. However, in later studies by different authors (157,159,160) it was found that the Mossbauer spectrum of $GdFe_2$ at 77° K was complex and the easy direction of magnetization could not be straightforwardly determined

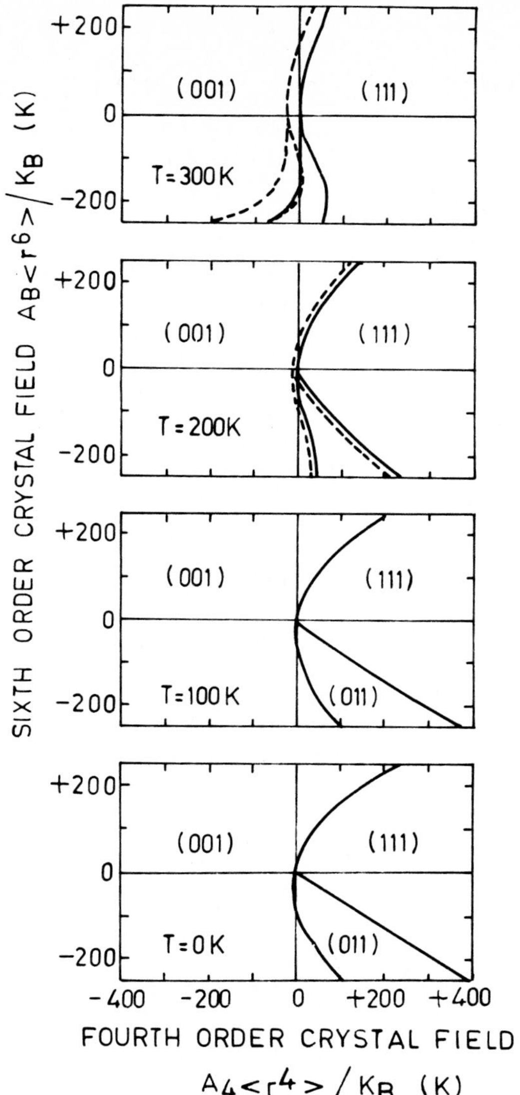

Fig. 3. Easy axis diagram for $SmFe_2$ *for four different temperatures and* $\mu_B H_{ex}/k_B = 130^o$ *K. The dashed lines indicate the transitions due to an additional bulk anisotropy* K_1/k_B *= -2 per formula unit (154).*

as one of the [100], [110] and [111] directions, observed previously (156). Such complex behaviour was attributed to competition between Fe-Fe and Fe-Gd exchange interactions.

A Mossbauer study of ^{57}Fe was carried out in $YbFe_2$, $NdFe_2$ and $PrFe_2$ Laves phase compounds with temperature range 4.2-300° K

(158,161-163). $YbFe_2$ is a ferrimagnet with a compensation temperature of 31(7)° K. Below 50° K it is magnetized along [100]; above this temperature the magnetization seems to deviate from [100]. Exchange interactions appear to dominate the crystalline electric field effects. In $NdFe_2$ two subspectra are observed at low temperature so the direction of magnetization is [110]. Around 160° K a rotation of magnetization from [110] towards [100] takes place with tipping of 10° with respect to [100]. This effect is attributed to dipolar fields.

The role of magnetostrictive effects may be taken into account in the interpretation of magnetic anisotropy in Mossbauer experiments on the RFe_2 compounds magnetized along [111] (143). These effects are apparently too small to account for the difference between the crystalline potentials observed in $YbFe_2$ on the one hand, and in $TmFe_2$ or $ErFe_2$ on the other. This difference seems to arise either from a nonconstancy of the true crystalline potential V_4+V_6 in the RFe_2 series or from anisotropic R-Fe exchange.

3.2.2 Rare Earth Iron Ternary Compounds

In magnetostrictive devices it would be highly desirable to reduce the magnetocrystalline anisotropy as much as possible while retaining a large magnetostriction. One possible technique for accomplishing this objective in to consider rare earth-iron pseudo-binary compounds of the form $R(Fe_{1-x}T_x)_2$ and $R^1_{x'}$, $R^2_{1-x'}$, Fe_2, where T and x and R^1, R^2 and x', respectively, are chosen to minimize the anisotropy. Here we present a summary of the Mossbauer and magnetic measurements in the ternary compounds.

$R(Fe_{1-x}T_x)_2$ and $R(Fe_{1-x}T_x)_3$ ternaries (R = rare earth and T = Aℓ, Mn, Co and Ni).

The magnetization and Mossbauer studies at ^{57}Fe nuclei have been reported (166-196) on $R(Fe_{1-x}T_x)_2$, where R = Ce, Gd, Tb, Dy, Ho, Er and Y and T = Aℓ, Mn, Co and Ni. The investigations were mainly confined to studies on the effects of changes in electron concentration on (a) the magnetic anisotropy (b) Mossbauer parameters and (c) the ordering temperatures. X-ray studies showed (166-169) that two structure types are formed for $R(Fe_xA\ell_{1-x})_2$. The compounds close to the RFe_2 or $RA\ell_2$ in $R(Fe_xA\ell_{1-x})_2$ are $MgCu_2$-type (cubic) C15 Laves phases for $1 \geq x \geq 0.67$ and $0.4 \geq x \geq 0$, and $MgZn_2$-type (hexagonal) C14 Laves phases for $0.57 \geq x \geq 0.47$, where R = Sc, Y, Pr,

Nd, Sm, Gd, Tb, Dy, Ho, Er, Tm and Lu. In the systems $R(Fe_{1-x}Mn_x)_2$ the Tb and Dy alloys possess the C15 cubic structure for all values of x. For Gd, the structure is C15 for x from 0 to 0.2 and 0.8 to 1.0; for the intermediate range x = 0.5 to 0.6 an unknown structure exists (176,177). For Er and Ho alloys, the structure is C15 cubic in the range x = 0 to 0.6 and C14 for the higher values of x.

The Curie temperature shows a linear decrease with increasing Aℓ concentration up to the phase boundary at x = 0.677 in $R(Fe_xA\ell_{1-x})_2$ (R = Gd, Dy and Ho). For hexagonal compounds only a slight variation of the Curie temperature with increasing Aℓ concentration is found. At high Aℓ concentration a small increase of T_c with increasing Aℓ values was observed. The small difference between the magnitude of the Curie temperatures for $R(Fe, A\ell)_2$ (R = Gd, Dy, Ho and Y) was attributed (168,169) to the domination of Fe-Fe exchange for $1 \geq x \geq 0.7$. For $Ho_{33.3}Fe_{66.6-x}A\ell_x$, x = 10, 20, 23.3 (iron-rich cubic), x = 26.6, 33.3, 38.3 (hexagonal), and x = 43.3, 50, 60 (aluminum-rich cubic) Mossbauer measurements at room temperature showed that the iron-rich cubic phase (x = 10) exhibits magnetic splitting (166,167) but that the other two iron-rich cubic phases do not. Presumably, they order magnetically at lower temperatures. Mossbauer spectra of the hexagonal phases exhibit magnetic splitting at temperatures near 4° K but not at room temperature. The aluminum rich cubic phases do not show magnetic ordering; in these phases the quadrupole coupling increases with increasing iron concentration, as does the electron density at the iron nuclei.

The results of Mossbauer measurements at ^{161}Dy and ^{57}Fe nuclei in $Dy(Fe_xA\ell_{1-x})_2$ in different concentrations are shown in Fig. 4. From Fig. 4(a) an increase in the magnetic hyperfine field and isomer shift of about 700 kOe and 1.5 mm/sec, respectively, is observed at the Dy nuclei in going from $DyAl_2$ to $DyFe_2$. The increase in the hyperfine field is partly due to the transfer of Dy-5d electrons to the iron site and partly due to the RKKY-polarization of conduction electrons by iron moments. The increase of the isomer shift is due to the removal of 5d-electrons which lowers the s-electron shielding and increases the electron density. In Fig. 4(b) a superposition of several different hyperfine fields and isomer shifts at ^{57}Fe nuclei are observed for intermediate concentrations. The effect is attributed to an appreciable amount of covalent, instead of long range metallic, bonds in these systems. It was concluded that the transferred d-electron density changes the spin and the charge density contributions

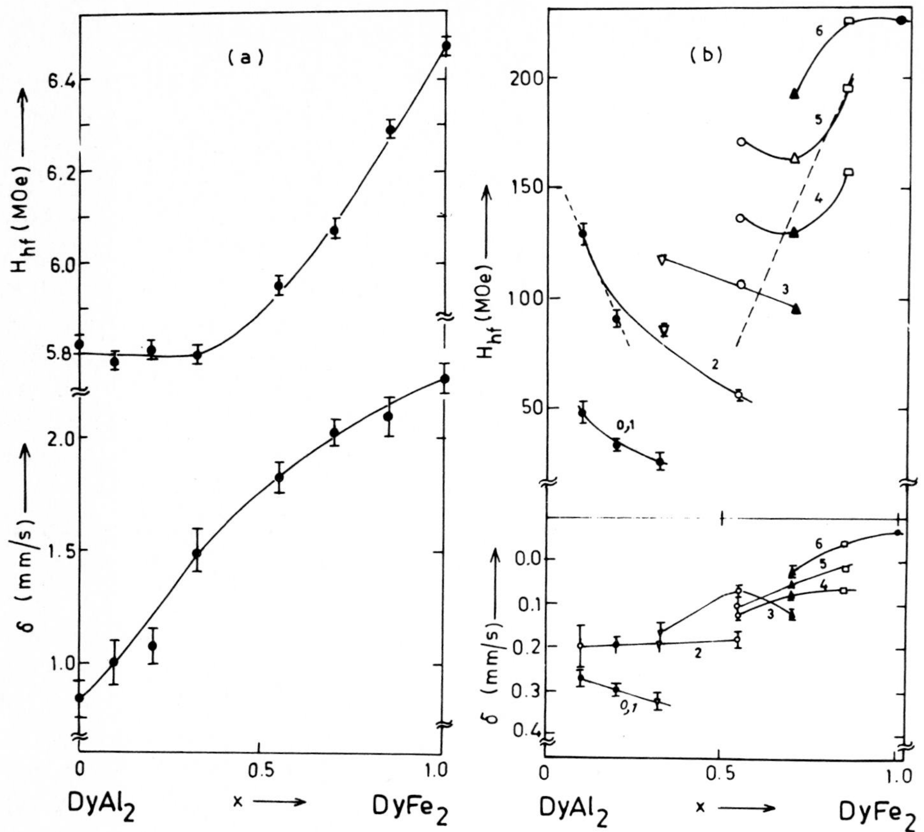

Fig. 4. Concentration dependence of hyperfine field H_{hf} and isomer shi shift (δ) at (a) ^{161}Dy nuclei and (b) ^{57}Fe nuclei in $Dy(Fe_xAl_{1-x})$ at 4.2° K. The parameters in (b) give the number of iron nearest neighbours (170).

at the iron site (170), increasing the isomer shift, as shown in the lowest curve of Fig. 4(b).

Mossbauer and magnetization (171) studies of $Tb(Fe_{1-x}Al_x)_2$ indicate that the total and sublattice magnetizations of this system vary with x. The Tb sublattice magnetization, in particular, varies quite rapidly for values of x near the point where the rhombohedral distortion vanishes (164).

The temperature dependence of Mossbauer and magnetic measurement in ternary compounds RFe_4Al_8 have been reported (65,172-174). The susceptibility and Mossbauer studies indicate the existence of two independent magnetic sublattices. The iron sublattice orders into an antiferromagnetic structure at about 120° K, whereas the rare earth sublattice orders (excluding those with La, Ce, Eu and Y) antiferromagnetically at about 20° K (65,174). The hyperfine field observed at 4.2° K at Fe nuclei in these compounds is approximately 110 kOe, about one third of the hyperfine field in pure Fe metal. This difference is attributed to the contribution of the core polarization to the hyperfine field, since the core polarization scales linearly with the magnetic moment (172,173).

Mossbauer measurements of $R(Fe_{1-x}Co_x)_2$ (R = Ce, Ho and Y) have been reported (178-180) at different temperatures and for various Co-concentrations. The Curie temperature decreases with increasing x, i.e., with increase of Co concentration. In Ce systems the Mossbauer spectra show magnetic ordering at 4.2° K in alloys with $0 < x < 0.94$. From the overall change in the hyperfine interaction at iron sites with concentration it was concluded that cerium atoms remain tetravalent. The variation of Curie temperature and distribution of hyperfine fields were attributed to changes of the iron moment as a function of its local environment (180). The Mossbauer spectra of $Ho(Fe_{1-x}Co_x)_2$ compounds show that the hyperfine field increases linearly with Co-concentration up to about 35% Co. This change is attributed to the filling of spin up and spin down d-sub bands by the extra electrons of the cobalt atom (178,179). The electric quadrupole splitting was found to increase with Co concentration and the angle θ between $\vec{q}$ and H_{eff} was observed to be 64° for $Ho(Fe_{.08}Co_{.92})_2$ and 58° for $HoFe_2$ so that the direction of magnetization deviates from the [001] direction (179).

Mossbauer investigations were carried out at different temperatures on compounds of the $R(^{57}Fe_xCo_{1-x})_2$ type with $0.01 < x < 0.05$ to study the possible variations of magnetic or structural properties of the RCo_2 compounds due to the presence of small amounts of iron (181-185). These findings indicate that the ^{57}Fe atoms randomly occupy Co sites and exert little effect on the magnetic properties. Slightly higher Curie temperatures (T_c) were observed by increasing the amount of Fe in $R(^{57}Fe_xCo_{1-x})_2$ for $0.01 \leq x \leq 0.05$ (R = Pr, Nd, Tb, Dy, Ho and Er). The Mossbauer spectra were characteristic of the [100] and [111] directions of magnetization for Pr and Dy, Tb and Er compounds, respectively. As the temperature increases the magnetic hyperfine fields decrease but retain the [100] or [111] directions of magnetization for the Pr and Dy or Tb and Er compounds, respectively. A quadrupole doublet is observed above T_c. The temperature dependence of ^{57}Fe hyperfine effective fields, H_{eff}, in Pr and Dy compounds (single iron site) and Tb and Er compounds (two iron sites, 3:1) is shown in Fig. 5. The steep decrease of H_{eff} in Pr, Er and Dy compounds near T_c suggests the existence of a first-order

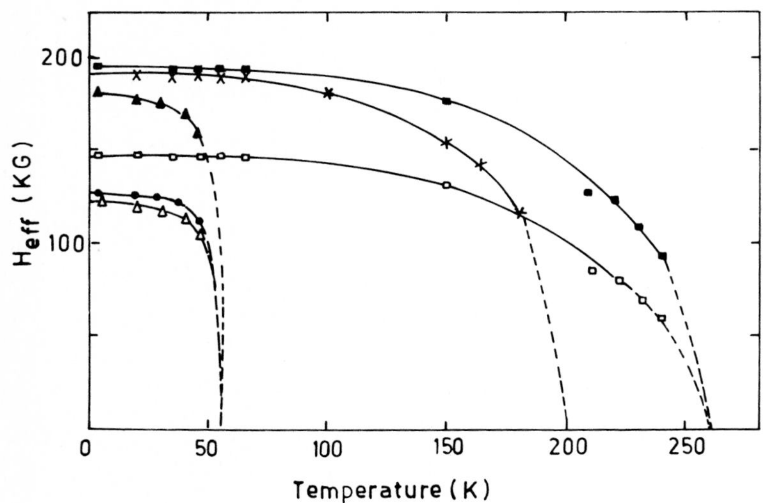

Fig. 5. The temperature dependence of the ^{57}Fe hyperfine effective fields, H_{eff}, ^{57}Fe doped in $PrCo_2$, $TbCo_2$, $DyCo_2$ and $ErCo_2$ (. and X represents Pr and Dy compounds, respectively, with $\vec{n}||$ to [100] cubic axis). For Tb and Er there are two inequivalent sites with intensity ratio 3:1 (□ and ■) and (▲ and △), respectively (183).

transition, while for the Tb compound, the smoother temperature dependence of H_{eff} indicates a second order transition at T_c (183). Spin-reorientation-like transitions were found in the Nd and Ho compounds (182,184). It was concluded that the R-Co interaction was not the only source of magnetocrystalline anisotropy but that additional contributions to the anisotropy free energy were present.

Crystallographic and Mossbauer investigations were reported for $Dy(Fe_{1-x}Co_x)_3$, $Er(Fe_{1-x}Co_x)_3$ and $Y(Fe_{1-x}Co_x)_3$ at different temperatures and concentrations (186,187). Spin reorientation was observed in Er and Y compounds from Mossbauer measurements. The variation of the ordering temperature with cobalt concentration showed a maximum at x = 0.3-0.5 for Dy and Y compounds. The difference in the ordering temperatures for both systems indicates that R-3d and R-R interactions were considerably smaller than 3d-3d exchange interactions. Moreover, the Fe-Co exchange is stronger than either Fe-Fe or Co-Co exchange forces in intermediate compositions.

The magnetic behaviour of $R(Fe_xNi_{1-x})_2$ (R = Ce, Gd, Dy, Ho and Y) has been studied by Mossbauer measurements (188-194). The Curie temperature decreases almost linearly in all the compounds with increasing Ni-concentration. Mossbauer studies indicate the disappearance of magnetic order at $x \approx 0.36$ in the $Ce(Fe_{1-x}Ni_x)_2$ system. The fitting of spectra for x = 0.06-0.12 at 4.2° K showed the easy direction of magnetization to be [111]. The variation of the ^{57}Fe isomer shift and the quadrupole splitting of $Ce(Fe_{1-x}Ni_x)_2$ alloys with composition exhibit much larger changes than the $Ce(Fe_{1-x}Co_x)_2$ system. This has been attributed to changes in the density of states at the Fermi level and to changes in the effective valency of cerium atoms with increasing nickel concentration (189). The ^{57}Fe hyperfine field at 4.2° K for Gd, Dy and Ho compounds decreases with increasing Ni content. The behaviour of Curie temperature and hyperfine field suggests the diminution of the Fe-Fe contribution to the exchange interaction (191,193). Mossbauer and x-ray studies were carried out on $Dy(Fe_xNi_{1-x})_3$ and $Y(Fe_xNi_{1-x})_3$ systems (195,196). The variation of Curie temperature and hyperfine field with Ni concentration was observed to be the same as in $R(Fe_xNi_{1-x})_2$ pseudobinaries. The ordering temperatures and Mossbauer parameters are widely different in Fe-Co and Fe-Ni ternaries of RFe_3 type, and suggest that the iron moments are more delocalized when the iron is partially replaced by nickel.

$R^1_xR^2_{1-x}Fe_2$ Ternaries

Mossbauer studies of $R^1_xR^2_{1-x}Fe_2$ compounds (R^1 = Ho, Dy and R^2 = Tb, Er and Tm) have been reported (197-203) at different temperatures between 4.2-300° K. The easy direction of magnetization was determined from Mossbauer spectra in terms of (x, T) spin-orientation diagrams (Fig. 6 (a, b)). Theoretical spin orientation diagrams were calculated assuming that the magnetocrystalline anisotropy was due to the anisotropy of the interaction between the 4f-electrons of the rare earth ions with the crystalline fields. The general features of the experimental results were reproduced by this theory. The contribution of the iron sublattice was also calculated and improved the agreement between the theory and experiment. Crystal field parameters, deduced using a point charge model, showed that a charge between +1 and +2 on the iron atom best describes

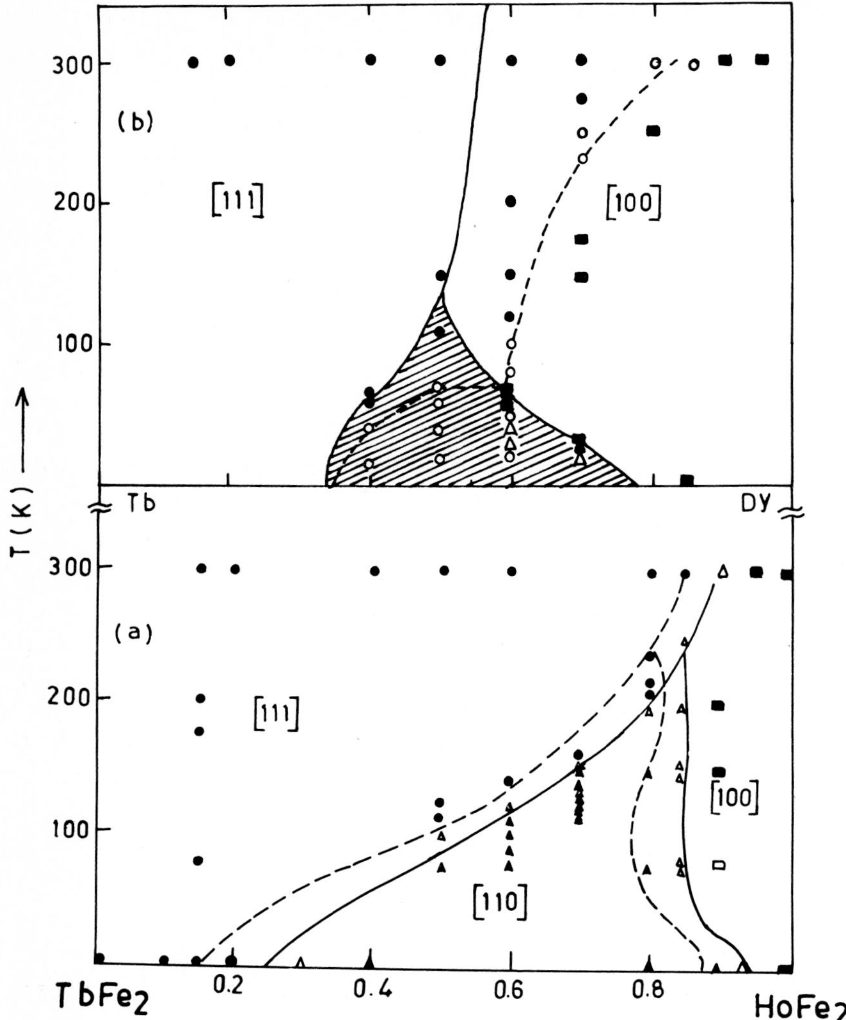

Fig. 6. Spin orientation diagram in (a) $Ho_xTb_{1-x}Fe_2$ (b) $Dy_xTb_{1-x}Fe_2$ systems. The solid and dashed lines are the calculated and experimental boundaries, respectively, between regions with [111], [100] and unusual (shaded area b) directions of easy magnetization (197,203).

the experimental features of the spin orientations.

The Mossbauer studies on $R^1_xR^2_{1-x}Fe_2$ compounds with R^1 and R^2 non S-state rare earths revealed that the magnetic anisotropy is due mainly to the interaction of the rare earth ions with the crystalline field. It is expected that for an S-state ion, e.g., Gd, the easy direction of magnetization will be determined only by the iron sublattice anisotropy or that of exchange interactions.

In $Gd_xY_{1-x}Fe_2$, the Curie temperature increased with increasing Gd concentration. As temperature and composition vary, the Mossbauer spectra show a change in the easy direction of magnetization. At 78° K the direction changes from [111] to [110] for $0 < x < 0.2$ and at 300° K a similar change takes place for $0.6 < x < 0.8$. The Gd-Fe exchange interaction seems to be responsible for determination of direction of magnetization at low temperatures. The increase of the ^{57}Fe hyperfine field with increasing Gd content is attributed to the change of the 4s conduction electrons (204). The ^{57}Fe Mossbauer measurements in $Er_xY_{1-x}Fe_2$ and $Yb_xY_{1-x}Fe_2$ showed that the direction of magnetization is along [111]. The hyperfine fields at the two inequivalent iron sites varied as a function of $\underline{x}$; the difference of 30 kOe between sites in both the compounds is attributed to additional polarization of the conduction electrons by the dipolar fields (205). In another communication (206), Mossbauer studies were reported on a series of ferrimagnetic $Tb_xY_{1-x}Fe_2$ compounds. Mossbauer results indicate a distortion (compression, stretching or displacement) of each tetrahedron for Fe atoms along the direction of magnetization. This distortion is indicated by the value of θ far from 70° for the intense component and near 0° for the less intense component. The quadrupole interaction and isomer shift are different for the two components. All of these results imply (206) that the iron sites corresponding to the two components of the spectrum are not equivalent chemically or crystallographically and are interpreted in terms of the iron atoms being displaced from the conventional position.

3.2.3 RFe_3, R_2Fe_{17} and R_6Fe_{23} Compounds

The RFe_3 intermetallic compounds crystallize in the rhombohedral $PuNi_3$ type structure, which belongs to the R $\bar{3}$ m space group. In this structure the rare earth atoms occupy two crystallographically inequivalent sites, a and c, in the ratio of 3:6; the iron atoms occupy three inequivalent sites, b, c, and h, in the ratio of 3:6:18. The point symmetries of the latter are $\bar{3}$m, 3m and m, respectively. In reality the situation is made more complex by the electric quadrupole and magnetic dipole interactions which can introduce additional inequalities within a given group of positions. For h positions with m point symmetry the situation is different. The EFG tensor is not axially symmetric, thus for different easy directions the h positions split into a maximum of three subgroups (h_1),

(h_2) and (h_3) occupied in the ratio of 6:6:6. Depending upon the easy axis orientation, the Mossbauer spectrum might consist of three, four or five patterns.

The ^{57}Fe Mossbauer studies of RFe_3 intermetallics (R = Sm, Gd, Tb, Dy, Ho, Er) have been reported over a wide range of temperature 4-800° K (187,207-217). The Mossbauer parameters are listed in Table 10a. The temperature dependence of the Mossbauer spectrum of $SmFe_3$ was interpreted as a superposition of three sextets for which the easy direction of magnetization is parallel to the c-axis and does not change with temperature. In $GdFe_3$ the Mossbauer spectrum showed the direction of magnetization to be along the a-axis at 4.2° K but to shift slightly at higher temperature. From the sign of the quadrupole spitting at 1(b) and 2(c) iron sites in $TbFe_3$, the magnetization is deduced to be along the crystallographic b-axis (210,214). At room temperature the Mossbauer analysis of $DyFe_3$ indicated that the magnetization was along the b-axis, but below the room temperature the easy axis does not point in any of the principal crystallographic axes (214,217). A rotation of the easy direction of magnetization is observed between 150° K and 200° K. In $HoFe_3$ a spin reorientation takes place at T $\sim$100° K; below this temperature the magnetization is parallel to the a-axis and above this temperature it is parallel to the b-axis. Arif et al (209) and Bowden and Day (215,216) observed a spin transition at T = 47° K in $ErFe_3$. From their Mossbauer and magnetization studies they found that the easy direction does not prefer to align along any individual crystalline axes. This behaviour was attributed to the rare earth sublattice magnetization and to the position of c and h iron atoms. However, Van der Kraan et al (214) and Gubben et al (187) reported that the easy direction of magnetization turns from the c-axis to the b-axis at T = 50° K with increasing temperature. They find that the dipolar contribution to the hyperfine fields in RFe_3 compounds does not play an important role, as observed by Arif et al (209). From the similarity in behaviour of the temperature dependence of the hyperfine fields and the isomer shifts associated with the different Fe sites, it was concluded that there is no substantial difference in the electronic configuration of these Fe atoms.

The rare earth-iron compounds of the type R_2Fe_{17} crystallise in the hexagonal (Th_2Ni_{17}) or rhombohedral (Th_2Zn_{17}) structure for heavy rare earth and light rare earth compounds, respectively. Most of the R_2Fe_{17} compounds behave as normal ferro-or ferrimagnets with the rare earth spin moment coupled parallel or anti-parallel

TABLE 10(a) : *Mossbauer parameters for various RFe_3 intermetallic compounds.*

		$SmFe_3$	$GdFe_3$	$TbFe_3$	$TbFe_3^*$	$DyFe_3$	$DyFe_3$	$HoFe_3$	$HoFe_3$	$HoFe_3^*$	$ErFe_3^*$	$ErFe_3$
	$T(^oK)$	393	4.2	295	77	295	77	79	138	77	77	25
b)	δ(mm/s)	+0.07	+0.31	+0.18	-0.25	+0.17	-0.18	+0.31	+0.24	-0.24	-0.11	+0.30
	ΔE(mm/s)	+0.62	-0.34	-0.47	--	-0.47	--	-0.42	-0.43	--	--	+0.36
	H_{eff}(kOe)	149	233	225	245	218	249	231	238	253	253.5	200
c)	δ(mm/s)	+0.07	+0.29	+0.20	-0.17	+0.18	-0.10	+0.29	+0.28	-0.17	-0.185	+0.21
	ΔE(mm/s)	+0.92	-0.48	-0.53	--	-0.50	--	-0.50	-0.50	--	--	+0.95
	H_{eff}(kOe)	226	249	232	245.5	223	245	240	242	246.5	227.5	268
h_1)	δ(mm/s)	+0.07	+0.30	+0.15	-0.16	+0.15	-0.17	+0.27	+0.25	-0.16	-0.16	+0.26
	ΔE(mm/s)	-0.05	-0.48	-0.13	--	-0.14	--	-0.38	-0.23	--	--	-0.08
	H_{eff}(kOe)	202	259	228	242.5	218	237.5	253	235	237.5	235.5	244
h_2)	δ(mm/s)	+0.07	+0.30	+0.15	-0.16	+0.15	-0.17	+0.29	+0.25	-0.16	--	--
	ΔE(mm/s)	-0.05	+0.27	-0.13	--	-0.14	--	+0.23	-0.23	--	--	--
	H_{eff}(kOe)	202	236	228	242.5	218	233.5	229	235	224.5	--	--
h_3)	δ(mm/s)	+0.07	+0.30	+0.19	-0.16	+0.16	-0.17	+0.29	+0.24	-0.16	--	--
	ΔE(mm/s)	-0.05	+0.27	+0.58	--	+0.44	--	+0.23	+0.49	--	--	--
	H_{eff}(kOe)	202	236	211	228.5	208	234.5	229	226	224.5	--	--
	$T_c(^oK)$	650	725	655	--	605	--	575	--	--	--	--
	$T_c(Moss. ^oK)$	---	--	--	643	--	--	--	--	577	533	--
	Ref.	214	214	214	209,210	214	209	214	214	209,210	209,210	213

* δ in this column is relative to $^{57}Co/Pd$ source and in all other columns relative to $Na_2Fe(CN)_5NO.2H_2O$.

TABLE 10(b) : *Mossbauer parameters of various R_6Fe_{23} compounds.*

Compound	H_{eff}(kOe)				Easy Direction	T (^oK)	Ref.
	b	f_1	f_2	d			
Ho_6Fe_{23}	352	310	267	283	[111]	4.2	4
	273	214	200	226	--	298	211
Er_6Fe_{23}	367	314	262	272	[111]	4.2	4
Tm_6Fe_{23}	364	315	255	270	--	4.2	4
Yb_6Fe_{23}	355	313	249	272	--	4.2	4
Yb_6Fe_{23}	370	309	253	270	[100]	16	226

to the Fe moments. There are four non-equivalent crystallographic sites for Fe (in Wyckoff's notation these are indicated by 4f, 6g, 12j, 12k, and 18f, 18h, 9d, 6c for the hexagonal and rhombohedral structure, respectively). Mossbauer and magnetization measurements in R_2Fe_{17} have been reported (218-224) for R = Nd, Pr, Tb, Ho, Dy, Er, Tm and Y. The Mossbauer spectra were resolved into subspectra corresponding to the four Fe sites. The hyperfine fields of the Fe atoms at the f sites were consistently ∿20% greater than the weighted average of the hyperfine fields (H_{av}) at the other three sites. In the magnetically ordered state all of these intermetallic compounds except Tm_2Fe_{17} have a basal plane anisotropy which is determined partially by the rare earth sublattice and partially by the 3d sublattice (218-220,222-224). Of special interest are the results on Tm_2Fe_{17} where the easy direction of magnetization changes at 72° K from parallel to the c-axis to parallel to the a-axis. This change is accompanied by a marked increase in the hyperfine fields (222). Below this transition temperature, the increase in the hyperfine field of the f site Fe atoms is rather large ($\Delta H_{eff} \approx$ 44 kOe). A small increase in the average hyperfine field at other Fe atoms ($\Delta H_{eff} \approx$ 14 kOe) is observed. The temperature dependence of the isomer shift at the f-site iron shows that it is systematically higher in Tm_2Fe_{17} than in Er_2Fe_{17} at temperatures below 70° K, whereas above 80° K it is the same for both compounds. Around 72° K, the isomer shift increases by ∿0.10 mm/s with decreasing temperatures in Tm_2Fe_{17}; i.e., the electron charge density at the f-site iron nucleus abruptly becomes smaller. The magnetization measurements also show a sharp peak near 72° K which is related to the change in magnetic anisotropy. It was concluded that the iron atoms, which form a part of the f site in R_2Fe_{17}, are normally responsible for basal plane anisotropy. Only in the case of Tm_2Fe_{17} does the strong anisotropy of this rare earth dominate at low temperatures.

The R_6Fe_{23} compounds have a cubic structure of the type Th_6Mn_{23} and space group Fm3m (O_h^5). The iron atoms occupy four crystallographic inequivalent sites in the ratio of 1:6:8:8. Mossbauer parameters in some of these compounds have been listed in Table 10(b). From the hyperfine field studies at various Fe sites in rare earth-iron compounds (RFe_3, R_2Fe_{17} and R_6Fe_{23}) the concentration dependence of the magnitude of the iron moment is found to be largely determined by the local environment (225).

3.3 Rare Earth Amorphous Compounds

An amorphous solid is one in which three dimensional periodicity is absent and which possesses only short range order. Amorphous materials can have quite different properties than those of crystalline solids. Amorphous alloys of rare earths with transition metals have been much studied. These materials have potential application in thermo-magnetic recording and in magnetic devices such as bubble domain memories. Many of the devices make use of Gd-Co based films which have both the necessary bulk anisotropy with the easy direction perpendicular to the plane of the film, and a compensation point near room temperature (227,228). It has become clear that a rich variety of complex magnetic structures can occur in amorphous solids (229-231). In this section we shall survey the information about the structure, bonding and magnetic properties of rare earth glasses and amorphous alloys deduced from Mossbauer studies.

Mossbauer studies of rare earth glasses have been very limited. We report those on thulium doped sodasilica glasses (232,233) and on europium silicate and phosphate glasses (234). In soda-silica glasses containing Tm, relatively narrow absorption lines and well resolved quadrupole splittings were observed, suggesting that the Tm^{3+} ion preferentially occupies sites of well defined symmetry which are similar to the case of Tm_2O_3. The exact symmetry of the thulium iron site is unknown. The strong temperature dependence of the quadrupole splitting shows the presence of a local CEF and a significant degree of short-range order (233).

In europium silicate and phosphate glasses, broad single line Mossbauer spectra are observed (234). The isomer shift for the glass, relative to Eu_2O_3, is 0.10(2) mm/sec at 300°, 77° and 4° K and the line width 3.9 mm/sec. The smallest isomer shift of Eu^{2+}, relative to Eu_2O_3, is about -10 mm/sec. It thus follows that europium is present in these glasses as Eu^{3+} and that the environment of Eu^{3+} in glasses is very similar to that of Eu^{3+} in the oxide. The broadening of the europium line in silicate and phosphate glasses was attributed to an unresolved quadrupole splitting and disorder in the glass structure. The recoilless fraction was smaller than in crystalline Eu_2O_3 at all temperatures. The most probable reason is that Eu^{3+} is less tightly bound in the glass than in the crystal. Such a reduction of the binding forces is consistent with a random network model of the glass structure with Eu^{3+} ions occupying interstitial positions.

It was proposed that in amorphous materials crystal fields also exist, although, no longer with cubic symmetry (235). The magnetic properties of the amorphous rare earth transition-metal compounds depend upon the nature of the crystal fields acting on the rare earth atoms. Mossbauer studies on amorphous RFe_2 (R = Tb, Dy, Ho and Er) alloys were reported by Sarkar et al (236-237) and showed that the crystalline electric field is well defined in these amorphous alloys, in agreement with the proposed model (235). X-ray (238) and neutron diffraction (239) showed no evidence of the cubic Laves structure in the crystalline phase or of any other microstructure; thus, the materials were topologically disordered. The data supported the simple picture of random close packing (RCP) of atomic spheres (235) since the radial distribution (238-240) function shows peaks at distances corresponding to the rare earth-rare earth separation, the iron-rare earth separation and the iron-iron separation. The iron-iron separation is much the same as in the Laves phase, but the iron-rare earth and rare earth-rare earth separations are greater in the amorphous phase.

Fig. 7 shows ^{57}Fe spectra of $HoFe_2$ in the amorphous and polycrystalline states. The measurements in the paramagnetic state were made at 295° K on the amorphous sample and at 700° K on the polycrystalline sample. Amorphous $HoFe_2$ is paramagnetic at room temperature, while the crystalline $HoFe_2$ is paramagnetic above 596° K. In spite of the difference in ordering temperature and the difference in interatomic distances in the amorphous and polycrystalline states, the two spectra are much the same. Measurement on $DyFe_2$ and $ErFe_2$ showed a well resolved doublet with quadrupole splitting of the same magnitude ($\sim$0.4 mm/sec) as in the corresponding crystalline phase (237).

Fig. 7 also shows the spectra for amorphous $HoFe_2$ at 230° K and 145° K. The partially resolved doublet, due to the quadrupolar interaction, persists from 295° K to 230° K, but at 145° K, the pattern changes into a broad spectrum, indicating that the Curie temperature for amorphous $HoFe_2$ is between 145° K and 230° K. Spectra for amorphous and polycrystalline $DyFe_2$ at room temperature is also shown. The amorphous sample displays the doublet of the paramagnetic regime, but the polycrystalline sample shows the typical six-line spectrum of the ^{57}Fe nucleus in a hyperfine exchange field. Room temperature measurements of $TbFe_2$ showed a broad absorption curve indicative of magnetic ordering.

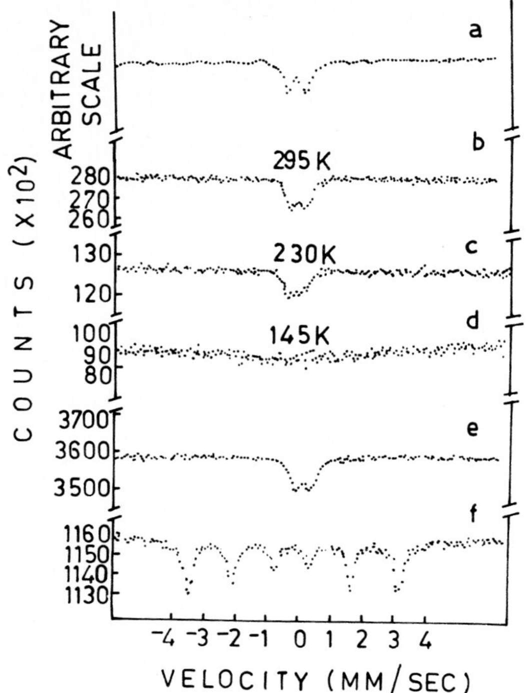

Fig. 7. Mossbauer absorption spectra of ^{57}Fe in $HoFe_2$ (a) paramagnetic polycrystalline sample (b) paramagnetic amorphous sample (c) amorphous sample approaching the magnetic transition and (d) in the magnetically ordered state. $DyFe_2$ at 295° K (e) amorphous (f) polycrystalline (237).

The above Mossbauer studies of ^{57}Fe in amorphous RFe_2 alloys reported an absence of resolved hyperfine structure below the Curie temperature T_c. However, in more recent studies Forester et al (241) found a broad but resolved distribution of the hyperfine fields in amorphous $DyFe_2$. Fig. 8 shows ^{57}Fe Mossbauer spectra of $DyFe_2$ at different temperatures. At low temperatures the mean value of $H_{eff}(Fe)$ is 225 kOe, virtually the same as for crystalline $DyFe_2$ (228 kOe) (145), although much smaller than for elemental iron, presumably due to electron transfer from the rare earth to the iron d-band. It was also observed that, whereas 12 Fe nearest neighbours surround rare earth atoms in crystalline RFe_2, this number is reduced on the average to about 6-7 in the amorphous state.

Mossbauer and magnetization studies have also been reported on amorphous $GdFe_2$ and YFe_2 (242-244). The Curie temperature (Table 8) of amorphous $GdFe_2$ is 500° K and drops sharply as one proceeds to the right in the series (decreasing R spin) approaching the zero

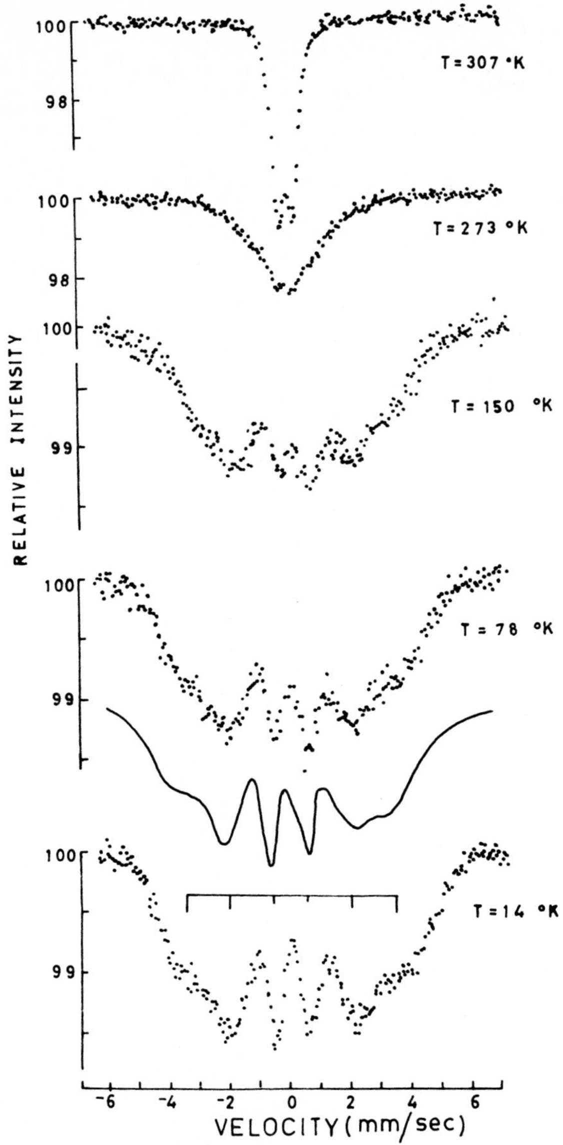

Fig. 8. Temperature dependence of ^{57}Fe Mossbauer spectra in amorphous $DyFe_2$ (241).

spin limit; YFe_2 exhibits <u>no long-range order</u>. This is in marked contrast to the analogous crystalline Laves phase compounds, which show a much weaker dependence on the rare earth spin.

Yttrium, although not a rare earth, represents a non-magnetic "pseudo rare earth" substitution which provides information on the

Fe-Fe exchange interaction. The temperature dependence of Mossbauer spectra of amorphous YFe_2 (242) shows behaviour characteristic of long-range magnetic order for $T < T_c = 55(5)^\circ$ K in contradiction to earlier magnetization studies (245). Above T $\sim 58^\circ$ K the spectra exhibit a quadrupole doublet which is characteristic of all RFe_2 amorphous alloys (236,237,241). The quadrupole splitting decreases monotonically from 0.495 $\pm$ .006 mm/sec at 80° K to 0.423 $\pm$ .006 mm/sec at 300° K (242). As the temperature is lowered below 58° K, a magnetic hyperfine splitting develops. The average magnetic hyperfine field $\bar{H}_{eff}(T)$ grows smoothly with decreasing T and closely follows an S = 1 Weiss molecular field dependence. The measured value $\bar{H}_{eff}$ = 233(6) kOe at 5° K is comparable with that in amorphous $GdFe_2$ ($\bar{H}_{eff}$ = 248 kOe), which has the largest hyperfine field of any heavy rare earth (RFe_2) amorphous alloy. For further understanding of the low temperature magnetic phase, spectra were taken in external magnetic fields up to 70 kOe in the Yttrium compound and, for comparison, spectra were taken on amorphous $GdFe_2$ under similar conditions (243). Like Gd, Y has no orbital anisotropy but Y, unlike Gd, carries no magnetic moment. In $GdFe_2$ at 20° K a field of 7 kOe, sufficient to overcome the sample demagnetizing field, completely aligns the moments in this low anisotropy ferrimagnet. For YFe_2 at 10° K, under small external magnetic fields, completely different and anomalous behaviour was observed; at 10 kOe the spectra virtually remained unchanged while at 20 kOe a slight broadening was observed. This behaviour is not characteristic of amorphous ferromagnetism but of spero-magnetism (246) (to be discussed later in detail). The Mossbauer spectrum at 163° K and in an external field of 70 kOe shows that the magnetic component collapses as the temperature is raised; the magnetic and non-magnetic components appear to maintain the same relative areas. It was concluded that amorphous YFe_2 displays characteristic spin-glass behaviour (242,243).

Magnetic structure and properties of rare earth transition-metal (RT_3) amorphous alloys,(R = Gd, Dy, Y and T = Fe, Co, Ni) have been studied by Mossbauer spectroscopy and magnetization measurements (247-251). Magnetic exchange interactions depend strongly on the distance between interacting atoms, bond angle for super-exchange and the number of interacting neighbours. These parameters have a fixed value on equivalent sites in crystalline materials, but no two atomic sites are equivalent in amorphous solids. This inequiva-lency of sites leads to a distribution in magnitude of the atomic moments. Moment distributions are narrowest in rare earth compounds and alloys because the 4f shell is well shielded from its neighbourhood. Moment distributions are best inferred from the hyperfine field distributions $P(H_{hf})$ measured by NMR or Mossbauer spectroscopy. The $P(H_{hf})$ distributions provide a detailed picture of the appearance of magnetism. Further information is obtained from the transition probabilities for Mossbauer transitions which yield $\langle \sin^2\psi \rangle$, where ψ is the angle between a spin and γ-direction imposed in the experiment.

The inequivalency of the sites also leads to a distribution in the exchange interaction due to the dependence of J on distance and bond angle for super-exchange. Further, if the coupling between pairs of spins is negative, then different exchange paths can favour different relative alignments of a pair. This effect may lower the ordering temperature and produce a random, non-collinear structure in amorphous lattices. With this information, the magnetic structures occurring in amorphous solids can be correlated with Mossbauer and magnetization measurements.

A magnetic one-subnetwork is defined as a group of atoms with similar magnetic interactions. Possible one-subnetwork structures are shown in Fig. 9(a). Ferromagnetism is conceptually the simplest

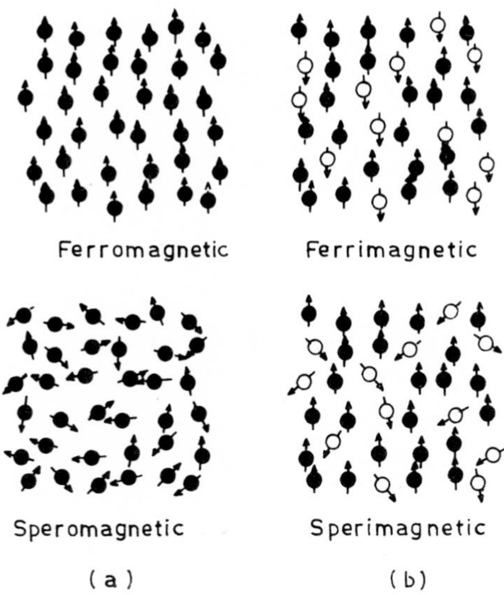

Fig. 9. Amorphous magnetic structure (a) one-sub-network (ferromagnet, speromagnetic) (b) two-sub-network (ferrimagnetic, sperimagnetic).

type of magnetic order which can exist in an amorphous solid. Amorphous antiferromagnetism is not so evident, though it is easy to picture two interpenetrating sublattices. A novel feature of amorphous magnetism is the possibility that the spins within a domain may be frozen into more or less random orientations (called speromagnetism-spins randomly distributed over all directions). Spins distributed over some directions with greater probability than for other directions results in an asperomagnetic structure. A speromagnet resembles a spin glass while an asperomagnet is a random ferromagnet.

The two-subnetworks are distinguishable groups of atoms with different magnetic interactions. Possible two-subnetwork structures are shown in Fig. 9(b). The ferrimagnetic structure is the direct analogue of crystalline ferrimagnetism except for the random positions of atoms belonging to the two-subnetwork. A sperimagnetic structure is one in which the moments of one or both sub-networks are frozen into random orientations (i.e., random ferrimagnet).

Mossbauer spectra of ^{161}Dy resonance were taken for amorphous and crystalline $DyFe_3$, $DyCo_3$ and $DyNi_3$ at 4.2° K (247,248). The overall splitting for amorphous and crystalline materials was similar but in the amorphous alloys some lines were broadened due to a distribution of magnetic and quadrupole interactions. Magnetic hyperfine field, quadrupole interaction, and ordering temperatures for these materials are listed in Table 5. From the hyperfine fields it appears that the Dy moments in the three alloys at 4.2° K are close to the free-ion value 10 μ_B. The Dy hyperfine field in the amorphous alloys is only reduced by about 1% compared to that of the crystalline alloys. This data confirms that the 4f shell of Dy is well shielded from the influence of neighbouring atoms so far as its moment is concerned.

The magnetic structure of these alloys cannot be collinear, because the net moments are incompatible with the sum or the difference of the Dy moment (10 μ_B) and three times the transition metal moments. In contrast $GdCo_3$ has a collinear ferrimagnetic structure (Fig. 10). The Fe moments and Co moments lie essentially in the plane with relative intensities of the $\Delta m = 0$ transition given by $I_0 = \frac{1}{2}\langle \sin^2\psi \rangle = 0.49(2)$. Moreover, by saturating the magnetization with a small external field and varying the angle ψ, the transition metal moments were shown to be ferromagnetically coupled. The Dy moments have almost no preferred orientation and $I_0 = \frac{1}{2}\langle \sin^2\psi \rangle = 0.37(4)$. The magnetic structure compatible with these results for $DyFe_3$ and $DyCo_3$ compounds in which the moments on

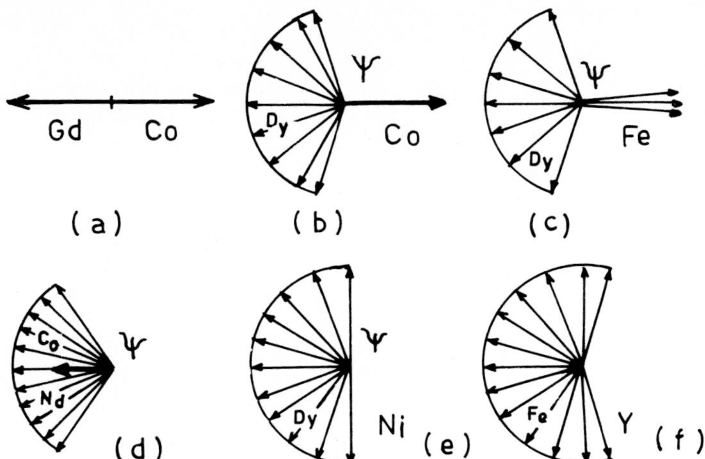

Fig. 10. Magnetic structures proposed for (a) Gd-Co (b,c) Dy-Co, Dy-Fe with ψ ≥ 70°, (d) Nd-Co with ψ ∼60° and (e,f) Dy-Ni and Y-Fe with ψ ∼90° is (a) ferrimagnetic (b,c,d) sperimagnetic (e,f) asperomagnetic.

one (or both) sublattices are directed at random in space is called sperimagnetic. The magnetic moment of 5 μ_B in the $DyNi_3$ compound may be explained by an asperomagnetic structure in which there is no moment on Ni and the Dy-Dy exchange is much weaker than the single ion anisotropy so that the Dy moments are uniformly distributed over a hemisphere.

Properties of amorphous thin films of Nd-Co, Nd-Fe, Gd-Fe and Gd-Co have been studied (244,251,252) by Mossbauer, x-ray diffraction and magnetization measurements. Magnetic films, prepared by evaporation or sputtering, exhibit uniaxial magnetic anisotropy. For Nd-Fe, magnetization data show that Nd contributed a much smaller fraction of its free ion moment to the net magnetization than in Nd-Co, assuming that the Fe subnetwork is taken to be collinear and ferromagnetic. It was therefore suggested that there may be some scatter in the orientations of the moments on the iron sub network. Further, the intensities of the Mossbauer lines are consistent with the iron moments being distributed in a cone of half angle $\psi \leq 50^\circ$ normal to the plane of the film. The relative intensity of the $\Delta m = 0$ transitions is $I_0 = \frac{1}{2}\langle\sin^2\psi\rangle = 0.15$. The proposed magnetic structures for Nd-Co and Nd-Fe are sperimagnets, like the Dy analogs, with the difference that the rare earth and transition-metal moments are coupled in the opposite sense (Fig. 10).

3.4 Rare Earth Chalcogenides

The results of systematic studies of the magnetic properties of rare earth chalcogenides have been reviewed in recent years (253-255). The Eu-monochalcogenides, EuO, EuS, EuSe and EuTe, are special members of a series of magnetic superconductors which are well described by the Heisenberg model of ferromagnetism (256,257). The relations between the interatomic exchange and hyperfine interactions, along with experimental results of microscopic studies, have been presented for Eu-monochalcogenides (257). In this section we shall review the recent Mossbauer studies on RX, $(Eu_{1-x}Sn_x)X$ and $Eu_xSr_{1-x}X$ (R = Eu, Sm, Tm and X = O, S, Se, Te) compounds.

Mossbauer studies in magnetically ordered europium-monochalcogenides (EuO, EuS, EuSe, EuTe and mixed series of $(Eu_{1-x}Sn_x)X$ and $Eu_{1-x}Sr_xS$ compounds) under high pressures and external high fields have been reported at various temperatures (258-268). The transferred and supertransferred hyperfine fields corresponding to the saturation values of the hyperfine fields have been discussed in terms of a two parameter approximation of the Heisenberg operator, limited to effective nearest and next nearest neighbour interactions. For sufficient dilution of the magnetic system a superparamagnetic behaviour is observed which supports the idea of fairly long range magnetic interaction.

Information about the exchange interactions is obtained from studies of hyperfine fields acting at the nuclear sites of the ions in magnetic substances. At the magnetic ions these fields consist of the contribution to the magnetic hyperfine field by the ion itself and of the transferred hyperfine field from the magnetic neighbours. Zinn (257) has discussed in detail the transferred hyperfine interactions in EuO, EuS, EuSe and EuTe. Typical Mossbauer spectra of EuO and EuTe at 4.2° K at various pressures is shown in Fig. 11 (a,b). For EuO, EuS and EuTe, a pressure induced increase of isomer shift δ was observed with an initial slope of $(\partial\delta/\partial p)_T$ = 5.0, 11.0 and 14.8 (10^{-3} mm/sec per kbar), respectively. The pressure induced variation of δ with lattice parameters for EuO, EuS, and EuTe shows nearly the same slope $(\partial\delta/\partial \ell n a)_T$ = -18(2), -20(1) and -20(3) mm/sec, respectively. This variation is explained on the basis of compression of closed inner shells primarily of the $5s^2$ shell (258,259).

The observed variation of B_{eff} with pressure (Fig. 11) was interpreted in terms of J_1 and J_2, the exchange integrals of the Eu ion with the first and second neighbour shell, respectively. The

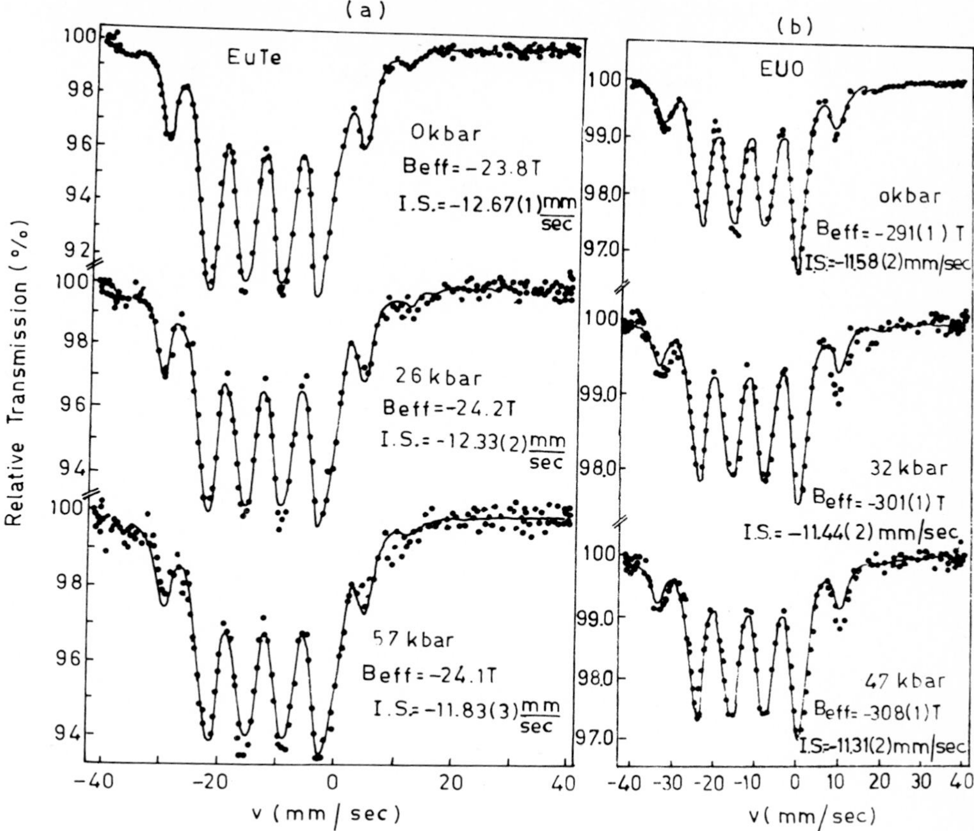

Fig. 11. Pressure dependence of Mossbauer spectra of (a) EuTe (b) EuO at 4.2° K (258,259).

transferred hyperfine field originates from the direct overlap between nearest neighbour Eu ions. The ferromagnetic exchange integral J_1 corresponds to this overlap. The supertransferred field results from the next nearest neighbour Eu shell ions via the p-orbitals of the chalcogen ligands. The supertransferred field should increase with the covalency of the ligands, as this is connected with the anti-ferromagnetic exchange integral J_2. With these considerations for the transferred hyperfine fields in the ferromagnetic and antiferromagnetic state, the variation of B_{eff} with pressure is explained as follows:

In EuO and EuS the increase of negative hyperfine field B_{eff} is mainly caused by the rapidly increasing J_1 overlap. This behaviour is attributed to an induced reduction due to overlap of positive spin-polarization of the $5s^2$ shell. The supertransferred fields are less sensitive to the lattice constant and contribute a constant amount to B_{eff}. In antiferromagnetic EuTe, all contributions from the first cation shell cancel each other. The transferred field remains practically unchanged.

High-pressure Mossbauer studies of ^{119}Sn and ^{151}Eu transferred hyperfine fields in EuS, EuSe and EuTe doped with small amounts of substitutional Sn^{2+} have been reported (267,268) and compared to data on pure Eu-chalcogenides. The high pressure (above 14 kbar) Mossbauer spectra of ^{119}Sn and ^{151}Eu of $(Eu_{0.99}Sn_{0.01})Se$ were similar to that of $(Eu_{0.986}Sn_{0.01})S$ (a ferromagnet). EuSe orders ferromagnetically at pressures above 14 kbar. The ^{151}Eu isomer shift δ increases with a slope of $(\partial\delta/\partial p)_T$ = 12(2) and 15(2) (10^{-3} mm/sec per kbar) for the $(Eu_{1-x}Sn_x)S$ and $(Eu_{1-x}Sn_x)Se$ systems, respectively. These results agree with the systematic behaviour of isomer shift found in pure EuO, EuS and EuTe (258) compounds. The $^{119}Sn^{2+}$ isomer shifts are nearly constant $(\partial\delta/\partial p)_T \leq 5 \times 10^{-4}$ mm/sec per kbar and indicate that any compression of the s-valence-electron shell must be counter-balanced by an increased shielding, probably coming from p-admixture. The variation of Curie temperatures of EuS and EuSe with lattice parameters shows a slight curvature. The increase in slope with decreasing lattice parameter can be attributed to the exponential increase of J_1 exchange with shrinking Eu-Eu distance. The variation of transferred hyperfine fields of Sn^{2+} in EuS and EuTe as a function of lattice constants is shown in Fig. 12. The dramatic increase of transferred hyperfine field in ferromagnetic EuS and EuSe with decreasing lattice constant is correlated with the volume dependence of the Curie temperature. The slope of B_{thf}, the transferred hyperfine field, in ferromagnetic EuS is about 10 times steeper than that in antiferromagnetic EuSe and EuTe. In the last two compounds B_{thf} is completely due to J_2 exchange. This behaviour is indicated in Fig. 12 by a dashed-dotted line which connects the antiferromagnetic B_{thf} of EuTe and EuSe and includes a linear extrapolation to EuS. The magnitudes of B_{thf} induced by pressure are: $(dB_{thf}/d\ell n V)_p$ = -17(2)T for EuTe and $(dB_{thf}/d\ell n V)_V$ = -45(4)T for EuS. These pressure induced changes of B_{thf} are a factor of 4.7 (EuTe) and 12.5 (EuS) larger than in the pure chalcogenides. This supports the view that the increase of the lattice parameter within the chalcogen series mainly reflects the effect of different anion radii.

Mossbauer and magnetization studies have been reported (260,265)

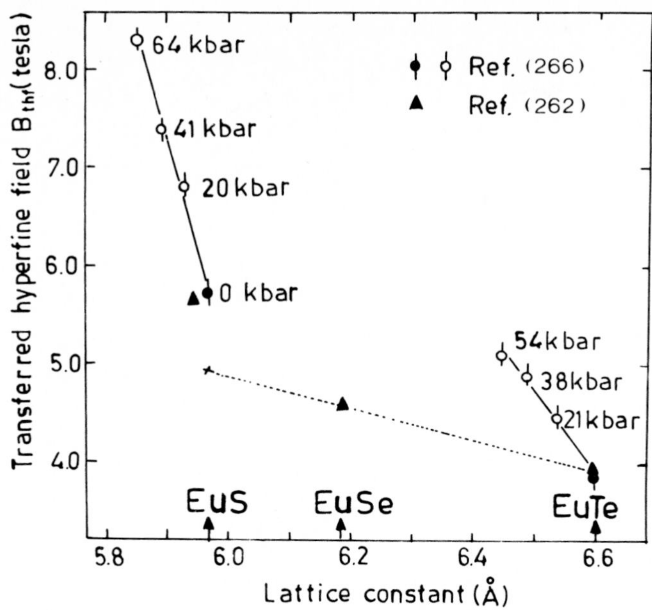

Fig. 12. Magnitudes of Sn^{2+} transferred hyperfine fields B_{thf} (extrapolated to 0° K) as a function of the lattice constants (268)

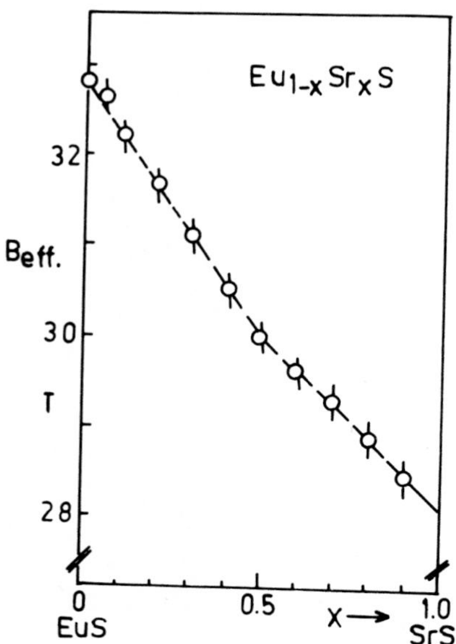

Fig. 13. Concentration dependence of the effective hyperfine field at ^{151}Eu in $Eu_{1-x}Sr_xS$ (260).

on $Eu_xSr_{1-x}S$. Substitution of magnetic Eu cations for diamagnetic ions like Sr^{2+} lowers the ordering temperature to 2.9° K in $Eu_{0.5}Sr_{0.5}S$ from 16.6° K in EuS. The Sr-rich samples show well resolved hyperfine spectra at temperatures below about 100 m° K but exhibit complex spectra at elevated temperatures. The concentration dependence of the effective hyperfine field, deduced from Mossbauer measurements, is shown in Fig. 13. The effective hyperfine field H_{eff} = -32.990(5)T for EuS and H_{eff} = -28.1(2)T for SrS.

From the analysis of the concentration dependence of the lattice constant, hyperfine fields and ferromagnetic ordering temperature, it was concluded that the model for the transferred magnetic hyperfine and exchange interactions, limited to nearest and next nearest neighbours, is a very poor approximation in Eu chalcogenides. The results strongly support the model which allows fairly long range transferred magnetic hyperfine and the exchange interactions.

Mossbauer studies of ^{149}Sm for SmS, SmSe and SmTe at room temperature and SmS under applied pressure up to 11 kbar, as well as for $Sm_{1-x}M_xS$ (M = Ce, La, Gd, Tm, Y) for different samarium concentrations have been reported (34,68,269). The electronic configuration and metal-insulator transition in samarium sulfide has been discussed. A sharp increase in isomer shift at 6 kbar pressure in SmS, and at x = 0.15 in $Sm_{1-x}Y_x$ (34), indicated a phase transition. In semiconducting SmS the observed isomer shift was found to be close to that expected for Sm^{2+}. In the metallic phase the value of the shift did not correspond to metallic Sm^{3+} but indicated that a substantial fraction of Sm ion remained as Sm^{2+}, i.e., in the $4f^6$ configuration. Samarium, in metallic SmS, has a higher divalent character when the transition is produced by doping with Y than when produced by pressure. Intermediate valence configurations were found in both systems. The isomer shift results of ^{153}Eu as a dilute probe in SmS doped with various di- and trivalent elements showed that in the semiconducting phase the conduction band is primarily of the 5d character with the d electrons localized on the trivalent doping ions. In the metallic phase, participation of the 6s electrons is believed to explain the band character (68). Mossbauer measurements at ^{169}Tm in semiconducting TmSe and TmTe have been reported (270,271). It was observed from the temperature dependence of hyperfine fields in TmSe that there is an unusual mixed valence state where interconfiguration fluctuations between the Tm^{3+} and Tm^{2+} electronic configurations occur on a very rapid time scale. Magnetic ordering is observed around 2.8° K. Analysis of

Mossbauer spectra below the magnetic ordering temperature in TmTe indicated that the Tm ion is principally in the Tm^{2+} state in this material. Futhermore, the spectrum of TmTe at 3° K was an asymmetric doublet due to electronic relaxation effects and characteristic of non-cubic site symmetry. The semiconductor to metal transition occuring in TmTe under pressure at room temperature has been associated with a phase containing Tm^{2+} ions which transforms into an isostructural phase containing smaller Tm^{3+} ions or mixed configurations (271).

REFERENCES

1. S. Offer, I. Nowik and S.G. Cohen, Chemical Applications of Mossbauer Spectroscopy, Academic, New York, 1968, p. 427.
2. A.F. Clifford, Developments in Applied Spectroscopy, Vol. 8, Plenum, New York, 1970, p. 255.
3. R.L. Cohen, Proc. 9th Rare Earth Research Conf., 1 (1971) 144.
4. K.H.J. Buschow, Rep. Progr. Phys., 40 (1977) 1179.
5. R.G. Barnes, Handbook on Physics and Chemistry of Rare Earths, Vol. 2, North Holland, Amsterdam, 1979, p. 387.
6. R.L. Cohen, J. Phys. (Paris) Colloq., 41 (1980) C1-333.
7. G.M. Kalvius, Hyperfine Interactions in Excited Nuclei, Gordon & Breach, New York, 1971, p. 523.
8. B.D. Dunlap, Mossbauer Effect Methodology, Vol. 7, Plenum, New York, 1971, p. 123.
9. B.D. Dunlap, Mossbauer Effect Data Index 1970, Plenum, New York, 1972, p. 25.
10. B.D. Dunlap and G.M. Kalvius, Mossbauer Isomer Shifts, North Holland, Amsterdam, 1978, p. 15.
11. E.R. Bauminger, G.M. Kalvius and I. Nowik, Mossbauer Isomer Shifts, North Holland, Amsterdam, 1978, p. 661.
12. G.K. Shenoy, (Preprint) Indo-U.S. Rare Earth Conference, India (1980) to be published.
13. A.E. Clark and H.S. Belson, Phys. Rev. B, 5 (1972) 3642.
14. J.M.D. Coey, J. Inorg. Nucl. Chem., 38 (1976) 1139.
15. U.F. Klein, G. Wortman and G.M. Kalvius, Solid State Commun., 18 (1976) 291.
16. E.R. Bauminger, A. Diamant, I. Felner, I. Nowik and S. Ofer, Phys. Rev. Lett., 34 (1975) 962.
17. H. Weiss and H. Langhoff, Phys. Lett., 69A (1979) 448.
18. E. Loh, C.L. Chien and J.C. Walker, Phys. Lett., 49A (1974) 357.
19. E. Loh, F.T. Parker and J.C. Walker, Hyperfine Interactions in Excited Nuclei, 3 (1971) 888.
20. E. Loh, F.T. Parker and J.C. Walker, U.S. Atomic Energy Commission Progress Report, AT (30-1)-2028 (1970) 138.
21. G.J. Bowden, R.K. Day and P. Martinson, J. Phys. F : Metal Phys., 8 (1978) 2051.
22. E. Callen and H.B. Callen, Phys. Rev., 139 (1965) 455.
23. R.A. Reese and R.G. Barnes, J. Appl. Phys. 40 (1969) 1493.
24. R.A. Reese and R.G. Barnes, Phys. Rev., 163 (1967) 465.
25. J. Stohr and W. Wagner, J. Phys. F : Metal Phys., 5 (1975) 812.
26. S. Hufner, P. Kienle, W. Weidemann and H. Eicher, Z. Physik, 182 (1965) 499.
27. H.P. Wit and L. Niesen, Hyperfine Interact., 1 (1976) 501.
28. R.L. Cohen, Phys. Rev., 169 (1968) 32.

29. P. Bonville, P. Imbert, G. Jehanno and F. Gonzalez-Jimenez, J. Phys. Chem. Solids, 39 (1978) 1273.
30. D.L. Uhrich and R.G. Barnes, Phys. Rev., 164 (1976) 428.
31. I. Nowik and E.R. Bauminger, Mossbauer Effect Data Index, 1975, Plenum, New York, 1976, p. 407 (and references therein).
32. E. Gerdau, H. Winkler and F. Sabathil, Hyperfine Interact., 4 (1978) 630.
33. W. Fischer, H. Huhnermann and K. Mandrek, Z. Phys., 254 (1972) 127.
34. J.M.D. Coey, S.K. Ghatak, M. Avignon and F. Holtzberg, Phys. Rev. B, 14 (1976) 3744.
35. J.M.D. Coey, S.K. Ghatak and F. Holtzberg, AIP Conf. Proc., 24 (1975) 38.
36. J.M.D. Coey and O. Massenet, Valence Instabilities and Narrow-Band Phenomena, Plenum, New York, 1976, p. 211.
37. I. Nowik, Valence Instabilities and Narrow-Band Phenomena, Plenum, New York, 1976, p. 261.
38. J.N. Chazalviel, M. Campagna, G.K. Wertheim and P.H. Schmidt, Solid State Commun., 19 (1976) 725.
39. P. Glentworth, A.L. Nichols, N.R. Large and R.J. Bullock, J. Chem. Soc. Dalton (1973) 969.
40. C.L. Chien, S. De Benedetti and F. De S. Barros, Phys. Rev. B, 10 (1974) 3913.
41. M. Loewenhaupt, Z. Physik, 267 (1974) 219.
42. K.H.J. Buschow, W.J. Huiskamp, H. Th. LeFever, F.J. Van Steenwijk and R.C. Thiel, J. Phys. F: Metal Phys., 5 (1975) 1625.
43. F.J. Van Steenwijk, H. Th. LeFever, R.C. Thiel and K.H.J. Buschow, Physica, 79B (1975) 604.
44. F.J. Van Steenwijk, W.J. Huiskamp, H. Th. LeFever, R.C. Thiel and K.H.J. Buschow, Physica, 86-88B (1977) 89.
45. K.H.J. Buschow and R.J. Van Steenwijk, Physica, 85B (1977) 122.
46. E. Baggio-Saitovitch, F.J. Litterst and H. Micklitz, J. Phys. (Paris) Colloq., 37 (1976) C6-529.
47. C.L. Chien and A.W. Sleight, Phys. Rev., B, 18 (1978) 2031.
48. J.M.D. Coey, O. Massenet, M. Kasaya and J. Etourneau, J. Phys. (Paris) Colloq., 40 (1979) C2-333.
49. H. de Graaf, W.J. Huiskamp, R.C. Thiel, H. Th. LeFever and K.H.J. Buschow, Physica, 98B (1979) 60.
50. F. Gotz, G. Czjzek, J. Fink, H. Schmidt, V. Oestreich and P. Fulde, J. Appl. Phys., 50 (1979) 7513.
51. D. Schroeer, Ch. S. Kuo and R.L. Lambe, Phys. Status. Solidi (b), 92 (1979) 565.
52. H. de Graaf, R.C. Thiel and K.H.J. Buschow, Physica, 100B (1980) 81
53. M.M. Abd-Elmeguid, H. Micklitz and K.H.J. Buschow, J. Phys. (Paris) Colloq., 41 (1980) C1-127.
54. K.H.J. Buschow, Rept. Progr. Phys., 42 (1979) 1373.
55. G.M. Kalvius, U.F. Klein and G. Wortman, J. Phys. (Paris) Colloq., 35 (1974) C6-139.
56. E. Gorlich, H.U. Hrynkiewicz, R. Kniec, K. Latka and K. Tomala, Phys. Status Solidi (b), 64 (1974) K147.
57. E.R. Bauminger, I. Felner, D. Levron, I. Nowik and S. Ofer, Phys. Rev. Lett., 33 (1974) 890.
58. E.R. Bauminger, I. Felner, D. Froindlich, D. Levron, I. Nowik, S. Ofer and R. Yankovsky, J. Phys. (Paris) Colloq., 35 (1974) C6-61
59. E.R. Bauminger, I. Felner, D. Levron, I. Nowik and S. Ofer, Proc. Int. Conf. Mossbauer Spectroscopy, Vol. I, Cracon, 1975, p. 69.
60. E.R. Bauminger, I. Felner, D. Levron, I. Nowik and S. Ofer, Solid State Commun., 18 (1976) 1073.
61. G. Wortmann, J. Moser and U.F. Klein, Phys. Lett., 55A (1976) 486.
62. J. Moser, U.F. Klein, G. Wortman and G.M. Kalvius, J. Phys. (Paris) Colloq., 37 (1976) C6-421.

63. J. Moser, U.F. Klein, G. Wortman and G.M. Kalvius, Physica, 86-88B (1977) 243.
64. E.R. Bauminger, I. Felner and S. Ofer, J. Magn. Magn. Mater., 7 (1978) 317.
65. I. Felner and I. Nowik, J. Phys. Chem. Solids., 39 (1978) 951.
66. I. Felner and I. Nowik, J. Phys. Chem. Solids, 40 (1979) 1035.
67. E.R. Bauminger, A. Diamant, I. Felner, I. Nowik, A. Mustachi and S. Ofer, J. Phys. (Paris) Colloq., 37 (1976) C6-49.
68. R.L. Cohen, I. Nowik, K.W. West and E. Bucher, Phys. Rev. B, 16 (1977) 4455.
69. G.J. Kemerink, D.O. Boerma, H. de Waard, J.D. de Wit and S.A. Drentje, J. Phys. (Paris) Colloq., 41 (1980) C1-435.
70. D. Mihaila-Tarabasanu, U. Wagner, F.E. Wagner and G.M. Kalvius, Int. Conf. Mossbauer Spectroscopy, Vol. 1, Cracow, 1975, p. 291.
71. E. Mohs, G.K. Wolf, U. Wagner and F.E. Wagner, Proc. Int. Conf. Mossbauer Spectroscopy, Vol. I, Bucharest, 1977, p. 83.
72. R. Yanovsky, E.R. Bauminger, I. Felner, I. Nowik and S. Ofer, Hyperfine Interact., 3 (1977) 263.
73. L. Niesen and S. Ofer, Hyperfine Interact., 4 (1978) 347.
74. H. Armon, E.R. Bauminger, A. Diamant, I. Nowik and S. Ofer, Phys. Lett., 44A (1973) 279.
75. D.B. Prowse and J.D. Cashion, Phys. Status Solidi (a), 22 (1974) 631.
76. J.W. Ross and J. Sigalas, J. Phys. F: Metal Phys., 5 (1975) 1973.
77. K. Tomala, G. Czjzek, J. Fink and H. Schmidt, Solid State Commun., 24 (1977) 857.
78. G. Czjzek, J. Fink, H. Schmidt and K. Tomala, Proc. Int. Conf. Mossbauer Spectroscopy, Vol. I, Bucharesl, 1977, p. 83.
79. F.J. Van Steenwijk, H. Th. LeFever, R.C. Thiel and K.H.J. Buschow, Physica, 92B (1977) 52.
80. D.C. Cook and J.D. Cashion, J. Phys. C: Solid State Phys., 12 (1979) 605.
81. L. Ben-Dor, I. Shilo and I. Felner, J. Solid State Chem., 24 (1978) 401.
82. D.C. Cook and J.D. Cashion, J. Phys. C: Solid State Phys., 9 (1976) L97.
83. G. Mennenga and L. Niesen, J. Phys. (Paris) Colloq., 41 (1980) C1-439.
84. I. Nowik, I. Felner, and M. Seh, J. Magn. Magn. Mater., 15-18 (1980) 1215.
85. J.K. Yakinthos and J. Chappert, Solid State Commun., 17 (1975) 979.
86. D.W. Forester and W.A. Ferrando, Phys. Rev. B, 13 (1976) 3991.
87. D.W. Forester and W.A. Ferrando, Phys. Rev. B, 14 (1976) 4769.
88. Ch. S. Kuo and D. Schroeer, Phys. Status Solidi (b), 87 (1978) 325.
89. J.M. Friedt, G.K. Shenoy and B.D. Dunlap, J. Phys. (Paris) Colloq., 40 (1979) C2-243.
90. F.J. Litterst, J. Tejada and G.M. Kalvius, J. Appl. Phys., 50 (1979) 7636.
91. R. Iraldi, V.N. Nguyen, J. Rossat-Mignod and F. Tcheou, Solid State Commun., 15 (1974) 1543.
92. V.G. Stankevich, I.I. Lukashevich, N.I. Filippov and I.A. Gladkikh, Sovt. Phys.--JETP, 39 (1974) 134.
93. D.M. Sweger, R. Segnan and J.J. Rhyne, Phys. Rev. B, 9 (1974) 3864.
94. M. Belakhovsky, J. Chappert and D. Schmitt, J. Phys. C: Solid State Phys., 10 (1977) L493.
95. R. Abbundi, J.J. Rhyne, D.M. Sweger and R. Segnan, Phys. Rev. B, 18 (1978) 3313.
96. H.P. Wit and L. Niesen, Proc. Conf. On Hyperfine Interactions, Uppsala, 1974, p. 246.
97. L. Niesen, H.P. Wit, P.J. Kikkert and H. de Waard, Proc. Conf. Mossbauer Effect, Cracow, 1975, p. 207.

98. L. Niesen and H.P. Wit, J. Phys. (Paris) Colloq., 37 (1976) C6-639.
99. L. Niesen, Hyperfine Interact., 2 (1976) 15.
100. H.P. Wit, L. Niesen, H. de Waard, Hyperfine Interact., 5 (1978) 233
101. P.J. Kikkert and L. Niesen, J. Phys. (Paris) Colloq., 41 (1980) C1-203.
102. J. Stohr, J.D. Cashion, W. Wagner and L. Asch, Solid State Commun., 18 (1976) 35.
103. W. Wagner, L. Asch and G.M. Kalvius, Hyperfine Interact., 4 (1978) 441.
104. W. Wagner, G.M. Kalvius, L. Asch and L.L. Hirst, J. Magn. Magn. Mater, 15-18 (1980) 54.
105. G.K. Shenoy, J. Stohr and G.M. Kalvius, Solid State Commun., 13 (1973) 909.
106. J. Stohr and G.K. Shenoy, Solid State Commun., 14 (1974) 583.
107. J. Stohr, W. Wagner and G.K. Shenoy, Phys. Lett., 47A (1974) 177.
108. J. Stohr and W. Wagner, J. Phys. F: Metal Phys., 5 (1975) 912.
109. J. Stohr, J.D. Cashion and W. Wagner, J. Phys. F: Metal Phys., 5 (1975) 1417.
110. C. Rettori and D. Davidov, Phys. Rev. B, 10 (1974) 4033.
111. L. Niesen and P.J. Kikkert, Proc. Conf. on Hyperfine Interactions, Uppsala, 1974, p. 160.
112. L. Niesen, P.J. Kikkert and H. de Waard, Hyperfine Interact., 3 (1977) 109.
113. C.W. Kimball, A.E. Dwight, G.M. Kalvius, B. Dunlap and M.V. Nevitt, Phys. Rev. B, 12 (1975) 819.
114. H.P. Wit, N. Teekens, L. Niesen and S.A. Drentje, Hyperfine Interact., 4 (1978) 674.
115. R.K. Day and J.B. Dunlop, J. Magn. Magn. Mater., 15-18 (1980) 651.
116. P. Ruden and H. Micklitz, J. Magn. Magn. Mater., 15-18 (1980) 993.
117. J. Hodges and G. Jehanno, J. Magn. Magn. Mater., 15-18 (1980) 51.
118. P. Bonville, J.A. Hodges, P. Imbert, G. Jehanno, R. Chevrel and M. Sergent, Rev. Phys. Appl., 15 (1980) 1139.
119. W. Schilling, G. Burger, K. Isebeck and H. Wenzl, in Vacancies and Interstitials in Metals, North Holland, Amsterdam, 1969, p. 255.
120. G.K. Shenoy, J. Stohr, W. Wagner, G.M. Kalvius and B.D. Dunlap, Solid State Commun., 15 (1974) 1485.
121. G.K. Shenoy, R. Poinsot, L. Asch, J.M. Friedt and B.D. Dunlap, Phys. Lett., 49A (1974) 429.
122. G.K. Shenoy, L. Asch, J.M. Friedt and B.D. Dunlap, J. Phys. (Paris) Colloq., 35 (1974) C6-425.
123. B.D. Dunlap and G.R. Davidson, M. Eibschutz, H.J. Guggenheim and R.C. Sherwood, J. Phys. (Paris) Colloq., 35 (1974) C6-429.
124. F. Gonzalez-Jimenez, P. Imbert and F. Hartmann-Bourtron, Phys. Rev. B, 9 (1974) 95.
125. J. Stohr, Phys. Rev. B, 11 (1975) 3559.
126. C. Borely, F. Gonzalez-Jimenez, P. Imbert and F. Varret, J. Phys. Chem. Solids, 36 (1975) 683.
127. C. Borely, F. Gonzalez-Jimenez, P. Imbert and F. Varret, J. Phys. Chem. Solids, 36 (1975) 605.
128. G.R. Davison, B.D. Dunlap, M. Eibschutz and L.G. van Uitert, Phys. Rev. B, 12 (1975) 1681.
129. R. Yanovsky, E.R. Bauminger, D. Leveron, I. Nowik and S. Ofer, Solid State Commun., 17 (1975) 1511.
130. I. Nowik, I. Felner and R. Yanovsky, J. Phys. (Paris) Colloq., 37 (1976) C6-431.
131. G.K. Shenoy, B.D. Dunlap, S. Dattagupta and L. Asch, Phys. Rev. Lett., 37 (1976) 539.
132. J.A. Hodges, P. Bonville, F. Hartmann-Boutron and P. Imbert, Proc. Int. Conf. Mossbauer Spectroscopy, Vol. I, Bucharest, 1977, p. 207.

133. S. Dattagupta, G.K. Shenoy, B.D. Dunlap and L. Asch, Phys. Rev. B, 16 (1977) 3893.
134. A.M. Afanas'ev, E.V. Onishchenko, L. Asch and G.M. Kalvius, Phys. Rev. Lett., 40 (1978) 816.
135 . P. Bonville, J.A. Hodges, P. Imbert and F. Hartmann-Boutron, Phys. Rev. B, 18 (1978) 2196.
136. J.W. Ross and E. Trone, J. Phys. F: Metal Phys., 8 (1978) 983.
137. B.D. Dunlap, G.K. Shenoy, J.M. Friedt, M. Meyer and G.J. McCarthy, J. Appl. Phys., 49 (1978) 1448.
138. J.A. Hodges and P. Imbert, J. Phys. (Paris) Colloq., 40 (1979) C2-253.
139. H. Winkler, P. Alwardt and E. Gerdau, J. Phys. (Paris) Colloq., 40 (1979) C2-250.
140. B.D. Dunlap, G.K. Shenoy and G.M. Kalvius, Phys. Rev. B, 10 (1974) 26.
141. B.D. Dunlap, G.K. Shenoy and L. Asch, Phys. Rev. B, 13 (1976) 18.
142. M. Tachiki, S. Maekawa, R. Treder and M. Levy, Phys. Rev. Lett., 34 (1975) 1579.
143. M. Rosen, H. Klimer, U. Atzmony and M.P. Dariel, Phys. Rev. B, 9 (1974) 254.
144. H. Klimer, M. Rosen, M.P. Dariel and U. Atzmony, Phys. Rev. B, 10 (1974) 2968.
145. G.J. Bowden, D. St. P. Bunbury, A.P. Guimaraes and R.E. Snyder, J. Phys. C: Solid State Phys., 1 (1968) 1376.
146. C.W. Kimball, A.E. Dwight, R.S. Preston and S.P. Taneja, AIP Conf. Proc., 18 (1974) 1242.
147. S.P. Taneja, A.E. Dwight, C.W. Kimball and R.S. Preston, Proc. Nucl. Phys. and Solid State Phys. Symposium, Bangalore (India), 16C (1973) 313.
148. D. Barb, E. Burzo and M. Morariu, 5th Int. Conf. on Mossbauer Spectroscopy, Bratislava, Vol. I, 1973, p. 37.
149. G.J. Bowden, J. Phys. F: Metal Phys., 3 (1973) 2206.
150. H. Schechter, D. Bukshpan and I. Nowik, Phys. Rev. B, 14 (1976) 3087.
151. D. Bukshpan, H. Schechter and I. Nowik, J. Magn. Magn. Mater., 7 (1978) 212.
152. D. Feder (Bukshpan) and I. Nowik, J. Magn. Magn. Mater., 12 (1979) 149.
153. S. Japa, K. Krop and M. Przybylski, J. Phys. (Paris) Colloq., 41 (1980) C1-199.
154. A.M. Van Deipan, H.W. de Wijn and K.H.J. Buschow, Phys. Rev. B, 8 (1973) 1125.
155. U. Atzmony, M.P. Dariel, E.R. Bauminger, D. Lebenbaum, I. Nowik and S. Ofer, Proc. 10th Rare Earth Conf., Vol. I, 1973, p. 605.
156. U. Atzmony and M.P. Dariel, Phys. Rev. B, 10 (1974) 2060.
157. J.N.J. Vander Velden, A.M. Van der Kraan, P.C.M. Gubbens and K.H.J. Buschow, Proc. Int. Conf. Mossbauer Spectroscopy, Vol. I, Cracow, 1975, p. 181.
158. C. Neyer, B. Srour, Y. Gros, F. Hartmann-Boutron and J.J. Capponi, J. Phys. (Paris), 38 (1977) 1449.
159. D.C. Creagh and S.H. Ayling, J. Mater. Sci., 13 (1978) 113.
160. T. Mizoguchi, Y. Tanaka, T. Tschida and Y. Nakamura, J. Phys. (Paris) Colloq., 40 (1979) C2-211.
161. C. Meyer, F. Hartmann-Boutron, Y. Gros, B. Srour and J.J. Capponi, J. Phys. (Paris) Colloq., 40 (1979) C5-191.
162. C. Meyer, Y. Gros. F. Hartmann-Boutron and J.J. Capponi, J. Phys. (Paris) 40 (1979) 403.
163. C. Meyer, F. Hartmann-Boutron, J.J. Capponi, J. Chappert and O. Massenet, J. Magn. Magn. Mater. 15-18 (1980) 1229.

164. A.E. Dwight and C.W. Kimball, Acta Cryst. B, 30 (1974) 2791.
165. D. Bukshpan and I. Nowik, J. Magn. Magn. Mater., 12 (1979) 162.
166. A.E. Dwight, C.W. Kimball, R.S. Preston, S.P. Taneja and L. Weber, Proc. 10th Rare Earth Conf., Vol. II, 1973, p. 1027.
167. A.E. Dwight, C.W. Kimball, R.S. Preston, S.P. Taneja and L. Weber, J. Less Common Metals, 40 (1975) 285.
168. R. Grossinger and W. Steiner, Phys. Status Solidi (a), 28 (1975) K135.
169. R. Grossinger, W. Steiner and K. Krec, J. Magn. Magn. Mater., 2 (1976) 196.
170. H. Maletta, G. Crecelius and W. Zinn, J. Phys. (Paris) Colloq., 35 (1974) C6-279.
171. R.S. Preston, S.P. Taneja, S.M. Drensky, A.E. Dwight, C.W. Kimball and L.R. Sill, AIP Conf. Proc., 34 (1976) 194.
172. K.H.J. Buschow and A.M. Van der Kraan, J. Phys. F: Metal Phys., 8 (1978) 921.
173. A.M. Van der Kraan and K.H.J. Buschow, Physica, 86-88 (B+C) (1977) 93.
174. I. Felner and I. Nowik, Solid State Commun., 28 (1978) 67.
175. W.E. Wallace, A.S. Ilyushin and D. Lopez, Proc. Int. Conf. Mossbauer Spectroscopy, Cracow, Vol. 2, 1975, p. 99.
176. A.S. Ilyushin and W.E. Wallace, J. Solid State Chem., 17 (1976) 131.
177. A.S. Ilyushin and W.E. Wallace, J. Solid State Chem., 17 (1976) 373
178. A.P. Guimaraes and D. St. P. Burnbury, J. Phys. F: Metal Phys., 3 (1973) 885.
179. A.M. Van der Kraan and P.C.M. Gubbens, J. Phys. (Paris) Colloq., 35 (1974) C6-469.
180. G. Longworth and I.R. Harris, J. Less Common Metals, 41 (1975) 175.
181. U. Atzmony, M.P. Dariel and G. Dublon, Phys. Rev. B, 14 (1976) 3715
182. U. Atzmony and G. Dublon, Physica, 86-88(B+C) (1977) 167.
183. G. Dublon and U. Atzmony, J. Phys. F: Metal Phys. 7 (1977) 1069.
184. U. Atzmony, M.P. Dariel and G. Dublon, Phys. Rev. B, 17 (1978) 396.
185. H.R. Corson, G.R. Hoy and B. Kolk, J. Phys. (Paris) Colloq., 40 (1979) C2-159.
186. S.K. Arif, D. St. P. Bunbury and G.J. Bowden, J. Phys. F: Metal Phys., 5 (1975) 1792.
187. P.C.M. Gubbens, A.M. Van der Kraan and K.H.J. Buschow, J. Phys. (Paris) Colloq., 40 (1979) C5-200.
188. E. Burzo, D. Barb and M. Bodea, Proc. Int. Conf. Mossbauer Spectroscopy, Vol. I, Cracow, 1975, p. 169.
189. I.R. Harris and G. Longworth, J. Less Common Metals, 45 (1976) 63.
190. G. Crecelius, H. Maletta, J. Hauck, J. Fink, G. Czjzek and H. Schmidt, J. Magn. Magn. Mater, 4 (1977) 40.
191. M. Bodea, D. Barb and E. Burzo, Proc. Int. Conf. Mossbauer Spectroscopy, Vol. I, Bucharest, 1977, p. 387.
192. E. Burzo, Phys. Rev. B, 17 (1978) 1414.
193. E. Burzo, Solid State Commun., 18 (1976) 1431.
194. S.K. Arif, I. Sigalas and D. St. P. Bunbury, Phys. Status Solidi (a), 41 (1977) 585.
195. S.K. Arif and D. St. P. Bunbury, Phys. Status Solidi (a), 33 (1976) 91.
196. S.C. Tsai, K.S.V.L. Narasimhan, G.J. Kunesh and R.A. Butera, J. Appl. Phys., 45 (1974) 3582.
197. U. Atzmony, M.P. Dariel, E.R. Bauminger, D. Lebenbaum, I. Nowik and S. Ofer, Phys. Rev. Lett., 28 (1972) 244.
198. U. Atzmony, M.P. Dariel, E.R. Bauminger, D. Lebenbaum, I. Nowik and S. Ofer, Perspectives in Mossbauer Spectroscopy, Plenum, New York, 1973, p. 11.

199. U. Atzmony, M.P. Darial, E.R. Bauminger, D. Lebenbaum, I. Nowik and S. Ofer, Phys. Rev. B, 7 (1973) 4220.
200. G. Dublon, U. Atzmony, M.P. Dariel and H. Shaked, Phys. Rev. B, 12 (1975) 4628.
201. U. Atzmony and M.P. Dariel, AIP Conf. Proc., 24 (1975) 662.
202. M. Rosen, H. Klimker, U. Atzmony and M.P. Dariel, J. Phys. Chem. Solids, 37 (1976) 513.
203. U. Atzmony, M.P. Dariel and G. Dublon, Phys. Rev. B, 15 (1977) 3565
204. M. Morariu, E. Burzo and D. Barb, J. Phys. (Paris) Colloq., 37 (1976) C6-615.
205. M.P. Dariel, U. Atzmony and D. Lebenbaum, Phys. Status Solidi, 59 (1973) 615.
206. R.S. Preston, A.E. Dwight, A.J. Fedro and C.W. Kimball, AIP Conf. Proc., 24 (1975) 660.
207. S.K. Arif, G.J. Bowden and D. St. P. Bunbury, Proc. 18th Ampere Congress, Amsterdam, 1974, p. 85.
208. S.K. Arif, D. St. P. Bunbury, G.J. Bowden and R.K. Day, J. Phys. F: Metal Phys., 5 (1975) 1037.
209. S.K. Arif, D. St. P. Bunbury, G.J. Bowden and R.K. Day, J. Phys. F: Metal Phys., 5 (1975) 1048.
210. S.K. Arif, D. St. P. Bunbury, G.J. Bowden and R.K. Day, J. Phys. F: Metal Phys., 5 (1975) 1785.
211. S.P. Taneja, A.E. Dwight and C.W. Kimball, Proc. Nucl. Phys. and Solid State Physics Symposium, Calcutta, India, Vol. 18C, 1975, p. 547.
212. A.M. Van der Kraan, J.N.J. Vander Veldon, P.C.M. Gubbens and K.H.J. Buschow, Proc. Int. Conf. Mossbauer Spectroscopy, Cracow, Poland, Vol. 1 (1975) 179.
213. A.M. Van der Kraan, P.C.M. Gubbens and K.H.J. Buschow, Phys. Status Solidi (a), 31 (1975) 495.
214. A.M. Van der Kraan, J.N.J. Vander Velden, J.H.F. Van Apeldoorn, P.C.M. Gubbens and K.H.J. Buschow, Phys. Status Solidi (a), 35 (1976) 137.
215. G.J. Bowden and R.K. Day, J. Phys. F: Metal Phys., 7 (1977) 181.
216. R.K. Day and G.J. Bowden, Physica, 86-88 (B+C) (1977) 71.
217. S. Japa, K. Krop, R. Radwanski and J. Wolinki, J. Phys. (Paris) Colloq., 40 (1979) C2-19.
218. P.C.M. Gubbens and K.H.J. Buschow, J. Appl. Phys., 44 (1973) 3739.
219. P.C.M. Gubbens and K.H.J. Buschow, Proc. Int. Conf. on Magn., Moscow, 1974, p. 60.
220. P.C.M. Gubbens, J.J. Van Loef and K.H.J. Buschow, J. Phys. (Paris) Colloq., 35 (1974) C6-617.
221. J.B.A.A. Elemans, P.C.M. Gubbens and K.H.J. Buschow, J. Less Common Metals, 44 (1976) 51.
222. P.C.M. Gubbens and K.H.J. Buschow, Phys. Status Solidi (a), 34 (1976) 729.
223. W. Steiner and R. Haferal, Phys. Status Solidi (a), 42 (1977) 739.
224. P.C.M. Gubbens, A.M. Van der Kraan and K.H.J. Buschow, Physica, 86-88B (1977) 199.
225. P.C.M. Gubbens, A.M. Van der Kraan and K.H.J. Buschow, Solid State Commun., 26 (1978) 107.
226. P.C.M. Gubbens, J.H.F. Van Apeldoorn, A.M. Van der Kraan and K.H.J. Buschow, J. Phys. F: Metal Phys., 4 (1974) 921.
227. P. Chaudhary, J.J. Cuomo and R.J. Gambino, I.B.M. J. Res. Develop., 11 (1973) 66.
228. R. Meyer, H. Jouve and J.P. Rebouillant, IEEE Trans. Magn., 11 (1975) 1335.
229. J.M.D. Coey, J. Chappert, J.P. Rebouillant and T.S. Wang, Phys. Rev. Lett., 36 (1976) 1061.
230. R. Arresse-Boggiano, J. Chappert, J.M.D. Coey, A. Lienard and J.P. Rebouillant, J. Phys. (Paris) Colloq., 37 (1976) C1-771.

231. R.C. Taylor, T.R. McGuire, J.M.D. Coey and A. Gangulee, J. Appl. Phys., 49 (1978) 2885.
232. C.R. Kurkjain and E.A. Sigety, Proc. 7th Int. Cong. of Glass, Brussels, Gordon and Breach, New York, 1964, p. 39.
233. D.L. Uhrich and R.G. Barnes, Phys. Chem. Glasses, 9 (1968) 184.
234. M.F. Taragin and J.C. Eisenstein, Phys. Rev. B, 9 (1970) 3490 and J. Non-crystalline Solids, 11 (1973) 395.
235. R. Harris, M. Plischke and M.J. Zuckermann, J. Phys. (Paris) Colloq., 35 (1974) C4-265.
236. D. Sarkar, R. Segnan and A.E. Clark, AIP Conf. Proc. No. 18, Pt. 1 (1974) 636.
237. D. Sarkar, R. Segnan, E.K. Cornall, E. Collen, R. Harris, M. Plischke and M.J. Zuckermann, Phys. Rev. Lett., 32 (1974) 542.
238. G.S. Cargill, AIP Conf. Proc., 24 (1974) 138.
239. J.J. Rhyne, S.J. Pickart and H.A. Alperin, Phys. Rev. Lett., 29 (1972) 1562.
240. J.J. Rhyne, AIP Conf. Proc., 29 (1975) 182.
241. D.W. Forester, R. Abbundi, R. Segnan and D. Sweger, AIP Conf. Proc., 24 (1975) 115.
242. W.P. Pala, D.W. Forester and R. Segnan, AIP Conf. Proc., 34 (1976) 322.
243. D.W. Forester, W.P. Pala and R. Segnan, Amorphous Magnetism II, Plenum, New York, 1977, pp. 135-143.
244. C. Vihoria, P. Lubitz and V. Ritz, J. Appl. Phys., 49 (1978) 4908.
245. J.J. Rhyne, J.H. Schelleng and N.C. Koon, Phys. Rev., 10 (1974) 4672.
246. J.M.D. Coey, J. Appl. Phys., 49 (3) Pt. II (1978) 1646.
247. R. Arrese-Boggiano, J. Chappert, J.M.D. Coey, A. Lienard and J.P. Rebouillat, J. Phys. (Paris) Colloq., 37 (1976) C6-771.
248. J.M.D. Coey, J. Chappert, J.P. Rebouillat and T.S. Wang, Phys. Rev. Lett., 36 (1976) 1061.
249. J. Chappert, R. Arrese-Boggiano and J.M.D. Coey, J. Magn. Magn. Mater., 7 (1978) 175.
250. J.P. Rebouillat, A. Lienard, J.M.D. Coey, R. Arrese-Boggiano and J. Chappert, Physica, (86-88)B (1977) 773.
251. R.C. Taylor and A. Gangulee, J. Appl. Phys., 48 (1977) 358.
252, Y. Nishihara, T. Katayamma, Y. Yamaguchi, S. Ogawa and T. Tsushima, Japanese J. Appl. Phys., 17 (1978) 1083.
253. E. Bucher, K. Andres, F.J. di Salvo, J.P. Maita, A.C. Gossard, A.S. Co-oper and G.W. Hall, Jr., Phys. Rev. B, 11 (1975) 500.
254. S. Methessel, New Developments in Semiconductors, Noordhoff Int. Pub. Leiden, Netherlands, 1973, p. 37.
255. P. Wachter, C.R.C. Critical Reviews in Solid State Sciences, 3 (1972) 189.
256. R.H. Swendsen, Phys. Rev. B, 5 (1972) 116 and Phys. Rev. B, 11 (1975) 1935.
257. W. Zinn, J. Magn. Magn. Mater., 3 (1976) 23.
258. U.F. Klein, G. Wortmann and G.M. Kalvius, J. Magn. Magn. Mater., 3 (1976) 50.
259. U.F. Klein, J. Moser, G. Wortmann and G.M. Kalvius, Physica, 86-88B (1977) 118.
260. G. Crecelius, H. Meletta, H. Pink and W. Zinn, J. Magn. Magn. Mater, 5 (1977) 150.
261. Ch. Sauer and W. Zinn, Physica, 86-88B (1977) 1031.
262. Ch. Sauer J. Magn. Magn. Mater., 3 (1976) 46.
263. N.A. Blum and R.B. Frankel, AIP Conf. Proc., 29 (1976) 416.
264. N. Bykovetz, Solid State Commun., 18 (1976) 143 and Ph.D. thesis, Univ. of Pennsylvania (1975).
265. H. Maletta and G. Crecelius, J. Phys. (Paris) Colloq., 37 (1976) C6-645.

266. F. Hulliger, M. Landolt, R. Schmelezer and I. Zarback, Solid State Commun., 17 (1975) 751.
267. J. Moser, G. Wortmann, N. Bykovertz and G.M. Kalvius, J. Magn. Magn. Mater., 12 (1979) 77.
268. M. Abd Elmeguid and G. Kaindl, J. Phys. (Paris) Colloq., 40 (1979) C2-310.
269. M. Eibschutz, R.L. Cohen, E. Buehler and J.H. Wernick, Phys. Rev. B, 6 (1972) 18.
270. B.B. Triplett, N.S. Dixon, P. Bool Chand, S.S. Hanna and E. Buchner, J. Phys. (Paris) Colloq., 12 (1974) C6-653.
271. B.B. Triplet, Y. Mahmud, N.S. Dixon, S.S. Hanna and E. Holtzberg, Phys. Lett., 67A (1978) 151.

CHAPTER 16

MOSSBAUER SPECTROSCOPIC STUDIES OF FERROELECTRIC COMPOUNDS

S.K. Date* and U. Gonser†

* Physical Chemistry Division, National Chemical Laboratory, Pune, 411008, India.

† Fachbereich Angewandte Physik, Universitat des Saarlandes, D66 Saarbrucken, West Germany.

1. INTRODUCTION

It was some sixty years ago that the phenomenon now known as ferroelectricity was first recognised in Rochelle salt (1). For the next thirty years, the phenomenon was thought to be a rarity in nature and attempts to understand it at the microscopic level were formulated in terms of specific crystal structures which supported a ferroelectric instability. In fact, a deeper understanding of the causes of the ferroelectric phase transition is one of the key problems of molecular science today (2). The microscopic breakthrough came in 1960 with the recognition of the fundamental relationship between lattice dynamics and ferroelectricity and more importantly of the existence of a soft mode behaviour (SMB) or instability at the ferroelectric phase transition (3). Those who are interested in ferroelectrics from the lattice dynamical point of view encounter the following question: what role is played by the so-called soft mode behaviour (condensation of the soft phonon) around the critical temperature. The SMB studies about the structural phase transitions in many materials, including well-known ferroelectrics and anti-ferroelectrics have lately been a subject of considerable interest and therefore of many detailed experimental investigations by various techniques such as neutron, x-ray, light and Brillouin scatterings, magnetic resonance (NMR, ESR, etc.) and Mossbauer spectroscopy (4). There are two basic reasons for the increased activities: On one side is the lattice dynamical theory of ferroelectric phase transitions proposed by Cochran (3,5) and further developed by Anderson and others (6) in the early sixties and on the other side is the potential technological applications with these materials.

The Mossbauer effect (ME), which is the resonance emission and absorption of gamma rays by atoms bound in solids without the excitation of phonons (i.e. the observation of zero phonon gamma rays), is relatively a new comer to the group of techniques used in the study of ferroelectric materials (7). The information to be obtained from ME in ferroelectrics comes from various paramter such as, isomer shift (IS), quadrupole splitting (ΔE_Q) of the observed resonance on the one hand and a study of the absorption or Mossbauer "f" fraction on the other. A review of the literature shows that during the last fifteen years, ME for Fe^{57} and Sn^{119} have been primarily used to study ferroelectric materials on the microscopic scale. This has been feasible due to the fundamental relations that exist between the dielectric constant and normal modes of vibrations of the crystal on one side and the Mossbauer "f" factor and the same normal modes of vibration on the other. Table 1 gives some typical examples of ferroelectric and antiferroelectric materials investigated by ME. Before the work for the present article was started, two short reviews have appeared in the literature (8,9) and we have used both of them thoroughly.

We will describe in brief the basic properties and classification of ferroelectrics in order to understand the characteristic behaviour of ME parameters, namely, the resonance fraction f(T), the quadrupole interaction (E_Q), and the center shift of the spectrum around the transition temperature. In the second part of our article, a brief summary of the lattice dynamical theory introduced by Cochran and further developed by Anderson and others will be outlined. As examples, we have chosen well-known systems such as

(i) $BaTiO_3$: Co^{57}, Fe^{57} and Sn^{119}

(ii) $PbTiO_3$: Co^{57}, Fe^{57}

(iii) $K_4Fe(CN)_6 \cdot 3H_2O$(KFCT)

(iv) $PbZrO_3$: Sn^{119} and $PbZrO_3$: Fe^{57}

(v) diatomic GeTe-SnTe system

(vi) Layer-type and mixed oxides

although there are many more interesting examples available in the literature. In the last part of our article, we will describe some of the applications of the ME to the study of photochromatic and photorefractive materials, such as, $SrTiO_3$: Fe, $LiNbO_3$: Fe systems.

TABLE 1

	Ferroelectric/Antiferroelectric Materials	Remarks
1.	Pervoskite oxides and others	
	$BaTiO_3$	
	$PbTiO_3$	
	$PbZrO_3$	ABO_3: Mossbauer impurity as probes for IS, ΔE_Q, f(T) study, Photochromatic and Photorefactive properties
	$LiNbO_3$	
	$LiTaO_3$	
	$BiFeO_3$	
	$SrTiO_3$	
2.	Mixed oxides/Solid solutions	
	$Pb(NbFe)O_3$	
	$La_{1-x}Sr_xFeO_3$	
	(Mn, Fe) YO_3	IS, ΔE_Q(T), f(T)
	$BiFeO_3$-$PbZrO_3$	
	$BiFeO_3$-$PbTiO_3$	
3.	Chalcogenides	
	Sn(Te, Se, S)	Center shift, f(T) studies
	SnTe-GeTe	
4.	Alums	
	$NH_4Fe(SO_4)_2.12H_2O$	ΔE_Q(T), f(T) dependence
	$NH_4(Al, Fe)(SO_4)_2.12H_2O$	
5.	Hydrogen bonded materials	
	$K_4Fe(CN)_6.3H_2O$	
	KH_2PO_4 and KD_2PO_4	Dynamics of hydrogen atoms
	$NH_4H_2PO_4$	

2. CLASSIFICATION OF FERROELECTRICS

Let us look into the basic properties and classification of ferroelectrics (1,4) in order to understand the characteristic behaviour of ME parameters around ferroelectric transition temperatures.

A ferroelectric crystal is one which exhibits spontaneous polarisation (i.e. the crystal is polarised in the absence of an external electric field) and further the direction of the spontaneous polarisation may be altered under the influence of an electric field. These crystals are a subclass of spontaneously polarised pyroelectric crystals. However, the polarisation is usually masked by the surface charges, which collect on the surface from the atmosphere. When the temperature of such a crystal is altered, the polarisation changes and this change can be experimentally observed. Pyroelectrics are, in turn, a subclass of piezoelectric crystals, i.e. these crystals become polarised under the influence of external stresses and this property is entirely a matter of crystal symmetry. We know that 20 out of 32 classes lack a center of symmetry. Out of these 20 classes, only ten have a unique polar axis and these ten classes belong to the pyroelectric group of crystals. We remember here once again that ferroelectrics are only those pyroelectrics whose direction of spontaneous polarisation is reversible by an applied field. This definition reveals an interesting difference between ferroelectricity (FE) and ferromagnetism (FM). In case of FM materials, whenever the moments are expontaneously aligned, they can also be reversed by means of an external field. The phenomenon of ferroelectricity, on the other hand, is interesting only when the moments are loosely aligned, i.e. when the interactions are so delicately balanced as to allow reversal of spontaneous polarisation. The ferroelectric properties of a ferroelectric material disappear above a critical temperature T_c, the so-called ferroelectric Curie temperature. Associated with the transition from the ferroelectric to the nonferroelectric phase are the anomalies in other properties such as dielectric constant, specific heat, crystal structure etc. The transition may be of the first or second order. It is also known from earlier studies that the spontaneous polarisation in the ferroelectric crystals is associated with the spontaneous electrostructural strains in the crystal and therefore in the ferroelectric phase, the crystal structure has a lower symmetry which changes at the transition temperature, T_c.

We summarize the criteria according to which classification of the FE materials have been or could be proposed.

(1) Crystal chemical classification: According to this classification these materials may be divided into two groups. The first group comprises hydrogen-bonded crystals, such as KH_2PO_4, Rochelle salt, tri-glycine sulphate. The second group includes the oxides such as $BaTiO_3$, $PbTiO_3$, $KNbO_3$, $LiNbO_3$ etc.

(2) Polarisation direction classification: Here again, FE materials can be divided into two groups: a group involving a single axis of spontaneous polarisation such as Rochelle salt, KH_2PO_4 etc. and another group comprising crystals which can polarise along several axes that are crystallographically equivalent in the non-polar phase such as $BaTiO_3$, ferroelectric alums etc.

(3) Centre of symmetry classification in the non-polar phase: A first group of ferroelectrics is characterised by a non-polar phase which is piezoelectric (non-centrosymmetrical) such as Rochelle salt, KH_2PO_4 and isomorphous compounds. A second group of ferroelectrics is characterised by a centrosymmetrical non-polar phase, e.g. $BaTiO_3$.

(4) Phase change classification: This involves classification according to the nature of the phase change occurring at the Curie point, i.e. the nature of the microscopic mechanism which gives rise to spontaneous polarisation of the crystal. The first group undergoes a transition of the order-disorder type, such as KH_2PO_4. A second group of compounds undergoes a displacive transition such as $BaTiO_3$ and other oxides.

In order-disorder transitions, permanent electric dipoles which are present above the transition temperature become ordered at the transition temperature. Such a transition is similar to a ferromagnetic transition and can be treated with the same types of ordering theories as are used in the magnetic case. We will not discuss them in detail here due to the scope of this review. On the other hand, a displacive ferroelectric transition is very different from a magnetic transition in that no electric dipoles exist above the transition temperature. These dipoles come into existence at the phase transition only, when the unit cell distorts in such a way that an ordered set of electric dipoles is created. The process leading to such a distortion of the crystal can be described in terms of a cancellation of long-range coulomb forces by short-range forces between the ions. This lattice dynamical theoretical approach was first suggested by Cohnran and developed further by many others. The cancellation of these forces leads to an anomaly in the frequency of the transverse optical (TO) vibrational mode of the crystal. This particular mode has a wave vector q equal to zero, which corresponds to entire Bravais sublattices moving as rigid units relative to each other. In other words, the temperature dependent TO vibrational mode decreases in frequency as the Curie temperature is approached from the higher temperature side. We will not go into details about the mathematical description of the lattice dynamical approach but wish to point out the characteristic features and their influence on some of the ME parameters due to the softening of the crystal lattice around T_c (10)

If W_T is the frequency of the optical mode of interest with wave vector q = O, we write

$$W_T^2 = G\ (T - T_O), \qquad (1)$$

where T_O is a temperature near the transition temperature T_c. Eqn. (1) along with the Lyddane-Sachs-Teller relation (LST)

$$\frac{W_L^2}{W_T^2} = \frac{\varepsilon_s}{\varepsilon_\infty} \qquad (2)$$

leads us to the experimentally determined Curie Law temperature dependence of the static dielectric constant ε_s. (W_L is the frequency of the longitudinal mode with q = O and ε_∞ is the high frequency dielectric constant).

In observing the ME spectrum of transmitted gamma ray intensity close to the ferroelectric phase transition, there are two quantities of interest to us. These are a change in the absorption and/or change in the shape and position of the line. Of these, the first is heavily dependent on the temperature dependence of the optical phonon spectrum while the second tells us about the changes in the local field gradients and symmetry of the lattice.

3. EXPERIMENTAL RESULTS

Early ME studies in ferroelectric $BaTiO_3$ and $PbTiO_3$

It has long been recognised that ferroelectricity in AbO_3 type compounds, of which $BaTiO_3$ is a typical member, is closely associated with (i) intense electric field at the B lattice site, (2) temperature dependent homogeneous transverse optical B-O-B mode (3) nature of B-O bond, and (4) displacement of the B ion in the surrounding oxygen octahedron. In view of the important role played by the B ion, several attempts have been made either directly or indirectly to investigate its environment. It has been well-established that $BaTiO_3$ exhibits a ferroelectric phase transition (cubic to tetragonal) around 120° C and in addition, it undergoes two more phase transitions, one at 5° C (tetragonal to orthorhombic) and the other at -90° C (orthorhombic to rhombohedral) at which the direction and magnitude of spontaneous polarisation and the dielectric constant change abruptly. All these three transitions have been investigated by Fe^{57} and Sn^{119} ME techniques. In the pioneering experiment Bhide and Multani (11)probed the lattice with Co^{57} impurity ions. Fig. 1 shows

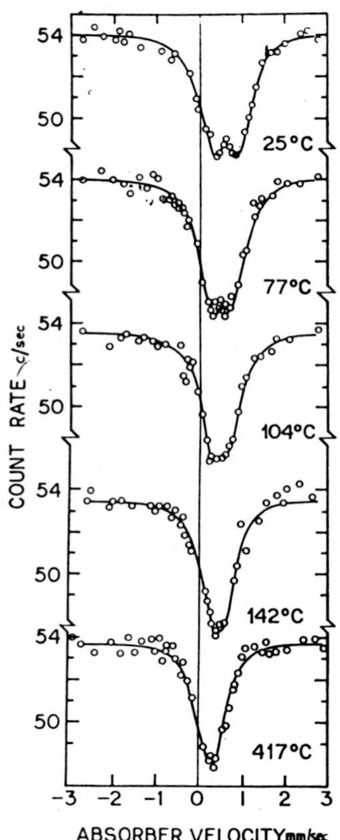

Fig. 1. Typical ME spectra for $BaTiO_3$:Co^{57} source matched against a standard 310 stainless steel absorber at different temperatures.

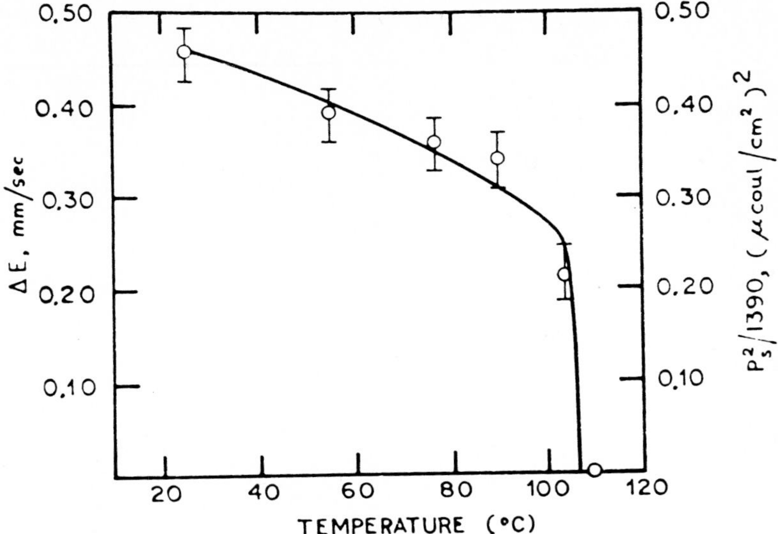

Fig. 2. Variation of the quadrupole splitting with temperature is represented by open circles. The solid line indicates, on a matching scale, the temperature dependence of the square of spontaneous polarisation.

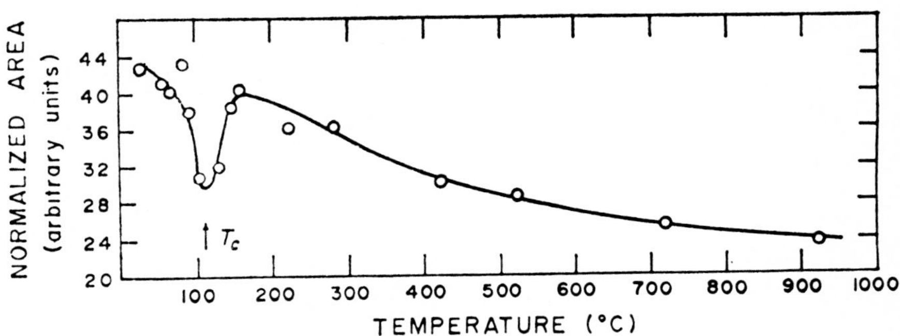

Fig. 3. Temperature variation of the normalized area under the Fe^{57} resonance

typical spectra obtained at different temperatures. In Fig. 2 and 3, the temperature dependence of ΔE_Q and the spontaneous polarisation P_S and the normalized area under the resonance are plotted. Fig. 4 shows the anomalous behaviour at the transition temperature of four different parameters, spontaneous polarisation, dielectric constant, normalised area under the resonance curve and percentage absorption

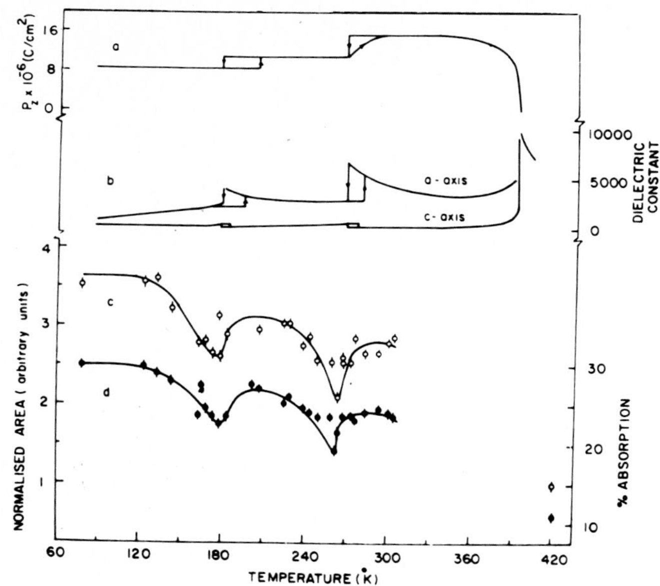

Fig. 4. Variation of (a) the spontaneous polarisation (b) dielectric constant (c) the normalized area under the Sn^{119} resonance and (d) percentage absorption as a function of temperature.

as derived by Sn^{119} ME studies as a function of temperature (12). It is clearly seen that all these quantities show anomalous temperature dependence near the transition temperature. It may be pointed out that the effective EFG experienced by the Co^{57} and Sn^{119} impurity ions is compatible with the perfect lattice local environment model suggested by these authors. In 1972 Bhide and Hegde (13) studied the ferroelectric phase transition in $PbTiO_3$ around 490° C by incorporating the Co^{57} atoms as microprobes substitutionally at Ti^{4+} lattice sites. Figures 5-11 exhibit the detailed analysis of their work. Figs. 5-6 show typical Mossbauer spectra for $PbTiO_3$:Co^{57} sources matched against 310 enriched stainless steel and single crystal $K_4[Fe(CN)_6].3H_2O$ absorbers at various temperatures. The variation of quadrupole splitting with temperature, which is similar to that observed in $BaTiO_3$, is shown in Fig. 7. From the systematic behaviour of impurity ions in ABO_3 type compounds it is presumed that Fe^{3+} ions occupy Ti^{4+} sites substitutionally and as expected the isomer shift and ΔE_Q values are characteristic of high spin ferric ions in distorted oxygen octahedral surroundings.

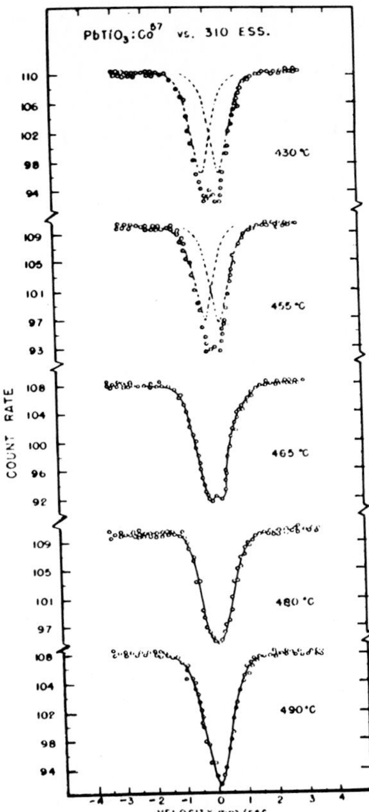

Fig. 5. Typical spectra for PbTiO :Co^{57} source matched against 310 enriched stainless steel absorber at various temperatures.

Isomer shift and center shift

The abrupt change in the center shift and resonance area at T_c have been analysed. Fig. 8 shows the variation of center shift against temperature. To understand the discontinuous changes at the transition temperature, we have to realise that the center shift is made up of two components : Isomer shift and second order Doppler shift (SOD)

$$\delta E_{total} = \delta E_{IS}(T) + \delta E_{SOD(T)} \tag{3}$$

Both of these contributing factors have either an explicit or an implicit temperature dependence. The change in δE_{IS} arises out of the

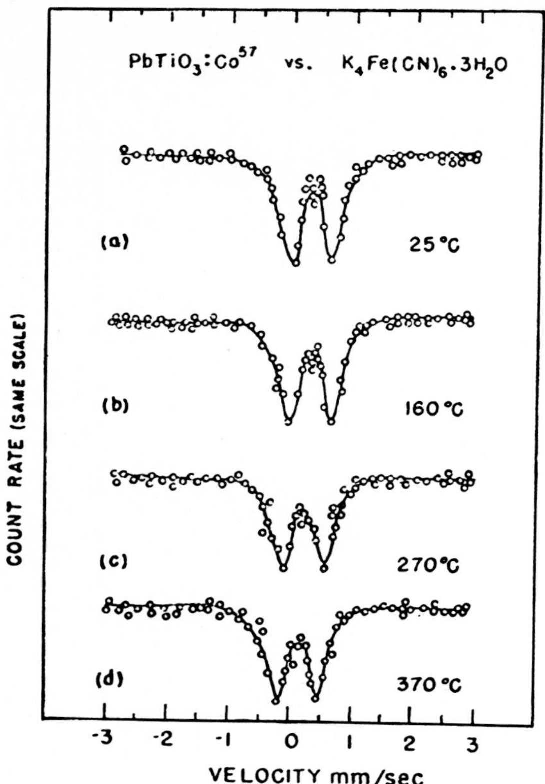

Fig. 6. Typical spectra for $PbTiO_3:Co^{57}$ source matched against $K_4Fe(CN)_6.3H_2O$ single crystal absorber at various temperatures.

thermal expansion as a consequence of increasing temperature. In the high temperature region, one can write

$$\delta E_{IS}(T) = \delta E_{IS}(0) + \alpha T \tag{4}$$

where α is a constant which can be evaluated from the knowledge of the thermal expansion and the pressure dependence of the isomer shift. The changes in $\delta E_{SOD}(T)$ are caused essentially by the variation of the mean square velocity as a function of temperature. In the region $T \geq \Theta_D/3$, one can write 1/4

$$\frac{\delta E_{SOD}(T)}{E_r} = \frac{3k_B T}{2Mc^2} \left(1 + \frac{\Theta D^2}{20T^2}\right) \tag{5}$$

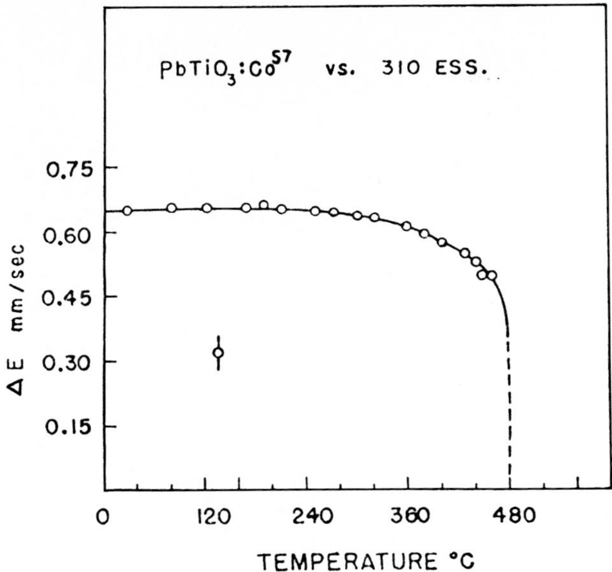

Fig. 7. Variation of quadrupole splitting (ΔE_Q) with temperature.

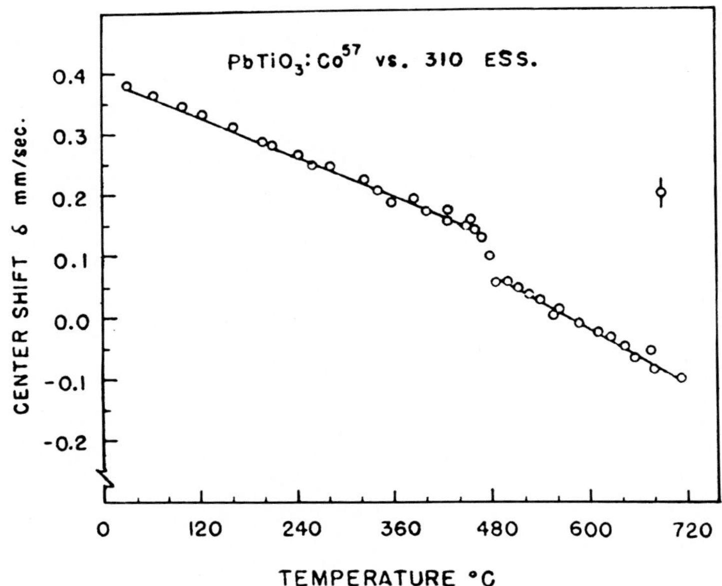

Fig. 8. Variation of center shift with temperature.

where E_r = gamma-ray energy

M = mass of the nucleus

k_B = Boltzmann constant

Θ_D, T = Debye and ambient temperatures.

If we combine Eqns. (4) and (5)

$$\frac{\delta E_{total}(T)}{E_\gamma} + \left(\frac{3k_\beta}{2Mc^2} - \frac{\alpha}{E_\gamma}\right)T = \frac{\delta E_{IS}(O)}{E_\gamma} - \frac{3k_\beta \Theta_D^2}{40\ Mc^2\ T}. \quad (6)$$

If we plot the left hand side of Eqn. (6) against reciprocal temperature, the intercept will give $\delta E_{IS}(O)/E_\gamma$ and the slope of the curve (Fig. 9) will yield the Debye temperature. The curve shows a discontinuous break of 0.075 $\pm$ 0.04 mm/sec. Assuming the volume changes discontinuously at T_c as available from the data of Shirane et al (15) and using the wellknown equation proposed by Pound et al (16) and Pipkorn et al (17)

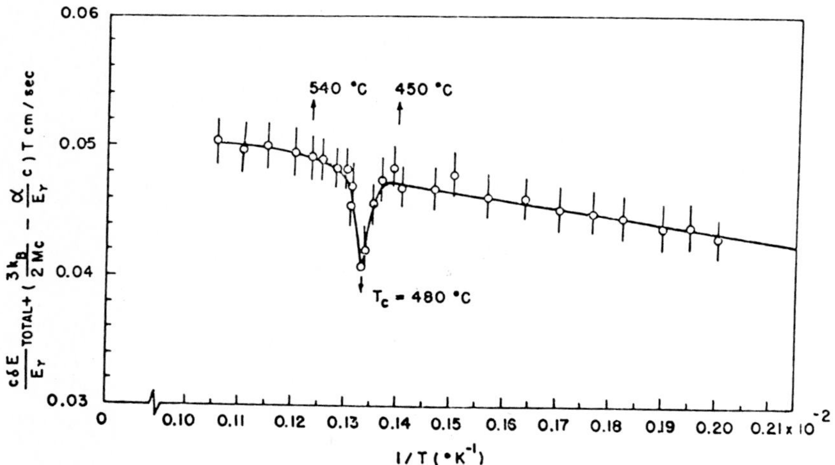

Fig. 9. Plot of LHS of Equation No. 6 against the reciprocal of temperature ($^oK^{-1}$) showing clearly an anomaly in the vicinity of T_c.

$$\frac{\delta E_{IS}}{E_\gamma} = (5.37 \pm 0.07) \times 10^{12}\left(1 - \frac{\Delta V}{V}\right), \quad (7)$$

it can be estimated that the sudden change in δE_{IS} would be about 0.01 mm/sec. Alternatively the sudden change may be due to the change in Debye temperature and/or to the disappearance of the soft mode.

Further detailed theoretical and experimental studies will throw more light on these lattice dynamical instabilities.

Soft mode behaviour

It has been suggested by Cochran (3) and Anderson (6) that ferroelectricity in the pervoskite ferroelectrics arises because of the temperature dependent optical mode which decreases in frequency as the temperature of the crystal approaches T_c from the high temperature side. In second order phase transition, the disappearance of the optical mode leads to the instability leading to a phase transition at T_c. However, in the first order phase transition, the instability sets in earlier because of the optic acoustic mode mixing. In ferroelectrics undergoind either of these phase transitions, ME studies provide a method for the investigation of the temperature dependent soft mode.

After stuitable corrections, the area A under the resonance curve can be shown to be proportional to the Lamb-Mossbauer factor (L-M factor). Assuming a monoatomic cubic lattice and the Debye model, the Lamb-Mossbauer factor f(T) can be expressed as

$$f(T) = \exp\{-\frac{E_\gamma^2}{2Mc^2}\cdot\frac{6}{k_\beta\Theta_D}[\frac{1}{4}+(\frac{T}{\Theta_D})^2\int_0^{\Theta_D/T}\frac{x\,dx}{e^x-1}]\}. \quad (8)$$

From this expression, it is possible to determine the Debye temperature of the lattice from the knowledge of the temperature dependence of L-M factor. Fig. 10 shows the area under the resonance, after corrections for background and normalisation for the off-resonance count rate, as a function of temperature. This variation shows anomalous variation of L-M factor in the vicinity of transition temperature. Following the analysis of Bhide and Hegde (13), Gleason and Walker (18), one can write

$$f(T) = \exp\{-\frac{E_\gamma^2}{2Mc^2}\;\frac{6}{k_\beta\Theta_D}[\frac{1}{4}+(\frac{T}{\Theta_D})^2\int_0^{\Theta_D/T}\frac{x\,dx}{e^{x-1}}]\}$$

$$\exp[-\;(\frac{2}{\exp\{h\;P(T-T_o)^{\frac{1}{2}}/k_\beta T\}-1}+1)$$

$$X(\frac{(hk)^2}{2Mh\;P(T-T_o)^{\frac{1}{2}}}\;\epsilon_{an}^2)]. \quad (9)$$

From this expression, it is immediately clear that, for a second order phase transition, f(T) becomes zero at T_C. However, for a

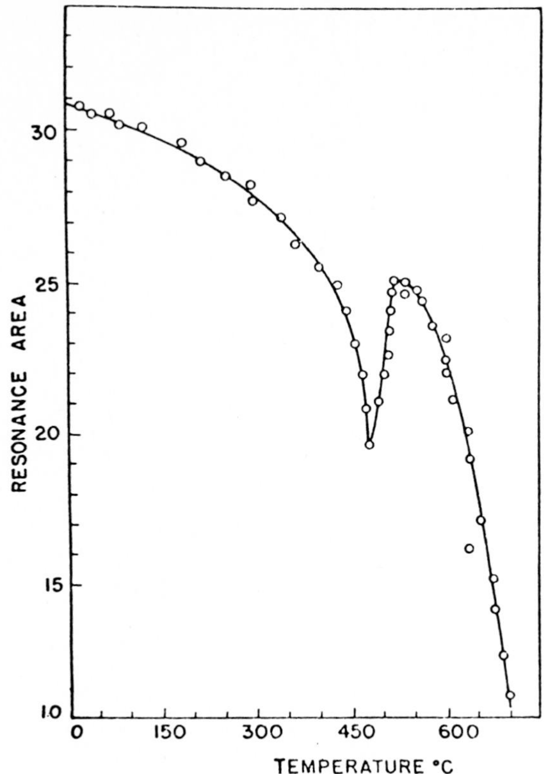

Fig. 10. Temperature variation of the normalized resonance area, showing a well-defined anomaly in the vicinity of T_c.

first order transition, the crystal becomes unstable at T_c even before T_o is reached and one expresses the temperature dependence of the soft mode as

$$W_{an}^2 = G(T - T_c) + G(T_c - T_o). \qquad (10)$$

It is seen clearly that f(T)-T variation will reach a minimum at T_c. Fig. 11 shows both the experimentally observed normalised area under resonance and the calculated values of f(T) in the case of the ferroelectric $PbTiO_3:Co^{57}$ system. Recently, Samuel and Sundaram (19) have suggested an improvement in the theory of the ME in ferroelectrics by (i) completing the formal expression of f(T) through a correction factor, (ii) replacing the polarisation vector with another characteristic of ferroelectric substances, namely,

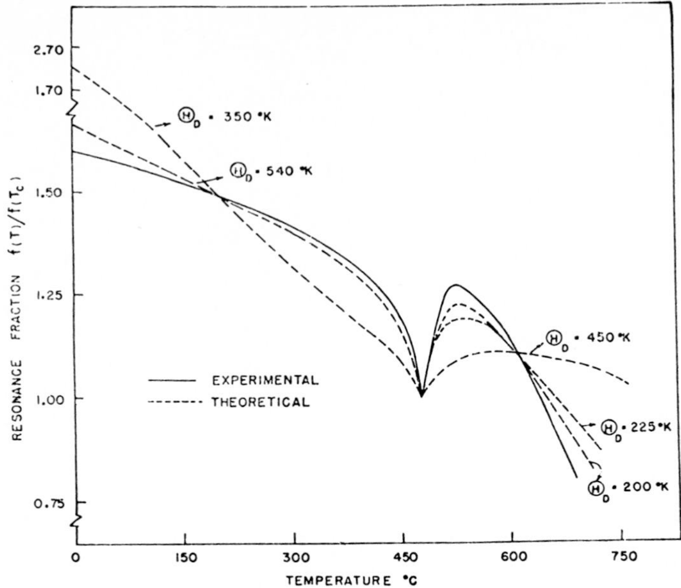

Fig. 11. Computer-fitted curves of the theoretical resonance fraction in the presence of Cochran-type optic mode with temperatures, normalized to that at T_c, with the corresponding normalized area under the resonance curve.

the fraction of ferroelectrically active modes, and (iii) examining analytically the possibility of an anomalous behaviour in the second order Doppler shift. The results of these calculations are: (i) The observed anomalous behaviour of f(T) is explained using a model consistent with lattice dynamical theory, (ii) It shows no anomalous behaviour in the second order Doppler shift. On the other hand, Wissel (20) has pointed out that definite theoretical predictions on the influence of a soft mode on the Lamb-Mossbauer factor cannot be made. The principal objections to the earlier theoretical expressions are: (i) Harmonic approximations are used to derive the expression, (ii) It is necessary to use the complete dispersion relation and its temperature dependence, (iii) Detailed knowledge of the lattice modes and their dependence on the wave vector is, in general, available neither from experiments nor from theory. In fortuitous circumstances, as Wissel (20) pointed out, the anomalous behaviour and change in lineshape may occur. Meissner and Binder (21) have also recently questioned the observation of an anomalous behaviour in the Lamb-Mossbauer factor, x-ray and neutron scattering experiments.

Recent work on $BaTiO_3$:Fe^{57} and Sn^{119} ME studies (22)

Since 1963 a number of papers on the observation of the anomalous behaviour of Mossbauer parameters, namely the "f" factor and "SOD" shift in ferroelectric $BaTiO_3$, have been published. However, the results are contradictory. As reported earlier, Bhide and Multani (11) described detailed experiments with Fe^{57} and Co^{57} doped $BaTiO_3$ in which a large f-factor anomaly at the Curie temperature was observed. Maguire and Rees (23) were unable to repeat this observation with samples prepared in the same manner. Using ceramic samples, Yarmarkin et al (24) concluded that a substitution at the Ti site by Fe^{57} is unlikely; instead several Ti and Fe containing phases are formed at the grain boundaries. The results obtained with Sn^{119} ME studies on $BaTiO_3$ ceramic samples are also controversial. While Bokov et al (25) reported anomalies at the Curie temperature, Mitrofanov et al (26) and Plotnikova et al (27) did not find any anomalous behaviour in various ME parameters.

The basic question in all these ME experiments with $BaTiO_3$ remains unanswered: what is the position and local environment of the impurity atoms, i.e. Co^{57}, Fe^{57} and Sn^{119} in $BaTiO_3$? More specifically do the impurity atoms substitute Ti and form a similar chemical bond in the lattice? Undoubtedly more conclusive are the results with melt grown crystals in Mossbauer Rayleigh scattering investigations (28-30).

It was our intention to repeat these experiments where contradictory results have been obtained, particularly experiments with crystals where the resonance atoms were already contained in the melt from which they had been grown (22). We have not observed any effect of phase transitions on the hyperfine parameters or f-factor in $BaTiO_3$. There are subtle differences in the spectra of $BaTiO_3$:Co^{57} sources, However, all the investigated $BaTiO_3$:Co^{57} sources show essentially spectra which do not depend strongly on the parameters of the thermal treatment. In spite of a very large number of prepared sources, we never succeeded in observing spectra as obtained by either Bhide and Multani (11) or by Maguire and Rees (23). Fig. 12 shows typical ME spectra of a $BaTiO_3$:Co^{57} source after an initial hydrogen firing matched against a standard single line absorber. A well resolved six line pattern is observed with H_{int}=365 kOe and δ=-0.03 mm/sec with an additional single line at the centre

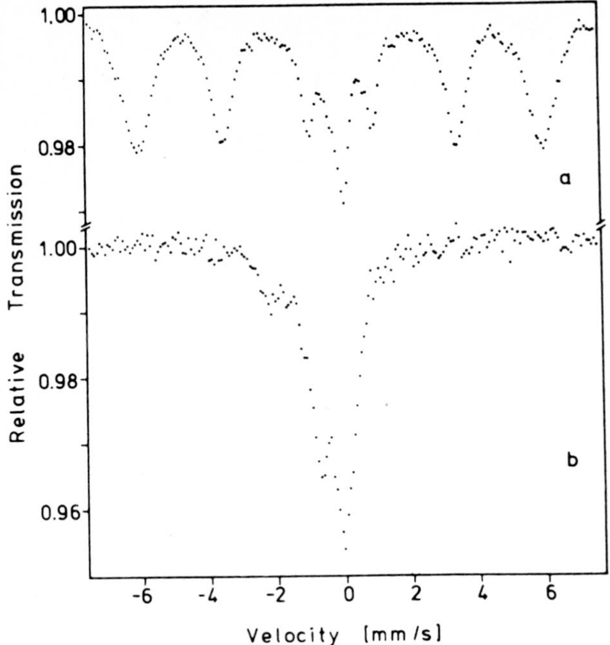

Fig. 12. ME spectra of a $BaTiO_3$:Co^{57} source at room temperature (a) after hydrogen firing, (b) after air firing and quenching.

(δ = 0.04 mm/sec). The relatively broad lines (ca ∿ 1 mm/sec) indicate a hyperfine field value which is somewhat larger than the value reported earlier for α-Fe(330 kOe). After air-firing and quenching the magnetic spectrum has disappeared. The new spectrum can be decomposed into two quadrupole doublets (parameters are δ_1 = -0.33 mm/sec, ΔE_{Q1} = 0.73 mm/sec, δ_2 = -0.99 mm/sec, ΔE_{Q2} = 2.02 mm/sec) which can be easily assigned to the two valence states of iron, Fe^{3+} and Fe^{2+}, on the basis of well-known isomer shift and quadrupole splitting systematics (31). Searching with great care in the neighbourhood of the Curie temperature, no anomalies or even significant changes in the spectra were detected. Similarly, no anomalies or significant changes in the Sn^{119} spectra were detected (Fig. 13). Our spectra can be regarded as a clear indication that Co or Fe atoms are not really distributed and incorporated into the $BaTiO_3$ host lattice, but are conglomerated wherever defects in the lattice offer sufficient room. In fact, in one out of six sources of $BaTiO_3$, Bhide and Multani (11) obtained a perfect lattice local

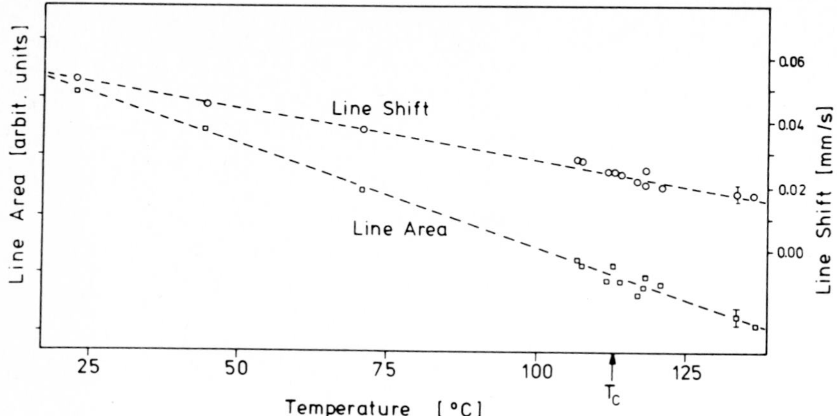

Fig. 13. Line shift and line area versus temperature of a 1.4 wt. % doped $BaTiO_3:Sn^{119}$ single crystal absorber.

environment for the probe atoms. Our detailed experimental results are summarised below:

(i) We have demonstrated the difficulties in substituting the probe atoms at Ti^{4+} site in the $BaTiO_3$ lattice. The commonly used thermal diffusion techniques, even up to 1400° C are not sufficient to incorporate the probe atoms into the host lattice.

(ii) Our $BaTiO_3:Co^{57}$ source spectra show clearly that there are precipitated phases (or there is a precipitated phase): ferromagnetic or paramagnetic Fe-Co alloys after hydrogen heat treatment or the corresponding oxide mixture after air firing.

(iii) However, it is possible to substitute Ti^{4+} by Fe^{3+} or Sn^{4+} when the crystals are grown from the melt.

(iv) Weakly Fe^{57} doped ceramic samples indicate that the iron atoms occupy partly Ti^{4+} substitutional sites although these samples exhibit defect structure.

(v) In no case has a change in the ME parameters been observed by variation of temperature through the Curie temperature of $BaTiO_3$. Recent theoretical studies (20,21) predict that no such effects exist, which is the principal observation in our careful experiments.

"f(T)" anomaly in antiferroelectrics

It has been suggested (32) that in antiferroelectrics the entire optical branch may be temperature dependent near the transition temperature, in contrast to ferroelectrics where the long wavelength part (k = 0) of the optical branch is temperature dependent. The behaviour of the temperature dependent frequency of transverse optical modes from the linearized model of a crystal is expressed as

$$W_T^2(k) = \gamma[T-\{T_c + T(k)\}]. \quad (11)$$

For an antiferroelectrics, the quantity T(k) has its maximum value for k near to the edge of the Brillouin zone; decreasing k, it decreases and for k = 0 it is zero. For this reason, a greater relative decrease in the spectral area of the Mossbauer spectrum can be expected in the neighbourhood of antiferroelectric compared to ferroelectric transition temperature. A suitable substance for studying the anomalous behaviour is $PbZrO_3$ with an addition of Fe, Co, Sn impurities. In the antiferroelectric phase of $PbZrO_3$ the symmetry is orthorhombic and it undergoes a phase transition to the cubic phase at 230° C, as evaluated by X-ray and neutron scattering techniques. In the vicinity of the transition temperature, dielectric constant reaches a maximum and above this temperature, the dielectric constant follows the Curie-Weiss law $\varepsilon = C/T - T_o$ with $C = 1.6 \times 10^5$ degrees. Around 230° C, strong electric fields can induce a ferroelectric phase. Recently, two Mossbauer studies using Fe^{57} and Sn^{119} impurities have been reported (33,34). Fig. 14, 15 show the area fraction of recoil-free resonance absorption for $PbZrO_3$. Fe^{57} Mossbauer studies show a relative drop in recoilless fraction of about 30% at 230° C. On the other hand, Sn^{119} measurements show a relative change in 'f' factor by 40%. This is greatly in excess of the previously measured figure of about 10% or less in the case of ferroelectric $BaTiO_3$. The softening of the transverse optical modes near the edge of Brillouin zone in antiferroelectrics is responsible for such a large anomalous behaviour. These results agree well with Dvorak's theoretical predictions (32).

Teisseron and Baudry (35) have studied the temperature dependence of electric field gradient acting on Ta^{181} probe atoms substituting at Zr^{4+} sites in lead zirconate by time differential perturbed angular corrections (TDPAC). The existence of an intermediate

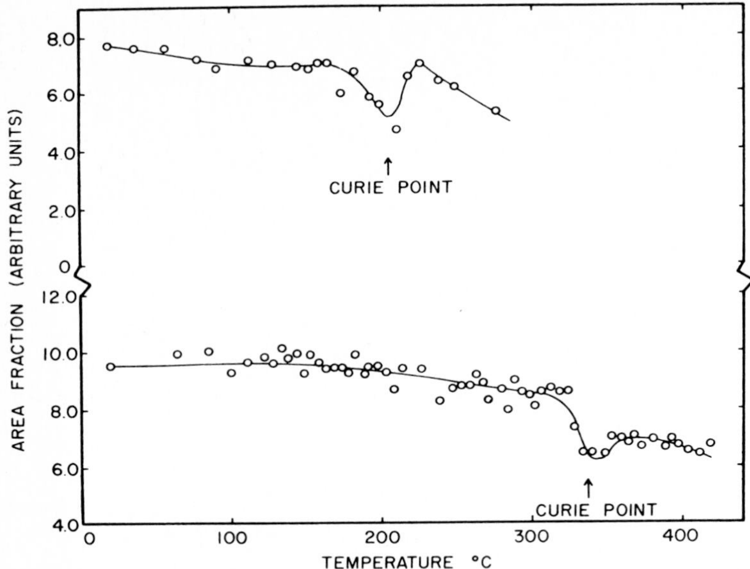

Fig. 14. Area under the absorption peak as a function of temperature in $PbZrO_3:Fe^{57}$ system.

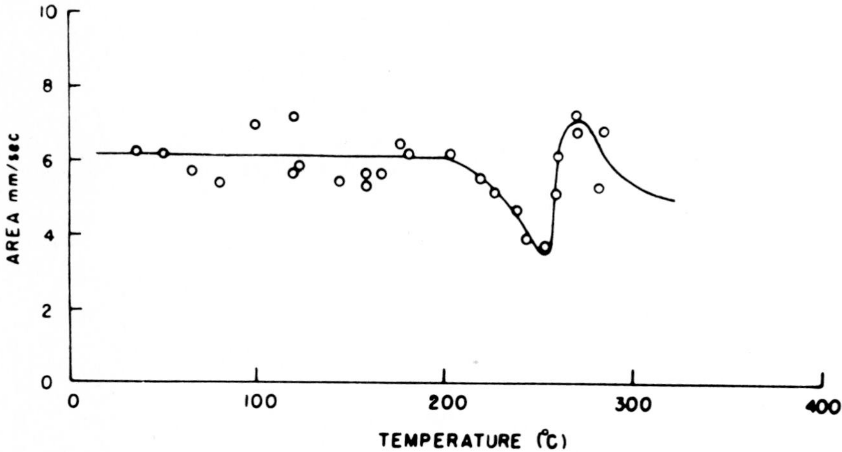

Fig. 15. Temperature variation of the normalized area, exhibiting the anomaly in the vicinity of the Curie temperature T_c in $PbZrO_3:Sn^{119}$ system.

ferroelectric phase between the antiferroelectric and paraelectric phases has been confirmed. Dynamical effects, SMB near the phase transition, especially in the paraelectric phase, are shown to be present. However, so far there are no meticulous ME experiments reported in the literature.

KFCT results

During last few years contradictory results have been obtained on the behaviour of the Mossbauer recoil-free fraction in $K_4[Fe(CN)_6]\cdot 3H_2O$ near its ferroelectric transition temperature. In the crystal structure, the water molecules are located in layers perpendicular to the [010] axis. Between the layers of water molecules, there are two layers of $Fe(CN)_6^{-4}$ ions and interspread potassium ions. The basic structure is either tetragonal or pseudotetragonal. This modification changes irreversibly into a ferroelectric monoclinic structure around 218° K. This structure shows spontaneous polarisation in [101] direction below 248° K. Hazony et al (36) measured the spectral area on single crystals of KFCT and found an anomalous increase related to the 'f' factor near T_c when the direction of transmitted gamma rays was close to the ferroelectric axis. On the other hand, Gleason and Walker (18) and Clauser (37) did not observe any changes in the spectral area of powder samples at the transition temperature. In earlier single crystal experiments, it was observed that the motion of $Fe(CN)_6^{-4}$ complexes resulting from the FE instability is perpendicular to [010] axis. Accordingly, the anomalous effect will vary with the direction of incidence. If the directional unstable mode coincides with the observed dielectric anomaly, we can argue in favour of a displacive transition. Recently, Montano et al (38) and Placido (39) remeasured the spectral area in an oriented single crystal of KFCT along its ferroelectric axis, i.e. [010] axis, and found an increase in the 'f' factor at the transition temperature. They demonstrated this behaviour on a simple model which assumes a change in orientation of the principal axes of the EFG tensor. During the reorientation, the direction of the principal Z axis passes through the direction parallel to k_{max} and an increase in area is expected, as experimentally observed. Associated with the anomaly, a change in the intensity ratio of the quadrupole doublet will cause an apparent "line shift". Fig. 16 shows the experimental results of resonance area versus temperature.

f(T) dependence in diatomic ferroelectrics

In their ME studies on the properties and phase transitions in

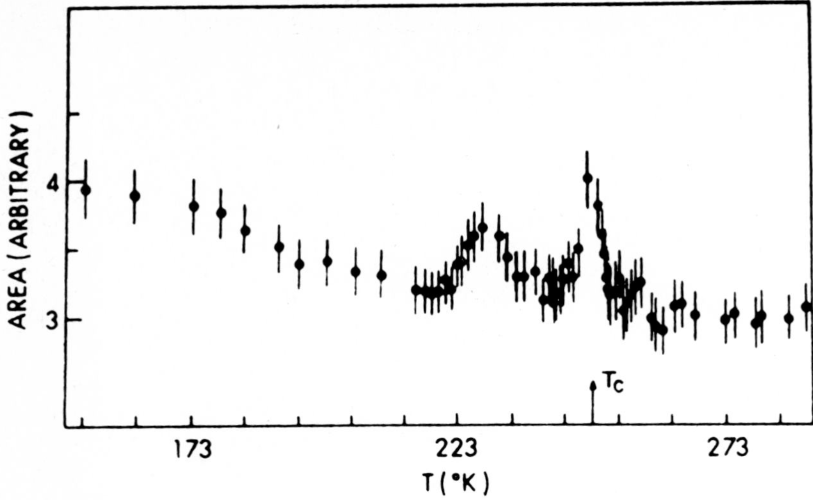

Fig. 16. Experimental area of KFCT versus temperature in the [101] direction.

the $Ge_xSn_{1-x}Te$ system, Rigamonti and Petrini (40) observed a critical behaviour of f(T) in the neighbourhood of the phase transition. It is well-known that, on cooling, GeTe shows a transformation from cubic to rhombohedral structure around 670° K. This phase transition is a displacive type and ferroelectric in character. The structure could be associated with a lattice instability against a transverse optical mode, along a [111] direction; the resulting polarisation cannot be detected through dielectric measurements due to the high conductivity of semiconducting GeTe. It was suggested, therefore, that GeTe would be an example of a ferroelectric diatomic "ionic" crystal. The GeTe-SnTe system forms a continuous series of solid solutions and X-ray data show that $Ge_xSn_{1-x}Te$ exhibit the cubic-rhombohedral transformation at a critical temperature which almost linearly decreases towards 0 °K as a function of tin concentration. On the other hand, SnTe is an example of a "paraelectric" compound whose cubic structure remains stable down to about 0 °K. Fig. 17 shows the decrease in the recoil-free fraction for $Ge_{0.1}Sn_{0.9}Te$ and $Ge_{0.3}Sn_{0.7}Te$ compounds near the transition temperature 130 °K and 270° K respectively.

Layer-type and mixed oxide ferroelectrics

Sultanov et al (41) have studied perovskite-like layer type ferro-

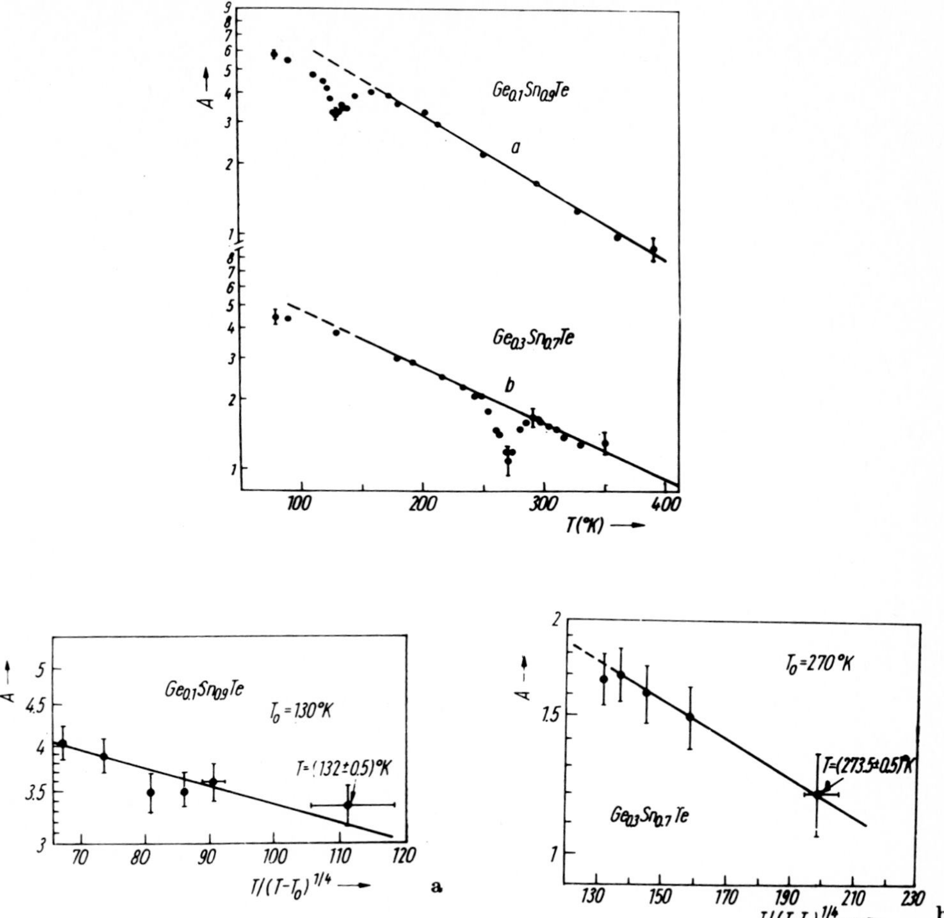

Fig. 17. Semi-Logarithmic plot of the area under Sn^{119} resonance versus temperature.

electrics $Pr_2Bi_4Ti_3Fe_2O_{18}$ and $La_2Bi_4Ti_3Fe_2O_{18}$ using Fe^{57} Mossbauer spectroscopy. These materials are classified as diffuse phase transition ferroelectrics. In such materials, different kinds of ions occupy equivalent crystallographic positions in the unit cell. It is assumed that the statistical distribution of these ions in equivalent sites is responsible for the diffuseness of the phase transition. These authors have reported on the temperature dependence of quadrupole splitting and the integral intensity of Mossbauer lines (the area under the absorption peaks). Fig. 18 shows typical ME spectra at various temperatures. The temperature dependence of the

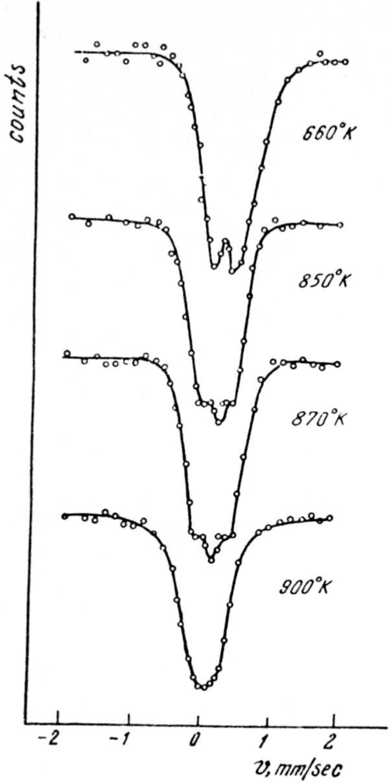

Fig. 18. Typical ME spectra of $Pr_2Bi_4Ti_3Fe_2O_{18}$ at different temperatures

quadrupole splitting is observed up to 870° K. It remains almost constant below 400° K and above this temperature it decreases almost linearly. The quadrupole splitting is proportional to the electric field gradient and the latter in its turn is closely proportional to the square of the spontaneous polarisation, P_s as observed in displace-type ferroelectrics. In the diffuse-type phase transitions, the P_s diminishes monotonically with the temperature and is not equal to zero at temperatures above the transition temperature. The temperature dependence of the line intensity, i.e. the area under the absorption peak normalised to that at 600° K, I_T/I_{600} is shown in Fig. 19. It is seen that the intensity decreases on approaching the Curie temperature from the paraelectric phase, passes through a broad minimum and then begins its usual increase with decreasing temperature. The minimum corresponds to the average Curie temperature of this ferroelectric material, 830° K, as estimated from dielectric

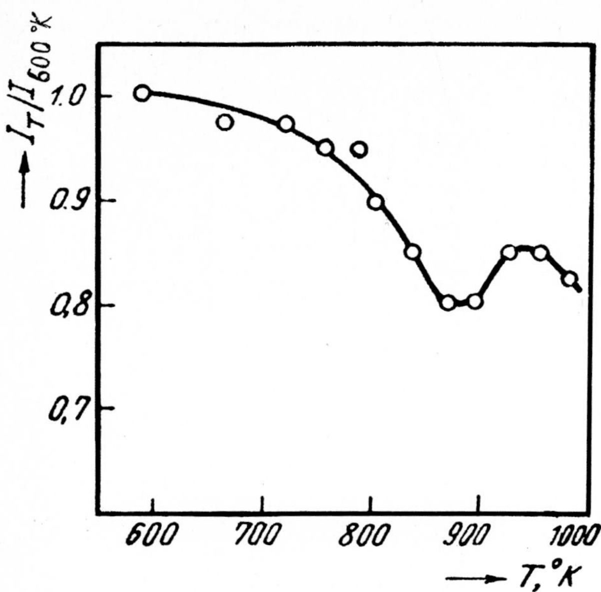

Fig. 19. Temperature dependence of Mossbauer line intensity relative to that at 600 K.

measurements. It is interesting to note that a large decrease in the Mossbauer line intensity occurs around 700° K representing the beginning of the diffuse phase transition range in the material. The authors have attributed the anomalous behaviour to the temperature dependent frequency of the transverse optical branch, a central importance in the lattice dynamical theory of ferroelectricity.

Canner et al (33) have studied antiferroelectric $PbZrO_3$ (PZ), ferroelectric $PbTiO_3$ (PT) and the closely related mixed system $PbTi_{0.2}Zr_{0.8}O_3$ (PTZ) by doping about 5-mol.% of $BiFeO_3$. In these systems, a very small difference was found in the isomer shift of these compounds (Fig. 20), independent of whether they were in the ferroelectric,antiferroelectric or cubic states, indicating that the bonding character is not affected by the transitions among the states. In the study of $PbTiO_3$-$BiFeO_3$ solid solutions, these authors observed a very small discontinuity in the curve of isomer shift versus temperature at T_c (Fig. 21). A dip in the Mossbauer fraction is observed at the Curie temperature (22 Fig.), in contrast to the anomalous cusp-like behaviour observed by Bhide and Hegde (11). Similar anomalous behaviour was also observed by them in their study of $PbZrO_3$-$BiFeO_3$ (PZ) and mixed oxide (PTZ) systems. Fig. 14

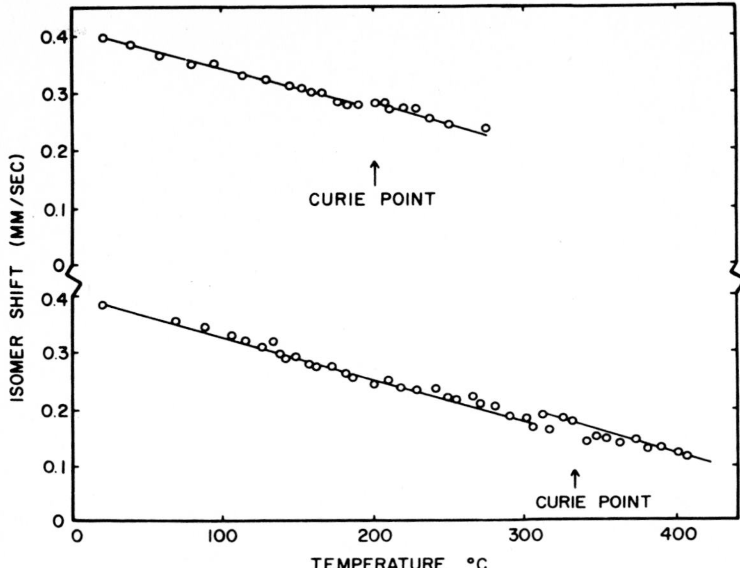

Fig. 20. A plot of isomer shift as a function of temperature for PZ (upper curve) and PTZ (lower curve).

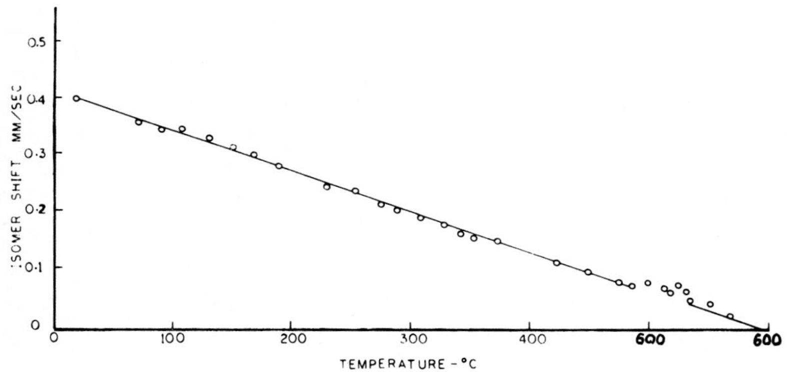

Fig. 21. Isomer shift with respect to iron metal as a function of temperature.

shows the temperature dependence of area for PZ and PTZ samples. The dip was pronounced (30%) and approximately equal for both systems and corresponds to low frequency lattice vibrational mode at the Brillouin-zone boundary.

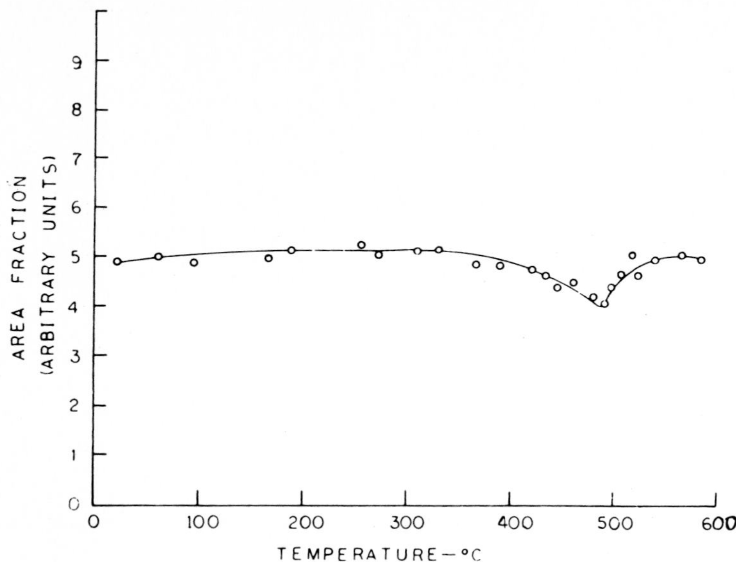

Fig. 22. Area under the absorption peak as a function of temperature.

Photochromic and Photorefractive ferroelectric materials

It is the characteristic of photochromic materials to have a change in colour reversibly under light illumination. The term "photorefractive" materials is used to refer to optically induced changes of refractive index in spontaneously polarised materials (42). Typically these materials are transparent or light coloured in the normal or thermally stable state and become more darkly coloured after irradiation with UV, red or blue light. The induced photochromic optical absorption decays thermally at room temperature during times ranging from few seconds to several days, depending on the material. The materials can also be returned to their original state by irradiation with visible light. Many applications, such as electro-optic modulators, memories and displays, utilizing ferroelectric photochromic materials are possible. In earlier years, transition metal ions doped in $SrTiO_3$, $CaTiO_3$, $BaTiO_3$ were studied with the aim of possible photochromic applications. During the seventies transition metal ion doped $LiNbO_3$ and $LiTaO_3$ were found to be more suitable for electro-optic photo-refractive applications. In fact, iron doped $LiNbO_3$ and $LiTaO_3$ have proved extraordinarily successful as high resolution holographical storage media (43,44). We have used Fe^{57} ME to probe microscopic defect structure (45,48) associated

with photorefractive processes responsible for the storage in $LiNbO_3$. As an important technological application, we will discuss in detail the Fe doped $LiNbO_3$ system.

The light induced change of the refractive index has been observed in photosensitive $LiNbO_3$: Fe crystals exposed to irradiation at impurity optical absorption frequency of about 4800 Å. The photochromic or photorefractive effect is based on the spatial modulation of photocurrents by non-uniform illumination of the crystal. The storage process in $LiNbO_3$ is due to the bulk photovoltaic effect and to the subsequent diffusion and drift of the photocarriers. The generation of photovoltaic and diffusion currents by non-uniform light intensity causes an electronic space charge distribution which modulates the index of refraction via the electro-optic effect (43). By doping with iron ions (or transition metal ions) in $LiNbO_3$, the storage process is enhanced by orders of magnitude indicating the essential role of impurities. In Fe doped $LiNbO_3$, the indicating impurities act as donor-acceptor traps via inter-valence exchange such as $Fe^{3+} \rightleftarrows Fe^{2+}$ which has been substantiated using chemical arguments and some indirect conclusions from earlier studies.

Typical ME spectra of oxidized $LiNbO_3$ single crystal absorbers at various temperatures (45) are shown in Fig. 23. At room temperature, the spectrum exhibits a non-Lorentzian absorption with wings extending over $\pm$ 10 mm/sec. At 77° K and 4.2° K, the spectra show clearly the expected paramagnetic hyperfine structure for Fe^{3+}-$^6S_{5/2}$ state alongwith the electron spin relaxation effects. The various Mossbauer parameters are characteristic of the high spin Fe^{3+} state indicating that the iron ions occupy a substitutionally unique site in $LiNbO_3$.

The ME spectra for the Fe doped $LiNbO_3$ single crystals reduced in an argon atmosphere (45) around 800° C are shown in Fig. 24. This reduction procedure is absolutely necessary for the fixation of the image stored in the crystal. The paramagnetic hyperfine structure of Fe^{3+} ions is predominant in the oxidized state while an asymmetric quadrupole split spectrum is observed in the reduced state. Fig. 25 shows the gradual conversion of the Fe^{3+} to Fe^{2+} state with its characteristic hyperfine interaction parameters (49). The assignment of the Fe^{2+} state is further substantiated by single crystal orientation experiments and magnetic field perturbation experiments (47,48) on polycrystalline samples. We have also shown that the changes in the electronic structure of impurity iron ions have a direct influence on the photorefractive sensitivity and the saturation value of the refractive index. After different treatments,

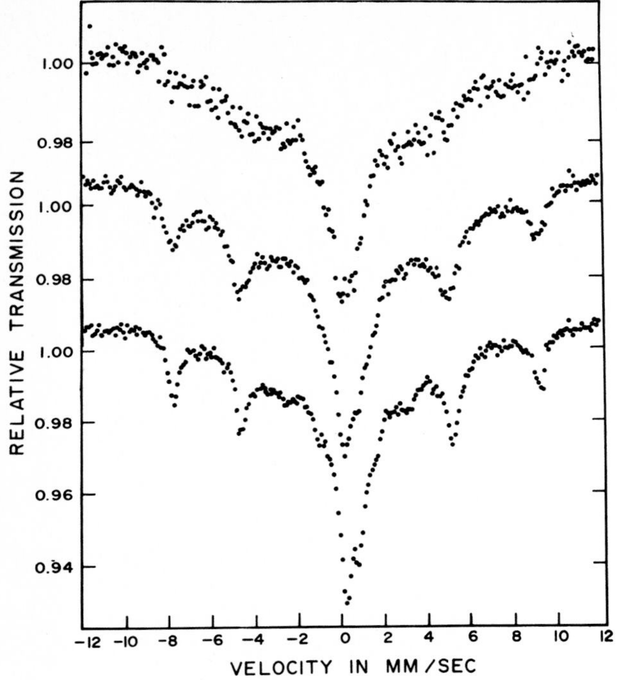

Fig. 23. Typical ME spectra of Fe^{3+} ions in lithium niobate at various temperatures: (a) 298° K, (b) 77° K and (c) 4.2° K.

the results are plotted in Fig. 26 which show the photorefractive sensitivity, relative percentage of Fe^{2+} and Fe^{3+} concentrations monitored in ME experiments, concentration of Fe^{3+} evaluated from ESR experiments versus optical absorption constant (50). With increasing conversion from Fe^{3+} to Fe^{2+}, the photorefractive sensitivity becomes saturated. Fig. 26 also shows the close correlation between the concentration of Fe^{2+} impurity ions and optical absorption band around 2.6 eV which is known to lead to an anisotropic charge transport upon optical excitation. Recently Lauer et al (51) and Engelmann and Gonser (52) reported a valence state change of iron impurities in $LiNbO_3$ after X-ray and UV irradiation. The $LiNbO_3$ crystals used earlier were irradiated using a copper X-ray source (60 KV, 40 mA and Fe filter) at 4.2° K. Fig. 27 shows the ME spectra of $LiNbO_3$:Fe single crystal before and after X-ray irradiation, the effect being the conversion of Fe^{3+} to Fe^{2+}. A reconversion is possible by annealing the crystal at temperature greater than 160° K for a sufficiently long time. During the UV irradiation experiments

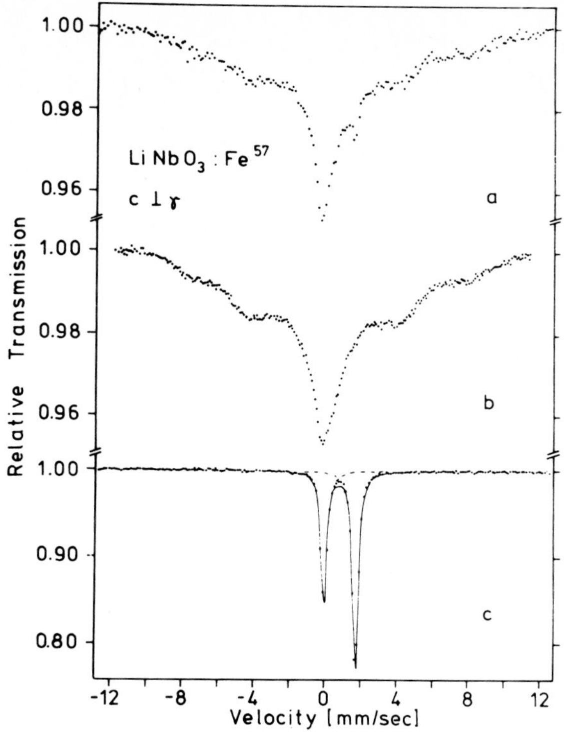

Fig. 24. ME spectra of a $LiNbO_3{:}Fe^{3+}$ single crystal absorber at 295° K (a) in "as grown state", (b) after annealing in air at 800° C for 24 hours and (c) after annealing in argon atmosphere at 800° C for 24 hours.

(λ = 360 nm), similar $Fe^{3+} \rightleftarrows Fe^{2+}$ conversion was observed at 4.2° K (Fig. 28).

The inter-valence exchange process responsible for the storage in $LiNbO_3$ can be followed by heating the crystal in Li_2CO_3 powder around 570° C for nearly three days (53). Fig. 29 shows ME spectrum of a crystal reduced with Li_2Co_3 powder. All hyperfine interaction parameters are almost equal to those obtained in argon reduction experiments. During the heating of Fe doped $LiNbO_3$ crystals in Li_2CO_3 powder, Li^{1+} ions diffuse into the lattice, filling all the Li vacancies for a greater part of the time. Dus to non-availability of lithium sites, iron ions may enter the other alternative site, i.e. Nb^{5+} is substitutionally occupied by Fe^{2+} ions. A complete summary of our results follows:

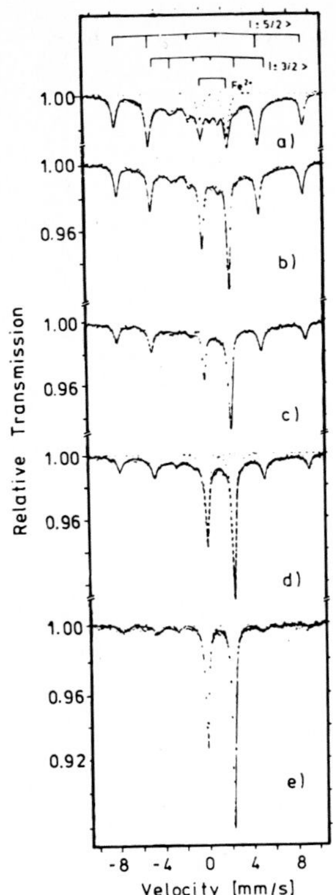

Fig. 25. ME spectra of a $LiNbO_3$-0.22 Wt% Fe_2O_3 absorber measured at 77° K after various annealing procedures at 1000° C: (a) as grown (b) 760 torr argon atmospher, 50 hrs. (c) 760 torr argon, additional 50 hrs. (d) 50 torr argon, additional 10 hrs. (e) 0.2 torr argon, additional 20 hrs.

(i) The ME spectra with single crystal of $LiNbO_3:Co^{57}$ (source) and $LiNbO_3$: Fe^{57} (absorber) revealed the existance of high spin Fe^{3+} (in oxidized state) and Fe^{2+} (in reduced state) valence states.

(ii) The Fe^{2+}/Fe^{3+} ratio could be changed by reducing or oxidizing heat treatments either in an argon or a lithium carbonate atmosphere.

(iii) For both Fe^{3+} and Fe^{2+} valence states, the principal axis of the electric field gradient tensor is found to be parallel to the optical axis with no drastic change in the surrounding local environment.

(iv) The UV and X-ray excitation of a single crystal of Fe doped $LiNbO_3$ produces a metastable Fe^{2+} state, anologous to the

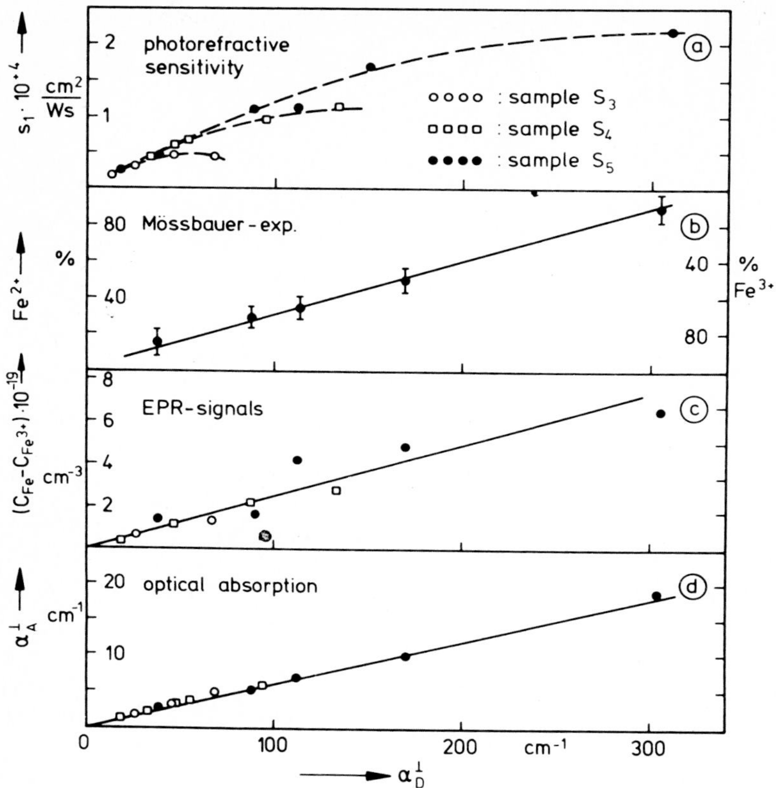

Fig. 26. Photorefractive sensitivity (a) relative percentage of Fe^{2+} *and* Fe^{3+} *concentration monitored in ME experiments (b) concentration of* Fe^{2+} *evaluated from EPR measurements (c) and optical absorption at 1.1 eV (d) versus D-band absorption introduced by different chemical treatments.*

storage process and the formation of a Fe^{2+} state with light irradiation at 4800 Å.

(v) A direct correlation has been observed between the changes in the local electronic environment, photorefractive sensitivity and the saturation value of the index of refraction.

$SrTiO_3$: Fe system

Strontium titanate, $SrTiO_3$, shows incipient ferroelectric behaviour at low temperatures, very much like $KTaO_3$. Unlike $KTaO_3$, however, it shows a non-ferroelectric phase transition involving oxygen-cage tilting, which occurs at about 105° K at atmospheric

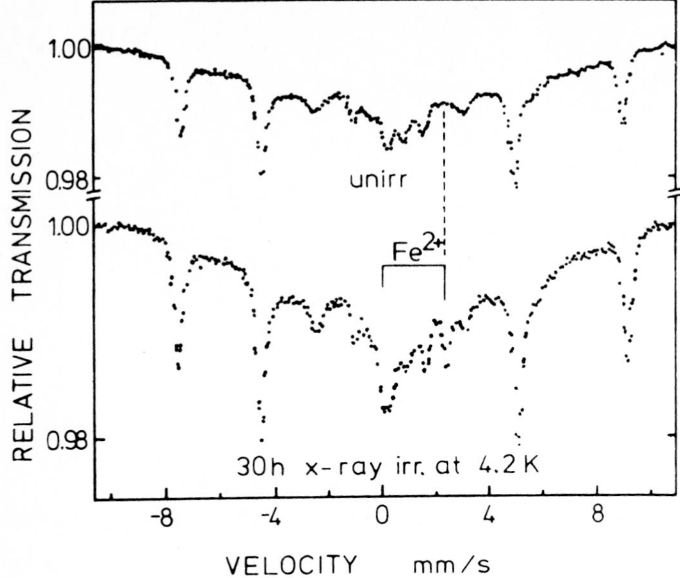

Fig. 27. Typical ME spectra of a $LiNbO_3:Fe^{3+}$ single crystal absorber (a) before irradiation, (b) after x-ray irradiation at 60 KV, 40 mA.

pressure. This phase transition was studied earlier by Bhide and Bhasin (54) using single crystal sources or polycrystalline absorber experiments. In source experiments, the authors observed a break (within the experimental error) in the curve of isomer shift against temperature. However, the assumption of substitution of Ti^{4+} by Co^{2+} ions seems unjustified, as seen clearly from our recent and more accurate experiments (22).

Recently the $SrTiO_3$: Fe system (55) has been studied in great detail due to its photochromic, thermochromic and cathodochromic properties. Several attempts have been made to understand the charge transfer processes involved in the colour changes in the crystal, which are, in turn, related to the changes in the impurity valence state and its associated electronic environment. In the $SrTiO_3$: Fe system, Fe^{4+}, Fe^{3+}, Fe^{2+} and associated defect structures have been identified (56). There is a unique possibility of observing an unusual ionic state, Fe^{5+} in cathodochromic studies which have not been observed so far via ME methods (57), although EPR studies (58) have identified the unusual ionic state. Both the photochromic and photorefractive processes in many **ferroelectric** materials are being continuously studied all over the world. It is hoped that the optical

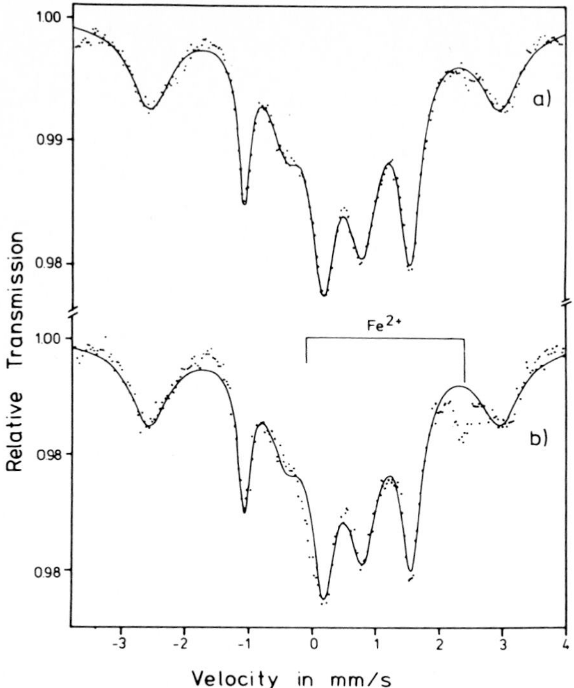

Fig. 28. Typical ME spectra of a single crystal absorber (a) before irradiation (b) after UV irradiation at λ = 360 nm.

storage applications will be developed in future.

4. CONCLUSIONS

The elucidation of the microscopic and macroscopic nature of ferroelectric properties are of great scientific and technological importance. Considerable efforts have been undertaken to shed light on ferroelectric materials near the transition temperature via ME studies. Mossbauer results close to the transition temperature are quite controversial. Some workers have found dramatic effects in the hyperfine parameters as well as in recoil-free fraction. Other workers could not find any effect at all. Thus, the question can be asked: why, in some cases, dramatic changes have been observed by Mossbauer spectroscopy or is Mossbauer spectroscopy an inadequate method of investigating ferroelectric properties? On the other hand, this technique has proved itself as an invaluable tool in understanding the microscopic mechanisms responsible for optical

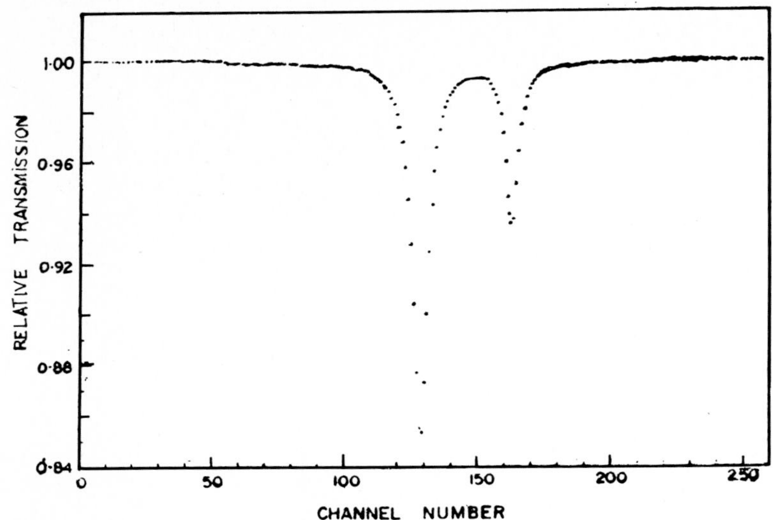

Fig. 29. Quadrupole split spectrum of a single crystal $LiNbO_3$: Fe absorber reduced in Li_2Co_3 powder at 570° C for 60 hours.

storage in ferroelectric materials.

ACKNOWLEDGEMENTS

It is a great pleasure to thank many of our co-workers involved in "Ferroelectrics" programme. In particular, we are indebted to Professor W. Keune and Dr. W. Wildner for their continued interest, co-operation and fruitful discussions during the last six years. One of us (SKD) thanks Alexander Von Humboldt Foundation for a visiting fellowship.

REFERENCES

1. For an excellent review, see F. Jona and G. Shirane, Ferroelectric Crystals, Pergamon Press, New York, 1962.
2. A.R. Von Hippel, J. Phys. Soc. Japan, 28 (Suppl) (1970) 1.
3. W. Cochran, Adv. Phys., 9 (1960) 387; ibid, 10 (1960) 401.
4. M.E. Lines and A.M. Glass, Principles and Applications of Ferroelectrics and Related Materials, Clarendon Press, Oxford, 1977.
5. W. Cochran, Phys. Rev., 166 (1968) 507.
6. P. Anderson, in G.I. Skanavi (Ed.), Fizika Dielectrikov, Akad. Nauk SSSR Fizicheskii Inst. im P.N. Lebedeva, Moscow, 1960.
7. G.K. Wertheim, Mossbauer Effect: Principles and Applications, Academic Press, New York, 1964.

8. A review paper was presented by one of us (SKD) at Intern. Spring School on Appl. Mossbauer Spectroscopy, Hunfeld, West Germany, 1975
9. D. Barb and S. Constantinescu, J. Phys. (Paris) (Suppl), C1 (1979) 1.
10. G.K. Shenoy, in S.G. Cohen and M. Pasternak (Eds.), Perspectives in Mossbauer Spectroscopy, Plenum Press, 1973, p. 141.
11. V.G. Bhide and M.S. Multani, Phys. Rev., 139 (1965) A1983. ibid, 149 (1966) 289.
12. V.G. Bhide and V.V. Durge, Solid State Commun., 10 (1972) 401.
13. V.G. Bhide and M.S. Hegde, Phys. Rev. B, 5 (1972) 3488.
14. G.K. Wertheim, D.N.E. Buchanan and H.J. Guggenheim Phys. Rev., B, 2 (1970) 1392.
15. G. Shirane et al, J. Phys. Soc. Japan, 6 (1961) 265.
16. R.V. Pound, G.B. Benedek and R. Drever, Phys. Rev. Lett., 1 (1961) 405.
17. D.N. Pipkorn et al, Phys. Rev., 135 (1964) A1604.
18. T.G. Gleason and J.C. Walker, Phys. Rev., 188 (1969) 893.
19. E.A. Samuel and V.S. Sundaram, Phys. Lett., 47A (1974) 421.
20. Ch. Wissel, Solid State Commun., 17 (1975) 1011.
21. G. Meissner and K. Binder, Phys. Rev. B, 12 (1975) 3948.
22. W. Wildner et al, Ferroelectrics, 23 (1980) 193. see also W. Wildner, Ph.D. Thesis, Universitat des Saarlandes, Saarbrucken, 1978.
23. H.G. Muguire and L.V.C. Rees, J. Phys. (Paris) Colloq., 33 (1972) C2-173.
24. V.K. Yarmarkin, B.A. Shustrov and V.A. Motornyi, Proc. Int. Conf. Mossbauer Spect., Cracow, 1975, p. 325.
25. V.A. Bokov et al, Proc. Int. Meeting on Ferroelectricity, Prague, 1966, p. 80.
26. K.P. Mitrofanov et al, Proc. Int. Meeting on Ferroelectricity, Prague, 1966, p. 87.
27. M.V. Plotnikova, K.P. Mitrofanov and Yu. N. Venevtsev, Ferroelectrics, 8 (1974) 535; see also JETP Lett., 17 (1973) 97.
28. D.A. O'Conner and E.R. Spicer, Phys. Lett., 29A (1969) 136.
29. D.A. O'Conner, Proc. Int. Conf. Mossbauer Spectroscopy, Cracow, 1975, p. 369.
30. C.N.W. Darlington, W.J. Fitzgerald and D.A. O'Conner, Phys. Lett., 54A (1975) 35.
31. Two recent books are available; (i) U. Gonser (Ed.), Mossbauer Spectroscopy, Springer Verlag, Berlin, 1975 and (ii) P. Gutlich, R. Link and A. Trautwein (Eds.), Mossbauer spectroscopy and Transition Metal Chemistry, Springer Verlag, Berlin, 1978.
32. V. Dvorak, Phys. Status Solidi, 14 (1966) K161.
33. J.P. Canner, C.M. Yagnik, R. Gerson and W.J. James, J. Appl. Phys., 42 (1971) 4708; see also C.M. Yagnik et al, J. Appl. Phys., 42 (1971) 395; ibid, J. Appl. Phys., 40 (1969) 4713.
34. A.P. Jain, S.N. Shringi and M.L. Sharma, Phys. Rev. B, 2 (1970) 2756.
35. G. Teisseron and A. Baudry, Phys. Rev., 118 (1975) 4518.
36. Y. Hazony, D.E. Earls and I. Lefkowitz, Phys. Rev., 166 (1968) 507.
37. M.J. Clauser, Phys. Rev. B, 1 (1970) 357.
38. P.A. Montano, H. Shechter and U. Shimony, Phys. Rev. B, 3 (1971) 858.
39. F. Placido, Ferroelectrics, 8 (1970) 537.
40. A. Rigamonti and G. Petrini, Phys. Status Solidi, 41 (1970) 591.
41. G.D. Sultanov, F.A. Mirishli and I.H. Ismailzade, Ferroelectrics, 5 (1973) 197; ibid, 8 (1974) 539.
42. See Reference 4, Ch. 12.
43. G.A. Alphonse and W. Philips, RCA Review, 37 (1976) 184. and references therein.
44. D.L. Staebler and W. Philips, Appl. Opt., 13 (1974) 788.

45. W. Keune, S.K. Date, I. Dezsi and U. Gonser, J. Appl. Phys. 46 (1975) 3914; see also Ferroelectrics, 13 (1976) 443.
46. S.K. Date, W. Keune and U. Gonser, Hyperfine Interact., 7 (1979) 369.
47. S.K. Date et al, Ferroelectrics, (1980).
48. S.K. Date et al, to be published in Hyperfine Interact.
49. S.K. Date et al, J. Phys. (Paris) Colloq., 37 (1976) C6-117.
50. H. Kurz et al, Appl. Phys. 12 (1979) 355.
51. J. Lauer, H.D. Pfannes and W. Keune, J. Phys. (Paris) Colloq., 40 (1979) C2-561.
52. H. Engelmann and U. Gonser, Ferroelectrics, 23 (1980) 97.
53. S.K. Date, P.S. Joag and A.S. Nigvekar, Nat. Acad. Sci. Lett., vol. 1 (1978).
54. V.G. Bhide and H.C. Bhasin, Phys. Rev., 159 (1967) 586; Phys. Rev., 172 (1968) 290.
55. B.W. Faughnan, D.L. Staebler and Z.J. Kiss, Appl. Solid State Sci., 2 (1974) 107.
56. C.T. Luiskutty and P.J. Ouseph, Solid State Commun., 13 (1973) 405. see also Phys. Status Solidi (b), 58 (1973) K171.
57. S.K. Date, unpublished work.
58. K.W. Blazey et al, Solid State Commun., 16 (1975) 589; ibid, 16 (1975) 1289.

Index